Biology of Plants

FIFTH EDITION

Vincent van Gogh (1853–1890) Dutch
Undergrowth with Two Figures, 1890

In late June, 1890, one month before his suicide in the
French town of Auvers-sur-Oise, Vincent van Gogh
described this painting in a letter to his brother: "the
undergrowth around poplars, violet trunks running across
the landscape, perpendicular like columns; the depths of
the woods are blue, and at the bottom of the big trunks
the grassy ground, full of flowers, white, pink, yellow and
green, long grass turning russet . . ." The bold colors and
slashed brushwork convey the passion and struggle
characteristic of Van Gogh's work.

Biology of Plants

FIFTH EDITION

Peter H. Raven

Missouri Botanical Garden and
Washington University, St. Louis

Ray F. Evert

University of Wisconsin, Madison

Susan E. Eichhorn

University of Wisconsin, Madison

WORTH PUBLISHERS

Biology of Plants

FIFTH EDITION

Copyright © 1971, 1976, 1981, 1986, 1992 by Worth Publishers, Inc.

All rights reserved.

Printed in the United States of America

Library of Congress Catalog Card Number: 91-72015

ISBN: 0-87901-532-2

Printing: 1 2 3 4 5 — 96 95 94 93 92

Development editor: Sally Anderson

Design: Malcolm Grear Designers

Art director: George Touloumes

Production editor: Elizabeth Mastalski

Production supervisor: Barbara Anne Seixas

Photographs: Leonora Morgan

Line art: Rhonda Nass (section-opening paintings), Carol Watkins, Jane Nass-Barnidge, Ellen Marie Dudley, Jana Fothergill, TSI Graphics Inc., and Demetrios Zangos

Typographer: New England Typographic Service, Inc.

Color separations and stripping: Creative Graphic Services Corporation

Printing and Binding: Von Hoffmann Press, Inc.

Cover: Vincent van Gogh, *Undergrowth with Two Figures* (detail), 1890; oil on canvas, 19½″ x 39¼″, bequest by Mary E. Johnston, 1967, to the Cincinnati Art Museum

WORTH PUBLISHERS
33 Irving Place
New York, New York 10003

Preface

For our ancestors, not long ago, a cleared field or felled trees meant civilization and survival. That perception has changed. With the extraordinary growth of the human population in this century, we now realize that it is the trees and plants that are threatened, as species are lost to cement, dam-created lakes, commercial forests, and deserts.

If we are not to destroy the plants critical to the air we breathe and the whole biosphere we depend upon, we must come to a better understanding of their function and role. Recent knowledge of such topics as the molecular and cellular biology of plant growth and development has vastly expanded our view. We are learning how to build sustainable systems in agriculture and forestry, thanks to this growing field—a field not of tree stumps but of knowledge.

As we approached the writing of this fifth edition of BIOLOGY OF PLANTS, we once again—urged on by our publisher—explored ways in which the book's length could be reduced. We sent questionnaires, spoke with teachers, and beseeched reviewers, only to discover, once again, that there was no consensus on the topics that should be eliminated. Given this lack of uniformity in botany courses, together with the value of being free to choose from a larger menu, so to speak, and the small price differential between this and much more limited presentations, we felt free to indulge our own strong inclination—to present you with an introductory textbook that we think does justice to the subject.

Despite its title, the book treats all the groups of organisms traditionally considered by university departments of botany: viruses, bacteria, the photosynthetic protists, fungi, and the plants themselves. We continue to define plants as consisting only of bryophytes and vascular plants, a coherent evolutionary line derived from ancestors that would be classified as green algae if we knew what they were. We do not include red and brown algae with the plants because they were independently derived from unicellular ancestors. We have also left the green algae in the kingdom Protista, an arbitrary decision but one that makes the kingdom Plantae easier to visualize. Fungi, of course, are treated as a wholly independent group. And, for this edition, we have acknowledged the fundamental differences of Archaebacteria and Eubacteria by placing them in separate kingdoms. Consequently, this book recognizes six kingdoms: Archaebacteria, Eubacteria, Protista, Fungi, Plantae, and Animalia.

While retaining the strengths of previous editions, we have made a number of significant changes. The most obvious is the move to full color throughout the book. Each of the life cycles, which were redrawn in two colors for the last edition, has now been rendered in full color, with special care given to the color selections in order to distinguish, unmistakably, between phases. We think you will find these illustrations, as well as the numerous other illustrations to which color has been added, even clearer and more informative than before.

We have worked to bring this edition up to date with the latest findings at all levels of organization, from molecular and cellular to whole plant to ecological systems. Virtually every chapter has been reorganized to some extent to provide an even more logical sequence.

Major organizational changes involve moving the Evolution chapter forward to directly follow Genetics, which is now integrated into a single chapter. Within the Genetics chapter, the material on nucleic acids and their transcription has been expanded in order to provide a solid background for the discussion in later chapters of the growth and development of plants at a cellular level.

Within the Diversity section (Section 3), bacteria and viruses are now discussed in a single chapter, and the two kingdoms of bacteria are clearly delineated. The chapter on fungi has been brought up to date with a considerable amount of recent information. In Chapter 15 on bryophytes, the three major groups are now recognized as divisions, consistent with a modern understanding of their structure and biochemical features. The bryophyte groups are regarded as plants with relatively generalized characteristics, not directly related to one another.

The greatly expanded discussion of ferns recognizes the major groups of ferns and carefully explains and illustrates their features. In addition, there is now a separate chapter on gymnosperms, broadening and deepening the treatment of this key group of seed plants. As a minor point, we have adopted a modern classification of the genera of lycopods (club mosses) consistent with the treatment that is to appear in *The Flora of North America*, which will soon be published by the Flora of North America Association. In Chapter 19, modern insights into the evolution of the flowering plants have been added, and the book has moved away from the traditional treatment of angiosperms as having been derived from ancestors with large magnolia-like flowers.

Within the section on the structure and development of the plant body, secondary growth in roots is now considered in an integrated chapter on root growth and development. Secondary growth in stems is covered in a separate chapter, which follows the chapter on primary growth and development in stems. A notable change for this edition is an extensive revision of the plant hormones chapter (Chapter 25). Special attention has been given to the molecular basis of hormone action, plant biotechnology, and genetic engineering (recombinant DNA) technology, which is providing biologists with an opportunity never before available—to transfer genetic traits between highly diverse organisms.

In the Ecology section, boxed essays have been added on the great Yellowstone fire, on the controversy surrounding the logging of ancient forests in the Pacific Northwest, and on the competition for light in plant communities.

The fifth edition of BIOLOGY OF PLANTS remains true to the book's original goals: to present basic information about plants in an accurate, accessible, well-illustrated, and interesting way. As in previous editions, we pay special attention to what we think of as the book's interlocking themes: (1) the plant body as the dynamic result of the processes of growth and development carried out by chemical interactions; (2) evolutionary relationships as the guide to understanding form and function in organisms; and (3) ecology as a way of emphasizing our dependence on plants to sustain our lives and those of all other living organisms. BIOLOGY OF PLANTS has been written and organized to allow for as much flexibility as possible in the sequence in which topics are covered in the classroom. Care has been taken to enhance the book's usefulness for the many courses in which less detail is needed and only parts of the book are assigned. We have tried to provide students with a clear and interesting introduction to botany and with an up-to-date reference to topics that may catch their imagination and into which they may wish to delve more deeply.

We strive to introduce our readers not only to what is known about plant biology but also to new developments and remaining unsolved problems. In so doing, we hope to enlist new talents in working toward the solutions on which all of our futures depend.

We greatly appreciate the enthusiastic support and constructive suggestions made by teachers who used the previous edition. In addition, we have received substantial assistance from: Michael R. Sussman, University of Wisconsin, who reviewed the material on membranes; Paul W. Ludden and Wayne Becker, both of the University of Wisconsin, who reviewed the chapters on energy; Anthony Bleecker, University of Wisconsin, Alan Jones, University of North Carolina, Karen Kindle, Cornell University, and Bernard O. Phinney, University of California, Los Angeles, who provided considerable help on the growth regulation and growth response chapter; Malgre Carreno, University of Wisconsin, who advised on rhizobial interactions in legumes; and Thomas J. Givnish and Hugh H. Iltis, both of the University of Wisconsin, for special help with the ecology section.

Others to whom we wish to express our thanks for reviewing parts of the manuscript or supplying information are:

Paul Ahlquist, *University of Wisconsin*
Frank J. Alfieri, *Long Beach State University*
Janice Antonovics, *Duke University*
Jerry Baskin, *University of Kentucky*
Wayne Becker, *University of Wisconsin*
John Beebe, *Calvin College*
David Bilderback, *University of Montana*
Meredith Blackwell, *Louisiana State University*
Andrew Blaustein, *Oregon State University*
Lois Brako, *Missouri Botanical Garden*
Rita Calvo, *Cornell University*
James Caponetti, *University of Tennessee*
Sherman Carlquist, *Rancho Santa Ana Botanic Garden*
Edward Clebsch, *University of Tennessee*
David S. Conant, *Lyndon State College*
John Corliss, *University of Maryland*
Thomas B. Croat, *Missouri Botanical Garden*
William Louis Culberson, *Duke University*
Jerry Davis, *University of Wisconsin, LaCrosse*
Roland Dute, *Auburn University*
James Ehleringer, *University of Utah*

John W. Einset, *Eni-Chem Americas, Inc.*
Peter K. Endress, *University of Zurich*
Robert C. Evans, *Rutgers University*
Sharon Eversman, *Montana State University*
Donald R. Farrar, *Iowa State University*
Priscilla Fawcett, *Waterloo, Iowa*
Gary L. Floyd, *Ohio State University*
Gerald J. Gastony, *Indiana University*
Janice Glime, *Michigan Technological University*
Linda E. Graham, *University of Wisconsin*
Earl Gritton, *University of Wisconsin*
Mason E. Hale, *Smithsonian Institution*
Ray Hammerschmidt, *Michigan State University*
D. L. Hawksworth, *C-A-B International Mycological Institute, Kew*
John Herr, *University of South Carolina*
Ronald Hoham, *Colgate University*
Max Hommersand, *University of North Carolina*
Judith Jernstedt, *University of California, Davis*
Arthur Kelman, *North Carolina State University*
Bryce Kendrick, *University of Waterloo*
Theodore Kozlowski, *University of California, Santa Barbara*
Dale Kristensen, *Queen's University*
Wayne Kussow, *University of Wisconsin*
Joseph LaPointe, *New Mexico State University*
David Layzall, *Queen's University*
John N.A. Lott, *McMaster University*
David J. Mclaughlin, *University of Minnesota, St. Paul*
Dorothy McMeekin, *Michigan State University*
Charles N. Miller, Jr., *University of Montana*
Brent D. Mishler, *Duke University*
Roy Moore, *University of Ulster at Coleraine, Northern Ireland*
Robbin Moran, *Missouri Botanical Garden*
R.G.E. Murray, *University of Western Ontario*
Lytton Musselman, *Old Dominion University*
William Newcomb, *Queen's University*
Knut Norstog, *Waterloo, Iowa*
Curt Peterson, *Auburn University*
Larry Peterson, *University of Guelph*
Tom Phillips, *University of Illinois*
Robert Platt, *Ohio State University*
Peter Quail, *Plant Gene Expression Center*
Beverly Roberts, *Department of Marine Research, Florida*
Emanuel D. Rudolph, *Ohio State University*
Paula Sanchini, *Coe College*
Frank S. Santamour, *U.S. National Arboretum, Washington, D.C.*
Rudolf Schmid, *University of California, Berkeley*

Steven R. Seavey, *Lewis and Clark College*
J. Kenneth Shull, *Appalachian State University*
Richard Storey, *Colorado College*
Boyd Strain, *Duke University*
Harry D. Thiers, *San Francisco State University*
Jennifer Thorsch, *University of California, Santa Barbara*
Sue Tolin, *Virginia Polytechnic Institute and State University*
Peter M. Vitousek, *Stanford University*
Warren H. Wagner, Jr., *University of Michigan*
John Webster, *University of Exeter*
John West, *University of California, Berkeley*
D. Reid Wiseman, *College of Charleston*
John Worrall, *University of British Columbia*
Eli Zamski, *The Hebrew University, Jerusalem, Rehovot*

We are grateful to Rhonda Nass for her superb paintings, to Carol Watkins and Jane Nass-Barnidge for their outstanding artwork, to Damian S. Neuberger for his excellent photomicroscopy, and to Ellen Marie Dudley and Jana Fothergill for their fine artistic contributions.

We also wish to thank the following people at the University of Wisconsin: Claudia Lipke, for several color photographs; Mark Wetter and Theodore Cochrane, for plant identification; Mohommad Fayyaz, Mary Bauschelt, and Glen Grieger, for greenhouse material; and William A. Russin, Derek Webb, and Megan Maguire, for general assistance. In addition, we wish to thank those people who helped us obtain the many fine photographs.

The three authors would like to express their appreciation to their spouses: to Tamra Engelhorn Raven, for her assistance and encouragement; to Mary Evert, for her patience and inspiration; and to Henry (Ike) Eichhorn, for his support and especially his forbearance during the preparation of the fifth edition.

Finally, we wish to express our sincere thanks to Sally Anderson, Betsy Mastalski, George Touloumes, Barbara Seixas, Demetrios Zangos, Patricia Lawson, Mary Mazza, Lisa Douglas, and the many other people at Worth Publishers who have made outstanding contributions to this book. This edition has benefited greatly from their unfailing commitment to the highest quality in the publishing of textbooks.

Peter H. Raven
Ray F. Evert
Susan E. Eichhorn
February, 1992

Contents in Brief

CHAPTER 1   An Introduction to Botany   1

SECTION 1   *The Plant Cell: Structure and Metabolism*   **14**

CHAPTER 2   Introduction to the Eukaryotic Cell   15

CHAPTER 3   The Molecular Composition of Cells   47

CHAPTER 4   The Movement of Substances into and out of Cells   62

CHAPTER 5   The Flow of Energy   74

CHAPTER 6   Respiration   86

CHAPTER 7   Photosynthesis   99

SECTION 2   *Genetics and Evolution*   **120**

CHAPTER 8   Genetics and Heredity   121

CHAPTER 9   The Process of Evolution   150

SECTION 3   *Diversity*   **170**

CHAPTER 10   The Classification of Living Things   171

CHAPTER 11   Bacteria and Viruses   186

CHAPTER 12   Fungi   208

CHAPTER 13   Protista I: Water Molds, Slime Molds, Chytrids, and Unicellular Algae   244

CHAPTER 14   Protista II: Red, Brown, and Green Algae   268

CHAPTER 15   Bryophytes   298

CHAPTER 16   Seedless Vascular Plants   317

CHAPTER 17   Gymnosperms   356

CHAPTER 18   Introduction to the Angiosperms   380

CHAPTER 19   Evolution of the Angiosperms   401

SECTION 4   *The Angiosperm Plant Body: Structure and Development*   **440**

CHAPTER 20   Early Development of the Plant Body   441

CHAPTER 21   Cells and Tissues of the Plant Body   453

CHAPTER 22   The Root: Structure and Development   470

CHAPTER 23   The Shoot: Primary Structure and Development   488

CHAPTER 24   Secondary Growth in Stems   520

SECTION 5   *Physiology of Seed Plants*   **544**

CHAPTER 25   Regulating Growth and Development: The Plant Hormones   545

CHAPTER 26   External Factors and Plant Growth   573

CHAPTER 27   Plant Nutrition and Soils   593

CHAPTER 28   The Movement of Water and Solutes in Plants   616

SECTION 6   *Ecology and the Human Prospect*   **636**

CHAPTER 29   The Dynamics of Communities and Ecosystems   637

CHAPTER 30   The Biomes   657

CHAPTER 31   Plants and People   686

APPENDIX A: Fundamentals of Chemistry   713

APPENDIX B: Metric Table and Temperature Conversion Scale   726

APPENDIX C: Classification of Organisms   728

APPENDIX D: Geologic Eras   735

GLOSSARY   737

ILLUSTRATION ACKNOWLEDGMENTS   763

INDEX   769

Contents

CHAPTER 1   **An Introduction to Botany**   1

The Evolution of Plants   1
The Evolution of Communities   6
The Appearance of Human Beings   10
Summary   12
Suggestions for Further Reading   13

SECTION 1   *The Plant Cell: Structure and Metabolism*   14

CHAPTER 2   **Introduction to the Eukaryotic Cell**   15

Prokaryotes and Eukaryotes   16
The Plant Cell   17
Essay: Origin of the Cell Theory   18
Plasma Membrane   18
Nucleus   19
Plastids   20
Mitochondria   23
Microbodies   25
Vacuoles   25
Essay: Viewing the Microscopic World   26

Ribosomes   27
Endoplasmic Reticulum   28
Golgi Apparatus   28
Cytoskeleton   29
Ergastic Substances   30
Flagella and Cilia   31
Essay: The Role of Actin and Myosin in Cytoplasmic Streaming in Giant Algal Cells   33
Cell Wall   33
Plasmodesmata   37
Cell Division   38
The Cell Cycle   38
Essay: The Role of Microtubules in Cell Division: Immunofluorescence Microscopy   41
Summary   44

CHAPTER 3   **The Molecular Composition of Cells**   47

Organic Molecules   47
Carbohydrates   48
Lipids   52
Essay: Radiocarbon Dating   53
Proteins   55
Nucleic Acids   59
Other Nucleotide Derivatives   61
Summary   61

CHAPTER 4   **The Movement of Substances into and out of Cells**   62

Principles of Water Movement   62
Essay: Imbibition   64
Structure of Cellular Membranes   66
Transport of Solutes across Membranes   68
Essay: Patch-Clamp Recording in the Study of Ion Channels   69

Endocytosis and Exocytosis   70
Transport via Plasmodesmata   71
Summary   73

CHAPTER 5   **The Flow of Energy**   **74**

The Laws of Thermodynamics   74
Oxidation-Reduction   77
Enzymes and Living Systems   78
Enzymes as Catalysts   79
Cofactors in Enzyme Action   81
Enzymatic Pathways   82
Regulation of Enzyme Activity   82
The Energy Factor: ATP   83
Summary   84

CHAPTER 6   **Respiration**   **86**

Glycolysis   87
The Aerobic Pathway   90
Anaerobic Pathways   97
Essay: Bioluminescence   97
Summary   98

CHAPTER 7   **Photosynthesis**   **99**

Overview of Photosynthesis   99
The Nature of Light   101
The Light-Dependent Reactions   103
The Light-Independent Reactions   109
Essay: The Carbon Cycle   115
Summary   117
Suggestions for Further Reading   118

SECTION 2   *Genetics and Evolution*   120

CHAPTER 8   **Genetics and Heredity**   **121**

Sexual Reproduction in Eukaryotes   121
Structure of Eukaryotic
Chromosomes   122
Meiosis   124
How Are Characteristics Inherited?   128

Mutations   134
The Determination of the
Phenotype   135
The Molecular Basis of Inheritance   136
The Nature of DNA   136
How Do Genes Work?   140
Regulating Gene Transcription   145
Summary   147
Suggestions for Further Reading   149

CHAPTER 9   **The Process of Evolution**   **150**

The Behavior of Genes in
Populations: The Hardy-Weinberg
Law   152
Responses to Selection   154
The Divergence of Populations   157
The Evolutionary Role of
Hybridization   160
Essay: Adaptive Radiation in
Hawaiian Tarweeds   164
The Origin of Major Groups of
Organisms   168
Summary   168
Suggestions for Further Reading   169

SECTION 3   *Diversity*   170

CHAPTER 10   **The Classification of
Living Things**   **171**

The Binomial System   172
The Major Groups of Organisms   175
Relationships within the Eukaryotes   176
Formal Classification of Organisms   178
Sexual Reproduction   181
Summary   185
Suggestions for Further Reading   185

CHAPTER 11   **Bacteria and Viruses**   **186**

Characteristics of Bacteria   187
Major Groups of Bacteria   190
Essay: Citrus Canker   200

Viruses   200
Summary   205
Suggestions for Further Reading   207

CHAPTER 12   **Fungi**   **208**

Biology of the Fungi   210
The Evolution of Fungi   211
Fungal Reproduction   212
The Major Groups of Fungi   212
Essay: Phototropism in a Fungus   214
Essay: From Pathogen to Symbiont:
Fungal Endophytes   220
Essay: Predaceous Fungi   223
Mycorrhizae   238
Summary   242
Suggestions for Further Reading   243

CHAPTER 13   **Protista I: Water Molds,**
Slime Molds, Chytrids,
and Unicellular Algae   **244**

Ecology of Unicellular Protista   245
Symbiosis and the Origin of the
Chloroplast   245
Characteristics of the Divisions   246
Division Oomycota   248
Essay: Hormonal Control of
Sexuality in a Water Mold   250
Division Chytridiomycota   251
Division Acrasiomycota   252
Division Myxomycota   255
Division Chrysophyta   257
Essay: Diatom Motility   261
The Dinoflagellates: Division
Pyrrhophyta   262
Essay: Red Tides   263
Essay: Mitosis in Dinoflagellates   265
The Euglenoids: Division
Euglenophyta   265
Summary   266
Suggestions for Further Reading   267

CHAPTER 14   **Protista II: Red, Brown,**
and Green Algae   **268**

Characteristics of Red, Brown, and
Green Algae   269
Red Algae: Division Rhodophyta   270
Brown Algae: Division Phaeophyta   274
Green Algae: Division Chlorophyta   279
Essay: Algae and Human Affairs   281
Essay: Symbiotic Green Algae   294
Summary   296
Suggestions for Further Reading   297

CHAPTER 15   **Bryophytes**   **298**

The Nature of Bryophytes   299

The Liverworts: Division
Hepatophyta   300
Essay: Spore Discharge in
Liverworts   306
The Hornworts: Division
Anthocerophyta   307
The Mosses: Division Bryophyta   308
Summary   316
Suggestions for Further Reading   316

CHAPTER 16   **Seedless Vascular Plants**   **317**

The Evolution of Vascular Plants   317
Organization of the Vascular
Plant Body   319
Reproductive Systems   322
The Divisions of Seedless
Vascular Plants   323
Division Rhyniophyta   325
Division Zosterophyllophyta   326
Division Trimerophyta   326
Division Psilotophyta   326
Division Lycophyta   330
Division Sphenophyta   338
Division Pterophyta   343
Essay: Coal Age Plants   346
Summary   353
Suggestions for Further Reading   355

CHAPTER 17   **Gymnosperms**   **356**

Progymnosperms   359
Extinct Gymnosperms   361
Living Gymnosperms   361
Summary   378
Suggestions for Further Reading   379

CHAPTER 18   **Introduction to the**
Angiosperms   **380**

The Flower   383
The Angiosperm Life Cycle   388
Essay: Hay Fever   397
Essay: Genetic Self-incompatibility   398
Summary   399
Suggestions for Further Reading   400

CHAPTER 19   **Evolution of the Angiosperms**   **401**

Relationships of the Angiosperms   401
Origin of the Angiosperms   405
Evolutionary Radiation of
the Angiosperms   406
Evolution of the Flower   408
Essay: Hot Pollination in the
Arum Lilies   416
The Evolution of Fruits   429
Biochemical Coevolution   435
Summary   438
Suggestions for Further Reading   439

SECTION 4   *The Angiosperm Plant Body: Structure and Development*   **440**

CHAPTER 20   Early Development of the Plant Body   441

The Mature Embryo and Seed   441
Formation of the Embryo   444
Requirements for Seed Germination   447
From Embryo to Adult Plant   448
Essay: Vegetative Reproduction: Some Ways and Means   450
Essay: Wheat: Bread and Bran   452
Summary   452

CHAPTER 21   Cells and Tissues of the Plant Body   453

The Tissue Systems   454
Tissues and Their Component Cells   455
Summary   467

CHAPTER 22   The Root: Structure and Development   470

Root Systems   470
Origin and Growth of Primary Tissues   472
Primary Structure   475
Effect of Secondary Growth on the Primary Body of the Root   480
Origin of Lateral Roots   483
Aerial Roots and Air Roots   483
Adaptations for Food Storage   485
Summary   487

CHAPTER 23   The Shoot: Primary Structure and Development   488

Origin and Growth of the Primary Tissues of the Stem   489
Primary Structure of the Stem   491
Relation between the Vascular Tissues of the Stem and the Leaf   496

Morphology of the Leaf   498
Structure of the Leaf   500
Grass Leaves   505
Development of the Leaf   506
Leaf Abscission   509
Essay: Leaf Dimorphism in Aquatic Plants   510
Transition between Vascular Systems of the Root and the Shoot   511
Development of the Flower   511
Essay: Plants, Air Pollution, and Acid Rain   512
Stem and Leaf Modifications   515
Essay: Convergent Evolution   518
Summary   519

CHAPTER 24   Secondary Growth in Stems   520

The Vascular Cambium   521
Effect of Secondary Growth on the Primary Body of the Stem   523
The Wood: Secondary Xylem   532
Summary   541
Suggestions for Further Reading   542

SECTION 5   *Physiology of Seed Plants*   **544**

CHAPTER 25   Regulating Growth and Development: The Plant Hormones   545

Auxin   547
Cytokinins   551
Ethylene   554
Abscisic Acid   555
Gibberellins   555
The Molecular Basis of Hormone Action   559
Plant Biotechnology   563
Essay: Arabidopsis thaliana: A Weed with Promise   568
Summary   571

CHAPTER 26 **External Factors and Plant Growth**   573

The Tropisms   573
Circadian Rhythms   576
Photoperiodism   578
Chemical Basis of Photoperiodism   581
Hormonal Control of Flowering   584
Dormancy   585
Cold and the Flowering Response   587
Nastic Movements   587
Generalized Effects of Mechanical Stimuli on Plant Growth and Development: Thigmomorphogenesis   589
Solar Tracking   590
Summary   591
Suggestions for Further Reading   592

CHAPTER 27 **Plant Nutrition and Soils**   593

General Nutritional Requirements   593
Functions of Inorganic Nutrients in Plants   595
The Soil   597
Essay: The Water Cycle   600
Nutrient Cycles   602
Nitrogen and the Nitrogen Cycle   602
Essay: Halophytes: A Future Resource?   608
The Phosphorus Cycle   611
Human Impact on Nutrient Cycles   612
Soils and Agriculture   612
Plant Nutrition Research   613
Essay: Compost   614
Summary   615

CHAPTER 28 **The Movement of Water and Solutes in Plants**   616

Movement of Water and Inorganic Nutrients through the Plant Body   616
Translocation: The Movement of Substances through the Phloem   629
Essay: Radioactive Tracers and Autoradiography in Plant Research 633
Summary   634
Suggestions for Further Reading   634

SECTION 6   *Ecology and the Human Prospect*   636

CHAPTER 29 **The Dynamics of Communities and Ecosystems**   637

Interactions between Organisms   638
Essay: Pesticides and Ecosystems   643
Nutrient Cycling   645
Trophic Levels   646
Development of Communities and Ecosystems   649
Essay: The Great Yellowstone Fire   652
Summary   655

CHAPTER 30 **The Biomes**   657

Life on the Land   657
Rainforests   662
Essay: Alexander von Humboldt   664
Savannas and Deciduous Tropical Forests   666
Deserts   668
Essay: How Does a Cactus Function?   671
Grasslands   672
Temperate Deciduous Forests   674
Essay: Competing for Light   676
Temperate Mixed and Coniferous Forests   678
Mediterranean Scrub   680
Taiga   681
Essay: Jobs versus Owls   682
Arctic Tundra   683
Summary   684

CHAPTER 31   **Plants and People**   **686**

   The Agricultural Revolution   687
   Essay: The Origin of Corn   693
   The Growth of Human Populations   698
   Agriculture in the Future   700
   Summary   709
   Suggestions for Further Reading   710

APPENDIX A: **Fundamentals of Chemistry**   **713**

   Atoms   713
   Bonds and Molecules   717
   Water and the Hydrogen Bond   720
   Water As a Solvent   721
   Acids and Bases   721
   Chemical Reactions   722

APPENDIX B: **Metric Table and Temperature
               Conversion Scale**   **726**

APPENDIX C: **Classification of Organisms**   **728**

APPENDIX D: **Geologic Eras**   **735**

GLOSSARY   737

ILLUSTRATION ACKNOWLEDGMENTS   763

INDEX   769

Biology of Plants

FIFTH EDITION

Life on earth depends on the ability of plants to capture the sun's energy. This reaction requires the green pigment chlorophyll, which is present, for example, in the leaves of this wild rose.

CHAPTER 1

An Introduction to Botany

"What drives life is . . . a little current, kept up by the sunshine." The Nobel laureate Albert Szent-Györgyi, with this simple sentence, summed up one of the greatest marvels of evolution—the photosynthetic process, which is necessary to life on this planet. When a particle of light strikes a molecule of chlorophyll, one of its electrons absorbs that energy and is boosted to a higher energy level. This excited electron is then transferred to an acceptor molecule, initiating a flow of electrons that, within a fraction of a second, converts the energy momentarily gained by the electron to chemical energy as the electron returns to its original energy level. This process is known as photosynthesis. With very few exceptions, all life on this planet is dependent upon the energy momentarily gained by such electrons.

Only a few types of organisms—plants, algae, and some bacteria—possess chlorophyll, which can, when embedded in the membranes of a living cell, carry out photosynthesis. Once light energy is trapped in chemical form, it becomes available as an energy source to all other organisms, including human beings. We are totally dependent upon photosynthesis, a process for which plants are exquisitely adapted.

The Evolution of Plants

Like all other living organisms, plants have had a long evolutionary history. The planet earth itself—an accretion of dust and gases swirling in orbit around the star that is our sun—is some 4.5 billion years old. The earliest known fossils are about 3.5 billion years old and consist of several kinds of small, relatively simple cells (Figure 1–1). These fossils have been found in some of the oldest rocks on earth.

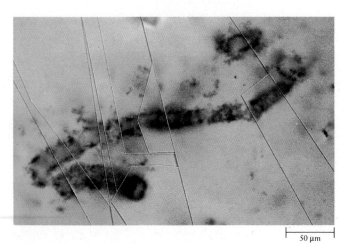

50 μm

1–1
The earliest known fossils, chainlike bacteria from ancient rocks in Australia, dated at 3.5 billion years of age. They are about a billion years younger than the earth itself, but there are few suitable older rocks in which to look for earlier evidence of life. More complex organisms—those with eukaryotic cellular organization—did not evolve until about 1.5 billion years ago. For at least 2 billion years, therefore, bacteria were the only forms of life on earth. These cells have been magnified 260 times.

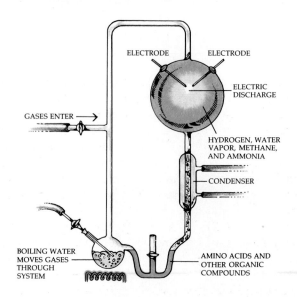

1–2
Stanley Miller, while a graduate student at the University of Chicago in the 1950s, used apparatus such as that diagrammed here, to simulate conditions he believed existed on the primitive earth. Hydrogen, methane, and ammonia were circulated continuously between a lower "ocean," which was heated, and an upper "atmosphere," through which an electric discharge was transmitted. At the end of 24 hours, about half of the carbon originally present in the methane gas was converted to amino acids and other organic molecules.

As events are reconstructed, these first cells were formed by a series of chance occurrences. Imagine the earth enveloped by swirling gases that spew from countless numbers of volcanoes. This early atmosphere seems to have consisted primarily of nitrogen gas, mixed with relatively large amounts of carbon dioxide and water vapor. These three molecules contain the chemical elements carbon, oxygen, hydrogen, and nitrogen, which make up about 98 percent of the material found in living organisms today.

Through the thin atmosphere, the rays of the sun beat down on the harsh, bare surface of the young earth, bombarding it with light, heat, and ultraviolet radiation. Molecules of gases such as hydrogen sulfide, ammonia, and methane also seem to have been present in the early atmosphere. In contrast, oxygen gas, which now makes up about 21 percent of our atmosphere, was not formed until living organisms evolved and began to carry out photosynthesis. Thus the first steps in the evolution of life took place in an **anaerobic** (without oxygen) atmosphere.

As the crust of the earth was cooling and stabilizing, violent storms raged, accompanied by lightning and the release of electrical energy. Radioactive substances in the earth emitted large quantities of energy, and molten rock and boiling water erupted from beneath the earth's surface. The energy in this vast crucible broke apart the simple gases of the atmosphere and re-formed them into larger and more complex molecules. Ultraviolet light bathed the surface of the earth, breaking these molecules and the gases, and causing the formation of still more new molecules.

Current theories propose that the compounds formed in the early atmosphere tended to be washed out of it by the driving rains, collecting in the oceans, which grew larger as the earth cooled. Using the same gases theorized to have existed then, scientists performed experiments in which they simulated the conditions presumed to have existed at that early time on earth (Figure 1–2). Under these experimental conditions, complex organic molecules, similar to those that form the fundamental building blocks of all life, were formed. On the early earth, the oceans, and probably small pools near volcanoes, became an increasingly rich mixture of such organic molecules.

Some organic molecules have a tendency to aggregate in groups. In the primitive oceans these groups probably took the form of droplets, similar to the droplets formed by oil in water. Such droplets of organic molecules appear to have been the forerunners of primitive cells, the first forms of life.

According to current theories, these organic molecules also served as the source of energy for the earliest forms of life. The primitive cells or cell-like structures were able to use these abundant compounds to satisfy their energy requirements. As they evolved and became more complex, these cells were increasingly able to control their own destinies. With this increasing complexity, they acquired the ability to grow, to reproduce, and to pass on their characteristics to subsequent generations.

Cells that satisfy their energy requirements by consuming the organic compounds produced by external sources are known as **heterotrophs** (Gk. *heteros,* "other" + *trophos,* "feeder"). A heterotrophic organism is one that is dependent on an outside source of organic molecules for its energy. This category of organisms today includes all living things classified as animals or fungi (Figure 1–3a) and many of the one-celled organisms—most bacteria and some protists.

As the primitive heterotrophs increased in number, they began to use up the complex molecules on which their existence depended—and which had taken millions of years to accumulate. Organic molecules in free solution (not inside a cell) became more and more scarce, and competition began. Under the pressure of this competition, cells that could make efficient use of the limited energy sources now available were more likely to survive than cells that could not. In the course of time, by the long, slow process of elimination of the

less fit, cells evolved that were able to make their own energy-rich molecules out of simple nonorganic materials. Such organisms are called **autotrophs,** "self-feeders." Without the evolution of these early autotrophs, life on earth would soon have come to an end.

The most successful of the autotrophs were those that evolved a system for making direct use of the sun's energy—that is, the process of photosynthesis (Figure 1–3b). The earliest photosynthetic organisms, although simple in comparison with plants, were much more complex than the primitive heterotrophs. To capture and use the sun's energy required a complex pigment system that could catch and hold the energy of light and, linked to this system, a way of fixing the energy in an organic molecule.

Evidence of the activities of photosynthetic organisms has been found in rocks 3.4 billion years old, about 100 million years after the first fossil evidence of life on earth. We can be almost certain, however, that both life and photosynthetic organisms evolved considerably earlier than the evidence suggests. In addition, there seems to be no doubt that heterotrophs evolved before autotrophs. With the arrival of autotrophs, the flow of energy in the biosphere (that is, the living world and its environment) came to assume its modern form: radiant energy channeled through the photosynthetic autotrophs to all other forms of life.

(a)

(b)

1–3

A modern heterotroph and a photosynthetic autotroph. (a) A fungus, Amanita muscaria, *a poisonous mushroom commonly known as fly agaric, growing on a forest floor in New York State.* Amanita, *which, like other fungi, absorbs its food (often from other organisms), is heterotrophic. (b) Large-flowered trillium* (Trillium grandiflorum), *one of the first plants to flower in spring in the deciduous woods of eastern and midwestern North America. Like most vascular plants, trilliums are rooted in the soil; photosynthesis occurs chiefly in the leaves of this autotrophic organism. The flowers are produced in well-lighted conditions before the leaves appear on the surrounding trees. The underground portions (rhizomes) of the plant live for many years and spread to produce new plants vegetatively under the thick cover of decaying leaves and other organic material on the forest floor. Trilliums also reproduce via seeds, which are dispersed by ants (see Figure 19–48).*

PHOTOSYNTHESIS AND THE EVOLUTION OF ATMOSPHERIC OXYGEN

As photosynthetic organisms increased in number, they changed the face of the planet. This biological revolution came about because one of the most efficient strategies of photosynthesis—the one employed by nearly all living autotrophs—involves splitting the water molecule (H_2O) and releasing its oxygen. Thus, as a result of photosynthesis, the amount of oxygen gas (O_2) in the atmosphere increased. This increase in oxygen level had two important consequences.

First, some of the oxygen molecules in the outer layer of the atmosphere were converted to ozone (O_3) molecules. When there is a sufficient quantity of ozone molecules in the atmosphere, it filters the ultraviolet rays—rays highly destructive to living organisms—from the sunlight that reaches the earth. By about 450 million years ago, organisms, protected by the ozone layer, could survive in the surface layers of water and on the land.

Second, the increase in free oxygen opened the way to a much more efficient utilization of the energy-rich carbon-containing molecules formed by photosynthesis —it enabled organisms to break down those molecules by respiration. As is discussed in Chapter 6, respiration yields far more energy than can be extracted by any anaerobic process.

Before the atmosphere accumulated oxygen and became **aerobic,** the only cells that existed were **prokaryotic**—simple cells that lacked a nuclear envelope and did not have their genetic material organized into complex chromosomes. Another name for prokaryotes is "bacteria," and all of the kinds of organisms that existed on earth prior to about 1.5 billion years ago were bacteria, some heterotrophic and others autotrophic. According to the fossil record, the increase of relatively abundant free oxygen was accompanied by the first appearance of **eukaryotic** cells—cells that have nuclear envelopes, complex chromosomes, and membrane-bound organelles. Eukaryotic organisms, in which the individual cells are usually much larger than those of the bacteria, appeared about 1.5 billion years ago and were well established and diverse by 1 billion years ago. All living systems, except for the bacteria, are composed of one or more eukaryotic cells.

THE SEA AND THE SHORE

Early in evolutionary history, the principal photosynthetic organisms were microscopic cells floating below the surface of the sunlit waters. Energy abounded, as did carbon, hydrogen, and oxygen, but as the cellular colonies multiplied, they quickly depleted the mineral resources of the open ocean. (It is this shortage of essential minerals that is the limiting factor in any modern plans to harvest the seas.) As a consequence, life began to develop more abundantly toward the shores, where the waters were rich in nitrates and minerals carried down from the mountains by rivers and streams and scraped from the coasts by the ceaseless waves.

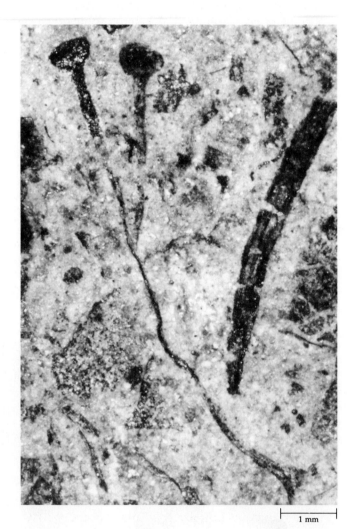

1–4

A fossil of Cooksonia, *one of the earliest and simplest plants known, from the late Silurian period (414–408 million years ago).* Cooksonia *consisted of little more than a branched axis with terminal sporangia, or spore-producing structures.*

The short, straight line at the bottom of this micrograph and those that follow provides a reference for size; a millimeter, abbreviated mm, is 1/10 centimeter. The same system is used to indicate distances on a road map.

The rocky coast presented a much more complicated environment than the open sea, and, in response to these evolutionary pressures, living organisms became increasingly complex in structure and more diversified. Not less than 650 million years ago, organisms evolved in which many cells were linked together to form an integrated, multicellular body (Figure 1–4). In these primitive organisms we see the early stages in the evolution of plants, fungi, and animals. Fossils of multicellular organisms are much easier to detect than those of simpler ones; thus, the history of life on earth is much better documented from the time of their first appearance.

On the turbulent shore, multicellular photosynthetic organisms were better able to maintain their position against the action of the waves, and, in meeting the challenge of the rocky coast, new forms developed. Typically, these new forms evolved relatively strong cell walls for support and specialized structures to anchor their bodies to the rocky surfaces (Figure 1–5). As these organisms increased in size, they were confronted with the problem of how to supply food to the dimly lit, more deeply submerged portions of their bodies, where photosynthesis was not taking place. As a result of these new pressures, specialized food-conducting tissues evolved that extended down the center of their bodies and connected the upper photosynthesizing parts with the lower, nonphotosynthesizing structures.

THE TRANSITION TO LAND

The body of the familiar plant can best be understood in terms of its long history and, in particular, in terms of the evolutionary pressures involved in the transition to land. The requirements of a photosynthetic organism are relatively simple: light, water, carbon dioxide for photosynthesis, oxygen for respiration, and a few minerals. On land, light is abundant, as are oxygen and carbon dioxide—both of which circulate more freely in air than in water—and the soil is generally rich in minerals. The critical factor, then, for the transition to land is water.

Land animals, generally speaking, are mobile and able to seek out water just as they seek out food. Fungi, though immobile, remain largely below the surface of the soil or within whatever damp organic material they feed upon. Plants utilize an alternative evolutionary strategy. **Roots** anchor the plant in the ground and collect the water required for maintenance of the plant body and for photosynthesis, while the **stems** provide support for the principal photosynthetic organs, the **leaves**. A continuous stream of water moves upward through the roots and stems, and then out through the leaves. The outermost layer of cells (the epidermis) of all of the aboveground portions of the plant that are ul-

1–5
Multicellular photosynthetic organisms anchored themselves to rocky shores early in the course of their evolution. These kelp, seen at low tide on the rocks at Botanical Beach on Vancouver Island, British Columbia, are brown algae (Phaeophyta), a group in which multicellularity evolved independently of other groups of organisms.

timately involved in photosynthesis are covered with a waxy **cuticle** that retards water loss. However, the cuticle also tends to prevent the necessary exchange of gases between the plant and the surrounding air. The solution to this dilemma is found in pairs of specialized epidermal cells (the guard cells), which regulate small openings between them called **stomata** (singular: **stoma**). The stomata open and close in response to environmental and physiological signals, thus helping the plant maintain a balance between its water losses

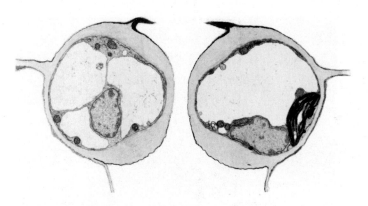

1–6

Transverse section of a mature stoma of a sugar beet (Beta vulgaris) leaf. The stomata, or small openings, in the aerial parts of the plant are regulated by the guard cells, which border the stomata. Each stoma is bordered by two guard cells.

and its oxygen and carbon dioxide requirements (Figure 1–6).

In younger plants and in those with a life span of one year (annuals), the stem is also a photosynthetic organ. In longer-lived plants (perennials), the stem may become thickened and woody and covered with cork, which, like the cuticle-covered epidermis, retards water loss. In both annuals and perennials, the stem serves to conduct, via the **vascular system** (conducting system), a variety of substances between the photosynthetic and nonphotosynthetic parts of the plant body. The vascular system has two major components: the **xylem,** through which water passes upward through the plant body; and the **phloem,** through which food manufactured in the leaves and other photosynthetic parts of the plant is transported throughout the plant body. It is this efficient conducting system that has given the main group of plants—the vascular plants—their name.

Plants, unlike animals, continue to grow throughout their lives. All plant growth originates in **meristems**—localized regions of perpetually embryonic tissues. Meristems located at the tips of all roots and shoots—**apical meristems**—are involved with the extension of the plant body. Thus the roots are continuously reaching new sources of water and minerals, and the photosynthetic regions are continuously extending toward the light. The type of growth that originates from apical meristems is known as **primary growth. Secondary growth,** which results in a thickening of stems, branches, and roots, originates from two lateral meristems—the **vascular cambium** and the **cork cambium.**

Plants also had to overcome new challenges in order to reproduce on land. These challenges were met at first mainly by drought-resistant spores, and later by the evolution of complex, multicellular structures in which the gametes, or reproductive cells, were held. In the **seed plants,** which include almost all familiar plants except the ferns, mosses, and liverworts, the young plant, or embryo, is enclosed within specialized coverings provided by the parent. There the embryo is protected from both drought and predators.

Thus, in summary, the vascular plant (Figure 1–7) is characterized by a root system that serves to anchor the plant in the ground and to collect water and inorganic ions from the soil; a stem or trunk that raises the photosynthetic parts of the plant body toward its energy source, the sun; and leaves, highly specialized photosynthetic organs. Roots, stems, and leaves are interconnected by a complicated and efficient vascular system for the transport of food and water. The reproductive cells of plants are enclosed within multicellular structures, and in seed plants the embryos are protected by resistant coverings. All of these characteristics are adaptations to a photosynthetic existence on land.

The Evolution of Communities

The invasion of the land by plants changed the face of the continents. Looking down from an airplane on one of the earth's great expanses of desert or on one of its mountain ranges, one can begin to imagine what the world looked like before the appearance of plants. Yet even in these regions, the traveler who goes by land will find an astonishing variety of plants punctuating the expanses of rock and sand. In those parts of the world where the climate is more temperate and the rains are more frequent, communities of plants dominate the land and determine its character. In fact, to a large extent, they *are* the land. Rainforest, savanna, woods, desert, tundra—each of these words brings to mind a portrait of a landscape (see Figure 1–8 on the next two pages). The main features of each landscape are its plants, enclosing us in a dark green cathedral in our imaginary rainforest, carpeting the ground beneath our feet with wild flowers in a meadow, moving in great golden waves as far as the eye can see across our imaginary prairie. Only when we have sketched these **biomes**—natural communities of wide extent, characterized by distinctive, climatically controlled groups of plants and animals—in terms of trees and shrubs and grasses can we fill in other features, such as deer, antelope, rabbits, or wolves.

How do vast plant communities, such as those seen on a continental scale, come into being? To some extent we can trace the evolution of the different kinds of plants and animals that populate these communities.

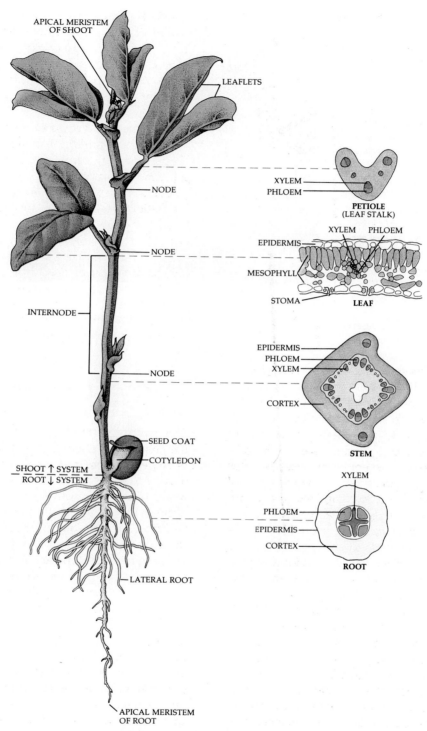

1–7

Diagram of a young broad bean (Vicia faba) plant, showing the principal organs and tissues of the modern vascular plant body. The organs—root, stem, and leaf—are composed of tissues, which are groups of cells with distinct structures and functions. Collectively, the roots make up the root system, and the stems and leaves together make up the shoot system of the plant. In the great majority of vascular plants, the shoot system is above the soil surface and the root system is below. Unlike roots, stems are divided into nodes and internodes. The node is the part of the stem at which one or more leaves are attached, and the internode is the part of the stem between two successive nodes. (In the broad bean, the first few foliage leaves are divided into two leaflets each.) Buds (embryonic shoots) commonly arise in the axils (upper angle between leaf and stem) of the leaves. Lateral, or branch, roots arise from the inner tissues of the roots. The vascular tissues—xylem and phloem— occur together and form a continuous vascular system throughout the plant body. (It is the mesophyll tissue of leaves that is specialized for photosynthesis.)

1–8
Some examples of the enormous diversity of biological communities on earth. (a) The temperate deciduous forest, which covers most of the eastern United States and southeastern Canada, is dominated by trees that lose their leaves in the cold winters. Here are paper birches and a red maple photographed in early autumn in the Adirondack Mountains of New York State. (b) Underlain with permafrost, Arctic tundra is a treeless biome characterized by a short growing season. Shown here is a tundra pond photographed in late summer in Denali National Park in Alaska. (c) Mediterranean climates are rare on a world scale. Cool, moist winters, during which the plants grow, are followed by hot, dry summers, during which the plants become dormant. Shown here is a pine-oak chapparal photographed on Mount Diablo in California. (d) Deserts typically receive less than 25 centimeters of rain per year. Here in the Sonoran desert in Arizona, the dominant plant is the giant saguaro cactus. Adapted for life in a dry climate, saguaro cactuses have shallow, wide-spreading roots and thick stems for storing water. (e) In Africa, as photographed here in Tanzania, savannas are inhabited by huge herds of grazing mammals, such as these Thomson's gazelles. The trees are acacias, one of which is seen here with a browsing giraffe. (f) The tropical rainforest, shown here in Costa Rica, is the richest, most diverse biome on earth, with at least two-thirds of all species of organisms on earth found there.

(a)

(d)

Even with accumulating knowledge, however, we have only begun to glimpse the far more complex pattern of development, through time, of the whole system of organisms that make up these various communities.

Such communities, along with the nonliving environment of which they are a part, are known as ecological systems, or **ecosystems.** Ecosystems are discussed later in greater detail. For now, it is sufficient to regard an ecosystem as a kind of corporate entity made up of transient individuals. Some of these individuals, the larger trees, live as long as several thousand years; others, the microorganisms, live only a few hours or even minutes. Yet the ecosystem as a whole tends to be

remarkably stable (although not static); once in balance, it does not change for centuries. Our grandchildren will someday walk along a woodland path once followed by our great-grandparents, and where they saw a pine tree, a mulberry bush, a meadow mouse, wild blueberries, or a towhee, these children, if this woodland still exists, will see roughly the same kinds of plants and animals in the same numbers.

An ecosystem functions as an integrated unit, although many of the organisms in the system compete for resources. Virtually every living thing, even the smallest bacterial cell or fungal spore, provides a food source for some other living organism. In this way, the

(b)

(c)

(e)

(f)

energy captured by green plants is transferred in a highly regulated way through a number of different types of organisms before it is dissipated. Moreover, interactions among the organisms themselves, and between the organisms and the nonliving environment, produce an orderly cycling of elements such as nitrogen and phosphorus. Energy must be added to the ecosystem constantly, but the elements are cycled through the organisms, returned to the soil, decomposed by soil bacteria and fungi, and recycled. These transfers of energy and the cycling of elements involve complicated sequences of events, and in these sequences each group of organisms has a highly specific role. As a conse-

quence, it is impossible to change a single element in an ecosystem without the risk of destroying the balance upon which the stability of the ecosystem depends.

At the base of productivity in virtually all ecosystems are the plants, algae, and photosynthetic bacteria that alone have the ability to capture energy from the sun and to manufacture organic molecules that they and all other kinds of organisms may require for life. There are roughly half a million kinds of organisms capable of photosynthesis, and at least eight or ten times that many heterotrophic organisms, which are completely dependent upon the photosynthesizers. For animals, including human beings, there are many kinds of molecules—including essential amino acids, vitamins, and minerals—that can be obtained only through plants or other photosynthetic organisms. Furthermore, the oxygen that is released into the atmosphere by photosynthetic organisms makes it possible for life to exist on the land and in the surface layers of the ocean. Oxygen is necessary for the energy-producing metabolic activities of the great majority of organisms, including photosynthetic organisms.

The Appearance of Human Beings

Human beings are relative newcomers to the world of living organisms (Figure 1–9). If the entire history of the earth were measured on a 24-hour time scale starting at midnight, cells would appear in the warm seas before dawn. The first multicellular organisms would not be present until well after sundown, and the earliest appearance of humans (about 2 million years ago) would be about half a minute before the day's end. Yet humans more than any other animal—and almost as much as the plants that invaded the land—have changed the surface of the planet, shaping the biosphere according to their own needs, ambitions, or follies.

The development of agriculture, starting at least 11,000 years ago, in time made it possible to maintain large populations of people in towns and cities. This development (reviewed in detail in Chapter 31) allowed specialization and the diversification of human culture. One of the characteristics of this culture is that it examines itself and the nature of other living things, including plants. Eventually, the science of biology developed within the human communities that had been made possible through the domestication of plants. That part of biology that deals with plants and, by tradition, with bacteria, fungi, and algae is called **botany,** or **plant biology.**

PLANT BIOLOGY

The study of plants has been pursued for thousands of years, but like all branches of science, it has become diverse and specialized only during the past three centuries. Until little more than a century ago, botany was a branch of medicine, pursued chiefly by physicians as an avocation or principal specialization. Today, however, it is an important scientific discipline that has many subdivisions: **plant physiology,** which is the study of how plants function, that is, how they capture and transform energy and how they grow and develop; **plant morphology,** the study of the form of plants;

1–9

The clockface of biological time. Life appears relatively early in the earth's history, sometime before 6:00 A.M. on a 24-hour scale. The first multicellular organisms do not appear until the twilight of that 24-hour day, and the genus Homo *is a very late arrival—less than a minute before midnight.*

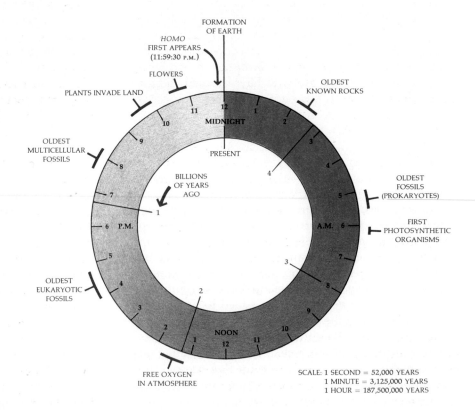

plant anatomy, the study of their internal structure; **plant classification,** also called taxonomy or systematics, the naming and classifying of plants; **cytology,** the study of cell structure, function, and life histories; **genetics,** the study of heredity and variation; **molecular biology,** the study of the structure and function of biological macromolecules; **ecology,** the study of the relationships between organisms and their environment; and **paleobotany,** the study of the biology and evolution of fossil plants.

In times past, all organisms were considered to be either plants or animals, and microscopic organisms were assigned to either the plant or animal kingdom as they were discovered. Fungi were considered to be plants, presumably because most of them do not move around and because their growth form is vaguely more like that of the familiar green plants than it is like that of the animals. The differences between bacteria and other living organisms are much more fundamental than those that separate other groups of organisms, as is explained in the following chapter and in greater detail in Chapter 11. (Viruses, also discussed in Chapter 11, are not really living organisms but are pieces of the genetic apparatus of other organisms; viruses replicate themselves by directing the metabolic processes of cells.)

Among the eukaryotes, there are many kinds of unicellular organisms that differ greatly among themselves. The heterotrophic eukaryotes, which have traditionally been called protozoa, were at one time grouped with the animals, whereas the autotrophic eukaryotes, which have traditionally been called algae, were grouped with the plants. The interrelations between the heterotrophic and autotrophic groups, however, are obvious to those who have studied them in detail; these two groups do not really represent different evolutionary lines. Instead, all basically unicellular (eukaryotic) organisms are now grouped in the kingdom Protista, which is considered in detail in Chapters 13 and 14. Among the algae, several evolutionary lines have become multicellular—the green, brown, and red algae—but algae are, for the most part, unicellular. On the other hand, all plants are multicellular; however, plants are not directly related to the multicellular algae, except for the green algae from which they evolved during their invasion of the land. Because of the unique features of plants—multicellular, terrestrial, nonmobile, photosynthetic organisms—they are recognized as a distinct kingdom, but with a narrower definition than has been traditional in the past.

Whereas plants obtain their food by photosynthesis (a few exceptional plants have lost this ability but are clearly derived from the rest), animals ingest their food, and fungi (discussed in Chapter 12) absorb it after secreting enzymes and digesting it externally. These three multicellular lines are each treated as distinct kingdoms of eukaryotes, namely, kingdom Plantae, kingdom Animalia, and kingdom Fungi. All other eukaryotes—a very diverse group—are assigned to the kingdom Protista. Because of clear molecular evidence—discussed further in Chapters 10 and 11—we are placing bacteria in two kingdoms, Eubacteria and Archaebacteria. This book therefore recognizes six kingdoms of organisms.

Included in this book are all organisms that have traditionally been studied by botanists: plants as well as bacteria, viruses, fungi, and autotrophic protists (algae). Only the animals have traditionally been the province of zoologists. Although we do not regard algae, fungi, bacteria, or viruses as plants, and shall not refer to them as plants in this book, they are included here because of tradition, and because they are normally considered as part of the botanical portion of the curriculum, just as botany itself used to be considered a part of medicine. *Virology, bacteriology, phycology* (the study of algae), and *mycology* (the study of fungi) are well-established fields in their own right, but they still fall loosely under the umbrella of botany.

BOTANY AND THE FUTURE

In this chapter, we have ranged from the beginnings of life on this planet to the evolution of plants and ecosystems to the development of agriculture and civilization. These broad topics are of interest to many people other than botanists, or plant biologists. The urgent efforts of botanists and agricultural scientists will be needed to feed the rapidly growing world population of human beings, as is discussed in Chapter 31. Extant plants, algae, and bacteria offer the best hope of providing a renewable source of energy for human activities, just as extinct plants, algae, and bacteria have been responsible for the massive accumulations of gas, oil, and coal on which our modern industrial civilization depends. In an even more fundamental sense, the role of plants, along with that of algae and photosynthetic bacteria, commands our attention. As the producers of energy-containing compounds in the global ecosystem, these organisms are the route by which all other living things, including ourselves, obtain energy, oxygen, and many other materials necessary for our continued existence. As a student of botany, you will be in a better position to assess the important ecological and environmental issues of the day and, by understanding, help to build a healthier world.

As we approach the beginning of the twenty-first century, it has become apparent that human beings, who will number well over 6 billion by the year 2000, are managing the earth with an intensity that would have been unimaginable a few decades ago. Every hour, manufactured chemicals fall on virtually every square centimeter of the planet's surface. The protective stratospheric ozone layer formed 450 million years ago has been seriously depleted by the use of chlorofluorocarbons (CFCs), and damaging ultraviolet rays penetrating the depleted layer have increased the incidence of skin cancer in people all over the world. Moreover, it has been estimated that, by the middle of the next century, the average temperature will have increased between 1° and 6°C, as the greenhouse effect —the trapping of heat radiating from the earth's surface out into space—is intensified through the increased amounts of carbon dioxide, nitrogen oxides, CFCs, and methane in the atmosphere, resulting from human activities. Most serious of all, a large portion of the total number of species of plants, animals, fungi, and microorganisms is disappearing during our lifetime —the victims of human exploitation of the earth. All of these trends are alarming, and they demand our utmost attention.

Marvelous new possibilities have been developed during the past few years for the better utilization of plants by people—these developments are discussed throughout this book. It is now possible to stimulate the growth of plants, to deter their pests, to control weeds in crops, and to form hybrids between plants with much more precision and far wider possibilities than ever before. The potential of these new discoveries is growing with every passing year, as additional discoveries are made and new applications are developed. For example, the methods of genetic engineering (discussed further in Chapter 25) now make it possible, in principle, to transfer either natural or synthetic genes from one plant or animal to another in order to produce certain characteristics. These methods, first applied in 1973, have already been the basis of billions of dollars in investments and increased hope for the future. Discoveries still to be made undoubtedly will far exceed our wildest dreams and go far beyond the facts that are available to us now.

As we turn to Chapter 2, in which our attention narrows to a cell so small it cannot be seen by the unaided eye, it is well to keep these broader concerns in mind. A basic knowledge of plant biology is useful in its own right and is essential in many fields of endeavor, but it is also increasingly relevant to some of society's most crucial problems and to the difficult decisions that will face us in choosing among the proposals for diminishing them. Our own future, the future of the world, and the future of all kinds of plants—as individual species and as components of the life-support systems into which we all have evolved—depend upon our knowledge. Thus, this book is dedicated not only to the botanists of the future, whether teachers or researchers, but also to the informed citizens, scientists and laypeople alike, in whose hands such decisions lie.

Summary

Only a few kinds of organisms—plants, algae, and some bacteria—have the capacity to capture energy from the sun and to fix it in organic molecules by the process of photosynthesis. Virtually all life on earth depends, directly or indirectly, on the products of this process.

The planet earth is about 4.5 billion years old. Initially, its atmosphere is thought to have consisted primarily of nitrogen gas, with relatively large amounts of water vapor and carbon dioxide. The four elements present in these gases—carbon, hydrogen, nitrogen, and oxygen—make up about 98 percent of the material found in all living organisms. In the turbulent early atmosphere, the gases were spontaneously recombined into new, larger molecules. Oxygen (which now makes up nearly 21 percent of the earth's atmosphere) was essentially absent until photosynthetic organisms began to produce it in quantity. As a result, ultraviolet rays (now largely blocked from the surface of the earth by ozone) bombarded the surface of the earth and assisted in the synthesis of molecules.

Heterotrophs, organisms that feed on organic molecules or other organisms, were the first to evolve. The oldest known fossils date back 3.5 billion years. Autotrophic organisms, those that could produce their own food by photosynthesis, evolved no less than 3.4 billion years ago. Until about 1.5 billion years ago, bacteria—the prokaryotes—were the only organisms that existed. Eukaryotes, with larger, much more complex cells, evolved at that time. Multicellular eukaryotes began to evolve at least 650 million years ago, and they began to invade the land about 450 million years ago.

Plants are basically a terrestrial group, one of several evolutionary lines that consist entirely or mostly of multicellular organisms. The other major groups are fungi, which absorb their food, and animals, which ingest it. The unicellular eukaryotic organisms are placed in the kingdom Protista, along with three smaller groups of multicellular eukaryotes—the red, brown, and green algae. All six of these multicellular lines evolved independently from unicellular protists.

Plants, which evolved from green algae, have achieved a number of specialized characteristics that suit them for life on land. These characteristics are best developed among the members of the dominant group known as the vascular plants. Among them are a waxy cuticle, penetrated by specialized openings known as stomata through which gas exchange takes place, and an efficient conducting system. This system consists of xylem, in which water and absorbed nutrients pass from the roots to the stems and leaves, and phloem, which transports the products of photosynthesis to all parts of the plant. Plants increase in length by primary growth and in girth by secondary growth, both of which take place in zones of rapid cell division known as meristems.

As plants have evolved, they have come to constitute biomes, great terrestrial assemblages of plants and animals. The interacting systems made up of biomes and their nonliving environments are called ecosystems. Human beings, which appeared about 2 million years ago, developed agriculture at least 11,000 years ago and have subsequently become the dominant ecological force on earth. Humans have used their knowledge of plants to foster their own development and will continue to do so with increasingly greater importance in the future.

Suggestions for Further Reading

Dickerson, R.E.: "Chemical Evolution and the Origin of Life," *Scientific American* 239(3):70–86, September 1978.

An excellent brief account of the chemical changes thought to have occurred during the evolution of life on earth.

Groves, David I., John S.R. Dunlop, and Roger Buick: "An Early Habitat of Life," *Scientific American* 245(4):64–73, October 1981.

A vivid description of a mud flat in Australia that may have sheltered microorganisms some 3.5 billion years ago.

Margulis, Lynn: *Early Life*, Science Books International, Inc., Boston, 1982.

An outstanding essay concerning the evolution of life on earth and its history from about 3.5 billion to about 700 million years ago; semipopular in style and easily read.

Milne, David, David Raup, John Billingham, Karl Niklas, and Kevin Padian (eds.): *The Evolution of Complex and Higher Organisms*, National Aeronautics and Space Administration Special Publication SP-478, U.S. Government Printing Office, Washington, D.C., 1985.

Traces the history of life on earth and the course of its evolution and is informative about directions for future research.

Schopf, J. William: *Earth's Earliest Biosphere: Its Origin and Evolution*, Princeton University Press, Princeton, N.J., 1983.*

A splendid collection of papers documenting the earliest history of earth prior to the evolution of eukaryotic cells.

* Available in paperback.

The Plant Cell: Structure and Metabolism

With the onset of cool autumn evenings, the leaflets of this sumac leaf have begun to show their fall coloring, due largely to red anthocyanin pigments being synthesized by the leaf for the first time. As these water-soluble pigments are synthesized, the chlorophyll pigments essential for photosynthesis break down, and the leaf gradually undergoes a transformation.

CHAPTER 2

Introduction to the Eukaryotic Cell

Cells are the structural and functional units of life. The smallest organisms are composed of single cells. The largest are made up of billions of cells, each of which still lives a partly independent existence. The realization that all organisms are composed of cells was one of the most important conceptual advances in the history of biology because it provided a unifying theme for the study of all living things. When studied at the cellular level, even the most diverse organisms are remarkably similar to one another, both in their physical organization and in their biochemical properties. The **cell theory** was formulated early in the nineteenth century, well before the presentation of Darwin's *theory of evolution,* but these two great unifying concepts are, in fact, closely related. In the similarities among cells, we catch a glimpse of a long evolutionary history that links modern organisms, including plants and ourselves, with the first cellular units that took shape on earth billions of years ago.

There are many different kinds of cells. Within our own bodies there are more than 100 distinct cell types. In a teaspoon of pond water, one can find several different types of one-celled organisms, and in an entire pond, there are probably several hundred clearly different kinds. Plants are composed of cells that are superficially quite different from those of our own bodies, and insects have many kinds of cells not found in plants or in vertebrates. Thus, one remarkable fact about cells is their diversity.

A second, even more remarkable fact about cells is their similarity. Every living cell is a self-contained and at least partially self-sufficient unit, and each is bounded by an outer membrane—the plasma membrane, or plasmalemma (often simply called the cell membrane)—that controls the passage of materials into and out of the cell and so makes it possible for the cell to differ biochemically and structurally from its surroundings. Enclosed within this membrane is the cytoplasm, which, in most cells, includes a variety of discrete bodies and various dissolved or suspended molecules. In addition, every cell contains DNA (deoxyribonucleic acid), which encodes the genetic information (see Chapter 8), and this code, with rare exceptions, is the same for every organism, whether bacterium, oak tree, or human.

Prokaryotes and Eukaryotes

Two fundamentally distinct groups of organisms can be recognized: **prokaryotes** and **eukaryotes.** These terms are derived from the Greek word *karyon*, meaning "kernel" (nucleus). The name *prokaryote* means "before a nucleus," and *eukaryote*, "with a good, or true, nucleus."

The prokaryotes are represented by the bacteria, including the cyanobacteria (see Chapter 11). Prokaryotic cells differ most notably from eukaryotic cells in that they lack nuclei; that is, their DNA is not surrounded by a membranous envelope (Figure 2–1). Nor is their DNA associated with proteins to form chromosomes. Bacteria have no specialized membrane-bound structures to perform specific functions, although they have some internal membranes that are involved with specific functions.

In eukaryotic cells the DNA, combined with protein, is located in chromosomes, which are in a nucleus bounded by two membranes called the nuclear envelope. Eukaryotic cells are further divided into distinct compartments that perform different functions (Figure 2–2).

Compartmentation in eukaryotic cells is accomplished by means of membranes, which, when seen

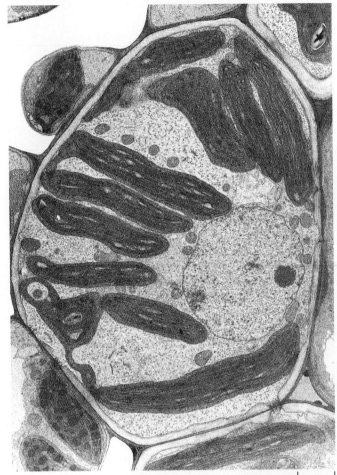

0.25 μm

with the aid of an electron microscope, look remarkably similar in various organisms. When suitably preserved and stained, these membranes may have a three-layered appearance, consisting of two dark layers separated by a lighter layer (Figure 2–3). The term "unit membrane" has been used to designate membranes that have such an appearance.

2–1

Escherichia coli, a bacterium that is a common, usually harmless inhabitant of the human digestive tract, is an example of a prokaryotic cell. Each of these rod-shaped organisms has a cell wall, a plasma membrane, and cytoplasm. The two cells in the center are the products of a recent cell division and have not yet separated completely. The DNA is found in the less granular area—the nucleoid— in the center of each cell. The densely granular appearance of the cytoplasm is largely due to the presence of numerous ribosomes (entities involved with protein synthesis).

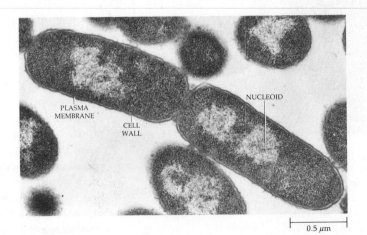

PLASMA MEMBRANE

CELL WALL

NUCLEOID

0.5 μm

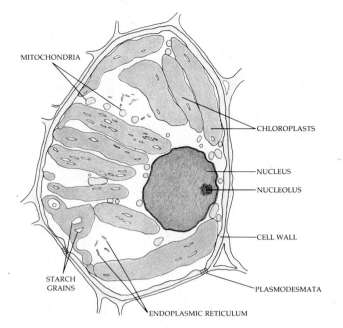

2-2

At left is a cross section of a plant cell as seen with an electron microscope. In the accompanying labeled drawing, some of the cell's components are identified.

This particular cell is a bundle-sheath cell from the leaf of a corn plant (Zea mays). The nucleus—site of the cell's genetic information—can be seen at the center right. A single, well-defined nucleolus is visible within the nucleus. Several chloroplasts (organelles involved in photosynthesis) and mitochondria (organelles involved in the conversion of food to usable energy) can also be seen in this cell. Some of the chloroplasts contain oval starch grains that are light in appearance.

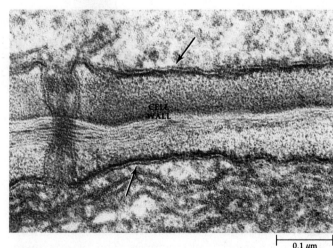

0.1 μm

2-3

Under high magnification, cellular membranes often have a three-layered (dark-light-dark) appearance, as seen in the plasma membranes (see arrows) on either side of the common wall between two cells from a corn (Zea mays) leaf.

The Plant Cell

The plant cell typically consists of a more or less rigid **cell wall** and a **protoplast**. The term *protoplast* is derived from the word **protoplasm**, which is used to refer to the contents of cells. A protoplast is the unit of protoplasm inside the cell wall.

A protoplast consists of **cytoplasm** and a **nucleus** (see Table 2–1). The cytoplasm includes distinct, membrane-bound entities (organelles such as plastids and mitochondria), systems of membranes (the endoplasmic reticulum and dictyosomes), and nonmembranous entities (such as ribosomes, actin filaments, and microtubules). The rest of the cytoplasm—the "cellular soup," or cytoplasmic matrix, in which the nucleus, various entities, and membrane systems are suspended

Table 2–1 *An Inventory of Plant Cell Components*

Cell wall	Middle lamella Primary wall Secondary wall Plasmodesmata	
Protoplast	Nucleus	Nuclear envelope Nucleoplasm Chromatin Nucleolus
	Cytoplasm	Ground substance Organelles bounded by two membranes: Plastids Mitochondria Organelles bounded by one membrane: Microbodies Vacuoles (tonoplast) Endomembrane system Endoplasmic reticulum Dictyosomes Vesicles Plasma membrane Cytoskeleton Microtubules Actin filaments Ribosomes
Ergastic substances*	Crystals Anthocyanins Starch grains Tannins Fats, oils, and waxes Protein bodies	

* May occur in cell wall, ground substance, and/or organelles.

Origin of the Cell Theory

In the seventeenth century, the English physicist Robert Hooke, using a microscope of his own construction, noted that cork and other plant tissues are made up of small cavities separated by walls. He called these cavities "cells," meaning "little rooms." The word did not take on its present meaning, however, until 150 years later.

In 1838, Matthias Schleiden, a German botanist, came to the conclusion that all plant tissues are organized in the form of cells. In the following year, zoologist Theodor Schwann extended Schleiden's observation to animal tissues and proposed a cellular basis for all life. The cell theory is of tremendous and central importance to biology because it emphasizes the basic similarity of all living systems and so brings an underlying unity to widely varied studies involving many different kinds of organisms.

In 1858, the cell theory took on broader significance when the great pathologist Rudolf Virchow generalized that cells can arise only from preexisting cells: "Where a cell exists, there must have been a preexisting cell, just as the animal arises only from an animal and the plant only from a plant. . . . Throughout the whole series of living forms, whether entire animal or plant organisms or their component parts, there rules an eternal law of continuous development."

In the broad perspective of evolution, Virchow's concept takes on an even larger significance. There is an unbroken continuity between modern cells—and the organisms that they compose—and the primitive cells that first appeared on earth at least 3.5 billion years ago.

 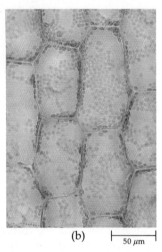

(a)   (b)   50 μm

2–4

An Elodea *cell in two planes of focus. (a) Focus on upper surface of the cell. (b) Focus on middle of the cell. The numerous disk-shaped structures are chloroplasts located in the cytoplasm along the wall. In surface view (a), the disk-shaped chloroplasts appear circular in outline. In (b) the chloroplasts have their broad surfaces facing the surface of the wall and appear to be elongated. Notice in (b) the absence of chloroplasts in the center of the cell—that is, within the vacuole.*

—is called the **ground substance**. (The term "cytosol" is often used to refer to the ground substance.) The cytoplasm is delimited from the cell wall by a single membrane, the **plasma membrane**. In contrast to most animal cells, plant cells develop one or more liquid-filled cavities, or **vacuoles**, within their cytoplasm. The vacuole is bounded by a single membrane called the **tonoplast**.

In a living plant cell, such as that of the pondweed *Elodea* (Figure 2–4), the cytoplasm is frequently in motion; the organelles, as well as various substances suspended in the ground substance, can be observed being swept along in an orderly fashion in the moving currents. This movement is known as **cytoplasmic streaming**, or **cyclosis**, and it continues as long as the cell is alive. Cytoplasmic streaming undoubtedly facilitates the exchange of materials within the cell and between the cell and its environment; however, it is not known whether this is a primary function of cytoplasmic streaming.

Plasma Membrane

Among the various membranes of the cell, the plasma membrane typically has the clearest dark-light-dark appearance in electron micrographs. The plasma membrane has several important functions: (1) it mediates the transport of substances into and out of the protoplast (see Chapter 4); (2) it coordinates the synthesis and assembly of cell wall microfibrils (cellulose); and (3) it translates hormonal and environmental signals involved in the control of cell growth and differentiation.

POLYSOME

ENDOPLASMIC RETICULUM

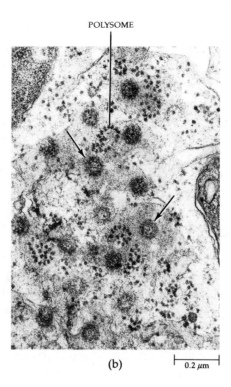

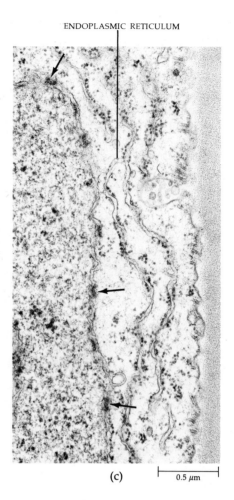

(a) 0.5 μm

(b) 0.2 μm

(c) 0.5 μm

2–5

Nuclear pores as revealed (a) in a freeze-etch preparation of the surface of the nuclear envelope of an onion (Allium cepa) root-tip cell, and (b) and (c) in electron micrographs of nuclei in parenchyma cells of the seedless vascular plant Selaginella kraussiana. *In (b) the pores are shown in surface view, whereas in (c) they are shown in sectional view (see arrows). Note the polysomes (coils of granules) on the surface of the nuclear envelope in (b) and the rough endoplasmic reticulum paralleling the nuclear envelope in (c). (See pages 27 and 28.)*

Nucleus

The nucleus is often the most prominent structure within the protoplast of eukaryotic cells. The nucleus performs two important functions: (1) it controls the ongoing activities of the cell by determining which protein molecules are produced by the cell and when they are produced, as we shall see in Chapter 8; and (2) it stores the genetic information, passing it on to the daughter cells in the course of cell division.

In eukaryotic cells, the nucleus is bounded by a pair of membranes called the **nuclear envelope**. The nuclear envelope, as seen with the aid of an electron microscope, contains a large number of circular pores that are 30 to 100 nanometers in diameter (see Figure 2–5); the inner and outer membranes are joined around each pore to form the margin of its opening. The pores are not merely holes in the envelope; each has a complicated structure. They provide a direct passageway through the nuclear envelope for the exchange of materials between the nucleus and the cytoplasm. In various places the outer membrane of the envelope may be continuous with the endoplasmic reticulum, a complex system of membranes that plays a central role in cell biosynthesis. The nuclear envelope may be considered a specialized, locally differentiated portion of the endoplasmic reticulum.

If the cell is treated by special staining techniques, thin threads and grains of **chromatin** can be distinguished from the **nucleoplasm**, or nuclear ground substance. Chromatin is made up of DNA combined with large amounts of proteins called histones. During the process of nuclear division, the chromatin becomes progressively more condensed until it takes the form of **chromosomes**. Chromosomes (chromatin) of nondividing nuclei apparently are attached at one or more sites to the nuclear envelope. The hereditary information is carried in the molecules of DNA contained within the chromosomes. The content of DNA per cell is much higher in eukaryotic organisms than in bacteria.

Different organisms vary in the number of chromosomes present in their somatic (body) cells. *Haplo-*

pappus gracilis, a desert annual, has 4 chromosomes per cell; cabbage, 20; bread wheat, 42; and one species of the fern *Ophioglossum,* about 1250. The reproductive cells, or gametes, however, have only half the number of chromosomes that is characteristic of the somatic cells of the organism. The number of chromosomes in the gametes is referred to as the **haploid** ("single") number, and that in the somatic cells is called the **diploid** ("double") number.

Often the only structures within a nucleus that are discernible with the light microscope are the spherical structures known as **nucleoli** (singular: **nucleolus**), one or more of which are present in each nondividing nucleus (Figure 2–2). In many diploid organisms, the nucleus contains one nucleolus to each haploid set of chromosomes. The nucleoli may fuse and then appear as one large structure. Each nucleolus contains high concentrations of RNA (ribonucleic acid) and proteins, along with large loops of DNA emanating from several chromosomes. The loops of DNA, known as nucleolar organizer regions, are the sites of formation of ribosomal RNA. The nucleolus, in fact, is the site of assembly of ribosomes.

Plastids

Together with vacuoles and cell walls, **plastids** are characteristic components of plant cells. Each plastid is bounded by an envelope consisting of two membranes. Internally, the plastid is differentiated into a system of membranes and a more or less homogeneous ground substance, the **stroma**. Mature plastids are commonly classified on the basis of the kinds of pigments they contain.

Chloroplasts, the sites of photosynthesis (see Chapter 7), contain chlorophylls and carotenoid pigments. The chlorophyll pigments are responsible for the green color of these plastids. In plants, chloroplasts are usually disk-shaped and measure between 4 and 6 micrometers in diameter. A single mesophyll ("middle of the leaf") cell may contain 40 to 50 chloroplasts; a square millimeter of leaf contains some 500,000. The chloroplasts are usually found with their broad surfaces parallel to the cell wall (see Figures 2–4b and 2–6).

The internal structure of the chloroplast is complex (Figure 2–7). The stroma is traversed by an elaborate system of membranes in the form of flattened sacs

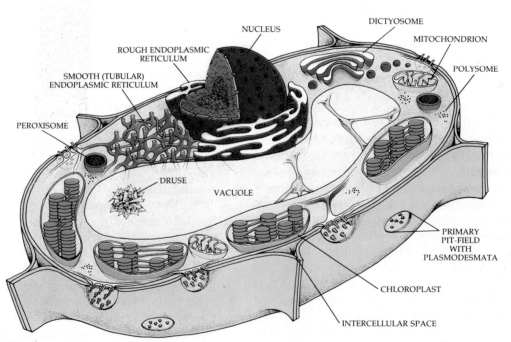

2–6

A three-dimensional diagram of a chloroplast-containing plant cell. Typically, the disk-shaped chloroplasts are located in the cytoplasm along the wall, with their broad surfaces facing the *surface of the wall. The center of this cell is occupied by a vacuole (bounded by the tonoplast), and it contains a druse (a crystal aggregate). In this cell, the nucleus lies in the cytoplasm along* *the wall, although in some cells it might appear suspended by strands of cytoplasm in the center of the vacuole, surrounded by a thin layer of cytoplasm.*

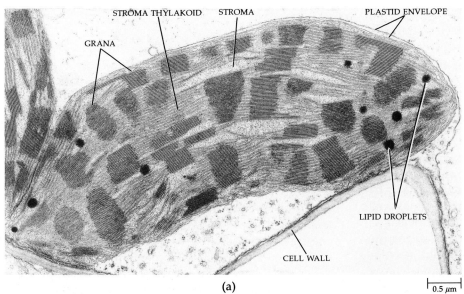

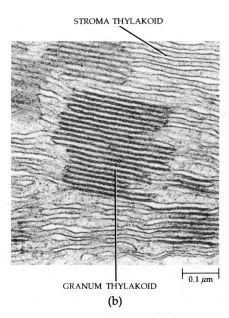

2–7

(a) *A chloroplast of a corn* (Zea mays) *leaf.* (b) *Detail showing grana composed* *of stacks of disklike thylakoids. The* *thylakoids of the various grana are inter-* *connected by other thylakoids, commonly* *called stroma thylakoids.*

called **thylakoids**. The thylakoids are believed to constitute a single, interconnected system. Chloroplasts are generally characterized by the presence of **grana** (singular: **granum**)—stacks of disklike thylakoids that resemble a stack of coins. The thylakoids of the various grana are connected with each other by thylakoids (the stroma thylakoids, or intergrana thylakoids) that traverse the stroma. Chlorophylls and carotenoid pigments are found embedded in the thylakoid membranes.

The chloroplasts of green algae and plants often contain starch grains and small lipid (oil) droplets. The starch grains are temporary storage products and accumulate only when the alga or plant is actively photosynthesizing (Figure 2–2). They may be lacking in the chloroplasts of plants that have been kept in the dark for as little as 24 hours but often reappear after the plant has been in the light for only 3 or 4 hours.

Chloroplasts are semiautonomous organelles that resemble bacteria in several ways. For example, like bacteria, chloroplasts contain one or more *nucleoids*—clear, grana-free regions containing DNA. The DNA of the plastid, like that of a bacterium, exists in circular form; moreover, it is not associated with histones. In addition, the ribosomes of both bacteria and chloroplasts are about two-thirds as large as the cytoplasmic ribosomes of eukaryotes. Also, protein synthesis on bacterial and chloroplast ribosomes is inhibited by antibiotics, such as chloramphenicol and streptomycin, that have no effect on eukaryotic ribosomes.

Table 2-2 *Measurements Used in Microscopy*

1 centimeter (cm)	= 1/100 meter = 0.4 inch*
1 millimeter (mm)	= 1/1,000 meter = 1/10 cm
1 micrometer (μm)**	= 1/1,000,000 meter = 1/10,000 cm
1 nanometer (nm)	= 1/1,000,000,000 meter = 1/10,000,000 cm
1 angstrom (Å)†	= 1/10,000,000,000 meter = 1/100,000,000 cm

or

1 meter = 10^2 cm = 10^3 mm = 10^6 μm = 10^9 nm = 10^{10} Å

* A metric-to-English conversion table is found in Appendix B.
** Micrometers were formerly known as microns (μ), and nanometers as millimicrons (mμ).
† The angstrom is not an accepted measurement in the International System of Units; in the past, however, it was widely used in microscopy.

The ability to form chloroplasts and the pigments associated with them involves the contribution of both nuclear and plastid DNA. Overall control clearly resides in the nucleus, however. Although some organelle proteins are encoded, or specified, by plastid DNA and synthesized in the plastid, most of the proteins in the plastids are encoded by nuclear DNA, synthesized in the ground substance, and then imported into the plastid.

Chloroplasts are the ultimate source of most of our food supplies and of our fuel. Chloroplasts are not only sites of photosynthesis; they are also involved in amino acid synthesis and fatty acid synthesis, and provide space for the temporary storage of starch.

Chromoplasts (Gk. *chroma*, "color") are also pigmented plastids (Figure 2–8). Of variable shape, chromoplasts lack chlorophyll but synthesize and retain carotenoid pigments, which are often responsible for the yellow, orange, or red colors of many flowers, old leaves, some fruits, and some roots. Chromoplasts may develop from previously existing green chloroplasts by a transformation in which the chlorophyll and internal membrane structure of the chloroplast disappear and masses of carotenoids accumulate, as occurs during the ripening of many fruits. The precise functions of chromoplasts are not well understood, although at times they act as attractants to insects and other animals with which they have coevolved, playing an essential role in the cross-pollination of flowering plants and the dispersal of fruits and seeds (see Chapter 19).

Leucoplasts (Figure 2–9) are nonpigmented plastids. Some synthesize starch (*amyloplasts*; Figure 2–10), but others are thought to be capable of forming a variety of substances, including oils and proteins. Upon exposure to light, leucoplasts may develop into chloroplasts.

Proplastids are small, colorless or pale green, undifferentiated plastids that occur in meristematic (dividing) cells of roots and shoots. They are the precursors of the other, more highly differentiated plastids such as chloroplasts, chromoplasts, or amyloplasts (Figure 2–11). If the development of a proplastid into a more highly differentiated form is arrested by the absence of light, it may form one or more **prolamellar bodies**, which are semicrystalline bodies composed of tubular membranes (Figure 2–12). Plastids containing prola-

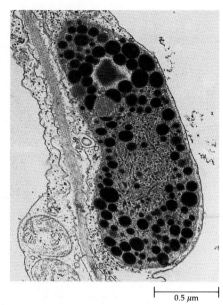

2–8
Chromoplast from a Forsythia *petal. The chromoplast contains numerous electron-dense lipid droplets in which the pigment responsible for the color of the petal is stored.*

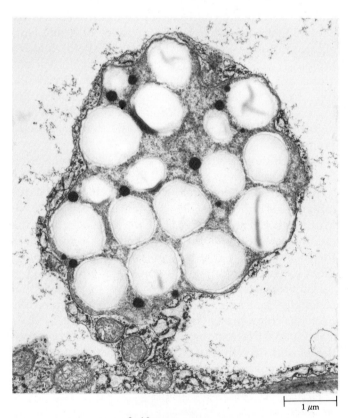

2–10
Amyloplast from the embryo sac of soybean (Glycine max). The round, clear bodies are starch grains. The smaller, dense bodies are lipid droplets.

2–9
Leucoplasts clustered around the nucleus in an epidermal cell (center) of a Zebrina leaf. The purple color is due to anthocyanins in vacuoles of the epidermal cells.

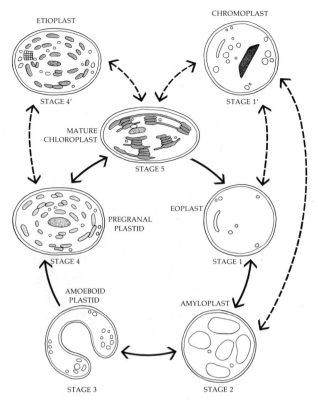

2–11

Plastid developmental cycle as proposed by J. M. Whatley. Stages 1–3: Eoplast (eo-, "early"), amyloplast, and amoeboid can be grouped together as nongreen proplastid stages. Stage 4: The pregranal plastid may be green or nongreen. In the absence of light, the pregranal stage may be represented by an etioplast (stage 4'). As indicated, chromoplasts (stage 1') may be formed from several types of plastid.

mellar bodies are called **etioplasts**. Etioplasts form in leaf cells of plants grown in the dark. During subsequent development of etioplasts into chloroplasts in the light, the membranes of the prolamellar bodies develop into thylakoids. In nature the proplastids in the embryos of seeds first develop into etioplasts; then, upon exposure to light, the etioplasts develop into chloroplasts. The various kinds of plastids are remarkable for the relative ease with which they can change from one type to another.

Plastids reproduce by fission, the process of dividing into equal halves, which is characteristic of bacteria. In meristematic cells, the division of proplastids roughly keeps pace with cell division. In mature cells, however, the greater proportion of the final plastid population may be derived from the division of mature plastids.

Mitochondria

Like plastids, **mitochondria** are bounded by two membranes (Figures 2–13 and 2–14). The inner membrane is extensively folded into pleats or shelf-like projections known as **cristae** (singular: crista), which greatly increase the surface area available to enzymes and the reactions associated with them. Mitochondria are generally smaller than plastids, measuring about half a micrometer in diameter, and there is great variability in length and shape. As a rule, mitochondria are barely visible with a light microscope, but they can readily be seen with the aid of an electron microscope.

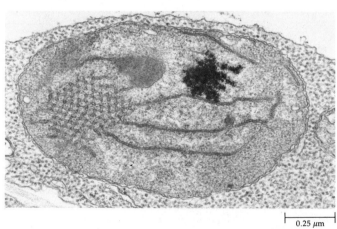

0.25 μm

2–12

Etioplast, with a semicrystalline prolamellar body composed of tubular membranes, in a leaf cell of tobacco (Nicotiana tabacum) *grown in the dark. When exposed to light, the membranes of the prolamellar body develop into thylakoids.*

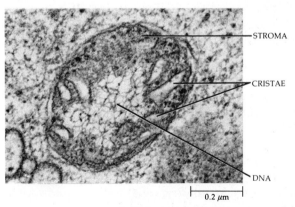

STROMA

CRISTAE

DNA

0.2 μm

2–13

Mitochondrion in a leaf cell of spinach (Spinacia oleracea), *in a plane of section revealing some strands of DNA in the nucleoid. The envelope of the mitochondrion consists of two separate membranes. The inner membrane folds inward to form cristae, which are embedded in a dense stroma. The small granules in the stroma are ribosomes.*

Mitochondria are the sites of respiration, a process involving the release of energy from organic molecules and its conversion to molecules of ATP (adenosine triphosphate), the chief immediate source of chemical energy for all eukaryotic cells. (These processes are discussed further in Chapter 6.) Most plant cells contain hundreds or thousands of mitochondria, the number of mitochondria per cell being related to the cell's demand for ATP.

With the aid of time-lapse photography, mitochondria can be seen to be in constant motion, turning and twisting and moving from one part of the cell to another; they also fuse and divide by fission. Mitochondria tend to congregate where energy is required. In cells in which the plasma membrane is very active in transporting materials into or out of the cell, they often can be found arrayed along the membrane surface. In motile, single-celled algae, mitochondria are typically clustered at the bases of the flagella, presumably providing energy for flagellar movement.

Mitochondria, like plastids, are semiautonomous organelles, containing the components necessary for the synthesis of some of their own proteins. The inner membrane of the mitochondrion encloses a liquid matrix that contains proteins, RNA, DNA, ribosomes similar to those of bacteria, and various solutes (dissolved substances). The DNA occurs as circular molecules in one or more clear areas, the nucleoids (Figure 2–13).

EVOLUTIONARY ORIGIN OF MITOCHONDRIA AND CHLOROPLASTS

On the basis of the close similarity between bacteria and the mitochondria and chloroplasts of eukaryotic cells, it seems probable that mitochondria and chloroplasts originated as bacteria that found shelter within larger heterotrophic cells. These larger cells were the forerunners of the eukaryotes. The smaller cells, which contained (and still contain) all the mechanisms necessary to trap and/or convert energy from their surroundings, donated these useful capacities to the larger cells. Cells with these respiratory and/or photosynthetic "assistants" had a clear advantage over their contemporaries and undoubtedly soon multiplied at their contemporaries' expense. All but a very few modern eukaryotes contain mitochondria, and all autotrophic eukaryotes also contain chloroplasts; both seem to have been acquired by independent symbiotic events. (Symbiosis is the close association between two or more dissimilar organisms that may be, but is not necessarily, beneficial to each.) The smaller cells—now established as symbiotic organelles within the larger cells—obtained protection from environmental extremes. As a consequence, eukaryotes were able to invade the land and acidic waters, where the prokaryotic cyanobacteria are absent but the eukaryotic green algae abound.

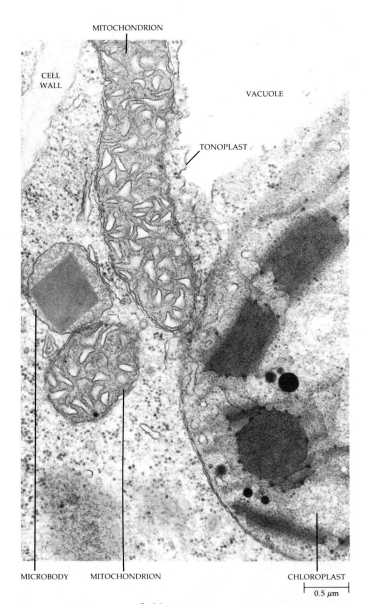

2–14
Organelles in a leaf cell of tobacco (Nicotiana tabacum). A microbody with a large crystalline inclusion and a single bounding membrane may be contrasted with the two mitochondria and the plastid, each of which is bounded by two membranes. Due to the plane of section, the double nature of the plastid envelope is apparent only in the lower portion of the micrograph. A single membrane, the tonoplast, separates the vacuole (the clear area at the upper right) from the rest of the cytoplasm.

Microbodies

Unlike plastids and mitochondria, which are bounded by two membranes, **microbodies** are spherical organelles bounded by a single membrane. They range in diameter from 0.5 to 1.5 micrometers. Microbodies have a granular interior and sometimes contain a crystalline body composed of protein (Figure 2–14). Inasmuch as they lack DNA and ribosomes, microbodies must import all of their proteins. Microbodies are generally associated with one or two segments of endoplasmic reticulum, the three-dimensional system of internal membranes within cells.

Some microbodies, called *peroxisomes*, play an important role in glycolic acid metabolism, which is associated with photorespiration (see page 112); in green leaves, peroxisomes are closely associated spatially with mitochondria and chloroplasts (Figure 2–14). Other microbodies, called *glyoxysomes,* contain enzymes necessary for the conversion of fats to carbohydrates during germination in many seeds.

Vacuoles

Together with the presence of plastids and a cell wall, the vacuole is one of the three characteristics that distinguish plant cells from animal cells. Vacuoles are membrane-bound regions within the cell that are filled with a liquid called **cell sap**. These organelles are surrounded by the tonoplast, or vacuolar membrane (Figures 2–6 and 2–14). Different kinds of vacuoles with distinct functions may be found in a single mature cell.

The immature plant cell typically contains numerous small vacuoles, which increase in size and fuse into a single vacuole as the cell enlarges. In the mature cell, as much as 90 percent of the volume may be taken up by the vacuole, with the cytoplasm consisting of a thin peripheral layer closely pressed against the cell wall (Figure 2–4). By filling such a large proportion of the cell with "inexpensive" vacuolar contents, plants not only save "expensive" (in terms of energy) nitrogen-consuming cytoplasm but also acquire a large surface between the thin layer of cytoplasm and the cell's external environment. Most of the increase in the size of the cell results from enlargement of the vacuoles. A direct consequence of this strategy is the development of turgor pressure and the maintenance of tissue rigidity, one of the principal roles of the vacuole and tonoplast (see Chapter 4).

The principal component of the cell sap is water, with other components varying according to the type of plant and its physiological state. Vacuoles typically contain salts and sugars, and some contain dissolved proteins. Sometimes the concentration of a particular

2–15
Druses, aggregates of crystals composed of calcium oxalate, in epidermal cells of a cotyledon of redbud (Cercis canadensis), *as seen with a scanning electron microscope.*

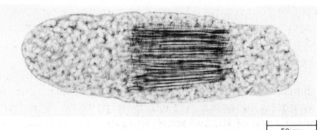

2–16
A bundle of raphides—needlelike crystals of calcium oxalate—in a vacuole of a leaf cell of the snake plant (Sansevieria). *The tonoplast is not discernible in this photomicrograph. The granular substance surrounding the crystals is cytoplasm.*

material in the vacuole is sufficiently great for it to form crystals. Calcium oxalate crystals, which can assume several different forms, are especially common (Figures 2–15 and 2–16). Cell sap is usually slightly acidic. Some cell sap, such as that of the vacuoles in citrus fruits, is very acidic—hence the tart, sour taste of the fruit.

Vacuoles are important storage compartments for various metabolites (products of metabolism), such as the reserve proteins in seeds and the malic acid in CAM plants (see page 116). They also remove toxic secondary products of metabolism, such as nicotine, an alkaloid, from the rest of the cytoplasm (see also Figure 19–49).

The vacuole is often a site of pigment deposition. The blue, violet, purple, dark red, and scarlet colors of plant cells are usually caused by a group of pigments known as the anthocyanins (see Chapter 19). Unlike most other plant pigments, the anthocyanins are readily soluble in water, and they are dissolved in the cell sap. They are responsible for the red and blue colors of many vegetables (radishes, turnips, cabbages), fruits (grapes, plums, cherries), and a host of flowers (cornflowers, geraniums, delphiniums, roses, and peonies).

Viewing the Microscopic World

Most cells can be seen only with the aid of a microscope. The units of measurement generally used for describing cells are micrometers and nanometers (see Table 2–2). Unaided, the human eye has a resolving power of about 1/10 millimeter, or 100 micrometers. This means that if one looks at two lines that are less than 100 micrometers apart, they will appear as a single line. Similarly, two dots less than 100 micrometers apart look like a single blurry dot. In order to resolve structures closer together than this, microscopes must be used.

The best light microscopes have a resolving power of 0.2 micrometer, or about 200 nanometers, and so improve on the naked eye about 500 times. It is theoretically impossible to build a light microscope that will do better than this. The limiting factor is the wavelength of light, which ranges from about 0.4 micrometer for violet light to 0.7 micrometer for red light.

Notice that resolving power and magnification are different; using the best light microscope, if one takes a photograph of two lines that are less than 200 nanometers apart, one can enlarge the image indefinitely, but the two lines will still blur together. By using more powerful lenses, one can increase the magnification, but this will not improve the resolution.

The Transmission Electron Microscope

With the transmission electron microscope (TEM), the limit of resolution imposed by the light microscope can be reduced because electrons have a much shorter wavelength than visible light. Theoretically, a TEM operating at an accelerating voltage of 100,000 volts should have a resolution of about 0.002 nanometer. All things considered, however,

with biological specimens, the resolution is more like 2 nanometers, which is still 100 times better than the resolution of the light microscope.

The TEM has its disadvantages. In order to produce a beam of electrons, air must be pumped out of the microscope to create a vacuum. Because the specimen to be examined must be put into the vacuum, in the path of the electron beam, it is not possible to examine it in a living condition. The specimen must therefore be preserved by fixation and embedded in plastic so that it can be sectioned by a special cutting instrument (ultramicrotome). Because electrons have very limited penetrating power, the block of plastic containing the specimen must be cut exeedingly thin (50 to 100 nanometers thick). Such thin sections are effectively two-dimensional slices of tissue, which fail to convey a three-dimensional perspective. Also, it is often difficult to determine what changes may have been produced in the material in the course of its preparation. The micrograph in Figure 2–2 is one of the many transmission electron micrographs found in this book.

The Scanning Electron Microscope

Although the resolution of the scanning electron microscope (SEM) is only about 10 nanometers, this instrument is a valuable tool for biologists. Unlike the TEM, which uses electrons that have passed through the specimen to form an image, the SEM uses electrons that are scattered or emitted from the surface of the specimen. The electron beam is focused into a fine probe, which is rapidly passed back and forth over the specimen; complete scanning from top to bottom usually takes only a few seconds. Variations in the surface of the specimen affect the pattern in which the

Sometimes the pigments are so brilliant that they mask the chlorophyll in the leaves, as in the ornamental red maple. Anthocyanins are also responsible for the brilliant red colors of some leaves in autumn. These pigments form in response to cold, sunny weather, when the leaves stop producing chlorophyll. As the chlorophyll that is present disintegrates, the newly formed anthocyanins are unmasked. In leaves that do not form anthocyanin pigments, the breakdown of chlorophyll in autumn may unmask the more stable yellow-to-orange carotenoid pigments already present in the chloroplasts. The most spectacular autumnal coloration develops in years when cool, clear weather prevails in the fall.

Vacuoles are also involved with the breakdown of

macromolecules and the recycling of their components within the cell. Entire cell organelles, such as mitochondria and plastids, may be deposited and degraded in vacuoles. Because of this digestive activity, vacuoles are comparable in function with the organelles known as *lysosomes* that occur in animal cells.

Vacuoles have long been considered to arise from the endoplasmic reticulum. When it was realized that vacuoles are analogous to the lysosomes of animal cells, attempts were initiated to determine whether the vacuoles of at least some plant cells had origins similar to those of animal-cell lysosomes. In animal cells, lysosome formation is associated with a specialized region of cytoplasm called "GERL," for Golgi associated–endoplasmic reticulum–lysosomal complex. GERLs have

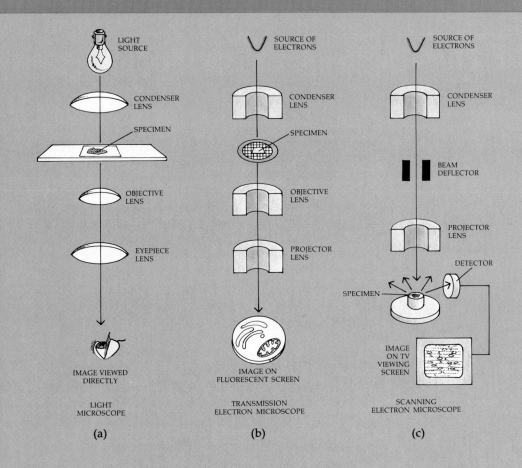

A comparison of (a) the light microscope, (b) the transmission electron microscope, and (c) the scanning electron microscope. The light microscope is shown upside-down, to emphasize its similarities with electron microscopes. The focusing lenses in the light microscope are glass or quartz; those in the electron microscopes are magnetic coils. In both the light microscope and the transmission electron microscope, the illuminating beam passes through the specimen; in the scanning electron microscope, it is deflected from the surface.

electrons are scattered from it; holes and fissures appear dark, and knobs and ridges appear light, imparting a three-dimensional image. Only surface features can be examined with the SEM; consequently, it is used to examine whole cells, tissues, and the surfaces of various structures. Figure 2–15 was prepared with a scanning electron microscope, as were others in this textbook.

now been found in several kinds of plant cells, including root-tip cells of radish and a euphorbia and cells of mung bean cotyledons. The Golgi apparatus is discussed on page 28.

Ribosomes

Ribosomes are small particles, only about 17 to 23 nanometers in diameter, consisting of about equal amounts of protein and RNA (see page 143). They are the sites at which amino acids are linked together to form proteins and are abundant in the cytoplasm of metabolically active cells. Ribosomes either occur freely in the cytoplasm or are attached to the endoplasmic re-

ticulum; free and membrane-bound ribosomes commonly are found in the same cell. Ribosomes are also found in nuclei. As mentioned previously, plastids and mitochondria contain ribosomes similar to those in prokaryotes.

Ribosomes actively involved in protein synthesis occur in clusters or aggregates called polyribosomes, or **polysomes**. Cells that are synthesizing proteins in large quantities often contain extensive systems of polysome-bearing endoplasmic reticulum. Polysomes are often attached to the outer surface of the nuclear envelope (Figure 2–5b). Membrane-bound ribosomes and free ribosomes are both structurally and functionally identical, differing from one another only in the proteins they are making at any given time.

Endoplasmic Reticulum

The **endoplasmic reticulum** is a complex three-dimensional membrane system of indefinite extent. In sectional view, the endoplasmic reticulum appears as two parallel membranes with a narrow, transparent space, or lumen, between them (Figure 2–5c). The form and abundance of the endoplasmic reticulum within a cell vary greatly depending on the cell type, its metabolic activity, and its stage of development. For example, in cells that secrete or store proteins, the endoplasmic reticulum has the form of flattened sacs, or **cisternae** (singular: cisterna), with numerous ribosomes attached to its outer surface. Endoplasmic reticulum bearing ribosomes is called *rough endoplasmic reticulum* (Figure 2–5c). The ribosomes are assembled into polysomes, and the rough endoplasmic reticulum is a major site of protein synthesis. In contrast, cells that secrete lipids have an extensive system of tubules that lack ribosomes. Endoplasmic reticulum lacking ribosomes is called *smooth endoplasmic reticulum*. Smooth endoplasmic reticulum is usually tubular in form. Both rough and smooth forms of endoplasmic reticulum may occur within the same cell; usually, numerous connections exist between them (Figure 2–6).

The endoplasmic reticulum appears to function as a communications system within the cell. In some electron micrographs, it can be seen to be continuous with the outer membrane of the nuclear envelope. In fact, these two structures together seem to form a single membrane system. When the nuclear envelope breaks down at cell division, its fragments are similar to portions of endoplasmic reticulum. It is easy to visualize the endoplasmic reticulum as a system for channeling materials—such as proteins and lipids—to different parts of the cell. In addition, the endoplasmic reticulum of adjacent cells is interconnected by way of cytoplasmic threads, called plasmodesmata, which traverse their common walls (see page 37).

The endoplasmic reticulum is the principal site of membrane synthesis within the cell. It appears to give rise to vacuolar and microbody membranes, as well as to the cisternae of dictyosomes in at least some plant cells.

Golgi Apparatus

The term **Golgi apparatus** is used to refer collectively to all of the **dictyosomes**, or *Golgi bodies*, of a cell. Dictyosomes are groups of flat, disk-shaped sacs, or cisternae, which are often branched into a complex series of tubules at their margins (Figure 2–17). Those in higher plant cells usually have four to eight cisternae stacked together.

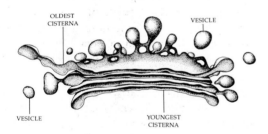

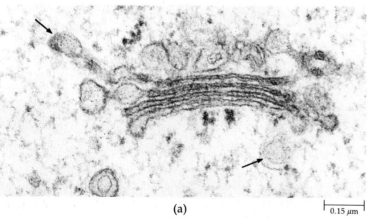

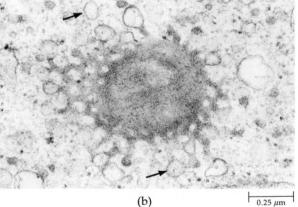

(a)   ⊢ 0.15 μm ⊣   (b)   ⊢ 0.25 μm ⊣

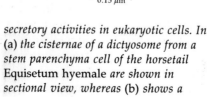

2–17

The dictyosome consists of a group of flat, membranous sacs associated with vesicles that apparently bud off from the sacs. It serves as a "packaging center" for the cell and is concerned with *secretory activities in eukaryotic cells. In (a) the cisternae of a dictyosome from a stem parenchyma cell of the horsetail* Equisetum hyemale *are shown in sectional view, whereas (b) shows a* *single cisterna in a surface view. The arrows in both micrographs indicate secretory vesicles along the margins of the cisternae.*

2–18

A diagrammatic representation of the endomembrane concept. This drawing illustrates the idea that new membranes are synthesized at the rough endoplasmic reticulum. Small transition vesicles pinch off the smooth-surfaced portion of the endoplasmic reticulum and carry membranes and enclosed materials to the forming face of the dictyosome. Secretory vesicles derived from cisternae at the maturing face of the dictyosome migrate to the plasma membrane and fuse with it, contributing new membrane to the plasma membrane and discharging their contents in the region of the wall.

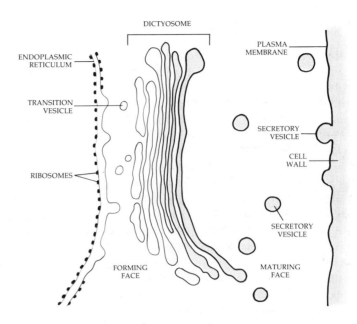

Customarily the two opposite poles or surfaces of a stack are referred to as forming (cis) and maturing (trans) faces. The membranes of the forming cisternae are structurally similar to membranes of the endoplasmic reticulum, while those toward the opposite pole, or maturing face, become progressively more like plasma membrane (Figure 2–18).

Dictyosomes are involved in secretion, and most higher plant dictyosomes are involved in cell wall synthesis. Noncellulosic cell wall polysaccharides synthesized by the dictyosomes collect in vesicles that are pinched off cisternae on the maturing faces. These secretory vesicles migrate to and fuse with the plasma membrane (Figure 2–18). The vesicles then discharge their contents to the cell exterior, and the polysaccharides become part of the cell wall. The products secreted by the dictyosomes are not always synthesized entirely by the dictyosomes. Some products are formed in part by other cellular components, such as the endoplasmic reticulum, and then are transferred to the dictyosomes, where they are modified prior to secretion. A good example is glycoproteins—carbohydrate-protein compounds—which are important cell wall constituents. The protein portion is synthesized on polysomes of rough endoplasmic reticulum, and the carbohydrate portion is synthesized in the dictyosome, where the glycoprotein is then assembled.

THE ENDOMEMBRANE CONCEPT

Membranes are dynamic, mobile structures that continuously change their shape and surface area. An exam-

ple of the mobility of cellular membranes is provided by the concept of the **endomembrane system**. According to this concept, the internal cytoplasmic membranes (with the exception of mitochondrial and plastid membranes) constitute a continuum, with the endoplasmic reticulum being the initial source of membranes. New cisternae are added to the dictyosomes from the endoplasmic reticulum via transition vesicles, while secretory vesicles derived from the dictyosomes eventually contribute to growth of the plasma membrane, which is also part of the system (Figure 2–18). Hence, the endoplasmic reticulum and dictyosomes are a functional unit, in which the dictyosomes act as the main vehicles for the transformation of endoplasmic-reticulum-like membranes into plasma-membrane-like membranes.

It is important to note that even in tissues undergoing very little growth or cell division, a turnover of membrane components is continuously taking place. Although the membranes persist, their constituent molecules are constantly being replaced.

Cytoskeleton

Virtually all eukaryotic cells possess a **cytoskeleton**, a complex network of protein filaments that extends throughout the ground substance and is intimately involved in many processes, including cell division, growth and differentiation, and the movement of organelles from one location to another within the cell. The cytoskeleton of plant cells consists of two types of protein filament: microtubules and actin filaments.

MICROTUBULES

Microtubules are long, thin cylindrical structures about 24 nanometers in diameter and of varying lengths. Each microtubule is built up of subunits of the protein called tubulin. The subunits are arranged in a helix to form 13 rows, or "protofilaments," around a hollow core (see Figure 3–19). All of the protofilaments are aligned in parallel with the same polarity; hence, the microtubule is a polar structure, for which we can designate plus and minus ends. Microtubules are dynamic structures that undergo regular sequences of breakdown and re-formation at specific points in the cell cycle (see Figure 2–40). Their assembly takes place at specific "nucleating sites" or at microtubule organizing centers, which, in plant cells, have a poorly defined amorphous structure.

Microtubules have many functions. In enlarging and differentiating cells, microtubules just inside the plasma membrane are involved in the orderly growth of the cell wall, especially through their control of the alignment of the cellulose microfibrils that are added by the cytoplasm to the cell wall (Figure 2–19). The direction of expansion of the cell is governed, in turn, by the alignment of cellulose microfibrils in the wall. Microtubules serve to direct dictyosome vesicles toward the developing wall. They make up the spindle fibers, which form in dividing cells, and play a role in chromosome movement and in the formation of the cell plate (the initial partition between dividing cells). In addition, microtubules are important components of flagella and cilia and are involved in the movement of these structures.

ACTIN FILAMENTS

Actin filaments, also called microfilaments and filamentous actin (F actin), like microtubules, are polar structures with distinct plus and minus ends. They are composed of a contractile protein called *actin*, which is similar to the actin of muscle, and occur as long filaments 5 to 7 nanometers wide. In addition to single filaments, bundles of actin filaments have been found in many plant cells (Figure 2–20).

Actin filaments have been implicated in a variety of roles in plant cells, including involvement in cell wall deposition, tip growth of pollen tubes, nuclear migration following cell division, and cytoplasmic streaming. They are spatially associated with microtubules and, like microtubules, are organized into distinct arrays that change in distribution during the cell cycle.

Ergastic Substances

Ergastic substances are passive products of the protoplast; some are storage products, others are waste products. These substances may appear and disappear at different times in the life of a cell. We have already discussed several examples of ergastic substances: starch grains, crystals, and anthocyanin pigments. Other examples are resins, gums, tannins, lipid droplets, and protein bodies (Figure 2–21). Ergastic substances are found in the cell wall, in the ground substance, and in organelles, including vacuoles.

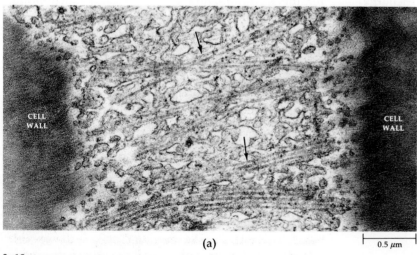

(a)

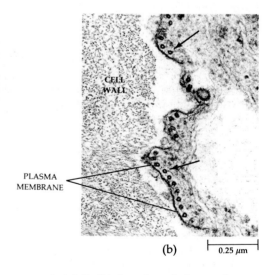

(b)

2–19

Microtubules (indicated by arrows) seen in (a) longitudinal and (b) transverse views in leaf cells of Botrychium virginianum, *a fern. A glancing section more or less parallel to and just inside the wall and plasma membrane is shown in (a). In (b) the microtubules can be seen to be separated from the wall by the plasma membrane.*

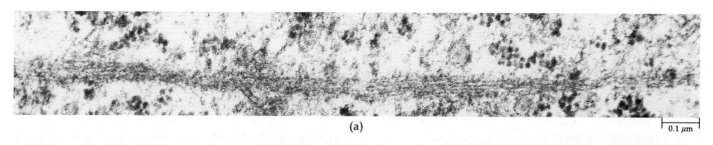

(a)

`0.1 μm`

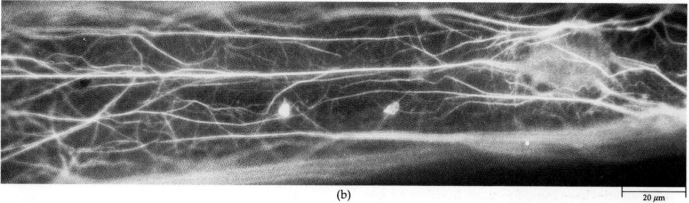

(b)

`20 μm`

2-20

(a) *A bundle of actin filaments as revealed in an electron micrograph of a* *leaf cell of corn (Zea mays). (b) Several bundles of actin filaments as revealed in a* *fluorescence micrograph of a stem hair of tomato (Lycopersicon esculentum).*

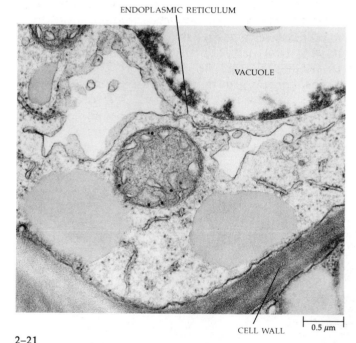

ENDOPLASMIC RETICULUM

VACUOLE

CELL WALL   `0.5 μm`

2-21

Cytoplasmic components in a parenchyma cell from the thickened stem, or corm, of the quillwort Isoetes muricata. Two lipid droplets are visible to the lower right and lower left of the mitochondrion (near the center of this micrograph). The dense material lining the vacuole is tannin. Both the lipid droplets and the tannin are ergastic substances. Above, and on either side of the mitochondrion, a cisterna of the endoplasmic reticulum is distended. Some vacuoles are thought to arise from the endoplasmic reticulum in this way.

Storage products, such as starch and sugars, are *primary metabolites*—substances that play a basic role in cell metabolism. Other ergastic substances, such as tannins, are *secondary metabolites* (secondary products) that play no apparent function in the plant's primary metabolism (see also Biochemical Coevolution, pages 435 through 438).

Flagella and Cilia

Flagella and **cilia** (singular: **flagellum** and **cilium**) are hairlike structures that extend from the surface of many different types of eukaryotic cells. They are relatively thin and constant in diameter (about 0.2 micrometer), but they vary in length from about 2 to 150 micrometers. By convention, those that are longer or are present alone or in small numbers are usually referred to as flagella, whereas those that are shorter or occur in greater numbers are called cilia. There is no definite distinction between these two types of structures, however, and we will use the term *flagellum* to refer to both.

In some algae and other protists, flagella are locomotor structures, propelling the organisms through the water. In plants, flagella are found only in the sex cells (gametes) and then only in plants that have motile sperm, including the mosses, liverworts, ferns, cycads, and maidenhair tree *(Ginkgo biloba)*. Some flagella, called tinsel flagella, bear one or two rows of minute,

TINSEL
FLAGELLUM

WHIPLASH
FLAGELLUM

| 2.5 μm |

2–22
*The two types of flagella are repre-
sented in this single cell of the colonial
organism* Synura petersenii, *a golden
alga. This organism has a long tinsel
flagellum (left) and a somewhat shorter
whiplash flagellum (right).*

lateral appendages, whereas others, called whiplash
flagella, lack such processes (Figure 2–22).

One of the most intriguing of the discoveries made
possible by the electron microscope is the internal
structure of flagella. Each flagellum has a precise internal organization (Figure 2–23). An outer ring of nine
pairs of microtubules surrounds two additional microtubules in the center of the flagellum proper. This basic
9-plus-2 pattern of organization is found in all flagella
of eukaryotic organisms.

The movement of a flagellum originates from within
the structure itself; flagella are capable of movement
after they have been detached from cells. The movement is produced by a sliding microtubule mechanism,
in which the outer pairs of microtubules move past one
another without contracting. As the pairs slide past one
another, their movement causes localized bending of
the flagellum. Sliding of the pairs of microtubules results from cycles of attachment and detachment of
enzyme-containing "arms" that join neighboring pairs
in the outer ring (Figure 2–23).

Flagella grow out of cylinder-shaped structures in
the cytoplasm known as *basal bodies,* which then form
the basal portion of the flagellum. Basal bodies have an
internal structure that resembles the structure of the
flagellum proper, except that the outer tubules in the
basal body occur in triplets rather than in pairs and the
two central tubules are absent.

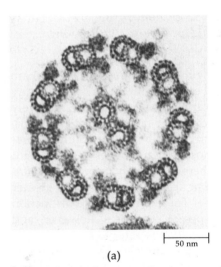

| 50 nm |

(a)

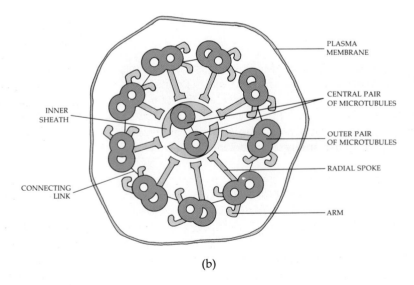

PLASMA
MEMBRANE

CENTRAL PAIR
OF MICROTUBULES

INNER
SHEATH

OUTER PAIR
OF MICROTUBULES

RADIAL SPOKE

CONNECTING
LINK

ARM

(b)

2–23
(a) *Transverse section through the
flagellum of the unicellular green alga*
Chlamydomonas, *showing the typical
9-plus-2 arrangement of microtubules*
*found in the flagella of almost all
eukaryotic cells.* (b) *Diagram of a flagellum in transverse section. The "arms,"*
*the radial spokes, and the connecting
links are formed from different types
of protein.*

The Role of Actin and Myosin in Cytoplasmic Streaming in Giant Algal Cells

Much of our understanding of the mechanism of cytoplasmic streaming comes from work on the giant internodal cells of green algae such as *Chara* and *Nitella* (see Chapter 14). In these cells, which measure 2 to 5 centimeters long, the chloroplast-containing layer of cytoplasm bordering the wall is stationary. Spirally arranged bundles of actin filaments extend for several centimeters along the cells, forming distinct "tracks" that are firmly attached to the stationary chloroplasts. The moving layer of cytoplasm occurs between the bundles of actin filaments and the tonoplast, and contains the nucleus, mitochondria, and other cytoplasmic components.

The generating force necessary for cytoplasmic streaming comes from an interaction between actin and **myosin,** a protein molecule with an ATPase-containing head that is activated by actin. ATPase is an enzyme that breaks down—hydrolyzes—ATP to release energy (see Chapter 5). Apparently, the organelles in the streaming cytoplasm are indirectly attached to the actin filaments by myosin molecules, which use the energy released by ATP hydrolysis to "walk" along the actin filaments, pulling the organelles with them. Streaming always occurs from the minus to the plus ends of the actin filaments, all of which are similarly oriented within a bundle.

(a) A track followed by the streaming cytoplasm in the algal cell. (b) A longitudinal section through part of the cell, showing the arrangement of stationary and streaming layers of cytoplasm. The proportions have been distorted in both diagrams for clarity.

(a)

CELL WALL   PLASMA MEMBRANE

ACTIN BUNDLES

TONOPLAST

VACUOLE

MITOCHONDRIA
IN STREAMING
CYTOPLASM

CHLOROPLASTS
IN STATIONARY
(b) CYTOPLASM

Cell Wall

The presence of a cell wall, above all other characteristics, distinguishes plant cells from animal cells. Its presence is the basis of many of the characteristics of plants as organisms. The cell wall limits the size of the protoplast and prevents rupture of the plasma membrane when the protoplast enlarges following the uptake of water by the cell.

Once regarded as merely an outer, inactive product of the protoplast, the cell wall is now recognized as having specific and essential functions. Cell walls contain a variety of enzymes and play important roles in the absorption, transport, and secretion of substances in plants. They may also serve as sites of lysosomal, or digestive, activity. In addition, the cell wall may play an active role in defense against bacterial and fungal pathogens by receiving and processing information from the surface of the pathogen and transmitting this information to the plasma membrane of the host cell.

Through gene-activated processes, the host cell may become resistant to attack through the production of *phytoalexins* (see page 643)—antibiotics that are toxic to the pathogens—or through the synthesis and deposition of substances such as lignin, which act as barriers to invasion. Certain cell wall polysaccharides, dubbed "oligosaccharins," may even function as hormones, regulating plant growth and development.

COMPONENTS OF THE CELL WALL

The most characteristic component of plant cell walls is **cellulose**, which largely determines their architecture. Cellulose is made up of repeating molecules of glucose attached end to end (see page 51). These long, thin cellulose molecules are united into **microfibrils** about 10 to 25 nanometers wide (Figure 2–24). Cellulose has crystalline properties (Figure 2–25) because of the orderly arrangement of cellulose molecules in certain

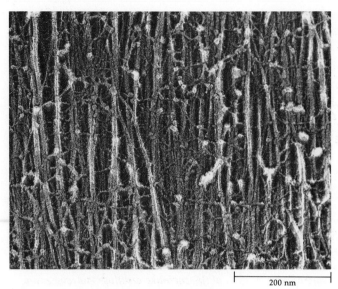

2–24
*Surface view of the primary wall of a
carrot* (Daucus carota) *cell, prepared
by a fast-freeze, deep-etch technique,
showing cellulose microfibrils cross-linked
by an intricate web of matrix molecules.
Compare this electron micrograph with
the schematic diagram in Figure 2–27.*

2–25
Stone cells (sclereids) *from the flesh of a
pear* (Pyrus communis) *seen in polarized
light. Clusters of such stone cells are
responsible for the gritty texture of this
fruit. The stone cells have very thick
secondary walls traversed by numerous
simple pits, which appear as lines in
the walls. The walls appear bright in
polarized light because of the crystalline
properties of their principal component,
cellulose.*

parts, the *micelles*, of the microfibrils (Figure 2–26). The microfibrils wind together to form fine threads that may coil around one another like strands in a cable. Each "cable," or **macrofibril**, measures about 0.5 micrometer in width and may reach 4 micrometers in length. Cellulose molecules wound in this fashion may have a strength exceeding that of an equivalent thickness of steel.

The cellulose framework of the wall is interpenetrated by a cross-linked *matrix* of noncellulosic molecules. Some of these molecules are polysaccharides called *hemicelluloses* and others are *pectic substances*, or *pectins*, which are closely related chemically to hemicelluloses (Figure 2–27).

Another important constituent of the walls of many kinds of cells is **lignin**, which, aside from cellulose, is the most abundant polymer found in plants. Physically, lignin is rigid, and it serves to add rigidity to the wall; it is commonly found in the walls of plant cells that have a supporting or mechanical function.

Cutin, **suberin**, and **waxes** are fatty substances commonly found in the walls of the outer, protective tissues of the plant body. Cutin, for example, is found in the walls of the epidermis, and suberin is found in those of the secondary protective tissue, cork. Both substances occur in combination with waxes and function largely to reduce water loss from the plant (see page 53).

CELL WALL LAYERS

Plant cell walls vary greatly in thickness, depending partly on the role the cells play in the structure of the plant and partly on the age of the individual cell. The cellulosic layers formed first make up the **primary wall**. The region of union of the primary walls of adjacent cells is called the **middle lamella** (also called intercellular substance). Many cells deposit additional wall layers. These form the **secondary wall**. If present, the secondary wall is laid down by the protoplast of the cell on the inner surface of the primary wall (Figure 2–28).

The Middle Lamella

The middle lamella is composed mainly of pectic substances. Frequently, it is difficult to distinguish the middle lamella from the primary wall, especially in cells that develop thick secondary walls. When a cell wall becomes lignified, this process characteristically begins in the middle lamella and then spreads to the primary wall and finally to the secondary wall.

The Primary Wall

The primary wall is deposited before and during the growth of the cell. In addition to cellulose, hemicelluloses, and pectin, primary walls contain enzymes and

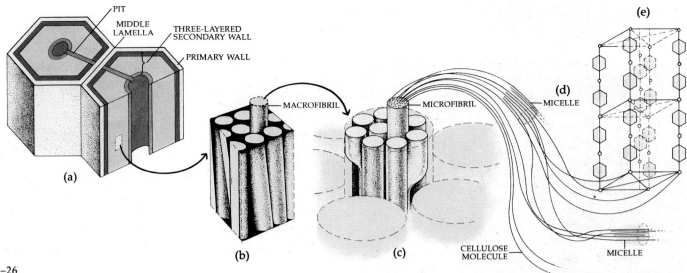

2-26
The detailed structure of a cell wall. (a) Portion of wall showing the middle lamella, primary wall, and three layers of secondary wall. Cellulose, the principal component of the cell wall, exists as a system of fibrils of different sizes. (b) The
largest fibrils, macrofibrils, can be seen with the light microscope. (c) With the aid of an electron microscope, the macrofibrils can be resolved into microfibrils about 10 nanometers wide. (d) Parts of the microfibrils, the micelles, are
arranged in an orderly fashion and impart crystalline properties to the wall. (e) A fragment of a micelle shows parts of the chainlike cellulose molecules in a lattice arrangement.

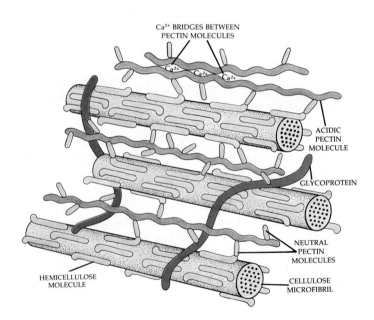

2-27
Schematic diagram showing how the cellulose microfibrils and matrix components (pectins, hemicelluloses, and glycoproteins) of the primary wall may be interconnected. Hemicellulose molecules are linked to the surface of the cellulose microfibrils by hydrogen bonds. Some of these hemicellulose molecules are cross-linked by neutral pectin molecules to acidic pectin molecules. Acidic pectin molecules are linked by calcium ions (Ca^{2+}). The glycoproteins are probably attached to the pectin molecules.

glycoproteins (Figure 2–27). Primary walls may also become lignified. The pectic polysaccharides are the most abundant matrix components of the primary walls of most flowering plants.

Actively dividing cells commonly have only primary walls, as do most mature cells involved with such metabolic processes as photosynthesis, respiration, and secretion. These cells—that is, living cells with primary walls but without secondary walls—are able to lose their specialized cellular form, divide, and differentiate into new types of cells. For this reason, it is principally cells with only primary walls that are involved in wound healing and regeneration in the plant.

Usually, primary cell walls are not of uniform thickness throughout but have thin areas that are called **primary pit-fields** (Figure 2–28). Cytoplasmic threads, or plasmodesmata, which connect the living protoplasts of adjacent cells, are commonly aggregated in the primary pit-fields.

The Secondary Wall

Although many plant cells have only a primary wall, in others a secondary wall is deposited by the protoplast inside the primary wall. Secondary wall formation occurs mostly after the cell has stopped growing and the primary wall is no longer increasing in surface area. Secondary walls are particularly important in specialized cells that have a strengthening function and in those involved in the conduction of water; in these cells, the protoplast often dies after the secondary wall has been laid down. Cellulose is more abundant in secondary walls than in primary walls, and pectic substances are lacking; the secondary wall is therefore rigid and not readily stretched. Enzymes and glycoproteins,

35

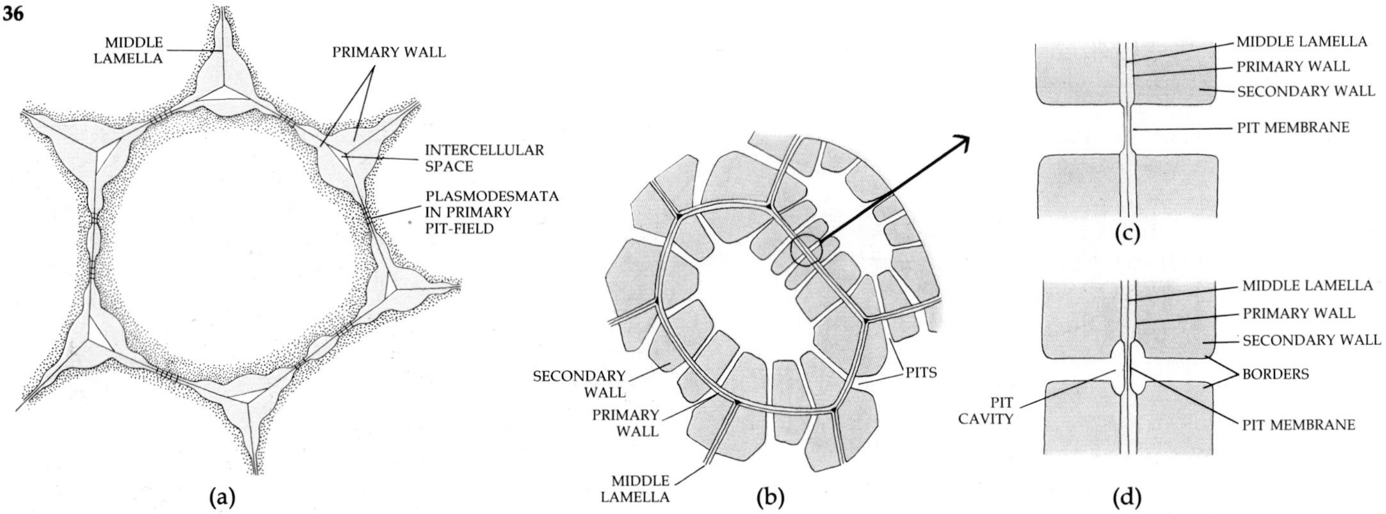

2–28

Primary pit-fields, pits, an plasmodes-mata. (a) Parenchyma cell with primary walls and primary pit-fields, thin areas in *the walls. As shown here, plasmodesmata commonly traverse the wall at the primary pit-fields. (b) Cells with secondary walls* *and numerous simple pits. (c) A simple pit-pair. (d) A bordered pit-pair.*

which are relatively abundant in primary cell walls, apparently are absent in secondary cell walls. The matrix of the secondary wall is composed of hemicellulose.

Frequently, three distinct layers—designated S_1, S_2, and S_3, for the outer, middle, and inner layer, respectively—can be distinguished in the secondary wall (Figure 2–29). The layers differ from one another in the orientation of their cellulose microfibrils. Such multiple-wall layers are found in certain cells of the secondary xylem, or wood. The laminated structure of such secondary walls greatly increases the strength of the walls. The cellulose microfibrils are laid down in a denser pattern than in the primary wall. Lignin is common in the secondary walls of cells found in wood.

When the secondary cell wall is deposited, it is not laid down over the primary pit-fields of the primary wall. Consequently, characteristic depressions, or **pits**, are formed in the secondary wall (Figure 2–28). In some instances, pits are also formed in areas where there are no primary pit-fields.

A pit in a cell wall usually occurs opposite a pit in the wall of an adjoining cell. The middle lamella and two primary walls between the two pits are called the **pit membrane**. The two opposite pits plus the membrane constitute a **pit-pair**. Two principal types of pits are found in cells that have secondary walls: **simple** and **bordered**. In bordered pits, the secondary wall arches over the **pit cavity**. In simple pits, there is no overarching (Figure 2–28). (See page 459 for additional information on the properties of pit membranes in tracheary elements.)

GROWTH OF THE CELL WALL

Cell walls grow both in thickness and in surface area. Extension of the wall is a complex process under the

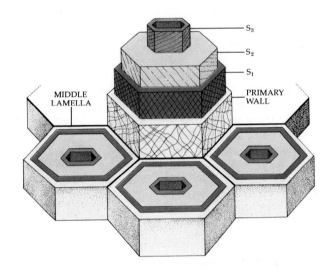

2–29

Diagram showing microfibril organization in primary walls and in three layers (S_1, S_2, S_3) of secondary wall.

close biochemical control of the protoplast. It requires loosening of wall structure—a phenomenon influenced by the hormone auxin (see Chapter 25)—as well as an increase in protein synthesis, in respiration to provide needed energy, and in uptake of water by the cell. Most of the new microfibrils are placed on top of those previously formed (layer upon layer), although some may be inserted into the existing wall structure.

In cells that enlarge more or less uniformly in all directions, the microfibrils are laid down in a random array, forming an irregular network. Such cells are found in the pith of stems, in storage tissues, and in tissue cultures. In contrast, in elongating cells, the microfibrils of the side walls are deposited in a plane at right angles (transverse) to the axis of elongation. As the wall increases in surface area, the orientation of the outer microfibrils becomes more nearly longitudinal, or parallel to the long axis of the cell.

The matrix substances—pectic substances and hemicelluloses—are carried to the wall in dictyosome vesicles, as are the glycoproteins. The type of matrix substance synthesized and secreted by a cell at any given time depends on the stage of development. For example, pectins are more characteristic of enlarging cells, whereas hemicelluloses predominate in cells that are no longer enlarging.

Cellulose synthesis is poorly understood, but it is now quite clear that (1) cellulose microfibrils are synthesized at the surface of the cell by an enzyme complex embedded in the plasma membrane, and that (2) the orientation of the microfibrils is controlled by microtubules lying just inside the plasma membrane.

Plasmodesmata

As mentioned previously, the protoplasts of adjacent plant cells are connected with one another by **plasmodesmata** (singular: **plasmodesma**). Although such structures have long been visible with the light microscope (Figure 2–30), they were difficult to interpret. It was not until they could be observed with an electron microscope that their nature was confirmed (Figure 2–31).

Plasmodesmata may occur throughout the cell wall, or they may be aggregated in primary pit-fields or in the membranes between pit-pairs. With the electron microscope, plasmodesmata appear as narrow canals (about 30 to 60 nanometers in diameter) lined by a

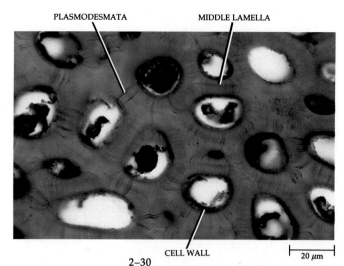

CELL WALL 20 µm

2–30
Light micrograph of plasmodesmata in the thick primary walls of persimmon (Diospyros) *endosperm, the nutritive tissue within the seed. The plasmodesmata appear as fine lines extending from cell to cell across the walls. The middle lamella appears as a light line between these cells.*

plasma membrane and traversed by a tubule of endoplasmic reticulum known as the *desmotubule*. Many plasmodesmata are formed during cell division as strands of tubular endoplasmic reticulum become trapped within the developing cell plate (see Figure 2–39). Plasmodesmata are also formed in the walls of nondividing cells. These structures provide a pathway for the transport of certain substances between cells (see page 71).

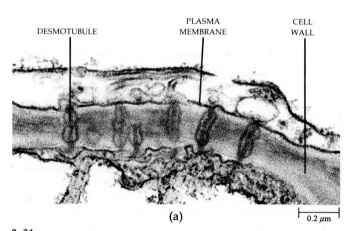

(a) 0.2 µm

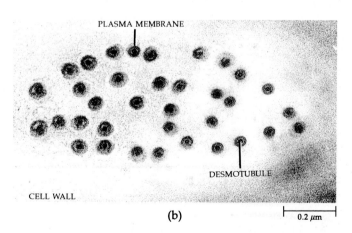

(b) 0.2 µm

2–31
Electron micrographs of plasmodesmata in the walls of corn (Zea mays) *leaf cells, as seen in longitudinal* (a) *and transverse* (b) *views. Plasmodesmata are narrow, plasma-membrane-lined canals in the walls, each traversed by a tubule of endoplasmic reticulum known as the desmotubule.*

37

Cell Division

In one-celled organisms, the cell grows by absorbing substances from the environment and using these materials to synthesize new structural and functional molecules. When such a cell reaches a certain size, it divides. The two daughter cells, each about half the size of the original mother cell, then begin growing again. In a one-celled organism, cell division may occur every day or even every few hours, producing a succession of identical organisms. In many-celled organisms, cell division, together with cell enlargement, is the means by which the organism grows. In all of these instances, the new cells are genetically and initially structurally similar to both the parent cell and one another.

Cell division in eukaryotes consists of two overlapping stages: **mitosis** and **cytokinesis**. Mitosis is the process by which a nucleus gives rise to two daughter nuclei, each morphologically and genetically equivalent to the other. Cytokinesis involves the division of the cytoplasmic portion of a cell and the separation of daughter nuclei into separate cells.

The Cell Cycle

Actively dividing cells pass through a regular sequence of events known as the cell cycle (Figure 2–32). Completion of the cycle requires varying periods of time, depending on both the type of cell and external factors, such as temperature or availability of nutrients. Commonly, the cycle is divided into *interphase* and the four phases of *mitosis*.

INTERPHASE

Interphase, the phase between successive mitotic divisions, was once regarded as the "resting phase" in the cell cycle. Nothing could be further from the truth, however, for interphase is a period of intense cellular activity.

Interphase can be divided into three phases, which are designated G_1, S, and G_2 (Figure 2–32). The G_1 phase (G stands for gap) occurs after mitosis. It is a period of intense biochemical activity, during which the cell increases in size, and the various organelles, internal membranes, and other cytoplasmic components increase in number. In order for the cycle to continue to the S phase, the G_1 phase must proceed past a critical point called the *restriction point*, or *Start*, at which time the cell undergoes an internal change that commits it to complete the subsequent steps of the cycle.

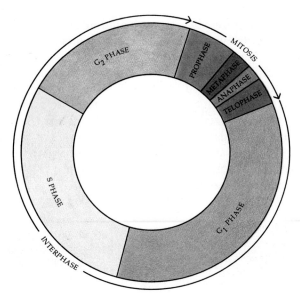

THE CELL CYCLE

2–32
The cell cycle can be divided into mitosis and interphase, the interval between successive mitotic divisions. Cells usually spend most of their time in interphase, which itself can be divided into three phases. The first of these (G_1) is a period of general growth and replication of cytoplasmic organelles. During the second phase (S), DNA synthesis takes place. The third (G_2) phase is a period of final preparation for mitosis. After the G_2 phase, mitosis begins.

The S (synthesis) phase—the period of DNA synthesis—apparently requires a signal from the cytoplasm before it can begin. The appearance of this *S-phase activator* presumably switches on DNA synthesis. During the S phase, many of the histones and other DNA-associated proteins are also synthesized. Following the S phase, the cell enters the G_2 phase, which continues until mitosis starts. During the G_2 phase, structures directly involved with mitosis, such as the components of the spindle fibers, are formed.

MITOSIS

Mitosis, or nuclear division, is a continuous process that is conventionally divided into four major phases: *prophase, metaphase, anaphase,* and *telophase* (Figures 2–33 through 2–35). These four phases make up the process by which the genetic material synthesized during the S phase is divided equally between two daughter nuclei.

Mitosis is triggered by a cytoplasmic factor that is present only in M-phase cytoplasm (M stands for mitosis), the *M-phase-promoting factor (MPF)*, which causes

2–33

Mitosis, a diagrammatic representation with four chromosomes. During early prophase, the chromosomes become visible as long threads scattered throughout the nucleus; the chromosomes shorten and thicken until each can be seen to consist of two threads (chromatids) attached to each other at their centromeres; a kinetochore develops on each centromere and, finally, the nucleolus and nuclear membrane disappear.

Metaphase begins with the appearance of the spindle in the area formerly occupied by the nucleus. During metaphase, the chromosomes migrate to the equatorial plane of the spindle. At full metaphase (shown here) the centromeres of the chromosomes lie on that plane.

Anaphase begins as the kinetochores on each chromosome separate. The sister chromatids, now called daughter chromosomes, then move to opposite poles of the spindle.

Telophase—more or less the reverse of prophase—begins when the daughter chromosomes have completed their migration.

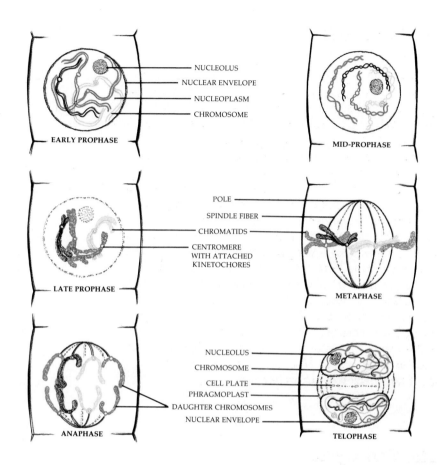

chromosome condensation. The rise and fall in the levels of MPF are correlated with fluctuations in the level of another protein called *cyclin*. Although cyclin is synthesized at a rapid and constant rate throughout the cell cycle, it is abruptly destroyed about halfway through the M phase. Thus, in each cycle, the concentration of cyclin rises steadily from zero and then precipitously drops back to zero again. Apparently, the surge in MPF activity is triggered by the increase in the cyclin concentration above a certain threshold.

Mitosis and cytokinesis depend on the coordination of many events and processes associated with the cell cycle—events and processes linked together as a dependent sequence.

Preprophase Band

One of the earliest signs that a plant cell is going to divide is the appearance of a narrow, ringlike band of microtubules just beneath the plasma membrane. (See the essay on page 41.) This relatively dense band of microtubules encircles the nucleus in a plane corresponding to the equatorial plane of the future mitotic spindle (see Figure 2–40). Inasmuch as it appears just before prophase, it is called the **preprophase band**. The preprophase band disappears after initiation of the mitotic spindle, long before initiation of the cell plate in late te-

lophase. Yet, as the cell plate forms, it grows outward to fuse with the cell wall of the mother cell precisely at the zone previously occupied by the preprophase band.

Prophase

As viewed with a microscope, the transition from the G₂ phase of interphase to **prophase** is not a clearly defined event. During prophase, the chromatin, which is diffuse in the interphase nucleus, gradually condenses into well-defined chromosomes. Initially, however, the chromosomes appear as elongated threads scattered throughout the nucleus. As prophase advances, the threads shorten and thicken, and as the chromosomes become more distinct, each can be seen to be composed of not one but two threads coiled about one another. During the preceding S phase, each chromosome duplicated itself; hence, each chromosome now consists of two sister **chromatids.** By late prophase, after further shortening, the two chromatids of each chromosome lie side by side and nearly parallel, joined to one another at their **centromeres** (Figure 2–36). The centromere, which consists of a specific DNA sequence needed to attach the DNA molecule to the mitotic spindle, divides the chromosome into two arms of lengths specific to each chromosome.

During prophase, a *clear zone* appears around the

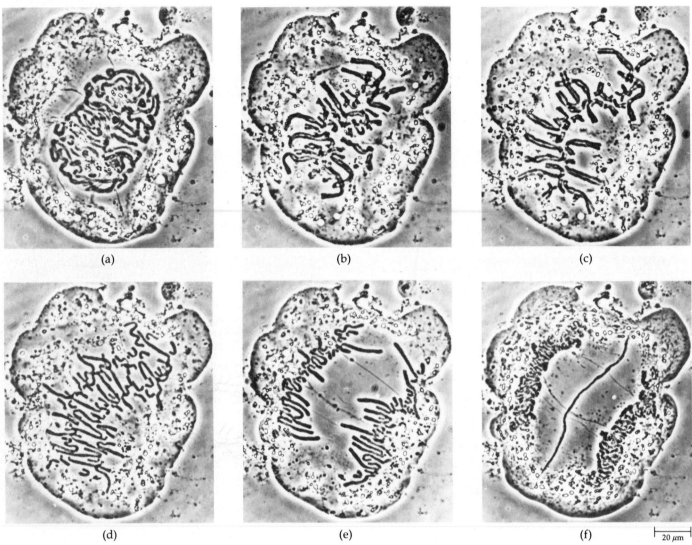

(a)   (b)   (c)

(d)   (e)   (f)   ⊢—⊣ 20 μm

2–34
Mitosis as seen with phase-contrast optics in a living cell of the African blood lily (Haemanthus katherinae). *The spindle is barely discernible in these cells, which have been flattened in order to show all of the chromosomes more clearly. (a) Late prophase: The chromosomes have condensed. A clear zone has developed around the nucleus. (b) Late prophase–early metaphase: The nuclear envelope has disappeared, and the ends of some of the chromosomes are protruding into the cytoplasm. (c) Metaphase: The chromosomes are arranged with their centromeres on the equatorial plane. (d) Mid-anaphase: The sister chromatids (now called daughter chromosomes) have separated and are moving to opposite poles of the spindle. (e) Late anaphase. (f) Telophase-cytokinesis: The daughter chromosomes have reached the opposite poles, and the two chromosome masses have begun the formation of two daughter nuclei. Cell plate formation is nearly completed.*

nuclear envelope (Figure 2–34a). Microtubules appear in this zone. The microtubules are initially randomly oriented, but by late prophase they are aligned parallel to the nuclear surface along the spindle axis. This is the earliest manifestation of mitotic spindle assembly.

Toward the end of prophase, the nucleolus gradually becomes indistinct and finally disappears. Either simultaneously or shortly afterward, the nuclear envelope breaks down, marking the end of prophase (Figure 2–37).

Metaphase

Metaphase begins as the **spindle**, a three-dimensional structure that is widest in the middle and tapers toward the poles, occupies the area formerly occupied by the nucleus. The spindle consists of *spindle fibers*, which are bundles of microtubules (see Figure 2–40c). Some of the spindle microtubules become attached to specialized protein complexes called **kinetochores**, which developed on each centromere during late pro-

The Role of Microtubules in Cell Division: Immunofluorescence Microscopy

Immunofluorescence microscopy has been used to reveal the location of microtubules at different stages of cell division. Unlike other procedures, immunofluorescence microscopy has the capacity to provide three-dimensional views. Here root-tip cells of onion (*Allium cepa*) were stained with specially prepared fluorescent antibodies to tubulin and photographed by fluorescence microscopy. The fluorescent antibodies, which became attached to the protein tubulin, revealed the location of the microtubules. (a) A preprophase band of microtubules (horizontal arrows) encircles the cell just before prophase.

Other microtubules (arrowheads) outline the nuclear envelope (not visible). (b) The preprophase band has disappeared, and the area it previously occupied (arrows) appears as a nonfluorescing band. Microtubules of the spindle extend from this region to the spindle poles. (c) In early telophase, after the chromosomes have separated, new microtubules form a phragmoplast, in which cell plate formation takes place. (d) As cell plate formation continues, the microtubules become concentrated where the cell plate is extending outward toward the walls of the dividing cell.

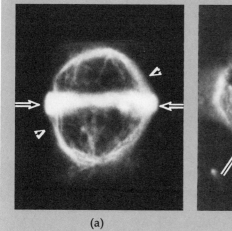
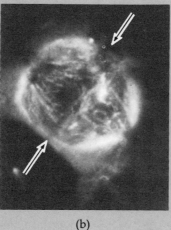
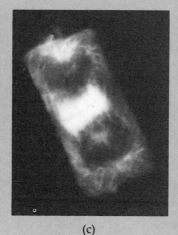

(a)   (b)   (c)   (d)

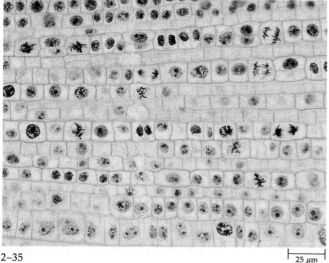

2–35

Photomicrograph of dividing cells in an onion (Allium) root tip. By comparing these cells with the phases of mitosis illustrated in Figures 2–33 and 2–34, you should be able to identify the mitotic phases shown in this micrograph.

25 μm

phase (Figure 2–36). These microtubules are called *kinetochore microtubules.* The remaining spindle microtubules, which extend from pole to pole, are called *polar microtubules.* The kinetochore microtubules extend to opposite poles from the sister chromatids of each chromosome. These early events of metaphase are referred to by some biologists as *prometaphase.*

Eventually, the kinetochore microtubules align the chromosomes midway between the spindle poles so that the kinetochores lie on the equatorial plane of the spindle. When the chromosomes have all moved to the equatorial plane, or metaphase plate, the cell has reached full metaphase. The chromatids are now in position to separate (Figure 2–34c).

Organization of the Mitotic Spindle

As mentioned previously, the assembly of microtubules takes place at specific sites known as microtubule organizing centers. In animals and many protists, the microtubule organizing center responsible for the formation of the mitotic spindle is the **centrosome,** a cloud

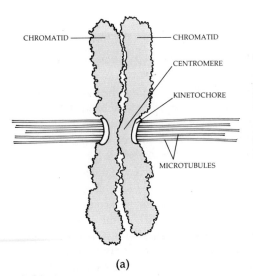

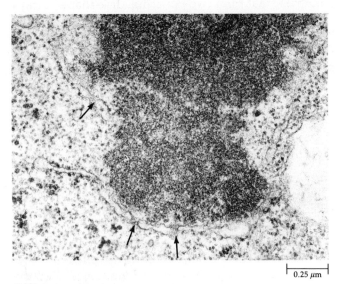

2–36

(a) *A diagram of a fully condensed chromosome. The chromosomal DNA was replicated during interphase, so that each chromosome now consists of two identical parts, called chromatids. The chromatids are joined together at their* centromeres. *During late prophase, a kinetochore developed on each centromere, facing opposite directions. Attached to the kinetochores are microtubules that form part of the spindle.* (b) *Electron micrograph of* a portion of a metaphase chromosome of a green alga, showing microtubules attached to the kinetochores of sister chromatids. The three-layered structure typical of most kinetochores is discernible here.

2–37

A portion of a late prophase nucleus of a parenchyma cell in a horsetail (Equisetum hyemale) stem. The arrows point to nuclear pores in fragments of the nuclear envelope, which is in the process of breaking down.

of amorphous material typically surrounding a pair of *centrioles*, structures identical to the basal bodies associated with flagella. During mitosis, a centrosome is found at each pole of the spindle, and the spindle microtubules can be seen emanating from them. In plant cells, the microtubule organizing centers provide for the assembly of the spindle, even though the spindle poles are poorly defined and centrioles are absent.

Anaphase

Anaphase (Figure 2–34d, e) begins abruptly with the simultaneous separation of all the sister chromatids at the centromeres. The sister chromatids are now called **daughter chromosomes.** As the kinetochores of daughter chromosomes move toward opposite poles, the arms of the chromosomes seem to drag behind. As anaphase continues, daughter chromosomes pull apart, with the tips of longer arms separating last.

The poleward movement of the daughter chromosomes during anaphase apparently is associated with the shortening of the kinetochore microtubules at the kinetochore end, the kinetochores seemingly "chewing" their way poleward along their microtubules. In the process, the daughter chromosomes are pulled toward the poles by the kinetochores. One opponent of this model for anaphase chromosome movement (he favors a role for actin filaments) has dubbed it the "Pac-Man" model, after the video game of the same name. Some separation of the daughter chromosomes may also come about by the spindle poles themselves moving farther apart. During this process, the polar microtubules increase in length.

By the end of anaphase, the two identical sets of daughter chromosomes have moved to opposite poles.

Telophase

During **telophase** (Figure 2–34f) the separation of the two identical sets of chromosomes is made final as nuclear envelopes are organized around each set (Figure 2–38). The membranes of these nuclear envelopes are derived from rough endoplasmic reticulum. The spindle apparatus disappears. In the course of telophase, the chromosomes become increasingly indistinct, elongating to become slender threads again. Nucleoli also re-form at this time. When these processes are completed and the chromosomes have once more disappeared from view, mitosis is completed and the two daughter nuclei have entered interphase.

The two daughter nuclei produced during mitosis are genetically equivalent to one another and to the nucleus that divided to produce them. This is important, for the nucleus is the control center of the cell, as is described in more detail in Chapter 8. It contains coded instructions that specify the production of proteins, many of which mediate cellular processes by acting as enzymes and some of which serve directly as structural elements in the cell. This hereditary blueprint must be exactly passed on to the daughter cells, and its precise distribution is ensured in eukaryotic organisms by the organization of chromosomes and their division during the process of mitosis.

The duration of mitosis varies with the tissue and the organism involved. However, prophase is the longest phase and anaphase is the shortest. In a root tip, the relative lengths of time for each of the four phases may be as follows: prophase, 1 to 2 hours; metaphase, 5 to 15 minutes; anaphase, 2 to 10 minutes; and telophase, 10 to 30 minutes. By contrast, the duration of interphase may range from 12 to 30 hours.

CYTOKINESIS

As noted previously, *cytokinesis*, which typically overlaps mitosis, is the process by which the cytoplasm is divided. In most organisms, cells divide by ingrowth of the cell wall, if present, and constriction of the plasma membrane, a process that cuts through the spindle fibers. In all plants (bryophytes and vascular plants) and in a few algae, cell division occurs by the formation of a **cell plate** (Figures 2–38 through 2–40).

In early telophase, an initially barrel-shaped system of fibrils called the **phragmoplast** forms between the two daughter nuclei. The fibrils of the phragmoplast, like those of the mitotic spindle, are composed of microtubules that form two opposing, disk-shaped arrays on either side of the division plane. As seen with the light microscope, small droplets appear across the plane and gradually fuse, forming the cell plate. The cell plate—preceded by the phragmoplast—grows outward until it reaches the wall of the dividing cell, completing the separation of the two daughter cells. With the electron microscope, the fusing droplets can be seen to be vesicles derived from the Golgi apparatus. The vesicles

2–38

In plant cells, separation of the daughter chromosomes is followed by formation of a cell plate, which completes the separation of the dividing cells. Here numerous Golgi vesicles can be seen fusing in an early stage of cell plate formation. The two groups of chromosomes on either side of the developing cell plate are at telophase. Arrows point to portions of the nuclear envelope reorganizing around the chromosomes.

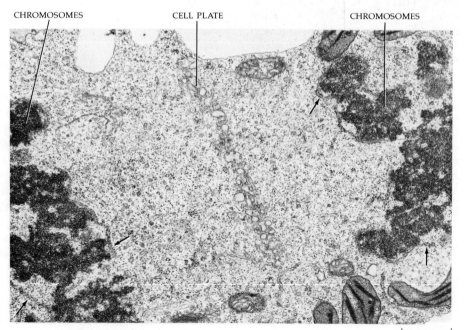

CHROMOSOMES   CELL PLATE   CHROMOSOMES

1 μm

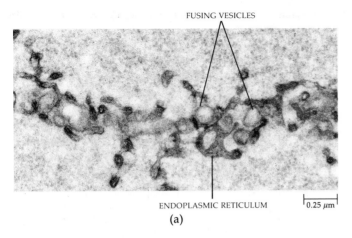

FUSING VESICLES

ENDOPLASMIC RETICULUM

0.25 μm

(a)

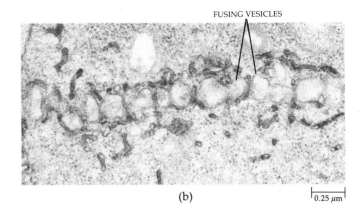

FUSING VESICLES

(b)

0.25 μm

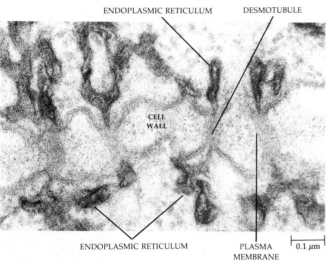

ENDOPLASMIC RETICULUM   DESMOTUBULE

CELL
WALL

ENDOPLASMIC RETICULUM   PLASMA
MEMBRANE

0.1 μm

(c)

2–39

*Progressive stages in cell plate formation
in root cells of lettuce* (Lactuca sativa), *showing association of the endoplasmic
reticulum with the developing cell plate
and the origin of plasmodesmata. (a) A
relatively early stage, with numerous
small, fusing vesicles of dictyosome
origin and loosely arranged elements of
tubular endoplasmic reticulum. (b) An
advanced stage, revealing a persistent
close relationship between the endoplasmic reticulum and fusing vesicles. Strands
of tubular endoplasmic reticulum become
trapped during cell plate consolidation.
(c) Mature plasmodesmata, which consist
of a plasma-membrane-lined canal and
a tubule, the desmotubule, of endoplasmic
reticulum.*

presumably contain pectic substances that form the middle lamella, and their membranes contribute to the formation of the plasma membrane on either side of the plate. Plasmodesmata are formed at this time, when segments of tubular endoplasmic reticulum are "caught" between fusing vesicles (Figure 2–39).

The developing cell plate eventually fuses with the mother cell wall precisely at the zone demarcated earlier by the preprophase band. Actin filaments have been found bridging the gap between the leading edge of the phragmoplast and the cell wall, possibly providing an explanation as to how the expanding phragmoplast "remembers" the site of the former preprophase band.

Following the formation of the middle lamella, each protoplast deposits a primary wall next to the middle lamella. In addition, each daughter cell deposits a new wall layer around the entire protoplast; this new wall is continuous with the wall at the cell plate (Figure 2–40). The original wall of the mother cell stretches and ruptures as the daughter cells enlarge.

Summary

All living matter is composed of cells. Cells are extremely varied, ranging in structure and function from independent, single-celled organisms to the highly specialized, interdependent cell types found in complex multicellular plants and animals. However, cells are remarkably similar in their basic structure. All cells are bounded by an outer membrane (known as the plasma membrane, or plasmalemma). Enclosed within this membrane are the cytoplasm and the hereditary information in the form of DNA. Collectively, the contents of the cell are called protoplasm.

Cells are of two fundamentally different types: prokaryotic and eukaryotic. Prokaryotic (bacterial) cells lack nuclei and membrane-bound organelles, and are represented today by bacteria, including the cyanobacteria. Eukaryotic cells have true nuclei and are compartmented, containing distinct structures that perform different functions.

INTERPHASE
(a)

PREPROPHASE
(b)

METAPHASE
(c)

TELOPHASE AND CYTOKINESIS
(d)

CYTOKINESIS
(e)

EARLY INTERPHASE
(f)

INTERPHASE
(g)

CELL ENLARGEMENT
(h)

2–40

Changes in the distribution of microtubules during the cell cycle and cell wall formation during cytokinesis. (a) During interphase, and in enlarging and differentiating cells, the microtubules lie just inside the plasma membrane. (b) Just before prophase, a ringlike band of microtubules, the preprophase band, encircles the nucleus in a plane corresponding to the equatorial plane of the future mitotic spindle. (c) During metaphase, the microtubules occur in the form of a spindle. (d) During telophase, microtubules are organized into a phragmoplast between the two daughter nuclei, and the cell plate forms at the equator of the phragmoplast. (e) As the cell plate matures in the center of the phragmoplast, the phragmoplast and developing cell plate grow outward until they reach the wall of the dividing cell. (f) During early interphase, microtubules radiate outward from the nuclear envelope into the cytoplasm. (g) Each sister cell forms its own primary wall. (h) With enlargement of the daughter cells (only the upper one is shown here), the mother cell wall is torn. In (g) and (h) the microtubules once more lie just inside the plasma membrane.

Eukaryotic cells are divided into compartments by membranes, which control the passage of materials into and out of the cell and into and out of organelles.

Plant cells typically have a cell wall and a protoplast, the unit of protoplasm inside the cell wall. The protoplast consists of the cytoplasm and a nucleus. The cytoplasmic matrix, or ground substance, of plant cells is frequently in motion, a phenomenon known as cytoplasmic streaming, or cyclosis. The plasma membrane separates the cytoplasm from the cell wall.

In addition to cell walls, plant cells are characterized by the presence of vacuoles within their cytoplasm. These organelles are cavities that are filled with cell sap, an aqueous solution of materials including a variety of salts, sugars, and other substances. Vacuoles play an important role in cell enlargement and the maintenance of tissue rigidity. In addition, many vacuoles are involved in the breakdown of macromolecules and the recycling of their components within the cell. The vacuole is bounded by a membrane called the tonoplast.

The nucleus, which is the control center of the cell, is often the most prominent structure within the protoplast. It is surrounded by a nuclear envelope composed of two membranes. Enclosed within the envelope is the chromatin, which becomes visible as distinct chromosomes during nuclear division. Chromatin and chromosomes consist of DNA and proteins.

Together with vacuoles and cell walls, plastids are characteristic components of plant cells. Each plastid is bounded by an envelope consisting of two membranes. Mature plastids are classified on the basis of the kinds of pigments they contain: chloroplasts contain chlorophylls and carotenoid pigments; chromoplasts contain carotenoid pigments; and leucoplasts are nonpigmented.

Like plastids, mitochondria are organelles bounded by two membranes; the inner one is folded to form an extensive inner membrane system, increasing the surface area available to enzymes and the reactions associated with them. Mitochondria are the principal sites of respiration in eukaryotic cells. Unlike plastids and mitochondria, microbodies are organelles bounded by a single membrane.

In addition to organelles, the cytoplasm of eukaryotic cells contains two membrane systems: the endoplasmic reticulum and the dictyosomes. The endoplasmic reticulum is an extensive three-dimensional system of membranes. It often has numerous ribosomes attached to it. Ribosomes are also found free in the cytoplasm. Ribosomes are the sites at which amino acids are linked together to form proteins. During protein synthesis, the ribosomes occur in clusters called polyribosomes, or polysomes.

Dictyosomes, or Golgi bodies, are groups of flat, disk-shaped membranous sacs from which numerous vesicles bud off. The dictyosomes serve as collection and packaging centers for complex carbohydrates and other substances, which are transported to the surface of the cell in the vesicles. The vesicles also serve as a source of plasma membrane material. The endoplasmic reticulum and dictyosomes, together with the plasma membrane, are part of a functional unit called the endomembrane system.

The ground substance of eukaryotic cells is permeated by the cytoskeleton, a complex network of protein filaments, of which there are two types in plant cells: microtubules and actin filaments. Microtubules are thin cylindrical structures of variable length and are composed of subunits of the protein tubulin. They play a role in mitosis, cell plate formation, the growth of the cell wall, and the movement of flagella. Actin filaments are contractile proteins composed of actin. These long filaments occur singly and in bundles, and they play a causative role in cytoplasmic streaming.

Flagella are hairlike structures that project from the surface of many different types of eukaryotic cells, serving as locomotor structures. All flagella of eukaryotic cells have the same highly characteristic 9-plus-2 internal structure; that is, an outer ring of nine pairs of microtubules surrounding an inner pair of microtubules in the center of the flagellum.

The cell wall is the major distinguishing feature of the plant cell. It determines the structure of the cell, the texture of plant tissues, and many important characteristics that distinguish plants as organisms. All plant cells have a primary wall. In addition, many have a secondary wall. The region between the primary walls of adjacent cells is called the middle lamella. Cellulose is the principal component of primary and secondary walls. The cellulose microfibrils occur in a cross-linked matrix of noncellulosic molecules, such as hemicelluloses and pectin. Lignin may also be present in cell walls but is especially characteristic of cells with secondary walls.

The protoplasts of adjacent cells are connected to one another by plasmodesmata, which provide a pathway for the transport of certain substances between cells.

Dividing cells pass through a cell cycle consisting of interphase and mitosis. Interphase can be further divided into three phases: a G_1 phase of general growth and replication of organelles; an S phase, during which the genetic material, DNA, is duplicated; and a G_2 phase, a period of final preparation for mitosis.

When the cell is in interphase, the chromosomes are in an uncoiled state and are difficult to distinguish from the nucleoplasm. During mitosis, the chromosomal material, or chromatin, condenses, and each chromosome can be seen to consist of two parallel threads—the chromatids—held together at their centromeres. The centromeres separate, and the duplicate chromatids—now called daughter chromosomes—move to opposite poles. The separation of the two identical sets of chromosomes is completed when new nuclear envelopes are formed around them. In this way, the genetic material is equally distributed between the two daughter nuclei.

Mitosis is generally followed by cytokinesis, the division of the cytoplasm. In plants and certain algae, the cytoplasm is divided by a cell plate that begins to form during mitotic telophase. Once the cytoplasm is divided, the protoplasts lay down new cell walls.

CHAPTER 3

The Molecular Composition of Cells

3–1

In photosynthesis, organisms capture light energy and use it to form organic molecules by combining the carbon and oxygen from carbon dioxide in the atmosphere with hydrogen obtained from water. These organic molecules provide the energy that powers living systems and are also used to build the larger structural molecules of which living organisms are composed. The sun-dappled plants shown here, among them broad-leaf members of the genus Heliconia, *are growing beside a waterfall at Misal-Ha in Chiapas, Mexico.*

As discussed in the previous chapter, cells are remarkably similar in basic structure, despite their variety. These similarities become even more striking when we examine cells at the molecular level. (You may find it helpful to read Appendix A, Fundamentals of Chemistry, before proceeding with this chapter.)

Although the earth and its atmosphere contain 92 different types of naturally occurring elements, only relatively few of them were selected in the course of evolution to form the complex, highly organized material that makes up the cells of living organisms. In fact, as shown in Table 3–1, about 99 percent (by fresh weight) of living matter is composed of only six elements. In cells, these elements are found chiefly in organic molecules and, in particular, in water.

Water makes up more than half of all living tissue and more than 90 percent of the weight of most plant tissues. (See Appendix A for a review of some important properties of water.) By contrast, ions such as potassium (K^+), sodium (Na^+), and calcium (Ca^{2+}) account for no more than 1 percent. Almost all the rest of the tissue, chemically speaking, is composed of organic molecules.

Organic Molecules

Organic molecules, by definition, are molecules that contain covalently bonded carbon backbones. In addition to carbon, nearly all organic molecules contain hydrogen, and most of them contain oxygen as well (Figure 3–1). Nitrogen and sulfur occur less frequently than oxygen, and only a small percentage of organic molecules contain other elements. Just as living matter is composed of relatively few elements, the atoms of these few elements are arranged to form relatively few organic molecules that, in various combinations, make up most of the dry weight of living organisms.

Table 3-1 *Atomic Composition of Three Representative Organisms (Percentage by Fresh Weight)*

Element	Human	Alfalfa	Bacterium
Carbon (C)	19.37%	11.34%	12.14%
Hydrogen (H)	9.31	8.72	9.94
Nitrogen (N)	5.14	0.83	3.04
Oxygen (O)	62.81	77.90	73.68
Phosphorus (P)	0.63	0.71	0.60
Sulfur (S)	0.64	0.10	0.32
C H N O P S total:	97.90%	99.60%	99.72%

SOURCE: H. J. Morowitz, *Energy Flow in Biology* (New York: Academic Press, 1968).

The four principal types of organic molecules found in living organisms are carbohydrates, lipids, proteins, and nucleic acids (Table 3–2). Lipids, proteins, and nucleic acids are relatively large structures with high molecular weights and are therefore called *macromolecules*. Large molecules that are made up of similar or identical subunits are known as **polymers** ("many parts"); the subunits are called **monomers** ("single parts").

Carbohydrates

Carbohydrates—the most abundant organic molecules in nature—are molecules that contain carbon combined with hydrogen and oxygen. They are the primary energy-storage molecules in most living things. Carbohydrates are formed from small molecules known as

sugars. There are three principal kinds of carbohydrates, classified according to the number of sugar molecules they contain: monosaccharides, disaccharides, and polysaccharides.

MONOSACCHARIDES AND DISACCHARIDES

Monosaccharides ("single sugars") are the simplest carbohydrates; they are made up of a chain of carbon atoms to which hydrogen atoms and oxygen atoms are attached in the proportion of one carbon atom to two hydrogen atoms to one oxygen atom. Monosaccharides can be described by the formula $(CH_2O)_n$, where $n = 3$ or some larger number. These proportions gave rise to the term *carbohydrate* ("hydrate of carbon") for sugars and for the larger molecules formed from sugar subunits. Examples of several common monosaccharides are shown in Figure 3–2. As the figure indicates, the five-carbon and six-carbon sugars can also exist in ring form; in fact, they are normally found in this form when dissolved in water.

Disaccharides consist of two monosaccharides joined together. As shown in Figure 3–3, the union is accomplished by the removal of a molecule of water from the pair of monosaccharide molecules, a process known as **condensation.** Such joined molecules can be broken apart by **hydrolysis**—the addition of a molecule of water at each linkage—to form the monosaccharide units again. Hydrolysis is an exergonic, or "downhill," reaction; the chemical bonding energy of its products is less than that of the original molecule (see Chapter 5). Thus, energy is released in such a reaction. Conversely, linking two monosaccharides together to form a disaccharide requires the input of energy.

Table 3-2 *Four Important Classes of Organic Molecules*

Organic Molecules	Functions	Components	Elemental Composition
Carbohydrates	Energy source, structural material, building blocks for other molecules	Simple sugars	Carbon, hydrogen, and oxygen
Lipids	Energy storage, structural components of membranes, barrier to water loss	Fatty acids and glycerol in fats and oils	Carbon, hydrogen, and oxygen
Proteins	Structural material, enzymes	Amino acids	Carbon, hydrogen, oxygen, nitrogen, and sulfur
Nucleic acids	Storage, transmission, translation of genetic information; protein synthesis	Nucleotides (nitrogenous bases, sugars, and phosphates)	Carbon, hydrogen, oxygen, nitrogen, and phosphorus

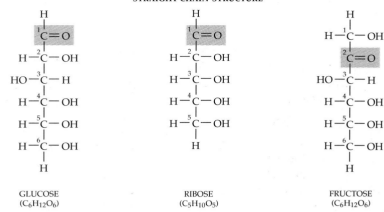

GLYCERALDEHYDE
($C_3H_6O_3$)

GLUCOSE
($C_6H_{12}O_6$)

RIBOSE
($C_5H_{10}O_5$)

FRUCTOSE
($C_6H_{12}O_6$)

3–2

Examples of biologically important monosaccharides. Five-carbon sugars (pentoses) and six-carbon sugars (hexoses) can exist in both chain and ring forms. The position occupied by each carbon atom has a number. By convention, the carbon atoms of the rings are not shown; their presence is assumed at each corner of the ring not occupied by another atom. The letters α (alpha) and β (beta) refer to the position of one of the –OH groups on the ring forms, as indicated in color. Note that groups shown with vertical bonds project above and below the plane of the ring. The lower portions of the rings (which project toward you) are thickened to suggest a three-dimensional orientation.

RING-FORM STRUCTURE

α-GLUCOSE
($C_6H_{12}O_6$)

β-RIBOSE
($C_5H_{10}O_5$)

β-FRUCTOSE
($C_6H_{12}O_6$)

The monosaccharide glucose is the form in which sugar is most often transported through animal systems. Sucrose, a disaccharide composed of glucose and fructose, is the form in which most sugar is transported in plants. Sucrose is common table sugar (cane or beet sugar).

The ultimate source of sugar in all plant cells is photosynthesis, in which energy from the sun is converted to the chemical bond energy needed to form a sugar molecule. When the sugar molecule is broken down in respiration, the energy of the chemical bonds is released.

3–3

The condensation reaction that produces the disaccharide sucrose, the form in which sugar is generally transported in plants. Sucrose is formed from an α-glucose unit and a β-fructose unit, joined in a 1 ⟶ 2 linkage (the bonding between the two rings involves the carbon 1 of glucose and the carbon 2 of fructose). In order to represent this bond on paper, we must rotate the structural formula for β-fructose 180° (right to left), which has the disconcerting effect of turning everything upside down to our eyes. In the three-dimensional world in which the molecules actually exist, however, formation of this 1 ⟶ 2 linkage creates no problems.

As you can see, the condensation reaction producing sucrose involves the removal of a molecule of water. Splitting sucrose back into its constituent monosaccharides requires the addition of a water molecule (hydrolysis). Formation of sucrose from glucose and fructose requires an energy input of 5.5 kilocalories per mole by the cell.

α-GLUCOSE   β-FRUCTOSE

H_2O

SUCROSE

$^6 CH_2OH$ ... AMYLOSE (a)

$^6 CH_2OH$ A BRANCH POINT IN AMYLOPECTIN (b)

(c)

3–4

In most plants, accumulated sugars are stored in the form of starch. Starch is composed of two different types of polysaccharides: (a) amylose and (b) amylopectin. A single molecule of amylose may contain 1000 or more α-glucose units with carbon 1 of one glucose ring linked to carbon 4 of the next in a long, unbranched chain that winds to form a helix (c). A molecule of amylopectin may contain 1000 to 6000 α-glucose units; short chains containing about 24 to 36 α-glucose units periodically branch off the main chain. (d) Perhaps because of their helical nature, starch molecules tend to cluster into grains. In this scanning electron micrograph of a single storage cell of potato (Solanum tuberosum), *the spherical structures are starch grains, which form in amyloplasts, one grain per amyloplast.*

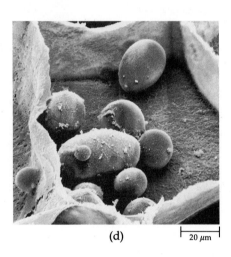

(d)   20 μm

POLYSACCHARIDES

Polysaccharides are macromolecules, polymers made up of monosaccharides (the monomers) linked together in long chains. Some polysaccharides are storage forms of sugar. **Starch,** which is built of many glucose molecules, is the principal storage polysaccharide in plants (Figure 3–4), and glycogen is the common storage form of sugar in fungi, bacteria, and animals. In some plants —most notably grasses that originated in temperate zones and also a few dicotyledons—the principal storage polysaccharides in leaves and stems are sucrose and polymers of fructose called **fructans**. Polysaccharides must be hydrolyzed before they can be used as energy sources or transported through living systems.

Polysaccharides are also important structural compounds. In plants, the principal structural polysaccharide is cellulose (Figure 3–5). Moreover, cellulose is the most abundant polysaccharide in nature. Although cellulose and starch are made of the same building materials, the arrangement of the long-chain molecules in cellulose makes it rigid, and so its biological role is very different from that of starch. Also, the bonds linking the glucose units in cellulose are different from those found in starch, and cellulose is not readily hydrolyzed by enzymes that break down other polysaccharides. In addition to cellulose, plant cell walls commonly contain two other types of polysaccharide, *pectins* (pectic polysaccharides) (Figure 3–6) and *hemicelluloses* (Figure 3–7).

Once glucose molecules are incorporated into the plant cell wall in the form of cellulose, they are no longer available to the plant as an energy source. In fact, only some bacteria, fungi, protozoa, and a very few animals (silverfish, for example) possess enzyme systems capable of breaking down cellulose. Other organisms, such as cattle (and other ruminants), termites, and cockroaches, are able to utilize cellulose as a source of energy only because of the microorganisms (which possess the necessary enzyme systems) that inhabit their digestive tracts.

3–5

(a) *In living systems, α-ring and β-ring forms of the glucose molecule occur in equilibrium. The molecules pass through the straight-chain form to get from one ring structure to another. Whereas starch consists of α-glucose monomers, cellulose (b) consists entirely of β-glucose monomers, joined in 1 → 4 linkages. (Note that the structural formulas for alternating β-glucose units have been rotated 180°, front to back, to show the bonding.) In cellulose, the –OH groups (in color), which project from both sides of the chain, form hydrogen bonds with –OH groups on neighboring chains, resulting in the formation of bundles of cross-linked parallel cellulose molecules (c).*

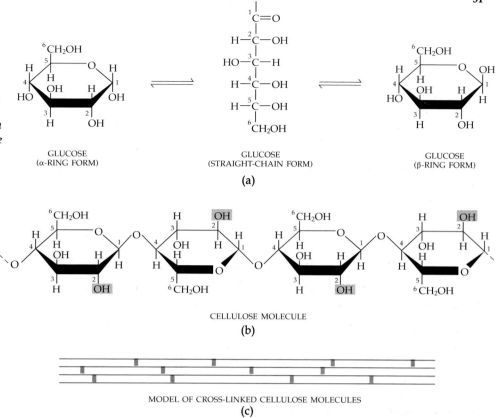

GLUCOSE
(α-RING FORM)

GLUCOSE
(STRAIGHT-CHAIN FORM)

GLUCOSE
(β-RING FORM)

(a)

CELLULOSE MOLECULE

(b)

MODEL OF CROSS-LINKED CELLULOSE MOLECULES

(c)

3–6

Pectic compounds are built up of residues of α-galacturonic acid (a), which is a derivative of glucose. Polymers of this sugar derivative are known as pectic acid (b). Calcium and magnesium salts of pectic acid make up most of the middle lamella, which binds adjacent plant cells. In pectin and protopectin, different proportions of the carboxyl-group hydrogen (shown in color) are replaced with methyl (–CH₃) groups. Protopectin is a common constituent of the cell wall and is less soluble than pectin, which is commonly found dissolved in plant juices.

α-GALACTURONIC
ACID

(a)

PECTIC ACID

(b)

GLUCOSE

XYLOSE

GALACTOSE

FUCOSE

3-7

Building-block sugars of the hemicellulose xyloglucan. The backbone of xyloglucan consists of a chain of glucose residues to which side chains of xylose are attached.

In dicotyledons (one of two major groups of flowering plants), the side chains may be elongated by linkage with galactose and fucose residues. The xyloglucans

play an important structural role in stabilizing the primary wall by hydrogen bonding to the cellulose microfibrils.

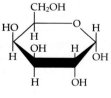

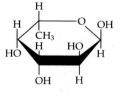

The emphasis in recent years on preserving the tropical rainforests of the world and on planting trees is related in part to the fact that enormous quantities of carbon are locked up in cellulose. This is carbon that might otherwise contribute to the level of carbon dioxide in the atmosphere and, hence, to the "greenhouse effect" and global warming. More serious, however, is the extinction of species of plants and other organisms that accompanies the destruction of the forest (see pages 12 and 665).

Chitin is another important structural polysaccharide; it is the principal structural component of fungal cell walls and also of the relatively hard outer coverings, or exoskeletons, of insects and crustaceans. The monomer of chitin is a six-carbon sugar to which a nitrogen-containing group has been added (see Figure 12–4).

Lipids

Lipids are fats and fatlike substances. They have two principal distinguishing characteristics: (1) they are generally hydrophobic ("water-fearing") and thus are insoluble in water (although they are soluble in other lipids and in other hydrophobic solvents), and (2) they contain a large number of carbon-hydrogen bonds and, as a consequence, release a larger amount of energy in oxidation than do other organic compounds. On oxidation, fats yield, on the average, about 9.3 kilocalories per gram, as compared to a yield of about 3.8 kilocalories per gram for carbohydrates. These two characteristics of lipids determine their roles as structural materials and as energy reserves.

FATS AND OILS

Fats and oils are storage forms of lipids called **triglycerides.** Fats are triglycerides that are solid at room temperature, while oils are liquid. In animals, fats are the most common form of energy-storage molecule. In plants, excess food energy most commonly is stored as starch, less often as oil (Figure 3–8). Cells synthesize fats from sugars.

Fats and oils have similar chemical structures (Figure 3–9). Each consists of three fatty acids joined to a glycerol molecule. Fatty acids are long hydrocarbon chains that carry a terminal carboxyl group, giving them the characteristics of a weak acid. The carboxyl end of each fatty acid forms a link with the glycerol molecule; as each link forms, a water molecule is released in a manner similar to that associated with disaccharide formation (Figure 3–3). In the reverse reaction, the fatty acids can be split from the glycerol by hydrolysis.

The physical nature of the triglyceride is determined by the chain lengths of the fatty acids and by whether the acids are **saturated** or **unsaturated.** In saturated fatty acids, all of the carbon atoms are linked to as many hydrogen atoms as possible. In unsaturated fatty acids, some of the carbon atoms are joined by double bonds; such carbon atoms are able to form additional bonds with other atoms (hence the term "unsaturated"). The presence of double bonds tends to lower

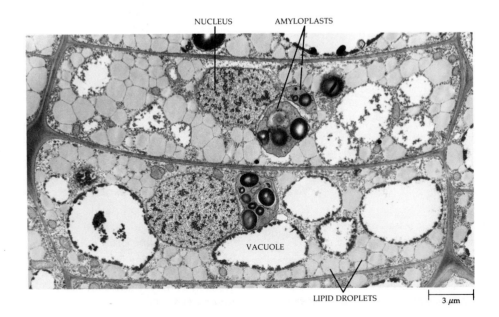

3–8
Two cambial cells from the fleshy, underground stem (corm) of the quillwort Isoetes muricata. *During winter these cells contain a large quantity of stored oil in the form of lipid droplets. In addition, carbohydrate in the form of starch grains is stored within amyloplasts. Several vacuoles can be seen in each of these cells. The dense material lining the vacuoles is tannin.*

the melting point of the molecule, and thus triglycerides containing unsaturated fatty acids tend to be liquid (oily) at room temperature. Examples found in plants are safflower oil, peanut oil, and corn oil. Triglycerides found in animals tend to have more saturated fatty acids and are thus solid (for example, lard and butter) at room temperature.

CUTIN, SUBERIN, AND WAXES

As mentioned in Chapter 2, **cutin** and **suberin** are unique, insoluble lipid polymers that are important structural components of many plant cell walls. (Suberin also contains a lignin-like component.) The major function of these polymers is to form a matrix in which **waxes**—long-chain lipid compounds—are embedded. Together the cutin-wax or suberin-wax combinations form barrier layers that help prevent the loss of water and other molecules from the aerial, or aboveground, parts of the plant. The waxes, in particular, constitute the major barrier.

Cutin, together with its embedded waxes, forms the **cuticle,** which covers the outer walls of epidermal cells (see Chapter 1). As seen in Figure 3–10, the cuticle consists of several layers. The first layer is composed of waxes deposited on the surface as epicuticular wax (Figure 3–11). Beneath the epicuticular wax is the cuticle proper, consisting of cuticular wax and cutin. This layer may be followed by one or more so-called cuticular layers consisting of cellulose, cutin, and wax. Finally, a layer of pectin may occur between the cuticle and the cell wall.

Radiocarbon Dating

All organic materials contain carbon. All of this carbon previously existed in the form of carbon dioxide and made its way into living organisms by means of photosynthesis. Most of the carbon atoms present in carbon dioxide have an atomic weight of 12 (^{12}C), but a fixed proportion of the atoms are carbon 14 (^{14}C), a radioactive isotope of carbon.

Carbon 14 is produced as a result of bombardment by high-energy particles from outer space and occurs in small amounts as heavy carbon dioxide. Plant cells use carbon dioxide to make carbohydrates and other organic molecules, accepting $^{14}CO_2$ almost as readily as $^{12}CO_2$. All animals are directly or indirectly dependent upon the products of plant metabolism for food; thus a fixed proportion of carbon atoms in the tissues of all living things is radioactive carbon 14. After death, carbon is no longer ingested and the proportions shift, with the radioactive carbon 14 decaying slowly and the amount of carbon 12 remaining the same. Carbon 14 has a half-life of 5730 years, so a fossil of this age should contain just half the carbon 14 of a living plant. Objects such as fossils or even structures made of wood or some other once-living material can be dated quite accurately by measuring the ratio of ^{14}C to ^{12}C.

Radiocarbon dating is particularly useful for studying archeological remains. This method of dating depends on the assumption that the proportion of ^{14}C to ^{12}C has remained constant in the atmosphere within the time span under study. Atmospheric nuclear testing has made it impossible for future archeologists to date current history in this way.

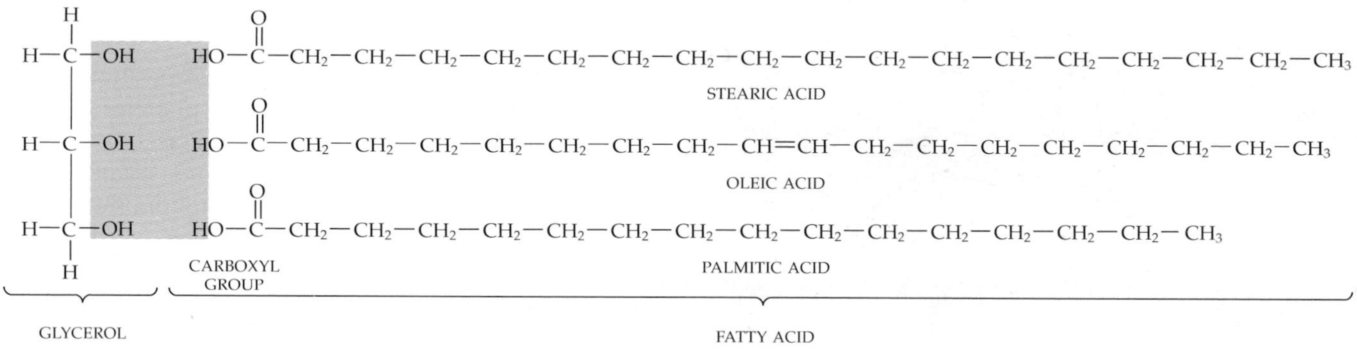

3–9

A fat molecule consists of three fatty acids joined to a glycerol molecule (hence the term "triglyceride"). The long hydrocarbon chain of each fatty acid terminates in a carboxyl (–COOH) group, which becomes covalently bonded to the glycerol molecule. Each bond is formed when a molecule of water (color) is removed (condensation). The physical properties of a fat—such as its melting point—are determined by the lengths of its fatty acid chains and by whether the chains are saturated or unsaturated. Three different fatty acids are shown here. Stearic acid and palmitic acid are saturated, and oleic acid is unsaturated, as you can see by the double bond in its hydrocarbon chain.

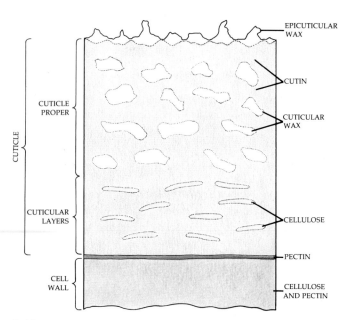

3–10

Diagram showing various layers of the cuticle on the outer wall of an epidermal cell. The outer layer, the epicuticular layer, is composed entirely of wax. The second layer, the cuticle proper, consists of wax and cutin, whereas the following layer or layers, the cuticular layers, are made up of cellulose and cutin. Wax may also be present in the cuticular layers. A layer of pectin may occur between the cuticle and cell wall, which itself is composed of cellulose and pectin.

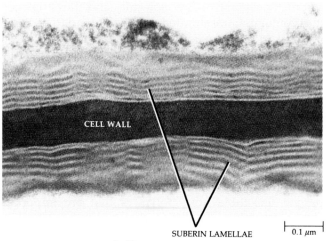

3–12

Electron micrograph showing the suberin lamellae in the walls between two cork cells of potato (Solanum tuberosum). *Note the alternating light and dark bands, which are interpreted as consisting of wax and suberin, respectively.*

Suberin, a major component of all cork cell walls, is also found in the Casparian strip of endodermal cells (see page 488) and the bundle-sheath cell walls in the leaves of grasses (see page 505). As seen with the electron microscope, suberized walls have a lamellar, or layered, appearance, with light bands alternating with dark bands (Figure 3–12). The light bands are believed to be composed of waxes and the dark bands of suberin. Some cuticles also have a lamellar appearance.

PHOSPHOLIPIDS

Closely related to the triglycerides are the phospholipids—various compounds in which glycerol is attached to only two fatty acids, with the third position on the glycerol molecule occupied by a phosphate group (Figure 3–13). Phospholipids are very important in cellular structure, particularly in the cellular membranes.

The phosphate end of the phospholipid molecule is hydrophilic ("water-loving") and thus is soluble in water, in contrast to the hydrophobic fatty acids. If phospholipids are added to water, they tend to form a film along its surface, with their polar heads under the water and their insoluble fatty acid chains (tails) protruding above the surface (Figure 3–14a). In the watery interior of the cell, phospholipids tend to align themselves in double layers, with the insoluble fatty acids oriented toward one another and the phosphate ends directed outward (Figure 3–14b). Such configurations are important in the structure of cellular membranes (see page 66).

3–11

Scanning electron micrograph of the upper leaf surface of Eucalyptus cloeziana, *showing deposits of epicuticular wax. Beneath these deposits is the cuticle, a wax-containing layer covering the outer walls of the epidermal cells. Waxes help protect exposed plant surfaces from water loss.*

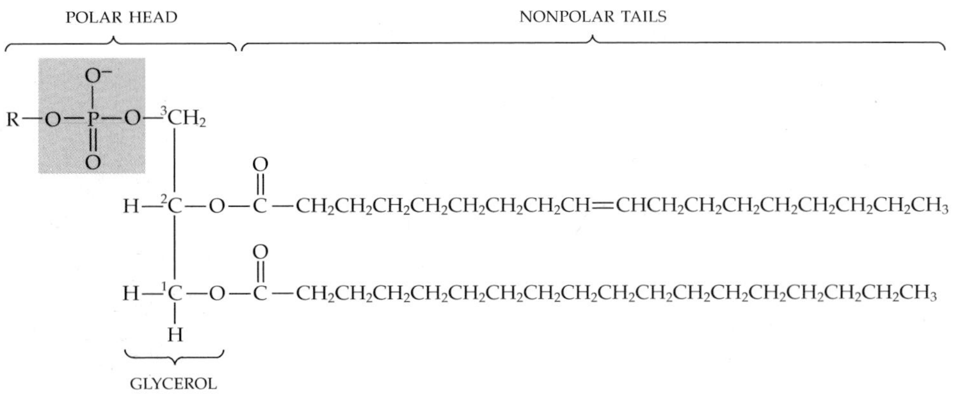

3–13

A phospholipid molecule consists of two fatty acids linked to a glycerol molecule, as in a fat, and a phosphate group (indicated by color) linked to the third carbon of the glycerol. The phosphate group usually contains an additional chemical group, indicated by the letter R. The fatty acid tails are nonpolar and uncharged and are therefore hydrophobic (insoluble in water); the polar head containing the phosphate and R group is hydrophilic (soluble in water).

POLAR HEAD   NONPOLAR TAILS

$$R-O-\underset{\underset{O}{\overset{\overset{O^-}{|}}{\underset{||}{P}}}}{}-O-^3CH_2$$

$$H-^2C-O-\overset{\overset{O}{||}}{C}-CH_2CH_2CH_2CH_2CH_2CH_2CH_2CH=CHCH_2CH_2CH_2CH_2CH_2CH_2CH_2CH_3$$

$$H-^1C-O-\overset{\overset{O}{||}}{C}-CH_2CH_2CH_2CH_2CH_2CH_2CH_2CH_2CH_2CH_2CH_2CH_2CH_2CH_2CH_2CH_3$$

$$H$$

GLYCEROL

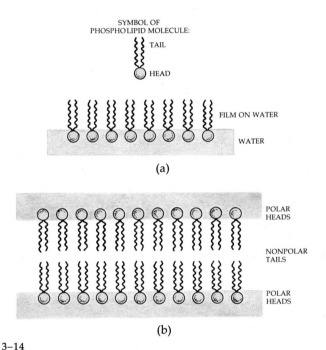

SYMBOL OF
PHOSPHOLIPID MOLECULE:

TAIL

HEAD

FILM ON WATER

WATER

(a)

POLAR HEADS

NONPOLAR TAILS

POLAR HEADS

(b)

3–14

(a) Because phospholipids have water-soluble heads and water-insoluble tails, they tend to form a thin film on a water surface, with their tails extending above the water. (b) Surrounded by water, phospholipids spontaneously arrange themselves in two layers, with their hydrophilic heads extending outward (into the water) and their hydrophobic tails inward (away from the water). This arrangement forms the structural basis of cellular membranes.

Proteins

Proteins are polymers of nitrogen-containing molecules known as **amino acids**. The same basic set of 20 amino acids is used in the formation of all proteins, but because protein molecules are large and complex—often containing several hundred amino acids—the number of different amino acid sequences and, hence, the possible variety of protein molecules are enormous. As shown below, the amino acid sequence in a protein is very important because it determines the structure and therefore the function of that protein. A single cell of the bacterium *Escherichia coli* can contain 600 to 800 different kinds of proteins at any one time, and the cell of a plant or animal probably has several times that number.

In plants, the largest concentration of proteins is found in some seeds, in which as much as 40 percent of the dry weight may be protein. These proteins appear to function as storage forms of amino acids that are used by the embryo when it resumes growth upon germination of the seed.

AMINO ACIDS

Every amino acid has the same basic structure, consisting of an amino group, a carboxyl group, and a hydrogen atom bonded to a central carbon atom. In addition, every amino acid has an "R" group—an atom or a group of atoms—bonded to the central carbon (Figure 3–15a, b). It is the R group that determines the identity of any particular amino acid. Figure 3–15c shows the complete structure of the 20 amino acids found in proteins. The amino acids are grouped according to their polarity and electrical charge, which are important in determining the properties of the various amino acids and, in particular, of the proteins formed from their combination.

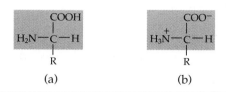

(a)   (b)

NONPOLAR (HYDROPHOBIC)

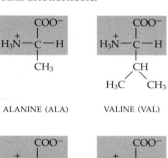

ALANINE (ALA)   VALINE (VAL)

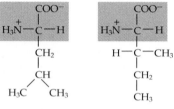

LEUCINE (LEU)   ISOLEUCINE (ILE)

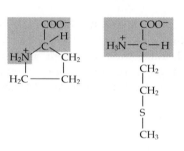

PROLINE (PRO)   METHIONINE (MET)

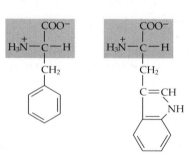

PHENYLALANINE (PHE)   TRYPTOPHAN (TRP)

POLAR, UNCHARGED (HYDROPHILIC)

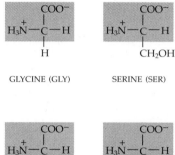

GLYCINE (GLY)   SERINE (SER)

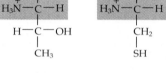

THREONINE (THR)   CYSTEINE (CYS)

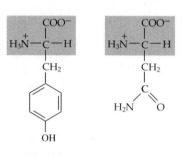

TYROSINE (TYR)   ASPARAGINE (ASN)

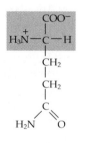

GLUTAMINE (GLN)

ACIDIC, NEGATIVELY CHARGED (HYDROPHILIC)

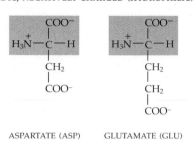

ASPARTATE (ASP)   GLUTAMATE (GLU)

BASIC, POSITIVELY CHARGED (HYDROPHILIC)

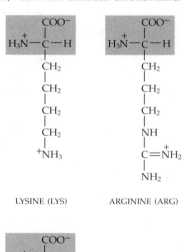

LYSINE (LYS)   ARGININE (ARG)

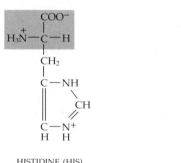

HISTIDINE (HIS)

(c)

3–15

(a) *The general formula of an amino acid. Every amino acid contains an amino group (–NH₂) and a carboxyl group (–COOH) bonded to a central carbon atom. A hydrogen atom and a side group (R) are also bonded to the same carbon atom. This basic structure is the same in all amino acids. The side group*

is different in each kind of amino acid. (b) At pH 7, both the amino and the carboxyl groups are ionized. (c) The 20 amino acids used in making proteins. As you can see, the essential structure is the same in all 20 molecules, but the side groups differ. Amino acids with uncharged polar side groups are relatively

hydrophilic and are usually on the outside of proteins, whereas the side chains on nonpolar amino acids tend to aggregate on the inside. Amino acids with acidic and basic side groups are very polar, and they are almost always found on the outside of protein molecules.

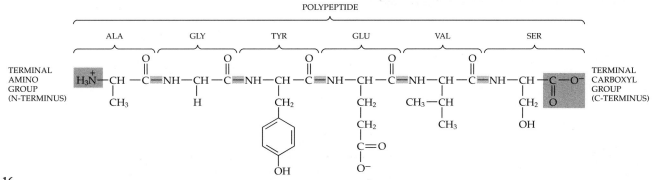

3–16

The links between amino acid residues are known as peptide bonds (blue). Peptide bonds are formed by the removal of a molecule of water (condensation).

The bonds always form between the carboxyl (–COO⁻) group of one amino acid and the amino (–NH₃⁺) group of the next. Consequently, the basic structure of

a protein is a long, unbranched molecule. This linear arrangement of the amino acids is known as the primary structure of the protein.

POLYPEPTIDES

A chain of amino acids is known as a **polypeptide.** As in the case of polysaccharides and triglycerides, a molecule of water is removed to form each link in the chain. The amino group of one amino acid always links to the carboxyl group of the next amino acid; this bond is known as the **peptide bond** (Figure 3–16). The amino acid subunits of a polypeptide are often referred to as amino acid residues. Because of the way in which the amino acids are bonded together, there is always a free amino group at one end of the chain (the N-terminus) and a free carboxyl group at the other (the C-terminus).

Polypeptide chains can also be cross-linked by disulfide bonds, which are covalent sulfur-to-sulfur bonds formed between two molecules of the amino acid cysteine. A disulfide bond may link two cysteine residues in the same polypeptide chain or in two different chains.

Proteins are large polypeptides; they have molecular weights ranging from 10^4 (10,000) to more than 10^6 (1,000,000). In comparison, water has a molecular weight of 18, and glucose has a molecular weight of 180. (The average molecular weight of an amino acid residue is about 120; if you remember this figure, you can rapidly calculate the approximate number of amino acids in a protein on the basis of its total molecular weight. The way in which molecular weights are determined is described in Appendix A.)

LEVELS OF PROTEIN ORGANIZATION

The sequence of amino acids in the polypeptide chain is referred to as the **primary structure** of the protein. As the polypeptide chain is assembled, regular hydrogen-bond interactions take place between the peptide bonds along the chain, giving rise to alpha helices and beta pleated sheets, which make up the protein's **secondary structure.** Certain combinations of alpha helices and beta pleated sheets may be folded to produce a so-

called globular protein molecule **(tertiary structure).** Finally, the polypeptide chains of a protein having two or more chains may interact to form a **quaternary structure.**

Primary Structure

The primary structure of a protein is simply the linear sequence of amino acids in the polypeptide chain (Figures 3–16 and 3–18a). Each kind of protein has a different primary structure—a unique protein "word" consisting of a unique sequence of amino acid "letters."

Secondary Structure

The most common secondary structure is the *alpha helix,* which resembles a spiral staircase (Figures 3–17a and 3–18b). The alpha helix is rigid and very uniform in its geometry, with a turn of the helix occurring every 3.6 amino acids. The rigid structure is maintained by hydrogen bonds across successive turns of the spiral, with each peptide bond regularly hydrogen-bonded to another peptide bond nearby in the chain.

In *beta pleated sheets,* there are no hydrogen bonds between peptide bonds within a chain. Instead, hydrogen bonding occurs between peptide bonds of adjacent polypeptide chains (Figure 3-17b).

Tertiary and Quaternary Structure

Polypeptide chains may also fold up to form globular structures—the tertiary structure of the protein (Figure 3–18c). Most biologically active proteins, such as enzymes, are globular. The specialized proteins of cellular membranes are also globular in structure. In addition, microtubules are composed of a large number of spherical subunits, each of which is a globular protein (Figure 3–19).

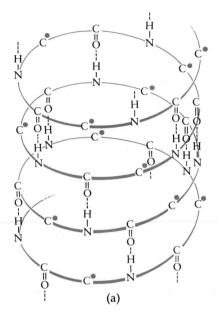

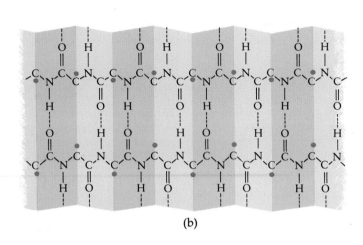

(a)

(b)

3–17

Protein secondary structures. (a) The alpha helix. In this representation, the R groups and H atoms are omitted from the carbons indicated by red dots. The helix is held in shape by hydrogen bonds, indicated by the dashed lines. The bonds form between the oxygen atom of the

carboxyl group of one peptide bond and the hydrogen of the amino group in another peptide bond that occurs four amino acids farther along the chain. Every peptide bond of the chain participates in such hydrogen bonding. (b) The

beta pleated sheet. The pleats are formed by hydrogen bonding between atoms of the backbone of the polypeptide; the R groups, which are attached to the carbons indicated by red dots, extend above and below the folds of the pleat.

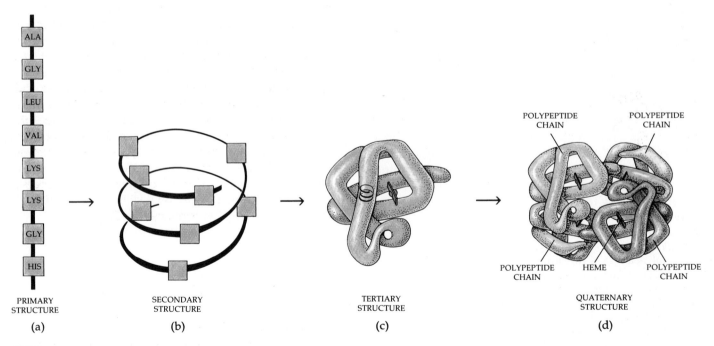

PRIMARY STRUCTURE (a)

SECONDARY STRUCTURE (b)

TERTIARY STRUCTURE (c)

QUATERNARY STRUCTURE (d)

3-18

The four levels of protein organization. (a) The primary structure of a protein consists of a linear sequence of amino acids linked together by peptide bonds.

(b) The polypeptide chain may coil into an alpha helix, one form of secondary structure. (c) The alpha helix folds up to form a three-dimensional, globular

structure, the tertiary structure. (d) The combination of several chains into a single functional molecule is the quaternary structure.

The *tertiary structure* of a protein is determined by its primary structure, or amino acid sequence. The tertiary structure is maintained by hydrogen bonds between the R groups of amino acids in adjacent loops of the chain and by interactions among the various amino acid R groups, based on their various charges (Figure 3–15c), and between the amino acid R groups and water molecules. These bonds are relatively weak and can be broken quite easily by physical or chemical changes in the environment, such as heat or increased acidity. This breakdown is called *denaturation.* The coagulation of egg white when an egg is cooked is a common example of protein denaturation. When proteins are denatured, the polypeptide chains are unfolded, causing a loss of the biological activity of the protein.

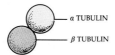

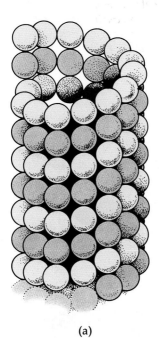

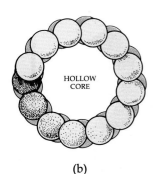

HOLLOW CORE

(a)   (b)

3–19

Microtubules play many roles in the biology of the cell (page 30). (a) Longitudinal view of a short portion of a microtubule, which is assembled from complexes of two proteins, alpha and beta tubulin, to form 13 vertical "protofilaments" around a hollow core. The subunits first come together to form a soluble dimer. The dimers then self-assemble into insoluble hollow tubules. (b) Transverse section of a microtubule, showing the profiles of the 13 protofilaments.

Organisms cannot live at extremely high temperatures because their enzymes and other proteins become unstable and nonfunctional due to denaturation.

Many proteins are composed of more than one polypeptide chain. These chains may be held to each other by hydrogen bonds, disulfide bridges, hydrophobic forces, attractions between positive and negative charges, or, most often, by a combination of these types of interactions. This level of organization of proteins, which involves the interaction of two or more polypeptides, is called a *quaternary structure* (Figure 3–18d).

ENZYMES

Enzymes are large, complex globular proteins that act as *catalysts,* substances that accelerate the rate of a chemical reaction by lowering the energy of activation (see Appendix A), but that remain unchanged in the process. Because they remain unaltered, catalysts can be used over and over again and so are typically effective in very small amounts.

In the laboratory, the rates of chemical reactions are usually accelerated (up to a point) by the application of heat, which increases the force and frequency of collisions among molecules. In nature, however, hundreds of different reactions are going on in the cell at the same time, and heat would speed up all these reactions indiscriminately. Moreover, heat would melt the lipids, denature the proteins, and have other generally destructive effects on the cell. Because of enzymes, cells are able to carry out chemical reactions at great speeds and at relatively low temperatures. If enzymes were not present, the reactions would occur anyway, but at a rate so slow that their effects would be negligible.

Enzymes are sometimes named by adding the ending *-ase* to the root of the name of the substrate (the reacting molecule or molecules). Amylase catalyzes the hydrolysis of amylose (starch), and sucrase catalyzes the hydrolysis of sucrose into glucose and fructose. Nearly 2000 different enzymes are now known, and each of them is capable of catalyzing some specific chemical reaction. The behavior of enzymes in biological reactions is further explained in Chapter 5.

Nucleic Acids

The information dictating the structures of the enormous variety of protein molecules found in living organisms is encoded in and translated by molecules known as **nucleic acids.** Just as proteins consist of long chains of amino acids, nucleic acids consist of long chains of **nucleotides.** A nucleotide, however, is a more

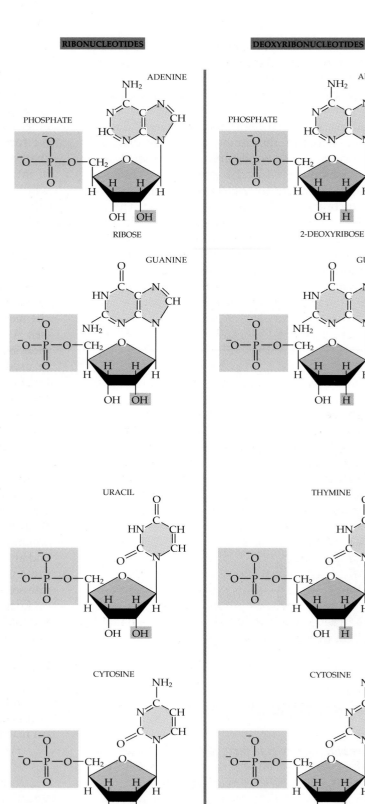

3–20

The building blocks of RNA and DNA. Each nucleotide building block contains a phosphate group, a sugar, and a nitrogenous base, which can be either a purine or a pyrimidine.

The purines in DNA and RNA are the same, but one type of DNA pyrimidine nucleotide contains thymine, whereas its RNA counterpart contains uracil. The only other difference between DNA and RNA is the presence of one more oxygen atom on the sugar (ribose) component of the RNA. As described in Chapter 8, however, the biological roles of the two nucleic acids are profoundly different.

complex molecule than an amino acid. As shown in Figure 3–20, it consists of three subunits: a phosphate group, a five-carbon sugar, and a nitrogenous base—a molecule that has the properties of a base and contains nitrogen.

The sugar subunit of a nucleotide may be either ribose or deoxyribose, which contains one less oxygen atom than ribose (Figure 3–20). Ribose is the sugar in the nucleotides that form **ribonucleic acid (RNA),** and deoxyribose is the sugar in the nucleotides that form **deoxyribonucleic acid (DNA).**

Five different nitrogenous bases are found in the nucleotides that are the building blocks of nucleic acids. Two of these, adenine and guanine, have a two-ring structure and are known as *purines.* The other three, cytosine, thymine, and uracil, have a single ring structure and are known as *pyrimidines.* Adenine, guanine, and cytosine are found in both DNA and RNA, while thymine is found only in DNA and uracil only in RNA. As we shall see in Chapter 5, adenine and ribose are also found in the nucleotides that are essential participants in the chemical reactions occurring within living systems.

Although their chemical components are very similar, DNA and RNA play very different biological roles. DNA is the primary constituent of the chromosomes of the cell and is the carrier of the genetic message. The primary function of RNA is to translate the genetic message present in DNA into proteins. Some RNA molecules function as enzymelike catalysts (sometimes referred to as ribozymes). The discovery of the structure and function of the nucleic acids is undoubtedly the greatest triumph thus far of the molecular approach to the study of biology. In Section 2, we shall trace the events leading to the key discoveries and shall consider in some detail the marvelous processes—the details of which are still being worked out—by which these molecules perform their functions.

Other Nucleotide Derivatives

Nucleotides and compounds derived from them serve a variety of functions within the cell. Two nucleotide derivatives of prime importance are ATP (adenosine triphosphate) and ADP (adenosine diphosphate) (see Figure 5–10). Almost all energy exchanges within the cell involve the transfer of phosphate groups; ATP and ADP are the most important molecules in phosphate transfer. The role of these energy-exchange molecules is discussed in Chapter 5.

Nucleotides play important roles in a variety of other energy-exchange molecules. One such compound is NAD$^+$ (nicotinamide adenine dinucleotide), which contains a nucleotide of adenine plus a phosphate (see Figure 5–8). Widespread use of the purine-sugar-phosphate combination in energy transfers may have to do with the fact that these comparatively large, charged molecules do not pass through membranes and thus cannot "escape" into and out of cells or membrane-bound organelles.

Summary

Living matter is composed of only a few of the naturally occurring elements. The bulk of living matter is water. Most of the rest of living material is composed of organic molecules—carbohydrates, lipids, proteins, and nucleic acids.

Carbohydrates serve as a primary source of chemical energy for living systems and as important structural elements in cells. The simplest carbohydrates are the monosaccharides, such as glucose and fructose. Monosaccharides can be combined to form disaccharides, such as sucrose, and polysaccharides, such as starch and cellulose. These molecules can usually be broken apart by the addition of a water molecule at each linkage, a chemical reaction known as hydrolysis.

Lipids are another source of energy and structural material for cells. Compounds in this group—fats, oils, cutin, suberin, waxes, and phospholipids—are generally insoluble in water.

Proteins are very large molecules composed of long chains of amino acids; these chains are known as polypeptides. The 20 different amino acids used to form proteins vary according to their R groups. From these amino acids, enormous numbers of different protein molecules are built. The principal levels of protein organization are: (1) primary structure, the linear amino acid sequence; (2) secondary structure, the alpha-helix or beta-pleated-sheet formation of the polypeptide chain; (3) tertiary structure, the folding of the alpha helices and beta pleated sheets into globular shapes; and (4) quaternary structure, which results from specific interactions between two or more polypeptide chains.

Enzymes act as catalysts in chemical reactions; most enzymes are globular proteins. Because of enzymes, cells are able to accelerate the rate of chemical reactions at relatively low temperatures.

Nucleotides are complex molecules consisting of a phosphate group, a five-carbon sugar, and a nitrogenous base. They are the building blocks of the nucleic acids deoxyribonucleic acid (DNA) and ribonucleic acid (RNA), which transmit and translate the genetic information. Some RNA molecules function as catalysts. Two nucleotide derivatives—ATP and ADP—are involved in most energy exchanges within the cell.

CHAPTER 4

The Movement of Substances into and out of Cells

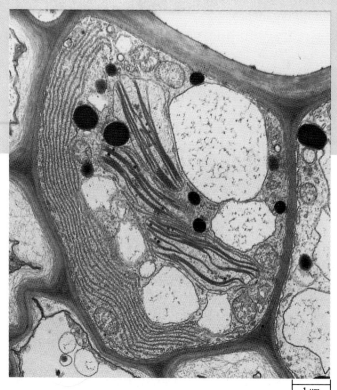

4-1

Electron micrograph of a leaf cell from a "grass fern," Vittaria guineensis. *In addition to the plasma membrane bordering the wall, numerous other membranes permeate the protoplast. Especially conspicuous here are the parallel arrays of endoplasmic reticulum (left), the internal membranes of the chloroplasts (center), and the tonoplasts, which form the boundaries of the vacuoles (clear areas containing flocculent material). Cellular membranes such as these have the ability to regulate the passage of substances across them.*

All cells are separated from their surroundings by a surface membrane—the plasma membrane. Eukaryotic cells are further divided internally by a variety of membranes, including the endoplasmic reticulum, dictyosomes, and the bounding membranes of organelles (Figure 4–1). These membranes are not impenetrable barriers, for cells are able to regulate the amount, kind, and often the direction of movement of substances that pass across their membranes. This ability is an essential capacity of living cells because few metabolic processes could occur at reasonable rates if they depended upon the concentration of necessary substances found in the cell's surroundings. In fact, one criterion by which we identify living systems is the difference between the concentration of a variety of substances in living matter and in the surrounding, nonliving environment.

Control of the exchange of substances across membranes depends on the physical and chemical properties of the membranes and of the ions or molecules that move through them. Water is the most important of the molecules moving into and out of cells.

Principles of Water Movement

The movement of water, whether in living systems or in the nonliving world, is governed by three basic processes: bulk flow, diffusion, and osmosis.

BULK FLOW

Bulk flow, the overall movement of water (or some other liquid), occurs in response to differences in the potential energy of water, usually referred to as **water potential.**

4–2

Water at the top of a falls, like a boulder on a hilltop, has potential energy. The movement of water molecules as a group, as from the top of the falls to the bottom, is referred to as bulk flow.

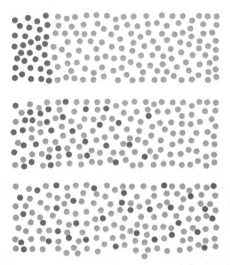

4–3

A diagram of diffusion, which is the result of the random movement of individual molecules. Such random movement produces a net movement from an area in which a substance has a high concentration to an area in which that substance has a lower concentration. Notice that as one type of molecule (indicated by red) diffuses to the right, the other (indicated by blue) diffuses in the opposite direction. The result is an even distribution of the two types of molecules. Can you see why the net movement of molecules will slow down as equilibrium (even distribution) is approached?

A simple example of water that has potential energy is water behind a dam or at the top of a waterfall (Figure 4–2). As this water runs downhill, its potential energy can be converted to mechanical energy by a waterwheel or to mechanical and then electrical energy by a hydroelectric turbine.

Pressure is another source of water potential. If we put water into a rubber bulb and squeeze the bulb, this water, like the water at the top of a waterfall, has water potential, and it will move to an area of lower water potential. Can we make the water that is running downhill run uphill by means of pressure? Yes, we can, but only so long as the water potential produced by the pressure exceeds the water potential produced by gravity. *Water moves from an area of higher water potential to one of lower water potential, regardless of the reason for the difference in potential.*

The concept of water potential is useful because it enables physiologists to predict the way in which water will move under various conditions. Water potential is usually measured in terms of the pressure required to stop the movement of water—that is, the hydrostatic pressure—under the particular circumstances involved. The units used to express this pressure are the bar and megapascal (MPa). (These are metric units of pressure. A bar is equal to the average pressure of the air at sea level—about 70.3 grams per square centimeter. One megapascal is equal to 0.1 bar.) By convention, the water potential of pure water is set at zero, so that the water potential of an aqueous solution will have a negative value (less than zero).

DIFFUSION

Diffusion is a familiar phenomenon. If a few drops of perfume are sprinkled in one corner of a room, the scent will eventually permeate the entire room even if the air is still. If a few drops of dye are put in one end of a tank of water, the dye molecules will slowly become evenly distributed throughout the tank (Figure 4–3). This process may take a day or more, depending on the size of the tank, the temperature, and the size of the dye molecules.

Why do the dye molecules move apart? If you could observe the individual dye molecules in the tank, you would see that each one moves individually and at random. Imagine a thin section through the tank, running from top to bottom. Dye molecules will move into and out of the section, some moving in one direction, some moving in the other. But you would see more dye molecules moving from the side of greater dye concentration. Why? Simply because there are more dye molecules at that end of the tank. For example, if there are more dye molecules on the left, more of them will move randomly to the right, even though there is an equal probability that any one dye molecule will move from right to left. Consequently, the overall (net) movement of dye molecules will be from left to right. Similarly, the net movement of water molecules is from right to left.

What happens when all the molecules are evenly distributed throughout the tank? Their even distribution does not affect the behavior of the molecules as individuals; they still move at random. But there are now as many molecules of dye and as many molecules of water on one side of the tank as on the other, and so there is no net direction of motion. There is, however, just as much individual motion as before, provided the temperature has not changed.

Substances that are moving from a region of higher concentration to a region of lower concentration are said to be moving *down or along a concentration gradient.* Diffusion occurs only down a concentration gradient. A substance moving in the opposite direction, toward a higher concentration of its own molecules, would be moving *against a concentration gradient,* which is analogous to being pushed uphill. The steeper the downhill gradient—that is, the larger the difference in concentration—the more rapid the net movement. Also, diffusion is more rapid in gases than in liquids and is more rapid at higher than at lower temperatures. Can you explain why?

Notice that there are two concentration gradients in our tank; the dye molecules are moving along one of them, and the water molecules are moving in the opposite direction along the other. The two kinds of molecules are moving independently of each other. In both cases, the movement is down a gradient. When the molecules have reached a state of equal distribution (that is, when there are no more gradients), they continue to move, but now there is no net movement in either direction. In other words, the net transfer of the molecules is zero; the system may be said to be in a state of **dynamic equilibrium.**

The concept of water potential is also useful in understanding diffusion. A high concentration of dye molecules in one region of the tank means a low concentration of water molecules there and, thus, low water potential. If the pressure is equal everywhere, water molecules, as they move down the concentration gradient, are moving from a region of higher water potential to a region of lower water potential. When dynamic equilibrium is reached, water potential is equal in all parts of the tank.

Imbibition

Water molecules exhibit a tremendous cohesiveness because of their polarity, that is, the difference in charge between one end of a water molecule and the other (see Appendix A). Similarly, because of this difference in charge, water molecules can cling (adhere) either to positively charged or to negatively charged surfaces. Many large biological molecules, such as cellulose, are polar and so attract water molecules. The adherence of water molecules is also responsible for the biologically important phenomenon called imbibition or, sometimes, hydration.

Imbibition (from the Latin *imbibere,* "to drink in") is the movement of water molecules into substances such as wood or gelatin, which swell, or increase in volume, as a result of the accumulation of water molecules. The pressures developed by imbibition can be astonishingly large. It is said that stone for the ancient Egyptian pyramids was quarried by driving wooden pegs into holes drilled in the rock face and then soaking the pegs with water. The swelling wood created a force that split the slab of stone. In living plants, imbibition occurs particularly in seeds, which may increase to many times their original size as a result. Imbibition is essential to the germination of the seed.

The net result of diffusion is that the diffusing substance eventually becomes evenly distributed. Briefly, **diffusion** may be defined as *the dispersion of substances by a movement of their ions or molecules, which tends to equalize their concentrations throughout the system.*

Cells and Diffusion

Except over very short distances, diffusion is essentially a slow process. It is efficient only if the concentration gradient is steep and the volume relatively small. For instance, the rapid spread of a perfume through the air is due primarily to air currents, not to diffusion. Similarly, in many cells, the transport of materials is speeded by active streaming of the cytoplasm. Cells also hasten diffusion by their own metabolic activities. For example, in a nonphotosynthesizing cell, oxygen is used up within the cell almost as rapidly as it enters, thereby maintaining a steep oxygen concentration gradient from outside to inside. Carbon dioxide is produced by the cell, and so a gradient from inside to outside the cell is maintained for carbon dioxide.

Similarly, within a cell, materials are often produced in one place and used in another. Thus, a concentration gradient can be established between two regions of the cell, and materials can diffuse down the gradient from the site of production to the site of use.

Most biological organic molecules are polar and therefore hydrophilic, and so cannot freely diffuse through the lipid barrier of cellular membranes. However, oxygen, which is soluble in lipids, diffuses freely through these same membranes. Carbon dioxide and water also diffuse in and out freely because, though polar, they are small, uncharged molecules.

OSMOSIS

While permitting passage of water, cellular membranes block the passage of most dissolved substances. (The dissolved substances are called **solutes,** and the water is the **solvent** of the solution.) Such membranes are known as **differentially permeable membranes;** diffusion of water through them is known as **osmosis.** Osmosis involves a net flow of water from a solution that has a high water potential to a solution that has a low water potential. In the absence of other factors that influence water potential (such as pressure), the movement of water by osmosis is from a region of lower solute concentration (and therefore of higher water concentration) into a region of higher solute concentration (and lower water concentration). The presence of solute decreases the water potential, creating a water potential gradient down which water moves. The water potential is affected not by the nature of the solute but only by how much solute the solution contains—the number of particles of solute (molecules or ions).

Osmosis results in a buildup of pressure as water molecules continue to diffuse across the membrane into a region of lower water concentration. If water is separated from a solution by a *semipermeable membrane* (a membrane that allows the ready passage of water but no passage of solute) in a system such as that shown in Figure 4–4, the water will move across the membrane and cause the solution to rise in the tube until an equilibrium is reached—that is, until the water potential is the same on both sides of the membrane. If enough pressure is applied on the solution in the tube by a piston, say, such as in Figure 4–4c, it is possible to prevent the net movement of water into the tube. The pressure that would have to be applied to the solution to stop water movement is called the *osmotic pressure.* The tendency of water to move across a membrane because of the effect of solutes on water potential is called the **osmotic potential** (also called solute potential).

Osmosis and Living Organisms

The movement of water across the plasma membrane in response to differences in water potential causes some crucial problems for living systems, particularly those in an aqueous environment. These problems vary according to whether the water potential of the cell or organism is higher than, similar to, or lower than the water potential of its environment. For example, the water potential of one-celled organisms that live in salt water is usually similar to the water potential of the medium they inhabit, which is one way of solving the problem.

Many types of cells live in environments with relatively high water potentials. In many freshwater single-celled organisms, such as *Euglena*, the water potential of the cell is lower than that of the surrounding medium; consequently, water tends to move into the cell by osmosis. If too much water moves into the cell, it can dilute the cell contents to the point of interfering with function and can eventually even rupture the plasma membrane. This is prevented by a specialized organelle known as a contractile vacuole, which collects water from various parts of the cell body and pumps it out of the cell with a rhythmic contraction.

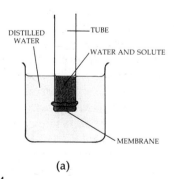

(a)

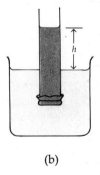

(b)

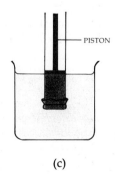

(c)

4–4

Osmosis and osmotic potential. (a) The tube contains a solution; the beaker contains distilled water. (b) The semipermeable membrane permits the passage of water but not solute. The movement of water across the membrane by osmosis and into the solution causes the solution *to increase in volume and thus to rise in the tube. The solution will rise until the osmotic potential—the tendency of water to move across the membrane into the region of lower water concentration—is counter-balanced by the pressure exerted* *by the column of solution, which is proportional to the height h and density of the solution. (c) The force that must be applied to the piston to oppose the rise of the solution in the tube is called the osmotic pressure.*

TURGOR

If a plant cell is placed in a solution with a relatively high water potential, the protoplast expands and the plasma membrane stretches and exerts pressure against the cell wall. However, the plant cell does not rupture because it is restrained by the relatively tough cell wall.

Plant cells tend to concentrate relatively strong solutions of salts within their vacuoles, and they can also accumulate sugars, organic acids, and amino acids. As a result, plant cells constantly absorb water by osmosis and build up their internal hydrostatic pressure. This pressure against the cell wall keeps the cell **turgid,** or stiff. Consequently, the hydrostatic pressure in plant cells is commonly referred to as turgor pressure. **Turgor pressure** is defined as *the pressure that develops in a plant cell as a result of osmosis and/or imbibition* (see the essay "Imbibition," page 64). Equal to and opposing the turgor pressure at any instant is the inwardly directed mechanical pressure of the cell wall, called the **wall pressure.**

Turgor in the plant is especially important in the support of nonwoody plant parts. As discussed in Chapter 2, most of the growth of a plant cell is the direct result of water uptake, with the bulk of the increase in size of the cell resulting from enlargement of the vacuoles. The hormone auxin presumably contributes to this water uptake by relaxing the cell wall and thus decreasing the resistance the wall exerts to turgor pressure (see Chapter 25).

Turgor is maintained in most plant cells because they generally exist in a medium with a relatively high water potential. However, if a turgid plant cell is placed in a solution with a relatively low water potential, water will leave the cell by osmosis, and the vacuole and protoplast will shrink, causing the plasma membrane to pull away from the cell wall (Figure 4–5). This phenomenon is known as **plasmolysis.** The process can be reversed if the cell is then transferred to pure water. Figure 4–6 shows *Elodea* leaf cells before and after plasmolysis. Although the plasma membrane and the tonoplast—the membrane surrounding the vacuole—are, with few exceptions, permeable only to water, the cell walls allow both solutes and water to pass freely through them. The loss of turgor by plant cells may result in *wilting,* or drooping of leaves.

Structure of Cellular Membranes

All of the membranes of the cell have the same basic structure, consisting of a *lipid bilayer* in which are embedded globular proteins, many of which extend across the bilayer and protrude on either side (Figure 4–7). The portion of these *transmembrane proteins* embedded in the bilayer is hydrophobic, whereas the portion or portions exposed on either side of the membrane are hydrophilic.

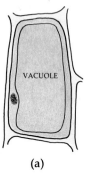
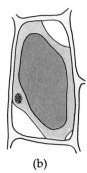
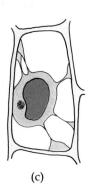

(a)   (b)   (c)

4–5

Plasmolysis in a leaf epidermal cell. (a) Under normal conditions, the protoplast is in close contact with the cell wall. (b) When the cell is placed in a relatively strong sugar solution, water passes out of the cell into the sugar solution and the protoplast contracts slightly. (c) When immersed in a stronger (more concentrated) sugar solution, the cell loses even larger amounts of water and the protoplast contracts still further.

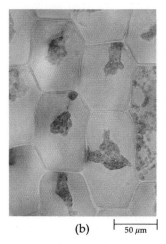

(a)   (b)   ├── 50 µm ──┤

4–6

Elodea leaf cells. (a) Turgid cells and (b) cells after being placed in a relatively strong sucrose solution. The cells in (b) are plasmolyzed.

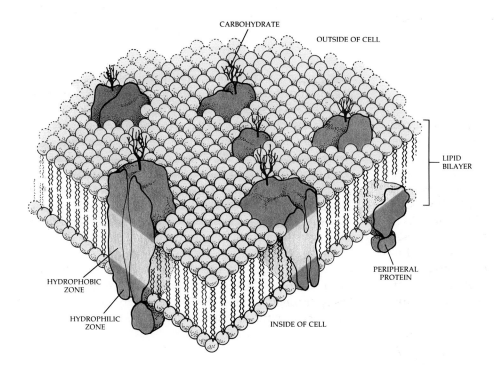

CARBOHYDRATE

OUTSIDE OF CELL

LIPID
BILAYER

PERIPHERAL
PROTEIN

HYDROPHOBIC
ZONE

HYDROPHILIC
ZONE

INSIDE OF CELL

4–7
Fluid-mosaic model of membrane structure. The membrane is composed of a bilayer (double layer) of lipid molecules —with their hydrophobic "tails" facing inward—and large protein molecules. The proteins traversing the bilayer are known as transmembrane proteins. Protruding from the bilayer, on the inner surface of the membrane, other proteins, called peripheral proteins, are attached to some of the transmembrane proteins. The portion of a transmembrane protein molecule embedded in the lipid bilayer is hydrophobic; the portion or portions exposed on either side of the membrane are hydrophilic. Short carbohydrate chains are attached to the outside of the plasma membrane. The whole structure is quite fluid, and hence the proteins can be thought of as floating in a lipid "sea."

The two surfaces of a membrane differ considerably in chemical composition. For example, there are two major types of lipids in the plasma membrane of plant cells—phospholipids (the more abundant) and sterols (particularly stigmasterol; not cholesterol, which is the major sterol in animal tissues)—and the two layers of the bilayer have different concentrations of each. Moreover, the transmembrane proteins have definite orientations within the bilayer, and the portions protruding on either side have different amino acid compositions and tertiary structures. Other proteins are also associated with membranes, including the so-called *peripheral proteins*, which are attached to some of the transmembrane proteins protruding from the bilayer on the inner surface of the membrane. Transmembrane proteins and other proteins tightly bound to the membrane are called *integral proteins*. On the outer surface, short-chain carbohydrates are attached to the protruding proteins. The carbohydrates, which form a coat on the outer surface of the membranes of some eukaryotic cells, are believed to play important roles in cell-to-cell adhesion processes and in the "recognition" of molecules (such as hormones, viruses, and antibiotics) that interact with the cell.

Two basic configurations have been identified among transmembrane proteins (Figure 4–8). One is a relatively simple rodlike structure consisting of an alpha helix embedded in the hydrophobic interior of the membrane, with less regular, hydrophilic portions extending on either side of it. The other configuration is found in large globular proteins with complex tertiary and quaternary structures that result from repeated "passes" through the membrane. In such "multipass" membrane proteins, the polypeptide chain usually crosses the lipid bilayer as a series of alpha helices.

Whereas the lipid bilayer provides the basic structure and impermeable nature of cellular membranes, the proteins are responsible for most membrane functions. Most membranes are composed of 40 to 50 percent lipid (by weight) and 60 to 50 percent protein, but the amounts and types of proteins in a membrane reflect its function. Membranes involved with energy transduction, such as the internal membranes of mitochondria and chloroplasts, consist of about 75 percent protein. Some of the proteins are enzymes that catalyze membrane-associated reactions, whereas others are carriers involved in the transport of specific molecules into and out of the cell or organelle. Still others act as

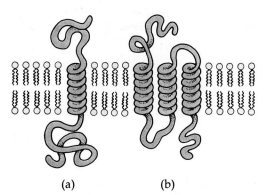

(a)   (b)

4–8
Two configurations of transmembrane proteins. Some extend across the lipid bilayer as a single alpha helix (a) and others—multipass proteins—as multiple alpha helices (b). The portions of protein protruding on either side of the membrane are hydrophilic; the intramembranous helical portions are hydrophobic.

receptors for receiving and transducing chemical signals from the cell's internal or external environment. Although some of the integral proteins appear to be anchored in place (perhaps to the cytoskeleton), the lipid bilayer is generally quite fluid. Some of the proteins float more or less freely in the bilayer, and they and the lipid molecules can move laterally within it, forming different patterns, or mosaics, that vary from time to time and place to place—hence the name "fluid-mosaic" for this model of membrane structure.

Transport of Solutes across Membranes

As mentioned previously, hydrophobic molecules (such as oxygen) and small, uncharged polar molecules (such as carbon dioxide and water) can permeate cell membranes freely by simple diffusion. The observation that hydrophobic molecules diffuse readily across plasma membranes provided the first evidence of the lipid nature of the membrane.

However, most substances required by cells are polar and require **transport proteins** to transfer them across membranes. Each transport protein is highly selective; it may accept one type of ion (such as Ca^{2+} or K^+) or molecule (such as a sugar or an amino acid) and exclude a nearly identical one. All transport proteins with known membrane orientations have proven to be multipass transmembrane proteins. These proteins provide the specific solutes they transport with a continuous pathway across the membrane without the solutes coming into contact with the hydrophobic interior of the lipid bilayer.

Transport proteins can be grouped into three broad classes: pumps, carriers, and channels (Figure 4–9). **Pumps** are driven by either chemical energy (ATP) or light energy, and in plant and fungal cells they typically are proton pumps (they are, in fact, the enzyme H^+-ATPase; ATPase is the ATP-hydrolyzing enzyme). Both carrier and channel proteins are driven by the energy of electrochemical gradients (see below). **Carriers** bind the specific solute being transported and undergo a conformational change in order to transport the solute across the membrane. **Channel proteins** form water-filled pores that extend across the membrane and, when open, allow specific solutes (usually inorganic ions of appropriate size and charge) to pass through them.

Another basis for the three categories of transport protein is the speed of transport. The number of solute molecules transported per protein per second is relatively slow with pumps (fewer than 500 per second), intermediate with carriers (500 to 10,000 per second), and most rapid with channels (10,000 to many millions per second).

If a molecule is uncharged, the direction of its transport is determined only by the difference in the concentration of the molecule on the two sides of the membrane (the **concentration gradient**). If a solute carries a net charge, however, both the concentration gradient and the total electrical gradient across the membrane (the membrane potential) influence its transport. Together, the two gradients constitute the **electrochemical gradient.** Plant cells typically maintain electrical gradients across the plasma membrane and the tonoplast. The ground substance is electrically negative relative to both the aqueous medium outside the cell and the solution (cell sap) inside the vacuole. Transport down a concentration gradient or an electrochemical gradient is called **passive transport.** Passive transport is exhibited by all channel proteins and some carrier proteins (Figure 4–9). Passive transport with the assistance of carrier proteins is called **facilitated diffusion.**

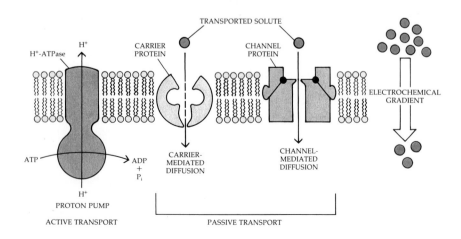

4–9
Diagram illustrating the three broad classes of transport proteins: pumps, carriers, and channel proteins. The proton pump (H^+-ATPase) is driven by the chemical bond energy of ATP. Because this process moves solutes (here H^+) against their concentration gradient, it is called active transport. Transport down a concentration gradient or electrochemical gradient by simple diffusion or by channel- or carrier-mediated diffusion is called passive transport.

Patch-Clamp Recording in the Study of Ion Channels

The membranes of both plant and animal cells contain ion channels, or channel proteins, which form pathways for passive ion movement. When activated, these ion channels become permeable to ions, allowing solute flow across the membrane. The opening and closing of ion channels is referred to as *gating*.

The first evidence for gated ion channels was obtained from electrophysiological experiments using intracellular microelectrodes. Such methods can be applied only to relatively large cells, however, and consequently the currents measured flow through many channels at a time; in addition, different types of channels may be open simultaneously. Plant cells pose another problem: when a microelectrode is inserted into the protoplast, it usually penetrates both the plasma membrane and the tonoplast, so that the data reflect the collective behavior of both membranes.

The patch-clamp technique has revolutionized the study of ion channels. With this technique, which involves the electrical analysis of a very small patch of membrane, it is possible to identify a single ion-specific channel in the plasma membrane or tonoplast and to study the transport of ions through that channel.

In patch-clamp experiments, a glass electrode (micropipette) with a tip diameter of about 1.0 micrometer is brought into contact with an isolated (wall-less) protoplast or isolated vacuole. When gentle suction is then applied, an extremely tight seal is formed between the micropipette and the membrane. When the patch is still attached to the cell, it is called the *cell-attached* or *on-cell configuration* (a). Other configurations include the *whole-cell configuration*, in which the patch within the rim of the micropipette is ruptured and contact is established between the contents of the cell and those of the micropipette (b); the *outside-out patch*, which is obtained from the whole-cell configuration by withdrawing the micropipette and allowing some of the membrane to reseal over the tip of the pipette (c); and the *inside-out patch*, which is obtained when the pipette is withdrawn from a cell so that only a small patch of membrane remains intact within the tip (d).

A detached patch has an advantage over the whole-cell configuration because with a detached patch it is easy to alter the composition of the solution on either side of the membrane to test the effect of different solutes on channel behavior. With either configuration, the electrical potential difference across the membrane, or membrane potential, can be controlled by an electrical circuit that both measures the voltage across the membrane and continuously adjusts it to match the voltage imposed by the experimenter.

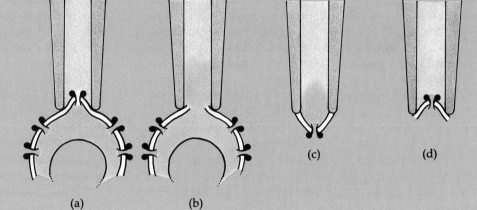

Different recording configurations used with patch-clamp technique: (a) *cell-attached (on-cell);* (b) *whole-cell;* (c) *outside-out;* (d) *inside-out.*

(a)   (b)   (c)   (d)

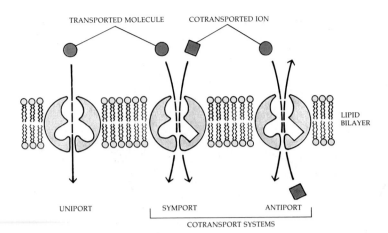

4–10

Diagram illustrating how carrier proteins function. A uniport carrier protein simply transfers a solute across the membrane. In the cotransport systems, the transport of one solute depends on the simultaneous or sequential transfer of a second solute, in either the same direction (symport) or the opposite direction (antiport).

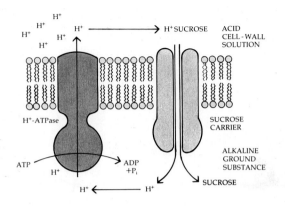

4–11

Primary and secondary active transport. The proton pump (H⁺-ATPase) generates a proton motive force (primary active transport), which energizes the uptake of various solutes across the membrane (secondary active transport). In this illustration, sucrose is being cotransported with H⁺ (H⁺-sucrose symport) across the plasma membrane from the apoplast (cell wall) into the cytoplasm.

All channel proteins and some carrier proteins are *uniports*, which are simply proteins that transport only one solute from one side of the membrane to the other. Other carrier proteins function as *cotransport systems*, in which the transfer of one solute depends on the simultaneous or sequential transfer of a second solute. The second solute may be transported in the same direction *(symport)* or in the opposite direction *(antiport)* (Figure 4–10). Neither simple diffusion nor passive transport is capable of moving solutes *against* a concentration gradient or an electrochemical gradient. The capacity to move solutes against a concentration gradient or an electrochemical gradient requires energy, and this process is called **active transport,** which is always mediated by carrier proteins. As indicated above, the proton pump in plant and fungal cells is energized by ATP and can be measured as an H⁺-ATPase located in the membrane. The enzyme generates a large electrical potential and a pH gradient, that is, a gradient of protons (hydrogen ions), that provide the driving force for solute uptake by all the H⁺-coupled cotransport systems. By this process, even neutral solutes can be accumulated to concentrations much higher than those outside the cell, simply by cotransporting with a charged molecule (for example, an H⁺). The energy-yielding first process (the pump) is referred to as "primary active transport," and the second process (cotransporters) as "secondary active transport" (Figure 4–11).

Endocytosis and Exocytosis

The transport proteins involved with the transfer of ions and polar molecules across the plasma membrane are unable to transport large molecules such as polysaccharides and proteins. Yet most animal cells and at least some plant cells are able both to ingest and to secrete such macromolecules through the formation of membrane-bound vesicles or saclike structures, which serve as vehicles for the substances being transported.

Uptake of macromolecules into the cell occurs by **endocytosis,** a process involving invagination of the plasma membrane. Two main types of endocytosis are recognized: phagocytosis and pinocytosis.

Phagocytosis ("cellular eating") involves the ingestion of relatively large, solid particles, such as bacteria or cellular debris, via large vesicles derived from the plasma membrane. Many one-celled organisms, such as amoebas, feed in this way. Among the organisms discussed in this book, the plasmodial slime molds and cellular slime molds exhibit phagocytosis (Figure 4–12). A unique example of phagocytosis in plants is found in the nodule-forming roots of legumes during the release of *Rhizobium* bacteria from infection threads. The bacteria are enveloped by portions of the plasma membrane of the root hairs (see page 604).

Pinocytosis ("cellular drinking") involves the inges-

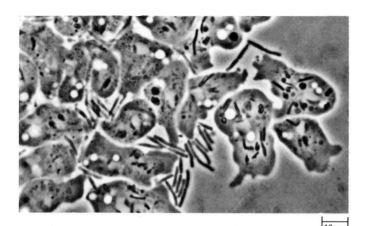

4–12

The amoeboid feeding stage of the cellular slime mold Dictyostelium au-reum. *These amoebas have been feeding upon the rod-shaped bacterium* Esche-richia coli *by phagocytosis. Notice the whole bacterial cells both inside and outside the amoebas.*

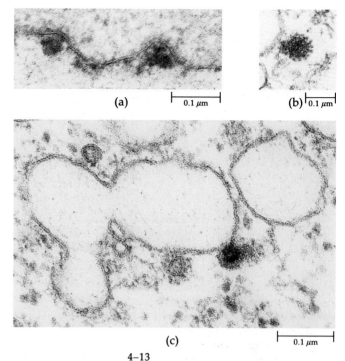

(a)   0.1 μm   (b) 0.1 μm

(c)   0.1 μm

4–13

*Evidence for endocytosis in corn (*Zea mays*) root-cap cells that have been ex-posed to a solution containing lead nitrate. (a) Granular deposits containing lead can be seen in two coated pits. (b) A coated vesicle with lead deposits. (c) Here, one of two coated vesicles has fused with a large dictyosome vesicle. This coated vesicle (dark structure) still contains lead deposits, but it appears to have cast off its coat, which is lying just to the right of it. The second coated vesicle is clearly intact.*

tion of fluid and solutes via small vesicles derived from the plasma membrane. Until very recently, pinocytosis was thought to be uncommon in plant cells. Studies using isolated protoplasts and intact tissues and the salts of heavy metals (such as lead) now indicate, how-ever, that an endocytic cycle similar to that of animal cells exists in plant cells (Figure 4–13). This cycle begins at specialized regions of the plasma membrane called *coated pits.* Coated pits are depressions of the plasma membrane coated with bristlelike structures on their cytoplasmic surfaces. Shortly after they are formed (within a minute or so), coated pits invaginate and pinch off to form *coated vesicles.* Within an even shorter time, the coated vesicles shed their coats and then fuse with some other membrane-bound structure (for exam-ple, dictyosomes or small vacuoles), releasing their con-tents in the process. The coats of the pits and vesicles contain several proteins, including *clathrin,* a protein complex composed of three large and three smaller poly-peptide chains that together form a three-legged structure (a triskelion). It is the arrangement of the tri-skelions on the cytoplasmic surface of the membranes that forms the characteristic coat on the surface of coated pits and vesicles. Smooth, uncoated vesicles of various size also appear to be involved in endocytosis in plant cells.

Many substances are also exported or secreted from cells in vesicles. This reverse endocytosis is referred to as **exocytosis.** One example cited in Chapter 2 is the role of dictyosome vesicles in cell wall formation. The vesicles, with their enclosed cell wall precursors, move to the surface of the cell, fuse with the plasma mem-brane, and expel their contents into the region of the developing cell wall. A second example cited in Chapter 2 involves the role of dictyosome vesicles in cell plate formation. Both smooth and coated vesicles have been implicated in exocytosis in plant cells.

Transport via Plasmodesmata

As described in Chapter 2, neighboring cells of the plant body are interconnected by narrow strands of cy-toplasm called plasmodesmata, which provide poten-tial pathways for the passage of substances from cell to cell. The term **symplast** is used to refer to the intercon-nected protoplasts and their plasmodesmata. The movement of substances from cell to cell by means of

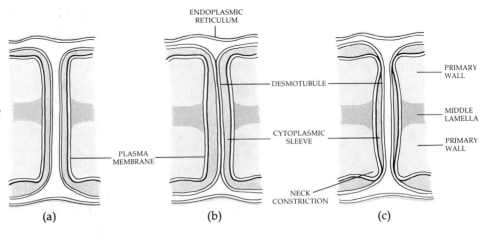

4–14

Diagrams illustrating possible variation in the structure of plasmodesmata. (a) Both the desmotubule and the cytoplasmic sleeve between the desmotubule and plasma membrane lining the plasmodesmatal canal are open pathways. (b) The cytoplasmic sleeve is open, but the desmotubule is closed. (c) The desmotubule is open, but the cytoplasmic sleeve is closed at both ends by neck constrictions.

the plasmodesmata is called **symplastic transport.** In contrast, the movement of substances in the cell wall continuum, or **apoplast,** surrounding the symplast is called **apoplastic transport.**

Plasmodesmata may provide a more efficient pathway between neighboring cells than the less direct, alternative route of plasma membrane, cell wall, and plasma membrane. It is believed that cells and tissues that are far removed from direct sources of nutrients can be supplied with nutrients either by simple diffusion or by bulk flow through plasmodesmata. In addition, some substances are believed to move through plasmodesmata to and from the xylem and phloem—the tissues concerned with long-distance transport in the plant body.

As mentioned in Chapter 2, a plasmodesma typically is traversed by a tubular strand of endoplasmic reticulum, called a *desmotubule,* which is continuous with the endoplasmic reticulum of the adjacent cells. In most electron micrographs, the desmotubule does not resemble the adjoining endoplasmic reticulum. It is much smaller in diameter and contains a central, rodlike structure. Considerable controversy has centered on the interpretation of the central rod, but today most inves-

tigators believe it represents the merger of the inner leaflets (inner portions of the bilayers) of the tightly appressed endoplasmic reticulum forming the desmotubule (Figure 4–14). If this interpretation is correct, the desmotubule lacks a lumen, and all transport via the plasmodesma would be restricted to the channel surrounding the desmotubule. This channel, called the *cytoplasmic sleeve* (formerly called the "cytoplasmic annulus"), appears to be subdivided into several narrower channels by proteinaceous particles. Interestingly, these narrower channels have about the same diameter as the "connexons" of gap junctions in animal cells, which form connecting channels between adjacent cells.

Evidence for transport between cells via plasmodesmata comes from studies involving fluorescent dyes and electrical currents. The dyes, which do not easily cross the plasma membrane, can be observed moving from the injected cells into neighboring cells and beyond (Figure 4–15). Such studies have revealed that most plasmodesmata can allow the passage of molecules up to molecular weights of 700 to 900 daltons (a dalton is the weight of a single hydrogen atom), values more than adequate for sugars and amino acids to move freely across these intercellular connections.

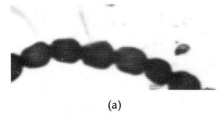

(a)

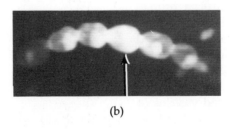

(b)

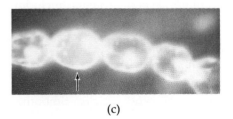

(c)

4–15

Hairs from stamens of Setcreasea purpurea *before (a) and after (b and c) injection of one of the hair cells with the fluorescent dye disodium fluorescein. When the dye was injected into the cytoplasm of the cell indicated by the arrow,*

it passed in both directions into the cytoplasm of its neighboring cells. (b) Two minutes after injection; (c) five minutes after injection. Note in (c) that the dye is present in the cytoplasm, which

lines the walls of the cell. Inasmuch as the plasma membrane is impermeable to the dye, the movement of dye from cell to cell must have occurred via plasmodesmata in the adjoining cell walls.

The passage of pulses of electrical current from one cell to another is monitored by means of receiver electrodes placed in neighboring cells. The extent of the electrical force detected is found to vary with the frequency of plasmodesmata and with the number of cells and length of cells between the injection and receiver electrodes.

Whether the plasmodesmata have any control over the movement of substances from cell to cell has not yet been determined, although some investigators have found what might be "valves" in some plasmodesmata.

Summary

The plasma membrane regulates the passage of materials into and out of the cell, a function that makes it possible for the cell to maintain its structural and functional integrity. This regulation depends on interaction between the membrane and the materials that pass through it.

Water is one of the principal materials passing into and out of cells. Water moves by bulk flow, diffusion, and osmosis. Water potential determines the direction in which the water moves; that is, the movement of water is from regions of high water potential to regions of lower water potential. Bulk flow is the overall movement of water molecules, as when water flows downhill or moves in response to pressure.

Diffusion involves the independent movement of molecules and results in net movement along a concentration gradient. Carbon dioxide and oxygen are two important molecules that move into and out of cells by diffusion across the membrane. Diffusion is most efficient when the distance involved is small and when the concentration gradient is steep. The rate of movement of substances within cells is increased by cytoplasmic streaming.

Osmosis is the diffusion of water through a differentially permeable membrane, that is, one that permits the movement of water but inhibits the passage of solutes. In the absence of other forces, the movement of water by osmosis is from a region of lower solute concentration (one of higher water potential) to a region of higher solute concentration (one of lower water potential). The turgor (rigidity) of plant cells is a result of osmosis.

Cellular membranes are composed of lipid bilayers in which globular proteins are suspended. Some of these proteins are transport proteins, which transfer ions and small polar molecules across the membrane. Three broad classes of transport proteins are recognized: pumps, carriers, and channels. Transport down a concentration gradient or an electrochemical gradient —passive transport—is exhibited by all channel proteins and by some carriers. Transport against an electrochemical gradient—active transport—is always mediated by carrier proteins and requires an input of chemical energy from ATP.

Controlled movement of large molecules into and out of a cell may occur by endocytosis or by exocytosis, processes in which substances are transported in vesicles. Endocytosis of solids is called phagocytosis, whereas endocytosis of dissolved molecules is called pinocytosis. Exocytosis provides a means of exporting materials (such as some cell wall components) from the cell.

Movement of substances between plant cells may also occur by way of narrow strands of cytoplasm, the plasmodesmata, which interconnect the protoplasts of neighboring cells. Such movement is known as symplastic transport. The movement of substances in the cell wall continuum is known as apoplastic transport.

See the end of Chapter 7 for a list of suggestions for further reading.

CHAPTER 5

The Flow of Energy

5–1
The energy of the summer sun is stored in these wheat plants, ready for harvest. Of the radiant energy of the sun striking the wheat field, about one percent is converted into stored chemical energy.

Life here on earth depends on the flow of energy from thermonuclear reactions taking place at the heart of the sun (Figure 5–1). The amount of radiant energy delivered by the sun to the earth is 13×10^{23} (the number 13 followed by 23 zeros) calories per year. About a third of this solar energy is immediately reflected back into space. Much of the remaining two-thirds is absorbed by the earth and converted to heat, but about one percent of the solar energy reaching the earth becomes, through a series of operations performed by the cells of plants and other photosynthetic organisms, the energy that drives nearly all the processes of life. Living systems change energy from one form to another, transforming the radiant energy of the sun into the chemical and mechanical energy used by living organisms (Figure 5–2).

These concepts of a vital relationship between plants and animals, and between energy and life, form part of the study of **thermodynamics**—the science of energy exchanges. Before we look at the details of how photosynthetic organisms use solar energy, let us first consider a few key thermodynamic principles and then explore the manner in which enzymes mediate many of the reactions and processes carried out by cells.

The Laws of Thermodynamics

Energy is an elusive concept. It is usually defined operationally, that is, by what it does rather than what it is: energy does work.

THE FIRST LAW

The development of the steam engine in the latter part of the eighteenth century changed scientific thinking about the nature of energy. Energy came to be asso-

5–2

An example of the flow of biological energy. The radiant energy of sunlight is produced by the fusion reactions taking place in the sun. Chloroplasts, present in all photosynthetic eukaryotic cells, capture this energy, which is stored briefly in ATP, and use it to convert water and carbon dioxide into carbohydrates, such as glucose, starch, and other foodstuff molecules. Oxygen is released into the air as a product of these photosynthetic reactions.

Mitochondria—organelles present in essentially all eukaryotic cells—break down these carbohydrates and capture their stored energy in ATP molecules. This process—cellular respiration—consumes oxygen and produces carbon dioxide and water, completing the cycle.

With each transformation, some energy is dissipated to the environment in the form of heat. Thus the flow of biological energy is one-way and can continue only so long as there is an input of energy from the sun.

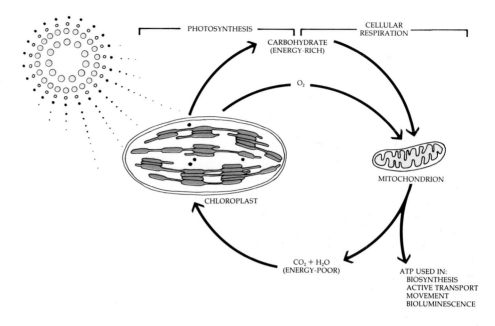

ciated with work, and heat and motion came to be seen as forms of energy: heat = **thermal energy,** and motion = **kinetic energy.** The way was opened for the formulation of the laws of thermodynamics. The **first law of thermodynamics** states, quite simply: *Energy can be changed from one form to another but cannot be created or destroyed.*

In the steam engine of a locomotive, for instance, the chemical energy in the fuel is converted to thermal energy when the fuel is burned (Figure 5–3). The thermal energy converts liquid water in the boiler to steam. The pressure of the steam causes the piston/rod to move; thus thermal energy is converted to kinetic energy. This kinetic energy is transferred to the wheels of the locomotive, and so the wheels begin to turn. Thus chemical energy (fuel) has been converted to kinetic energy (the moving train).

If you were to compare the amount of energy released from the burning fuel with the amount used to turn the wheels, you would see that the former is larger than the latter. At first glance, it looks as if energy has been destroyed and the first law violated, but closer examination shows that the kinetic energy of the train is not the only energy produced. Some of the original chemical energy in the fuel has gone into heat of friction as the wheels roll along the track, and some has left the boiler through the exhaust valve. Comparing the amount of energy released from the fuel with the

5–3

In this locomotive, energy stored in the fossil fuel coal is released by burning the coal to produce thermal energy, which converts liquid water to steam. The steam is then used to drive the piston/ rod, which causes the wheels to turn. Thus, the chemical energy stored in the coal has been converted to kinetic energy. The excess heat in the boiler is dissipated through an exhaust valve.

sum of the kinetic energy of the moving train plus the heat energy of friction plus the energy in the exhaust, you will find that the two amounts are very close to each other, as the *first law* says they must be.

The notion of **potential energy** also resulted from eighteenth-century studies on the steam engine. A barrel of oil or a ton of coal could be assigned a certain amount of potential energy, expressed in terms of the amount of heat the oil or coal would liberate when burned. The efficiency of the conversion of the potential energy to usable energy depends on the design of the energy-conversion system. Most engines work with less than 25 percent efficiency.

Although these concepts were formulated in terms of engines running on heat energy, they apply to other systems as well. For example, a boulder pushed to the top of a hill gains potential energy. Given a little push (the energy of activation), the boulder rolls down the hill again, converting the potential energy to kinetic energy and to heat produced by friction. As mentioned previously, water can also possess potential energy (page 63). As it moves by bulk flow from the top of a waterfall or over a dam, it can turn waterwheels that turn gears, for example, to grind corn. Thus the potential energy of water, in this system, is converted to the mechanical energy of the wheels and gears and to heat, which is produced by the movement of the water itself as well as by the turning wheels and gears.

Light is another form of energy, as is electricity. Light can be changed to electrical energy, and electrical energy can be changed to light (say, by letting it flow through the tungsten wire in a light bulb).

The first law of thermodynamics states that in energy exchanges and conversions, wherever they take place and whatever they involve, the sum of the energy of the products of the reaction plus the energy released in the reaction is equal to the energy possessed by the initial reactants.

THE SECOND LAW

In a biological sense, the **second law of thermodynamics** is the more interesting one. It predicts the direction of all events involving energy exchanges; thus it has been called "time's arrow." One way of stating the second law is: *In all energy exchanges and conversions, if no energy leaves or enters the system under study, the potential energy of the final state will always be less than the potential energy of the initial state.* The second law is in keeping with everyday experience (Figure 5–4). Of its own accord, a boulder will roll downhill but never uphill. Heat will flow from a hot object to a cold one and never the other way. Our cells can process carbohydrates enzymatically to yield carbon dioxide and water but—since we cannot capture the energy of the

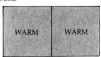

INITIAL STATE   FINAL STATE
COPPER BLOCKS

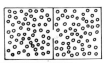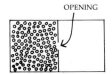

HEAT FLOWS FROM HOT TO COLD BODY

OPENING

GAS MOLECULES FLOW FROM ZONE OF HIGH PRESSURE TO ZONE OF LOW PRESSURE

ORDER BECOMES DISORDER

5–4
Some illustrations of the second law of thermodynamics. In each case, a concentration of energy—in the hot copper block, in the gas molecules under pressure, and in the neatly organized books—is dissipated. In nature, processes tend toward randomness, or disorder. Only an input of energy can reverse this tendency and reconstruct the initial state from the final state. Ultimately, however, disorder will prevail, since the total amount of energy in the universe is finite.

sun as plants do—our enzymatic processes cannot produce carbohydrates from carbon dioxide and water.

A process in which the potential energy of the final state is less than that of the initial state is one that yields energy (otherwise it would be in violation of the first law). An energy-yielding process is an **exergonic** ("energy-out") reaction. In contrast, **endergonic** ("energy-in") reactions are energy-requiring reactions. As the second law predicts, only exergonic reactions can take place spontaneously. In order to proceed, endergonic reactions require an energy input that is greater than the difference in energy between products and reactants.

One important factor in determining whether or not a reaction is exergonic is ΔH, the change in heat content of the system (Δ stands for change, H for heat content). As noted in Appendix A, the energy change that occurs when glucose, for instance, is oxidized can be measured in a calorimeter and expressed in terms of ΔH. The oxidation of a mole of glucose yields 673 kilocalories. (A mole is the amount of a substance, in grams, that equals its molecular weight. For example, the atomic weight of carbon is 12 and the atomic weight of oxygen is 16, so the molecular weight of CO_2 is 44. One mole of CO_2 is therefore 44 grams of CO_2.)

$$C_6H_{12}O_6 + 6O_2 \longrightarrow 6CO_2 + 6H_2O$$

$$\Delta H = -673 \text{ kcal/mole}$$

Generally speaking, an exergonic chemical reaction is also an exothermic reaction—that is, it gives off heat and thus has a negative ΔH. However, there are exceptions. One of the most dramatic is found with a substance known as dinitrogen pentoxide, which decomposes spontaneously and with explosive force to nitrogen dioxide and oxygen, and in so doing absorbs heat:

$$2N_2O_5 \longrightarrow 4NO_2 + O_2$$

$$\Delta H = +26.18 \text{ kcal/mole}$$

In short, another factor besides the gain or loss of heat can determine the direction of a process. This factor, called **entropy**, is a measurement of the disorder, or randomness, of a system. Let us return to water as an example. The change from ice to liquid water and the change from liquid water to water vapor are both endothermic processes—a considerable amount of heat is absorbed from the surroundings as they occur. Yet, under the appropriate conditions, these changes proceed spontaneously. The key factor in these processes is the increase in entropy. In the case of the change of ice into water, a solid is being turned into a liquid, and some of the hydrogen bonds that hold the water molecules together in a crystal (ice) are being broken. As the water turns to vapor, the rest of the hydrogen bonds are ruptured as the individual water molecules separate, one by one. In each case, the disorder of the system has increased.

The notion that there is more disorder associated with more numerous and smaller objects than with fewer, larger ones is in keeping with our everyday experience. If there are 20 papers on a desk, the possibilities for disorder are greater than if there are 2 or even 10. If each of the 20 papers is cut in half, the entropy of the system—the capacity for randomness—increases. The relationship between entropy and energy is also a commonplace idea. If you were to find your room tidied up and your books in alphabetical order on the shelf, you would recognize that someone had been at work—that energy had been expended. To organize the papers on a desk similarly requires the expenditure of energy.

Now let us return to the question of the energy changes that determine the course of chemical reactions. As discussed, both the change in the heat content of the system (ΔH) and the change in entropy (which is symbolized as ΔS) contribute to the overall change in energy. This total change—which takes into account both heat and entropy—is called the **free-energy change** and is symbolized as ΔG, after the American physicist Josiah Willard Gibbs (1839–1903), who was one of the first to integrate all of these ideas.

Keeping ΔG in mind, let us examine once again the oxidation of glucose. The ΔH of this reaction is -673 kcal/mole. The ΔG is -686 kcal/mole. Thus, the entropy factor has contributed 13 kcal/mole to the free-energy change of the process. In this case, change in heat content and in entropy both contribute to the lower energy state of the products of the reaction.

The relationship among ΔG, ΔH, and entropy is given in the following equation:

$$\Delta G = \Delta H - T\,\Delta S$$

This equation states that the free-energy change is equal to the change in heat content (a negative value in exothermic reactions, remember) minus the change in entropy multiplied by the absolute temperature T. In exergonic reactions, ΔG is always negative, but ΔH may be negative, positive, or zero. Since T is always positive, the greater the change in entropy, the more negative ΔG will be; that is, the more exergonic the reaction will be. Therefore it is possible to state the second law in another, simpler way: *In general, all natural processes are exergonic.*

BIOLOGY AND THE SECOND LAW

The laws of thermodynamics are of crucial importance to biology because they are the organizing principles under which a number of different kinds of processes can be unified. Also, as we shall see in the following chapter, they permit a kind of biochemical bookkeeping.

The most interesting implication of the second law, as far as biology is concerned, is the relationship between entropy and order. Living systems are continuously expending large amounts of energy to maintain order. Stated in terms of chemical reactions, living systems continuously expend energy to maintain a position far from equilibrium. If equilibrium were to occur, the chemical reactions in a cell would, for all practical purposes, stop, and no further work could be done. At equilibrium, a cell would be dead.

Oxidation-Reduction

Chemical reactions are essentially energy transformations in which energy stored in chemical bonds is transferred to other, newly formed chemical bonds. In such transfers, electrons shift from one energy level to another (see Appendix A). In many reactions, electrons pass from one atom or molecule to another. These reactions, known as oxidation-reduction (or redox) reactions, are of great importance in living systems. The *loss* of an electron is known as **oxidation,** and the atom or molecule that loses the electron is said to be oxidized. The reason electron loss is called oxidation is that oxygen, which attracts electrons very strongly, is most often the electron acceptor.

Reduction involves the *gain* of an electron. Oxidation and reduction occur simultaneously; the electron lost by the oxidized molecule is accepted by another molecule, which is thus reduced; hence the term "redox reactions."

Redox reactions may involve only a solitary electron, as when sodium loses an electron and becomes oxidized to Na^+, and chlorine gains an electron and is reduced to Cl^-. Often, however, the oxidation of organic molecules involves the removal of electrons and hydrogen ions (protons), and the reduction of such molecules involves the gain of electrons and protons. For example, when glucose is oxidized, electrons and hydrogen ions are lost by the glucose molecule and are gained by oxygen to form water:

$$C_6H_{12}O_6 + 6O_2 \longrightarrow 6CO_2 + 6H_2O + energy$$

The electrons move to lower energy levels as the atoms rearrange to form the product molecules, and energy is released.

Conversely, during photosynthesis, electrons and hydrogen ions are transferred from water to carbon dioxide, thereby reducing the carbon dioxide, eventually forming glucose:

$$6CO_2 + 6H_2O + energy \longrightarrow C_6H_{12}O_6 + 6O_2$$

In this case, the electrons are forced to a higher energy level by the input of energy that is required for the reaction to take place.

In living systems, the energy-capturing reactions (photosynthesis) and the energy-releasing reactions (glycolysis and respiration) are oxidation-reduction reactions. As we have seen, the complete oxidation of a mole of glucose releases 686 kilocalories of free energy (that is, $\Delta G = -686$ kcal/mole). Conversely, the reduction of carbon dioxide to form a mole of glucose stores 686 kilocalories of free energy in the chemical bonds of glucose. If the energy released during the oxidation of glucose were to be released all at once, most of it would be dissipated as heat. Not only would it be of no use to the cell, but the resulting temperature increase would destroy the cell. Enzymatic mechanisms have evolved in living systems, however, that regulate these chemical reactions—and a multitude of others—in such a way that energy is stored in particular chemical bonds from which it can be released stepwise in small amounts as required by the cell.

Enzymes and Living Systems

In any living system, thousands of different chemical reactions occur, many of them simultaneously. The sum of all these reactions is referred to as **metabolism** (from the Greek *metabolē*, meaning "change"). If one merely listed the individual chemical reactions, it would be difficult to understand a cell's metabolic activities. Fortunately, there are some guiding principles that lead one through the maze of cell metabolism. First, virtually all the chemical reactions that take place in a cell involve **enzymes**—the catalysts and regulators for the metabolic processes of living systems. Second, biochemists group these reactions in ordered series of steps called pathways, with each pathway often containing a dozen or more sequential reactions. Each pathway serves a specific function in the overall life of the cell or organism. Furthermore, certain pathways have many steps in common, such as those that are concerned with the synthesis of the different amino acids or the various nitrogenous bases. Some pathways converge; for example, the pathway by which fats are broken down to yield energy leads to the same pathway by which glucose is broken down to yield energy.

Many living systems have unique pathways. Plant cells expend energy building cellulose walls, an activity not engaged in by animal cells. It is not surprising that the distinctive differences among cells and organisms are reflected not only in their forms and functions but also in their biochemistry. What is surprising, however, is that much of the metabolism of even the most diverse of organisms is exceedingly similar; the differences in many of the metabolic pathways of humans, oak trees, mushrooms, and jellyfish are very slight. Some pathways are found in virtually all living systems.

The magnitude of the chemical work carried out by a cell is illustrated by the fact that, for the most part, the thousands of different molecules found within a cell are synthesized there. The total of chemical reactions involved in this synthesis is called **anabolism.** Anabolic reactions usually involve an increase in atomic order (a decrease in entropy) and are almost always energy-requiring (endergonic). Cells are also constantly involved in the breakdown of larger molecules; these activities, known collectively as **catabolism,** involve a decrease in atomic order (an increase in entropy) and are usually energy-liberating (exergonic). Catabolic reactions serve two purposes: (1) they release the energy required for anabolism and other work of the cell, and (2) they serve as a source of raw materials for anabolic processes. Hence, both aspects of metabolism—anabolism and catabolism—are essential for normal cellular activities.

Living systems carry out this multitude of chemical activities under carefully controlled conditions. Most of their chemical reactions are carried out within living cells, and thousands of different kinds of molecules are intermingled. Temperatures cannot be high; otherwise, many of the fragile structures on which life depends would be destroyed. How is all this complex chemical work accomplished? The question can be answered in a single word: enzymes. Without enzymes, biochemical reactions would take place so slowly that, for all practical purposes, they would not occur at all, and the activities we associate with life would cease to exist.

Enzymes as Catalysts

Enzymes are the catalysts of biological reactions. They resemble other catalysts in that they are not used up in the course of the reaction and so can be used over and over again. Enzymes differ from other catalysts, however, in that they are very selective in their action. Some enzymes catalyze a reaction with only a single set of reactants. In enzyme-catalyzed reactions, the reactant (or reactants) on which an enzyme operates is called its **substrate.** The selectivity an enzyme exhibits in choosing a substrate is known as its **specificity.**

Enzymes enormously accelerate the rate at which reactions occur. For instance, the reaction of carbon dioxide with water,

$$CO_2 + H_2O \rightleftharpoons H_2CO_3$$

can occur spontaneously, as it does in the oceans. In the human body, however, this reaction is catalyzed by an enzyme, carbonic anhydrase. Carbonic anhydrase is one of the most effective enzymes known—each enzyme molecule leads to the production of about 6×10^5 (600,000) molecules of the product, carbonic acid, per second. The catalyzed reaction is about 10^7 times faster than the uncatalyzed one.

THE ACTIVE SITE

Most enzymes are complex globular proteins consisting of one or more polypeptide chains (see Chapter 3). They are folded to form a groove or pocket into which the reacting molecule or molecules—the substrate—fit and where the reactions occur. This portion of the enzyme is known as the **active site** (Figure 5–5). The active site is formed by the very exact folding of the polypeptide chain. The relationship between the active site and the substrate is very precise.

The active site not only has a precise three-dimensional shape but also has exactly the correct array of charged and uncharged, or hydrophilic and hydrophobic, areas on its binding surface. If a particular portion of the substrate has a negative charge, the correspond-

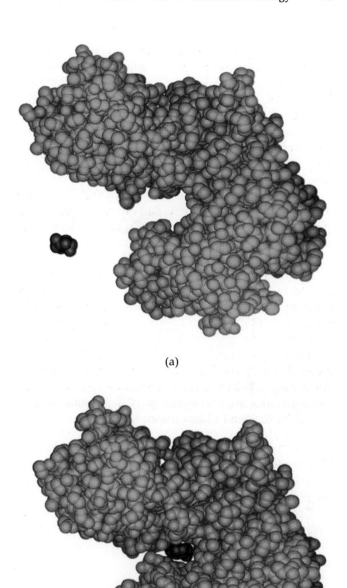

(a)

(b)

5–5
(a) *Space-filling models of the yeast enzyme hexokinase (green) and one of its substrates, glucose (red). Such models, which are produced by computer techniques, show the three-dimensional shape of the molecules. (b) Here the glucose molecule is shown colliding with the enzyme and binding to the active site, which appears as a cleft in the side of the hexokinase molecule. In the absence of glucose, hexokinase has an open cleft. When glucose is bound by hexokinase, the cleft is partially closed.*

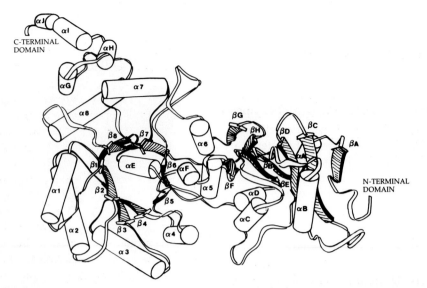

5-6

Schematic diagram of the structure of one of the two subunits of ribulose 1,5-bisphosphate carboxylase/oxygenase (Rubisco) from the photosynthetic bacterium Rhodospirillum rubrum. (Alpha helices are represented by cylinders and beta pleated sheets by arrows.) Rubisco is the key enzyme in photosynthetic carbon dioxide fixation, catalyzing the carboxylation of ribulose 1,5-bisphosphate, the first step in the Calvin cycle (see Chapter 7). This enzyme also catalyzes the oxygenation of ribulose 1,5-bisphosphate, the first step in photorespiration.

The subunit, which consists of 466 amino acid residues, is a "two-domain" protein, with a smaller, N-terminal domain and a larger, C-terminal domain. The N-terminal domain consists of residues 1–137 and contains beta strands βA–βE, which form a mixed five-stranded beta pleated sheet. Helix A is on one side of the beta sheet, and helices B and C on the other side. A region of extended chain after helix D makes the connection to the C-terminal domain, which consists of residues 138–466.

The two subunits of the enzyme

interact tightly to form the functional Rubisco molecule of Rhodospirillum rubrum. The core of this binding area consists of interactions between the two C-terminal domains and between two regions from the N-terminal domain of one subunit and regions from the C-terminal domain of the second subunit. These subunit interactions have functional importance because some of the residues involved occur in or close to the active-site region; hence, each active site of the enzyme is built up from residues of both subunits.

ing feature on the active site has a positive charge, and so on. Thus, the active site not only confines the substrate molecule but also orients it in the correct direction.

The amino acids involved in the active site need not be adjacent to one another on the polypeptide chains. In fact, in an enzyme with quaternary structure, these amino acids may even be on different polypeptide chains (Figure 5–6). The amino acids are brought together at the active site by the precise folding of the polypeptide chains in the molecule.

THE INDUCED-FIT HYPOTHESIS

Studies of enzyme structure have suggested that the binding that takes place between enzyme and substrate alters the conformation of the enzyme, thus inducing an even closer fit between the active site and the reactants. It is believed that this *induced fit* may place some strain on the reacting molecules and so further facilitate the reaction (Figure 5–7).

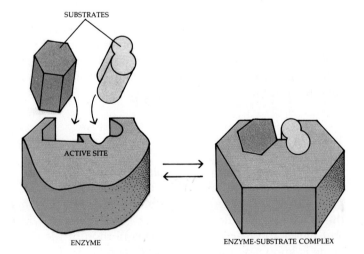

5-7

A model of the induced-fit hypothesis of enzyme action. The active site is believed to be flexible and thus adjustable in its shape, or conformation, to that of the substrate molecule. This induces a particularly close fit between the active site and the substrate and may also put some strain on the substrate molecule.

Cofactors in Enzyme Action

The catalytic activity of some enzymes appears to depend only on their structure as proteins. Most enzymes, however, require one or more nonprotein components, known as **cofactors,** without which the enzymes fail to function. In such cases, the protein portion of the enzyme is called an *apoenzyme.*

IONS AS COFACTORS

Certain ions are cofactors for particular enzymes. For example, magnesium ion (Mg^{2+}) is required in most enzymatic reactions involving the transfer of a phosphate group from one molecule to another. The two positive charges on the magnesium ion hold the negatively charged phosphate group in position. Other ions, such as Ca^{2+} and K^+, play similar roles in other reactions. In some cases, ions serve to hold the folds of the enzyme protein together.

COENZYMES AND VITAMINS

Nonprotein organic cofactors can also play a crucial role in enzyme-catalyzed reactions. Such cofactors are called **coenzymes.** For example, in some oxidation-reduction reactions, electrons are passed along to a molecule that serves as an electron acceptor. There are several different electron acceptors in any given cell, and each is tailor-made to hold the electron at a slightly different energy level. As an example, let us look at just one, nicotinamide adenine dinucleotide (NAD^+), which is shown in Figure 5–8. At first glance, NAD^+ looks complex and unfamiliar, but if you look at it more closely, you will find that you recognize most of its component parts. The two ribose (five-carbon sugar) units are linked by a *pyrophosphate bridge.* One of the ribose units is attached to the nitrogenous base adenine. The other is attached to another nitrogenous base, nicotinamide. A nitrogenous base plus a sugar is called a *nucleoside,* and a nucleoside plus a phosphate is called a **nucleotide.** A molecule that contains two of these combinations is called a dinucleotide.

The nicotinamide ring is the active end of NAD^+, that is, the part that accepts electrons. Nicotinamide is a vitamin—niacin. Vitamins are compounds that are required in small quantities by many living organisms; humans and other animals cannot synthesize vitamins and so must obtain them in their diets. When nicotinamide is present, our cells can use it to make NAD^+. Many vitamins are coenzymes or parts of coenzymes.

Nicotinamide adenine dinucleotide, like many other

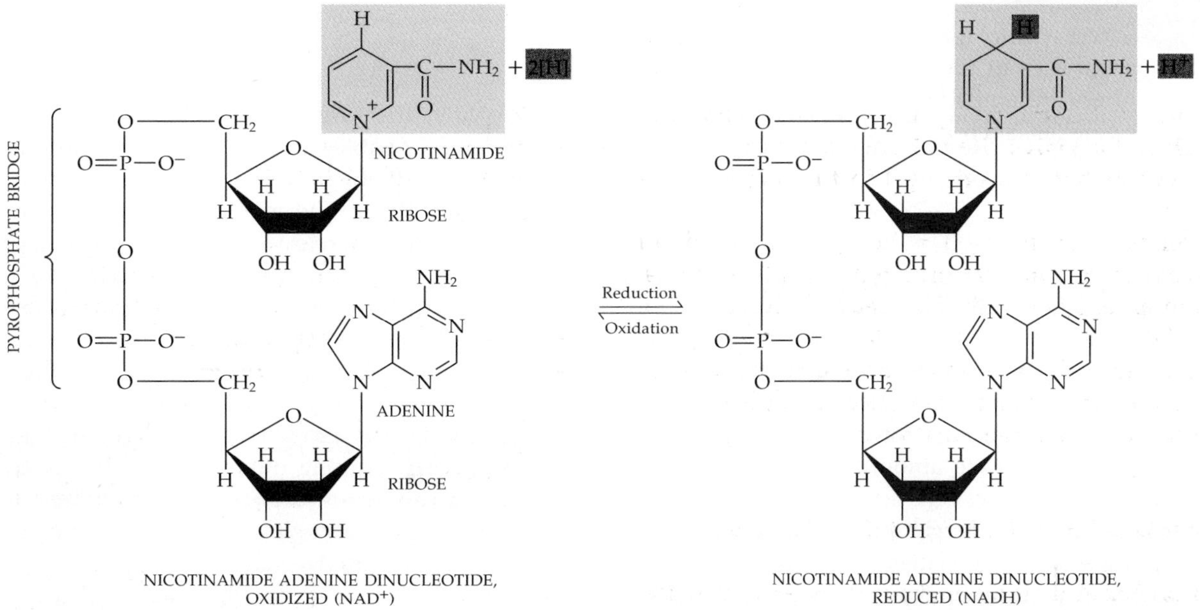

NICOTINAMIDE ADENINE DINUCLEOTIDE, OXIDIZED (NAD^+)

NICOTINAMIDE ADENINE DINUCLEOTIDE, REDUCED (NADH)

5–8

Nicotinamide adenine dinucleotide, an electron acceptor, in its oxidized form, NAD^+, and in its reduced form, NADH. Nicotinamide is a derivative of nicotinic acid, one of the essential B vitamins.

Notice how the bonding within the nicotinamide ring shifts as the molecule changes from the oxidized to the reduced form, and vice versa. Reduction of NAD^+ to NADH requires two electrons and one

hydrogen ion (H^+). The two electrons, however, generally travel as components of two hydrogen atoms; thus, there is one hydrogen ion "left over" when NAD^+ is reduced.

coenzymes, is recycled. That is, NAD$^+$ is regenerated when NADH + H$^+$ passes its electrons to another electron acceptor. Thus, although this coenzyme is involved in many cellular reactions, the actual number of NAD$^+$ molecules required is relatively small.

Some enzymes use cofactors that remain attached to the protein. Such cofactors are referred to as *prosthetic groups*. Examples of prosthetic groups are the iron-sulfur clusters of ferredoxins (page 109) and the pyridoxal phosphate (vitamin B$_6$) of some transaminases.

Enzymatic Pathways

Enzymes are involved in ordered series of steps—the pathways we referred to earlier. Consequently, an organism carries out its chemical activities with remarkable efficiency. First, there is little accumulation of waste products, since each product tends to be used in the next reaction along the pathway. A second advantage of such sequential reactions is understandable when one considers that most chemical reactions can go in either direction; that is, they are reversible (see Appendix A). If each product along a series of reactions is used by the next reaction almost as rapidly as it is formed, the tendency for reversal of the reactions is minimized. Moreover, if the eventual end product is also used up rapidly, the whole series of reactions will move toward completion. A third advantage is that the groups of enzymes making up a common pathway can be segregated within the cell. Some are found in membrane-bound sacs in the ground substance of the cytoplasm. Others are embedded in the membranes of specialized organelles, such as mitochondria and chloroplasts.

Regulation of Enzyme Activity

Another remarkable feature of the metabolic activity of cells is the extent to which each cell regulates the synthesis of the products necessary for its well-being, making them in the appropriate amounts and at the rates required. At the same time, cells avoid overproduction, which would waste both energy and raw materials. The availability of reactant molecules or of cofactors is a principal factor in limiting enzyme action, and for this reason most enzymes probably work at a rate well below their maximum.

Temperature affects enzymatic reactions. An increase in temperature increases the rate of enzyme-catalyzed reactions, but only to a point. As can be seen in

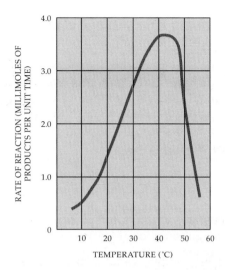

RATE OF REACTION (MILLIMOLES OF PRODUCTS PER UNIT TIME)

TEMPERATURE (°C)

5–9

The effect of temperature on the rate of an enzyme-controlled reaction. The concentrations of enzyme and reacting molecules (substrate) were kept constant. The rate of the reaction, as in most metabolic reactions, approximately doubles for every 10°C rise in temperature up to about 40°C. Above this temperature, the rate decreases as temperature increases, and at about 60°C the reaction stops altogether, presumably because the enzyme is denatured.

Figure 5–9, the rate of most enzymatic reactions approximately doubles for each 10°C rise in temperature in the range 10° to 40°C but then drops off very quickly after about 40°C. The increase in reaction rate occurs because of the increased energy of the reactants; the decrease in the reaction rate above about 40°C occurs as the enzyme molecule itself begins to vibrate, disrupting the relatively weak forces that hold it in its specifically active shape. This degradation of the enzyme molecule is known as *denaturation* (page 59).

The pH of the surrounding solution also affects enzyme activity. Among other factors, the conformation of an enzyme depends on the attraction and repulsion between negatively charged (acidic) and positively charged (basic) amino acids. As the pH changes, these charges change, and so the shape of the enzyme changes until it is so drastically altered that the enzyme is no longer functional. More important, probably, the charges of the active site and substrate are changed so that the binding capacity is affected. Some enzymes are frequently found at a pH that is not their optimum, suggesting that this discrepancy may be not an evolutionary oversight but a way of controlling enzyme activity.

Living systems also have more precise ways of turning enzyme activity on and off. Some enzymes are produced in an inactive form and are activated, usually by another enzyme, only when they are needed. Specific mechanisms by which enzymes may be induced or repressed are discussed in Chapter 8.

The Energy Factor: ATP

All of the biosynthetic (anabolic) activities of the cell (and many other activities as well) require energy. A large proportion of this energy is supplied by a single molecule, **adenosine triphosphate (ATP),** a nucleotide derivative that is the cell's chief energy currency.

At first glance, ATP (Figure 5–10) also appears to be a complex molecule. However, as with NAD$^+$, you will find that its component parts are familiar. ATP is made up of adenine, the five-carbon sugar ribose, and three phosphate groups. These three phosphate groups, with strong negative charges, are linked to each other by **phosphoanhydride bonds** and linked to the ribose by a **phosphoester bond.**

To understand the role of ATP, let us briefly review the concept of the chemical bond and bond energies.

Because a chemical bond is a stable configuration of electrons, reacting molecules must possess a certain amount of energy in order to collide with sufficient force to overcome their mutual repulsion and to weaken existing chemical bonds, allowing the formation of new bonds. This energy is the energy of activation (see Appendix A). Because of enzymes, which reduce the required energy of activation to a level already possessed by a significant proportion of the reacting molecules, the reactions essential to life can proceed at an appropriate rate. As we have seen, however, the *direction* in which a reaction proceeds is determined by the free-energy change ΔG. Only if the reaction is exergonic (negative ΔG) will it proceed to any signficant extent. Yet many cellular reactions, including anabolic reactions such as the formation of a disaccharide from two monosaccharide molecules, are endergonic (positive ΔG). In such reactions, the electrons forming the chemical bonds of the product are at a higher energy level than the electrons in the bonds of the starting materials—that is, the potential energy of the product is greater than the potential energy of the reactants, an apparent violation of the second law of thermodynamics. Cells circumvent this difficulty by **coupled reactions** in which endergonic reactions (or

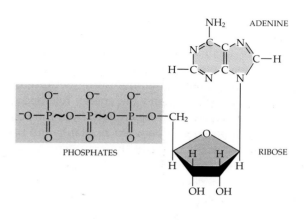

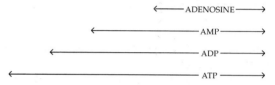

5–10
The structure of adenosine triphosphate (ATP), adenosine diphosphate (ADP), and adenosine monophosphate (AMP). A phosphoester bond links the first phosphate group to the ribose of adenosine, whereas phosphoanhydride bonds, designated by the symbol ~, link the second and third phosphate groups to the molecule. At pH 7, the phosphate groups are fully ionized. In order to represent more accurately its position in the ATP molecule, we have rotated the adenine 180° (right to left) from the orientation depicted in Figure 5–8.

transport processes, such as the active transport of a substance against a concentration gradient) are linked to exergonic reactions that provide a surplus of energy, making the net process exergonic and thus able to proceed spontaneously. The molecule that most frequently supplies energy in such coupled reactions is ATP.

Because of its internal structure, the ATP molecule is well suited to this role in living systems. Energy is released from the ATP molecule when one phosphate group is removed by hydrolysis, producing a molecule of ADP (adenosine diphosphate):

$$ATP + H_2O \longrightarrow ADP + phosphate$$

In the course of this reaction, some 7.3 kcal/mole of ATP are released. Removal of a second phosphate group produces AMP (adenosine monophosphate) and releases an equivalent amount of chemical energy:

$$ADP + H_2O \longrightarrow AMP + phosphate$$

The covalent bonds—that is, the phosphoanhydride bonds—linking these two phosphates to the rest of the molecule are symbolized in Figure 5–10 by the symbol ~ and were for many years called high-energy bonds. However, this designation is misleading because these bonds are easily broken, and the energy released during the reaction does not arise entirely from the bonds. The difference in energy between reactants and products is also partly the result of the arrangement of the electron orbitals of the ATP or ADP molecules. The phosphate groups each carry a negative charge and so tend to repel each other. When a phosphate group is removed, the molecule undergoes a change in electron configuration that results in a structure with less energy.

In most reactions that take place within a cell, the terminal phosphate group of ATP is not simply removed but is transferred to another molecule. This addition of a phosphate group to a molecule is known as **phosphorylation;** the enzymes that catalyze such transfers are known as *kinases*. This reaction transfers some of the released energy to the phosphorylated compound, which, thus energized, can participate in some other metabolic reaction.

Let us look at a simple example of energy exchange involving ATP in the formation of sucrose in sugarcane. Sucrose is formed from the monosaccharides glucose and fructose; under standard thermodynamic conditions, sucrose synthesis is strongly endergonic, requiring an input of 5.5 kilocalories for each mole of sucrose formed:

$$\text{glucose} + \text{fructose} \rightleftharpoons \text{sucrose} + H_2O$$

However, the synthesis of sucrose in sugarcane, when coupled with the breakdown of ATP, is actually exergonic. During the series of reactions involved in the formation of sucrose, a phosphate group is transferred to a glucose molecule and to a fructose molecule, thus energizing each of them; hence we have the overall equation:

$$\text{glucose} + \text{fructose} + 2ATP + 2H_2O \longrightarrow$$
$$\text{sucrose} + H_2O + 2ADP + 2 \text{ phosphate}$$

In this reaction, 5.5 kilocalories are utilized in the formation of sucrose, and the overall energy difference in products and reactants is about 9.1 kilocalories. Thus, the sugarcane plant is able to form sucrose by coupling the breakdown of two molecules of ATP to the synthesis of a covalent bond between glucose and fructose.

Where does the ATP originate? As we shall see in the next chapter, energy released in the cell's catabolic reactions, such as in the breakdown of glucose, is used to "recharge" the ADP molecule to ATP. Of course, the energy released in these catabolic reactions is originally derived from the sun as radiant energy that is converted, during photosynthesis, into chemical energy. Some of this chemical energy is stored in the ATP molecule before being converted to chemical bond energies of other organic molecules. Thus the ATP/ADP system serves as a universal energy-exchange system, shuttling between energy-releasing and energy-requiring reactions.

Summary

Life on this planet is dependent on the flow of energy from the sun. A small fraction of this energy, captured in the process of photosynthesis, is converted to the energy that drives the many other metabolic reactions associated with living systems and from which living systems derive their order and organization.

In photosynthesis, the energy of the sun is used to forge high-energy carbon-carbon and carbon-hydrogen bonds; then, in respiration, these bonds are broken down to carbon dioxide and water, and energy is released. Some of this energy is used to drive cellular processes, but as in machines, some energy is lost in each energy-conversion step.

Living systems thus operate in accord with the laws of thermodynamics. The first law of thermodynamics states that energy can never be created or destroyed, although it can be changed from one form to another. The potential energy of the initial state (or reactants) is equal to the potential energy of the final state (or products) plus the energy released in the process or reaction. The second law of thermodynamics states that, in the course of energy conversions, if no energy enters or leaves a system, the potential energy of the final state will always be less than the potential energy of the initial state. The difference in energy between the initial and final states is known as the free-energy change and is symbolized as ΔG. Exergonic (energy-yielding) reactions have a negative ΔG. Factors that determine the ΔG include ΔH, the change in heat content; ΔS, the change in entropy; and the absolute temperature T:

$$\Delta G = \Delta H - T \Delta S$$

Energy transformations in cells involve the transfer of electrons from one energy level to another and, often, from one atom or molecule to another. Reactions involving the transfer of electrons from one molecule to another are known as oxidation-reduction reactions. An atom or molecule that loses electrons is oxidized; one that gains electrons is reduced.

Metabolism is the total of all the chemical reactions that take place in cells. Reactions resulting in the breakdown, or degradation, of molecules are known collectively as catabolism. Biosynthetic reactions—the building of new molecules—are called anabolism. Metabolic reactions take place in ordered series of steps called pathways. Each step in a pathway is controlled by a particular enzyme.

Enzymes serve as highly specific catalysts; they enormously increase the rate at which reactions take place but are not used up in the process. Enzymes are large protein molecules folded in such a way that particular groups of amino acids form an active site. The reacting molecules, known as the substrates, fit precisely into this active site. Many enzymes require cofactors, which may be simple ions, such as Mg^{2+} or K^+, or nonprotein organic molecules, such as NAD^+. The latter are known as coenzymes.

Enzyme-catalyzed reactions are under tight metabolic control. The rate of enzymatic reactions is affected by temperature, pH, and other environmental factors.

ATP supplies the energy for most of the activities of the cell. The ATP molecule consists of the nitrogenous base adenine, the five-carbon sugar ribose, and three phosphate groups. The phosphate groups are linked by two covalent bonds that are easily broken, each yielding some 7.3 kcal/mole. Cells are able to carry out endergonic reactions and processes by coupling them with exergonic reactions that provide a surplus of energy. Such coupled reactions usually involve ATP or related triphosphate compounds.

Respiration

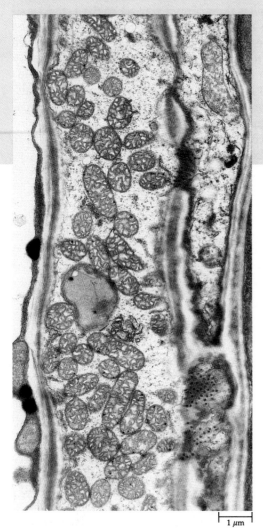

6–1

*Mitochondria from a leaf cell of barley
(Hordeum vulgare). Mitochondria are
the sites of cellular respiration, by which
chemical energy is transferred from
carbon-containing compounds to ATP.
Most of the ATP is produced on the
surfaces of the cristae by enzymes that
form a part of the structure of these
membranes. (A solitary plastid is visible
in the center of this micrograph.)*

Respiration is the process by which the chemical energy of carbohydrates is transferred to ATP—the universal energy-carrier molecule—and is thus made available for the immediate energy requirements of the cell (Figure 6–1). In the following pages, we will describe in some detail how a cell breaks down carbohydrates and captures and stores much of the released energy in the phosphoanhydride bonds of ATP. The process is described in detail because it provides an excellent illustration both of the chemical principles described in the previous chapter and of the way in which cells perform biochemical work.

As mentioned in Chapter 3, energy-yielding carbohydrate molecules are generally stored in plants as sucrose or starch. A necessary preliminary step to respiration is the hydrolysis of these storage molecules to monosaccharides. Respiration itself is generally considered to begin with glucose, an end product of the hydrolysis of both sucrose and starch.

Glucose can be used as a source of energy under both aerobic (with O_2) and anaerobic (without O_2) conditions. However, maximum energy yields for oxidizable organic compounds are generally achieved only under aerobic conditions. Consider, for example, the overall reaction for the complete oxidation of glucose:

$$C_6H_{12}O_6 + 6O_2 \longrightarrow 6CO_2 + 6H_2O + energy$$

With oxygen as the ultimate electron acceptor, this reaction is highly exergonic (energy-yielding), with a ΔG of −686 kcal/mole. This reaction represents the process of **respiration.** (When energy is extracted from organic compounds without the involvement of oxygen, the process is called **fermentation,** which is discussed later in this chapter.)

Respiration involves three distinct stages: glycolysis, the Krebs cycle, and the electron transport chain. In **glycolysis,** the six-carbon glucose molecule is broken down to a pair of three-carbon molecules of pyruvic acid or pyruvate. (Pyruvic acid dissociates, producing pyruvate and a hydrogen ion. Pyruvic acid and pyruvate exist in dynamic equilibrium, and the two terms can be used interchangeably.) In the **Krebs cycle,** the pyruvate molecules are further broken down to carbon dioxide and water, and the resulting electrons are passed to the **electron transport chain.**

As the glucose molecule is oxidized, some of its energy is extracted in a series of small, discrete steps and is stored in the phosphoanhydride bonds of ATP.

In accord with the second law of thermodynamics, some of this chemical energy is dissipated as heat energy. In birds, mammals, and to some extent other vertebrates, the heat generated by cellular respiration is conserved by various mechanisms so that the body temperature of the organism generally remains above that of the environment. In plants, respiration normally has virtually no effect on the temperature of the plant body. However, in the flowers and inflorescences of some plants, an alternative electron transport chain produces little ATP, and most of the free energy present in the respiratory substrate—usually starch—is liberated in the form of heat. Floral heat production (thermogenicity) is under tight biological control, as evidenced by the fact that the same upper temperature is reached regardless of ambient (surrounding) temperature. Familiar examples of such thermogenic plants are the Eastern skunk cabbage (*Symplocarpus foetidus*) and the houseplants *Dieffenbachia, Philodendron,* and *Monstera,* all members of the arum family (Araceae) (see the essay on pages 416 and 417). Probably the most spectacular case of thermogenicity is found in the voodoo lily (*Sauromatum guttatum*) (Figure 6–2). An "explosive" release of heat serves as a "volatilizer" for pungent floral odors that attract insect pollinators, such as carrion beetles and flesh-feeding flies.

Glycolysis

Glycolysis (from *glyco,* meaning "sugar," and *lysis,* meaning "splitting") occurs in a series of nine steps, each catalyzed by a specific enzyme. This series of reactions is carried out by virtually all living cells, from bacteria to the eukaryotic cells of plants and animals. Glycolysis is an anaerobic process that occurs in the ground substance of the cytoplasm. Biologically, glycolysis may be considered a primitive process, in that it most likely arose before the appearance of atmospheric oxygen and before the origin of cellular organelles.

The glycolytic pathway is shown in detail in Figure 6–3. As the series of reactions is discussed, notice how the carbon skeleton of the molecule is disassembled as its atoms are rearranged step by step. It is not intended that you memorize these steps; simply follow them

APPENDIX

6–2

In the voodoo lily (Sauromatum guttatum), thermogenicity may result in a temperature increase as great as 14°C and an oxygen consumption as high as that of a hummingbird in flight. An "explosive" release of heat serves as a "volatilizer" for pungent floral odors that attract insect pollinators, such as carrion beetles and flesh-feeding flies. The primary site for the production of both heat and odor is the long, naked appendix that extends upward out of the floral chamber, in which the pollinators are temporarily trapped. A day later, a second "explosion" within the floral chamber stimulates the insects to activity at the time when the male flowers are shedding their pollen; this sequence of events results in pollination of the female flowers. The inflorescence then shrivels, and the insects are free to escape and cross-pollinate other inflorescences.

closely. Note especially the formation of ATP from ADP and the formation of NADH from NAD⁺. ATP and NADH represent the cell's net energy harvest from this process. In Chapter 7, we will see that reactions 4, 5, and 6 also occur in the Calvin cycle. This repetition illustrates a principle of biochemical evolution: pathways do not arise entirely anew; rather, a few new reactions are added to an existing set to make a "new" pathway.

Step 1   The first step in glycolysis requires an input of energy in the form of ATP. The enzyme hexokinase catalyzes the transfer of the terminal phosphate group of ATP to glucose, producing glucose 6-phosphate in an energy-yielding reaction.

Step 2   In this step, the molecule is rearranged, again with the help of a specific enzyme (phosphoglucoisomerase). The six-sided ring characteristic of glucose becomes a five-sided fructose ring. (As shown in Figure 3–2, glucose and fructose have the same types and numbers of atoms ($C_6H_{12}O_6$) but differ in the arrangement of these atoms.) This reaction can proceed in either direction. It is pushed forward by the accumulation of glucose 6-phosphate from step 1 and is pulled forward by the disappearance of fructose 6-phosphate as it enters step 3.

Step 3   This step, which is similar to step 1, results in the attachment of a phosphate to the first carbon of the fructose 6-phosphate molecule, which produces fructose 1,6-bisphosphate, that is, fructose with phosphates in the 1 and 6 positions. The conversion of the glucose molecule to the higher-energy fructose 1,6-bisphosphate is accomplished at the expense of two molecules of ATP. Thus far, no energy has been recovered, but the overall yield will more than compensate for this initial investment.

Step 4   This is the cleavage step of glycolysis. The six-carbon molecule is split into two interconvertible three-carbon molecules—glyceraldehyde 3-phosphate and dihydroxyacetone phosphate. However, because the glyceraldehyde 3-phosphate is used up in subsequent reactions, all of the dihydroxyacetone phosphate is eventually converted to glyceraldehyde 3-phosphate. Thus, the products of subsequent steps must be counted twice to account for the fate of each glucose molecule. With the completion of step 4, the preparatory reactions that require an input of ATP energy are complete.

Step 5   Next, glyceraldehyde 3-phosphate molecules are oxidized—that is, hydrogen atoms with their electrons are removed—and NAD^+ is converted to NADH. This is the first reaction from which the cell harvests energy. Energy from this oxidation reaction is also used to attach an additional phosphate group to what is now the 1 position of each of the three-carbon

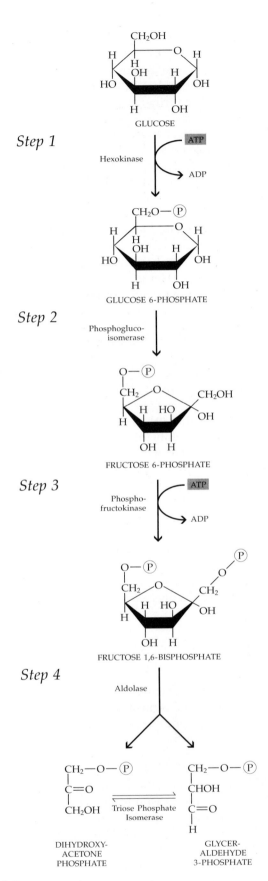

6–3
The steps of glycolysis. $\circled{P}$ = *phosphate group.*

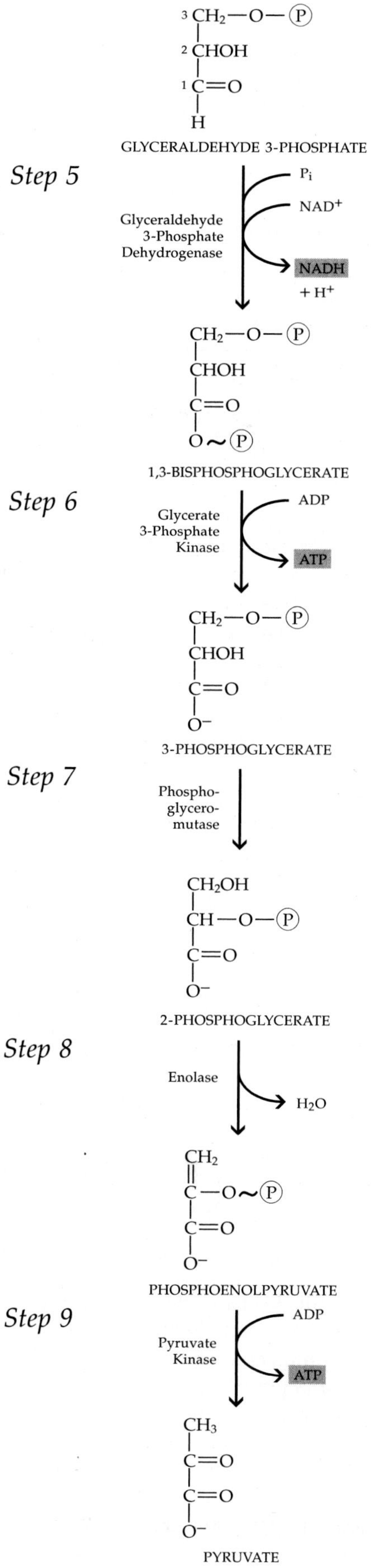

^{3}CH$_2$—O—(P)

^{2}CHOH

^{1}C=O

H

GLYCERALDEHYDE 3-PHOSPHATE

Step 5

Glyceraldehyde
3-Phosphate
Dehydrogenase

P$_i$

NAD$^+$

NADH

+ H$^+$

CH$_2$—O—(P)

CHOH

C=O

O~(P)

1,3-BISPHOSPHOGLYCERATE

Step 6

Glycerate
3-Phosphate
Kinase

ADP

ATP

CH$_2$—O—(P)

CHOH

C=O

O$^-$

3-PHOSPHOGLYCERATE

Step 7

Phospho-
glycero-
mutase

CH$_2$OH

CH—O—(P)

C=O

O$^-$

2-PHOSPHOGLYCERATE

Step 8

Enolase

H$_2$O

CH$_2$

C—O~(P)

C=O

O$^-$

PHOSPHOENOLPYRUVATE

Step 9

Pyruvate
Kinase

ADP

ATP

CH$_3$

C=O

C=O

O$^-$

PYRUVATE

molecules. (The designation P$_i$ indicates inorganic phosphate that is available as a phosphate ion in solution in the cytoplasm.)

Step 6 The bond energy of phosphate released from the 1,3-bisphosphoglycerate molecule is used to recharge a molecule of ADP (a total of two molecules of ATP are formed per molecule of glucose). This is a highly exergonic reaction (that is, ΔG has a large negative value) and pulls all the previous reactions forward. The formation of ATP by the enzymatic transfer of a phosphate group from a metabolic intermediate to ADP is referred to as **substrate-level phosphorylation.**

Step 7 The remaining phosphate group is transferred from the 3 position to the 2 position of the glycerate molecule.

Step 8 In this step, a molecule of water is removed from the three-carbon compound, and as a consequence of the rearrangement of electrons in the molecule, a high-energy phosphorylated compound (phosphoenolpyruvate) is formed.

Step 9 The phosphate group is transferred to a molecule of ADP, forming another molecule of ATP (again, a total of two molecules of ATP are formed per molecule of glucose). This is also a highly exergonic reaction, driving the reaction sequence in the forward direction.

SUMMARY OF GLYCOLYSIS

The complete sequence of glycolysis begins with one molecule of glucose (Figure 6–4). Energy enters the sequence at steps 1 and 3 by the transfer of a phosphate group from an ATP molecule—one at each step—to the sugar molecule. The six-carbon molecule splits at step 4, and after this point the sequence yields energy. At step 5, two molecules of NAD$^+$, in being reduced to two molecules of NADH, store some of the energy from the oxidation of glyceraldehyde 3-phosphate. At steps 6 and 9, two molecules of ADP take energy from the system, forming additional phosphoanhydride bonds to become two molecules of ATP.

Glycolysis (from glucose to pyruvate) can be summarized by the overall equation

glucose + 2NAD$^+$ + 2ADP + 2P$_i$ $\longrightarrow$
2 pyruvate + 2NADH + 2H$^+$ + 2ATP + 2H$_2$O

Thus, one glucose molecule is converted to two molecules of pyruvate. The net harvest—the energy yield—is two molecules of ATP and two molecules of NADH per molecule of glucose. Two moles of pyruvate have a total energy content of about 546 kilocalories, com-

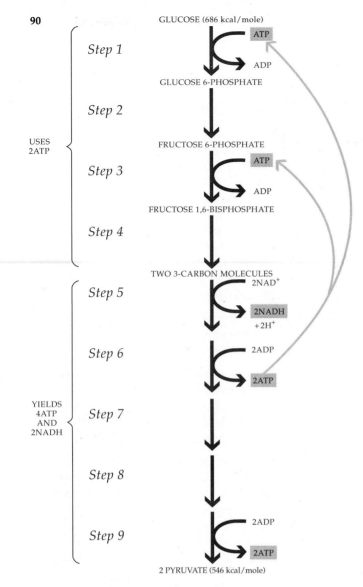

GLUCOSE (686 kcal/mole)

Step 1

ATP

ADP

GLUCOSE 6-PHOSPHATE

Step 2

USES
2ATP

FRUCTOSE 6-PHOSPHATE

Step 3

ATP

ADP

FRUCTOSE 1,6-BISPHOSPHATE

Step 4

TWO 3-CARBON MOLECULES

2NAD$^+$

Step 5

2NADH
+2H$^+$

Step 6

2ADP

2ATP

YIELDS
4ATP
AND
2NADH

Step 7

Step 8

Step 9

2ADP

2ATP

2 PYRUVATE (546 kcal/mole)

6–4

A summary of glycolysis. Two molecules of ATP and two molecules of NADH represent the energy yield. Most of the energy stored in the original glucose molecule remains in the two molecules of pyruvate.

pared with 686 kilocalories stored in a mole of glucose; hence, a large portion of the energy stored in the original glucose molecule is still present in the two pyruvate molecules.

Note that glycolysis is, in essence, an internal oxidation-reduction reaction. Compare the original substrate for glycolysis (glucose) with the product molecules (pyruvate). The $-CH_3$ (methyl group) of pyruvate arises from carbons 1 and 6 of the glucose molecule, and these two carbons are more reduced in pyruvate than in glucose. In contrast, the $-COO^-$ (carboxylate group) of pyruvate arises from the two middle carbons (carbons 3 and 4) of glucose, and these carbons are more oxidized in pyruvate than in the glucose molecule from which they were derived.

Also note that the two molecules of NADH can yield additional ATP molecules in the mitochondria when used as electron donors to the electron transport chain of the aerobic pathway (see page 94).

The Aerobic Pathway

Pyruvate is a key intermediate in cellular energy metabolism, and it can be utilized in one of several pathways. Which pathway it follows depends in part upon the conditions under which metabolism takes place and in part upon the specific organism involved and, in some cases, the particular tissue in the organism. The principal environmental factor that determines which pathway is followed is the availability of oxygen.

In the presence of oxygen, pyruvate is oxidized completely to carbon dioxide, and glycolysis is but the initial phase of respiration. This aerobic pathway results in the complete oxidation of glucose and a much greater ATP yield than can be achieved by glycolysis alone. These reactions take place in two stages—the Krebs cycle and the electron transport chain—both occurring within mitochondria in eukaryotic cells.

Mitochondria are surrounded by two membranes, the inner one convoluted inwardly into folds called cristae (see Figure 6–14). Within the innermost compartment, surrounding the cristae, is the matrix, a dense solution containing enzymes, coenzymes, water, phosphates, and other molecules involved in respiration. Thus, a mitochondrion resembles a self-contained chemical factory. The outer membrane allows most small molecules to move in or out freely, but the inner one permits the passage of only certain molecules, such as pyruvate and ATP, while it prevents the passage of others. Some of the enzymes of the Krebs cycle are found in solution within the matrix. Other Krebs-cycle enzymes and the components of the electron transport chain are built into the surfaces of the cristae.

A PRELIMINARY STEP: THE OXIDATION OF PYRUVATE

Before entering the Krebs cycle, pyruvate is both oxidized and decarboxylated. In the course of this exergonic reaction, a molecule of NADH is produced from NAD$^+$. The original glucose molecule has now been oxidized to two acetyl (CH_3CO) groups; two molecules of CO_2 have been liberated; and two molecules of NADH have been formed from NAD$^+$ (Figure 6–5).

Each acetyl group is then temporarily attached to coenzyme A (CoA)—a large molecule, a portion of which is a nucleotide and a portion of which is pan-

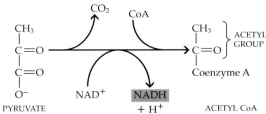

6–5

The three-carbon pyruvate molecule is oxidized and decarboxylated to form the two-carbon acetyl group, which is attached to coenzyme A to form acetyl CoA. The oxidation of the pyruvate molecule is coupled to the production of NADH from NAD+. Acetyl CoA is the form in which carbon atoms derived from glucose enter the Krebs cycle.

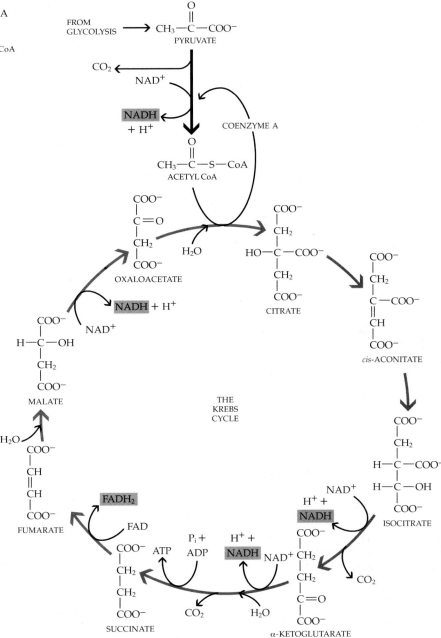

6–6

In the course of the Krebs cycle, two carbons enter as the acetyl group and two carbons are oxidized to carbon dioxide; the hydrogen atoms are passed to electron carriers (NAD+ and FAD). As in glycolysis, a specific enzyme is involved at each step.

tothenic acid, one of the B-complex vitamins. The combination of the acetyl group and CoA is known as acetyl CoA (Figure 6–5).

Fats and amino acids can also be converted to acetyl CoA and enter the respiratory sequence at this point. A triglyceride molecule is first hydrolyzed to glycerol and three fatty acids. Then, beginning at the carboxyl end, two-carbon acetyl groups are successively removed as acetyl CoA from the fatty acids. A molecule such as palmitic acid (see Figure 3–9), which contains 16 carbon atoms, yields eight molecules of acetyl CoA. Amino acids can also be converted to other Krebs-cycle intermediates such as α-ketoglutarate, oxaloacetate, and fumarate, and thereby enter the cycle.

THE KREBS CYCLE

The Krebs cycle is named in honor of Sir Hans Krebs, who was largely responsible for its elucidation. Krebs postulated this metabolic pathway in 1937 and later received a Nobel Prize in recognition of his brilliant work. The Krebs cycle is also called the tricarboxylic acid (TCA) cycle because it is initiated with the formation of an organic acid (citrate) that has three carboxylic acid groups.

The Krebs cycle always begins with acetyl CoA, its only real substrate. Upon entering the Krebs cycle (Figure 6–6), the two-carbon acetyl group is combined with a four-carbon compound (oxaloacetate) to produce a

RIBOFLAVIN

CH_3

CH_3

+ 2[H]

CH_3

CH_3

Reduction

Oxidation

CH_2

$(HCOH)_3$

CH_2

CH_2

$(HCOH)_3$

CH_2

O

$O=P—O^-$

O

$O=P—O^-$

O

$O=P—O^-$

O

$O=P—O^-$

NH_2

NH_2

ADENINE

RIBOSE

H   H

H   H

OH OH

H   H

H   H

OH OH

FLAVIN ADENINE DINUCLEOTIDE,
OXIDIZED (FAD)

FLAVIN ADENINE DINUCLEOTIDE,
REDUCED (FADH₂)

6–7

Flavin adenine dinucleotide, an electron acceptor, in its oxidized form (FAD) and its reduced form (FADH₂). Riboflavin is a vitamin (vitamin B₂) made by all plants and many microorganisms. It is a pigment and, in its oxidized form, is bright yellow. A related electron acceptor, flavin mononucleotide (FMN), consists of riboflavin with one phosphate group. It accepts electrons from NADH in the electron transport chain.

six-carbon compound (citrate). (The coenzyme A is released to combine with a new acetyl group.) In the course of the cycle, two of the six carbons are oxidized to CO_2 and oxaloacetate is regenerated—thus literally making this series of reactions a cycle. Each turn around the cycle uses up one acetyl group and regenerates one molecule of oxaloacetate, which is then ready to begin the Krebs cycle again. In the course of these steps, some of the energy released by the oxidation of the carbon atoms is used to convert ADP to ATP (one molecule per cycle), and some is used to convert NAD^+ to NADH (three molecules per cycle). In addition, some of the energy is used to reduce a second electron carrier —the coenzyme flavin adenine dinucleotide (FAD) (Figure 6–7). One molecule of FADH₂ is formed from FAD in each turn of the cycle. Oxygen is not directly involved in the Krebs cycle; the electrons and protons removed in the oxidation of carbon are all accepted by NAD^+ and FAD:

$$\text{oxaloacetate} + \text{acetyl CoA} + \text{ADP} + P_i + 3NAD^+ + \text{FAD} \longrightarrow$$
$$\text{oxaloacetate} + 2CO_2 + \text{CoA} + \text{ATP} +$$
$$3NADH + 3H^+ + FADH_2$$

The Krebs cycle is summarized in Figure 6–8.

THE ELECTRON TRANSPORT CHAIN

The glucose molecule is now completely oxidized. Some of its energy has been used to produce ATP from

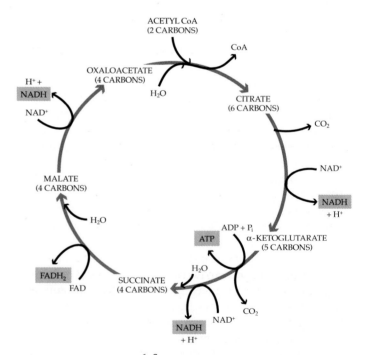

6–8

A summary of the Krebs cycle. One molecule of ATP, three molecules of NADH, and one molecule of FADH₂ represent the energy yield of each acetyl group passing through the cycle.

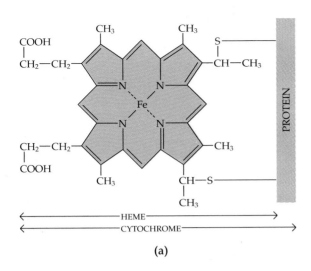

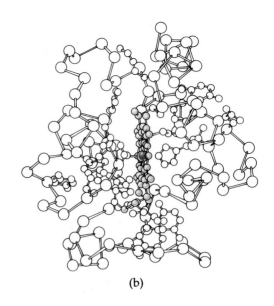

(a)

(b)

6–9

Cytochromes are molecules that partici-pate in electron transfer in mitochondria. (a) Each cytochrome contains an atom of iron held in a nitrogen-containing (porphyrin) ring. The porphyrin ring *with its atom of iron is known as heme. Each iron atom accepts one electron and is reduced from Fe^{3+} to Fe^{2+}. The cytochrome shown here is cytochrome c.* *(b) The overall structure of the cyto-chrome c molecule, showing the posi-tion of the heme group (color) within the globular protein.*

ADP in substrate-level phosphorylation. Most of the energy, however, still remains in the electrons removed from the carbon atoms as they were oxidized. These electrons were passed to the electron carriers NAD^+ and FAD and are at a high energy level. In the course of the electron transport chain, they are passed "down-hill" to oxygen, and the energy released is used to form ATP from ADP. This process is known as **oxidative phosphorylation.**

The electron carriers of the electron transport chain of the mitochondria differ from NAD^+ and FAD in their chemical structures. Some of them belong to a class of compounds known as *cytochromes*—protein molecules with an iron-containing porphyrin ring, or heme group, attached (Figure 6–9). Each cytochrome differs in its protein structure and in the energy level at which it holds the electrons. In their reduced form, cy-tochromes carry a single electron without a proton.

Non-heme iron proteins—the *iron-sulfur proteins*—are another component of the electron transport chain. The iron of these proteins is not attached to a porphyrin ring; instead, the iron atoms are attached to sulfides and to the sulfurs of cysteine residues of the protein chain (Figure 6–10). Like cytochromes, the iron-sulfur proteins carry an electron but not a proton.

The most abundant components of the electron transport chain are *quinone molecules* (Figure 6–11). Unlike cytochromes and iron-sulfur proteins, quinones carry both an electron and a proton—the equivalent of a hydrogen atom. By alternating electron transfer be-tween components that carry an electron only and those that carry a hydrogen atom, protons can be shuttled across the mitochondrial membrane. For ex-ample, every time a quinone molecule accepts an elec-

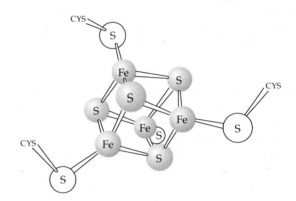

6–10

Postulated arrangement of iron and sulfur atoms in the center of an iron-sulfur protein. The iron-sulfur center shown here consists of four iron atoms and eight sulfur atoms, four of which are contributed by four cysteine residues of the protein (polypeptide) chain of the enzyme. Iron-sulfur proteins are involved in electron transfer.

tron from a cytochrome, it also picks up a proton (H^+) from the surrounding medium. When the quinone gives up its electron to the next carrier, such as a cy-tochrome, the proton is again released into the me-dium. Because the electron carriers are arranged in the membrane so that protons are picked up on one side of the membrane and released on the other, a proton gra-dient is generated across the membrane. The impor-tance of this gradient is discussed later. Most of these quinone molecules are not attached to proteins and are thought to be able to move across the membrane.

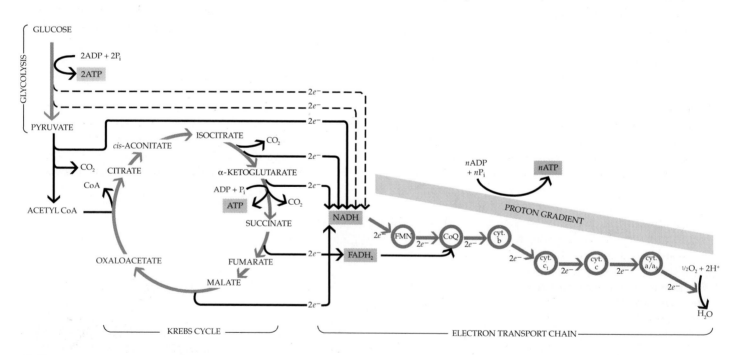

6–11

Oxidized and reduced forms of the quinone ubiquinone. Ubiquinone (the oxidized form of coenzyme Q, see Figure 6–12) can be reversibly reduced to

ubiquinol. Ubiquinone accepts both electrons and protons from a donor in the electron transport chain. The name

"ubiquinone" is a reflection of the ubiquity of this compound (it occurs in virtually all cells).

At the top of the electron transport chain are the electrons held by NADH and FADH$_2$. For every molecule of glucose oxidized, the yield of the Krebs cycle was two molecules of FADH$_2$ and six molecules of NADH, and the oxidation of pyruvate to acetyl CoA yielded two molecules of NADH. Recall that an additional two molecules of NADH were produced in glycolysis; in the presence of oxygen, electrons from these NADH molecules are transported into the mitochondrion. Electrons from all NADH molecules are trans-

ferred to the electron acceptor flavin mononucleotide (abbreviated FMN; see Figure 6–7), the first component of the electron transport chain. Electrons from FADH$_2$ molecules are transferred to an electron acceptor farther down the electron transport chain.

As electrons flow along the electron transport chain from a higher to a lower energy level, the released energy is harnessed and used to generate the proton gradient that, in turn, drives the formation of ATP from ADP + P$_i$ (Figure 6–12). At the end of the chain, the

6–12

A summary of respiration. Glucose is first broken down to pyruvate, with a yield of two ATP molecules and the reduction (dashed arrows) of two NAD$^+$ molecules. Pyruvate is oxidized to acetyl CoA, and one molecule of NAD$^+$ is reduced (note that this and subsequent reactions must be counted twice for each glucose molecule; this electron passage is

indicated by solid arrows). In the Krebs cycle, the acetyl group is oxidized and the electron acceptors, NAD$^+$ and FAD, are reduced. NADH and FADH$_2$ then transfer their electrons to the electron transport chain, which consists mostly of a series of cytochromes. As these cytochromes pass the electrons downhill, the energy released is used to form ATP from

ADP, as shown in Figure 6–14. Three molecules of ATP are produced for each pair of electrons passed from NADH to oxygen, and two for each pair passed from FADH$_2$ to oxygen. The electron transport chain is shown here in abbreviated form, with some of its intermediates deleted.

electrons are accepted by oxygen and combine with protons (hydrogen ions) to produce water. Each time one pair of electrons passes from NADH to oxygen, enough protons are "pumped" across the membrane to generate three molecules of ATP. Each time a pair of electrons passes from $FADH_2$, which holds them at a slightly lower energy level than NADH, enough protons are "pumped" to form two molecules of ATP.

THE MECHANISM OF OXIDATIVE PHOSPHORYLATION: CHEMIOSMOTIC COUPLING

For many years, the mechanism of oxidative phosphorylation—that is, the way in which ATP is formed from ADP and phosphate as electrons pass down the electron transport chain—was a puzzle. A major breakthrough occurred in 1961, when the British biochemist Peter Mitchell (Nobel laureate in chemistry, 1978) proposed that the process is driven by a proton gradient (a differential H^+ concentration across a membrane) produced across the inner mitochondrial membrane during electron transport. Continuing studies have revealed many details of this mechanism, which is known as **chemiosmotic coupling.**

According to this ingenious concept, protons are pumped out of the mitochondrial matrix to the outer mitochondrial compartment as electrons from NADH are passed along the electron transport chain, which forms a part of the inner mitochondrial membrane. Each pair of electrons crosses the membrane three times as it moves from one electron carrier to the next (and finally to oxygen). This triple passage generates an electrochemical gradient—a chemical gradient (of protons) and an electrical gradient (of charge)—that drives

the protons back into the matrix through specialized channels in the inner mitochondrial membrane.

It is now known that the channels through which the protons flow back into the matrix are provided by a large enzyme complex known as **ATP synthase.** This enzyme complex consists of two major components, or factors, F_O and F_1 (Figure 6–13a). The F_O is embedded in the inner mitochondrial membrane, traversing it from outside to inside and serving as the channel through which protons flow. The F_1 is attached to the F_O on the matrix side of the membrane. The F_1 components appear as protruding knobs in electron micrographs (Figure 6–13b). F_1 alone cannot make ATP from ADP and phosphate in its isolated form, but it can hydrolyze ATP to ADP and phosphate, thus functioning as an ATPase. However, its usual function when attached to the F_O component in the intact mitochondrion is just the reverse: as protons flow down the electrochemical gradient from the outside into the matrix, passing through the F_O component and then the F_1 component, the free energy that is released powers the synthesis of ATP from ADP and phosphate.

Figure 6–14 summarizes the chemiosmotic-coupling mechanism as it occurs in oxidative phosphorylation. The term "chemiosmotic" reflects the fact that the production of ATP in oxidative phosphorylation includes both chemical processes and transport across a differentially permeable membrane.

Control of Oxidative Phosphorylation

Electrons continue to flow down the electron transport chain only if ADP is available for conversion to ATP. Thus, oxidative phosphorylation is regulated by the "law of supply and demand." When the energy requirements of the cell decrease, fewer molecules of

6–13

(a) *Diagram of the ATP synthase complex. The F_O portion is embedded in the inner mitochondrial membrane and extends across it, and the F_1 component extends into the mitochondrial matrix. The F_1 component is a complex of three α and three β subunits, and the F_O component is composed of smaller c subunits, which surround a larger subunit (not shown). The two components are attached to each other through a neck consisting of subunits designated b and δ. (b) In this electron micrograph, the knobs protruding from the membrane of these vesicles are the F_1 portions of ATP synthase complexes. The vesicles are formed from broken mitochondrial membrane, and the knobs are protruding out into the medium.*

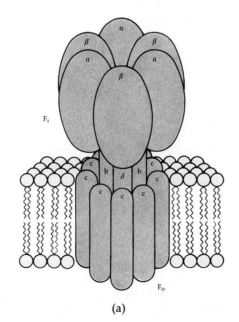

(a)

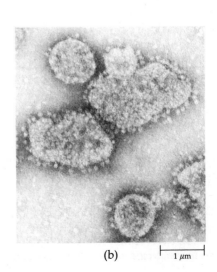

(b) ⊢——⊣ 1 μm

6–14

(a) *Three-dimensional diagram of a mitochondrion. (b) Detail of a crista, showing the F_1 components of the ATP synthase complexes lining the inner membrane on the matrix side. (c) According to the chemiosmotic-coupling hypothesis, protons are pumped out of the mitochondrial matrix as electrons are passed down the electron transport chain, which forms part of the inner mitochondrial membrane. The inward movement of protons down the electro-chemical gradient as they pass back across the inner membrane through the ATP synthase complex provides the energy for synthesis of ATP from ADP and phosphate. The exact number of protons pumped out of the matrix as each electron pair moves down the chain remains to be determined, as does the number that must flow through ATP synthase for each molecule of ATP formed.*

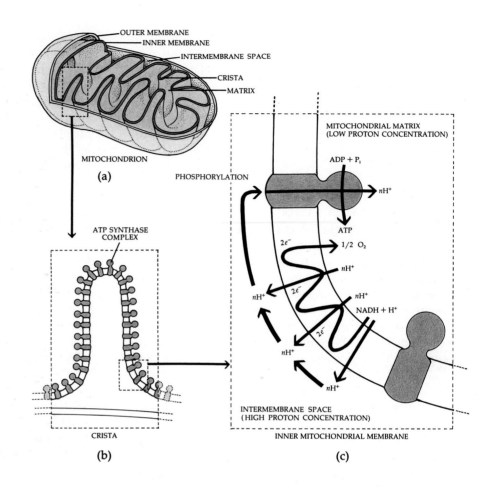

OVERALL ENERGY HARVEST

ATP are used, fewer molecules of ADP become available, and the electron flow is decreased.

We are now in a position to see how much of the energy originally present in the glucose molecule has been recovered in the form of ATP. The "balance sheet" for ATP yield given in Table 6–1 may help you keep track of the discussion that follows.

Glycolysis in the presence of oxygen yields 2 molecules of ATP directly plus 2 molecules of NADH. The electrons held by these 2NADH molecules are transported across the mitochondrial membrane at a "cost" of 2ATP molecules. Thus the net yield from reoxidation of the 2NADH molecules is 4 molecules of ATP.

The conversion of pyruvate to acetyl CoA yields 2 molecules of NADH (inside the mitochondrion) for each molecule of glucose and so produces 6 molecules of ATP.

The Krebs cycle yields, for each molecule of glucose, 2 molecules of ATP, 6 molecules of NADH, and 2 molecules of $FADH_2$, for a total of 24 molecules of ATP.

As the balance sheet shows (Table 6–1), the complete yield from a single molecule of glucose is 36 mole-

Table 6–1   *Energy Yield from One Molecule of Glucose*

Glycolysis:	2 ATP 2 NADH ⟶ 4 ATP	⟶ 6 ATP
Pyruvate ⟶ acetyl CoA:	1 NADH ⟶ 3 ATP	(×2) ⟶ 6 ATP
Krebs cycle:	1 ATP 3 NADH ⟶ 9 ATP 1 $FADH_2$ ⟶ 2 ATP	(×2) ⟶ <u>24 ATP</u>
		Net yield 36 ATP

cules of ATP. All but 2 of the 36 molecules of ATP have come from reactions in the mitochondrion, and all but 4 involve the oxidation of NADH or $FADH_2$ along the electron transport chain.

The total difference in free energy (ΔG) between the reactants (glucose and oxygen) and the products (carbon dioxide and water) is -686 kilocalories per mole. Approximately 39 percent of this, or 263 kilocalories per mole (7.3×36), has been captured in the terminal phosphoanhydride bonds of the 36 ATP molecules (Figure 6–15), an efficiency of about 39 percent.

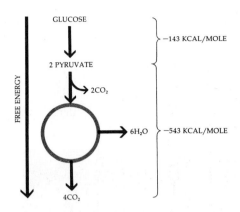

6–15

The energy changes in the oxidation of glucose. The complete respiratory sequence proceeds with an energy drop of 686 kilocalories per mole of glucose. Of this, about 39 percent (263 kilocalories) is conserved in 36 ATP molecules. By contrast, in the anaerobic pathways only 2 ATP molecules are produced, representing only about 2 percent of the available energy of glucose.

The ATP molecules, once formed, are exported across the membrane of the mitochondrion by a shuttle system that simultaneously brings in one molecule of ADP for each ATP exported.

Anaerobic Pathways

In eukaryotic cells (as well as most bacteria), pyruvate usually follows the aerobic pathway and is completely oxidized to carbon dioxide and water. However, in the absence of oxygen, pyruvate is not the end product of glycolysis. The NADH produced during the oxidation of glyceraldehyde 3-phosphate must be reoxidized to NAD$^+$. Without this reoxidation, glycolysis would soon stop because the cell would run out of NAD$^+$ as an electron acceptor.

In many bacteria, fungi, protists, and animal cells, this oxygenless, or anaerobic, process may result in the formation of lactate: this process is therefore called **lactate fermentation.** In yeast and most plant cells, pyruvate is broken down to ethanol (ethyl alcohol) and carbon dioxide, and this anaerobic process is therefore called **alcoholic fermentation.** In this process, the reoxidation of NADH is preceded by the removal of carbon dioxide (decarboxylation) (Figure 6–16).

Thermodynamically, lactate fermentation and alcoholic fermentation are similar. In both, the NADH is

Bioluminescence

In most organisms, the energy stored in ATP is used for cellular work, such as biosynthesis, transport, and movement. In some, however, the chemical energy is sometimes reconverted to light energy. Bioluminescence is probably an accidental by-product of energy exchanges in most luminescent organisms, such as the fungus *Mycena lux-coeli* shown here, photographed by its own light. In some luminescent organisms, such as fireflies, in which the flashes of light serve as mating signals, bioluminescence serves a useful function.

reoxidized, and the energy yield for glucose breakdown is limited to the 2 ATPs produced during glycolysis. The complete, balanced equation for the fermentation of glucose can be written as follows:

$$\text{glucose} + 2\text{ADP} + 2\text{P}_i \longrightarrow \begin{Bmatrix} 2 \text{ lactate } or \\ 2 \text{ ethanol} + 2\text{CO}_2 \end{Bmatrix} +$$

$$2\text{ATP} + 2\text{H}_2\text{O}$$

During alcoholic fermentation, approximately 7 percent of the total available energy of the glucose molecule—about 52 kilocalories per mole—is released, with about 93 percent remaining in the two alcohol molecules. However, if one considers the efficiency with which the anaerobic cell conserves much of those 52 kilocalories as ATP (7.3 kilocalories per mole of ATP, or 14.6 kilocalories per mole total), the efficiency of energy conservation is about 26 percent. As noted in Chapter 5, the efficiency of human-made machines seldom exceeds 25 percent, and that of respiration is only 39 percent.

The fact that glycolysis does not require oxygen suggests that the glycolytic sequence evolved early, before free oxygen was present in the atmosphere. Presumably, primitive one-celled organisms used glycolysis (or something very much like it) to extract energy from the

6–16
(a) The steps by which pyruvate is converted anaerobically to ethanol. In the first step, carbon dioxide is released. In the second, NADH is oxidized and acetaldehyde is reduced. Most of the energy of the glucose remains in the alcohol, which is the principal end product of the sequence. However, by regenerating NAD^+, these steps allow glycolysis to continue, with its small but sometimes vital yield of ATP. (b) The consequences of anaerobic glycolysis. Yeast cells, which are present on the grapes, mix with the juice when the grapes are crushed. Storing the mixture under anaerobic conditions causes the yeast to break down the glucose in the grape juice to alcohol. In modern wine-making, pure yeast cultures are added to relatively sterile grape juice for fermentation, rather than relying on the yeasts carried on the grapes.

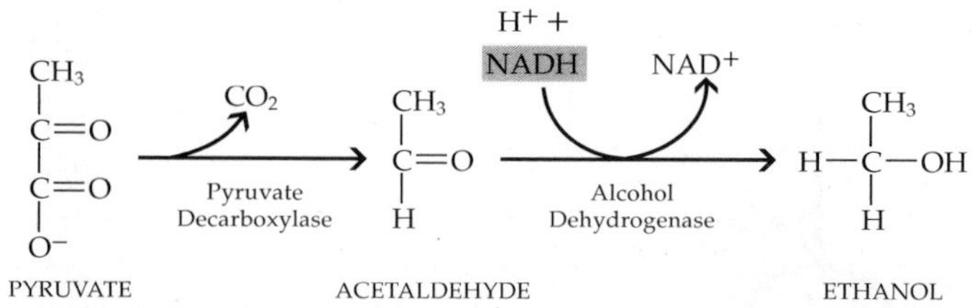

(a)

(b)

organic molecules they absorbed from their watery surroundings. Although the anaerobic pathways generate only two molecules of ATP for each glucose processed, this low yield was and is adequate for the needs of many organisms.

Summary

Respiration, or the complete oxidation of glucose, is the chief source of energy in most cells. As the glucose is broken down in a series of small enzymatic steps, some of the energy released is packaged in the form of terminal phosphoanhydride bonds in ATP and the rest is lost as heat.

The first phase in the breakdown of glucose is glycolysis, in which the six-carbon glucose molecule is split into two three-carbon molecules of pyruvate; two molecules of ATP and two of NADH are formed. This reaction occurs in the cytoplasmic ground substance.

In the course of respiration, the three-carbon pyruvate molecules are broken down within the mitochondrion to two-carbon acetyl groups, which then enter the Krebs cycle as acetyl CoA. In the Krebs cycle, the acetyl group is broken apart in a series of reactions to yield two additional molecules of carbon dioxide, one molecule of ATP, and four molecules of reduced electron carriers (three NADH and one $FADH_2$).

The final stage of respiration is the electron transport chain, which involves a series of electron carriers and enzymes embedded in the inner membrane of the mitochondrion. Along this series of electron carriers, the high-energy electrons accepted by NAD^+ during glycolysis and by NAD^+ and FAD during the Krebs cycle pass "downhill" to oxygen. The large quantity of free energy released during the passage of electrons down the electron transport chain powers the pumping of protons (H^+ ions) out of the mitochondrial matrix. This creates an electrochemical gradient of H^+ across the inner membrane of the mitochondrion. When the protons pass back into the matrix through the ATP synthase complex, the energy that is released is used to form ATP from ADP and phosphate. In the course of the breakdown of the glucose molecule, 36 molecules of ATP are gained, most of them in the mitochondrion.

In the absence of oxygen, pyruvate may be converted either to lactate (in many bacteria, fungi, and animal cells) or to ethanol and carbon dioxide (in yeasts and most plant cells). These anaerobic processes—called fermentation—yield 2 ATPs for each glucose molecule (2 pyruvate molecules).

CHAPTER

Photosynthesis

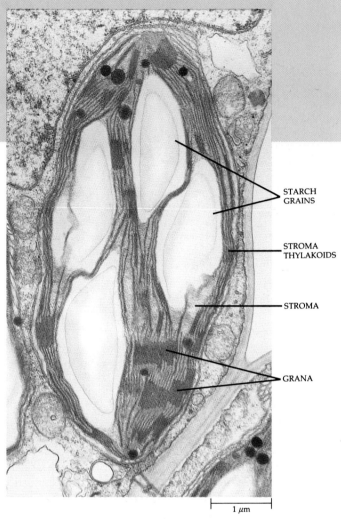

STARCH GRAINS

STROMA THYLAKOIDS

STROMA

GRANA

1 μm

7–1
A chloroplast from a mesophyll cell of the pigweed (Amaranthus retroflexus) *leaf. The chloroplast is the site of photosynthesis in eukaryotic organisms. The light-capturing reactions occur in thylakoids, where the chlorophylls and other pigments are found. Many of the thylakoids characteristically occur in disklike stacks called grana. The series of reactions by which this energy is used to synthesize carbon-containing compounds occurs in the stroma, the material surrounding the thylakoids. During periods of intense photosynthesis, some of the photosynthate is stored temporarily in the chloroplast as grains of starch. At night, sucrose is produced from the starch and exported from the leaf.*

In the previous chapter, we described the breakdown of carbohydrates to yield the energy required for the many kinds of activities carried out by living systems. In the pages that follow, we will complete the circle by describing the way in which light energy from the sun is captured and converted to chemical energy.

This process—photosynthesis—is the route by which virtually all energy enters our biosphere. Each year more than 100 billion metric tons of sugar are produced worldwide by photosynthetic organisms. The importance of photosynthesis, however, extends far beyond the sheer weight of this product. Without this flow of energy from the sun, channeled largely through the chloroplasts of eukaryotic cells (Figure 7–1), the pace of life on this planet would swiftly diminish and then, following the inexorable second law of thermodynamics, would virtually cease altogether.

Overview of Photosynthesis

The importance of photosynthesis in the economy of nature was not recognized until comparatively recent times. Aristotle and other Greeks, observing that the life processes of animals were dependent upon the food they ate, thought that plants derived their food from the soil.

More than 300 years ago, in one of the first carefully designed biological experiments ever reported, the Belgian physician Jan Baptista van Helmont (ca. 1577–1644) offered the first experimental evidence that soil alone does not nourish the plant. Van Helmont grew a small willow tree in an earthenware pot, adding only water to the pot. At the end of 5 years, the willow had increased in weight by 74.4 kilograms, whereas the soil had decreased in weight by only 57 grams. On the basis of these results, van Helmont concluded that all

99

the substance of the plant was produced from the water and none from the soil! Van Helmont's conclusions, however, were too broad.

Toward the end of the eighteenth century, the English clergyman/scientist Joseph Priestley (1733–1804) reported that he had "accidently hit upon a method of restoring air that had been injured by the burning of candles." On 17 August 1771, Priestley "put a [living] sprig of mint into air in which a wax candle had burned out and found that, on the 27th of the same month, another candle could be burned in this same air." The "restorative which nature employs for this purpose," he stated, was "vegetation." Priestley extended his observations and soon showed that air "restored" by vegetation was not "at all inconvenient to a mouse." Priestley's experiments offered the first logical explanation of how air remained "pure" and able to support life despite the burning of countless fires and the breathing of many animals. When he was presented with a medal for his discovery, the citation read in part: "For these discoveries we are assured that no vegetable grows in vain . . . but cleanses and purifies our atmosphere." Today we would explain Priestley's experiments simply by saying that plants take up the CO_2 produced by combustion or exhaled by animals, and that animals inhale the O_2 released by plants.

Later, the Dutch physician Jan Ingenhousz (1730–1799) confirmed Priestley's work and showed that the air was "restored" only in the presence of sunlight and only by the green parts of plants. In 1796, Ingenhousz suggested that carbon dioxide is split in photosynthesis to yield carbon and oxygen, with the oxygen then released as gas. Subsequently, the proportion of carbon, hydrogen, and oxygen atoms in sugars and starches was found to be about one atom of carbon per molecule of water (CH_2O), as the word "carbohydrate" indicates. Thus, in the overall reaction for photosynthesis,

$$CO_2 + H_2O + \text{light energy} \longrightarrow (CH_2O) + O_2$$

it was generally assumed that the carbohydrate came from a combination of water molecules and the carbon atoms of carbon dioxide and that the oxygen was released from the carbon dioxide. This entirely reasonable hypothesis was widely accepted; but, as it turned out, it was quite wrong.

The investigator who upset this long-held theory was C. B. van Niel of Stanford University. Van Niel, then a graduate student, was investigating the activities of different types of photosynthetic bacteria. One particular group of such bacteria—known as the purple sulfur bacteria—reduces carbon to carbohydrates during photosynthesis but does not release oxygen. The purple sulfur bacteria require hydrogen sulfide for their photosynthetic activity. In the course of photosynthesis, globules of sulfur accumulate inside the bacterial cells (Figure 7–2). Van Niel found that the following re-

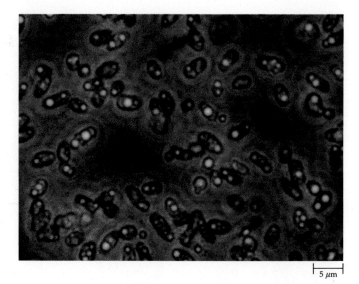

5 µm

7–2
Purple sulfur bacteria. In these cells, hydrogen sulfide plays the same role that water plays in the photosynthetic process of plants. The hydrogen sulfide is oxidized to elemental sulfur, which accumulates in globules, visible within these cells.

action takes place during photosynthesis in these bacteria:

$$CO_2 + 2H_2S \xrightarrow{\text{Light}} (CH_2O) + H_2O + 2S$$

This finding was simple and did not attract much attention until van Niel made a bold extrapolation. He proposed the following generalized equation for photosynthesis:

$$CO_2 + 2H_2A \xrightarrow{\text{Light}} (CH_2O) + H_2O + 2A$$

In this equation, H_2A represents an oxidizable substance, such as hydrogen sulfide or free hydrogen. In algae and green plants, however, H_2A is water (Figure 7–3). In short, van Niel proposed that water, *not* carbon dioxide, was the source of the oxygen in photosynthesis.

This brilliant speculation, proposed in the early 1930s, was proven years later, when investigators used a heavy isotope of oxygen ($^{18}O_2$) to trace the oxygen from water to oxygen gas:

$$CO_2 + 2H_2{}^{18}O \xrightarrow{\text{Light}} (CH_2O) + H_2O + {}^{18}O_2$$

Thus, in the case of algae and green plants, in which water serves as the electron donor, the complete, balanced equation for the production of glucose by photosynthesis becomes:

$$6CO_2 + 12H_2O \xrightarrow{\text{Light}} C_6H_{12}O_6 + 6O_2 + 6H_2O$$

7–3

*The bubbles on the leaves of this sub-
mersed pondweed,* Elodea, *are bubbles
of oxygen, one of the products of photosynthesis. Van Niel was the first to
propose that the oxygen produced in
photosynthesis came from the splitting
of water rather than the breakdown
of carbon dioxide.*

About 200 years ago, as noted above, light was discovered to be required for the process we now call photosynthesis. In fact, photosynthesis occurs in two
stages, only one of which actually requires light. Evidence for this two-stage process was first presented in
1905 by the English plant physiologist F. F. Blackman,
as the result of experiments in which he measured the
individual and combined effects of changes in light intensity and temperature on the rate of photosynthesis.
These experiments showed that photosynthesis has
both a light-dependent stage and a light-independent
stage.

In Blackman's experiments, the light-independent
reactions increased in rate as the temperature was increased, but only up to about 30°C, after which the rate
began to decrease. From this evidence it was concluded
that these reactions were controlled by enzymes, since
this is the way enzymes are expected to respond to temperature (see Figure 5–9). This conclusion has since
been proven to be correct.

In the first stage of photosynthesis—the light-
dependent stage—light energy is used to form ATP
from ADP and to reduce electron-carrier molecules, notably the coenzyme $NADP^+$. In the second stage of
photosynthesis—the light-independent stage—the energy of ATP is used to link carbon dioxide covalently to
an organic molecule, and the reducing power of
NADPH is then used to reduce the newly bound carbon atom to the oxidation level of the carbon atoms in a
simple sugar. In the process, the chemical energy of
ATP and NADPH is converted into forms suitable for
transport and storage, and, at the same time, a carbon
skeleton is generated from which all other organic molecules can be built. This conversion of CO_2 into organic
compounds is known as **carbon fixation.**

The Nature of Light

Over 300 years ago, the English physicist Sir Isaac
Newton (1642–1727) separated light into a spectrum of
visible colors by letting it pass through a prism. By this
experiment, Newton showed that white light is actually
made up of a number of different colors, ranging from
violet at one end of the spectrum to red at the other.
Their separation is possible because light of different
colors is bent (diffracted) at different angles in passing
through a prism.

In the nineteenth century, through the genius of the
British physicist James Clerk Maxwell (1831–1879), it
came to be known that what we experience as light is
but a very small part of a vast continuous spectrum of
radiation, the electromagnetic spectrum (Figure 7–4).
As Maxwell showed, all the radiations included in this
spectrum travel in waves. The wavelengths—that is,
the distances from one wave peak to the next—vary
from those of X-rays, which are measured in nanometers, to those of low-frequency radio waves, which are
measured in kilometers. The shorter the wavelength,
the greater the energy. Within the spectrum of visible
light, red light has the longest wavelength and violet
has the shortest. Another feature that these radiations
have in common is that, in a vacuum, they all travel at
the same speed—300,000 kilometers per second.

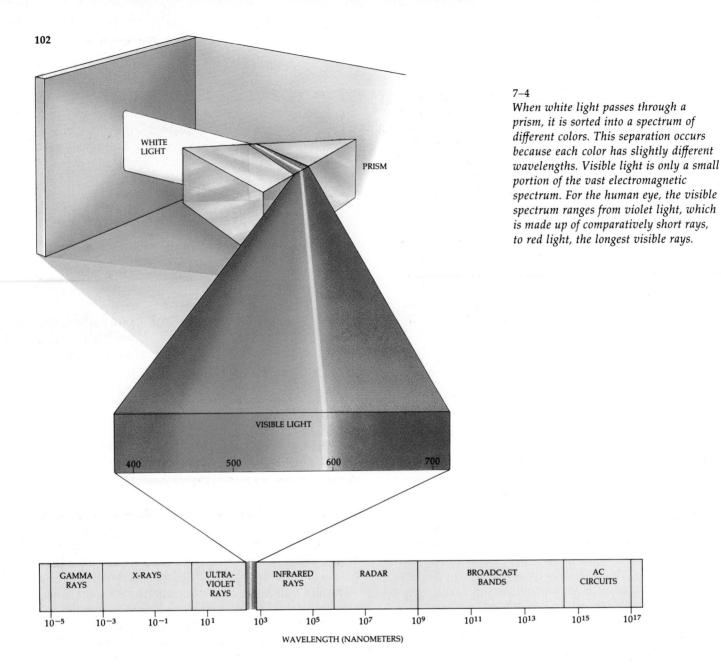

WHITE
LIGHT

PRISM

VISIBLE LIGHT

400   500   600   700

GAMMA RAYS	X-RAYS	ULTRA-VIOLET RAYS	INFRARED RAYS	RADAR	BROADCAST BANDS	AC CIRCUITS

10^{-5} 10^{-3} 10^{-1} 10^{1} 10^{3} 10^{5} 10^{7} 10^{9} 10^{11} 10^{13} 10^{15} 10^{17}

WAVELENGTH (NANOMETERS)

7–4
When white light passes through a prism, it is sorted into a spectrum of different colors. This separation occurs because each color has slightly different wavelengths. Visible light is only a small portion of the vast electromagnetic spectrum. For the human eye, the visible spectrum ranges from violet light, which is made up of comparatively short rays, to red light, the longest visible rays.

By 1900 it had become clear, however, that the wave model of light was not adequate. The key observation, a very simple one, was made in 1888: when a zinc plate is exposed to ultraviolet light, it acquires a positive charge. The metal, it was soon deduced, becomes positively charged because the light energy dislodges electrons, forcing them out of the metal atoms. Subsequently it was discovered that this photoelectric effect, as it is known, can be produced in all metals. Every metal has a critical wavelength for the effect; the light (visible or invisible) must be of that particular wavelength or shorter (must be more energetic) for the effect to occur.

With some metals, such as sodium, potassium, and selenium, the critical wavelength is within the spectrum of visible light, and as a consequence, visible light striking the metal can set up a moving stream of electrons

(an electric current). Exposure meters, television cameras, and the electric eyes that open doors at supermarkets and airline terminals all operate on this principle of turning light energy into electrical energy.

The wave model of light would lead one to predict that the brighter the light—that is, the stronger, or more intense, the beam—the greater the force with which the electrons would be dislodged. Whether or not light can eject the electrons of a particular metal, however, depends on the wavelength of the light, not on its intensity. A very weak beam of the critical wavelength is effective, whereas a stronger (brighter) beam of a longer wavelength is not. Furthermore, increasing the intensity of light increases the number of electrons dislodged but not the velocity at which they are ejected from the metal. To increase the velocity, one must use a shorter wavelength of light. Nor is it necessary for energy to accumulate in the metal. With even a dim beam of the critical wavelength, an electron may be emitted the instant the light hits the metal.

To explain such phenomena, the particle model of light was proposed by Albert Einstein in 1905. According to this model, light is composed of particles of energy called **photons,** or quanta of light. The energy of a photon (a quantum of light) is not the same for all kinds of light but is, in fact, inversely proportional to the wavelength—the longer the wavelength, the lower the energy. Photons of violet light, for example, have almost twice the energy of photons of red light, the longest visible wavelength.

The wave model of light permits physicists to describe certain aspects of its behavior mathematically, whereas the photon model permits another set of mathematical calculations and predictions. These two models are no longer regarded as opposing one another; rather, they are complementary, in the sense that both are required for a complete description of the phenomenon we know as light.

THE FITNESS OF LIGHT

Light, as Maxwell showed, is only a tiny band in a continuous spectrum. From the physicist's point of view, the difference between light and darkness—so dramatic to the human eye—is only a few nanometers of wavelength, or, expressed differently, a small amount of energy. Why is it that this tiny portion of the electromagnetic spectrum is responsible for vision, for phototropism (the curving of an organism toward light), for photoperiodism (the seasonal changes that take place in an organism with the changing length of day and night), and also for photosynthesis, on which all life depends? Is it an amazing coincidence that all these biological activities depend on these same wavelengths?

George Wald of Harvard University, one of the foremost experts on the subject of light and life, says no. He thinks that if life exists elsewhere in the universe, it is probably dependent upon this same fragment of the vast spectrum of radiation. Wald bases this conjecture on two points. First, living things are composed of large, complicated molecules held in special configurations and relationships to one another by hydrogen bonds and other weak bonds. Radiation of even slightly higher energy (shorter wavelength) than that of violet light can break these bonds and so disrupt the structure and function of the molecules. Radiation with a wavelength below 200 nanometers drives electrons out of atoms to create ions; hence, it is called ionizing radiation. The energy of radiation of wavelengths longer than those of the visible band is largely absorbed by water, which makes up the bulk of all living things. When such radiation is absorbed by organic molecules,

it causes them to increase their motion (increasing heat) but does not trigger changes in their structure. Only those radiations within the range of visible light are capable of exciting molecules—that is, of raising electrons to higher energy levels—and so of producing chemical and, ultimately, biological changes.

The second reason that the visible band was "chosen" by living things is that it, above all, is what is available. Most of the radiation reaching the earth from the sun is within this range. Higher-energy wavelengths are screened out by the oxygen and ozone high in the atmosphere. Much infrared radiation is screened out by water vapor and carbon dioxide before it reaches the earth's surface.

This is an example of what has been termed "the fitness of the environment"; the suitability of the environment for life and the suitability of life for the physical world are exquisitely interrelated. If they were not, life could not exist.

The Light-Dependent Reactions

THE ROLE OF PIGMENTS

The first step in the conversion of light energy to chemical energy is the absorption of light. A **pigment** is any substance that absorbs visible light. Some pigments absorb all wavelengths of light and so appear black. However, most absorb only certain wavelengths and transmit or reflect the wavelengths they do not absorb. **Chlorophyll,** the pigment that makes leaves green, absorbs light principally in the violet and blue wavelengths and also in the red; because it reflects green light, it appears green. The light absorption pattern of a pigment is known as the **absorption spectrum** of that substance.

One line of evidence that chlorophyll is the principal pigment involved in photosynthesis is the similarity between its absorption spectrum and the action spectrum for photosynthesis (Figure 7–5). An **action spectrum** demonstrates the relative effectiveness of different wavelengths of light for a specific light-requiring process, such as photosynthesis, flowering, or phototropism. Similarity between the absorption spectrum of a pigment and the action spectrum of a light-requiring process is considered as evidence that that particular pigment is responsible for that particular process.

When pigments absorb light, electrons are temporarily boosted to a higher energy level. As the electrons return to a lower energy level, there are three possible consequences: (1) the energy may be dissipated as heat; (2) the energy may be reemitted almost instantaneously as light energy of longer wavelength, a phenomenon

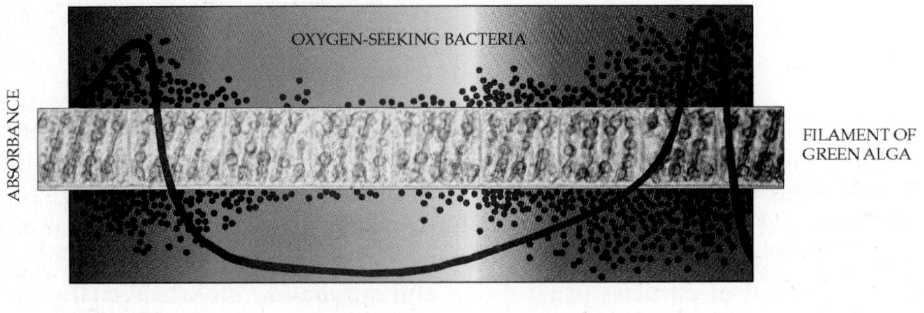

ABSORBANCE

OXYGEN-SEEKING BACTERIA

FILAMENT OF GREEN ALGA

7–5

Results of an experiment performed in 1882 by T. W. Engelmann revealed the action spectrum of photosynthesis in the filamentous alga Spirogyra. *Like investigators working more recently, Engelmann used the rate of oxygen production to measure the rate of photosynthesis. Unlike his successors, however, he lacked sensitive electronic devices for detecting oxygen. As his oxygen indicator, he chose bacteria that are attracted by oxygen. In place of the mirror and diaphragm usually used to illuminate objects under view in his microscope, he substituted a "microspectral apparatus," which, as its name implies, produced a tiny spectrum of colors that it projected upon the slide under the microscope. Then he arranged a filament of algal cells parallel to the spread of the spectrum. The oxygen-seeking bacteria congregated mostly in the areas where the violet and red wavelengths fell upon the algal filament. As you can see, the action spectrum for photosynthesis that Engelmann revealed in his experiment paralleled the absorption spectrum of chlorophyll (as indicated by the curve). He concluded that photosynthesis depends on the light absorbed by chlorophyll. This is an example of the sort of experiment that scientists refer to as "elegant," for it was not only brilliant but also simple in design and conclusive in its results.*

known as fluorescence; or (3) the energy may be captured by the formation of a chemical bond, as it is in photosynthesis.

If chlorophyll is isolated and placed in a test tube and light is permitted to strike it, the chlorophyll will fluoresce. In other words, the pigment molecules absorb light energy, thereby raising the electrons momentarily to a higher energy level, but the electrons then fall back again to a lower energy level. As they fall to a lower energy level, they release much of this energy as light. None of the light absorbed by isolated chlorophyll molecules can be converted to any form of energy useful to living systems. Chlorophyll can convert light energy to chemical energy only when the chlorophyll molecules are associated with certain proteins and embedded in the specialized membranes of the thylakoids.

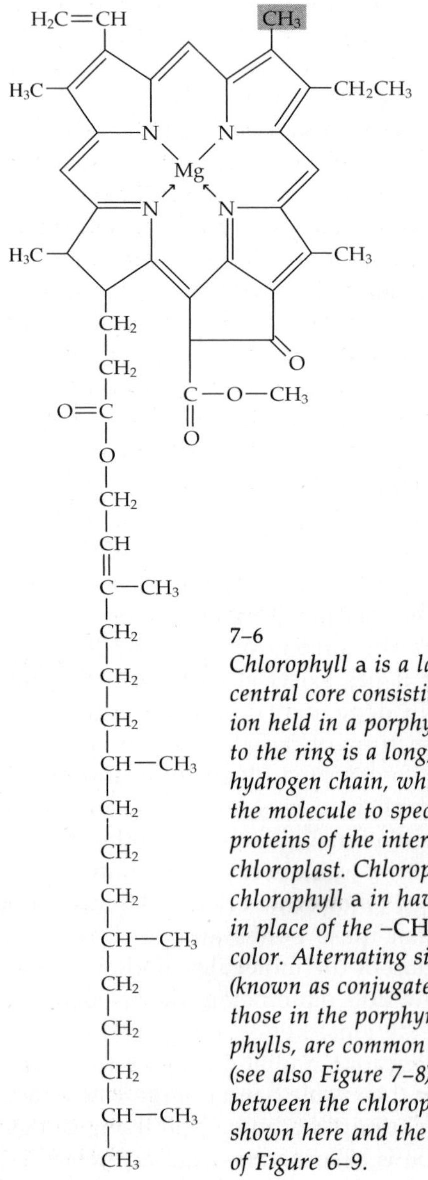

THE PHOTOSYNTHETIC PIGMENTS

Pigments that participate in photosynthesis include the chlorophylls, the carotenoids, and the phycobilins.

There are several different kinds of chlorophyll, which differ from one another both in the details of their molecular structure and in their specific absorption properties. Chlorophyll *a* (Figure 7–6) occurs in all photosynthetic eukaryotes and in the cyanobacteria, and it is considered to be essential for the type of photosynthesis carried out by organisms of these groups.

The vascular plants, bryophytes, green algae, and euglenoid algae also contain chlorophyll *b*. Chlorophyll *b* is an **accessory pigment**—a pigment that serves to broaden the range of light that can be used in photo-

7–6

Chlorophyll a *is a large molecule with a central core consisting of a magnesium ion held in a porphyrin ring. Attached to the ring is a long, insoluble carbon-hydrogen chain, which serves to anchor the molecule to specific, hydrophobic proteins of the internal membranes of the chloroplast. Chlorophyll* b *differs from chlorophyll* a *in having a –CHO group in place of the –CH$_3$ group indicated in color. Alternating single and double bonds (known as conjugated bonds), such as those in the porphyrin ring of chlorophylls, are common among pigments (see also Figure 7–8). Note the similarity between the chlorophyll* a *molecule shown here and the cytochrome molecule of Figure 6–9.*

synthesis (Figure 7–7). When a molecule of chlorophyll *b* absorbs light, the energy is transferred to a molecule of chlorophyll *a*, which then transforms it into chemical energy during the course of photosynthesis. In the leaves of green plants, chlorophyll *b* generally constitutes about one-fourth of the total chlorophyll content.

Chlorophyll *c* takes the place of chlorophyll *b* in some groups of algae, most notably the brown algae (see Chapter 14) and the diatoms (see Chapter 13). The photosynthetic bacteria (other than cyanobacteria) contain either *bacteriochlorophyll* (in purple bacteria) or *chlorobium chlorophyll* (in green sulfur bacteria). These bacteria cannot extract electrons from water and thus do not evolve oxygen. Chlorophylls *b* and *c* and the photosynthetic pigments of purple bacteria and green sulfur bacteria are simply chemical variations of the basic structure shown in Figure 7–6.

Two other classes of pigments that are involved in the capture of light energy are the **carotenoids** and the **phycobilins.** The energy absorbed by these accessory pigments must be transferred to chlorophyll *a*; like chlorophylls *b* and *c*, these accessory pigments cannot substitute for chlorophyll *a* in photosynthesis.

Carotenoids are red, orange, or yellow fat-soluble pigments found in all chloroplasts and in cyanobacteria. Like the chlorophylls, the carotenoid pigments of chloroplasts are embedded in the thylakoid membranes. Two groups of carotenoids—**carotenes** and **xanthophylls**—are normally present in chloroplasts (xanthophylls contain oxygen in their molecular structure and carotenes do not). The beta-carotene found in plants is the principal source of the vitamin A required by humans and other animals (Figure 7–8). In green leaves, the color of the carotenoids is masked by that of the much more abundant chlorophylls.

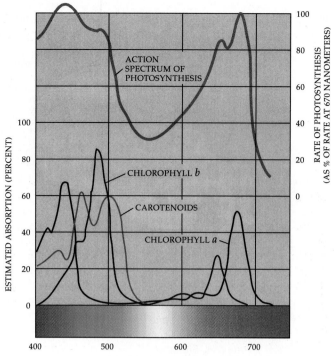

7–7
The lower curves show the absorption spectra for chlorophyll a, chlorophyll b, and the carotenoids in the chloroplast, and the upper curve shows the action spectrum for photosynthesis. The action spectrum of photosynthesis indicates that chlorophyll a, chlorophyll b, and the carotenoids absorb the light used in photosynthesis.

The third major class of accessory pigments, the phycobilins, are found in the cyanobacteria and in the chloroplasts of the red algae. Unlike the carotenoids, the phycobilins are water-soluble.

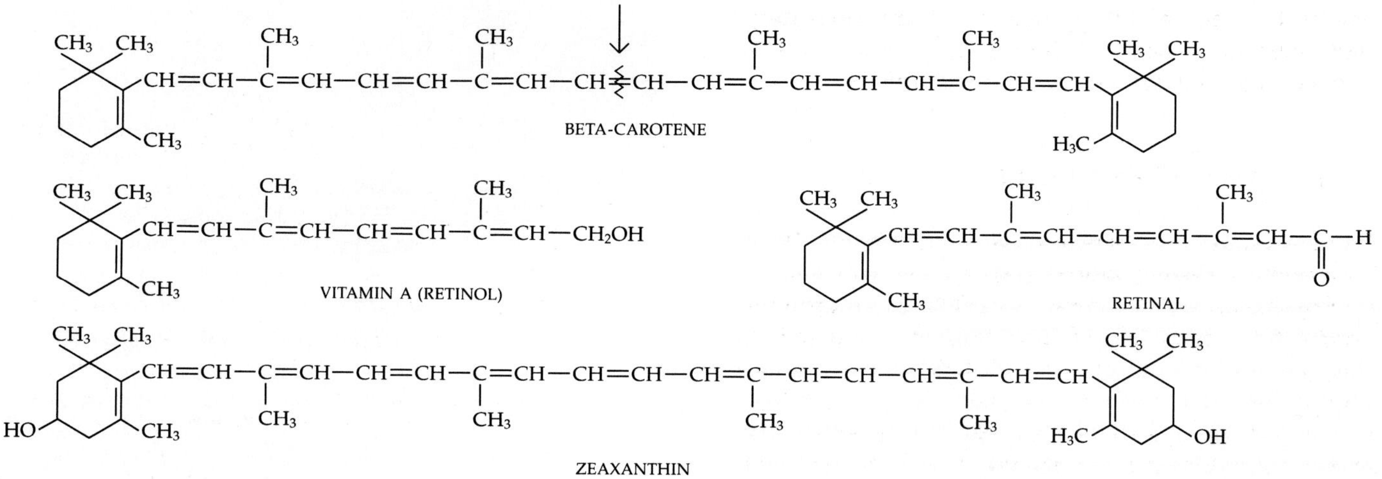

BETA-CAROTENE

VITAMIN A (RETINOL)

RETINAL

ZEAXANTHIN

7–8
Related carotenoids. Cleavage of the beta-carotene molecule at the point indicated by the arrow yields two molecules of vitamin A (retinol). Oxidation of vitamin A yields retinal, the pigment involved in vision. Note the conjugated bonds (alternating single and double bonds) in the carbon chains. Zeaxanthin (a xanthophyll) is the pigment responsible for the yellow color of corn (Zea mays) kernels.

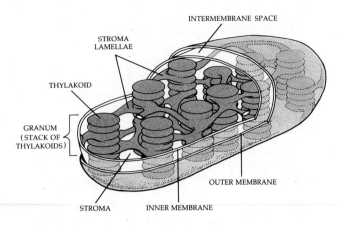

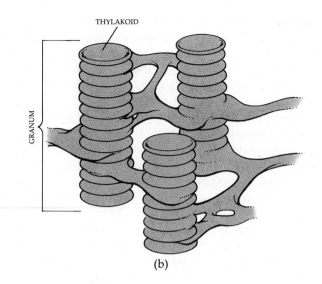

7–9

Diagrams depicting (a) *the three-dimensional structure of a chloroplast and* (b) *the arrangement of the pigment-containing thylakoid membranes. Stacks of disklike thylakoids, the grana, are interconnected by thylakoids (stroma thylakoids, or lamellae) that traverse the stroma.*

THE PHOTOSYSTEMS

In the chloroplast (Figures 7–1 and 7–9), the chlorophyll and other pigment molecules are embedded in the thylakoids in discrete units of organization called **photosystems** (Figure 7–10). Each photosystem includes an assembly of about 250 to 400 pigment molecules and consists of two closely linked components: a *reaction center protein-pigment complex* and an *antenna protein complex*. Within the photosystems, the chlorophyll molecules are bound to specific chlorophyll-binding proteins and held in place to allow efficient capture of light energy.

All of the pigments within a photosystem are capable of absorbing photons, but only one pair of chlorophyll molecules per photosystem can actually use the energy in the photochemical reaction. This special pair of chlorophyll molecules is situated at the core of the **reaction center** of the photosystem. The other pigment molecules, called **antenna pigments** because they act as an antennalike network for the gathering of light, are located in the antenna protein complex. In addition to chlorophyll, varying amounts of carotenoid pigments are also located in each antenna complex.

Light energy absorbed by a pigment molecule anywhere in the network is transferred from one pigment molecule to the next until it reaches the reaction center, with its special pair of chlorophyll molecules. When either of these two chlorophyll molecules absorbs the energy, one of its electrons is boosted to a higher energy

7–10

Internal surface of a thylakoid, as viewed by the freeze-fracture technique. The particles embedded within the membrane are thought to be the structural units of the photosystems involved in the light reactions.

level and is transferred to an acceptor molecule to initiate electron flow. The chlorophyll molecule is thus oxidized and positively charged.

According to present evidence, there are two different kinds of photosystems. In Photosystem I, the chlorophyll molecules of the reaction center are a form of chlorophyll *a* known as P_{700}. The "P" stands for pigment and the subscript "700" designates the optimal absorption peak in nanometers, the 700 nanometers being the wavelength of light. The reaction center of Photosystem II also contains a special form of chlorophyll *a*. Its optimal absorption peak is at 680 nanometers, and accordingly it is called P_{680}. Much of our understanding of the structure of Photosystem II in plants comes from studies of similar complexes in photosynthetic bacteria. Especially notable are the studies of Johann Deisenhofer, Robert Huber, and Hartmut Michel, who determined the precise chemical structure of the reaction center in the purple bacterium *Rhodopseudomonas viridis* by X-ray crystallography, providing us with the first good glance at how pigment molecules are arranged to capture light energy (Figure 7–11). Deisenhofer, Huber, and Michel were awarded the Nobel Prize for chemistry in 1988 for their work.

In general, Photosystem I and Photosystem II work together simultaneously and continuously. As we shall see, however, Photosystem I can operate independently.

A MODEL OF THE LIGHT-DEPENDENT REACTIONS

Figure 7–12 shows the present model of how the two photosystems work together. According to this model, light energy enters Photosystem II, where it is trapped by molecules of P_{680} in the reaction center, either directly or indirectly via one or more of the pigment molecules. When a P_{680} molecule is excited, its energized electron is transferred to an acceptor molecule designated as "Q," probably a quinone (page 94). By an as yet poorly understood reaction, the electron-deficient P_{680} molecule is able to replace its electrons (the electrons actually move in pairs) by extracting them from a water molecule. When the electrons pass from the water molecule to the P_{680} molecules, the water molecule is dissociated into protons and oxygen gas. This light-dependent oxidative splitting of water molecules is called **photolysis.** The enzyme that catalyzes the photolysis of water is located on the inside of the thylakoid membrane. Thus, the photolysis of water molecules contributes to the generation of a proton gradient across the membrane. Manganese is an essential cofactor for the oxygen-evolving mechanism.

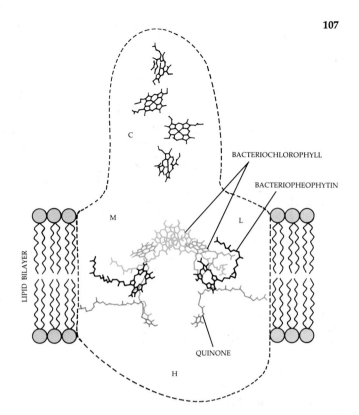

7–11

The arrangement of the reaction center of the purple bacterium Rhodopseudomonas viridis *as determined by X-ray crystallography. The reaction center consists of four protein subunits: C, L, M, and H. The C (for cytochrome) subunit is bound to the outer surface of the membrane. The L and M subunits, which make up the pigment system, span the lipid bilayer and contain a total of four bacteriochlorophyll molecules, two bacteriopheophytin molecules, and two quinones. The H subunit is bound to the cytoplasmic surface of the membrane. The arrangements of bacteriochlorophylls in the L and M subunits enable one pair of bacteriochlorophylls to absorb light energy and produce an excited electron. That electron then is quickly transferred through intermediate bacteriochlorophylls and bacteriopheophytin molecules to a bound quinone. The electron "hole" in the original pigment molecule is filled by an electron from the cytochrome, and the electron on the bound quinone is transferred to a mobile quinone resulting in a source of chemical reducing power. The reduced quinone then can reduce NAD^+ to NADH, which in turn can reduce $NADP^+$ to NADPH. As in eukaryotic plant cells, the NADPH is a source of reducing power. A similar series of reactions is believed to occur in Photosystem II of plants.*

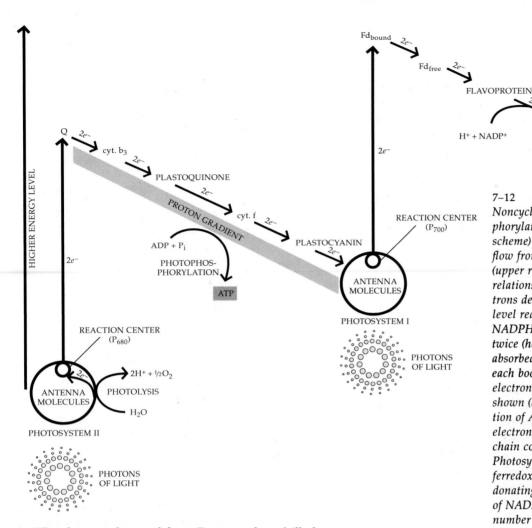

7–12

Noncyclic electron flow and photophosphorylation. This zigzag scheme (the Z scheme) shows the pathway of electron flow from water (lower left) to $NADP^+$ (upper right), as well as the energy relationships. To raise the energy of electrons derived from water to the energy level required to reduce $NADP^+$ to NADPH, the electrons must be boosted twice (heavy red lines) by photons absorbed in Photosystems I and II. After each boosting step, the high-energy electrons flow downhill via the routes shown (black arrows). Photophosphorylation of ADP to yield ATP is coupled to electron flow in the electron transport chain connecting Photosystem II to Photosystem I (see Figure 7–13). Bound ferredoxin (Fd_{bound}, upper right) is shown donating its electrons for the reduction of $NADP^+$ to NADPH. Ferredoxin has a number of roles in the chloroplast, including acting as a source of reducing power in amino acid biosynthesis and fatty acid biosynthesis.

The electrons boosted from P_{680} pass downhill along an electron transport chain to Photosystem I. The components of this electron transport chain resemble those of the electron transport chain of respiration; cytochromes, iron-sulfur proteins, and quinones are involved. In addition, photosynthetic electron transport involves chlorophyll and the copper-containing protein plastocyanin. The electron transport chain between the

7–13

The chemiosmotic coupling mechanism of photophosphorylation. In this process protons are "pumped" across the thylakoid membrane, from the outside (stroma) to the inside (thylakoid space), as a result of the arrangement of the electron carriers in the membrane. The proton concentration in the thylakoid space builds up partly from the splitting of water there and partly because of the oxidation of plastoquinone (PQ) at the inner face of the membrane. As the protons flow down the gradient from the thylakoid space back into the stroma, ADP is phosphorylated to ATP by ATP synthase. For every three protons that pass through the ATP synthase, one ATP is synthesized. The C of CF_O and CF_1 stands for "chloroplast."

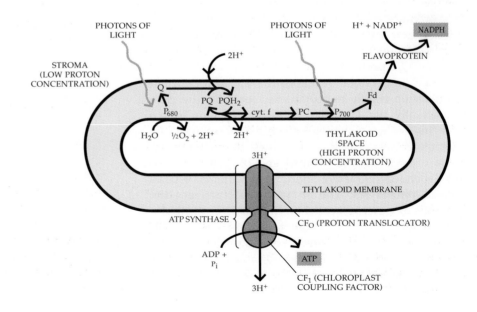

two photosystems is arranged so that ATP can be formed from ADP and phosphate by a mechanism similar in many ways to the oxidative phosphorylation that takes place in mitochondria (page 23). In chloroplasts, the process is called **photophosphorylation** (Figure 7–13).

In Photosystem I, light energy boosts the electrons from a P_{700} molecule to an electron acceptor called bound ferredoxin ("Fe" for iron, and "redoxin" to indicate that it carries out reduction-oxidation, or redox, reactions), which is an iron-sulfur protein. The electrons then are passed downhill through a series of intermediates (including free ferredoxin and a flavoprotein) to the coenzyme $NADP^+$, resulting in the reduction of the $NADP^+$ to NADPH and oxidation of the P_{700} molecule. The electrons removed from the P_{700} molecule are replaced by the electrons that have moved down the electron transport chain from Photosystem II. Two photons must be absorbed by Photosystem II and two by Photosystem I in order to reduce one molecule of $NADP^+$ to NADPH.

Thus, in the light there is a continuous flow of electrons from water to Photosystem II to Photosystem I to $NADP^+$. This unidirectional flow of electrons from water to $NADP^+$ is called **noncyclic electron flow,** and the production of ATP that occurs during noncyclic electron flow is called **noncyclic photophosphorylation.**

The free-energy change (ΔG) for the reaction

$$H_2O + NADP^+ \longrightarrow NADPH + H^+ + \tfrac{1}{2}O_2$$

is 51 kilocalories per mole. The energy of the equivalent mole of photons of 700-nanometer light (the equivalent of a mole of photons is called an *einstein*) is 40 kilocalories per einstein. Because four photons are required to boost two electrons to the level of NADPH, 160 kilocalories are available. Approximately one-third of the available energy is captured as NADPH. The total energy harvest from noncyclic electron flow (based on the passage of 12 pairs of electrons from H_2O to $NADP^+$) is 12ATP and 12NADPH.

Cyclic Photophosphorylation

As mentioned previously, Photosystem I can work independently of Photosystem II (Figure 7–14). In this process, called **cyclic electron flow,** electrons are boosted from P_{700} to bound ferredoxin, the electron acceptor. Instead of the electrons being passed downhill to $NADP^+$, they are shunted to an acceptor in the electron transport chain that connects Photosystems I and II, and then they pass downhill through that chain back into the reaction center of Photosystem I. ATP is produced in the course of this passage, and inasmuch as this process involves a cyclic flow of electrons, it is

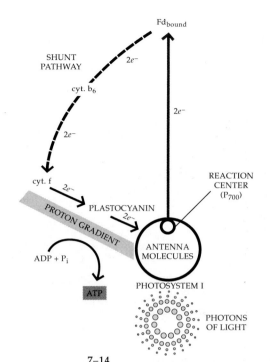

7–14

Cyclic photophosphorylation involves only Photosystem I. ATP is produced from ADP by the same process shown in Figure 7–13, but oxygen is not released and $NADP^+$ is not reduced.

called **cyclic photophosphorylation.** It is believed that the most primitive photosynthetic mechanisms worked in this way, and this is apparently the way in which some bacteria carry out photosynthesis. Photosynthetic eukaryotes are also able to synthesize ATP by cyclic electron flow. However, no water is split, no O_2 is evolved, and no NADPH is formed. The only product is ATP.

Cyclic electron flow and photophosphorylation are believed to occur when the cell has enough reducing power in the form of NADPH but requires additional ATP for other metabolic needs. Thus, the chloroplast is able to adapt to changing conditions by varying the proportion of cyclic and noncyclic electron flow.

The Light-Independent Reactions

In the second stage of photosynthesis, the chemical energy harvested by the light-dependent reactions is used to reduce carbon. Carbon is available to photosynthetic cells in the form of carbon dioxide. For algae and cyanobacteria, this carbon dioxide is found dissolved in the surrounding water. In most plants, carbon dioxide reaches the photosynthetic cells through special openings, called stomata, which are found in the plants' leaves and green stems (Figure 7–15). Because the plastid is the primary assimilatory organelle of the plant

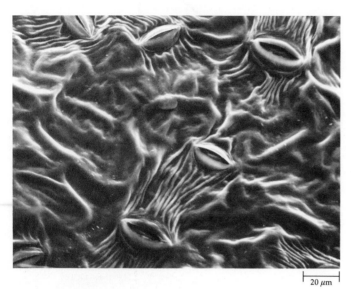

7–15
Scanning electron micrograph showing stomata on the lower surface of a cottonwood (Populus) leaf. Carbon dioxide reaches the photosynthetic cells through the stomata.

$20\ \mu m$

RIBULOSE 1,5–
BISPHOSPHATE
(RuBP)

Ribulose 1,5–
Bisphosphate
Carboxylase
(Rubisco)

2 MOLECULES OF
3-PHOSPHOGLYCERATE
(PGA)

7–16
Melvin Calvin and his collaborators, Andrew A. Benson and James A. Bassham, briefly exposed photosynthesizing algae to radioactive carbon dioxide ($^{14}CO_2$). They found that the radioactive carbon is covalently linked to a molecule of ribulose 1,5–bisphosphate (RuBP). The resulting six-carbon compound then immediately splits to form two molecules of 3-phosphoglycerate (PGA). The radioactive carbon atom, indicated here in color, next appears in one of the two molecules of PGA. This is the first step in the Calvin cycle.

cell, the light-independent, or carbon-fixing, reactions of photosynthesis compete with the nitrogen and sulfur assimilatory pathways for the energy (ATP and NADPH) produced by the light-dependent reactions.

THE CALVIN CYCLE: THE THREE-CARBON PATHWAY

The reduction of carbon occurs in the stroma of the chloroplast by means of a series of reactions known as the Calvin cycle (named after its discoverer, Melvin Calvin, who received a Nobel Prize for his work on the elucidation of this pathway). The Calvin cycle is analogous to the Krebs cycle (page 91) in that, by the end of each turn of the cycle, the starting compound is regenerated. The starting (and ending) compound in the Calvin cycle is a five-carbon sugar with two phosphate groups—**ribulose 1,5-bisphosphate (RuBP)**. The process begins when carbon dioxide enters the cycle and is "fixed" (bonded covalently) to RuBP. The resultant six-carbon compound immediately splits to form two molecules of 3-phosphoglycerate, or PGA (Figure 7–16). (Each PGA molecule contains three carbon atoms; hence the designation of the Calvin cycle as the C_3, or three-carbon, pathway. The six-carbon intermediate has never been isolated.)

RuBP carboxylase (commonly called **"Rubisco"**),

the enzyme catalyzing this crucial initial reaction, is quite abundant in chloroplasts, making up more than 15 percent of the total chloroplast protein. (It is said to be the most abundant protein in the world. Can you say why?)

The complete cycle is diagrammed in Figure 7–17. As in the Krebs cycle, each step is catalyzed by a specific enzyme. At each full turn of the cycle, a molecule of carbon dioxide enters the cycle and is reduced, and a molecule of RuBP is regenerated. Six revolutions of the cycle, with the introduction of six atoms of carbon, are necessary to produce a six-carbon sugar, such as glucose. The overall equation for the production of a molecule of glucose is

$$6CO_2 + 12NADPH + 12H^+ + 18ATP \longrightarrow$$
$$1\ glucose + 12NADP^+ + 18ADP + 18P_i + 6H_2O$$

The immediate product of the cycle is **glyceraldehyde 3-phosphate,** the primary molecule transported from the chloroplast to the ground substance of the cell. This same triose-phosphate ("triose" means a three-carbon sugar) is formed when the fructose 1,6-bisphosphate molecule is split at the fourth step in glycolysis, and it is interconvertible with another triose-phosphate, dihydroxyacetone phosphate (page 88). Using the energy provided by hydrolysis of the phosphate bonds, the first four steps of glycolysis can be reversed to form glucose from glyceraldehyde 3-phosphate.

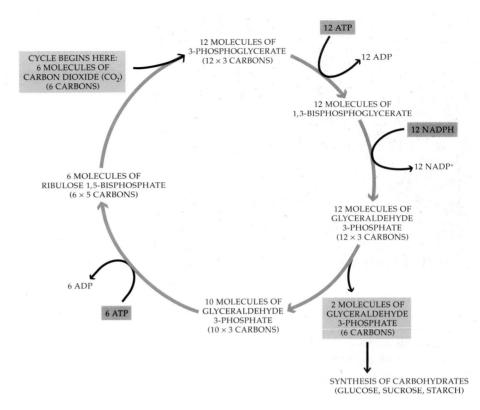

7–17

A summary of the Calvin cycle. At each full turn of the cycle, one molecule of carbon dioxide enters the cycle. Six turns are summarized here—the number required to make two molecules of glyceraldehyde 3-phosphate, and thus one molecule of glucose. Six molecules of ribulose 1,5-bisphosphate (RuBP), a five-carbon compound, are combined with six molecules of carbon dioxide, yielding twelve molecules of 3-phosphoglycerate, a three-carbon compound. These are reduced to twelve molecules of glyceraldehyde 3-phosphate. Ten of these three-carbon molecules are combined and rearranged to form six five-carbon molecules of RuBP. The "extra" molecules of glyceraldehyde 3-phosphate represent the net gain from the Calvin cycle; they serve as the starting point for the synthesis of sugars, starch, and other cellular components. The energy that "drives" the Calvin cycle is the ATP and NADPH produced by the light-dependent reactions of Figure 7–12.

The Products of Photosynthesis

Although glucose commonly is represented as a product of photosynthesis in summary equations, in reality, very little free glucose is generated in photosynthesizing cells. Rather, most of the fixed carbon is converted either to **sucrose,** the major transport sugar in plants, or to **starch,** the major storage carbohydrate in plants (Chapter 3).

Much of the glyceraldehyde 3-phosphate produced by the Calvin cycle is exported to the ground substance, where it can be rapidly converted to glucose 1-phosphate and fructose 6-phosphate. The glucose 1-phosphate then is converted to the nucleotide-sugar UDP-glucose (uridine diphosphate-glucose), which is linked with the fructose 6-phosphate to form sucrose 6-phosphate, the immediate precursor of sucrose. Removal of the phosphate by hydrolysis leaves sucrose.

Most of the glyceraldehyde 3-phosphate that remains in the chloroplast is converted to starch, which is stored temporarily during the light period as grains in the stroma (Figure 7–1). Through a series of reactions, the glyceraldehyde 3-phosphate is converted to glucose 1-phosphate, which then is used to produce the nucleotide-sugar ADP-glucose, the immediate precursor of starch. The glucose is added directly to the growing starch grain by the enzyme starch synthetase. At night, sucrose is produced from starch for export from the leaf. Carbon derived from starch appears to be transported from the chloroplast into the ground substance as glucose rather than as triose-phosphate.

THE FOUR-CARBON PHOTOSYNTHETIC PATHWAY

The Calvin cycle is not the only carbon-fixing pathway used in the light-independent reactions. In some plants, the first product of CO_2 fixation to be detected is not the three-carbon molecule PGA, but rather the four-carbon molecule oxaloacetate (which is also an intermediate in the Krebs cycle). Plants that employ this pathway (along with the Calvin cycle) are commonly called **C$_4$ plants** (for four-carbon), as distinct from the **C$_3$ plants,** which use *only* the Calvin cycle. (The C$_4$ pathway is also referred to as the Hatch-Slack pathway after two Australian plant physiologists who played key roles in its elucidation.)

The oxaloacetate is formed when carbon dioxide is fixed to phosphoenolpyruvate (PEP) in a reaction catalyzed by the enzyme PEP carboxylase, which is found in the ground substance of the C$_4$ plant cell (Figure 7–18). The oxaloacetate is then reduced to malate or converted, with the addition of an amino group, to the amino acid aspartate in the chloroplast of the same cell. These steps occur in mesophyll cells. The next step is a surprise: the malate (or aspartate, depending on the

7-18

Carbon dioxide fixation by the C_4 pathway. Carbon dioxide is "fixed" to phosphoenolpyruvate (PEP) by the enzyme PEP carboxylase. Depending on the species, the resulting oxaloacetate is either reduced to malate or converted to aspartate through the addition of an amino group (–NH$_2$). The malate or aspartate moves into the bundle-sheath cells where CO_2 is released for use in the Calvin cycle.

species) moves from the mesophyll cells to bundle-sheath cells surrounding the vascular bundles of the leaf, where it is decarboxylated to yield CO_2 and pyruvate. The CO_2 then enters the Calvin cycle and reacts with RuBP to form PGA and other intermediates of the cycle. Meanwhile, the pyruvate returns to the mesophyll cells, where it reacts with ATP to form more molecules of PEP (Figure 7–19). Hence, the anatomy of the leaves of C_4 plants imparts a spatial separation between the C_4 pathway and the Calvin cycle.

The two primary carboxylating enzymes use different forms of the carbon dioxide molecule as a substrate. RuBP carboxylase uses CO_2, whereas PEP carboxylase uses the hydrated form of carbon dioxide, the bicarbonate ion (HCO_3^-), as its substrate. RuBP carboxylase is found in the chloroplast, whereas PEP carboxylase occurs in the cytoplasmic ground substance.

Typically, the leaves of C_4 plants are characterized by an orderly arrangement of the mesophyll cells around a layer of large bundle-sheath cells, so that together the two form concentric layers around the vascular bundle (Figure 7–20). This wreathlike arrangement has been termed Kranz anatomy. (*Kranz* is the German word for "wreath.") In some C_4 plants, the chloroplasts of the mesophyll cells have well-developed grana while those of the bundle-sheath cells have either poorly developed grana or none at all (Figure 7–21). In addition, when photosynthesis is occurring, the bundle-sheath chloroplasts commonly form larger and more numerous starch grains than the mesophyll chloroplasts.

Efficiency of C_4 Plants

Fixation of CO_2 in C_4 plants has a larger energy cost than in C_3 plants. For each molecule of CO_2 fixed in the C_4 pathway, a molecule of PEP must be regenerated at the cost of two phosphate groups of ATP. Also, C_4 plants need five ATPs altogether to fix one molecule of CO_2, whereas C_3 plants need only three. One might well ask why C_4 plants should have evolved such a seemingly clumsy and energetically expensive method of providing carbon dioxide to the Calvin cycle.

Photosynthesis in C_3 plants is always accompanied by **photorespiration,** a process that *consumes* oxygen and *releases* CO_2 in the presence of light (Figure 7–22). Photorespiration is a wasteful process. Unlike mitochondrial respiration, photorespiration is not accompanied by oxidative phosphorylation; hence, it yields no ATP. In addition, photorespiration diverts some of the reducing power generated in the light-dependent reactions from the biosynthesis of glucose into the reduction of oxygen. Under normal atmospheric conditions, as much as 50 percent of the carbon fixed in photosynthesis by a C_3 plant may be reoxidized to CO_2 during photorespiration. Photorespiration is very active in C_3 plants, seriously limiting their efficiency, but it is nearly absent in C_4 plants.

The major substrate oxidized by photorespiration in C_3 plants is glycolate. Glycolate is oxidized in the peroxisomes (page 25) of photosynthetic cells and is formed by the oxidative breakdown of RuBP by RuBP carboxylase, the identical enzyme that fixes CO_2 into PGA in the C_3 pathway. How is this possible?

RuBP carboxylase can promote the reaction of RuBP with *either* CO_2 or O_2 (Figure 7–22). In fact, the full name of the enzyme is *RuBP carboxylase/oxygenase,* in recognition of this dual activity. When the CO_2 concentration is high and that of O_2 is relatively low, RuBP carboxylase fixes CO_2 to RuBP to yield PGA. When the CO_2 concentration is low and that of O_2 is relatively high, the enzyme also acts as an oxygenase, combining

7–19
The pathway for carbon fixation in the corn plant (Zea mays), a C_4 *plant.* CO_2 *is first fixed in mesophyll cells as oxaloacetate, which is rapidly converted to malate. The malate is then transported to bundle-sheath cells, where the* CO_2 *is released. The* CO_2 *thus released enters the Calvin cycle, ultimately yielding sugars and starch. Pyruvate returns to the mesophyll cell for regeneration of phosphoenolpyruvate (PEP).*

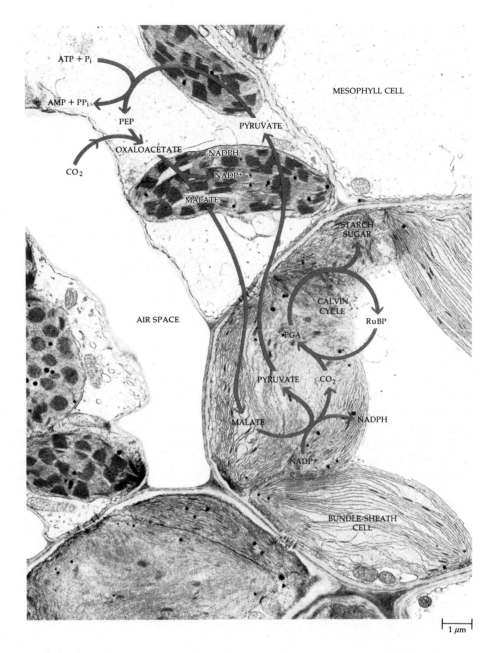

7–20
Transverse section of a portion of a corn (Zea mays) leaf. As is typical of C_4 *plants, the vascular bundles (composed of xylem and phloem) are surrounded by large, chloroplast-containing bundle-sheath cells that are, in turn, surrounded by a layer of mesophyll cells. The* C_4 *pathway takes place in the mesophyll cells; the Calvin cycle occurs in the bundle-sheath cells.*

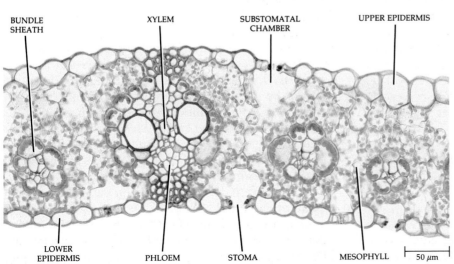

7–21

Electron micrograph showing portions of chloroplasts in a mesophyll cell (above) and a bundle-sheath cell (below) of a corn (Zea mays) leaf. Compare the well-developed grana of the mesophyll cell chloroplast with the poorly developed grana of the bundle-sheath cell chloroplast. Note the plasmodesmata in the wall between these two cells. The intermediates of photosynthesis move from one cell to the other via the plasmodesmata in this C_4 plant.

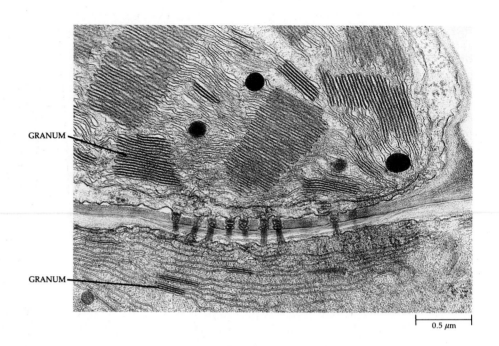

GRANUM

GRANUM

0.5 μm

7–22

Reactions catalyzed by RuBP carboxylase/oxygenase. Reaction 1—fixation of CO_2 in the Calvin cycle—is favored by high CO_2 and low oxygen concentrations. Reaction 2 also occurs to a significant extent, especially in the presence of low CO_2 and high oxygen concentrations (normal atmospheric conditions). The phosphoglycolate is converted to glycolate, the substrate for photorespiration, which takes place in peroxisomes. Reaction 2 lessens the efficiency of photosynthesis since only one molecule of PGA is formed from RuBP rather than two, as in Reaction 1.

RuBP and oxygen to yield phosphoglycolate and PGA, instead of the two PGA molecules normally formed in the carboxylation. The phosphoglycolate is then converted to glycolate—the substrate oxidized during photorespiration.

High CO_2 and low O_2 concentrations limit photorespiration. Consequently, C_4 plants have a distinct advantage over C_3 plants because CO_2 fixed by the C_4 pathway is essentially "pumped" from the mesophyll cells into the bundle-sheath cells, thus maintaining a high $CO_2:O_2$ ratio at the site of the action of RuBP carboxylase (Figure 7–16). This high $CO_2:O_2$ ratio favors the carboxylation of RuBP. In addition, since both the Calvin cycle and photorespiration are localized in the inner, bundle-sheath layer of cells, any CO_2 liberated by photorespiration into the outer, mesophyll layer can be refixed by the C_4 pathway that operates there. The CO_2 liberated by photorespiration can thus be prevented from escaping from the leaf. Moreover, compared with C_3 plants, C_4 plants are superior utilizers of available CO_2; this is in part due to the fact that the enzyme PEP carboxylase is not inhibited by O_2. As a result, the net photosynthetic rates (that is, total photosynthetic rate minus photorespiratory loss) of C_4 grasses, such as corn (*Zea mays*), sugarcane (*Saccharum officinale*), and sorghum (*Sorghum vulgare*), can be two to three times the rates of C_3 grasses, such as wheat (*Triticum aestivum*), rye (*Secale cereale*), oats (*Avena sativa*), and rice (*Oryza sativa*) under the same environmental conditions.

Because C_4 plants evolved primarily in the tropics, they are especially well adapted to high light intensities, high temperatures, and dryness. The optimal tem-

The Carbon Cycle

By photosynthesis, living systems incorporate carbon dioxide from the atmosphere into organic, carbon-containing compounds. In respiration, these compounds are broken down into carbon dioxide and water. These processes, viewed on a world-wide scale, result in the carbon cycle. The principal photosynthesizers in this cycle are plants, phytoplankton, marine algae, and cyanobacteria. They synthesize carbohydrates from carbon dioxide and water and release oxygen into the atmosphere. About 100 billion metric tons of carbon per year are bound into carbon compounds by photosynthesis.

Some of the carbohydrates are used by the photosynthesizers themselves. Plants release carbon dioxide from their roots and leaves, and marine algae and cyanobacteria release carbon dioxide into the water where it maintains an equilibrium with the carbon dioxide of the air. Some 500 billion metric tons of carbon are "stored" as dissolved carbon dioxide in the seas, and some 700 billion metric tons in the atmosphere. Some of the carbohydrates are used by animals that feed on live plants, on algae, and on one another, releasing carbon dioxide. An enormous amount of organic carbon is contained in the dead bodies of plants and other organisms as well as in discarded leaves and shells, feces, and other waste materials. All these materials settle into the soil or sink to the ocean floors, where they are used consumed by decomposers—small invertebrates, bacteria, and fungi. Carbon dioxide is also released by these processes into the reservoir of the air and oceans. Another, even larger store of carbon lies below the surface of the earth in the form of coal and oil, deposited there some 300 million years ago.

In the course of years and averaged over the entire earth, the natural processes of photosynthesis and respiration essentially balance one another. Over the long span of geologic time, the carbon dioxide concentration of the atmosphere has varied, but for the last 10,000 years—at least up to the Industrial Revolution—it has remained relatively constant. By volume, carbon dioxide is a very small proportion of the atmosphere, only about 0.035 percent. It is important, however, because carbon dioxide, along with water vapor and some other gases, including methane, absorbs infrared radiation from the sun. Since the start of the Industrial Revolution, the amount of carbon dioxide in the atmosphere has been increasing, owing mainly to our use of fossil fuels, such as coal, oil, and natural gas, and to our destruction of forests.

Although models of global warming differ greatly, most scientists agree that the earth's average temperature may increase between 1° and 6°C by the middle of the next century. As a consequence of continued global warming, climates will change to the extent that the cultivation of certain crops will become difficult or impossible in many areas where they are currently grown, and ultimately the polar ice caps will melt, resulting in the flooding of coastal areas. One of the difficulties in predicting the exact effects of global warming has to do with our inability to predict changes in precipitation. In any event, the consequences are projected to be serious, and attention must be given to the problem by finding alternatives to fossil fuels and, to some extent, by planting forests.

Other gases, such as methane, chlorofluorocarbons (CFCs), and nitrogen oxides, also affect global warming significantly. Methane is produced not only from natural sources but also as a result of human activities.

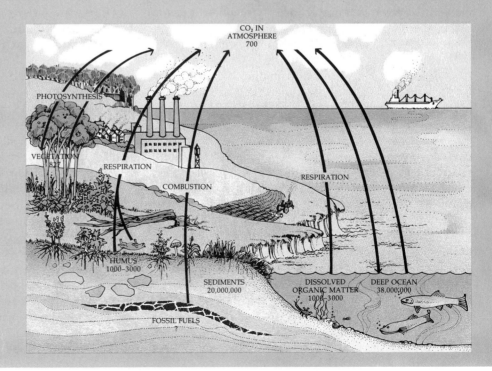

The carbon cycle. The arrows indicate the movement of carbon atoms. The numbers are all estimates of the amount of carbon stored, expressed in billions of metric tons. The amount of carbon released by respiration and combustion has, it is believed, begun to exceed the amount fixed by photosynthesis.

7–23

The numbers shown here indicate the percentages of the total number of grass species having the C$_4$ pathway in each of 32 local floras in North America. The highest percentages are found in those regions with the highest temperatures during the growing season.

perature range for C$_4$ photosynthesis is much higher than that for C$_3$ photosynthesis, and C$_4$ plants flourish even at temperatures that would eventually be lethal to many C$_3$ species. Because of their more efficient use of carbon dioxide, C$_4$ plants can attain the same photosynthetic rate as C$_3$ plants but with smaller stomatal openings and, hence, with considerably less water loss. Analyses of the geographic distribution of C$_4$ species in North America have shown that they are generally most abundant in climates with high temperatures. Differences appear to exist, however, between monocots and dicots in the particular type of high temperature favored. For example, C$_4$ grasses are most abundant in regions having the highest temperatures during the growing season (Figure 7–23); in contrast, C$_4$ dicots are most abundant in regions having the greatest aridity during the growing season.

A striking illustration of different growth patterns in C$_4$ plants is found in our lawns, which, in the cooler parts of the country at least, consist mainly of C$_3$ grasses, such as Kentucky bluegrass (*Poa pratensis*) and creeping bent (*Agrostis tenuis*). Crabgrass (*Digitaria sanguinalis*), which all too often overwhelms these dark green, fine-leaved grasses with patches of its yellowish green, broader leaves, is a C$_4$ grass that grows much

more rapidly in the heat of summer than the temperate C$_3$ grasses mentioned above.

All of the plants now known to utilize C$_4$ photosynthesis are flowering plants, including at least 19 families, 3 of monocots and 16 of dicots; however, no family has been found that contains only C$_4$ species. This pathway has undoubtedly arisen independently many times in the course of evolution.

CRASSULACEAN ACID METABOLISM

Crassulacean acid metabolism, commonly designated **CAM,** has evolved independently in many succulent plants, including the cacti (Cactaceae) and stonecrops (Crassulaceae). Plants are considered to have CAM if their photosynthetic cells have the ability to fix CO$_2$ in the dark via the activity of PEP carboxylase. The malic acid so formed is stored in the vacuole. During the following light period, the malic acid is decarboxylated, and the CO$_2$ is transferred to RuBP of the Calvin cycle *within the same cell.* Thus, CAM plants, like C$_4$ plants, utilize both C$_4$ and C$_3$ pathways; they differ from the C$_4$ plants, however, in that there is a temporal separation of the two pathways in the CAM plants, rather than a spatial one as in the C$_4$ plants.

CAM plants are largely dependent upon nighttime accumulation of carbon dioxide for their photosynthesis because their stomata are closed during the day, retarding water loss. This is obviously advantageous in the conditions of high light intensity and water stress under which most CAM plants live. If all atmospheric CO_2 uptake in a CAM plant occurs at night, the water-use efficiency of that plant can be many times greater than that of a C_3 or C_4 plant. During periods of prolonged drought, some CAM plants can keep their stomata closed both night and day, maintaining low metabolic rates by refixing respiratory CO_2 through the dark accumulation of malic acid followed by CO_2 fixation via the Calvin cycle the following day.

Among the vascular plants, CAM is more widespread than C_4 photosynthesis. It has been reported in at least 23 families of flowering plants, mostly dicots, including such familiar houseplants as the maternity plant (*Kalanchoë daigremontiana*), wax plant (*Hoya carnosa*), and snake plant (*Sansevieria zeylanica*), a monocot. Not all CAM plants are highly succulent; two examples of less succulent ones are the pineapple (*Ananas comosus*) and Spanish "moss" (*Tillandsia usneoides*), both members of the family Bromeliaceae (monocots). Some nonflowering plants have also been reported to show CAM activity, including the bizarre gymnosperm *Welwitschia mirabilis* (see Figure 17–36), the aquatic species of quillwort (*Isoetes*), and some ferns. *Welwitschia*, however, fixes CO_2 predominantly the way a C_3 plant does.

ADAPTIVE SIGNIFICANCE OF PHOTOSYNTHETIC MECHANISMS

From our discussions of C_3, C_4, and CAM photosynthesis, it may have become apparent that the photosynthetic mechanism is important but is not the only factor that determines where a plant lives. All three mechanisms have advantages and disadvantages, and a plant can compete successfully only when the benefits of its type of photosynthesis outweigh other factors. For example, although C_4 plants generally tolerate higher temperatures and drier conditions than C_3 species, C_4 plants may not compete successfully at temperatures below 25°C. This is, in part, because they are more sensitive to cold than C_3 species. In addition, as discussed, CAM plants, which are best adapted to very arid conditions, conserve water by closing their stomata during the day. This practice, however, severely reduces their ability to take in and fix CO_2. Hence, CAM plants grow slowly and compete poorly with C_3 and C_4 species under less extreme conditions. Thus, to some extent, each photosynthetic type of plant is the victim of its own mechanism.

Summary

In photosynthesis, light energy is converted to chemical energy and carbon is "fixed" into organic compounds. The generalized equation for this reaction is

$$CO_2 + 2H_2A + \text{light energy} \longrightarrow (CH_2O) + H_2O + 2A$$

in which H_2A represents water or some other substance that can be oxidized (that is, from which electrons can be removed), and (CH_2O) represents carbohydrate.

The first step in photosynthesis is the absorption of light energy by pigment molecules. The pigments involved in photosynthesis in eukaryotes include the chlorophylls and the carotenoids, which are packed in the thylakoids of chloroplasts as photosynthetic units called photosystems. Light absorbed by pigment molecules boosts their electrons to a higher energy level. Because of the way the pigment molecules are arranged in the photosystems, they are able to transfer this energy to special pigment molecules at the reaction centers. Most photosynthetic organisms contain two photosystems, Photosystem I and Photosystem II.

Not all of the photosynthetic reactions require light. The series of reactions requiring light are referred to as the "light-dependent reactions," and those not requiring light are called the "light-independent reactions."

In the currently accepted model of the light-independent reactions, light energy enters Photosystem II, where it is trapped by the P_{680} chlorophyll molecules of the reaction center. Electrons are boosted uphill from P_{680} to an electron acceptor. As the electrons are removed from P_{680}, they are replaced by low-energy electrons from water molecules, and oxygen is produced. Pairs of electrons then pass downhill to Photosystem I along an electron transport chain; this passage generates a proton gradient that drives the synthesis of ATP from ADP and phosphate (photophosphorylation). Meanwhile, light energy absorbed in Photosystem I is passed to the P_{700} chlorophyll molecules of the Photosystem I reaction center. The energized electrons are ultimately accepted by the coenzyme molecule $NADP^+$, and the electrons removed from P_{700} are replaced by the electrons from Photosystem II. The energy yield from the light-dependent reactions is stored in the molecules of NADPH and in the ATP formed by photophosphorylation. Photophosphorylation also occurs in cyclic electron flow, a process that does not require Photosystem II.

Like oxidative phosphorylation in mitochondria, photophosphorylation in chloroplasts is a chemiosmotic process. As electrons flow down the electron transport chain from Photosystem II to Photosystem I, protons are pumped from the stroma into the thylakoid space, creating a gradient of potential energy. As protons flow down this gradient from the thylakoid space back into the stroma, passing through an ATP synthe-tase, ATP is formed.

In the light-independent reactions, which take place in the stroma of the chloroplast, the NADPH and ATP produced in the light-dependent reactions are used to reduce carbon dioxide to organic carbon. This reduction is accomplished by means of the Calvin cycle. In the Calvin cycle, a molecule of carbon dioxide is combined with the starting compound, a five-carbon sugar called ribulose 1,5-bisphosphate (RuBP), to form two mole-cules of the three-carbon compound 3-phosphoglycer-ate (PGA). At each turn of the cycle, one carbon atom enters the cycle. Six turns of the cycle produce two mol-ecules of the three-carbon molecule, glyceraldehyde 3-phosphate. At each turn of the cycle, RuBP is regenerated. Although glucose commonly is repre-sented as an end product of photosynthesis, most of the fixed carbon is converted to either sucrose or starch.

Plants in which the Calvin cycle is the only carbon-fixation pathway, and in which the first detectable product of CO_2 fixation is PGA, are called C_3 plants. In the so-called C_4 plants, carbon dioxide is initially fixed to phosphoenolpyruvate (PEP) to yield oxaloacetate, a four-carbon compound. The oxaloacetate is rapidly converted to either malate or aspartate, which transfers the CO_2 to RuBP of the Calvin cycle. The Calvin cycle in C_4 plants occurs in bundle-sheath cells, whereas the C_4 pathway takes place in the mesophyll cells. The C_4 plants are more efficient utilizers of CO_2 than C_3 plants, in part because PEP carboxylase is not inhibited by O_2, thus enabling C_4 plants to capture CO_2 with minimal water loss. Also, photorespiration, a process that con-sumes oxygen and releases CO_2 in the light, is active in C_3 plants but nearly absent in C_4 plants.

Crassulacean acid metabolism (CAM) is found in many succulent plants. In CAM plants, the fixation of CO_2 by PEP carboxylase into C_4 compounds occurs at night, when the stomata are open. The C_4 compounds are stored overnight, and then, during the daytime, when the stomata are closed, the fixed CO_2 is trans-ferred to RuBP of the Calvin cycle. The Calvin cycle and the C_4 pathway occur within the same cells in CAM plants; hence, these two pathways, which are spatially separated in C_4 plants, are temporally sepa-rated in CAM plants.

Suggestions for Further Reading

Alberts, Bruce, Dennis Bray, Julian Lewis, Martin Raff, Keith Roberts, and James D. Watson: *Molecular Biology of the Cell,* 2d ed., Garland Publishing, Inc., New York, 1989.

A large book covering both the molecular biology of the cell and the be-havior of cells in multicellular animals and plants. A truly modern ap-proach to the cell, this book is beautifully illustrated. Written for introductory courses in cell biology.

Asimov, I.: *Life and Energy,* Avon Books, New York, 1962.

An elementary but elegant description of the energetic basis of life by one of the greatest science writers of our time.

Baker, D.A., and J.L. Hall: *Solute Transport in Plant Cells and Tissues,* Longman Scientific & Technical, Harlow, Essex, England, 1988.

A multiauthored book emphasizing aspects of transport physiology cur-rently receiving attention by researchers in the field and covering a broad range of topics.

Becker, Wayne M., and David W. Deamer: *The World of the Cell,* 2d ed., Benjamin/Cummings, Redwood City, Calif., 1991.

An outstanding general cell-biology book, includes topics in bioenergetics and cellular energy metabolism; intended for undergraduates.

Conant, James B. (ed.): *Harvard Case Histories in Experimental Science,* vol. 2, Harvard University Press, Cambridge, Mass., 1964.

Case No. 5, "Plants and the Atmosphere," edited by Leonard K. Nash, de-scribes early work on photosynthesis. The narrative, often presented in the words of the investigators, illuminates the historical context in which these initial discoveries were made.

Cram, Jane M., and Donald J. Cram: *The Essence of Organic Chemis-try,* Addison-Wesley Publishing Co., Inc., Reading, Mass., 1978.

An elementary textbook emphasizing the features of organic chemistry most relevant to life processes.

"Frontiers in Biology: The Cell Cycle," *Science* 246(4930):603–640, November, 1989.

A series of six articles dealing with various aspects of the cell cycle; de-scribes experimental work that has provided evidence for various control mechanisms.

Gabriel, Mordecai L., and Seymour Fogel (eds.): *Great Experiments in Biology,* Prentice Hall, Inc., Englewood Cliffs, N.J., 1955.*

Presents many of the fundamental discoveries of biology as seen firsthand through the eyes of their discoverers. The examples are well chosen, and their value is greatly enhanced by accompanying explanatory notes and chronological tables of key developments in various areas of biology.

Govindjee, and William J. Coleman: "How Plants Make Oxygen," *Scientific American*, February 1990, pages 50–58.

An interestingly written article about the "water-oxidizing clock" used by photosynthesizing cells to split water molecules.

Gunning, B.E.S., and A.W. Robards (eds.): *Intercellular Communications in Plants: Studies on Plasmodesmata*, Springer-Verlag, New York, 1976.

An outstanding series of reviews of plasmodesmatal structure, distribution, and probable function by leading specialists in the field.

Gunning, B.E.S., and M.W. Steer: *Plant Cell Biology: An Ultrastructural Approach*, reprint edition published by M.W. Steer, 1986.

A collection of plant electron micrographs and accompanying explanations.

Kluge, M., and I.P. Ting: "Crassulacean Acid Metabolism: Analysis of an Ecological Adaptation," *Ecological Studies*, vol. 30, Springer-Verlag, New York, 1978.

A comprehensive treatment of all aspects of Crassulacean acid metabolism, with special emphasis on CAM as an ecological adaptation. The significance of CAM in agriculture is also considered.

Lehninger, Albert L., David L. Nelson, and Michael M. Cox: *Principles of Biochemistry*, 2d ed., Worth Publishers, Inc., New York, 1992.

This introductory text is outstanding both for its clarity and for its consistent focus on the living cell. There are numerous medical and practical applications throughout.

Monson, Russell K.: "On the Evolutionary Pathways Resulting in C_4 Photosynthesis and Crassulacean Acid Metabolism (CAM)," *Advances in Ecological Research* 19:57–110, 1989.

An excellent review of the literature comparing CO_2 assimilation by C_3, C_4, and CAM plants and C_3–C_4 intermediates.

Prescott, David M.: *Cells: Principles of Molecular Structure and Function*, Jones and Bartlett Publishers, Boston, 1988.

An up-to-date yet concise textbook of cell biology, written for a first course at the undergraduate level. Chapter 4 is devoted to energy flow and metabolism.

Robinson, David G.: *Plant Membranes: Endo- and Plasma Membranes of Plant Cells*, John Wiley & Sons, Inc., New York, 1985.

An up-to-date monograph on the structure and function of plant membranes.

Salisbury, Frank B., and Cleon W. Ross: *Plant Physiology*, 4th ed., Wadsworth Publishing Co., Inc., Belmont, Calif., 1991.

A detailed and useful review of the entire subject.

Stryer, Lubert: *Biochemistry*, 3d ed., W.H. Freeman and Company, San Francisco, 1988.

A good introduction to cellular energetics.

Stumpf, P.K., and E.E. Conn (eds.): *The Biochemistry of Plants: A Comprehensive Treatise*, vol.1, *The Plant Cell*, N.E. Tolbert (ed.), Academic Press, New York, 1980.

A multiauthored book considering the structure and function of plant cells. The first chapter is introductory and discusses the cell in general. Each of the remaining chapters is devoted to a different subcellular component.

Ting, I.P., and M. Gibbs (eds.): "Crassulacean Acid Metabolism," *Proceedings of the Fifth Annual Symposium in Botany*, University of California, Riverside, American Society of Plant Physiology, printed at the Waverly Press, Baltimore, Md., 1982.*

A series of papers addressing questions on the relationship between structure and function of CAM plants in natural habitats and on the evolution of CAM.

Zelitch, Israel: "Photosynthesis and Plant Productivity," *Chemical and Engineering News* 57(6):28–48, 1979.

A concise but thorough description of the major areas of research in photosynthesis. Discusses a number of research areas with direct relevance to agriculture.

* Available in paperback.

Genetics and Evolution

Corn, or maize (Zea mays), is one of the three most important crop plants in the world. Corn has long been an important organism for genetic studies. The different colors of some of the kernels in this painting are the result of genetic transposition, the study of which brought American geneticist Barbara McClintock a Nobel Prize.

CHAPTER 8

Genetics and Heredity

Ever since people first started to look at the world around them, they have puzzled and wondered about heredity. Why is it that the offspring of all living things —whether dandelions, dogs, aardvarks, or oak trees— always resemble their parents and never some other species? Why does a child have her mother's eyes or her father's chin or, even more puzzling, her grandfather's nose?

Questions such as these were recorded as far back as the writings of the Greeks, and they were probably old even then. Such questions have always been important. Throughout history, biological inheritance has been a major factor in determining the distribution of wealth, power, land, and privilege. From the point of view of biology, heredity is at the heart of any definition we may have of life.

Much of our present knowledge of heredity rests directly on certain principles discovered by Gregor Mendel (Figure 8–1). Although Mendel never saw a chromosome, he developed principles that demonstrate the existence of genes, chromosomes, and the process of meiosis. In this chapter we shall examine the physical basis of heredity, that is, meiosis, discuss some aspects of Mendel's work, and then consider the molecular basis of inheritance.

Sexual Reproduction in Eukaryotes

One of the major characteristics of eukaryotes as a group is that they undergo sexual reproduction, a process that does not occur in prokaryotes (bacteria). Although some eukaryotes do not reproduce sexually, it is evident that most such organisms have lost the capacity to do so during the course of their evolutionary history.

8–1
Gregor Mendel (1822–1884), shown here standing at the right, holding a fuchsia, carried out his studies in the garden of an Austrian monastery. His work in the field of genetics, which was not understood and was therefore largely ignored during his lifetime, was rediscovered in 1900.

Sexual reproduction involves a regular alternation between meiosis and fertilization. **Meiosis** is the process of nuclear division in which the chromosome number is reduced from the diploid (2*n*) to the haploid (*n*) number. During meiosis, the nucleus of a diploid cell undergoes two divisions, one of which is a reduction division. These divisions result in the production of four daughter nuclei, each containing one-half the number of chromosomes of the original nucleus. In plants, meiosis occurs during the production of spores in flowers, cones, or similar structures. **Fertilization,** or syngamy, is the process by which two haploid cells (gametes) fuse to form a diploid zygote. Fertilization thus reestablishes the diploid chromosome number.

Diploid organisms have two sets of chromosomes, one derived from each parent. Corresponding chromosomes, which form pairs during the course of meiosis, are called homologous chromosomes, or **homologs.** The interaction of products indirectly derived from genes on each of these sets of chromosomes determines the genetic characteristics of the diploid organism.

Structure of Eukaryotic Chromosomes

The chromosomes of eukaryotes are made up of DNA and protein, which together form chromatin. Each eukaryotic chromosome consists of a single threadlike molecule of double-stranded DNA, which is continuous throughout its entire length. If the DNA double helix (see Figure 8–15) were stretched out, it would extend for several centimeters. The most abundant of the DNA-binding proteins are the **histones,** which are present in such enormous quantities that their total mass is about equal to that of the DNA. The histones play an important role in packing the DNA into the nucleus in an orderly manner.

Because they contain very high concentrations of the amino acids arginine and lysine, the histones are positively charged. In contrast to the histones, the double-stranded DNA, with its exposed phosphate groups, is negatively charged. Hence, the histones bind tightly around the DNA, forming repeating structural units called **nucleosomes,** which give the chromatin the "beads-on-a-string" appearance seen in electron micrographs (Figure 8–2a) after treatments that "relax" the native structure of the DNA. Each nucleosome consists of an octamer (eight-molecule complex) of four different kinds of histones, with the DNA helix wound around the octamer twice in a path resembling a spring (Figure 8–2b). Within such nucleosomes, the DNA is efficiently packaged and protected from enzymatic degradation.

Between the nucleosomes are strings of *spacer,* or *linker, DNA* with nonhistone proteins attached to them. The nonhistone proteins that occur in chromosomes are very diverse. They include regulators of the copying, or transcription, of specific genes, enzymes involved in the modification of nucleic acids and proteins, and proteins that dictate the structure of chromosomes and nuclei.

Some portions of chromosomes are packaged in a highly condensed state called **heterochromatin.** The rest of the chromatin, which is not condensed during interphase, is called **euchromatin.** The condensed organization of heterochromatin prevents transcription of mRNA; indeed, the expression of euchromatic genes is often inhibited if they are located near a heterochromatic region.

During early meiosis or mitosis, the euchromatin condenses. DNA initially aggregates into a series of tight "coils" separated by uncoiled regions. These coiled regions, or **chromomeres,** are much larger than nucleosomes and can be seen with a light microscope (Figure 8–2c). Even larger coils are formed at certain stages of nuclear division (Figure 8–3b).

8–2

(a) *Electron micrograph of nucleosomes and connecting threads of DNA from a chicken erythrocyte (red blood cell). The nucleosomes—the structures that look like beads on a string—are each approximately 10 nanometers in diameter.*

(b) *Each nucleosome consists of an octamer of four different kinds of histones (two each of H2A H2B, H3, and H4), around which the DNA helix is wound twice. Nucleosomes are separated from each other by spacer, or linker, DNA. Histone H1 is thought to be associated with the spacer DNA between successive nucleosomes.*

(c) *Photomicrograph of meiotic chromosomes in a tarweed (Hemizonia pungens). Here the beadlike structures visible along the paired chromosomes are not nucleosomes but chromomeres, which are regions of aggregated chromatin that are larger than nucleosomes and therefore visible in a light microscope.*

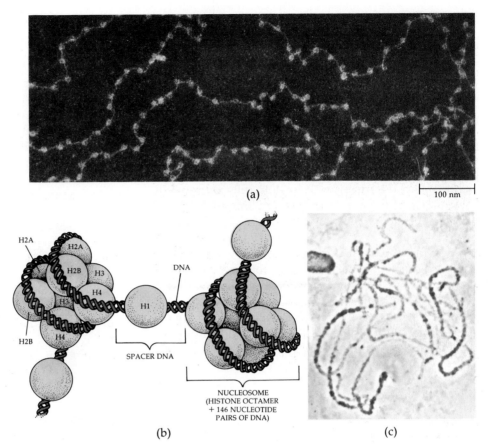

(a)

100 nm

H2A
H2A
H2B
H3
H4
H3
H4
H2B
DNA
H1
SPACER DNA
NUCLEOSOME
(HISTONE OCTAMER
+ 146 NUCLEOTIDE
PAIRS OF DNA)

(b)

(c)

(a)

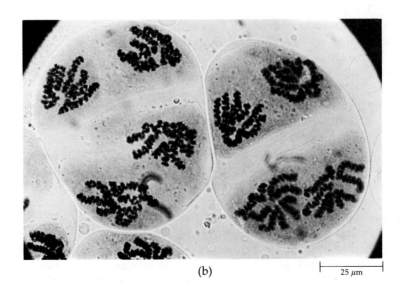

(b)

25 μm

8–3

(a) *Wake-robin (Trillium erectum) flowers in the early spring.* (b) *The second anaphase of meiosis during microspore formation in T. erectum. Separation of the clearly visible chromosomes, which contain the genetic material, DNA, is nearly complete. Each of the newly forming nuclei will possess only half the number of chromosomes that were present in the original nucleus at the beginning of meiosis.*

Meiosis

Meiosis occurs only in specialized diploid cells and only at particular times in the life cycle of a given organism. Through meiosis and cytokinesis, a single diploid cell gives rise to four haploid cells—either gametes or spores. A **gamete** is a cell that unites with another gamete to produce a diploid **zygote.** The zygote may then divide, either meiotically, producing four unicellular haploid cells, or mitotically, eventually forming a multicellular diploid organism. If haploid cells are formed, they may function as (1) independent, unicellular, haploid organisms or (2) gametes, in which case they may eventually unite with one another (fertilization). If a multicellular diploid organism is formed, it will, in most cases, eventually produce either haploid spores or gametes by meiosis. A **spore** is a cell that can develop into an organism without uniting with another cell. Spores often divide mitotically, producing multicellular organisms that are entirely haploid and that eventually give rise to gametes by mitosis (see Figure 10–10, page 184).

FIRST MEIOTIC DIVISION

Meiosis consists of two successive nuclear divisions. Refer to Figure 8–7 on page 126 to help keep track of the processes described in the following paragraphs.

In **prophase I** (prophase of the first meiotic division), the chromosomes—present in the diploid number—first become visible as long, slender threads. As in mitosis (see Chapter 2), the chromosomes have already duplicated during the preceding interphase. Consequently each chromosome, at the beginning of prophase I, consists of two identical chromatids attached at the centromere. At this early stage of meiosis, however, each chromosome appears to be single rather than double.

Before the individual chromatids become apparent, the homologous chromosomes pair with each other. The pairing is very precise, beginning at one or more sites along the length of the chromosomes and proceeding in a zipperlike fashion, such that the same portions of the homologous chromosomes lie next to one another. Each homolog is derived from a different parent and is made up of two identical chromatids. Thus, a homologous pair consists of four chromatids. The pairing of homologous chromosomes is a necessary part of meiosis; the process cannot occur in haploid cells because such homologs are not present. The pairing is called **synapsis,** and the associated pairs of homologous chromosomes are called **bivalents.**

During the course of prophase I, the paired threads become more and more tightly contracted, and consequently the chromosomes shorten and thicken. With

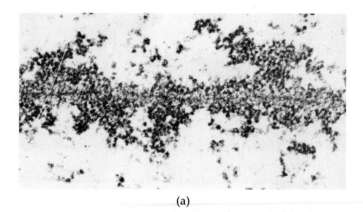

(a)

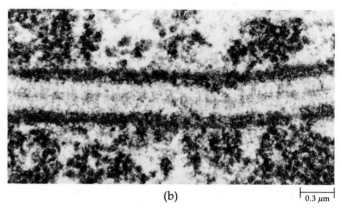

(b) $\overline{\quad}$ 0.3 μm

8–4
(a) *Portion of a chromosome of* Lilium *early in prophase I, prior to pairing. Note the dense axial core. This core, which consists mainly of proteins, may arrange the genetic material of the chromosome in preparation for pairing and genetic exchange.* (b) *Synaptonemal complex in a bivalent of* Lilium.

the aid of an electron microscope, it is possible to identify a densely staining axial core, consisting mainly of proteins, in each chromosome (Figure 8–4a). During mid-prophase, the axial cores of a pair of homologous chromosomes approach each other to within 0.1 micrometer, forming a **synaptonemal complex** (Figure 8–4b).

In favorable material, it can now be seen that each axial core is double; that is, each bivalent is made up of four chromatids, two per chromosome. During the time when the synaptonemal complex exists, portions of the chromatids break apart and rejoin with corresponding segments from their homologous chromatids. This **crossing-over** results in chromatids that are complete but which have a different representation of genes than originally. Figure 8–5 shows the visible evidence that crossing-over has taken place—the X-like configuration called a **chiasma** (plural: chiasmata).

As prophase I proceeds, the synaptonemal complex ceases to exist. Eventually, the nuclear envelope breaks down. The nucleolus usually disappears as RNA synthesis is temporarily suspended. Finally, the homologous chromosomes appear to repulse one another. Their chromatids are held together at the chiasmata, however. These chromatids separate very slowly. As they separate, some of the chiasmata slip toward the end of the chromosome arm. One or more chiasmata may occur in each arm of the chromosome, or only one in the entire bivalent; the appearance of a particular bivalent can vary widely, depending on the number of chiasmata present (Figure 8–5).

In **metaphase I,** the spindle—an axis of microtubules similar to that which functions in mitosis—becomes conspicuous (Figure 8–6). As meiosis proceeds, individual microtubules become attached to the centromeres of the chromosomes of each bivalent. These paired chromosomes then move to the equatorial plane of the cell, where they line up randomly in a configuration characteristic of metaphase I. The centromeres of the paired chromosomes line up on opposite sides of the equatorial plane; in contrast, in mitotic metaphase, as we have seen, the centromeres of the individual chromosomes line up directly on the equatorial plane.

Anaphase I begins when the homologous chromosomes separate and begin to move toward the poles. (Notice again the contrast with mitosis. In mitotic anaphase, the centromeres separate and the sister chromatids separate.) In meiotic anaphase I, the centromeres do not separate and the sister chromatids remain together; it is the homologs that separate. Because of the exchanges of chromatid segments that result from crossing-over, however, the chromatids are not identical, as they were at the onset of meiosis.

In **telophase I,** the coiling of the chromosomes relaxes and the chromosomes become elongated and once again indistinct. New nuclear envelopes are formed from the endoplasmic reticulum, as telophase gradually changes to interphase. Finally, the spindle disappears, the nucleolus is reformed, and protein synthesis commences again. In many organisms, however, no interphase intervenes between meiotic divisions I and II: in these organisms, the chromosomes pass more or less directly from telophase I to prophase II of the second meiotic division.

SECOND MEIOTIC DIVISION

At the beginning of the second meiotic division, the chromatids are still attached by their centromeres. This division resembles a mitotic division: the nuclear envelope (if one re-formed during telophase I) becomes disorganized, and the nucleolus disappears by the end of **prophase II.** At **metaphase II,** a spindle again becomes obvious, and the chromosomes—each consisting of two chromatids—line up with their centromeres on the equatorial plane. At **anaphase II,** the centromeres separate and are pulled apart, and the newly separated chromatids, now called daughter chromosomes, move to opposite poles (see Figure 8–3). At **telophase II,** new

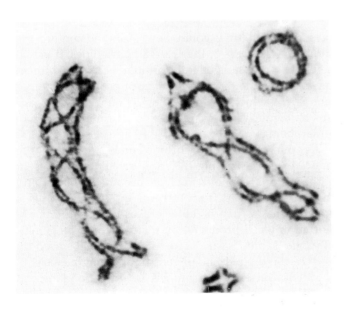

8–5

Variations in the number of chiasmata can be seen in the paired chromosomes of a grasshopper, Chorthippus parallelus.

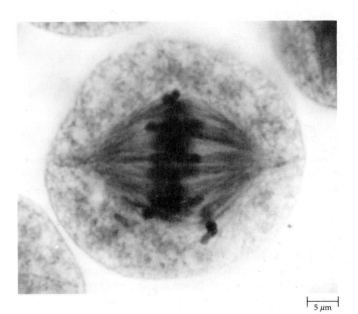

5 μm

8–6

The spindle in a pollen mother cell of wheat (Triticum aestivum) *during metaphase I of meiosis.*

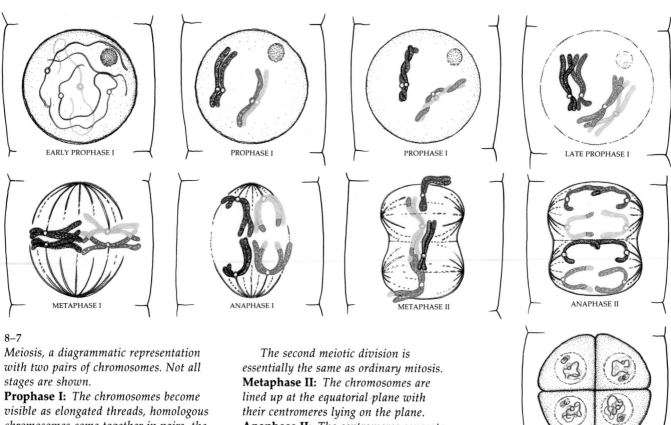

EARLY PROPHASE I PROPHASE I PROPHASE I LATE PROPHASE I

METAPHASE I ANAPHASE I METAPHASE II ANAPHASE II

LATE TELOPHASE II

8–7

Meiosis, a diagrammatic representation with two pairs of chromosomes. Not all stages are shown.
Prophase I: *The chromosomes become visible as elongated threads, homologous chromosomes come together in pairs, the pairs coil round one another, and the paired chromosomes become very short.*
Metaphase I: *The paired chromosomes move into position on the metaphase plate with their centromeres evenly distributed on either side of the equatorial plane of the spindle.*
Anaphase I: *The paired chromosomes separate and move to opposite poles.*

The second meiotic division is essentially the same as ordinary mitosis.
Metaphase II: *The chromosomes are lined up at the equatorial plane with their centromeres lying on the plane.*
Anaphase II: *The centromeres separate, and the chromatids separate and move toward opposite poles of the spindle.*
Telophase II: *The chromosomes have completed their migration. Four new nuclei, each with the haploid number of chromosomes, are formed.*

Meiosis in crested wheat grass (Agropyron cristatum), $n = 7$, is shown opposite.

nuclear envelopes and nucleoli are organized, and the contracted chromosomes relax as they fade into an interphase nucleus.

The stages of meiosis are summarized in Figure 8–7.

THE CONSEQUENCES OF MEIOSIS

The end result of meiosis is that the genetic material present in the diploid nucleus has *divided twice*. Therefore, each cell has only half as many chromosomes as the original diploid nucleus. But more important are the genetic consequences of the process. At metaphase I, the orientation of the bivalents is random; that is, the chromosomes derived from one parent are randomly divided between the two new nuclei. In addition, because of crossing-over, each chromosome usually contains segments that have been derived from both parents. If the original diploid cell had two pairs of homologous chromosomes, $n = 2$, there are four possible

ways in which they could be distributed among the haploid cells. If $n = 3$, there are 8 possibilities; if $n = 4$, there are 16. The general formula is 2^n. In human beings, $n = 23$, and so the number of possible combinations is 2^{23}, which is equal to 8,388,608. Many organisms have much higher numbers of chromosomes than human beings.

As the number of chromosomes increases, the chance of reconstituting the same set that was present in the original diploid nucleus becomes smaller and smaller. Quite apart from this, the existence of at least one chiasma in each bivalent makes it almost impossible that any cell produced by meiosis could be genetically the same as any one of those that fused to produce the diploid line of cells undergoing meiosis.

Meiosis differs from mitosis in three fundamental ways (Figure 8–8):

1. Although the genetic material is replicated only one time, during the interphase preceding prophase I in meiosis, there are two nuclear divisions involved, producing a total of four nuclei.

Early Prophase I. Chromosomes appear as threads. Each thread is actually double-stranded, composed of two identical chromatids.

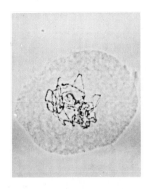

Prophase I. Homologous chromosomes pair. This is a crucial point of difference between meiosis and mitosis. Chiasmata are visible.

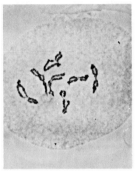

Metaphase I. Bivalents are now lined up randomly at the equatorial plane of the cell, with their centromeres evenly distributed on either side of the plane.

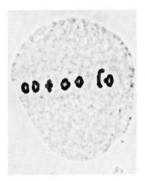

Anaphase I. The homologous chromosomes have separated from each other and are moving toward opposite poles of the cell.

Late Telophase I. The chromosomes are regrouped at each pole, and the cell is dividing to form two cells.

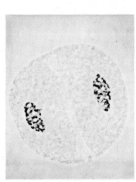

Prophase II. The chromosomes are reappearing. Each still consists of two chromatids. Because of crossing-over, the chromatids are no longer identical to each other.

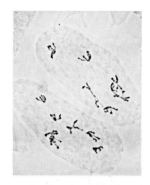

Metaphase II. The chromosomes are lined up at the equatorial plane of the cell, with their centromeres on the plane.

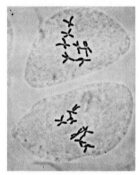

Anaphase II. The centromere of each chromosome has divided and the chromatids—now chromosomes—are moving toward opposite poles.

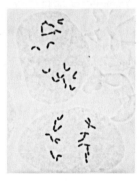

Telophase II. The chromosomes have now completely separated and new cell walls are forming.

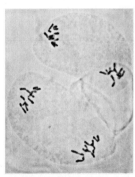

Tetrad. New plasma membranes and cell walls form as the process of cytokinesis is completed. These four haploid cells will become pollen grains.

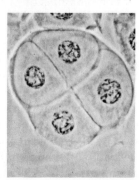

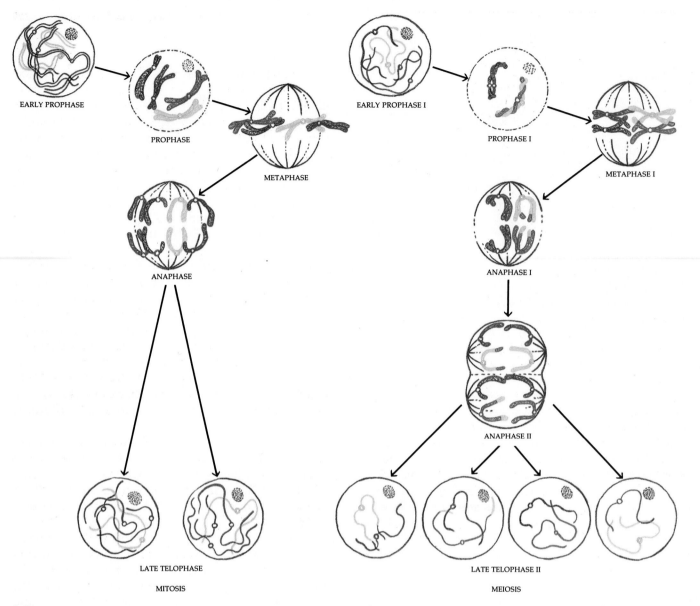

EARLY PROPHASE

PROPHASE

METAPHASE

ANAPHASE

LATE TELOPHASE

MITOSIS

EARLY PROPHASE I

PROPHASE I

METAPHASE I

ANAPHASE I

ANAPHASE II

LATE TELOPHASE II

MEIOSIS

8–8

A comparison of mitosis and meiosis.

2. Each of the four nuclei produced in meiosis is haploid, containing only one-half the number of chromosomes present in the original diploid nucleus from which it was produced.

3. The nuclei produced by meiosis contain entirely new combinations of chromosomes.

In meiosis, nuclei *different* from the original nucleus are produced, in contrast with mitosis, which produces nuclei with chromosome complements *identical* to those of the original nucleus. The genetic and evolutionary consequences of the behavior of chromosomes in meiosis are profound. Because of meiosis and fertilization, the populations of diploid organisms that occur in nature are far from uniform; instead, they consist of individuals that differ from one another in many characteristics.

How Are Characteristics Inherited?

Figure 8–9 is a plot of the distribution of ear length in a particular variety of corn, but a curve of this shape could just as well represent the distribution of weight among a random assortment of pinto beans or height among oak trees. Environmental factors, such as rain in a cornfield, might affect the actual measurements involved in preparing such a graph, but they rarely affect the shape of the curve.

A bell-shaped curve of this sort is known as a ''normal'' curve. The pattern of variation is said to be *continuous*. The smooth, symmetrical shape of the curve indicates that populations cannot be divided into a series of sharply contrasting forms. Human height follows a similar pattern of inheritance. Much of the vari-

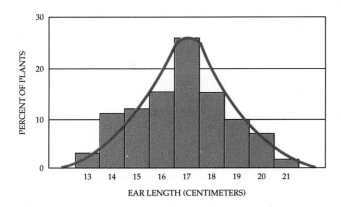

8–9

Distribution of ear length of the Black Mexican variety of corn (Zea mays). This is an example of a phenotypic characteristic that is determined by the interaction of a number of genes. Such characteristics show continuous variations; if these variations are plotted as a curve, the curve is bell-shaped, with the mean, or average, falling in the center of the curve.

ation that is seen in nature is of exactly this kind, and the early scholars who tried to understand how the characteristics of plants and animals were inherited were baffled by this variation. In plants, the most important components of agricultural yield—dimensions, weight, and height—all exhibit continuous variation. We know now that patterns of this sort are produced by the combined interaction of many genes, and that the patterns can be analyzed, at least in principle, by determining the individual roles of these genes.

MENDELIAN GENETICS

The branch of genetics that deals with relatively clearcut traits and their inheritance is generally termed Mendelian genetics, in recognition of the work of Gre-

gor Mendel. Mendel carried out his studies from 1856 to 1863 in the garden of the Augustinian monastery of St. Thomas in Brünn, then in Austria. Because national boundaries have shifted, this city is now in Czechoslovakia; it is currently called Brno. Mendel carefully selected his principal experimental subject, the garden pea *(Pisum sativum),* and concentrated on discrete, qualitative characteristics, such as differences in flower color and seed shape.

Mendel made large numbers of experimental crosses and analyzed the numerical relationships among the progenies. Importantly, he studied not only the offspring of the first generation but also those of subsequent generations and their crosses. Moreover, before he made a cross between two different kinds of peas, he first obtained true-breeding lines for each of the traits in which he was interested; this was a simple matter, since self-pollination occurs automatically in the garden pea (Figure 8–10). Earlier students of heredity had not used such well-defined traits, nor had they handled their experimental material so carefully; their results, consequently, were ambiguous and difficult to analyze.

The interpretation of Mendel's results was surprisingly clear. Table 8–1 lists seven traits of the pea plants that he used in his experiments. Crosses between individuals that differ in one single trait, such as those Mendel carried out, are called **monohybrid** crosses (those that involve two traits are called **dihybrid** crosses). When Mendel crossed plants with contrasting characteristics, he observed that, in every case, one of the alternative characteristics could not be seen in the first generation (F_1). For example, the seeds of all the progeny of the cross between yellow-seeded plants and green-seeded plants were as yellow as those of the yellow-seeded parent. Mendel called the characteristic for yellow seeds, as well as the other characteristics that were seen in the F_1 generation, **dominant.** He called the

8–10

In a flower, the pollen develops in the anthers, and the egg cells develop in the ovules. Pollination occurs when pollen grains are trapped on the stigma. The pollen grains germinate and grow down to the ovules, where fertilization occurs. In most species, this process involves pollen from one plant reaching the stigma of another plant, a phenomenon known as cross-pollination. The fertilized egg, or zygote, develops into an embryo within the ovule, and the ovule matures to form the seed.

In the flower of the garden pea, the stigma and anthers are enclosed com-

pletely by petals. Unlike the flowers of many other species of plants, pea flowers do not open completely, and pollination usually results in self-fertilization. In his crossbreeding experiments, Mendel pried open the bud before the pollen matured and removed the anthers with forceps. Then he pollinated the flower artificially by dusting the stigma with pollen collected from another plant that had the characteristic(s) in which he was interested.

In this drawing, several petals and anthers have been removed, and the ovary has been cut open to reveal the ovules.

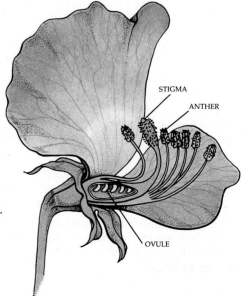

STIGMA

ANTHER

OVULE

Table 8–1   *Mendel's Pea-Plant Experiments*

Trait	Dominant	Recessive	F_2 Generation Dominant	F_2 Generation Recessive
Seed form (7)	Round	Wrinkled	5474	1850
Seed color (1)	Yellow	Green	6022	2001
Flower site (4)	Axial	Terminal	621	207
Flower color (1)	Red	White	705	224
Pod form (4)	Inflated	Tight	882	299
Pod color (5)	Green	Yellow	428	152
Stem length (4)	Tall	Dwarf	787	277

The number following the name of the trait indicates the chromosome on which the gene for that particular trait is located (see page 132 for discussion).

characteristics that did not appear in the first generation **recessive.** When plants of the F_1 generation were allowed to self-pollinate, the recessive characteristic reappeared in the F_2 generation in ratios of approximately 3 dominant to 1 recessive (Table 8–1). Thus, the characteristics were still present in the F_1 generation, but masked.

These results can be easily understood in terms of meiosis, a phenomenon that was unknown at the time Mendel was carrying out his studies. The characteristics

of diploid organisms are determined by interactions between alleles. An **allele** is one of two or more alternative forms of the same gene. Alleles occupy the same site, or **locus,** on homologous chromosomes. Hence, each diploid cell has two alleles for each gene, one on each of the homologous chromosomes.

Consider a cross between a white-flowered plant and a red-flowered plant. The allele for white flower color, which is recessive, is indicated by the lowercase letter *w.* The contrasting allele for red flower color, which is dominant, is indicated by the capital letter *W.* In the true-breeding strains of garden pea with which Mendel worked, white-flowered individuals had the genetic constitution, or **genotype,** *ww.* Red-flowered individuals had the genotype *WW.* Individuals such as these, which have two identical alleles at a particular locus on their homologous chromosomes, are said to be **homozygous.** When plants with these contrasting characteristics are crossed, every individual in the F_1 generation receives a *W* allele from the red-flowered parent and a *w* allele from the white-flowered parent, and thus has the genotype *Ww.* Such an individual is said to be **heterozygous** for the gene for flower color.

In the course of meiosis, a heterozygous individual will form two kinds of gametes, *W* and *w,* which will be present in equal proportions. As indicated in Figure 8–11a, the *W* and *w* gametes derived, respectively, from the two parents, will recombine to form one *WW* indi-

8–11

(a) *A cross between a pea plant with two dominant alleles for red flowers (WW) and a pea plant with two recessive alleles for white flowers (ww). The phenotype of the offspring in the F₁ generation is red, but note that its genotype is Ww. The F₂ generation is shown in the diagram below. The W allele, being dominant, determines the appearance, or phenotype, of the flower. Only when the offspring receives a w allele from each parent, producing a ww genotype, does the recessive trait (white flowers) appear. The ratio of dominant-to-recessive phenotypes in the F₂ generation is thus 3 to 1, as expected. The genotypic ratio is 1 to 2 to 1. (b) A testcross between a pea plant with red flowers and one with white flowers. Although the red-flowered plant is phenotypically identical to the WW plant shown in (a), results of the testcross reveal that it is heterozygous (Ww) for this gene; only a heterozygous parental genotype could produce the genotypes observed in the progeny. In this case, the genotypic ratio is 1 Ww to 1 ww.*

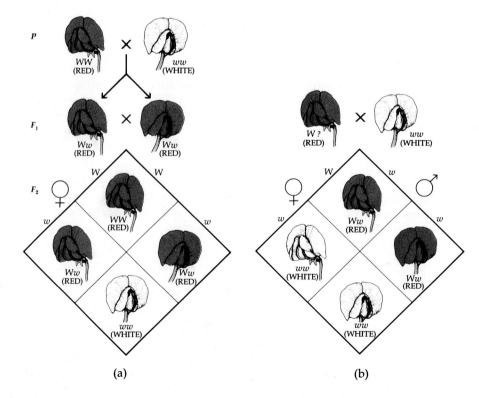

(a)   (b)

vidual, one *ww* individual, and two *Ww* individuals, on average, for every four offspring produced. In terms of their appearance, or **phenotype,** the heterozygous *Ww* individuals will be red-flowered; they are indistinguishable in this respect from the homozygous *WW* individuals. The products of the allele from the red-flowered parent are sufficient to mask those of the allele from the white-flowered parent. This, then, is the basis for the 3-to-1 phenotypic ratios that Mendel observed (Table 8–1).

How can one tell whether the genotype of a plant with red flowers is *WW* or *Ww*? As shown in Figure 8–11b, one can tell by crossing such a plant with a white-flowered plant and counting the progeny of the cross. Mendel performed just this kind of experiment, which is known as a **testcross**—the crossing of an individual showing a dominant characteristic with a second individual that is homozygous recessive for that trait.

THE PRINCIPLE OF SEGREGATION

The principle that was established by these experiments is the **principle of segregation,** which is sometimes known as Mendel's first law. According to this principle, hereditary traits are determined by discrete factors (now called genes) that appear in pairs, one of each pair being inherited from each parent. During meiosis, the pairs of factors are separated, or *segregated.* Hence, each gamete that is produced by an offspring at maturity contains only one member of each pair that the offspring possesses. This concept of discrete factors explained how a characteristic could persist from generation to generation without blending with other characteristics, as well as how it could seemingly disappear and then reappear in a later generation. The development of the principle of segregation, therefore, was of the utmost importance for understanding both genetics and evolution.

INCOMPLETE DOMINANCE

In the previous examples, the action of the dominant allele, when it was present, masked the existence of the second, recessive allele. Dominant and recessive characteristics are not always so clear-cut, however. In cases of **incomplete dominance,** the phenotype of the heterozygote is intermediate between the phenotype of the parent homozygotes, since the action of one allele does not completely mask the action of the other.

For example, in snapdragons, a cross between a red-flowered plant and a white-flowered plant produces a plant that has pink flowers. When the F_1 generation is intercrossed, the characteristics segregate again, the result in the F_2 generation being one red-flowered (ho-

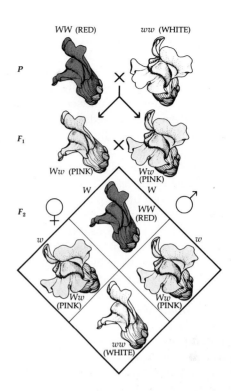

8–12
When a snapdragon with red flowers and one with white flowers are crossed, the resulting progeny have pink flowers (Ww). When the pink-flowered plants are crossed, the results show that, although the gene products blend in the heterozygote, the genes themselves remain discrete and segregate according to the Mendelian ratio 1:2:1 in both genotype and phenotype.

mozygous) plant to two pink-flowered (heterozygous) plants to one white-flowered (homozygous) plant (Figure 8–12). Thus, cases of incomplete dominance also conform to Mendel's principle of segregation.

INDEPENDENT ASSORTMENT

In inheritance patterns involving more than one gene, certain differences in the patterns depend on whether the genes are located relatively close together on the same chromosome or on different chromosomes. We shall consider first the relatively simple situation in which the genes are located on different chromosomes and then the more complicated case in which they are located on the same chromosome.

Mendel studied hybrids that involved two pairs of contrasting characteristics and are called dihybrid crosses. For example, he crossed strains of garden peas in which one parent had seeds that were round and yellow and the other parent had seeds that were wrinkled and green. As Table 8–1 shows, the alleles for round seeds and yellow seeds are both dominant, and the wrinkled and green alleles are recessive. All the plants of the F_1 generation had seeds that were round and yellow. When the F_1 seeds were planted and the flowers allowed to self-pollinate, 556 F_2 plants were produced. Of these, 315 plants showed the two dominant characteristics—round and yellow—in their seeds, and 32 combined the recessive characteristics, wrinkled and green. All the rest of the plants produced seeds that were unlike those of either parent: 101 were wrinkled and yellow, and 108 were round and green. Totally new combinations of characteristics had appeared.

Figure 8–13 shows the basis for such results. In a cross involving two pairs of dominant and recessive alleles, with each pair on a different chromosome, the ratio of distribution of phenotypes is 9:3:3:1. The 9 represents the proportion of the F_2 progeny that shows both dominant alleles; 3 represents the proportions of the two possible combinations of dominant and recessive alleles; and 1 is the proportion that shows both recessive alleles. In the example shown, one parent carries both dominant alleles and the other carries both recessive alleles. Suppose that each parent carried one recessive and one dominant allele. Would the results be the same? If you are not sure of the answer, try making a diagram of the possibilities, as was done in Figure 8–13. Diagrams of the sort presented in that figure are called Punnett squares; they are named after the man who developed them, the English geneticist Reginald Crundall Punnett. Punnett was one of the scientists who rediscovered Mendel's work early in the twentieth century.

From these experiments, Mendel formulated his second law, the **principle of independent assortment.** This law states that the inheritance of a pair of factors for one trait is independent of the simultaneous inheritance of factors for other traits; in other words, such factors assort independently, as though no other factors were present. This independent assortment results directly from the random manner in which the centromeres of the bivalents line up on either side of the equatorial plane at metaphase I.

LINKAGE

When Mendel was performing his experiments, the existence of chromosomes was unknown. Knowing that genes are located on chromosomes, one can readily see that if two different genes are located relatively close together on the same chromosome pair, they generally will not segregate independently. In this case, a 9:3:3:1 ratio will not be obtained in the F_2 generation.

Either through good fortune or by careful selection

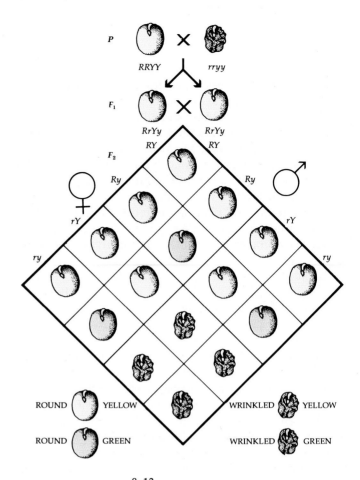

8–13
Independent assortment. In one of Mendel's experiments, he crossed a plant having round (RR) and yellow (YY) seeds with a plant having wrinkled (rr) and green (yy) seeds. The seeds of the F_1 generation were round and yellow. In the F_2 generation, however, as shown in the diagram, the recessive traits reappeared. Furthermore, they appeared in new combinations. In a cross such as this, involving two pairs of alleles on different chromosomes, the expected phenotypic ratio in the F_2 generation is 9:3:3:1. The genotypic ratio is 1 rryy to 2 rrYy to 1 rrYY to 2 Rryy to 4 RrYy to 2 RrYY to 1 RRyy to 2 RRYy to 1 RRYY.

of the traits he studied, Mendel avoided the phenomenon of linkage, which he probably would not have been able to explain. The garden pea has seven pairs of chromosomes, and the seven major traits studied by Mendel are located on four different chromosomes, as shown in Table 8–1 (page 130). In addition, the traits located on the same chromosome are so far apart that, with a single exception, they segregate independently of one another. The genes for pod form and stem length, which are on chromosome 4, are so close to one another that Mendel should have observed the effects of linkage between these two features. He made no report of such a dihybrid cross, however, and so we do not know whether he made one or not.

The linkage of genes was first discovered by the English geneticist William Bateson and his coworkers in 1905, while they were studying the genetics of the sweet pea (*Lathyrus odoratus*). These scientists crossed a doubly homozygous recessive strain of sweet peas that had red petals and round pollen grains with a second strain (resembling the wild form of the species) that had purple petals and long pollen grains. From this cross, they obtained the following characteristics in the F_2 generation:

284	purple	long
21	purple	round
21	red	long
55	red	round

If the genes for flower color and pollen shape were on the same chromosome, there should have been only two types of progeny. If they were on two different chromosomes, then the ratio of the four classes should have been 216:72:72:24, or 9:3:3:1. Clearly, the genes of the "parental" types were being held together, as if they were *linked* to one another.

The explanation for these results is that the two genes are "linked" on one chromosome but are sometimes exchanged between homologous chromosomes during crossing-over. It is now known that crossing-over—the breakage and rejoining of chromosomes that results in the appearance of chiasmata—occurs in prophase I of meiosis (see Figure 8–7). The greater the distance between two genes on a chromosome, the greater the chance that a crossover will occur between them. The closer together two genes are, the greater will be their tendency to assort together in meiosis and, thus, the greater is their "linkage." Maps can be constructed, based on the amount of crossing-over that occurs between the genes; and such **genetic maps** provide an approximation of the actual positions of the genes on the chromosomes.

POLYGENIC INHERITANCE

The first experiment that illustrated how many genes can interact to produce a continuous pattern of variation in plants (in this case, wheat) was carried out by the Swedish scientist H. Nilson-Ehle. Table 8–2 shows the phenotypic effects of various combinations of two genes, each with two alleles, which act together to control the intensity of color in wheat kernels. By this and similar studies, we are beginning to learn how many genes may interact to produce the complex characteristics of plants and animals. The principles first developed by Mendel well over a century ago can thus be applied to an analysis of the sorts of characteristics that had puzzled earlier students of heredity.

Table 8–2 *The Genetic Control of Color in Wheat Kernels*

Parents	$R_1R_1R_2R_2$ $\times$ $r_1r_1r_2r_2$		
	(Dark red) (White)		
F_1	$R_1r_1R_2r_2$ (Medium red)		

F_2		Genotype		Phenotype	
1		$R_1R_1R_2R_2$		Dark red	
2 2 } 4		$R_1R_1R_2r_2$ $R_1r_1R_2R_2$		Medium-dark red Medium-dark red	
4 1 1 } 6		$R_1r_1R_2r_2$ $R_1R_1r_2r_2$ $r_1r_1R_2R_2$		Medium red Medium red Medium red	15 red to 1 white
2 2 } 4		$R_1r_1r_2r_2$ $r_1r_1R_2r_2$		Light red Light red	
1		$r_1r_1r_2r_2$		White	

Mutations

The studies on independent assortment described above depend on the existence of differences between the alleles of a gene. How do such differences arise? The first answer to this question was provided by the Dutch scientist Hugo de Vries (Figure 8–14).

THE ORIGIN OF THE CONCEPT OF MUTATION

In 1901, de Vries studied the inheritance of characteristics in a kind of evening primrose (*Oenothera glazioviana*) that was abundantly naturalized on the coastal dunes of Holland. In this plant, he found that, although the patterns of heredity were generally well ordered and predictable, a characteristic occasionally appeared that had not been observed previously in either parental line. De Vries hypothesized that this new characteristic was the phenotypic expression of a change in a gene. Moreover, according to his hypothesis, the changed gene would then be passed along just as the other genes were. De Vries spoke of this hereditary change in one of the alleles of a gene as a **mutation** and of the organism carrying it as a *mutant.*

Ironically, only two of roughly 2000 changes in the evening primrose that were observed by de Vries were what we would term mutations today. All of the rest were due to new genetic combinations or to the presence of extra chromosomes rather than actual abrupt changes in any particular gene.

THE KINDS OF MUTATIONS

Today, any change in the genetic message of an organism is called a mutation. Such changes may involve either alterations in the coding sequence itself, or changes in the way in which the genetic message is organized. There are five principal kinds of mutations.

Point Mutations

Point mutations involve only one or a few nucleotides. They may arise from chemical or physical damage to the DNA. **Mutagens,** such as ionizing radiation, ultraviolet radiation, or various kinds of chemicals, generally cause point mutations; in human beings and other animals, this is often how cancer starts. Point mutations may also arise by very-low-frequency mispairing during the replication of DNA.

Deletions

Small sections of the chromosome may be deleted, as by X-rays. This usually results in a change in the characteristics of the organism. Many deletions apparently occur in nature as the result of crossing-over between

8–14

(a) *Hugo de Vries, shown standing next to* Amorphophallus titanum, *a member of the same family as the calla lily. The plant, a native of the Sumatran jungles, has one of the most massive inflorescences, or flower clusters, of any of the angiosperms. This picture was taken in the arboretum of the Agricultural College at Wageningen, Holland, in 1932.* (b) Oenothera glazioviana, *the evening primrose. DeVries reported on mutations he observed in this organism.*

(a)   (b)

different parts of the same chromosome, at sites where duplications occur.

Position Effects

Genes do not necessarily occur in a fixed position on the chromosomes but may move around. In bacteria, **plasmids**—small circular molecules of DNA separate from the main chromosome—may enter that chromosome at places where they share a common base sequence. (The consequences of such a process for bacterial and viral evolution are discussed in Chapter 11, and the consequences for genetic engineering in Chapter 25.) Both in bacteria and in eukaryotes, genes may move as small, mobile portions of the chromosome called **transposons,** which migrate from one chromosomal position to another at random. In either case, the changed position of the genes involved may disrupt the action of their new neighbors, or vice versa, and lead to effects that we recognize as mutations. The proximity of genes to areas of heterochromatin or other regions that control gene expression may likewise contribute to these effects.

Inversions and Translocations

During the course of evolution of particular groups of organisms, a chromosomal segment may break free and then reenter its chromosome with its sequence of genes oriented in the direction opposite to the original one. Such a change in chromosomal sequence is called an inversion. Another sort of change that is frequent in certain groups of plants involves the breaking off of a segment of one chromosome and the reattachment of that segment on another chromosome. Such a change in position is called a translocation; translocations are often reciprocal. In the case of either an inversion or a translocation, the genes on the chromosomal segment involved may express themselves differently in their new environment, thus resulting in a mutation. The potential patterns of recombination are also altered by these changes, with important consequences for plants.

Changes in Chromosome Number

Mutationlike effects may also be associated with changes in chromosome number, which occur spontaneously and rather frequently but are usually eliminated promptly. Whole chromosomes may, under certain circumstances, be added to or subtracted from the basic set, or whole sets of chromosomes may be added in the phenomenon of polyploidy, the duplication of whole sets of chromosomes. We shall discuss the evolutionary importance of polyploidy in plants in Chapter 9. In any of these cases, associated alterations of the phenotype may occur.

THE EVOLUTIONARY EFFECTS OF MUTATIONS

When a mutation occurs in a predominantly haploid organism, such as the fungus *Neurospora* or a bacterium, the phenotype associated with this mutation is immediately exposed to the environment. If favorable, the mutation tends to increase in the population as a result of natural selection; if unfavorable, as is usually the case, it is quickly eliminated from the population. Other mutations may be more nearly neutral and then may persist by chance alone, but most are either negative or positive in their effect on the organism in which they occur. In a diploid organism, the situation is very different. Each chromosome and every gene is present in duplicate, and a mutation on one of the homologs, even if it would be unfavorable in a double dose, may have much less effect or even be advantageous when present in a single dose. For this reason, such a mutation may persist in the population. The mutant gene may eventually alter its function, or the selective forces on the population may change in such a way that the effects of the mutant gene become advantageous.

Although most mutations are harmful, the capacity to mutate is extremely important, for it allows the individuals in a species to vary and to adapt to changing conditions. Mutations thus provide the basis for evolutionary change. Mutations in eukaryotes occur spontaneously at a rate of about 5×10^{-6} per locus per cell division (that is, 1 mutant gene at a given locus per 200,000 cell divisions); this, together with recombination, provides abundant variation of the sort that is necessary for evolutionary change through natural selection.

The Determination of the Phenotype

The appearance of any organism—its phenotype—is the result of many complex, interacting biochemical processes. Some of these processes occur during the course of development of the organism, and some continue at any stage in its life. In plants especially, these interactions never cease. Simple changes at the level of DNA can have effects upon the final appearance of an organism that are both complex and largely unpredictable. When the science of genetics was relatively new, genes were described in terms of their more obvious phenotypic effects, that is, as purple-petaled, hairy, and so forth. It was eventually found, however, that single genes could control whole complexes of traits, a phenomenon known at **pleiotropy,** and that most traits are controlled by the combined effects of several or many genes, a phenomenon known as **epistasis.** In epistatic interactions, one gene modifies the phenotypic expression of another, nonallelic gene. In fact, no gene ever acts in isolation; its effects are always modified by the internal environment, which is, of course, the end product of the interaction of thousands of genes.

We have come a long way in our understanding of the mechanism of heredity and of the ways in which the characteristics of organisms are produced. Much more remains to be learned, however, and the field will remain one of central importance to the science of botany for many years to come.

The Molecular Basis of Inheritance

By the end of the nineteenth century, biologists knew that heredity is associated with the nucleus in cells and, in particular, with the chromosomes. It was not until the early 1950s, however, that DNA was generally accepted as the substance of which genes are made. Before then, many biologists believed that only proteins had the capacity to carry the necessary genetic information.

Some say that the twentieth century will be remembered as the time in history when human beings first reached the moon, but others predict that it will be best known as the age in which we discovered the nature of DNA and thus began to unravel the secrets of heredity.

The Nature of DNA

In 1951, the American geneticist James D. Watson went to England, where he arranged to work with Francis Crick of the Cavendish Laboratory in Cambridge. Watson and Crick were among the scientists who were convinced that DNA, rather than protein, is the genetic material.

Watson and Crick had two types of information with which to work as they set out to determine how the DNA molecule is put together. First, they knew that for DNA to be the genetic material, it had to meet at least four requirements:

1. It must carry genetic information from cell to cell and from generation to generation; furthermore, it must carry a great deal of information.

2. It must copy itself, for the chromosome does this before every cell division; moreover, it must replicate itself with great precision.

3. On the other hand, a gene must sometimes change, or mutate.

Table 8–3 *Composition of DNA in Several Species**

Source	Purines		Pyrimidines	
	Adenine	Guanine	Cytosine	Thymine
Human being	30.4	19.6	19.9	30.1
Ox	29.0	21.2	21.2	28.7
Salmon sperm	29.7	20.8	20.4	29.1
Wheat germ	28.1	21.8	22.7	27.4
Escherichia coli	24.7	26.0	25.7	23.6
Sheep liver	29.3	20.7	20.8	29.2

* In moles per 100 gram-atoms, percent; after Chargaff, Erwin, *Essays on Nucleic Acids*, 1963.

4. It must have some mechanism for "reading out" the stored information and translating it into action in the living individual.

Watson and Crick were well aware that the DNA molecule could be the genetic material only if it could be shown to have the size, configuration, and complexity required to code the tremendous store of information needed by living things and to make exact copies of this code.

The second type of information that Watson and Crick used came from previous biochemical studies of the DNA molecule. Among these data were the following:

1. The molecule is very large and also long and thin.

2. The three components (a nitrogenous base, a sugar, and a phosphate) are arranged in nucleotides, as shown in Figure 3–20 on page 60.

3. According to X-ray diffraction studies conducted by Rosalind Franklin and Maurice Wilkins at King's College in London, the long molecule of DNA is composed of regularly repeating units that appear to be arranged in a spiral or, more accurately, a helix.

4. The data shown in Table 8–3 indicate that the ratio of nucleotides containing adenine to those containing thymine is 1 to 1, as is the ratio of nucleotides containing guanine to those containing cytosine.

Watson and Crick did not perform experiments in the usual sense but, rather, assembled all the facts then known about DNA into a meaningful whole. They worked with the accumulated data, with measurements provided by the X-ray diffraction photographs of Franklin and Wilkins, and with a tin model that they attempted to construct in a way consistent with the physical and chemical data about DNA (Figure 8–15a).

(a)

8-15

(a) Watson (left) and Crick with their tin model of DNA. (b) A representation of the Watson-Crick concept of the DNA molecule. On the left, the molecule is drawn in side view with the axis indicated by the vertical rod. The molecule consists of two polynucleotide chains coiled together to form a double helix with an overall diameter of 2 nanometers (20 angstroms). The molecule coils to the right.

The two chains that make up the double helix are composed of nucleotides containing deoxyribose sugar residues (S) in which the sugar of each nucleotide is linked by a phosphate group (P) to the sugar of the adjacent nucleotide. The regular sequence of sugars and phosphates forms the backbone of the molecule. The sugars of each chain project into the cylinder formed by the coiling of the two chains.

The nitrogenous base pairs (indicated by heavy horizontal lines) are flat molecules occupying the central area in the cylinder (the dashed rectangles in the cross sections on the right). The base pairs, represented by thymine (T) and adenine (A) at level A and by cytosine (C) and guanine (G) at level B, are linked by hydrogen bonds. The broken circular line in the cross sections indicates the outer edge of the double helix when the molecule is viewed end on.

The bases are stacked above one another at intervals of 3.4 angstroms and are rotated 36° with each step. Thus, there are 10 base pairs per complete turn of the helix. As a result of this rotation, the successive side views of the base pairs appear as lines of varying lengths, depending on the viewing angle.

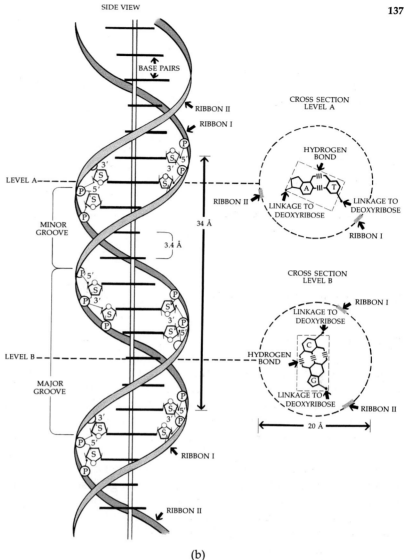

(b)

THE DOUBLE HELIX

By piecing together the various data, Watson and Crick were able to deduce that DNA is not a single-stranded helix, as are many proteins, but is instead a huge, entwined double helix. Imagine the banister of a spiral staircase, which forms a single helix. If a ladder were twisted into the shape of a helix, keeping the rungs perpendicular to the sides, a crude model of a double helix would be formed.

In a DNA molecule, the two sides are made up of alternating deoxyribose sugar molecules and phosphate groups (Figures 8–15b and 8–16). The rungs of the ladder are formed by the nitrogenous bases—adenine (A), thymine (T), guanine (G), and cytosine (C)—one base for each sugar-phosphate, with two bases forming each rung. The paired bases—the rungs of the ladder—are joined together by hydrogen bonds and are always purine-pyrimidine combinations (Figure 8–16). For this reason, the ratio of purines to pyrimidines in a molecule of DNA is always 1 to 1.

8–16

The double-stranded structure of a portion of the DNA molecule. Each nucleotide consists of a sugar (deoxyribose), a phosphate group, and a nitrogenous base (a purine or a pyrimidine). Note the repetitive sugar-phosphate-sugar-phosphate sequence that forms the backbone of the molecule. Each phosphate group is attached to the 5′ carbon of one sugar and to the 3′ carbon of the sugar in the adjacent nucleotide, so that the strand has a 5′ end and a 3′ end. The bridges formed by the phosphate groups between the nucleotides in the two strands run in opposite directions; that is, the strands are antiparallel. (Compare this figure with Figure 8–15b on the preceding page.)

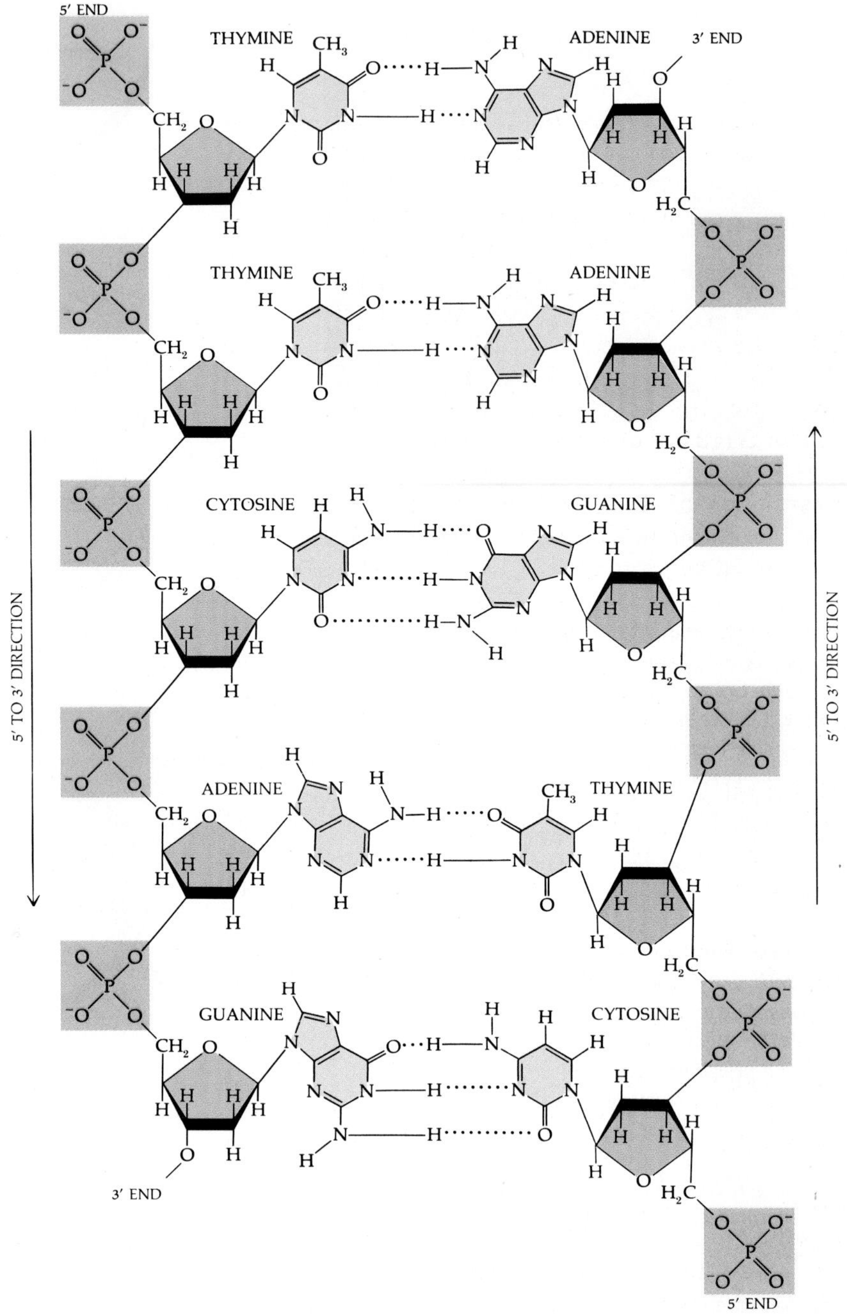

Watson and Crick noticed that the nucleotides along any one strand of the double helix could be assembled in any order, such as ATGCGTACATT and so on. Because a DNA molecule may be several thousand nucleotides long, a wide variety of arrangements is possible. The number of paired bases ranges from about 5000 for the simplest virus up to an estimated 5,000,000,000 in the 46 chromosomes of human beings. In a single human cell, the DNA—which, if extended in a single thread, would be about 1.5 meters long—contains an amount of information equal to some 600,000 printed pages averaging 500 words each, the equivalent of a library of about 1000 books. In short, the DNA molecule does indeed have the capacity to store the necessary genetic information.

THE MOLECULE THAT COPIES ITSELF

The most exciting discovery came when Watson and Crick set out to construct the matching DNA strand. In doing this, they encountered an interesting and important restriction: not only could purines not pair with purines and pyrimidines not pair with pyrimidines, but adenine could pair only with thymine, and guanine could pair only with cytosine. Only these two combinations of nitrogenous bases form the correct hydrogen bonds; adenine forms two hydrogen bonds with thymine, and guanine forms three hydrogen bonds with cytosine.

Now look again at Table 8–3. The Watson-Crick model explains the base composition of DNA in a simple, logical way. Perhaps the most important property of the model is that the two strands are *complementary*, that is, each strand contains a sequence of bases that is complementary to the sequence of bases in the other strand. When the DNA molecule reproduces itself, it simply "unzips" down the middle; the nitrogenous bases break apart at the hydrogen bonds (Figure 8–17). The two strands separate, and new strands form along each old one. If a T (thymine) is present on the old strand, only an A (adenine) in the new strand can pair with it; a G (guanine) will pair only with a C (cytosine), and so on. In this way, each strand forms a complementary copy of itself, and two exact replicas of the original molecule are produced. The age-old question of how hereditary information is duplicated and passed on for generation after generation was, in principle, answered by the discovery of the structure of DNA.

In what could be considered one of the great understatements of all time, Watson and Crick wrote in their original brief publication: "It has not escaped our notice that the specific pairing we have postulated immediately suggests a possible copying mechanism for the genetic material." In 1962, nine years after their original hypothesis was published, Watson, Crick, and Wilkins shared a Nobel Prize in recognition of their landmark studies.

Subsequently, it has been shown that because the DNA molecule is very long, double-stranded, and helical, the replication process is complex, requiring many enzyme-catalyzed steps. Initiation of **DNA replication** always begins at a specific nucleotide sequence known as *origin of replication*. Initiation requires special initiator proteins and enzymes known as *helicases*, which break the hydrogen bonds linking the complementary bases at the origin of replication, opening up the helix so replication can occur. The synthesis of new strands is catalyzed by enzymes known as **DNA polymerases**. When viewed with an electron microscope, the localized regions of replication appear as Y-shaped structures, which are called **DNA replication forks.**

DNA replication is further complicated by the configuration of the sugar-phosphate-sugar attachments in the DNA molecule. In each strand, the phosphate group that joins two deoxyribose molecules is attached to one sugar at the 5′ position (carbon 5 of deoxyribose) and to the adjacent sugar at the 3′ position (the third carbon in the ring of deoxyribose). This configuration gives each strand a 5′ end and a 3′ end. Moreover, the two strands run in opposite directions; that is, the direction from the 5′ end to the 3′ end of each strand is opposite. The strands are said to be **antiparallel** (Figure 8–16).

DNA polymerase synthesizes new DNA strands only in the 5′ to 3′ direction. Along one of the original strands, therefore, DNA can be synthesized continuously in the 5′ to 3′ direction as a single unit. This

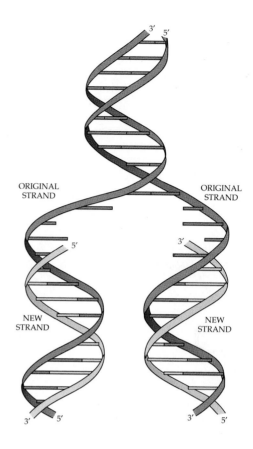

8–17
Replication of the DNA molecule as predicted by the Watson-Crick model. The strands separate down the middle as the paired bases separate at the hydrogen bonds. Each of the original strands then serves as a template along which a new, complementary strand forms from nucleotides available in the cell. Subsequent research has resulted in modification of some of the details of this process, as we shall see shortly, but the underlying principle is unchanged.

newly synthesized DNA is known as the **leading strand.** Because of DNA's antiparallel nature, the other strand is synthesized, again in the 5′ to 3′ direction, as a series of fragments that are each individually synthesized in a direction opposite to the overall direction of replication. As the fragments, known as *Okazaki fragments*, lengthen, they are connected to the growing strand. The overall result is a strand that grows in the same direction as the leading strand, but at a slightly slower rate—hence, it is known as the **lagging strand.** The complex process of DNA replication is summarized in Figure 8–18.

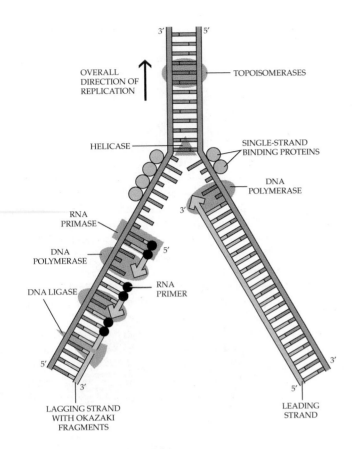

OVERALL
DIRECTION OF
REPLICATION

TOPOISOMERASES

HELICASE

SINGLE-STRAND
BINDING PROTEINS

DNA
POLYMERASE

RNA
PRIMASE

DNA
POLYMERASE

DNA LIGASE

RNA
PRIMER

LAGGING STRAND
WITH OKAZAKI
FRAGMENTS

LEADING
STRAND

8–18

A summary of DNA replication. The two strands of the DNA double helix separate, and new complementary strands are synthesized in the 5′ to 3′ direction, using the original strands as templates. The beginning of a new strand is provided by a primer formed of nucleotides of RNA. Synthesis of the leading strand is continuous, while synthesis of the lagging strand is discontinuous. The short DNA molecules of the lagging strand are called Okazaki fragments. The synthesis of the RNA primers is catalyzed by an enzyme known as RNA primase. The synthesis of each Okazaki fragment ends when the DNA polymerase runs into the RNA primer attached to the 5′ end of the previous fragment. Following replacement of the RNA primer of the previous Okazaki fragment by DNA nucleotides, the fragment is joined to the growing strand by the enzyme DNA ligase. Enzymes called topoisomerases prevent DNA tangling during replication, and the separated strands of the replication fork are stabilized by single-strand binding proteins.

How Do Genes Work?

Watson and Crick disclosed the chemical nature of the gene and suggested the way in which it duplicated itself. However, one question remained unanswered: How does the nucleotide sequence in DNA specify the sequence of amino acids in a protein?

The search for the answer to this question led to **ribonucleic acid (RNA),** the sister molecule of DNA (discussed in Chapter 3). RNA's involvement was long suspected because cells that are synthesizing large amounts of protein invariably contain large amounts of RNA. Moreover, unlike DNA, which is found mostly in the nucleus, RNA is found mostly in the cytoplasm, where protein synthesis takes place.

As it turned out, not one but three kinds of RNA play roles as intermediates in the steps that lead from DNA to proteins: messenger RNA (mRNA), transfer RNA (tRNA), and ribosomal RNA (rRNA).

TRANSCRIBING RNA FROM DNA

Molecules of RNA are synthesized by a process known as **RNA transcription,** which is similar to DNA replication (Figures 8–19 and 8–20). In eukaryotic cells, there are different enzymes, called *RNA polymerases,* for transcription of mRNA, tRNA, and rRNA. Each single-stranded RNA molecule is formed along one strand of the DNA helix by the same base-pairing copying principle that applies to the formation of a new strand of DNA. The presence of adenine in the parent DNA strand leads to the insertion of a uracil in the RNA strand. Specific nucleotide sequences of DNA, called *promoters,* are the start signals for RNA synthesis, and other sequences, called *terminators,* are the stop signals for RNA synthesis. Like a strand of DNA, each RNA molecule has a 5′ end and a 3′ end, and, as in the synthesis of DNA, the nucleotides are added one at a time to the 3′ end of the growing RNA strand.

MESSENGER RNA AND THE GENETIC CODE

Messenger RNA transcripts are working copies of the genetic information. Incorporating the instructions encoded in DNA, mRNA dictates the sequence of amino acids in proteins.

In prokaryotic cells, most proteins are encoded by a single continuous segment of DNA sequence from which a functional mRNA molecule is copied (Figure 8–21). In eukaryotic cells, however, most genes have coding sequences, called **exons,** interrupted by noncoding sequences called **introns** (Figure 8–21). During transcription, the entire length of the gene is copied, resulting in a very long RNA molecule, the *primary tran-*

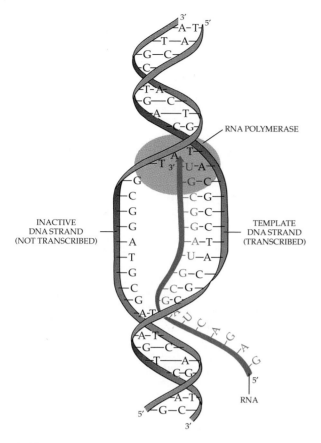

8–19

A schematic representation of RNA transcription. At the point of attachment of the enzyme RNA polymerase, the DNA opens up, and, as the RNA polymerase moves along the DNA molecule, the two strands of the molecule separate. Nucleotide building blocks are assembled in RNA in a 5' to 3' direction. Note that the RNA strand is complementary—not identical—to the template strand from which it is transcribed; its sequence is, however, identical to that of the inactive (untranscribed) DNA strand, except for the replacement of thymine (T) by uracil (U).

8–20

A bacterial gene in action. In the micrograph, you can see several different messenger RNA (mRNA) strands (shown in color in the diagram) being transcribed simultaneously from the same DNA template. The longest one, at the left, was the first one synthesized. As each mRNA strand peels off the DNA molecule, ribosomes attach to the mRNA, translating its encoded information into protein. You can also see molecules of RNA polymerase, the enzyme that catalyzes the transcription of RNA from DNA. The RNA polymerase molecule at the far right is approximately at the point where transcription begins. The protein molecules are not visible in this micrograph.

script. Before this molecule leaves the nucleus, the intron sequences are removed by special "RNA processing enzymes," and the remaining exon sequences are spliced together. After this process, called *RNA splicing,* has been completed, the RNA molecule passes from the nucleus into the cytoplasm as mRNA.

In eukaryotes, therefore, the mRNA that serves as a template for protein synthesis is actually spliced together from segments of the mRNA precursor (primary transcript) that is transcribed directly from chromosomal DNA. The chain of information in both bacteria and eukaryotes flows from gene to mRNA molecule to protein—the "central dogma" of genetics. In eukaryotes, however, the processing of transcripts adds an extra level of complication: gene ⟶ primary transcript ⟶ mRNA ⟶ protein.

Messenger RNA molecules are large, ranging in size from a few hundred to about 10,000 nucleotides. During protein synthesis, each sequence of three nucleotides—called a **codon**—specifies a single amino acid. The so-called **genetic code,** by which the sequence of nucleotides in mRNA is translated into amino acids in the course of protein synthesis, consists of 64 codons (Figure 8–22). Of the 64 codons, 61 specify particular

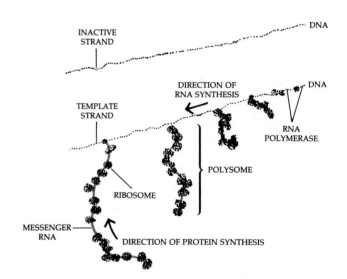

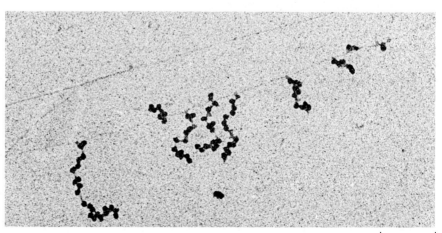

0.25 μm

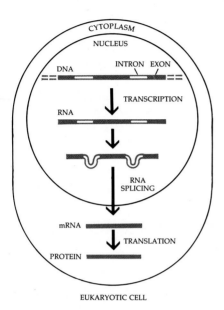

DNA
TRANSCRIPTION
mRNA
TRANSLATION
PROTEIN

PROKARYOTIC CELL

CYTOPLASM

NUCLEUS

DNA   INTRON EXON

TRANSCRIPTION

RNA

RNA
SPLICING

mRNA

TRANSLATION

PROTEIN

EUKARYOTIC CELL

8–21
The transfer of information from DNA to protein by means of mRNA is simpler in prokaryotic cells than in eukaryotic cells. In eukaryotic cells, the coding regions (exons) of the DNA are interrupted by noncoding regions (introns). The introns must be removed enzymatically and the exons spliced together in order to form the mRNA.

	SECOND LETTER				
	U	C	A	G	
U	UUU ⎫ phe UUC ⎭ UUA ⎫ leu UUG ⎭	UCU ⎫ UCC ⎪ ser UCA ⎪ UCG ⎭	UAU ⎫ tyr UAC ⎭ UAA stop UAG stop	UGU ⎫ cys UGC ⎭ UGA stop UGG trp	U C A G
C	CUU ⎫ CUC ⎪ leu CUA ⎪ CUG ⎭	CCU ⎫ CCC ⎪ pro CCA ⎪ CCG ⎭	CAU ⎫ his CAC ⎭ CAA ⎫ gln CAG ⎭	CGU ⎫ CGC ⎪ arg CGA ⎪ CGG ⎭	U C A G
A	AUU ⎫ AUC ⎪ ile AUA ⎭ AUG met	ACU ⎫ ACC ⎪ thr ACA ⎪ ACG ⎭	AAU ⎫ asn AAC ⎭ AAA ⎫ lys AAG ⎭	AGU ⎫ ser AGC ⎭ AGA ⎫ arg AGG ⎭	U C A G
G	GUU ⎫ GUC ⎪ val GUA ⎪ GUG ⎭	GCU ⎫ GCC ⎪ ala GCA ⎪ GCG ⎭	GAU ⎫ asp GAC ⎭ GAA ⎫ glu GAG ⎭	GGU ⎫ GGC ⎪ gly GGA ⎪ GGG ⎭	U C A G

FIRST LETTER / THIRD LETTER

8–22
The genetic code, consisting of 64 codons (triplet combinations of bases) and their corresponding amino acids (see Figure 3–15, page 56). Of the 64 codons, only 61 specify particular amino acids. The other three codons are "stop signals," which cause the chain to terminate. Since 61 triplets code 20 amino acids, there obviously must be "synonyms"; for example, leucine has six codons. Most of the synonyms differ only in the third nucleotide. The code is shown here as it would appear in the mRNA molecule.

amino acids; the other three are termination (stop) codons. Since proteins contain 20 different amino acids, some amino acids must be specified by more than one codon; hence, the genetic code is said to be *degenerate*.

One of the most remarkable discoveries of molecular biology is that the genetic code is identical in all organisms, with a very few exceptions that concern only small details. Genes from bacteria can function perfectly, under the proper circumstances, in mammalian cells. Also, plant genes, at the proper stage of their expression, can be introduced into bacteria, where the plant genes can then synthesize their own products. Not only does this observation provide a striking demonstration of the fact that all life on earth has had a common origin, it also provides the basis for the techniques of genetic engineering, which hold such promise for human progress in the future (see Chapter 25).

TRANSFER RNA

Transfer RNA molecules are sometimes called "the dictionary of the language of life" because of the role they play in the translation of mRNA into the language of proteins. Transfer RNA molecules are relatively small, consisting of about 80 nucleotides that form a long, single strand that folds back on itself (Figure 8–23). There are more than 20 different tRNA molecules in every cell, at least one for each of the 20 amino acids found in proteins.

Each tRNA molecule has two important attachment sites. One of these sites, the **anticodon,** binds to the codon on an mRNA molecule, and the other, at the 3' end of the tRNA molecule, attaches to a particular amino acid. Attachment of tRNA molecules to their amino acids is brought about by enzymes known as aminoacyl-tRNA synthetases. There are at least 20 different aminoacyl-tRNA synthetases, one or more for each amino acid. These enzymes are key elements in the translation of the genetic message; they determine

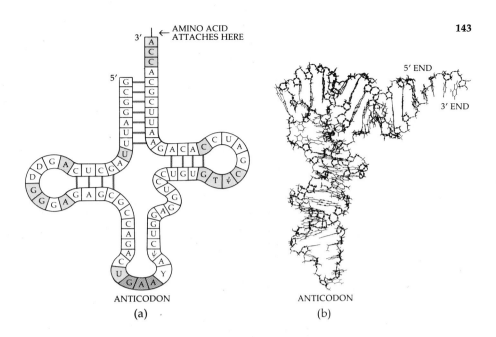

8–23

(a) *The two-dimensional structure of a tRNA molecule. Such molecules consist of about 80 nucleotides linked together in a single chain. The chain always terminates in a CCA sequence. An amino acid can link to its specific tRNA at this end. Some nucleotides are the same in all tRNAs; these are shown in gray. The other nucleotides vary according to the particular tRNA. The symbols D, γ, ψ, and T represent unusual modified nucleotides characteristic of tRNA molecules. The significance of these unusual nucleotides is that they prevent certain hydrogen bonds from forming while promoting other cross-chain associations, thus determining the final folding of the molecule.*

Each portion of the tRNA molecule appears to have a unique function. The acceptor end is where the amino acid is bound. The right-hand loop, known as the TψC loop, is the same in every tRNA molecule; it probably governs the binding of tRNA to ribosomes. The left-hand loop, the DHU loop, in contrast, differs from one tRNA molecule to another; it may be involved in determining which amino acid is enzymatically added to the acceptor end. Some of the nucleotides are hydrogen-bonded to one another, as indicated by the colored lines. The unpaired nucleotides at the bottom of the diagram (indicated in color) are known as the anticodon. They serve to "plug in" the tRNA molecule to an mRNA codon.

(b) *The molecule folds over on itself, producing this three-dimensional structure. This is a photograph of a model.*

8–24

Ribosomes, in both prokaryotes and eukaryotes, consist of two subunits, one large and one small. Each subunit is composed of specific rRNA and protein molecules. Studies of rates of sedimentation in the ultracentrifuge have revealed that the size and density of both the subunits and the whole ribosome differ in prokaryotes and eukaryotes. (a) The subunits of the Escherichia coli *ribosome have sedimentation values of 50S and 30S; these values, which are a function of both molecular weight and the shape of the molecule, are not additive. The prokaryotic 30S and 50S ribosomal subunits combine to form a ribosome that has a sedimentation value of only 70S. (b) and (c) Two views of the three-dimensional structure of the* E. coli *ribosome, as revealed by electron micrographs.*

which amino acid will be associated with which tRNA (and thus with which anticodon).

RIBOSOMAL RNA AND RIBOSOMES

As the name implies, **ribosomal RNA** is associated with **ribosomes,** which are large complexes of RNA and protein molecules. Ribosomes consist of two subunits, one small and the other large, each composed of specific rRNA molecules and proteins (Figure 8–24). Functionally, ribosomes are large protein-synthesizing machines on which tRNA molecules position themselves in precise relationship to the mRNA molecules so as to read accurately the genetic message encoded in the mRNA. Each ribosome has two sites involved in protein synthesis: an *A (aminoacyl) site,* to which the incoming tRNA, with its amino acid, binds, and a *P (peptidyl) site,* to which the growing polypeptide chain is attached.

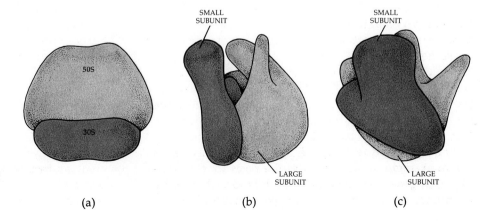

TRANSLATING mRNA INTO PROTEIN

The synthesis of protein is known as **translation** because it involves the transfer of information from one language (nucleotides) to another (amino acids). Typically, protein synthesis consumes more energy than any other biosynthetic process. The major stages in translation are initiation, elongation of the polypeptide chain, and chain termination (Figure 8–25).

Initiation begins when the smaller ribosomal unit attaches to a strand of mRNA near its 5′ end, exposing its first codon, the *initiation codon*. The first tRNA then pairs with the initiation codon of the mRNA in an antiparallel fashion: the initiation codon usually is (5′)–AUG–(3′) and the tRNA anticodon is (3′)–UAC–(5′). The initiator tRNA, which binds to the AUG codon, carries a modified form of the amino acid methionine (*N*-formyl methionine, or fMet), the amino acid that starts the polypeptide chain. The combination of the small ribosomal subunit, mRNA, and the initiator tRNA is known as the *initiation complex*. The larger ribosomal subunit then attaches to the smaller subunit, resulting in the fMet–tRNA being locked in place at the P site and the A site being made available for an incoming aminoacyl tRNA. The energy for this step is provided by the hydrolysis of guanosine triphosphate (GTP).

At the beginning of the **elongation** stage, the sec-

8–25

Three stages in protein synthesis. (a) Initiation. The smaller ribosomal subunit attaches to the 5′ end of the mRNA molecule. The first tRNA molecule, bearing the modified amino acid fMet, plugs into the AUG initiation codon on the mRNA molecule. The larger ribosomal subunit locks into place, with the tRNA occupying the P (peptidyl) site. The A (aminoacyl) site is vacant. The initiation complex is now complete. (b) Elongation. A second tRNA with its attached amino acid moves into the A site, and its anticodon plugs into the mRNA. A peptide bond is formed between the two amino acids brought together at the ribosome. At the same time, the bond between the first amino acid and its tRNA is broken. The ribosome moves along the mRNA chain in a 5′ to 3′ direction, and the second tRNA, with the dipeptide attached, is moved to the P site from the A site as the first tRNA is released from the ribosome. A third tRNA moves into the A site, and another peptide bond is formed. The growing peptide chain is always attached to the tRNA that is moving from the A site to the P site, and the incoming tRNA bearing the next amino acid always occupies the A site. This step is repeated over and over until the polypeptide is complete. (c) Termination. When the ribosome reaches a termination codon (in this example, UGA), the polypeptide is cleaved from the last tRNA and the tRNA is released from the P site. The A site is occupied by a release factor that triggers the dissociation of the two subunits of the ribosome.

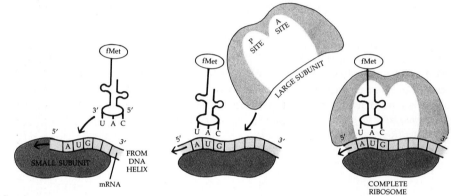

(a) INITIATION

(b) ELONGATION

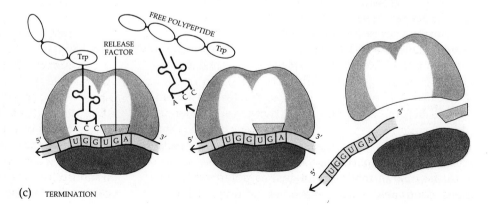

(c) TERMINATION

ond codon of the mRNA is positioned opposite the vacant ribosomal A site. A tRNA with an anticodon complementary to the second mRNA codon plugs into the mRNA and, with its amino acid, occupies the A site of the ribosome. With both P and A sites occupied, an enzyme, **peptidyl transferase,** which is part of the larger ribosomal subunit, forges a peptide bond between the two amino acids, attaching the first amino acid (fMet) to the second. The first tRNA is released from the ribosome and reenters the cytoplasmic tRNA pool. The ribosome moves one more codon down the mRNA molecule; consequently, the second tRNA, to which the fMet and second amino acid are now attached, is transferred from the A to the P position. A third aminoacyl tRNA moves into the A position opposite the third codon on the mRNA, and the step is repeated. The P position accepts the tRNA bearing the growing polypeptide chain; the A position accepts the tRNA bearing the new amino acid that will be added to the chain. As the ribosome moves along the mRNA strand, the initiator portion of the mRNA molecule is freed, and another ribosome can form an initiation complex with it. A group of ribosomes translating the same mRNA molecule is known as a **polyribosome,** or **polysome** (Figure 8–26).

The cyclic translation process just described continues until **termination** occurs when one of three possible *stop codons* (UAG, UAA, or UGA) is encountered on the mRNA. There are no tRNAs that recognize these codons, and so no tRNAs will enter the A site in response to them. Instead, cytoplasmic proteins called *release factors* bind directly to any stop codon at an A site on the ribosome, and the completed polypeptide chain is freed.

POLYPEPTIDE TARGETING AND SORTING

The polypeptides encoded by nuclear genes are synthesized by a process that begins in the ground substance of the cell. A mechanism is required, therefore, to assure that each of the polypeptides eventually is directed to the right cellular "compartment." This mechanism is called **polypeptide** (or **protein**) **targeting** and **sorting.**

Shortly after polypeptide synthesis begins on ribosomes in the ground substance, one of two divergent pathways is followed. Those ribosomes involved in the synthesis of polypeptides destined for the endoplasmic reticulum (or for endoplasmic reticulum–derived membranes) become associated with the membrane of the endoplasmic reticulum early in the translational process. These polypeptides are transferred across the membrane of the endoplasmic reticulum (or are inserted into it, in the case of integral proteins) as polypeptide synthesis continues. Those ribosomes involved with the

8–26
Clusters of ribosomes reading the same mRNA strand. Such groups are called polyribosomes, or polysomes.

synthesis of polypeptides destined for the nucleus, mitochondria, chloroplasts, or peroxisomes remain free in the ground substance. The polypeptides released from these free ribosomes either remain in the ground substance or are taken up by the appropriate cellular component.

The synthesis of polypeptides destined for the endoplasmic reticulum begins with a **signal sequence** (or *signal peptide*) of hydrophobic amino acids at the amino terminus. It is the signal sequence that causes the mRNA–ribosome complex to bind to the endoplasmic reticulum. The growing polypeptide chain then is transported across the membrane into the lumen of the endoplasmic reticulum. After it has crossed the membrane, the signal sequence is cleaved from the polypeptide chain by a *signal peptidase.* Growth of the polypeptide chain continues and the polypeptide gradually assumes its final three-dimensional configuration within the lumen. When synthesis is complete, the ribosomal subunits separate from the mRNA, and the newly formed polypeptide is sequestered in the lumen of the endoplasmic reticulum.

Regulating Gene Transcription

The transcription of genes into mRNA is under elaborate cellular control. Even in bacteria, several interlocking systems operate to regulate which genes are transcribed and to what extent. Eukaryotes appear to share many of the properties of bacterial gene regulation, although less is known about the details of control mechanisms in eukaryotes.

Because there has been intensive research on some bacterial genes, we know a great deal about how their transcription is regulated. Perhaps the best-understood case is the lactose *(lac)* system of the bacterium *Esche-*

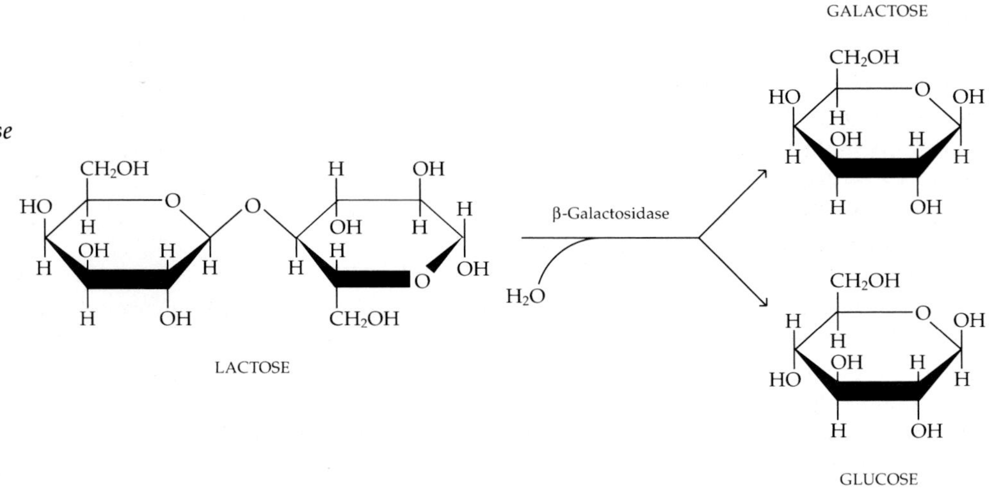

8–27

In the bacterium Escherichia coli, *the splitting of a molecule of lactose into galactose and glucose requires the enzyme β-galactosidase. β-Galactosidase is an inducible enzyme; that is, its production is regulated by an inducer—in this case, its substrate, the disaccharide lactose.*

GALACTOSE

β-Galactosidase

H_2O

LACTOSE

GLUCOSE

richia coli. Normally, these bacteria do not encounter the disaccharide lactose in their environment, so they do not synthesize the enzymes necessary to metabolize it. If lactose is added to the bacterial growth medium, the cells begin to produce large amounts of the enzyme β-galactosidase, which splits the disaccharide into glucose and galactose, thus permitting the lactose to be metabolized (Figure 8–27). In short, the presence of lactose induces the production of the enzyme molecules needed to break it down. Such enzymes are said to be **inducible.**

Many cases of such induction of enzymes by energy-rich substrates are known. In other cases, the enzymes involved in the synthesis of particular amino acids or other metabolites are not produced by the cell when that amino acid is present. Such enzymes are said to be **repressible.** This repression of enzyme synthesis by key metabolites involves the same control system as induction.

THE OPERON

Francois Jacob and Jacques Monod, working in France on studies for which they received a Nobel Prize in 1965, developed the concept of the **operon** to explain how bacterial cells regulate enzyme biosynthesis. An operon (Figure 8–28) consists of a promoter, one or more **structural genes** (that is, genes that code for proteins, often enzymes that work sequentially in a particular reaction pathway), and another DNA sequence known as an **operator.** The operator is a sequence of nucleotides located between the promoter and the structural gene or genes; the operator may overlap the promoter, the adjacent structural gene, or both. The *lac* operon of *E. coli* contains the gene for β-galactosidase and two other genes involved in lactose metabolism.

Transcription of the structural genes often depends on the activity of still another gene, the **regulator,** which may be located anywhere on the bacterial chro-

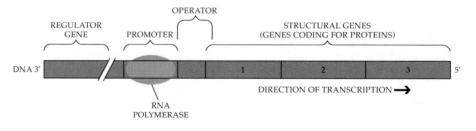

8–28

A schematic representation of an operon. An operon consists of a promoter, an operator, and structural genes (that is, genes that code for proteins, often enzymes that work sequentially in a particular reaction pathway). The

promoter, which precedes the operator, is the binding site for RNA polymerase. The operator is the site at which a repressor protein can bind; it may overlap the promoter, the first structural gene (as shown here), or both. Another gene

involved in operon function is the regulator, which codes for the repressor. Although the regulator may be adjacent to the operon, in most cases it is located elsewhere on the bacterial chromosome.

mosome. This gene codes for a protein called the **repressor,** which binds to the operator. When a repressor is bound to the operator, it obstructs the promoter. As a consequence, RNA polymerase either cannot bind to the DNA molecule or, if bound, cannot begin its movement along the molecule. The result in either case is the same: no mRNA transcription occurs. However, when the repressor is removed, transcription may begin. Evidence for the existence of the regulator gene was derived from studies of *E. coli* cells that could not stop making β-galactosidase. In these cells, a mutation in the regulator gene for the *lac* operon provided the essential clue that such a gene existed in normal cells.

The capacity of the repressor to bind to the operator and thus to block protein synthesis depends, in turn, on another molecule that functions as an **effector.** Depending on the operon, an effector can either activate or inactivate the repressor for that particular operon (Figure 8–29). For example, when lactose is present in the growth medium, the first step in its metabolism produces a closely related sugar, allolactose, that binds to and inactivates the repressor, removing it from the operator of the *lac* operon. As a consequence, RNA polymerase can begin its movement along the DNA molecule, transcribing the structural genes of the operon into mRNA. In the case of the tryptophan *(trp)* operon, the presence of the amino acid activates the repressor, which then binds to the operator and blocks the synthesis of the unneeded enzymes (Figure 8–29). Both allolactose and tryptophan—as well as the molecules that interact with the repressors of other operons—are allosteric effectors, which exert their effects by causing an alteration in the configuration of the repressor molecule.

Some 75 different operons have now been identified in *E. coli*, comprising 260 structural genes. Some are, like the *lac* operon, inducible, while others are, like the *trp* operon, repressible. Note, however, that both inducible and repressible systems are examples of negative control, since both involve repressors that turn off transcription.

Summary

Sexual reproduction involves two events: the reduction of the diploid chromosome number through meiosis and its reestablishment through fertilization.

Meiosis results in the production of either gametes or spores. A gamete is a haploid cell that fuses with another gamete, producing a diploid zygote. A spore is a cell that, without fusing with another cell, develops into a mature haploid organism.

Meiosis involves two sequential nuclear divisions and results in a total of four nuclei (or cells), each of which has the haploid number of chromosomes.

In the first meiotic division, the homologous chromosomes pair lengthwise. The chromosomes are double, each consisting of two chromatids; chiasmata form between the chromatids of the homologs. These chiasmata are the visible evidence of crossing-over—the exchange of chromatid segments between homologous chromosomes. The bivalents line up at the equatorial plane in a random manner (but with the centromeres of the paired chromosomes on either side of the plane), so that the chromosomes from the female parent and those from the male parent are completely reassorted during anaphase I. This reassortment, together with crossing-over, ensures that each of the products of meiosis differs from the parental set of chromosomes and from each other. In this way, meiosis permits the expression of the variability that is stored in the diploid genotype.

In the second meiotic division, the chromosomes divide as in mitosis.

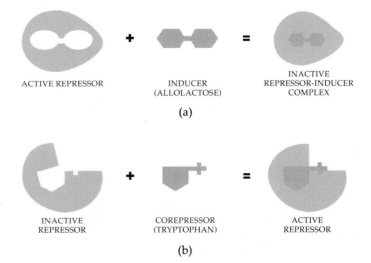

8–29

In an operon system, the synthesis of protein is regulated by interactions involving either a repressor and an inducer or a repressor and a corepressor. (a) In inducible systems, such as the lac *operon, the repressor molecule is active until it combines with the inducer (in this case, allolactose). (b) In repressible systems, such as the* trp *operon, the repressor is not active until it combines with the corepressor.*

ACTIVE REPRESSOR + INDUCER (ALLOLACTOSE) = INACTIVE REPRESSOR-INDUCER COMPLEX

(a)

INACTIVE REPRESSOR + COREPRESSOR (TRYPTOPHAN) = ACTIVE REPRESSOR

(b)

The genetic makeup of an organism is known as its genotype; its outward characteristics constitute its phenotype. In diploid organisms—a category that includes most familiar plants and animals—every gene commonly is present twice. The alternative members for each gene are known as alleles. The phenotype of diploid organisms with respect to a simple trait like those studied by Mendel is determined by the interaction of the two alleles that are present at the same locus on homologous chromosomes. Both alleles may be alike (ho-

mozygous), or they may be different (heterozygous). Consequently, mutations in diploid organisms are more difficult to detect than those that occur in haploid organisms, but they may be more important in an evolutionary sense. There are a number of different kinds of mutations: point mutations, deletions, position effects (including plasmids and transposons), inversions, translocations, and changes in chromosome number.

Although both alleles are present in the genotype, only one may be detected in the phenotype. The allele

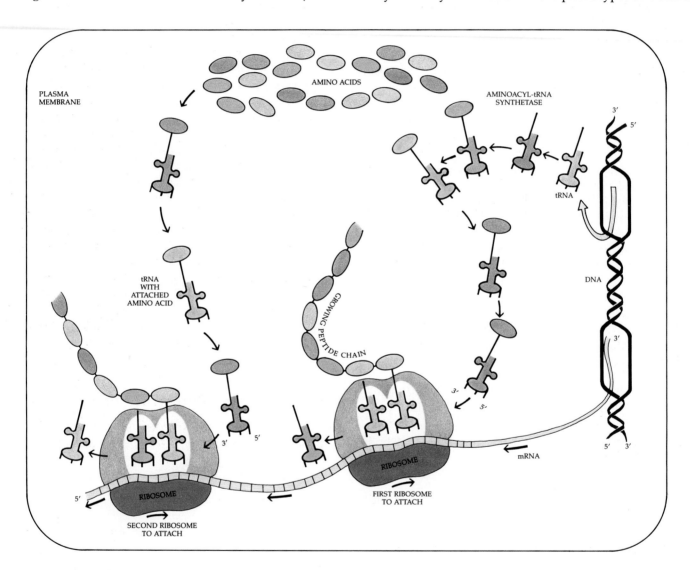

8–30

A summary of protein sythesis in a bacterial cell. At least 32 different kinds of tRNA molecules are transcribed from the DNA of the bacterial cell. These molecules are so structured that each can be attached at one end (by an aminoacyl-tRNA synthetase) to a specific amino acid. Each contains an anticodon that is complementary to an mRNA codon for that particular amino acid.

The process begins when an mRNA strand is transcribed from a DNA template. At the point of attachment to the ribosome, the matching tRNA molecule, with its amino acid, plugs in momentarily to the codon in the mRNA. As the ribosome moves along the mRNA strand, a tRNA linked to its particular amino acid fits into place and the first

tRNA molecule is released, leaving behind its amino acid, now enzymatically linked to the second amino acid by a peptide bond. As the process continues, the amino acids are brought into line one by one, following the exact order originally dictated by the DNA from which the mRNA was transcribed.

that is expressed in the phenotype is the dominant allele. The one that is concealed in the phenotype is the recessive allele. When two organisms that are each heterozygous for a given pair of alleles are crossed, the ratio of dominant to recessive in the phenotype of the offspring is 3:1. If the action of one allele is insufficient to mask the action of its alternative form (incomplete dominance), the heterozygotes are distinguishable, and the phenotypic ratio is 1:2:1.

A given characteristic is usually controlled by the interaction of the products of more than one gene, which is the major reason that pre-twentieth-century scholars found it so difficult to analyze the principles of heredity. The pattern of segregation for most features is continuous and is controlled by many genes. Some of these genes affect other genes, a phenomenon known as epistasis. Other genes affect more than one feature of the organism, a phenomenon known as pleiotropy.

The molecular key to understanding the process of heredity was provided by the early 1950s with the discovery of the role of DNA. A variety of evidence strongly indicated that the genetic information resided in DNA, and the description by James Watson and Francis Crick in 1953 of the double-helical structure of DNA, with its complementary base-pairing, provided a molecular model. This led to rapid advances in our knowledge, and the intimate workings of the cell's genetic machinery can now be described in surprising detail.

DNA is replicated by the synthesis of a complementary copy of each of the two strands of the double helix. A variety of enzymes act in concert to unravel DNA coils, unwind the double helix locally, and add new bases to each of the two separated strands.

The genetic information in the DNA is not expressed directly but is transferred via messenger RNA (mRNA). The long molecules of mRNA are assembled by complementary base-pairing along one strand of the DNA helix and then pass to the cytoplasmic ribosomes. This process, called transcription, is under rigid genetic control. Each sequence of three nucleotides in the mRNA molecule is the codon for a specific amino acid.

The genetic code has been deciphered, that is, it is now known which amino acid is called for by a given mRNA codon. Of the 64 possible triplet combinations of the four-lettered nucleotide code, 61 specify particular amino acids and three are termination codons. With 61 combinations coding for 20 amino acids, there is more than one codon for many amino acids.

Not all gene information represents information about protein sequence. Much of the genetic information in eukaryotic nuclear mRNA is transcribed from DNA segments called introns; these mRNA segments are excised from the mRNA before it reaches the cytoplasm. The remaining mRNA segments, transcribed from portions of the DNA known as exons, are spliced together in the nucleus before moving into the cytoplasm.

At the ribosomes, mRNA is combined with a series of small molecules called transfer RNAs (tRNAs), which are bound to specific amino acids. Each tRNA has a three-base sequence (anticodon) that is complementary to a codon of the mRNA. The mRNA-designated tRNA molecule fits its complementary three-base anticodon onto the next codon of the mRNA base sequence and, guided by the ribosome, transfers its specific amino acid to the end of the growing peptide chain. The amino acid is then bound to the chain by a peptide bond formed by the enzymes in the ribosome. This process (protein synthesis) is called translation because it involves the transfer of information from one language (nucleotides) to another (amino acids).

A principal means of genetic regulation in bacteria is the operon system. An operon is a linear sequence of genes coding for a group of functionally related proteins plus the promoter and operator. The structural genes of the operon are transcribed as a single mRNA molecule. Transcription from the operon is controlled by the promoter and operator sequences, which are adjacent to the structural genes and bind specific proteins. The *lac* operon is an example of an inducible operon. It is turned from "off" to "on" when an inducer binds to and inactivates the repressor. Other operons, such as the *trp* operon, are repressible. These are turned from "on" to "off" by the action of a corepressor that binds to an inactive repressor. This activates the repressor, which then binds to the operator. Both induction and repression are forms of negative regulation.

Suggestions for Further Reading

Judson, H.F.: *The Eighth Day of Creation*, Simon and Schuster, New York, 1979.

An exciting book about the discovery of the principles of heredity.

Mlot, C.: "On the Trail of Transfer RNA Identity," *BioScience* 39(11): 756–759, 1989.

Progress in the quest to learn how a single amino acid is recognized in protein synthesis.

Stahl, F.: "Genetic Recombination," *Scientific American*, February 1987, pages 90–101.

An analysis of crossing over at the molecular level.

Suzuki, D., A. Griffiths, Jeffrey H. Miller, and R. Lewontin: *An Introduction to Genetic Analysis*, 4th ed., W. H. Freeman and Company, New York, 1989.

Excellent genetics text, very widely used.

Watson, J., et al.: *Molecular Biology of the Gene*, 4th ed., The Benjamin/Cummings Publishing Company, Menlo Park, Calif., 1987.

The definitive text on DNA structure and function.

CHAPTER 9

The Process of Evolution

9–1

Charles Darwin in 1840, four years after he returned from his voyage on H.M.S. Beagle. *In his later book,* The Voyage of the *Beagle, Darwin made the following comments about his selection for the voyage: "Afterwards, on becoming very intimate with Fitz Roy [the captain of the* Beagle], *I heard that I had run a very narrow risk of being rejected on account of the shape of my nose! He . . . was convinced that he could judge of a man's character by the outline of his features; and he doubted whether anyone with my nose could possess sufficient energy and determination for the voyage. But I think he was afterwards well satisfied that my nose had spoken falsely."*

In 1831, as a young man of 22, Charles Darwin (Figure 9–1) set forth on a five-year voyage as ship's naturalist on a British navy ship, HMS *Beagle.* The book he wrote about the journey, *The Voyage of the* Beagle, is not only a classic work of natural history but also provides us with insight into the experiences that led directly to Darwin's proposal of his theory of evolution by natural selection.

At the time of Darwin's historic voyage, most scientists—and nonscientists as well—still believed in the theory of "special creation." According to this idea, each of all the many different kinds of living organisms was created (or otherwise came into existence) in its present form. Some scientists, such as Jean Baptiste de Lamarck (1744–1829), had proposed theories of evolution but could not convincingly explain the mechanism by which the process occurred.

Darwin's theory was able to bring about a great intellectual revolution simply because the mechanism for evolution he presented was so convincing that there was no longer room for reasonable scientific doubt. Particularly important in the genesis of Darwin's ideas were his experiences during a stay of some five weeks in the Galápagos Islands, an archipelago that lies in equatorial waters about 950 kilometers off the western coast of South America (Figure 9–2). There he made two particularly important observations. First, he noted that the plants and animals found on the islands, although distinctive, were similar to those on the nearby South American mainland. If each kind of plant and animal had been created separately and was unchangeable, as was then generally believed, why did the plants and animals of the Galápagos not resemble those of Africa, for example, rather than those of South America? Or, indeed, why were they not utterly unique, unlike organisms anywhere else on earth? Second, people familiar with the islands pointed out variations that occurred from island to island in such organisms as the giant tortoises. Sailors who took these

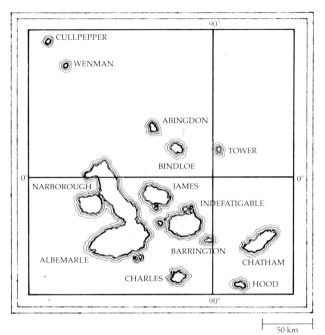

50 km

9–2

The Galápagos Islands are a small cluster of volcanic islands some 950 kilometers off the coast of Ecuador. Since they came into existence (starting a few million years ago), they have been colonized from time to time by plant and animal voyagers accidentally swept by wind or water from the mainland. Some of those organisms have managed to survive, reproduce, and adapt themselves to these bleak islands.

tortoises on board and kept them as sources of fresh meat on their sea voyages were able to tell from which island a particular tortoise had come simply by noting its appearance. If the Galápagos tortoises had been specially created, why did they not all look alike?

Darwin began to wonder if all the tortoises and other strange animals and plants of the Galápagos might not have been derived at different times from organisms that existed on the mainland of South America. Once they reached this remote archipelago, these organisms might have spread slowly from island to island, changing bit by bit in response to local conditions and eventually becoming distinct races that could be differentiated easily by a human observer.

In 1838, after returning from his historic voyage, Darwin read a book entitled *Essay on the Principles of Population*, by Thomas Malthus, an English clergyman. Published in 1798, eleven years before Darwin's birth, the book sounded an early warning about the explosive growth of the human population. Darwin saw that Malthus's reasoning was theoretically correct, not only for the human population but also for all other popula-

tions of organisms. For example, a single breeding pair of elephants—which have the slowest rate of reproduction of all animals—could give rise to a population of some 19 million individuals in 750 years, if all of their progeny were to survive long enough to reproduce. Despite this, the number of elephants on earth remains relatively constant; where there were two individuals 750 years ago, there are, in general, only two at the present time. But what determines which two elephants, out of a possible 19 million, are the surviving progeny?

In 1842, when he was 33 years old, Charles Darwin wrote out his overall argument for evolution by natural selection and then continued to enlarge and refine his manuscript for many years. The independent discovery of these principles by another English naturalist, Alfred Russel Wallace (1823–1913), who sent Darwin an essay on the subject in 1858, induced Darwin finally to publish his book, *On the Origin of Species by Means of Natural Selection, or the Preservation of Favoured Races in the Struggle for Life*. It appeared in November 1859 and immediately was recognized as one of the most influential and important books of all time.

In his book, Darwin named the process by which surviving progeny are chosen **natural selection** and presented extensive evidence that this mechanism is the principal means by which evolution occurs. He likened natural selection to **artificial selection,** the process by which breeders of domesticated plants and animals deliberately change the characteristics of the strains or races in which they are interested. They do this by allowing only those individuals with desirable characteristics to breed. Darwin recognized that wild organisms are also variable. Some individuals have characteristics that enable them to produce more progeny than others under the prevailing environmental conditions; as they do so, their characteristics gradually become more common in the population than the characteristics of individuals that produce fewer progeny. In the long run, this tendency can be counted on to produce slow but steady changes in the frequencies of different characteristics in populations; it is the process that leads to evolution.

In artificial selection, breeders can concentrate their efforts on one or a few characteristics of interest, such as fruit size or animal weight. In natural selection, however, the entire organism must be "fit" in terms of the total environment in which it lives; in other words, it is the entire phenotype that is subject to natural selection.

Evolutionary changes that result in the differentiation of major groups of organisms, such as plants or fungi, obviously require long periods of time; thus it is no coincidence that the conclusions of the geologist Charles Lyell, who demonstrated that the earth was much older than previously thought, had a profound influence on Darwin. In developing his theory of evolution, Darwin needed such an ancient earth as a stage on which to view the unfolding of the diversity of living things. The many distinctive kinds of fossils that were being discovered—fossils that, with increasing age, became more and more different from living organisms—also provided evidence important to the development of the theory of evolution. Darwin's emphasis on natural selection as the primary mechanism by which evolution occurs soon was accepted by scientists generally. In order to understand in genetic terms how it occurs, we need first to consider the ways in which genes behave in populations—the subject of our next section.

The Behavior of Genes in Populations: The Hardy-Weinberg Law

In the nineteenth century, when most biologists believed in some sort of blending inheritance (in which parental characteristics were presumed to blend together in the progeny), it was difficult to understand why rare characteristics did not simply become so diluted that they eventually disappeared. Darwin was unable to solve this problem because in his day so little was known about the mechanisms of heredity; as far as we know, he was unaware of Mendel's findings, and he certainly never referred to them. In fact, it was not until 1900, 16 years after Mendel's death, that scientists again took up the kinds of studies that he had initiated.

When early geneticists did so, they immediately were troubled by the fact that dominant genes did not drive out recessive ones, which were not expressed in heterozygous individuals. The question is clearly important to understanding evolution, because it is the genetic variability of populations that provides the "raw material" on which the process of natural selection operates. Its answer, although it was not immediately obvious, lay in a proper understanding of the particulate nature of the gene. The appropriate calculations were simultaneously presented in 1908 by G. H. Hardy, an English mathematician, and G. Weinberg, a German physician.

The **Hardy-Weinberg law,** as it is now known, states that, in a large population in which random mating occurs and in the absence of forces that change the proportions of alleles (discussed below), the original ratio of dominant alleles to recessive alleles will be retained from generation to generation.

For example, consider the alleles of a single gene— say, gene *A*. Let us make up an artificial population in which half of the individuals are homozygous *AA* and the other half are homozygous *aa* (the alternative allele). Figure 9–3 shows that the proportion of *AA* individuals (or *aa*, or *Aa*) in the third (or fourth or fifth) generation will, on the average, be the same as it was in the second generation.

For studies in population genetics, the elements of the Hardy-Weinberg law are usually stated in algebraic terms, with the fractions used in Figure 9–3 expressed as decimals. A given gene may have two, three, or more alleles present alternatively at a single locus, although thus far we have considered only situations in which there are two. When this is the case, the frequency (p) of the dominant allele plus the frequency (q) of the recessive allele must together equal the frequency of the whole; $p + q = 1$. (The frequency of an allele is simply the proportion of that allele in a population in relation to all alleles of the same gene.) This summation of the frequencies is equivalent to saying that, if there are only two alleles, *A* and *a*, of a particular gene and if half (0.50) of the alleles in the gene pool are *A*, then the other half must be *a*. Similarly, if 99 out of 100 (0.99) are *A*, then 1 out of 100 (0.01) is *a*.

How then do we find the relative proportions of individuals that are *AA*, *Aa*, and *aa*? These proportions can be calculated, keeping in mind that the frequency of *A* is p and the frequency of *a* is q, by multiplying the frequency of *A* in males by that in females (p^2), *A* in males by *a* in females (pq), *a* in males by *A* in females (pq); and *a* in males by *a* in females (q^2). The sum of these frequencies is $p^2 + 2pq + q^2$. This equals, as you may remember from the binomial theorem, $(p + q)^2$, and, since $p + q = 1$, then $(p + q)^2 = 1$. So, if half (0.50) of the gene pool is *A* and half is *a*, the proportion of *AA* (p^2) is 0.25, the proportion of *Aa* ($2pq$) is 0.50, and the proportion of *aa* (q^2) is 0.25.

$$p^2 + 2pq + q^2 = (0.50)^2 + 2(0.50)(0.50) + (0.50)^2$$
$$= 0.25 + 2(0.25) + 0.25$$
$$= 0.25 + 0.50 + 0.25$$
$$= 1$$

These proportions agree with those found by Mendel in the second (F_2) generation of crosses between homozygous dominant (for example, *AA*) and homozygous recessive (*aa*) individuals. What happens, however, in subsequent generations? As can be seen in Figure 9–4, the genotypic frequencies remain constant indefinitely. Similarly, the allele frequencies remain constant at p (for allele *A*) = 0.5 and q (for allele *a*) = 0.5, so that the Hardy-Weinberg law can be viewed as predicting a state of "genetic equilibrium."

We know that many populations do not exist in a state of equilibrium, however, for if they did there would be no changes in the representation of individual alleles and thus no evolution. Four factors operate to produce deviations from the Hardy-Weinberg equilibrium.

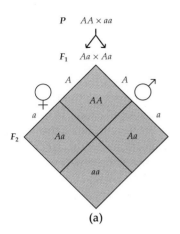

(a)

9-3

(a) *If two populations, one homozygous for a dominant allele (AA) and one homozygous for a recessive one (aa), interbreed, all of the first generation (F_1) resemble the parents with the dominant alleles, even though the progeny are heterozygous (Aa). When the F_1 generation interbreeds, however, the Mendelian (phenotypic) 3:1 ratio will appear in the F_2 generation, which, on the average, will be 1/4 AA, 1/2 Aa, and 1/4 aa.*

Random breeding of the F_2 generation results in the following probabilities of producing AA individuals in the F_3 generation: (b) If AAs crossbreed, the chances of producing AAs are 4/4 (because all

the offspring in the F_3 would be AA); this total must be multiplied by 1/16 (because the probability of AA breeding with AA is 1/4 $\times$ 1/4) for a total of 1/16. (c) If AAs breed with Aas, the chances of producing AAs are 1/2 $\times$ 1/8, or 1/16. (d) If Aas cross with AAs, the chances of producing AAs are again 1/16. (e) If Aas crossbreed, the chances of producing AAs are again 1/16. Thus, in the third generation, AA individuals, on the average, number 1/16 + 1/16 + 1/16 + 1/16, or 1/4, which is the same as in the second generation. In short, sexual recombination alone does not change the proportions of different alleles.

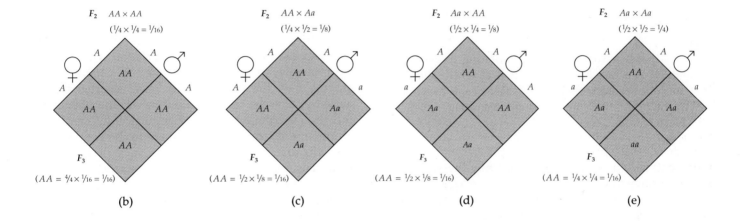

(b)   (c)   (d)   (e)

9-4

Possible combinations of gametes in a population consisting of AA, Aa, and aa individuals, illustrating the principle expressed in the Hardy-Weinberg law. The frequency of allele A is represented by p, that of allele a by q.

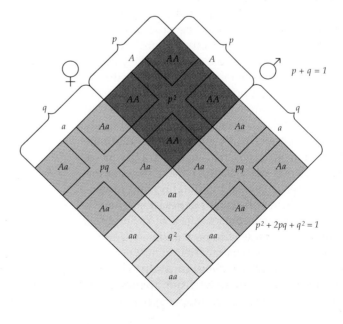

Population Size

The Hardy-Weinberg law operates only when there is a large population. In a small population, the random loss of one or more individual genotypes—by failure to breed, for example—can lead to the elimination of one or more of the alleles from the population.

Migration

Further distortion is caused by migration into or out of a particular population. If individuals with particular genetic characteristics enter or leave a population in a proportion different from that in which such individuals are already represented, the frequencies of particular alleles and genotypes will obviously change. Migration—by the dispersal of seeds, for example—is as characteristic of plants as it is of animals.

Mutation

If an allele of a particular gene is mutating to another allelic form at a rate higher than that at which the reverse mutation is occurring, the frequencies of those alleles in the population are obviously changing.

Selection

Selection is simply a term used to describe the nonrandom reproduction of genotypes. In any variable population, some individuals leave more progeny than others. As a result, certain alleles become more frequent in the population and others become less frequent. Selection is the major factor that causes deviations from the Hardy-Weinberg equilibrium, and selection is also the primary reason for evolutionary change. Mutations are, of course, the basis for variability in populations, and changes in populations occur as a result of selection acting on the variability provided by mutations.

Although selection is sometimes thought of as a creative force, it is important to remember that it is merely a description of events that have already taken place. When the proportion of a certain allele is higher in one generation than it was in a preceding generation, and in the absence of an alternative explanation, it is said that selection has taken place. Selection does not actually *cause* the changes that occur. By continuously eliminating certain alleles from a population under a given set of conditions, selection *channels* the pattern of variation in characteristics that is caused by mutation and recombination. As a result, in subsequent generations there is a proportionately higher representation of individuals that can survive better under the prevailing conditions. In unusual or marginal environments, genotypes that were poorly represented throughout most of the range of a species may be increased greatly by the process of selection.

In the case of a deleterious recessive allele, even one that causes the death of homozygous individuals, the lower the frequency of the allele in the population, the less it will be acted on by selection. Such a relationship occurs because the proportion of recessive alleles in homozygotes (the only state in which the allele is subject to selection) decreases precipitously as the frequency of the allele decreases in the population. In short, the lower the frequency of a recessive allele, the less it is exposed to the action of selection. As the frequency of the recessive allele drops, the progress toward its removal from the population also slows down.

Responses to Selection

GENETIC FACTORS

The response of a population to selection is affected by the principles of genetics, discussed in Chapter 8. In general, only the phenotype is being selected, in the sense that it is the relationship between the phenotype and the environment that determines how many offspring an individual will contribute to the next generation. Because nearly all characteristics of individual organisms are determined by the interactions of many genes, phenotypically similar individuals can have very different genotypes.

When some feature, such as tallness, is strongly selected for, there is an accumulation of alleles that contribute to this feature and an elimination of those alleles that work in the opposite direction. But selection for a polygenic characteristic is not simply the accumulation of one set of alleles and the elimination of others. Gene interactions, such as epistasis and pleiotropy (see page 135), are of fundamental importance in determining the course of selection in a population.

Another feature of importance in determining the nature of selective change is the necessity of producing an organism that "works." Strong selection for one feature may be limited in its effects because it leads to the accumulation of so many other undesirable features in the population that the organism might no longer be able to survive and reproduce. Such effects occur primarily because of linkage within the genotype.

PHENOTYPIC FACTORS

Other kinds of limits to selection are imposed by the fact that an organism must meet conflicting environmental demands. For example, in harsh alpine environments, plants must grow and photosynthesize rapidly if they are to store enough carbohydrates to survive the long, severe winters. Therefore, they must respond with renewed metabolic activity to the first sign of favorable growing conditions in spring. However, if their response is so finely tuned that they respond even to a temporary thaw in the middle of winter, they will be eliminated from the population. Desert plants are in a similar situation; their seeds must not germinate until conditions are appropriate for their survival.

An interesting example of the way in which the different features of an organism interact was detected among tropical trees of the pea family (Fabaceae) by Daniel Janzen of the University of Pennsylvania. Some of the species Janzen studied have small, numerous seeds, whereas others have larger, fewer seeds. Larger seeds are advantageous in that they provide a more abundant supply of food to the germinating seedlings, but they are also more easily preyed upon by insects, especially seed beetles of the family Bruchidae. Such beetles are often so abundant that they destroy the entire seed crop in a given year, and so present a very serious barrier to reproduction for the plants.

On investigation, Janzen found that in the legumes that had large seeds, the seeds were poisonous and thus avoided predation by beetles. Although the smaller seeds of other species were attractive to the beetles, they evidently escaped predation more efficiently because of their size. Therefore, it is reasonable to conclude, as Janzen did, that seed poisons and small seeds represent alternative responses to a common selective force—predation by seed beetles—and that either response would allow the particular species in which it had evolved to survive.

In the light of Janzen's reasonable explanation of the situation in tropical legumes, it is interesting to note that the seeds and fruits of plants on oceanic islands are often larger than those of related species on the mainland (Figure 9–5). The kinds of pests that attack the plants on the mainland are often absent on the islands. It seems likely that this provides the opportunity for the plants to produce larger fruits and seeds, ensuring a better start for their seedlings because they contain greater amounts of stored food. This isolation from pests is somewhat comparable to the way poisons protect the seeds of the mainland tropical legumes, in that it has apparently allowed large seeds to evolve in the island plants.

CHANGES IN NATURAL POPULATIONS

Under certain circumstances, the characteristics of populations may change rapidly. Particularly during the past few centuries, the influence of human beings in many areas has been so great that some populations of other organisms have had to adjust rapidly in order to survive. Evolutionary biologists have been particularly interested in examples of such rapid changes because the principles involved are presumed to be the same as those that govern changes in populations generally.

In a Maryland study, the plants growing in the ungrazed part of a pasture were found to be taller than those in the grazed part. Such differences clearly could have originated directly from the effects of the grazing. In fact, when samples of the plant species involved were taken from grazed and ungrazed parts of the pasture and grown together in a garden, it was found that some of the differences were genetic. Thus plants of white clover (*Trifolium repens*), Kentucky bluegrass (*Poa pratensis*), and orchard grass (*Dactylis glomerata*) from the grazed part of the pasture remained shorter than those from the ungrazed part. This allowed the deduction that shorter plants of these species, perhaps because they were missed during intensive grazing, were at a selective advantage under conditions of grazing, and that the alleles responsible for short stature had increased in the grazed portion of the pasture. A second example of this sort is illustrated in Figure 9–6.

In Wales, the tailings and dumps around a number of abandoned lead mines are rich in lead (up to 1 percent) and zinc (up to 0.03 percent)—substances that are highly toxic to most plants. Because of the presence of these metals, the tailings and dumps are often nearly completely bare. Observing that one species of grass, *Agrostis tenuis*, was colonizing the mine soil, scientists

9–5
Island species, isolated from their natural enemies, are often able to produce larger fruits and seeds. Shown here are two related species of Zanthoxylum, *members of the citrus family (Rutaceae); one of them is found in Hawaii, the other on the Asian mainland.*

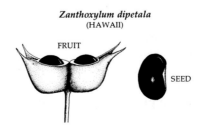

Zanthoxylum dipetala
(HAWAII)

FRUIT

SEED

Zanthoxylum ailanthoides
(ASIAN MAINLAND)

FRUIT

  SEED

(b)

(a)

9–6
*Prunella vulgaris is a common herb
of the mint family; it is widespread in
woods, meadows, and lawns in temperate
regions of the world. Most populations
consist of erect plants, such as those
shown in (a), which grow in open, often
somewhat moist, grassy places throughout
the cooler regions of the world. Populations found in lawns, however, always
consist of prostrate plants, such as those
shown in (b), growing in Berkeley, California. Erect plants of* P. vulgaris *cannot
survive in lawns because they are damaged by mowing and do not have the
capability for resprouting low branches
from the base, which would be necessary
for survival. When lawn plants are
grown in an experimental garden, some
remain prostrate whereas others grow
erect. The prostrate habit is determined
genetically in the first group and
environmentally in the second group.*

took some of the plants from these areas and others
from nearby pastures, and grew them together either in
normal soil or in mine soil. In the normal soil, the mine
Agrostis plants were definitely slower growing and
smaller than the pasture plants. On the mine soil, however, the mine plants grew normally but the pasture
plants did not grow at all. Half the pasture plants in the
mine soil were dead in three months and had mis-
shapen roots that were rarely more than 2 millimeters
long. But a few of the pasture plants (3 of a sample of
60) showed some resistance to the effects of the metal-
rich soil. They were doubtless genetically similar to the
plants originally selected in the development of the
lead-resistant strain of *Agrostis*. The mine was no more
than 100 years old, and so the lead-resistant strain had
developed in a relatively short period of time. The resistant plants were selected from the genetically variable plants found in adjacent habitats and were then
constituted, through the agency of natural selection, as
a distinct strain.

ASEXUAL REPRODUCTION AND EVOLUTION

In ways that we shall explore later in this book, new
plants are often able to arise directly from existing ones.
This is what happens, for example, when ivy touches
the soil and grows roots and when a chrysanthemum or
iris plant is divided. Reproduction of this sort, which is
called asexual reproduction, is widespread in nature
also. From an evolutionary standpoint, it is important
to note that the new plants are the products of mitosis
and thus are genetically identical to their parent.

Many kinds of plants reproduce both sexually and
asexually, so that some individuals are produced following genetic recombination, while others are produced following mitosis (Figure 9–7). Since the latter
are genetically identical to their parents, they may

9–7
Violets reproduce both sexually and asexually. The large flowers are cross-pollinated by insects, and the seeds may be carried some distance from the parent plant by ants. The smaller flowers, closer to the ground, are self-pollinating and never open. Seeds from these flowers drop close to the parent plant and produce plants that are genetically similar to the parent. Presumably such plants are better able (on the average) to grow successfully near the parent. Both of these forms of reproduction are sexual and involve genetic recombination. Asexual reproduction in violets occurs when horizontally creeping stems (either stolons or rhizomes) produce new, genetically identical plants close to the parent.

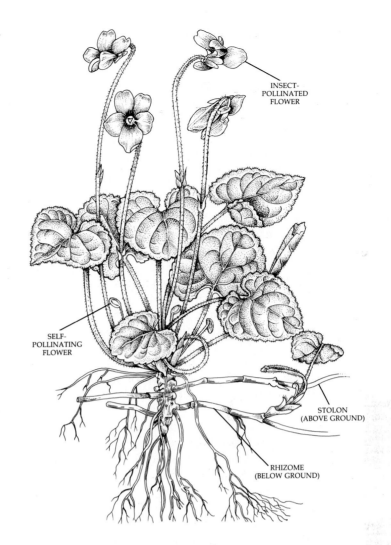

INSECT-
POLLINATED
FLOWER

SELF-
POLLINATING
FLOWER

STOLON
(ABOVE GROUND)

RHIZOME
(BELOW GROUND)

flourish locally if their genotypes are especially favored in a particular habitat. Lacking recombination and genetic variability, however, they are not able to adjust as readily to changing conditions as are plants of the same species produced sexually. Sexual reproduction was characteristic of the first plants, and asexual reproduction is, in every case, secondarily derived from it.

The Divergence of Populations

The world is physically and biologically complex, and individual species of plants almost always have discontinuous distributions. Their habitats are certainly discontinuous, whether they be lakes, streams, the tops of mountains, sunny spots in the woods, or a certain kind of soil. Given this observation, the movement of alleles between separated populations is more or less limited, and the populations then respond to the selective pressures of their local situations in different ways. Even though normal *gene flow* (the movement of alleles from one population to another as a consequence of the migration of individuals) may provide a source of variability for a distant population, it cannot completely counteract the effects of local selective forces.

The distances necessary to isolate populations effec-

tively from other populations of the same species vary with the dispersibility of the plants in question, but such distances are often quite small. For several kinds of insect-pollinated plants that grow in temperate regions, a gap of only 300 meters may effectively isolate two populations. Rarely will more than 1 percent of the pollen that reaches a given individual come from this far away. In plants whose pollen is spread by the wind, very little falls more than 50 meters from the parent plant under normal circumstances. The chance of pollen from such a plant reaching a receptive stigma on a plant that is farther away decreases very rapidly with increasing distance. This is not to say, of course, that two pine trees separated by 50 meters are out of genetic contact but, rather, that the sort of progeny that each produces will be affected much more by the demands of its local environment than by alleles introduced from a distance.

Any two separated populations diverge from one another because they are responding to different selective forces. If they regain contact, they may merge reproductively, or the differences that have accumulated between them may lead to some degree of genetic isolation. Genetic isolation, or reproductive isolation, can come about in a variety of ways, as is discussed later in this chapter.

ECOTYPIC AND CLINAL VARIATION

Developmental plasticity is the tendency of individuals to vary over time in response to different environmental conditions, or for genetically identical organisms to differ in response to different environmental stimuli. Such plasticity is much greater in plants than it is in animals because the open system of growth that is characteristic of plants can more easily be modified in various ways that produce striking differences in the way a particular genotype is expressed.

Environmental factors, as every gardener knows, can cause profound differences in the phenotypes of many species of plants. Leaves that develop in the shade, for example, may be thinner and larger than those that develop in the sun. Such differences so impressed early botanists that they were not sure whether the differences between plant populations growing in different habitats were genetically determined or simply resulted directly from different environmental factors.

This question was first attacked experimentally in the 1920s by the Swedish botanist Göte Turesson, who transplanted to his experimental gardens members of distinctive populations of certain plant species that he found in southern Sweden. In the great majority of the 31 species he investigated, the differences he had observed when the plants were growing in their original habitats were genetically controlled. The distinctive plants of these species maintained these differences regardless of the habitats in which they were grown. In a very few species, the differences he observed were the result of the direct action of the environment and were affected by changes in habitat. Differences in plant stature, season of flowering, color of leaves, and so on were usually genetically controlled. Turesson termed such genetic races, differentiated with respect to particular habitats, **ecotypes.**

Of particular interest in the study of ecotypes is the work of Jens Clausen, David Keck, and William Hiesey, which was carried out under the auspices of the Carnegie Institution of Washington at the Institution's Department of Plant Biology, Stanford, California. These scientists dealt experimentally with a number of species of plants from the western United States. They established transplant gardens at three localities in California: (1) Stanford, near sea level, with predominant winter rainfall, (2) Mather, on the western slope of the Sierra Nevada, at about 1400 meters elevation, with cold, snowy winters and hot, mostly dry summers, and (3) Timberline, near the crest of the Sierra Nevada at roughly the same latitude as the other two stations but at 3050 meters elevation, with very cold, snowy winters and short, cool, rather dry summers (Figure 9–8). Working mainly with plants that could be propagated asexually, so that genetically identical individuals could be grown at all three locations, Clausen and his colleagues expanded on Turesson's experiments.

The habitats of the western United States are more sharply distinct than those Turesson had studied in southern Sweden, and thus it is not surprising that so many plant species of the western United States have sharply defined ecotypes. One species studied by the Carnegie group was the perennial herb *Potentilla glandulosa*. A close relative of the strawberry (*Fragaria*), this herb ranges through a wide variety of climatic zones in California, and native populations occur near each of the three experimental stations mentioned above. When *P. glandulosa* plants from the different locations were grown side by side in gardens set up at the three stations, a number of ecotypic differences between the original strains became apparent. There appeared to be four distinct ecotypes of *P. glandulosa*, each with morphological characteristics that are strongly correlated with particular physiological responses critical to the survival of the ecotype in its native environment.

9–8

Transplant gardens of the Carnegie Institution, all at approximately the same latitude in central California. (a) Stanford, near sea level; (b) Mather, in the central Sierra Nevada at about 1400 meters elevation; and (c) Timberline, at 3050 meters elevation.

(a)

For example, the Coast Ranges ecotype consists of plants that grew actively in both winter and summer when cultivated at Stanford, which lies within their native range. Plants of this ecotype survived at Mather, outside their native range, even though they were subjected to about five months of cold winter weather. At Mather they became winter-dormant, but they stored enough food during their growing season to carry them through this long, unfavorable season. At Timberline, plants of the Coast Ranges ecotype almost invariably died during the first winter; the short growing season at this high elevation did not permit them to store enough food to survive the long winter. Other plants that occur in California's Coast Ranges produce ecotypes that have physiological responses comparable to those of *P. glandulosa*. Indeed, strains of unrelated plant species that occur together naturally at a given place often are more similar to one another physiologically than they are to other populations of their own species.

The physiological and morphological characteristics of ecotypes, as in *P. glandulosa*, usually have a complex genetic basis, involving dozens (or, in some cases, perhaps even hundreds) of genes. Sharply defined ecotypes are characteristic of regions, like the western United States, where the breaks between adjacent habitats are sharply defined. On the other hand, when the environment changes more gradually from one habitat to another, the characteristics of the plants that occur in that region may do likewise. A gradual change of this kind in the characteristics of populations of an organism is called a **cline.**

Clines are frequently encountered in organisms that live in the sea, where the temperature often rises or falls very gradually with changes in latitude. They are also characteristic of organisms that occur in such areas as the eastern United States, where rainfall gradients may extend over thousands of kilometers. When populations of plants are sampled along a cline, the differences are often proportional to the distance between the populations.

PHYSIOLOGICAL DIFFERENTIATION

In order to understand why ecotypes flourish where they do, we must understand the physiological basis for their ecotypic differentiation. For example, Scandinavian strains of goldenrod (*Solidago virgaurea*) from shaded habitats and exposed habitats show experimental differences in their photosynthetic response to light intensity during growth. The plants from shaded environments grew rapidly under low light intensities, whereas their growth rate was markedly retarded under high light intensities. In contrast, plants from exposed habitats grew rapidly under conditions of high light intensity but much less well at low light levels.

In another experiment, arctic and alpine populations of the widespread herb *Oxyria digyna* were studied, using strains from an enormous latitudinal gradient extending southward from Greenland and Alaska to the mountains of California and Colorado. Plants of northern populations had more chlorophyll in their leaves, as well as higher respiration rates at all temperatures, than plants from farther south. High-elevation plants from near the southern limits of the species' range carried out photosynthesis more efficiently at high light intensities than did low-elevation plants from farther north. Thus the respective races could function better in their own habitats, marked, for example, by high light intensities in high mountain habitats, lower intensities in the far north, and so forth. The existence of *O. digyna* over such a wide area and such a wide range of ecological conditions is made possible, in part, by differences in metabolic potential among its constituent populations.

(b)

(c)

REPRODUCTIVE ISOLATION

A population of plants tends to respond to selection as an integrated unit. As individual populations of a species become more and more different from one another, the characteristics that originally allowed them to interbreed with one another may change along with characteristics of more immediate importance—those that allow them to grow successfully in their different habitats. For this reason, plants from widely different populations may be unable to produce hybrids (hybrids are the offspring of genetically dissimilar parents) or, if they do, the hybrids may not be fertile. In general, the greater the difference in appearance of two populations, the less likely it is that they will be able to form hybrids.

When two populations become reproductively isolated—unable to form fertile hybrids with one another —they have no effect on each other's subsequent evolution, at least in a genetic sense. Such isolation is therefore one of the more crucial steps in the evolutionary divergence of populations and the formation of species.

Reproductive isolation has often been used as a basis for defining the category *species*. This criterion is not generally applicable, however, because the very different-looking, ecologically distinct species in some groups of plants—particularly such long-lived plants as trees and shrubs—often can form fertile hybrids with one another. In contrast, hybrids between species of herbaceous plants are often sterile, or can be formed only with difficulty. In such short-lived plants, both perennials and annuals, even the individual populations may be reproductively isolated. Because of these differences, it is apparent that reproductive isolation is not a general characteristic of plant species and should not be used to define species of plants generally.

When subjected to similar selective pressures, populations of annual plants apparently change much more rapidly than populations of long-lived plants. The speed with which the characteristics of annuals change seems to be related to their short life cycles and leads to the differences between populations of annuals and other relatively short-lived plants just mentioned. In areas where annual plants can grow successfully, such as deserts or summer-dry regions (California, for example), they may constitute a third or more of the total number of plant species present.

A number of factors, in addition to hybrid sterility or the inability to hybridize at all, tend to keep different species of plants distinct when they occur together. Some are factors that prevent the formation of hybrids in the first place. For example, two plant species capable of forming fertile hybrids may occur in the same area but in different habitats. In the eastern United States, scarlet oak (*Quercus coccinea*) can be found growing together with black oak (*Q. velutina*) over a very wide area, and the two species, which are wind-pollinated, form fertile hybrids readily in cultivation. Nevertheless, hybrids between these two species are rare in nature. In general, scarlet oaks are found in relatively moist, low areas with acidic soil, whereas black oaks are found in drier, well-drained habitats. Only where the environment has been disturbed—as by burning or cutting of the trees—do hybrids become more common, apparently because new kinds of habitats are created to which some hybrid individuals may be better suited than either of their parents.

Among the other mechanisms that can prevent the formation of hybrids between species that occur together are seasonal differences in time of flowering. If two species do not flower together, they will not hybridize in nature even when they grow side by side. Alternatively, they may be pollinated by different kinds of insects or other animals, and pollen may rarely be transferred between them. They will hybridize only when a pollinator mistakenly visits the "wrong" flower.

CLUSTERS OF SPECIES

The evolutionary processes we have been discussing lead to the production of groups of related species in many different geographic areas. The clusters of species of animals on the Galápagos Islands, especially the tortoises we mentioned earlier and the Galápagos finches, are famous because of their role in the development of Darwin's theory of evolution. The development of such patterns is known as **adaptive radiation**. (See "Adaptive Radiation in Hawaiian Tarweeds," pages 164–165.)

Differentiation on islands is particularly striking because, in the absence of competition, organisms seem more likely to produce forms that are highly unusual in comparison with those of related species that occur on the continents. In island localities, the characteristics of plants and animals may change more rapidly than on the mainland, and features that are never encountered elsewhere may arise. Similar clusters of species may also arise in mainland areas, of course, and may involve spectacular differentiation.

The Evolutionary Role of Hybridization

Even if species hybridize only rarely in nature, the hybrids that they do produce may be important because

9–9

Sycamores (Platanus), *which are called plane trees in Britain, offer examples of well-differentiated populations that have retained the ability to hybridize. Present-day species of this genus have been isolated from one another for at least 50 million years in widely scattered localities. One of these, the oriental plane (P. orientalis), is native from the eastern Mediterranean region to the Himalayas. This handsome tree has been widely cultivated in southern Europe since Roman times, but it cannot be grown in northern Europe away from the moderating influence of the sea. After the discovery of the New World, one of the North American species, P. occidentalis, was brought into cultivation in the colder portions of northern Europe, where it flourished. About 1670, these two very different trees hybridized spontaneously where they were cultivated together in England, producing the intermediate and fully fertile London plane, Platanus × hybrida. This hybrid, which is capable of growing in regions with cold winters, is now grown as a street tree throughout the temperate regions of the world.*

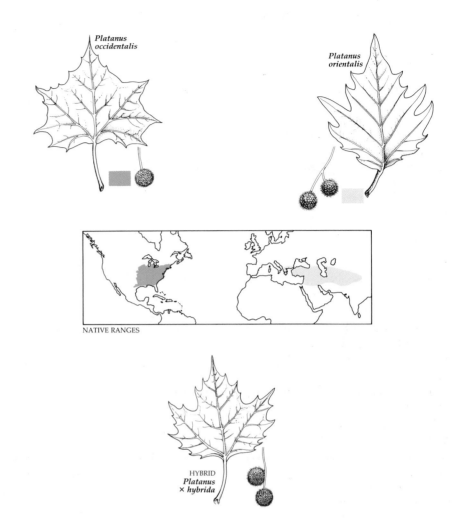

of the way in which such hybrid individuals recombine the characteristics of their parents. The environment can change repeatedly, and individuals of hybrid origin often contain new genetic combinations that are better suited to the new environment than those of either parent; the hybrids may even be able to colonize some habitat where neither parent can grow.

When the habitats of the parental species are found in the same area but are sharply distinct (as in the case of the scarlet oak and the black oak), there may be little opportunity for the establishment of hybrids, but where the habitats intergrade or are disturbed, the situation may be very different. Here the recombination of genetic material that was originally characteristic of the two differentiated species may have a greater potential for producing well-adapted offspring than would changes within a single population (Figure 9–9). The characteristics of a population of hybrids may become stabilized if the hybrids are better adapted to some par-

ticular environmental conditions than either of their parents.

Hybridization between species is an important evolutionary mechanism in many groups of plants. In some genera, the recombination of genetic material between species seems to be the chief means of developing new species that can grow in diverse habitats. These genera primarily comprise woody trees or shrubs, such as *Eucalyptus*, oaks (*Quercus*), manzanitas (*Arctostaphylos*), and mountain lilacs (*Ceanothus*; Figure 9–10). Clusters of species that interact in this way tend to be most common on islands or in ecologically diverse areas, such as California. Another example, in this case involving pollinating agents, is shown in Figure 9–11.

The establishment and stabilization of hybrid populations depend on the fertility of the hybrids. Even if the hybrids are sterile, however, they may still be able to propagate themselves, either asexually or by regaining their fertility through polyploidy.

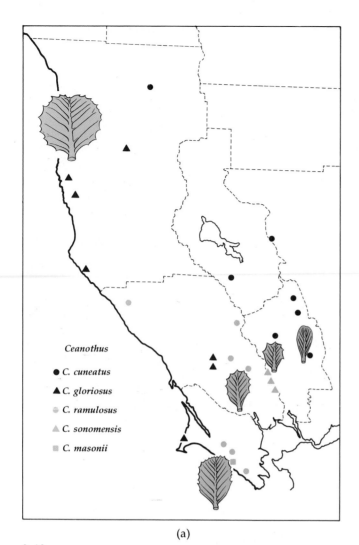

(a)

9–10

The establishment of hybrid populations is an important evolutionary mechanism in many groups of woody plants, including the mountain lilacs, Ceanothus, *shown here. (a) Map showing a portion of central California (with county lines; note San Francisco Bay at the bottom), a geologically complex region. Two relatively widespread and distinct species of mountain lilac, the coastal C. gloriosus and the interior C. cuneatus (which ranges far to the east out of the area depicted), have produced three series of hybrid populations; each is variable but stabilized and is able to grow better than either parent in the areas in which they occur together. Leaves from representative populations are shown on the map. (b) Segregation of the extremes in the progeny of an artificial cross between the parental species. In habit, some of the intermediates resemble the named intermediate species of hybrid origin shown on the map to the left. (c) Flowering branch of C. gloriosus. (d) Flowering branch of C. cuneatus.*

9–11
Richard Straw, then of California State University, Los Angeles, hypothesized that hybridization accounts for the origin of Penstemon spectabilis, *a flowering plant found in the mountains of southern California. One of the parental species,* P. grinnellii, *has broad, two-lipped, pale blue flowers that are pollinated mainly by large bees, such as carpenter bees. Another species,* P. centranthifolius, *has long, slender red flowers that are visited mostly by hummingbirds. Their suspected hybrid derivative,* P. spectabilis, *shown here, is ecologically and morphologically intermediate between the two, having rose-purple flowers. It is pollinated by a specialized family of pollen-gathering wasps, which visit neither of the parental species.*

(a)

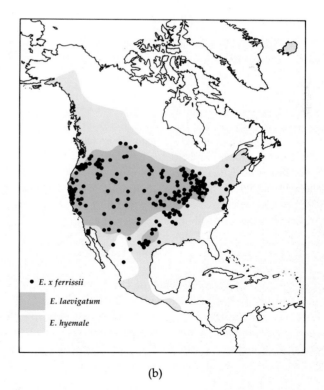

(b)

9–12
One of the most abundant and vigorous of the horsetails (see Figure 16–23, page 338) found in North America is Equisetum × ferrissii, *a completely sterile hybrid of* E. hyemale *and* E. laevigatum. *Horsetails propagate readily from small fragments of underground stems, and the hybrid maintains itself over its wide range through such vegetative propagation. (a) Stems of* E. × ferrissii, *with strobili. (b) Range of* E. × ferrissii *and those of its parental species.*

Adaptive Radiation in Hawaiian Tarweeds

Some spectacular groups of native plants and animals occur in the Hawaiian Islands. On this archipelago, whose major islands arose from the sea in isolation over millions of years, the habitats are exceedingly diverse, and the distances to mainland areas from which plants and animals migrated to colonize the islands is great. Those organisms that did reach the Hawaiian Islands often changed greatly in their characteristics as they occupied the varied habitats that were available to them. The process by which such changes occur is called *adaptive radiation.*

The striking Hawaiian silversword alliance, which includes 28 species in three closely related genera of the sunflower family, Asteraceae, is the most remarkable example of adaptive radiation in plants. These species belong to the tarweed subtribe Madiinae, most of whose members are found in California and adjacent regions. The silverswords belong to the genus *Argyroxiphium,* which is closely related to two other genera also found only in Hawaii, *Dubautia* and *Wilkesia.* The species of these genera range in habit from small, mat-forming shrubs and rosette plants to large trees and climbing vines. They grow in habitats as diverse as exposed lava, dry scrub, dry woodland, moist forest, wet forest, and bogs. In these habitats, the annual precipitation ranges from less than 40 centimeters to more than 1230 centimeters. The rainiest of these habitats is among the wettest places on earth.

This enormous variation in habitats is paralleled by significant variation in leaf sizes and shapes among these plants. For example, species of *Dubautia* that grow in bright, dry habitats usually have very small leaves, while those that inhabit the shaded understories of wet forests have much larger ones. The silversword, *A. sandwicense,* which grows on the dry alpine slopes of Haleakala Crater on the island of Maui, has leaves that are covered with a dense, silvery mat of hairs. These hairs apparently provide protection from intense solar radiation and aid in the conservation of moisture. The leaves of the closely related greensword, *A. grayanum,* which also occurs on Maui but in wet forest and bog habitats, lacks these hairs.

Important physiological differences likewise characterize the species of Hawaiian tarweeds, which must meet very different kinds of environmental challenges. Robert Robichaux of the University of Arizona has shown, for example, that the species of *Dubautia* that occur in dry habitats have a much greater tolerance to water stress than those that occur in moist to wet habitats. The species that grow in dry habitats have leaves with more elastic cell walls and under dry conditions are able to maintain higher turgor pressures than their relatives.

Despite their great diversity in appearance and the very different habitats in which they grow, all 28 species of these

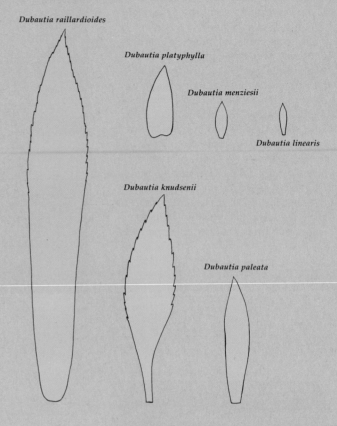

Silhouettes of leaves of six species of Dubautia, *showing the great differences in leaf shape and size between species that grow in wet habitats (larger leaves) and those that grow in dry ones.*

three genera are very closely related to one another. Any two of them can be hybridized, as far as we know, and all of the hybrids are at least partly fertile. Additionally, experimental hybrids between Hawaiian and Californian tarweeds have recently been produced, underscoring their very close relationship. Further studies of these relationships are being carried out by Gerald Carr of the University of Hawaii, Donald Kyhos of the University of California, Davis, and Bruce Baldwin of Duke University. The entire group of three genera appears certainly to have evolved in isolation on the Hawaiian Islands following the arrival of a single, original colonist from western North America.

The silversword, Argyroxiphium sandwicense, a remarkable plant that grows on the
exposed, upper cinder slopes of Haleakala
Crater on the island of Maui, where the plants
are exposed to high levels of solar radiation
and low humidities.

Dubautia reticulata. *Species of the genus
Dubautia (which includes 21 of the estimated
28 species of Hawaiian tarweeds) range from
trees and shrubs to lianas and small, matted
plants that are scarcely woody. Dubautia
reticulata, shown here, grows in the wet forests
of Maui, where it may reach a height of 8
meters or more and develop a trunk with a
diameter of nearly 0.5 meter.*

Wilkesia gymnoxiphium. *This bizarre,
yuccalike plant occurs only on the island
of Kauai, where it is restricted to dry scrub
vegetation along the fringes of Waimea Canyon. Kauai is the oldest of the main islands of
the Hawaiian archipelago. Robert Robichaux,
shown here, is studying the physiological
ecology of this fascinating group of plants.*

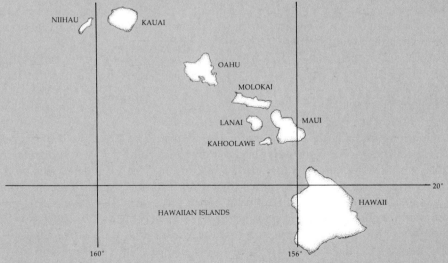

Dubautia scabra. *This low, matted, herbaceous member of the genus is found in moist
to wet habitats on several of the islands in
the Hawaiian archipelago. It is believed to be
the progenitor of several additional species on
the younger islands of the group.*

The Hawaiian Islands. *The oldest of the main
islands, Kauai, includes some rocks that are as
much as six million years old; the youngest,
Hawaii, is still being formed. The Hawaiian
archipelago is gradually moving northwest
with the Pacific Plate, the older islands gradually being eroded below the surface of the*
sea and the younger ones continuously
being formed, apparently as they pass over a
"hot spot", which is a thin spot in the earth's
crust through which lava erupts. Thus we
know that there were islands in approximately
the present position of Hawaii much more
than six million years ago.

A S E X U A L   R E P R O D U C T I O N   I N   H Y B R I D S

Even if hybrids are sterile, they may become widespread if they are able to reproduce asexually (Figure 9–12). In some groups of plants, sexual reproduction is combined with frequent asexual reproduction, so that recombination occurs but successful genotypes can be multiplied exactly (see Figure 9–7).

An outstanding example of such a system is the extremely variable Kentucky bluegrass (*Poa pratensis*), which in one form or another occurs throughout the cooler portions of the Northern Hemisphere. Occasional hybridization with a whole series of related species has produced hundreds of distinctive races of this grass, each characterized by a form of vegetative reproduction called **apomixis,** in which seeds are formed but the embryos are asexually produced from the parent—to which they are identical genetically. In apomictic species, or in other species in which vegetative reproduction is well developed, the individual strains may be particularly successful in specific habitats. In addition, such vegetatively propagated strains do not require outcrossing (cross-pollination between individuals of the same species) and so often do well in situations, such as high mountains, where pollination by insects may be uncertain.

The hundreds of species of hawthorn (*Crataegus*) and blackberry (*Rubus*) found in the eastern half of the United States and adjacent Canada are apomictic derivatives of groups of species in which occasional hybrids occur. For these genera, the wholesale destruction of forests and other natural habitats by humans has led to conditions favorable to the establishment of many new genotypes that would otherwise have had no place in the original vegetation.

P O L Y P L O I D Y

Cells or individuals that have more than two sets of chromosomes are called **polyploids.** Polyploid cells arise at a low frequency as the result of a "mistake" in mitosis in which the chromosomes divide but the nucleus does not. In this way, a cell with twice the usual number of chromosomes is produced. If such a cell then goes through interphase and divides, it can give rise, either sexually or asexually, to a new individual that will have twice the number of chromosomes of its parents or parent. Polyploid individuals also can be produced experimentally by the use of colchicine, a drug that inhibits the formation of the spindle in mitosis by its disruptive effects on microtubule formation.

In a polyploid, the range of variation is often narrowed considerably from that in a related diploid because in the polyploid each gene is present in multiple copies. For instance, for polyploids having twice the usual complement of chromosomes (and assuming a frequency for the recessive allele of 0.50), the proportion of homozygous recessive individuals in a population (assuming random segregation) is 1/16, or a frequency of $(0.50)^4$, rather than the 1/4, or $(0.50)^2$, found in a diploid population. Polyploid races or species are often self-pollinated, which further reinforces their decreased variability. Some polyploids are better suited to dry habitats or colder temperatures than their diploid relatives, whereas others do better in certain types of soils. In ways such as these, polyploids may be adapted to colonize habitats near or beyond the tolerance limits of their diploid ancestors.

Polyploid plants occur at low frequencies in many natural populations. When such polyploids hybridize, which they often are able to do more easily than their diploid counterparts, a stable polyploid derivative, fertile from the start, may be formed. Less commonly, polyploids of hybrid origin have arisen by chromosome doubling in sterile diploid hybrids; such chromosome doubling sometimes functions to restore fertility.

This less common mode was involved in one of the earliest well-documented cases of polyploidy—the origin of a polyploid hybrid between the radish (*Raphanus sativus*) and the cabbage (*Brassica oleracea*). Both genera belong to the mustard family, Brassicaceae, and they are closely related. Both species have 18 chromosomes in their vegetative cells and regularly form 9 pairs of chromosomes at meiotic metaphase I. The hybrid between them, which was obtained experimentally with some difficulty, had 18 unpaired chromosomes in meiosis and thus was completely sterile. In the polyploid that appeared spontaneously among these hybrid plants, there were 36 chromosomes in the vegetative cells, and 18 pairs of chromosomes were formed regularly in meiosis. In other words, the polyploid hybrid had all the chromosomes of the radish and the cabbage in its cells, and these paired regularly among themselves, functioning normally. Consequently, the polyploid had a relatively high fertility.

A number of polyploids originated as weeds in habitats associated with the activities of human beings, and sometimes these polyploids have been spectacularly successful. One of the best known is a salt-marsh grass of the genus *Spartina* (Figure 9–13). One native species, *S. maritima*, occurs in marshes along the coasts of Europe and Africa. A second species, *S. alterniflora*, was introduced into Great Britain from eastern North America in about 1800, spreading from the localities where it was first planted and forming large but local colonies.

In Britain, the native *S. maritima* is short in stature whereas *S. alterniflora* is much taller, frequently growing to 0.5 meter and occasionally to 1 meter or even more in height. Near the harbor at Southampton, in

(a)

(b)

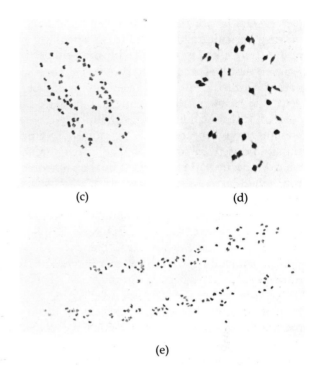

(c)

(d)

(e)

9–13

Polyploidy has been investigated extensively among grasses of the genus Spartina, *which grow in salt-marsh habitats along the coasts of North America and Europe. (a) This salt marsh is on the coast of Great Britain. (b) A* Spartina *hybrid. (c)* Spartina maritima, *the native European species of salt-marsh*

grass, has 2n = 60 chromosomes, shown here from a cell in meiotic anaphase I. (d) Spartina alterniflora *is a North American species with 2n = 62 chromosomes (there are 30 bivalents and 2 unpaired chromosomes, shown here from a cell in meiotic metaphase I). (e) A vigorous polyploid, S. anglica, arose*

spontaneously from a hybrid between these species and was first collected in the early 1890s. This polyploid, which has 2n = 122 chromosomes, shown here from a cell in meiotic anaphase I, is now extending its range through the salt marshes of Great Britain and other temperate countries.

southern England, both the native species and the introduced species existed side by side throughout the nineteenth century. In 1870, botanists discovered a sterile hybrid between these two species that reproduced vigorously by rhizomes. Of the two parental species, *S. maritima* has a somatic chromosome number of $2n = 60$ and *S. alterniflora* has $2n = 62$; the hybrid, owing perhaps to some minor meiotic misdivision, also has $2n = 62$. This sterile hybrid, which was named *Spartina × townsendii*, still persists. About 1890, a vigorous seed-producing polyploid, named *S. anglica*, was derived naturally from it. This fertile polyploid, which had a diploid chromosome number of $2n = 122$ (one chromosome pair was evidently lost), spread rapidly along the coasts of Great Britain and northwestern France. It is often planted to bind mud flats, and such use has contributed to its further spread.

One of the most important polyploid groups of plants is the genus *Triticum*, the wheats. The most commonly cultivated crop in the world, bread wheat, *T. aestivum*, has $2n = 42$ chromosomes. Bread wheat orig-

inated at least 8000 years ago, probably in central Europe, following the natural hybridization of a cultivated wheat with $2n = 28$ chromosomes and a wild grass of the same genus with $2n = 14$ chromosomes. The wild grass probably occurred spontaneously as a weed in the fields where wheat was being cultivated. The hybridization that gave rise to bread wheat probably occurred between polyploids that arose from time to time within the populations of the two ancestral species.

It is likely that the desirable characteristics of the new, fertile, 42-chromosome wheat were easily recognized, and it was selected for cultivation by the early farmers of Europe when it appeared in their fields. One of its parents, the 28-chromosome cultivated wheat, had itself originated following hybridization between two wild 14-chromosome species in the Near East. Species of wheat with $2n = 28$ chromosomes are still cultivated at present, along with their 42-chromosome derivative. Such 28-chromosome wheats are the chief grains used in macaroni products because of the desirable agglutinating properties of their proteins.

New strains produced by artificial hybridization can improve agricultural performance. Particularly promising is *Triticosecale*, a group of man-made hybrids between wheat *(Triticum)* and rye *(Secale);* these hybrids often are called "triticale," based on an earlier, incorrect name. Some of these hybrids, combining the high yield of wheat with the ruggedness of rye, are more resistant to wheat rust—an economically important fungal disease (see pages 235–238). This characteristic could be particularly important in the subtropical and tropicalhighlandregionsoftheworld,whererustisthechief factor limiting wheat production. *Triticosecale* is now widely cultivated and gaining rapidly in popularity in France, and other areas (see page 702). The most important strain of this new crop has $2n = 42$ chromosomes. It was derived by chromosome doubling following hybridization of 28-chromosome wheat with 14-chromosome rye.

In nature, polyploids are selected by the environment rather than by human beings. Polyploidy is a major evolutionary mechanism. Events similar to those in the history of wheat must have taken place well over 100,000 times just to account for the present representation of polyploids in the flora of the world, in which polyploids amount to more than half of the total species of plants. Among these polyploids are many of our most important crops—not only wheat but also cotton, tobacco, sugarcane, bananas, potatoes, and safflower, to mention just a few. To this list can be added many of our most attractive garden flowers, such as chrysanthemums, pansies, and daylilies.

The Origin of Major Groups of Organisms

As knowledge about the ways in which species originate has become better developed, evolutionary scientists have turned to the question of the origin of genera and other higher taxonomic groups of organisms. As suggested by the essay on Hawaiian tarweeds (pages 164–165), genera originate through the same kinds of evolutionary processes that are responsible for the origin of species. If a particular species has a distinctive adaptation to a habitat that is very different from the habitat in which its progenitor grows, the adapted species may come to be very different from that progenitor. It may eventually give rise to many new species and come to constitute a novel evolutionary line. In time, it may become so distinct that this line may be classified as a new genus, family, or even class of organisms. No special mechanisms are necessary to account for the origin of taxonomic groups above the level of species—only discontinuity in habitat, range, or way of life.

Much has been written about whether evolution is uniformly a gradual process, as it has often been viewed in the past, or whether it takes place in spurts: long periods of gradual change or of no change at all, punctuated by periods of rapid change. The latter model is called the **punctuated equilibrium** model of evolution. Some of the proponents of this model have asserted that macroevolution—the process whereby higher taxonomic groups originate—takes place by principles different from those characteristic of microevolution, the sort of gradual change to which we have devoted most of this chapter. All of the kinds of features that distinguish higher taxonomic categories, however, also distinguish some species, and so the process of evolution seems to be of one kind only.

Summary

Natural selection is the process by which organisms with more favorable characteristics with respect to their environment leave more surviving progeny. Natural selection occurs because of genetically determined variations in natural populations. It takes place as organisms progressively occupy different kinds of habitats. Reproductive success (fitness) can be determined only in relation to the habitat in which a particular population occurs.

The base line for the genetics of populations is the Hardy-Weinberg law. It states that, in a large population in which there is random mating and in the absence of forces that change the proportions of alleles, the ratio of dominant to recessive alleles remains constant from generation to generation.

Four principal factors that can cause changes in the proportions of alleles and deviations from the Hardy-Weinberg equilibrium (necessary for evolution to occur) are population size, migration, mutation, and selection. Of these, selection is the most consequential. It is defined as the nonrandom reproduction of genotypes, some being favored over others in the particular environments in which they occur.

Recessive alleles are relatively inaccessible to selection in diploid organisms. The less frequent they become, the higher is the proportion of such alleles that are hidden (masked) in heterozygotes.

A population's response to selective pressures is complex for a number of reasons. Only the phenotype is accessible to selection, and similar phenotypes can result from very different combinations of alleles. Because of epistasis and pleiotropy, single alleles cannot be selected in isolation; selection affects the whole genotype.

Populations adjust to particular environments and form sharply defined units, or ecotypes, if the lines between the environments are sharply drawn. If environments gradually intergrade, populations of plants may form clines.

While populations are changing, they also diverge in those factors that allow them to intercross successfully. After a period of isolation, two populations may be more or less incompatible or may produce sterile hybrids. Related species of relatively long-lived plants, such as trees and shrubs, are less apt to be reproductively isolated than are related species of annuals and other short-lived plants.

Hybrid populations derived from two species are common in plants, especially trees and shrubs. This is particularly true in environments with relatively few species, such as oceanic islands, where adjustment to environmental changes may be especially critical, and in regions with sharply defined environmental breaks and diverse climates, such as California. Such hybrids may reproduce and play an important role in plant evolution.

Even if the hybrids between two species are sterile, they may be propagated by apomixis or other forms of vegetative reproduction or become fertile following doubling in chromosome number (polyploidy).

Carried out over time, the same processes responsible for the evolution of species give rise to genera and other major groups.

Suggestions for Further Reading

Anderson, Edgar: *Plants, Man and Life,* University of California Press, Berkeley, 1967.*

A fascinating account of the evolution of certain groups of plants that reflects the personality of one of the greatest twentieth-century students of botany.

Berra, Tim M.: *Evolution and the Myth of Creationism,* Stanford University Press, Stanford, Calif., 1990.*

A basic guide to the facts of the evolution debate.

Dawkins, Richard: *The Blind Watchmaker,* W. W. Norton & Company, New York, 1986.

Subtitled "Why the evidence of evolution reveals a universe without design," this excellent book will help you to understand the field more clearly.

Futuyma, Douglas: *Evolutionary Biology,* 2d ed., Sinauer Associates, Inc., Sunderland, Mass., 1986.

An excellent text, providing a clear explanation of the process of evolution.

Gould, Stephen Jay: *Wonderful Life,* W. W. Norton & Company, New York, 1989.*

A truly marvelous essay on evolution by an outstanding student of the field. Highly recommended.

Grant, Verne: *Plant Speciation,* Columbia University Press, New York, 1971.

This comprehensive treatment of plant evolution provides a thorough introduction to most aspects of the field.

Minkoff, E.C.: *Evolutionary Biology,* Addison-Wesley Publishing Company, Inc., Reading, Mass., 1983.

Excellent account of evolution, simpler than that of Futuyma.

Moore, John: "Science as a Way of Knowing: Evolutionary Biology," *American Zoologist,* 23 (1):1–68, 1983.

A concise exposition of evolutionary theory, presented clearly as a series of hypotheses and deductions.

Stebbins, G. Ledyard: *Processes of Organic Evolution,* 3d ed., Prentice Hall, Inc., Englewood Cliffs, N.J., 1977.*

A concise review of the field by one of its outstanding practitioners.

* Available in paperback.

Diversity

Yellow lady's slipper (Cypripedium calceolus), *an orchid, blooming in early summer in a Wisconsin woods. Orchidaceae is the largest family of flowering plants, with some 20,000 species, the* majority of which grow in the tropics. As is true of most orchids, especially those of temperate climates, the lady's slipper is endangered.

CHAPTER 10

The Classification of Living Things

At least 10 million different kinds of living organisms share our biosphere. We humans differ from these other organisms both in the degree of our curiosity and in our power of speech. As a consequence of these two characteristics, we have long sought to inquire about other creatures and to exchange information about them. In order to do this, it was necessary to give the organisms names.

Most familiar organisms have been given common names, but even for the simplest of purposes, such names may be inadequate. Sometimes the names are misleading, particularly when we are exchanging information with people from different parts of the world. A sycamore or a bluebell in Great Britain may or may not be the same as the plants that are called sycamores and bluebells in North America. A pine in Europe or the United States is not the same as a pine in Australia. A yam in the southeastern United States is totally different from the vegetable known as a yam a few hundred kilometers away in the West Indies. When different languages are involved, the problems become hopelessly complex. For these reasons, biologists designate organisms with Latin names that are officially recognized by international organizations of botanists, bacteriologists, and zoologists.

These formal Latin names originated from informal systems of naming plants. Different kinds of organisms have long been given names corresponding to categories such as "oaks," "roses," or "dandelions." In medieval times, when an interest in communicating information about organisms was growing, Latin was the language of scholarship. Hence, the Latin names for these "kinds" of organisms were adopted as standard and were widely disseminated in books printed with the newly discovered movable type. The names were often those that the Romans had used; in other cases, they were newly invented ones or names that were put into a Latin form. These "kinds" eventually came to be called **genera** (singular: **genus**), and the individual members of these genera, such as red oaks or willow oaks, were called "species."

At first, the species were identified by descriptive Latin phrases consisting of one to many words; these phrase names are called "polynomials." The first word in a polynomial was the name of the genus to which the plant belonged. Thus, all oaks were identified by polynomials beginning with the word *Quercus*, and all roses with polynomials that started with the word *Rosa*. The ancient Latin names for these plants have continued to be used to designate genera.

The Binomial System

A simplification in the system of naming living things was made by the eighteenth-century Swedish professor and naturalist Carl Linnaeus (Figure 10–1), whose ambition was to name and describe all of the known kinds of plants, animals, and minerals. In 1753, Linnaeus published a two-volume work, *Species Plantarum* ("The Kinds of Plants"). In this work, Linnaeus used polynomial designations for all species of plants as a means of describing and naming them. He regarded these poly-

nomials as the proper names for the species included in these volumes, but, in adding an important innovation, he laid the foundation for the binomial system of nomenclature—the one we still use today. In the margin of the *Species Plantarum*, next to the "proper" polynomial name of each species, Linnaeus wrote a single word that, when combined with the generic name, formed a convenient "shorthand" designation for the species. For example, for catnip, which was formally named *Nepeta floribus interrupte spicatus pedunculatis* (meaning "*Nepeta* with flowers in an interrupted pedunculate spike"), Linnaeus wrote the word "cataria" (meaning "cat-associated") in the margin of the text, thus calling attention to a familiar attribute of the plant. He and his contemporaries soon began calling this species *Nepeta cataria*, which is still its official designation today.

The convenience of this new system was obvious, and the cumbersome polynomial names were soon replaced by binomial ("two-term") names. The earliest binomial name applied to a particular species has priority over other names applied to the same species later. The rules governing the application of scientific names to plants are embodied in the *International Code of Botanical Nomenclature*.

A species name, such as *Nepeta cataria*, consists of two parts, the first of which is the generic name. A generic name may be written alone when one is referring to the entire group of species making up that genus. For instance, Figure 10–2 shows three species of the violet genus, *Viola*. If a species is discovered to have been placed in the wrong genus initially and must then be transferred to another genus, the second part of its name—the **specific epithet**—moves with it to the new genus. If there is already a species in that genus that has that particular specific epithet, however, an alternative name must be found.

Each species has a *type specimen*, usually a dried plant specimen housed in a museum, which is designated either by the person who originally named that species or by a subsequent author if the original author failed to do so. The type specimen serves as a basis for comparison with other specimens in determining whether they are members of the same species or not.

A specific epithet is meaningless when written alone; thus *biennis* is an epithet used in conjunction with dozens of different generic names. For example, *Artemisia biennis*, a kind of wormwood, and *Lactuca biennis*, a species of wild lettuce, are two very different members of the sunflower family, and *Oenothera biennis*, an evening primrose, belongs to a different family. Because of the danger of confusing names, a specific epithet is always preceded by the name or the initial letter of the genus that includes it: for example, *Oenothera biennis* or *O. biennis*. Names of genera and species are printed in italics or are underlined when written or typed.

10–1

Carl Linnaeus (1707–1778), the naturalist who devised the binomial system for naming organisms. Linnaeus believed that each living thing corresponded more or less closely to some ideal model and that by classifying all living things, he was revealing the grand pattern of creation.

(a)

(b)

(c)

10-2
Three members of the violet genus. (a) The long-spurred violet, Viola rostrata, *which grows in temperate regions of eastern North America as far west as the Great Lakes. (b)* Viola quercetorum, *a yellow-flowered violet of California and southern Oregon. (c) Pansy,* Viola

tricolor *var.* hortensis, *an annual, cultivated strain of a mostly perennial species that is native to western Europe. These photographs indicate the kinds of differences in flower color and size, leaf shape and margin, and other features*

that distinguish the species of this genus, even though there is an overall similarity between all of them. There are about 500 species of the genus Viola; *most of them are found in temperate regions of the Northern Hemisphere.*

Some species consist of two or more races, which are called subspecies or varieties. All of the subspecies or varieties of one species bear an overall resemblance to one another but exhibit one or more important differences. As a result of these subdivisions, although the binomial name is still the basis of classification, the names of some plants and animals may consist of three parts. Names of subspecies and varieties are also written in italics or underlined, and the subspecies or variety that includes the type specimen of the species repeats the name of the species. Thus the peach tree is *Prunus persica* var. *persica*, whereas the nectarine is *Prunus persica* var. *nectarina*. The repeated *persica* in the name of the peach tree tells us that the type specimen of the species *P. persica* belongs to this variety.

WHAT IS A SPECIES?

Groups of populations that resemble one another relatively closely and other groups of populations less closely are called **species**, but the application of this term differs widely from one group of organisms to another. The word "species" itself has no special connotation; it simply means "kind" in Latin. The patterns of variation that occur in different groups of organisms as a result of the evolutionary processes discussed in Chapter 9 differ greatly from one another, so that the term "species" cannot be applied in a uniform way. For example, genetic recombination is unknown in some groups (e.g., algae related to *Euglena*, and most bacteria), while in other groups, outcrossing is prevalent and interspecific hybridization is frequent. Despite such differences, the term "species" provides a convenient way to talk about and catalog organisms.

OTHER TAXONOMIC GROUPS

Linnaeus (and earlier scientists) recognized the plant, animal, and mineral kingdoms, and the kingdom is still the most inclusive unit used in biological classification. Scientists employ several additional taxonomic categories between the levels of genus and kingdom. Thus, genera are grouped into **families,** families into **orders,** orders into **classes,** classes into **divisions,** and divisions into **kingdoms.** The groups botanists call divisions are called phyla (singular: phylum) by zoologists, an unfortunate difference that has historical roots.

Regularities in the form of the names for the different categories make it possible to recognize them as names at that level. For example, names of plant families end in -aceae, with a very few exceptions. Older names are allowed as alternatives for a few families, such as Fabaceae, the pea family, which may also be called Leguminosae; Apiaceae, the parsley family, also known as Umbelliferae; and Asteraceae, the sunflower family, also known as Compositae. Names of plant orders end in -ales. No scientific names except those of genera, species, subspecies, and varieties are written in italics or underlined.

Sample classifications of corn (*Zea mays*) and the commonly cultivated edible mushroom (*Agaricus bisporus*) are given in Table 10-1 on page 174.

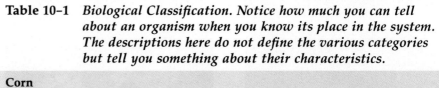

Table 10–1   *Biological Classification. Notice how much you can tell about an organism when you know its place in the system. The descriptions here do not define the various categories but tell you something about their characteristics.*

Corn

Category	Name	Description
Kingdom	Plantae	Organisms that are terrestrial, have chlorophyll *a* and *b* contained in chloroplasts, and show structural differentiation.
Division	Anthophyta	Vascular plants with seeds and flowers; ovules enclosed in an ovary, pollination indirect; the angiosperms.
Class	Monocotyledones	Embryo with one cotyledon; flower parts usually in threes; many scattered vascular bundles in the stem.
Order	Commelinales	Monocots with fibrous leaves; reduction and fusion in flower parts.
Family	Poaceae	Hollow-stemmed monocots with reduced greenish flowers; fruit a specialized achene (caryopsis); the grasses.
Genus	*Zea*	Robust grasses with separate staminate and carpellate flower clusters; caryopsis fleshy.
Species	*Zea mays*	Corn.

Edible Mushroom

Category	Name	Description
Kingdom	Fungi	Nonmotile, multinucleate, heterotrophic, absorptive organisms in which chitin predominates in the cell walls.
Division	Basidiomycota	Dikaryotic fungi that form a basidium bearing four spores (basidiospores); the basidiomycetes.
Class	Hymenomycetes	Basidiomycetes that produce basidiomata, or "fruiting bodies," and club-shaped, aseptate basidia that line gills or pores.
Order	Agaricales	Fleshy fungi with radiating gills or pores.
Family	Agaricaceae	Agaricales with gills.
Genus	*Agaricus*	Dark-spored soft fungi with a central stalk and gills free from the stalk.
Species	*Agaricus bisporus*	The common edible mushroom.

The Major Groups of Organisms

In Linnaeus's time, all organisms were considered to be either plants or animals. Animals moved, ate things, breathed, and grew until they were adults. Plants did not move, eat, or breathe; they were not observed to feed on other organisms and seemed to be able to grow indefinitely.

As new groups of organisms were discovered, they were classified either as plants or as animals. Thus fungi and bacteria were grouped with the plants, and protozoa were grouped with animals. Eventually, however, other organisms were discovered, such as *Chlamydomonas*, a swimming green alga that moves *and* manufactures its own food. Organisms of this sort could not be classified easily as either plants or animals, and by the 1930s the traditional division of living organisms into two kingdoms had clearly become inadequate and little more than a historical curiosity.

Unfortunately, no universally accepted alternative has been proposed, and scientists disagree about how many kingdoms of organisms should be recognized and what kinds of organisms should be included in each. Moreover, the old division into plants and animals is still widely reflected in the organization of college and university science departments, research projects, and textbooks (including this one). That is why we include algae, fungi, and bacteria in this text, in addition to plants.

PROKARYOTES

As discussed in Chapter 2, it is now evident that the most fundamental division in the living world is that between the prokaryotes and the eukaryotes (Figure 10–3). The prokaryotes, or bacteria, are classified in two kingdoms, Archaebacteria and Eubacteria, the members of which differ from one another fundamentally in the sequences of bases in their ribosomal RNAs (rRNAs). Prokaryotes differ far more from eukaryotes than do plants and animals from one another, and the differences between the two kingdoms of prokaryotes are more significant than the differences that separate any of the kingdoms of eukaryotes. Because of their morphological simplicity, prokaryotes are classified largely on the basis of their biochemical features, which are enormously diverse.

Prokaryotes (discussed in detail in Chapter 11) do not have membrane-bound cellular organelles, microtubules, or the complex 9-plus-2 flagella that are characteristic of eukaryotes. Their genetic material consists of a single, circular molecule of DNA, which is not associated with histone proteins. Although several mechanisms leading to genetic recombination are known in

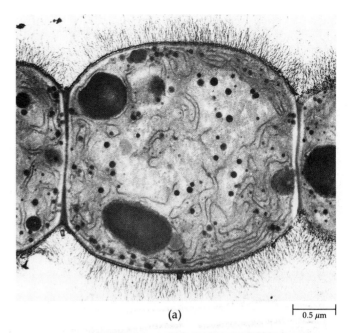

(a)

0.5 μm

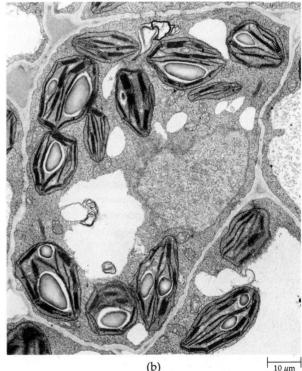

(b)

10 μm

10–3
Electron micrographs of (a) *a prokaryote, the cyanobacterium* Anabaena, *and* (b) *a eukaryotic cell, from a sugar beet* (Beta vulgaris) *leaf. Note the greater complexity of the eukaryotic cell, with its conspicuous nucleus and chloroplasts, and its much larger size (note the scale markers).*

bacteria, such recombination occurs relatively infrequently and only in certain groups. When it does occur, it is accomplished by means other than sexual reproduction.

VIRUSES

As also discussed in Chapter 11, viruses are segments of DNA or RNA that have achieved the ability to use the machinery of other cells to reproduce themselves. Viruses have the ability to assemble a protein coat around themselves, which protects them as they move from one host to another.

EUKARYOTES

All eukaryotes have a definite nucleus that is bounded by a double membrane, the nuclear envelope. Within the nuclear envelope are complex chromosomes in which the DNA is associated with histone proteins. These chromosomes divide regularly by mitosis. The flagella and cilia of eukaryotes have the characteristic 9-plus-2 pattern of microtubules described in Chapter 2. Microtubules are also components of the cytoplasm of eukaryotes, which contain many membranes. Organelles, including mitochondria, occur in the cells of all eukaryotes, and vacuoles, which are bounded by a single membrane, or tonoplast, also occur widely in eukaryotic organisms, especially plants.

In addition, many eukaryotes exhibit two important features not found in prokaryotes: integrated multicellularity and sexual reproduction. Bacterial cells sometimes remain together in filamentous or even three-dimensional masses following their division, but there are usually no protoplasmic connections between the individual cells and hence no overall integration of the entire filament or mass. In plants, the protoplasts of contiguous cells are connected by plasmodesmata, which traverse their cell walls. In animals, there are no cell walls, and the cells are separated primarily by their plasma membranes.

Relationships within the Eukaryotes

The system of classification used in this book divides organisms into six kingdoms, with the prokaryotes divided into the kingdoms Eubacteria and Archaebacteria and the eukaryotes divided into kingdoms Protista, Fungi, Plantae, and Animalia (Figure 10-4). As mentioned above, Eubacteria and Archaebacteria are sharply distinct from each other, with accumulating evidence suggesting that Archaebacteria are more closely related to the eukaryotes than are Eubacteria. Among the eukaryotes, on the other hand, the relationships are more complex and the divisions between kingdoms much less clear-cut. The predominantly unicellular phyla and divisions (equivalent terms used in different groups of organisms) and some of the multicellular lines associated with them are grouped in the kingdom Protista, from which the three kingdoms consisting essentially of multicellular organisms—Plantae, Animalia, and Fungi—have been derived.

THE ORIGINS OF MULTICELLULARITY

Multicellular organisms originated from unicellular ancestors many times in the history of the eukaryotes. In each case, the ancestors, if they are known, are classified as members of the kingdom Protista. Although the plant, animal, and fungi kingdoms consist almost entirely of multicellular organisms, there are a few unicellular fungi (the yeasts) that were probably originally derived from multicellular ancestors. Plants, animals, and fungi were certainly derived from different groups of unicellular protists. The three groups differ from each other fundamentally in their mode of nutrition: plants manufacture their food, animals ingest food, and fungi secrete digestive enzymes, digesting their food externally and then absorbing it.

In addition to the three major multicellular groups of organisms that we recognize as kingdoms, there are three other divisions—the red algae, green algae, and brown algae—in which multicellular organisms are well represented. The brown algae are exclusively multicellular, the red algae are almost entirely multicellular, and the green algae consist of large numbers of both unicellular and multicellular organisms. Since the members of these divisions are photosynthetic and since at least some of the organisms included are multicellular, it has been argued by some scientists that these divisions should be included in the plant kingdom. Since the red algae and brown algae share no immediate common ancestor with the green algae and plants, however, classifying all of these groups as plants is clearly incorrect from an evolutionary point of view. The green algae include the ancestors of the plants, but these algae are sharply distinct from any living plants (for one thing, they are aquatic). For these reasons, we prefer to recognize the plants as a distinct kingdom and to retain the red algae, brown algae, and green algae in the kingdom Protista. Although these three algal divisions are unusual in including so many multicellular organisms, most divisions and phyla of protists have at least some members that are multicellular.

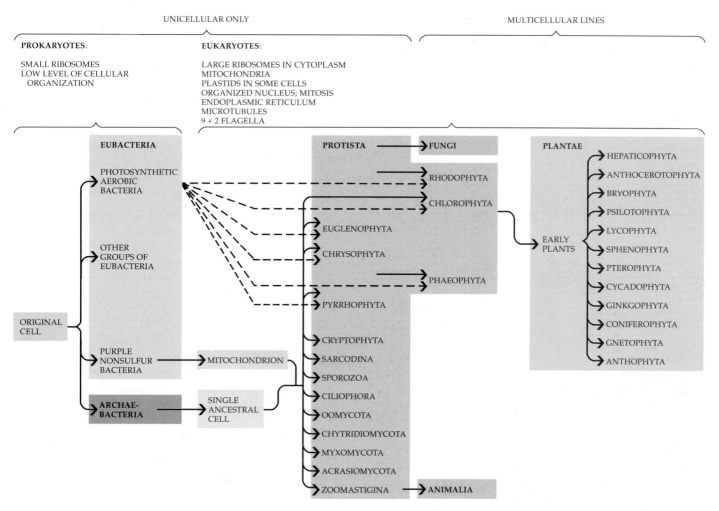

10–4
One possible scheme of the evolutionary relationships among organisms. The solid lines indicate phylogenetic relationships, and the dashed lines indicate the establishment of symbiotic relationships. All eukaryotic organisms were derived from a single line of cells in which mitochondria, originally symbiotic bacteria, were incorporated very early. From this single line of cells, the diverse assemblage of one-celled organisms known as the Protista developed. Those protists that developed symbiotic relationships with photosynthetic bacteria gave rise to the several different lines of modern algae. From particular single-celled protists, the fungi, plants, and animals evolved.

MITOCHONDRIA AND CHLOROPLASTS

One of the most significant features in the evolution of eukaryotes has been their acquisition of mitochondria and chloroplasts. As explained in Chapter 2, it is virtually certain that these complex organelles had independent, symbiotic origins, as is evident in their structure and the nature of their hereditary apparatus. Mitochondria most closely resemble the group of Eubacteria called the purple nonsulfur bacteria (see Chapter 11), a group of anaerobic (oxygen-avoiding) bacteria. In these bacteria, the plasma membrane is elaborately folded throughout the cytoplasm, resembling the cristae that occur in the interior of mitochondria. There are also biochemical similarities between mitochondria and the purple nonsulfur bacteria.

Chloroplasts seem to represent the descendents of photosynthetic, aerobic (oxygen-requiring) bacteria. The different groups of photosynthetic protists (algae) are apparently not directly related to one another; they seem to have acquired their chloroplasts independently at various times in the distant past. Hypotheses about the nature of these chloroplasts will be discussed in Chapters 11, 13, and 14.

Table 10–2   *Classification of Living Organisms Included in This Book. (See Appendix C for summary descriptions of these groups.)*

Prokaryotes

Kingdom Eubacteria	Eubacteria	
Kingdom Archaebacteria	Archaebacteria	

Eukaryotes

Kingdom Protista	Heterotrophic protists	Division Oomycota (water molds)
		Division Chytridiomycota (chytrids)
		Division Acrasiomycota (cellular slime molds)
		Division Myxomycota (plasmodial slime molds)
	Photosynthetic protists ("algae")	Division Chrysophyta (diatoms and chrysophytes)
		Division Pyrrophyta (dinoflagellates)
		Division Euglenophyta (euglenoids)
		Division Rhodophyta (red algae)
		Division Phaeophyta (brown algae)
		Division Chlorophyta (green algae)
Kingdom Fungi	Fungi	Division Zygomycota (zygomycetes)
		Division Ascomycota (ascomycetes)
		Division Basidiomycota (basidiomycetes)
Kingdom Plantae	Bryophytes	Division Hepaticophyta (liverworts)
		Division Anthocerotophyta (hornworts)
		Division Bryophyta (mosses)
	Vascular plants	
	Seedless vascular plants	Division Psilotophyta (psilotophytes)
		Division Lycophyta (lycophytes)
		Division Sphenophyta (horsetails)
		Division Pterophyta (ferns)
	Seed plants	Division Cycadophyta (cycads)
		Division Ginkgophyta (ginkgo)
		Division Coniferophyta (conifers)
		Division Gnetophyta (gnetophytes)
		Division Anthophyta (angiosperms)
		Class Dicotyledones (dicots)
		Class Monocotyledones (monocots)

Formal Classification of Organisms

The following is a synopsis of the six-kingdom system of classification used in this book (see Table 10–2, which does not include the kingdom Animalia).

PROKARYOTES:
KINGDOM EUBACTERIA AND
KINGDOM ARCHAEBACTERIA

Prokaryotes lack nuclear envelopes, plastids, mitochondria, microtubules, and 9-plus-2 flagella. The members of both kingdoms, Eubacteria and Archae-bacteria, exhibit solitary unicellular or colonial unicellular organization (Figure 10–5) and very rarely form protoplasmic connections between the cells. Most are heterotrophic, absorbing their nutrients, but some are chemosynthetic or photosynthetic. Reproduction is predominantly by cell division, although genetic recombination occurs in several groups. They are either nonmotile or motile by simple flagella or gliding.

The majority of bacteria are classified as Eubacteria. The Archaebacteria include methane-producing bacteria (Figure 10–6), sulfate reducers, extreme halophiles (bacteria that grow in very salty environments), and extreme thermophiles (bacteria that occur in very hot habitats).

Eubacteria differ from Archaebacteria in the base

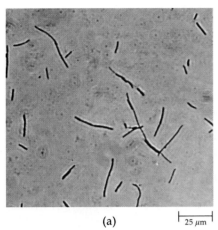

(a) ⊢ 25 μm ⊣

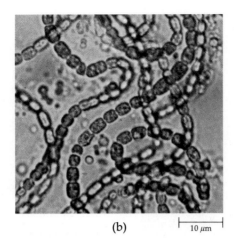

(b) ⊢ 10 μm ⊣

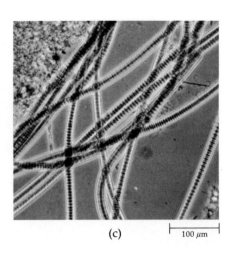

(c) ⊢ 100 μm ⊣

10–5

Eubacteria. This kingdom comprises most of the bacteria, which are prokaryotic organisms. (a) Lactobacillus acidophilus, bacteria that sour milk. (b) A gelatinous colony of the cyanobacterium Nostoc. The cyanobacteria are one of the most abundant groups of photosynthetic bacteria. (c) Oscillatoria, another cyanobacterium, in which the growth form is filamentous.

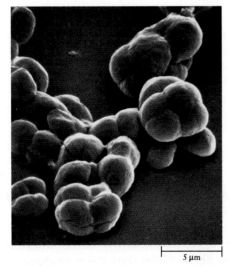

⊢ 5 μm ⊣

10–6

Archaebacteria. Methane-producing bacteria, such as the cells shown here, produce methane from carbon dioxide and hydrogen gas. They are anaerobes and can therefore live only in the absence of oxygen—a condition prevailing on the young earth but occurring today only in isolated environments, such as the muck and mud at the bottom of swamps.

sequences in their rRNAs, in the fact that muramic acid is a component of the cell walls of Eubacteria but not of Archaebacteria, and in other biochemical characteristics of fundamental importance. Based on several biochemical features, including rRNA base sequences, the Archaebacteria appear to be more closely related to eukaryotes than are the Eubacteria. Both kingdoms of prokaryotes are discussed in Chapter 11.

KINGDOM PROTISTA

Protista (Figure 10–7) are here considered to comprise all organisms traditionally regarded as protozoa (one-celled "animals"), as well as all algae except for the group of bacteria now called "cyanobacteria" that are sometimes misleadingly called "blue-green algae." (The term "algae" is an informal one used for photosynthetic eukaryotes other than plants and for the cyanobacteria; nearly all algae are aquatic.) Also included in the kingdom Protista are some heterotrophic groups of organisms, including the water molds and their relatives (division Oomycota), the chytrids (division Chytridiomycota), the cellular slime molds (division Acrasiomycota), and the plasmodial slime molds (division Myxomycota)—four groups of organisms that have traditionally been placed with the fungi.

The reproductive cycles of protists are varied but typically involve both cell division and sexual reproduction. Protists may be motile by 9-plus-2 flagella or cilia or by amoeboid movement, or they may be nonmotile. The predominantly unicellular groups of protists included in this book are discussed in Chapter 13, and the three major groups of algae—green algae, brown algae, and red algae—are discussed in Chapter 14. The heterotrophic protists known as protozoa—in other words, those that have historically been treated as animals instead of as fungi—are not included in this text.

(a)

(b)

(c)

(d)

(e)

10–7

Protista. (a) Plasmodium of a plasmodial slime mold, Physarum, *growing on a leaf. (b)* Postelsia palmiformis, *the "sea palm" (division Phaeophyta), growing on exposed intertidal rocks off Vancouver Island, British Columbia. (c) Volvox, a motile colonial green alga (division Chlorophyta). (d) Sebdenia polydactyla, a red alga (division Rhodophyta). (e) A pennate diatom (division Chrysophyta) showing the intricately marked shell characteristic of this group.*

In summary, the kingdom Protista includes a very heterogeneous assemblage of unicellular, colonial, and multicellular eukaryotes that do not have the distinctive characteristics of the animals, plants, or fungi.

KINGDOM ANIMALIA

The animals are multicellular organisms with eukaryotic cells lacking cell walls, plastids, and photosynthetic pigments. Nutrition is primarily ingestive with digestion in an internal cavity but in some forms is absorptive, and a number of groups lack an internal digestive cavity. The level of organization and tissue differentiation in complex animals far exceeds that of the other kingdoms, particularly with the evolution of complex sensory and neuromotor systems. The motility of the organism (or, in sessile forms, of its parts) is based on contractile fibrils. Reproduction is predominantly sexual. Animals are not discussed in this book, except in relation to some of their interactions with plants and other organisms.

KINGDOM FUNGI

Fungi (Figure 10–8) are nonmotile, filamentous eukaryotes that lack plastids and photosynthetic pigments

10–8

Fungi. (a) Red blanket lichen (Herpo-thallon sanguineum) growing on a tree trunk. (b) A flower fly (Syrphus) that has been killed by the fungus Ento-mophthora muscae, *a zygomycete. (c) A white coral fungus (family Clavariaceae). (d) Mushrooms (genus probably My-cena), with dew droplets, growing in a rainforest in Peru. (e) An earthball,* Scleroderma aurantium.

and absorb their nutrients from either dead or living organisms. The fungi have traditionally been grouped with plants, but there is no longer any doubt that the fungi are in fact an independent evolutionary line. Aside from their filamentous growth habit, fungi have little in common with any of the groups that have been classified as algae. The cell walls of fungi include a matrix of chitin. The structures in which fungi form their spores are often complex. Fungal reproductive cycles, which also can be quite complex, typically involve both sexual and asexual processes. Fungi are discussed in Chapter 12.

KINGDOM PLANTAE

Plants—the three divisions of bryophytes (mosses, liverworts, and hornworts) and the nine divisions of vascular plants—constitute a kingdom of photosynthetic organisms adapted for a life on the land. Their ancestors were specialized green algae. All plants are multicellular and are composed of vacuolate eukaryotic cells with cellulosic cell walls. Their principal mode of nutrition is photosynthesis, although a few plants have become heterotrophic. Structural differentiation occurred during the evolution of plants on land, with trends toward the evolution of organs specialized for photosynthesis, anchorage, and support (Figure 10–9). In more complex plants, such organization has produced specialized photosynthetic, vascular, and covering tissues. Reproduction in plants is primarily sexual, with cycles of alternating haploid and diploid generations; in the more advanced members of the kingdom, the haploid generation (gametophyte) has been reduced during the course of evolution. The bryophytes are discussed in Chapter 15 and the vascular plants in Chapters 16–19.

(a)

(b)

(c)

(d)

(e)

(f)

(g)

(h)

(i)

(j)

10–9

Plants. (a) Sphagnum, *the peat moss (division Bryophyta), forms extensive bogs in cold and temperate regions of the world.* (b) Marchantia *is by far the most familiar of the thallose liverworts (division Hepaticophyta). It is a widespread, terrestrial genus that grows on moist soil and rocks.* (c) Diphasiastrum digitatum, *a "club moss" (division Lycophyta).* (d) *Wood horsetail,* Equisetum sylvaticum (*division Sphenophyta).* (e) *Bulblet fern,* Cystopteris bulbifera (*division Pterophyta).* (f) *The dandelion,* Taraxacum officinale, *and* (g) *fishhook cactus,* Mammillaria microcarpa, *are dicots (class Dicotyledones, division Anthophyta).* (h) *Foxtail barley,* Hordeum jubatum, *and* (i) Cymbidium *orchids are monocots (class Monocotyledones, division Anthophyta).* (j) *Sugar pine,* Pinus lambertiana, *and incense cedar,* Calocedrus decurrens, *are both conifers (division Coniferophyta).*

Sexual Reproduction

Prokaryotes have several means of achieving genetic recombination. As described in more detail in Chapter 11, a portion of the bacterial genetic material may be transferred from one cell to another, but there is no mechanism comparable to meiosis by which it can then be regularly transmitted along with the genetic material of the recipient cell. The only method for regular transmission of the new genetic information in prokaryotes is the incorporation of the new fragment into the DNA molecule of the recipient bacterium, a system that is neither precise nor readily repeatable.

Sexual reproduction, which involves a regular alternation between meiosis and fertilization, has a high selective advantage because it is the primary mechanism that produces and maintains variability in natural populations of eukaryotes. Variability, as we saw in Chapter 9, is the basic trait that makes possible the adjustment of living organisms to their environment through the process of evolution.

THE EVOLUTION OF DIPLOIDY

The first eukaryotic organisms were probably haploid and asexual, but once sexual reproduction was established among them, the stage was set for the evolution of diploidy. It seems likely that this condition first arose when two haploid cells combined to form a diploid zygote; such an event probably took place repeatedly.

Presumably the zygote then divided immediately by meiosis (zygotic meiosis), thus restoring the haploid condition (Figure 10–10a). In organisms with this simple kind of life cycle (primitive eukaryotes and fungi), the zygote is the only diploid cell.

By "accident"—an accident that occurred in a number of separate evolutionary lines—some of these zygotes divided mitotically instead of meiotically and, as a consequence, produced an organism that was composed of diploid cells, with meiosis occurring later. In animals, this delayed meiosis (gametic meiosis) results in the production of gametes—eggs and sperm. If they come together, these gametes fuse, an event that immediately restores the diploid state (Figure 10–10b). Therefore, in animals, gametes are the only haploid cells.

In plants, meiosis (sporic meiosis) results in the production of spores, not gametes. Spores are cells that can divide directly by mitosis to produce a multicellular haploid organism; this is in contrast to gametes, which can develop only following fusion with another gamete. Multicellular haploid organisms that appear in alternation with diploid forms are found in plants; in some brown, red, and green algae; and in two closely related genera of chytrids and one or more other groups of protists not discussed in this book. Such organisms exhibit the phenomenon known as **alternation of generations** (Figure 10–10c). Among the plants, the haploid, gamete-producing generation is called the

10–10

Diagrams of the principal types of life cycles. In these diagrams, the diploid phase of the cycle takes place below the broad bar, and the haploid phase occurs above it. The four white arrows signify the products of meiosis; the single white arrow represents the fertilized egg, or zygote.

(a) In zygotic meiosis, the zygote divides by meiosis to form four haploid cells, which divide by mitosis to produce either more haploid cells or a multicellular individual that eventually gives rise to gametes by differentiation. This type of life cycle is found in Chlamydomonas *and a number of other algae and in the fungi.*

(b) In gametic meiosis, the haploid gametes are formed by meiosis in a diploid individual and fuse to form a diploid zygote that divides to produce another diploid individual. This type of life cycle is characteristic of most animals and some protists (Oomycota), as well as the brown alga Fucus.

(c) In sporic meiosis, the sporophyte, or diploid individual, produces haploid spores as a result of meiosis. These spores do not function as gametes but undergo mitotic division. This gives rise to multicellular haploid individuals (gametophytes), which eventually produce gametes that fuse to form diploid zygotes. These zygotes, in turn, differentiate into diploid individuals. This kind of life cycle, known as alternation of generations, is characteristic of the plants and many algae. A similar sort of life cycle is found in the chytrid Allomyces *and one other closely related genus and in a few other groups of protists not included in this book.*

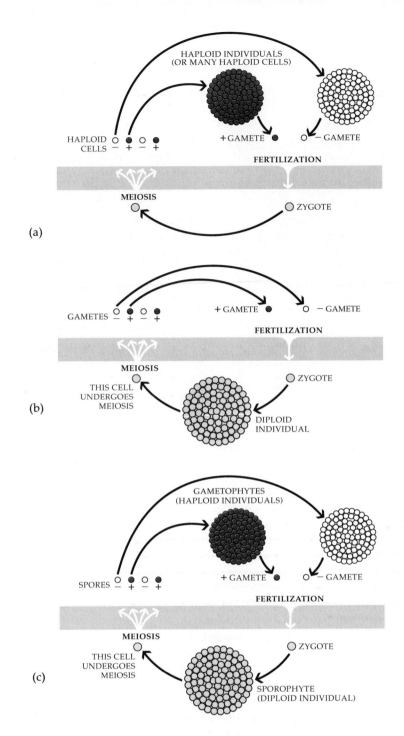

(a)

(b)

(c)

gametophyte and the diploid, spore-producing generation is called the **sporophyte.** This same terminology is used for the algae and sometimes other groups as well.

In some algae—most of the red algae, many of the green algae, a few of the brown algae—the haploid and diploid forms are the same in external appearance. Such types of life cycles are said to exhibit an alternation of **isomorphic** generations.

In contrast to the life cycles discussed above, there are some in which the haploid and diploid forms are not identical. During the history of these groups, mutations occurred that were expressed in only one generation, although the alleles were, obviously, present in both the diploid and the haploid generations. In life

cycles of this kind, the gametophyte and sporophyte became notably different from one another, and an alternation of **heteromorphic** generations originated. Such life cycles are characteristic of plants and some brown and red algae.

In the bryophytes (mosses, liverworts, and hornworts), the gametophyte is dominant; it is nutritionally independent from and usually larger than the sporophyte, which may be more complex structurally. In the vascular plants, on the other hand, the sporophyte dominates; it is much larger and more complex than the gametophyte, which is nutritionally dependent on the sporophyte in nearly all groups.

Diploidy permits the storage of more genetic information and so perhaps allows a more subtle expression of the organism's genetic background in the course of development. This may be the reason that the sporophyte is the large, complex, and nutritionally independent generation in vascular plants. One of the clearest evolutionary trends in this group, which predominates in most terrestrial habitats, is the increasing dominance of the sporophyte and suppression of the gametophyte. Among the flowering plants, the female gametophyte is a microscopic body that consists of only seven cells, and the male gametophyte consists of only three cells. Both of these gametophytes are nutritionally dependent on the sporophyte.

Summary

Biologists have developed methods of naming and classifying organisms that permit them to designate an organism very precisely, an essential factor in scientific communication. The classification of an organism also reveals its relationship to other living things.

Organisms are designated scientifically by a name that consists of two words—a binomial. The first word in the binomial is the name of the genus (plural: genera), and the second word, the specific epithet, combined with the name of the genus, completes the name of the species. Species are sometimes subdivided into subspecies or varieties. Genera are grouped into families, families into orders, orders into classes, and classes into divisions. Divisions are grouped into kingdoms, the kingdom being the largest unit used in classification of the living world. In this text, living organisms are grouped into six kingdoms, the first two of which consist of prokaryotes, the other four of eukaryotes: (1) Eubacteria, a kingdom of prokaryotes (bacteria); (2) Archaebacteria, another kingdom of bacteria; (3) Protista, the protozoa, eukaryotic algae, chytrids, slime molds, and water molds; (4) Animalia, multicellular organisms that are not photosynthetic; (5) Fungi, the fungi; and (6) Plantae, the bryophytes and vascular plants, photosynthesizing organisms that are terrestrial and more complex than the algae.

Mitochondria, which are characteristic of all eukaryotes, probably originated from bacteria similar to the purple nonsulfur bacteria, whereas chloroplasts appear to have originated from at least three different groups of photosynthetic bacteria. The possession of chloroplasts in different groups of eukaryotes, therefore, does not signify a direct relationship between them. Multicellularity originated independently a number of times in different groups of protists. Some of the multicellular lines, including the red algae, brown algae, and green algae, are retained among the protists in the six-kingdom system of classification used in this book, whereas three major multicellular lines—animals, plants, and fungi—are recognized as separate kingdoms.

In the evolution of organisms, diploidy evolved subsequent to the process of sexual reproduction. In primitive eukaryotes and all fungi, the zygote formed by fertilization divides immediately by meiosis. From early cycles of this sort, more complex life cycles involving diploid phases were derived on a number of occasions, when the zygote divided by mitosis. If the haploid cells produced by meiosis function immediately as gametes, the result is the type of life cycle found in animals and in some groups of protists. If the haploid cells divide by mitosis, as in many algae, all plants, and two genera of chytrids (Protista), they are considered spores; the diploid generation that gives rise to such spores is called the sporophyte. The haploid generation, which develops from the spores by mitosis, is called the gametophyte; it eventually produces gametes by mitosis. If the gametophyte and sporophyte in a particular life cycle are approximately equal in size and complexity, the alternation of generations is said to be isomorphic; if they differ widely in size and complexity, it is said to be heteromorphic.

Suggestions for Further Reading

Bold, Harold G., C.S. Alexopoulos, and T. Delevoryas: *Morphology of Plants and Fungi,* 4th ed., Harper & Row, Publishers, Inc., New York, 1980.

A well-illustrated and ample treatment of the diversity of plants, algae, and fungi.

Gledhill, D.: *The Names of Plants,* Cambridge University Press, Cambridge, England, 1985.

An outstanding concise discussion of the rules and procedures by which plants are named, with a good glossary.

Jones, S.B., and Arlene E. Luchsinger: *Plant Systematics,* 2d ed., McGraw-Hill Book Company, New York, 1986.

A good, practical overall treatment of the field.

Margulis, Lynn: *Symbiosis in Cell Evolution,* W. H. Freeman and Company, New York, 1981.

A fascinating discourse on the origin of eukaryotic cells by serial symbiotic events.

Margulis, Lynn, and Karlene V. Schwartz: *Five Kingdoms: An Illustrated Guide to the Phyla of Life on Earth,* 2d ed., W. H. Freeman and Company, New York, 1988.

An excellent and concise account of the diversity of life on earth.

Scagel, Robert F., et al.: *Nonvascular Plants: An Evolutionary Approach,* Wadsworth Publishing Co., Inc., Belmont, Calif., 1982.

An exhaustive, thoroughly illustrated review of the diversity of bryophytes, fungi, bacteria, and those groups of protists treated in this book.

CHAPTER 11

Bacteria and Viruses

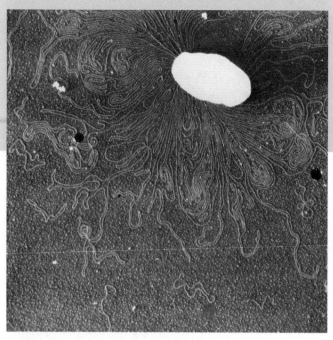

11–1

Because of the way in which it was prepared, this cell of the common colon bacterium Escherichia coli *is spilling out its DNA. The bacterial chromosome consists of a single giant molecule of DNA containing some 400,000 base pairs. The smaller circles of DNA visible on the left and bottom of the micrograph are plasmids with approximately 5000 base pairs. (The black dots are holes in the plastic film on which the preparation was made; the scattered white dots may be dust particles or pieces of the cell.)*

Of all organisms, bacteria—the prokaryotes—are the simplest in structure, smallest physically, and most abundant in terms of numbers of individuals (Figure 11–1). They make up the very distinctive kingdoms Eubacteria and Archaebacteria. Bacteria are metabolically highly diverse, and several groups are photosynthetic. Their cells are very small, usually 1 to 10 micrometers in diameter, compared with 10 to 100 micrometers for most eukaryotic cells, and this difference in diameter reflects an enormous difference in volume. The organization of a prokaryotic cell, as we saw in Chapter 2, is much simpler than that of any eukaryote. Even though individual bacteria are quite small, the total weight of all the bacteria in the world is estimated to exceed that of all other living organisms combined; in the sea, for example, bacteria make up an estimated 90 percent of the total combined weight of all organisms. In a single gram of fertile agricultural soil, there may be 2.5 billion bacteria, 400,000 fungi, 50,000 algae, and 30,000 protozoa! About 2500 species of bacteria are currently recognized, but many more await discovery.

For at least 2 billion years, bacteria were the only form of life on earth. The earth is about 4.5 billion years old, and the oldest known sedimentary rocks, from Greenland, are about 3.9 billion years old. The oldest known fossils, 3.5 billion years in age, are from Western Australia (see Figure 1–1, page 2), while others almost as old have been found in South Africa. Both the nature of some of these fossils and the chemical composition of the rocks in which they are found indicate that photosynthesis occurred at that early time. In contrast to these ancient records of bacteria, the oldest eukaryotes are about 1.5 billion years old.

In this chapter, we shall consider not only bacteria but also viruses, which are even smaller than bacteria. A virus consists primarily of a genome that replicates itself within a host cell by directing the machinery of that cell to synthesize viral nucleic acids and proteins.

Since no virus can grow or replicate on its own, viruses are not regarded as living organisms in the usual sense of "living." Most viruses evidently originated as pieces of the genetic machinery of bacteria.

Characteristics of Bacteria

BACTERIAL STRUCTURE AND CLASSIFICATION

Bacteria lack an organized nucleus surrounded by a nuclear envelope (see Chapter 2). They do not have complex chromosomes such as those of the eukaryotes but, rather, a large, circular molecule of DNA—the so-called "bacterial chromosome." Included in the cytoplasm are a number of ribosomes and granular inclusions. The DNA is found in a less granular area called the **nucleoid.** Bacteria do not reproduce sexually, although genetic recombination occurs occasionally among the Eubacteria. Unlike eukaryotes, bacteria are never truly multicellular. Although some bacteria form filaments or masses of cells, these cells are connected to each other either because their cell walls fail to separate completely following cell division or because they are held together within a common gelatinous capsule or mucilaginous sheath. Plasmodesmata between such linked cells are extremely rare, occurring only in a few species of cyanobacteria. Consequently, there is no continuity or communication between bacterial cells comparable to that occurring in multicellular eukaryotes.

Bacteria lack organelles surrounded by membranes, but they do have other structures that play similar roles. The plasma membrane often has folds and convolutions that extend into the interior of the cell (Figure 11–2). These internal membranes increase the surface area to which enzymes are bound and serve to separate different enzymatic functions. In some photosynthetic bacteria, such as the cyanobacteria (formerly misleadingly called "blue-green algae"), the photosynthetic pigments are located on internal membranes. In others, these pigments are located in discrete spherical bodies called chromatophores.

The cell walls of bacteria consist of a matrix of disaccharides cross-linked by short chains of amino acids (peptides). In one group, the gram-positive bacteria (those that retain the purple dye crystal violet when subjected to the gram staining procedure), the cell wall ranges from 15 to 80 nanometers thick, depending on the species. The cell walls of gram-positive bacteria consist of a single macromolecule of **peptidoglycan,** a type of compound not present in any eukaryote. In gram-negative bacteria (which do not retain crystal violet), an additional layer consisting of large molecules of lipopolysaccharides (polysaccharides with attached lipids) encases the peptidoglycan layer. The cell wall of gram-negative bacteria—that is, the peptidoglycan layer—is only about 10 nanometers thick.

The adhesive properties of individual bacterial cells are determined by a layer of tangled fibers of polysaccharides called the **glycocalyx.** Although bacteria often do not form a glycocalyx when they are grown in laboratory cultures, in nature this structure may play a key role in the initiation and progression of bacterial infection by mediating the adhesion of the bacteria to different substrates. By means of enzymatic reactions that occur in the glycocalyx, for example, certain bacteria are able to colonize the intact enamel surfaces of teeth as well as many other seemingly resistant substrates.

A gelatinous layer—the **capsule**—is often formed outside a bacterial cell wall. The capsule is apparently secreted through the wall by the bacterial protoplast. Like the wall, the capsule is composed of polysaccharides, but it is not as firmly bound to the bacterium and can be removed by vigorous washing.

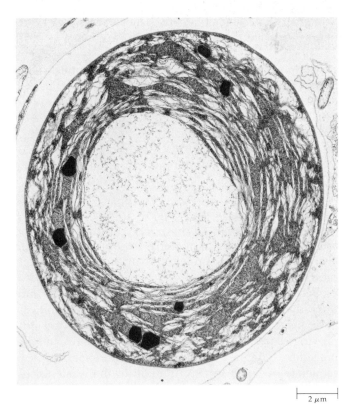

2 μm

11–2
A single cell of the bacterium Prochloron, *showing the extensive internal membrane structure.* Prochloron *is a photosynthetic bacterium with chlorophylls* a *and* b *and carotenoids, the same pigments found in the green algae and plants.*

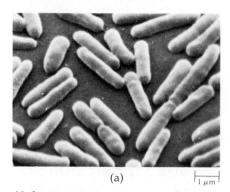

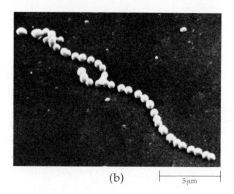

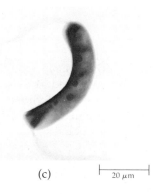

| (a) | ⊢—1 μm | (b) | ⊢—5 μm | (c) | ⊢—20 μm |

11–3

The three major forms of bacteria: (a) bacilli, (b) cocci, and (c) spirilla. Cell shape is a relatively constant feature in most species of bacteria. Bacilli include the microorganisms that cause lockjaw (Clostridium tetani), as well as the familiar Escherichia coli. They also are

responsible for many plant diseases, including fire blight of pears and apples (Erwinia amylovora) and bacterial wilt of tomatoes, potatoes, and bananas (Pseudomonas solanacearum). Among the cocci are Diplococcus pneumoniae,

the cause of bacterial pneumonia; Streptococcus lactis, a common milk-souring agent; and Nitrosococcus nitrosus, soil bacteria that oxidize ammonia to nitrites. The spirilla, which are less common, are helically coiled.

Bacteria vary greatly in form and cellular composition, but cell shape is nevertheless a useful basis for bacterial classification. Straight, rod-shaped bacteria are known as **bacilli,** spherical ones as **cocci,** and long, coiled ones as **spirilla** (Figure 11–3). Bacilli or cocci may adhere to form chains, and such behavior is characteristic of particular genera. The actinomycetes are one group that characteristically form chains (Figure 11–4b).

Certain kinds of bacteria have very slender, rigid, helical flagella, each several times longer than the cell to which it is attached. Lacking microtubules, these flagella differ greatly from those of eukaryotes (page 31). The bacterial flagella, which rotate very rapidly, enable the bacterium to swim with a jerky motion. They are 3 to 12 micrometers long and only 10 to 20 nanometers thick—usually too fine to be seen with a light microscope (Figure 11–5). Pili, which are found in certain bacteria, are similar to flagella but shorter and thinner (Figure 11–6); certain pili may serve a function during the exchange of genetic material in some bacteria. Pili also play an important role in the attachment of bacteria to surfaces.

Because the structure of bacteria is so simple and their metabolism so diverse, biochemical differences have been of key importance in establishing and improving bacterial classification. As we saw in the preceding chapter, fundamental differences in the sequence of bases in rRNAs indicate the existence of two very distinctive kingdoms of bacteria, Eubacteria and Archaebacteria. Other biochemical differences separate the very diverse members of these two kingdoms, including the presence of muramic acid, known in no other organism, in the cell walls of Eubacteria. It appears increasingly likely that Archaebacteria are more closely related to eukaryotes than are Eubacteria, and many interesting discoveries in this area are being made as detailed investigations proceed.

For the purposes of our discussion here, we shall first consider the general features of all bacteria—the prokaryotes—together and then discuss the special features of Eubacteria and Archaebacteria separately.

BACTERIAL ECOLOGY

Bacteria occur in virtually all habitats, and because of their unusual metabolic characteristics, some bacteria can survive in environments that support no other form of life. Most bacteria are heterotrophs—organisms that cannot make organic compounds from simple inorganic substances, but must use other organisms as a source of these compounds. The largest group of heterotrophic bacteria is the **saprobes,** which obtain their nourishment from dead organic matter. Saprobic bacteria and fungi are responsible for the decay and recycling of organic material in the soil. A number of other kinds of bacteria are autotrophs—organisms that manufacture organic compounds from inorganic precursors either by photosynthesis (photoautotrophs) or by chemosynthesis (chemoautotrophs).

Chemoautotrophic bacteria obtain the energy needed for their synthetic reactions in a unique fashion: from the oxidation of reduced inorganic compounds containing nitrogen, sulfur (see Figure 11–10d), and iron or from the oxidation of gaseous hydrogen. The inorganic compounds that these bacteria oxidize provide a source of energy, as does light for photosynthetic organisms. Their carbon source, however, is the same: carbon dioxide.

Some bacteria are **obligate anaerobes;** they live only in the absence of oxygen. Others are **facultative anaerobes;** they can survive and grow without oxygen but grow more vigorously in its presence. Respiration

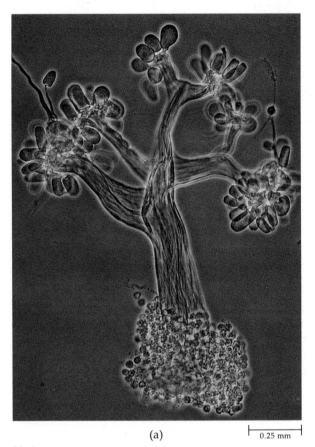

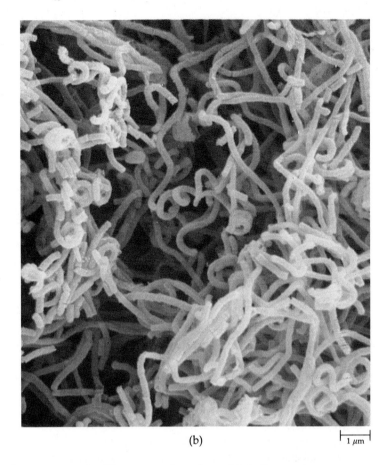

(a) 0.25 mm

(b) 1 μm

11–4
Two of the diverse types of bacteria. (a) Myxobacteria, or gliding bacteria, have a pattern of organization similar to that of the slime molds (see Chapter 13). This is a micrograph of a fruiting body of Chondromyces crocatus. *Each fruiting body, which may contain as many as 1 million cells, consists of a central stalk that branches to form clusters of single-cell spores. Normally, the myxobacteria are rods that glide together and eventually differentiate into the kinds of bodies shown here. (b) Actinomycetes are abundant in soil, where they are largely responsible for the "moldy" odor of damp soil and decaying organic material.* Streptomyces scabies, *the cause of potato scab disease, is shown here.*

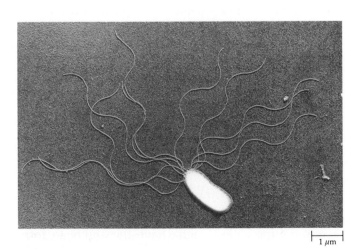

1 μm

11–5
Flagella on Pseudomonas marginalis, *a common bacterium that is widespread in soils. It causes a soft-rot disease found in many fleshy and leafy vegetables.*

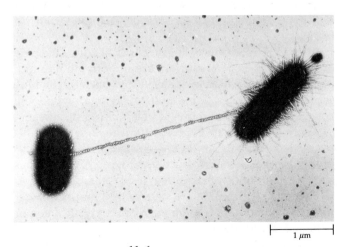

1 μm

11–6
Pili in Escherichia coli. *A conjugation pilus connects the two cells, and many shorter pili can be seen on the bacterial cell at the right.*

(which requires oxygen) yields much more energy than fermentation, as discussed in Chapter 6. Some bacteria are able to grow at very high temperatures, such as around deep-sea vents, where the temperature may approach 360°C; others grow near the freezing point of water and can survive indefinitely when it is much colder.

Bacteria play a vital role in the functioning of the world ecosystem. The autotrophic bacteria make major contributions to the global carbon balance; the role of certain other groups of bacteria in fixing atmospheric nitrogen is likewise of major significance (see page 196). Through the action of decomposers, materials incorporated into the bodies of once-living organisms are released and made available for successive generations. More than 90 percent of the CO_2 production in the biosphere, other than that associated with human activities, results from the metabolic activity of bacteria and fungi. The abilities of certain bacteria to decompose synthetic substances, such as petroleum, pesticides, and dyes, may lead to their widespread use in cleaning up spills and dumps when the techniques of using these bacteria are better developed.

As one example of the way in which bacteria participate in global biological processes, consider the element sulfur. Sulfur is made available to plants (which cannot utilize elemental sulfur) by chemoautotrophic bacteria, such as some strains of *Thiobacillus*, which oxidize elemental sulfur to sulfate:

$$2S + 2H_2O + 3O_2 \longrightarrow 4H^+ + 2SO_4^{2-}$$

Sulfates are accumulated by plants, and the sulfur is incorporated into proteins. The degradation of proteins (see Chapter 27) liberates the sulfur-containing amino acids, among others, and a number of bacteria are capable of breaking down these amino acids to release hydrogen sulfide (H_2S). Sulfates are also reduced to H_2S by certain soil microorganisms, such as *Desulfovibrio*.

In addition to their other ecological roles, bacteria are important as agents of disease in both animals and plants. Human diseases caused by bacteria include tuberculosis, cholera, anthrax, gonorrhea, whooping cough, bacterial pneumonia, Legionnaire's disease, typhoid fever, botulism, syphilis, diphtheria, and tetanus. More than 200 species of bacteria are associated with plant diseases, some of which are described later in this chapter; many of these diseases are highly destructive.

Industrially, bacteria are sources of a number of important antibiotics: streptomycin, aureomycin, neomycin, and tetracycline, for example, are produced by actinomycetes. Bacteria are also widely used commercially for the production of drugs and other substances, such as vinegar, various amino acids, and enzymes. The production of almost all cheeses involves bacterial fermentation of lactose into lactic acid, which coagu-

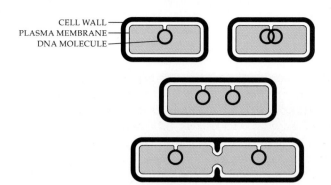

11–7
A schematic diagram of the mode of attachment of bacterial DNA to the plasma membrane, which leads to the distribution of one DNA molecule to each daughter cell. As you can readily see by comparison with Figure 11–1, the relative length of the DNA is actually much greater than shown here.

lates milk proteins. The same kinds of bacteria that are responsible for the production of cheese are also responsible for the production of yogurt and for the lactic acid that preserves sauerkraut and pickles.

BACTERIAL REPLICATION

The chief mode of reproduction in bacteria is by **fission:** each cell increases in size and divides into two cells. In this process, the plasma membrane and cell wall grow inward, eventually dividing the cell in half (Figure 11–7). Chains of bacteria are formed when the newly formed cell wall only partially splits. During fission, the single, circular, double-stranded DNA molecule that contains the hereditary information—the bacterial chromosome—is believed to be attached to a point on the interior surface of the plasma membrane. After the DNA molecule replicates, two identical circular DNA molecules are attached side by side to the plasma membrane. New plasma membrane and cell wall material is formed between these two points of attachment as the cell divides, and the membrane eventually pinches in between the two points, providing each daughter cell with an identical DNA molecule.

Major Groups of Bacteria

A few groups of Eubacteria have the ability to form thick-walled **endospores,** which enable these bacteria to survive for extended periods when growth is not possible. Endospores are resistant to dehydration and heat, some being able to withstand boiling water for 1 hour. They are formed by the division of a bacterial cell into two or more portions. A very thick coat is formed around each spore, enclosing a DNA molecule. The survival capacity of these spores is incredible; they may germinate to produce new individuals after being held for decades or even centuries under conditions unfavorable to growth.

Genetic recombination in bacteria follows the transfer of a portion of a DNA molecule from one cell to another. These fragments may be transmitted by direct contact (conjugation; see Figure 11–6), be carried into a cell by a virus (transduction), or enter a cell in solution as "naked" DNA (transformation). In the new cell, these fragments either may act in concert with the chromosomal DNA molecule to produce messenger RNA or may actually be incorporated into the host chromosome. In nearly all bacterial cells, there are small, circular fragments of DNA—**plasmids** (see Figure 11–1)—in addition to the chromosomal DNA. Some plasmids can be integrated into and then replicated with the bacterial chromosome. The transfer of plasmids by conjugation or other methods is important in the spread of characteristics, such as antibiotic resistance, from one species of bacterium to another. Genetic engineering is mainly based on the use of bacterial plasmids (see Chapter 25).

For most bacteria, in which genetic recombination is not known, mutation is the major source of variability that allows the bacteria to adjust to new conditions. Since some species of bacteria divide every 12.5 minutes under optimal conditions, mutations can rapidly become established in a population. Even when genetic recombination does occur, mutation still provides the basic variability, as it does for all organisms.

We shall now consider in more depth some of the major groups of prokaryotes under the headings of the two kingdoms, Archaebacteria and Eubacteria. The groups selected for individual treatment under the Eubacteria are those we consider to be of special evolutionary and ecological importance. These include the cyanobacteria, mycoplasmas, photosynthetic bacteria, nitrogen-fixing bacteria, and bacteria that cause diseases in plants.

ARCHAEBACTERIA

The Archaebacteria are diverse but have in common a series of biochemical characteristics of fundamental importance, including base sequences in rRNAs, lack of muramic acid in their cell walls, and lipid composition of their plasma membranes. Archaebacteria are diverse metabolically and, to some extent, morphologically also. Genetic recombination has not been observed in the members of this kingdom.

One of the leading groups of Archaebacteria consists of the methanogenic bacteria, or **methanogens**

(see Figure 10–6). These bacteria, which have long been recognized as a distinct group since they are the only microorganisms that produce a hydrocarbon (methane) as their major metabolic product, are strictly anaerobic. They are common in sewage-treatment plants, in bogs, in the ocean depths, and in the digestive tracts of cattle and other ruminant mammals (those that chew a cud), where they make possible the digestion of cellulose. Most of the natural gas reserves we now use as fuel were produced by the activities of methanogenic bacteria in the past. Some methanogens produce methane from CO_2 and H_2, obtaining their own energy in the process. Three divisions of methanogenic Archaebacteria have been recognized, and there are probably more still to be discovered or described properly.

One of the Archaebacteria, *Methanosarcina*, is capable of fixing atmospheric nitrogen, a property that is otherwise confined to Eubacteria and which will be discussed later.

A second major group of Archaebacteria consists of extreme **halophiles:** bacteria that occur everywhere in nature where the salt concentration is high—in places such as salt and soda lakes, as well as in basins where seawater is left to evaporate, yielding table salt (Figure 11–8). Halophiles grow best at temperatures from 35 to 50°C. One member of this group is *Halobacterium halobium*, which has in its plasma membrane patches of purple pigment, bacteriorhodopsin, that are in effect light-driven pumps. These pigments enable the bacterium to pump out protons, much as in the oxidative phosphorylation that occurs in mitochondria (see Chapter 6), and to harness the proton gradient produced for the synthesis of ATP—a source of energy. The members of this group are therefore autotrophic, but the system that makes them autotrophic differs radically from the systems found in all other groups of autotrophic organisms.

A third major group of Archaebacteria consists of extremely **thermophilic** ("heat-loving"), mostly sulfur-metabolizing bacteria. These bacteria may be either autotrophic, obtaining their energy by forming hydrogen sulfide (H_2S) from elemental sulfur or by comparable pathways, or heterotrophic. They grow best at temperatures from 70 to 105°C and so occur around hot springs and in similar places. The genus *Pyrococcus*, for example, occurs around hot-water vents at the bottom of the Mediterranean Sea.

A fourth group of Archaebacteria consists of a single genus, *Thermoplasma*, containing a single species. *Thermoplasma* resembles the mycoplasmas (page 195) in lacking a cell wall and being very small; individuals vary from spherical (0.3 to 2 micrometers in diameter) to filamentous. *Thermoplasma* was confused with the true mycoplasmas, which are Eubacteria, until its rRNA base sequences were examined by C. R. Woese of the University of Illinois and his colleagues; their results

11–8
Halophilic archaebacteria growing in
evaporating ponds of seawater, as seen in
an aerial view taken near San Francisco
Bay, California. The ponds yield table
salt, as well as other salts of commercial
value. As the water evaporates and
the salinity increases, the halophiles
multiply, forming massive growths, or
"blooms," that turn the seawater pink or
reddish purple.

showed clearly that it is an extremely reduced member of the Archaebacteria. *Thermoplasma* is only known to occur in acidic, self-heating coal refuse piles in southern Indiana and western Pennsylvania, in areas within the piles where the temperatures range from 32 to 80°C—the sort of very unusual habitat where Archaebacteria seem to thrive.

Archaebacteria often grow in places that resemble habitats characteristic of the early earth. In view of their overall distinctiveness, it has been suggested that the methanogenic bacteria might have originated more than 3 billion years ago, when there was an atmosphere rich in CO_2 and H_2, and that they might have persisted in certain habitats similar to those in which they first evolved.

EUBACTERIA

Eubacteria, with many more major evolutionary lines, families, and genera than Archaebacteria, are highly diverse. Virtually all characteristics discussed in this chapter pertain to Eubacteria, with the exception of the distinctive features of Archaebacteria just discussed.

Cyanobacteria

The cyanobacteria deserve special emphasis because of their great ecological importance, especially in the global carbon and nitrogen cycles, as well as their evolutionary significance. Photosynthetic cyanobacteria have chlorophyll *a*, together with carotenoids and other, unusual accessory pigments known as **phycobilins.** There are two kinds of phycobilins: **phycocyanin,**

a blue pigment, and **phycoerythrin,** a red one. Within the cells of cyanobacteria are numerous layers of membranes, often parallel to one another (see Figure 11–16a), and masses of ribosomes. These membranes are photosynthetic thylakoids that resemble those found in chloroplasts, which, in fact, correspond in size to the entire cyanobacterial cell. The main storage product of the cyanobacteria, as in all bacteria, is glycogen. Cyanobacteria resemble chloroplasts and may have given rise to at least some eukaryotic chloroplasts symbiotically. In biochemical detail, cyanobacteria are especially similar to the chloroplasts of red algae (see page 270).

Most cyanobacteria have a mucilaginous sheath, or coating, which is often deeply pigmented, particularly in species that sometimes occur in terrestrial habitats. The colors of the sheaths in different species include light gold, yellow, brown, red, emerald green, blue, violet, and blue-black. Despite their former name ("blue-green algae"), therefore, only about half of the species of cyanobacteria are actually blue-green in color.

Although more than 7500 species of cyanobacteria have been described and given names, experimental studies suggest that there may actually be as few as 200 distinct, free-living nonsymbiotic species. Like other bacteria, cyanobacteria sometimes grow under extremely inhospitable conditions, from the water of hot springs to the frigid lakes of Antarctica, where they sometimes form luxuriant mats 2 to 4 centimeters thick in water beneath more than 5 meters of permanent ice. However, cyanobacteria are absent in acidic waters, where eukaryotic algae are often abundant. The greenish color of some polar bears in zoos is due to the presence of colonies of cyanobacteria that develop within the hollow hairs of their fur.

Layered chalk deposits called stromatolites (Figure 11–9), which have a continuous geologic record covering 2.7 billion years, are produced when colonies of cyanobacteria bind calcium-rich sediments. Today, stromatolites are formed only in a few places, such as shallow pools in hot, dry climates. Their abundance in the fossil record is evidence that such environmental conditions were prevalent in the past, when cyanobacteria played the decisive role in elevating the level of free oxygen in the atmosphere of the early earth.

Cyanobacteria often form filaments and may grow in large masses 1 meter or more in length. Some cyanobacteria are unicellular, a few form branched filaments, and a very few form plates or irregular colonies (Figure 11–10; see also Figure 10–5b and c). A cyanobacterial cell (except for heterocysts, discussed below) may divide, and the resulting subunits may then separate to form new colonies. In addition, the filaments may break into fragments called **hormogonia.** As in other filamentous or colonial bacteria, the cells of cyanobacteria are usually joined only by their walls or by mucilaginous sheaths, so that each cell leads an independent life; small plasmodesmata have been found, however, in some cyanobacteria.

Some filamentous cyanobacteria are motile, gliding and rotating around the longitudinal axis. Short segments that break off from a cyanobacterial colony may glide away from their parent colony at rates as rapid as 10 micrometers per second. This movement may be connected with the extrusion of mucilage through small pores in the cell wall, together with the production of contractile waves in one of the surface layers of the wall. Some cyanobacteria exhibit intermittent jerky movements.

The cells of cyanobacteria living in freshwater or marine habitats—especially those that inhabit the surface layers of the water, in the community of microscopic organisms known as the **plankton**—commonly contain bright, irregularly shaped structures called gas vacuoles. These vacuoles provide and regulate the buoyancy of the organisms, thus allowing them to float at certain levels in the water. When numerous cyanobacteria become unable to regulate their gas vacuoles properly—for example, because of extreme fluctuations of temperature or oxygen supply—they may float to the surface of the body of water and form visible masses called "blooms." Some cyanobacteria that form blooms secrete chemical substances that are toxic to other organisms, causing large numbers of deaths. The Red Sea apparently was given its name because of the blooms of planktonic species of *Trichodesmium*, a red cyanobacterium.

Many marine cyanobacteria occur in limestone (calcium carbonate) or lime-rich substrates, such as coralline algae (see page 272) and the shells of mollusks. Some freshwater species, particularly those that grow

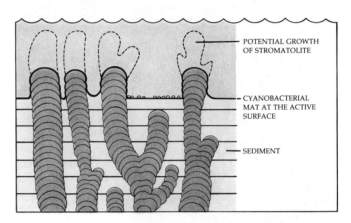

POTENTIAL GROWTH OF STROMATOLITE

CYANOBACTERIAL MAT AT THE ACTIVE SURFACE

SEDIMENT

11–9

Stromatolites are produced when flourishing colonies of cyanobacteria bind calcium carbonate into domed structures, like the ones shown in the diagram and photograph, or others with a more intricate form. Such structures are abundant in the fossil record but today are being formed only in a few, highly suitable environments, such as that shown here on the tidal flats of Hamlin Pool in Western Australia.

in hot springs, often deposit thick layers of lime in their colonies.

Many genera of cyanobacteria can fix nitrogen (see page 196). In filamentous cyanobacteria, the nitrogen fixation often occurs within **heterocysts,** which are specialized, enlarged cells (Figures 11–11 and 11–12). The wall of a heterocyst is identical to that of any other cyanobacterial cell, but outside the wall, the heterocyst forms a bilayered envelope. The outer portion of this envelope consists of a polysaccharide and the inner layer of glycolipids. Within a heterocyst, the cell's internal membranes are reorganized into a concentric or reticulate pattern. Since heterocysts lack photosystem II

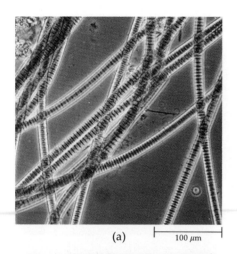

(a) |—————| 100 µm

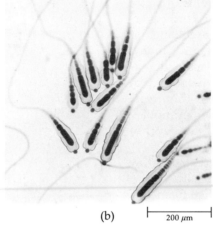

(b) |———| 200 µm

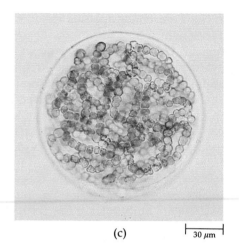

(c) |—| 30 µm

(d) |———| 50 µm

11–10

Three common genera of cyanobacteria and a related form. (a) Oscillatoria, *in which the only form of reproduction is by means of fragmentation of the filament. (b)* Calothrix, *a filamentous form with a basal heterocyst (see page 193).* Calothrix *is capable of forming akinetes —enlarged cells that develop a resistant outer envelope—just above the heterocysts. (c) A gelatinous "ball" of* Nostoc commune, *containing numerous filaments. These cyanobacteria occur frequently in freshwater habitats. (d)* Thiothrix, *a genus that lacks chlorophyll but is apparently closely allied to the cyanobacteria.* Thiothrix *obtains energy by the oxidation of* H_2S. *The chains of cells, each filled with particles of sulfur, are attached to the substrate at the base (center of micrograph) and so form a characteristic rosette.*

|—| 5 µm

11–11

Cell division in Anabaena. *This electron micrograph shows a chain of cells held together along incompletely separated walls, as well as a single cell (at the right) undergoing cell division. The first cell on the right end of the chain is a heterocyst. The sort of cell division shown here, in which the margins of the cell grow inward, is characteristic of all organisms except plants and a few genera of algae, in which a cell plate is formed.*

(see Chapter 7), the cyclic photophosphorylation occurring in these cells does not result in the evolution of oxygen. The oxygen that is present is either rapidly reduced by hydrogen, a by-product of nitrogen fixation, or expelled through the wall of the heterocyst. Nitrogenase, the enzyme that catalyzes the nitrogen-fixing reactions, is sensitive to the presence of oxygen, and nitrogen fixation is an anaerobic process. The formation of heterocysts in *Nostoc* and other genera is inhibited by the presence of ammonia or nitrates, but when the levels of nitrogen-containing substances fall below a specific threshold, heterocysts begin to appear.

In addition to the heterocysts, some cyanobacteria form resistant spores called **akinetes** (Figures 11–10b and 11–12), enlarged cells around which thickened envelopes develop. Like the endospores formed by other Eubacteria, akinetes are resistant to heat and drought and thus allow the cyanobacterium to survive during unfavorable periods.

Cyanobacteria may occur as symbionts within the bodies of some sponges, amoebas, flagellated protozoa, diatoms, green algae that lack chlorophyll, other cyanobacteria, mosses, liverworts, vascular plants, and oomycetes, in addition to their familiar role as the photosynthetic partner in many lichens (see Chapter 12). Some symbiotic cyanobacteria lack a cell wall, in which

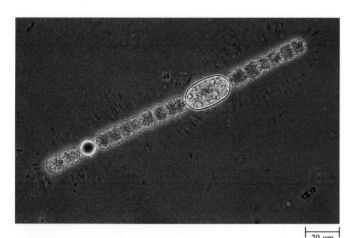

11–12
The filaments of Anabaena, *which is a nitrogen-fixing cyanobacterium, are composed of barrel-shaped cells held in a gelatinous matrix. Nitrogen fixation takes place within specialized cells called heterocysts (third cell from the left). Like* Calothrix *(see Figure 11–10b),* Anabaena *forms akinetes (large oval body toward the right). Electron micrographs of* Anabaena *also can be seen in Figures 11–11 and 11–16.*

20 μm

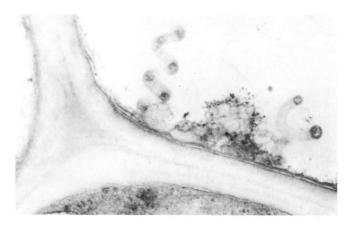

11–13
Spiroplasmas in cells of stunted corn (Zea mays) plants.

case they function as chloroplasts. The symbiotic cyanobacterium divides at the same time as the host cell by a process similar to chloroplast division. As mentioned above and to be reviewed in Chapters 13 and 14, bacteria similar to cyanobacteria appear to have given rise to at least some chloroplasts.

Mycoplasmas

Mycoplasmas include the smallest known cells—they can be as small as 0.1 micrometer in diameter. Each mycoplasmal cell consists of an external plasma membrane, DNA, RNA, ribosomes, soluble proteins, sugars, and lipids. Mycoplasmas lack cell walls. An individual mycoplasma probably contains fewer than 650 genes, which is about one-fifth the number found in *Escherichia coli* and other common Eubacteria.

Aside from *Thermoplasma*, which superficially resembles mycoplasmas but is in fact a reduced member of the kingdom Archaebacteria, there are five genera of mycoplasmas, with 50 of the 60 recognized species assigned to the genus *Mycoplasma*. The analyses of C. R. Woese and his colleagues have shown that these five genera were probably derived from a single group of Eubacteria and are therefore directly related to one another.

Mycoplasmas have been identified in more than 200 plant species and associated with more than 50 plant diseases, many with symptoms of yellowing (see Figure 11–14b) and stunting. Among these plant-pathogenic mycoplasmas are the spiroplasmas—long, thin, helical mycoplasmas that are motile in liquid and less than 0.2 micrometer in diameter (Figure 11–13). Some have been cultured on artificial media, including *Spiroplasma citri*, the agent of stubborn and little-leaf diseases of citrus, and other plant pathogens as well. Spiroplasmas exhibit vigorous whirling and flexing movements when observed in liquid culture. Proof of pathogenicity—that the organisms isolated from diseased tissues actually caused the disease—has been obtained only with a few spiroplasmas, however.

In vascular plants, mycoplasmas are generally confined to the elements of the phloem known as sieve tubes. Most mycoplasmas are believed to move passively from one sieve-tube member (see Chapter 21) to another through the sieve-plate pores, as the sugar solution is transported in the phloem (Figure 11–14a). Motile spiroplasmas, however, may also be able to move actively in phloem tissue.

Photosynthetic Bacteria

There are at least five groups of photosynthetic Eubacteria: (1) cyanobacteria (see pages 192–195), (2) green sulfur bacteria, (3) purple sulfur bacteria (see Figure 7–2, page 100), (4) purple nonsulfur bacteria, and (5) *Prochloron*. Like plants, photosynthetic bacteria contain chlorophyll. Like photosynthetic eukaryotes, the cyanobacteria and *Prochloron* contain chlorophyll *a*. The chlorophylls present in the other groups of photosynthetic bacteria differ in several ways from chlorophyll *a*, but they all have the same basic structure (see Figure 7–6, page 104).

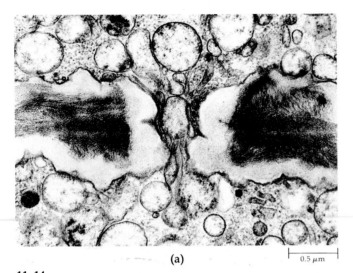

(a)

0.5 μm

(b)

11–14
(a) *Mycoplasma-like organisms apparently traversing a sieve-plate pore in a young inflorescence of a coconut palm (Cocos nucifera) affected by lethal yel-* *lowing disease.* (b) *A devastated grove of coconut palms—now looking like telephone poles—in Jamaica. Lethal yel-* *lowing has been responsible for the death of palms of many genera in southern Florida and elsewhere.*

The colors characteristic of photosynthetic bacteria are associated with the presence of several accessory pigments that function in photosynthesis. In the two groups of purple bacteria, these pigments are yellow and red carotenoids; in the cyanobacteria, as we have seen, they are phycobilins.

In the green sulfur and purple sulfur bacteria, sulfur compounds play the same role in photosynthesis that water plays in organisms containing chlorophyll *a*. That is,

$$CO_2 + 2H_2S \xrightarrow{\text{light}} (CH_2O) + H_2O + 2S$$

As discussed in Chapter 7, an understanding of the course of photosynthesis in the purple sulfur bacteria led C. B. van Niel to propose the generalized equation of photosynthesis:

$$CO_2 + 2H_2A \xrightarrow{\text{light}} (CH_2O) + H_2O + 2A$$

in which H_2A is a generalized hydrogen donor.

In the photosynthetic purple nonsulfur bacteria, other compounds, including alcohols, fatty acids, and keto acids, serve as electron (hydrogen) donors for the photosynthetic reaction. The bacterium that became symbiotic early in the history of eukaryotes, giving rise to mitochondria, probably was a member of this group. This conclusion is based both on the similar metabolic features of mitochondria and purple nonsulfur bacteria and on comparisons of the base sequences in their rRNAs.

Because of their requirement for H_2S or a similar substrate, the photosynthetic sulfur bacteria are able to grow only in habitats that contain large amounts of decaying organic material, recognizable by its sulfurous odor. In these bacteria, elemental sulfur may accumulate as deposits within the cell (see Figure 11–10d).

The fifth group of photosynthetic bacteria consists of the single genus *Prochloron* (see Figure 11–2). As far as is known, *Prochloron* lives only along tropical seashores, as a symbiont within colonial sea squirts (ascidians). *Prochloron* is one of a very few genera of prokaryotes to contain the same photosynthetic pigments as green algae and plants—namely, chlorophylls *a* and *b*, together with carotenoids. Despite these similarities, however, *Prochloron* is unlikely to be the actual ancestor of green algal chloroplasts because the rRNA base sequences of *Prochloron* resemble those of cyanobacteria more closely than those of green algal chloroplasts.

Nitrogen-Fixing Bacteria and the Nitrogen Cycle

The conversion of nitrogen gas, which constitutes about 80 percent of the atmosphere, to **nitrates**—a form in which the nitrogen is available for biological reactions—is **nitrogen fixation,** an ecological process of fundamental importance. Of all living organisms, only a few genera of bacteria are capable of nitrogen fixation. Outstanding among them are the symbiotic bacteria *Rhizobium* and *Bradyrhizobium* (see Chapter 27), which form nodules on the roots of legumes. In addition, a group of filamentous bacteria known as actino-mycetes are involved in nitrogen fixation, forming nodules on the roots of several groups of trees and shrubs and a few perennial herbs, including alders (*Alnus*), wax myrtles (*Myrica*), and mountain lilacs (*Ceanothus*). Actinomycetes are efficient nitrogen fixers that effectively contribute to the accumulation of nitrogen in soils (Figure 11–15).

In addition to *Rhizobium, Bradyrhizobium,* and the actinomycetes, which fix nitrogen within nodules that form on the roots of plants, certain nitrogen-fixing bacteria are regularly associated with the roots and leaves of plants, where they utilize the carbohydrate-rich exudates from the plants and in turn provide nitrogen in a usable form.

Many free-living bacteria also play important roles in nitrogen fixation. In the ocean, for example, species of the cyanobacterium *Trichodesmium* account for about a quarter of the total nitrogen fixed there, an enormous amount.

Symbiotic cyanobacteria are likewise very important in nitrogen fixation. The ones that occur as the photosynthetic partner in many lichens, for example (see Chapter 12), enhance the capacity of the lichens to colonize bare areas, where fixed nitrogen is in especially short supply. In the warmer parts of Asia, rice can often be grown continuously on the same land without the addition of fertilizers because of the presence of nitrogen-fixing cyanobacteria in the rice paddies. Here the cyanobacteria, especially members of the genus *Anabaena* (Figure 11–16), often occur in association with the small, floating water fern *Azolla*, which forms masses on the paddies. Because of the nearly obligate association of *Azolla* with *Anabaena* (see page 607),

(a)

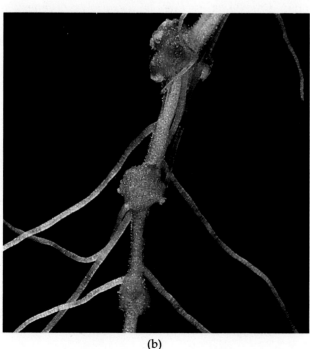

(b)

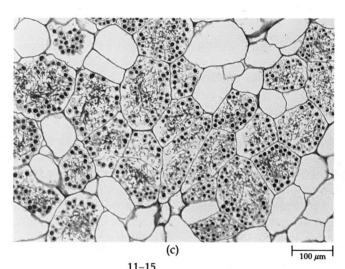

(c)

$\vdash\!\!-\!\!-\!\dashv$ 100 μm

11–15

Alders form symbiotic relationships with actinomycetes of the genus Frankia, *which fix nitrogen. Three aspects of the relationship are shown here, all involving red alder,* Alnus rubra, *as studied by John Torrey and his colleagues at the Harvard Forest in Massachusetts.* (a) *Alder seedlings grown in different soil mixtures after inoculation with* Frankia. *The seedlings at the left were not inoculated and had no nodules; those at the right were inoculated and formed abundant nodules.* (b) *Nodules on alder root.* (c) *Micrograph of a section of a root nodule lobe, stained with toluidine blue, showing cortical cells filled with* Frankia. *The bacteria fill the center of the infected cells, and the swollen terminal ends of the filaments, called vesicles, fill the periphery. The vesicles are the probable site of the N₂-fixing enzyme, nitrogenase.*

11–16
(a) *A cell of the cyanobacterium* Ana-baena azollae, *showing the major features visible with the electron microscope. The gelatinous sheath of this cell has been destroyed in preparing the specimen. This organism is associated with the common floating water fern* Azolla *and is responsible for nitrogen fixation in these plants (see Figure 10–3a and Chapter 27). (b) Women planting rice in a paddy in Perak, Malaysia. In Southeast Asia, rice can often be grown on the same land continuously without the addition of fertilizers because of* Anabaena azollae, *which grows in the tissues of* Azolla *growing in the rice paddies.*

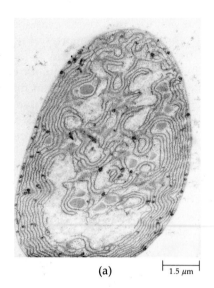

(a) ⊢——⊣ 1.5 μm

(b)

paddies covered with *Azolla* remain rich in fixed nitrogen. *Anabaena* carries out its nitrogen fixation in heterocysts (see page 194), as do other genera of free-living cyanobacteria, such as *Nostoc*.

The nitrogen contained within proteins is returned to the nitrogen cycle by a series of steps carried out by a variety of microorganisms. Certain microorganisms break down proteins into peptides, which are subsequently broken down into their constituent amino acids. Many organisms use a process called *ammonification* to break down amino acids, with the consequent production of ammonium ions (NH_4^+; see Chapter 27). Ammonia (NH_3), which is in equilibrium with ammonium ions, can be converted to nitrite ions (NO_2^-), by the chemoautotrophic bacteria *Nitrosomonas* and *Nitrococcus*, and the nitrites are oxidized to nitrates (NO_3^-) by the bacterium *Nitrobacter*. The conversion of ammonia to nitrites and nitrates constitutes the process of **nitrification.** This process releases energy, which is used by these chemoautotrophs to reduce carbon dioxide to carbohydrate. Several other kinds of bacteria are capable of reversing the process, converting nitrates to nitrites and, ultimately, ammonia. **Denitrification,** the conversion of nitrates to nitrogen gas or nitrous oxide, results in the loss of nitrogen from the soil and thus completes the nitrogen cycle (see Figure 27–8).

Plant-Pathogenic Bacteria

Many economically important diseases of plants are caused by Eubacteria, which contribute substantially to the one-eighth of crops worldwide that are lost to disease. Almost all kinds of plants can be affected by bacterial diseases, and many of these diseases can be extremely destructive (Figure 11–17).

Virtually all plant-pathogenic bacteria are gram-negative, rod-shaped bacteria. They are parasites—symbionts that are harmful to their hosts. The symptoms caused by plant-pathogenic bacteria are quite varied, with the most common appearing as spots of various sizes on stems, leaves, flowers, and fruits (Figure 11–18). Almost all such bacterial spots are caused by members of two closely related genera, *Pseudomonas* and *Xanthomonas* (see the essay on page 200).

11–17
This population of giant saguaro cactus (Carnegiea gigantea) at an elevation of 1000 meters on the southern slope of the Santa Catalina Mountains near Tucson, Arizona, was attacked by the bacterium Erwinia carnegieana *after being weakened by severe freezing conditions in January 1962.*

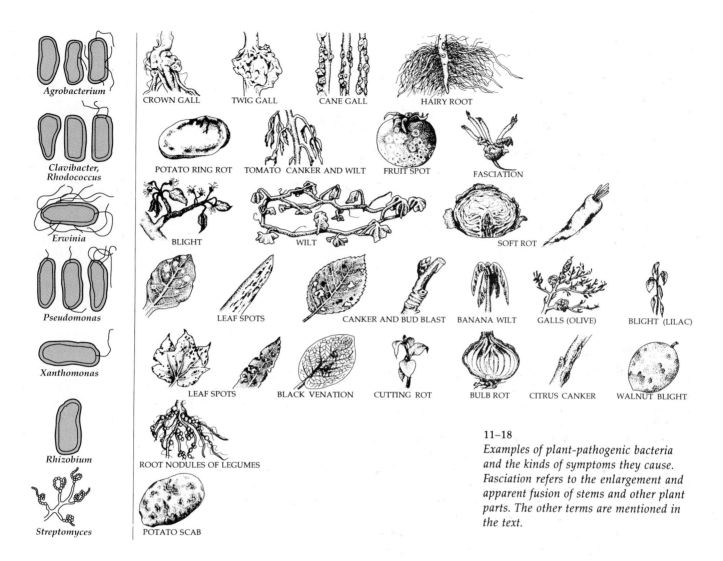

11–18
Examples of plant-pathogenic bacteria and the kinds of symptoms they cause. Fasciation refers to the enlargement and apparent fusion of stems and other plant parts. The other terms are mentioned in the text.

Some of the most destructive diseases of plants—such as blights, soft rots, and wilts—are also caused by bacteria. Blights are characterized by rapidly developing necroses (dead, discolored areas) on stems, leaves, and flowers. Fire blight in apples and pears, caused by *Erwinia amylovora*, is a widespread, economically important disease that can kill young trees within a single season. Fire blight was so destructive to pear trees that by the 1930s, their cultivation on a commercial scale had been eliminated from almost all parts of the United States except the Northwest. Bacterial soft rots occur most commonly in the fleshy storage organs of vegetables, such as potatoes or onions, and in fleshy fruits, such as tomatoes and eggplants. The most destructive soft rots are caused by bacteria of the genus *Erwinia*, with heavy losses occurring in the post-harvest period.

Bacterial vascular wilts mainly affect herbaceous plants. The bacteria invade the vessels of the xylem, where they multiply, interfering with the movement of water and inorganic nutrients and resulting in the wilting and death of the plants. The bacteria commonly degrade portions of the vessel walls and can even cause the vessels to rupture. Once the walls have ruptured, the bacteria then spread to the adjacent parenchyma tissues, where they continue to multiply. In some bacterial wilts, the bacteria ooze to the surface of the stems or leaves through cracks formed over cavities filled with cellular debris, gums, and bacteria. More commonly, however, the bacteria do not reach the surface of the plant until the plant has been killed by the disease. Among the most important examples are bacterial wilt of alfalfa and bean plants (each caused by different species of *Clavibacter*); bacterial wilt of cucurbits, such as squashes and watermelons (caused by *Erwinia tracheiphila*); and black rot of crucifers, such as cabbage (caused by *Xanthomonas campestris*). The most economically important wilt disease of plants, however, is caused by *Pseudomonas solanacearum*. It affects at least

Citrus Canker

One of the most dramatic examples of a human reaction to a bacterial disease of plants began in August 1984, when diseased citrus seedlings were discovered at a nursery in Avon Park, Florida. The diagnosis was that the seedlings had a form of citrus canker, a disease caused by one of the more than 100 disease-causing groups of the bacterium *Xanthomonas campestris*. Some strains of this bacterium cause lesions on the stems, leaves, and fruits of citrus trees, spread easily through an infected grove, and are transported widely on crates, clothing, and transported plants, and in many other ways.

The appearance of citrus canker greatly alarmed Florida citrus producers and agricultural authorities because of the devastating effects of an Asiatic form of *X. campestris* that had caused great destruction to the citrus groves in the state from 1915 to 1927, when it was thought to have been eliminated. More than 3 million citrus trees were destroyed during that 12-year period. Following the detection of citrus canker in 1984, over 7 million citrus seedlings were destroyed within four months in an effort to control the disease, and major funds were committed to research.

Unfortunately, the major eradication program that was begun in 1984 may have been based on a costly error. Although many cultures of *X. campestris* are maintained in culture collections, we do not know enough about the characteristics of many of them to determine their pathogenicity accurately. In retrospect, it appears that the bacterial strain that appeared in 1984 has leaf spotting as its primary effect. There is no evidence that it destroys fruits and entire trees like the strain that is believed to have disappeared from Florida in 1927. Such destructive strains were discovered at a few localities along the western coast of Florida in 1987, but they seem not to have been the ones that appeared in 1984 and set off the major eradication program.

Ultimately, the State of Florida burned approximately 20 million citrus trees during the 1980s, fearful that citrus canker might wipe out their $1 billion-a-year citrus industry. Believing that State authorities confused a relatively harmless, leaf-spotting strain of *X. campestris* with the much more destructive Asian strains, many nursery owners are now suing the State of Florida for reimbursement for plants that were seized and destroyed. Thus the threat of a new outbreak of citrus canker has been extremely costly both in terms of funds expended for destruction of nursery stock and in terms of legal claims against the State. This example illustrates not only how important it is to make a proper diagnosis of disease agents but also how difficult it is to distinguish different strains of a pathogen such as *X. campestris*.

---

44 different genera of plants, including such major crops as bananas, peanuts, tomatoes, potatoes, eggplants, and tobacco, to name a few. This disease occurs worldwide in tropical, subtropical, and warm temperate areas.

The members of another genus of bacteria, *Agrobacterium*, cause crown gall in plants, "gall" being a general term for swelling. Because of their role in genetic engineering, these bacteria will be discussed in Chapter 25.

Viruses

A virus consists primarily of a genome that replicates itself only within a specific host cell by directing the machinery of that cell to synthesize viral nucleic acids and proteins (Figure 11–19). Since no virus can grow or replicate on its own outside of a cell, viruses are not considered to be alive in the usual sense. Virus genomes are composed of highly organized sequences of nucleic acids, either DNA or RNA—double-stranded or single-stranded—depending on the virus. All viruses (but not viroids, described later) have a protein covering, called a capsid, which encloses the nucleic acid, and some viruses also have a lipid-rich envelope surrounding the capsid. The protein or lipoprotein coat determines the surfaces to which a particular kind of virus will adhere and protects the nucleic acid as the virus passes from one host to another.

Viruses are important because they are so widespread in the cells of plants and other organisms, and because they frequently cause disease. In humans, viruses are responsible for many infectious diseases, including smallpox, chicken pox, measles, mumps, influenza, colds (often complicated by secondary bacterial infections), infectious hepatitis, polio, rabies, herpes, and the deadly AIDS (acquired immune deficiency syndrome). They also cause many diseases in pet, farm, and wild animals and wild and cultivated plants; some viruses that infect plants are closely related to those that affect animals. Viruses cause diseases by disrupting the normal functioning of the host cells and tissues.

The cells of every kind of organism may at times contain one or more distinctive forms of virus; there

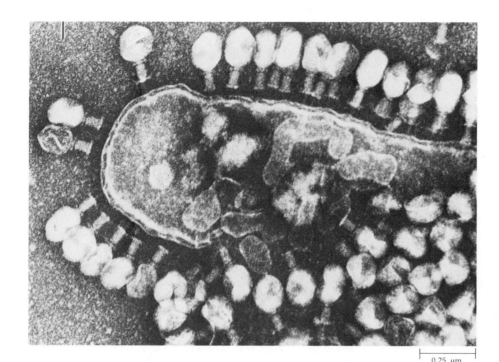

Bacterial viruses infecting a cell of Escherichia coli. *Some of the viruses have discharged their DNA into the bacterial cell, which has begun to break down.*

0.25 μm

may be literally millions of different viruses. Over 400 plant viruses have been described; they fall into about 25 distinct groups. Viruses that infect bacteria are called **bacteriophages** ("bacteria eaters") or simply **phages** (Figure 11–19). Such viruses, and others, have been widely used in molecular biological experiments. Viruses also serve as gene-transfer systems; as such, they have become indispensable tools of modern biotechnology (see Chapter 25).

THE NATURE OF VIRUSES

The existence of viruses was first recognized when the causative agent of tobacco mosaic disease was found to pass through the porcelain filters that were commonly used to trap bacteria. Viruses range in size from about 15 nanometers to more than 300 nanometers in diameter. Thus they are comparable in size to large molecules; a hydrogen atom is about 0.1 nanometer in diameter, and a protein molecule is at most a few nanometers thick.

When a virus multiplies within a host cell, it essentially "takes over" the biosynthesis of that cell, using its own nucleic acids to "command" the host to produce more virus particles, thereby competing with the genetic material of the host cell in regulating cell functions. Since viruses are restricted to certain hosts, the responses of individual plant species exposed to an unknown virus can be used to aid in the identification of the virus.

Until the 1930s, viruses were considered to be extremely small bacteria. Evidence against this point of view began to accumulate in 1933, when Wendell Stanley, a biochemist working at the Rockefeller Institute for Medical Research, prepared an extract from diseased tobacco plants that contained the tobacco mosaic virus and purified it. In the presence of high salt concentrations, the purified virus precipitated in the form of crystals, thus behaving more like a chemical than an organism. When these needlelike crystals were redissolved and the solution applied to a tobacco leaf, the characteristic symptoms of tobacco mosaic disease appeared. Thus the virus had retained its ability to infect tobacco plants even after it had been crystallized, unlike any known organism.

Most plant viruses, like the tobacco mosaic virus, are RNA viruses, having an RNA genome and containing no DNA. Many animal viruses are DNA viruses, containing no RNA. In sharp contrast to viruses, all independent living cells contain both RNA and DNA. Viruses also lack plasma membranes, cytoplasm, and ribosomes, as well as the enzymes necessary for protein synthesis and energy production.

Two groups of plant viruses, the caulimoviruses and geminiviruses, have DNA as their genetic material. Geminiviruses are small, spherical particles that almost always exist in pairs. Bean golden mosaic is a plant disease caused by a geminivirus; it is spread from plant to plant by whiteflies. Another such disease is maize streak virus, which is spread by leafhoppers and has the smallest known genome of any virus.

VIRAL STRUCTURE AND REPLICATION

The structures of a large number of viruses have now been elucidated by electron microscopy. The tobacco mosaic virus, for example, is a rodlike particle about 300 nanometers long and 15 nanometers in diameter, with a single strand of RNA over 6000 bases long (Figure 11–20). The capsid consists of over 2000 identical protein molecules arranged with a helical symmetry (Figure 11–21). The most common shape of a virus particle, however, is an icosahedron, a 20-sided figure (Figure 11–22). Many plant viruses have such a shape, including Tulare apple mosaic (Figure 11–23a), tobacco ringspot, cucumber mosaic, and cowpea mottle viruses. Other plant viruses are bullet-shaped (Figure 11–23b) or appear as long, flexible threads (Figure 11–23c).

In the course of its replication, a virus sheds its capsid and, if one is present, its envelope, freeing its nucleic acid. Within its host, the RNA or DNA of the virus then multiplies by taking over the genetic machinery of a cell, thus producing the nucleic acids and proteins necessary to reassemble additional virus particles. In an RNA virus, such as tobacco mosaic virus, the viral RNA directs the formation of a complementary strand of RNA, which then serves as the template for the production of new viral RNA molecules. The viral RNA, which is single-stranded, acts as mRNA, utilizing the ribosomes of its host cell and directing the synthesis of enzymes and viral capsid and envelope proteins.

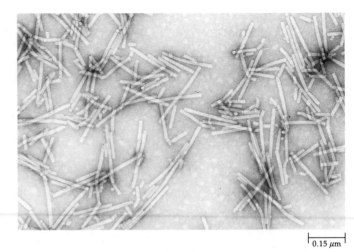

0.15 μm

11–20
Tobacco mosaic virus (TMV) particles as seen with the electron microscope.

PLANT VIRAL DISEASES

Well over 1000 kinds of plant diseases are known to be caused by viruses, with more than 400 different kinds of viruses known to be involved. The effects of some of these diseases are illustrated in Figures 11–24 and 11–25. Some viruses occur primarily in the parenchymal tissues, where they may reduce the numbers of, or sometimes eliminate, any chloroplasts that are present. Other viruses replicate in the sucrose-rich fluid circu-

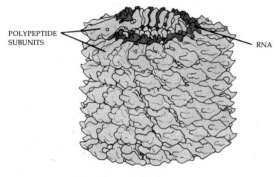

POLYPEPTIDE SUBUNITS

RNA

11–21
A diagram of a portion of a tobacco mosaic virus (TMV) particle. This virus has a central core of RNA and a protein coat, or capsid, composed of 2200 identical polypeptide molecules, each containing 158 amino acid residues, folded into irregularly shaped subunits. The RNA fits into a groove at the narrower end of the subunit and is represented here by the red chain exposed at the top.

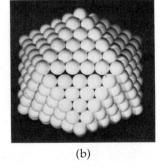

(a)   25 nm   (b)

11–22
(a) An adenovirus, one of the many kinds of viruses that cause colds in humans. This virus is an icosahedron. Each of its 20 sides is an equilateral triangle made up of protein subunits. There is a total of 252 subunits. (b) A model of an adenovirus, consisting of 252 tennis balls.

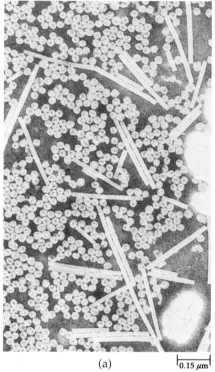

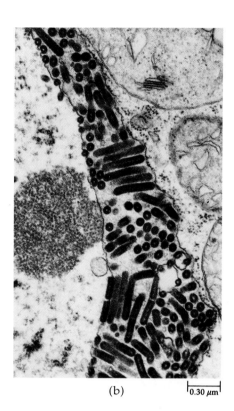

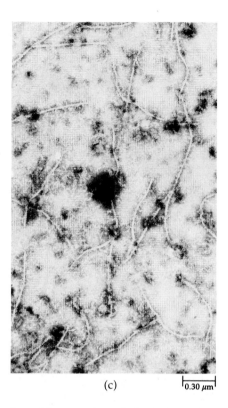

(a) $\overline{0.15\ \mu m}$ (b) $\overline{0.30\ \mu m}$ (c) $\overline{0.30\ \mu m}$

11–23

Plant viruses; all of these are RNA viruses. (a) A mixture of Tulare apple mosaic virus and tobacco mosaic virus. Tulare apple mosaic virus is icosahedral, whereas tobacco mosaic virus is rigid and elongated. (b) Particles of a rhabdovirus

in a cell from a pepper plant (Capsicum frutescens); this is a typical bacillus-shaped virus. The rhabdovirus particles are seen here in the space between the inner and outer membranes of the nuclear envelope. As the virus migrates

from this space through the inner nuclear membrane, it acquires a portion of that membrane, which then surrounds the particle. (c) Particles of a flexible, elongated virus in a cell from a Christmas cactus (Zygocactus truncatus).

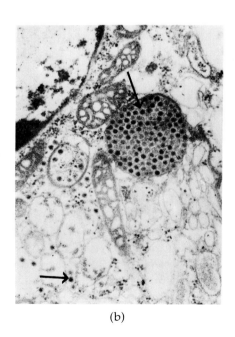

(a)   (b)   (c) $\overline{0.5\ \mu m}$

11–24

(a) Tumors produced by the wound tumor virus in sweet clover (Melilotus alba). (b) Wound tumor virus particles (located by arrows) visible in an electron

micrograph of a cell of the host plant. (c) The virus is transmitted by the clover leafhopper (Agallia constricta). This electron micrograph shows an epidermal

cell. Viruses are being produced in the honeycomblike area at the upper left. Individual viruses can be seen in the dark area below.

11–25
*Streaked flowers of Rembrandt tulips.
The streaking is caused by a viral infection transmitted directly from plant to
plant.*

lating in the phloem, a tissue that they may ultimately destroy. Viral diseases of plants are almost always transmitted from plant to plant by invertebrate animals, such as insects or nematodes. Sucking insects, such as aphids or leafhoppers, regularly transmit viruses from one plant to another as they move about sucking fluids from the phloem or merely from the epidermal cells (see Chapter 28). A few kinds of plant viruses are able to multiply in the cells of the animal vector as well as in those of their host plant.

As noted earlier, except for the geminiviruses and caulimoviruses, plant viruses are RNA viruses. With a few exceptions, among them the comoviruses (which have capsids made of two kinds of protein), the capsids of RNA plant viruses consist of many copies of a single protein. The most striking characteristic of plant RNA viruses is the frequent occurrence of divided genomes, with more than one molecule of RNA being present, all packaged within a separate but identical protein capsid. However, other viruses have two to four different-sized RNA molecules, each contained within an individual capsid, so that the virus particles may be heterogeneous in shape or density. The ways in which these individual RNA molecules interact in promoting viral infectivity differ from group to group.

Necroses and mosaic diseases are among the common plant diseases caused by viruses. In mosaic dis-

eases, light green or yellow areas—ranging in size from small flecks to large stripes—appear on the leaves or other green areas. Sometimes the whole infected plant may be lighter green than normal. Almost all mosaic diseases are caused by viruses. The yellow blotches or borders on the leaves of some prized horticultural varieties may be caused by viruses; the variegated appearance of some flowers is the result of viral infections that are passed on from generation to generation (Figure 11–25).

Viral diseases greatly reduce the productivity of many different kinds of crops around the world, and much effort is being expended to find effective ways to control these diseases. The production of virus-free plants by meristem culture is one highly effective control method. A small growing tip (meristematic tissue), which is usually free of viruses in most plants, is dissected away from the parent plant and induced to form a whole new individual. Meristem culture has been used to eliminate viruses and to greatly increase yields in several crops, including potatoes and rhubarb (see Figure 25–22).

Another approach to the control of plant viruses has focused on the transmission of viruses within the host plant. All viruses move either by short-distance cell-to-cell transmission through plasmodesmata or by long-distance spread through the vascular system. Cell-to-cell spread is often facilitated by one or more viral gene products that make possible the passage of unusually large particles through the plasmodesmata (see Figures 2–31 and 4–14). Genetically engineered plants containing specific viral genes that do not allow the passage of viral particles, thus preventing cell-to-cell transmission, will be increasingly useful as crops.

VIROIDS AND OTHER INFECTIOUS PARTICLES

Viroids are the smallest known agents of infectious disease. They are much smaller than the smallest viral genomes, they lack protein coats, and they do not code for any protein or genetic product, other than more RNA. Although the existence of similar particles is suspected in some animals, viroids have been isolated only from plants. They consist of small, single-stranded molecules of RNA that replicate autonomously in susceptible cells. Viroids have been identified as the agents responsible for some economically important plant diseases. One type of viroid, for example, has resulted in the death of millions of coconut trees in the Philippines over the last half century, and another nearly destroyed the chrysanthemum industry in the United States in the early 1950s.

The first viroid to be characterized—potato spindle

11–26
Electron micrograph of potato spindle tuber viroid (arrows) mixed with portions of the double-stranded DNA molecule of a bacteriophage. This micrograph illustrates the tremendous difference in size between a viroid and a virus.

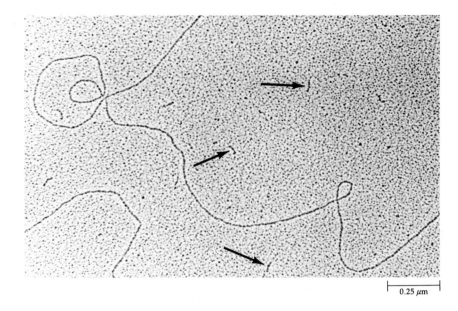

0.25 μm

tuber viroid, or PSTV—was identified by Theodor O. Diener of the U.S. Department of Agriculture in 1971. Potatoes infected with PSTV are elongated (spindle-shaped) and gnarled; they sometimes have deep crevices in their surfaces. PSTV is one of the largest viroids of the dozen or so known; its RNA, which may exist either as a closed circle or as an open, hairpin-shaped structure, is 359 bases long. In both forms, complementary base pairs are joined by hydrogen bonds, resulting in a structure resembling double-stranded RNA. Under the electron microscope, both forms of PSTV appear as rods about 50 nanometers long (Figure 11–26). The PSTV genome is only about one-tenth the size of that of the simplest virus. Viroids are found almost exclusively in the nuclei of infected cells; they replicate by forming a complementary strand that acts as a template, much like virus replication. Presumably, host enzymes catalyze this process.

Because of their location in the nucleus and their apparent inability to act as messenger RNAs, it has been suggested that viroids may cause their symptoms by interfering with gene regulation in the infected host cells. Certain plant proteins found in healthy cells are present in significantly increased quantities in infected cells.

THE ORIGIN OF VIRUSES

Because of the simplicity of viral genomes and viral structure, some earlier investigators mistakenly thought that viruses might represent the direct descendants of the self-replicating units from which the first cells evolved. This line of descent is unlikely, for viruses exist only by virtue of their ability to utilize the genetic machinery of their host cells. They compete with the nucleic acids of these cells and take over the host genetic and metabolic activities in directing the formation of new viral particles. It is clear, therefore, that viruses came into being after the evolution of cells in which the genetic code was already established.

Viruses probably originated independently on many occasions during the history of life on earth. They can evolve with astonishing rapidity in response to strong selection pressures. It is likely that novel viruses are still evolving today from both bacteria and eukaryotes.

Summary

Bacteria—the prokaryotes—are the smallest and structurally simplest organisms. They lack an organized nucleus and membrane-bound cellular organelles. Bacteria are placed in the kingdoms Eubacteria and Archaebacteria. Bacteria do not reproduce sexually, but rather by fission. Most of their genetic material is incorporated into a single circular molecule of double-stranded DNA, which replicates before cell division; often, additional small circles of double-stranded DNA, known as plasmids, are also present. Except for mycoplasmas, all bacteria have rigid cell walls. In the Eubacteria, this wall is mainly composed of peptidoglycan. Gram-negative bacteria, which have walls that do not retain the dye crystal violet, have a thick lipopolysaccharide coating over the peptidoglycan layer, the actual cell wall. All bacteria are unicellular, although their cells may adhere to one another after they divide, forming chains of cells (filaments) or other seemingly multicellular structures. Plasmodesmata are very rare, occurring only in some cyanobacteria.

The Archaebacteria, which include the methane-producers, differ from the Eubacteria in the composition of their cell walls, in their ribosomal RNA base sequences, and in other features. They survive in unusual habitats resembling those that existed early in the earth's history. Archaebacteria are more similar to eukaryotes in several fundamental features than are Eubacteria.

Most species of bacteria are heterotrophic; they share with the fungi the role of decomposers in the world ecosystem, and they are enormously abundant. As a group, the bacteria are extremely versatile metabolically; besides heterotrophs, there are photosynthetic (photoautotrophic) and chemoautotrophic bacteria. Some bacteria are aerobic, others are obligate anaerobes, and still others are facultative anaerobes. A number of genera play important roles in the cycling of nitrogen, sulfur, and carbon. Many bacteria are important pathogens, in both plants and animals. One distinctive group of bacteria, the mycoplasmas, which lack a cell wall and are very small, include a number of disease-causing organisms.

Photosynthetic bacteria can be classified into at least five distinct groups, two of which, the cyanobacteria and *Prochloron*, contain chlorophyll *a*, the same molecule that occurs in all photosynthetic eukaryotes. Bacteria seem clearly to have been involved in the symbiotic origin of chloroplasts. Early in the history of eukaryotes, a symbiotic event similar to the one involved in the origin of chloroplasts seems to have given rise to mitochondria; purple nonsulfur bacteria apparently were the group involved in this event.

Chemosynthetic bacteria derive their energy from the oxidation of inorganic molecules. Some of the genera that oxidize proteins and amino acids convert ammonium ion to nitrites, and other genera are able to convert the nitrites to nitrates. In addition, many cyanobacteria, both symbiotic and free-living, are able to fix atmospheric nitrogen; filamentous cyanobacteria do so within specialized cells called heterocysts, from which oxygen is excluded. Most of the conversion of atmospheric nitrogen to nitrates, however, is carried out by bacteria of the genera *Rhizobium* and *Bradyrhizobium*, which form nodules on the roots of plants of the pea family (Fabaceae, or Leguminosae), and by actinomycetes that form nodules on the roots of certain other plants. Without bacteria, the nitrogen cycle would not occur, and life on earth as we know it would not be possible.

Bacterial cells may be rod-shaped (bacilli), spherical (cocci), or coiled (spirilla). If the cell wall does not divide completely following cell fission, the daughter cells may adhere in groups, in filaments, or in solid masses. Bacteria may have flagella and be motile; the rotation of the flagella moves the bacteria through the water. Bacteria may also have shorter rodlike structures known as pili. Some bacteria move by gliding.

Viruses are nonliving fragments of genomes that have evolved both from bacteria and from eukaryotes. They contain either RNA or DNA, surrounded by an outer protein coat, or capsid, and sometimes also by a lipid-containing envelope. Viruses are comparable in size to large macromolecules. Many are spherical, with icosahedral (20-sided) symmetry, and a number of others are rod-shaped, with helical symmetry.

Viruses are responsible for many diseases of humans and other animals, and also for more than 1000 kinds of plant diseases. Important among these are necroses and mosaic diseases.

Viruses cannot reproduce outside living cells and seem to have originated as escaped portions of the genetic apparatus of either bacteria or eukaryotes, which broke loose, acquired the ability to synthesize a protein-rich covering, and began to function within the cells of organisms. When they replicate, viruses take over the genetic machinery of the host cell.

RNA viruses may either act as messenger RNA in the host cells or else direct the synthesis of a complementary strand of RNA that acts as a template in producing viral RNA. DNA viruses simply compete with the DNA of the host cell. Of the viruses that cause diseases in plants, all but two groups are RNA viruses. Most viral diseases are transmitted from plant to plant by insects or nematodes. Within their host plants, viruses either move from cell to cell by way of the plasmodesmata or are spread over longer distances through the vascular system.

Viroids, which cause certain plant diseases, consist of small molecules of RNA. Unlike viruses, viroids lack protein coats. They are thought to interfere with gene regulation in infected host cells, where they occur mainly within the nucleus.

Suggestions for Further Reading

Agrios, George N.: *Plant Pathology*, 3d ed., Academic Press, Inc., New York, 1988.

A systematic account of the diseases of plants and the organisms that cause them.

Atlas, Ronald M.: *Microbiology: Fundamentals and Applications*, 2d ed., Macmillan Publishing Company, New York, 1988.

An excellent account of all aspects of the basic biology of bacteria, viruses, and microscopic eukaryotes.

Brock, Thomas D., and Michael T. Madigan: *Biology of Microorganisms*, 6th ed., Prentice Hall, Inc., Englewood Cliffs, N.J., 1991.

An interesting and often entertaining presentation of microbiology, including algae and protozoa, with an emphasis on the whole cell and its ecology.

Cooper, J.I., and F.O. MacCallum: *Viruses and the Environment*, Chapman and Hall, New York, 1984.*

An excellent introduction to the ecology of viruses.

Dickinson, C.H., and J.A. Lucas: *Plant Pathology and Plant Pathogens*, 2d ed., Halsted Press, John Wiley & Sons, Inc., New York, 1982.

A well-written introduction to plant diseases, a subject of considerable general interest and economic importance.

Diener, T.O.: "The Viroid—A Subviral Pathogen," *American Scientist* 71:481–489, 1983.

An excellent review of the properties of these fascinating RNA particles.

Galloway, John: "Nature's Second-Favourite Structure," *New Scientist* 117:36–39, March 31, 1988.

A fascinating short essay on shapes in nature, including those of some viruses.

Lynch, J.: "Microbes Are Rooting for Better Crops," *New Scientist* 118:45–49, April 28, 1988.

By studying the communities of microorganisms that live around a plant's roots, scientists hope to find ways to improve crop yield.

Margulis, Lynn: *Early Life*, Science Books International, Publishers, Portola Valley, Calif., 1982.

An excellent account of the evolutionary history of bacteria.

Matthews, R.E.F.: *Plant Virology*, 3d ed., Academic Press, Inc., New York, 1991.

A new edition of the most comprehensive general text currently available.

Poole, Philip: "Microbes on the Move," *New Scientist* 125:38–41, March 3, 1989.

Studies of the orientation and mechanisms of movement in bacteria are receiving wide attention.

Postgate, John: "Microbial Happy Families," *New Scientist* 121:40–44, January 21, 1989.

A good account of the ways in which ribosomal RNA sequences and other biochemical features are applied to the classification of bacteria.

Sagan, Dorion, and Lynn Margulis: *Garden of Microbial Delights: A Practical Guide to the Subvisible World*, Harcourt Brace Jovanovich, Publishers, Boston, 1988.

An impressive account of the diversity and interest of microorganisms, the dominant and oldest forms of life on earth.

Shapiro, James A.: "Bacteria and Multicellular Organisms," *Scientific American*, June 1988, pages 82–89.

Bacteria, while essentially unicellular, differentiate into various cell types and form highly regular colonies that appear to be guided by sophisticated temporal and spatial control systems.

Sonea, S., and M. Paniset: *A New Bacteriology*, Jones and Bartlett Publishers, Inc., Boston, 1983.*

A marvelous short book that presents an informative view of bacteria as living organisms.

* Available in paperback.

CHAPTER 12

Fungi

12–1
Mycelia of several basidiomycetes under the bark of a fallen tree. The activities of fungi and bacteria make it possible for the organic material that is incorporated into the bodies of organisms to be recycled in the ecosystem.

Fungi are as distinct from algae and from plants as they are from animals. They are discussed in this text because they have traditionally been grouped with plants and because of their complex interrelationships with plants. However, the only characteristics that these two groups share, other than those common to all eukaryotes, is a somewhat filamentous, or elongate, multicellular growth form. (A few fungi—the yeasts—are unicellular.) Fungi have no motile cells at any stage of their life cycles and no direct evolutionary connection with the plants. Clearly the two groups were derived independently from different groups of single-celled eukaryotes. Therefore, we treat fungi as a distinct kingdom—the kingdom Fungi.

Traditionally, the fungi have been considered to include not only the related divisions covered in this chapter but also the heterotrophic protists discussed in Chapter 13. Because there is little evidence for a direct relationship between the fungi and the several groups of protists that have traditionally been included with them, the two categories are separated in this book for clarity of discussion. No group of protists seems a likely ancestor of the fungi, even though that ancestor, were it known, would certainly be considered a member of the kingdom Protista.

The fungi, together with the heterotrophic bacteria and a few other groups of heterotrophic organisms, are the decomposers of the biosphere (Figure 12–1). Their activities are as necessary to the continued existence of the world, as we know it, as are those of the food producers. Decomposition releases carbon dioxide into the atmosphere and returns nitrogenous compounds and other materials to the soil, where they can be used again—recycled—by plants and eventually by animals. It is estimated that, on the average, the top 20 centimeters of fertile soil may contain nearly 5 metric tons of fungi and bacteria per hectare (2.47 acres). Some 500 known species of fungi, representing a number of dis-

12–2
The common mold Rhizopus *growing on strawberries.*

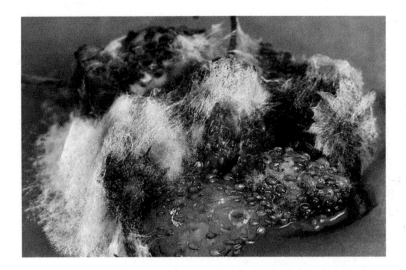

tinct groups, are marine, breaking down organic material in the sea just as their relatives do on land.

As decomposers, fungi often come into direct conflict with human interests. A fungus makes no distinction between a rotten tree that has fallen in the forest and a fence post; the fungus is just as likely to attack one as the other. Equipped with a powerful arsenal of enzymes that break down organic products, fungi are often nuisances and are sometimes highly destructive. Fungi attack cloth, paint, cartons, leather, waxes, jet fuel, petroleum, wood, paper, insulation on cables and wires, photographic film, and even the coating of the lenses of optical equipment—in fact, almost any conceivable substance. Although individual species of fungi are highly specific to particular substrates, as a group they attack virtually anything. Everywhere, they are the scourge of food producers, distributors, and sellers alike, for they grow on bread, fresh fruits (Figure 12–2), vegetables, meats, and other products. Fungi reduce the nutritional value, as well as the palatability, of such foodstuffs. They also produce toxins, some of which—the aflatoxins—are highly carcinogenic, showing their effects at concentrations as low as a few parts per billion.

The importance of fungi as commercial pests is enhanced by their ability to grow under a wide range of conditions. Thus some strains of *Cladosporium herbarum*, which attacks meat in cold storage, can grow at a temperature as low as −6°C. In contrast, one species of *Chaetomium* grows optimally at 50°C and survives even at 60°C.

The qualities that make fungi such important pests can also make them commercially valuable. Certain yeasts are useful because they produce substances such as ethanol and carbon dioxide, which play a central role in baking, brewing, and winemaking. Other fungi provide the distinctive flavors and aromas of specific kinds of cheese. The commercial use of fungi in industry is

growing, and many antibiotics—including penicillin, the first such substance to be used widely—are produced by fungi. Dozens of different kinds of fungi are eaten regularly by humans, and some of them are cultivated commercially. The ability of fungi to break down substances is leading to their investigation in relation to waste clean-up programs.

A striking example of the potential value of compounds derived from fungi is cyclosporine, a "wonder drug," which was isolated from the soil-inhabiting fungus *Tolypocladium inflatum*; it became available in 1979. Cyclosporine suppresses the immune reactions that cause rejection of organ transplants, thus reducing the likelihood of transplant rejection. It does not, however, have the undesirable side effects of other drugs used for this purpose. Only the advent of this remarkable drug made it possible to resume organ transplants, which had essentially been abandoned, in the early 1980s. Everyone is now aware of the success that followed.

The kinds of relationships between fungi and other organisms are extremely diverse. For example, about four-fifths of all vascular plants form mutually beneficial associations (called mycorrhizae) between their roots and fungi; these associations play a critical role in plant nutrition (see page 238). As another example, lichens (see page 223) are symbiotic associations between fungi and either algal or cyanobacterial cells; lichens occupy some extremely hostile habitats. Many fungi attack living organisms, rather than dead ones, and sometimes do so in surprising ways (see the essay "Predaceous Fungi" on page 223). They are the most important single cause of plant diseases; well over 5000 species of fungi attack economically valuable crop and garden plants, as well as many wild plants. Certain fungi attack living trees, causing enormous losses to the world's timber crop. Other fungi cause serious diseases in humans and domestic animals.

Some 77,000 distinct species of fungi (including lichens) have been described, and many more await discovery. There may ultimately prove to be as many species of fungi as there are of plants, although far fewer fungi have been named thus far.

Biology of the Fungi

Fungi are primarily terrestrial. Although some fungi are unicellular, most are filamentous, and structures such as mushrooms consist of a great many such filaments, packed tightly together (Figure 12–3). Fungal filaments are known as **hyphae,** and a mass of hyphae is called a **mycelium** (see Figure 12–1). Growth of hyphae occurs at their tips, but proteins are synthesized throughout the mycelium. The hyphae grow rapidly, and an individual fungus may produce more than a kilometer of new mycelium within 24 hours. (The words "mycelium" and **mycologist**—a scientist who studies fungi —are derived from the Greek word *myketos,* meaning "fungus.")

The cell walls of plants and many protists are built on a framework of cellulose microfibrils, which is interpenetrated by a matrix of noncellulosic molecules, such as hemicelluloses and pectic substances. In fungi, the

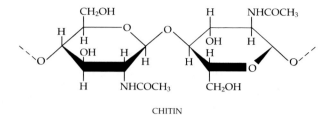

CHITIN

12–4
Chitin consists of repeating units of β–1,4–linked N-acetylglucosamine. Similar linkage is found between the glucose units of cellulose and between the polysaccharide units of bacterial cell walls, suggesting that such a linkage forms a particularly strong polysaccharide structure. Chitin is characteristic of the cell walls of fungi, as well as the skeletons of insects and other arthropods.

cell wall is composed primarily of another polysaccharide—**chitin**—which is the same material found in the hard shells, or exoskeletons, of arthropods, such as insects, arachnids, and crustaceans (Figure 12–4). Chitin is more resistant to microbial degradation than is cellulose.

With their rapid growth and filamentous form, fungi have a relationship to their environment that is very different from that of any other group of organisms. The surface-to-volume ratio of fungi is very high, so they are in as intimate a contact with the environment as are the bacteria. Usually no part of a fungus is more than a few micrometers from its external environment, being separated from it only by a thin cell wall and the plasma membrane. With its extensive mycelium, a fungus can have a profound effect on its surroundings—for example, in binding soil particles together. Hyphae of the same species often fuse, even when they have grown from different spores, thus increasing the intricacy of the network.

An aspect of this intimate relationship between fungus and environment is that all parts of the fungus are metabolically active; quiescent layers of tissue, such as heartwood, are absent in fungi. The enzymes and other substances secreted by fungi have an immediate effect on the surroundings and are important for the maintenance of the fungus itself. Fungi absorb food mostly at or near the growing tips of their hyphae.

All fungi are heterotrophic. In obtaining their food, they function either as saprobes (living on organic materials from dead organisms), as parasites, or as mutualistic symbionts. Some fungi, mainly yeast, can release energy by fermentation, such as in the production of ethyl alcohol from glucose. Glycogen is the primary

12–3
A mushroom, such as this individual of Hygrocybe aurantiosplendens *growing in a redwood forest in central California, is made up of densely packed hyphae, collectively known as mycelium. Mushrooms reproduce by spores, which are formed on structures that line the gills under the cap.*

The Evolution of Fungi

The first fungi were probably unicellular eukaryotic organisms that apparently no longer have any living counterparts. From these organisms were derived the **coenocytic** fungi, in which many nuclei are found in a common cytoplasm. (*Coenocytic* means "contained in a common vessel," or multinucleate, the nuclei not separated by walls.) The living coenocytic fungi are classified as members of the division Zygomycota, commonly known as the zygomycetes.

In members of the division Ascomycota, called ascomycetes, and the division Basidiomycota, or basidiomycetes, the mycelia are septate—divided by cell walls—but the septa, or cross walls, are perforated by a central pore (Figure 12–6). In the ascomycetes, however, the nuclei pass readily through the pores in the hyphae, and the mycelia are therefore functionally coenocytic. Evolutionary relationships between the three divisions of fungi are difficult to establish; there is no evidence of a close connection between the zygomycetes and the other two groups, which evidently share a common ancestor.

Fungi may be among the oldest eukaryotes, but their early history is poorly understood. The oldest fossils that resemble fungi occur in strata about 900 mil-

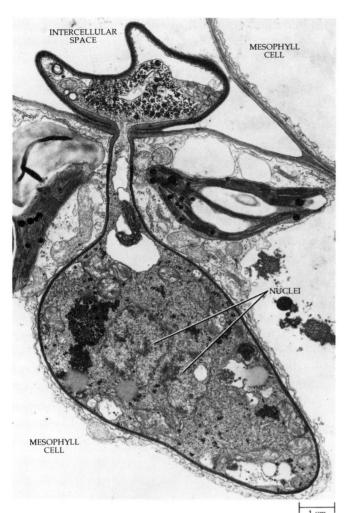

12–5
Electron micrograph of a haustorium of Melampsora lini, *a rust fungus, growing in a leaf cell of flax* (Linum usitatissimum). *In the intercellular space at the top of the micrograph is the haustorial mother cell. The narrow penetration hypha leads to the large, bulbous haustorium within the lower mesophyll cell.*

storage polysaccharide in some fungi, as it is in animals and bacteria; lipids serve an important storage function in other fungi.

Specialized hyphae, known as **rhizoids,** anchor some kinds of fungi to the substrate. Parasitic fungi often have similar specialized hyphae, called **haustoria** (singular: haustorium), which absorb nourishment directly from the cells of other organisms (Figure 12–5).

All fungi have cell walls and produce spores. Fungi are nonmotile throughout their life cycle, lacking cells with cilia or flagella. Many fungal spores are dispersed by wind or water, while others are carried from place to place by insects or other arthropods.

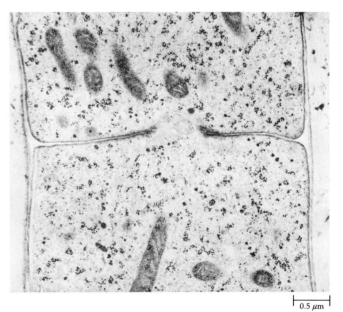

12–6
Electron micrograph of a septum between two cells in the ascomycete Gibberella acuminata. *The large globular structures are mitochondria, and the tiny dark granules are ribosomes. This specimen was thin-sectioned through the central pore region of a septum in the final stages of formation.*

lion years old, but the oldest that have been identified with certainty as fungi are from the Ordovician period, 450 to 500 million years ago (see Appendix D). Coenocytic fungi resembling zygomycetes were associated with the earliest vascular plants in the Silurian period, some 400 million years ago (see page 323), and all three divisions of fungi were certainly in existence by the close of the Carboniferous period, 300 million years ago. The great majority of fungi are terrestrial, and the invasion of land by plants and animals, which began about 430 million years ago, seems to have been the event that led to their evolutionary diversification.

Fungal Reproduction

The reproductive structures of fungi are separated from the hyphae by septa. These reproductive structures are called **gametangia** if they are directly involved in the production of gametes, and **sporangia** (zygomycetes) or **conidiogenous cells** (ascomycetes and some basidiomycetes) if they are involved in the production of asexual spores. Meiosis typically follows the formation of the zygote in all fungi; in other words, meiosis in fungi is zygotic (see Figure 10–10a, page 184).

Nonmotile spores are the characteristic means of reproduction in fungi. Some of the spores are dry and very small; they can remain suspended in the air for long periods, thus being carried to great heights and for great distances. This property helps to explain the very wide distributions of many species of fungi. Other spores are slimy and stick to the bodies of insects and other arthropods, which may then spread them from place to place. The spores of most fungi are propelled ballistically into the air (see the essay "Phototropism in a Fungus" on page 214). The bright colors and powdery textures often seen in association with many types of molds are produced by the spores.

MITOSIS AND MEIOSIS

One of the most characteristic features of the fungi involves nuclear division. In fungi, the processes of meiosis and mitosis are different from those that occur in plants, animals, and many protists. In many fungi, the nuclear envelope does not disintegrate and re-form but is constricted near the midpoint between the two daughter nuclei; in others, it breaks down near the mid-region. In most fungi, the spindle forms within the nuclear envelope; in some basidiomycetes, however, it appears to form within the cytoplasm and move into the nucleus. All fungi lack centrioles, but they form unique structures called **spindle pole bodies,** which appear at the spindle poles. This unique combination of

features indicates that fungi are not directly related to other living eukaryotes; they clearly deserve their present status as a separate kingdom.

The Major Groups of Fungi

DIVISION ZYGOMYCOTA

Most zygomycetes live on decaying plant and animal matter in the soil; some are parasites of plants, insects, or small soil animals; a few occasionally cause severe infections in humans and domestic animals. There are approximately 765 described species of zygomycetes. Most of them have coenocytic hyphae, within which the cytoplasm can often be seen streaming rapidly. Zygomycetes can usually be recognized by their profuse, rapidly growing hyphae, but some of them also exhibit a unicellular, yeastlike form of growth under certain environmental conditions. Asexual reproduction by means of spores produced in more or less specialized sporangia borne on the hyphae is almost universal in zygomycetes.

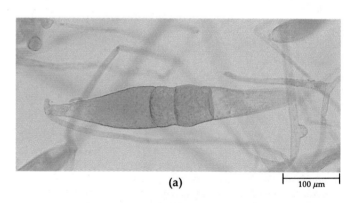

(a) |———| 100 μm

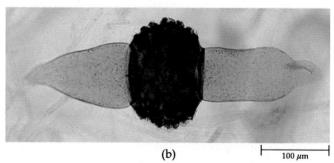

(b) |———| 100 μm

12–7
Rhizopus stolonifer, *black mold.*
(a) *Gametangia, the gamete-producing structures, are in the process of fusing to produce a zygosporangium.* (b) *A zygosporangium, or sexual resting structure. Such a zygosporangium contains one to several diploid nuclei, the zygotes.*

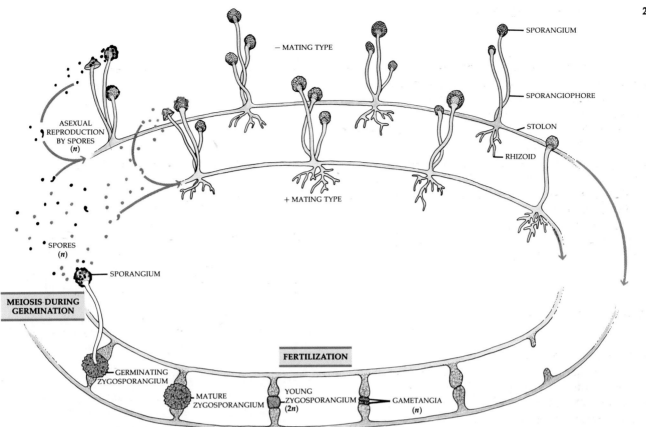

12–8

In Rhizopus stolonifer, *as in most other zygomycetes, asexual reproduction by means of haploid spores is the chief mode of reproduction. Less frequently, sexual reproduction occurs. In this* common species, it involves genetically differentiated mating strains, which have traditionally been labeled + and − types. (Although the mating strains are morphologically indistinguishable from one another, they are shown here in two colors.) The zygosporangium in Rhizopus develops a thick, rough, black coat and becomes dormant, often for several months.

The terms "Zygomycota" and "zygomycetes" refer to the chief characteristic of the division—the production of thick-walled sexual structures called **zygosporangia** (Figure 12–7), which often remain dormant for long periods. When two appropriate individuals are in close proximity, hormones are produced that cause outgrowths near their hyphal tips to come together and develop into gametangia. Some species are **homothallic** (self-fertile) and can produce sexual structures from a single genetic strain; many others, like *Rhizopus* (Figure 12–8), are **heterothallic** and require a combination of + and − strains for sexual reproduction.

In any event, the gametangia become separated from the rest of the fungal body by the formation of septa (see Figure 12–8). The walls between the two touching gametangia dissolve, and the two multinucleate protoplasts come together. A thick-walled zygosporangium is produced following the fusion of two multinucleate gametangia. Inside the zygosporangium, the gametes—which are simply nuclei—fuse, forming one or more diploid nuclei, or zygotes. (In zygomycetes, the zygosporangium has often been called a zygospore, but since it contains several to many diploid nuclei, themselves zygotes, we prefer the terminology presented here.)

Meiosis occurs at the time of germination of the zygosporangium, which cracks open, producing a sporangium that is similar to the asexually produced sporangium. Haploid spores, produced asexually within the sporangium, germinate, and the cycle begins again.

One of the best-known and most familiar members of this division is *Rhizopus stolonifer*, a black mold that forms cottony masses on the surface of moist, carbohydrate-rich foods and similar substances that are exposed to air (see Figure 12–2). This organism is also a serious pest of stored fruits and vegetables. The life cycle of *R. stolonifer* is illustrated in Figure 12–8. The mycelium of *Rhizopus* is composed of several distinct kinds of haploid hyphae. Most of the mycelium consists of rapidly growing, coenocytic hyphae, which grow through the substrate, absorbing nutrients. From them, arching hyphae called **stolons** are formed. The stolons form rhizoids wherever their tips come into contact with the substrate. From each of these points, a sturdy, erect branch arises, which is called a **sporangiophore** because it produces a spherical sporangium at its apex. Each sporangium begins as a swelling, into which a number of nuclei flow; the sporangium is eventually isolated by the formation of a septum. The protoplasm within is cleaved, and a cell wall forms around each

Phototropism in a Fungus

In *Pilobolus*, a zygomycete 5 to 10 millimeters tall that grows on dung, the sporangia are shot toward the light. The sporangiophore of this fungus grows toward the light so that all light rays entering the subsporangial swelling, which is essentially a lens, converge on a basal photoreceptive area. Light focused elsewhere promotes maximum growth of the sporangiophore on the side away from the light, causing the sporangiophore to bend toward the light. The high turgor pressure of the sap in the vacuole of the subsporangial swelling eventually causes it to split, which blasts the sporangium

off to a distance of 2 meters or more at an initial velocity that may approach 50 kilometers per hour! Considering that the sporangia are only about 80 micrometers in diameter, this is an enormous distance. After the sporangium has been fired off, the sporangiophore collapses. The sporangium adheres where it lands, and if this is on a blade of grass, the sporangium may be eaten by an herbivore. It then passes through the digestive tract of the herbivore unharmed and is deposited in the dung to begin the cycle anew.

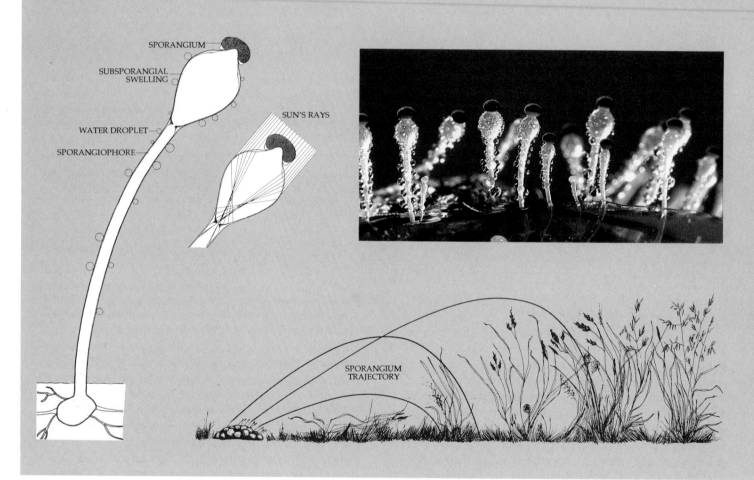

SPORANGIUM

SUBSPORANGIAL SWELLING

WATER DROPLET

SPORANGIOPHORE

SUN'S RAYS

SPORANGIUM TRAJECTORY

spore. The sporangium wall becomes black as it matures, giving the mold its characteristic color. Each spore, when liberated, can germinate to produce a new mycelium.

One of the most important groups of zygomycetes includes *Glomus* and related genera, which always grow in intimate association with the roots of plants, forming mycorrhizae, which will be discussed on page 238. Another group of zygomycetes that has great ecological significance, the order Entomophthorales, is parasitic on insects and other small animals (see Figure

10–8b). Species of this order, most of which reproduce by means of a terminal, asexual spore that is discharged at maturity, are being increasingly used in the biological control of insect pests of crops.

The trichomycetes, a third group of zygomycetes, have an intriguing relationship with arthropods. These fungi are found in aquatic insect larvae, millipedes, crayfish, and even crustaceans from thermal vents. Trichomycetes are seldom seen because they occur in the gut of the host organism, where some are thought to provide vitamins to the animal.

DIVISION ASCOMYCOTA

The ascomycetes comprise about 30,000 described species, including a number of familiar and economically important fungi. Most of the blue-green, red, and brown molds that cause food spoilage are ascomycetes, including the salmon-colored bread mold *Neurospora*, which has played an important role in the development of modern genetics. Ascomycetes are the cause of a number of serious plant diseases, including the powdery mildews, which primarily attack leaves; chestnut blight (caused by the fungus *Crypthonectria parasitica*, accidentally introduced into North America from northern China); and Dutch elm disease (caused by *Ophiostoma ulmi*, a fungus native to Europe). Many yeasts are also ascomycetes, as are the edible morels and truffles (Figure 12–9). Thousands of additional species of ascomycetes—some undoubtedly of great economic importance—await discovery and scientific description.

Ascomycetes, with the exception of the unicellular yeasts, are filamentous when they are growing. In general, their hyphae are septate. The septa are perforated, however (see Figure 12–6), and the cytoplasm and nuclei can move through the septal pores. The hyphal cells of the vegetative mycelium may be either uninucleate or multinucleate. Some ascomycetes are homothallic, others heterothallic.

The life cycle of an ascomycete is diagrammed in Figure 12–10; this diagram should be consulted frequently to understand the following material. In most of the species of this division, asexual reproduction is by the formation of specialized spores known as **conidia** (a Greek word meaning "fine dust"). Conidia, which are usually multinucleate, are formed from conidiogenous cells (Figure 12–11); such cells are usually borne at the tips of modified hyphae called **conidiophores** ("conidia bearers"). Unlike zygomycetes, ascomycetes do not produce sporangia within which a number of spores are produced following mitosis.

Sexual reproduction in ascomycetes always involves the formation of an **ascus** (plural: asci), a saclike structure within which haploid **ascospores** are formed following meiosis. Both asci and ascospores are unique structures; they distinguish the ascomycetes from all other fungi (Figure 12–12). Ascus formation usually occurs within a complex structure composed of tightly interwoven hyphae—the **ascoma** (plural: ascomata; formerly known as the ascocarp). Many ascomata are macroscopic. An ascoma may be open and more or less

(a)

(b)

(c)

12–9

Ascomycetes. (a) A morel, Morchella esculenta. *The true morels are among the choicest edible fungi. Mushroom gatherers look for them when the oak leaves are "the size of a mouse's ear." Morels were first grown successfully in culture in 1983 but have not yet been developed into a commercial crop.*

(b) *Scarlet cup,* Sarcoscypha coccinea, *a beautiful fungus with an open ascoma (apothecium). (c) The highly prized, edible ascoma of a black truffle,* Tuber melanosporum. *In the truffles, this spore-bearing structure is produced below ground and remains closed, liberating its ascospores only when the ascoma decays*

or is broken open by digging animals. Truffles are mycorrhizal (see page 238), mainly on oaks and hazelnuts, and are searched for by specially trained dogs and pigs. Recently, they have been cultivated commercially on a small scale by inoculating the roots of seedling host plants with their spores.

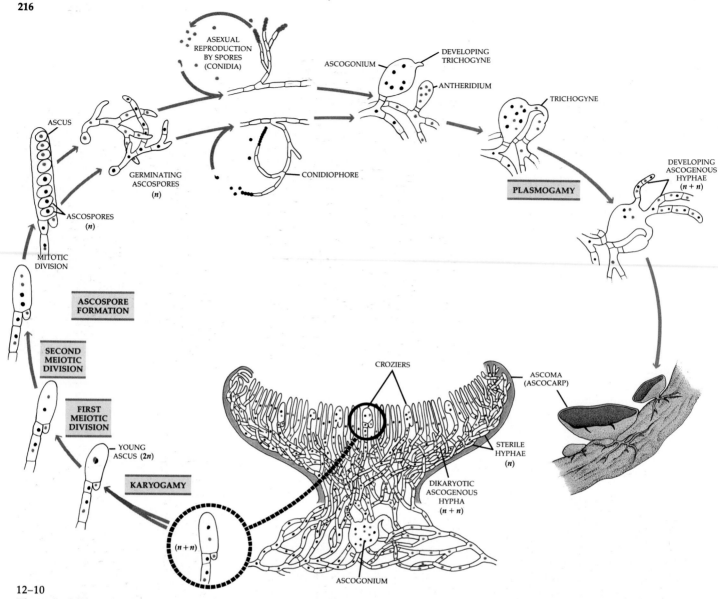

ASEXUAL REPRODUCTION BY SPORES (CONIDIA)

ASCOGONIUM

DEVELOPING TRICHOGYNE

ANTHERIDIUM

TRICHOGYNE

ASCUS

CONIDIOPHORE

PLASMOGAMY

DEVELOPING ASCOGENOUS HYPHAE (n + n)

GERMINATING ASCOSPORES (n)

ASCOSPORES (n)

MITOTIC DIVISION

ASCOSPORE FORMATION

SECOND MEIOTIC DIVISION

FIRST MEIOTIC DIVISION

YOUNG ASCUS (2n)

KARYOGAMY

(n + n)

CROZIERS

ASCOMA (ASCOCARP)

STERILE HYPHAE (n)

DIKARYOTIC ASCOGENOUS HYPHA (n + n)

ASCOGONIUM

12–10

The typical life cycle of an ascomycete. Asexual reproduction occurs by way of specialized spores, known as conidia, which are usually multinucleate. Sexual reproduction involves the formation of asci and ascospores. Karyogamy is fol- lowed immediately by meiosis in the ascus, resulting in the production of ascospores.

cup-shaped (an apothecium; see Figure 12–9b), closed and spherical (a cleistothecium; Figure 12–12b), or spherical to flask-shaped with a small pore through which the ascospores escape (a perithecium; Figure 12–12c). The asci usually develop on the inner surface of the ascoma. The layer of asci is usually called the **hymenium,** or hymenial layer (Figure 12–13).

In the life cycle of an ascomycete (see Figure 12–10, top left corner), the mycelium is initiated with the ger- mination of an ascospore on a suitable substrate. Soon after, the mycelium begins to reproduce asexually by forming conidiophores. Many crops of conidia are pro- duced during the growing season, and it is the conidia that are primarily responsible for the reproduction of the fungus.

Sexual reproduction, involving ascus formation, occurs on the same mycelium that produces conidia; it is preceded by the formation of multinucleate gametan- gia called antheridia and ascogonia. The male nuclei of the **antheridium** pass into the **ascogonium** via the **trichogyne,** which is an outgrowth of the ascogonium. **Plasmogamy,** the fusion of the two protoplasts, has now taken place. The male nuclei may then pair with the genetically different female nuclei within the com- mon cytoplasm, but they *do not fuse with them yet.* **Ascogenous hyphae** now begin to grow out of the ascogonium. As these special hyphae develop, compat- ible pairs of nuclei migrate into them. Cell division in the developing ascogenous hyphae occurs in such a way that the resultant cells are invariably **dikaryotic** (containing two compatible haploid nuclei) rather than **monokaryotic** (containing one nucleus each).

12–11
Conidia are the characteristic asexual spores of ascomycetes; they are usually multinucleate. In these electron micrographs, stages in the formation of conidia in Nomuraea rileyi, *which infects the velvetbean caterpillar, are shown.*
(a) Scanning electron micrograph of conidia at various stages of development.
(b) Transmission electron micrograph of conidia. In this species, the new conidia are produced at the base of the chain.

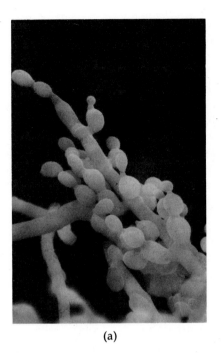

(a)

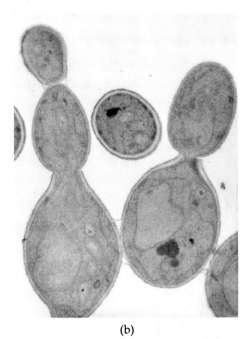

(b)

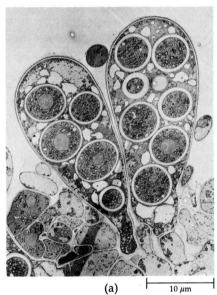

(a)   10 μm

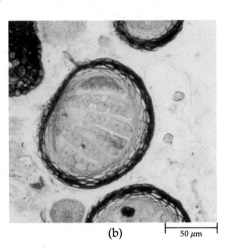

(b)   50 μm

(c)   100 μm

12–12
(a) An electron micrograph showing two asci of Ascodesmis nigricans *in which ascospores are maturing. (b) Ascoma of* Erysiphe aggregata, *showing the enclosed asci and ascospores. This completely enclosed type of ascoma is called a cleistothecium. (c) An ascoma of* Coniochaeta, *showing the enclosed asci and ascospores. Note the small pore at the top. This sort of ascoma, with a small opening, is known as a perithecium.*

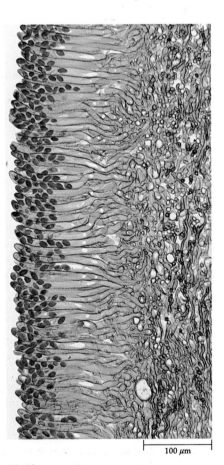

100 μm

12–13
*A stained thin section through the hymenial layer of a morel (*Morchella*), showing asci with ascospores.*

The asci form at the tips of the dikaryotic, ascogenous hyphae. The apical cell of the dikaryotic hypha grows to form a hook, or **crozier.** In this hooked cell, the two nuclei divide in such a way that their spindle fibers are parallel to each other. Two of the daughter nuclei are close to one another at the top of the hook; one of the remaining two daughter nuclei is near the tip, and the other is near the basal septum of the hook. Two septa are then formed; these divide the hook into three cells, of which the middle one becomes the ascus. It is in this middle cell that **karyogamy** occurs: the two nuclei fuse to form a diploid nucleus (zygote), the only diploid nucleus in the life cycle of the ascomycetes (with the exception of the parasexual cycles discussed on page 221). Soon after karyogamy, the young ascus begins to elongate. The diploid nucleus then undergoes meiosis, which is generally followed by one mitotic division, producing an eight-spored ascus. These haploid nuclei are then cut off in segments of the cytoplasm to form ascospores. In most ascomycetes, the ascus becomes turgid at maturity and finally bursts, releasing its ascospores explosively into the air in the cup fungi and some of the perithecium-forming species. The ascospores are generally propelled about 2 centimeters from the ascus, but some species propel them as far as 30 centimeters. This initiates their airborne dispersal.

Unicellular Ascomycetes: The Yeasts

Yeasts are primarily unicellular; in fact, the term is generally applied to fungi with a unicellular growth form. There are at least 60 genera of yeasts, with approximately 500 known species. The unicellular growth form characteristic of yeasts occurs in some members of all three divisions of fungi, but many of the organisms that we call yeasts, including a number of those with the greatest economic importance, are ascomycetes. At least a quarter of the genera of yeasts, however, are basidiomycetes, members of the third division of fungi, which we shall discuss on page 228.

Yeasts reproduce by several different mechanisms, each of which produces a new daughter cell. Some form of budding is most common (Figure 12–14a); thus, each yeast cell can be regarded as a conidiogenous cell. Some species of yeasts are capable of assuming mycelial growth under suitable environmental conditions. Many important animal pathogens grow as yeasts at body temperature, exhibiting hyphal growth at lower temperatures when they function as saprobes. Other yeasts remain unicellular under all conditions.

Some yeasts multiply vegetatively only in the haploid condition; at times, morphologically unaltered vegetative cells may function as gametes, fusing to form a diploid cell, the **fusion cell,** which functions as an ascus (Figure 12–14b). Such a life cycle is predominantly haploid. In other yeasts, such as the familiar *Saccharomyces cerevisiae*, which is used in making bread, wine, and beer, the fusion cell multiplies asexually to form a population of diploid cells, which may ultimately undergo meiosis. Such yeasts have both haploid and diploid multiplying generations. In other yeasts, the ascospores fuse in pairs immediately after they are formed; the ascospore is the only haploid cell in the life cycle, which is predominantly diploid.

Yeasts are important to humans because of their ability to ferment carbohydrates, breaking down glucose to produce ethyl alcohol and carbon dioxide. Thus yeasts are utilized by vintners as a source of ethanol, by bakers as a source of carbon dioxide, and by brewers as a source of both substances. Many domestically useful strains of yeast have been developed by selection and breeding, and the techniques of genetic engineering are now being used to improve these strains further through the addition of useful genes from other organisms. In modern winemaking, pure strains of yeasts are added to relatively sterile grape juice; historically, and even now in some areas of the world, wild yeasts that are present on the grapes are used to produce wine (see Figure 6–16, page 98). Some of the flavors of wine come directly from the grape, but most arise from the

12–14
Yeasts. (a) *Budding cells of bread yeast,* Saccharomyces cerevisiae. (b) *Asci with ascospores of* Schizosaccharomyces octosporus.

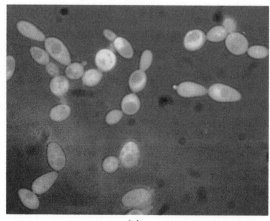

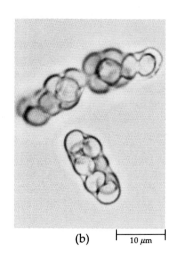

(a)   (b) |— 10 μm —|

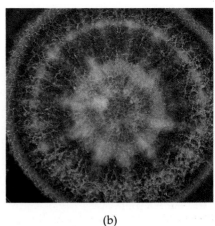

12–15
Penicillium *and* Aspergillus—*two of the common genera of Deuteromycota.* (a) *A culture of* Penicillium notatum, *the original penicillin-producing fungus, showing the distinctive colors produced during growth and spore development.* (b) *A culture of* Aspergillus fumigatus, *a fungus that causes respiratory disease in humans. Notice the concentric growth pattern produced by successive "pulses" of spore production.*

(a)

(b)

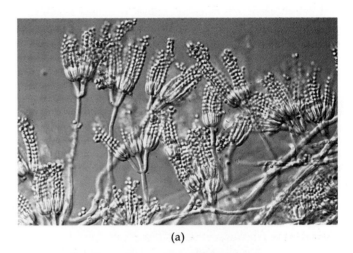

(a)

direct action of the yeast. In brewing beer, the medium is sterilized by heating prior to fermentation, and pure yeast cultures are added to it. As mentioned above, most of the yeasts important in the production of wine, cider, sake, and beer are strains of a single species, *Saccharomyces cerevisiae*, although other species also play a role. Most lager beer, for example, is made using *S. carlsbergensis*. *Saccharomyces cerevisiae* is now virtually the only species used in baking bread (Figure 12–14). Some species of yeast are important as human pathogens, causing such diseases as thrush and cryptococcosis, the latter caused by a basidiomycete that has the growth habit of a yeast when it grows as a human pathogen.

A number of yeasts, especially *S. cerevisiae*, have become important laboratory organisms for genetic research. Like *Escherichia coli* among the bacteria, this yeast has become the organism of choice for studies of the metabolism, genetics, and development of eukaryotic cells, and for chromosome studies. In the future, such detailed knowledge, coupled with the ease of manipulating the genetics of yeasts growing on defined media, will doubtless greatly increase the importance of yeasts for scientific investigations and industrial applications.

Deuteromycota

The deuteromycetes, or conidial fungi, are a miscellaneous assemblage of about 17,000 distinct species of fungi in which the sexual reproductive features are either not known or not used as the basis of classification (Figures 12–15 and 12–16). They have often been called "Fungi Imperfecti" because they can be considered imperfect in their lack of sexual stages, but in fact many of them are abundant, flourishing organisms—so that the term "imperfect" is somewhat misleading.

(b)

12–16
The conidiogenous cells anu conidiophores of deuteromycetes are used in their classification. (a) Penicillium (*brushlike*) *and* (b) Aspergillus (*tightly clumped and arising from the swollen top of the conidiophore*). *Note the long chains of small, dry conidia.*

From Pathogen to Symbiont: Fungal Endophytes

The leaves and stems of plants are often riddled with fungal hyphae; such fungi are called **endophytes,** a general term used for a plant or fungus growing within a plant. Although some fungal endophytes cause disease symptoms in the plants they inhabit, others produce no such effects. Instead, they protect host plants from their natural enemies: herbivores and, in some cases, pathogenic microbes.

In many species of grasses, endophytic fungi infect the flowers of the host and proliferate in the seeds; eventually, a substantial mass develops throughout the stems and leaves of the mature grass plant, the fungal hyphae growing between the host cells. A good example of such a relationship is provided by tall fescue, *Festuca arundinacea,* a grass that covers more than 15,000 square kilometers (35 million acres) of lawns, fields, and pastures in the United States, especially in the East and Midwest. Tall fescue plants that are free of fungus provide good forage for livestock, but cattle feeding on infected plants become lethargic and cease grazing, often panting and drooling excessively. If the animals are not moved to other forage, they become feverish, gain weight slowly, produce little milk, have difficulty in conceiving, develop gangrene, and eventually die, if they have not been slaughtered by that time for their salvage value. In the 1970s, scientists at the University of Kentucky discovered that these symptoms were associated with the endophytic ascomycete *Sphacelia typhina.* In relationships with fungal endophytes, which are common in grasses, the host gains protection from its enemies (including, unfortunately for us, cattle!) and the fungus obtains all of its nutritional needs from the grass. The deterrent effects on herbivores occur because the fungi produce alkaloids—bitter, nitrogen-rich compounds that are abundant in some plants—which have physiological effects on humans and other animals (see page 436).

Cattle can distinguish between infected grasses and those that are free of fungi; given a choice, they will readily choose the latter over the former. If only infected plants are available, however, cattle will eat them, although in low quantities, to avoid starvation. Tall fescue plants infected by *Sphacelia* also appear to be more resistant to the attacks of insects and other invertebrate herbivores than are uninfected plants. Since they grow more vigorously than uninfected plants, infected plants are extremely difficult to eliminate once established in pastures. This poses an important problem: in the southeastern United States alone, annual losses in beef due to this fungus may amount to $500 million.

The endophytic fungi that infect hosts other than grasses are not usually transmitted through the seeds of their host plants; their spores simply are blown from plant to plant or are carried by insects. Like the fungal endophytes of grasses, however, these fungi often protect their hosts from attack by insects, microbial pathogens, and grazing animals and occur in most individuals of some species. In many of these relationships, the fungi may infect only the vegetative parts of the host and remain metabolically inactive for long periods; if the plant tissues are damaged by herbivores, the fungi may grow rapidly and produce toxins, protecting both the

The asexual phase of the life cycle of a fungus, the **anamorph,** is the only phase known for most deuteromycetes. In some of the fungi classified as deuteromycetes, the **teleomorph,** or sexual phase of the life cycle, apparently has been lost in the course of evolution; in others, it may simply not have been discovered yet; and in still others, it may simply not be used as the principal basis of classification because of the close resemblance between the conidium-producing features and those of other deuteromycetes. Thus for some species of the well-known deuteromycete genera *Aspergillus* and *Penicillium,* the teleomorph is known, but the species are usually still classified as members of these genera because of their overall resemblance to the other species. It is for all of these reasons that mycologists regard the deuteromycetes as a miscellaneous assemblage of fungi, not equivalent to the three fungal divisions.

On the basis of their overall characteristics, most deuteromycetes are clearly ascomycetes; they reproduce only by means of conidia. A few deuteromycetes are basidiomycetes or zygomycetes, the former marked by the septa and clamp connections that are characteristic of the basidiomycetes (see Figure 12–28, page 231). New molecular techniques are now supplementing traditional morphological features in the evaluation of the relationship between particular deuteromycetes and their teleomorph-forming relatives.

Many fungi exhibit the phenomenon of **heterokaryosis,** in which genetically different nuclei occur together in a common cytoplasm. The nuclei may differ from one another because of mutation or because of the fusion of genetically distinct hyphae. Since genetically different nuclei may occur in different proportions in different parts of a mycelium, these sectors may have different properties.

Particularly among the deuteromycetes, genetically distinct haploid nuclei occasionally fuse. Within the resulting diploid nuclei, the chromosomes may associate, recombination may follow, and genetically novel haploid nuclei may be formed. This phenomenon, which is

Claviceps purpurea, which causes the plant disease known as ergot, is seen here growing on stalks of rye (Secale cereale).

plants and new sites for fungal infection. Thousands of species of fungi may be involved in these relationships, almost all of them ascomycetes. The selective advantages of such associations for both the host plant and the fungus are evident, and it is a simple matter to see how plant pathogens have been transformed into members of mutually beneficial symbiotic associations on numerous occasions.

Some of the alkaloids produced by another fungus, *Claviceps purpurea*, which infects rye (*Secale cereale*) and other grasses, are identical to those produced by *Sphacelia*. The rye-infecting fungus causes a plant disease called ergot. Domestic animals and people who eat the infected grain develop a disease called ergotism, which is often accompanied by gangrene, nervous spasms, psychotic delusions, and convulsions. It occurred frequently during the Middle Ages, when it was known as St. Anthony's fire. In one epidemic in the year 994, more than 40,000 people died. As recently as 1951, there was an outbreak of ergotism in a small French village, in which 30 people became temporarily insane, believing that they were pursued by demons or snakes; five of the villagers died. Some of the ergot alkaloids have the property of enhancing muscle constriction and have been used for more than 400 years to hasten uterine contraction during childbirth; others dilate the veins, decreasing blood pressure, and are used in the treatment of migraine. Because of these useful medical properties, efforts are now under way to produce improved strains of *C. purpurea* and to grow them commercially.

known as **parasexuality,** was discovered in *Aspergillus nidulans*, a deuteromycete. Within the hyphae of this common fungus, there is one diploid nucleus, on the average, for every 1000 haploid nuclei. Parasexual cycles are characteristic of certain groups of fungi, such as the deuteromycetes, adding considerably to their genetic and evolutionary flexibility.

Among the deuteromycetes are many organisms of great economic importance. For example, a number of the most important plant pathogens are members of this artificial group. **Anthracnose** diseases of plants, which cause lesions and blackening, are generally caused by members of this group. For example, an often fatal disease of dogwoods (*Cornus florida*) that was detected over a wide area of the eastern United States in the late 1980s is caused by a deuteromycete of the genus *Discula*.

Other members of the group produce effects that are valued by humans. For example, certain members of the genus *Penicillium* give some types of cheese the

appearance, flavor, odor, and texture so highly prized by gourmets. One such mold, *P. roquefortii*, was first found in caves near the French village of Roquefort. Legend has it that a peasant boy left his lunch, a fresh piece of mild cheese, in one of these caves and on returning several weeks later found it blue-marbled, tart, and fragrant. Only cheeses from the area around these particular caves are permitted to bear the name Roquefort, although there are many other blue cheeses—Danish, Stilton, Gorgonzola—which are ripened by the same fungus. Another species of *Penicillium*, *P. camembertii*, gives Camembert and Brie cheeses their special qualities.

In the Orient, soy paste (miso) is produced by fermenting soybeans with *Aspergillus oryzae*; soy sauce (shoyu) is produced by fermenting soybeans with a mixture of *A. oryzae* and *A. soyae*, though lactic acid bacteria are also actively involved in producing the final product. (Tofu, or bean curd, and tempeh, a similar protein-rich food of tropical Asia, are produced following fermentation of soybeans with species of *Mucor* and *Rhizopus*, respectively; these fungi are zygomycetes.) *Aspergillus oryzae* is also important for the initial steps in brewing sake, the traditional alcoholic beverage of Japan; the yeast *Saccharomyces cerevisiae* is important later in the process. Citric acid is produced commercially in large amounts from colonies of *Aspergillus* grown under very acidic conditions.

Antibiotics are substances produced by one living organism that inhibit the growth of other living organisms (such as bacteria) and so may be therapeutically useful to humans and other organisms. Many important antibiotics are produced by deuteromycetes. The first antibiotic was discovered by Sir Alexander Fleming, who noted in 1928 that a strain of *Penicillium* that had contaminated a culture of *Staphylococcus* growing on a nutrient agar plate had completely halted the growth of the bacteria. Ten years later, Howard Florey and his associates at Oxford University purified penicillin and later came to the United States to promote the large-scale production of the drug. The demand during World War II was so great that the production of penicillin was increased from a few million units per month in 1942 to more than 700 billion units in 1945. Penicillin is effective in curing a wide variety of bacterial diseases caused by gram-positive bacteria, including pneumonia, scarlet fever, syphilis, gonorrhea, diphtheria, and rheumatic fever. Many of the substances that we use as antibiotics undoubtedly also play significant ecological roles in nature, allowing the organisms that produce them to outcompete others.

Not all substances produced by deuteromycetes are useful to us. For example, the trichothecenes, which are produced by a number of genera of this group, are potent inhibitors of protein synthesis in eukaryotes and have been suspected as agents of biological warfare. Another group of toxic chemicals, which are potent causative agents for liver cancer in humans, are the aflatoxins. The aflatoxins are secondary metabolites produced by certain strains of *Aspergillus flavus* and *A. parasiticus*. These fungi frequently grow in stored food products, especially corn, wheat, and peanuts; in tropical countries, aflatoxins have been estimated to contaminate at least 25 percent of the food.

In recent years, aflatoxins have been detected occasionally in corn harvested in the United States, despite strong efforts to detect and destroy contaminated corn. Drought conditions in 1988 were apparently especially favorable to the growth of the fungus, and the U.S. Department of Agriculture blocked the export of more then 4 million bushels of corn in the first four months of that year alone. Improved testing methods are being developed to assist in the location of contaminated corn and corn products.

One group of deuteromycetes, the dermatophytes (from the Greek words *dermatos*, skin + *phyton*, plant), are the cause of ringworm, athlete's foot, and other fungal skin diseases. Such diseases are especially prevalent in the tropics. The pathogenic stages of these fungi are asexual, but most of these organisms have now been correlated with species of ascomycetes; nevertheless, they continue to be classified on the basis of their disease-causing forms. Athlete's foot flourishes under warm, wet conditions and usually disappears rapidly if a person switches from boots or shoes to sandals and keeps his or her feet dry. Although there are typically about 2 million bacteria per square centimeter on the sole of the foot, competing with the dermatophytes for the available nutrients, the fungi can outcompete them because they grow directly into the epidermal cells and also poison the bacteria with the antibiotics they secrete. Dermatophytes are much more active than bacteria in metabolizing keratin, the tough, fibrous protein to which the contents of epidermal cells are converted as they are sloughed off. Eventually, however, the bacteria may develop resistant strains and then proliferate, causing severe infection. During World War II, more soldiers had to be sent back from the South Pacific because of skin infections than because of wounds received in battle. Many fungi are associated with particular diseases; one of these, *Candida albicans*, which is yeastlike under certain conditions, causes thrush and other infections of the mucous membranes.

In recent years, fungal diseases have become much more prevalent and more adequately diagnosed in humans than formerly. Hundreds of thousands of fungal spores are inhaled by everyone each day, but, until recently, most of these had little effect. Since the 1950s, however, drugs routinely used to facilitate the acceptance of organ transplants, as well as chemotherapy, large doses of steroids, and other common medical treatments, have resulted in large populations of people whose immune systems have been partially or almost completely suppressed, and who cannot overcome infections. People with AIDS are subject to opportunistic fungal attack for the same reason.

Since 1980, at least 100 fungi, not previously connected with disease in humans, have been identified as pathogenic in people with suppressed immune systems. For example, *Fusarium oxysporum*, a common fungal pathogen of plants, has infected and destroyed the lungs of dozens of patients. *Aspergillus flavus* produces an enzyme that digests internal tissues, creating holes that soon are occupied by large masses of fungal hyphae. With at least a dozen new fungal diseases being identified each year, research into methods of controlling them is being intensified. New antifungal antibiotics are urgently needed.

Predaceous Fungi

Among the most highly specialized of the fungi are the predaceous fungi; they have developed a number of mechanisms for capturing small animals they use as food. Although microscopic fungi with such habits have been known for many years, recently it has been learned that a number of species of gilled fungi also attack and consume nematodes (small roundworms). The oyster mushroom, *Pleurotus ostreatus*, for example, grows on decaying wood (a, b). Its hyphae secrete a substance that anesthetizes nematodes, after which the hyphae envelope and penetrate these tiny worms; the fungus apparently uses them primarily as a source of nitrogen, thus supplementing the low levels of nitrogen that are present in wood.

Some of the microscopic deuteromycetes secrete on the surface of their hyphae a sticky substance in which passing protozoa, rotifers, small insects, or other animals become stuck (c). More than 50 species of this group trap or snare nematodes. In the presence of the roundworms, the fungal hyphae produce loops that swell rapidly, closing the opening like a noose when a nematode rubs against its inner surface. Presumably the stimulation of the cell wall increases the amount of osmotically active material in the cell, causing water to enter the cells and increase their turgor pressure; the outer wall then splits and a previously folded inner wall expands as the trap closes.

(a)

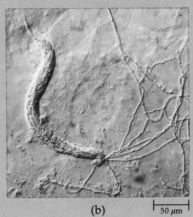
(b) ⊢ 50 μm ⊣

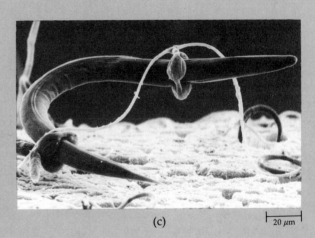
(c) ⊢ 20 μm ⊣

(a) *The oyster mushroom,* Pleurotus ostreatus. *(b) Hyphae of the oyster mushroom, which produce a substance that anesthetizes, converging on the mouth of an immobilized* nematode. *(c) The predaceous deuteromycete* Arthrobotrys dactyloides *has trapped a nematode. The trap consists of rings, each comprising three cells, which when triggered* swell rapidly to about three times their original size in 0.1 second and strangle the nematode. Once the worm has been trapped, fungal hyphae grow into its body and digest it.

Lichens

Lichens are mutualistic symbiotic associations between ascomycetes and certain genera of green algae or cyanobacteria.* The photosynthetic members of these associations provide the energy-rich carbon compounds for both partners, and they are protected from environmental extremes by their fungal partners, which pass on to them mineral nutrients that reach the fungi from the external environment. Because of this symbiotic relationship, lichens are able to live in some of the harshest habitats on earth (Figure 12–17). Lichen-forming fungi constitute about 13,500 species of varied appearance, different from the roughly 30,000 named species of ascomycetes. Some 40 genera of photosynthetic organisms are found in combination with these ascomycetes. The most frequent are the green algae *Trebouxia, Pseudotrebouxia,* and *Trentepohlia* and the cyanobacterium *Nostoc;* about 90 percent of all lichens have one of these four genera as their photosynthetic component. A number of lichens incorporate both a green alga and a cyanobacterium.

*There are also about a dozen species of basidiomycetes that form associations with algae, but they are closely related to free-living basidiomycetes and do not resemble the other lichen-forming fungi.

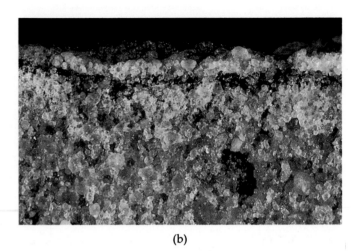

(a)

(b)

12–17

(a) *In this seemingly lifeless, dry region of Antarctica, lichens live just beneath the exposed surface of the sandstone.* (b) *In the fractured rock, the colored bands are distinct zones biologically. The black and white zones are formed by a lichen, while the lower green zone is produced by a nonlichenized unicellular green alga. The air temperatures in this part of Antarctica rise almost to freezing in the summer and probably fall to −60°C in winter.*

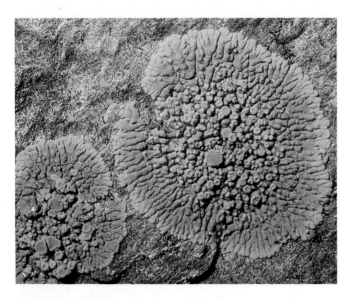

12–18

A crustose ("encrusting") lichen, Caloplaca sacicola, *growing on a bare rock surface in central California.*

Lichens are extremely widespread in nature; they occur from arid desert regions to the Arctic and grow on bare soil, tree trunks, sunbaked rocks, fence posts, and windswept alpine peaks all over the world (Figures 12–18 to 12–20). Some lichens are so tiny that they are almost invisible to the unaided eye; others, like the reindeer "mosses," may cover kilometers of land with ankle-deep growth. One species, *Verrucaria serpuloides*, is a permanently submerged marine lichen. Lichens are often the first colonists of newly exposed rocky areas. In Antarctica, there are more than 350 species of lichens (see Figure 12–17) but only two species of vascular plants; seven species of lichens actually occur within 4° of the South Pole! The activities of lichens in forming the first soil on rock surfaces initiate biological succession (see page 649) in these areas. Those lichens in which the primary or secondary photosynthetic partner is a cyanobacterium are of special importance because they also contribute fixed nitrogen to the soil; such lichens are the primary factor in supplying nitrogen in many regions.

The colors of lichens range from white to black, through shades of red, orange, brown, yellow, and green, and these organisms contain many unusual chemical compounds. Many lichens are used as sources of dyes; for example, in early days the characteristic color of Harris tweed resulted from the treatment of the wool with a lichen dye. Many lichens have also been used as medicines, perfume odor-holding bases, or minor sources of food.

Lichens have long been the subject of biological investigations because of the intriguing nature of the associations between a fungus and its included alga or cyanobacterium. The fungal partner apparently plays the major role in determining the form of the lichen; recently, however, it has been demonstrated that a single fungus, when associated with different kinds of algae, can produce morphologically very different individuals that have traditionally been placed in different genera.

(a)

(b)

(c)

12–19

(a) *Crustose and foliose ("leafy") lichens growing on the gravestone of Roland Thaxter (1858–1932), renowned mycologist from Harvard University.* (b) *Parmelia perforata, a foliose ("leafy") lichen growing around a hummingbird's* nest on a dead tree branch in Mississippi. (c) *Old-man's beard (Usnea hirta), a hanging lichen that often occurs in masses on the limbs of trees. Although it is superficially similar to Usnea and also occupies a similar ecological niche,* "Spanish moss," which is common throughout the southern United States, is not a lichen but a flowering plant—a member of the pineapple family (Bromeliaceae).

(a)

(b)

(c)

12–20

Several fruticose ("shrubby") lichens. (a) *Goldeye lichen, Teloschistes chrysophthalmus.* (b) *British soldiers, Cladonia cristatella, are 1 to 2 centimeters tall.* (c) *Cladonia subtenuis, commonly called "reindeer moss," is actually a lichen. Lichens of this group, which are abundant in the Arctic,* concentrated radioactive substances following atmospheric nuclear tests. Reindeer feeding on the lichens concentrated these radioactive substances still further and passed them on to the humans and other animals that consumed them or their products, especially milk and cheese.

12–21
A cross section of the lichen Lobaria
verrucosa. *The simplest lichens consist
of a crust of fungal hyphae entwining
colonies of algae. In the more complex
lichens, however, the hyphae and the
algal cells are organized in a thallus with
a definite growth form and characteristic
internal structure. The lichen shown here
has four distinct layers: (1) the upper
cortex, a protective surface of heavily
gelatinized fungal hyphae; (2) the algal
layer, which consists of algal cells and
loosely interwoven, thin-walled hyphae;
(3) the medulla, which is a thick layer
of loosely packed, weakly gelatinized
hyphae. This layer, which makes up about
two-thirds of the thickness of the thallus,
appears to serve as a storage area, with
enlarged cells in the fungal hyphae; and
(4) the lower cortex, which is thinner
than the upper one and is covered with
fine projections (rhizines) that attach the
lichen to its substrate.*

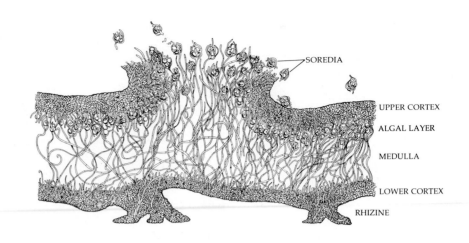

SOREDIA

UPPER CORTEX

ALGAL LAYER

MEDULLA

LOWER CORTEX

RHIZINE

Most of the algae or cyanobacteria found in lichens also occur as free-living species, whereas the lichen fungi are generally found in nature only in lichens.

Lichens commonly reproduce by simple fragmentation, by the production of special powdery propagules known as **soredia** (Figure 12–21), or by small outgrowths known as **isidia**. Fragments, soredia, and isidia all contain both fungal hyphae and algae or cyanobacteria; they act as small dispersal units to establish the lichen in new localities. The fungal component of the lichen may also form ascomata, which are similar to those of other ascomycetes, except that in lichens the ascomata may endure and produce spores slowly but continuously over a number of years. These spores form new lichens when they germinate and come into contact with the appropriate green algae or cyanobacteria.

Biology of the Lichens

How can the lichens survive under environmental conditions so severe that they exclude any other form of life? At one time, it was thought that the secret of the lichen's success was that the fungal tissue protected the alga or cyanobacterium from drying out. Actually, one of the chief factors in lichen survival seems to be the fact that they dry out very rapidly. Lichens are frequently very desiccated in nature, with a water content ranging from only 2 to 10 percent of their dry weight.

When the lichen dries out, photosynthesis ceases; in this state of "suspended animation," even blazing sunlight or great extremes of heat or cold can be endured by some species of lichens. Cessation of photosynthesis depends, in large part, on the fact that the upper cortex of the lichen becomes thicker and more opaque when dry, cutting off the passage of light energy. A wet lichen may be damaged or destroyed by light intensities or temperatures that do not harm a dry lichen.

When a lichen is wetted by rain, it absorbs 3 to 35 times its own weight in water in a very short time. If a dry, brittle, lichen is submerged in water, it becomes soft and pliable within a few minutes. This is the simple physical process of imbibition—the lichen takes up water just as blotting paper does.

The lichen reaches its maximum vitality, as judged by the rate of photosynthesis, after it has been soaked with water and has begun to dry. Its rate of photosynthesis reaches a peak when the water content is 65 to 90 percent of the maximum the organism can hold; below this level, if the lichen continues to lose water, the rate of photosynthesis decreases. In many environments, the water content of the lichen varies markedly in the course of a day, with most photosynthesis taking place only during a few hours, usually in the early morning after wetting by fog or dew. As a consequence, lichens have an extremely slow rate of growth, their radius increasing at rates ranging from 0.1 to 10 millimeters a year. Calculated on this basis, some mature lichens have been estimated to be as much as 4500 years old. They achieve their most luxuriant growth along seacoasts and on fog-shrouded mountains.

Lichens apparently absorb some mineral nutrients from their substrate (this is suggested by the fact that particular species are characteristically found on particular kinds of rocks or soil or tree trunks), but most of their mineral nutrients are derived from the surrounding air and from rainfall. Lichens clearly play an important role in the function of ecosystems. Because they have no means of excreting the elements that they absorb, however, some lichens are particularly sensitive to toxic compounds; the toxins cause the chlorophyll in their algal or cyanobacterial cells to deteriorate. Lichens are very sensitive indicators of the toxic components of polluted air, and they are being used increasingly to monitor atmospheric pollutants, especially around cities. Lichens are particularly sensitive to sulfur dioxide, probably because it quickly destroys their limited supply of chlorophyll. Since those lichens that include cyanobacteria are especially sensitive to sulfur dioxide, air pollution can substantially limit nitrogen fixation in natural communities, causing long-range changes in soil fertility. Both the state of health of the lichens and their chemical composition are used as indicators of the health of their environment. For example, analysis of lichens can trace the amounts of heavy metals and other pollutants around industrial sites. Fortunately, many lichens have the ability to bind heavy metals outside their cells and thus escape damage themselves.

When nuclear tests were being conducted in the atmosphere, lichens were used to monitor the fallout. Now, lichens provide a useful way to monitor the contamination by radioactive substances following events such as the explosion of the Chernobyl nuclear power plant in 1986.

The Nature of the Relationship between Fungi and Photosynthetic Organisms

What is the relationship between the two components of a lichen? First of all, it is important to remember that lichens are **polyphyletic**—they have been derived many times independently from different kinds of ascomycetes—so that the nature of the symbiosis doubtless varies. Nonetheless, a number of features, which we shall discuss now, are common to all lichens. The fungus clearly obtains organic carbon from the alga or cyanobacterium, since the lichen behaves like a photosynthetic organism, dependent only on light, air and mineral nutrients. In fact, the movement of organic carbon from the included alga or cyanobacterium to the fungus has been established by tracer experiments using ^{14}C-labeled carbon dioxide. In those lichens that include cyanobacteria such as *Nostoc*, nitrogen fixation by the bacteria and the transfer of nitrogen compounds to the fungus are also important. In the lichen, the fungal hyphae form a close network around the included algal or cyanobacterial cells. Haustoria, which are very common in lichens, penetrate the photosyn-

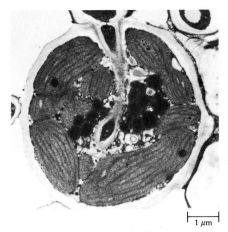

12–22
A haustorium of the fungal component of the lichen Strigula elegans, *penetrating the green algal component,* Cephaleuros virescens. *The densely packed chloroplasts are prominent in the algal cell.*

thetic cells (Figure 12–22); other specialized hyphae, called **appressoria,** lie along the surface of the photosynthetic cells, penetrating them by means of specialized pegs.

The association with a fungus profoundly affects the nature of the metabolic output from the included photosynthetic cells. For example, green algae secrete large quantities of two particular alcohols, D-sorbitol and D-ribitol, only when the algae are associated with fungi to form lichens. Conversely, lichen-forming fungi produce large numbers of extracellular secondary metabolic products called **lichen substances** or **lichen acids,** which sometimes amount to as much as 10 percent of the dry weight of a lichen. More than 350 lichen substances have been identified; they are mostly weak phenolic acids, similar to the constituents of tannins. Some of them are pigmented, having yellow, orange, or red colors, a few of which are used as natural dyes. Lichen substances are useful in lichen identification and classification because they differ enough to separate species. The biological function of these substances remains unknown, although they may play a role in deterring potential herbivores.

We can separate the algal or cyanobacterial and fungal components of a particular lichen and grow them alone in pure culture. Under such conditions, the fungus grows very slowly in compact colonies that look quite unlike the actual lichen. The fungus requires a large number of complex carbohydrates for growth and does not ordinarily form spore-producing bodies. In contrast, isolated lichen algae or cyanobacteria grow

12–23
Scanning electron micrographs of early stages in the interaction between fungal and algal components of British soldier lichens (Cladonia cristatella) in laboratory culture. The photosynthetic component in this lichen (see Figure 12–20b) is Trebouxia, *a green alga.*
(a) An algal cell surrounded by fungal hyphae. (b) Penetration of algal cells by fungal haustoria (arrow). (c) Mixed groups of fungal and algal components developing into the mature lichen.

(a)   25 µm

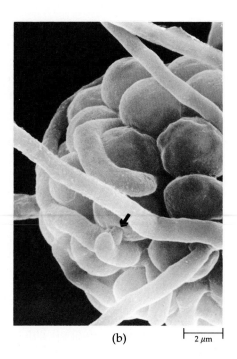

(b)   2 µm

(c)   20 µm

12–24
(a) Corn smut is a familiar plant disease in which the fungus Ustilago maydis *produces black, dusty-looking masses of spores in ears of corn. These are cooked and eaten in Mexico and Central America, where they are regarded as a delicacy.* Ustilago *is a smut, a member of the class Teliomycetes of the division Basidiomycota. (b) Blister rust (Kuehneola uredinis) on a blackberry leaf photographed in San Mateo County, California.*

(a)

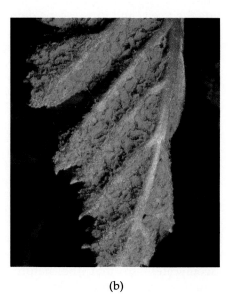

(b)

more rapidly when they are free-living than when they are in the lichen association. In some ways, it is appropriate to think of the lichen partnership not as mutualism but rather as controlled parasitism of the alga or cyanobacterium by the fungus; nevertheless, a lichen does amount to more than the sum of its components, and in that sense it is a distinct "organism." When grown together in culture, the fungus seems first to bring its photosynthetic partner under control and then to establish the characteristic appearance of the mature lichen (Figure 12–23).

DIVISION BASIDIOMYCOTA

Last of the three divisions of fungi to be discussed, the division Basidiomycota includes the most familiar fungi. Among the 16,000 distinct species of this division are the mushrooms, toadstools, stinkhorns, puffballs, and shelf fungi, as well as two important groups of plant pathogens, the rusts and smuts (Figure 12–24). Basidiomycetes play a central role in the decomposition of plant litter, often constituting two-thirds of the living biomass (other than animals) in the soil. Even though

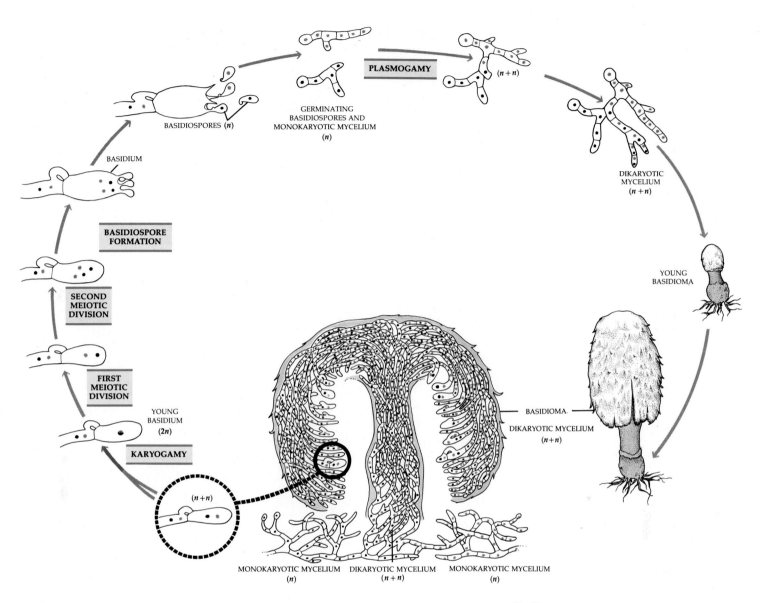

PLASMOGAMY

$(n+n)$

GERMINATING
BASIDIOSPORES AND
MONOKARYOTIC MYCELIUM
(n)

BASIDIOSPORES (n)

BASIDIUM

DIKARYOTIC
MYCELIUM
$(n+n)$

**BASIDIOSPORE
FORMATION**

**SECOND
MEIOTIC
DIVISION**

YOUNG
BASIDIOMA

**FIRST
MEIOTIC
DIVISION**

YOUNG
BASIDIUM
$(2n)$

KARYOGAMY

$(n+n)$

BASIDIOMA
DIKARYOTIC MYCELIUM
$(n+n)$

$(n+n)$

MONOKARYOTIC MYCELIUM   DIKARYOTIC MYCELIUM   MONOKARYOTIC MYCELIUM
(n) $(n+n)$ (n)

12–25
*Life cycle of a mushroom (division
Basidiomycota, class Hymenomycetes).
Monokaryotic mycelia are produced from
basidiospores, and dikaryotic mycelia
from them, often following the fusion of
different mating types, in which case the
mycelia are heterokaryotic. Dikaryotic
mycelia form the basidioma, within
which basidia form on the hymenia that
line the gills, ultimately releasing up to
billions of basidiospores.*

the larger basidiomycetes are the best-known fungi,
many more species undoubtedly await discovery.

A diagram of the life cycle of basidiomycetes will
provide a convenient reference point as we proceed
with our discussion of this important group (Figure 12–
25). Basidiomycetes are distinguished from other fungi
by their production of **basidiospores,** which are borne
outside a club-shaped spore-producing structure called
the **basidium** (plural: basidia) (Figure 12–26). In nature, most basidiomycetes reproduce primarily through
the formation of basidiospores.

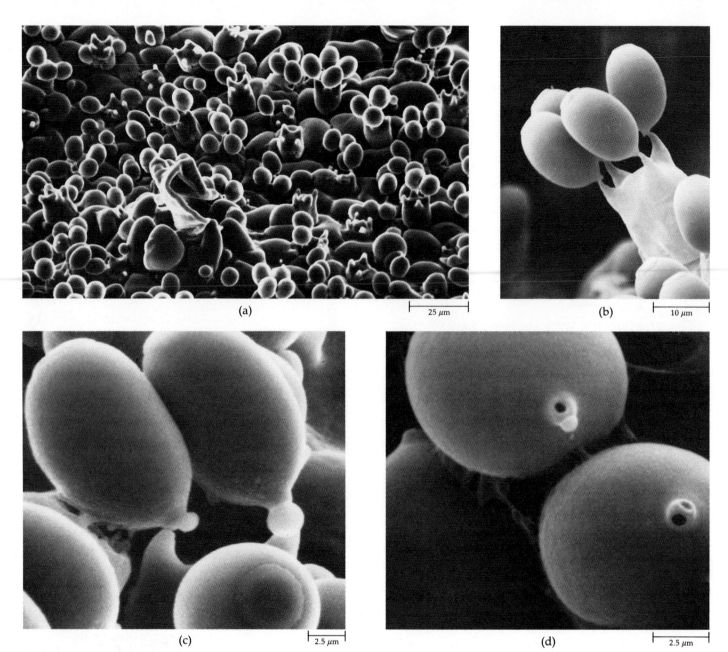

(a) 25 μm

(b) 10 μm

(c) 2.5 μm

(d) 2.5 μm

12–26

Scanning electron micrographs of basidiospores of the inky cap mushroom, Coprinus cinereus. (a) *The hymenium showing numerous basidia frozen at the time of basidiospore release.* (b) *The top of a basidium, with an elongate basidiospore attached asymmetrically to each of its four sterigmata.* (c) *Close-up of two basidiospores. A droplet forms at the base of each basidiospore prior to its release from the sterigma. The basidiospore on the right is shown at a later stage than the one on the left.* (d) *Detail of the base of two basidiospores. On the right, two pores are visible, one where the basidiospore was attached to the sterigma, the other through which the droplet was secreted. In the basidiospore on the left, the droplet is still in place.*

The mycelium of the basidiomycetes is always septate, but the septa are perforated. The pore of the septum has an inflated, barrel-shaped margin, called a **dolipore;** such dolipore septa are important characteristics of basidiomycetes, allowing most members of the division to be recognized from their hyphae alone. On either side of the dolipore, the cytoplasm is delimited by differentiated membrane caps called **parenthesomes** because, in profile, they resemble a pair of parentheses (Figure 12–27). Parenthesomes are absent in rusts and smuts.

In most species of basidiomycetes, the mycelium passes through two distinct phases—monokaryotic and dikaryotic—during the life cycle of the fungus. When it germinates, a basidiospore produces a mycelium that may be multinucleate initially; septa are soon formed and the mycelium is divided into monokaryotic (uninucleate) cells. Commonly, the dikaryotic mycelium is produced by fusion of monokaryotic hyphae from different mating types (in which case it is heterokaryotic), or it may arise from a single monokaryotic mycelium when septa are not formed after nuclear division (in which case it is homokaryotic). In either case, the result is formation of a dikaryotic (binucleate) mycelium since karyogamy does not immediately follow plasmogamy.

The apical cells (those at the apex, or tip) of the dikaryotic mycelium usually divide by the formation of clamp connections (Figure 12–28). These clamp connections, which ensure the allocation of one nucleus of each type to the daughter cells, are highly characteristic of the basidiomycetes.

The mycelium that forms the **basidiomata** (singular: basidioma; formerly called basidiocarps)—fleshy, spore-producing bodies, such as mushrooms and puffballs—is dikaryotic. The formation of the basidiomata may require light. As it forms the basidiomata, the dikaryotic mycelium becomes differentiated into specialized hyphae that play different functions within the basidiomata.

The basidiomycetes can be divided into three classes: Hymenomycetes, Gasteromycetes, and Teliomycetes. The Hymenomycetes include all fungi that produce basidiospores exposed early *on* a basidioma, such as mushrooms and shelf fungi. The Gasteromycetes include those fungi, such as puffballs, that form spores enclosed *in* a basidioma. The Teliomycetes, the rusts and smuts, do not form a basidioma; instead, these fungi produce their spores in masses. The basidia of Hymenomycetes are formed on a distinct fertile layer, the hymenium—the source of the group's name—and not from scattered, single, dispersed cells, as in some Teliomycetes and some Gasteromycetes. The basidiomata that are characteristic of the Hymenomycetes and Gasteromycetes are analogous to the ascomata of the ascomycetes.

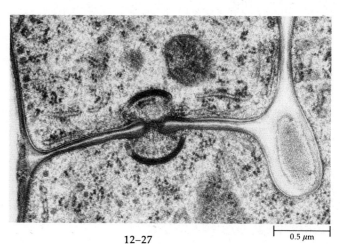

12–27
A dolipore septum, shown in an electron micrograph of Auricularia auricula, *a common soil-inhabiting basidiomycete. Such septa are characteristic of the dikaryotic hyphae of basidiomycetes. Each dolipore septum is perforated by a pore. Parenthesomes are visible on either side of the dolipore.*

0.5 μm

12–28
(a) *In the basidiomycetes, dikaryotic hyphae characteristically are distinguished by the presence of clamp connections over the septa. Clamp connections are formed during cell division; they presumably ensure the proper distribution of the two genetically distinct types of nuclei in the basidioma.* (b) *Electron micrograph of a clamp connection and characteristic septa in a hypha of* Auricularia auricula.

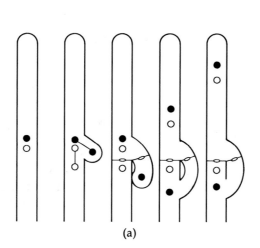

(a)

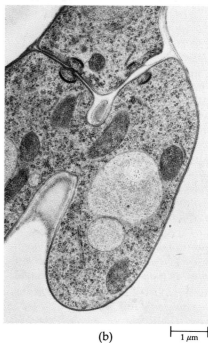

(b)

1 μm

(a)

(b)

(c)

(d)

12–29

Basidiomycetes—members of the class Hymenomycetes. (a) *The fly agaric,* Amanita muscaria. *The mushrooms are at various stages of growth; one has been picked to show the gills. Among the characteristics of this genus of mushrooms, many members of which are poisonous, are the spots on the cap, the ring on the stalk, and the cup, or volva, around its base.* (b) Leccinum subglabripes, *a bolete. In the boletes, the gills found in most kinds of mushrooms are replaced by tubes ending in pores.* (c) *A shelf fungus,* Ganoderma applanatum. *Shelf fungi are responsible for most wood rot.* (d) *An edible spine fungus,* Hericium coralloides. *The hymenium, an outer spore-bearing layer of basidia, is borne on the surface of the downwardly directed teeth.*

Class Hymenomycetes

The class Hymenomycetes contains the edible and poisonous mushrooms (see Figure 12–3), coral fungi, tooth fungi, and shelf or bracket fungi (Figure 12–29c); they produce their basidiospores in a hymenium enclosed in a basidioma. Some Hymenomycetes have club-shaped, nonseptate (internally undivided) basidia, usually bearing four basidiospores, each on a minute projection called a **sterigma** (plural: sterigmata) (Figures 12–26 and 12–30). Others—the jelly fungi (Figure 12–31)—have septate (internally divided) basidia like those of the Teliomycetes.

The structure that one recognizes as a mushroom or toadstool is a basidioma (see Figure 12–25). ("Mushroom" is sometimes popularly used to designate the edible forms of basidiomata, and "toadstool" is used to designate the inedible ones, but mycologists do not recognize such a distinction and use only the term "mushroom." In this book, all such forms are referred to as mushrooms; this does not mean that they are all edible.) The mushroom generally consists of a **cap,** or **pileus,** that sits atop a **stalk,** or **stipe.** The masses of hyphae in the basidiomata usually form distinct layers. Early in its development—the "button" stage—the mushroom may be covered by a membranous tissue that ruptures as the mushroom enlarges. In some genera, remnants of this tissue are visible as patches on the upper surface of the cap and as a sheath, or **volva,** at the base of the stipe (see Figure 12–29a). In many Hymenomycetes, the lower surface of the cap consists of radiating strips of tissue called **gills** (see Figure 12–30), which are lined with the hymenium. In other members of the class, the hymenium is located elsewhere; for example, in the spine fungi (see Figure 12–29d) the hymenium covers downwardly directed pegs, or spines.

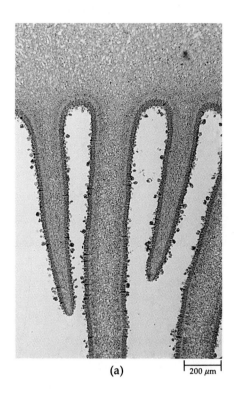

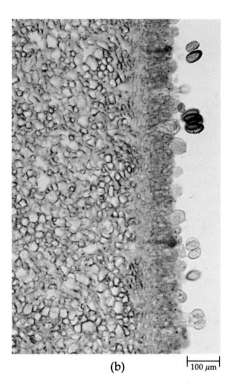

(a)   ⊢—200 μm—⊣   (b)   ⊢100 μm⊣

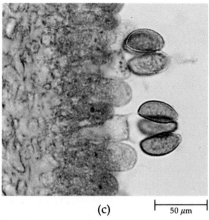

(c)   ⊢—50 μm—⊣

12–30
Stained sections through the gills of
Coprinus, *a common mushroom, at progressively higher magnifications. The
hymenial layer is stained darker in each
of these preparations. (a) Outlines of
some of the gills. (b) Developing basidia
and basidiospores are shown in a section
through the hymenial layer. (c) Nearly
mature basidiospores attached to basidia
by sterigmata.*

In the shelf fungi and boletes (see Figure 12–29b, c),
the hymenium lines vertical tubes that open as pores.

In Hymenomycetes, basidia form in hymenia; each
basidium develops from a terminal cell of a dikaryotic
hypha. Soon after the young basidium enlarges, karyogamy occurs. This is followed almost immediately by
meiosis of each diploid nucleus, resulting in the formation of four haploid nuclei (see Figure 12–25). Each of
the four nuclei then migrates into a sterigma, which enlarges at its tip to form a uninucleate, haploid basidiospore. The reproductive capacity of a single mushroom
is tremendous, with billions of spores being produced
by a single basidioma.

In relatively uniform habitats, such as lawns and
fields, the mycelium from which mushrooms are produced spreads underground, growing downward and
outward and forming a ring of mushrooms on the surface; this ring may grow to as large as 30 meters in diameter. In an open area, the mycelium expands evenly
in all directions, dying at the center and sending up
spore-producing bodies at the outer edges, where it
grows most actively because this is the area in which
the nutritive material in the soil is most abundant. (This
is exactly what happens when a fungus is inoculated at
the center of a Petri plate.) As a consequence, the
mushrooms appear in rings, and as the mycelium
grows, the rings become larger. Such circles of mushrooms are known in European folk legends as "fairy
rings" (Figure 12–32).

12–31
*A jelly fungus growing on a dead tree
limb in the Amazon forest of Brazil. Jelly
fungi are the only Hymenomycetes that
form septate basidia.*

The best known basidiomycetes are the gill fungi. *Agaricus campestris*, the common field mushroom, is particularly abundant. The closely related *A. bisporus* is one of the few mushrooms cultivated commercially. It is now grown in more than 70 countries, and the value of the world crop exceeds $14 billion. Together with the Oriental shiitake mushroom, *Lentinus edodes*, *A. bisporus* makes up about 86 percent of the world's mushroom crop. Other mushrooms are also being cultivated, and some are gathered in large quantities in nature.

Not all mushrooms are edible, however; the gill fungi also include many that are poisonous. For example, the genus *Amanita* includes the most poisonous of all mushrooms, as well as some that are edible. Even one bite of the "destroying angel," *A. virosa*, can be fatal. Some basidiomycetes contain chemicals that cause humans who eat them to hallucinate (Figure 12–33).

Class Gasteromycetes

The Gasteromycetes produce basidiospores inside basidiomata, which are completely enclosed for at least part of their development. This class probably evolved from the Hymenomycetes. Among the more familiar Gasteromycetes are the stinkhorns, earthstars, false truffles, bird's-nest fungi, and puffballs (Figure 12–34). Stinkhorns (Figure 12–34b) have a remarkable morphology. They develop underground in tough sacs. At maturity, they differentiate into a lower receptacle derived from the sac, an elongating stalk, and a gleba, which is a convoluted, external hymenium. The gleba forms an unpleasant-smelling, sticky mass of spores that attracts flies and beetles, which disperse the spores.

12–32
A "fairy ring" formed by the mushroom Marasmius oreades. *Some fairy rings are estimated to be up to 500 years old. Because of the exhaustion of key nutrients, the grass immediately inside such a ring is often stunted and lighter green than the grass outside the ring.*

The puffballs are familiar Gasteromycetes. At maturity, the interior of a puffball dries up, and it releases a cloud of spores when struck (Figure 12–34a). A single puffball may produce several trillion basidiospores. The bird's-nest fungi (Figure 12–34c) begin their development like puffballs, but the disintegration of much of their internal structure leaves them looking like miniature bird's nests.

12–33
Mushrooms figure prominently in the religious ceremonies of several groups of Indians in southern Mexico and Central America; the Indians eat certain basidiomycetes for their hallucinogenic qualities. (a) *One of the most important of these mushrooms is* Psilocybe mexicana, *shown here growing in a pasture near Huautla de Jiménez, Oaxaca, Mexico.* (b) *The shaman María Sabina is shown eating* Psilocybe *in the course of a midnight religious ceremony.* (c) *Psilocybin is the chemical responsible for the colorful visions experienced by those who eat these "sacred" mushrooms; it is a structural analogue of LSD and mescaline (see Figure 19–51a, page 437).*

(a)

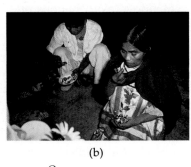

(b)

PSILOCYBIN

(c)

(a)

(b)

(c)

(d)

12–34
Basidiomycetes, members of the class Gasteromycetes. (a) Puffballs, Lycoperdon ericetorum. Raindrops cause the thin outer layer of the puffball to dimple, forcing out a puff of air, mixed with spores, through the opening. (b) Stinkhorn, Phallus impudicus. The basidiospores are released in a foul-smelling, sticky mass at the top of the fungus. Flies, such as the individuals of Mydaea urbana *shown here, visit it for food and spread the spores, which adhere to their legs and bodies in great numbers. (c) White-egg bird's-nest fungus,* Crucibulum laeve. *The round structures of the "nests" of these fungi contain the basidiospores, which are splashed out and dispersed by raindrops. (d) Earthstar,* Geastrum saccutum, *showing one fully opened individual and two others in earlier stages of development. The outer layer is folded back in this genus, raising the spore mass above the dead leaves.*

smuts (see Figure 12–24) produce their spores in masses, which are smaller in rusts and are called **sori.** As plant pathogens, the rusts and smuts are of tremendous economic importance, causing billions of dollars of damage to crops throughout the world each year.

The life cycles of many rusts are complex, and these pathogens are a constant challenge to the plant pathologist whose task it is to keep them under control. Until recently, the rusts were thought to be obligate parasites on vascular plants, but now several species are grown in artificial culture. Some smuts are also able to complete their development under laboratory conditions.

An example of a rust life cycle is provided by *Puccinia graminis,* the cause of black stem rust of wheat. It is one of some 7000 species of rusts. Numerous strains of *P. graminis* exist; in addition to wheat, they parasitize other cereals such as barley, oats, and rye, and various species of wild grasses. *Puccinia graminis* is a continuous source of economic loss for the wheat grower. In a single year, the losses in Minnesota, North Dakota, South Dakota, and the prairie provinces of Canada amounted to nearly 8 million metric tons. As early as 100 A.D., Pliny described wheat rust as "the greatest pest of the crops." Today plant pathologists combat black stem rust largely by breeding resistant wheat varieties, but mutation and recombination in the rust make any advantage short-lived. Parasexual processes resulting in somatic recombination are one factor in the production of new infective strains of wheat rust.

Puccinia graminis is **heteroecious;** that is, it requires two different hosts to complete its life cycle (see Figure 12–35 on the following two pages). **Autoecious** para-

Class Teliomycetes

The class Teliomycetes consists of two large orders of fungi: the rusts (order Uredinales) and the smuts (order Ustilaginales). The members of this class differ from other basidiomycetes in that they do not form basidiomata. They do, however, form dikaryotic hyphae and basidia, which are septate like those of the jelly fungi (members of the class Hymenomycetes). The rusts and

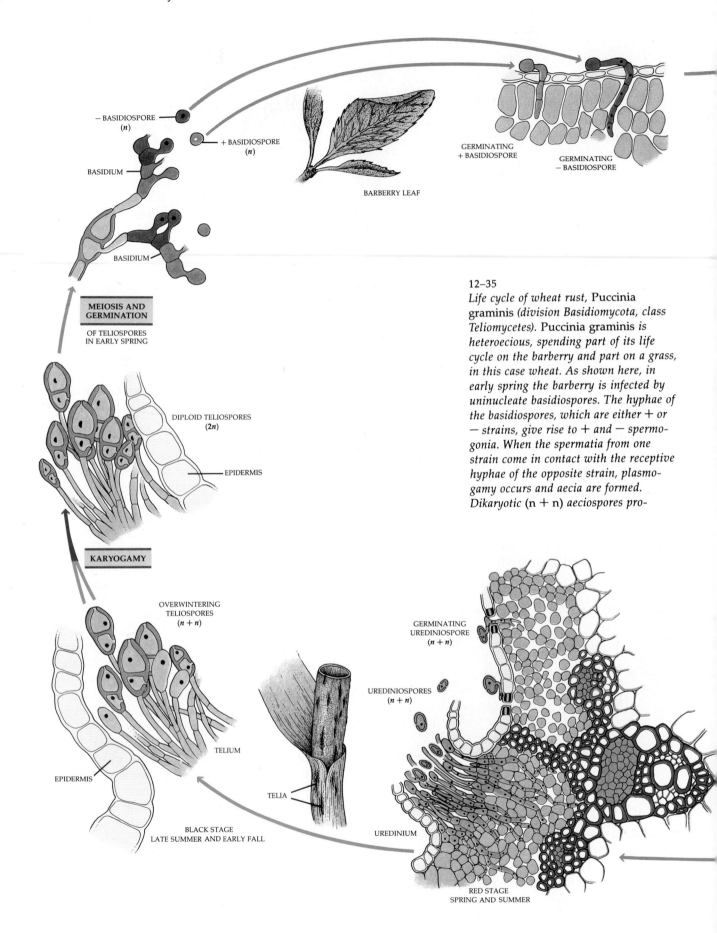

− BASIDIOSPORE
(n)

+ BASIDIOSPORE
(n)

BASIDIUM

BASIDIUM

BARBERRY LEAF

GERMINATING
+ BASIDIOSPORE

GERMINATING
− BASIDIOSPORE

**MEIOSIS AND
GERMINATION**

OF TELIOSPORES
IN EARLY SPRING

DIPLOID TELIOSPORES
(2n)

EPIDERMIS

KARYOGAMY

OVERWINTERING
TELIOSPORES
(n + n)

GERMINATING
UREDINIOSPORE
(n + n)

UREDINIOSPORES
(n + n)

EPIDERMIS

TELIUM

TELIA

BLACK STAGE
LATE SUMMER AND EARLY FALL

UREDINIUM

RED STAGE
SPRING AND SUMMER

12–35
Life cycle of wheat rust, Puccinia
graminis *(division Basidiomycota, class
Teliomycetes).* Puccinia graminis *is
heteroecious, spending part of its life
cycle on the barberry and part on a grass,
in this case wheat. As shown here, in
early spring the barberry is infected by
uninucleate basidiospores. The hyphae of
the basidiospores, which are either + or
− strains, give rise to + and − spermogonia. When the spermatia from one
strain come in contact with the receptive
hyphae of the opposite strain, plasmogamy occurs and aecia are formed.
Dikaryotic (n + n) aeciospores pro-*

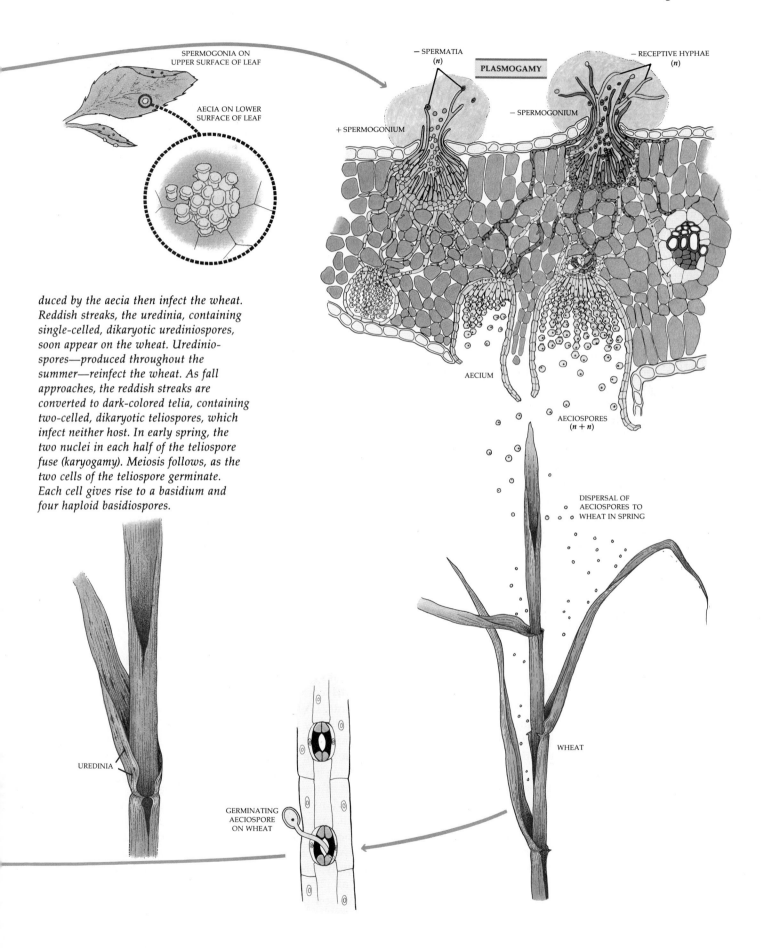

SPERMOGONIA ON
UPPER SURFACE OF LEAF

AECIA ON LOWER
SURFACE OF LEAF

— SPERMATIA
(*n*)

PLASMOGAMY

— RECEPTIVE HYPHAE
(*n*)

+ SPERMOGONIUM

— SPERMOGONIUM

AECIUM

AECIOSPORES
(*n* + *n*)

DISPERSAL OF
AECIOSPORES TO
WHEAT IN SPRING

duced by the aecia then infect the wheat.
Reddish streaks, the uredinia, containing
single-celled, dikaryotic urediniospores,
soon appear on the wheat. Urediniospores—produced throughout the
summer—reinfect the wheat. As fall
approaches, the reddish streaks are
converted to dark-colored telia, containing
two-celled, dikaryotic teliospores, which
infect neither host. In early spring, the
two nuclei in each half of the teliospore
fuse (karyogamy). Meiosis follows, as the
two cells of the teliospore germinate.
Each cell gives rise to a basidium and
four haploid basidiospores.

UREDINIA

GERMINATING
AECIOSPORE
ON WHEAT

WHEAT

sites, in contrast, require only one host. *Puccinia graminis* can grow indefinitely on its grass host, but there it reproduces only asexually. In order for sexual reproduction to take place, the rust must spend part of its life cycle on barberry *(Berberis)* and part on a grass. One method of attempting to eliminate this rust has been to eradicate barberry bushes. For example, the crown colony of Massachusetts passed a law ordering "whoever . . . hath any barberry bushes growing in his or their land . . . shall cause the same to be extirpated or destroyed on or before the thirteenth day of June, A.D. 1760."

Infection of the barberry occurs in the spring, when uninucleate basidiospores infect the plant by forming haploid mycelia, which first develop spermogonia, primarily on the upper surfaces of the leaves. The form of *P. graminis* that grows on barberry consists of separate + and − strains, so that the rust's basidiospores and the spermogonia derived from them are either + or −. Each spermogonium is a flask-shaped pustule lined by cells that form sticky, uninucleate cells called **spermatia.** The mouth of the spermogonium is surrounded by a brush of orange, stiff, unbranched, pointed hairs, the **periphyses,** which hold droplets of sugary, sweet-smelling nectar. The nectar, which is attractive to flies, contains the spermatia. Among the periphyses, branched **receptive hyphae** are also to be found. Flies visit the spermogonia and feed on the nectar. In moving from one spermogonium to another, they transfer spermatia. If a + spermatium of one spermogonium comes in contact with the − receptive hypha of another spermogonium, or vice versa, plasmogamy occurs and dikaryotic hyphae are produced. Aecial initials are produced from the dikaryotic hyphae that extend downward from the spermogonium. **Aecia** are then formed primarily on the lower surface of the leaf, where they produce chains of **aeciospores.** The dikaryotic aeciospores must then infect the wheat; they will not grow on barberry.

The first external manifestation of infection on the wheat is the appearance of rust-colored, linear streaks on the leaves and stems (the red stage). These streaks are **uredinia,** containing unicellular, dikaryotic **urediniospores.** Urediniospores are produced throughout the summer and reinfect the wheat; they are also the primary means by which the wheat rust has spread throughout the wheat-growing regions of the world. In late summer and early fall, the red-colored sori gradually darken and become **telia** with two-celled dikaryotic **teliospores** (the black stage). The teliospores are overwintering spores, which infect neither wheat nor barberries. In early spring, prior to germination, the two haploid nuclei of each cell fuse to form a diploid nucleus (karyogamy). With the onset of germination,

meiosis takes place, apparently in the short, curved basidia that emerge from the two cells of the teliospore. Septa are formed between the resultant nuclei, which then migrate into the sterigmata and develop into basidiospores. Thus, the year-long cycle is completed.

In certain regions, the life cycle of wheat rust can be short-circuited through the persistence of the uredinial state when actively growing plant tissues are continuously available. In the North American plains, urediniospores from winter wheat in the southwestern states and Mexico drift north to southern Manitoba. Later generations scatter westward to Alberta, and finally there is a southern drift at the end of summer, apparently moving along the eastern flank of the Rocky Mountains, and so back to the wintering grounds. Under such circumstances, wheat rust does not depend on barberries to persist, so that barberry eradication was not an effective tool for controlling wheat rust in this area. In contrast, the spread of urediniospores from the south is prevented in Eurasia, where there are extensive east-west mountain ranges, and barberries *are* necessary for survival.

Mycorrhizae

Certain fungi play a crucial role in the mineral nutrition of vascular plants. For many forest trees, if seedlings are grown in a sterile nutrient solution and then transplanted to grassland soils, they grow poorly and may eventually die from malnutrition (Figure 12–36). However, if a small amount (for example, 0.1 percent by volume) of forest soil containing the appropriate fungi is added to the soil around the roots of the seedlings, they will grow normally. The restoration of normal growth is caused by the functioning of **mycorrhizae** ("fungus roots"), which are intimate and mutually beneficial symbiotic associations between roots and fungi.

Mycorrhizae occur in most groups of vascular plants. Only a few families of flowering plants characteristically lack mycorrhizae; these include the mustard family (Brassicaceae) and the sedge family (Cyperaceae). The clumped, very fine roots of Proteaceae, the protea family, appear to play a major role similar to that of mycorrhizae in other plants.

Many plants seem to grow normally when they are well supplied with essential elements (see Chapter 27), especially phosphorus, even if mycorrhizae are lacking; however, if the essential elements are present in limited supply, plants that lack mycorrhizae grow poorly or not at all. The ability of mycorrhizae to absorb and transport soil phosphorus was shown in experiments using

12–36

Mycorrhizae and tree nutrition. Nine-month-old seedlings of white pine (Pinus strobus) were raised for two months in a sterile nutrient solution and then transplanted to prairie soil. The seedlings on the left were transplanted directly. The seedlings on the right were grown for two weeks in forest soil containing fungi before being transplanted to the grassland soil.

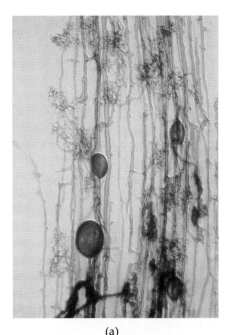

(a)

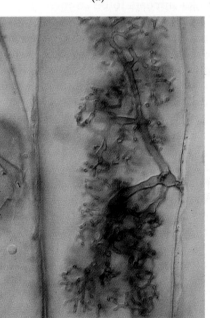

(b)

12–37

Endomycorrhizae. Glomus versiforme, a zygomycete, is shown here growing in association with the roots of leeks, Allium porrum. (a) Arbuscules (highly branched structures) and vesicles (dark, oval structures). Arbuscules predominate in young infections, with vesicles becoming common later. (b) Arbuscules growing inside a leek root cell.

radioactive ^{32}P. Increased absorption of zinc, manganese, and copper—three other essential nutrients—has likewise been demonstrated. Since the hyphal network of mycorrhizae extends several centimeters out from colonized roots, the plants are able to obtain nutrients from a much larger volume of soil than would be possible otherwise. The fungus clearly benefits from the association, obtaining carbohydrates from the host plant.

ENDOMYCORRHIZAE

There are two major types of mycorrhizae: **endomycorrhizae** and **ectomycorrhizae.** Of these two, endomycorrhizae are by far the more common, occurring in about 80 percent of all vascular plants. The fungal component is a zygomycete, with fewer than 200 species involved in such associations worldwide. Thus, endomycorrhizal relationships are not highly specific. The fungal hyphae penetrate the cortical cells of the plant root, where they form either minute, highly branched, treelike structures called **arbuscules** (Figure 12–37) or swellings called **vesicles.** (Vesicles are often called intermatrical spores and may also occur between host cells.) Endomycorrhizae are often called vesicular-arbuscular mycorrhizae (abbreviated as V/A mycorrhizae). Most, or perhaps all, exchange between plant and fungus takes place at the arbuscules. The hyphae extend out into the surrounding soil for several centimeters and thus greatly increase the amount of phosphates and other essential nutrients absorbed.

Endomycorrhizae are of particular importance in the tropics, where soils tend to be positively charged and thus to retain phosphates so tightly that this nutrient is available only in very limited supplies for plant growth. (In contrast, most temperate soils are negatively charged.) Since the impoverished farmers of the tropics are often unable to afford fertilizers, endomycorrhizae play an especially critical role in making phosphates available to crops in these regions. A better understanding of mycorrhizal relationships is certainly one key to lessening the amount of fertilizer that must be applied to agricultural lands to secure satisfactory crops, and to managing rangelands and other natural plant communities effectively. The commercial applications of selected strains of endomycorrhizae to crops to increase yields appears to be an increasingly attractive possibility. These fungi are also particularly important in some ecologically devastated regions, such as coal slag heaps, where plant growth is especially difficult. Unfortunately, endomycorrhizal fungi cannot be grown successfully in pure culture, and thus they can be multiplied only in host plant "pot cultures."

ECTOMYCORRHIZAE

Ectomycorrhizae (Figure 12–38) are characteristic of certain groups of trees and shrubs that are found primarily in temperate regions. These groups include the beech family (Fagaceae), which includes the oaks; the willow, poplar, and cottonwood family (Salicaceae); and the pine family (Pinaceae), as well as certain groups of tropical trees that form dense stands of only one or a few species. The trees that grow at timberline in different parts of the world—such as pines in the northern mountains, *Eucalyptus* in Australia, and southern beech (*Nothofagus;* see Figure 19–10, page 408)—often are ectomycorrhizal. The ectomycorrhizal association apparently makes the trees more resistant to the harsh, cold, dry conditions that occur at the limits of tree growth.

In ectomycorrhizae, the fungus surrounds but does not penetrate living cells in the roots. The hyphae grow between the cells of the root cortex, forming a characteristic structure, the **Hartig net** (Figure 12–39), which eventually surrounds many of the cortical cells. The presence of the Hartig net seems to prolong the life of the individual cells and of the root as a whole. The Hartig net functions as the interface between the fungus and the plant. In addition to this structure, the

12–38

Ectomycorrhizal rootlets from a western hemlock (Tsuga heterophylla). *In such ectomycorrhizae, the fungus commonly forms a sheath of hyphae, called a fungal mantle, around the root. Hormones secreted by the fungus cause the root to branch. This growth pattern and the hyphal sheath impart a characteristic branched and swollen appearance to the ectomycorrhizae. The narrow strands extending from the mycorrhizae are mycelial strands that function as extensions of the root system.*

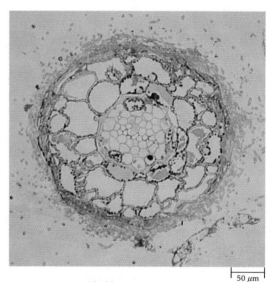

50 μm

12–39

Transverse section of an ectomycorrhizal rootlet of Pinus. *The hyphae of the fungus form a mantle around the root and also penetrate between the cortical cells, where they form the characteristic Hartig net.*

roots are surrounded by a mantle of hyphae, from which mycelial strands extend into the surrounding soil (see Figure 12–38). In roots with ectomycorrhizae, root hairs are often absent, their function apparently carried out by the fungal hyphae. Ectomycorrhizae are mostly formed with basidiomycetes, including many genera of mushrooms, but some ectomycorrhizae involve associations with ascomycetes, including truffles. At least 5000 species of fungi are involved in ectomycorrhizal associations, often with a high degree of specificity.

OTHER KINDS OF MYCORRHIZAE

Two other kinds of mycorrhizae are those characteristic of the heather family (Ericaceae) and a few closely related groups, and those associated with the orchids (Orchidaceae). In Ericaceae, the fungus may constitute up to 80 percent of the mycorrhizae by weight. It forms a prominent web of hyphae over the root surface, from which fine lateral hyphae penetrate the cortical cells of the plant. Although it is energetically expensive for the plant to support such a large fungal biomass, the relationship seems to play an important role in allowing Ericaceae to colonize the kinds of infertile, acidic soils where they are especially frequent. Both ascomycetes and basidiomycetes are involved in mycorrhizal associations with Ericaceae.

The mycorrhizae on the roots of Ericaceae seem to function primarily in enhancing nitrogen, rather than phosphorus, uptake by the plants, a factor of special importance in acidic soils. They also seem to increase the tolerance of some Ericaceae to heavy-metal poisoning, always a risk in acid conditions, whether natural or associated with acid precipitation.

In nature, orchid seeds germinate only in the presence of suitable fungi. This mycorrhizal association is unique in that the fungus, which is internal, also supplies its host with carbon, at least when the host is a seedling. The fungi involved in such associations are mostly basidiomycetes, with more than 100 species involved.

MYCORRHIZAE AND THE HISTORY OF VASCULAR PLANTS

A study of the fossils of early plants has revealed that endomycorrhizal associations were as frequent then as they are in their contemporary descendants (Figure 12–40). This finding led K. A. Pirozynski and D. W. Malloch to suggest that the evolution of mycorrhizal associations may have been a critical step in allowing colonization of the land by plants. Given the poorly de-

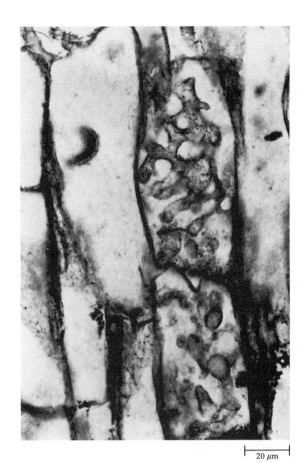

12–40
Silicified roots of a gymnosperm from the Triassic period of Antarctica, showing well-developed arbuscules. Endomycorrhizae were characteristic of the earliest plants for which fossils are known.

veloped soils that would have been available at the time of the first colonization, the role of mycorrhizal fungi (probably zygomycetes) may have been of crucial significance, particularly in facilitating the uptake of phosphorus and other nutrients. A similar relationship has been demonstrated among contemporary plants colonizing extremely nutrient-poor soils, such as slag heaps: those individuals with endomycorrhizae have a much better chance of surviving. Thus it may have been not a single organism but rather a symbiotic association of organisms, comparable to a lichen, that initially invaded the land.

Summary

Fungi , together with the heterotrophic bacteria, are the principal decomposers of the biosphere, breaking down organic products and recycling carbon, nitrogen, and other components to the soil and air. Fungi are rapidly growing, nonphotosynthetic organisms that characteristically form filaments called hyphae, which may be either septate or aseptate. In most fungi, the hyphae are highly branched, forming a mycelium. Parasitic fungi often have specialized hyphae (haustoria) by means of which they extract organic carbon from the living cells of other organisms.

Fungi, which are almost all terrestrial, reproduce by means of spores, which are usually wind-dispersed. No motile cells are formed at any stage of their life cycle. Glycogen is the primary storage polysaccharide. The primary component of fungal cell walls is chitin.

Most fungi are saprobes; that is, they live on organic material from dead organisms. Many fungi are economically important to humans as destroyers of foodstuffs and other organic materials. The kingdom also includes yeasts, *Penicillium* and other producers of antibiotics, cheese molds, and edible mushrooms.

The zygomycetes (division Zygomycota) mostly have coenocytic mycelia. Their asexual spores are generally formed in sporangia—saclike structures in which the entire contents are converted to spores. The zygomycetes are so named because they form zygosporangia during sexual reproduction.

Division Ascomycota, the members of which are commonly called ascomycetes, has some 30,000 named, distinct species, more than any other group of fungi. The distinguishing characteristic of ascomycetes is the ascus—a saclike structure in which the meiotic spores, known as ascospores, are formed. In the ascomycete life cycle, the protoplasts of male and female gametangia fuse, and the female gametangia produce specialized hyphae that are dikaryotic (each compartment containing a pair of haploid nuclei). The ascus forms at the tip of a dikaryotic hypha. Ascospores are generally forcibly expelled. The asci are incorporated into complex spore-producing bodies called ascomata (formerly known as ascocarps). Asexual reproduction may also take place by spore formation; typically, the ascomycetes form asexual spores known as conidia. Yeasts are single-celled ascomycetes in which asexual reproduction occurs by budding, a form of conidia formation. Sexual reproduction occurs by the formation of asci not enclosed in spore-producing bodies.

An artificial group of fungi—Deuteromycota—includes many thousands of species with no known sexual cycle. Most of these deuteromycetes are actually ascomycetes, but some are basidiomycetes.

Lichens are symbiotic partnerships between ascomycetes (and a few basidiomycetes) and either green algae or cyanobacteria. The fungi obtain their food from their photosynthetic partners, surrounding the partners with hyphae and penetrating their cells. A lichen is morphologically and physiologically different from either organism as it exists separately. The ability of the lichen to survive under adverse environmental conditions is related to its ability to withstand desiccation and remain dormant when dry.

The division Basidiomycota, the members of which are called basidiomycetes, includes many of the largest and most familiar fungi. They include mushrooms, puffballs, shelf fungi, and others, as well as the rusts and smuts, which are important plant pathogens. The distinguishing characteristic of the basidiomycetes is their production of basidia. Like the ascus, the basidium is produced at the tip of a dikaryotic hypha and is the structure in which meiosis occurs. Each basidium typically produces four basidiospores, and these are the principal means of reproduction in the basidiomycetes. Basidiospores differ from ascospores in that the former are formed outside the basidium, but, like ascospores, they are forcibly propelled away from the basidium at maturity.

In the classes Hymenomycetes (which includes the mushrooms, or gill fungi) and Gasteromycetes (which includes the puffballs and their relatives), the basidia are incorporated into complex spore-producing bodies called basidiomata (formerly called basidiocarps). The basidia in Hymenomycetes are borne externally, often lining gills or tubes, and open by pores; the basidia in Gasteromycetes are borne internally for at least part of their development. The members of the class Teliomycetes, the rusts and smuts, do not form basidiomata; they have septate basidia, as do the jelly fungi (members of the class Hymenomycetes). Gasteromycetes and hymenomycetes other than the jelly fungi form nonseptate basidia.

Mycorrhizae—symbiotic associations between plant roots and fungi—characterize all but a few families of vascular plants. Endomycorrhizae, in which the fungal partners are zygomycetes, occur in about 80 percent of all kinds of vascular plants. In such associations, the fungus actually penetrates the cortical cells of the host. In the second major type of mycorrhizal association, ectomycorrhizae, the fungus does not penetrate the host cells but forms a feltlike layer over its roots and a network around the cortical cells. Mostly basidiomycetes, but also some ascomycetes, are involved in ectomycorrhizal associations. Mycorrhizal associations are important in obtaining phosphorus and other nutrients for the plant and in providing organic carbon for the fungus. Such associations were characteristic of the first plants to invade the land.

Suggestions for Further Reading

Ahmadjian, Vernon, and Surindar Paracer: *Symbiosis: An Introduction to Biological Associations*, University Press of New England, Hanover, N.H., 1986.

An outstanding account of lichens and other symbiotic associations, so important in understanding biology.

Ayres, Peter, and Nigel Paul: "Weeding with Fungi," *New Scientist*, September 1, 1990, pages 36–39.

A herbicide made from fungi, which can devastate weeds, could be the ideal means of weed control.

Bessette, Alan, and Walter J. Sundberg: *A Quick Reference Guide to the Mushrooms of North America*, Macmillan Publishing Company, New York, 1987.**

An outstanding field guide, well illustrated with photographs, to the common mushrooms likely to be found in North America.

Carefoot, G.L., and E.R. Sprott: *Famine on the Wind, Man's Battle Against Plant Disease*, Rand McNally & Company, Skokie, Ill., 1967.

A highly informative account of important fungal diseases that have influenced the history of many countries.

Christensen, Clyde M.: *Molds, Mushrooms and Mycotoxins*, University of Minnesota Press, Minneapolis, 1975.

An engagingly written account of some fungi and their importance to humans.

Cooke, W.B.: *The Fungi of Our Mouldy Earth*, Beihefte zur Nova Hedwigia, heft 85, J. Cramer, Berlin, 1986.

Fine and lively account of the characteristics of fungi, including chapters on their collection, preparation, isolation, and identification.

Hale, Mason E.: *The Biology of Lichens*, 3d ed., University Park Press, Baltimore, Md., 1983.

An outstanding, concise summary of all aspects of the morphology, physiology, systematics, ecology, and economic uses and applications of the lichens.

Hawksworth, D.L., B.C. Sutton, and G.C. Ainsworth: *Ainsworth & Bisby's Dictionary of the Fungi*, 7th ed., Commonwealth Mycological Institute, Kew, Surrey, England, 1983.

A comprehensive account of all generic and higher-level names and terms used for the fungi; very useful as a reference.

Kendrick, Bryce: *The Fifth Kingdom*, Mycologue Publications, 331 Daleview Place, Waterloo, Ontario, Canada N2l 5M5, 1985.**

A beautifully written, comprehensive, and fascinating guide to the fungi, illuminating all aspects of their biology.

McKnight, Kent H., and Vera B. McKnight: *A Field Guide to Mushrooms of North America*, Peterson Field Guide Series, Houghton Mifflin Company, Boston, 1987.**

A beautifully illustrated guide that includes most of the common and edible species found in North America.

Matossian, Mary K.: *Poisons of the Past: Molds, Epidemics, and History*, Yale University Press, New Haven, Conn., 1989.

A highly original account of the effects that food poisoning resulting from molds on bread may have had on human history.

"Microbes for Hire," *Science 85*, July/August 1985.

An outstanding series of articles on the actual and potential industrial uses of microorganisms, including yeasts and other fungi.

Pirozynski, K.A., and D.L. Hawksworth, eds.: *Coevolution of Fungi with Plants and Animals*, Academic Press, Inc., San Diego, 1988.

A fine collection of papers on all aspects of fungal symbiosis and coevolution.

Smith, D.C., and A.E. Douglas: *The Biology of Symbiosis*, Edward Arnold, Baltimore, Md., 1987.**

An excellent account of biological symbiosis, including several chapters on symbiotic relationships involving the fungi.

*Available in paperback.

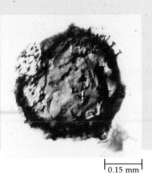

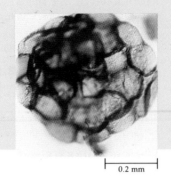

l0.15 mm⊣   l0.2 mm⊣

Protista I: Water Molds, Slime Molds, Chytrids, and Unicellular Algae

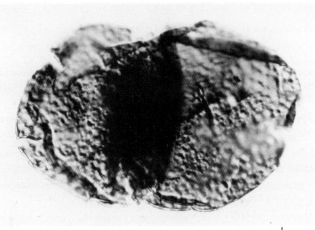

⊢0.1 mm⊣

13–1
Three very different kinds of acritarchs that occurred together about 700 million years ago in the seas of what is now the Grand Canyon region of Arizona. Acritarchs, which first appeared in the fossil record about 1.5 billion years ago, were evidently eukaryotes, as indicated by their size and by the complexity of their cell walls. For hundreds of millions of years, these organisms appear to have been the only eukaryotes. Acritarchs, which were very diverse, were probably planktonic autotrophs. Some closely resemble the resting zygotes of dinoflagellates (see page 264), while others resemble certain green algae.

Eukaryotes appeared in the fossil record about 1.5 billion years ago (Figure 1–1); by that time, bacteria had been in existence for more than 2 billion years. For hundreds of millions of years, all eukaryotes were unicellular. The first known fossils of multicellular eukaryotes are about 630 million years old, although the earlier existence of multicellular organisms is indicated by what appear to be worm burrows that are dated at about 700 million years old. For at least 800 million years of their time on earth, all eukaryotes were members of what is regarded today as the kingdom Protista.

A number of evolutionary lines of multicellular organisms have originated from unicellular ancestral eukaryotes, and there are some multicellular members in almost every division of protists. Four of these wholly or partly multicellular groups are photosynthetic. Among them, the plants, which arose from multicellular green algae, are so distinct that they are treated as a separate kingdom (kingdom Plantae) and are discussed in Chapters 15–19. The remaining three groups—the green, brown, and red algae—which are not directly related to one another, are placed in separate divisions in the kingdom Protista and are covered in Chapter 14. In this chapter, we shall describe the seven predominantly unicellular divisions of protists that, in the past, were treated as plants or fungi—the water molds, slime molds, chytrids, and unicellular algae. Those groups of protists that have traditionally been regarded solely as Protozoa—such as ciliates and amoebas—are not included in this book since they are not normally studied by botanists.

Of the seven divisions of protists discussed in this chapter, four consist of heterotrophic organisms. They have traditionally been the concern of the *mycologist*, or student of fungi. Most scientists now interpret the available evidence as indicating that there is no direct relationship between these heterotrophic protists and the fungi. Nevertheless, they continue to be described by terminology used for the fungi. For example, these organisms are called molds, a traditional word for some fungi, and their filaments are called hyphae, as in the fungi.

The other three divisions covered in this chapter, along with the three divisions discussed in the following chapter, consist principally of photosynthetic organisms. They are studied by *phycologists*, or students of algae, and also by *protozoologists*, students of protozoa.

Until recently, little evidence has been available to indicate direct evolutionary relationships between any of the ten divisions of the kingdom Protista discussed in this book. Over the past few years, however, comparisons of base sequences in nucleic acids, such as chloroplast DNA, have begun to indicate such relationships. In general, the divisions included here are linked by their basic eukaryotic structure and their lack of the distinguishing features of plants, animals, or fungi.

0.2 mm

13–2
Marine phytoplankton. The organisms shown here are dinoflagellates and filamentous and unicellular diatoms.

Ecology of Unicellular Protista

In all bodies of water, minute photosynthetic cells and tiny animals occur as suspended **plankton** (from the Greek word for "wanderer"). The planktonic algae and the cyanobacteria, which together constitute the **phytoplankton,** are the beginning of the food chain for the heterotrophic organisms that live in the ocean and in bodies of fresh water. Phytoplankton, the photosynthetic constituent of the plankton, is usually composed of single-celled organisms. Some of these are quite simple in appearance, while others have intricate and delicately detailed forms that are sometimes joined into colonies or filaments (Figure 13–2). The smallest members of the phytoplankton—**nanoplankton**—are so small that they normally pass through a plankton net, which has mesh openings of 40 to 76 micrometers. The nanoplankton includes algae and cyanobacteria; these organisms are incredibly abundant, and they make important contributions to marine and freshwater productivity.

The heterotrophic plankton—**zooplankton**—consists mainly of tiny crustaceans, the larvae of many different phyla of animals, and many heterotrophic protists and bacteria. In the sea, most small and some large fish, as well as most of the great whales, feed on the plankton, and still larger fish feed on the smaller ones. In this way, the "great meadow of the sea," as the phytoplankton is sometimes called, can be likened to the meadows of the land, serving as the source of nourishment for heterotrophic organisms.

Of the four heterotrophic divisions discussed in this chapter, two consist primarily of aquatic organisms:

Oomycota, the water molds and related organisms, and Chytridiomycota, the chytrids. The other two divisions —Acrasiomycota, the cellular slime molds, and Myxomycota, the plasmodial slime molds—are primarily terrestrial. The water molds and chytrids have flagellated stages in their life cycles; these organisms occur as parasites or as saprobes, consumers of organic material derived from other organisms. The cellular and plasmodial slime molds, in contrast, feed actively on bacteria and small bits of organic debris. The most "animal-like" of the organisms included in this book, the slime molds are conventionally claimed as "protozoa" by zoologists.

Symbiosis and the Origin of the Chloroplast

In all organisms that produce oxygen during photosynthesis—the cyanobacteria, eukaryotic algae, and plants —chlorophyll *a* is involved in the process. The chlorophyll *a* is found in chloroplasts of all these organisms except cyanobacteria, which lack chloroplasts. The similarity of the chloroplasts in these various groups of autotrophs suggests a common origin, and yet the groups differ so greatly in their characteristics that they could not have originated following a single symbiotic event (Chapter 2). In other words, the autotrophic groups are not directly related to one another.

The ease and frequency with which symbiotic events occur can be seen in the wide variety of such relationships that occur today. The modern examples in-

13–3

A symbiotic alga, the unicellular green alga Chlorella, *which is discussed in more detail in Chapter 14. (a) Each bell-shaped cell of the protozoan* Vorticella *contains numerous cells of the symbiotic alga. (b) Electron micrograph of a* Vorticella *containing cells of* Chlorella. *Each algal cell is found in a separate vacuole (perialgal vacuole) bounded by a single membrane. The protozoan provides the algae with protection and shelter, while the algae probably produce a carbohydrate that serves as nourishment for their host cell.*

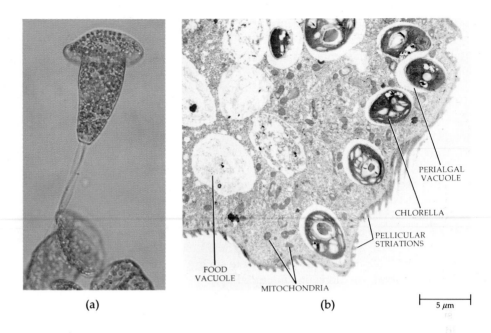

PERIALGAL
VACUOLE

CHLORELLA

PELLICULAR
STRIATIONS

FOOD
VACUOLE

MITOCHONDRIA

(a)   (b)   5 μm

clude symbioses of algae with invertebrate animals, fungi (mostly in lichens), other algae, vertebrates, liverworts, and a few kinds of vascular plants (in which the algae grow in cavities in stems, petioles, and leaves). Among invertebrates, for example, some 150 genera belonging to eight different phyla are known to have algal symbionts growing within their cells (Figure 13–3). Symbiotic dinoflagellates (photosynthetic protists) are responsible for much of the productivity of coral reefs, and other symbiotic associations are of equal importance ecologically.

Photosynthetic protists have traditionally been regarded as related to one another; it was realized only recently that their common ability to manufacture food had been acquired mostly by independent evolutionary events. Thus, autotrophic eukaryotes other than plants have traditionally been grouped together as "algae," with filamentous, heterotrophic eukaryotes considered "fungi" and nonfilamentous, heterotrophic eukaryotes considered "protozoa." The members of these informal groups, however, are not necessarily related. Some algae are more closely related to some protozoa than they are to other algae, and vice versa. For example, the euglenoids are virtually identical overall to some groups of the protozoans known as zoomastigotes (phylum Zoomastigina), but many of the euglenoids are photosynthetic, with chloroplasts closely resembling those of the green algae. Ancestors of the euglenoids apparently acquired these chloroplasts by ingesting green algae, a group to which the euglenoids are not directly related. Euglenoid chloroplasts are surrounded by three membranes, those of green algae by two; the outer membrane of the euglenoid chloroplasts evidently represents the plasma membrane of the green algae the euglenoids originally ingested. Other groups of protists, such as the dinoflagellates, apparently also acquired their chloroplasts in this way.

Characteristics of the Divisions

The heterotrophic protists considered in this book (four divisions) are highly diverse. The Chytridiomycota (chytrids) and Oomycota (oomycetes) have coenocytic (multinucleate and aseptate) hyphae, which can be extensive and highly branched in some oomycetes. Oomycetes produce motile asexual spores—zoospores—with two flagella, one tinsel and one whiplash (see Figure 2–22, page 32); both flagella may be lateral, or both may be apical. Chytrid zoospores, in contrast, have a single, posterior whiplash flagellum. The Myxomycota spend most of their nonreproductive phases either as streaming masses of cytoplasm, called plasmodia, or as encysted masses, which provide a means of escaping drought. The Acrasiomycota are amoeba-like organisms that do not form flagellated cells at any stage of their life cycle. The principal features of the heterotrophic and autotrophic divisions of protists discussed in this book (including those in Chapter 14) are summarized in Table 13–1.

The autotrophic divisions of protists also differ widely from one another in the nature of their flagella (when present) and in their biochemical characteristics, such as pigmentation, food reserves, and cell wall components. The names of some of the autotrophic divisions are derived from the colors of the predominant accessory pigments, which mask the grass-green color of the chlorophylls. Accessory pigments contribute en-

Table 13-1 *Comparative Summary of Characteristics in Ten Divisions of Protista*

Division	Number of Species	Photosynthetic Pigments	Carbohydrate Food Reserve	Flagella	Cell Wall Component	Habitat
Oomycota (water molds)	580	None	Glycogen	2: 1 tinsel, 1 whiplash; in reproductive cells only	Cellulose	Aquatic or need water
Chytridiomycota (chytrids)	575	None	Glycogen	1, posterior, whiplash; in reproductive cells only	Chitin, other polymers	Freshwater or marine
Acrasiomycota (cellular slime molds)	70	None	Glycogen	None (amoeboid)	Cellulose	Terrestrial
Myxomycota (plasmodial slime molds)	500	None	Glycogen	2, whiplash; in reproductive cells only	None	Terrestrial
Chrysophyta (chrysophytes and diatoms)	6700	Chlorophylls *a* and *c*; carotenoids, mainly fucoxanthin	Chrysolaminarin	None or 1 or 2, apical, whiplash or tinsel, equal or unequal	None, or cellulose, with silica scales in some; silica in diatoms	Marine and freshwater
Pyrrhophyta (dinoflagellates)	2100	Chlorophylls *a* and *c*; carotenoids, mainly peridinin	Starch	None or 2, lateral, tinsel	Cellulose, other materials	Marine and freshwater
Euglenophyta (euglenoids)	1000	Chlorophylls *a* and *b*; carotenoids	Paramylon	1 (to 3), apical tinsel (one row of hairs)	No cell wall; have a mostly proteinaceous pellicle	Mostly freshwater
*Rhodophyta (red algae)	4000	Chlorophyll *a*, phycobilins, carotenoids	Floridean starch	None	Sulfated galactons, cellulose, calcium carbonate in many	Marine, some freshwater; many tropical species
*Phaeophyta (brown algae)	1500	Chlorophylls *a* and *c*; carotenoids, mainly fucoxanthin	Laminarin, mannitol	2, lateral; tinsel forward, whiplash behind; in reproductive cells only	Cellulose matrix with alginic acids (polysaccharides)	Almost all marine; flourish in cold ocean waters
*Chlorophyta (green algae)	7000	Chlorophylls *a* and *b*; carotenoids	Starch	None or 2 (or more); apical or lateral; equal; whiplash	Polysaccharides, sometimes cellulose	Mostly freshwater, but some marine

*Divisions consisting largely or entirely of multicellular organisms; discussed in Chapter 14.

ergy absorbed from light, primarily to photosystem II. A rich diversity of storage products is found among these organisms, with most having distinctive carbohydrate food reserves, often in addition to lipids (see below). Reflecting the heterotrophic ancestry of these divisions is the fact that some genera of chrysophytes, dinoflagellates, and euglenoids regularly ingest solid particles of food, just as do the members of many divisions that did not acquire chloroplasts symbiotically during the course of evolution.

Different groups of protists accumulate a rich variety of carbohydrates and lipids through photosynthesis. Heterotrophic protists store their food as glycogen, as do fungi and animals. In green algae, starch is the food reserve; it accumulates within the chloroplasts (as it does in plants, which are derived from the green algae); in the members of all other divisions, the food reserves are accumulated in the cytoplasm outside the chloroplasts. In dinoflagellates, the food reserves are starch; in red algae, they consist of **floridean starch,** a unique molecule that resembles the amylopectin portion of true starch and thus is more like glycogen than like starch. In brown algae, the food reserve is another glucose-containing polysaccharide, **laminarin,** which has linkages between the glucose units that are different from the linkages in starch. In the Chrysophyta, the main storage product is **chrysolaminarin,** which is a more highly polymerized form of laminarin. Many brown algae also store mannitol, an alcohol derived from the sugar mannose. The euglenoids synthesize paramylon, a polysaccharide with a helical configuration; they differ sharply from all other divisions in this respect.

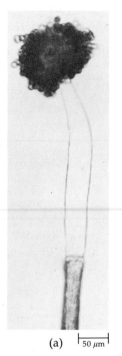
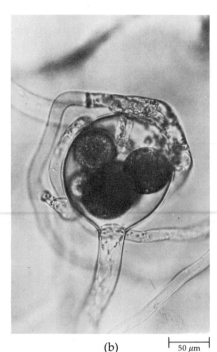

(a) |‑ 50 μm ‑| (b) |‑ 50 μm ‑|

13–4

Achlya ambisexualis, an oomycete that reproduces both asexually and sexually. (a) Empty sporangium with zoospores encysted about its opening, a distinctive feature of Achlya. (b) Sex organs, showing fertilization tubes extending from the antheridium through the wall of the oogonium to the eggs.

Division Oomycota

The division Oomycota, with about 580 species, is a distinctive group, commonly called oomycetes. The cell walls of these organisms are composed largely of cellulose or cellulose-like polymers, thus differing markedly from the chitin cell walls of the fungi. The chromosomes of oomycetes resemble those of most other eukaryotic organisms and differ greatly from the highly condensed chromosomes found in fungi. Meiosis and mitosis resemble the same processes in other eukaryotes; centrioles are present. For all of these reasons, biologists have rejected the traditional linkage of oomycetes with fungi; the two groups seem not to be directly related. The oomycetes range from unicellular to highly branched, coenocytic, and filamentous forms.

Most species of oomycetes can reproduce both sexually and asexually. Sexual reproduction is oogamous and involves an **oogonium,** which contains one to

many eggs, and an **antheridium,** containing numerous male nuclei (Figure 13–4). Fertilization results in the formation of a thick-walled zygote, the **oospore,** which serves as a resting spore; the oospore is the structure for which this division is named. As mentioned previously, asexual reproduction among the oomycetes is by means of motile zoospores, which have two flagella—one tinsel and one whiplash.

One large group of the division Oomycota is aquatic. The members of this group—the so-called water molds—are abundant in fresh water and are easy to isolate from it. Most of them are saprobic, but a few are parasitic, including species that cause diseases of fish and fish eggs.

In some water molds, such as *Saprolegnia* (Figure 13–5), sexual reproduction can occur with male and female sex organs borne on the same individual; in other words, these water molds are **homothallic.** Other water molds, such as some species of *Achlya* (Figure 13–4), are **heterothallic**—male and female sex organs are borne by different individuals, or, if they are borne on

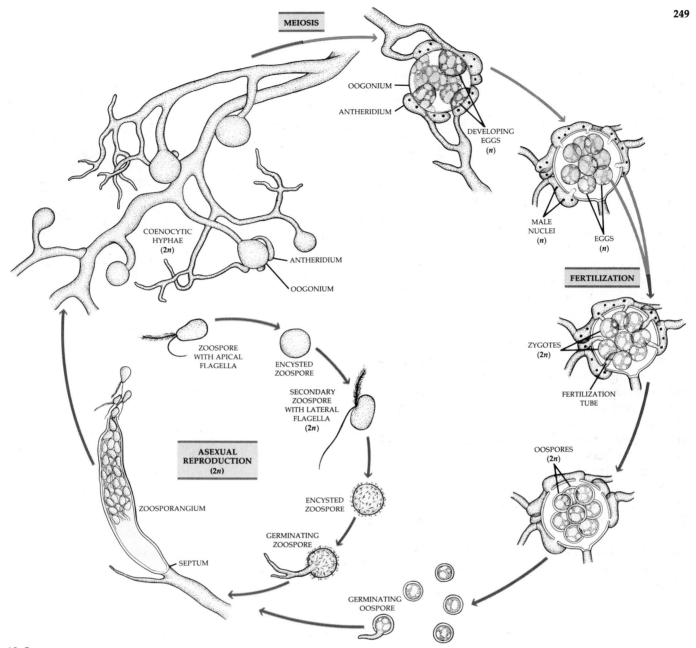

13–5

Life cycle of Saprolegnia, *an oomycete. The mycelium of this water mold is diploid. Reproduction is mainly asexual. Biflagellated zoospores released from a sporangium swim for a while and then encyst. Each eventually gives rise to a secondary zoospore, which also encysts and then germinates to produce a new mycelium.*

During sexual reproduction, oogonia and antheridia are formed, in this species, on the same hypha. Meiosis occurs within these structures. The oogonia are enlarged cells in which a number of spherical eggs are produced. The antheridia develop from the tips of other filaments of the same individual and produce numerous male nuclei. In mating, the antheridia grow toward the oogonia and develop fertilization tubes, which penetrate the oogonia, as seen in Achyla *in Figure 13–4b.*

Male nuclei travel down the fertilization tubes to the female nuclei and fuse with them. Following each nuclear fusion, a thick-walled zygote—the oospore—is produced. On germination, the oospore develops into a hypha, which eventually produces a sporangium, beginning the cycle anew.

one individual, that individual is genetically incapable of fertilizing itself. Both *Saprolegnia* and *Achyla* can reproduce sexually and asexually.

Another group of oomycetes is primarily terrestrial, although the organisms still form motile zoospores when open water is available. Among this group—the Peronosporales—are several forms that are economically important. As the noted mycologist C. J. Alexopoulos said, "At least two of them have had a hand—or should we say hypha!—in shaping the economic history of an important portion of mankind."

One of these is *Plasmopara viticola*, which causes

Hormonal Control of Sexuality in a Water Mold

Achyla ambisexualis is a heterothallic water mold; it produces male and female sex organs on different individuals. The late John R. Raper of Harvard University studied the hormonal control of sexuality in these organisms. The female hyphae secrete a substance that induces the initial development of antheridia on the hyphae of the male organism. (a) Appearance of undifferentiated hyphae before the addition of this substance—characterized and named antheridiol by Alma Barksdale of The New York Botanical Garden. (b) Antheridial branches produced within two hours of the addition of crystalline antheridiol. (c) Antheridial branches growing toward a plastic particle containing antheridiol.

After the developing antheridia appear, the male organism secretes its own hormone, called oogoniol, which induces the formation of oogonia on the hyphae of the female organism. With the appearance of young oogonia, the antheridial hyphae are attracted to them, and distinct, mature antheridia develop. Raper attributed this latter response to a substance he called "hormone C," presumably secreted by the female, but subsequent research suggested that antheridiol elicits this developmental reaction. After the antheridia are formed, there is a differentiating response in the female oogonia, presumably hormonally induced, which leads to the maturation of the oogonium and its enclosed female gametes (eggs). Thus, even in protists, sexual reproduction can involve a highly coordinated sequence of events that are hormonally controlled.

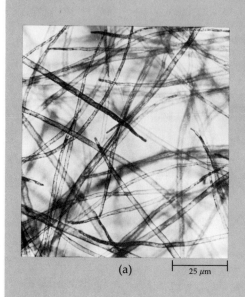

(a)   25 μm

(b)   25 μm

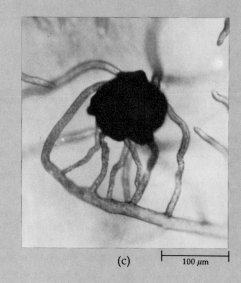

(c)   100 μm

downy mildew in grapes. Downy mildew was accidentally introduced into France in the late 1870s on grape stock from the United States that had been imported because of its resistance to other diseases. The mildew soon threatened the entire French wine industry. It was eventually brought under control by a combination of good fortune and skillful observation. French vineyard owners in the vicinity of Medoc customarily put a disagreeable mixture of copper sulfate and lime on vines growing along the roadside to discourage passersby from picking the grapes. A professor from the University of Bordeaux who was studying the problem of downy mildew noticed that these plants were free from symptoms of the disease. After conferring with the vineyard owners, the professor prepared his own mixture of chemicals—the Bordeaux mixture—which was made generally available in 1882. The Bordeaux mixture was the first chemical used in the control of a plant disease.

Another economically important member of this group is the genus *Phytophthora* (which means "plant destroyer"). *Phytophthora*, with about 35 species, is a particularly important plant pathogen that causes widespread destruction of many crops, including cacao, pineapples, tomatoes, rubber, papayas, onions, strawberries, apples, soybeans, tobacco, and citrus. A widespread member of this genus, *P. cinnamomi*, which occurs in soil, has killed or rendered unproductive millions of avocado trees in southern California and elsewhere. It has also destroyed tens of thousands of hectares of valuable eucalyptus timberland in Australia. The zoospores of *P. cinnamomi* are attracted to the

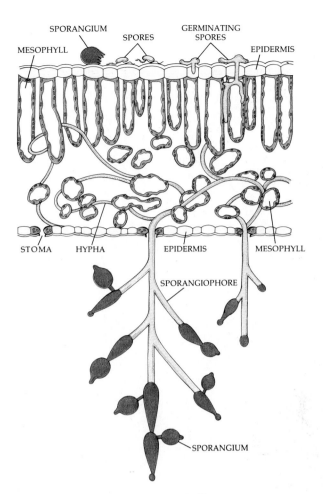

SPORANGIUM SPORES GERMINATING SPORES
MESOPHYLL EPIDERMIS

STOMA HYPHA EPIDERMIS MESOPHYLL

SPORANGIOPHORE

SPORANGIUM

13–6
Phytophthora infestans, the cause of the late blight of potatoes. The cells of the potato leaf are shown in green. In the presence of water and low temperatures, either zoospores are released from the sporangia and swim to the germination site (as shown here) or the sporangia germinate directly through a germ tube.

plants they infect by chemicals exuded by the roots. This oomycete also produces resistant spores that can survive for up to six years in moist soil. Extensive breeding efforts are now under way with avocados and other susceptible crops to produce strains resistant to this oomycete.

The best-known species of *Phytophthora*, however, is *P. infestans* (Figure 13–6), the cause of the late blight of potatoes, which produced the great potato famine of 1846–47 in Ireland. The population of Ireland, which had risen from 4.5 million people in 1800 to about 8.5 million in 1845, fell as a result of this famine to 6.5 million people in 1851. About 800,000 people starved to death, and great numbers of people emigrated, many of them to the United States. Virtually the entire Irish potato crop was wiped out in a single week in the summer of 1846. This was a disaster for the Irish peasants, who depended almost entirely on potatoes for their food, each adult consuming between four and six kilograms of potatoes every day—the amount required to supply sufficient quantities of protein to keep alive.

Another member of this division, blue mold *(Peronospora hyoscyami)*, devastated tobacco crops in the United States and Canada in 1979, causing an estimated loss of almost a quarter of a billion dollars. Blue mold spreads by means of airborne, multinucleate spores. Epidemics such as this will become increasingly abundant in the future as the genetic variability of major crops is narrowed, a problem we shall consider in Chapter 31.

Division Chytridiomycota

The Chytridiomycota, or chytrids, are a predominantly aquatic group of protists, consisting of about 575 species. These organisms are varied in form, in the nature of their sexual interactions, and in their life histories. The cell walls of some chytrids include chitin, although other polymers are also found; primarily because of this feature, some scientists believe that chytrids might be the group from which fungi arose. Preliminary analyses of rRNA likewise indicate that there may be a direct evolutionary relationship between chytrids and fungi, although the matter needs further investigation. Meiosis and mitosis in chytrids resemble these processes in most other eukaryotes, and chytrids differ fundamentally from fungi in these respects.

Almost all chytrids are coenocytic, with few septa at maturity. They are distinguished primarily by their characteristic motile cells (zoospores and gametes), which have a single, posterior, whiplash flagellum. Some chytrids are simple, unicellular organisms that do not develop a mycelium. In them, the whole organism is transformed into a reproductive structure at the appropriate time. Other chytrids have slender rhizoids that extend into the substrate and serve as an anchor (Figure 13–7). Some species are parasites of algae, protozoa, and aquatic oomycetes and of the spores, pollen grains, or other parts of plants; other species are saprobic on such substrates as dead insects.

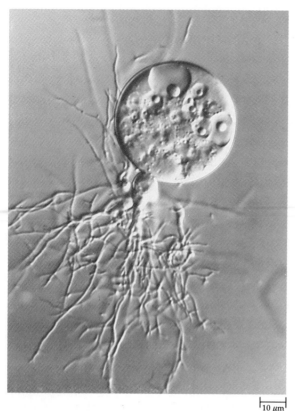

13–8 (on facing page)
In the life cycle of the chytrid Allomyces arbusculus, *there is an alternation of isomorphic generations. The haploid and diploid individuals are indistinguishable until they begin to form reproductive organs. The haploid individuals (gametophytes) produce approximately equal numbers of colorless female gametangia and orange male gametangia. The male gametes, which are about half the size of the female gametes, are attracted by sirenin, a hormone produced by the female gametes. The zygote loses its flagella and germinates to produce a diploid individual. This sporophyte forms two kinds of sporangia: (1) asexual sporangia—colorless, thin-walled structures that release diploid zoospores—which in turn germinate and repeat the diploid generation, and (2) sexual sporangia—thick-walled, reddish-brown structures that are able to withstand severe environmental conditions. After a period of dormancy, meiosis occurs in these sexual, resistant sporangia, resulting in the formation of haploid zoospores. These zoospores develop into gametophytes, which produce gametangia at maturity.*

$\overline{10 \ \mu m}$

13–7
Chytridium confervae, *a common chytrid, as seen with the aid of differential interference contrast system optics. Note the slender rhizoids extending downward.*

One of the chytrids in the genus *Coelomomyces* is an obligate parasite of mosquito larvae and other flies. This genus exhibits a life cycle comparable with that of the rust fungi, in that it has an alternation of hosts between small aquatic crustaceans known as copepods and mosquito larvae. *Coelomomyces* is being investigated as a possible biological control for mosquitoes.

Other chytrids are much more complex in their structure and reproduction. Consider, for example, the life cycle of the genus *Allomyces*. Some species have an alternation of isomorphic generations like that shown in Figure 13–8, whereas in other species the alternating generations are heteromorphic—the haploid and diploid individuals do not resemble one another closely. Alternation of generations is characteristic of plants and of many algae but is otherwise found only in *Allomyces*, in one other closely related genus of chytrids (see Figure 10–10, page 184), and in a very few other heterotrophic protists not considered in this book. In terms of its life cycle, morphology, and physiology, *Allomyces* is one of the best-known protists; its sex hormones have been investigated extensively.

Division Acrasiomycota

The cellular slime molds—a group of about 70 species in six genera—are probably more closely related to the amoebas (phylum Rhizopoda) than to any other group. These molds are common inhabitants of most litter-rich soils, where they usually exist as free-living amoeba-like cells, or **myxamoebas.** These myxamoebas feed on bacteria by surrounding and engulfing them (see Figure 4–12, page 71). Unlike fungi, with which they were once grouped, the cellular slime molds have cellulose-rich cell walls and undergo normal mitosis, in which the nuclear envelope breaks down. In addition, again unlike the fungi, cellular slime molds have centrioles.

Dictyostelium discoideum, the first cellular slime mold to be discovered (1933), reproduces by cell division and shows little morphological differentiation until it exhausts the available supply of bacteria. In response to starvation, the cellular slime molds produce spores, but not in a simple way. Individual cells first aggregate to form a motile sluglike mass. The myxamoebas

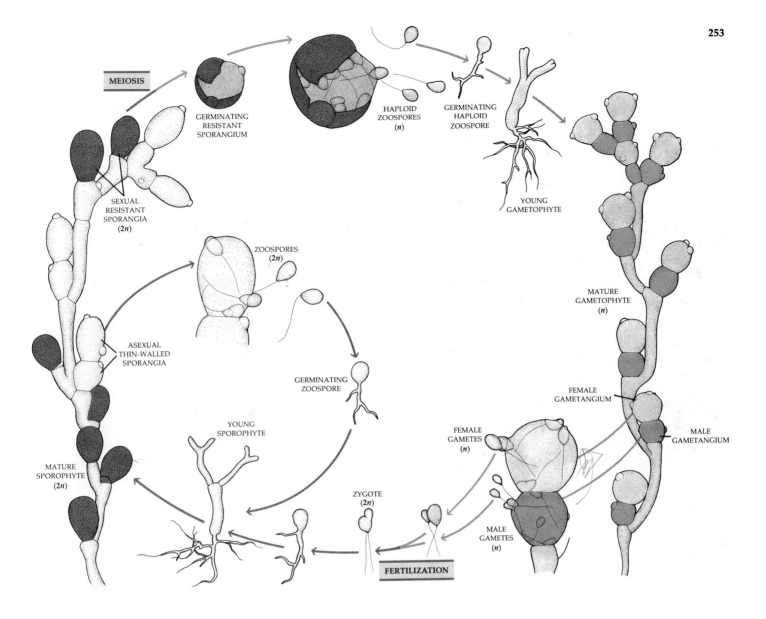

MEIOSIS

GERMINATING
RESISTANT
SPORANGIUM

HAPLOID
ZOOSPORES
(n)

GERMINATING
HAPLOID
ZOOSPORE

YOUNG
GAMETOPHYTE

SEXUAL
RESISTANT
SPORANGIA
(2n)

MATURE
GAMETOPHYTE
(n)

ZOOSPORES
(2n)

ASEXUAL
THIN-WALLED
SPORANGIA

GERMINATING
ZOOSPORE

FEMALE
GAMETANGIUM

YOUNG
SPOROPHYTE

FEMALE
GAMETES
(n)

MALE
GAMETANGIUM

MATURE
SPOROPHYTE
(2n)

ZYGOTE
(2n)

MALE
GAMETES
(n)

FERTILIZATION

retain their individuality in the mass, which usually contains 10,000 to 125,000 individuals. This mass, called a **pseudoplasmodium** (Figure 13–9), or slug, migrates to a new place before differentiating and releasing spores. It moves by the contraction of molecules of a protein called myosin (the same one that is responsible for muscle contraction in animals) in the extracellular matrix around the pseudoplasmodium. Because it moves to a new area before forming spores, *Dictyostelium* avoids releasing new spores into habitats depleted of bacteria.

The aggregation of myxamoebas is initiated when one or more starved cells begin to secrete cyclic adenosine monophosphate (cAMP) into the substrate. The cAMP diffuses away from the cells, establishing a concentration gradient along which the surrounding cells move toward the cells that are secreting the cAMP. The secreting cells, in turn, are stimulated to emit a new "pulse" of cAMP after a period of about five minutes. At least three waves of cells are recruited in this fashion. As the cells accumulate at the aggregation center,

their plasma membranes become sticky, which causes them to adhere to each other and results in the formation of a pseudoplasmodium surrounded by a cellulose sheath.

Myxamoebas of *Dictyostelium discoideum* are very sensitive to light and heat and migrate toward light and from cooler to warmer places. The myxamoebas produce ammonia (NH_3); the higher the temperature and the light intensity, the greater the amount of ammonia produced. The movement of the pseudoplasmodia is most rapid at intermediate concentrations of ammonia, a restriction that causes the pseudoplasmodia to move toward warmer, lighter places while avoiding those that are too hot or too bright.

The eventual developmental fate of an individual cell is determined by its position in the aggregation, which appears to be controlled by which stage of the cell cycle the individual myxamoeba was in when aggregation began. Cells that divide between about 1.5 hours before and 40 minutes after the onset of starvation enter the aggregation last. When migration ceases,

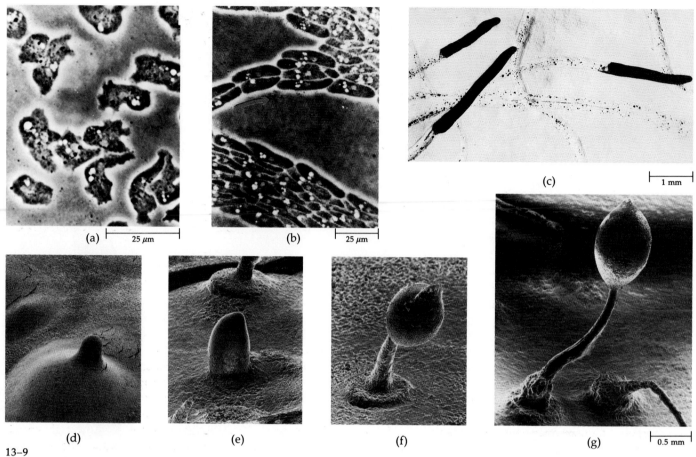

13–9

Life cycle of Dictyostelium discoideum. *(a) The feeding stage of the amoebas. The light gray area in the center of each cell is the nucleus, and the white areas are contractile vacuoles. (b) Amoebas aggregating. The direction in which the* stream is moving is indicated by an arrow. (c) Migrating pseudoplasmodia, each formed of many amoebas. Each sluglike mass deposits a thick slime sheath, which collapses behind it. (d–g) At the end of the migration, the pseudo- *plasmodium gathers itself into a mound and begins to rise vertically, differentiating into a stalk and mass of spores, as seen in these scanning electron micrographs.*

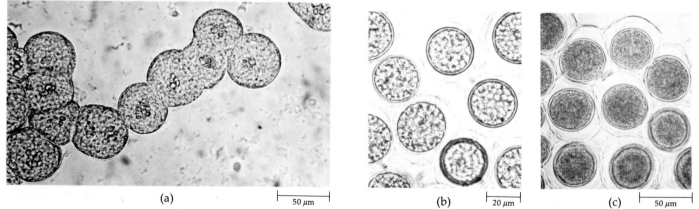

13–10

The process of macrocyst formation in Dictyostelium mucoroides. *(a) Each zygote, or giant cell, is beginning to engulf* the amoebas around it. (b) Giant cells have engulfed all of the amoebas, and each has laid down a cellulose wall. *(c) Mature macrocysts; at this stage they appear homogeneous.*

the anterior cells become the dead, cellulose-coated stalk cells of the developing spore-bearing body. The posterior cells of the pseudoplasmodium then move to the top of the stalk and become dormant spores. Eventually, the basal disk and stalk cells die, and the spores are dispersed. If spores fall on a warm, damp surface, they germinate. Each spore releases a single myxamoeba, and the cycle is repeated.

Reproduction involving asexual spores is common in the cellular slime molds. Sexual reproduction also occurs frequently and involves structures known as **macrocysts.** In *Dictyostelium mucoroides,* for example, the macrocysts are somewhat flattened, irregularly circular to ellipsoidal, multicellular structures from 25 to 50 micrometers in diameter (Figure 13–10). During the formation of macrocysts, pairs of haploid amoebas first fuse, forming zygotes; subsequently, these zygotes attract and then engulf amoebas that are nearby. The macrocysts are formed by smaller aggregations of amoebas than those involved in the formation of slugs, and the aggregations are rounded in outline, rather than being elongate. A thin membrane laid down by the amoebas encircles each macrocyst. Later, a heavy cell wall, rich in cellulose, is laid down around the whole group within the membrane. Within the macrocyst, the zygote—the only diploid cell of the life cycle —undergoes meiosis and several mitotic divisions before germination and the release of new, haploid amoebas.

Division Myxomycota

This division comprises the plasmodial slime molds, or myxomycetes, a group of about 500 species that seems to have no direct relationship to the cellular slime molds or any other group. When conditions are appropriate, the plasmodial slime molds exist as thin, streaming masses of protoplasm that creep along in amoeboid fashion. Lacking a cell wall, this "naked" mass of protoplasm is called a **plasmodium.** As these plasmodia travel, they engulf and digest bacteria, yeast cells, fungal spores, and small particles of decaying plant and animal matter. The successful culture of plasmodia on media that do not contain particulate matter suggests that plasmodia can also obtain food by direct absorption.

The plasmodium may grow to weigh as much as 20 to 30 grams, and because slime molds are spread out thinly, this amount can cover an area of up to several square meters (Figure 13–11). The plasmodium contains many nuclei but is not divided by cell walls. As it grows, the nuclei divide repeatedly and synchronously; that is, all the nuclei in a plasmodium divide at the

13–11
Plasmodium of a plasmodial slime mold,
Physarum, *growing on a tree trunk.*

same time. Centrioles are present, and mitosis is normal, although the chromosomes are very small.

Typically, the moving plasmodium is fan-shaped, with flowing protoplasmic tubules that are thicker at the base of the fan and spread out, branch, and become thinner toward their outer ends. The tubules are composed of slightly solidified protoplasm through which more liquefied protoplasm flows rapidly. The foremost edge of the plasmodium consists of a thin film of gel separated from the substrate by only a plasma membrane and a slime sheath (Figure 10–7a).

Plasmodial growth continues as long as an adequate food supply and moisture are available. Generally, when either of these is in short supply, the plasmodium migrates away from the feeding area. At such times, plasmodia may be found crossing roads or lawns, climbing trees, or in other unlikely places. In many species, when the plasmodium stops moving, it divides into a large number of small mounds. The mounds are similar in size and volume, and so their formation is probably controlled by chemical effects within the plasmodium. The life cycle of a typical plasmodial slime mold is summarized in Figure 13–12. Each mound produces a mature sporangium (Figure 13–13), usually borne at the tip of a stalk; this sporangium is often extremely ornate. Meiosis occurs in the young diploid spores after cleavage and wall formation, and four nuclei are produced. Three of the four nuclei disintegrate, leaving the spore with a single haploid nucleus. In some members of this group, discrete sporangia are not produced, and the entire plasmodium may develop either into a **plasmodiocarp** (Figure 13–13a), which retains the former shape of the plasmodium, or into an **aethalium** (Figure 13–13b), in which the plasmodium forms a large mound that is essentially a single large sporangium.

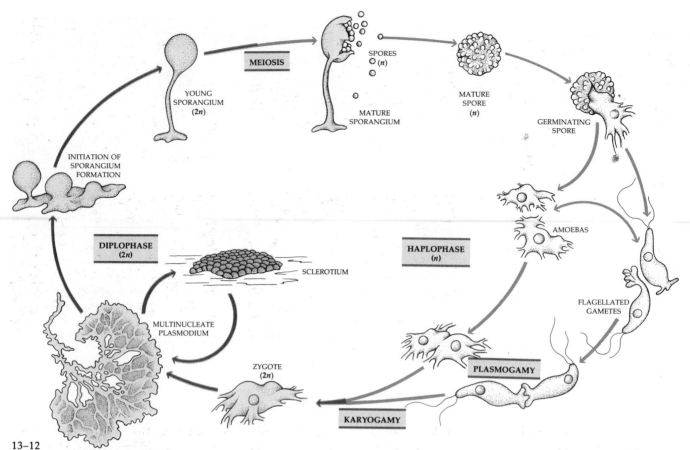

MEIOSIS

SPORES
(*n*)

YOUNG
SPORANGIUM
(2*n*)

MATURE
SPORANGIUM

MATURE
SPORE
(*n*)

GERMINATING
SPORE

INITIATION OF
SPORANGIUM
FORMATION

AMOEBAS

DIPLOPHASE
(2*n*)

SCLEROTIUM

HAPLOPHASE
(*n*)

FLAGELLATED
GAMETES

MULTINUCLEATE
PLASMODIUM

ZYGOTE
(2*n*)

PLASMOGAMY

KARYOGAMY

13–12

Life cycle of a typical myxomycete. Sexual reproduction in the plasmodial slime molds consists of three distinct phases: plasmogamy, karyogamy, and meiosis. Plasmogamy is the union of two protoplasts, which may be amoebas or flagellated gametes derived from germinated spores; it brings two haploid nuclei together in the same cell. Karyo-

gamy is the fusion of these two nuclei, resulting in the formation of a diploid zygote and the initiation of the so-called diplophase of the life cycle. The plasmodium is a multinucleate, free-flowing mass of protoplasm that can pass through a silk cloth or a piece of filter paper and remain unchanged. Plasmodia

often form hardened sclerotia in nature and are able to survive dry periods well in this condition. In any event, the active plasmodium may ultimately form sporangia. Within the developing sporangia, meiosis restores the haploid condition and thus initiates the haplophase of the life cycle.

When habitats dry out, plasmodia can quickly form an encysted stage, the **sclerotium.** Sclerotia can be seen easily on firewood piles because these organisms are often brightly colored in shades of yellow and orange. They are of great importance for the survival of plasmodial slime molds, especially in fast-drying habitats such as dead cacti or soil in deserts, where these organisms are abundant.

The spores of myxomycetes are also resistant to environmental extremes and can be very long-lived; some have germinated after being kept in the laboratory for more than 60 years. Thus spore formation in this group seems to make possible not only genetic recombination but also survival under adverse conditions.

Under favorable conditions, the spores split open and the protoplast slips out. The protoplast may remain amoeboid, or it may develop one or two whiplash flagella; the amoeboid and flagellated stages are readily

interconvertible. The amoebas feed by the ingestion of bacteria and organic material and multiply by mitosis and cell cleavage. If the food supply is used up or conditions are otherwise unfavorable, an amoeba may cease moving about, become round, and secrete a thin wall to form a microcyst. These microcysts can remain viable for a year or more, resuming activity when favorable conditions return.

After a period of growth, plasmodia appear in the amoeba population. Their appearance is governed by a number of factors, including cell age, environment, density of amoebas, and cAMP, which play a role similar to that discussed for the cellular slime mold *Dictyostelium discoideum* (see pages 252–255). One method of plasmodium formation is by the fusion of gametes, which are usually genetically different from one another and are ultimately derived from different haploid spores. These gametes are simply some of the amoebas

13–13
*Spore-producing structures in the
division Myxomycota.* (a) *Plasmodiocarp
of* Hemitrichia serpula. (b) *Aethalia of*
Lycogala *growing on bark.* (c) *Sporangium of* Arcyria nutans. (d) *Sporangia
of* Stemonitis splendens.

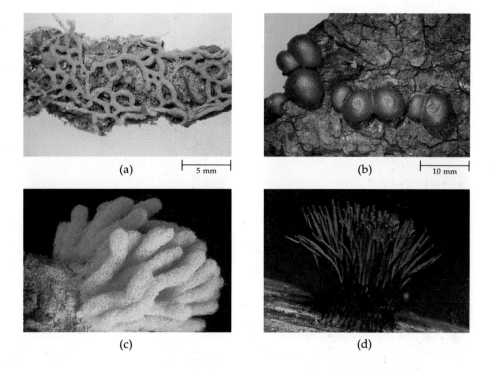

or flagellated cells that now have a new role. In some species and strains, however, the plasmodium is known to form directly from a single amoeba; such plasmodia are usually haploid, like the amoebas from which they arose.

Division Chrysophyta

Chrysophytes are mostly autotrophic, unicellular organisms that are abundant in fresh and salt water throughout the world. They have chlorophylls *a* and *c*, the color of which is largely masked (in the members of the classes Chrysophyceae and Bacillariophyceae) by the abundance of the golden-brown accessory pigment fucoxanthin, a carotenoid. The carbohydrate food reserve of Chrysophyta is a substance called chrysolaminarin. The cells of chrysophytes may be naked and lack a cell wall, or they may have cell walls made up primarily of cellulose. Some members of the division bear prominent silica scales, while others have shells, usually made up of interwoven cellulosic fibrils, that may be impregnated with minerals.

Chrysophyte chloroplasts are biochemically similar to those of brown algae (See Chapter 14) and some dinoflagellates. Another biochemical similarity between the chrysophytes and brown algae is that both groups store food as the polysaccharide laminarin (or its more polymerized form, chrysolaminarin) outside the chloroplast; in addition, the brown algae store mannitol. Both chrysophytes and brown algae have unequal flagella of similar structure (see Figure 2–22, page 32). On the other hand, the presence of algin in the cell walls of brown algae and its absence in chrysophytes would seem to be evidence against a close relationship between the two groups. Macromolecular comparisons will presumably lead to the solution of this problem soon.

THE GOLDEN ALGAE: CLASS CHRYSOPHYCEAE

This class of Chrysophyta—the golden algae—consists of about 500 species (Figure 13–14). Until recently, they were thought to be primarily freshwater organisms, but it is now known that they are abundant members of the nanoplankton and contribute greatly to the productivity of the marine plankton.

Many Chrysophyceae lack a clearly defined cell wall but have silica scales or skeletal structures, which may be superficial or internal and often are exceedingly elaborate. Most members of this class are flagellated, unicellular organisms (see Figure 2–22), although a few lack flagella and some are amoeboid. Except for the presence of their chloroplasts, these amoeboid cells are indistinguishable from amoebas (phylum Rhizopoda), and the two groups may be closely related. A few Chrysophyceae actually ingest bacteria and other or-

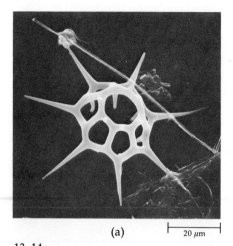

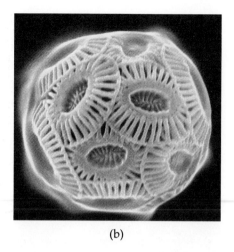

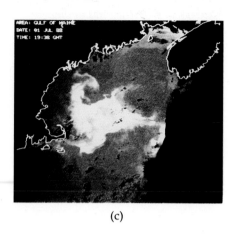

(a)   (b)   (c)

20 μm

13–14

Marine golden algae. (a) A scanning electron micrograph of the siliceous skeleton of Distephanus speculum, *a cold-water species. In life, the skeleton of* Distephanus *is encased in an amoeboid protoplast that contains many small chloroplasts. (b)* Emiliania huxleyi, *a coccolithophorid. This is the most widespread and abundant of the estimated 150 species of this group of extremely minute golden algae. Although these nanoplankton are quite abundant, they are so small* *that they are not caught by conventional plankton nets; they also dissolve in acid fixatives, so they are difficult to find and to study. The plates of the coccolithophorids consist of calcium carbonate. (c) A bloom consisting almost entirely of* Emiliania huxleyi, *appearing white against the gray of the water in the Gulf of Maine. The darkest areas are land (outlined in white) or clouds. The bloom is located east of Cape Cod; Nova Scotia* *appears in the upper right corner. Such blooms have been observed to grow to cover nearly 40,000 square kilometers in the course of three weeks. Within the blooms, there are approximately 3500 cells and 120,000 detached plates per milliliter. The image was recorded by the advanced very-high-resolution radiometer on a National Oceanic and Atmospheric Administration (NOAA) satellite on July 1, 1988.*

ganic particles; for example, individuals of the genus *Dinobryon* each consume about 36 bacteria per hour and are the major consumers of bacteria in some of the cooler lakes of North America. In the members of this class, an individual cell usually contains one or two large chloroplasts; a large granule of chrysolaminarin often accumulates near the posterior end of the cell. Reproduction in most golden algae is asexual and involves zoospore formation. Some Chrysophyceae are colonial.

THE DIATOMS: CLASS BACILLARIOPHYCEAE

The diatoms are mostly unicellular organisms; they are exceedingly important components of the phytoplankton. As such, they are a primary source of food for aquatic animals in both marine and freshwater habitats. It is estimated that there may be at least 5600 living species of diatoms. Counting the great number of extinct species, the diatoms have accounted for at least 40,000 species overall. There are often tremendous numbers of individuals in small areas; thus, over 30 to 50 million individuals of the freshwater genus *Achnanthes* may occur on 1 square centimeter of a submerged rock in the streams of North America. Large numbers of species likewise may occur together. In two small samples of mud from the ocean near Beaufort, North Carolina, for example, 369 species of diatoms were identified. Most species of diatoms occur in plankton, but some are bottom dwellers or grow on other algae or plants.

Diatoms differ from other Chrysophyta in lacking flagella, except for some male gametes, and in their unique shells. These thin double shells, or frustules, are made of polymerized, opaline silica ($SiO_2 \cdot nH_2O$); the two halves (valves) fit together, one upon the other. Electron microscopy has shown that fine tracings on the diatom shells are actually composed of a large number of minute, intricately shaped depressions, pores, or passageways that connect the living protoplasm within the shell to the outside environment (Figure 13–15). These ornamentations are remarkably similar to those on the scales of Chrysophyceae, which may have been the evolutionary antecedent of diatom frustules. The most conspicuous features within the protoplast of diatoms are the brownish plastids that contain chlorophylls *a* and *c* as well as fucoxanthin;

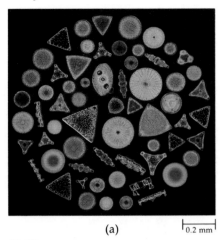

(a) |0.2 mm|

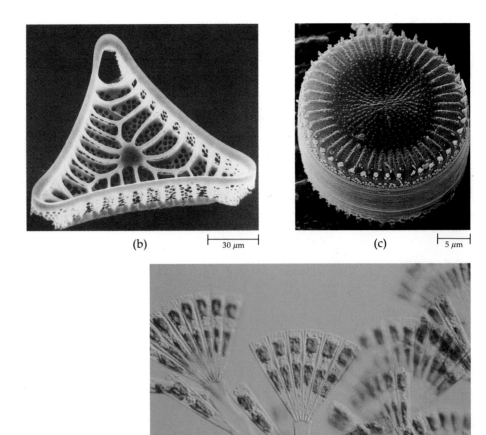

(b) |30 μm| (c) |5 μm|

13–15
(a) *A selected array of marine diatoms, as seen with a light microscope. Scanning electron micrographs of (b) one valve of an* Entogonia *shell and (c)* Cyclotella meneghiniana, *a centric diatom that occurs in brackish water. (d)* Licmophora flasellata, *a stalked pennate diatom, as seen with a light microscope.*

(d)

such plastids closely resemble those of Chrysophyceae. Diatoms reproduce primarily asexually by cell division (Figure 13–16).

On the basis of symmetry, two types of diatoms are recognized: the **pennate** diatoms, which are bilaterally symmetrical (Figure 13–15d), and the **centric** diatoms, which are radially symmetrical (Figure 13–15c). Centric diatoms, which have a larger surface-to-volume ratio than pennate ones and consequently float more easily, are more abundant in large lakes and marine habitats. When sexual reproduction occurs in the centric diatoms, it is oogamous. The male gametes may have a single tinsel flagellum; they are the only flagellated cells found in the diatoms at any stage of the life cycle. In the pennate diatoms, sexual reproduction is isogamous, and both male and female gametes are nonflagellated.

Although most species of diatoms are autotrophic, many can become heterotrophic, absorbing organic carbon; these heterotrophic species are primarily pennate diatoms that live on the bottom of the sea in relatively shallow habitats. A few diatoms are obligate heterotrophs; they lack chlorophyll and are no longer capable of providing their own food through photosynthesis. On the other hand, some diatoms, lacking their characteristic shells, live symbiotically in large marine protozoa (order Foraminifera) and provide organic carbon to their hosts.

The silica shells of diatoms have accumulated over millions of years, forming the fine, crumbly substance known as diatomaceous earth, which is used as an abrasive in silver polish and as a filtering and insulating material. In the Santa Maria, California, oil fields there is a subterranean deposit of diatomaceous earth that is 900 meters thick; near Lompoc, California, more than 270,000 metric tons of diatomaceous earth are quarried annually for industrial use.

Diatoms first became abundant in the fossil record some 100 million years ago, during the Cretaceous period. Many of the fossil species are identical to those still living today, which indicates an unusual persistence through geological time.

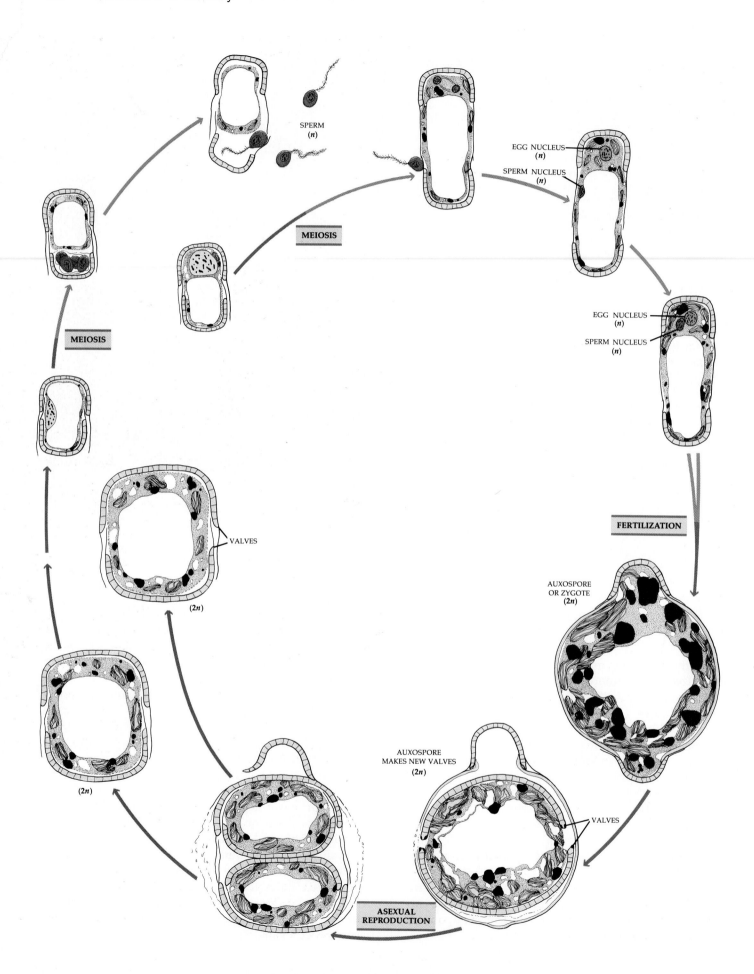

SPERM
(*n*)

MEIOSIS

MEIOSIS

EGG NUCLEUS
(*n*)

SPERM NUCLEUS
(*n*)

EGG NUCLEUS
(*n*)

SPERM NUCLEUS
(*n*)

VALVES

FERTILIZATION

(2*n*)

AUXOSPORE
OR ZYGOTE
(2*n*)

(2*n*)

AUXOSPORE
MAKES NEW VALVES
(2*n*)

VALVES

(2*n*)

ASEXUAL
REPRODUCTION

13–16

Reproduction in a centric diatom. Reproduction in the diatoms is mainly asexual, occurring by cell division. Each diatom is contained within a silica "box," with one of the valves of the box, or shell, enclosing the other. When cell division takes place, each daughter shell (bottom left) receives one of the valves of the previous shell (bottom right) and constructs a second valve. The existing valve always forms the large part (the lid) of the silica box, with the new valve fitting inside it. Thus one cell of each new pair tends to be smaller than the cell from which it is derived. In some species, the shells are expandable and are enlarged by the growing protoplasm within them. In other species, however, the shells are more rigid. Thus, individual cells become smaller through successive cell divisions. When the individuals of these species have decreased in size to about 30 percent of the maximum diameter, sexual reproduction may occur (top left). Certain cells function as male gametangia; they each produce four sperm through meiosis. Other cells function as female gametangia; in these, three of the four products of meiosis are non-functional, so that one egg is produced per cell. This is an example of gametic meiosis (see Figure 10–10b, page 184). After fertilization, the resulting auxospore, or zygote, expands to the full size characteristic of the species. The walls formed by the auxospores are often different from those of the asexually reproducing cells of the same species. Once the auxospore is mature, it divides and produces new shells identical in all their intricate markings to those of the parent cells.

Diatom Motility

Despite their lack of flagella or other locomotor organelles, many species of pennate diatoms are motile. Their locomotion results from a rigorously controlled secretion that occurs in response to a wide variety of physical and chemical stimuli. All motile diatoms seem to possess, along each shell, a fine groove called the raphe, which is basically a pair of pores connected by a complex slit in the siliceous wall of the diatom. Many nonmotile diatoms are attached to one another, with their shells arranged in long filaments. The raphe apparently evolved as a locomotor device by modification of the apical pores, which secrete the substances that unite the nonmotile diatoms into filaments.

A diatom moves in response to an external stimulus—such as mechanical disturbance, light, heat, or a toxic chemical—by initiating contractions in contractile bundles that lie adjacent to the raphe system. These contractions move dehydrated crystalline bodies to reservoir areas adjacent to pores in the raphe, and the crystalline bodies are then discharged into the pores. Here they take up water and expand into twisting fibrils. The fibrils move along the raphe until they touch a surface. They immediately adhere to anything they touch and then contract. If the object to which they adhere is large enough, the diatom moves toward it, depositing a trail of secreted material analogous to the slime trail of a snail. If the object is small, it is transported along the raphe, and the diatom remains stationary. Even motile diatoms are usually at rest. Each can move only a limited distance, since only a limited supply of the required crystalline bodies is available at any one time.

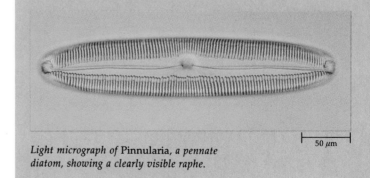

50 μm

Light micrograph of Pinnularia, *a pennate diatom, showing a clearly visible raphe.*

THE YELLOW-GREEN ALGAE: CLASS XANTHOPHYCEAE

The organisms of this class, which consists of about 600 species, are yellow-green. They resemble golden algae in having chlorophyll *c* but differ from them in their lack of fucoxanthin. Yellow-green algae are mostly nonmotile organisms, although some are amoeboid or flagellated, as are the gametes.

One of the better-known members of the class is *Vaucheria*, the "water felt," a coenocytic, little-branched, filamentous alga. *Vaucheria* reproduces both asexually, by the formation of large, compound zoospores that are multiflagellated, and sexually, by oogamy (Figure 13–17). *Vaucheria* is widespread in freshwater, brackish, and marine habitats and often is found on mud that is alternately immersed in water and exposed to air.

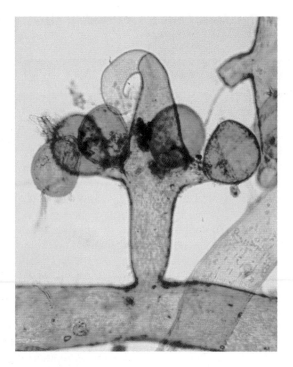

13–17
Vaucheria, *the "water felt," is a coenocytic, filamentous member of the division Chrysophyta.* Vaucheria *is oogamous, producing oogonia (spherical structures) and antheridia (the curved portion at the top).*

13–18
The "armor" of some dinoflagellates consists of cellulose plates in vesicles inside the plasma membrane. The vesicles of genera that appear to be naked may or may not contain cellulose plates .

The Dinoflagellates: Division Pyrrhophyta

Most dinoflagellates are unicellular biflagellates. About 2100 species are known, many of them abundant and highly productive members of the marine plankton; others occur in fresh water. In the dinoflagellates, the flagella beat within two grooves; one groove encircles the body like a belt, and the second groove is perpendicular to the first. The beating of the flagella in their respective grooves causes the dinoflagellate to spin like a top as it moves. The encircling flagellum is ribbonlike. There are also numerous nonmotile dinoflagellates, some of which are nonflagellated. Some genera of dinoflagellates ingest solid food particles, obtaining part or all of their nutrition in this way. A few dinoflagellates are highly toxic (see the essay "Red Tides" on the next page).

Many of the dinoflagellates are bizarre in appearance, with stiff cellulosic plates forming a wall (theca), which often looks like a strange helmet or part of an ancient coat of armor (Figures 13–18 and 13–19). The plates of the wall are in vesicles inside the plasma membrane, not outside like the cell wall of most algae.

Most dinoflagellates contain chlorophylls *a* and *c*, which are generally masked by carotenoid pigments, including peridinin, which is similar to fucoxanthin; their chloroplasts were probably derived from ingested chrysophytes. Other dinoflagellates have green, blue-green, or other types of chloroplasts; these were likely derived following the ingestion of pigmented cells of various groups and the subsequent establishment of stable symbioses. The carbohydrate food reserve in dinoflagellates is starch. Some of the species do not contain chlorophyll and are heterotrophic. Some dinoflagellates are capable of ingesting other cells. Colorless dinoflagellates obtain their nutrients by ingesting other cells, dissolved organic carbon, or small, particulate organic material.

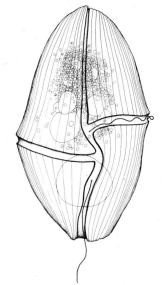

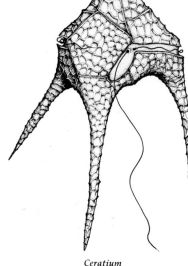

Gymnodinium costatum   *Ceratium*   *Exuviaella*   *Gymnodinium neglectum*

Red Tides

In late August 1987, the west coast of Florida was ravaged by a major red tide incident—one of dozens that are known to have occurred over the past 150 years. Hundreds of thousands of dead fish littered the beaches (a), and millions of tourist dollars were lost. The organism responsible for most red tides and their associated effects in the Gulf of Mexico is the dinoflagellate *Gymnodinium breve*. Following a period of rapid dinoflagellate reproduction, a toxic bloom formed in which *G. breve* was so abundant that the sea was colored reddish brown. Environmental factors that favor such blooms include warm surface temperatures, high nutrient content in the water, low salinity (which often occurs during rainy periods), and calm seas. Thus rainy weather followed by sunny weather in the summer months is often associated with red tide outbreaks.

Two months later, *G. breve* invaded estuaries along the North Carolina coast, areas where it had never before been observed. The dinoflagellates became so abundant that the waters turned yellowish, devastating the tourist industry in this region. At least 41 cases of respiratory, gastrointestinal, or neurological illness were reported in people who swam in these waters—all associated with the toxins produced by the dinoflagellate bloom. *Gymnodinium* is thought to have traveled north from Florida in the Gulf Stream; with it came schools of a fish called menhaden, which consumed large amounts of the dinoflagellates. Bottle-nosed dolphins ate the menhaden, and fully half of the dolphin population in the western Atlantic died. The fish had not been injured by the toxic dinoflagellates in their guts, but the dolphins, eating the fish whole, had been poisoned. In their weakened con-

dition, dolphins proved easy prey for bacterial and viral diseases.

When shellfish, such as mussels and clams, ingest toxic dinoflagellates or other organisms, they accumulate and concentrate the toxins. Depending on the species of toxic organisms the shellfish have consumed, they themselves may become dangerously toxic to the people who eat them. Along the Atlantic Coast, fisheries are commonly closed in summer, and people regularly suffer from poisoning after consuming mussels, oysters, scallops, or clams taken from certain regions. On the other hand, lobsters, crabs, shrimp, and cleaned fish are safe to eat even during red-tide conditions.

Other dinoflagellates are responsible for the formation of red tides in different areas. Thus *Gonyaulax tamarensis* is the organism involved along the northeastern Atlantic coast, from the Canadian Maritime provinces to southern New England, while *Gymnodinium catenella* causes red tides at times along the Pacific Coast from Alaska to California. About 20 species of dinoflagellates have been identified as producing toxins that kill birds and mammals, render shellfish toxic, or produce a widespread tropical fish-poisoning disease called ciguatera.

The poisons produced by some dinoflagellates, such as *G. catenella*, are extraordinarily powerful nerve toxins. The chemical nature and biological activity of most of these toxins are relatively well known. On the other hand, the factors that cause the red tides are poorly understood and are under active study at a number of research centers.

(a)

(b)   5 μm

(a) *Fishes littering a beach in Florida during the red tide incident of 1982.* (b) *Gymnodinium breve, the unarmored dinoflagellate responsible for the outbreaks of red tide along* *the west coast of Florida. The curved, transverse flagellum lies in a groove encircling the organism. Its longitudinal flagellum, only a portion of which is visible, extends from the* *middle of the organism toward the lower left. The apical groove at the top is an identifying characteristic of* Gymnodinium.

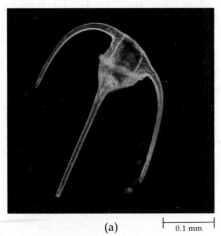

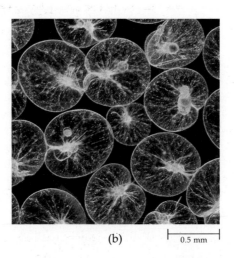

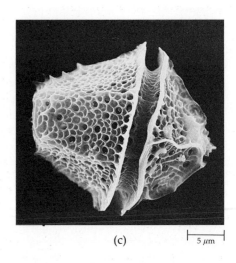

(a)   ⊢ 0.1 mm ⊣   (b)   ⊢ 0.5 mm ⊣   (c)   ⊢ 5 μm ⊣

13–19
(a) Ceratium tripos, *an armored dinoflagellate.* (b) Noctiluca scintillans, *a bioluminescent marine dinoflagellate.*

(c) Gonyaulax polyedra, *the dinoflagellate responsible for the spectacular red*

tides along the coast of southern California.

Dinoflagellates occur as symbionts in many other kinds of organisms, including sponges, jellyfish, sea anemones, tunicates, corals, octopuses and squids, gastropods, turbellarians, and certain protists. In the giant clams of the family Tridachnidae, the dorsal surface of the inner lobes of the mantle may appear chocolate-brown as a result of the presence of symbiotic dinoflagellates. When they are symbionts, the dinoflagellates lack thecae and appear as golden spherical cells called **zooxanthellae** (Figure 13–20).

Zooxanthellae are primarily responsible for the photosynthetic productivity that makes possible the growth of coral reefs in tropical waters, which are notoriously nutrient-poor. Coral tissues may contain as many as 30,000 symbiotic dinoflagellates per cubic millimeter, primarily within cells in the lining of the gut of the coral polyps. Here the dinoflagellates produce glycerol instead of starch, and the glycerol is used directly for the nutrition of the corals. Since the algae require light for photosynthesis, the corals that contain them grow mainly in ocean waters less than 60 meters deep. Many of the variations in the shapes of coral are related to the light-gathering properties of different geometrical arrangements, somewhat similar to the ways in which various branching patterns of trees serve to expose their leaves to sunlight.

The chief method of reproduction of dinoflagellates is by longitudinal cell division, with each daughter cell receiving one of the flagella and a portion of the theca and then constructing the missing parts in a very intricate sequence. Some nonmotile species form zoospores. In some of these species, only the zoospore has the typical dinoflagellate structure, whereas the mature forms have cells with no flagella or may even be joined together in chainlike colonies. Sexual reproduction has been found in a number of species of dinoflagellates; it is generally isogamous, but anisogamy occurs in a few species. Dinoflagellate zygotes, which are called hystrichospheres, form thick, chemically inert, ornamented cell walls; they are similar to some acritarchs (Figure 13–1).

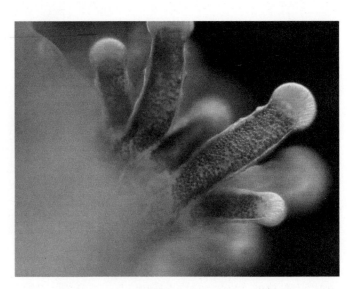

13–20
Zooxanthellae, the symbiotic form of dinoflagellates, shown here in a tentacle of a coral animal. These symbionts are responsible for much of the productivity of coral reefs.

Mitosis in Dinoflagellates

Dinoflagellates have a unique type of mitosis in which some of the features of cell division in bacteria may have been retained. In the dinoflagellate cell, as shown in (a), the chromosomes are always visible and do not condense prior to mitosis. There are extremely large amounts of DNA in each cell. The chromosomes are attached to the nuclear envelope, which persists during mitosis. Dinoflagellate chromosomes have a much lower ratio of protein to DNA than other eukaryotic chromosomes, and they may have evolved from bacterial chromosomes independent of the evolution of other eukaryotic chromosomes.

Cytoplasmic channels, shown extending through the center of (b), invade the dividing nucleus at the time of mitosis. The chromosomes remain attached to the nuclear envelope and are carried on the sides of these channels, which contain bundles of microtubules similar to those of a eukaryotic spindle apparatus. The microtubules are all oriented in one direction and presumably regulate the separation of the portions of the nuclear envelope with its attached chromosomes. The organism illustrated in both (a) and (b) is *Crypthecodinium cohnii*.

At least two species of dinoflagellates have binucleate cells, in which one nucleus is called **dinokaryotic** and the other is eukaryotic. In this second nucleus—which was presumably obtained by the host as a result of the invasion of a chrysophyte-like symbiotic organism—the chromosomes do not become condensed at any stage during the cell cycle, indicating that they are not identical with "typical" eukaryotic chromosomes.

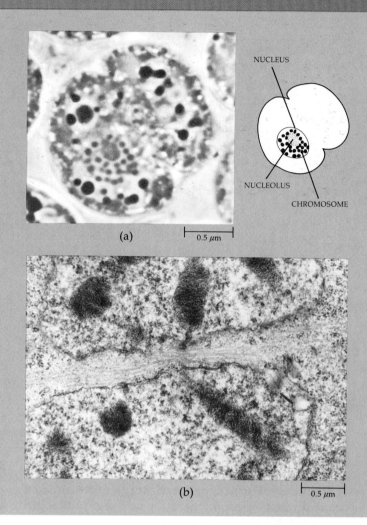

(a)   0.5 µm

NUCLEUS

NUCLEOLUS

CHROMOSOME

(b)   0.5 µm

The Euglenoids: Division Euglenophyta

Most of the estimated 1000 species of euglenoids occur in fresh water and especially in water that is rich in organic matter. Various species of euglenoids range from less than 10 micrometers to more than 500 micrometers (0.5 millimeter) long; they are quite variable in form. All are unicellular except the colonial genus *Colacium*.

The similarities between the chloroplasts of the euglenoids and those of the green algae—both possess chlorophylls *a* and *b* together with several carotenoids—suggest that individual euglenoids ingested green algae and subsequently reduced the green algal cells, forming a stable relationship with the former green algal chloroplasts. Euglenoids store their carbohydrate food reserves in the form of paramylon, a polysaccharide that is not found in any other group of organisms.

About a third of the approximately 40 genera of euglenoids have chloroplasts and are autotrophic; the remaining genera are heterotrophic and either ingest or absorb their food. Euglenoids probably should be regarded as one of the groups of zoomastigote protozoa (phylum Zoomastigina).

Euglenoids reproduce by cell division, with the individual cells remaining motile throughout the process. The nuclear envelope remains intact during mitosis, as it does in dinoflagellates, most green algae, most fungi, and ciliated protozoa. Euglenoid chromosomes, like those of dinoflagellates, remain condensed during interphase and throughout the mitotic cycle. Sexual reproduction is not known to occur in euglenoids.

The division Euglenophyta takes its name from *Euglena*, a common genus that is often used for laboratory studies of cell biology. Many species of *Euglena* are elongated, as shown in Figure 13–21. The cell is complex and contains numerous small chloroplasts. A long

13–21
(a) Euglena, *showing two large storage bodies of paramylon and the helically arranged proteinaceous strips of the pellicle.* (b) *The structure of* Euglena, *as interpreted from electron micrographs.*

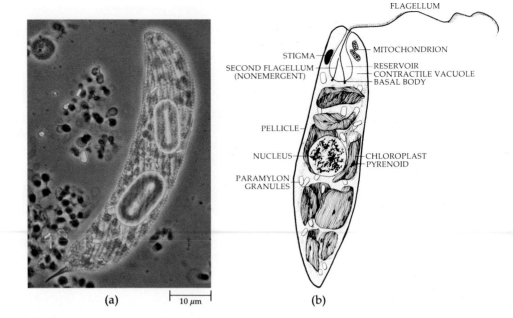

(a)

|← 10 μm →|

(b)

FLAGELLUM

STIGMA

MITOCHONDRION

SECOND FLAGELLUM
(NONEMERGENT)

RESERVOIR
CONTRACTILE VACUOLE
BASAL BODY

PELLICLE

NUCLEUS

CHLOROPLAST
PYRENOID

PARAMYLON
GRANULES

emergent flagellum with very fine hairs is present along one side, as well as a short, nonemergent flagellum. The emergent flagellum is usually held in front of the cell like a spinning lasso.

In *Euglena*, the flagella are attached at the base of the flask-shaped opening—the **reservoir**—at the anterior end of the cell. The contractile vacuole collects excess water from all parts of the cell and discharges it into the reservoir. The cell is delimited by a plasma membrane, inside which is a series of flexible, interlocking, mostly proteinaceous strips that are arranged helically. These strips, together with the plasma membrane, form a structure called the **pellicle.** Unlike the stiff walls of plant cells, the flexible pellicle permits *Euglena* to change its shape, providing an alternative means of locomotion for mud-dwelling forms.

Summary

The four divisions of heterotrophic protists that have often been treated as fungi are Oomycota, Chytridiomycota, Acrasiomycota, and Myxomycota. They have a more or less typical mitotic cycle, and centrioles are present. The first two groups are primarily aquatic; the latter two are terrestrial.

Members of the division Oomycota range from unicellular forms to highly branched coenocytic filamentous forms. The zoospores of this group have two flagella—one tinsel and one whiplash. The cell walls are composed largely of cellulose or cellulose-like polymers. Sexual reproduction involves a large, immobile egg and small, motile male nuclei. Two of the partly terrestrial members of this division are *Phytophthora*—a very important causative agent of plant diseases, including the blight that caused the potato famine of 1846–47 in Ireland—and *Plasmopara viticola*, which causes downy mildew of grapes.

The Chytridiomycota, or chytrids, are unicellular or coenocytic organisms that are aquatic; in most species that have been examined, the cell walls contain chitin. The motile spores and gametes have a single posterior whiplash flagellum. Although mitosis in the chytrids is unspecialized, they may be related to the group that gave rise to fungi. *Allomyces* and one related genus are the only nonphotosynthetic organisms known to have an alternation of generations similar to that of plants and green algae.

The cellular slime molds, division Acrasiomycota, are a small group of amoeba-like organisms that aggregate together at one stage of their life cycle to form pseudoplasmodia, or slugs. Cyclic AMP plays a major role in the aggregation of the individual amoebas into these "slugs," which undergo complex patterns of differentiation that have led to their being used as models in this field. Flagellated cells are not known. Sexual reproduction occurs in at least some members of this division; it takes place by means of macrocysts.

The plasmodial slime molds, division Myxomycota, may exist as streaming, multinucleate masses of protoplasm called plasmodia, which are usually diploid. These plasmodia ultimately form sporangia in which diploid spores are formed. Meiosis occurs within each of the spores, and three of the resulting nuclei disintegrate, leaving one haploid nucleus in each spore. Under favorable conditions, the spores split open, producing amoebas, which may become flagellated. These amoebas or flagellated cells may function as gametes. Plasmodium formation often, but not always, follows fusion of the gametes.

Three divisions of photosynthetic, primarily unicellular protists are discussed. Two of these—Chrysophyta and Pyrrhophyta—have similar chloroplasts that contain chlorophylls *a* and *c* and fucoxanthin; their chloroplasts, as well as those of the brown algae (division Phaeophyta), may have had a common origin and may have become established in some Pyrrhophyta and Chrysophyta following ingestion of organisms that had such chloroplasts. Different Pyrrhophyta, however, have different kinds of chloroplasts and probably acquired them by ingesting other kinds of organisms. In an analogous example, the chloroplasts of the division Euglenophyta, organisms that are probably best regarded as a group of zoomastigote protozoa, seem to have come from the green algae, division Chlorophyta, following the ingestion of such organisms.

Chrysophyta are important components of freshwater and marine phytoplankton. A number of the members of one class, Chrysophyceae, are small algae that are found in great quantities in the nanoplankton. They are very important contributors to photosynthetic productivity in the sea. A second class, the diatoms, class Bacillariophyceae, consists of at least 5600 living species of unicellular organisms with unique double silica cell walls. Diatoms have contributed in a major way to the marine and freshwater phytoplankton for the past 100 million years. The only flagellated cells in the life cycle of diatoms are the male gametes, which have been observed in only a few species. A third class of Chrysophyta, the Xanthophyceae, consists of organisms that lack the pigment fucoxanthin; they are called yellow-green algae.

The dinoflagellates (division Pyrrhophyta) are unicellular biflagellates that are mostly marine. Dinoflagellates are characterized by two flagella, which beat in different planes, causing the organism to spin. Stiff cellulose plates, frequently bizarre in appearance, are often present in vesicles beneath the plasma membrane. In their symbiotic form, in which they are called zooxanthellae, dinoflagellates are major contributors to the productivity of coral reefs; they also occur widely as symbionts in many other marine animals.

Euglenophyta, the euglenoids, are a small group of organisms, most of which occur in fresh water and are unicellular. They contain chlorophylls *a* and *b* and store carbohydrates as an unusual polysaccharide called paramylon. Euglenoids lack a cell wall but have a flexible series of mostly proteinaceous strips, which, together with the plasma membrane, make up the pellicle. The cells contain a contractile vacuole and bear flagella; chloroplasts occur in about a third of the approximately 40 genera. Sexual reproduction is unknown in euglenoids.

Suggestions for Further Reading

Bold, Harold C., and Michael J. Wynne: *Introduction to the Algae: Structure and Reproduction*, 2d ed., Prentice Hall, Inc., Englewood Cliffs, N.J., 1985.

A detailed reference work on the algae that contains a wealth of information on all groups; taxonomically oriented.

Corliss, John O.: "The Kingdom Protista and Its 45 Included Phyla," *BioSystems* 17:87–126, 1984.

A modern arrangement of all phyla/divisions of protists, with comments on the possible phylogenetic relationships among the major groups.

Gray, W.D., and Constantine J. Alexopoulos: *Biology of the Myxomycetes*, Ronald Press Co., John Wiley & Sons, Inc., New York, 1968.

A well-written book about the biology of the plasmodial slime molds, emphasizing ultrastructural, biochemical, and physiological aspects.

Lembi, Carole A., and J. Robert Waaland: *Algae and Human Affairs*, Cambridge University Press, Cambridge, England, 1988.

Excellent collection of papers on positive and negative interactions between algae (as well as cyanobacteria) and humans.

Margulis, Lynn, John O. Corliss, Michael Melkonian, and David J. Chapman (eds.): *Handbook of Protoctista*, Jones and Bartlett Publishers, Boston, 1990.

The definitive account of the organisms included in this and the next chapter.

Raper, Kenneth B.: *The Dictyostelids*, Princeton University Press, Princeton, N.J., 1984.

An excellent discussion of the growth, morphogenesis, and systematics of cellular slime molds.

Ross, Ian K.: *Biology of the Fungi: Their Development, Regulation, and Associations*, McGraw-Hill Book Company, New York, 1979.

An outstanding account of the fungi, biologically oriented and emphasizing the divisions treated in this chapter.

Round, F.E.: *The Ecology of Algae*, Cambridge University Press, New York, 1981.

A comprehensive treatment of freshwater and marine algal ecology, organized according to principles and well written.

Round, F.E., R.M. Crawford, and D.G. Mann: *The Diatoms: Biology and Morphology of the Genera*, Cambridge University Press, New York, 1990.

In this book, an excellent and comprehensive account of the biology of diatoms precedes a detailed treatment of the individual genera.

Sharnoff, Sylvia D.: "Beauties from a Beast: Woodland Jekyll and Hydes," *Smithsonian* 22:98–103, July 1991.

A delightful account of slime molds, accompanied by breathtaking color photographs.

Taylor, F.J.R. (ed.): *The Biology of Dinoflagellates*, Blackwell Scientific Publications, Oxford, England, 1987.

The most recent comprehensive treatment of major aspects of the biology of these unique protists.

CHAPTER 14

Protista II: Red, Brown, and Green Algae

14–1
Algae on the rocks at low tide, photographed at Las Cruces in central Chile.

The open sea, the shore, and the land are the three life zones that make up our biosphere. Of these zones, the sea and the shore are the more ancient. Here algae play a role comparable to the role played by plants in the far younger terrestrial world (Figure 14–1). Often, algae are also dominant in freshwater habitats—ponds, streams, and lakes—where they may be the most important contributors to the productivity of these ecosystems. Everywhere they grow, algae play an ecological role comparable to that of the plants in land habitats.

Along the rocky shore can be found the larger, more complex algae, or seaweeds, typically arranged in fairly distinct visible bands or layers in relation to tide levels (Figure 14–2). Their structural complexity reflects their ability to survive in this challenging life zone, where twice each day they are subject to great fluctuations of humidity, temperature, salinity, and light, in addition to the pounding action of the surf and the abrasive action of suspended sand particles churned up by the waves.

Anchored offshore beyond the zone of waves, algae provide shelter for a rich diversity of microscopic organisms, as well as for larger fishes and invertebrate animals that feed on the microorganisms and on each other. These extensive algal growths may be so dense as to be called forests. For example, off the coast of California, there are huge beds of giant kelps—brown algae whose broad, 15-meter-long fronds are buoyed upward on sinuous stalks 30 meters or more from their holdfasts on the bottom. Many large carnivores, including sea otters and tuna, find food and refuge in these kelp beds, which are also harvested by humans for food and industrial products.

The three divisions of algae that are wholly or partly multicellular—the larger algae—form the subject of this chapter. These are the algae that form masses along the coasts and at times float free in the open ocean, as in the Sargasso Sea, a wide tract of ocean in the north

14–2
Rocks along the coast of Tatoosh Island, Washington, showing the different layers of algal species from lower to higher tide levels.

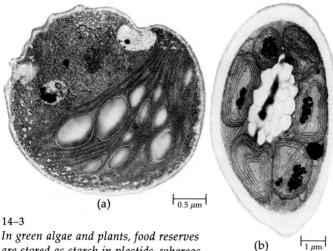

(a)   0.5 μm

(b)   1 μm

14–3
In green algae and plants, food reserves are stored as starch in plastids, whereas other photosynthetic eukaryotes store their food reserves in the cytoplasm.
(a) Chlorella, a green alga, with large, light-colored starch grains in its plastids.
(b) Masses of floridean starch grains surrounded by plastids are seen in the cytoplasm of Batrachospermum moniliforme, *a red alga.*

Atlantic. One of these divisions, the green algae (division Chlorophyta), also includes many unicellular algae, some of which are important contributors to the productivity of the plankton, both in the sea and in fresh water. The exceedingly diverse green algae are of special interest because they are almost certainly the group that gave rise to the plants, an event that occurred more than 430 million years ago. Although there is probably no existing green alga that exactly resembles the ancestor of plants, many retain some of its probable features.

In addition to the green algae, which inhabit marine, freshwater, and even terrestrial habitats, there are two other divisions of primarily marine, multicellular algae: red algae (division Rhodophyta) and brown algae (division Phaeophyta). In addition to their abundant marine representatives, there are a few freshwater genera of red algae and brown algae. All brown algae are multicellular, whereas a few red algae are unicellular.

Characteristics of Red, Brown, and Green Algae

As discussed in the preceding chapter, the chloroplasts of red, brown, and green algae differ in their biochemical features. In red algae, the chloroplasts contain chlorophyll *a* and phycobilins. These chloroplasts are similar to cyanobacteria, an ancient group from which they were almost certainly derived following the incorporation of cyanobacteria as symbiotic elements in the cells of ancestral red algae. Green algae, in contrast, have chloroplasts containing chlorophylls *a* and *b* and carotenoids, the same substances found in the chloroplasts of euglenoids and plants, which are descended from green algae. Although the chloroplasts of green algae bear some biochemical resemblance to the chloroplasts of *Prochloron* (see Figure 11–2, page 187), the accumulating evidence is equivocal as to whether they are related directly to the chloroplasts of red algae (that is, cyanobacteria) or not. Finally, the golden-brown chloroplasts of the brown algae, which contain chlorophylls *a* and *c*, might be related either directly to cyanobacteria, which would have changed biochemically after becoming symbiotic, or to some unknown (or possibly extinct) group of bacteria.

The members of the three divisions considered in this chapter also differ greatly in many other characteristics, some of which are summarized in Table 13–1 (page 247). Like the plants, green algae store their food reserves, which are starch, within their chloroplasts (Figure 14–3a). All other groups of photosynthetic eukaryotes store their food reserves outside the chloroplasts (Figure 14–3b). The food reserves of red algae are floridean starch, a polymer that resembles the amylo-pectin portion of true starch, not the amylose part, and thus resembles glycogen more closely than starch. The food reserves of brown algae are stored as laminarin—another polymer of glucose, in which the units have a linkage that differs from that found in starch—and as the alcohol mannitol.

Although cellulose is found in the cell walls of some red, brown, and green algae, the cell walls differ greatly in constitution, both within and between these groups. In addition, there are many other differences. The red algae lack flagella at all stages of their life cycles. The flagellated reproductive cells of the brown algae, like those of the chrysophytes, have unequal flagella—a tinsel flagellum directed forward and a trailing whip-lash flagellum (see Figure 2–22, page 32). In sharp contrast, the flagella of green algae (when present) are almost always equal whiplash flagella.

When algal cells divide, the plasma membranes gen-erally pinch inward from the margin of the cell (fur-rowing), as occurs in animals, fungi, and protists. However, cell plates like those of plants occur in a few genera of filamentous green algae. Most algal cells have centrioles, which become the basal bodies for the fla-gella of any motile cells that are produced. In the red algae, however, there are evidently analogous struc-tures called **polar rings.**

In general, the multicellular members of the divi-sions discussed here do not have a complex array of tis-sues like those found in vascular plants. Certain kelps (brown algae), however, have a central conducting strand of cells that resemble the sieve elements of vas-cular plants. The reproductive structures of algae are generally single cells, not multicellular structures with sterile protective jackets, such as are found in plants (see Chapter 15). However, some of the green algae have reproductive structures of intermediate complex-ity that may provide an indication of the sorts of inter-mediate stages that occurred in the evolution of plant reproduction.

Diverse types of life cycles occur in red, brown, and green algae. Different green algae have gametic, sporic, and zygotic meiosis (see Figure 10–10, page 184), whereas red algae may have either sporic or zygotic meiosis. Brown algae have either gametic or sporic meiosis but not zygotic meiosis, which does, however, occur in the chrysophytes, a related division.

The three divisions of algae discussed in this chapter are ancient groups, differing sharply in their character-istics. Fossils similar to unicellular green algae occur in the Bitter Springs Formation of central Australia, which is about 900 million years old. Fossils of red algae, at least 700 million and perhaps more than 1.1 billion years old, were reported in 1990, and the unicellular antecedents of the three divisions described in this chapter doubtless evolved much earlier. By the Cam-brian period, at least 550 million years ago, large

siphonous (coenocytic) green algae had evolved. Other multicellular lines of green algae appeared throughout the Paleozoic era; for example, the oldest stoneworts, representatives of the most distinct branch of green algae, date from the late Silurian period, about 420 mil-lion years ago. From about the same time come fossils of green algae such as *Parka*, which apparently was similar to the modern green alga *Coleochaete* (see page 285). Those green algae and red algae that had calcar-eous cell walls are best represented in the fossil record, for obvious reasons; brown algae are poorly repre-sented as fossils.

Red Algae: Division Rhodophyta

Red algae are particularly abundant in tropical and warm waters, although many are found in the cooler regions of the world; there are about 4000 species. Fewer than 100 different species of red algae occur in fresh water, but in the sea the number of species is greater than that of brown and green algae—the other two divisions including seaweeds—combined. Red algae usually grow attached to rocks or to other algae; there are a few floating forms, and a few that are either unicellular or colonial. Structurally, the majority of red algae are complex (Figures 14–4 and 14–5).

The chloroplasts of red algae contain phycobilins, water-soluble accessory pigments that mask the color of chlorophyll *a* and give red algae their distinctive color. These pigments are particularly well suited to the absorption of the green and blue-green light that pene-trates into deep water, where red algae are well repre-sented. Biochemically and structurally, the chloroplasts of red algae closely resemble the cyanobacteria from which they were almost certainly derived directly fol-lowing an endosymbiotic event.

Red algae are unusual among the algae, and unique among the major groups of aquatic organisms, in hav-ing no flagellated cells, as already mentioned. In con-trast to green and brown algae, the red algae have polar rings in place of centrioles. Many red algae produce un-usual toxic terpenoids that may assist in deterring her-bivores.

The cell walls of most red algae include a rigid inner component consisting of microfibrils, which may be cellulose or another polysaccharide, and a mucila-ginous matrix, usually a sulfated polymer of galactose, such as agar or carrageenan (see the essay "Algae and Human Affairs" on page 281). It is this mucilaginous matrix that gives red algae their characteristic flexible, slippery texture.

In addition, certain red algae deposit calcium car-bonate in their cell walls. Many of these algae are espe-cially tough and stony, and constitute the family Corallinaceae, the coralline algae. Coralline algae are

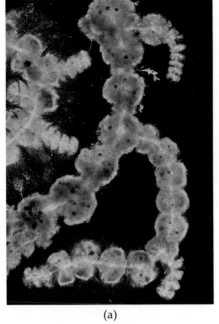

(a)

(b)

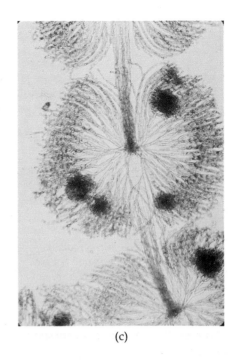

(c)

14–4

(a) *The simple, filamentous red alga* Batrachospermum moniliforme. *The soft, gelatinous, branched axes of this freshwater red alga are most frequently* found in cold streams, ponds, and lakes, where they occur throughout the world. (b) Batrachospermum sirodotia, *show-* *ing the whorls of lateral branches.* (c) *Two carposporophytes of* Batrachospermum.

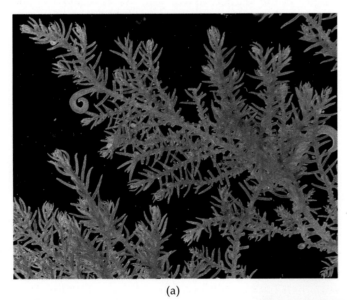

(a)

(b)

14–5

(a) *In* Bonnemaisonia hamifera, *the basically filamentous structure of the red algae is clearly evident. The branched filaments of this red alga are hooked, enabling it to cling to other seaweeds.* (b) *Jointed coralline algae in a tidal pool in California.* (c) *The reef-building, unjointed coralline red alga* Porolithon craspedium. (d) *Irish moss (*Chondrus crispus), *an important source of carrageenan.*

(c)

(d)

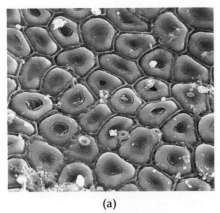

(a)

(b)

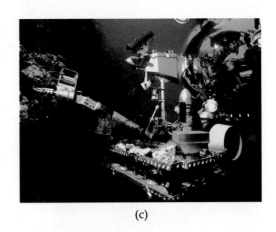

(c)

14–6

(a) *Scanning electron micrograph of an unidentified species of crustose red alga from a depth of 268 meters on a seamount in the Bahamas, nearly 100 meters below the lowest limits established for any other photosynthesizing organism. At this depth, the light intensity was estimated* as *0.0005 percent of its value at the ocean surface. The alga formed patches about a meter across, covering about 10 percent of the rock surface. When tested in the laboratory, this alga was found to be about 100 times more efficient than its* shallow-water relatives in capturing and using light energy. (b) Purple crustose coralline algae from the same seamount. (c) The research submersible Johnson Sea Link I. The deepest records of algae were made with the aid of this vessel.*

common throughout the oceans of the world, growing on stable surfaces that receive enough light (Figure 14–5b, c). An encrusting coralline alga has been found at 268 meters, a greater depth than was previously recorded for any photosynthetic organism (Figure 14–6). This is more than 100 meters below the depth to which sunlight normally penetrates.

Coralline algae play an important role in building coral reefs. Indeed, the productivity of such reefs, and their ability to grow in relatively infertile tropical waters, depends directly on the coralline algae and the symbiotic dinoflagellates, or zooxanthellae, found within the coral animals themselves (see Figure 13–20, page 264). The reef-building coralline algae are usually unjointed (Figure 14–5c), whereas many of the other coralline algae are conspicuously jointed (Figure 14–5b). Some coralline algae form crusts on rocks (Figure 14–6). Coralline algae are an ancient group, as we have seen, with possible fossils known from as early as the Precambrian period, more than 700 million years ago.

Most red algae are composed of filaments, which are often densely interwoven and are held together by the mucilage of the intercellular matrix. Simpler forms, such as the freshwater red alga *Batrachospermum*, are composed of highly branched filaments (see Figure 14–4). Growth in filamentous red algae is initiated by a single, dome-shaped apical cell, which cuts off segments sequentially to form an axis. This axis, in turn, forms whorls of lateral branches (Figure 14–4b). Many red algae are multiaxial—that is, made up of many coherent filaments forming a three-dimensional body. In such forms, the filaments are interconnected secondarily, forming a network within a common mucilaginous matrix. In many red algae, pit connections are a distinctive feature of the cell walls. These are lens-shaped plugs held in the walls by their equatorial grooves (Figure 14–7). A few genera, such as *Porphyra*, have cells densely packed together into one- or two-layered sheets. Although there are only a few single-celled red algae, these include one of the most extensively studied members of the division, *Porphyridium*.

The life history of most red algae consists of three phases: (1) a haploid gametophyte; (2) a diploid phase, called a **carposporophyte;** and (3) another diploid

14–7

A pit connection in the red alga Palmaria. Pit connections are distinct, lens-shaped plugs that form between the cells of red algae as the cells divide. These connections are also formed frequently between the cells of adjacent filaments that come into contact with each other, linking together the bodies of red algae. Pit connection cores are protein, and their outer cap layers are, at least in part, polysaccharide.

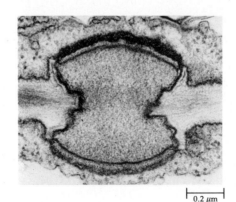

├─────┤ 0.2 μm

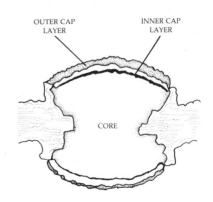

OUTER CAP LAYER

INNER CAP LAYER

CORE

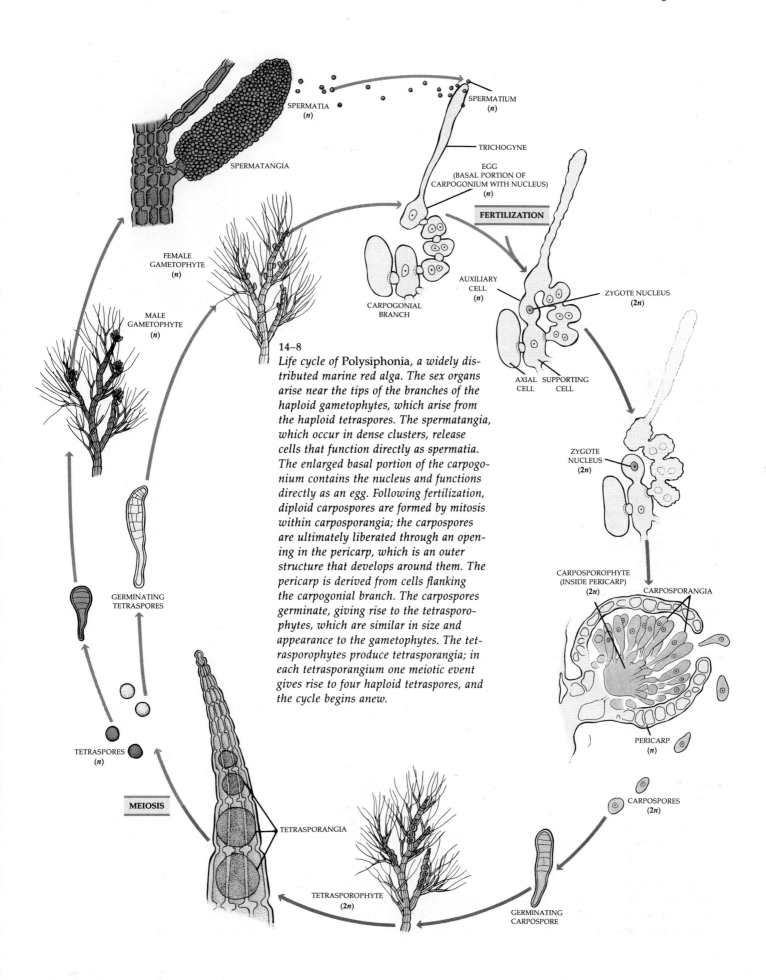

14–8
Life cycle of Polysiphonia, *a widely distributed marine red alga. The sex organs arise near the tips of the branches of the haploid gametophytes, which arise from the haploid tetraspores. The spermatangia, which occur in dense clusters, release cells that function directly as spermatia. The enlarged basal portion of the carpogonium contains the nucleus and functions directly as an egg. Following fertilization, diploid carpospores are formed by mitosis within carposporangia; the carpospores are ultimately liberated through an opening in the pericarp, which is an outer structure that develops around them. The pericarp is derived from cells flanking the carpogonial branch. The carpospores germinate, giving rise to the tetrasporophytes, which are similar in size and appearance to the gametophytes. The tetrasporophytes produce tetrasporangia; in each tetrasporangium one meiotic event gives rise to four haploid tetraspores, and the cycle begins anew.*

SPERMATIA
(*n*)

SPERMATIUM
(*n*)

SPERMATANGIA

TRICHOGYNE

EGG
(BASAL PORTION OF
CARPOGONIUM WITH NUCLEUS)
(*n*)

FERTILIZATION

FEMALE
GAMETOPHYTE
(*n*)

AUXILIARY
CELL
(*n*)

ZYGOTE NUCLEUS
(2*n*)

MALE
GAMETOPHYTE
(*n*)

CARPOGONIAL
BRANCH

AXIAL
CELL

SUPPORTING
CELL

ZYGOTE
NUCLEUS
(2*n*)

GERMINATING
TETRASPORES

CARPOSPOROPHYTE
(INSIDE PERICARP)
(2*n*)

CARPOSPORANGIA

TETRASPORES
(*n*)

PERICARP
(*n*)

MEIOSIS

CARPOSPORES
(2*n*)

TETRASPORANGIA

TETRASPOROPHYTE
(2*n*)

GERMINATING
CARPOSPORE

phase, called a **tetrasporophyte** (Figure 14–8). Male algae (gametophytic generation) bear spermatangia, which release nonmotile spermatia, male gametes that are carried by water currents. The female sex organ, the **carpogonium,** develops a long, hairlike outgrowth, the trichogyne, which is similar in structure and function to the trichogynes of the ascomycetes (see Chapter 12). When a spermatium contacts a trichogyne, the two fuse, creating the pore through which the male nucleus passes. The male nucleus then migrates to the egg nucleus and fuses with it.

In the simpler red algae, such as the filamentous *Batrachospermum*, the diploid carposporophyte generation (Figure 14–4c) develops directly from the fertilized carpogonium. *Batrachospermum* lacks a tetrasporophyte, and meiosis occurs directly in the apical cell of the diploid carposporophyte. In the more advanced red algae, the diploid nucleus formed by the fusion of the sperm nucleus with the egg within the carpogonium is transferred to a second cell, the auxiliary cell, from which the carposporophyte develops. *Polysiphonia* provides an example of this kind of life cycle (Figure 14–8).

In most red algae, the gametophyte and tetrasporophyte generations resemble one another closely and are therefore said to be isomorphic, as in *Polysiphonia*. Coralline algae, for example, have isomorphic life cycles. In them and in other red algae with life cycles of this kind, the diploid **carpospores,** formed within the carposporangia, are released and develop into free-living tetrasporophytes. The tetrasporophyte produces still another kind of spore-producing structure, the **tetrasporangium.** Each tetrasporangium undergoes meiosis, giving rise to four haploid tetraspores. On germination, each tetraspore produces a gametophyte. An increasing number of heteromorphic life cycles are also being discovered, in which the tetrasporophytes either are microscopic and filamentous or else consist of a thin crust that is tightly attached to the rock substrate.

Brown Algae: Division Phaeophyta

The brown algae, an almost entirely marine group, include the most conspicuous seaweeds of temperate waters. Although there are only about 1500 species, the brown algae dominate rocky shores throughout the cooler regions of the world (Figure 14–9). The larger brown algae of the order Laminariales, a number of which form extensive beds offshore, are called kelps. In clear water, brown algae flourish from low-tide level to a depth of 20 to 30 meters; on gently sloping shores, they may extend 5 to 10 kilometers from the coastline. Even in the tropics, where the brown algae are less

(a)

common, there are immense floating masses of *Sargassum* (Figure 14–10) in such areas as the Sargasso Sea (so named because of the abundance of *Sargassum)* in the Atlantic Ocean northeast of the Caribbean islands.

Brown algae range in size from microscopic forms to the largest of all seaweeds. Their algal cells typically contain numerous disk-shaped, golden-brown plastids that have biochemical and structural similarities to the plastids of chrysophytes (see Chapter 13), with which they may have had a common origin. In addition to chlorophylls *a* and *c*, the chloroplasts of brown algae also contain various carotenoids, including an abundance of the xanthophyll fucoxanthin, which gives the members of this division their characteristic dark-brown or olive-green color.

The simplest brown algae are branched, filamentous organisms, such as *Ectocarpus* (Figure 14–11), in which growth takes place by means of **intercalary meristems,** meristems within the filaments. In more specialized filamentous brown algae, the filaments are united within a common matrix to form a solid body. In the most complex types, a three-dimensional cellular structure is formed, composed of a tissue like the parenchyma of plants.

Large kelps, such as *Laminaria*, are differentiated into regions known as the **holdfast, stipe,** and **blade,** with the meristematic region located between the blade and stipe (see Figure 14–9b). The pattern of growth resulting from this type of meristematic activity is particularly important in the commercial utilization of *Macrocystis*, which is harvested along the California coast. When the older blades are harvested at the surface by kelp-cutting boats, *Macrocystis* is able to regen-

(b)

(c)

14–9

Brown algae. (a) Bull kelp (Durvillea antarctica) *exposed at low tide off a rocky coast in New Zealand. (b) Detail of the kelp* Laminaria, *showing holdfasts, stipes, and the bases of several fronds. (c) Rockweed* (Fucus vesiculosus) *densely covers many rocky shores that are exposed at low tide. When submerged, the air-filled bladders on the blades carry them up toward the light. Photosynthetic rates of frequently exposed marine algae are one to seven times as great in air as in water, whereas the rates are higher in water for those rarely exposed. This difference accounts in part for the vertical distribution of seaweeds in intertidal areas.*

BLADE

FLOAT
(AIR-FILLED
BLADDER)

STIPE

(a)

(b)

14–10

(a) The brown alga Sargassum *has a complex pattern of organization.* Sargassum, *like* Fucus, *is a member of the order Fucales and has a life cycle like that shown in Figure 14–14. (b) Two species of this genus, which lack sexual reproduction, form the great free-floating masses of the Sargasso Sea.*

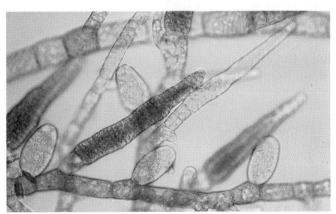

14–11

Ectocarpus, *a brown alga that has simple branched filaments. This micrograph of* E. siliculosus *shows unilocular sporangia (the short, rounded, light-colored structures) and plurilocular sporangia (the longer, dark-colored structures), which are borne on sporophytes. Meiosis takes place within the unilocular sporangia.* Ectocarpus *occurs in shallow water and estuaries throughout the world, from cold Arctic and Antarctic waters to the tropics.*

erate from its deeper portions. Giant kelps, such as *Macrocystis* and *Nereocystis*, may be more than 60 meters long; they grow very rapidly so that a considerable amount of material is available for harvest. One of the most important products derived from the kelps is a mucilaginous intercellular material called algin, which is important as a stabilizer and emulsifier for some foods and for paint, and as a coating for paper.

The internal structure of kelps is complex. Some of them have, in the center of the stipe, elongated cells that are modified for food conduction, resembling phloem cells in the vascular plants (Figure 14–12). These elongated kelp cells have sieve plates and are able to conduct food material rapidly, at rates as high as 60 centimeters per hour, from the blades at the water surface to the poorly illuminated stipe and holdfast regions far below. Lateral translocation from the outer photosynthetic layers to the inner cells takes place in many relatively thick kelps. Mannitol is the primary material that is translocated, along with amino acids.

The rockweed *Fucus* (see Figure 14–9c) is a dichotomously branching brown alga that has **air bladders** near the ends of its blades. The pattern of differentiation of *Fucus* otherwise resembles that of the kelps. *Sargassum* (see Figure 14–10) is related to *Fucus*. Some species of *Sargassum* remain attached, while in others the individuals form floating masses in which the holdfasts have been lost; both forms occur within certain species. *Fucus* and *Sargassum* grow by means of repeated divisions from a single apical cell and not from a meristem located within the body, such as is characteristic of the kelps.

The life cycles of most brown algae involve an alternation of generations and, therefore, sporic meiosis (see Figure 10–10c, page 184). The gametophytes of the more primitive brown algae produce multicellular reproductive structures called **plurilocular gametangia,** which may function as male or female gametangia or produce flagellated haploid spores that give rise to new gametophytes. The diploid sporophytes produce both **plurilocular** and **unilocular sporangia** (see Figure 14–11). The plurilocular sporangia form diploid zoospores that produce new sporophytes. Meiosis takes place within the unilocular sporangia, producing haploid zoospores that germinate to produce gametophytes.

In *Ectocarpus,* the gametophyte and sporophyte are similar in size and appearance (isomorphic). Many of the larger brown algae, including the kelps, undergo an alternation of heteromorphic generations—a large sporophyte and a microscopic gametophyte, as in the common kelp *Laminaria* (Figure 14–13). In *Laminaria,* the unilocular sporangia are produced on the surface of the mature blades. Half of the zoospores that the sporangia produce grow into male gametophytes and half into female gametophytes. The plurilocular gametangia borne on these gametophytes, according to one theory, have become modified during the course of their evolution into one-celled antheridia, each of which releases a single sperm, and one-celled oogonia, each with a single egg. The fertilized egg in *Laminaria* remains attached to the female gametophyte and develops into a new sporophyte. In several genera of brown algae, the female gametes attract the male gametes by secreting olefinic hydrocarbons (open-chain hydrocarbons with one or more double bonds).

Fucus has a unique life cycle (see Figure 14–14, page 278), in which, according to one interpretation, meiosis is gametic and leads, following one to several mitotic divisions, to the production of the gametes (see Figure 10–10b, page 184). Zygotic meiosis, which characterizes two large classes of green algae, is not known in brown algae.

14–12

Some brown algae, such as the giant kelp Macrocystis integrifolia, *have evolved sieve tubes comparable to those found in vascular plants. (a) Cross section showing a sieve plate. (b) Longitudinal section of part of a stipe, with sieve tubes. In this view, the sieve plates appear thickened because they are covered with the wall substance callose.*

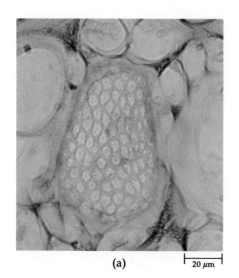

(a) |⎯⎯⎯| 20 μm

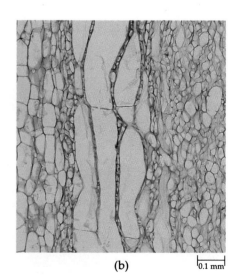

(b) |⎯⎯| 0.1 mm

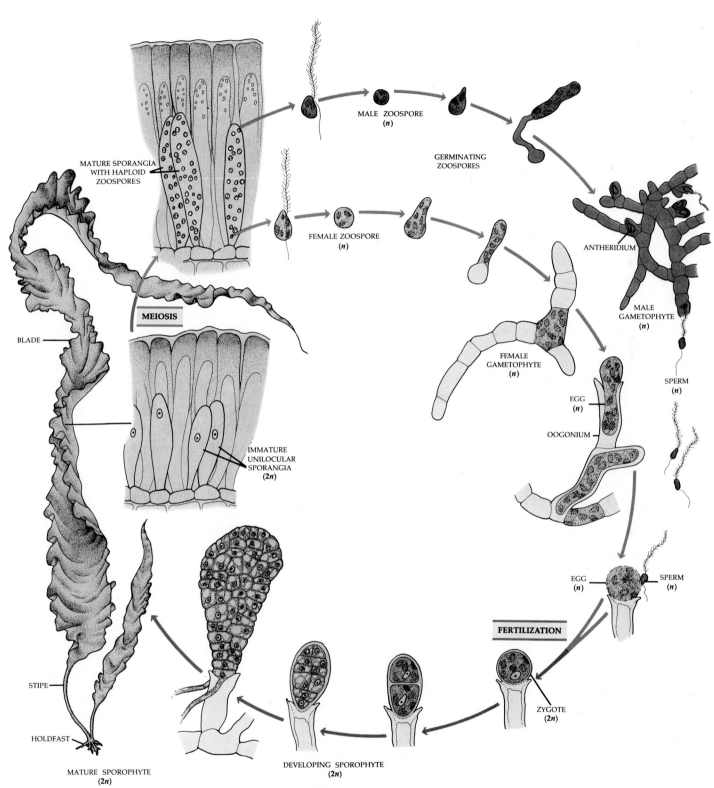

14–13

Life cycle of the kelp Laminaria, *an example of sporic meiosis. Like most of the brown algae,* Laminaria *has an alternation of heteromorphic generations in which the sporophyte is conspicuous.*

Motile haploid zoospores are produced in the sporangia following meiosis. From these zoospores grow the microscopic, filamentous gametophytes, which in turn produce the motile sperm and nonmotile

eggs. In the simpler brown algae, the sporophyte and gametophyte are often similar; they have an alternation of isomorphic generations.

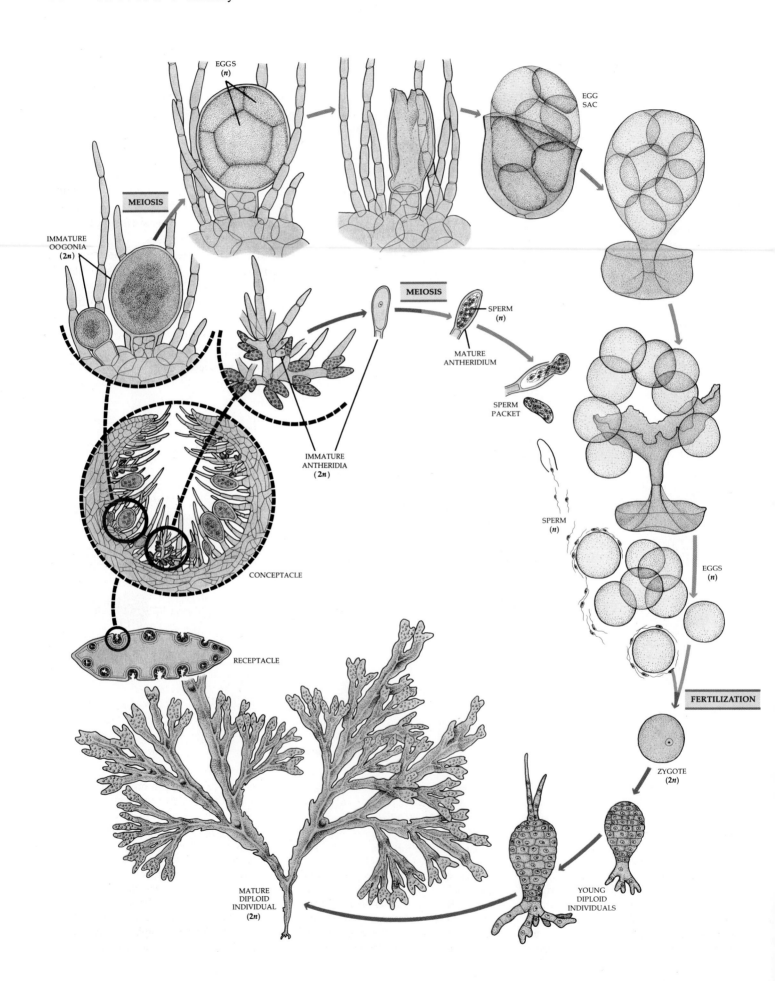

EGGS
(n)

EGG
SAC

MEIOSIS

IMMATURE
OOGONIA
(2n)

MEIOSIS

SPERM
(n)

MATURE
ANTHERIDIUM

SPERM
PACKET

IMMATURE
ANTHERIDIA
(2n)

SPERM
(n)

CONCEPTACLE

EGGS
(n)

RECEPTACLE

FERTILIZATION

ZYGOTE
(2n)

MATURE
DIPLOID
INDIVIDUAL
(2n)

YOUNG
DIPLOID
INDIVIDUALS

14–14 *(On facing page)*
In Fucus, *gametangia are formed in specialized hollow chambers (conceptacles), which are found in fertile areas (receptacles) at the tips of the branches of diploid individuals. There are two types of gametangia—oogonia and antheridia. Meiosis is followed immediately by mitosis to give rise to 8 eggs per oogonium and 64 sperm per antheridium. Eventually the eggs and sperm are set free in the water, where fertilization takes place. Meiosis is gametic, and the zygote grows directly into the new diploid individual.*

Green Algae: Division Chlorophyta

The green algae are the most diverse of all the algae, both in form and in life history. The group comprises at least 7000 species. Although most green algae are aquatic, they are found in a wide variety of habitats, including on the surface of snow (Figure 14–15), on tree trunks, in the soil, and in symbiotic associations with lichens, protozoa, and hydras. Some green algae—such as species of the unicellular genera *Chlamydomonas* and *Chloromonas,* found growing on the surface of snow, and *Trentepohlia,* a filamentous alga that grows on rocks and tree trunks or branches—produce large amounts of carotenoids as a shield against intense light. These accessory pigments give the algae an orange, red, or rust color. Most aquatic green algae are found in fresh water, but many groups are marine. Many green algae are microscopic, but some of the marine species are large; *Codium magnum* of Mexico, for example, sometimes attains a breadth of 25 centimeters and a length of more than 8 meters.

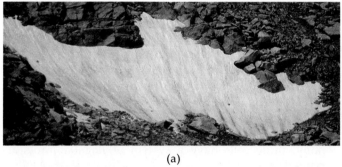

(a)

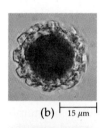

(b) ⊢ 15 μm ⊣

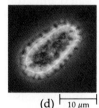

(d) ⊢ 10 μm ⊣

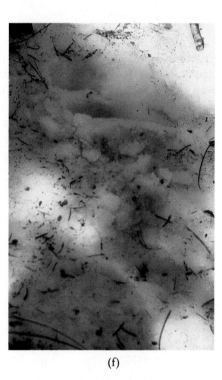

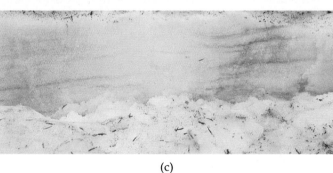

(c)

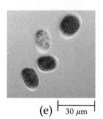

(e) ⊢ 30 μm ⊣

(f)

14–15
Snow algae. Snow algae are unique because of their tolerance to temperature extremes, acidity, high levels of irradiation, and minimal nutrients for growth. (a) In many parts of the world, as in alpine areas from northern Mexico to Alaska, the presence of large numbers of snow algae produces "red snow" during the summer. This photograph was taken near Cedar Breaks National Monument in Utah. (b) Dormant zygote of the snow alga Chlamydomonas nivalis. *The red color results from carotenoids that serve to protect the chlorophyll in the zygote following fertilization. (c) Green snow occurs just below the surface, usually near tree canopies in alpine forests. It is widespread, occurring as far south as Arizona and as far north as Alaska and Quebec. This photograph was taken at Cayuse Pass, Mt. Ranier National Park, Washington. (d) Resting zygote of the alga* Chloromonas brevispina, *found in green snow. (e) Three brilliant orange resting zygotes of* Chloromonas granulosa, *the alga responsible for orange snow. The single yellow zygote will change to orange as it matures. (f) Orange snow, associated with trees in forested alpine regions, receives more irradiation than green snow and less then red snow. It is less common than the other two kinds of colored snow, occurring from Arizona and New Mexico to Alaska. This photograph was taken at Bill Williams Mountain in Arizona.*

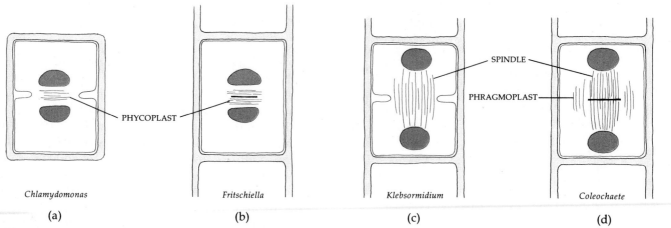

14–16

Cell division in two classes of the division Chlorophyta. In the class Chlorophyceae, (a) and (b), the mitotic spindle is absent (nonpersistent) and the daughter nuclei, which are relatively near one another, are separated by a

phycoplast. (a) Cell division by furrowing; (b) cell division by cell-plate formation. In the class Charophyceae, (c) and (d), the mitotic spindle is persistent and the daughter nuclei are relatively far apart. (c) Cell division by furrowing; (d) phrag-

moplast present and cell division by cell-plate formation. Ulvophyceae, like Charophyceae, also have a persistent spindle, but the spindle does not proliferate to form a phragmoplast.

Chlorophyta resemble plants in several important characteristics. They contain chlorophylls *a* and *b*; store starch, their food reserve, inside plastids (green algae and plants are the only two groups to do so); and have firm cell walls composed, in some genera, of polysaccharides such as cellulose with a matrix of hemicelluloses and pectic substances incorporated into the cellulose. In addition, the microscopic structure of the flagellated reproductive cells in some green algae resembles that of plant sperm. Many other biochemical details, such as the production of phytochrome (see Chapter 26) in at least one genus of green algae and the production of lignin-like substances by another, also indicate that there is a very close relationship between the two groups. For these reasons, green algae are believed to be the group from which plants originated.

Detailed investigations with the electron microscope have added greatly to the overall impression of diversity among the green algae. The members of this division represent several evolutionary lines that have been derived independently from unicellular, flagellated, green ancestors. Although the characteristics of their chloroplasts are uniform, the green algae are otherwise highly diverse; they may or may not have had a common ancestor.

CELL DIVISION IN GREEN ALGAE

The members of the largest class of green algae, the freshwater Chlorophyceae, have a unique mode of cell division involving a **phycoplast** (Figure 14–16). In

these algae, the daughter nuclei move toward one another as the nonpersistent mitotic spindle collapses, and a new system of microtubules, the phycoplast, develops parallel to the plane of cell division. Presumably, the role of the phycoplast is to ensure that the cleavage furrow will pass between the two daughter nuclei. The nuclear envelope persists throughout mitosis. In motile cells, the flagella are inserted at the anterior end. Internally, these cells have a system of flagellar roots arranged in a cross-shaped pattern of four narrow bands of microtubules (Figure 14–17). These bands of microtubules originate at or near the anterior basal bodies.

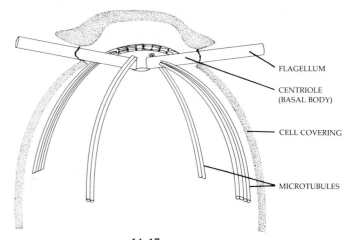

14–17

Diagram of the cross-shaped arrangement of microtubules associated with the flagellar centrioles that is characteristic of green algae of the class Chlorophyceae.

Algae and Human Affairs

People of various parts of the world, especially in the Far East, eat both red and brown algae. Kelps ("kombu") are eaten regularly as vegetables in China and Japan; they are sometimes cultivated but are mainly harvested from natural populations. *Porphyra* ("nori"), a red alga, is eaten by many inhabitants of the north Pacific Basin and has been cultivated in Japan, Korea, and China for centuries. The nori industry in this region is valued at more than $1 billion annually. Various other red algae are eaten on the islands of the Pacific and also on the shores of the North Atlantic. Seaweeds are generally not of high nutritive value as a source of carbohydrates because humans, like most other animals, lack the enzymes necessary to break down most of the materials in cell walls, such as cellulose and the protein-rich intercellular matrix. Seaweeds do, however, provide necessary salts, as well as a number of important vitamins and trace elements, and so are valuable supplementary foods. Some green algae, such as *Ulva*, or sea lettuce, are also eaten as greens.

In many northern temperate regions, kelp is harvested for its ash, which is rich in sodium and potassium salts and is therefore valued for industrial processes. Kelp is also harvested regionally and used directly for fertilizer.

Alginates, which are a group of substances derived from kelps such as *Macrocystis*, are widely used as thickening agents and colloid stabilizers in the food, textile, cosmetic, pharmaceutical, paper, and welding industries. Off the west coast of the United States, *Macrocystis* kelp beds can be harvested several times a year by cropping them just below the water surface.

One of the most useful direct commercial applications of any alga is the preparation of agar, which is made from a mucilaginous material extracted from the cell walls of a number of genera of red algae. Agar is used to make the capsules that contain vitamins and drugs, as a dental-impression material, as a base for cosmetics, and as a culture medium for bacteria and other microorganisms. Purified agarose is the gel often used in electrophoresis in biochemical experimentation. Agar is also employed as an antidrying agent in bakery goods, in the preparation of rapid-setting jellies and desserts, and as a temporary preservative for meat and fish in tropical regions. Agar is produced in many parts of the world, but Japan is the principal manufacturer. A similar algal colloid called carrageenan is used in preference to agar for the stabilization of emulsions such as in paints, cosmetics, and dairy products. In the Philippines, the red alga *Eucheuma* is cultivated commercially as a source of carrageenan.

(a)

(b)

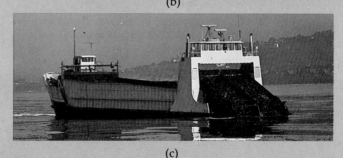

(c)

(a) *A forest of giant kelp* (Macrocystis pyrifera) *growing off the coast of California.* (b) *Harvesting the seaweed* Nudaria *by hand from submerged ropes in Japan.* (c) *A kelp harvester operating in the nearshore waters of California. Cutting racks at the rear of the ship are lowered 3 meters below the water's surface, and the ship moves backward through the kelp canopy. The cut kelp is moved via conveyor belts to a collecting bin on board the ship.*

In other green algae, the spindle persists during cell division, and in some of these algae, a **phragmoplast,** resembling that present in plants, is formed. The microtubules in a phragmoplast are oriented perpendicular to the plane of cell division (Figure 14–16). The spindle remains until it is "broken," either by the growing cell plate—which originates in the central region of the cell and grows outward to its margins—or by furrowing. This mode of cell division is apparently ancestral to that involving a phycoplast, a structure that evolved within the green algae. The mode of cell division is of great importance in the classification of the green algae.

CLASSIFICATION OF GREEN ALGAE

A small class of green algae, the unicellular Micromonadophyceae, which are scaly or naked flagellates, retains many of what seem to have been the ancestral features of the division. Contemporary phycologists divide green algae into four classes, each thought to be a separate evolutionary line. One of these classes, Pleurastrophyceae (see the essay "Symbiotic Green Algae" on page 294), is a small group that we shall not consider in detail. The other three classes—Charophyceae, Ulvophyceae, and Chlorophyceae—are large and include many familiar organisms.

The three major classes of green algae differ from one another in a number of fundamental features. In Chlorophyceae, which have a phycoplast, the daughter nuclei approach one another as the phycoplast forms. In Charophyceae, the nuclear envelope disintegrates at the start of mitosis, as it does in plants; in Chlorophyceae and Ulvophyceae, the envelope persists throughout mitosis. Each of these classes includes unicellular flagellated cells, which may be whole unicellular organisms (Chlorophyceae) or reproductive cells (Charophyceae, Ulvophyceae). The original members of each class were certainly unicellular flagellates; multicellular organisms originated independently in each class.

Charophyceae and Chlorophyceae occur mainly in fresh water, whereas Ulvophyceae are predominantly marine. The motile cells formed by members of these classes differ greatly; those of Charophyceae are asymmetrical, whereas those of the other two classes are either radially symmetrical or nearly so. The flagella of Charophyceae are subapical (inserted near, but not exactly at, the apex), and they extend at right angles from the cell. In contrast, the flagella of Chlorophyceae and Ulvophyceae are apical and directed forward.

Internally, the cells of Charophyceae possess a system of microtubules in the form of a flat, broad band (Figure 14–18). This band originates in a multilayered structure near the laterally directed basal bodies and ra-

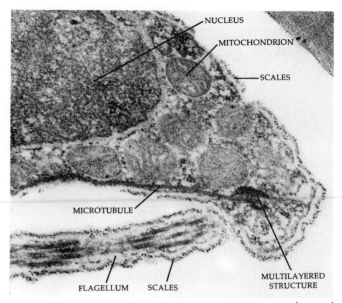

14–18
Electron micrograph of the anterior portion of a motile sperm of the green alga Coleochaete *(class Charophyceae), showing the multilayered structure near the mitochondria. The multilayered structure is also characteristic of plants and is one of the features linking plants with Charophyceae, their ancestors. As seen here, a layer of microtubules extends from the multilayered structure down into the posterior end of the cell and serves as a cytoskeleton for these wall-less cells. The flagellar and cellular membranes are covered by a layer of small, diamond-shaped scales.*

diates downward into the cell. The same kind of structure is also found in the sperm cells of plants. In addition, Charophyceae contain the photorespiratory enzyme glycolate oxidase in peroxisomes; this enzyme is otherwise known to occur in peroxisomes only in plants, although it occurs outside peroxisomes in a few green algae that are not members of the Charophyceae. On the basis of these and additional similarities, Charophyceae are considered more closely related to the plants than are the other classes of green algae, and the living form most similar to the ancestor of plants should be sought among the Charophyceae.

Sexual reproduction in Charophyceae and Chlorophyceae always involves the formation of a dormant zygote (zygospore) and zygotic meiosis (see Figure 10–10a, page 184). In contrast, sexual reproduction in Ulvophyceae often involves alternation of generations and sporic meiosis, and dormant zygotes are rare. In some siphonous (coenocytic) groups of Ulvophyceae, mainly found in warm tropical seas, gametic meiosis is characteristic.

CLASS CHAROPHYCEAE

Charophyceae consist of unicellular, few-celled, filamentous, and parenchymatous genera. Their relationship with one another is revealed by many fundamental structural and biochemical similarities. These include the asymmetrical flagellated cells, with lateral or subapical flagella that extend at right angles from the cells; zygotic meiosis and the production of dormant zygospores; and the presence of a multilayered structure in motile cells. The Charophyceae are also similar in the details of their cell division: all Charophyceae have a persistent mitotic spindle, and some of them develop a phragmoplast that aids in the formation of the cell plate.

Spirogyra (Figure 14–19) is a well-known genus of unbranched, filamentous Charophyceae, which often forms frothy or slimy floating masses in bodies of fresh water. Each filament is surrounded by a watery sheath that is slimy to the touch. The name *Spirogyra* refers to the helical arrangement of the one or more ribbonlike chloroplasts found within each uninucleate cell. The chloroplasts contain numerous **pyrenoids,** which are differentiated portions of the chloroplast. Pyrenoids are common among the algae but are lacking in the chloroplasts of plants except for members of the small division Anthocerophyta, the hornworts. Research has shown that, in two other genera of green algae, the enzyme RuBP carboxylase (see page 110) is concentrated in the pyrenoids. Pyrenoids may also be associated with the conversion of sugars to starch; deposits of starch are usually found surrounding the pyrenoids.

Asexual reproduction occurs in *Spirogyra* by cell division and fragmentation; there are no flagellated cells at any stage of the life cycle, but, as noted earlier, flagellated reproductive cells do occur in many genera of Charophyceae. During sexual reproduction in *Spirogyra*, a conjugation tube forms between two filaments; fertilization may occur in the tube, or one of the nuclei may migrate into the other filament, where fertilization takes place.

The desmids are a large group of freshwater green algae related to *Spirogyra*; like *Spirogyra*, they lack flagellated cells. Some desmids are filamentous, but most are unicellular. The cell construction of desmids is unusual in that the cell wall is in two sections with a narrow constriction—the isthmus—between them (Figure 14–20). Cell division in desmids, as in *Spirogyra*, involves the formation of a persistent spindle and furrowing. Desmids, of which there are thousands of species, frequent peat bogs and ponds that are poor in nutrients; they are important elements in the food chain in these habitats.

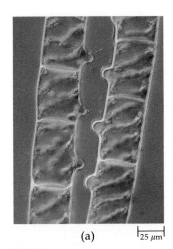  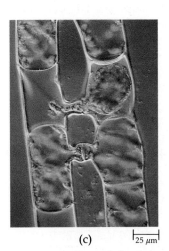 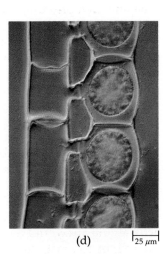

(a) $\vdash\!\!25\,\mu m\!\!\dashv$ (b) $\vdash\!\!25\,\mu m\!\!\dashv$ (c) $\vdash\!\!25\,\mu m\!\!\dashv$ (d) $\vdash\!\!25\,\mu m\!\!\dashv$

14–19

(a) *and* (b) *Sexual reproduction in* Spirogyra *follows the formation of conjugation tubes between the cells of adjacent filaments.* (c) *The contents of the cells of the* − *strain (on the left) pass through these tubes into the cells of the* + *strain.* (d) *Fertilization occurs within these cells; the resulting zygote develops a thick, resistant cell wall and is termed a zygospore. The vegetative filaments of* Spirogyra *are haploid, and meiosis occurs during germination of the zygospores, as it does in all Charophyceae.*

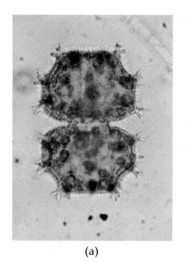
(a)

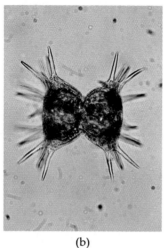
(b)

(c)

(d)

Plantlike Charophyceae

The two orders of green algae that we shall consider in connection with the ancestry of plants—the Coleochaetales and the Charales—are those that resemble plants most completely in the details of their cell division; that is, they have a phragmoplast and a nuclear envelope that disintegrates at the start of mitosis. In addition, they are oogamous, like plants. Neither group, however, is the direct ancestor of the plants because each is too specialized in certain respects. Plants were probably derived from an extinct member of the Charophyceae.

The order Coleochaetales includes branched filamentous genera and discoid (disk-shaped) ones. *Coleochaete*, which grows on the surface of submerged freshwater plants, is particularly complex (Figure 14–21; see also Figure 14–18). These algae reproduce asexually by zoospores; their zygotes are protected by a layer of sterile cells, which forms after fertilization. The vegetative cells are uninucleate, with one large chloroplast and one or more pyrenoids. Cell division may occur at the apices of the filaments or at the margins of the disk, depending on the form of the alga.

14–20

Desmids. The desmids are a group of thousands of species of unicellular, freshwater Charophyceae in which each cell is deeply constricted, giving the members of this group a very characteristic appearance. (a) Euastrum armatum. (b) Staurastrum brasiliense. (c) *A stained individual of* Micrasterias thomasiana, *showing mucilage filament extruded from an apical pore; this mucilage attaches the desmid to its substrate and also functions when the desmid moves.* (d) *Cell division in* M. thomasiana. *The smaller half of each daughter individual will grow to the size of the larger half, forming a full-sized desmid.*

14–21

(a) Coleochaete, *collected from the stem of an aquatic flowering plant growing in shallow water in a lake.* (b) *Individuals of this species of* Coleochaete *consist of a parenchymatous disk, which is generally one cell thick. The large cells are zygotes, which are protected by a cellular covering. The hair cells extending from the disk are ensheathed at the base; "Coleochaete" means "sheathed hair." These hairs are thought to discourage aquatic animals from feeding on the alga.*

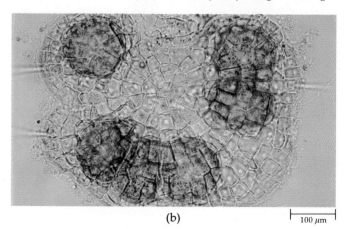
(a)   0.25 mm

(b)   100 μm

Coleochaete has been singled out for many years as a possible ancestor of the plants. Like a few other Charophyceae, *Coleochaete* resembles plants in having (1) a phragmoplast, (2) the photorespiratory enzyme glycolate oxidase localized in peroxisomes, (3) a multilayered structure associated with the flagella of reproductive cells (see Figure 14–18), and (4) multicellular antheridia. In addition, there are certain fossils from about the time when plants first evolved that resemble *Coleochaete* (Figure 14–22). Like the Charales, *Coleochaete* undergoes zygotic meiosis, a type of life history completely different from that in any plant, and so could not have been the literal ancestor of the plants. Nonetheless, its features, which include a number of structural and biochemical adaptations that are useful in the terrestrial environment, suggest the way in which the evolution of plants probably occurred.

The stoneworts, order Charales (Figure 14–23), are distinctive green algae found in fresh or brackish water. Some of them have heavily calcified cell walls and thus are well represented as fossils. There are about 250 living species. In their growth patterns, the stoneworts are complex; they exhibit apical growth, like plants, as well as differentiation into nodal and internodal regions, but they generate a filamentous axis. Whorls of short branches arise in the nodal regions. The sperm of stoneworts are produced in multicellular antheridia that are more complex than those found in any other group of protists. Their eggs are borne in oogonia of simpler construction than the antheridia. Sperm are the only flagellated cells in the stonewort life cycle.

14–22
Parka decipiens. This fossil is from the late Devonian period, about 380 million years ago. Parka *resembles* Coleochaete *in shape, tissue structure, and chemistry. This organism shows many features that suggest it may be related to the group of aquatic organisms that eventually gave rise to plants.*

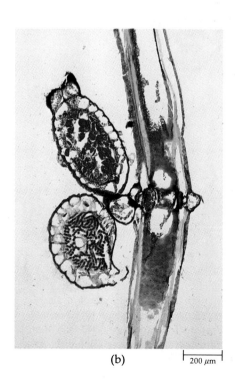

(a)   (b)   ⊢ 200 μm ⊣

14–23
(a) Chara, *a stonewort (class Charophyceae) that grows in shallow waters of temperate lakes. Its determinate growth pattern is clearly evident.* (b) *A stained segment of* Chara *showing gametangia. The top structure is an oogonium, and that below is an antheridium.*

CLASS ULVOPHYCEAE

Among the three major classes of green algae, the Ulvophyceae are the only ones that are primarily marine. Their flagellated cells are scaly or naked like those of Charophyceae, but they are nearly radially symmetrical and have apical, foreward-directed flagella like those of Chlorophyceae. The flagellated cells of Ulvophyceae may have two, four, or many flagella, like Chlorophyceae; those of Charophyceae are always biflagellated. Ulvophyceae have a closed mitosis, in which the nuclear envelope persists; the spindle is persistent through cytokinesis. They are the only group of green algae that have an alternation of generations, with sporic meiosis; unlike the other two classes that are discussed in detail in this chapter, Ulvophyceae rarely form dormant zygospores.

Ulvophyceae may be few-celled, filamentous, flat sheets of cells, or coenocytic. All filamentous and complex marine green algae that have been examined in detail share the characteristics of this class, which, unlike other classes of green algae, probably originated in the sea.

One evolutionary line of Ulvophyceae consists of filamentous algae with large, multinucleate septate cells. One of the genera of this group, *Cladophora* (Figure 14–24), is widespread in both fresh and salt water. Its filaments commonly grow in dense mats, which are either free-floating or attached to rocks and vegetation; the filaments elongate and branch near the ends. Each cell contains many nuclei and a single, peripheral, netlike chloroplast with many pyrenoids. Marine species of *Cladophora* have an alternation of isomorphic genera-

tions; most of the freshwater species do not have an alternation of generations, apparently having lost this characteristic during the course of their evolutionary adaptation from marine to freshwater habitats.

A second member of the evolutionary line that includes *Cladophora* is *Ulothrix zonata*, an alga common in cold-water streams, ponds, and lakes. The filaments of this species are attached to stones or other objects by means of a holdfast (Figure 14–25). All cells of the filament are essentially similar in appearance and contain a nucleus, a single, ringlike chloroplast, and a pyrenoid. Asexual reproduction occurs through formation of zoospores with four flagella; sexual reproduction occurs by means of biflagellated isogametes. Meiosis takes place prior to the germination of the zygote (zygotic meiosis); the filament is haploid. Although several groups of green algae that were once classified as species of *Ulothrix* are actually members of the Chlorophyceae (see Figure 14–37), *Ulothrix zonata* belongs to the Ulvophyceae.

A second kind of growth habit among the Ulvophyceae is that of *Ulva*, commonly known as sea lettuce (Figure 14–26). This familiar alga is common along temperate seashores throughout the world. Individuals of *Ulva* consist of a glistening, flat **thallus** (a simple, relatively undifferentiated vegetative body), which is two cells thick and a meter or more long in exceptionally large individuals. The thallus is anchored to the substrate by a holdfast produced by protuberances of the basal cells. Each cell of the thallus contains a single nucleus and chloroplast. *Ulva* is anisogamous and has an alternation of isomorphic generations like that of many other Ulvophyceae (Figure 14–27).

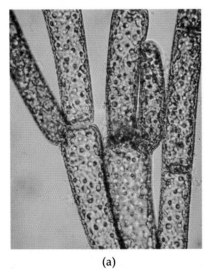

(a) (b) $\overline{\text{25 }\mu m}$ (c) $\overline{\text{25 }\mu m}$ (d)

14–24

Cladophora, *a member of the class Ulvophyceae, is widespread in marine and freshwater habitats. The marine species, like most Ulvophyceae, have an* alternation of generations, while the freshwater species do not. (a) Branched filaments of Cladophora. (b) Part of an individual cell, showing the netlike chloroplast. (c) Commencement of branching at the apical end of a cell. (d) An individual of Cladophora growing in a sluggish stream in California.

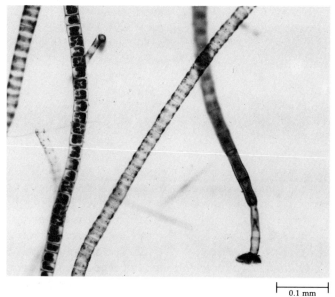

0.1 mm

14–25
Ulothrix zonata, *an unbranched, filamentous member of the class Ulvophyceae. The filament on the left with dense contents consists of sporangia in which zoospores are forming. The other filaments are vegetative. A holdfast can be seen at the lower end of the filament on the right.*

14–26
Sea lettuce, Ulva, *a common member of the class Ulvophyceae, which grows on rocks, pilings, and similar places in shallow seas worldwide.*

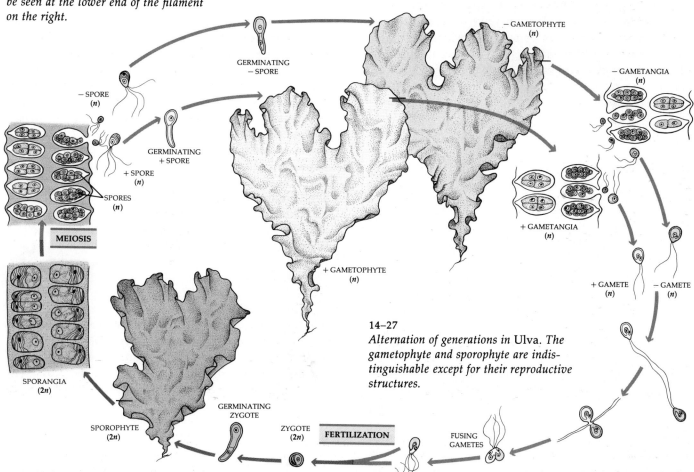

GERMINATING
– SPORE

– SPORE
(n)

GERMINATING
+ SPORE

+ SPORE
(n)

SPORES
(n)

MEIOSIS

SPORANGIA
(2n)

SPORANGIA
(2n)

SPOROPHYTE
(2n)

GERMINATING
ZYGOTE

ZYGOTE
(2n)

FERTILIZATION

FUSING
GAMETES

– GAMETOPHYTE
(n)

– GAMETANGIA
(n)

+ GAMETANGIA
(n)

+ GAMETOPHYTE
(n)

+ GAMETE
(n)

– GAMETE
(n)

14–27
Alternation of generations in Ulva. *The gametophyte and sporophyte are indistinguishable except for their reproductive structures.*

(a)

(c)

(b)   ├── 10 mm ──┤

14–28
Three genera of siphonous green algae (class Ulvophyceae). (a) A species of Codium, *abundant along the Atlantic coast. (b)* Ventricaria, *common in tropical waters; individuals are often about the size of a hen's egg. (c)* Aceta-bularia, *the "mermaid's wine glass," a mushroom-shaped siphonous green alga. The siphonous green alga in the background is* Dasycladus. *This photograph was taken in the Bahamas.*

The siphonous marine algae, characterized by very large, branched, coenocytic cells that are rarely septate, constitute a third evolutionary line of Ulvophyceae. These algae, which are very diverse, develop as a result of repeated nuclear division without the formation of cell walls (Figure 14–28). Cell walls are produced only in the reproductive phases of the siphonous green algae.

Codium magnum, mentioned on page 279, is a member of this group. Another, *Ventricaria (Valonia),* which is common in tropical waters, has been widely used in studies of cell walls and in physiological experiments requiring large amounts of cell sap. *Ventricaria* appears to be unicellular but is actually a large multinucleate vesicle containing separate rhizoids and young branches (Figure 14–28b); it grows to the size of a hen's egg. Another well-known siphonous green alga is *Acetabularia* (Figure 14–28c), which has been widely used in experiments on the genetic basis of differentiation. The siphonous green algae are primarily diploid; the gametes are the only haploid cells in the life cycle.

The chloroplasts of some siphonous green algae, including *Codium,* are symbiotic in the bodies of sea slugs (nudibranchs), marine mollusks that lack shells. The sea slugs eat the algae, and the algal chloroplasts persist in the cells that line the animal's respiratory chamber. In the presence of light, these chloroplasts carry on photosynthesis so efficiently that individuals of the nudibranch *Placobranchus ocellatus* are reported to evolve more oxygen than they consume.

Halimeda and related genera of siphonous green algae are notable for their calcified cell walls. When these algae die and disintegrate, they play a major role in the generation of the white carbonate sand that is so characteristic of tropical waters. A number of genera of algae, including *Halimeda* (Figure 14–29), contain secondary metabolites that significantly reduce feeding by herbivorous fishes. These metabolites apparently act as chemical defenses for this widespread tropical alga, which is often the most abundant photosynthetic organism in the reef communities where it occurs. The growth pattern of *Halimeda* is unusual: its fronds expand and grow rapidly at night, building up toxic compounds that deter herbivores, which are active mainly during the day. Within an hour of sunrise, the fronds turn green because of the migration of chloroplasts from the lower portions of the thallus and begin to photosynthesize rapidly.

14–29
Halimeda, *a siphonous green alga that is often dominant in reefs in warmer waters throughout the world. This alga produces distasteful compounds that retard grazing by fishes and other marine herbivores.*

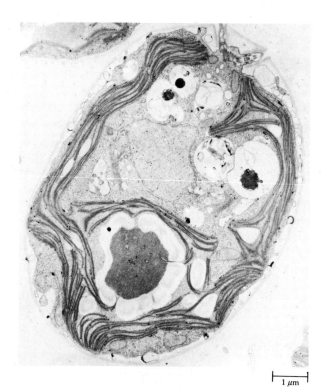

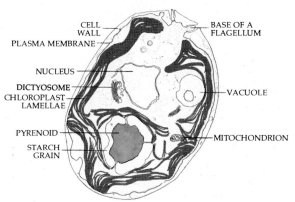

14–30
Chlamydomonas, *a unicellular green alga. Only the bases of the flagella can be seen in this electron micrograph.*

Toxic compounds, apparently playing defensive roles similar to those deduced for the toxins in *Halimeda,* are widely distributed among the algae, with more being discovered every year.

CLASS CHLOROPHYCEAE

Most of the green algae belong to this diverse group, which employs a mode of cell division involving a phycoplast. The uniqueness of this characteristic indicates that no other groups of organisms have arisen from members of this class. Chlorophyceae include flagellated and nonflagellated unicellular algae, motile and nonmotile colonial algae, filamentous algae, and parenchymatous algae. The members of this class live mainly in fresh water, although a few unicellular, planktonic species occur in coastal marine waters. A few Chlorophyceae are essentially terrestrial, occurring in habitats such as on snow (Figure 14–15), in soil, or on wood.

Motile Unicellular Chlorophyceae

Among the least complex of the Chlorophyceae are its unicellular, biflagellated members. The best known of these organisms is the genus *Chlamydomonas,* which is one of the most common freshwater green algae. Individuals of this genus are small (usually less than 25 micrometers long), green, and rounded or pear-shaped (Figure 14–30). The organism moves rapidly, with a characteristic darting motion that results from the beating of the two equal whiplash flagella protruding from its smaller, anterior pole. The flagella propel the alga through the water.

Each cell of *Chlamydomonas* has a single massive chloroplast containing a red pigment body—the **stigma,** or eyespot—which may be a shading device associated with a site of light perception. The photoreceptor in *Chlamydomonas* is homologous with rhodopsin, a visual pigment that is ubiquitous in multicellular animals. It is suspected, but has not yet been proved, that the visual pigment in the eyespots of dinoflagellates is also rhodopsin. These findings suggest a very ancient origin for the pigment. Individuals of *Chlamydomonas* are able to move toward light of appropriate intensities.

The chloroplast of *Chlamydomonas* also contains a roughly spherical pyrenoid. The uninucleate protoplast is surrounded by a thin glycoproteinaceous cell wall rich in hydroxyproline, inside which is the plasma membrane. There is no cellulose in the cell wall of *Chlamydomonas*. At the anterior end of the cell, there are two contractile vacuoles, which collect excess water and ultimately discharge it from the cell.

Under some environmental conditions, cells of *Chlamydomonas* become nonmotile. Cells in this state are usually nonflagellated, and their walls become gelatinous. When conditions change, the flagella may reappear and the cells again become free-swimming.

Chlamydomonas reproduces both sexually and asexually. During asexual reproduction, the haploid nucleus usually divides by mitosis to produce up to 16 daughter cells within the parent cell wall. Each cell then secretes a wall around itself and develops flagella. The cells secrete an enzyme that breaks down the parental wall, and the daughter cells can then escape, although fully formed daughter cells are often retained for some time within the parent cell wall. In ancient flagellates, such aggregations of daughter cells may have been the forerunners of colonial organisms.

Sexual reproduction, which occurs in some species of *Chlamydomonas*, involves the fusion of individuals belonging to different mating types (Figure 14–31). The vegetative cells are induced to form gametes by nitrogen starvation. The gametes, which resemble the vegetative cells, first become aggregated in clumps. Within these clumps, pairs are formed that stick together first by their flagellar membranes and later by a slender

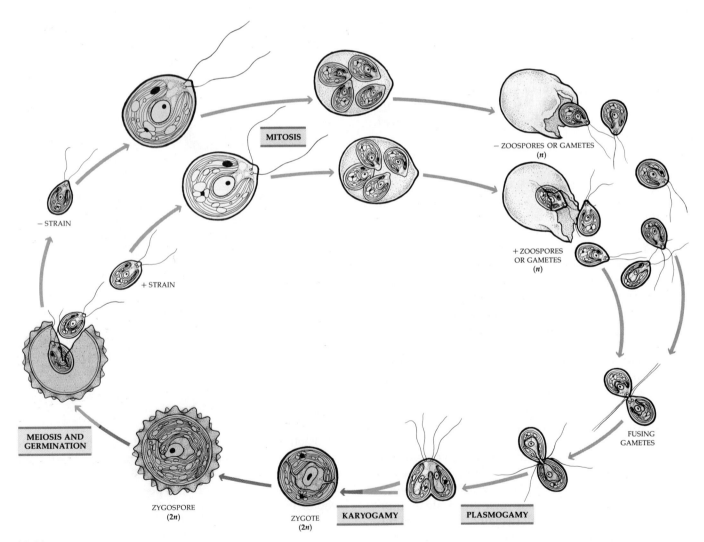

MITOSIS

− ZOOSPORES OR GAMETES
(*n*)

+ ZOOSPORES OR GAMETES
(*n*)

− STRAIN

+ STRAIN

FUSING GAMETES

MEIOSIS AND GERMINATION

ZYGOSPORE
(*2n*)

ZYGOTE
(*2n*)

KARYOGAMY

PLASMOGAMY

14–31

Life cycle of Chlamydomonas. *Sexual reproduction occurs when gametes of different mating types come together, cohering at first by their flagellar membranes and then by a slender protoplasmic thread—the conjugation tube. The protoplasts of the two cells fuse completely (plasmogamy), followed by the union of their nuclei (karyogamy). A thick wall is then formed around the diploid zygote. After a period of dormancy, meiosis occurs and four haploid cells emerge. Asexual reproduction of the haploid individuals by cell division is the most frequent mode of reproduction.*

14–32
Types of sexual reproduction, based on gamete form; each is found in at least one species of Chlamydomonas. *Isogamy—the gametes are equal in size and shape. Anisogamy—one gamete, conventionally termed male, is smaller than the other. Oogamy—one gamete, usually the larger, is nonmotile.*

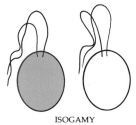
ISOGAMY

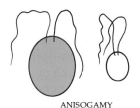
ANISOGAMY

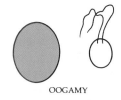
OOGAMY

protoplasmic thread that connects them at the base of their flagella. As soon as this protoplasmic connection is formed, the flagella become free, and one or both pairs of flagella propel the partially fused gametes through the water. The two gametes fuse completely, forming the zygote. Soon the four flagella shorten and eventually disappear, and a thick cell wall forms around the diploid zygote. This thick-walled, resistant zygote (zygospore) then undergoes a period of dormancy. Meiosis occurs at the end of the dormant period, resulting in the production of four haploid cells, each of which develops two flagella and a cell wall. These cells can either divide asexually or mate with a cell of another mating strain to produce a new zygote. Thus, *Chlamydomonas* exhibits zygotic meiosis (see Figure 10–10a, page 184), and the haploid phase is the dominant phase in its life cycle.

In most species of *Chlamydomonas*, the cells of the two mating types (conventionally designated + and −) are isogamous—identical in size and structure. In anisogamous species of *Chlamydomonas*, the female gametes are larger than the male gametes but are still motile; in oogamous species, the female gametes are nonmotile (Figure 14–32). Thus, the entire range of differences between gametes that occurs among algae is exhibited in the single genus *Chlamydomonas*.

Chlamydomonas traditionally has been considered primitive among the green algae, but it shares many advanced features with other Chlorophyceae and is not similar in detail to the ancestor of all green algae. *Chlamydomonas*-like cells do, however, resemble cells in what must have been some of the specialized evolutionary lines within its class, some of which will be discussed in the following four groups: (1) nonmotile unicellular genera, (2) nonmotile colonial genera, (3) motile colonial genera, and (4) filamentous and parenchymatous genera.

Nonmotile Unicellular Chlorophyceae

Chlorella is a unicellular green alga that lacks flagella, eyespots, and contractile vacuoles; individuals are rounder and often smaller than those of *Chlamydomonas* (Figure 14–33). In nature, *Chlorella* is widespread

in both fresh and salt water and in soil. Each *Chlorella* cell contains a single cup-shaped chloroplast, with or without a pyrenoid, and a single minute nucleus. The only known method of reproduction in *Chlorella* is asexual, each haploid cell dividing mitotically two or three times to give rise to four or eight nonmotile cells.

Chlorella, the first alga to be grown in culture, was used extensively in the studies that revealed some of the basic steps involved in photosynthesis. The ease with which it is maintained in culture makes it an ideal experimental organism. Currently, it is under investigation as a potential food source for humans. Pilot farms have been established in the United States, Germany, Japan, and Israel. The Japanese have been able to pro-

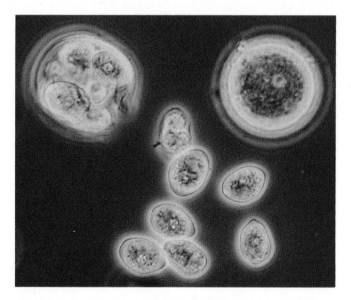

14–33
Chlorococcum echinozygotum, *a close relative of* Chlorella. Chlorococcum *is the commonest soil alga. At the upper left is a cell filled with asexual zoospores, which have formed mitotically within the cell. The smaller cells in the middle are biflagellate zoospores (the flagella are not visible). A nonmotile vegetative cell is at the upper right.*

cess *Chlorella* into a tasteless white powder that is rich in vitamins and protein and can be mixed with flour for the preparation of baked goods. *Chlorella* has also been investigated as a means of producing energy; in these experimental systems, the alga is grown together with a bacterium that converts to lipids the starch produced by the alga. Such systems can be used on barges or platforms in the open ocean, or even in space, which is one reason for the current interest in their development.

Nonmotile Colonial Chlorophyceae

Hydrodictyon, the "water net," is a nonmotile colonial member of the Chlorophyceae (Figure 14–34). Under favorable conditions, it accumulates in massive aggregates in ponds, lakes, and gentle streams. Each colony consists of many cylindrical cells arranged in the form of a large hollow cylinder. Initially uninucleate, each cell eventually becomes multinucleate. At maturity, each cell contains a large central vacuole and peripheral cytoplasm containing the nuclei and a large reticulate chloroplast with numerous pyrenoids. *Hydrodictyon* reproduces asexually through the formation of uninucleate, biflagellated zoospores. Eventually, the zoospores form into groups of four to nine (most typically six) within the cylindrical parent cell, lose their flagella, and form daughter colonies. Sexual reproduction in *Hydrodictyon* is isogamous, and meiosis is zygotic, as in all sexually reproducing Chlorophyceae.

Other genera of nonmotile colonial Chlorophyceae are important members of the phytoplankton in temperate ponds and lakes during summer.

Motile Colonial Chlorophyceae

In the motile colonial genera of Chlorophyceae, *Chlamydomonas*-like cells adhere in colonies that are propelled by the beating of the flagella of the individual cells (Figure 14–35). The members of this group have often been called the "volvocine line" after the genus *Volvox,* the largest, most complex organism exhibiting this type of colonial existence (Figure 14–36). In the colonies of some algae of this type, the cells are connected by cytoplasmic strands that provide for the integration of the whole organism. The simplest member of this group is *Gonium* (Figure 14–35a, b). The *Gonium* colony consists of separate cells held together within a gelatinous matrix. Each colony is made up of 4, 8, 16, or 32 cells (depending on the species) arranged in a slightly curved, shield-shaped disk. The flagella of each cell beat separately, pulling the entire colony forward. Each cell in *Gonium* can divide to produce a new colony.

A closely related colonial organism is *Pandorina,* which forms a tightly packed ovoid or ellipsoid colony, usually consisting of 16 or 32 cells held together within

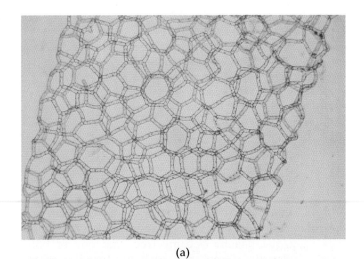

(a)

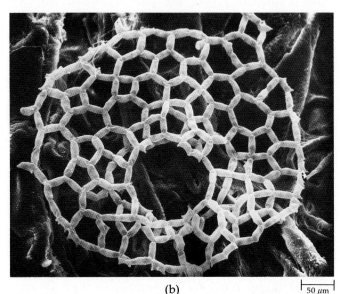

(b)   ⊢50 μm⊣

14–34
(a) *A stained portion of the "water net,"* Hydrodictyon, *a colonial member of the Chlorophyceae.* (b) *Scanning electron micrograph of a flattened young colony of* H. reticulatum.

a matrix (Figure 14–35c). The colony is polar, the eyespots being larger in the cells at one end of the colony. Each cell has two flagella, and because all the flagella point outward, *Pandorina* rolls through the water like a ball. When the cells attain their maximum size, the colony sinks to the bottom of the pond and each cell divides to form a daughter colony. The daughter colonies remain together until all have developed flagella. The parental matrix then breaks open like Pandora's box (which suggested the organism's name), releasing new daughter colonies.

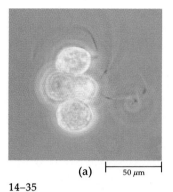 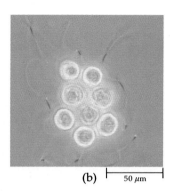 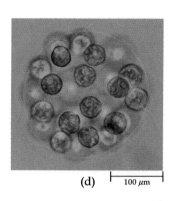

(a) ⊢ 50 μm ⊣ (b) ⊢ 50 μm ⊣ (c) ⊢ 100 μm ⊣ (d) ⊢ 100 μm ⊣

14–35

Several colonial Chlorophyceae. (a) and (b) Two views of Gonium, *(c)* Pandorina, *and (d)* Eudorina. *In these algae, cells similar to those of* Chlamydomonas *adhere in a gelatinous matrix to form multicellular colonies propelled by the beating of the flagella of the individual cells. Varying degrees of cellular specialization are found in different genera.*

14–36

Two spheroids (individuals) of Volvox carteri. *The spheroid at the left is asexual, with 16 large reproductive cells that divide to produce a new generation, which is normally asexual also. Here mitotic division has already occurred to produce juvenile spheroids, each of which will eventually digest a passageway out of the parent colony and swim away. When heat-shocked, females produce egg-bearing sexual female individuals (shown at the right) and males produce sperm-laden sexual male individuals. The zygotes are heat- and desiccation-resistant. The approximately 2000 small cells at the periphery of the transparent sphere are the* Chlamydomonas-like *somatic cells; in* V. carteri, *somatic cells are not interconnected in mature colonies as they are in some members of the genus.*

⊢ 0.5 mm ⊣

Eudorina is another spherical colony made up of green flagellates, the number of which is 32, 64, or 128 in the most common species (Figure 14–35d). It differs from *Gonium* and *Pandorina* in that some of the *Eudorina* cells, which are smaller than the others and are located in the anterior part of the colony (relative to the direction of movement), are incapable of reproducing to form new colonies. Here we see a beginning of specialization of function.

The most spectacular of these colonial green algae, however, is *Volvox* (Figure 14–36). *Volvox* consists of a hollow sphere made up of a single layer of 500 to 60,000 vegetative, biflagellated cells and a small number of reproductive cells. The number of cells in the sphere varies from species to species. When *Volvox* whirls through the water, it appears like a spinning universe of individual stars fixed in an invisible firmament.

Volvox exhibits polarity; that is, it has anterior and posterior poles. The flagella of each cell beat in such a way as to spin the entire colony clockwise around its axis as the colony moves forward (usually toward the light at most stages of the life cycle).

Most of the cells of the spherical *Volvox* colony are strictly vegetative, and the few that do engage in asexual reproduction are usually confined to the posterior hemisphere in a characteristic pattern. In some species of *Volvox*, the cells that will fulfill the reproductive function are not clearly distinguishable in juvenile spheroids, but they later become apparent as they undergo successive cycles of growth and division. Ultimately, these dividing cells give rise to new derivative spheroids, which "hatch" from the parental spheroid by releasing an enzyme that dissolves the transparent parental matrix. In more advanced members of the genus, however, asexual reproductive cells are set aside early in each reproductive cycle and are at all times structurally and functionally distinct from the vegetative, or somatic, cells. In these species, such as *V. carteri* (Figure 14–36), there is a true division of labor between two mutually dependent cell types. The nonmotile reproductive cells depend on the motile nonreproductive somatic cells to carry them to the surface, where they can obtain adequate light and CO_2 for photosynthesis. Thus, the more advanced species of *Volvox* are not simply colonial like the other species of the genus. They are instead multicellular and differentiated, as are plants and animals.

Sexual reproduction in *Volvox* is always oogamous, but reproductive patterns differ substantially among the various species. In some species, both eggs and sperm are produced within a single spheroid. In other species, members of a single genetically uniform clone (a culture derived asexually from a single individual) become sexually reproducing spheroids—either male or female (spheroids that include both sexes are unknown). In the more advanced species, however, sex is genetically determined, and each clone is either male or female. In all species that have been studied, sexual reproduction is synchronized within the population of colonies by a sexual inducer molecule, a glycoprotein with a molecular weight of about 30,000. This molecule is produced by a spheroid that has itself become sexual by some other, as yet poorly understood, mechanism. One male colony of *V. carteri* may produce enough inducer to induce sexual reproduction in over half a billion other colonies.

Volvox is one of the simplest multicellular organisms to show clear division of labor, and its sexual inducers are among the most potent biologically active substances known. Thus the genus (particularly *V. carteri*) has become the focus of a great deal of research in recent years. Like all of its relatives, *Volvox* is haploid; therefore, mutant genes are not masked by dominant alleles, and mutations affecting development can be readily detected. Hundreds of strains with specific, heritable developmental defects have been isolated and are being studied with the hope of learning how specific genes regulate cellular differentiation.

To summarize, several indicators of increasing specialization in the members of the colonial Chlorophyceae are evident. First, there is an increase in the cell number and in the size of the colonies. Second, there is increasing specialization in cell morphology and function. Finally, there is increasing sexual specialization paralleling that within the genus *Chlamydomonas*. Thus, *Gonium* and *Pandorina* have isogamous reproduction, whereas *Eudorina* and *Volvox* are oogamous. Nevertheless, this line clearly represents an evolutionary "dead end," in that it has not given rise to a more complex group of organisms.

Filamentous and Parenchymatous Chlorophyceae

The filamentous and parenchymatous Chlorophyceae, some of them members of the orders Chaetophorales and Oedogoniales, include algae that have the most complex structures found in the class. Their cells are

Symbiotic Green Algae

Many algae, including green algae, are symbiotic in other organisms. Most of the symbiotic green algae resemble the genus *Chlorella* (see Figure 14–3a); they can be found living within many freshwater protozoa, sponges, hydra, and some flatworms. Most of these algae reproduce by simple cell division and divide with the cells of their hosts, with which they remain in intimate contact.

The green alga *Tetraselmis convolutae*, a member of the class Pleurastrophyceae, is found mostly in the subepidermal cells of the marine flatworm *Convoluta roscoffensis*. Within the cells of the flatworm, *Tetraselmis* has no cell wall and is irregularly shaped; its plasma membrane, greatly increased in surface area by fingerlike projections, is in more or less direct contact with the vacuolar membrane of the host cell. When removed from the flatworm and cultured, *Tetraselmis* has a cell wall, four flagella, and an eyespot, all of which are missing when it is symbiotic. Another example of symbiosis involving green algal chloroplasts is discussed on page 288.

often specialized with respect to particular functions or positions in the algal body and, like plant cells, are sometimes connected by plasmodesmata. Although many genera of these groups, like plants, have a cell plate (Figure 14–37b), they also have a phycoplast, like all Chlorophyceae, and thus cannot be ancestral to plants. In addition, the nuclear envelope persists through the mitotic cycle, as it does in all Chlorophyceae.

One filamentous member of the Chlorophyceae, *Stigeoclonium* (Figure 14–37), superficially resembles *Ulothrix zonata* (Figure 14–25) but has all of the characteristics of Chlorophyceae. The two genera, together with other Chlorophyceae now recognized as distinct, were once considered to be the same. They are now placed in different classes, however, based on detailed studies with the electron microscope. *Stigeoclonium*, unlike *Ulothrix*, has branched filaments.

Fritschiella is a terrestrial alga that resembles *Stigeoclonium* in most of its microscopic features but is more complex. The body of *Fritschiella* consists of subterranean rhizoids, a parenchymatous prostrate system near the soil surface, and two kinds of erect branches (Figure 14–38). *Fritschiella* grows on damp surfaces, such as tree trunks, moist walls, and leaf surfaces. It has, like the plants, become adapted to a terrestrial existence, and it exhibits some of the features that were probably also found in ancestral plants.

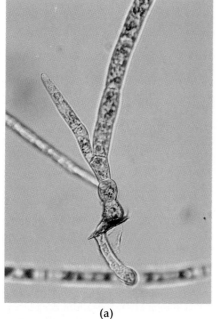

(a)

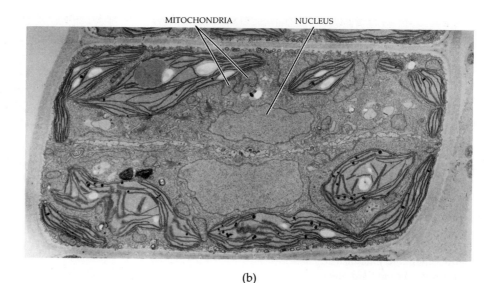

MITOCHONDRIA   NUCLEUS

(b)

14–37
Stigeoclonium, *a branched, filamentous member of the Chlorophyceae that was formerly considered to belong to the genus* Ulothrix (Ulvophyceae). *(a) Early branching in a young individual, with a dark holdfast at its base and a rhizoid extending downward from just above the*

holdfast. (b) Cell division in Stigeoclonium *occurs by cell-plate formation, which is nearly complete at the stage of cytokinesis shown in this electron micrograph. The cell plate can be seen as a narrow band running horizontally through the middle of the micrograph.*

The nuclei are close together during cytokinesis because the spindle collapses at telophase of the preceding mitosis. The phycoplast is not visible at this relatively low magnification.

Oedogonium is an unbranched filamentous alga; individuals are attached to the substrate by a holdfast. *Oedogonium* has very unusual features, especially in regard to cell division. The cells of this genus are uninucleate, with a netlike chloroplast around the periphery. As a cell of *Oedogonium* divides, a doughnut-shaped ring of wall material forms near its upper (apical) end. Following division of the nucleus, one of the daughter nuclei migrates into the upper end of the cell, and the wall ruptures violently and precisely at this ring. The ring material is then drawn out into a cylinder, forming the wall of the upper daughter cell, which is nearer the apex of the filament. A new cross wall forms in the plane of the phycoplast, just past the edge of the ruptured parental wall. As a consequence of ring expansion, the rims of the original mother cell bend outward, giving rise to the characteristic "caps," or annular scars (Figure 14–39). These scars reflect the number of divisions that have occurred in this cell.

Asexual reproduction in *Oedogonium* takes place by means of zoospore formation, with a single zoospore produced per cell. Each zoospore has a crown of about 120 flagella. Sexual reproduction is oogamous (Figure 14–40). Each antheridium produces two multiflagellated sperm, and each oogonium produces a single egg. Meiosis in *Oedogonium* is zygotic, as in all Chlorophyceae.

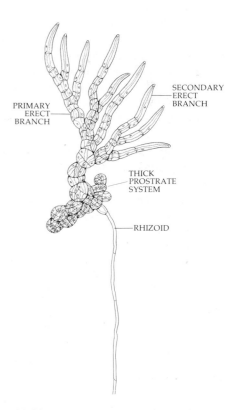

PRIMARY ERECT BRANCH

SECONDARY ERECT BRANCH

THICK PROSTRATE SYSTEM

RHIZOID

14–38
Fritschiella, *a terrestrial member of the Chlorophyceae. In its adaptation to a terrestrial habitat,* Fritschiella *has independently evolved some of the features that are characteristic of plants.*

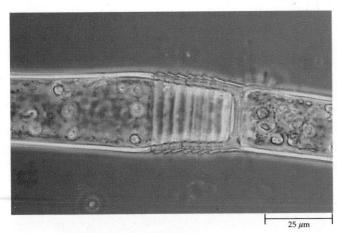

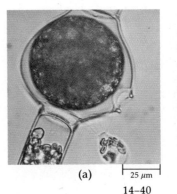
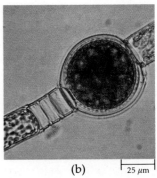

(a)   25 μm   (b)   25 μm

14–39
Oedogonium, *an unbranched, filamen-*
tous member of the Chlorophyceae. A
section of vegetative filament showing
annular scars.

14–40
Sexual reproduction in Oedogonium *is*
oogamous. Each oogonium produces a
single egg, whereas each antheridium
produces two multiflagellated sperm. (a)
An oogonium of Oedogonium cardiacum,
with a large, round egg and, at the lower
right, a sperm cell. (b) An oogonium of
Oedogonium foveolatum *with a young*
zygote and, to the left, a chain of empty
antheridia.

Summary

The three wholly or partly multicellular divisions of algae discussed in this chapter are the red algae (division Rhodophyta), brown algae (division Phaeophyta), and green algae (division Chlorophyta). The members of these divisions are largely aquatic; the brown and red algae are of great importance in marine habitats, and the green algae are often abundant in freshwater habitats, particularly in the warmer months. A number of genera of green algae also occur in the sea, and a few genera of red and brown algae live in fresh water. In all of these habitats, the members of these divisions of algae play an ecological role comparable to that of the plants in terrestrial habitats.

The various divisions of algae appear to have had their origins in symbiotic relationships between non-photosynthetic, protozoa-like eukaryotic cells and photosynthetic bacteria. The chloroplasts of red algae are biochemically and structurally very like the cyanobacteria from which they were probably derived. Whether the biochemically distinct chloroplasts of other algal divisions evolved from the chloroplasts of the red algae or were derived directly from groups of photosynthetic bacteria with different characteristics is being investigated currently.

Rhodophyta (red algae) are a large group particularly common in warmer marine waters. They nearly always grow attached to a substrate, and some grow at great depths (down to 268 meters). They contain phycobilins, which give them their characteristic colors, and chlorophyll *a.*

Phaeophyta (brown algae) include the largest and most complex of the marine algae. In many types the vegetative body is well differentiated into holdfast, stipe, and blade. Some have food-conducting tissues that approach, in their complexity, those of the vascular plants. The chloroplasts of brown algae contain chlorophylls *a* and *c,* together with abundant quantities of fucoxanthin, which gives them their olive-green to dark-brown color. The sporophyte is usually larger than the gametophyte.

Chlorophyta (green algae) are the largest and most diverse of the three divisions considered here, and the one from which the plants probably evolved. Their chloroplasts, which contain chlorophylls *a* and *b* and carotenoids, are the storage sites for starch; in these respects, they are identical to plant chloroplasts. The ancestral green algae appear to have been scaly, flagellated unicellular organisms. The division comprises at least five classes, of which we have considered three in detail.

Charophyceae and Chlorophyceae occur primarily in fresh water, whereas Ulvophyceae occur primarily in marine habitats. All three classes include unicellular, few-celled, filamentous, and parenchymatous genera, and Chlorophyceae include motile and nonmotile colonial genera as well. The motile spores of Charophyceae are asymmetrical and have lateral or subapical flagella that project at right angles, whereas those of the other two classes are radially symmetrical and have apical flagella.

In Chlorophyceae, the mitotic spindle collapses at telophase, and a phycoplast develops. A phycoplast is a system of cytokinetic microtubules oriented parallel to the plane of cell division. In Charophyceae and Ulvophyceae, the mitotic spindle persists during telophase. In addition, in some Charophyceae, there is a phragmoplast, a system of microtubules oriented at right angles to the plane of cell division, as in plants.

Sexual reproduction in Charophyceae and Chlorophyceae always involves the production of a dormant zygospore and is characterized by zygotic meiosis; sexual reproduction in Ulvophyceae often involves alternation of generations (sporic meiosis), and dormant zygotes are rare.

Suggestions for Further Reading

Bold, Harold C., and Michael J. Wynne: *Introduction to the Algae: Structure and Reproduction*, 2d ed., Prentice Hall, Inc., Englewood Cliffs, N.J., 1985.

A detailed reference work on algae, which contains a wealth of information on all groups; taxonomically oriented.

Bremer, Kåre: "Summary of Green Plant Phylogeny and Classification," *Cladistics* 1(4):369–385, 1985.

A careful attempt to trace the relationships of photosynthetic organisms; illustrates well the kinds of reasoning used to deduce evolutionary pathways.

Chapman, A.R.O.: *Biology of Seaweeds: Levels of Organization*, University Park Press, Baltimore, Md., 1979.*

A biologically oriented view of the green, brown, and red marine algae; the discussion ranges from cellular ultrastructure and function to community ecology.

Dring, M.J.: *The Biology of Marine Plants*, Edward M. Arnold, Publishers, Ltd., London, 1982.*

A concise introduction to experimental studies of growth patterns in algae, in relation to their genetics, evolution, and ecology.

Duffy, J. Emmett and Mark E. Hay: "Seaweed Adaptations to Herbivory," *BioScience* 40(5):368–375, 1990.

The chemical, structural, and morphological defenses of seaweeds are often adjusted to spatial or temporal patterns of attack.

Esser, Karl: *Cryptograms: Cyanobacteria, Algae, Fungi, Lichens*, Cambridge University Press, New York, 1982.

A good account of the groups enumerated in the title of the book, including a comprehensive manual of laboratory techniques for these organisms.

Graham, Linda E.: "The Origin of the Life Cycle of Land Plants," *American Scientist* 73:178–186, 1985.

This paper presents an outstanding discussion of the way in which a simple modification in the life cycle of an extinct green alga could have been responsible for the origin of plants.

Lee, R.E.: *Phycology*, 2d ed., Cambridge University Press, Cambridge, England, 1989.

Excellent account of all groups of algae.

Lembi, Carole A., and J. Robert Waaland: *Algae and Human Affairs*, Cambridge University Press, Cambridge, England, 1988.

Excellent collection of papers on interactions between algae and humans.

Lobban, Christopher S., Paul J. Harrison, and Mary Jo Duncan: *The Physiological Ecology of Seaweeds*, Cambridge University Press, New York, 1986.

An outstanding synthesis of the physiology of seaweeds and its relationship to their distribution.

Lobban, Christopher S., and Michael J. Wynne (eds.): *The Biology of Seaweeds*, Blackwell Scientific Publications/University of California Press, Berkeley, Calif., 1982.

An excellent advanced treatise on the structure, reproduction, ecology, physiology, biochemistry, and uses of seaweeds.

Saffo, Mary Beth: "New Light on Seaweeds," *BioScience* 37(9):654–664, 1987.

Recent studies have forced a reassessment of the role of light-harvesting pigments in depth zonation of seaweeds.

Scagel, R.F., et al.: *Nonvascular Plants: An Evolutionary Survey*, Wadsworth Publishing Co., Inc., Belmont, Calif., 1982.

An excellent synoptical treatment of all groups of photosynthetic organisms except vascular plants.

South, G.R., and A. Whittick: *Introduction to Phycology*, Blackwell Scientific Publications, Oxford, England, 1987.

A modern account of the algae, well written and illustrated.

Van Der Meer, John P.: "The Domestication of Seaweeds," *BioScience* 33:172–176, 1983.

An interesting analysis of the way in which the intelligent selection of new strains and improved management of crops could lead to much better yields of seaweeds.

*Available in paperback.

CHAPTER 15

Bryophytes

15–1
Dense growth of the moss Fissidens *on limestone rocks in a waterfall. This photograph was taken in a nature reserve just west of Yalta on the Crimean Peninsula in the Ukraine.*

A number of lines of evidence reviewed in the last chapter indicate that plants, an essentially terrestrial group, evolved from some group of ancient green algae. Like green algae, plants have chlorophyll *a* as their primary photosynthetic pigment and chlorophyll *b* and carotenoids as accessory pigments. A unique characteristic shared by plants and green algae is that starch, their primary carbohydrate food reserve, is deposited within the chloroplasts; in other eukaryotic photosynthetic organisms, the food reserves are deposited outside the chloroplasts. Cellulose is the principal component of plant cell walls and is an important cell wall component in some green algae also. Finally, plants form a phragmoplast and a cell plate during the process of cell division. The only other living organisms that share this characteristic are a few genera of green algae. Since all plants possess these characteristics, it seems likely that they are all descended from a remote common ancestor—or perhaps from several different, but related, green algae—that successfully invaded the land. A moss, one of the kinds of plants treated in this chapter, is shown in Figure 15–1.

Apparently, the ancestors of the plants had a well-developed alternation of heteromorphic generations, for this type of life cycle is universal among the living members of the kingdom. All plants are oogamous. In addition, all plants have embryos; their zygotes begin to divide while they are held within, and are nutritionally dependent on, the gametophyte generation. Green algae lack embryos, and in most green algae the development of the zygote is essentially independent of the female gametophyte. Dependent embryos certainly evolved early in the history of the plants and may well have been present, along with alternation of generations, in their common ancestor. The gametophytes of this ancestor probably bore multicellular gametangia like those of all plants today. However, plant gametangia, which differ from each other in the details of their development, may have evolved independently in different evolutionary lines. The earliest plants probably also had spores encased by walls containing the protective and highly resistant biopolymer sporopollenin, as all plants do today.

The organisms that were successful in making the transition from water to land developed features that prevented them from drying out. One of these features was the development of a sterile jacket layer around the sperm-producing and egg-producing cells of the male and female gametangia, called **antheridia** and **archegonia**, respectively. Plant embryos are protected within the archegonium. A sterile jacket layer was also formed around the spore-producing cells of the **sporangia**. The aerial parts of most plants are covered with a waxy layer, the cuticle, which helps prevent drying out. This cuticle is closely correlated with the presence of stomata, the specialized pores that function primarily in the regulation of gas exchange.

We do not know what kind of conducting tissue, if any, may have been present in the ancestor of the plants. Some fossils from as early as the Ordovician period, however, have elongate cells that resemble tracheids. There is also evidence of lignin-like substances in the walls of such cells from as early as the Silurian period. The first land plants were probably endomycorrhizal, with a zygomycete as the fungal partner (see Figure 12–40, page 241); mycorrhizal associations are found in all groups of living plants.

The oldest known fossils that resemble plants are from the early part of the Silurian period, about 430 million years ago (see Figure 1–4, page 4). Well-developed gametophytes are known from the Devonian period, some as much as 400 million years old (Figure 15–2). Because of the existence of these fossils, we may assume that the hypothetical common ancestor of the plants, a relatively complex green alga (see Figure 14–22, page 285), invaded the land more than 430 million years ago. Spore tetrads, fragments of what seem to have been specialized conducting cells, and cuticle-like fragments from the Ordovician period, about 450 million years ago, suggest the possibility of an even earlier origin for plants.

The Nature of Bryophytes

Three divisions of plants—the liverworts (division Hepatophyta), hornworts (division Anthocerophyta), and mosses (division Bryophyta)—have traditionally been called "bryophytes," and the three have been contrasted as a group with the divisions of vascular plants. In "bryophytes," the gametophytes are always nutritionally independent of the sporophytes, whereas the sporophytes are permanently attached to the gametophytes and vary in their dependence on them. In other words, the gametophyte is the conspicuous and dominant generation in the three "bryophyte" divisions. Among the other divisions of plants, the vascular plants, the sporophyte is the conspicuous and dominant generation.

Most scientists now agree that "bryophyte" should be understood as a collective term for three quite distinct divisions of relatively unspecialized plants. Liverworts are the most distinct and simplest of plants, lacking both stomata and, usually, also specialized conducting cells (water-conducting strands have been reported in the gametophytes of a very few genera). Liverworts seem clearly to constitute an evolutionary

15–2
(a) *Longitudinal section through a gametophyte from the Rhynie chert of Scotland. This plant lived in the early Devonian period about 400 million years ago. The arrows indicate the position of antheridia on the upper surface of the gametophyte. (b) Mature sperm are shown in a section through an antheridium. In their size and complexity, such gametophytes resemble those of the bryophytes. It is not known whether any of the early Devonian sporophytes from the same rock formations are related to these gametophytes or not.*

 (a) 1 mm

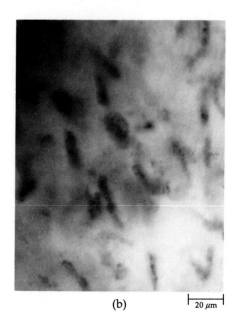 (b) 20 µm

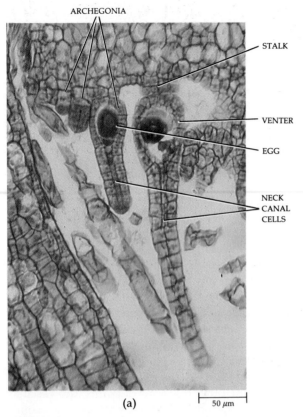

ARCHEGONIA

STALK

VENTER

EGG

NECK
CANAL
CELLS

(a)   50 μm

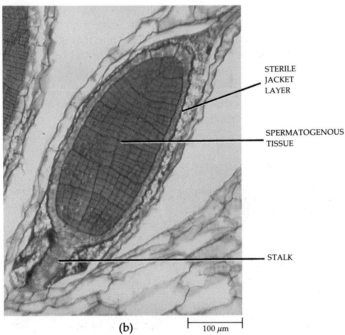

STERILE
JACKET
LAYER

SPERMATOGENOUS
TISSUE

STALK

(b)   100 μm

15–3
Gametangia of Marchantia, *a liverwort.*
(a) Details of several archegonia at
different stages of development. (b) One
developing antheridium. In Marchantia,
the archegonia and antheridia are borne
on different gametophytes.

branch separate from all other plants, and they might even have been derived independently from the green algae. Hornworts lack specialized conducting cells but have stomata, a characteristic they share with all other plants except liverworts. Mosses have specialized water-conducting and food-conducting cells in both their gametophytes and their sporophytes, although such cells are not found in all genera.

The three divisions that traditionally have been considered bryophytes share a number of unspecialized features. For example, their gametophytes are usually attached to the substrate by elongate single cells or filaments of cells called **rhizoids.** These rhizoids generally serve only to anchor the plants, since absorption of water and inorganic ions commonly occurs directly and rapidly throughout the gametophyte.

In these three divisions of bryophytes, as in the less specialized groups of vascular plants, the sperm must swim through water to reach the egg, which is located inside an archegonium. Archegonia are flask-shaped, with a long neck and swollen basal portion, the **venter,** which encloses a single egg (Figure 15–3a). The central cells of the neck, the **neck canal cells,** disintegrate when the archegonium is mature, resulting in a fluid-filled tube through which the sperm swim to the egg. The elongate or spherical antheridium is commonly stalked and consists of a sterile jacket layer, one cell thick, surrounding numerous **spermatogenous cells** (Figure 15–3b). Each spermatogenous cell, in liverworts, hornworts, and mosses, forms a single biflagellated sperm. After fertilization, the zygote is retained within the venter of the archegonium, where it develops into an embryo. For a period, the cells of the venter divide and enlarge, and the venter keeps pace with the growth of the young sporophyte within the archegonium; the enlarged archegonium is called a **calyptra.** The subsequent details of development differ among the liverworts, hornworts, and mosses.

The Liverworts: Division Hepatophyta

Liverworts, or hepatics, are a group of about 6000 species of small plants that are generally inconspicuous, although they may form relatively large masses in favorable habitats—often moist, shaded soil or rocks, tree trunks, or branches; a few kinds of liverworts grow in water. The name "liverwort" dates from the ninth century, when it was thought, because of the liver-shaped outline of the gametophyte in some genera, that these plants might be useful in treating diseases of the liver. According to the medieval "Doctrine of Signatures," the outward appearance of a body signaled the

15–4

(a) *Transverse section of the gametophyte of* Marchantia, *a thallose liverwort. Numerous chloroplast-bearing cells are evident in the upper layers, and there are several layers of colorless cells below them, as well as rhizoids that anchor the plant body to the substrate. Pores permit the exchange of gases in the air-filled chambers that honeycomb the photosynthetic dorsal layer. The specialized cells that surround each pore are usually arranged in four or five superimposed circular tiers of four cells each, and the whole structure is barrel-shaped. Under dry conditions, the cells of the lowermost tier, which usually protrude into the chamber, become juxtaposed; while under moist conditions, they separate. Thus the pores serve a function similar to that of the stomata of vascular plants. (b) A scanning electron micrograph of a pore on the dorsal surface of a gametophyte of* Marchantia.

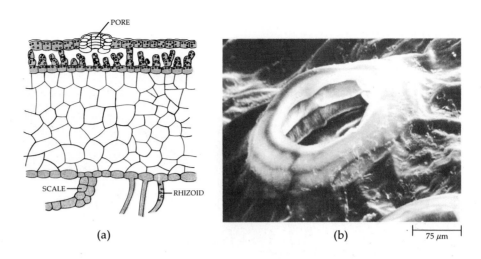

(a) (b) ⊢ 75 μm ⊣

possession of special properties. The Anglo-Saxon ending "-wort" (originally *wyrt*) means "herb"; it appears as a part of many plant names in the English language.

Generally understood to lack specialized conducting tissue, a cuticle, and stomata, liverworts are the simplest of all living plants. A few unusual liverworts have simple water-conducting tissue, the nature of which should be investigated in more detail. Regardless of the ultimate interpretation of the conducting cells in liverworts, it seems quite clear that the members of this division represent an evolutionary line distinct from all other plants, based on the unique patterns of growth and development in both their gametophytes and their sporophytes.

As in other groups of primitive plants, liverwort sperm often swim to the vicinity of the archegonia through a continuous film of water. Following fertilization, the zygotes develop into sporophytes that are generally less complex than those of mosses. The sporangia, or capsules, exhibit diverse mechanisms for the release of spores (see the essay "Spore Discharge in Liverworts" on page 306).

Most liverwort gametophytes develop directly from spores, but some genera form a filament of cells first, from which the mature gametophyte develops. In some groups of liverworts, the gametophytes are flattened, dorsiventral **thalli** (singular: **thallus;** a term traditionally used for undifferentiated plant bodies). The gametophytes of thallose liverworts grow from an apical

meristem. The gametophytes of most liverworts, however, are leafy. They grow from a single apical cell, which resembles an inverted pyramid with a base and three sides. Daughter cells are cut off from the sides of this single cell. Liverwort rhizoids are single-celled, unlike those of mosses, which are made up of several cells.

THALLOSE LIVERWORTS

Thallose liverworts are a diverse group of nonleafy hepatics. They can be found on moist, shaded banks and in other suitable habitats, such as flowerpots in a cool greenhouse. In a number of genera of thallose liverworts, the thallus is many cells thick—about 30 at the midrib and approximately 10 in the thinner portions—and is sharply differentiated into a thin, chlorophyll-rich upper (dorsal) portion and a thicker, colorless lower (ventral) portion (Figure 15–4a). The lower surface bears two kinds of rhizoids, as well as rows of scales. The upper surface is divided into raised regions, each of which marks the limits of an underlying air chamber and has a large pore that leads to this chamber (Figure 15–4b).

One of the most familiar liverworts is *Marchantia*, a widespread, terrestrial genus that grows on moist soil and rocks (see Figure 15–6 on page 304). Its dichotomously branching gametophytes are mostly one to a few centimeters long; its gametangia are restricted to specialized structures called **gametophores.** The gametophytes of *Marchantia* are unisexual, and the male and female gametophytes can be readily identified by the distinct structures they bear. The antheridia are borne on disk-headed stalks called **antheridiophores,** whereas the archegonia are borne on umbrella-headed stalks called **archegoniophores** (Figure 15–6). The life cycle of *Marchantia* is illustrated in Figure 15–5. In this genus, the sporophyte generation consists of a foot, a short seta, or stalk, and a capsule, or sporangium (Fig-

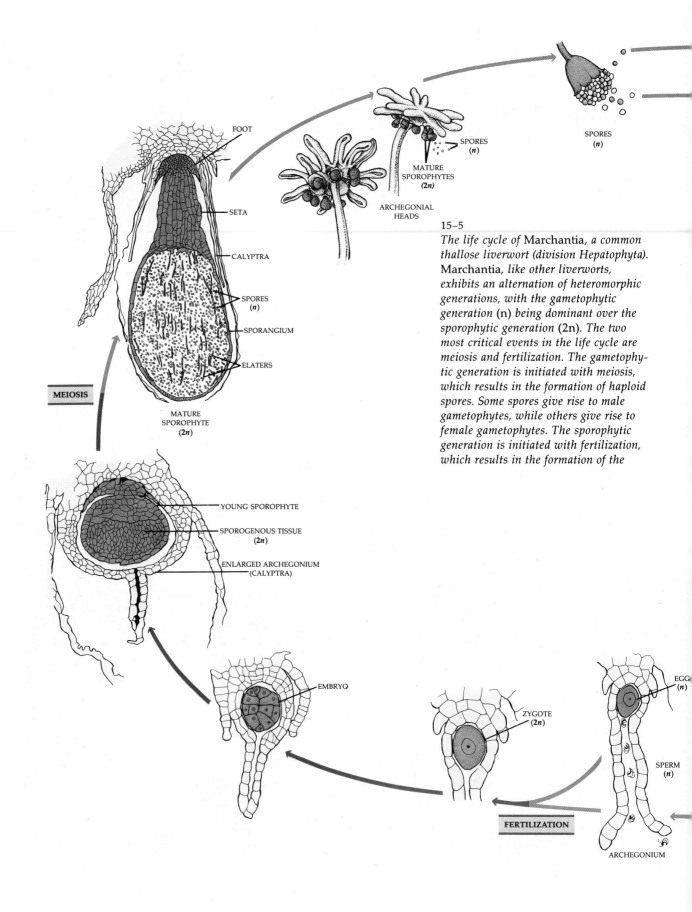

FOOT

SETA

CALYPTRA

SPORES
(*n*)

SPORANGIUM

ELATERS

MEIOSIS

MATURE
SPOROPHYTE
(*2n*)

SPORES
(*n*)

MATURE
SPOROPHYTES
(*2n*)

ARCHEGONIAL
HEADS

SPORES
(*n*)

15–5

The life cycle of Marchantia, *a common thallose liverwort (division Hepatophyta).* Marchantia, *like other liverworts, exhibits an alternation of heteromorphic generations, with the gametophytic generation (n) being dominant over the sporophytic generation (2n). The two most critical events in the life cycle are meiosis and fertilization. The gametophytic generation is initiated with meiosis, which results in the formation of haploid spores. Some spores give rise to male gametophytes, while others give rise to female gametophytes. The sporophytic generation is initiated with fertilization, which results in the formation of the*

YOUNG SPOROPHYTE

SPOROGENOUS TISSUE
(*2n*)

ENLARGED ARCHEGONIUM
(CALYPTRA)

EMBRYO

ZYGOTE
(*2n*)

EGG
(*n*)

SPERM
(*n*)

FERTILIZATION

ARCHEGONIUM

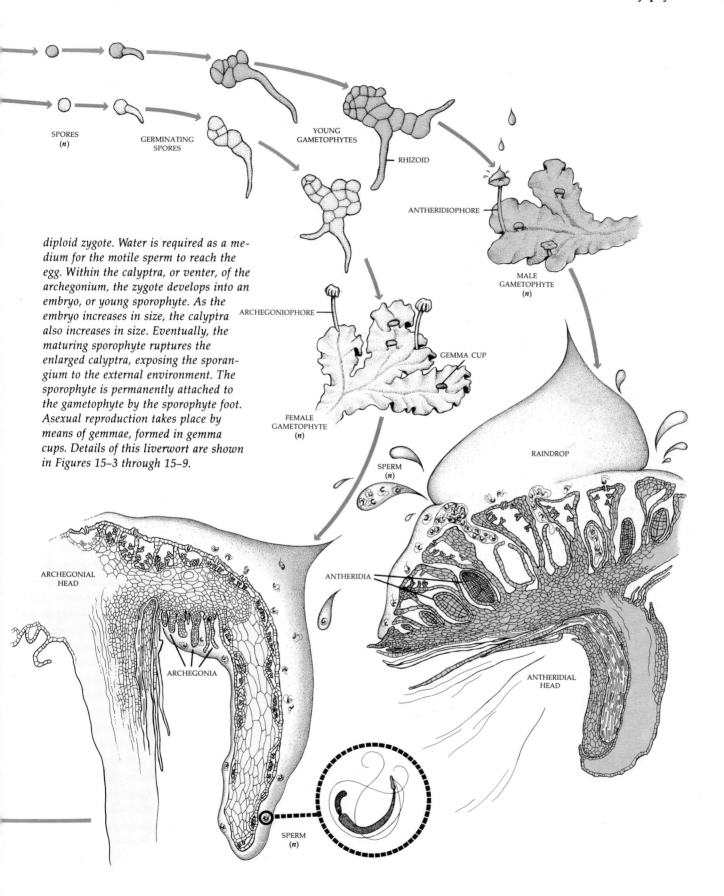

SPORES
(n)

GERMINATING
SPORES

YOUNG
GAMETOPHYTES

RHIZOID

ANTHERIDIOPHORE

MALE
GAMETOPHYTE
(n)

ARCHEGONIOPHORE

GEMMA CUP

FEMALE
GAMETOPHYTE
(n)

SPERM
(n)

RAINDROP

diploid zygote. Water is required as a medium for the motile sperm to reach the
egg. Within the calyptra, or venter, of the
archegonium, the zygote develops into an
embryo, or young sporophyte. As the
embryo increases in size, the calyptra
also increases in size. Eventually, the
maturing sporophyte ruptures the
enlarged calyptra, exposing the sporangium to the external environment. The
sporophyte is permanently attached to
the gametophyte by the sporophyte foot.
Asexual reproduction takes place by
means of gemmae, formed in gemma
cups. Details of this liverwort are shown
in Figures 15–3 through 15–9.

ARCHEGONIAL
HEAD

ANTHERIDIA

ARCHEGONIA

ANTHERIDIAL
HEAD

SPERM
(n)

15–6
Gametophytes of Marchantia. *The antheridia* (a) *and archegonia* (b) *are elevated on stalks above the thallus.*

(a)

(b)

ure 15–7). In addition to spores, the mature sporangium contains elongate cells called **elaters,** which have helically arranged hygroscopic (moisture-absorbing) wall thickenings (Figures 15–7 and 15–8; see also the essay "Spore Discharge in Liverworts" on page 306). The walls of these cells are sensitive to slight changes in humidity, and, after the capsule dehisces into a number of petal-like segments, the cell walls undergo a twisting action that aids in spore dispersal.

Fragmentation is the principal means of asexual reproduction in liverworts. Another widespread means of asexual reproduction in both liverworts and mosses is the production of **gemmae**—multicellular bodies that can give rise to new gametophytes. In *Marchantia*, the gemmae are produced in special cuplike structures—called gemma cups—located on the dorsal surface of the gametophyte (Figure 15–9). The gemmae are dispersed primarily by splashes of rain.

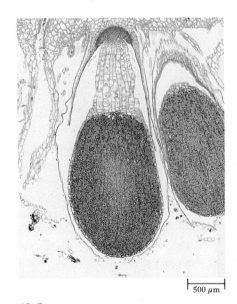

500 μm

15–7
A nearly mature sporophyte of Marchantia. *Note the elaters—helical, filamentous structures that assist in the dispersal of spores—within the spore-filled capsule.*

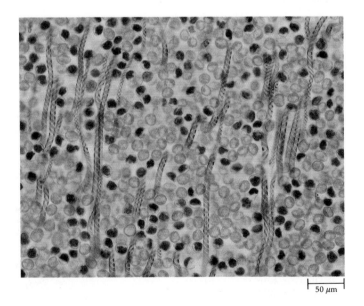

50 μm

15–8
Mature spores (red spheres) and elaters (green strands) from a capsule of Marchantia.

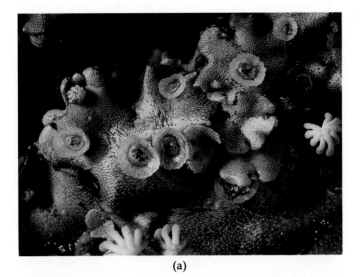

(a)

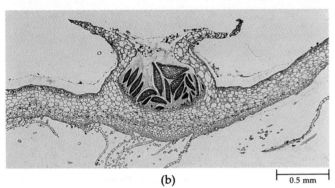

(b)

0.5 mm

15–9
(a) *Gametophytes of* Marchantia, *with gemma cups containing gemmae. The gemmae are splashed out by the rain and may then grow into new gametophytes, each genetically identical to the parent plant from which it was derived by mitosis. (b) Longitudinal section of a gemma cup.*

LEAFY LIVERWORTS

The leafy liverworts are a diverse group that includes more than 4000 of the 6000 species of division Hepatophyta (Figure 15–10). The leafy liverworts are especially abundant in the tropics and subtropics, in regions of heavy rainfall or high humidity (Figure 15–11), but they are also well represented in temperate regions. The plants are usually well branched and form small mats.

Liverwort leaves, like moss leaves, generally consist of only a single layer of nondifferentiated cells. The leaves of many genera are arranged in two rows, with a third row of reduced leaves along the lower surface of the gametophyte. (In mosses, the leaves are usually equal and spirally arranged around the stem; unlike liverwort leaves, moss leaves often have a thickened midrib.) Liverwort leaves are often two-lobed, and each lobe grows from two distinct apical growing points. In *Frullania*, a common liverwort that grows on bark, the leaves consist of a large dorsal lobe and a small, helmet-shaped ventral lobe (Figure 15–10c).

In the leafy liverworts, the antheridia generally occur in a packetlike swelling, the **androecium,** which develops on the lower portion of a modified leaf. The developing sporophyte, as well as the archegonium from which it develops, is characteristically surrounded by a tubular sheath, the **perianth** (Figure 15–10c).

15–10
Leafy liverworts. (a) Clasmatocolea puccionana, *showing the characteristic arrangement of the leaves.* (b) *The end of a branch of* C. humilis. *The capsule and the long stalk of the sporophyte are visible.* (c) *A portion of a branch of* Frullania, *showing the characteristic arrangement of its leaves. The antheridia are contained within the androecium. The archegonium and developing sporophyte are contained within the perianth.*

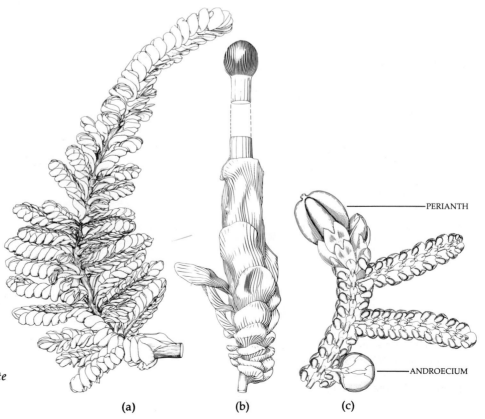

PERIANTH

ANDROECIUM

(a)   **(b)**   **(c)**

Spore Discharge in Liverworts

There are three methods of spore discharge in liverworts:
(a) Most liverworts resemble *Cephalozia* in the way their spores are dislodged. The capsule containing the elaters with their attached spores dries and splits open. Exposed to air, the elaters dry, initially coiling and then, upon reaching a critical point, suddenly expanding. As a result, the spores are dislodged from the elaters and released from the capsule.
(b) In *Marchantia*, the drying spore capsules, borne on stalks, split open, exposing the elaters to air. As the elaters dry, they twist, helping to disperse spores from the capsule.
(c) A less common method of spore discharge is seen in *Frullania*. As it dries, the sporangium peels open in four sections. The elaters are initially attached at both ends to the sporangium. Upon drying, one end of each elater snaps loose from the center of the sporangium, spreading the spores.

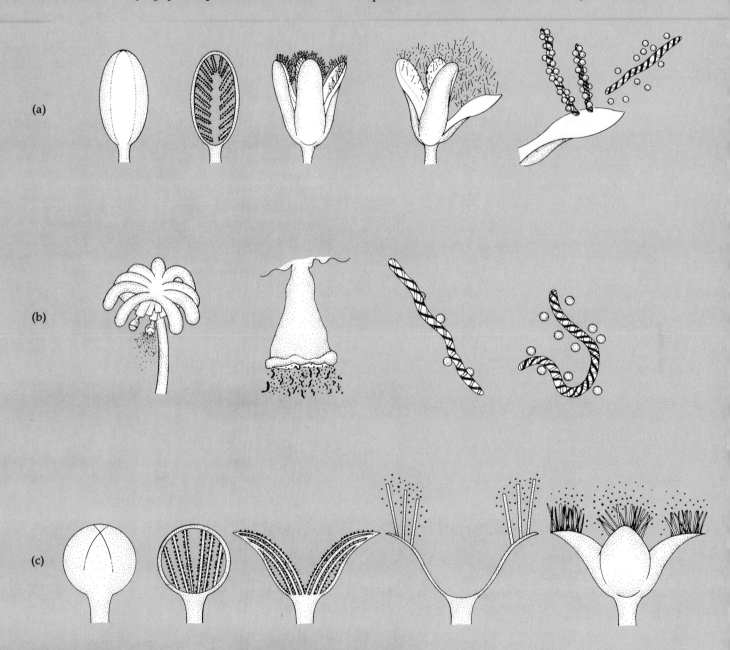

(a)

(b)

(c)

15–11
A leafy liverwort growing on the leaf of an evergreen tree in the rainforest of the Amazon Basin, near Manaus, Brazil.

The Hornworts: Division Anthocerophyta

Hornworts constitute a small division of unique plants that may be more closely related to green algae such as *Coleochaete* (see page 284) than to any other group. The members of the genus *Anthoceros* are the most familiar of the six genera (consisting of approximately 100 species of hornworts). The gametophytes of hornworts (Figure 15–12a) superficially resemble those of the thallose liverworts, but there are many features that indicate a relatively distant relationship. For example, each hornwort cell usually has a single large chloroplast, as in many algae, rather than the numerous, small, discoid chloroplasts found in the cells of other plants. Each chloroplast contains a pyrenoid, making the resemblance to green algae even more striking. These features might lead one to believe that hornworts had an evolutionary origin separate from that of other plants, but the presence of stomata, which apparently

(a)

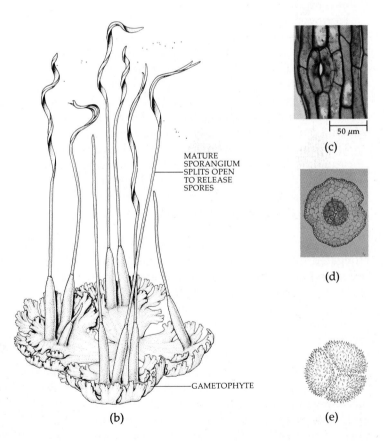

MATURE SPORANGIUM SPLITS OPEN TO RELEASE SPORES

GAMETOPHYTE

(b)

50 μm

(c)

(d)

(e)

15–12
Anthoceros, a hornwort. (a) *A gametophyte with attached sporophytes.* (b) *When mature, the sporangium splits,* and the spores are released. (c) *A stoma; stomata are abundant on the sporophytes of the hornworts, which are green and* photosynthetic. (d) *Developing spores and* (e) *mature spores.*

15–13
Anthoceros. (a) *Longitudinal section of a sporophyte, showing its foot embedded in the tissue of the gametophyte.* (b) *Longitudinal section of a sporangium, showing tetrads of spores with elater-like structures among them. The column in the central lower part of the sporangium consists of tissue that may function in conduction.*

MERISTEMATIC
REGION OF
SPOROPHYTE

FOOT

GAMETOPHYTE

STOMA

(a) 500 μm

(b) 100 μm

remain open until late in the organism's development, is a characteristic that hornworts share with all plants other than liverworts.

Hornwort gametophytes have a strong dorsiventral orientation; they are often rosettelike and are usually about 1 to 2 centimeters across. Individuals of *Anthoceros* have extensive internal cavities filled with mucilage, rather than with air, as in the gametophytes of thallose liverworts. These mucilage-filled cavities are often inhabited by cyanobacteria of the genus *Nostoc*, which fix nitrogen and supply it to their host plants.

The gametophytes of some species of *Anthoceros* are unisexual; others are bisexual. The antheridia and archegonia are sunken on the dorsal surface of the gametophyte, with the antheridia clustered in chambers. Numerous sporophytes may develop on the same gametophyte.

The sporophyte of *Anthoceros*, which is an upright elongated structure, consists of a foot and a long, cylindrical sporangium (Figures 15–12b and 15–13b). Very early in its development, a meristem, or zone of actively dividing cells, develops between the foot and sporangium; this meristem remains active as long as conditions are favorable for growth. As a result, the sporophyte continues to elongate for a prolonged period of time. It is green, having several layers of photosynthetic cells. It is also covered with a cuticle and has stomata (Figure 15–12c). Maturation of the spores and, ultimately, dehiscence of the sporangium begin near its tip and spread toward its base as the spores mature (Figure 15–12b). Among the spores are sterile, elongate structures, often multicellular, that resemble the liverwort elaters (Figure 15–13b). The dehiscing sporangium splits longitudinally into ribbonlike halves.

The Mosses: Division Bryophyta

Mosses constitute a diverse group of some 9500 species of small plants. They are often abundant in relatively moist areas, where a variety of species and a luxuriance of individuals can be found (Figure 15–1). These plants sometimes dominate the terrain to the exclusion of other plants over large areas of the far north and far south, as well as on rocky slopes above timberline in mountainous regions. A significant number of moss species are even able to withstand the long periods of severely cold temperatures on the Antarctic continent. Like the lichens, mosses are remarkably sensitive to air pollution, especially sulfur dioxide, and in highly polluted areas they are often absent or represented by only a few species. A number of species of mosses are found in deserts, and some form extensive masses on dry, exposed rocks where temperatures can become very high. Many mosses can remain alive for years in a dry condition, recovering growth quickly if moistened; other mosses are aquatic and die within a day or so if kept dry. A few species are even found on seaside, wave-splashed rocks, although none is truly marine.

Many groups of organisms contain members that are commonly called "mosses"—reindeer "mosses" are lichens, club "mosses" and Spanish "moss" are vascular plants, and sea "moss" and Irish "moss" are algae. The genuine mosses, however, are members of the division Bryophyta, which consists of three classes: Bryidae (the "true" mosses), Sphagnidae (the peat mosses), and Andreaeidae (the granite mosses). These groups are very distinct from one another, differing in many important features.

The gametophytes of all mosses are represented by two distinct phases: the **protonema** (plural: **protonemata;** from the Greek words *protos,* "first," and *nema,* "thread"), which arises directly from a germinating spore, and the **leafy gametophyte.** In the true mosses, the cells of the protonema occur in a single layer, and the branching filament resembles a filamentous green alga (Figure 15–14). Leafy gametophytes develop from minute budlike structures on the protonemata. In a few genera of mosses the protonema is persistent and assumes the major photosynthetic role, while the leafy shoots of the gametophyte are minute. Protonemata, which are characteristic of all mosses, are also found in some liverworts, but not in hornworts.

In the true mosses, the gametophyte is leafy and usually upright, rather than dorsiventrally flattened as it is in the leafy liverworts. It grows from an apical initial cell that resembles an inverted, three-sided pyramid, similar to that of the leafy liverworts. Although three ranks of leaves are produced initially, subsequent twisting of the axis results in displacement of these ranks to give the appearance of a spiral arrangement. In some genera (*Fontinalis,* an aquatic moss, for example) the original, three-ranked condition of the leaves is still obvious in the mature gametophytes.

Moss gametophytes, which exhibit varying degrees of complexity, range from as little as 0.5 millimeter to 50 centimeters or more in length. All have multicellular rhizoids, and the leaves are normally only one cell thick except at the midrib (which is lacking is some genera). In many mosses, the stems of the gametophytes and sporophytes have a central strand of water-conducting cells known as **hydroids.** Hydroids are elongate cells with inclined end walls that are thin and highly permeable to water, making them preferential pathways for water and solutes. They resemble the tracheary elements of vascular plants in lacking a living protoplast; consequently, hydroids are empty at maturity. Unlike tracheary elements, however, hydroids lack specialized wall thickenings (see page 458). In some moss genera, sieve elements (food-conducting cells, also known as **leptoids**) surround the strand of hydroids. The sieve elements, like hydroids, are elongate cells. Unlike hydroids, however, the sieve elements have living protoplasts at maturity. In general, the mature sieve elements of mosses resemble those of seedless vascular plants; both have degenerate nuclei and inclined end walls with small pores. It is likely that the water-conducting and food-conducting cells of mosses and vascular plants had a common origin, judged by their similarities in structure and function.

In mosses, the antheridia are often clustered within leafy structures called splash cups (Figure 15–15). The sperm from several to many antheridia are discharged

15–14
Protonema of a moss with a budlike structure. Protonemata are the first stage of the gametophyte generation of mosses and some liverworts. They often resemble filamentous green algae.

into a drop of water within such a cup and then are dispersed widely as a result of the action of raindrops falling into the cup. Insects may also carry drops of water, rich in sperm, from plant to plant. Since sperm can often be dispersed over relatively great distances in these ways, isolated female gametophytes often form sporophytes abundantly.

15–15
Leafy male gametophytes of the moss Polytrichum piliferum, *showing the mature antheridia clustered together in heads. The sperm are discharged into drops of water held within these leafy heads and are then splashed out of them by raindrops, sometimes reaching the vicinity of archegonia on other gametophytes.*

15–16

The two common growth forms found in the gametophytes of different genera of mosses. (a) "Cushiony" form, in which the gametophytes are erect and have few branches, shown here in Polytrichum juniperinum. *Spore capsules atop long, slender setae—sporophytes—can be seen rising above the gametophytes. (b) "Feathery" form, with matted, creeping gametophytes, shown here in* Thuidium delicatulum.

(b)

(a)

Two patterns of growth are common among moss gametophytes (Figure 15–16). In the first, the "cushiony" mosses, the gametophytes are erect and little-branched, usually bearing terminal sporophytes. In the second, the gametophytes are much-branched and "feathery," and the plants are creeping; the sporophytes are borne laterally. This second type of growth pattern is found, for example, in those mosses that hang in masses from the branches of trees in moist regions.

At maturity, gametangia are produced by most leafy gametophytes, either at the tip of the main axis or on a lateral branch. In some genera, the gametophytes are unisexual (Figure 15–17), but in other genera both archegonia and antheridia are produced by the same plant.

Moss sporophytes, which usually take 6 to 18 months to reach maturity in temperate species, are borne on the gametophytes. The capsules, or sporangia, are generally elevated on a **seta**, or stalk, which may

15–17

Gametangia of Mnium, *a moss. (a) Longitudinal section through an archegonial head showing the pink-stained archegonia surrounded by sterile structures called paraphyses. (b) Longitudinal section through an antheridial head showing antheridia surrounded by paraphyses.*

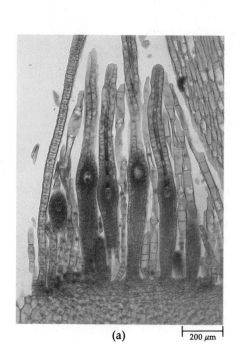

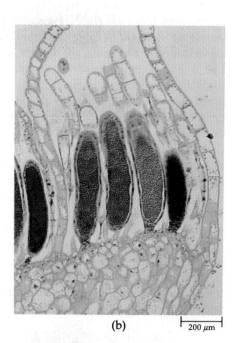

(a)   200 μm

(b)   200 μm

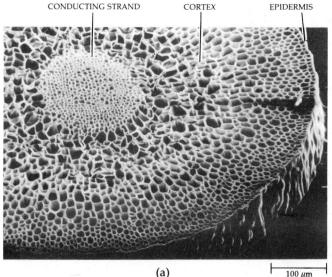

CONDUCTING STRAND CORTEX EPIDERMIS

(a)   100 μm

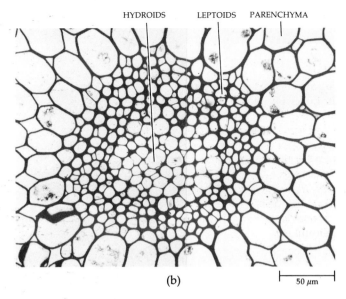

HYDROIDS LEPTOIDS PARENCHYMA

(b)   50 μm

15–18

Conducting strands in the seta of a sporophyte of the moss Dawsonia superba. *(a) General organization of the seta as seen in transverse section with the scanning electron microscope. (b) Transverse section showing central column of hydroids surrounded by a sheath of leptoids (sieve elements) and parenchyma of the cortex. (c) Longitudinal section of a portion of central strand, showing (from left to right) hydroids, leptoids, and parenchyma.*

HYDROID LEPTOID PARENCHYMA

(c)   20 μm

reach 15 to 20 centimeters in length in a few species; some mosses lack a seta entirely. The setae of most mosses contain a central strand of hydroids, which in some genera is surrounded by sieve elements (Figure 15–18). A short foot at the base of the seta is embedded in the tissue of the gametophyte, and the seta usually elongates early in the development of the sporophyte. Stomata are normally present on moss sporophytes; some moss stomata, however, consist only of a single, doughnut-shaped guard cell, which is a very different structure from that found in vascular plants.

Generally, the cells of the young and maturing sporophyte contain chloroplasts and carry out photosynthesis. When a moss sporophyte is mature, however, it gradually loses its ability to photosynthesize and turns yellow, then orange, and finally brown. The calyptra, derived from the archegonium, is commonly lifted upward with the capsule as the seta elongates. Prior to spore dispersal, the protective calyptra falls off, and the lid, or **operculum,** of the capsule bursts off, revealing a ring of teeth—the **peristome**—surrounding the opening (see Figure 15–20 on page 314). The teeth of the peristome are formed by the splitting, along a zone of weakness, of a cellular layer near the apex of the capsule. In most mosses, the teeth uncurl slowly when the air is relatively dry and curl up again when it is moist. The movements of the teeth ex-

pose the spores, which are gradually released. The peristome is a characteristic of the class Bryidae and is lacking in the other two classes of mosses. In mosses, each capsule sheds up to 50 million haploid spores, each of them capable of giving rise to a new gametophyte. A representative true moss life cycle is shown in Figure 15–19.

Asexual reproduction in the true mosses usually takes place by fragmentation. Virtually any part of the gametophyte is capable of regeneration, including the sterile parts of the sex organs. Many species produce gemmae, which are able to give rise to new gametophytes.

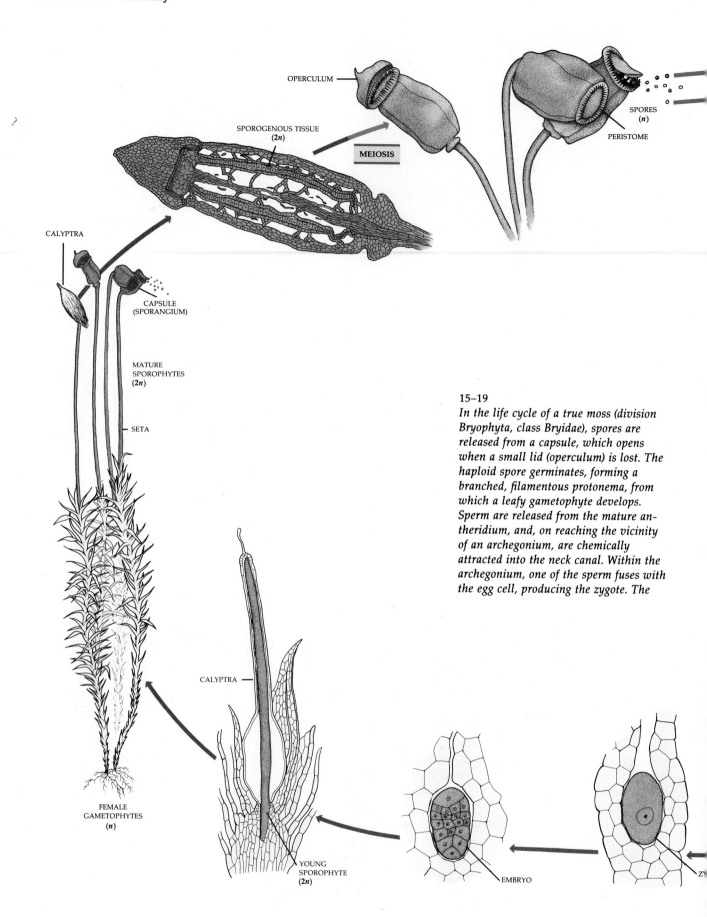

OPERCULUM

SPOROGENOUS TISSUE
(2n)

MEIOSIS

SPORES
(n)

PERISTOME

CALYPTRA

CAPSULE
(SPORANGIUM)

MATURE
SPOROPHYTES
(2n)

SETA

15–19
In the life cycle of a true moss (division Bryophyta, class Bryidae), spores are released from a capsule, which opens when a small lid (operculum) is lost. The haploid spore germinates, forming a branched, filamentous protonema, from which a leafy gametophyte develops. Sperm are released from the mature antheridium, and, on reaching the vicinity of an archegonium, are chemically attracted into the neck canal. Within the archegonium, one of the sperm fuses with the egg cell, producing the zygote. The

CALYPTRA

FEMALE
GAMETOPHYTES
(n)

YOUNG
SPOROPHYTE
(2n)

EMBRYO

ZY

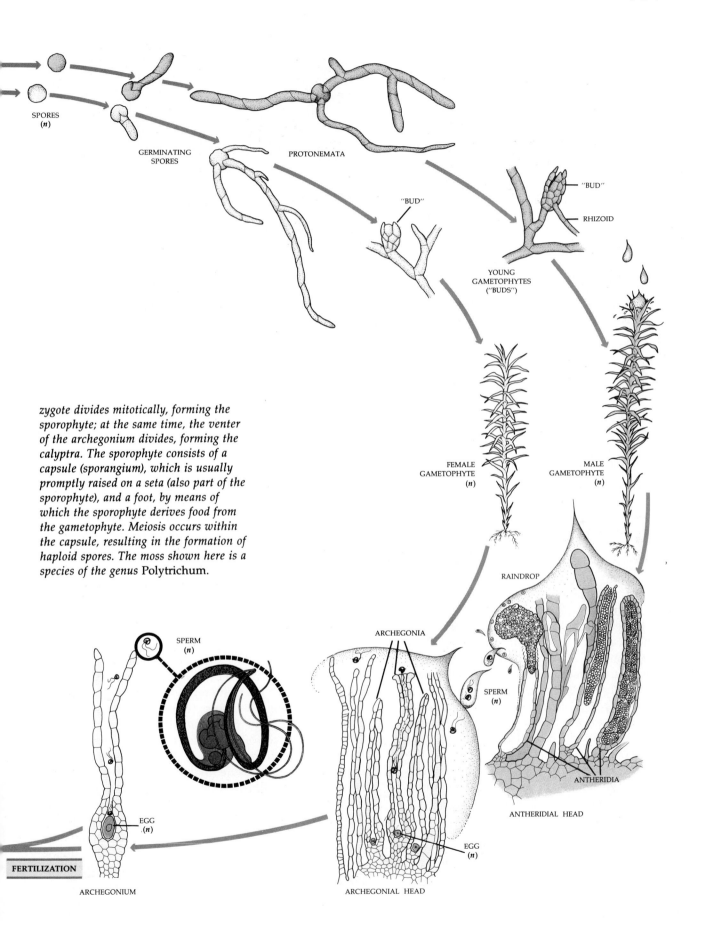

SPORES
(*n*)

GERMINATING
SPORES

PROTONEMATA

"BUD"

"BUD"

RHIZOID

YOUNG
GAMETOPHYTES
("BUDS")

zygote divides mitotically, forming the
sporophyte; at the same time, the venter
of the archegonium divides, forming the
calyptra. The sporophyte consists of a
capsule (sporangium), which is usually
promptly raised on a seta (also part of the
sporophyte), and a foot, by means of
which the sporophyte derives food from
the gametophyte. Meiosis occurs within
the capsule, resulting in the formation of
haploid spores. The moss shown here is a
species of the genus Polytrichum.

FEMALE
GAMETOPHYTE
(*n*)

MALE
GAMETOPHYTE
(*n*)

RAINDROP

SPERM
(*n*)

ARCHEGONIA

SPERM
(*n*)

ANTHERIDIA

ANTHERIDIAL HEAD

EGG
(*n*)

EGG
(*n*)

FERTILIZATION

ARCHEGONIUM

ARCHEGONIAL HEAD

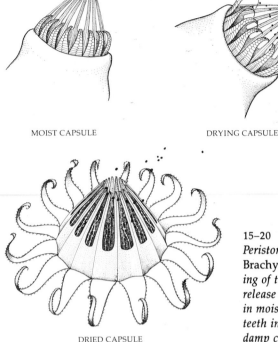

MOIST CAPSULE   DRYING CAPSULE

DRIED CAPSULE

(a)

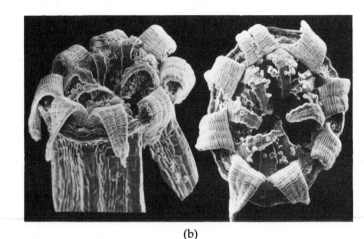

(b)

15–20

Peristome teeth in true mosses. (a) Brachythecium *has a peristome consisting of two rings of teeth, which open to release the spores in response to changes in moisture. The outer set of peristome teeth interlock with the inner set under damp conditions. As the capsule dries out,* the outer teeth pull away, allowing the dispersal of spores by the wind. (b) Scanning electron micrographs of the peristome teeth of two capsules of Orthotrichum, *showing the inner teeth curved inward and the outer teeth curved outward in dry conditions.*

THE PEAT MOSSES: CLASS SPHAGNIDAE

The approximately 350 species of peat mosses, all belonging to the genus *Sphagnum* (Figure 15–21), constitute a very distinct group, which clearly diverged from the main line of moss evolution a very long time ago. The stems of the gametophyte of *Sphagnum* bear clusters of branches, often five at a node, which are more densely tufted near the apex of the stem, forming a moplike head. These gametophytes form large, bright green or occasionally reddish clumps in boggy ground. The gametophytes of *Sphagnum* arise from a protonema that is platelike instead of filamentous and forms buds along its margins, from which the leafy gametophytes arise. The leaves of *Sphagnum* lack midribs, and the mature plants lack rhizoids. In the boggy places where they grow, *Sphagnum* plants are nearly always turgid and thus erect; they are also usually densely packed together. The distinctive leaves of this genus consist of large, dead cells surrounded by narrow green, or occasionally red-pigmented, living cells (Figure 15–21b). The dead cells contain pores and thickenings and readily become filled with water, so that the water-holding capacity of the peat mosses is up to 20 times its dry weight. By comparison, cotton absorbs only four to six times its dry weight.

Because of their superior absorptive qualities, *Sphagnum* mosses were used in Europe from the 1880s onward as dressings for wounds and boils; however, cotton dressings have been used almost exclusively for such purposes since World War I, probably because of their cleaner appearance. Gardeners mix peat moss with soil to increase the water-holding capacity of the soil and make it more acidic.

Sphagnum sporophytes (Figure 15–21a) are also distinctive. The red to blackish-brown capsules are nearly spherical and are raised on a stalk, the **pseudopodium,** which is part of the gametophyte and may be up to 3 millimeters long. The sporophyte has a very short seta. Spore discharge in *Sphagnum* is spectacular (Figure 15–21c). At the top of the capsule is a disk-shaped operculum, separated from the rest of the capsule by a circular groove. As the capsule matures, its internal tissues shrink and air is drawn in, presumably through the stomata, which cannot close. When the capsule wall dries, the air is trapped within. Contraction of the maturing capsule results in increased internal pressure, and the operculum is eventually blown off with an audible click. This happens in warm, sunny conditions. When the operculum is blown off, the escaping gas carries a cloud of spores out of the capsule. The most distinctive features of the class Sphagnidae are the lack of a peristome and the peculiar morphology of the gametophyte.

15–21

A peat moss, Sphagnum. (a) A gametophyte, with many attached sporophytes.
Some of the capsules, such as the two at
the front and to the left, have already
discharged their spores. (b) Structure of a
leaf. The interior, consisting of large,
dead cells, is surrounded by smaller living
cells, rich in chloroplasts. (c) Dehiscence
of a capsule. As the capsule dries, it is
filled with air. The stomata, through
which the air enters, close in the process
of drying; with further shrinking, the air
pressure inside the capsule blows off the
operculum, releasing a cloud of spores.

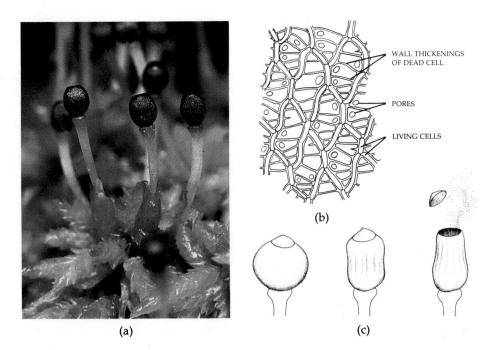

(a)

WALL THICKENINGS
OF DEAD CELL

PORES

LIVING CELLS

(b)

(c)

Ecology of *Sphagnum*

The peat mosses form extensive peat bogs, which are a
conspicuous feature of cold and temperate regions
throughout the world. The mosses contribute to the
acidity of their own environment by releasing H^+ ions;
in the center of the bogs the pH is often less than 4—
very unusual for a natural environment. Peat is formed
from the accumulation and compression of the mosses
themselves, as well as the sedges, reeds, grasses, and
other plants that grow among them. In Ireland and
some other northern regions, dried peat is widely used
as industrial fuel, as well as for domestic heating. At a
conservative estimate, peat bogs cover at least 1 percent
of the world's land surface, an enormous area equivalent to that of half of the United States.

THE GRANITE MOSSES: CLASS ANDREAEIDAE

The genus *Andreaea* consists of about 100 species of
small, blackish-green or dark reddish-brown tufted
rock mosses (Figure 15–22), which in their own way are
as peculiar as *Sphagnum*. The members of *Andreaea*
occur in mountainous or Arctic regions, often on granite rocks—which is the reason for their common name.
Although the gametophyte closely resembles that of
the true mosses, it arises from a thick structure with
many lobes rather than from a filamentous one. The
sporophyte lacks a true seta and is elevated above the
leaves on a stalk of gametophytic tissue, or pseudopodium, as in the peat mosses. The minute capsules of

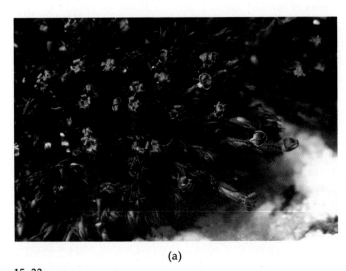

(a)

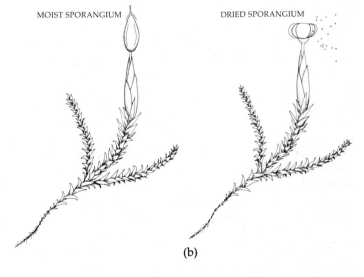

MOIST SPORANGIUM

DRIED SPORANGIUM

(b)

15–22

(a) Andreaea rothii *growing on granite
rock in Devon, England.* (b) The

Andreaea *sporangium contracts and
splits as it dries out, allowing the spores*

to fall out under dry conditions.

Andreaea are marked by four vertical lines of weaker cells along which the capsule splits. The capsule remains intact above and below the dehiscence lines. The resulting four valves are very sensitive to the humidity of the surrounding air, opening widely when it is dry—the spores can be carried widely by the wind under such circumstances—and closing when it is moist. This mechanism of spore discharge, by means of slits in the capsule, is different from that of any other moss (see Figure 15–22).

Andreaeobryum, a second genus of granite mosses, was discovered in Alaska in 1976. It differs from *Andreaea* in that its sporophyte has a true seta and its capsule splits to the apex.

have, at least in some groups, both specialized conducting tissue and stomata, resembling the remaining divisions of plants in these respects. The conducting tissue of mosses, when present, consists of hydroids, water-conducting cells, and leptoids, or sieve elements. A number of other structural and biochemical differences underscore the differences among these three divisions. For example, hornwort sporophytes are unique in having a basal meristem, which adds tissue to the sporangium over a period of months. On germinating, the spores of mosses produce a filamentous or platelike gametophyte known as a protonema, from which leafy gametophytes usually arise. Protonemata are found in some liverworts but are lacking in hornworts.

Summary

Plants seem to have been derived from a charophycean green alga, since the two groups share many unique features: chlorophyll *a* as the primary photosynthetic pigment, with chlorophyll *b* and carotenoids as accessory pigments; starch deposited within the chloroplasts; cellulosic cell walls; and cell division by forming a phragmoplast and cell plate. From the common ancestor of plants—which almost certainly displayed an alternation of heteromorphic generations, formed embryos, and had multicellular gametangia—a number of evolutionary lines were derived.

The bryophytes consist of three divisions of structurally rather simple, small plants in which the gametophytes are always nutritionally independent of the sporophytes, whereas the sporophytes are permanently attached to the gametophytes and vary in their dependence on them. Thus the gametophyte is the dominant generation. The male sex organs, antheridia, and female sex organs, archegonia, both have sterile jacket layers. Each archegonium contains a single egg. Numerous sperm are produced by each antheridium; the sperm are free-swimming and biflagellated. The sporophyte develops within the archegonium but usually grows out of it in mosses and hornworts, while remaining permanently attached to it. In mosses and some liverworts, the sporophyte is differentiated into a foot, a seta, and a capsule, or sporangium. Hornwort and moss sporophytes become more or less nutritionally independent of the gametophyte, whereas those of liverworts usually remain completely dependent on their gametophyte.

Among the "bryophytes," liverworts (division Hepatophyta) lack specialized conducting tissue (with a few possible exceptions) and stomata; hornworts (division Anthocerophyta) also lack specialized conducting tissue but have stomata; mosses (division Bryophyta)

Suggestions for Further Reading

Conard, Henry S., and Paul L. Redfearn, Jr.: *How to Know the Mosses and Liverworts*, 2d ed., William C Brown Co., Dubuque, Iowa, 1979.*

A good beginner's guide for use in identifying many of the more common bryophytes. It includes a profusely illustrated key and an excellent glossary.

Crum, Howard: *A Focus on Peatlands and Peat Bogs*, University of Michigan Press, Ann Arbor, 1988.

An outstanding book on peat bogs and the mosses and other organisms that form them.

Flowers, Seville: *Mosses: Utah and the West*, Brigham Young University Press, Provo, Utah, 1973.

An excellent reference on local flora, with an outstanding introduction, covering the morphology and ecology of mosses, methods of collecting and preserving them, and many other matters of interest.

Paolillo, Dominick J.: "The Swimming Sperms of Land Plants," *BioScience* 31:367–373, 1981.

A fascinating account of the similarities and differences between the sperm of different groups of plants.

Richardson, D.H.S.: *The Biology of Mosses*, John Wiley & Sons, Inc., New York, 1981.*

A well-written and well-illustrated short book on the mosses, covering all aspects of the group.

Schofield, W.B.: *Introduction to Bryology*, Macmillan Publishing Company, New York, 1985.

An outstanding analysis of all groups of mosses, liverworts, and hornworts, including general chapters on subjects such as physiology, ecology, and geography.

Smith, A.J.E. (ed.): *Bryophyte Ecology*, Chapman and Hall, New York, 1982.

An excellent collection of articles describing current research on bryophyte ecology.

*Available in paperback.

CHAPTER 16

Seedless Vascular Plants

16–1
Spores of the horsetail Equisetum
arvense, *as seen in a false-color scanning electron micrograph. The spores of*
Equisetum *form inside sporangia, which
are grouped within cone-like strobili at
the tips of fertile shoots (see Figure
16–26). Shown here are moist spores
wrapped tightly by thickened bands,
known as elaters, which are attached to
the spore walls. As the spores dry, the
elaters uncoil, helping to disperse the
spores from the sporangium.*

Plants, like all living things, had aquatic ancestors, and
the story of plant evolution is inseparably linked with
their progressive occupation of the land. In this
chapter, we shall first discuss the general features of
vascular plant evolution—features linked with life on
land—and then describe the features of the seedless
vascular plants, those with the most generalized features. This chapter will tell the story of the club mosses,
horsetails, ferns, and other groups of vascular plants
that retain some of the simpler features that were characteristic of the earliest members of this great group of
plants.

The Evolution of Vascular Plants

One of the key events in the early invasion of the land
by plants was the development of spores with durable
protective walls that allowed the spores to tolerate dry
conditions. As a result spores could be dispersed over
the surface of the land by wind. As plant bodies became larger, structures that allowed for more efficient
release and dispersal of the spores evolved (Figure 16–1).
An additional important evolutionary advance that accompanied this increase in size was the origin of the
cuticle, with a cutin framework and embedded waxes
that act as the primary barrier to the passage of water.
Stomata must have evolved at the same time as the cuticle to allow the exchange of gases needed for photosynthesis and respiration.

Relatively early in the history of plants, the evolution of efficient conducting systems, consisting of
xylem and phloem, solved the problem of water and
food conduction throughout the plant—a serious problem for any large organism growing on land. Many botanists believe that the acquisition of the ability to

16–2

Simple primitive vascular plants emerged from the water to colonize the barren landscape during the early Devonian period, some 408 to 387 million years ago. Shown here, from left to center, are Cooksonia, Rhynia, *and* Zosterophyl- *lum, all of which lacked leaves. By the beginning of the mid-Devonian period (387 to 374 million years ago), a more diverse plant community had evolved, with larger and more complex types of plants. Examples seen here on the right, from back to front, are* Psilophyton, *a robust trimerophyte;* Drepanophycus, *an early lycophyte; and* Protolepidodendron, *a lycophyte. The leaves of the lycophytes were microphylls. All of these early vascular plants lacked seeds.*

synthesize lignin, which is incorporated into the cell walls of supporting and water-conducting cells, was a pivotal step in the evolution of plants. The aboveground and belowground parts of early plants differed little from one another structurally, but ultimately the ancient plants gave rise to more specialized plants. These plants consisted of roots, which function in absorption and anchorage, and of shoots (stems and leaves), which provide a system well suited to the demands of life on land—namely, the acquisition of energy from sunlight, of carbon dioxide from the atmosphere, and of water. The morphological diversity of early plants was extensive; different forms were adapted to life in different terrestrial environments, in which the efficiency of photosynthesis was maximized by the morphological differentiation.

As early plants became increasingly suited to life on land, their gametophytic generation underwent a progressive reduction in size and became increasingly more protected by and nutritionally dependent on the sporophyte. Finally, seeds evolved in one evolutionary line. Seeds are structures that protect the embryonic sporophyte during its dispersal from the parent and during a following dormant period, and then nourish the sporophyte during the period of germination and the establishment of the young plant as an independent individual. In such ways are seed plants enabled to withstand unfavorable environmental conditions.

As discussed in Chapter 15, the common ancestor of the different divisions of plants was probably a relatively complex, multicellular green alga (Figure 14–22, page 285) that invaded the land about 450 million years ago, in the Ordovician period, probably as part of an endomycorrhizal relationship (Figure 12–40). An alga resembling *Coleochaete* (Figure 14–21, page 284) appears to be the best living model for this ancestor. Individuals of *Coleochaete* are haploid except for the zygote (they have zygotic meiosis). Many botanists now favor the hypothesis that the complex sporophytic generation, as it exists in plants, evolved independently from the diploid phase that is typical of many algal life cycles. Vascular plants were present in the Silurian period (438 to 408 million years ago); for example, well-preserved strands that are evidently vascular tissue have been found in North American rocks as much as 430 million years old. Vascular plants became numerous and diverse by the Devonian period (408 to 360 million years ago; Figure 16–2).

Because of their adaptations for existence on land, the vascular plants are dominant in terrestrial habitats. There are nine divisions with living representatives, including about 250,000 living species. In addition, there are several divisions that consist entirely of extinct vascular plants. In this chapter, we shall describe some of the evolutionary advances of the vascular plants and discuss seven divisions of seedless vascular plants, three of them extinct. In Chapters 17 to 19, we shall discuss the seed plants, which include five divisions with living representatives.

Organization of the Vascular Plant Body

The sporophytes of early vascular plants were dichotomously branched (evenly forked) axes that lacked roots and leaves. With evolutionary specialization, morphological and physiological differences arose between various parts of the plant body, bringing about the differentiation of roots, stems, and leaves—the organs of the plant (Figure 16–3). Collectively, the roots make up the **root system,** which anchors the plant and absorbs water and minerals from the soil. The stems and leaves together make up the **shoot system,** with the stems raising the specialized photosynthetic organs—the leaves—toward the sun and the vascular system con-

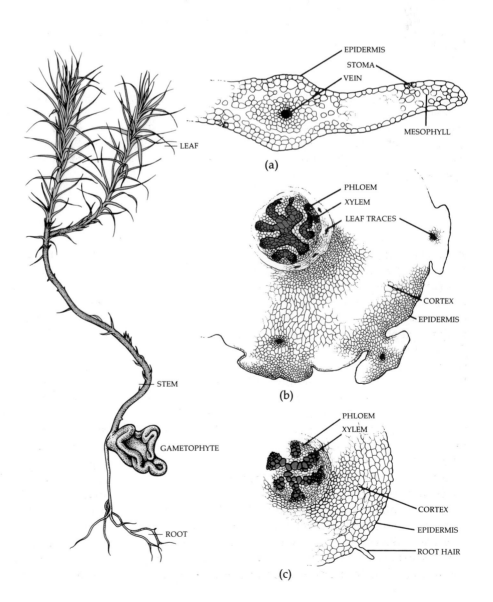

16–3
A diagram of a young sporophyte of the club moss Lycopodium clavatum, *which is still attached to its subterranean gametophyte. The tissues of each organ are shown in transverse sections of (a) leaf, (b) stem, and (c) root. In all three organs, the dermal tissue system is represented by the epidermis, and the vascular tissue system, consisting of xylem and phloem, is embedded in the ground tissue system. The ground tissue in the leaf—in* Lycopodium, *a microphyll—is represented by the mesophyll, and in the stem and root by the cortex, which surrounds a solid strand of vascular tissue, or protostele. The leaf is specialized for photosynthesis, the stem for support of the leaves and for conduction, and the root for absorption and anchorage.*

ducting water and minerals to the leaves and the end products of photosynthesis away from them.

The different kinds of cells of the plant body are organized into tissues, and the tissues are organized into still larger units called tissue systems. Three tissue systems—**dermal, vascular,** and **ground**—occur in all organs of the plant; they are continuous from organ to organ and reveal the basic unity of the plant body. The dermal tissue system makes up the outer, protective covering of the plant. The vascular tissue system comprises the conductive tissues—xylem and phloem—and is embedded in the ground tissue system (Figure 16–3). The principal differences in the structures of root, stem, and leaf lie primarily in the relative distribution of the vascular and ground tissue systems, as will be discussed in Section 4.

PRIMARY AND SECONDARY GROWTH

Primary growth may be defined as the growth that occurs relatively close to the tips of roots and stems. It is initiated by the apical meristems and is primarily involved with extension of the plant body—often the vertical growth of a plant. The tissues arising during primary growth are known as **primary tissues;** the part of the plant body composed of these tissues is called the **primary plant body.** Primitive vascular plants, and many contemporary ones as well, consist entirely of primary tissues.

In addition to primary growth, many plants undergo additional growth that thickens the stem and root; such growth is termed **secondary growth.** It results from the activity of lateral meristems, one of which, the **vascular cambium,** produces **secondary vascular tissues: secondary xylem** and **secondary phloem** (see Figure 24–6, page 523). The production of secondary vascular tissues is commonly supplemented by the activity of a second lateral meristem, the **cork cambium,** which forms a **periderm,** composed mostly of cork tissue. The periderm replaces the epidermis as the dermal tissue system of the plant. The secondary vascular tissues and periderm make up the **secondary plant body.** Secondary growth appeared in the middle Devonian period, about 380 million years ago, in several unrelated groups of vascular plants.

TRACHEARY ELEMENTS

Sieve elements, the conducting cells of the phloem, have soft walls and often collapse after they die, so they are rarely well preserved as fossils. In contrast, **tracheary elements,** the conducting cells of the xylem, have rigid, persistent walls and are sometimes well preserved. Because of their various wall patterns, the

tracheary elements provide valuable clues to the interrelationships of the different groups of vascular plants.

In fossil vascular plants from the Silurian and Devonian periods, the tracheary elements are elongate cells with long, tapering ends. Such tracheary elements, called **tracheids,** were the first type of water-conducting cell to evolve; they are the only type of water-conducting cell in most vascular plants other than angiosperms. Tracheids not only provide channels for the passage of water and minerals but also provide support for stems. Water-conducting cells are rigid, mostly because of the lignin in their walls. This rigidity made it possible for plants to evolve an upright habit and eventually for some of them to become trees.

Tracheids are more primitive (less specialized) than **vessel members**—the principal water-conducting cells in angiosperms. Vessel members apparently evolved independently from tracheids in several groups of vascular plants, including the angiosperms, the gnetophytes (page 375), several unrelated species of ferns, and certain species of *Selaginella* (a lycophyte) and *Equisetum* (a sphenophyte). The evolution of vessel members in these different groups provides an excellent example of convergent evolution—the independent development of similar structures by unrelated or only distantly related organisms (see "Convergent Evolution," page 518).

STELES

The primary vascular tissues—primary xylem and primary phloem—and the pith, if present, make up the central cylinder, or **stele,** of the stem and root in the primary plant body. Several types of steles are recognized, among them the protostele, the siphonostele, and the eustele.

The **protostele**—the most primitive type of stele—consists of a solid strand of vascular tissue in which the phloem either surrounds the xylem or is interspersed within it (Figures 16–3 and 16–4a). It is found in the extinct groups of seedless vascular plants discussed below, as well as in the Psilotophyta and Lycophyta and in the juvenile stems of some other living groups. In addition, it is the type of stele found in most roots.

The **siphonostele** is characterized by a central column of ground tissue, the **pith,** which is surrounded by the vascular tissue (Figure 16–4b). The phloem may form only outside the cylinder of xylem or on both sides of it. In the siphonosteles of ferns, the departure from the stem of the vascular strands leading to the leaves—the **leaf traces**—generally is marked by gaps —**leaf gaps**—in the siphonostele (as in Figure 16–4c). These leaf gaps are filled with parenchyma cells just like those that occur within and outside the vascular tissue of the siphonostele.

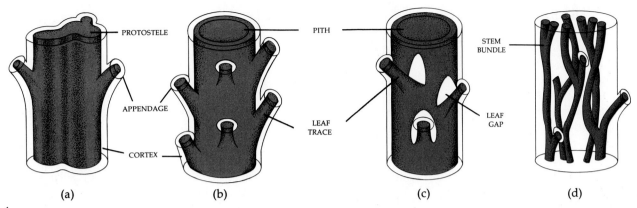

16-4
(a) *A protostele, with diverging appendages, the evolutionary precursors of leaves. (b) A siphonostele with no leaf gaps; the vascular traces leading to the* *leaves simply diverge from the solid cylinder. This sort of siphonostele is found in* Selaginella, *among other plants. (c) A siphonostele with leaf gaps, commonly* *found in seedless vascular plants. (d) A eustele. Siphonosteles and eusteles appear to have evolved independently from protosteles.*

If the vascular cylinder consists of a system of discrete strands around a pith, as it does in almost all seed plants, the stele is called a **eustele** (Figure 16-4d). Comparative studies of living and fossil vascular plants have suggested that the eustele of seed plants evolved directly from a protostele. Eusteles appeared first among the progymnosperms, a group of spore-bearing plants that are discussed in Chapter 17 (pages 359–361). Siphonosteles evidently evolved independently from protosteles. This relationship indicates that none of the groups of seedless vascular plants with living representatives gave rise to any living seed plants.

ORIGINS OF ROOTS AND LEAVES

Although the fossil record reveals little information on the origins of roots as we know them today, they must have evolved from the lower, often subterranean, portions of the axis of primitive vascular plants. For the most part, roots are relatively simple structures that seem to have retained many of the primitive structural characteristics no longer present in the stems of modern plants.

Leaves are the principal lateral appendages of the stem. Regardless of their ultimate size or structure, they arise as protuberances (leaf primordia) from the apical meristem of the shoot. There are two fundamentally distinct kinds of leaves, microphylls and megaphylls.

Microphylls are usually relatively small leaves; they contain only a single strand of vascular tissue (Figure 16-5a). Microphylls are associated with stems possessing protosteles and are characteristic of the lycophytes. The leaf traces leading to microphylls are not associated with leaf gaps. Even though the name *microphyll*

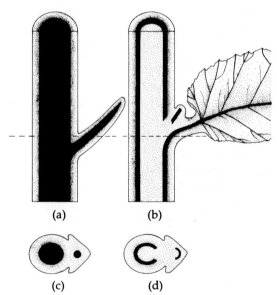

16-5
Longitudinal sections through a stem with a protostele and a microphyll (a) and a stem with a siphonostele and a megaphyll (b), emphasizing the nodes, or regions where the leaves are attached. Transverse sections through the nodes are shown below, (c) and (d). Note the presence of pith and a leaf gap in the stem with a siphonostele and their absence in the stem with a protostele. Microphylls are characteristic of lycophytes, while megaphylls are found in all other vascular plants.

322

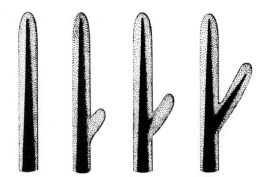

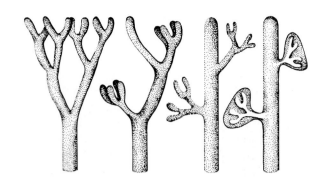

16–6
According to one widely accepted theory, microphylls (left) evolved as outgrowths of the main axis of the plant. Megaphylls (right) evolved by fusion of branch systems.

means "small leaf," some species of *Isoetes* have fairly long leaves. In fact, certain Carboniferous and Permian lycophytes had microphylls a meter or more in length.

Microphylls probably evolved as superficial lateral outgrowths of the stem (Figure 16–6). At first these structures were small, scalelike or spinelike outgrowths devoid of vascular tissue. Gradually, rudimentary leaf traces developed, which initially extended only to the base of the outgrowth. Finally, the leaf traces extended into the appendage, resulting in formation of the primitive microphyll.

Most **megaphylls,** as the name implies, are larger than most microphylls. With few exceptions, megaphylls are associated with stems that have either siphonosteles or eusteles, and the leaf traces leading to them are associated with leaf gaps (Figure 16–5b). Unlike the microphylls, the blade, or **lamina,** of most megaphylls has a complex system of veins.

Megaphylls are believed to have evolved from entire branch systems by a series of steps similar to that shown in Figure 16–6. The earliest plants had a leafless, dichotomously branching axis; unequal branching resulted in more aggressive branches "overtopping" the weaker ones. This was followed with a flattening out, or "planation," of the subordinated lateral branches. The final step was fusion, or "webbing," of the separate lateral branches to form a primitive lamina.

Reproductive Systems

All plants are oogamous and have an alternation of heteromorphic generations (Figure 16–7). In seed plants, the gametophyte is nutritionally dependent on the sporophyte. Oogamy is clearly favored in plants, since only one of the kinds of gametes must navigate across the hostile environment outside the plant. In vascular plants, the sporophyte is the dominant phase of the life cycle; it is larger and structurally much more complex than the gametophyte. In liverworts, hornworts, and mosses, as we saw in Chapter 15, the gametophyte is almost always larger than the sporophyte, which is at least partly nutritionally dependent on the gametophyte.

HOMOSPORY AND HETEROSPORY

Early vascular plants produced only one kind of spore as a result of meiosis; such vascular plants are said to be **homosporous.** Among living vascular plants, homospory is found in the psilotophytes, sphenophytes, some of the lycophytes, and almost all ferns. Upon germination, such spores almost always produce bisexual gametophytes—that is, gametophytes that bear both antheridia and archegonia.

Heterospory—the production of two types of spores in two different kinds of sporangia—is found in some of the lycophytes, in a few ferns, and in all seed

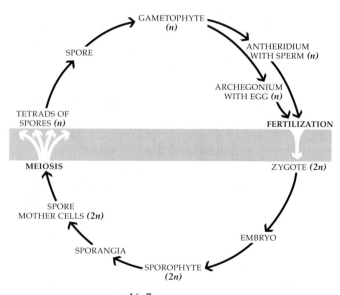

16–7
Generalized life cycle of a vascular plant.

plants. Heterospory arose many times in unrelated groups during the evolution of vascular plants. It was common as early as the Devonian period, with the earliest record from about 390 million years ago. The two types of spores are called **microspores** and **megaspores,** and they are produced in **microsporangia** and **megasporangia,** respectively. The two types of spores are defined on the basis of function and not necessarily relative size. Microspores give rise to male gametophytes (microgametophytes), and megaspores give rise to female gametophytes (megagametophytes). Both of these types of unisexual gametophytes are much reduced in size as compared with the gametophytes of homosporous vascular plants. In heterosporous plants, the gametophytes develop within the spore wall; in homosporous plants, they develop outside the spore wall. In *Equisetum,* although all of the spores are identical morphologically, they are physiologically differentiated, some giving rise to unisexual gametophytes.

GAMETOPHYTES AND GAMETES

The relatively large gametophytes of homosporous plants are independent of the sporophyte for their nutrition, although the subterranean gametophytes of some species—such as those of *Psilotum* and several genera of club mosses (Lycopodiaceae)—are heterotrophic, depending on endomycorrhizal fungi for their nutrients. Other genera of club mosses, like most ferns and the horsetails, have free-living, photosynthetic gametophytes. In contrast, the gametophytes of many heterosporous vascular plants, and especially those of the seed plants, are dependent on the sporophyte for their nutrition.

The evolution of the gametophyte of vascular plants is characterized by a progressive reduction in size and complexity, and angiosperm gametophytes are the most reduced of all (see page 388). Archegonia are found in all seedless vascular plants and most gymnosperms but are absent in all angiosperms; antheridia are also present in seedless vascular plants but are absent in all seed plants. In the seedless vascular plants, the motile sperm swim through water to the archegonium. These plants must therefore grow in habitats where water is at least occasionally plentiful.

The seed—a unique structure in which the embryo is shed from the parent plant enclosed within a resistant coat together with a supply of food that aids in the establishment of the embryo—evolved in the common ancestor of the seed plants. Seeds appeared at least 360 million years ago, at about the end of the Devonian period. Heterospory occurred in three genera of fossil lycophytes and one genus of fossil sphenophytes, but it is likely that the megaspores were shed, instead of being retained and developing on the parent plant as they are in seed plants. The evolution of the seed—the characteristic feature of the seed plants—and the major features of the divisions in which seeds occur will be discussed further in Chapters 17 to 19.

The Divisions of Seedless Vascular Plants

Three divisions of seedless vascular plants—the Rhyniophyta, Zosterophyllophyta, and Trimerophyta—flourished in the Devonian period and became extinct by about 360 million years ago, the end of that period. Of these, the rhyniophytes appeared at least 420 million years ago, in the mid-Silurian period. All three divisions consisted of seedless plants that were relatively simple in structure. A fourth division of seedless vascular plants, Progymnospermophyta, or progymnosperms, will be discussed in Chapter 17 because the members of this group may have been ancestral to the seed plants, both gymnosperms and angiosperms (Figure 16–8). In addition to these extinct divisions, we shall discuss in this chapter the Psilotophyta, Lycophyta, Sphenophyta, and Pterophyta, the four divisions of seedless vascular plants that have living representatives.

The overall pattern of diversification of plants may be interpreted in terms of the successive rise to dominance of four major plant groups; these four groups largely replaced the groups that were dominant earlier. In each instance, numerous species evolved in the groups that were rising to dominance. The major groups involved were:

(1) Early vascular plants—characterized by a simple and presumably primitive morphology. These included the rhyniophytes, zosterophyllophytes, and trimerophytes (Figure 16–9). Primitive vascular plants were dominant from the mid-Silurian period until the mid-Devonian period, from about 420 to 370 million years ago (see Figure 16–2).

(2) Ferns, lycophytes, sphenophytes, progymnosperms. These more complex groups were dominant from the late Devonian period through the Carboniferous period (see Figures 16–10, 17–1), from about 380 to about 290 million years ago (see the essay "Coal Age Plants" on pages 346 and 347).

(3) Seed plants arose starting in the late Devonian period, at least 360 million years ago, and evolved many new lines by the Permian period. Gymnosperms dominated land floras throughout most of the Mesozoic era until about 100 million years ago.

(4) Flowering plants, which appeared in the fossil record about 127 million years ago. This division became dominant within 20 to 30 million years and has remained dominant ever since.

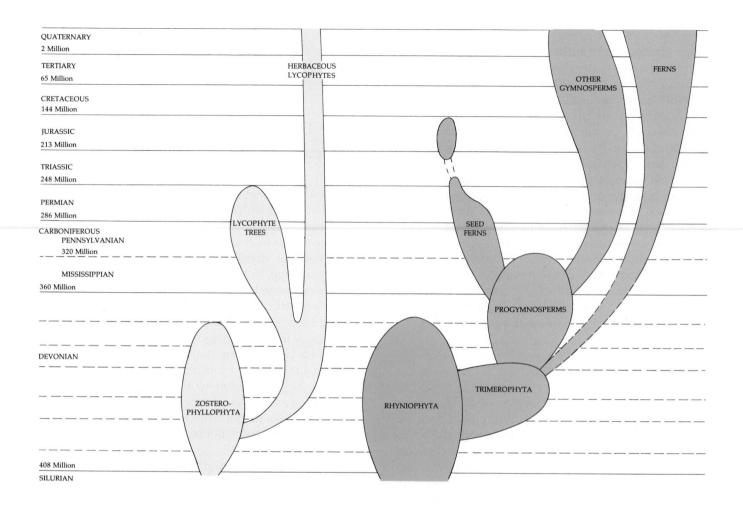

QUATERNARY 2 Million
TERTIARY 65 Million
CRETACEOUS 144 Million
JURASSIC 213 Million
TRIASSIC 248 Million
PERMIAN 286 Million
CARBONIFEROUS PENNSYLVANIAN 320 Million
MISSISSIPPIAN 360 Million
DEVONIAN
408 Million
SILURIAN

HERBACEOUS LYCOPHYTES

OTHER GYMNOSPERMS

FERNS

LYCOPHYTE TREES

SEED FERNS

PROGYMNOSPERMS

TRIMEROPHYTA

ZOSTERO-PHYLLOPHYTA

RHYNIOPHYTA

16–10

*Reconstruction of a Carboniferous period
swamp forest. See also Figure 17–1.*

16–8 (On the facing page)
Possible course of evolution of vascular plants from the zosterophyllophytes and rhyniophytes. The length of Devonian time is exaggerated, and the possible relationships with the angiosperms, or flowering plants, are not indicated. What the ancestral groups of the Psilotophyta and Sphenophyta may have been is not known. Only a few genera of fossil plants have been described that appear to be intermediate between the rhyniophytes, which appeared in the Silurian period (about 420 million years ago), and the zosterophyllophytes, which appeared in the fossil record about 10 million years later. The exact nature of the relationships between these two groups, and of their common ancestor, remains to be established.

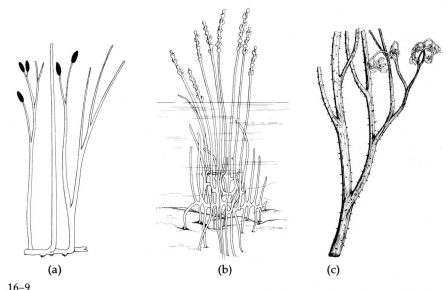

(a) (b) (c)

16–9

Early vascular plants. (a) Rhynia major is a rhyniophyte, one of the simplest of the known vascular plants. The axis was leafless and dichotomously branched. The sporangia, which were terminal, apparently released their spores by splitting longitudinally. (b) In Zosterophyllum and the other zosterophyllophytes, the sporangia, which were aggregated into a terminal spike, split along definite slits that formed around the outer margin. The zosterophyllophytes were larger than the rhyniophytes, but like the latter, they were mostly dichotomously branched plants that were either naked, spiny, or

toothed. (c) The trimerophytes were larger and more complex plants with a strong central axis with smaller side branches. The side branches were dichotomously branched and often had terminal masses of paired sporangia that tapered at both ends. The best-known genera included in this group are Psilophyton *and* Trimerophyton. *A reconstruction of P. princeps is shown here. Individuals of Rhynia major ranged up to about 0.5 meter tall, while some of the trimerophytes were up to 1 meter or more in height. See also Figure 16–2.*

Division Rhyniophyta

The earliest known vascular plants that we understand in detail belong to the division Rhyniophyta, a group that dates back to the mid-Silurian period, at least 420 million years ago. The group became extinct in the mid-Devonian period (about 380 million years ago). Earlier vascular plants were probably similar; their remains go back at least another 15 million years. Rhyniophytes were seedless plants, consisting of simple, dichotomously branching axes with terminal sporangia. Their plant bodies were not differentiated into roots, stems, or leaves, and they were homosporous. The name of the division comes from the good representation of these primitive plants in chert near the village of Rhynie, in Scotland.

Cooksonia, a rhyniophyte that is believed to have inhabited mud flats, has the distinction of being the oldest known vascular plant (see Figures 1–4, 16–2). Its slender, leafless aerial stems ranged up to about 6.5

centimeters long; their sporangia were globose. Tracheids have been identified in macerated bits of the axis. *Cooksonia* had become extinct by the early Devonian period, about 390 million years ago.

The best-known rhyniophyte is *Rhynia* (Figure 16–9a). Probably a marsh plant, its leafless, dichotomously branching aerial stems were attached to a rhizome (underground stem) with tufts of hairlike extensions. The aerial stems of *Rhynia*, which were about 20 centimeters long and 3 millimeters thick, were covered with a cuticle, bore stomata, and served as the photosynthetic organs.

The internal structure of *Rhynia* was similar to that of many of today's vascular plants. A single layer of superficial cells—the epidermis—surrounded the photosynthetic tissue of the cortex, and the center of the axis consisted of a solid strand of xylem surrounded by one or two layers of cells, which may or may not have been phloem cells. Apparently, the first xylem cells to mature occupied a central portion of the strand, and the last to mature were peripheral.

Division Zosterophyllophyta

The fossils of a second division of extinct seedless vascular plants—the Zosterophyllophyta—have been found in strata from the early to late Devonian period, from about 408 to 370 million years ago. Like the rhyniophytes, the zosterophyllophytes were leafless and dichotomously branched. It is possible that the group was aquatic. The aerial stems were covered with a cuticle, but only the upper ones contained stomata, indicating that the lower branches may have been embedded in mud. In *Zosterophyllum*, it has been suggested that the lower branches frequently produced lateral branches that forked into two axes, one that grew upward, the other downward (Figure 16–9b). The downward-growing branches may have functioned like a root, permitting the plant to spread outwardly from the center by providing support. The zosterophyllophytes are so named because of their general resemblance to the modern seagrass *Zostera*, marine angiosperms that superficially resemble grasses.

Unlike those of the rhyniophytes, the globose or kidney-bean-shaped sporangia of the zosterophyllophytes were borne laterally on short stalks. These plants were homosporous. The internal structure of the zosterophyllophytes was essentially similar to that of the rhyniophytes, except that in the Zosterophyllophytes the first xylem cells to mature were located around the periphery of the xylem strand and the last to mature were located in the center.

The zosterophyllophytes were almost certainly the ancestors of the lycophytes. The sporangia of early lycophytes were borne laterally, like those of the zosterophyllophytes, and the xylem of both divisions differentiated centripetally.

Division Trimerophyta

The division Trimerophyta probably evolved directly from the rhyniophytes; trimerophytes seem to represent the ancestral stock of the ferns, the progymnosperms, and perhaps the horsetails as well. The trimerophytes, which were larger and more complex plants than the rhyniophytes or zosterophyllophytes (Figure 16–9c), first appeared in the early Devonian period about 395 million years ago and had become extinct by the end of the mid-Devonian, about 20 million years later—a relatively short period of existence.

Although evolutionarily more advanced than the rhyniophytes, trimerophytes still lacked leaves. The main axis formed lateral branch systems that dichotomized several times. The trimerophytes, like the rhyniophytes and zosterophyllophytes, were homosporous. Some of their smaller branches terminated in elongate sporangia, while others were entirely vegetative. Besides their more complex branching pattern, the trimerophytes had a more massive vascular strand than the rhyniophytes. Together with a wide cortex composed of thick-walled cells, the large vascular strand probably was capable of supporting a fairly large plant, perhaps up to a meter in height. As in the rhyniophytes, the xylem of the trimerophytes differentiated centrifugally. The name of the division comes from the Greek words *tri*, *meros*, and *phyton*, meaning "three-parted plant," because of the three-parted branching of the secondary branches in the genus *Trimerophyton*.

Division Psilotophyta

The Psilotophyta include two genera, *Psilotum* and *Tmesipteris*, both living. *Psilotum* is tropical and subtropical in distribution. In the United States, it occurs in Florida, Louisiana, Arizona, Texas, Hawaii, as well as Puerto Rico, and it is a common greenhouse weed. *Tmesipteris* is restricted in distribution to Australia, New Caledonia, New Zealand, and other regions of the South Pacific. Both genera are very simple plants that resemble the rhyniophytes in their basic structure.

Psilotum is unique among living vascular plants in that it lacks both roots and leaves. The sporophyte consists of a dichotomously branching aerial portion with small scalelike outgrowths and a branching underground portion, or a system of rhizomes with many rhizoids. An endomycorrhizal zygomycete is present in the outer cortical cells of the rhizomes. *Psilotum* has a protostele (Figure 16–11).

Psilotum is homosporous, the spores being produced in sporangia borne on the ends of short, lateral branches. Upon germination, the spores give rise to bisexual gametophytes, which resemble portions of the rhizome (Figure 16–12). Like the rhizome, the subterranean gametophyte contains a symbiotic fungus. In addition, some gametophytes contain vascular tissue. The sperm of *Psilotum* are multiflagellated and require water to swim to the egg. Initially, the sporophyte is attached to the gametophyte by a foot, a structure that absorbs nutrients from the gametophyte. Eventually the sporophyte becomes detached from the foot, which remains embedded in the gametophyte. The life cycle of *Psilotum* is illustrated in Figure 16–14, pages 328–329.

Tmesipteris grows as an epiphyte on tree ferns and other plants (Figure 16–13) and in rock crevices. The leaflike appendages of *Tmesipteris* are larger than the scalelike outgrowths of *Psilotum*, but in other respects *Tmesipteris* is essentially similar to *Psilotum*.

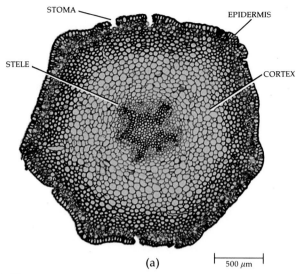

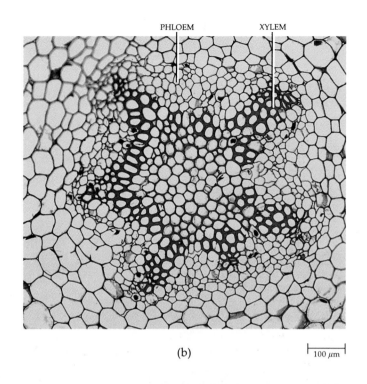

STOMA   EPIDERMIS

STELE   CORTEX

(a)

500 μm

PHLOEM   XYLEM

(b)

100 μm

16–11
Psilotum nudum. (a) *Transverse section of stem, showing mature tissues.* (b) *Detail of protostele, showing xylem and phloem.*

16–12
(a) *The subterranean gametophyte of* Psilotum nudum. *The gametophytes of psilotophytes are bisexual; that is, they bear both antheridia and archegonia.* (b) Psilotum nudum, *known locally as the moa plant, growing on a 1955 lava flow on the island of Hawaii. The yellow sporangia are clearly evident.*

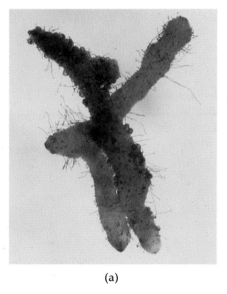

(a)

(b)

16–13
(a) Tmesipteris parva *growing on the trunk of the tree fern* Cyathea australis *in New South Wales, Australia.* (b) Tmesipteris lanceolata, *in New Caledonia, an island of the southwest Pacific.*

(a)

(b)

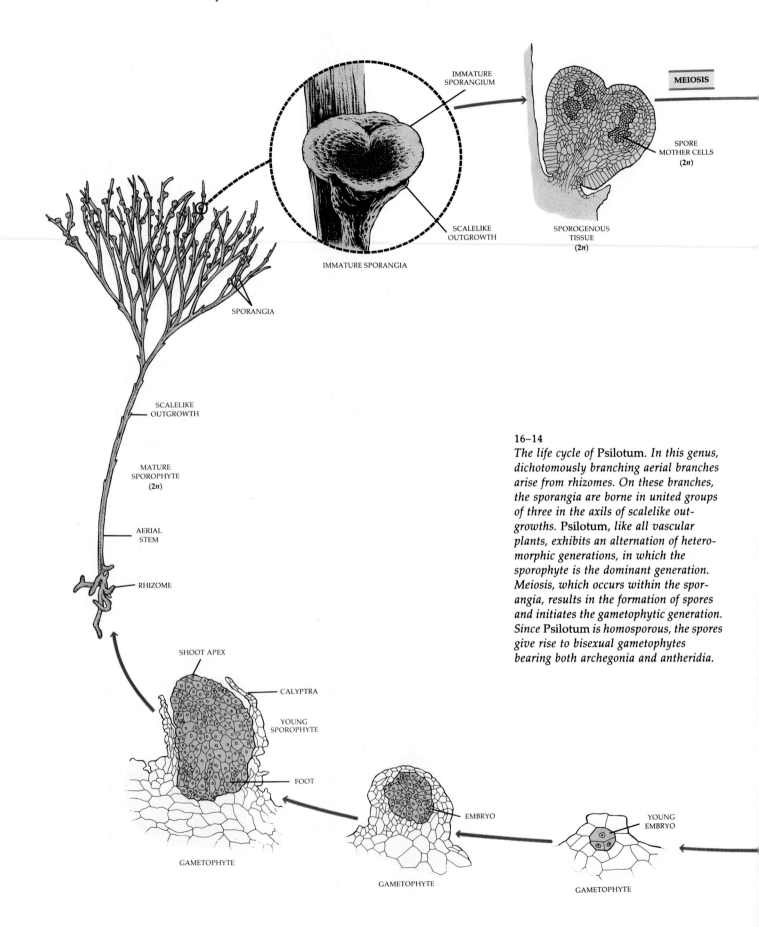

IMMATURE SPORANGIUM

SCALELIKE OUTGROWTH

IMMATURE SPORANGIA

MEIOSIS

SPORE MOTHER CELLS (2n)

SPOROGENOUS TISSUE (2n)

SPORANGIA

SCALELIKE OUTGROWTH

MATURE SPOROPHYTE (2n)

AERIAL STEM

RHIZOME

16–14

The life cycle of Psilotum. *In this genus, dichotomously branching aerial branches arise from rhizomes. On these branches, the sporangia are borne in united groups of three in the axils of scalelike outgrowths.* Psilotum, *like all vascular plants, exhibits an alternation of heteromorphic generations, in which the sporophyte is the dominant generation. Meiosis, which occurs within the sporangia, results in the formation of spores and initiates the gametophytic generation. Since* Psilotum *is homosporous, the spores give rise to bisexual gametophytes bearing both archegonia and antheridia.*

SHOOT APEX

CALYPTRA

YOUNG SPOROPHYTE

FOOT

GAMETOPHYTE

EMBRYO

GAMETOPHYTE

YOUNG EMBRYO

GAMETOPHYTE

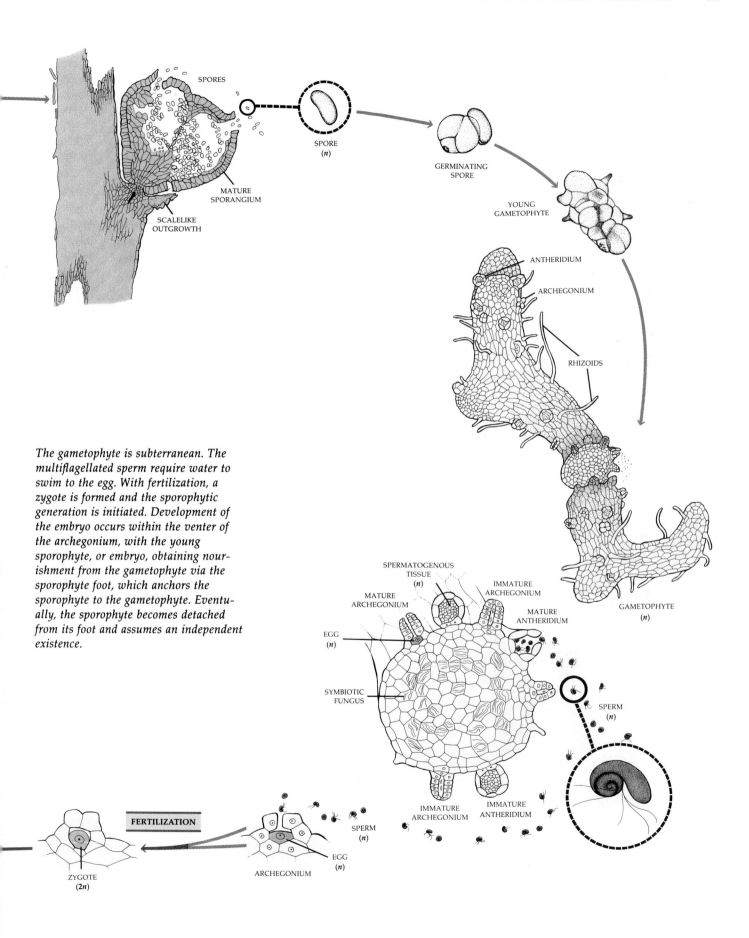

SPORES

SPORE
(*n*)

MATURE
SPORANGIUM

SCALELIKE
OUTGROWTH

GERMINATING
SPORE

YOUNG
GAMETOPHYTE

ANTHERIDIUM

ARCHEGONIUM

RHIZOIDS

GAMETOPHYTE
(*n*)

*The gametophyte is subterranean. The
multiflagellated sperm require water to
swim to the egg. With fertilization, a
zygote is formed and the sporophytic
generation is initiated. Development of
the embryo occurs within the venter of
the archegonium, with the young
sporophyte, or embryo, obtaining nourishment from the gametophyte via the
sporophyte foot, which anchors the
sporophyte to the gametophyte. Eventually, the sporophyte becomes detached
from its foot and assumes an independent
existence.*

SPERMATOGENOUS
TISSUE
(*n*)

IMMATURE
ARCHEGONIUM

MATURE
ARCHEGONIUM

MATURE
ANTHERIDIUM

EGG
(*n*)

SYMBIOTIC
FUNGUS

SPERM
(*n*)

IMMATURE
ARCHEGONIUM

IMMATURE
ANTHERIDIUM

FERTILIZATION

SPERM
(*n*)

EGG
(*n*)

ARCHEGONIUM

ZYGOTE
(2*n*)

Division Lycophyta

The 10 to 15 living genera and approximately 1000 living species of Lycophyta are the representatives of an evolutionary line that extends back to the Devonian period. The progenitors of the lycophytes were almost certainly early zosterophyllophytes (see Figure 16–9b). There are a number of orders of lycophytes, and at least three of the extinct ones included small to large trees. The three orders of living lycophytes, however, consist entirely of herbs; each order includes a single family. All lycophytes, living and fossil, possess microphylls, and this type of leaf is highly characteristic of the division. Tree lycophytes were among the dominant plants of the coal-forming forests of the Carboniferous period (see the essay "Coal Age Plants" on pages 346 and 347). Most lines of woody lycophytes became extinct before the end of the Paleozoic era, 248 million years ago.

LYCOPODIACEAE

Perhaps the most familiar living lycophytes are the club mosses, family Lycopodiaceae (see Figure 10–9c, page 182). All but two genera of living lycophytes belong to this family, most of the members of which were formerly grouped in the collective genus "Lycopodium." Seven of these genera are represented in the United States and Canada, but most of the estimated 400 species in the family are tropical. The boundaries of the predominantly tropical genera of this family are poorly understood, and as many as 15 genera may ultimately be recognized. Lycopodiaceae extend from Arctic regions to the tropics, but they rarely form conspicuous elements in any plant community. Most tropical species, many of which belong to the genus *Phlegmarius*, are epiphytes and thus rarely seen, but several of the temperate species form mats that may be evident on forest floors. Because they are evergreen, they are most noticeable in winter.

The sporophyte of most genera of Lycopodiaceae consists of a branching rhizome from which aerial branches and adventitious roots arise. Both stem and root are protostelic (Figure 16–15). The microphylls of Lycopodiaceae are usually spirally arranged, but they appear opposite or whorled in some members of the group. Lycopodiaceae are homosporous; the sporangia occur singly on the upper surface of fertile microphylls called **sporophylls**—modified leaves or leaflike organs

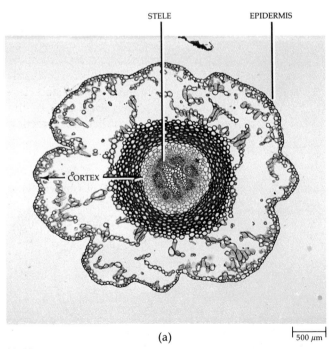

(a)   500 μm

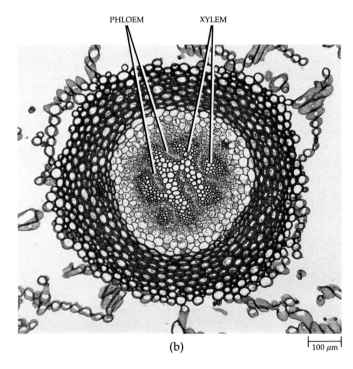

(b)   100 μm

16–15

(a) *Transverse section of* Diphasiastrum complanatum *stem, showing mature tissues.* (b) *Detail of protostele of* D. complanatum, *showing xylem and phloem. Both the stem and root of Lycopodiaceae show this protostelic organiza-* tion. *Note the large air spaces in the cortex. See also Figure 16–3.*

16–16
Huperzia lucidula *is a representative of those genera of Lycopodiaceae that lack strobili; the sporangia (small yellow structures along stem) are interspersed among sterile microphylls.*

that bear sporangia. In *Huperzia* (Figure 16–16) and *Phlegmarius*, the sporophylls are similar to ordinary microphylls and are interspersed among the sterile microphylls. In the other genera of Lycopodiaceae that occur in the United States and Canada, including *Diphasiastrum* and *Lycopodium* (Figure 10–9c), the nonphotosynthetic sporophylls are grouped into **strobili,** or cones, at the ends of the aerial branches.

Upon germination, the spores of Lycopodiaceae give rise to bisexual gametophytes that, depending on the genus, are either green, irregularly lobed masses (*Lycopodiella*, *Palhinhea*, and *Pseudolycopodiella* among the genera that occur in the United States and Canada) or subterranean, nonphotosynthetic, mycorrhizal structures (*Diphasiastrum*, *Huperzia*, *Lycopodium*, and *Phlegmarius* among the genera represented in the United States and Canada; see Figure 16–17). The development and maturation of archegonia and antheridia in a gametophyte of Lycopodiaceae may require

from 6 to 15 years, and these gametophytes may even produce a series of sporophytes in successive archegonia as they continue to grow.

Water is required for fertilization in Lycopodiaceae. The biflagellated sperm swim through water to the archegonium and then down its neck to the egg. Following fertilization, the zygote develops into an embryo, which grows within the venter of the archegonium. The young sporophyte may remain attached to the gametophyte for a long time, but eventually it becomes independent. The life cycle of *Lycopodium clavatum*, representative of those Lycopodiaceae that have a subterranean, mycorrhizal gametophyte and form a strobilus, is illustrated in Figure 16–17 on pages 332–333.

Among the genera of Lycopodiaceae that occur in the United States and Canada, *Huperzia* (Figure 16–16), the fir mosses, consists of seven species; *Lycopodium* (Figure 16–17), the tree club mosses, of five; *Diphasiastrum*, the club mosses and running pines, of 11; and *Lycopodiella* of six. These genera, and the others now recognized in Lycopodiaceae, differ from one another in many fundamentally important characteristics, including the arrangement of the sporophylls, the presence of rhizomes and organization of the vegetative body, the nature of the gametophyte, and basic chromosome numbers. They clearly should no longer be grouped together in a single genus, *Lycopodium*, an arrangement that conceals the great differences among them.

SELAGINELLACEAE

Among the living genera of lycophytes, *Selaginella*, the only genus of the family Selaginellaceae, has the most species, about 700. Most of them are tropical in distribution. Many grow in moist places; a few inhabit deserts, becoming dormant during the driest part of the year. Among the latter is the so-called resurrection plant, *S. lepidophylla*, which occurs in Texas, New Mexico, and Mexico (Figure 16–18a on page 334).

Basically, the herbaceous sporophyte of *Selaginella* is similar to that of some Lycopodiaceae; it bears microphylls, and its sporophylls are arranged in strobili (Figure 16–18b). Unlike Lycopodiaceae, however, *Selaginella* has a small, scalelike outgrowth, called a ligule, near the base of the upper surface of each microphyll and sporophyll. The stem and root are protostelic (Figure 16–19 on page 334).

Whereas Lycopodiaceae are homosporous, *Selaginella* is heterosporous, with unisexual gametophytes. Each sporophyll bears a single sporangium on its upper surface. Megasporangia are borne by **megasporophylls,** and microsporangia are borne by **microsporophylls.** Both kinds of sporangia occur in the same strobilus.

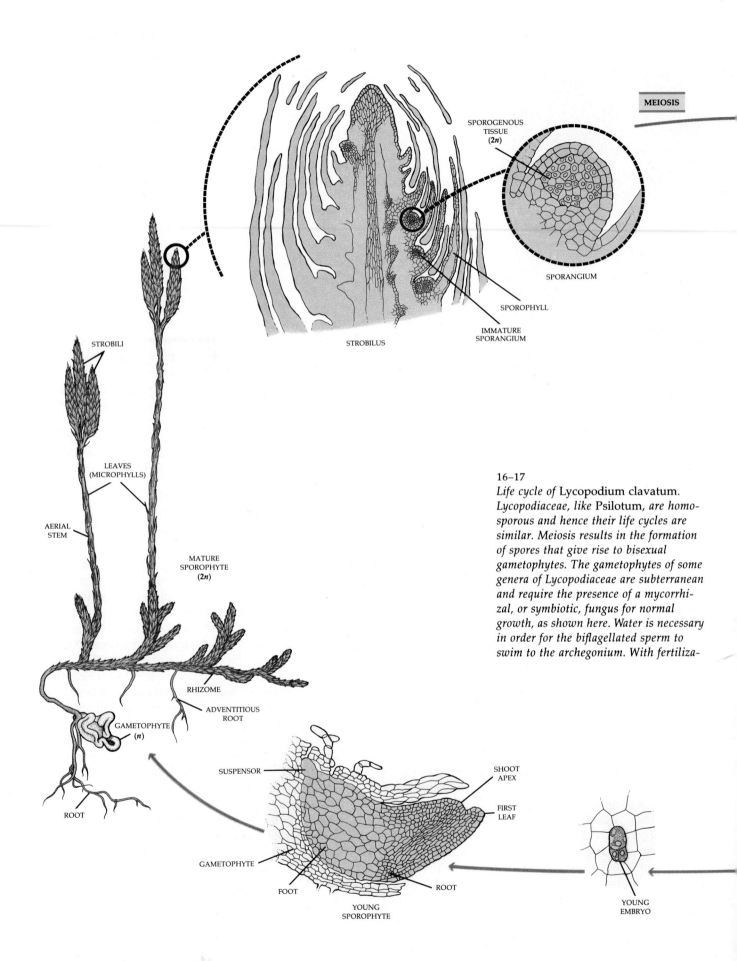

MEIOSIS

SPOROGENOUS
TISSUE
(2n)

SPORANGIUM

SPOROPHYLL

IMMATURE
SPORANGIUM

STROBILUS

STROBILI

LEAVES
(MICROPHYLLS)

AERIAL
STEM

MATURE
SPOROPHYTE
(2n)

RHIZOME

ADVENTITIOUS
ROOT

GAMETOPHYTE
(n)

ROOT

SUSPENSOR

SHOOT
APEX

FIRST
LEAF

GAMETOPHYTE

FOOT

ROOT

YOUNG
SPOROPHYTE

YOUNG
EMBRYO

16–17
Life cycle of Lycopodium clavatum.
Lycopodiaceae, like Psilotum, *are homosporous and hence their life cycles are
similar. Meiosis results in the formation
of spores that give rise to bisexual
gametophytes. The gametophytes of some
genera of Lycopodiaceae are subterranean
and require the presence of a mycorrhizal, or symbiotic, fungus for normal
growth, as shown here. Water is necessary
in order for the biflagellated sperm to
swim to the archegonium. With fertiliza-*

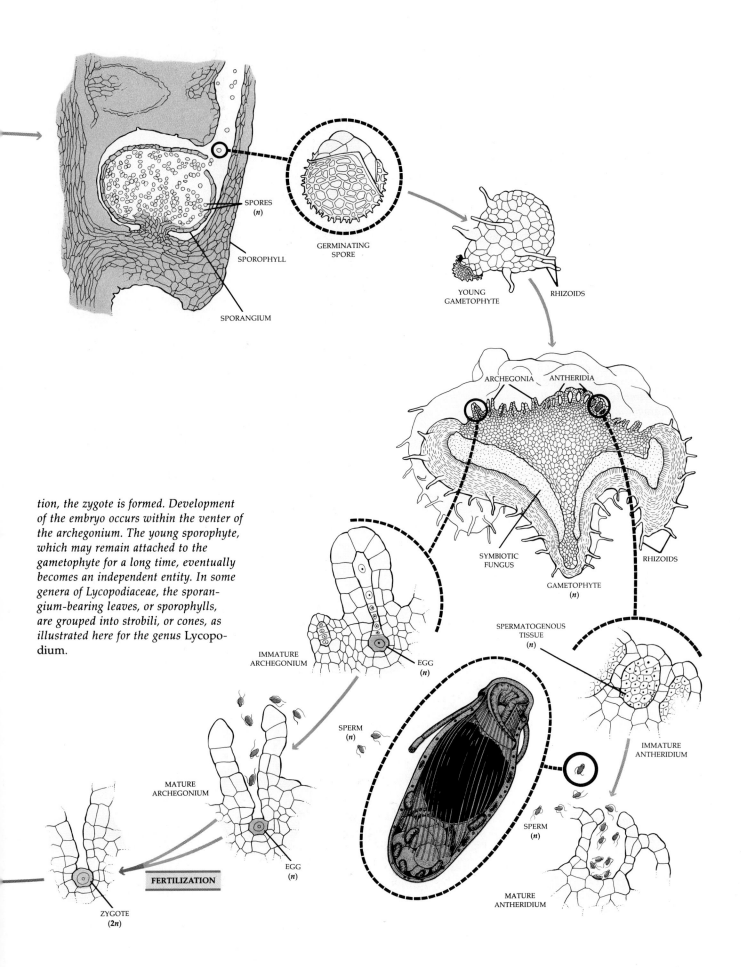

tion, the zygote is formed. Development of the embryo occurs within the venter of the archegonium. The young sporophyte, which may remain attached to the gametophyte for a long time, eventually becomes an independent entity. In some genera of Lycopodiaceae, the sporangium-bearing leaves, or sporophylls, are grouped into strobili, or cones, as illustrated here for the genus Lycopodium.

SPORES (*n*)

SPOROPHYLL

SPORANGIUM

GERMINATING SPORE

YOUNG GAMETOPHYTE

RHIZOIDS

ARCHEGONIA   ANTHERIDIA

SYMBIOTIC FUNGUS

RHIZOIDS

GAMETOPHYTE (*n*)

SPERMATOGENOUS TISSUE (*n*)

IMMATURE ARCHEGONIUM

EGG (*n*)

IMMATURE ANTHERIDIUM

SPERM (*n*)

MATURE ARCHEGONIUM

EGG (*n*)

SPERM (*n*)

FERTILIZATION

MATURE ANTHERIDIUM

ZYGOTE (**2***n*)

(a)

(b)

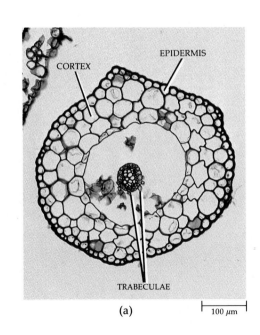

(c)

16–18
(a) Selaginella lepidophylla, *the resurrection plant, a plant that becomes completely dried out when water is not available but quickly revives following a rain. This plant was growing in Big Bend National Park, in Texas.* (b) Selaginella rupestris *with strobili.* (c) Selaginella willdenovii, *from the Old World tropics. Shade-loving, it climbs to 7 meters and has peacock-blue leaves with a metallic sheen. Note the clearly evident rhizomes.*

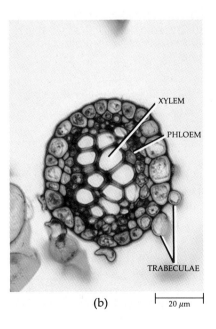

16–19
Selaginella. (a) *Transverse section of stem, showing mature tissues. The protostele is suspended in the middle of the hollow stem by elongate cortical cells (endodermal cells), called trabeculae. Only a portion of each trabecula can be seen here.* (b) *Detail of protostele.*

CORTEX   EPIDERMIS

TRABECULAE

(a)   100 μm

XYLEM

PHLOEM

TRABECULAE

(b)   20 μm

The male gametophytes (microgametophytes) in *Selaginella* develop within the microspore; they lack chlorophyll. At maturity the male gametophyte consists of a single prothallial, or vegetative, cell and an antheridium, which gives rise to many biflagellated sperm. The microspore wall must rupture in order for the sperm to be liberated.

During development of the female gametophyte (megagametophyte), the megaspore wall ruptures, and the gametophyte protrudes through the rupture to the outside. This is the portion of the female gametophyte in which the archegonia develop. It has been reported that the female gametophytes sometimes develop chloroplasts, although most *Selaginella* gametophytes derive their nutrition from food stored within the megaspores.

Water is required in order for the sperm to swim to the archegonia and fertilize the eggs. Commonly fertilization occurs after the gametophytes have been shed from the strobilus. During development of the embryos in both Lycopodiaceae and *Selaginella*, a structure called a **suspensor** is formed. Although inactive in Lycopodiaceae and some species of *Selaginella*, in other species of *Selaginella* the suspensor serves to thrust the developing embryo deep within the nutrient-rich tissue of the female gametophyte. Gradually, the developing sporophyte emerges from the gametophyte and becomes independent.

The life cycle of *Selaginella* is illustrated in Figure 16–22 on pages 336–337.

ISOETACEAE

The only member of the family Isoetaceae is *Isoetes*, or quillwort. Plants of *Isoetes* may be aquatic, or they may grow in pools that become dry at certain seasons. The sporophyte of *Isoetes* consists of a short, fleshy underground stem (corm) bearing quill-like microphylls on its upper surface and roots on its lower surface (Figure 16–20). In *Isoetes*, each leaf is a potential sporophyll.

Like *Selaginella*, *Isoetes* is heterosporous. The megasporangia are borne at the base of megasporophylls, and the microsporangia are borne at the base of microsporophylls, similar to the megasporophylls but located nearer the center of the plant (Figure 16–21). A ligule is found just above the sporangium of each sporophyll.

One of the distinctive features of *Isoetes* is the presence of a specialized cambium that adds secondary tissues to the corm. Externally the cambium produces only parenchyma tissue, whereas internally it produces a peculiar vascular tissue consisting of sieve elements, parenchyma cells, and tracheids in varying proportions.

16–20
Isoetes storkii. *View of the sporophyte showing quill-like leaves, stem, and roots.*

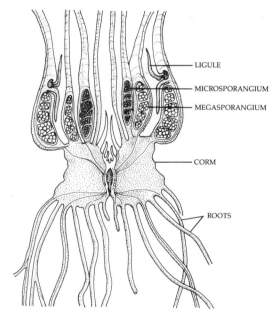

16–21
A diagram of a vertical section of an Isoetes *plant. Leaves are borne on the upper surface, and roots on the lower surface, of a short, fleshy underground stem. Some leaves (megasporophylls) bear megasporangia, and others (microsporophylls) bear microsporangia. The microsporophylls are located nearer the center of the plant.*

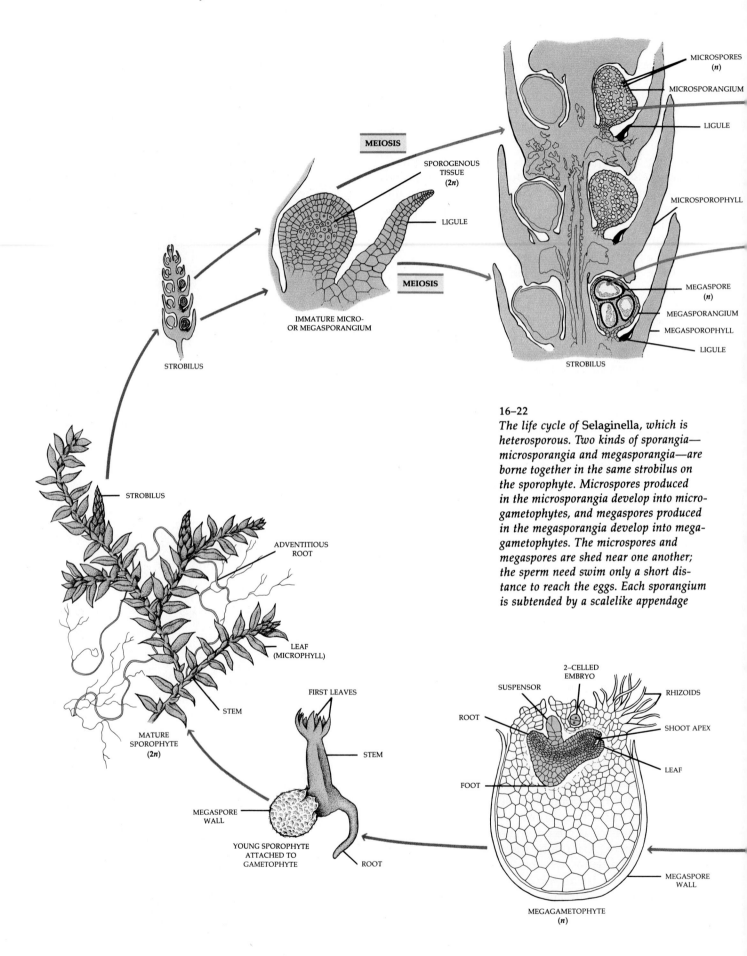

MICROSPORES (n)

MICROSPORANGIUM

LIGULE

SPOROGENOUS TISSUE (2n)

LIGULE

MEIOSIS

MICROSPOROPHYLL

MEIOSIS

MEGASPORE (n)

MEGASPORANGIUM

MEGASPOROPHYLL

LIGULE

IMMATURE MICRO-
OR MEGASPORANGIUM

STROBILUS

STROBILUS

16–22
The life cycle of Selaginella, *which is heterosporous. Two kinds of sporangia— microsporangia and megasporangia—are borne together in the same strobilus on the sporophyte. Microspores produced in the microsporangia develop into micro- gametophytes, and megaspores produced in the megasporangia develop into mega- gametophytes. The microspores and megaspores are shed near one another; the sperm need swim only a short dis- tance to reach the eggs. Each sporangium is subtended by a scalelike appendage*

STROBILUS

ADVENTITIOUS ROOT

LEAF (MICROPHYLL)

STEM

MATURE SPOROPHYTE (2n)

FIRST LEAVES

STEM

MEGASPORE WALL

YOUNG SPOROPHYTE ATTACHED TO GAMETOPHYTE

ROOT

2–CELLED EMBRYO

SUSPENSOR

RHIZOIDS

ROOT

SHOOT APEX

LEAF

FOOT

MEGASPORE WALL

MEGAGAMETOPHYTE (n)

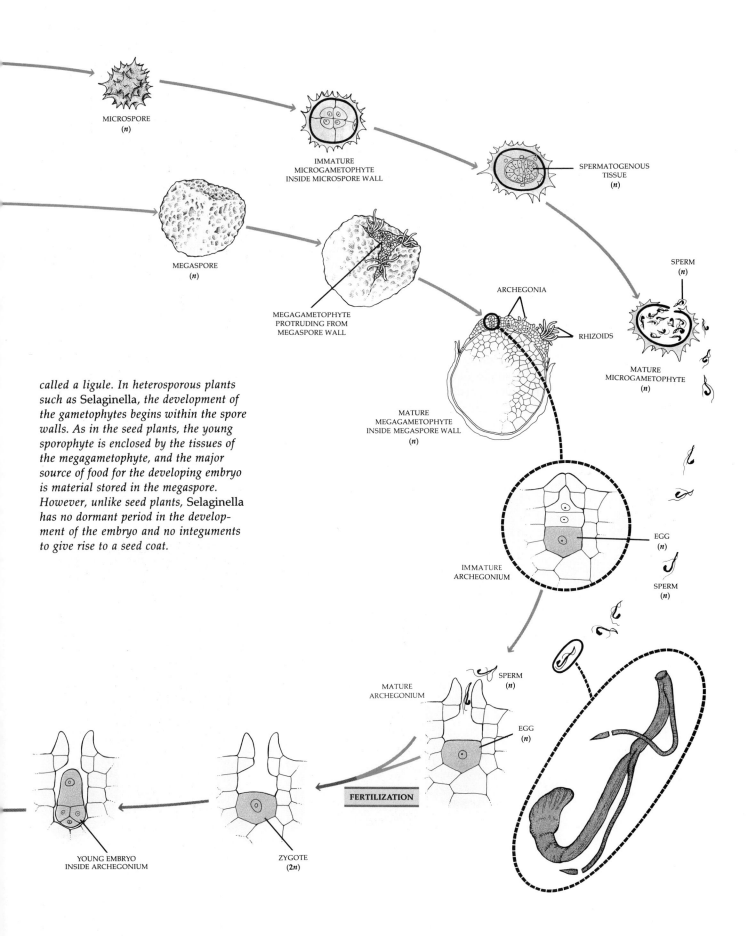

MICROSPORE
(*n*)

IMMATURE
MICROGAMETOPHYTE
INSIDE MICROSPORE WALL

SPERMATOGENOUS
TISSUE
(*n*)

MEGASPORE
(*n*)

MEGAGAMETOPHYTE
PROTRUDING FROM
MEGASPORE WALL

ARCHEGONIA

SPERM
(*n*)

RHIZOIDS

MATURE
MICROGAMETOPHYTE
(*n*)

MATURE
MEGAGAMETOPHYTE
INSIDE MEGASPORE WALL
(*n*)

*called a ligule. In heterosporous plants
such as* Selaginella, *the development of
the gametophytes begins within the spore
walls. As in the seed plants, the young
sporophyte is enclosed by the tissues of
the megagametophyte, and the major
source of food for the developing embryo
is material stored in the megaspore.
However, unlike seed plants,* Selaginella
*has no dormant period in the develop-
ment of the embryo and no integuments
to give rise to a seed coat.*

IMMATURE
ARCHEGONIUM

EGG
(*n*)

SPERM
(*n*)

MATURE
ARCHEGONIUM

SPERM
(*n*)

EGG
(*n*)

FERTILIZATION

YOUNG EMBRYO
INSIDE ARCHEGONIUM

ZYGOTE
(2*n*)

Some species of *Isoetes* from high elevations in the tropics have the unique characteristic of obtaining their carbon for photosynthesis from the sediment in which they grow rather than from the atmosphere. The leaves of these plants lack stomata, have a thick cuticle, and carry on essentially no gas exchange with the atmosphere. Like at least some of the other species of *Isoetes* in which the plants dry out for part of the year, these species have CAM photosynthesis (page 116).

Division Sphenophyta

Like the Lycophyta, the Sphenophyta extend back to the Devonian period. The sphenophytes reached their maximum abundance and diversity later in the Paleozoic era, about 300 million years ago. During the late Devonian and Carboniferous periods, they were represented by the calamites (see page 346), a group of trees that reached 18 meters or more in height, with a trunk that could be more than 45 centimeters thick. Today the Sphenophyta are represented by a single herbaceous genus, *Equisetum* (Figure 16–23), which consists of 15 species. Since *Equisetum* is essentially identical to *Equisetites*, a plant that appeared about 300 million years ago, in the Devonian period, *Equisetum* may be the oldest surviving genus of plants on earth.

The species of *Equisetum* are known as the horsetails; they are widespread in moist or damp places, by streams, and along the edge of woods. Horsetails are easily recognized because of their conspicuously jointed stems and rough texture. The small, scalelike leaves, which have a simple structure but are probably reduced megaphylls, are whorled at the nodes. When present, the branches arise laterally at the nodes and alternate with the leaves. The internodes (the portions of the stems between successive nodes) are ribbed, and the ribs are tough and strengthened with siliceous deposits—technically, opals—in the epidermal cells. Horsetails have been used to scour pots and pans, particularly in colonial and frontier times, and have thus earned the name "scouring rushes." The roots are adventitious, arising at the nodes of the rhizomes.

The aerial stems of *Equisetum* arise from branching underground rhizomes, and, although the plants may die back during unfavorable seasons, the rhizomes are perennial. The aerial stem is complex anatomically (Figure 16–24). At maturity, its internodes contain a hollow pith surrounded by a ring of smaller canals called carinal canals. Each of these smaller canals is associated with a strand of primary xylem and primary phloem.

Equisetum is homosporous. Sporangia are borne in groups of five to 10 along the margins of small umbrella-like structures known as **sporangiophores** (sporangia-bearing branches), which are clustered into strobili at the apex of the stem (see Figures 16–23a and

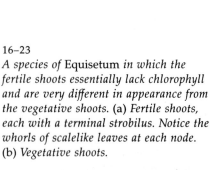

16–23

A species of Equisetum *in which the fertile shoots essentially lack chlorophyll and are very different in appearance from the vegetative shoots. (a) Fertile shoots, each with a terminal strobilus. Notice the whorls of scalelike leaves at each node. (b) Vegetative shoots.*

(a)

(b)

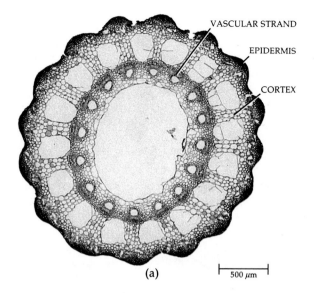

VASCULAR STRAND

EPIDERMIS

CORTEX

(a)   500 μm

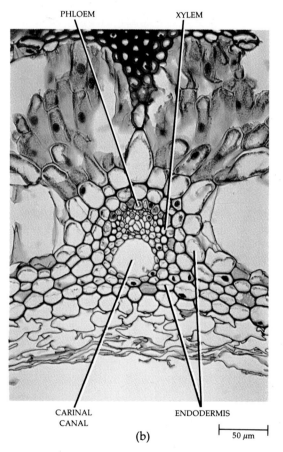

PHLOEM   XYLEM

CARINAL   ENDODERMIS
CANAL

(b)   50 μm

16–24
Stem anatomy of Equisetum. *(a)*
Transverse section of stem, showing
mature tissues. (b) Detail of vascular
strand, showing xylem and phloem.

16–26). The fertile stems of some species do not contain much chlorophyll. In these species, the fertile stems are sharply distinct from the vegetative stems, often appearing before the latter early in the spring (see Figure 16–23). In other species of *Equisetum,* the strobili are borne at the tips of otherwise vegetative stems (see Figure 10–9d, page 182). When the spores are mature, the sporangia contract and split along their inner surface, releasing numerous spores. Elaters, which arise from the outer layer of the spore wall, coil when moist and uncoil when dry, thus playing a role in spore dispersal (Figures 16–1 and 16–26).

The gametophytes of *Equisetum* are green and free-living, most being about the size of a pinhead; they become established mainly on mud that has recently been flooded and is rich in nutrients. The gametophytes (Figure 16–25), which reach sexual maturity in three to five weeks, are either bisexual or male. In bisexual gametophytes, the archegonia develop before the antheridia; this developmental pattern increases the probability of cross-fertilization. The sperm are multiflagellated and require water to swim to the eggs. The eggs of several archegonia on a single gametophyte may be fertilized and develop into embryos, or young sporophytes.

The life cycle of *Equisetum* is illustrated in Figure 16–26 on pages 340–341.

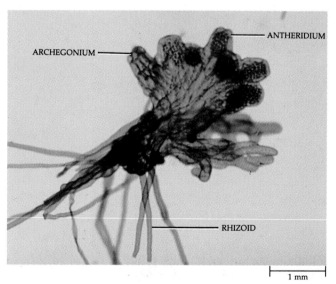

ARCHEGONIUM

ANTHERIDIUM

RHIZOID

1 mm

16–25
Equisetum. *Bisexual gametophyte*
showing male and female gametangia.
Compare this gametophyte with that
shown in Figure 15–2, a 400-million-year-
old plant fossil from Scotland.

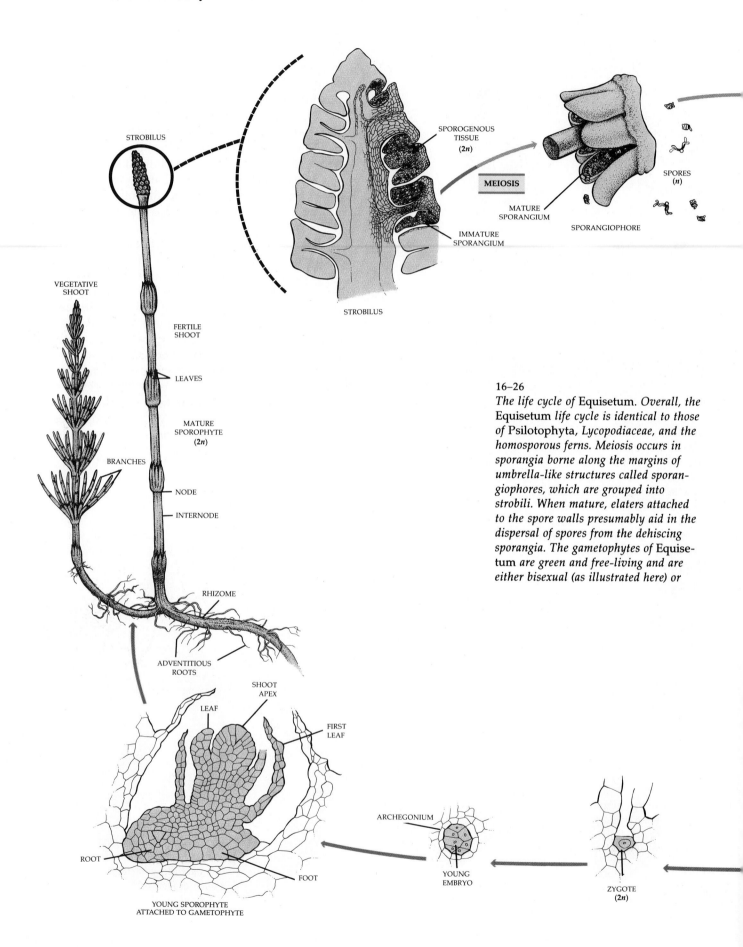

STROBILUS

SPOROGENOUS
TISSUE
(2n)

MEIOSIS

MATURE
SPORANGIUM

IMMATURE
SPORANGIUM

SPORES
(n)

SPORANGIOPHORE

STROBILUS

VEGETATIVE
SHOOT

FERTILE
SHOOT

LEAVES

MATURE
SPOROPHYTE
(2n)

BRANCHES

NODE

INTERNODE

RHIZOME

ADVENTITIOUS
ROOTS

SHOOT
APEX

LEAF

FIRST
LEAF

ARCHEGONIUM

ROOT

FOOT

YOUNG
EMBRYO

ZYGOTE
(2n)

YOUNG SPOROPHYTE
ATTACHED TO GAMETOPHYTE

16–26
The life cycle of Equisetum. *Overall, the*
Equisetum *life cycle is identical to those
of* Psilotophyta, Lycopodiaceae, *and the
homosporous ferns. Meiosis occurs in
sporangia borne along the margins of
umbrella-like structures called sporangiophores, which are grouped into
strobili. When mature, elaters attached
to the spore walls presumably aid in the
dispersal of spores from the dehiscing
sporangia. The gametophytes of* Equisetum *are green and free-living and are
either bisexual (as illustrated here) or*

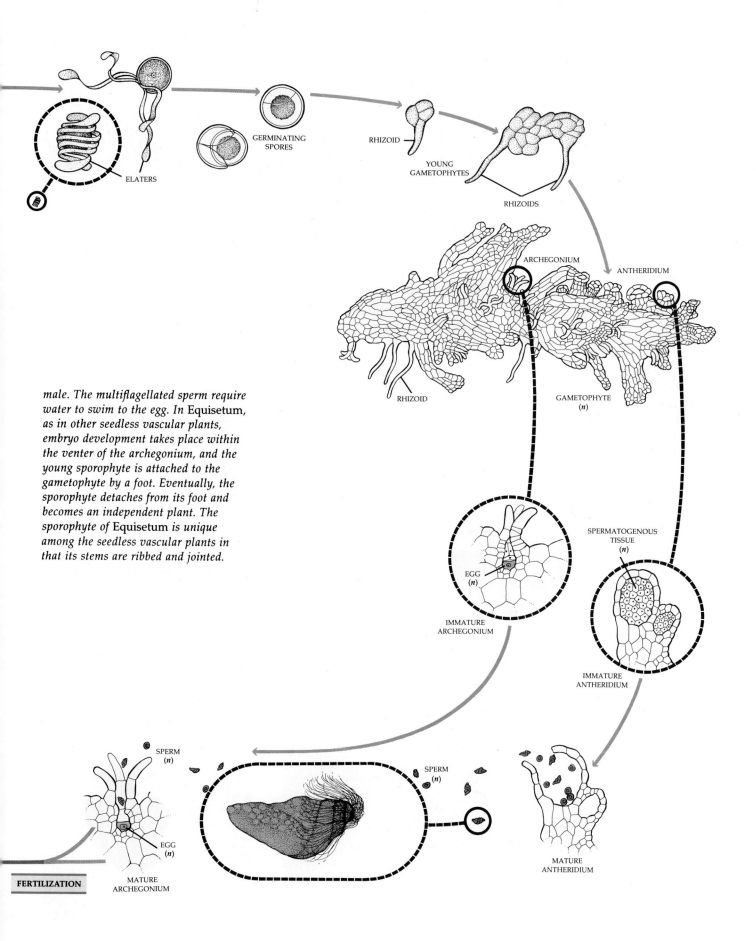

ELATERS

GERMINATING
SPORES

RHIZOID

YOUNG
GAMETOPHYTES

RHIZOIDS

ARCHEGONIUM

ANTHERIDIUM

RHIZOID

GAMETOPHYTE
(*n*)

male. The multiflagellated sperm require water to swim to the egg. In Equisetum, as in other seedless vascular plants, embryo development takes place within the venter of the archegonium, and the young sporophyte is attached to the gametophyte by a foot. Eventually, the sporophyte detaches from its foot and becomes an independent plant. The sporophyte of Equisetum is unique among the seedless vascular plants in that its stems are ribbed and jointed.

SPERMATOGENOUS
TISSUE
(*n*)

EGG
(*n*)

IMMATURE
ARCHEGONIUM

IMMATURE
ANTHERIDIUM

SPERM
(*n*)

SPERM
(*n*)

EGG
(*n*)

MATURE
ANTHERIDIUM

FERTILIZATION

MATURE
ARCHEGONIUM

(a)

(b)

16–27

The diversity of ferns, as illustrated by a few genera of the largest order of ferns, Filicales. (a) *A tree fern,* Cyathea, *at Monteverde, Costa Rica.* (b) Lindsaea, *Volcán Barba, Costa Rica.* (c) *Plagio-gyria, with distinct fertile and vegetative leaves, Volcán Poás, Costa Rica.* (d) Elaphoglossum, *with thick, undivided leaves, near Cuzco, Peru.* (e) Asplenium septentrionale, *a small fern that occurs all around the Northern Hemisphere, growing on metal-rich soil near a lead-silver mine in Wales.* (f) Polypodium polypodioides, *growing as an epiphyte on a juniper trunk in Florida.*

(c)

(d)

(e)

(f)

Division Pterophyta

Ferns have been relatively abundant in the fossil record from the Carboniferous period (see "Coal Age Plants," pages 346 and 347, and Figure 17–1) to the present. Today, ferns number about 11,000 species; they are the largest group of plants other than the flowering plants, and the most diverse (Figure 16–27 on the facing page).

The diversity of ferns is greatest in the tropics, where about three-fourths of the species are found. Here, not only are there many species of ferns, but ferns are abundant in many plant communities. Only about 380 species of ferns occur in the United States and Canada, whereas about 1000 occur in the small tropical country of Costa Rica in Central America; Costa Rica is about the size of West Virginia. Approximately a third of all species of tropical ferns grow upon the trunks or branches of trees as epiphytes (Figure 16–27f).

In both form and habitat, ferns exhibit great diversity (Figure 16–27). Some ferns are very small and have undivided leaves. *Lygodium*, a climbing fern, has leaves with a long, twining rachis (an extension of the leaf stalk, or petiole) that may be up to 30 meters or more in length. Some tree ferns (Figure 16–27a), such as those of the genus *Cyathea*, have been recorded to reach heights of more than 24 meters and to have leaves 5 meters or more in length. Although the trunks of such tree ferns may be 30 centimeters or more thick, their tissues are entirely primary in origin. Most of this thickness is the fibrous root mantle; the true stem is only four to six centimeters in diameter. The herbaceous genus *Botrychium* (see Figure 16–29a) is the only living fern known to form a vascular cambium.

In terms of the structure and method of development of their sporangia, ferns may be classified as either **eusporangiate** or **leptosporangiate** (Figure 16–28). The distinction between these two types of sporangia is of fundamental importance for understanding relationships among vascular plants. In a **eusporangium,** the parent cells, or initials, are located at the surface of the tissue from which the sporangium is produced. These initials divide by the formation of walls parallel to the surface, resulting in the formation of an inner and an outer series of cells. The outer cell layer, by further divisions in both planes—parallel to (periclinal divisions) and perpendicular to (anticlinal di-

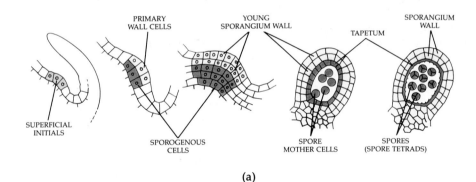

(a)

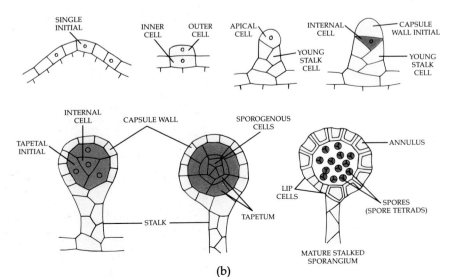

16–28

Development and structure of the two principal types of fern sporangia. (a) The eusporangium originates from a series of superficial parent cells, or initials. It develops a wall two or more layers thick (although at maturity the inner wall layers may be crushed) and a high number of spores. (b) The leptosporangium originates from a single initial cell, which first produces a stalk and then a capsule. Leptosporangia give rise to a relatively small number of spores.

(b)

visions) the surface—builds up the several-layered wall of the sporangium. The inner layer gives rise to a mass of irregularly oriented cells from which the spore mother cells ultimately arise. In many eusporangia, the inner wall layers are stretched and compressed during the course of development, so that the walls may apparently consist of a single layer of cells at maturity. Eusporangia, which are larger than leptosporangia and contain many more spores, are characteristic of all of the divisions of vascular plants we have considered to this point and of the more primitive ferns, including the members of the order Ophioglossales, which we shall discuss further below.

In contrast to the multicellular origin of eusporangia, **leptosporangia** arise from a single superficial initial cell, which divides transversely or obliquely (Figures 16–28 and 16–35). The inner of the two cells produced by this division may either contribute cells that produce a large part of the sporangial stalk or remain inactive and play no role in the further development of the sporangium, which is the more common condition. By a precise pattern of divisions, the outer cell ultimately gives rise to an elaborate, stalked sporangium, with a globose capsule that has a wall that is one cell thick. Within this wall is a nutritive structure two cell layers thick called the **tapetum.** The inner mass of the leptosporangium eventually differentiates into spore mother cells, which undergo meiosis to produce four spores each. After it nourishes the young dividing cells within the sporangium, the tapetum is deposited around the spores, creating ridges, spines, and other types of surface features that are often characteristic for individual families and genera. The spores are exposed following

the development of a crack in the so-called **lip cells** of the sporangium. The sporangia are stalked, and each contains a special layer of unevenly thick-walled cells called an **annulus.** As the sporangium dries out, contraction of the annulus causes tearing in the middle of the capsule. The sudden explosion and snapping back of the annulus to its original position then result in a catapult-like discharge of the spores. In eusporangia, the stalks are more massive and there is no annulus.

Most living ferns are homosporous; among them, heterospory is restricted to two orders of water ferns (see Figure 16–36), which will be discussed further below. A number of extinct ferns were heterosporous.

We shall now consider as examples three very different kinds of ferns: (1) the orders Ophioglossales and Marattiales, as examples of eusporangiate ferns; (2) the Filicales, or leptosporangiate ferns; and (3) the water ferns, orders Marsileales and Salviniales, the heterosporous ferns.

EUSPORANGIATE FERNS: ORDERS OPHIOGLOSSALES AND MARATTIALES

Of the three genera of the order Ophioglossales, *Botrychium,* the grape ferns (Figure 16–29a), and *Ophioglossum,* the adder's tongues (Figure 16–29b), are widespread in the North Temperate region. In both of these genera, a single leaf is typically produced each year from the rhizome. Each leaf consists of two parts: (1) a vegetative portion, or blade, which is deeply dissected in *Botrychium* and undivided in most species of

16–29
Representatives of the two genera of Ophiglossales that occur in North America. (a) Botrychium parallelum. *In the genus* Botrychium, *the lower, vegetative portion of the leaf is divided.* (b) *A species of* Ophioglossum. *In* Ophioglossum, *the lower portion of the leaf is undivided. In both genera, the erect, fertile, upper part of the leaf is sharply distinct.*

(a)

(b)

Ophioglossum and (2) a fertile segment. In *Botrychium*, the fertile segment is dissected in the same way as the vegetative portion and bears two rows of eusporangia on the outermost segments. In *Ophioglossum*, the fertile portion is undivided and bears two rows of eusporangia.

The gametophytes of *Botrychium* and *Ophioglossum* are subterranean, tuberous, elongate structures with numerous rhizoids; they have endophytic fungi and resemble the gametophytes of Psilotophyta. In *Botrychium*, the gametophytes usually possess a dorsal ridge in which the antheridia are embedded, with the archegonia generally located along the sides of the ridge. In the nature of their gametophytes, the structure of their leaves, and several other anatomical details, the Ophioglossales are sharply distinct from other living ferns. One member of this group, *Ophioglossum reticulatum*, has the highest chromosome number known in any living organism, with a diploid complement of about 1260 chromosomes.

The only other order of ferns that has eusporangia, the tropical Marattiales, is an ancient group with a fossil record that extends back to the Carboniferous period. The members of this order resemble more familiar groups of ferns more closely than they do Ophioglossales. *Psaronius*, the extinct tree fern illustrated in Figure 17–1 (page 357), was a member of this order. The six living genera of Marattiales include about 200 species.

FILICALES

Nearly all familiar ferns are members of the larger order Filicales, with at least 10,500 species. About 35 families and 320 genera are recognized in the order. Filicales differ from Ophioglossales and Marattiales in being leptosporangiate, and from the water ferns, which we shall discuss next, in being homosporous.

All ferns other than Ophioglossales and Marattiales, in fact, are leptosporangiate, and very few have the subterranean gametophytes with endophytic fungi that are characteristic of the Ophioglossales and Marattiales. Clearly, leptosporangia and the other distinctive features of most ferns are specialized characteristics, since most of the features that distinguish Ophioglossales and Marattiales are shared with other groups of primitive plants. We may therefore presume that Ophioglossales and Marattiales descended from an evolutionary branch that arose early in the history of ferns and that these two orders illustrate some of the features of the common ancestor of this group.

Most garden and woodland ferns of temperate regions have siphonostelic rhizomes (Figure 16–30). The rhizomes produce new sets of leaves each year. The fern embryo produces a true root, but this soon withers, and the rest of the roots are adventitious, arising from the rhizomes near the bases of the leaves. The leaves, or **fronds,** are megaphylls and represent the most conspicuous part of the sporophyte. Their high surface-to-

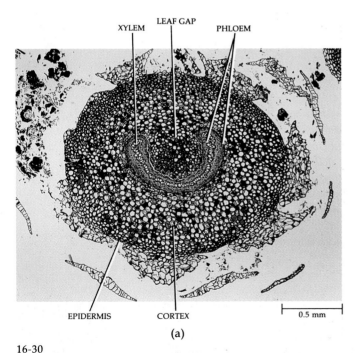

(a)

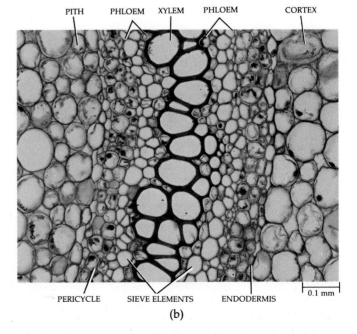

(b)

16-30

The anatomy of fern rhizomes. (a) Adiantum, or maidenhair fern. Transverse section of a rhizome, showing the siphono- *stele. Note the wide leaf gap. (b) Transverse section of part of the vascular region of a rhizome of the tree fern Dicksonia.* *The phloem is composed mainly of sieve elements; the xylem is composed entirely of tracheids.*

Coal Age Plants

The amount of carbon dioxide used in photosynthesis is about 100 billion metric tons annually, about a tenth of the total carbon dioxide present in the atmosphere. The amount of carbon dioxide returned as a result of oxidation of these living materials is about the same, differing only by 1 part in 10,000. This very slight imbalance is caused by the burying of organisms in sediment or mud under conditions in which oxygen is excluded and decay is only partial. This accumulation of partially decayed plant material is known as peat (page 315). The peat may eventually become covered with sedimentary rock and is thus placed under pressure. Depending on time, temperature, and other factors, peat may become compressed into soft or hard coal, one of the so-called fossil fuels.

During certain periods in the earth's history, the rate of fossil-fuel formation was greater than at other times. One such time was the Carboniferous period, which extended from 360 to 286 million years ago (see Figures 16–10 and 17–1). The lands were low, covered by shallow seas or swamps, and, in what are now temperate regions of Europe and North America, conditions were favorable for year-round growth. These regions were tropical to subtropical, with the equator then arcing across the Appalachians, over northern Europe, and through the Ukraine. Five groups of plants dominated the swamplands, and three of them were seedless vascular plants—lycophytes, sphenophytes (calamites), and ferns. The other two were gymnospermous types of seed plants—seed ferns (Pteridospermales) and cordaites (Cordaitales).

(a) *One of the dominant trees of the late Carboniferous period was the lycophyte* Lepidodendron, *some of which grew to heights of 40 meters or more.*

Lycophyte Trees

Two-thirds of the "Age of Coal" in the late Carboniferous period (Pennsylvanian) was dominated by lycophyte trees, most of which grew to heights of 10 to 35 meters and were sparsely branched (a). After the plant attained over half of its total height, the trunk branched dichotomously. Successive branching produced progressively smaller branches until, finally, the tissues of the branch tips lost their ability to grow further. The branches bore long microphylls. The tree lycophytes were largely supported by a massive periderm surrounding a relatively small amount of xylem.

Like *Selaginella* and *Isoetes*, lycophyte trees were heterosporous, and their sporophylls were aggregated into cones. Some of these trees produced seedlike structures.

As the swamplands began to dry up and the climate in Euramerica began to change toward the end of the Carboniferous period, the lycophyte trees vanished almost overnight, geologically speaking. Herbaceous lycophytes essentially similar to *Lycopodium* and *Selaginella* existed in the Carboniferous period, and representatives of some of them have survived to the present; there are fifteen living genera.

Calamites

Calamites, or giant horsetails, were plants of treelike proportions, reaching heights of 18 meters or more (see Figure 17–1). Like the plant body of *Equisetum*, that of the calamites consisted of a branched aerial portion and an underground rhizome system. In addition, the leaves and branches were whorled at the nodes. Even the stems were remarkably similar to those of *Equisetum*, except for the presence of secondary xylem in the calamites, which accounted for most of the great diameter of the stems (trunks up to one-third of a meter in diameter). The calamites, of which there were a number of genera, are now regarded as belonging to the same order as the living genus *Equisetum*.

The fertile appendages, or sporangiophores, of the calamites were aggregated into cones. Although most were homosporous, a few giant horsetails were heterosporous. Unlike most of the lycophyte trees, the giant horsetails survived the Carboniferous period and were abundant in the Permian.

(b) One of the most interesting of all gymnosperm groups is the seed ferns, a large group of primitive seed-bearing plants that appeared in the late Devonian period and flourished for about 125 million years. The remnants of these bizarre plants are common in rocks of Carboniferous age and have been well known to paleobotanists for a century or more. Their vegetative parts are so fernlike that for many years they were grouped with the ferns. This drawing is a reconstruction of the Carboniferous seed fern Medullosa noei. The plant was about 5 meters tall.

Ferns

Many of the ferns represented in the fossil record are recognizable as members of today's primitive fern families. The "Age of Ferns" in the late Carboniferous period was dominated by tree ferns such as *Psaronius*, one of the Marattiales—a eusporangiate group. Up to 8 meters tall, *Psaronius* had a stele that expanded toward the apex; the stele was covered below with adventitious roots, which played the key role in supporting the plant. The stem of *Psaronius* ended in an aggregate of large, pinnately compound fronds (see Figure 17–1).

Seed Plants

The two remaining plant groups that dominated the tropical lowlands of Euramerica were the seed ferns and the cordaites. Remnants of seed ferns are common in rocks of Carboniferous age (b). Their large, pinnately compound fronds were so fernlike that these plants were long regarded as ferns. Then in 1905, F. W. Oliver and D. H. Scott demonstrated that these plants bore seeds and so were gymnosperms. Many species were small, shrubby, or scrambling plants.

(c) Tip of young branch of the primitive conifer Cordaites, *with long, straplike leaves.*

Other probable seed ferns were tall, woody trees. The fronds of seed ferns were borne at the top of their stem, or trunk, with microsporangia and seeds borne on them. The seed ferns survived into the first part of the Mesozoic era. It has been suggested that the seed ferns and the ferns evolved from a common ancestor, but Figure 16–8 shows a more probable hypothesis.

The cordaites were widely distributed during the Carboniferous period both in swamps and in drier environments. Although some members of the order were shrubs, many were tall (15 to 30 meters), highly branched trees that formed extensive forests. Their long (up to 1 meter), straplike leaves were spirally arranged at the tips of the youngest branches (c). The center of the stem was occupied by a large pith, and a vascular cambium gave rise to a complete cylinder of secondary xylem. The root system, located at the base of the plant, also contained secondary xylem. The plants bore pollen-bearing cones and seed-bearing, conelike structures on separate branches. The cordaites were also abundant during the Permian period (286 to 248 million years ago), the drier and colder period that followed the Carboniferous period.

In Conclusion

The dominant tropical coal-swamp plants of the Carboniferous period in Euramerica—the lycophyte trees—became extinct prior to the Permian, a time of increasing tropical drought and extensive temperate to polar glaciation. Only the herbaceous relatives of the tree lycophytes and horsetails of the Carboniferous period continued to flourish and exist today, as do several families of ferns that appeared in the Carboniferous period. Both the seed ferns and the cordaites eventually disappeared. Only one group of Carboniferous gymnosperms, the conifers (not a dominant group at the time), survived and went on to produce new types during the Permian period. The living conifers are discussed in detail in Chapter 17.

16–31

"Fiddleheads" of the ostrich fern (Matteuccia struthiopteris). *Ostrich fern fiddleheads are gathered commercially in upper New England and New Brunswick; they are marketed fresh, canned, and frozen. These fiddleheads taste somewhat like crisp asparagus. The fiddleheads of some ferns are considered toxic, but those of the ostrich fern, which should be picked when they are less than 15 centimeters long, are apparently perfectly safe for human consumption. The ostrich fern is the most widely planted fern around house foundations in the eastern United States and Canada.*

volume ratio allows them to capture sunlight much more effectively than the microphylls of the lycophytes. The ferns are the only seedless vascular plants to possess megaphylls, other than perhaps the very reduced leaves of *Equisetum.* Commonly, the fronds are compound; that is, the lamina is divided into leaflets, or **pinnae,** which are attached to the **rachis,** an extension of the leaf stalk, or petiole. In nearly all ferns, the young leaves are coiled; they are commonly referred to as "fiddleheads" (Figure 16–31). This type of leaf development is known as **circinate vernation.** Uncoiling of the fiddlehead results from more rapid growth on the lower than on the upper surface of the leaf early in development and is mediated by the hormone auxin (see page 547), produced by the young pinnae on the inner side of the fiddlehead. This type of vernation protects the delicate embryonic leaf tip during development. Both fiddleheads and rhizomes are usually clothed with either hairs or scales, which are epidermal outgrowths; the characteristics of these structures are important in fern classification.

The sporangia of Filicales, all of which are homosporous, occur on the margins or lower surfaces of the leaves, on specially modified leaves, or on separate stalks. The sporangia commonly occur in clusters called **sori** (singular: **sorus;** Figures 16–32 and 16–33), which

16–32

Sori are clusters of sporangia found on the undersides of leaves of ferns. (a) In Dennstaedtia punctilobula *and other ferns of this genus, the sori are bare. (b) In the bracken fern* (Pteridium aquilinum) *shown here, as well as in the maidenhair ferns* (Adiantum), *the sori are located along the margins of the leaf blades, which are rolled back over them. (c) In the evergreen wood fern* (Dryopteris marginalis), *the sori, which are also located near the surface of the leaf blades, are completely covered by kidney-shaped indusia. (d) In* Onoclea sensibilis, *the sori are enfolded by globular lobes of the pinna (leaflet) and therefore not visible. After overwintering, the lobes separate slightly, and the spores are released early in the spring, often over the snow.*

(a)

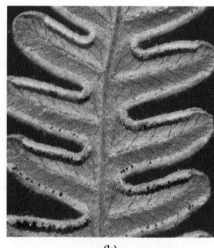
(b)

(c)

(d)

may appear as yellow, orange, brownish, or blackish lines, dots, or broad patches on the lower surface of a frond. In many genera, the young sori are covered by specialized outgrowths of the leaf, the **indusia** (singular: **indusium**), which may shrivel when the sporangia are ripe and ready to shed their spores. The shape of the sorus, its position, and the presence or absence of an indusium are important characteristics in the taxonomy of the Filicales.

The spores of Filicales ferns give rise to free-living, bisexual gametophytes, which are often found in moist places, such as the sides of pots in greenhouses. The gametophyte typically develops soon into a flat, heart-shaped, usually membranous structure, the **prothallus,** with numerous rhizoids on its central lower surface. Both antheridia and archegonia develop on the ventral surface of the prothallus. The antheridia generally appear first and occur primarily among the rhizoids. The archegonia are then formed near the notch, an indentation at the anterior end of the gametophyte. The difference in timing of appearance of the two kinds of gametangia promotes outcrossing in ferns. Water is required for the multiflagellated sperm to swim to the eggs in both heterosporous and homosporous ferns.

Some fern species, especially those that grow in dry, rocky habitats or in deserts, reproduce by spores that

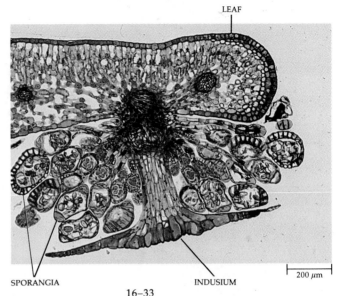

LEAF

SPORANGIA   INDUSIUM   200 μm

16–33
Cyrtomium falcatum, *a homosporous fern. Transverse section of a leaf, showing a sorus on the lower surface. The sporangia are in different stages of development and are protected by an umbrella-like indusium.*

(a)

(b)

(c)

16–34
In some ferns from widely scattered parts of the world, the gametophytes reproduce asexually and persist; sporophytes are not formed, either in the field or in the laboratory. These photographs show two of the three fern genera known to exhibit this habit in the eastern United States. (a) Typical habitat of persistent gametophytes of Vittaria *and* Trichomanes, *Ash Cave, Hocking County, Ohio. (b)* Trichomanes *gametophyte, Lancaster County, Pennsylvania. (c)* Vittaria *gametophyte, Franklin County, Alabama.*

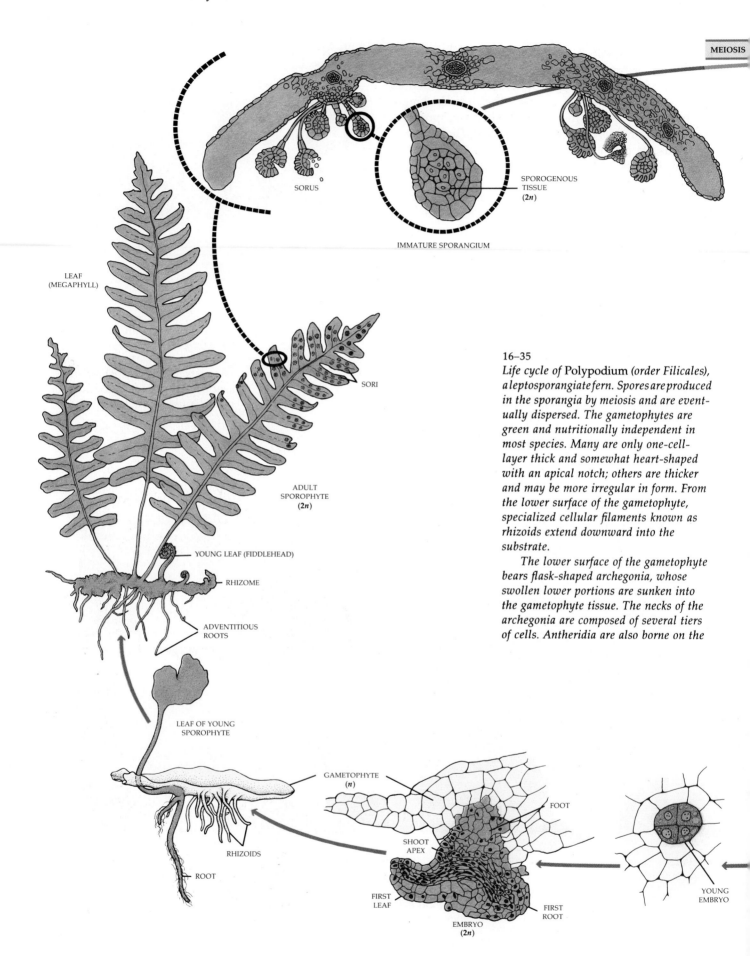

MEIOSIS

SORUS

SPOROGENOUS
TISSUE
(2n)

IMMATURE SPORANGIUM

LEAF
(MEGAPHYLL)

SORI

ADULT
SPOROPHYTE
(2n)

YOUNG LEAF (FIDDLEHEAD)

RHIZOME

ADVENTITIOUS
ROOTS

LEAF OF YOUNG
SPOROPHYTE

RHIZOIDS

ROOT

GAMETOPHYTE
(n)

FOOT

SHOOT
APEX

FIRST
LEAF

FIRST
ROOT

EMBRYO
(2n)

YOUNG
EMBRYO

16–35
Life cycle of **Polypodium** *(order Filicales),
a leptosporangiate fern. Spores are produced
in the sporangia by meiosis and are eventually dispersed. The gametophytes are
green and nutritionally independent in
most species. Many are only one-cell-
layer thick and somewhat heart-shaped
with an apical notch; others are thicker
and may be more irregular in form. From
the lower surface of the gametophyte,
specialized cellular filaments known as
rhizoids extend downward into the
substrate.*

*The lower surface of the gametophyte
bears flask-shaped archegonia, whose
swollen lower portions are sunken into
the gametophyte tissue. The necks of the
archegonia are composed of several tiers
of cells. Antheridia are also borne on the*

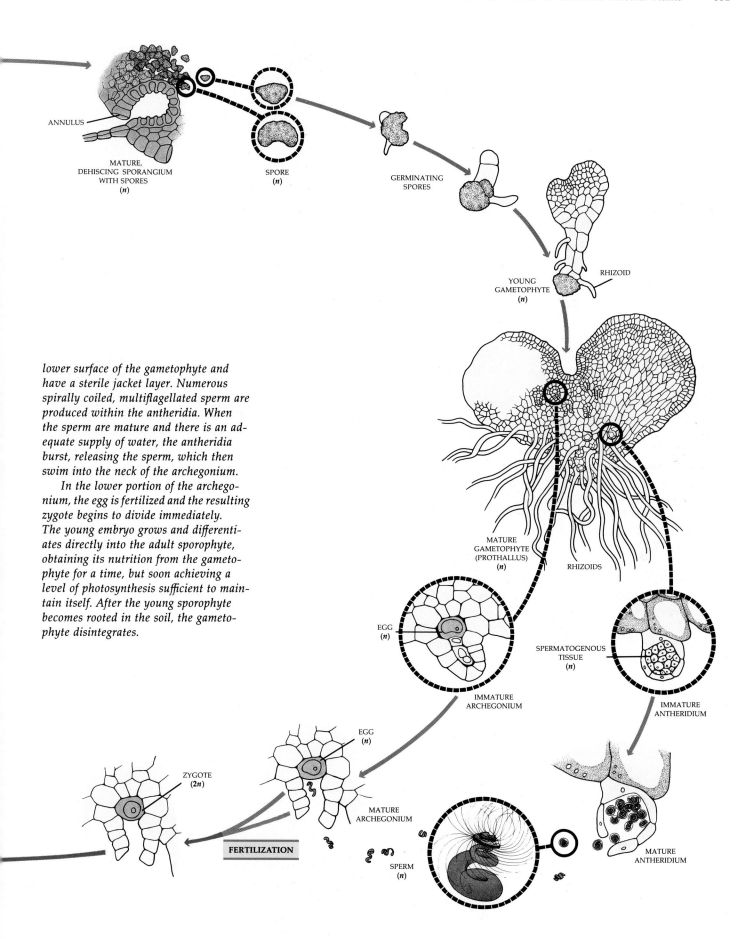

ANNULUS

MATURE,
DEHISCING SPORANGIUM
WITH SPORES
(n)

SPORE
(n)

GERMINATING
SPORES

YOUNG
GAMETOPHYTE
(n)

RHIZOID

lower surface of the gametophyte and
have a sterile jacket layer. Numerous
spirally coiled, multiflagellated sperm are
produced within the antheridia. When
the sperm are mature and there is an adequate supply of water, the antheridia
burst, releasing the sperm, which then
swim into the neck of the archegonium.

In the lower portion of the archegonium, the egg is fertilized and the resulting
zygote begins to divide immediately.
The young embryo grows and differentiates directly into the adult sporophyte,
obtaining its nutrition from the gametophyte for a time, but soon achieving a
level of photosynthesis sufficient to maintain itself. After the young sporophyte
becomes rooted in the soil, the gametophyte disintegrates.

MATURE
GAMETOPHYTE
(PROTHALLUS)
(n)

RHIZOIDS

EGG
(n)

SPERMATOGENOUS
TISSUE
(n)

IMMATURE
ARCHEGONIUM

IMMATURE
ANTHERIDIUM

ZYGOTE
(2n)

EGG
(n)

MATURE
ARCHEGONIUM

FERTILIZATION

SPERM
(n)

MATURE
ANTHERIDIUM

they produce asexually. Instead of the normal 64 haploid spores, their sporangia contain only 16 or 32 unreduced spores—that is, spores that have not undergone meiosis and therefore have the same number of chromosomes as the frond that produced them. The unreduced spores germinate, producing gametophytes that are normal in every respect except that neither archegonia nor antheridia are produced. A new leaf (frond) vegetatively proliferates from the notch region of the heart-shaped gametophyte, where the archegonia would normally be, and grows into a typical fern sporophyte. Thus, there is no difference in chromosome number between the two generations that alternate in the life cycle of these ferns. This type of life cycle, which involves the formation of a sporophyte without the union of gametes, is called **apogamy.** Apogamous ferns complete their gametophyte generation much more rapidly than is the case in ferns that reproduce sexually. For this reason, apogamy is thought to be favored in dry habitats, because the duration of the delicate gametophyte stage—the stage most susceptible to desiccation—is shortened.

Remarkably, the strap-shaped or filamentous gametophytes of some species of ferns, including three genera with six tropical species found in the southern Appalachians, persist without ever producing sporophytes; nor have they as yet been induced to produce sporophytes in the laboratory (Figure 16–34). They reproduce by vegetative outgrowths called gemmae that fall off and are blown away to found new colonies. These ferns appear to be distinct from the other, sporophyte-producing species of their respective genera, as judged by differences in their enzymes, and probably should be treated as distinct species. Such situations are common in mosses and are being discovered in ferns much more widely than was previously expected.

Early in its development, the embryo, or young sporophyte, receives nutrients from the gametophyte through a foot. Development is rapid, and the sporophyte soon becomes an independent plant, at which time the gametophyte disintegrates.

The life cycle of one of the Filicales is shown in Figure 16–35 on pages 350–351.

(a)

16–36
Water ferns. The two very distinct orders of water ferns are the only living heterosporous ferns. (a) Marsilea polycarpa, with its leaves floating on the surface of the water, photographed in Venezuela. (b) Marsilea, showing the germination of a sporocarp, with chains of sori. Each sorus contains a series of megasporangia and microsporangia. (c) Salvinia, with two floating leaves and one dissected, submerged leaf at each node. These two genera are representatives of the orders Marsileales and Salviniales, respectively.

WATER FERNS

The water ferns constitute two orders, Marsileales and Salviniales, which are structurally very different from each other and were almost certainly derived from different terrestrial ancestors. All water ferns are heterosporous, and they are the only living heterosporous ferns. There are five genera of water ferns. The slender rhizomes of the three genera of Marsileales, including *Marsilea* (which has about 50 species), grow in mud, on damp soil, or often with the leaves floating on the surface of water (Figure 16–36a). The leaves of *Marsilea* resemble those of a four-leaf clover. Drought-resistant, bean-shaped reproductive structures called **sporocarps,** which may remain viable even after 100 years of dry

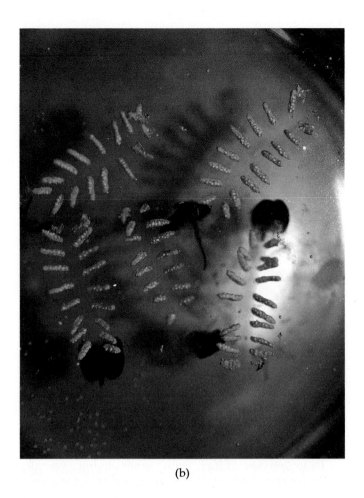

(b)

(c)

storage, germinate when placed in water to produce chains of sori, each bearing series of megasporangia and microsporangia (Figure 16–36b). The extremely specialized gametophytes and the heterospory of Marsileales are the primary reasons that we recognize these ferns as a distinct order.

The two genera of Salviniales, *Azolla* (see Figure 27–15, page 608) and *Salvinia* (Figure 16–36c), are small plants that float on the surface of water. Both genera produce their sporangia in sporocarps that are quite different in structure from those of Marsileales. In *Azolla*, the tiny, crowded, bilobed leaves are borne on slender stems. A pouch that forms on the upper, photosynthetic lobe of each leaf is inhabited by colonies of *Anabaena azollae*, one of the cyanobacteria; the lower, smaller lobe of each leaf is often nearly colorless. Because of the nitrogen-fixing abilities of the *Anabaena*, *Azolla* is important in maintaining the fertility of rice paddies and of certain natural ecosystems. The undivided leaves of *Salvinia*, which are up to 2 centimeters long, are borne in whorls of three on the floating rhizome. One of the three leaves hangs down below the surface of the water and is highly dissected, resembling

a mass of whitish roots. These "roots," however, bear sporangia, which reveals that they are actually leaves. The two upper leaves float on the water; they are covered by hairs that protect their surface from getting wet, and the leaves float back to the surface if they are temporarily submerged.

Summary

Vascular plants contain xylem and phloem and exhibit an alternation of generations in which the sporophyte is the dominant and nutritionally independent phase.

The plant bodies of many vascular plants consist entirely of primary tissues. Today, secondary growth is confined largely to the seed plants, although it occurred in several unrelated fossil groups of seedless vascular plants. The primary vascular tissues and associated ground tissues exhibit three basic arrangements: (1) the protostele, which consists of a solid core of vascular tissue; (2) the siphonostele, which contains a pith surrounded by vascular tissue; and (3) the eustele, which consists of a system of strands surrounding a pith, with the strands separated from one another by ground tissue.

Roots evolved from the underground portions of the primitive plant body. Leaves originated in more than one way. Microphylls, single-veined leaves whose leaf traces are not associated with leaf gaps, evolved as superficial lateral outgrowths of the stem. They are associated with protosteles and are characteristic of the lycophytes. Megaphylls, leaves with complex venation and leaf traces associated with leaf gaps, evolved from branch systems. Megaphylls are associated with siphonosteles and eusteles.

Vascular plants are either homosporous or heterosporous. Homosporous vascular plants produce only one type of spore, which gives rise usually to a bisexual gametophyte. Heterosporous plants produce microspores and megaspores, which germinate and give rise to male gametophytes and female gametophytes, respectively. The gametophytes of heterosporous plants are much reduced in size, compared with those of homosporous plants. In the history of the vascular plants, heterospory has evolved a number of times. There has been a long, continuous evolutionary trend toward a reduction in the size and complexity of the gametophyte, which culminated in the angiosperms. Seedless vascular plants have archegonia and antheridia; archegonia have been lost in all but a few gymnosperms, and both archegonia and antheridia have been lost in all angiosperms.

Vascular plants go back at least 430 million years; the earliest ones about which we have many structural details belong to the division Rhyniophyta, the oldest fossils of which are from the mid-Silurian period, about 420 million years ago. The plant bodies of the rhyniophytes and other contemporary plants were simple, dichotomously branching axes lacking roots and leaves. With evolutionary specialization, morphological and physiological differences arose between various parts of the plant body, bringing about the differentiation of root, stem, and leaf.

The living seedless vascular plants are classified in four divisions: the Psilotophyta (*Psilotum* and *Tmesip-* *teris*), the Lycophyta (which includes Lycopodiaceae, *Selaginella,* and *Isoetes*), the Sphenophyta *(Equisetum)*, and the Pterophyta (ferns). Most of the seedless vascular plants are homosporous. Heterospory is exhibited by *Selaginella, Isoetes,* and the water ferns.

The life cycles of the seedless vascular plants are essentially similar to one another: an alternation of heteromorphic generations in which the sporophyte is dominant and free-living. The gametophytes of the homosporous species are bisexual, producing both antheridia and archegonia, and are independent of the sporophyte for their nutrition. Those of heterosporous species are unisexual, much reduced in size, and, except for the few genera of heterosporous ferns, dependent on stored food derived from the sporophyte for their nutrition. All of the seedless vascular plants have motile sperm, and the presence of water is necessary for them to swim to the eggs.

Two orders of ferns (Ophioglossales and Marattiales) form eusporangia, like those of other seedless vascular plants; in eusporangia, the walls are several cell layers thick, and a number of cells participate in the initial stages of sporangial development. Other ferns—Filicales and the two orders of water ferns—form leptosporangia, specialized structures in which the wall consists of a single layer of cells and which develop from a single initial cell.

Psilotophytes differ from other living vascular plants in their lack of leaves (with the possible exception of *Tmesipteris*) and roots. Lycophytes are characterized by microphylls, associated with protosteles; the members of the other divisions have megaphylls, associated with siphonosteles or eusteles.

Two of the four divisions of seedless vascular plants that contain living representatives, the lycophytes and sphenophytes, extend back to the Devonian period. Among the seedless vascular plants, only the ferns, which first appear in the fossil record in the Carboniferous period, are represented by a large number of living species, about 11,000.

Five groups of vascular plants dominated the swamplands of the Carboniferous period ("Age of Coal"), and three of them were seedless vascular plants—lycophytes, sphenophytes, and ferns. The other two were gymnosperms—the seed ferns and the cordaites.

Suggestions for Further Reading

Banks, Harlan P.: *Evolution and Plants of the Past*, Wadsworth Publishing Co., Inc., Belmont, Calif., 1970.*

An introduction to plant evolution as it is revealed by the fossil record.

Camus, Josephine M., A. Clive Jermy, and Barry A. Thomas: *A World of Ferns*, Natural History Museum Publications, London, 1991.

Outstandingly beautiful account of the ferns.

Gensel, Patricia G., and Henry N. Andrews: *Plant Life in the Devonian*, Praeger Publishers, New York, 1984.

This advanced book traces what is known about the earliest vascular plants at the time of their most active evolution.

Gensel, Patricia G., and Henry N. Andrews: "Evolution of Early Land Plants," *American Scientist* 75:478–489, 1987.

An excellent and beautifully illustrated article on the first plants.

Gifford, Ernest M., and Adriance S. Foster: *Morphology and Evolution of Vascular Plants*, 3d ed., W.H. Freeman and Company, New York, 1989.

A well-organized, general account of the vascular plants, containing much interpretative material.

Lellinger, David B.: *A Field Manual of the Ferns & Fern-Allies of the United States and Canada*, Smithsonian Institution Press, Washington, D.C., 1985.

A handsomely illustrated and comprehensive guide to these fascinating plants.

Radford, Albert E., et al.: *Vascular Plant Systematics*, Harper & Row, Publishers, Inc., New York, 1974.*

A useful source book, including glossaries, techniques, bibliographies, indices, and useful discussions of all the various aspects of plant systematics—the scientific study of the kinds and diversity of plants, and of the relationships among them.

Stewart, Wilson N.: *Paleobotany and the Evolution of Plants*, Cambridge University Press, New York, 1983.

An excellent and well-illustrated account of all groups of fossil plants.

*Available in paperback.

17–1

A reconstruction of an Upper Carboniferous swamp forest dominated by the lycophyte tree Lepidodendron, forming the forest stand from left to center. Shown here are unbranched (resembling bottle brushes) and sparsely branched juvenile trees, as well as tall, many-branched canopy trees. Sigillaria, a lycophyte adapted to drier sites, is seen on the higher ground at the right (tree with tufts of long leaves). Other seedless trees include Calamites (the giant horsetails; far left and center foreground) and Psaronius (tree ferns with tapering trunks; far right). Seed plants also thrived, among

CHAPTER 17

Gymnosperms

them Medullosa *(a seed fern with bifurcated fronds; center and far right)* and Cordaites, *both mangrove-like forms (with roots in the water; ground level on the right) and tree forms (tall trees; far right).*

One of the most dramatic innovations to arise during the evolution of the vascular plants was the seed. Seeds seem to be one of the factors responsible for the dominance of seed plants in today's flora—a dominance that has become progressively greater over a period of several hundred million years (Figure 17–1). The reason is simple: the seed has survival value. The protection that a seed affords the enclosed embryo, and the stored food that is available to that embryo at the critical stages of germination and establishment, give seed plants a great selective advantage over their free-sporing relatives and ancestors.

All seed plants are heterosporous. In most of them, the highly reduced megagametophyte is retained within the original wall of the megaspore that gave rise to it (walls may not be formed around the megaspore nuclei in angiosperms). In angiosperms, more than one megaspore nucleus may participate in the formation of the megagametophyte (see page 390). In all seed plants, the developing megagametophyte is also retained within the megasporangium, which is fleshy and called the **nucellus** (plural: nucelli) in seed plants. The megasporangia of seed plants, unlike those of seedless heterosporous plants, are enveloped by one or two additional layers of tissue, the **integuments**. The integuments completely enclose the megasporangium except for an opening at the apex called the **micropyle**. This entire structure—the nucellus plus the integument(s)—is known as the **ovule** (Figure 17–2).

Following fertilization, the ovule develops into a seed, with the integument(s) developing into its **seed coat.** In other words, it is the ovule that develops into a seed; indeed, a seed is often called a mature ovule. In most modern seed plants, an embryo, or young sporophyte, develops within the seed before dispersal. Perhaps the development of the embryo before dispersal gives the seed a better chance of survival in cold and harsh conditions, and the Permian period, when conifers, cycads, and *Ginkgo* arose, was certainly a period of climatic extremes. In addition to a megaspore or embryo and a seed coat, all seeds contain stored food, which further increases the young plant's chances of survival.

The oldest known seedlike structures are from the late Devonian period, some 360 million years ago (Figure 17–3). During the next 50 million years, a wide array of seed-bearing plants evolved, including the seed ferns, which had seedlike structures, as well as the cordaites and conifers (see "Coal Age Plants" on pages 346–347). The earliest vascular plants, including the members of the three extinct divisions described in Chapter 16, had very wide distributions compared with those of their seed-bearing descendants. The reason for this is that seeds are less widely dispersed than spores. With the advent of plants that bore seeds, the floras—assemblages of plants—found in different regions of the world became more distinct from one another.

The seed plants, all of which possess megaphylls, include five divisions with living representatives: the Cycadophyta, the Ginkgophyta, the Coniferophyta, the Gnetophyta, and the Anthophyta. The first four of these divisions comprise the gymnosperms; the division Anthophyta comprises the angiosperms. Before beginning our discussion of seed plants, we shall briefly examine one more group of seedless vascular plants—the progymnosperms. They are discussed here, rather than in Chapter 16, because they are the likely progenitors of gymnosperms.

The oldest fossils in which angiosperms (division Anthophyta)—overwhelmingly the most successful vascular plants at the present time—have been identified with certainty are only about 127 million years old, from the early part of the Cretaceous period. Although

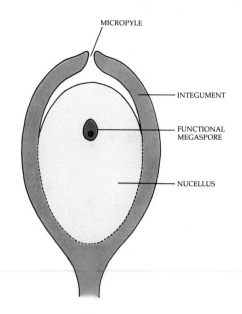

17–2
Longitudinal transection of an ovule, showing the arrangement of the integument, the megasporangium (nucellus), and the megaspore. The fertilization of ovules results in their maturation into seeds; in other words, seeds are mature, fertilized ovules.

the angiosperms actually must be somewhat older than our current understanding of the fossil record indicates, they are still relative newcomers in the broad picture of vascular plant evolution. Because they are such a large and important group, the angiosperms are considered in detail in Chapters 18 and 19.

The gymnosperms do not constitute an evolutionary line that is equivalent to the angiosperms; rather, the gymnosperms represent a series of evolutionary lines of

17–3
(a) *Reconstruction of a fertile branch of the late Devonian plant* Archeosperma arnoldii, *showing its seedlike structures. The cupules—cup-shaped structures that partly enclose the megasporangia—are arranged in pairs, and each cupule contains two flask-shaped seedlike structures about 4 millimeters long. The apex of each was dissected into lobes. (b) Diagram showing the position of the megaspore. (c) A megaspore released by maceration. This fossil, from Pennsylvania, is the oldest known seedlike structure—about 360 million years in age.*

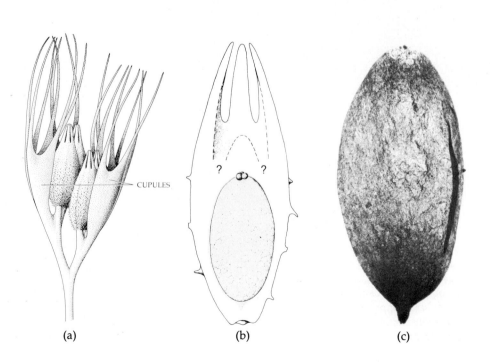

(a)   (b)   (c)

seed-bearing plants that lack the distinctive characteristics of the angiosperms. There are about 720 species of living gymnosperms, as compared with some 235,000 species of angiosperms—although individual gymnosperm species are often dominant over wide areas.

Progymnosperms

In the late Paleozoic era, there existed a group of plants called the progymnosperms (division Progymnospermophyta), which had characteristics intermediate between those of the trimerophytes and those of the seed plants. Although the progymnosperms reproduced by means of freely dispersed spores, they produced secondary xylem remarkably similar to that of gymnosperms (Figure 17–4): the progymnosperms were unique among woody Devonian plants in producing secondary phloem also. The progymnosperms and Paleozoic ferns may have evolved from the more ancient trimerophytes (Figure 16–9c), from which they differed primarily by having more elaborate branch systems and correspondingly more complex vascular systems. Another possibility is that the ferns evolved from the progymnosperms. Since the trimerophytes gave rise to the progymnosperms and since the ferns evolved from one of these groups, the uncertainties concern the timing of

17–5
Reconstruction of a portion of the branch system of Triloboxylon ashlandicum, *an* Aneurophyton-*type progymnosperm. The main axis bears vegetative branches at the top and bottom and fertile organs with sporangia in between.*

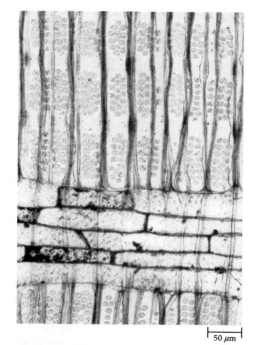

_{50 μm}

17–4
Radial view of the secondary xylem, or wood, of the progymnosperm Callixylon newberryi. *This fossil wood, with its regular series of pitted tracheids, is remarkably similar to that of certain gymnosperms.*

the origin of the ferns and the nature of their particular ancestors.

In the progymnosperms, the most important evolutionary advance over the trimerophytes is the presence of a bifacial vascular cambium—that is, one that produces both secondary xylem and secondary phloem. Vascular cambia of this type are characteristic of seed plants and apparently evolved first in the progymnosperms.

One kind of progymnosperm, the *Aneurophyton* type, which occurred in the Devonian period approximately 360 to 380 million years ago, featured three-dimensional branching (Figure 17–5) and had protosteles. The organization of these plants was similar to that of some early seed ferns, leading some paleobotanists to suggest that the branch systems of the *Aneurophyton*-type progymnosperm may have been the precursors of the fernlike leaves of early seed ferns.

17–6
Reconstruction of the progymnosperm
Archaeopteris, *which is common in the*
fossil record of eastern North America.
Specimens of Archaeopteris *attained*
heights of 20 meters or more, and some
of them seem to have formed forests.

17–7
Reconstruction of a frondlike lateral
branch system of the progymnosperm Archaeopteris macilenta. *Fertile leaves can*
be seen bearing maturing sporangia on
centrally located primary branches.

A second major kind of progymnosperm, the *Ar-chaeopteris* type, also appeared in the Devonian period, about 370 million years ago, and extended into the early Mississippian period, about 340 million years ago (Figure 17–6). This group is considered advanced because the lateral branch systems in these plants were flattened in one plane and bore laminar structures considered to be leaves (Figure 17–7). Eusteles apparently evolved in this group of progymnosperms and provided a strong linkage between this group and the living seed plants. The larger branches of *Archaeopteris*-type progymnosperms had a pith. Although most progymnosperms were homosporous, some species of *Archaeopteris* were heterosporous.

Some fossil logs of *Archaeopteris* are up to a meter or more in diameter and 10 meters long, indicating that at least some species of this group were large trees. The way in which fossils of the group occur suggests that *Archaeopteris* dominated extensive forests in some regions. As the reconstruction in Figure 17–6 suggests, individuals of *Archaeopteris* may have resembled conifers in their branching patterns, but this is not certain.

Morphological evidence that has been amassed during the past several decades strongly supports the contention that the gymnosperms evolved from the progymnosperms upon the evolution of the seed in what now seems to have been the common ancestor of all seed plants. At more or less the same time, seedlike structures (see Figure 17–3) evolved in another group of progymnosperms to give rise to the seed ferns. It is by no means certain, however, from which group of progymnosperms seed plants evolved.

Extinct Gymnosperms

Two groups of extinct gymnosperms—the seed ferns (division Pteridospermophyta) and the Cordaitales, primitive conifers—were discussed and illustrated in Chapter 16 (see "Coal Age Plants" on pages 346–347). As indicated in Figure 16–8, the seed ferns, which were a very diverse group, probably evolved separately from the progymnosperms and were not part of the evolutionary line leading to modern seed plants. According to the fossil record, seeds appeared at least 360 million years ago, in the late Devonian period, with plants that seem to have been seed ferns present at least 10 million years earlier.

The seedlike structure illustrated in Figure 17–3 is from a seed fern. Based on our current understanding of the relationships of these plants, it is virtually certain that none of them could have given rise to any modern seed plant.

Another group of extinct gymnosperms—the division Cycadeoidophyta—consisted of plants with palm-like leaves, somewhat resembling the living cycads (pages 374, 375). The cycadeoids (also called Bennettitales) were apparently members of the same evolutionary line as the living seed plants. The cycadeoids lived at the same time as the superficially similar cycads but were distinct from the cycads in having bisexual strobili and in other respects. Cycadeoids lived during the Jurassic and Cretaceous periods.

Living Gymnosperms

As noted earlier, there are four divisions of gymnosperms with living representatives: Cycadophyta (cycads), Ginkgophyta (maidenhair tree, ginkgo), Coniferophyta (conifers), and Gnetophyta (gnetophytes). The name *gymnosperm*, which literally means "naked seed," points to one of the principal characteristics of the plants belonging to these four divisions: their ovules and seeds are exposed on the surface of sporophylls and analogous structures. The four divisions of gymnosperms that have living representatives represent the achievement of a particular stage in evolutionary progression by various descendants of the progymnosperms.

With few exceptions, the female gametophyte of gymnosperms produces several archegonia. As a result,

more than one egg many be fertilized, and several embryos may begin to develop within a single ovule—a phenomenon known as **polyembryony.** In most cases, only one embryo survives and relatively few fully developed seeds contain more than one embryo.

In the seedless vascular plants, water is required for the motile, flagellated sperm to reach and fertilize the eggs. In the gymnosperms, water is not required as a medium of transport of the sperm to the eggs. Instead, the partly developed male gametophyte, the **pollen grain,** is transferred bodily· (usually passively, by the wind) to the vicinity of a female gametophyte within an ovule. This process is called **pollination.** After pollination, the male gametophyte produces a tubular outgrowth, the **pollen tube.**

In the conifers and gnetophytes, the sperm are nonmotile, and the pollen tubes convey them directly to the archegonia. In the cycads and *Ginkgo*, fertilization is transitional between the condition found in ferns and other seedless plants, in which free-swimming sperm occur, and the condition found in other seed plants, which have nonmotile sperm. The male gametophytes of cycads and *Ginkgo* are largely haustorial, absorbing nutrients from the ovule as they grow, and although a pollen tube is eventually produced, it does not penetrate the archegonium. The pollen tube may grow for several months in the tissue of the nucellus before reaching a cavity above the female gametophyte. At that time, the pollen tube bursts in the vicinity of the archegonium, releasing multiflagellated, swimming sperm cells. The sperm then swim to an archegonium, and one of them **fertilizes** the egg. With the development of sperm-conveying pollen tubes, seed plants no longer were dependent on the presence of free water to assure fertilization—a necessity for all seedless plants.

CONIFERS

By far the most numerous and widespread of the gymnosperm divisions living today, the Coniferophyta comprise some 50 genera with about 550 species. The tallest vascular plant, the redwood (*Sequoia sempervirens*) of coastal California and southwestern Oregon, is a conifer. Redwood trees attain heights of up to 117 meters and trunk diameters in excess of 11 meters. The conifers, which also include pines, firs, and spruces, are of great commercial value; their stately forests provide the wealth of vast regions of the North Temperate zone (see "Jobs versus Owls" on page 682). During the early Tertiary period, some genera were more widespread than they are now, and they dominated huge expanses on all of the northern continents.

The history of the conifers extends back at least to the late Carboniferous period, some 290 million years ago; as we have seen, the cordaites were primitive conifers. The leaves of modern conifers have many drought-resistant features, which may be related to the diversification of the division during the relatively dry and cold Permian period (286 to 248 million years ago). At that time, increasing worldwide aridity must have favored structural adaptations such as those of conifer leaves.

Pines

The pines (genus *Pinus*) include perhaps the most familiar of all gymnosperms (Figure 17–8); they dominate broad stretches of North America and Eurasia and are widely cultivated even in the Southern Hemisphere. There are about 90 species of pines, all of which are characterized by an arrangement of the leaves that is unique among living conifers. Pine leaves are needle-like. In the seedlings, they are spirally arranged and borne singly on the stems (Figure 17–9). After a year or two of growth, a pine begins to produce its leaves in bundles, or fascicles, each of which contains a specific number of long, needlelike leaves—from one to eight, depending on the species (Figure 17–10). These fascicles, wrapped at the base by a series of short, scalelike leaves, are actually short shoots in which the activity of the apical meristem is suspended. Thus, a fascicle of needles in a pine is morphologically a **determinate** (restricted in growth) branch. Under unusual circumstances, the apical meristem within a fascicle of needles in a pine may be reactivated and grow into a new shoot with **indeterminate** growth, or sometimes may even produce roots and grow into an entire pine tree (Figure 17–11).

17–8
Longleaf pines, Pinus palustris, *growing in North Carolina.*

(a)

(b)

17–9
(a) *Seedlings of longleaf pine,* Pinus palustris, *in Georgia, showing the long leaves in fascicles, or bundles, of three.*

(b) *A seedling of pinyon pine,* Pinus edulis, *showing juvenile leaves and a young taproot system. The mature leaves*

of this species are borne in fascicles of two needles each.

17–10

Bristlecone pine, Pinus longaeva, *in the White Mountains of California. Branch showing clusters of five needles, a mature ovulate cone on the right, and a young ovulate cone on the left. The individual needles of this species may remain functional for up to 45 years. It is also the longest-lived tree (see Figure 24–26).*

17–11

One-year-old Monterey pines (Pinus radiata) grown from rooted fascicles of needles. This experiment demonstrates that a fascicle of pine needles is actually a short shoot in which the activity of the apical meristem has been suspended but can be regenerated.

The leaves of pines, like those of many other conifers, are impressively suited for growth under arid conditions (Figure 17–12). The epidermis is covered with a thick cuticle; beneath the epidermis are one or more layers of compactly arranged, thick-walled cells—the hypodermis. The stomata are sunken below the surface of the leaf. The mesophyll, or ground tissue of the leaf, consists of parenchyma cells with conspicuous wall ridges that project into the cells like the pieces of a jigsaw puzzle. Commonly, the mesophyll is penetrated by two or more resin ducts. One vascular bundle, or two bundles side by side, are found in the center of the leaf. The veins are surrounded by transfusion tissue, composed of living parenchyma cells and short, nonliving tracheids. The transfusion tissue is believed to conduct materials between the mesophyll and the vascular bundles. An endodermis surrounds the transfusion tissue, so that the transfusion tissue and mesophyll are not in direct contact with each other.

Most pine species retain their needles for two to four years, and the overall photosynthetic balance of a given plant depends on the health of several years' crop of needles. In bristlecone pine *(Pinus longaeva),* the longest-lived tree (see Figures 17–10 and 24–26), the needles are retained for up to 45 years and remain photosynthetically active the entire time. Because the leaves of pines and other evergreens function for more than one season, they are exposed to damage by drought, freezing, or air pollution for much longer than

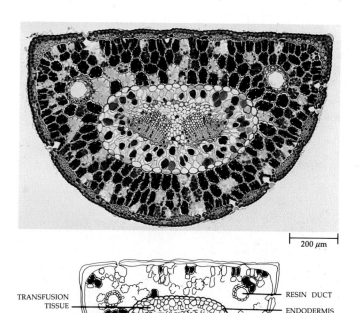

200 µm

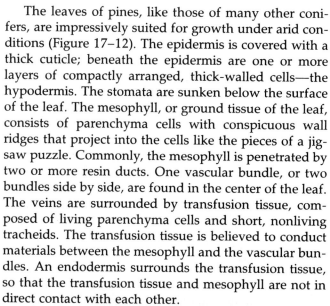

TRANSFUSION TISSUE
XYLEM
PHLOEM
MESOPHYLL
STOMA
RESIN DUCT
ENDODERMIS
VASCULAR BUNDLE
HYPODERMIS
EPIDERMIS

17–12

Pinus. *Transverse section of needle, showing mature tissues.*

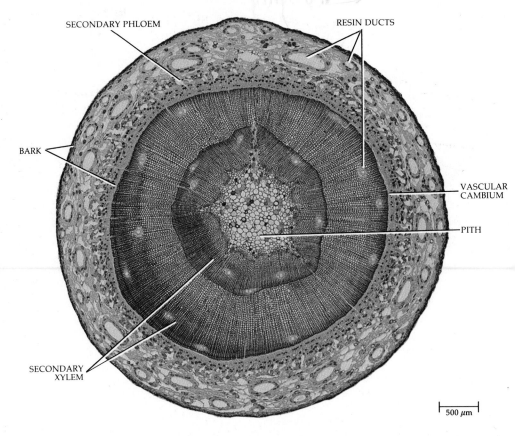

SECONDARY PHLOEM

RESIN DUCTS

BARK

VASCULAR CAMBIUM

PITH

SECONDARY XYLEM

500 μm

17–13
Pinus. *Cross section of stem, showing secondary xylem and secondary phloem separated from one another by vascular cambium. All of the tissues outside the vascular cambium, including the phloem, make up the bark.*

17–14
Monterey pine, Pinus radiata. Microsporangiate cones shedding pollen, which is blown about by the wind. Some of the pollen reaches the vicinity of the ovules in the ovulate cones and then germinates, producing pollen tubes and eventually bringing about fertilization.

the leaves of deciduous plants and are more often injured.

In the stems of pines and other conifers, secondary growth begins early and leads to the internal formation of substantial amounts of secondary xylem (Figure 17–13). Secondary xylem is produced toward the inside of the vascular cambium, and secondary phloem is produced toward the outside. The xylem of conifers consists primarily of tracheids; the phloem consists of sieve cells, which are the typical food-conducting cells of gymnosperms and seedless vascular plants (see page 461). Both kinds of tissues are traversed radially by narrow rays. With the initiation of secondary growth, the epidermis is eventually replaced with a periderm, which has its origin in the outer layer of cortical cells. As secondary growth continues, subsequent periderms are produced by active cell division deeper in the bark.

Reproduction in Pines

The microsporangia and megasporangia in pines and most other conifers are borne in separate cones on the same tree. Ordinarily, the microsporangiate cones are borne on the lower branches of the tree and the megasporangiate, or ovulate, cones are borne on the upper branches; in some pines, they are borne on the same branch, with the ovulate cones closer to the ends. Since the pollen is not normally blown straight upward, the ovulate cones are usually pollinated by pollen from another tree, thus enhancing outcrossing.

Microsporangiate cones in the pines are relatively small, usually 1 to 2 centimeters long (Figure 17–14).

The microsporophylls (Figure 17–15) are spirally arranged and more or less membranous; each bears two microsporangia. A young microsporangium contains many microsporocytes, or microspore mother cells; in early spring the microsporocytes undergo meiosis, each giving rise to four haploid microspores. Each microspore develops into a winged pollen grain, consisting of two **prothallial cells,** a **generative cell,** and a **tube cell** (Figure 17–16). This four-celled pollen grain is the immature male gametophyte. It is at this stage that the pollen grains are shed in enormous quantities; some are carried by the wind to the ovulate cones.

The ovulate cones of pines are much larger and more complex in their structure than are the pollen-bearing cones (Figure 17–17). The ovuliferous scales (cone scales), which bear the ovules, are not simply megasporophylls; instead, they are entire modified determinate branch systems properly known as **seed-scale complexes.** Each seed-scale complex consists of the ovuliferous scale—which bears two ovules on its upper surface—and a subtending sterile bract (Figure 17–18). The scales are arranged spirally around the axis of the cone. (The ovulate cone is, therefore, a compound structure, whereas the microsporangiate cone is a simple one, in which the microsporangia are directly attached to the microsporophylls.) Each ovule consists of a multicellular nucellus (the megasporangium) surrounded by a massive integument with an opening, the micropyle, facing the cone axis (Figure 17–18). Each megasporangium contains a single megasporocyte, or megaspore mother cell, which ultimately undergoes meiosis, giving rise to a linear series of four megaspores. However, only one of these megaspores is functional; the three nearest the micropyle soon degenerate.

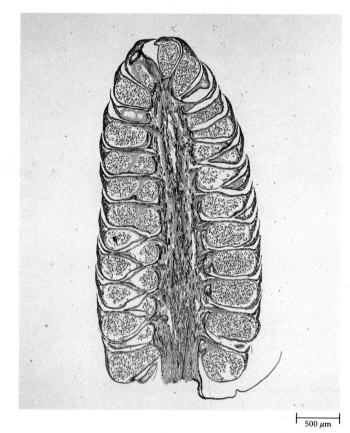

500 μm

17–15
Pinus. *Longitudinal view of a pollen-producing cone, showing microsporophylls and microsporangia containing mature pollen grains.*

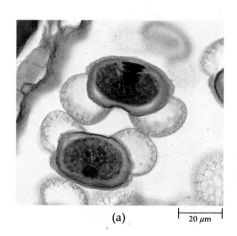

(a) 20 μm

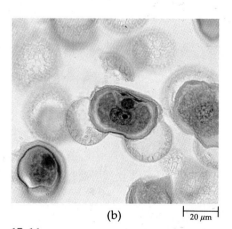

(b) 20 μm

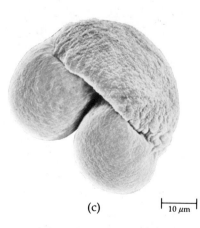

(c) 10 μm

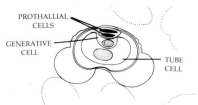

PROTHALLIAL CELLS
GENERATIVE CELL
TUBE CELL

17–16
Pinus. (a) *Pollen grains with enclosed immature male gametophytes. Each gametophyte consists of two prothallial cells, a relatively small generative cell, and a relatively large tube cell.* (b) *A somewhat older pollen grain. Here the prothallial cells, which have no apparent function, have degenerated.* (c) *Scanning electron micrograph of a pine pollen grain, with its two bladder-shaped wings. When the pollen grain germinates, the pollen tube emerges from the lower end of the grain between the wings.*

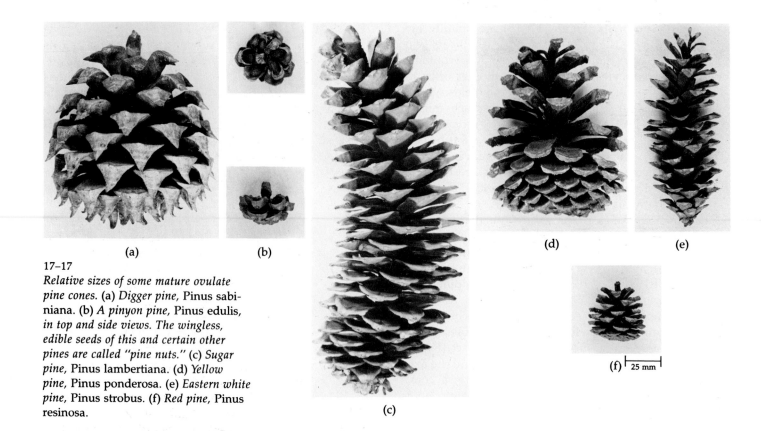

(a)   **(b)**   **(c)**   **(d)**   **(e)**

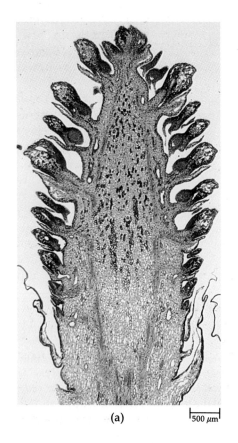

17–17
*Relative sizes of some mature ovulate
pine cones. (a) Digger pine,* Pinus sabiniana. *(b) A pinyon pine,* Pinus edulis,
*in top and side views. The wingless,
edible seeds of this and certain other
pines are called "pine nuts." (c) Sugar
pine,* Pinus lambertiana. *(d) Yellow
pine,* Pinus ponderosa. *(e) Eastern white
pine,* Pinus strobus. *(f) Red pine,* Pinus
resinosa.

(f) ⊢ 25 mm ⊣

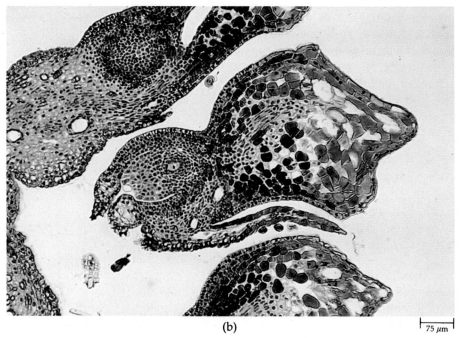

(a) ⊢ 500 μm ⊣   **(b)** ⊢ 75 μm ⊣

17–18
Pinus. *(a) Longitudinal view of young
female cone, or strobilus, showing its
complex structure. (b) Detail of a portion
of the strobilus. Note the megasporocyte*   *(megaspore mother cell) surrounded by
the nucellus. (See also the pine life cycle,
Figure 17–24.)*

17–19
Pinkish ovulate cones of lodgepole pine
(Pinus contorta) at the tip of a tall cen-
tral branch in late spring of their first
year, the time at which pollination
occurs. One-year-old ovulate cones
are visible at the base of this branch.
Yellowish, pollen-bearing microsporan-
giate cones cluster around the shorter
branches.

Pollination in pines occurs in the spring; the pollen adheres to a drop of sticky fluid about the micropyle. At this stage, the scales of the ovulate cone are widely separated (Figure 17–19). As the micropylar fluid evaporates, the pollen grain is drawn into the micropyle and comes in contact with the nucellus. After pollination, the scales grow together and help protect the developing ovules. Shortly after the pollen grain comes in contact with the nucellus, it germinates, forming a pollen tube. At this time, meiosis has not yet occurred in the megasporangium. About a month after pollination, the four megaspores are produced, only one of which develops into a megagametophyte. The development of the megagametophyte is sluggish; it often does not begin until some six months after pollination, and it may then require another six months for completion. In the early stages of megagametophyte development, mitosis proceeds without immediate cell wall formation. About 13 months after pollination, when the female gametophyte contains some 2000 free nuclei, cell wall formation begins. Then, approximately 15 months after pollination occurs, archegonia, usually two or three in number, differentiate at the micropylar end of the megagametophyte. The stage is now set for fertilization.

About 12 months earlier, the pollen grain germinated, producing a pollen tube that slowly digested its way through the tissues of the nucellus toward the developing megagametophyte. About a year after pollination, the generative cell of the four-celled male gametophyte undergoes division, giving rise to two kinds of cells—a **sterile cell** (stalk cell) and a **spermatogenous cell** (body cell). Subsequently, before the pollen tube reaches the female gametophyte, the spermatogenous cell divides, producing two sperm. The male gametophyte, or germinating pollen grain, is now mature. Seed plants do not form antheridia.

Some 15 months after pollination, the pollen tube reaches the egg cell of an archegonium, where it discharges much of its cytoplasm and both of its sperm into the egg cytoplasm (Figure 17–20). One sperm nucleus unites with the egg nucleus, and the other degenerates. Commonly, the eggs of all archegonia are fertilized and begin to develop into embryos (the phenomenon of polyembryony). Only one embryo usually develops fully, but about 3 to 4 percent of pine seeds contain more than one embryo and produce two or three seedlings upon germination.

During early embryogeny, four tiers of cells are produced near the lower end of the archegonium. Each of the four cells of the uppermost tier (that is, the tier farthest from the micropylar end of the ovule) begins to form an embryo. Simultaneously the four cells of the tier below the embryos, the suspensor cells, elongate greatly and force the four developing embryos into the female gametophyte. Thus, a second type of polyembryony is found in the pine life cycle. Once again, however, usually only one of the embryos develops fully. During embryogeny, the integument develops into a seed coat (see below).

The conifer seed is a remarkable structure, for it consists of a combination of two sporophytic generations—the seed coat (and remnants of the nucellus) and the embryo—and one gametophytic generation (Figure 17–21). The gametophyte serves as reserve food or nutritive tissue. The seed coat and embryo are diploid, and the female gametophyte is haploid. The embryo consists of a hypocotyl-root axis, with a root cap and apical meristem at one end and an apical meristem and several (generally eight) cotyledons, or seed leaves, at the other. The integument consists of three layers, of which the middle layer becomes hard and serves as the seed coat.

The seeds of pines are often shed from the cones during the autumn of the second year following the initial appearance of the cones and pollination. At maturity, the cone scales separate; the winged seeds of most species flutter through the air and are sometimes carried considerable distances by the wind. In some species of pines, such as lodgepole pines *(Pinus contorta),* the scales do not separate until the cones are subjected

17–20
Pinus. *Fertilization: union of a sperm nucleus with the egg nucleus. The second sperm nucleus (below) is nonfunctional; it will eventually disintegrate.*

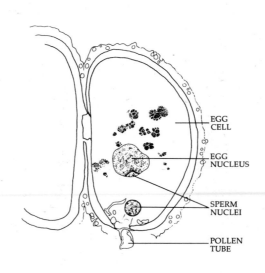

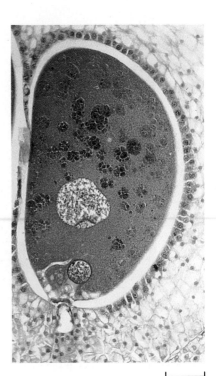

100 μm

to extreme heat (see the essay "The Great Yellowstone Fire" on page 652). When a forest fire sweeps rapidly through the pine grove and burns the parent trees, most of the fire-resistant cones are only scorched. These cones open, releasing the seed crop accumulated over many years, and reestablish the species. In other species of pines, including the limber pine (*Pinus flex-ilis*), whitebark pine (*Pinus albicaulis*), and pinyon pines of western North America, as well as a few similar species of Eurasia, the wingless, large seeds are harvested, transported, and stored for later eating by large, crowlike birds called nutcrackers. The birds miss many of the seeds they store, aiding in the dispersal of the pines.

The pine life cycle is summarized in Figure 17–24.

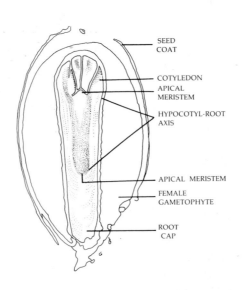

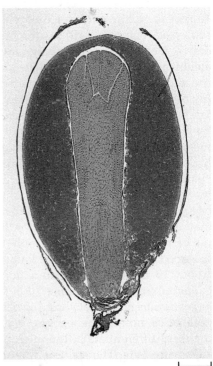

17–21
Pinus. *Longitudinal section of seed. The seed coat and embryo represent successive sporophyte (2n) generations, with a gametophyte generation intervening.*

500 μm

(a)

(b)

17–22
Two genera of the pine family (Pinaceae).
(a) *Ovulate cone of balsam fir (Abies balsamea). The upright cones, which are 5 to 10 centimeters long, do not fall to the ground whole, as they do in*

the pines; rather, they shatter and fall apart while still attached, scattering the winged seeds. (b) European larch (Larix decidua). The leaves of larch are needlelike, as in the pines; they are borne

singly both on long shoots and on short branch shoots and are spirally arranged. Unlike most conifers, the larches are deciduous; that is, they shed their leaves at the end of each growing season.

OTHER CONIFERS

Although other conifers (Figures 17–22 through 17–29) lack the needle clusters of pines and may also differ in a number of relatively minor details of their reproductive systems, the living conifers form a fairly homogeneous group. In most conifers other than the pines, the reproductive cycle takes only a year; that is, the seeds are produced in the same season as the ovules are pollinated. In such conifers, the time between pollination and fertilization usually ranges from three days to three or four weeks, instead of 15 months or so.

Among the important genera of conifers other than pines are the firs (*Abies*; Figure 17–22a), spruces (*Picea*), hemlocks (*Tsuga*), Douglas firs (*Pseudotsuga*), cypresses (*Cupressus*; Figure 17–23), and junipers (*Juniperus*; Figure 17–25 on page 372). In the yews (family Taxaceae), the ovules are not borne in cones but are solitary and surrounded by a fleshy, cuplike structure—the **aril** (Figure 17–26a on page 372). A number of other unique genera of conifers are found primarily in the Southern Hemisphere. Some of them, such as the Norfolk Island pine (*Araucaria heterophylla*) and the monkey-puzzle tree (*A. araucana*), are frequently cultivated in areas where the climate is mild; seedlings of Norfolk Island pines are also grown as houseplants.

17–23
In the subglobose cones of the cypresses, the scales are crowded together, as in this gowen cypress (Cupressus goveniana). The small trees of this species— only about 6 meters tall at maturity—are extremely local, being found only near Monterey, California.

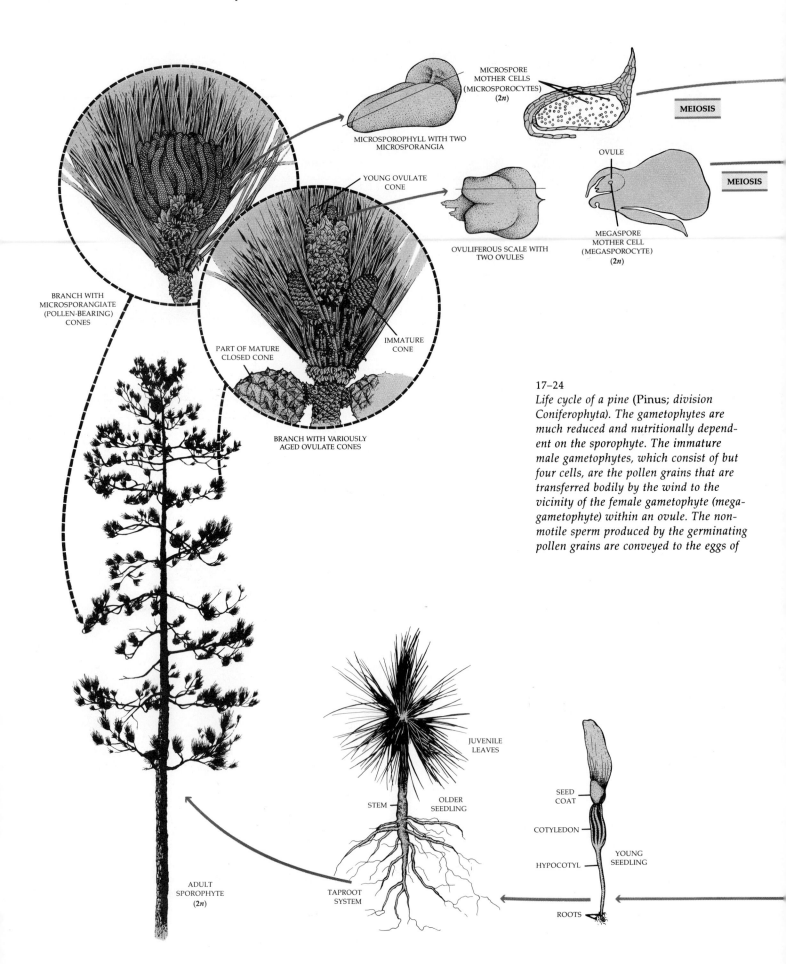

MICROSPORE
MOTHER CELLS
(MICROSPOROCYTES)
(2n)

MEIOSIS

MICROSPOROPHYLL WITH TWO
MICROSPORANGIA

YOUNG OVULATE
CONE

OVULE

MEIOSIS

OVULIFEROUS SCALE WITH
TWO OVULES

MEGASPORE
MOTHER CELL
(MEGASPOROCYTE)
(2n)

BRANCH WITH
MICROSPORANGIATE
(POLLEN-BEARING)
CONES

PART OF MATURE
CLOSED CONE

IMMATURE
CONE

BRANCH WITH VARIOUSLY
AGED OVULATE CONES

17–24

Life cycle of a pine (Pinus; division Coniferophyta). The gametophytes are much reduced and nutritionally dependent on the sporophyte. The immature male gametophytes, which consist of but four cells, are the pollen grains that are transferred bodily by the wind to the vicinity of the female gametophyte (megagametophyte) within an ovule. The nonmotile sperm produced by the germinating pollen grains are conveyed to the eggs of

JUVENILE
LEAVES

STEM

OLDER
SEEDLING

SEED
COAT

COTYLEDON

HYPOCOTYL

YOUNG
SEEDLING

ADULT
SPOROPHYTE
(2n)

TAPROOT
SYSTEM

ROOTS

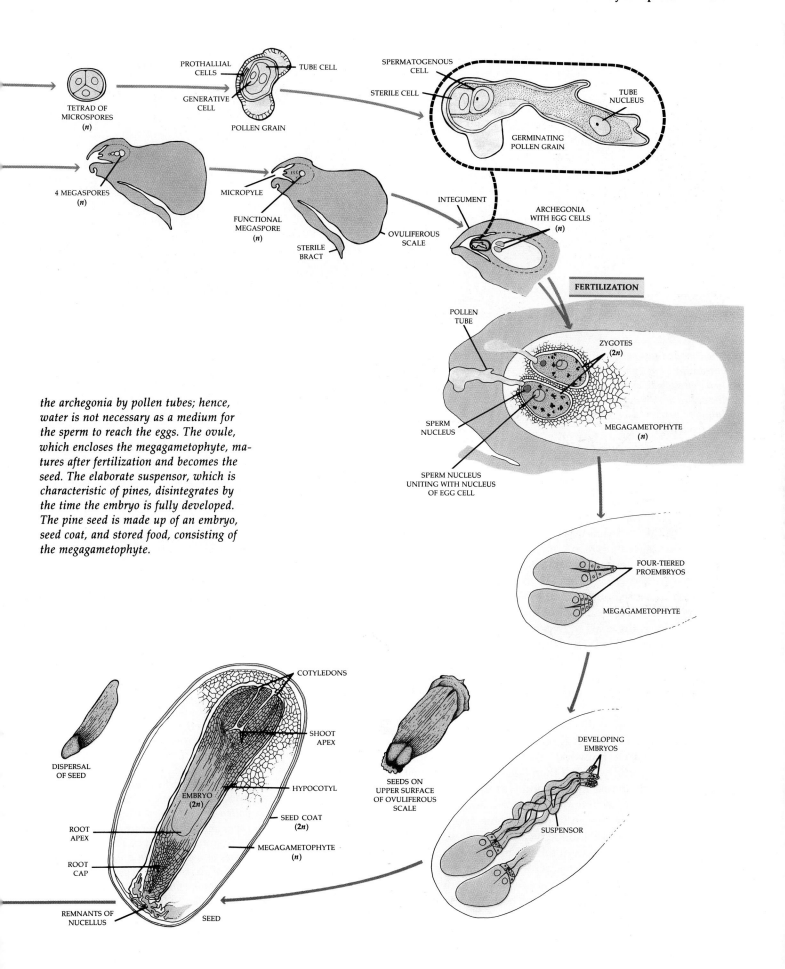

PROTHALLIAL CELLS

TUBE CELL

GENERATIVE CELL

TETRAD OF MICROSPORES (*n*)

POLLEN GRAIN

SPERMATOGENOUS CELL

STERILE CELL

TUBE NUCLEUS

GERMINATING POLLEN GRAIN

4 MEGASPORES (*n*)

MICROPYLE

FUNCTIONAL MEGASPORE (*n*)

STERILE BRACT

OVULIFEROUS SCALE

INTEGUMENT

ARCHEGONIA WITH EGG CELLS (*n*)

FERTILIZATION

POLLEN TUBE

ZYGOTES (2*n*)

SPERM NUCLEUS

SPERM NUCLEUS UNITING WITH NUCLEUS OF EGG CELL

MEGAGAMETOPHYTE (*n*)

the archegonia by pollen tubes; hence, water is not necessary as a medium for the sperm to reach the eggs. The ovule, which encloses the megagametophyte, matures after fertilization and becomes the seed. The elaborate suspensor, which is characteristic of pines, disintegrates by the time the embryo is fully developed. The pine seed is made up of an embryo, seed coat, and stored food, consisting of the megagametophyte.

FOUR-TIERED PROEMBRYOS

MEGAGAMETOPHYTE

COTYLEDONS

SHOOT APEX

DISPERSAL OF SEED

EMBRYO (2*n*)

HYPOCOTYL

ROOT APEX

SEED COAT (2*n*)

ROOT CAP

MEGAGAMETOPHYTE (*n*)

SEEDS ON UPPER SURFACE OF OVULIFEROUS SCALE

DEVELOPING EMBRYOS

SUSPENSOR

REMNANTS OF NUCELLUS

SEED

17–25

The common juniper (Juniperus communis) *has spherical ovulate cones like those of the cypresses, but in juniper the scales are fleshy and fused together. Juniper "berries" give gin its distinctive taste and aroma.*

(a)

(b)

17–26

The conifers of the yew family (Taxaceae) have seeds that are surrounded by a fleshy cup—the aril. The arils attract birds and other animals, which eat them and thus spread the seeds. (a) Members of the genus Taxus, *the yews—which occur around the Northern Hemisphere— produce red, fleshy ovulate structures. (b) Sporophylls and microsporangia of the pollen-bearing cones of a yew. Ovulate cones and pollen-bearing cones are found on separate individuals. The seeds and leaves of yews contain a toxic substance and are an important cause of poisoning by plants in children in the United States, although fatalities are extremely rare.*

One of the most interesting groups of conifers is the family Taxodiaceae, which appeared in the fossil record about 150 million years ago and is represented today by widely scattered species that are the remnants of populations that were much more widespread during the Tertiary period (Figure 17–29). One of the most remarkable of these is the coast redwood, *Sequoia sempervirens,* the tallest living plant. The famous "big tree," *Sequoiadendron giganteum,* which forms spectacular, widely scattered groves along the western slope of the Sierra Nevada of California, and the bald cypresses *(Taxodium)* of the southeastern United States and Mexico (Figure 17–27) also belong to this family.

Like most of the living genera of Taxodiaceae, *Metasequoia,* the dawn redwood, was much more widespread in the Tertiary period than it is now. *Metasequoia* occurred widely across Eurasia and was the most abundant conifer in western and arctic North America from the late Cretaceous period to the Miocene epoch (from about 90 to about 15 million years ago). It survived both in Japan and in eastern Siberia until a few million years ago. The genus *Metasequoia* was first described from fossil material by the Japanese paleobotanist Shigeru Miki in 1941 (Figure 17–29). Three years later the Chinese forester Tsang Wang, from the Central Bureau of Forest Research of China, visited the village of Mo-

17–27
The bald cypress (Taxodium distichum) is a deciduous member of the redwood family that grows in the swamps of the southeastern United States. In this autumn photograph, the leaves have begun to change color. Spanish "moss" (Tillandsia usneoides), actually a flowering plant related to the pineapple, is seen hanging in masses from the branches of these trees.

17–28
The dawn redwood (Metasequoia glyptostroboides). This tree, growing in Hubei Province in central China, is more than 400 years old.

17–29
Fossil branchlet of Metasequoia, about 50 million years old. The accompanying map shows the geographical distribution of some living and fossil members of the redwood family (Taxodiaceae).

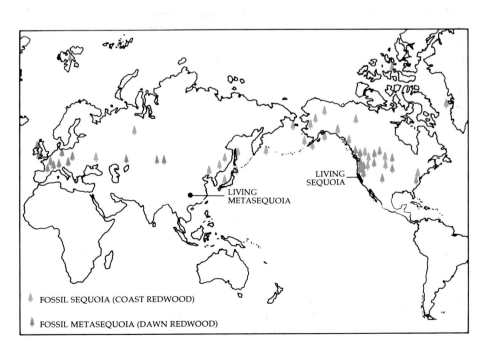

LIVING
METASEQUOIA

LIVING
SEQUOIA

🌲 FOSSIL SEQUOIA (COAST REDWOOD)

🌲 FOSSIL METASEQUOIA (DAWN REDWOOD)

tao-chi in remote Sichuan Province in south-central China. There he discovered a huge tree of a kind he had never seen before. The natives of the area had built a temple around the base of the tree. Tsang collected specimens of the tree's needles and cones, and studies of these samples revealed that the fossil *Metasequoia* had "come to life." In 1948, paleobotanist Ralph Chaney of the University of California at Berkeley led an expedition down the Yangtze River and across three mountain ranges to valleys where a thousand dawn redwoods were growing, the last remnant of the once great *Metasequoia* forest. Thousands of seeds were subsequently distributed widely, and this "living fossil" can now be seen growing in parks and gardens all over the world.

OTHER LIVING GYMNOSPERMS

Cycads

The other groups of living gymnosperms are remarkably diverse and scarcely resemble one another at all. Among them are the cycads, division Cycadophyta, which are palmlike plants found mainly in tropical and subtropical regions. These unique plants, which appeared at least 320 million years ago during the Car-

17–30
Male and female plants of Zamia pumila, *the only species of cycad native to the United States. The stems are mostly or entirely underground and, along with the storage roots, were used by Native Americans as food and as a source of starch. The two large gray cones in the foreground are ovulate cones; the smaller brown cones are microsporangiate cones.*

boniferous period, were so numerous in the Mesozoic era, along with the superficially similar cycadeoids, that this period is often called the "Age of Cycads and Dinosaurs." Living cycads comprise 11 genera, with about 140 species. *Zamia pumila*, which occurs commonly in the sandy woods of Florida, is the only cycad native to the United States (Figure 17–30).

Most cycads are fairly large plants; some reach 18 meters or more in height. Many have a distinct trunk that is densely covered with the bases of shed leaves. The functional leaves characteristically occur in a cluster at the top of the stem; thus the cycads resemble palms. (Indeed, a common name for some cycads is "sago palms.") Unlike palms, however, cycads exhibit true, if sluggish, secondary growth from a vascular cambium; the central portion of their trunks consists of a great mass of pith. Cycads are often highly toxic, with both neurotoxins and carcinogenic compounds abundant. They harbor cyanobacteria and make important contributions of fixed nitrogen to the areas where they occur.

The reproductive units of cycads are more or less reduced leaves with attached sporangia that are loosely or tightly clustered into conelike structures near the apex of the plant. The pollen and seed cones of cycads are borne on different plants (Figure 17–31). Their sperm are motile (Figure 17–32).

Recently, the role of insects in the pollination of cycads has received increasing emphasis. Beetles of several groups have frequently been found to be associated with the male cones, and less frequently with the female cones, of the members of several genera of cycads. For example, weevils (family Curculionidae) of the genus *Rhopalotria* carry out their entire life cycles upon and within male cones of *Zamia* and also visit the female cones. Pollen-eating beetles, although not weevils, have certainly been present throughout the history of the cycads; it seems reasonable to assume that there has been a long relationship between the members of the two groups. It is less certain that wind pollination plays an important role in the pollination of cycads, despite the general assumption that it does so.

Ginkgo

The maidenhair tree, *Ginkgo biloba*, is easily recognized by its fan-shaped leaves with their openly branched, dichotomous (forking) pattern of veins (Figure 17–33). It is an attractive and stately, but slow-growing, tree, which may reach a height of 30 meters or more. The leaves on the numerous spur shoots of *Ginkgo* are more or less entire, whereas those on the long shoots and seedlings are deeply lobed. Unlike most other gymnosperms, *Ginkgo* is deciduous; its leaves turn a beautiful golden color before falling in autumn.

(a)

(b)

17–31
(a) Encephalartos ferox, *a cycad native to Africa. Shown here is an ovulate plant with female cones.* (b) *An ovulate plant of* Cycas siamensis. *Several megasporophylls have been removed to reveal seeds on the upper surfaces of others.*

Ginkgo biloba is the sole living survivor of a genus that has changed little for more than 80 million years and is the only living member of the division Ginkgophyta. The living species shares features with other genera of gymnosperms that range back to the early Permian period, some 280 million years ago. There are probably no wild stands of *Ginkgo* anywhere in the world, but the tree was preserved in temple grounds in China and Japan. Introduced into other parts of the world, it has been an important feature of the parks and gardens of the temperate regions of the world for about 200 years. *Ginkgo* is especially resistant to air pollution and so is commonly cultivated in urban parks.

Like the cycads, *Ginkgo* bears its ovules and microsporangia on different individuals. The ovules of *Ginkgo* are borne in pairs on the end of short stalks and ripen to produce fleshy-coated seeds in autumn. In *Ginkgo*, fertilization within the ovules may not occur until after they have been shed from their parent tree. The male gametophyte forms an extensively branched, haustorium-like system that develops from an initially unbranched pollen tube. Attached to this system is a saclike structure that contains the tube nucleus, not enclosed within a tube cell in *Ginkgo*, and several internal cells.

Gnetophyta

The gnetophytes comprise three living genera and about 70 species of very unusual gymnosperms: *Gnetum, Ephedra,* and *Welwitschia.*

Gnetum, a genus of about 30 species, consists of trees and climbing vines with large, leathery leaves that closely resemble those of dicotyledons (Figure 17–34). It is found throughout the moist tropics.

17–32
Sexual reproduction in cycads and Ginkgo is unusual in combining motile sperm with pollen tubes. (a) *The sperm of the cycad* Zamia pumila, *shown here, swim by virtue of an estimated 40,000 flagella.* (b) *They are transported to the vicinity of the egg cells in the ovule by means of a pollen tube.*

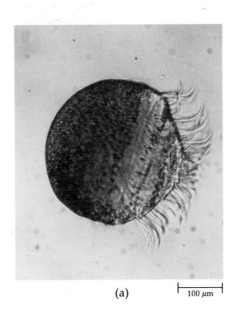

(a) |——| 100 μm

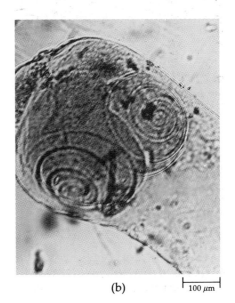

(b) |——| 100 μm

(a)

(b)

17–33

(a) Ginkgo biloba, *the maidenhair tree.*
This tree was given its English name
because of the resemblance between its
leaves and the leaflets of maidenhair
fern. (b) Ginkgo *leaves and fleshy seeds*
attached to spur shoot.

(a)

(b)

(c)

17–34

The large, leathery leaves of the tropical
gnetophyte Gnetum *resemble those of*
certain dicots. The species of Gnetum
grow as shrubs or woody vines in trop-
ical or subtropical forests. (a) *A function-*
ally microsporangiate inflorescence. (b)
Microsporangiate inflorescence and
leaves. (c) *Fleshy seeds with leaves.* (b)
and (c) *were photographed in the*
Amazon Basin of southern Venezuela.

(a)   (b)   (c)

17–35

Ephedra is the only one of the three living genera of Gnetophyta found in the United States. (a) *A male shrub of Ephedra viridis, in California. It is a densely branched shrub that has scalelike leaves, like other members of the genus.*

(b) *Microsporangiate strobili of* E. viridis. *Note the scalelike leaves on the stem.* (c) Ephedra trifurca, *in Arizona, with microsporangiate strobili.* (d) *Female plant of* E. viridis *with seeds.*

(d)

Most of the approximately 35 species of *Ephedra* are profusely branched shrubs with inconspicuous, small, scalelike leaves (Figure 17–35). With its small leaves and apparently jointed stems, *Ephedra* superficially resembles *Equisetum*. Most species of *Ephedra* inhabit arid or desert regions of the world.

Welwitschia is probably the most bizarre vascular plant (Figure 17–36). Most of the plant is buried in sandy soil. The exposed part consists of a massive, woody, concave disk that produces only two strap-shaped leaves, with the cone-bearing branches arising from meristematic tissue on the margin of the disk. *Welwitschia* grows in the coastal desert of southwestern Africa, in Angola, Namibia, and South Africa.

Although the genera of Gnetophyta are clearly related to one another and are appropriately placed together, they differ greatly in their characteristics. These genera do, however, have many angiosperm-like features, such as the similarity of their strobili to some angiosperm inflorescences, the presence of very similar vessels in their xylem, and the lack of archegonia in *Gnetum* and *Welwitschia*. For decades, scientists have debated whether the similarities between gnetophytes and angiosperms reflect descent from an immediate common ancestor or parallel evolution.

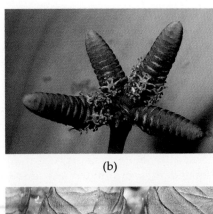

(b)

(c)

Recently, it has become clear, on the basis of careful comparisons and macromolecular analyses, that gnetophytes are the group most closely related to the angiosperms. In 1990, it was reported that double fertilization—a process involving the fusion of a second sperm nucleus with a nucleus of the female gametophyte—also occurs frequently in *Ephedra*. Thus double fertilization, hitherto considered a unique characteristic of the angiosperms, may actually have evolved in the common ancestor of angiosperms and gnetophytes. None of the living gnetophytes, however, could possibly be an ancestor of any angiosperm—each of the three living genera of gnetophytes has its own unique specializations. Interestingly, the reproductive structures of at least some species of all three genera of gnetophytes produce nectar and are visited by insects. Wind pollination is clearly important, but insects also play a role in the pollination of these unique plants.

17–36
The gnetophyte Welwitschia mirabilis, found only in the Namib Desert of Namibia and adjacent regions of southwestern Africa. Welwitschia *produces only two leaves, which continue to grow for the life of the plant. As growth continues, the leaves break off at the tips and split lengthwise; thus older plants appear to have numerous leaves. (a) A large, seed-producing plant. (b) Microsporangiate strobili. (c) Ovulate strobili; the insect is a fire bug, sucking sap from the strobili.* Welwitschia *is dioecious.*

Summary

Seed plants consist of five divisions with living representatives. One of these, the overwhelmingly successful angiosperms (division Anthophyta), is characterized by a unique set of reproductive features. The remaining four, which lack these specializations, are collectively termed the gymnosperms.

In addition to producing seeds, all seed plants bear megaphylls. The prerequisites of seed formation include heterospory; retention of a single megaspore (or, in angiosperms, of the products of a single meiotic division); development of the embryo, or young sporophyte, within the megagametophyte; and integuments. All seeds consist of a seed coat derived from the integument(s), an embryo, and stored food, which may be transferred to the embryo before the seed is mature. In gymnosperms, the stored food is provided by the haploid female gametophyte.

The oldest known seedlike structures occur in strata of the late Devonian period, about 360 million years old. The likely progenitors of the gymnosperms and angiosperms are the progymnosperms, a Paleozoic group of seedless vascular plants, which may also have given rise separately to the seed ferns and to the ferns. Among the major extinct groups of gymnosperms were the seed ferns (division Pteridospermophyta), a diverse group, and the cycadeoids (division Cycadeoidophyta), which resembled cycads but had bisexual cones.

Living gymnosperms comprise four divisions: Cycadophyta, Ginkgophyta, Coniferophyta, and Gnetophyta. Their life cycles are essentially similar: an alternation of heteromorphic generations with large, independent sporophytes and greatly reduced gametophytes. The ovules (megasporangia plus integuments) are exposed on the surfaces of the megasporophylls or analogous structures. At maturity the female gametophyte of most gymnosperms is a multicellular structure with several archegonia. The male gametophytes develop as pollen grains. Antheridia are lacking in all seed plants. In gymnosperms, the male gametes, or sperm, arise directly from the spermatogenous cell. Except for the cycads and *Ginkgo*, which have flagellated sperm, the sperm of seed plants are nonmotile.

In seed plants, free water is not necessary in order for the sperm to reach the eggs; instead, the sperm are conveyed to the eggs by a combination of pollination and pollen-tube formation. Pollination in seed plants is the transfer of pollen from microsporangium to megasporangium. Subsequently, one sperm of the male gametophyte (germinated pollen grain) unites with the egg, which in most gymnosperms is located in an archegonium. The second sperm has no apparent function (except perhaps in *Ephedra*), and it disintegrates. After fertilization in seed plants, each ovule develops into a seed.

The conifers (division Coniferophyta) are the largest and most widespread division of living gymnosperms, with about 50 genera and some 550 species. They dominate many plant communities throughout the world, with pines, firs, spruces, and other trees familiar over wide stretches of the north. Living cycads (division Cycadophyta) consist of 11 genera and about 140 species, mainly tropical but extending away from the equator in warmer regions. Cycads are palmlike plants with trunks and sluggish secondary growth. There is only one living species of the division Ginkgophyta, the maidenhair tree *(Ginkgo biloba)*, which is known only from cultivation. The three genera of the division Gnetophyta are the closest living relatives of the angiosperms (division Anthophyta), with which they share many characteristics.

Suggestions for Further Reading

Beck, Charles B. (ed.): *Origin and Evolution of Gymnosperms*, Columbia University Press, New York, 1988.
An excellent series of contemporary papers on every aspect of gymnosperm evolution.

Bierhorst, David W.: *Morphology of Vascular Plants*, Macmillan Publishing Company, New York, 1971.
An encyclopedic and profusely illustrated treatise on the morphology of vascular plants.

Gifford, Ernest M., and Adriance S. Foster: *Morphology and Evolution of Vascular Plants*, 3d ed., W.H. Freeman and Company, New York, 1989.
A well-organized, general account of the vascular plants, containing much interpretive material.

Niklas, Karl J.: "Aerodynamics of Wind Pollination," *Scientific American*, July 1987, pages 90–95.
The complex aerodynamics of pollen grains and the structures on which they land exert a high degree of control in the apparently random process of pollen capture.

Norstog, Knut: "Cycads and the Origin of Insect Pollination," *American Scientist* 75:270–279, 1987.
The extent of insect pollination in these ancient plants is just beginning to be understood.

Taylor, Thomas N.: "Reproductive Biology in Early Seed Plants," *BioScience* 32:23–28, 1982.
A good review of the reproductive features of early seed plants.

Introduction to the Angiosperms

18–1

A giant eucalyptus, or red tingle (Eucalyptus jacksonii), *growing in the Valley of the Giants in southwestern Australia. Note the man standing in the burnt-out base of this enormous angiosperm.*

The angiosperms make up much of the visible world of modern plants. Trees, shrubs, lawns, gardens, fields of wheat and corn, wildflowers, fruits and vegetables on the grocery shelves, the bright splashes of color in a florist's window, the geranium on a fire escape, duckweed and water lilies, eel grass and turtle grass, saguaro cacti and prickly pears—wherever you are, flowering plants are there also.

Angiosperms make up the division Anthophyta, which includes about 235,000 species and is thus, by far, the largest division of photosynthetic organisms. In their vegetative features, angiosperms are enormously diverse. In size, they range from species of *Eucalyptus*, trees well over 100 meters tall with trunks nearly 20 meters in girth (Figure 18–1), to some duckweeds, which are simple, floating monocots often scarcely 1 millimeter long (Figure 18–2). Some angiosperms are vines that climb high into the canopy of the tropical rainforest, while others are epiphytes that grow in that canopy. Many angiosperms, such as cacti, are adapted for growth in extremely arid regions. For more than 100 million years, the flowering plants have dominated the land.

In terms of their evolutionary history, the angiosperms are a group of seed plants with special characteristics: flowers, fruits, and distinctive life-cycle features that differ from those of other organisms. In this chapter, we shall outline these characteristics and place them in perspective; in the following chapter, we shall discuss the evolution of the angiosperms. In Section 4, we shall consider in some detail the structure and development of the angiosperm plant body (sporophyte).

The division Anthophyta includes two classes: the Monocotyledones (monocots), with about 65,000 species (Figure 18–3), and the Dicotyledones (dicots), with about 170,000 species (Figure 18–4). The similarities between these two groups are far greater than the differences; nonetheless, the two classes are clearly recog-

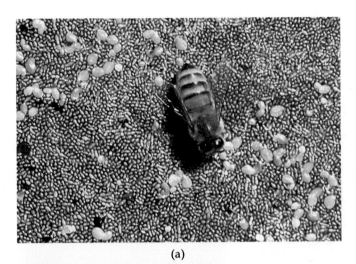

(a)

(c)

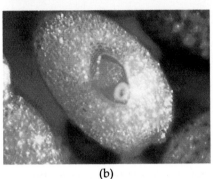

(b)

18–2

The duckweeds (family Lemnaceae) are the smallest flowering plants; their structural features mark them as extremely reduced relatives of the aroid family (Araceae), the family that includes calla lilies (Zantedeschia) and Philodendron. (a) A honeybee resting on a dense floating mat of three species of duckweed. The larger plants are Lemna gibba, *about 2 to 3 millimeters long; the smaller ones* are two species of Wolffia, *up to 1 millimeter long. (b) A flowering plant of W. borealis, with a circular concave stigma (looking like a tiny doughnut) and a minute anther just above it, both protruding from a central cavity. The whole plant is less than 1 millimeter long. (c) Flowering plant of L. gibba; two stamens and a style protrude from a pocket on the upper surface of the plant.*

(a)

(b)

(c)

18–3

Monocots. (a) A member of the palm family, the coconut palm (Cocos nucifera), growing in Tehuanatepec, Oaxaca, Mexico. The fruit of coconut is a drupe, not a nut (see Chapter 19). (b) Flowers and fruits of the banana plant (Musa × paradisiaca). The banana flower has an inferior ovary, and the tip of the fruit bears a large scar left by the fallen flower parts. (c) Rice (Oryza sativa) is a member of the grass family.*

(a)

(b)

(c)

18–4

Dicots. (a) The water lily (Nymphaea odorata). The very fragrant flower of this species contains numerous petals and stamens and is regular, or radially symmetrical. The genus Nymphaea is widely distributed in tropical and temperate regions throughout the world.

(b) Saguaro cactus (Carnegiea gigantea). The cacti, of which there are about 2000 species, are almost exclusively a New World family. The thick, fleshy stems, which store water, contain chloroplasts and have taken over the photosynthetic

function of the leaves. (c) Round-lobed hepatica (Hepatica americana), which flowers in deciduous woodlands in the early spring. The flowers have no petals but have six to ten sepals and numerous spirally arranged stamens and carpels.

nizable. Monocots include such familiar plants as the grasses, lilies, irises, orchids, cattails, and palms. The dicots include almost all the familiar trees and shrubs (other than the conifers) and many of the herbs (nonwoody plants). The major differences between monocots and dicots are summarized in Table 18–1.

In terms of their mode of nutrition, all but a few angiosperms are free-living, but both parasitic and saprophytic forms do exist (Figure 18–5). These latter plants mostly or completely lack chlorophyll. It has recently been shown that many, if not all, angiosperms considered to be saprophytes (and thus obtaining their nutrition from decaying organic matter) have obligate relationships with mycorrhizal fungi that are associated with a second plant, in this case, a green, actively growing photosynthetic angiosperm. The fungus forms a bridge that transfers carbohydrates from the photosynthetic plant to the "saprophyte." Thus the recipient plants might more accurately be considered parasites, at least in part. In addition to such plants, there are about 2800 species of parasitic dicots and about 200 species of parasitic monocots; dicot parasites include the mistletoes, *Cuscuta* (Figure 18–5a), and *Rafflesia* (Figure 18–5c). Interestingly, there is also a single species of parasitic gymnosperm, *Parasitaxus ustus* of New Caledonia. Parasitic flowering plants form specialized

Table 18–1 *Main Differences between Monocots and Dicots*

Characteristic	Dicots	Monocots
Flower parts	In fours or fives (usually)	In threes (usually)
Pollen	Basically tricolpate (having three furrows or pores)	Basically monocolpate (having one furrow or pore)
Cotyledons	Two	One
Leaf venation	Usually netlike	Usually parallel
Primary vascular bundles in stem	In a ring	Complex arrangement
True secondary growth, with vascular cambium	Commonly present	Absent

18–5

Parasitic and saprophytic angiosperms. These plants have little or no chlorophyll; they obtain their food as a result of the photosynthesis of other plants. (a) Dodder (Cuscuta salina), a parasite that is bright orange or yellow. Dodder is a member of the morning glory family (Convolvulaceae). (b) Indian pipe (Monotropa uniflora), a "saprophyte" that obtains its food from the roots of other plants via the fungal hyphae associated with its roots. (c) The world's largest flower, Rafflesia arnoldii, on Mount Sago, Sumatra. Plants of this genus are parasitic on the roots of a member of the grape family (Vitaceae). All of these plants are dicots. There are more than 3000 species of parasitic angiosperms, representing 17 families.

(a)

(b)

(c)

absorptive organs called haustoria, by means of which they penetrate the tissues of their hosts.

The Flower

The flower is a determinate shoot that bears sporophylls (Figure 18–6). The name "angiosperm" is derived from the Greek words *angeion*, meaning "vessel," and *sperma*, meaning "seed." The definitive structure of the flower is the carpel—the "vessel." A carpel is a specialized, longitudinally folded megaphyll. It contains the ovules, which develop into seeds after fertilization.

Flowers may be clustered in various ways into aggregations called **inflorescences** (Figures 18–7 and 18–8). The stalk of an inflorescence or of a solitary flower is known as a **peduncle,** while the stalk of an individual flower in an inflorescence is called a **pedicel.** The part of the flower stalk to which the flower parts are attached is termed the **receptacle.** Like any other shoot tip, the receptacle consists of nodes and internodes. In the flower, the internodes are very short, and consequently the nodes are very close together.

Many flowers include two sets of sterile appendages, the **sepals** and **petals,** which are attached to the receptacle below the fertile parts of the flower, the **stamens** and **carpels.** The sepals arise below the petals, and the stamens arise below the carpels. Collectively, the sepals form the **calyx,** and the petals form the **corolla.** Together, the calyx and corolla constitute the **perianth.** The sepals and petals are essentially leaflike in structure. Commonly the sepals are green and the petals are brightly colored, although in many flowers the members of both whorls are similar in color.

The stamens—collectively the **androecium** ("house of man")—are microsporophylls. In all but a few living angiosperms, the stamen consists of a slender stalk, or **filament,** upon which is borne a two-lobed **anther** containing four microsporangia, or pollen sacs.

The carpels—collectively the **gynoecium** ("house of woman")—are megasporophylls that are folded lengthwise, enclosing one or more ovules. A given flower may contain one or more carpels. If more than one carpel is present, they may be separate or fused together, in part or entirely. Sometimes the individual carpel or the group of fused carpels is called a pistil. The word "pistil" comes from the same root as "pestle," the instrument with a similar shape that pharmacists use for grinding substances into a powder in a mortar.

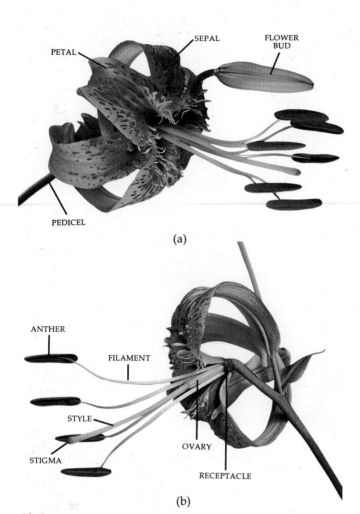

(a)

(b)

18–6

*Parts of a lily (Lilium henryi) flower.
(a) An intact flower. In some flowers,
such as lilies, the sepals and petals are
similar to one another, and the perianth
parts may then be referred to as tepals.
Note that the sepals are attached to the
receptacle below the petals. (b) A partly
dissected flower with two tepals and two
stamens removed to reveal the ovary.
The gynoecium consists of the ovary,
style, and stigma. The stamen consists of
the filament and anther. Note that the
sepals, petals, and stamens are attached
to the receptacle below the ovary, which
is made up, in the lily flower, of three
fused carpels. Such a flower is said to
be hypogynous.*

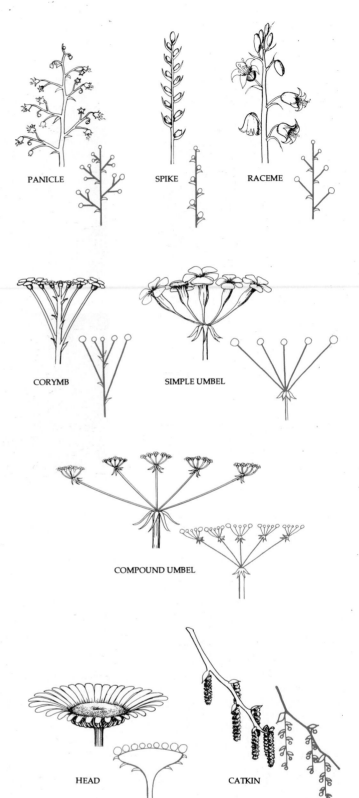

18–7

*Illustrations of some of the common
types of inflorescences found in the angiosperms, accompanied by simplified
diagrams (in color).*

(a)

(b)

(c)

(d)

(e)

18–8
Inflorescences of (a) *shooting star* (Dodecatheon pauciflorum), (b) *butter-and-eggs*
(Linaria vulgaris), (c) *lupine* (Lupinus
diffusus), (d) *bluebells* (Mertensia virginica), *and* (e) *water hemlock* (Cicuta
maculata). *Using Figure 18–7 as a guide,
can you identify the types of inflorescences
shown here?*

In most flowers, the individual carpels or group of fused carpels are differentiated into a lower part, the **ovary,** which encloses the ovules, and an upper part, the **stigma,** which receives the pollen. In many flowers a more or less elongated structure—the **style**—connects the stigma with the ovary. If the carpels are fused, there may be a common style or stigma, or each carpel may retain a separate style and stigma. The common ovary of such fused carpels is generally (but not always) partitioned into two or more locules—chambers of the ovary that contain the ovules. The number of locules is usually related to the number of carpels in the gynoecium.

The portion of the ovary where the ovules originate and to which they remain attached until maturity is called the **placenta.** The arrangement of the placentae (the placentation) and, consequently, of the ovules varies among different groups of flowering plants (Figure 18–9). In some flowers, the placentation is **parietal;** that is, the ovules are borne on the ovary wall or on extensions of it. In other flowers, the ovules are borne on a central column of tissue in a partitioned ovary with as many locules as there are carpels. This is **axile** placentation. In still others, the placentation is **free central,** the ovules being borne on a central column of tissue not connected by partitions with the ovary wall. And finally, in some flowers a single ovule occurs at the very base of a unilocular ovary. This is **basal** placentation. These differences are important in the classification of the flowering plants.

There are many variations on this basic flower structure in different kinds of plants. The majority of flowers include both stamens and carpels, and such flowers are said to be **perfect.** If either stamens or carpels are missing, the flower is **imperfect,** and depending on the part that is present, the flower is said to be either **staminate** or **carpellate** (or pistillate) (Figure 18–10). If both staminate and carpellate flowers occur on the same plant, as in corn (maize) and the oaks, the species is said to be **monoecious** (from the Greek words *monos,* "one," and *oecos,* "house"). If staminate and carpellate flowers are found on separate plants, the species is said to be **dioecious** ("two houses"), as in the willows and hemp *(Cannabis sativa).*

Any one of the floral whorls—sepals, petals, stamens, or carpels—may be lacking in the flowers of certain groups. Flowers with all four floral whorls are called **complete** flowers. If any whorl is lacking, the flower is said to be **incomplete.** Thus an imperfect flower is also incomplete, but not all incomplete flowers are imperfect.

The particular arrangement of the floral parts may be spiral on a more or less elongated receptacle, or similar parts—such as petals—may be attached in a whorl at a node. The parts may be united with other members of the same whorl *(coalescence)* or with members of

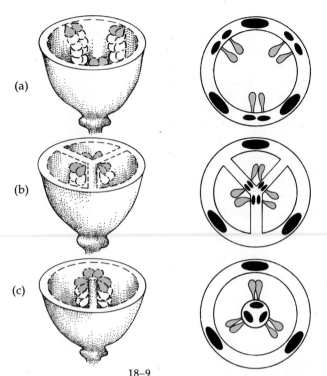

(a)

(b)

(c)

18–9

Types of placentation. (a) Parietal. (b) Axile. (c) Free central. Basal placentation is not shown here.

18–10

Staminate and carpellate flowers of a tan-bark oak (Lithocarpus densiflora). Most members of the family Fagaceae, including the true oaks (Quercus), are monoecious, meaning that the staminate and carpellate flowers are separate but are borne on the same tree.

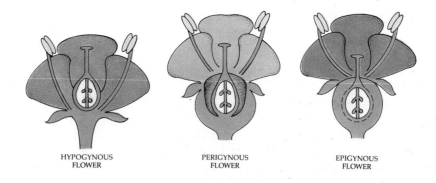

HYPOGYNOUS FLOWER   PERIGYNOUS FLOWER   EPIGYNOUS FLOWER

18–11

Types of flowers in common families of dicots, showing differences in the position of the ovary. In Ranunculaceae, for example, the sepals, petals, and stamens are attached below the ovary and there is no fusion; such flowers are said to be hypogynous. Many Rosaceae have superior ovaries, with the sepals, petals, and stamens attached below. Such flowers are said to be perigynous because these flower parts are fused together into a cup-shaped extension. Apiaceae, for example, have inferior ovaries; that is, the sepals, petals, and stamens appear to be attached above the ovaries. Such flowers are said to be epigynous.

other whorls *(adnation).* An example of adnation is the union of stamens with the corolla, which is fairly common. When the floral parts of the same whorl are not joined, the prefixes *apo-* (meaning "separate") or *poly-* may be used to describe the condition. When the parts are coalesced, either *syn-* or *sym-* is used. For example, in an aposepalous or polysepalous calyx, the sepals are not joined; in a synsepalous calyx, they are joined.

In addition to this variation in arrangement of flower parts (spiral or whorled), the level of insertion of the sepals, petals, and stamens on the floral axis varies in relation to the level of the ovary or ovaries (Figure 18–11). If the sepals, petals, and stamens are attached to the receptacle below the ovary, the ovary is said to be **superior** and the flower is said to be **hypogynous** (Figure 18–6). In some flowers with superior ovaries, the sepals, petals, and stamens are fused together to form a cup-shaped extension of the receptacle called the **hypanthium.** Such flowers are said to be **perigynous** (Figure 18–12); the sepals, petals, and stamens appear to arise from the margin of the cup. In other flowers the sepals, petals, and stamens apparently grow from the top of the ovary, which is **inferior.** Such flowers are said to be **epigynous** (Figure 18–13).

Finally, mention should be made of symmetry in flower structure. In some flowers, the corolla is made up of petals of similar shape that radiate from the center of the flower and are equidistant from each other; they are radially symmetrical. Such flowers are said to be **regular,** or actinomorphic (from the Greek root *actinos,* "star"). In other flowers, one or more members of at least one whorl are different from other members of the same whorl; they are bilaterally symmetrical. These flowers are said to be **irregular,** or zy-

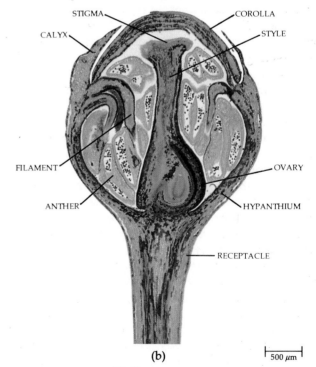

(a)

(b)

500 μm

18–12

Cherry (Prunus) flowers, (a) and (b), exhibit perigyny—their sepals, petals, and stamens are attached to a hypanthium (a cup-shaped extension of the receptacle). In (b) the stamens are bent and crowded in the hypanthium because the flower has not yet opened.

18–13
Apple (Malus sylvestris) *flowers,* (a) *and*
(b), *exhibit epigyny—their sepals, petals,*
and stamens apparently arise from the
top of the ovary. In (b) *the flower is very*
nearly open, but the stamens are not yet
erect.

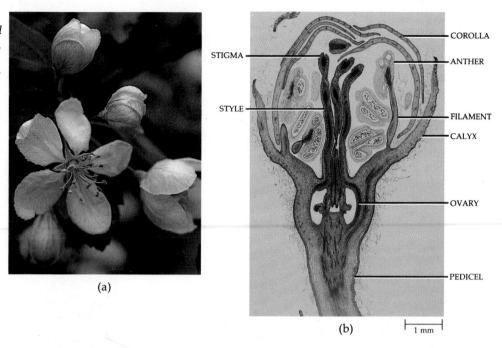

(a)

COROLLA

STIGMA

ANTHER

STYLE

FILAMENT

CALYX

OVARY

PEDICEL

(b)

1 mm

gomorphic (Greek *zygos,* a "yoke" or "pair"; for an
example, see Figure 18–8b, c).

The Angiosperm Life Cycle

Angiosperm gametophytes are very much reduced in
size—more so than those of any other heterosporous
plants, including the gymnosperms. The mature microgametophyte consists of only three cells; the mature
megagametophyte, which is held for its entire existence
within the tissues of the sporophyte, consists of only
seven cells in most species. Both antheridia and archegonia are lacking. Pollination is indirect: pollen is deposited on the stigma, after which the pollen tube con-
veys two nonmotile sperm to the female gametophyte.
After fertilization, the ovule develops into a seed,
which is enclosed in the ovary. At the same time, the
ovary (and sometimes additional structures associated
with it) develops into a fruit.

MICROSPOROGENESIS AND
MICROGAMETOGENESIS

Microsporogenesis is the formation of microspores
within the microsporangia, or pollen sacs, of the
anther. Microgametogenesis is the development of the
microspore into the microgametophyte, or pollen grain.

After it has first differentiated, the anther consists of
a uniform mass of cells, except for the partly differentiated epidermis. Eventually, four groups of fertile, or
sporogenous, cells become discernible within the anther;
each group is surrounded by several layers of sterile
cells. The sterile cells develop into the wall of the pollen
sac, which includes nutritive cells that supply food to
the developing microspores. The nutritive cells constitute the **tapetum,** the innermost layer of the pollen sac
wall (Figure 18–14). The sporogenous cells become microsporocytes, which divide meiotically, each diploid
microsporocyte giving rise to a tetrad of haploid microspores. Microsporogenesis is completed with formation
of the single-celled microspores.

During meiosis, each nuclear division may be followed immediately by cell wall formation, or the four
microspore protoplasts may be walled off simultaneously after the second meiotic division. The first condition is common in monocots, the second in dicots.
Subsequently, the major features of the pollen grains
are established (Figure 18–15). The pollen grains develop a resistant outer wall, the **exine,** and a cellulosic
inner wall, the **intine.** The exine is composed of a very
resistant substance known as **sporopollenin,** which apparently is derived partly from the tapetum and partly
from the microspores. The intine, which is composed of
cellulose and pectin, is laid down by the microspore
protoplasts.

Microgametogenesis in angiosperms is uniform and
begins when the uninucleate microspore divides mitotically, forming two cells within the original spore wall.

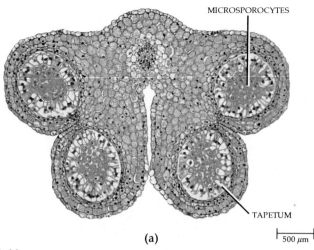

MICROSPOROCYTES

TAPETUM

(a)   500 μm

(b)   500 μm

18–14
*Two transverse sections of lily (Lilium)
anthers. (a) Immature anther, showing
the four pollen sacs containing micro-*
*sporocytes surrounded by the tapetum. (b)
Mature anther containing pollen grains.
The partitions between adjacent pollen*
*sacs break down prior to dehiscence, as
shown here.*

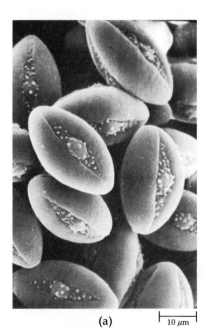

(a)   10 μm

(b)   100 μm

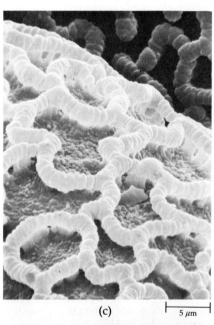

(c)   5 μm

18–15
*The wall of the pollen grain serves to
protect the male gametophyte on its often
hazardous journey between the anther
and the stigma. The outer layer, or exine,
is composed chiefly of a substance known
as sporopollenin, which appears to be a
polymer composed chiefly of carotenoids.
The exine, which is remarkably tough
and resistant, is often elaborately
sculptured.*

    *The sculpturing of pollen grain walls
is very precise and distinctly different
from one species to another, as revealed
in these scanning electron micrographs*
*of pollen grains. (a) Pollen grains of the
horse chestnut (Aesculus hippocastanum). The pore of each grain through
which a pollen tube may emerge is visible in the furrow. (b) Pollen grains of
a lily (Lilium longiflorum). (c) Detail
of the surface of a lily (L. longiflorum)
pollen grain. (d) Pollen grain of the
western ragweed (Ambrosia psilosta-
chya). The pollen of ragweed is a
primary cause of hay fever. Spiny pollen
grains such as these are common among
members of the sunflower family, Asteraceae, of which ragweed is a member.*

(d)

389

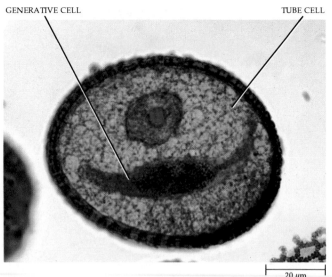

GENERATIVE CELL   TUBE CELL

20 μm

18–16
Mature pollen grain of Lilium, *contain-*
ing a two-celled male gametophyte. The
spindle-shaped generative cell will divide
mitotically, giving rise to two sperm; the
larger tube cell, which contains the gen-
erative cell, will form the pollen tube.
The round structure above the generative
cell is the tube cell nucleus, with its
red-stained nucleolus.

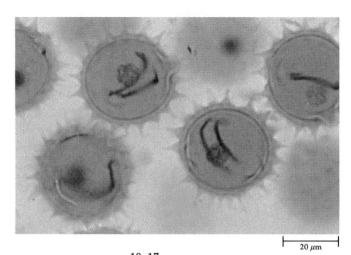

20 μm

18–17
Mature pollen grains—three-celled
male gametophytes of the telegraph plant,
Silphium *(family Asteraceae). Prior to*
pollination, each pollen grain contains
two filamentous sperm cells, which are
suspended in the cytoplasm of the larger
tube cell. In other words, the pollen of
Silphium *is shed at the three-celled*
stage, whereas that of Lilium, *shown*
in the preceding figure, is shed at the
two-celled stage.

One cell is called the **tube cell,** and the other is called
the **generative cell** (Figure 18–16). In about two-thirds
of flowering plant species, the microgametophyte is in
this two-celled stage at the time the pollen grains are
liberated by dehiscence of the anther. In the remainder,
the generative nucleus divides prior to the release of
the pollen grains, giving rise to two male gametes, or
sperm (Figure 18–17). Mature pollen grains are packed
with starch or oils, depending on the group, and are a
very nutritious source of food for animals.

Pollen grains, like the spores of seedless vascular
plants, vary considerably in shape and size, ranging
from less than 20 micrometers to over 250 micrometers
in diameter. They also differ in the number and arrangement of the pores through which the pollen tubes
ultimately grow. Nearly all families, many genera, and
a fair number of species of flowering plants can be
identified solely by their pollen grains, on the basis of
such characteristics as size, number of pores, and
sculpturing. In contrast to the large portions of plants
—such as leaves, flowers, and fruits—the pollen
grains, because of the chemical nature of the exine, are
extremely well represented in the fossil record. Thus
pollen provides a valuable index to the kinds of plants
and the nature of the climate that prevailed in the past.

Pollen grains and spores of seedless vascular plants
both have a sporopollenin wall, and both are the product of meiosis. Pollen grains, however, have two or
three nuclei when shed, as a result of mitosis; spores
have only one. The spores germinate from a suture
centrally located in the body, whereas pollen grains

germinate through their pores. The distinctive structure
and arrangement of these pores often allow pollen
grains to be distinguished from spores morphologically.

MEGASPOROGENESIS AND MEGAGAMETOGENESIS

Megasporogenesis is the process of megaspore formation within the nucellus (megasporangium). Megagametogenesis is the development of the megaspore into
the megagametophyte (embryo sac).

The ovule is a relatively complex structure, consisting of a stalk—the **funiculus**—bearing a nucellus enclosed by one or two integuments. Depending on the
species, one to many ovules may arise from the placentae, or ovule-bearing regions of the ovary wall. Initially,
the developing ovule is entirely nucellus, but soon it
develops one or two enveloping layers, the integuments, with a small opening, the micropyle, at one end
(Figure 18–18).

Early in the development of the ovule, a single
megasporocyte arises in the nucellus (Figure 18–18a).
The diploid megasporocyte divides meiotically (Figure
18–18b) to form four haploid megaspores, which are
generally arranged in a linear tetrad. With this, megasporogenesis is completed. Of the four megaspores,
three usually disintegrate and the one farthest from the
micropyle survives and develops into the megagametophyte.

18–18
Lilium. *Some stages in development of an ovule and embryo sac. (a) Two young ovules, each with a single, large megasporocyte. Integuments have not begun to develop. (b) Ovule now has integuments. The megasporocyte is in the first prophase of meiosis. (c) Ovule with eight-nucleate embryo sac (only six of the nuclei can be seen here, two at the chalazal end and four at the micropylar end). The polar nuclei have not yet migrated to the center of the sac.*

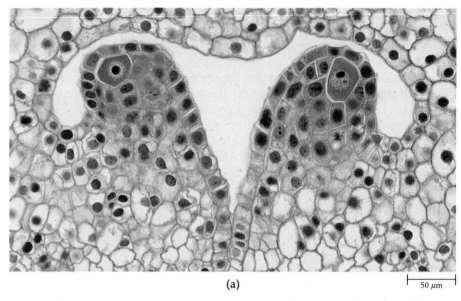

(a)

50 μm

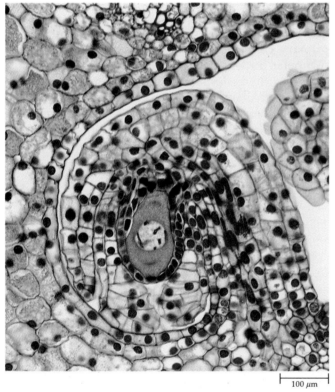

100 μm

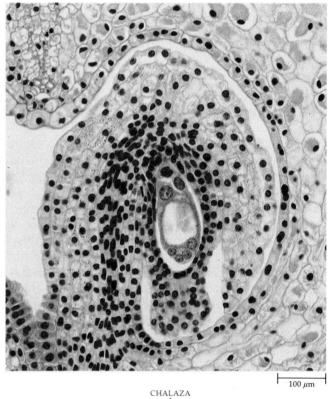

100 μm

CHALAZA

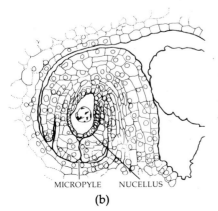

MICROPYLE   NUCELLUS

(b)

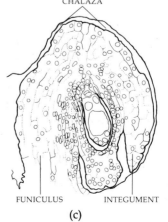

FUNICULUS   INTEGUMENT

(c)

The functional megaspore soon begins to enlarge at the expense of the nucellus, and the nucleus of the megaspore divides mitotically. Each of the resulting nuclei divides mitotically, followed by yet another mitotic division of the four resultant nuclei. At the end of the third mitotic division, the eight nuclei are arranged in two groups of four, one group near the micropylar end of the megagametophyte and the other at the opposite, or **chalazal,** end. One nucleus from each group migrates into the center of the eight-nucleate cell; these two nuclei are then known as the **polar nuclei.** The three remaining nuclei at the micropylar end become organized as the **egg apparatus,** consisting of an **egg cell** and two cellular **synergids.** Cell wall formation also occurs around the three nuclei left at the chalazal end, forming the so-called **antipodals.** The **central cell,** containing the polar nuclei, remains binucleate. The eight-nucleate, seven-celled structure is the mature female gametophyte, or **embryo sac** (Figure 18–18c).

The pattern of embryo-sac development just described is the most common one. Other patterns of embryo-sac development occur in about a third of the species of angiosperms that have been investigated.

POLLINATION AND FERTILIZATION

With **dehiscence** (shedding of contents) of the anthers, the pollen grains are transferred to the stigmas by a variety of vectors (see Chapter 19); the process whereby this transfer occurs is called pollination. Once in contact with the stigma, the pollen grains take up additional water; the water enters the pollen grains from the cells of the stigma surface along a water potential gradient. Following this hydration, the pollen grain germinates, forming a pollen tube. If the generative cell has not already divided, it soon does so, forming the two sperm. The germinated pollen grain, with its tube nucleus and two sperm, constitutes the mature microgametophyte (Figure 18–19).

The stigma and style are modified both structurally and physiologically to facilitate the germination of the pollen grain and growth of the pollen tube. The surface of many stigmas is essentially glandular tissue (*stigmatic tissue*), which secretes a sugary solution. The stigmatic tissue is connected with the ovule by *transmitting tissue,* which serves as a path through the style for the growing pollen tubes. Some styles contain open canals, which are lined with transmitting tissue. In such styles, the pollen tubes grow either along or among the cells of the lining. In most angiosperms, however, the styles are solid, with one or more strands of transmitting tissue extending from the stigma to the ovules. The pollen tubes grow either between the cells of the transmitting tissue or within their thick walls, depending on the kind of plant.

Commonly, the pollen tube enters the ovule through the micropyle and penetrates one of the synergids, which begins to degenerate soon after pollination has taken place but before the pollen tube has reached the embryo sac. The two sperm and the tube nucleus are then released into the synergid through the subterminal pore that develops in the pollen tube. Ultimately, one sperm nucleus enters the egg cell and the other enters the central cell, where it unites with the two polar nuclei (Figure 18–20). Recall that in most gymnosperms, only one of the two sperm is functional; one unites with the egg and the other degenerates. The involvement of both sperm in this process—the union of one with the egg and the other with the polar nuclei—is called **double fertilization;** it represents an unusual characteristic of the angiosperms, otherwise known only in *Ephedra* (division Gnetophyta), in which no endosperm is formed. In double fertilization in angiosperms, both sperm nuclei fuse with cells of the megagametophyte, one with the egg to form a diploid zygote, and the other with the polar nuclei to form the **primary endosperm nucleus.** Because there are often two polar nuclei, the latter process is called **triple fusion,** and the endosperm is often triploid. When there are more than two polar nuclei, the endosperm may have a higher multiple of the plant's basic chromosome number. The tube nucleus disintegrates during the

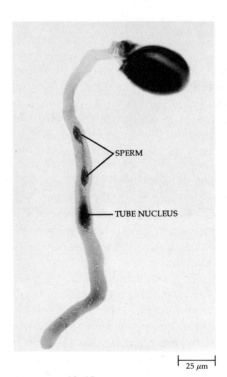

SPERM

TUBE NUCLEUS

25 μm

18–19
Mature male gametophyte of Solomon's seal (Polygonatum). The sperm and the tube nucleus can be seen in the pollen tube.

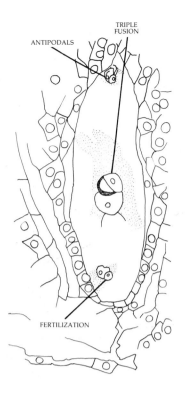

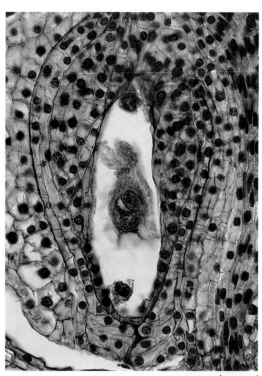

18–20
Lilium. *Double fertilization. Union of sperm and egg nuclei—"true" fertilization—can be seen in the lower half of the micrograph. Triple fusion of the other sperm nucleus and the two polar nuclei has taken place above.*

ANTIPODALS

TRIPLE FUSION

FERTILIZATION

50 μm

process of double fertilization, and the remaining synergid and antipodals disintegrate during the same process or early in the course of differentiation of the embryo sac.

DEVELOPMENT OF SEED AND FRUIT

In double fertilization, several processes are initiated: the primary endosperm nucleus divides, forming the **endosperm;** the zygote develops into an embryo; the integuments develop into a seed coat; and the ovary wall and related structures develop into a fruit.

In contrast to the embryogeny of the majority of gymnosperms, which begins with a free nuclear stage, embryogeny in angiosperms resembles that of the seedless vascular plants in that the first nuclear division of the zygote is accompanied by cell wall formation. In the early stages of development, the embryos of dicotyledons and monocotyledons undergo similar sequences of cell division, both becoming spherical bodies. It is with the formation of the cotyledons that a distinction first appears between dicot and monocot embryos: the dicot embryo develops two cotyledons, whereas the monocot embryo forms only one. The details of angiosperm embryogeny are presented in Chapter 20.

Endosperm formation begins with the mitotic division of the primary endosperm nucleus and usually is initiated prior to the first division of the zygote. In some angiosperms, a variable number of free nuclear divisions precede cell wall formation (nuclear-type endosperm formation); in other species the initial and subsequent mitoses are followed by cytokinesis (cellu-

lar-type endosperm formation). Although endosperm development may occur in a variety of ways, the function of the resulting tissue remains the same: to provide essential food materials for the developing embryo and, in many cases, the young seedling as well. In the seeds of some groups of angiosperms, the nucellus proliferates into a food-storage tissue known as **perisperm.** Some seeds may contain both endosperm and perisperm, as do those of the beet *(Beta)*. In many dicots and some monocots, however, most or all of these storage tissues are absorbed by the developing embryo before the seed becomes dormant, as in peas or beans. The embryos of such seeds commonly develop fleshy, food-storing cotyledons. The principal food materials stored in seeds are carbohydrates, proteins, and lipids.

Angiosperm seeds differ from those of gymnosperms in the origin of their stored food. In gymnosperms, the stored food is provided by the female gametophyte; in angiosperms, it is provided, at least initially, by endosperm, which is neither gametophytic nor sporophytic tissue.

Concomitantly with development of the ovule into a seed, the ovary (and sometimes other portions of the flower or inflorescence) develops into a fruit. As this occurs, the ovary wall, or **pericarp,** often thickens and becomes differentiated into distinct layers—the exocarp (outer layer), the mesocarp (middle layer), and the endocarp (inner layer), or into exocarp and endocarp only. These layers are generally more conspicuous in fleshy fruits than in dry ones. Fruits are discussed in greater detail in Chapter 19.

An angiosperm life cycle is summarized in Figure 18–21.

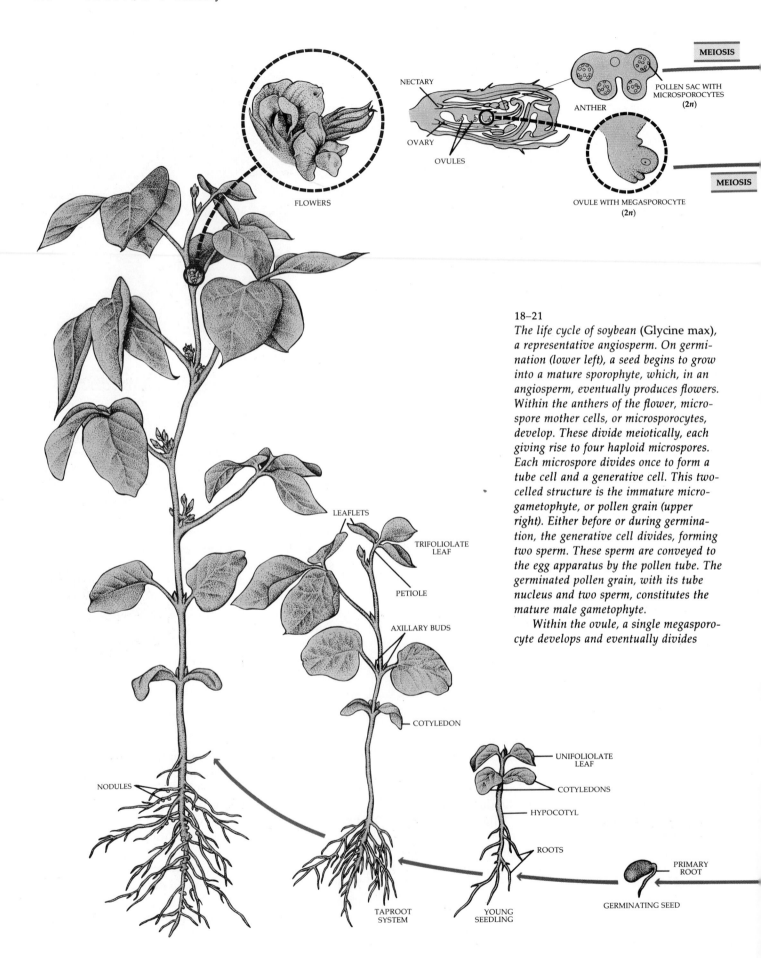

MEIOSIS

NECTARY

POLLEN SAC WITH
MICROSPOROCYTES
(2*n*)

ANTHER

OVARY

OVULES

FLOWERS

MEIOSIS

OVULE WITH MEGASPOROCYTE
(2*n*)

LEAFLETS

TRIFOLIOLATE
LEAF

PETIOLE

AXILLARY BUDS

COTYLEDON

UNIFOLIOLATE
LEAF

COTYLEDONS

HYPOCOTYL

NODULES

ROOTS

PRIMARY
ROOT

TAPROOT
SYSTEM

YOUNG
SEEDLING

GERMINATING SEED

18–21

*The life cycle of soybean (Glycine max),
a representative angiosperm. On germination (lower left), a seed begins to grow
into a mature sporophyte, which, in an
angiosperm, eventually produces flowers.
Within the anthers of the flower, microspore mother cells, or microsporocytes,
develop. These divide meiotically, each
giving rise to four haploid microspores.
Each microspore divides once to form a
tube cell and a generative cell. This two-
celled structure is the immature microgametophyte, or pollen grain (upper
right). Either before or during germination, the generative cell divides, forming
two sperm. These sperm are conveyed to
the egg apparatus by the pollen tube. The
germinated pollen grain, with its tube
nucleus and two sperm, constitutes the
mature male gametophyte.*

*Within the ovule, a single megasporo-
cyte develops and eventually divides*

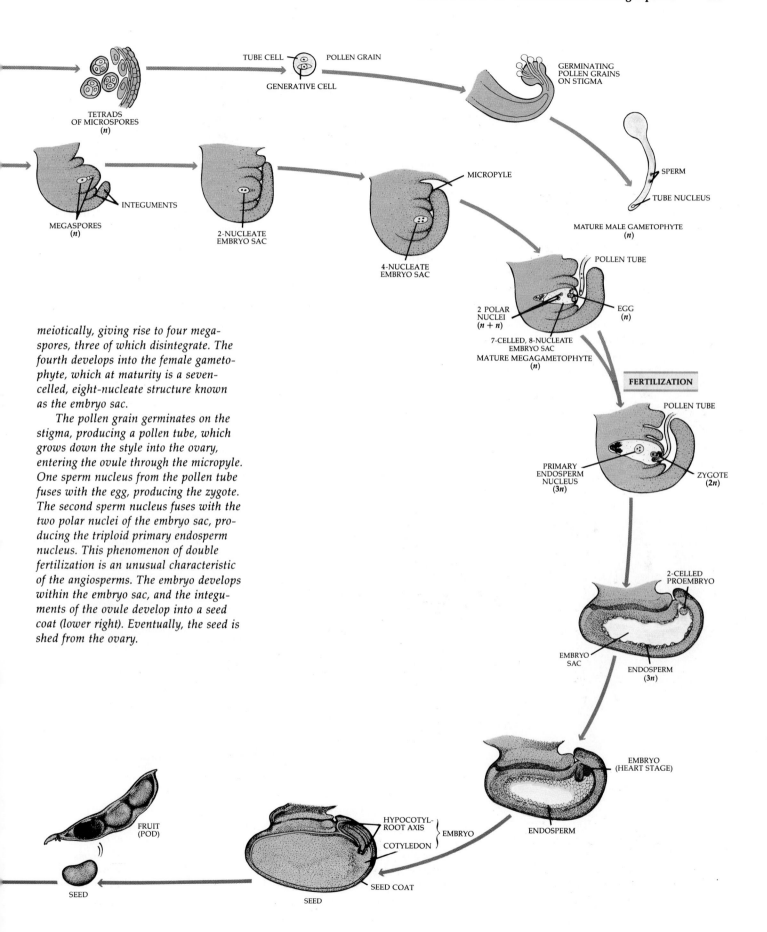

TUBE CELL POLLEN GRAIN

GENERATIVE CELL

TETRADS
OF MICROSPORES
(*n*)

GERMINATING
POLLEN GRAINS
ON STIGMA

SPERM

TUBE NUCLEUS

MATURE MALE GAMETOPHYTE
(*n*)

INTEGUMENTS

MEGASPORES
(*n*)

2-NUCLEATE
EMBRYO SAC

MICROPYLE

4-NUCLEATE
EMBRYO SAC

POLLEN TUBE

2 POLAR
NUCLEI
(*n* + *n*)

EGG
(*n*)

7-CELLED, 8-NUCLEATE
EMBRYO SAC
MATURE MEGAGAMETOPHYTE
(*n*)

FERTILIZATION

POLLEN TUBE

PRIMARY
ENDOSPERM
NUCLEUS
(3*n*)

ZYGOTE
(2*n*)

2-CELLED
PROEMBRYO

EMBRYO
SAC

ENDOSPERM
(3*n*)

EMBRYO
(HEART STAGE)

HYPOCOTYL-
ROOT AXIS

COTYLEDON

EMBRYO

ENDOSPERM

FRUIT
(POD)

SEED

SEED COAT

SEED

meiotically, giving rise to four megaspores, three of which disintegrate. The
fourth develops into the female gametophyte, which at maturity is a seven-
celled, eight-nucleate structure known
as the embryo sac.

The pollen grain germinates on the
stigma, producing a pollen tube, which
grows down the style into the ovary,
entering the ovule through the micropyle.
One sperm nucleus from the pollen tube
fuses with the egg, producing the zygote.
The second sperm nucleus fuses with the
two polar nuclei of the embryo sac, producing the triploid primary endosperm
nucleus. This phenomenon of double
fertilization is an unusual characteristic
of the angiosperms. The embryo develops
within the embryo sac, and the integuments of the ovule develop into a seed
coat (lower right). Eventually, the seed is
shed from the ovary.

FACTORS PROMOTING OUTCROSSING

Outcrossing (in plants, outcrossing is made possible by cross-pollination between individuals of the same species) is of critical importance for all eukaryotic organisms. It should therefore come as no surprise that angiosperms have evolved a variety of mechanisms that serve to promote the transfer of pollen from one plant to another. Flowers of most kinds of angiosperms have both stamens and carpels, but one or the other of these whorls is lacking in the individual flowers of dioecious and monoecious species (see Figure 18–10 for an example). In dioecious plants, such as willows, the pollen must pass from one individual to another to achieve fertilization; outcrossing is clearly complete. In monoecious plants, such as oaks, birches, and ragweed, the staminate and carpellate flowers are distinct but occur together on the same individual. Since the staminate and carpellate flowers of a single individual may mature at different times, their separation often leads to outcrossing.

Some seed plants other than angiosperms are also monoecious or dioecious, illustrating the importance of this condition in promoting outcrossing. Among the dioecious gymnosperms are *Ginkgo*, cycads, and junipers (conifers of the genus *Juniperus*, often misleadingly called "cedars" in North America). In all living seed plants other than angiosperms, the ovule- and pollen-producing parts occur in different structures, as in different kinds of cones.

Another way in which angiosperms promote outcrossing is through **dichogamy**, a condition in which the stamens and carpels reach maturity at different times—even though they occur together in the same flowers. Those plants in which the stamens of an individual flower mature before the stigmas become receptive are said to be **protandrous** (see Figure 18–24a), and those in which the stigmas become receptive before the stamens mature are said to be **protogynous**. As a result of these conditions, a given flower may be either effectively staminate or effectively carpellate at one time. Dichogamy is widespread among flowering plants (Figure 18–22).

(a)

(b)

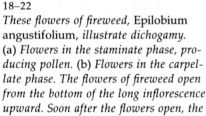

18–22

These flowers of fireweed, Epilobium angustifolium, *illustrate dichogamy.* (a) *Flowers in the staminate phase, producing pollen.* (b) *Flowers in the carpellate phase. The flowers of fireweed open from the bottom of the long inflorescence upward. Soon after the flowers open, the* anthers begin to shed pollen; about two days later, the style, which has been reflexed to one side, swings up into the center of the flower, the stigma opens, and the flower becomes carpellate. By this time, the anthers have completed shedding their pollen. As a result, the lowest open flowers on a stem that has been blooming for more than a few days are always carpellate and the upper ones staminate. Bees work up the stems of individual plants, carrying pollen to the carpellate flowers, and then fly to the lowest flowers of the next plant.

Hay Fever

In the temperate areas of the Northern Hemisphere, it is estimated that between 10 and 18 percent of all people will suffer from hay fever, which can be highly debilitating, at some point in their lives. Some of the proteins that occur in hollow spaces in pollen grain walls, and which can be released immediately following contact with a moist surface—as the grain swells and cracks—are responsible. Among them are proteins that can act as very powerful allergens and antigens in the human body, provoking strong reactions by the immune system. These are probably the incompatibility factors used in plant-to-plant recognition. The proteins may also be released in tiny particles of the tapetum, smaller than pollen grains, which may become air-borne as the anther splits open.

Wind-borne pollen, such as that of grasses, birch trees, and ragweed, is important as an agent of hay fever, probably mainly because it is shed in large amounts and thus more likely to reach susceptible victims than are the often larger pollen grains of insect-pollinated plants. The quantity of pollen inhaled seems to be the most important factor, on the whole. Some wind-borne pollen that is shed in huge amounts, however, such as that of corn (maize) and pines, rarely causes any difficulty. The scent of certain flowers can also cause reactions that resemble hay fever, perhaps in part at least by increasing the sensitivity of the nasal membranes.

When new kinds of plants, such as rapeseed, come into widespread cultivation, they often become important causes of hay fever. In the arid southwestern United States, the irrigation of large areas for lawns and golf courses, and the introduction of many kinds of weeds to the area, have made hay fever common where it was once virtually unknown.

The incidence of hay fever has been rising rapidly for more than 50 years, even though the pollen count is actually falling in many areas. Part of the increase is related to better detection of the problem, but there is clearly a genuine increase in hay fever. The answer to this problem no doubt lies in a better understanding of immune reactions.

Another strategy by which plants achieve outcrossing is through physical separation of the stamens and stigma(s) within a flower (Figure 18–23). In this way, pollen derived from the stamens of a given flower will rarely reach the stigma(s) of that flower, even if the stamens and stigmas mature at the same time. Such a condition, then, greatly increases the likelihood of the pollen's being dispersed to another flower.

Genetic self-incompatibility also occurs widely in the angiosperms, with at least some, and often many, members of almost every plant family being genetically self-incompatible. In many genetically self-incompatible plants, the pollen from a given individual either does not germinate or else dies after growing a short distance into the stigma of the same individual or that of an individual that is genetically identical for the self-incompatibility factors. In other genetically self-incompatible plants, the zygotes that result from self-fertilization, or fertilization by a gamete that has identical self-compatibility factors, are inviable. If a plant is genetically self-incompatible, it will be outcrossed even if its stamens and stigma come into contact regularly and mature at the same time (see the essay ''Genetic Self-incompatibility'' on page 398).

18–23
This flower of Easter lily (Lilium longi-florum) illustrates the wide separation of stigma and anthers that is characteristic of many plants.

Genetic Self-incompatibility

By recombining genes from different individuals, sexual reproduction produces variability in natural populations, giving these populations the potential for adaptation to change in their environment through progressive evolutionary modification. Plants that regularly self-pollinate lose some of this potential. Early in the history of the flowering plants, mechanisms evolved in many families that make cross-pollination almost mandatory, even when the flowers are perfect or when both staminate and carpellate flowers are present on the same plant.

Two basic mechanisms that promote cross-pollination are found among modern families of flowering plants. In the more common one, present in such economically important groups as grasses and legumes, the behavior of the pollen grain is determined by its own (haploid) genotype. If the grain's DNA carries at the incompatibility locus a gene that is identical to the gene at either of the two corresponding loci in the diploid stigma and style, then the entry of the pollen tube is barred. If it carries at this locus a gene that is different from either of those present in the stigmatic tissue, then the grain is accepted.

In the second kind of incompatibility system, found in the crucifer family (Brassicaceae) and the composite family (Asteraceae), the behavior of the pollen is determined by the genetics of the parent plant that produced the pollen, not by the genes of the pollen grain itself; in other words, the control system is based on the correspondence between two kinds of diploid tissue. In both systems, the kind of "match" that occurs at the incompatibility locus determines acceptance or rejection.

Although much remains to be learned about the physiology of these mechanisms, it is known that they depend upon "recognition" reactions between specific incompatibility proteins carried by the pollen grains and matching proteins produced in the stigmas or styles. In grasses, the reaction often occurs on the stigma surface. The scanning electron micrograph (a) shows part of the stigma of a flower of orchard grass (*Dactylis glomerata*). The stigma has numerous papilla of another grass species, rye (*Secale cereale*), in section. Outside the cuticle lie two other layers, an inner one composed of mucilaginous pectic materials and an outer section. Outside the cuticle lie two other layers, an inner one composed of mucilaginous pectic materials and an outer proteinaceous layer. The incompatibility response is shown when the tip of the pollen tube comes in contact with the outer layer, or very shortly thereafter. The micrographs taken with the fluorescence microscope (c and d) show stigmas of foxtail grass (*Alopecurus pratensis*) stained with a fluorescent stain for the wall polysaccharide known as callose. In (c), the pollination has been a compatible one, and the tube is seen growing toward the ovary after penetrating the stigma. In (d), the pollination has been incompatible: after the tube tip touched the surface protein layer, growth of the tube was halted and it filled with callose, which shows that it was rejected.

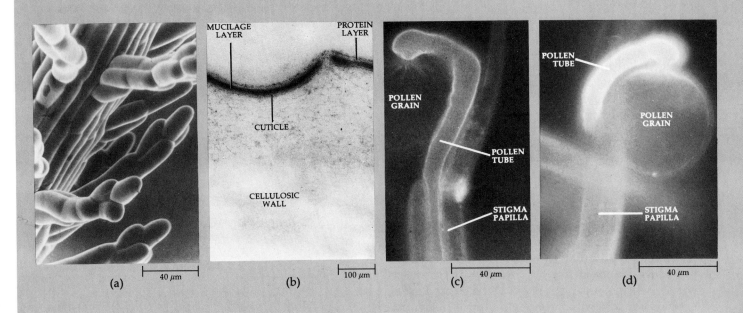

(a)

(b)

18–24

These two annual herbs are species of the genus Clarkia, of the evening primrose family, Onagraceae. They grow in the foothills of California and are closely related, as shown by careful comparison of their nucleic acid components. (a) Clarkia cylindrica is an outcrossing species; although, like all other members of the genus, it is genetically self-compatible, it is protandrous, the eight anthers opening and shedding their pollen approximately two days before the stigma is receptive. In addition, as you can see in this photograph, the stigma is widely separated from the anthers within the flower. (b) Sharply contrasting is C. heterandra, in which the flowers are much smaller and paler. This species has only four stamens, and their anthers shed pollen directly onto the stigma.

OUTCROSSING VERSUS SELF-POLLINATION

In Chapter 19, we shall examine many of the specific plant-animal relationships that result in the more-or-less accurate transfer of pollen between flowers. In most gymnosperms, pollen is dispersed by the wind, as it is in a number of angiosperms. In submersed angiosperms, pollen is dispersed either through or on the surface of the water. All of these methods lead to outcrossing.

Many angiosperms, however, have adopted self-pollination as a regular mode of reproduction, despite the advantages of outcrossing. In temperate regions, for example, more than half of all species of flowering plants are self-pollinated. Self-pollinating plants often have smaller and less conspicuous flowers than outcrossing ones (Figure 18–24), but then, the former have no need to attract animal visitors. In some self-pollinating plants, pollination occurs in the bud, after which the bud may or may not open; it often simply falls off, leaving a ripening ovary behind. In others, pollination occurs only after the buds open—although pollen may also be dispersed to some degree by animals that might visit such flowers.

Self-pollination can be advantageous under certain circumstances. Populations of self-pollinating plants normally have higher proportions of genetically similar individuals than populations in which outcrossing predominates. Depending on their genotype, many or all of the self-pollinated individuals in a given population may be well suited to a particular habitat; for example, they may, like many weeds, occur in similar, disturbed habitats throughout their range. A second, rather obvious advantage of self-pollination is the lack of dependency on animals or other vectors to achieve pollination. Whatever happens, short of physical damage or extreme drought, the self-pollinated plants will produce seed. This second advantage explains why self-pollinated plants are often relatively well represented in places where flower-visiting animals may be rare, as on high mountains or in the far north and south.

Summary

Angiosperms, or flowering plants, constitute the division Anthophyta, which, in turn, is divided into two large classes—Monocotyledones (65,000 species) and Dicotyledones (170,000 species). Flowering plants resemble gymnosperms in their essential reproductive details but differ in a number of fundamental ways. In angiosperm seeds, for example, the stored food is provided by a unique, usually triploid tissue called endosperm. The ovules are held within the megasporophylls (the carpels). The distinctive reproductive feature of the angiosperms, the flower, is characterized by the presence of carpels.

The flower is a determinate shoot that bears sporophylls. Individual flowers may have up to four whorls of appendages; from the outside in, they are the sepals (collectively, the calyx), the petals (collectively, the corolla), the stamens (collectively, the androecium), and the carpels (collectively, the gynoecium). The sepals and petals are sterile, with the sepals often green and protective, covering the flower in bud, and the petals often colored and serving a function in attracting insects. Individual stamens are often divided into a stalk, or filament, and an anther, containing four pollen sacs. Carpels are usually differentiated into a swollen lower part, the ovary, and a slender upper part, the style, terminating in the receptive stigma. One or more of these whorls may be missing in individual flowers.

Pollination in angiosperms takes place by the transfer of pollen from anther to stigma. The male gametes, or sperm, of angiosperms are transmitted by means of a pollen grain, which is an immature male gametophyte. At the time of dispersal, such a gametophyte may contain either two or three cells; initially, there is a tube cell and a generative cell, the latter dividing before or after dispersal to give rise to two sperm. The female gametophyte of an angiosperm is called an embryo sac; in many angiosperms, embryo sacs have eight nuclei at maturity, one of which is the egg (the number of cells varies in different groups). Both sperm function in angiosperm fertilization: one unites with the egg, producing a diploid zygote, and the other unites with usually two or sometimes more polar nuclei, giving rise to the primary endosperm nucleus. That nucleus divides, producing a unique kind of nutritive tissue in the seed (it may be absorbed by the embryo before the seed is mature), the endosperm. Outside the angiosperms, double fertilization is known only in *Ephedra* (division Gnetophyta), which lacks endosperm.

The ovaries (sometimes with some associated floral parts) develop into fruits, which enclose the seeds. The fruit is one of the principal characteristics of this great division of plants.

Outcrossing in angiosperms is promoted by dioecism, in which flowers are either staminate or carpellate and occur on different plants, and by monoecism, in which staminate and carpellate flowers are distinct but occur on the same plant. Outcrossing is also promoted by dichogamy, in which the stamens and carpels of a given flower mature at different times, or simply by the physical separation of the stamens and stigma(s) within a flower. Genetic self-incompatibility makes self-fertilization impossible even if the stamens and stigmas mature at the same time and come into contact with each other; this condition is common among the angiosperms.

Self-pollination is characteristic of many angiosperms, including more than half of the species that occur in temperate regions. Pollination often occurs within individual flowers, sometimes before they open (and they may *never* open, but simply fall off). Self-pollination results in the production of a high proportion of genetically similar individuals, which may be well adapted to a particular habitat.

Suggestions for Further Reading

Buckles, Mary Parker: *The Flowers Around Us: A Photographic Essay on Their Reproductive Structures*, University of Missouri Press, Columbia, Mo., 1985.

An attractive book, dealing well with the diversity of flowers and fruits.

Gamlin, Linda: "The Big Sneeze," *New Scientist* 126:37–41, June 2, 1990.

Interesting article on trends in understanding hay fever.

Gifford, Ernest M., and Adriance S. Foster: *Morphology and Evolution of Vascular Plants*, 3d ed., W.H. Freeman and Company, New York, 1989.

A well-organized, general account of the vascular plants, including excellent information on angiosperms.

King-Hele, Desmond: "Chronicle of the Lustful Plants," *New Scientist* 122:57–61, April 22, 1989.

This comic scientific poem written by a doctor-inventor and published 200 years ago created a sensation in the literary world: early investigations of sex in plants.

Lawrence, Susan V.: "Recent Advances in Hay Fever Research Are Nothing to Sneeze At," *Smithsonian*, September 1984, pages 100–111.

A fascinating account of some of the accidental effects of pollen, which are troublesome to so many of us.

CHAPTER 19

Evolution of the Angiosperms

(a)

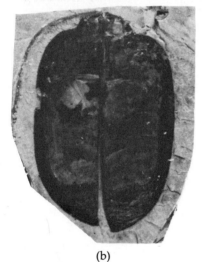

(b)

19–1
(a) *A longhorn beetle (family Cerambycidae), laden with pollen, visiting the flower of a member of the parsley family (Apiaceae) in the mountains of northeastern Arizona. (b) Carapace of a beetle from the same strata as* Archaeanthus, *an extinct angiosperm from 95 to 98 million years ago (see Figure 19–12). The evolution of the flowering plants is, to a large extent, the story of increasingly specialized relationships between flowers and their insect pollinators, in which beetles played an important early role.*

In a letter to a friend, Charles Darwin referred to the apparently sudden appearance of the angiosperms in the fossil record as "an abominable mystery." In the early fossil-bearing strata, about 400 million years old, one finds simple vascular plants, such as rhyniophytes and trimerophytes. Then there is a Devonian and Carboniferous proliferation of ferns, lycophytes, sphenophytes, and progymnosperms, which were dominant until about 300 million years ago. The early seed plants first appeared in the late Devonian period and led to the gymnosperm-dominated Mesozoic floras. Finally, during the first half of the Cretaceous period, toward the end of the Mesozoic era, the angiosperms appeared, gradually achieving worldwide domination in the vegetation by about 90 million years ago (Figure 19–1). By about 75 million years ago, many modern families and some modern genera of this division already existed (Figure 19–2).

Why did the angiosperms rise to world dominance and then continue to diversify to such a spectacular extent? In this chapter, we shall attempt to answer this question, centering our discussion on the relationships of the angiosperms, the origin of the division, the evolution of the flower, the evolution of fruits, and the role of certain chemical substances in angiosperm evolution. All five topics will help to illustrate the reasons for the enormous success of the group.

Relationships of the Angiosperms

Since the time of Darwin, scientists have attempted to understand the ancestry of the angiosperms. One approach has been to search for their possible ancestors in the fossil record. In this effort, particular emphasis has been placed on assessing the ease with which the

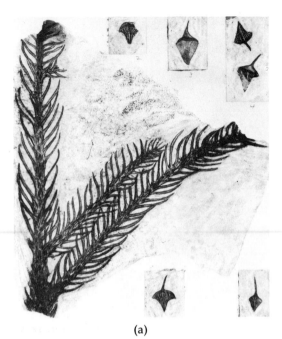

(a)

(b)

(c)

19-2

Fossils of a gymnosperm (a) and two angiosperms (b, c) from late Cretaceous deposits (about 70 million years ago) in Wyoming. (a) Twig and individual cone scales of Araucarites longifolia, *an extinct species of conifer belonging to a family (*Araucariaceae*) now restricted to the Southern Hemisphere. A familiar living member of this family is the Norfolk Island pine (*Araucaria heterophylla*). (b) Leaf of a fan palm,* Sabalites montana, *an extinct species distantly related to the palmettos of the southeastern United States. (c) Leaf of* Viburnum marginatum, *an extinct species belonging to a widespread genus of shrubs generally known as arrow-woods.*

ovule-bearing structures of various gymnosperms could be transformed into a carpel. Recently, a different approach has been suggested, which, instead of searching for ancestral groups, simply attempts to define the major natural groups of seed plants and to understand their interrelationships. The degree of relationship among these groups is then used to estimate how recently they diverged from a common ancestor. A leading scientist who has been pursuing this line of investigation is Peter Crane, of the Field Museum of Natural History in Chicago. He supplied much of the information on which this discussion is based.

The simplest situation in which to discuss phylogenetic (evolutionary) relationships is one involving only three different kinds (or groups) of plants. Ferns, cycads, and pines, for example, all possess xylem composed of tracheids, and this suggests that at some point in the distant past they shared a common ancestor that also possessed tracheids. The question can then be asked, Which two of these three plants have a more recent common ancestor? For three plants, there are only three possibilities from which to choose, and they can be conveniently summarized by three branching diagrams (Figure 19–3). Cycads and pines both have seeds and wood (secondary xylem), neither of which occurs in ferns, and this supports the idea that cycads and pines share a more recent common ancestor than either of these groups does with ferns (Figure 19–3a). Features that are thought to have been inherited from a common ancestor, such as the seeds, wood, and tracheids in this example, are said to be **homologous.** If an additional group of seed-bearing, woody plants is introduced into the picture—for example, the Douglas firs *(Pseudotsuga)*—more restricted homologies need to be found to clarify its relationship to cycads and pines. Pines and

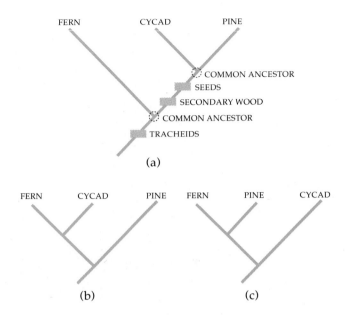

(a)

(b)   (c)

19–3
Three representations of possible patterns of relationship between ferns, cycads, and pines. Version (a) is probably correct and is based on the characteristics indicated.

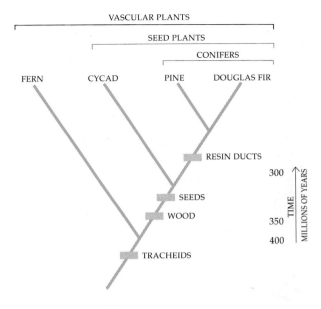

19–4
Phylogenetic relationships between ferns, cycads, pines, and Douglas firs, indicating the shared characteristics that support this particular pattern of relationship and the approximate times of divergence of these groups.

Douglas firs are similar in possessing resin ducts in their leaves, which suggests that they share a more recent common ancestor than either of these groups does with the cycads (Figure 19–4). As the analysis is extended by incorporating more similarities and more groups of plants, conflict frequently arises between conclusions about relationships that were derived either by studying more limited groups or by using fewer characteristics. In such cases, the most likely explanation (that is, the one favored by the largest number of homologies) is the one that should be adopted. The decision may then be tested and sometimes modified as additional features and additional groups of plants are included in the analysis.

Estimating relationships in this way leads to a pattern of groups nested within groups, which is termed **hierarchy.** Each group, or **clade,** is defined by one or more homologies and includes all the descendants of a single hypothetical common ancestor. The branching diagrams used to express the relationships between such plant groups are called **cladograms.** Vascular plants are a clade, within which seed plants form another clade, within which the conifers form yet another (Figure 19–4). The branching diagram also predicts the sequence in which homologies might be expected to appear in the fossil record. In our simple example, paleobotanical evidence conforms with these predictions: tracheids first appear at least 400 million years ago,

seeds and wood about 350 million years ago, and resin ducts about 300 million years ago.

Certain kinds of similarities that may at first sight appear to indicate relationships do not play a part in estimating evolutionary relationships with a cladogram. For example, the ovulate cones of cycads, pines, and Douglas firs look similar, but the similarity is only superficial. In cycads, each cone scale is a simple, modified, seed-bearing leaf; in conifers, however, each cone scale is a complex structure, as we discussed in Chapter 17. The cones of cycads and conifers, therefore, do not appear to have evolved from a common ancestor and could not be used to define a clade; they are said to be **analogous.** Their superficial similarity is due to **convergent evolution** (see the essay "Convergent Evolution" on page 518). In contrast, the compound ovulate cones of pines and Douglas firs are homologous and could be used to define a conifer clade.

A further potential source of confusion in assessing similarities results from adopting too narrow a view of a particular homology. For example, lycopsids, *Equisetum*, and ferns have free spores (that is, they all shed their spores instead of retaining them on the parent sporophyte) but this feature could not be used to define a clade because the spores of these plants are also homologous with part of the seed in seed plants. A group defined by its possession of free spores would be incomplete and would not include all the decendants of a

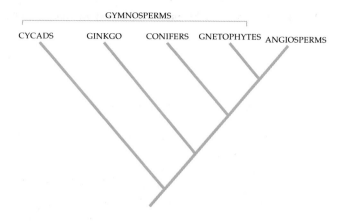

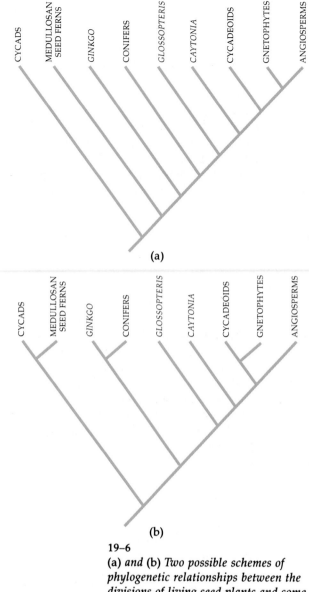

19–5

Phylogenetic relationships between the four divisions of living gymnosperms and the angiosperms.

(a)

(b)

19–6

(a) and (b) Two possible schemes of phylogenetic relationships between the divisions of living seed plants and some extinct groups. The medullosan seed ferns, Glossopteris, and Caytonia were all seed ferns, which, along with the other extinct groups, were discussed in Chapter 18. These cladograms were prepared by Peter Crane of the Field Museum of Natural History, Chicago.

single common ancestor. That same common ancestor would also have given rise to seed plants. In this example, there are two homologies: possession of spores and possession of megaspores modified into seeds; these define two clades, one of which is a subset of the other. Those vascular plants that form free spores do not form a clade by themselves.

Using the kind of reasoning just explained, which is known as **phylogenetic analysis** or **cladistics,** the problem of recognizing the ancestors of flowering plants (which may not yet have been discovered or may not have been preserved in the fossil record) is set aside. Instead, the objective is to define the major clades of seed plants and to assess their relationships in terms of relative recency of common ancestry.

Living seed plants provide the best place to begin the analysis because they are generally much better understood than their fossil relatives. One suggested cladogram of living seed plants is given in Figure 19–5. It is clear from this diagram that the only features that could define the gymnosperms would also have had to include the flowering plants. In effect, the gymnosperms are merely what is left of the clade of seed plants after the angiosperm clade has been removed from it.

After an analysis of living plants, the next stage is to incorporate what is known about fossil seed plants. Many fossil plants can be placed in groups that are known from their living members, but others are representatives of major clades that are now extinct. These extinct clades can provide important clues to recognizing homologies and may therefore be extremely useful in clarifying ideas on evolutionary relationships. Two proposed cladograms expressing the relationships between the major kinds of living and fossil seed plants are given in Figure 19–6. These ideas are certain to be refined by future research, especially as more is discov-

ered about some of the major extinct plant groups, but they do indicate how the flowering plants are currently thought to fit into the general pattern of evolution. It is significant, for example, that the cycadeoids, cycad-like plants that many have linked with angiosperms, and gnetophytes first appear in the fossil record in the Triassic period, about 225 million years ago; this seems to place some constraints on the possible earliest date of the appearance of the angiosperms. Biochemical methods are now being employed to define the relationships of angiosperms more clearly.

Origin of the Angiosperms

The unique characteristics of the angiosperms clearly arose in a seed plant, quite likely a shrub that lacked flowers, closed carpels, and fruits. The earliest known fossil angiosperm remains are pollen grains some 127 million years old, from the early Cretaceous period. Each of these early pollen grains has a single pore; the grains are similar to (but distinguishable from) fern spores and gymnosperm pollen. It is likely that most angiosperms that existed more than 127 million years ago had pollen that could not be distinguished from that of gymnosperms, which is one reason angiosperms have not yet been detected earlier in the fossil record.

The earliest known flowers, reported in 1990, were found near Melbourne, Australia (Figure 19–7). The flowers of this 120-million-year-old fossil angiosperm, which are subtended by bracts, are carpellate, and the leaves are simple. A perennial herb that perhaps grew in marshy places, this archaic dicot may have been similar to existing black peppers (Piperaceae) and resembles the kind of dicots that have been considered likely candidates to be ancestral forms. If most ancient angiosperms shared its herbaceous habit, they would have formed fossils only rarely—the leaves tend to shrivel while still attached to such plants—and

thus the plants would be difficult to detect in the fossil record.

Conservatively, the angiosperms are estimated to be at least 130 to 140 million years old. On the basis of the degree of difference between some macromolecules, scientists have estimated the time of divergence between monocots and dicots as from 200 to 230 million years ago. This very old date, however, appears to contradict the fossil evidence and needs to be evaluated further.

By no less than 120 million years ago, the three-pored pollen that is characteristic of all but the most primitive dicots had appeared in the fossil record. However old they may be as a group, the angiosperms did not become highly diverse until after that time, with most of the orders and families first appearing in Africa or South America. By approximately 90 million years ago, many existing orders and families of angiosperms had appeared (Figures 19–2, 19–8), and the flowering plants had achieved dominance throughout the Northern Hemisphere. During the subsequent 10 million years, they achieved dominance in the Southern Hemisphere as well. How did this come about, and what do we know about the origins of the division?

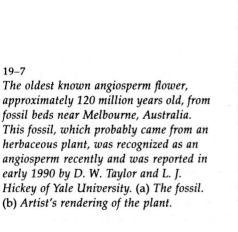

19–7

The oldest known angiosperm flower, approximately 120 million years old, from fossil beds near Melbourne, Australia. This fossil, which probably came from an herbaceous plant, was recognized as an angiosperm recently and was reported in early 1990 by D. W. Taylor and L. J. Hickey of Yale University. (a) The fossil. (b) Artist's rendering of the plant.

(a) 5 mm

(b) 5 mm

(a) |‒‒‒‒| 0.5 mm

(b) |‒‒| 0.2 mm

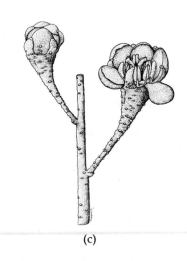

(c)

19–8
Silvianthemum suecicum, *a relatively specialized angiosperm that occurred in the late Cretaceous period in southern Sweden, about 83 million years ago. Well preserved in clay, these flowers are minute, perfect, and radially symmetrical, with five free sepals and five free petals,* three fused carpels, disc-shaped nectaries, and minute, numerous seeds. Like two other genera that occurred with it, Silvianthemum *is broadly related to the saxifrage family, Saxifragaceae, and especially to some of its woody relatives.* (a) *and* (b) *Two views of the* fossil flowers. (c) *Reconstruction of a flower and bud. Else Marie Friis and her coworkers are making outstanding contributions to our knowledge of fossil angiosperms.*

Evolutionary Radiation of the Angiosperms

Paleobotanist Daniel Axelrod, of the University of California at Davis, suggested on the basis of his analysis of the fossil record and the current occurrence of primitive angiosperms that the early evolution of the division may have taken place away from the lowland basins where fossil deposits are commonly found. In the hills and uplands of the tropics, the angiosperms could have evolved into a diverse array of forms without their remains being preserved in the fossil record. Only when they subsequently invaded the lowlands, commencing about 125 million years ago, would their presence and their diversity become evident.

The early flowering plants possessed many adaptive traits that made them particularly resistant to drought and cold. Among these were tough leaves, commonly reduced in size; vessel members (efficient water-conducting cells); and a tough, resistant seed coat that protected the young embryo from drying out. These features are not found in all of the flowering plants, nor are they restricted to the angiosperms, but they certainly have played major roles in the success of the division.

A number of other factors seem to have been important in the early and continuing success of the an-

giosperms, and we shall discuss some of them in more detail in this chapter. The evolution of sieve-tube members presumably made possible the more efficient conduction of sugars throughout the plant in the phloem, just as vessel members are more efficient than tracheids in the xylem. Perhaps of even greater importance, the precise systems of pollination and seed dispersal that became characteristic of the more advanced flowering plants allowed them to exist as widely scattered individuals in many different kinds of habitats. The enormous chemical diversity of the angiosperms, which includes the many kinds of defenses against diseases and herbivores, has likewise been of great importance, as we shall see below. These and other features bear directly on the fact that, compared with other plants, angiosperms reproduce rapidly and efficiently.

One important innovation in angiosperms was the evolution of the deciduous habit. There are some deciduous gymnosperms, including larches and bald cypresses, but very few in relation to the overall size of the group. The deciduous habit seems to have evolved first in tropical areas that experienced periodic drought. Descendants of these deciduous plants then spread to the north, where parts of the year were so cold that no water was available for growth. Other modifications of habit in the angiosperms, such as the evolution of herbaceous perennials and eventually of annuals, allowed their survival in conditions more extreme than those in which their woody progenitors could grow. All of these adaptations gave angiosperms a selective advantage in a world in which the climate was becoming progressively more stressful, that is, during the past 50 million years or so—the most important period of angiosperm evolution.

About 127 million years ago—when angiosperm pollen first became part of the fossil record—Africa and South America were directly connected with each other and with Antarctica, India, and Australia in a great southern supercontinent called Gondwanaland (Figure 19–9). Africa and South America began to separate at about this time, forming the southern Atlantic Ocean, but they did not move completely apart in the tropical regions until about 90 million years ago. India began to move northward at about the same time, colliding with Asia about 45 million years ago and causing the Himalayas to be thrust up in the process. Australia began to separate from Antarctica about 55 million years ago, but their separation did not become complete until more recently.

Within the central regions of West Gondwanaland, formed by what are now the continents of South America and Africa, arid-to-subhumid habitats, in which angiosperms and other kinds of organisms would have been challenged to produce new forms, were well represented. With the final separation of these two continents, at about the time the angiosperms became abundant in the fossil record worldwide, the world climate changed greatly. This was especially the case in these equatorial regions, which became milder, with fewer extremes of temperature and humidity. These changes are thought to have been related to the evolutionary success of the early angiosperms, which increasingly dominated floras all over the earth.

When India and Australia were in the far south, they were covered with vegetation typical of a cool temperate climate. Certain unique groups of plants and animals are shared by southern South America and southeastern Australia–Tasmania at the present time, having achieved their ranges by migration across Antarctica long before the onset of full glaciation there, which occurred about 20 million years ago (Figure 19–10). As Australia moved northward during the past 55 million years, it reached the great zone of aridity flanking the tropics, and the sorts of arid plant communities that are now so widespread on the Australian continent expanded greatly. As it neared Asia, the northern edge of Australia reached truly tropical climates for the first time and was invaded by the plants and animals of tropical Asia. To a large extent, however, the plants and animals characteristic of Asia and those characteristic of

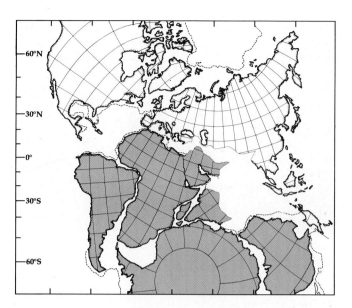

19–9
Relationship between the earth's southern land areas at the time of the first appearance of the angiosperms in the fossil record (127 million years ago). In the middle of the Cretaceous period, about 100 million years ago, South America was still directly connected with Africa, Madagascar, and India and, via Antarctica, with Australia. These combined land masses, indicated here in color, formed the supercontinent of Gondwanaland.

Australia have remained geographically distinct, and the line separating the areas where each is predominant is known as "Wallace's Line," after Alfred Russel Wallace, an early naturalist-explorer of these regions. (Wallace is also well remembered as the young man who, simultaneously with Darwin, proposed the theory of evolution by natural selection.) The original plants and animals of cool-temperate Australia have survived in the southeastern corner of the continent, in Tasmania, and in New Zealand, which separated from Australia–Antarctica about 80 million years ago and moved to the northeast.

The original flora and fauna of India did not fare so well, and only a few remnants of them survive today. India moved much farther than Australia, crossing the southern arid zone, the tropics, and the northern arid zone in the process. As a result, nearly all of India's original plants and animals became extinct; they were eventually replaced by desert, tropical, and mountain flora and fauna from Eurasia.

19-10
This forest of southern beech (Nothofagus menziesii), *growing in Fiordland,
South Island, New Zealand, is a relict
from the cool temperate forest that
extended from southern South America
across Antarctica to Australia and New
Zealand from about 80 million years ago
until perhaps 30 million years ago.
Increasingly large gaps have developed
between the land masses throughout that
entire period of time up to the present.*

Evolution of the Flower

Cretaceous fossil flowers, with few exceptions, fall into
two groups. Members of the first group resemble the
120-million-year-old Australian fossil discussed above
(see Figure 19–7). They have relatively few flower
parts, and their flowers are often unisexual. They may
be compared with, or considered ancestral to, existing
families such as Chloranthaceae, Piperaceae (the pepper family), and Platanaceae (the sycamore family).
Fossils up to about 110 million years old resemble the
sycamore family, which appears to be one of the oldest
extant groups of angiosperms.

The second group of ancient angiosperm fossils
consists of plants with large, robust, bisexual flowers
with many free parts, arranged spirally on an elongate
axis, as in *Magnolia* (Figure 19–11). An example of
this second group is *Archaeanthus linnenbergeri*, the
only known member of an extinct family of angiosperms (Figure 19–12). The stout flowers of *Archaeanthus* had an elongate axis with 100 to 130 spirally
arranged carpels, each eventually producing 10 to 18
seeds. There were three outer perianth parts and six to
nine inner ones; the stamens were numerous and spirally arranged. *Archaeanthus*, which inhabited what
were then the subtropical coastal plains of Kansas, was
probably deciduous, and it seems to have been a small
tree or shrub.

THE PARTS OF THE FLOWER

The extant families of archaic angiosperms often show
considerable variation both in the numbers of floral
parts and in the arrangement of these parts. Presumably this sort of variation also occurred in the angiosperms of the early Cretaceous period and explains the
wide divergence in floral types found then. Many more
specialized, modern families of angiosperms, in contrast, are highly constant in their floral features, with
little variation evident among their members.

What are some of the floral features that were characteristic of early angiosperms and are still present in
archaic members of the division today? We shall consider these features in relation to the different whorls of
the flower that were discussed and described in
Chapter 18.

The Perianth

In archaic angiosperms and their Cretaceous ancestors,
the perianth, if present, was never sharply divided into
calyx and corolla; there was always a gradual transition
between these whorls, as in modern magnolias and
water lilies. In some angiosperms, including the water

(a)

(b)

(c)

19–11

Flowers and fruits of the southern magnolia (Magnolia grandiflora). The cone-shaped receptacle bears numerous spirally arranged carpels from which curved styles emerge. Below the styles in (a) and (b) are the cream-white stamens. (a) The anthers have not yet shed their pollen, whereas the stigmas are receptive. In other words, the species is protogynous. (b) The floral axis of a second-day flower, showing stigmas that are no longer receptive and stamens that are shedding pollen. (c) Fruit, showing carpels and bright red seeds, each protruding on a slender stalk.

(a)

(b)

(c)

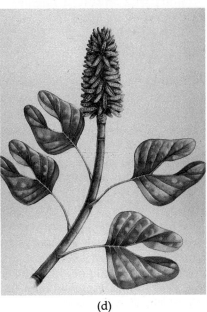

(d)

19–12

Archaeanthus linnenbergeri, an extinct angiosperm that generally resembles living magnolias. (a) Fossil reproductive axis. (b) Fossil leaf. (c) Reconstruction of a flowering branch. (d) Reconstruction of a fruiting branch. The reconstructions are based on the studies of David Dilcher of the University of Florida and were drawn by Megan Rohn. Careful studies by Dilcher and his students are revealing a great deal about the nature of early angiosperms and their flowers.

lilies, petals appear to have been derived from sepals; in other words, the petals can be viewed as modified leaves that have become specialized for attracting pollinators. In most members of the division, however, the petals were probably derived originally from stamens that lost their sporangia and then became specially modified for their new role. Most petals, like stamens, are supplied by just one vascular strand; in contrast, sepals are normally supplied by the same number of vascular strands as the leaves of the same plant (often three or more). Within sepals and petals alike, the vascular strands usually branch; so that the number of strands that enter them cannot be determined from the number of veins in the main body of the structures.

Petal fusion has occurred during the evolution of many groups of angiosperms, resulting in the familiar tubular corolla that is characteristic of many families. When a tubular corolla is present, the stamens often fuse with it and appear to arise from it. In a number of evolutionarily advanced families, the sepals are similarly fused into a tube.

The Androecium

Although the number of stamens in archaic angiosperms is highly variable, they are often broad and colored, playing an obvious role in attracting floral visitors. The stamens of a number of archaic angiosperms secrete scents, which also serve to attract insects. In contrast, the stamens of most contemporary angiosperms are usually thin (for example, see Figures 18–4 and 18–6), playing little or no role in the overall attractiveness of the flower to visitors.

In some specialized flowers the stamens become fused. Stamens may be fused together into columnar structures, as they are in the members of the pea, melon, mallow (Figure 19–13d), and sunflower families, or they may be fused with the corolla, as they are in the phlox, snapdragon, and mint families.

In some evolutionarily advanced families, some of the stamens have become secondarily sterile; that is, they have lost their sporangia and have become transformed into specialized structures, such as nectaries—glands that secrete nectar, a sugary fluid that attracts pollinators and provides food for them. (It is important to note, however, that most nectaries are not modified stamens but arose instead from other parts of the flower.) As we saw above, stamens also played a major role in the evolution of the petals in most angiosperms.

The Carpel

Generalized carpels are leaflike, with no specialized areas for the entrapment of pollen grains. In archaic plants, including the oldest angiosperm fossils that we

(a)

know, the carpels are always free from one another; in a few angiosperms, they are incompletely closed, although pollination is always indirect—the pollen does not contact the ovules. In modern angiosperms, the carpels are often fused, and they are sharply differentiated into stigmas, styles, and ovaries. There is much variation in the arrangement of the ovules among contemporary groups of angiosperms and often fewer ovules than in the more generalized and archaic families.

EVOLUTIONARY TRENDS AMONG FLOWERS

Insect pollination quite probably triggered the early evolution of angiosperms, with indirect pollination fostering competition between many pollen grains as they grew through the stigmatic tissue. The flowers of the earliest members of the division probably were bisexual, but unisexual flowers appeared in many families from the earliest times. The perianth consisted of leaflike members, but this feature was lost in many different groups and became differentiated into sepal-like and petal-like members early in the history of many other evolutionary lines. As we mentioned earlier, angiosperms from the Cretaceous period were variable in the number of their floral parts, with different representatives of the division achieving both high and low numbers. As the evolution of angiosperms proceeded, and as relationships with pollinators became more tightly linked, floral patterns became more stereotyped, with the following four trends evident (Figure 19–13):

(b)

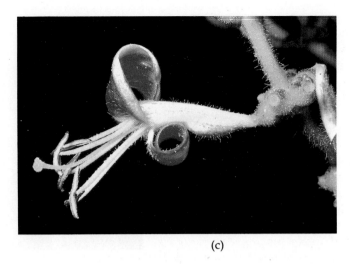

(c)

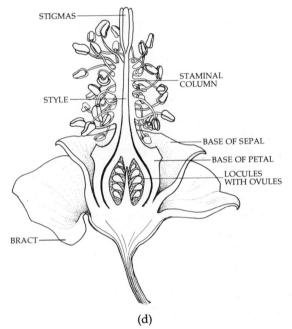

(d)

19–13

Examples of specialized flowers. (a) Wintergreen, Chimaphila umbellata. *The sepals (not visible) and petals are reduced to five each, the stamens to ten, and the five carpels are fused into a compound gynoecium with a single stigma. (b) Lotus,* Nelumbo lutea. *The undifferentiated sepals and the numerous petals and stamens are spirally arranged; the carpels are embedded in a flat-topped receptacle. (c) Chaparral honeysuckle,* Lonicera hispidula. *The ovary is inferior and has two or three locules; the sepals are reduced to small teeth at its apex. The petals are fused into a corolla tube in the zygomorphic (bilaterally symmetrical) flower, and the five stamens, which trude from the tube, are attached to its inner wall. The style is longer than the stamens, and the stigma is elevated above them; a pollinator visiting this flower would contact the stigma first, so that, if it were carrying pollen from another flower, it would deposit that pollen on the stigma before reaching the anthers. Fruits of this species are shown in Figure 19–46c. (d) A diagram of a cotton (Gossypium) flower, of the mallow family, showing the column of stamens fused around the style.*

1. From flowers with many parts that are indefinite in number, flowers have evolved with few parts that are definite in number.

2. The number of kinds of parts has been reduced from four in the primitive flower to three, two, or sometimes one in more advanced flowers. The shoot has become shortened so that the original spiral arrangement of parts is no longer evident. The floral parts have become fused.

3. The ovary has become inferior rather than superior in position.

4. The radial symmetry (regularity), or actinomorphy, of the primitive flower has given way to bilateral symmetry (irregularity), or zygomorphy, in the more advanced flowers.

19-14

Composites (family Asteraceae). (a) A diagram showing the organization of the head of a member of this family. The flowers are subordinated to the overall display of the head, which functions as a single large flower in attracting pollinators. (b) Thistle, Cirsium pastoris. *Members of the thistle tribe have only disk flowers. This particular species of thistle, with its bright red flowers, is regularly visited by hummingbirds, which are its primary agents of pollination. (c) Agoseris, a wild relative of the dandelion,* Taraxacum. *In the inflorescences of the chicory tribe (the group of composites to which dandelions and their relatives belong), there are no disk flowers; the marginal ray flowers, however, are often enlarged. (d) Sunflower,* Helianthus annuus.

EXAMPLES OF SPECIALIZED FAMILIES

Among the most evolutionarily specialized of the flowers are those of the family Asteraceae (Compositae), which are dicots, and those of the family Orchidaceae, which are monocots. In number of species, these are the two largest families of angiosperms.

Asteraceae

In the Asteraceae (the composites), the epigynous flowers are relatively small and closely bunched together into a head. Each of the tiny flowers has an infe-rior ovary composed of two fused carpels with a single ovule in one locule (Figure 19–14).

In composite flowers, the stamens are reduced to five in number and are usually fused to one another (coalescent) and to the corolla (adnate). The petals, also five in number, are fused to one another and also to the ovary, and the sepals are absent or reduced to a series of bristles or scales known as the **pappus.** The pappus often serves as an aid to dispersal by wind, as it does in the familiar dandelion, a member of the Asteraceae (Figure 19–14c; see also Figures 19–42 and 19–43). In other members of this family, such as beggar-ticks (*Bidens*), the pappus may be barbed, serving to attach the fruit to a passing animal and thus to enhance its chances of being dispersed from place to place. In many members of the family Asteraceae, each head includes two types of flower: (1) disk flowers, which make up the central portion of the aggregate, and (2) ray flowers, which are arranged on the outer periphery. The ray flowers are often carpellate, but sometimes they are completely sterile. In some members of the Asteraceae, such as sunflowers, daisies, and black-eyed Susans, the fused corolla of each ray flower forms a long strap-shaped ''petal.''

In general, the composite head has the appearance of a single large flower. Unlike many single flowers, however, the head matures over a period of days, with the individual flowers opening serially in a centripetal spiral. As a consequence, the ovules in a given head may be fertilized by a number of different pollen donors. The success of this plan as an evolutionary strategy is attested to by the great abundance of the members of the Asteraceae, which, with about 22,000 species, is the second largest family of flowering plants.

Orchidaceae

Another successful flower plan is that of the orchids (Orchidaceae), which, unlike the composites, are mon-ocots. There are probably at least 24,000 species of orchids, the largest family of flowering plants. Most of the species are tropical, with only about 140 native to the United States and Canada, for example. In the orchids, the three carpels are fused and, as in the composites, the ovary is inferior (Figure 19–15). Unlike the composites, however, each orchid ovary contains many thousands of minute ovules; consequently, each pollination event may result in the production of a huge number of seeds. Usually only one stamen is present (in one subfamily, the lady-slipper orchids, there are two), and this stamen is characteristically fused with the style and stigma into a single complex structure—the **column**. The entire contents of an anther are held together and dispersed as a unit—the **pollinium** (see Figure 19–26b). The three petals are modified so that the two lateral ones form wings and the third forms a cuplike lip that is often very large and showy. The sepals, also three in number, are often colored and similar to petals in appearance. The flower is always bilaterally symmetrical, and is often bizarre in appearance.

Among the orchids are some species with flowers the size of a pinhead and others with flowers more than 20 centimeters in diameter. Several genera contain saprophytic species; two Australian species grow entirely underground, their flowers appearing in cracks in

(a)

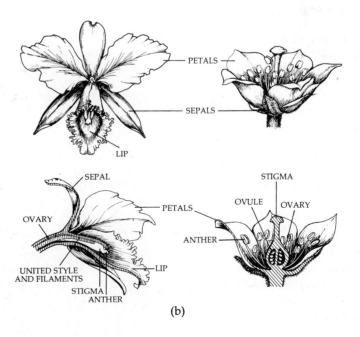

(b)

19–15

(a) *An orchid of the genus* Cattleya. *Orchids are one of the most specialized families of monocots. (b) A comparison of the parts of an orchid flower, shown on the left, with those of a radially symmetrical flower, shown on the right. The* ''lip'' *is a modified petal that serves as a landing platform for insects.*

(a)

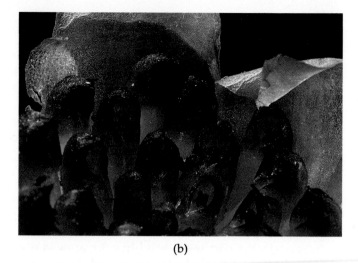
(b)

19–16

Rhizanthella, *a Western Australian orchid, grows entirely underground. Cracks in the soil form during the dry season, revealing the flowers, which* never show above the soil surface. (a) A view from above, with leaves and debris cleared away, of the parted bracts of Rhizanthella, *through which the plant's* pollinators (flies) enter. (b) A dozen flowers of Rhizanthella *are seen here surrounded by the protective bracts.*

the ground, where they are pollinated by flies (Figure 19–16). In the commercial production of orchids, the plants are cloned by making divisions of meristematic tissue, and thousands of identical plants can be produced rapidly and efficiently (see page 563). There are more than 60,000 registered hybrids of orchids, many of them involving two or more genera. The seed pods of orchids of the genus *Vanilla* are the natural source of the popular flavoring of the same name (Figure 19–17).

THE AGENTS OF EVOLUTION

Plants, unlike most animals, cannot move from place to place to find food or shelter or to seek a mate. In general, plants must satisfy those needs by growth responses and by the structures that they produce. Many angiosperms, however, have evolved a set of features that, in effect, allows them directed mobility in seeking a mate; this set of features is embodied in the flower.

19–17

Vanilla, *an orchid that is the source of the flavoring of the same name. Originally used by the Aztecs in what is now Mexico, vanilla is now cultivated primarily on Madagascar and other islands in the western Indian Ocean and elsewhere in the Old World. Vanilla is extracted from the dried, fermented seed pods of this orchid. Chocolate is a blend of cacao and vanilla. Synthetic vanilla flavoring (vanillin) is now used as the source of about 95 percent of all vanilla consumed.* (a) Flowers of the vanilla orchid (Vanilla planifolia). (b) Hand pollination of vanilla plants in Mexico; this procedure is carried out, even in wild plants, to ensure · a good crop of the seed pods from which vanilla is extracted.

(a)

(b)

By attracting insects and other animals with their flowers, and by directing the behavior of these animals so that cross-pollination (and therefore cross-fertilization) will occur at a high frequency, the angiosperms have transcended their rooted condition and, in a sense, become just as mobile as animals in this one respect. How was this achieved?

The earliest seed-bearing plants—various groups of gymnosperms—were pollinated passively. Large amounts of pollen were blown about by the wind, reaching the vicinity of the ovules only by chance. The ovules, which were borne on the leaves or within cones, exuded sticky drops of sap from their micropyles. These drops served to catch the pollen grains and to draw them to the micropyle, just as they do now in modern gymnosperms. As in most modern cycads (see page 374) and in some gnetophytes, insects feeding on the pollen and other flower parts began returning to these new-found sources of food and thus transferred pollen from plant to plant. Such a system is more efficient than passive pollination by the wind; it allowed much more accurate pollination with many fewer pollen grains involved.

The more attractive the plants were to the insects (see Figure 19–1), the more frequently they would be visited and the more seed they could produce. Any changes in the phenotype that made such visits more frequent or more efficient offered a selective advantage. Several important evolutionary developments followed. For example, plants that had flowers that provided special sources of food for their pollinators had a selective advantage. In addition to edible flower parts, pollen, and sticky fluid around the ovules, plants evolved floral nectaries, which, as previously mentioned, secrete the nutritious, sugary **nectar** that provides a source of energy for insects and other animals.

Attraction of insects to the naked ovules of gymnosperms sometimes resulted in the loss of some of the ovules to the insects. The evolution of a closed carpel, therefore, gave certain plants—the ancestors of the angiosperms—a reproductive, and thus a selective, advantage. Further changes in the shape of the flower, such as the evolution of the inferior ovary, may have been additional means of protecting the ovules from being eaten by insects and other animals, thus providing a further reproductive advantage.

Another important evolutionary development was the appearance of the bisexual flower. The presence of both carpels and stamens in a single flower (in contrast, for instance, to the separate microsporangiate and megasporangiate cones of living conifers) offers a selective advantage by making each visit by a pollinator more effective. The pollinator can both pick up and deliver pollen at each stop.

In the early part of the Tertiary period, 40 to 60 million years ago, such specialized groups of flower-visiting insects as bees and butterflies—which had been evolving with the angiosperms for about 50 million years at that point—became even more abundant and diverse. The rise and diversification of these groups of insects were directly related to the increasing diversity of angiosperms. In turn, the insects profoundly influenced the evolutionary course of angiosperms and contributed greatly to their diversification.

If a given plant species is pollinated by only one or a few kinds of visitors, selection favors specialization related to the characteristics of these visitors. Many of the modifications that have evolved in flowers were special adaptations that promoted constancy of a specific type of visitor to that particular kind of flower. Some of the special modifications of flowers that came about during the course of their evolution in response to specific pollinators will be described in the following pages.

Beetle-Pollinated Flowers

A number of modern species of angiosperms are pollinated solely or chiefly by beetles (Figure 19–18; see also the essay "Hot Pollination in the Arum Lilies," pages

19–18

Flower-visiting beetles. (a) *A pollen-eating beetle,* Asclera ruficornis, *at the open, bowl-shaped flowers of round-leaved hepatica (Hepatica americana) in spring in the woods of eastern North America. All species of this family (Oedemeridae) are obligate pollen-feeders as adults.* (b) *A cetoniine scarabid beetle,* Eupoecila australasiae, *visiting the flowers of* Angophora woodsiana, *near Brisbane, Australia. The cetoniine scarabs have membranous mouthparts used for drawing up nectar.*

(a)   (b)

Hot Pollination in the Arum Lilies

The familiar philodendrons that grow in your living room, the skunk cabbages of the swamps, and the arum lilies in your garden share a feature that is unique among plants: their inflorescences heat up greatly as a prelude to pollination. These three plants are members of the Araceae (aroids, or arum lilies), a large family of plants with about 106 genera and approximately 3000 species, the great majority of them tropical (a). In the members of this family, the flowers are small and unisexual. These flowers are densely crowded together on an erect column called a **spadix,** which is surrounded by a modified leaf—a sheath—called the **spathe.** The flowers of all but a few aroids emit strong, foul odors, which attract the flies and beetles that pollinate them. The flowers of some aroids differ from the norm; for example, many members of the very large genus *Anthurium* have pleasant scents and are pollinated by bees.

In many aroids, when the inflorescences are ready for pollination, they suddenly become very warm. In the common household philodendron, *Philodendron scandens* (b), for example, the temperature rises to 46°C even when the temperature of the surrounding air is as low as 4°C. The

heating occurs because the plants oxidize large amounts of stored food, mainly fat, sometimes consuming in one day as much as a quarter of the total weight of the spadix. Philodendrons are pollinated by beetles, which aggregate in large numbers on the inflorescences at dusk, gorging themselves on the pollen and flower parts, and mating. The beetles leave the next morning, heavily dusted with pollen, which they carry to the inflorescence they visit the next night. During the process of heat production in *Philodendron* and other aroids, the inflorescences swell greatly, apparently because of the increased metabolic activity. Interestingly, the substance that triggers the heating reaction seems to be salicylic acid, the active ingredient in aspirin, which was first derived from willow *(Salix)* bark.

This unusual feature probably has evolved partly because the flies that visit the inflorescences of many aroids are normally attracted by rotting organic matter, such as animal carcasses or dung, which becomes warm during the process of decomposition. Heat causes extremely bad-smelling substances such as amines, indoles, and skatoles to be dispersed more efficiently. Some aroids gain additional

(a)

(b)

(a) Monstera deliciosa, *a commonly cultivated tropical member of the Araceae. This species occurs as a native plant in the forests of Central America, from southern Mexico to Panama; the genus ranges widely throughout tropical America.* (b) The household philodendron, Philodendron scandens, *which is widespread as a native plant in the lowlands of Latin America.*

advantages from the heat they produce. For example, the eastern skunk cabbage, *Symplocarpus foetidus* (d), which occurs in the northeastern United States and adjacent Canada, flowers in later winter and early spring, when temperatures are often below freezing (see also Figure 6–2 on page 87). The inflorescences become 20° to 25°C warmer than the surrounding air and consequently are able to grow up through the snow that is often covering the ground. Eastern skunk cabbage is pollinated mainly by flies, but western skunk cabbage, *Lysichiton americanum* (c), is pollinated mostly by small, actively flying rove beetles (Staphylinidae).

Among the most impressive of the aroids is the devil's tongue, *Amorphophallus titanum* (e), which has an inflorescence that grows as tall as 2.5 meters; this plant occurs in the moist forests of Sumatra. The inflorescences, which last three to four days, have a greenish-purple spathe and a yellow-red spadix. The odor of a flowering devil's tongue is said to resemble rotting fish mixed with burnt sugar and is so powerful that some people have been said to faint after inhaling it. *Amorphophallus titanum* is pollinated by large carrion beetles.

(d)

(c)

(e)

(c) *Western skunk cabbage,* Lysichiton americanum. (d) Eastern skunk cabbage, *Symplocarpus foetidus.* (e) *Devil's tongue,* Amorphophallus titanum.

19–19
The foul-scented and often dark-colored flowers of many members of the milkweed family (Asclepiadaceae), such as those of this African succulent plant, Stapelia schinzii, are pollinated by carrion flies, as are the Araceae discussed in the essay on pages 416 and 417.

19–20
Bees have become as highly specialized as the flowers with which they have co-evolved. Their mouthparts have become fused into a sucking tube containing a tongue. The first segment of each of the three pairs of legs has a patch of bristles on its inner surface. Those of the first and second pairs are pollen brushes that gather the pollen that sticks to the bee's hairy body. On the third pair of legs, the bristles form a pollen comb that collects pollen from these brushes and from the abdomen. From the comb, the pollen is forced up into pollen baskets, concave surfaces fringed with hairs on the upper segment of the third pair of legs. Shown here is a honeybee (Apis mellifera) foraging in a flower of rosemary (Rosmarinus officinalis). In the rosemary flower, the stamens and stigma arch upward out of the flower, and both come into contact with the hairy back of any visiting bee of the proper size; here the anthers can be seen depositing white pollen grains on the bee.

416–417). The flowers of beetle-pollinated plants are either large and borne singly, such as those of magnolias, some lilies, California poppies, and wild roses, or small and aggregated in an inflorescence, such as those of dogwoods, elders, spiraeas, and many species of the parsley family (Apiaceae) (see Figure 19–1a). Members of some 16 families of beetles are frequent visitors to flowers, although, as a rule, these beetles derive most of their nourishment from other sources, such as sap, fruit, dung, and carrion. In beetles, the sense of smell is much more highly developed than the visual sense, and beetle-pollinated flowers are often white or dull in color, but they typically have strong odors (Figure 19–19). These odors are usually fruity, spicy, or similar to the foul odors of fermentation and are thus distinct from the sweeter odors of flowers that are pollinated by bees, moths, and butterflies. Some beetle-pollinated flowers secrete nectar; in others, the beetles chew directly on the petals or on specialized food bodies (pads or clusters of cells on the surfaces of the various floral parts), as well as eating the pollen. Many beetle-pollinated flowers have inferior ovaries, with the ovules well buried in the floral tissues, out of reach of the chewing jaws of the beetles on which they depend for their pollination.

Bee-, Wasp-, and Fly-Pollinated Flowers

Bees are the most important group of flower-visiting animals, being responsible for the pollination of more species of plants than the members of any other animal group. Modern families of bees have existed for at least 80 million years, and they subsequently became diverse along with the evolutionary radiation of the angiosperms. Both male and female bees live on nectar, and

19–21
A sweat bee (family Halictidae) gathering pollen from the stamens of a cactus (Echinocereus) *in Baja California, Mexico. The tall structures in the center of the flower are the stigmas.*

the females also collect pollen to feed the larvae. Bees have mouthparts, body hairs, and other appendages with special adaptations that make them suitable for collecting and carrying nectar and pollen (Figure 19–20). As Karl von Frisch and other investigators of insect behavior have shown, bees can learn quickly to recognize colors, odors, and outlines. The portion of the light spectrum that is visible to most insects, including bees, is somewhat different from the portion visible to humans. Unlike human beings, bees perceive ultraviolet as a distinct color, but not red, which therefore tends to merge with the background.

Many kinds of bees—especially solitary bees, which constitute a majority of species of the group (Figure 19–21)—are highly constant in their visits to flowers, confining their visits to one or a few plant species. Such constancy increases the efficiency of the particular species of bee—or of an individual bee—while it is visiting the flowers of one plant species. In relation to this specialization, bee species with narrowly restricted foraging habits often feature conspicuous morphological and physiological adaptations, such as coarse bristles in their pollen-collecting apparatus (if they visit plants with large pollen grains) or elongated mouthparts (if they take nectar from plants with long tubed flowers). When they are constant to this degree, bees exert a powerful evolutionary force for specialization of the plants that they visit. There are some 20,000 species of bees, the great majority of which visit flowers for food.

Bee flowers—that is, flowers that coevolved with bees—have showy, brightly colored petals, which are usually blue or yellow. They often have distinctive patterns by which bees can efficiently recognize them. Such patterns may include "honey guides," special markings that indicate the position of the nectar (Figure 19–22). Bee flowers are never pure red, and, as special photographic techniques have shown, they often have distinctive markings that are normally invisible to humans (Figure 19–23).

In bee flowers, the nectary is characteristically situated at the base of the corolla tube, where it is accessible only to such specialized organs as the mouthpiece of bees and not, for example, to the chewing mouthparts of beetles. Bee flowers characteristically have a "landing platform" of some sort (see Figures 19–15 and 19–22).

Bumblebees are among the most familiar flower visitors in the North Temperate zone (Figure 19–24). They are social bees that live in colonies. The queens (sexual females) overwinter, and, when they emerge in the spring, lay the eggs to establish a new colony. Bumblebees cannot fly until their wing muscles reach a temperature of about 32°C; to maintain this temperature, they must forage constantly on flowers with a copious supply of nectar. Many plants of the cool parts of North America and Eurasia, including lupines, larkspurs, and fireweed, are regularly pollinated by bumblebees throughout their ranges.

19–22
"Honey guides" on the flowers of the foxglove (Digitalis purpurea) *serve as distinctive signals to insect visitors. The lower lip of the fused corolla serves as a landing platform of the kind that is commonly found in bee flowers.*

(a)

(b)

19–23
The color perception of most insects is somewhat different from that of human beings. To a bee, for example, ultraviolet light (which is invisible to humans) is seen as a distinct color. These photographs show a flower of marsh marigold (Caltha palustris) (a) *in natural light, showing the solid yellow color as the flower appears to humans, and* (b) *in ultraviolet light. The portions of the flower that appear light in* (b) *reflect both yellow and ultraviolet light, which combine to form a color known as "bee's purple," whereas the dark portions of the flower absorb ultraviolet and therefore appear pure yellow when viewed by a bee. (See also page 429.)*

Some of the evolutionarily more advanced flowers, in particular the orchids, have developed complex passageways and traps that force the bees that visit them to follow a particular route into and out of the flower. This ensures that both anther and stigma come into contact with the bee's body at a particular point and in the proper sequence (see Figure 19–20).

An even more bizarre pollination strategy has been adopted by orchids of the genus *Ophrys*. The flower resembles a female bee, wasp, or fly (Figure 19–25). The males of these insect species emerge early in the spring, before the females. The orchids bloom early in the spring as well, and the male insects attempt to copulate with the orchid flower. During the course of its "sexual" visit, a pollinium may be deposited on the insect's body; when it visits another flower of the same species, the pollinium may be caught in the appropriate grooves on the stigma and thus bring about the pollination of that flower.

(a)

(b)

19–24
Bumblebees (Bombus). *These social bees are important pollinators of many genera of plants throughout the cooler parts of the Northern Hemisphere, and they have been introduced into regions where they are not native for the purpose of pollinating such plants as white clover* (Trifolium repens). (a) *A bumblebee gathering pollen from the flower of a California poppy* (Eschscholzia californica). (b) *Portion of the underground nest of a bumblebee, showing the cells in which the wormlike larvae complete their development. The bumblebees provision these cells with pollen and regurgitated nectar, which they obtain from flowers. Although a colony of bumblebees may visit a wide variety of flowers during the course of a season, an individual bee often visits only the flowers of one kind of plant on a single trip away from the nest.*

19–25
The beelike flowers of the orchid Ophrys speculum *attract male bees, which are so deceived by the resemblance to female bees of their species that they attempt to copulate with the flowers. In doing so, they often pick up a pollen sac (pollinium) from a flower and may then carry it to another flower of the same species. This orchid was photographed in Sardinia.*

An additional range of flowers with different characteristics is pollinated by flies of various kinds, including mosquitoes. These insects feed on nectar but do not gather pollen or store food for their larvae. Examples of flowers pollinated by mosquitos and flies are shown in Figures 19–19 and 19–26.

Flowers Pollinated by Moths and Butterflies

Flowers that coevolved with butterflies and diurnal moths (those that are active during the day rather than at night) are similar in many respects to bee flowers, mainly because butterflies, moths, and bees are all guided to flowers by a combination of sight and smell (Figure 19–27). Some species of butterflies, however, are able to perceive red as a distinct color, and some butterfly-pollinated flowers are red and orange.

(a)

(c)

19–26

Pollination by mosquitoes and other flies.
(a) Some small-flowered orchids, such as
Habenaria elegans, *in which the flowers*
are white or green and relatively incon-
spicuous, are visited and pollinated by mos-
quitoes in North Temperate and Arctic
regions. The mosquitoes obtain nectar
from the flowers. (b) A female mosquito

of the genus Aedes *with an orchid pol-*
linium attached to its head. Other small-
flowered orchids, such as those of the
genus Spiranthes, *are pollinated by bees.*
*(c) A fly on a flower of a lily (*Zigadenus
fremontii*). Notice the conspicuous yellow*
nectaries.

(b)

19–27

*Copper butterfly (*Lycaena gorgon*)*
sucking nectar from the flowers of a daisy.
The long sucking mouthparts of moths
and butterflies are coiled up at rest and
extended when feeding. They vary in
length from species to species: only a few
millimeters long in some of the smaller

moths, they are 1 to 2 centimeters long
in many butterflies, 2 to 8 centimeters
long in some hawkmoths of the North
Temperate Zone, and as long as 25
centimeters in a few kinds of tropical
hawkmoths.

19–28

Yucca moth (Tegeticula yucasella) *scraping pollen from a yucca flower. The female moth visits the creamy white flowers by night and gathers pollen, which she rolls into a tight little ball and carries in her specialized mouthparts to another flower. In the second flower, the moth pierces the ovary wall with her long ovipositor and lays a batch of eggs among the ovules. She then packs the sticky mass of pollen through the openings of the stigma. Moth larvae and seeds develop simultaneously, with the larvae feeding on the developing yucca seeds. When the larvae are fully developed, they gnaw their way through the ovary wall and lower themselves to the ground, where they pupate until the yuccas bloom again. It is estimated that only about 20 percent of the seeds are usually eaten.*

Most moths are nocturnal, and the typical moth-pollinated flower—as seen, for example, in several species of tobacco (*Nicotiana*)—is white or pale in color and has a heavy fragrance, a sweet penetrating odor that often is emitted only after sunset. Well-known flowers that are pollinated by moths include the yellow-flowered species of evening primrose (*Oenothera;* see Figure 8–14b, page 134) and the pink-flowered amaryllis (*Amaryllis belladonna*).

The nectary of a moth or butterfly flower is often located at the base of a long, slender corolla tube or a spur and is usually accessible only to the long sucking mouthparts of moths and butterflies. Hawkmoths, for instance, do not usually enter flowers, as bees do, but hover above them, inserting their long mouthparts into the floral tube. Consequently, hawkmoth flowers do not have the landing platforms, traps, and elaborate internal structural modifications seen in some of the bee flowers. Most moth-flower relationships, which typically involve smaller moths that do not use nearly as much energy as the hawkmoths, also involve shorter corolla tubes and often smaller flowers, over which the moths scramble. One of the most specialized moth-flower relationships is shown in Figure 19–28.

Bird-Pollinated Flowers

Some birds regularly visit flowers to feed on nectar, floral parts, and flower-inhabiting insects; many of these birds also serve as pollinators. In North and South America, the chief pollinators among the birds are hummingbirds (Figure 19–29); in other parts of the

19–29

A male Anna's hummingbird (Calypte anna) *at a flower of the scarlet monkey-flower* (Mimulus cardinalis) *in southern California. Note the pollen on the bird's forehead, which is in contact with the stigma of the flower.*

19–30
A collared sunbird (Anthreptes collarii) *perching and feeding on a bird-of-paradise* (Strelitzia reginae) *flower in South Africa.*

(a)

world, flowers are visited regularly by representatives of other specialized bird families (Figure 19–30).

Bird flowers have a copious, thin nectar (some actually drip with nectar when the pollen is mature) but usually have little odor because the sense of smell is poorly developed in birds. However, birds do have a keen color sense that is much like our own; it is not surprising, therefore, that most bird flowers are colorful, with red and yellow ones being the most common (see Figure 19–14b). Bird-pollinated flowers include red columbine (Figure 19–31a), fuchsia, passion flower, eucalyptus, hibiscus, poinsettia (Figure 19–31b,c), and many members of the cactus, banana, and orchid families.

(b)

(c)

19–31
Examples of bird-pollinated flowers. (a) Columbine (Aquilegia canadensis). *Alternating perianth segments are modified into nectar-filled tubes. Hummingbirds visit these hanging flowers, taking nectar from them on the wing; the nectar* of columbine flowers is inaccessible to most other kinds of animals. (b) and (c) Poinsettias (Euphorbia pulcherrima). In this familiar plant, a native of Mexico, the flowers are small, greenish, and clustered, but each cluster has a large, yellow *nectary from which abundant nectar flows. Ants are seen feeding on the nectar in (b). Modified upper leaves, bright red in color, attract hummingbirds to the clusters of flowers.*

Typically, these flowers are large or they form parts of large inflorescences, features that can be correlated with their importance as visual stimuli and their ability to hold large amounts of nectar.

Bird and other animal pollinators usually restrict their visits to the flowers of a particular plant species, at least for a short period of time. This is not the only factor promoting outcrossing, however. The pollinator must not confine its visits to a single flower or to the flowers of a single plant. When flowers are visited regularly by large animals with a high rate of energy expenditure, such as birds, hawkmoths, or bats, the flowers must produce large amounts of nectar to support the metabolic requirements of the animals and keep them coming back. On the other hand, if an abundant supply of nectar is available to animals with a lower rate of energy expenditure, such as small bees or beetles, these visitors will tend to remain at a single flower and, being satisfied there, will not move on to the flowers of other plants where they might bring about outcrossing. Consequently, species that are regularly pollinated by animals with a high rate of energy consumption, such as hummingbirds, have tended to evolve flowers with the nectar held in tubes or otherwise unavailable to smaller animals with lower rates of energy consumption. Similarly, the color red is a signal to birds but not to most insects. Birds, like ourselves, do not respond very strongly to odor clues. Thus, odorless, red flowers, being inconspicuous to insects, tend not to attract them, an adaptation that is advantageous in view of the copious production of nectar by such flowers.

Bat-Pollinated Flowers

Flower-visiting bats are found in tropical areas of both the Old World and the New World, and more than 250 species of bats—about a quarter of the total number of bat species—include at least some nectar, fruit, or pollen in their diet. Those species of bats that derive all or most of their nourishment from flowers have slender, elongated muzzles and long, extensible tongues, sometimes with a brushlike tip, and their front teeth are often reduced in size or are missing altogether.

Bat flowers are similar in many respects to bird flowers, being large, strong flowers that produce copious nectar (Figure 19–32). Because bats feed at night, bat flowers are typically dully colored, and many of them only open at night. Many bat-pollinated flowers are tubular or structurally modified to protect their nectar in other ways. Some bat-pollinated flowers—and fruits that are regularly dispersed by bats—hang down on long stalks below the foliage, where the bats can fly more easily; others are borne on the trunks of trees. Bats are attracted to the flowers largely through their

19–32
By thrusting its face into the tubular corolla of a flower of an organ-pipe cactus (Stenocereus thurberi), *this bat,* Leptonycteris curasoae, *is able to lap up nectar with its long, bristly tongue. Some of the pollen clinging to the bat's face and neck is transferred to the next flower it visits. This species of bat, which is one of the more specialized nectar-feeding bats, migrates from central and southern Mexico to the deserts of the southwestern United States during late spring and early summer. Here it feeds on the nectar and pollen of organ-pipe and saguaro cacti and on the flowers of agaves.*

sense of smell, and bat-pollinated flowers characteristically have either very strong fermenting or fruitlike odors, or musty scents like those produced by bats to attract one another. Bats fly from tree to tree, eating pollen and other flower parts, and carrying pollen from flower to flower on their fur. At least 130 genera of angiosperms are pollinated by bats or have their seeds dispersed by bats or both. Included among bat-pollinated angiosperms are such economically important plants as bananas, mangoes, kapok, and sisal.

Some bats derive a significant portion of their dietary protein from the pollen they consume. As yet another example of coevolution, the pollen of the flowers they visit has been found to contain significantly higher levels of protein than does the pollen of insect-pollinated flowers.

(a)

(b)

(c)

19–33

Unlike most angiosperms, the grasses have wind-pollinated flowers. Corn (Zea mays) has (a) staminate inflorescences (tassels) at the top of the stem and (b) ovulate inflorescences, with long protruding stigmas (the "silk" on the ears of corn), lower on the stem. (c) Grasses characteristically have enlarged, feathery stigmas

that efficiently catch the wind-blown pollen shed by the hanging anthers, as seen here in a grass of the genus Agropyron. (d) Scanning electron micrograph of a pollen grain of corn, showing the smooth pollen wall found in most wind-pollinated plants and the single aperture characteristic of monocots.

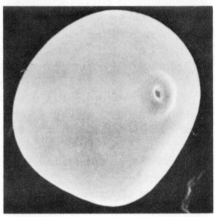

(d)

Wind-Pollinated Flowers

About a century ago, botanists considered wind-pollinated angiosperms to be the most primitive members of the division; they thought that all other kinds of angiosperms had evolved from them. The conifers—thought by some scientists at the time to have been the direct ancestors of the angiosperms—have small, drab-colored, odorless, unisexual cones and are pollinated by the wind. Similarly, the many examples of wind-pollinated flowers have dull colors, are relatively odorless, and do not produce nectar; their petals are either small or absent, and the sexes are often separated. However, studies of other characteristics of these wind-pollinated angiosperms—in particular, their specialized wood—have convinced most botanists that all wind-pollinated angiosperms evolved not from the conifers but from insect-pollinated angiosperms.

According to current interpretations of the evidence, wind-pollinated angiosperms originated independently from several different ancestral stocks. They are best represented in temperate regions and are relatively rare in the tropics. In temperate climates, many individual

trees of the same species are often found close together, and the dispersal of pollen by wind can occur readily in early spring, when the trees are leafless. In the tropics, on the other hand, many more kinds of trees are found in a given area, and the distance to the nearest individual of the same species may be quite great. Furthermore, in many tropical communities the trees are evergreen, and the dispersal of pollen by wind is more difficult than in temperate deciduous forests when the trees are leafless. Under these circumstances, pollination by insects and other animals that have the ability to seek out other individuals of the same plant species, sometimes over relatively great distances (20 kilometers or more, in some cases), is much more efficient than wind pollination.

Because wind-pollinated angiosperms do not depend on insects to transport their pollen from place to place, they devote no energy to the production of nutritious rewards for insect visitors. Wind pollination is very inefficient, however, and it is successful only where a large number of individuals of the same species grow fairly close together. Nearly all wind-borne pollen falls to the ground within 100 meters of the parent plant. Thus, if the individual plants are widely scattered, the chance that a pollen grain will reach a receptive stigma is very slim. Many wind-pollinated plants are dioecious (having male and female flowers on separate plants), such as the willows; monoecious (having separate male and female flowers on the same plant), such as the oaks (see Figure 18–10, page 386); or genetically self-incompatible, such as many grasses. Even though their pollen moves about somewhat randomly, these plants have evolved devices that foster a high degree of outcrossing.

Wind-pollinated flowers usually have well-exposed stamens that can easily lose their pollen to the wind. In some, the anthers are suspended from long filaments hanging from the flower (Figures 19–33 and 19–34). The abundant pollen grains, which are generally smooth and small, do not adhere to one another as do the pollen grains of insect-pollinated species. The large stigmas are characteristically exposed, and they often have branches or feathery outgrowths adapted for intercepting wind-borne pollen grains. Most wind-pollinated plants have ovaries with single ovules (and hence single-seeded fruits) because each pollination event consists of the meeting of one pollen grain with one stigma and leads to the fertilization of one ovule for each flower. Thus, each oak flower produces only a single acorn and each grass flower produces only a single grain, but plants with very small flowers tend to have, in compensation, multiple inflorescences (Figures 19–33 through 19–35).

19–35
Most common species of trees in temperate regions are wind-pollinated. The staminate flowers of the paper birch (Betula papyrifera) *hang down in catkins—flexible, thin tassels several centimeters long. These catkins are whipped by passing breezes, and the pollen, when mature, is scattered about by the wind.*

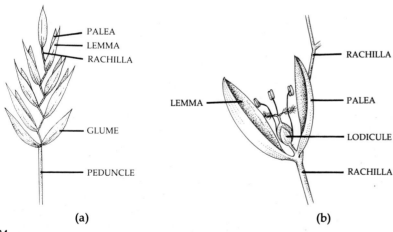

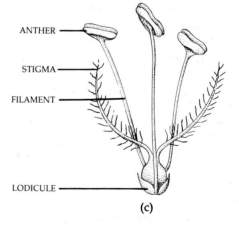

(a)

(b)

(c)

19–34
Grass flowers (florets) usually develop in clusters. (a) As a cluster matures, a single pair of dry, chaffy bracts—the glumes—separate a little, exposing the elongating spikelet, with from one to many florets (depending on the species of grass) attached to a central axis, or

rachilla. (b) Each floret is surrounded by two distinctive bracts of its own, the palea and the lemma. These are forced apart, exposing the inner parts of the flower (c), by the swelling of the lodicules—small, rounded bodies at the base of the carpel—and are spread wide when

the grass is in flower. The stamens, usually three in number, have slender filaments and long anthers, and the stigmas are typically long and feathery and so are efficient at intercepting the wind-borne pollen.

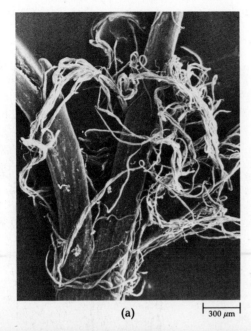

(a)

⊢ 300 μm ⊣

(b)

19–36
Unique pollination systems in submerged aquatic angiosperms. (a) The slender, branching stigmas of the sea-nymph, Amphibolis, seen here in a scanning electron micrograph, have captured many grains of the threadlike pollen released by staminate plants of this species. (b) Tiny staminate flowers of eel grass, Vallisneria, float into the depression in the surface-tension layer of the water that surrounds the much larger carpellate flower. The individual pollen grains are visible on the staminate flowers.

Water-Pollinated Flowers

Very few angiosperms—about 79 families and 380 genera—are submerged aquatic plants, living in marine and freshwater habitats. In 18 of these genera, the pollen is either transported under water or floats from one plant to another across the surface of the water. In some of these plants, pollen grains are either threadlike (filiform), thus increasing their chances of coming into

contact with receptive stigmas, or linked into chains, with a similar effect (Figure 19–36a). The filiform pollen of some aquatic genera is unusual in lacking an exine. In other genera of submerged aquatic plants, the pollen is dispersed across the surface of the water. For example, in the freshwater eel grass *Vallisneria*, the entire staminate flower is released from beneath the water surface and floats to the top. After the flower reaches the surface, its three stamens become erect and function like a sail. Floating just below the tension zone at the water surface, the carpellate flower creates a small depression into which the staminate flowers fall (Figure 19–36b).

In contrast to these plants with very specialized pollination systems, which are clearly advanced in an evolutionary sense, most aquatic angiosperms are either wind-pollinated or insect-pollinated, like their terrestrial ancestors. Their flowers are held above the surface of the water.

FLOWER COLORS

Color is one of the most conspicuous features of angiosperm flowers—a characteristic by which the members of the division are easily recognized. The varied colors of different kinds of flowers evolved in relation to their pollination systems and, in general, are advertisements for particular kinds of animals, as we have just seen.

The pigments that are responsible for the colors of angiosperm flowers are generally common in vascular plants other than angiosperms. It is the way in which they are concentrated in angiosperm flowers, and particularly in their corollas, that is a special characteristic of the flowering plants. Surprisingly, all flower colors are produced by a small number of pigments. Many red, orange, or yellow flowers owe their color to the presence of carotenoid pigments similar to those that occur in leaves (and in all plants, in green algae, and in some other organisms as well). The most important pigments in floral coloration, however, are **flavonoids,** which are compounds in which two six-carbon rings are linked by a three-carbon unit. Flavonoids probably occur in all angiosperms, and they are sporadically distributed among the members of other groups of plants. In leaves, flavonoids block far-ultraviolet radiation (which is destructive to nucleic acids and proteins). They usually selectively admit light of blue-green and red wavelengths, which are important for photosynthesis.

Pigments belonging to one major class of flavonoids, the **anthocyanins,** are major determinants of flower color (Figure 19–37). Most red and blue plant pigments are anthocyanins, which are water soluble and are found in vacuoles. By contrast, the carotenoids are fat soluble and are found in plastids. The color of an anthocyanin pigment depends on the acidity of the cell

PELARGONIDIN · CYANIDIN · DELPHINIDIN

19–37

Three anthocyanin pigments, the basic pigments on which flower colors in many angiosperms depend: pelargonidin (red), cyanidin (violet), and delphinidin (blue). Related compounds known as flavonols are yellow or ivory, and the carotenoids are red, orange, or yellow. Betacyanins (betalains) are red pigments that occur in one group of dicots. Mixtures of these different pigments, together with changes in cellular pH, produce the entire range of flower color in the angiosperms. Changes in flower color provide "signals" to pollinators, telling them which flowers have opened recently and are more likely to provide food.

sap of the vacuole; for example, cyanidin is red in acid solution, violet in neutral solution, and blue in alkaline solution. In some plants, the flowers change color after pollination, usually because of the production of large amounts of anthocyanins, and then become less conspicuous to insects.

The **flavonols,** another group of flavonoids, are very commonly found in leaves and also in many flowers. A number of these compounds are colorless or nearly so, but they may contribute to the ivory or white hues of certain flowers.

For all flowering plants, different mixtures of flavonoids and carotenoids (as well as changes in cellular pH) and differences in the structural, and thus the reflective, properties of the flower parts produce the characteristic colors. The bright fall colors of leaves come about when large quantities of colorless flavonols are converted into anthocyanins as the chlorophyll breaks down. In the all-yellow flowers of the marsh marigold (*Caltha palustris*), the ultraviolet-reflective outer portion is colored by carotenoids, whereas the ultraviolet-absorbing inner portion is yellow to our eyes because of the presence of a yellow chalcone, one of the flavonoids. To a bee or other insect, the outer portion of the flower appears to be a mixture of yellow and ultraviolet, a color called "bee's purple," whereas the inner portion appears pure yellow (see Figure 19–23). Most, but not all, ultraviolet reflectivity in flowers is related to the presence of carotenoids, and thus ultraviolet patterns are more common in yellow flowers than in others.

In the goosefoot, cactus, and portulaca families and in other members of the order Chenopodiales (Centro-

spermae), the reddish pigments are not anthocyanins or even flavonoids but a group of more complex aromatic compounds known as **betacyanins** (betalains). The red flowers of *Bougainvillea* and the red color of beets are due to the presence of betacyanins. No anthocyanins occur in these plants, and the families characterized by betacyanins are closely related to one another.

The Evolution of Fruits

Just as flowers have evolved in relation to their pollination by many different kinds of animals and other agents, so have fruits evolved for dispersal in many different ways. Fruit dispersal, like pollination, is a fundamental aspect of the evolutionary radiation of the angiosperms. Before we consider this subject in more detail, however, we must present some basic information about fruit structure.

A fruit is a mature ovary, which may or may not include some additional flower parts. A fruit in which such additional parts are retained is known as an **accessory fruit.** Although fruits usually have seed within them, some—**parthenocarpic fruits**—may develop without seed formation. The cultivated strains of bananas are familiar examples of this exceptional condition.

Fruits are generally classified as simple, multiple, or aggregate, depending on the arrangement of the carpels from which the fruit develops. **Simple fruits** develop from one carpel or from several united carpels. **Aggregate fruits,** such as those of magnolias, raspberries, and strawberries, consist of a number of separate carpels of one gynoecium. The individual parts of aggregate fruits are known as **fruitlets;** they can be seen, for example, in the magnolia fruit shown in Figure 19–11c. **Multiple fruits** consist of the gynoecia of more than one flower. The pineapple, for example, is a multiple fruit consisting of an inflorescence with many previously separate ovaries fused on the axis on which the flowers were borne (the other flower parts being squeezed between the expanding ovaries).

Simple fruits are by far the most diverse of the three groups. When ripe, they may be soft and fleshy, dry and woody, or papery. There are three main types of fleshy fruits—berries, drupes, and pomes. In **berries**—examples of which are tomatoes, dates, and grapes—there may be one to several carpels, each of which is typically many-seeded. The inner layer of the fruit wall

(a)

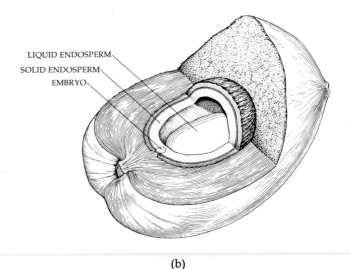

LIQUID ENDOSPERM
SOLID ENDOSPERM
EMBRYO

(b)

19–38
The coconut palm (Cocos nucifera) *has a very wide range on ocean shores throughout the world because its fruits are able to float for long periods and then germinate when they reach land. (a) A coconut germinating on the beach in* *Florida. (b) Diagram of a coconut fruit. The coconut milk is liquid endosperm; cell walls form around the nuclei in the liquid endosperm, which becomes solid by the time of germination. When coconuts are transported commercially, their* *husks are usually removed first, so that in temperate countries people usually see only the stony inner shell of the fruit surrounding the seed.*

is fleshy. In **drupes,** there may also be one to several carpels, but each carpel usually contains only a single seed. The inner layer of the fruit is stony and usually tightly adherent to the seed. Peaches, cherries, olives, and plums are familiar drupes. Coconuts are drupes whose outer layer is fibrous rather than fleshy, but in temperate regions we usually see only the coconut seed with the adherent stony inner layer of the fruit (Figure 19–38). **Pomes** are highly specialized fleshy fruits that are characteristic of one subfamily of the rose family. The pome is derived from a compound inferior ovary in which the fleshy portion comes largely from the enlarged base of the perianth. The inner portion, or endocarp, of a pome resembles a tough membrane, as you know from eating apples and pears, the two most familiar examples of this kind of fruit.

Dry simple fruits are classified as either dehiscent (Figures 19–39 and 19–40) or indehiscent (Figure 19–41). In **dehiscent fruits,** the tissues of the mature ovary wall (the pericarp) break open, freeing the seeds. In **indehiscent fruits,** on the other hand, the seeds remain in the fruit after the fruit has been shed from the parent plant.

There are several kinds of dehiscent simple dry fruits. The **follicle** is derived from a single carpel that splits down one side at maturity, as in the columbines and milkweeds (Figure 19–39a). Follicles were also characteristic of the extinct, middle Cretaceous plant *Archaeanthus* (see Figure 19–12), and they are also found in magnolias (see Figure 19–11c). In the pea family (Fabaceae), the characteristic fruit is a **legume.** Legumes resemble follicles, but they split along both sides (Figure 19–40). In the mustard family (Brassicaceae), the fruit is called a **silique** and is formed of two fused carpels. At maturity, the two sides of the fruit split off,

leaving the seeds attached to a persistent central portion (Figure 19–39c). The most common sort of dehiscent simple dry fruit is the **capsule,** which is formed from a compound ovary in plants with either a superior or an inferior ovary. Capsules shed their seeds in a variety of ways. In the poppy family (Papaveraceae), the seeds are often shed when the capsule splits longitudinally, but in some members of this family they are shed through holes near the top of the capsule (Figure 19–39b).

19–40 *(On the facing page)*
The legume, a kind of fruit that is usually dehiscent, is the characteristic fruit of the pea family, Fabaceae (also called Leguminosae). With about 18,000 species, Fabaceae is one of the largest families of flowering plants. Many members of the family are capable of nitrogen fixation because of the presence of nodule-forming bacteria of the genus Rhizobium *on their roots (see page 602). For this reason, these plants are often the first colonists on relatively infertile soils, as in the tropics, and they may grow rapidly there. The seeds of a number of plants of this family, such as peas, beans, and lentils, are important foods. (a) Legumes of the garden pea,* Pisum sativum. *(b) Legumes of* Albizzia polyphylla, *growing in Madagascar; each seed is in a separate compartment of the fruit. (c) Legume of* Griffonia simplicifolia, *a West African tree. The two valves of the legume are split apart, revealing the two seeds within.*

19–39
Dehiscent fruits. (a) Bursting follicles of a milkweed (Asclepias). (b) In some members of the poppy family (Papavera-ceae), such as poppies (genus Papaver), the capsule sheds its seeds through pores near the top of the fruit. (c) Plants of the mustard family (Brassicaceae) have a characteristic fruit known as a silique, in which the seeds arise from a central partition, and the two enclosing valves fall away at maturity.

(a)

CAPSULE
(*Papaver somniferum*)

(b)

SILIQUE
(*Brassica rapa*)

(c)

(a)

(b)

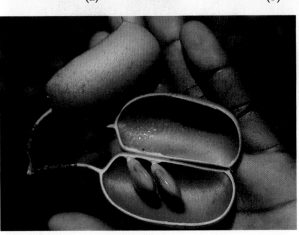

(c)

SAMARA
(*Fraxinus*)

(a)

(b)

19–41
Indehiscent fruits. (a) The samara, a winged fruit characteristic of ashes (Fraxinus) and elms (Ulmus), retains its single seed at maturity. Samaras are dispersed by wind. (b) Cypselas of burdock (Arctium), which attach them-selves to passing animals and are dispersed in that way. Burdock is a member of the family Asteraceae.

Indehiscent simple dry fruits are found in many different plant families (Figure 19–41). Most common is the **achene,** a small, single-seeded fruit in which the seed lies free in the cavity except for its attachment by the funiculus. Achenes are characteristic of the buttercup family (Ranunculaceae) and the buckwheat family (Polygonaceae). Winged achenes, such as those found in elms and ashes, are commonly known as **samaras** (Figure 19–41). The achenelike fruit that occurs in grasses (Poaceae) is known as a **caryopsis,** or grain; in it, the seed coat is firmly united to the fruit wall. In the Asteraceae, the complex, achenelike fruit is derived from an inferior ovary; technically, it is called a **cypsela** (Figure 19–41b; see also Figure 19–43). Acorns and hazelnuts are examples of **nuts,** which resemble achenes but have a stony fruit wall and are derived from a compound ovary. Finally, in the parsley family (Apiaceae) and the maples (Aceraceae), as well as a number of other, unrelated groups, the fruit is a **schizocarp,** which splits at maturity into two or more one-seeded portions (Figure 19–42a).

DISPERSAL OF FRUITS AND SEEDS

Just as flowers have evolved according to the characteristics of the pollinators that visit them regularly, so have fruits evolved in relation to their dispersal agents. In both coevolutionary systems, there have, in general, been many changes in relation to different dispersal agents within individual families and a great deal of convergent evolution toward similar-appearing structures with similar functions. We shall review some of the adaptation of fruits here, in relation to their dispersal agents.

Wind-Borne Fruits and Seeds

Some plants have light fruits or seeds that are dispersed by the wind (Figures 19–39a, 19–41a, 19–42). The dustlike seeds of all members of the orchid family, for example, are wind-borne. Other fruits have wings, which are sometimes formed from perianth parts, that allow them to be blown from place to place. In the schizocarps of maples, for example, each carpel develops a long wing (Figure 19–42a). The two carpels separate and fall when mature. Many members of the Asteraceae—dandelions, for example—develop a plumelike pappus, which aids in keeping the light fruits aloft (Figures 19–42b and 19–43). In some plants, the seed itself, rather than the fruit, bears the wing or plume; the familiar butter-and-eggs (*Linaria vulgaris*) has a winged seed, and both fireweed (*Epilobium*) and milkweed (*Asclepias;* see Figure 19–39a) have plumed seeds. In willows and poplars (family Salicaceae), the seed coat is covered with woolly hairs. In tumbleweeds

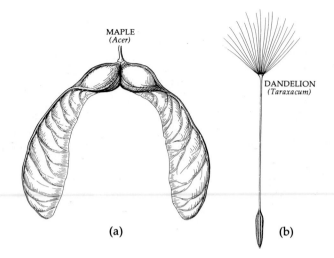

19–42
Wind-dispersed fruits. (a) In maples (Acer), each half of the schizocarp has a long wing. (b) The fruits of the dandelion (Taraxacum) and many other composites have a modified calyx, called the pappus, which is adherent to the mature cypsela and may form a plumelike structure that aids in wind dispersal.

19–43
The familiar small, indehiscent fruits of dandelions, which are technically known as cypselas (but often loosely called achenes), spread by their plumelike, modified calyx (the pappus). This photograph shows the fruiting heads of a plant of the genus Agoseris, which is closely related to the dandelions.

19–44

*In tumbleweeds (Salsola), the whole
plant breaks off and is blown across open
country, scattering its seeds as it tumbles
along. So many tumbleweeds blew into
Mobridge, South Dakota, on November 8,
1989, that the town looked like this. The
tumbleweeds are natives of Eurasia, but
they are widely naturalized as weeds in
North America and elsewhere.*

(a)

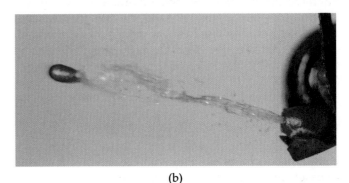

(b)

(Salsola), the whole plant (or a portion of it) is blown
along by the wind, scattering seeds as it moves (Figure
19–44).

Other plants shoot their seeds aloft. In touch-me-
not (Impatiens), the valves of the capsules separate suddenly, throwing seeds for some distance. In the witch
hazel (Hammamelis), the endocarp contracts as the fruit
dries, discharging the seeds so forcefully that they
sometimes travel as far as 15 meters from the plant.
Another example of self-dispersal is shown in Figure
19–45. In contrast to these active methods of dispersal,
the seeds or fruits of many plants simply drop to the
ground and are dispersed more or less passively (or are
dispersed by agents that operate only sporadically, such
as floods).

19–45

*Dwarf mistletoe (Arceuthobium), a
parasitic dicot that is the most serious
cause of loss of forest productivity in
the western United States. (a) A plant
growing on a pine branch in California.
(b) Seed discharge. Very high hydrostatic
pressure builds up in the fruit and shoots
the seeds as much as 15 meters laterally.
The seeds have an initial velocity of
about 100 kilometers per hour. This is
one of the ways in which the seeds are
spread from tree to tree, although they
are also sticky and can be carried from
one tree to another over much longer
distances by adhering to the feet or
feathers of birds.*

Water-Borne Fruits and Seeds

The fruits and seeds of many plants, especially those
growing in or near water sources, are adapted for floating, either because air is trapped in some part of the
fruit or because the fruit contains tissue that includes
large air spaces. Some fruits are especially adapted for
dispersal by ocean currents; notable among these is the
coconut (see Figure 19–38), which is why almost every
newly formed Pacific atoll quickly acquires its own coconut tree. Rain is also a common means of fruit and
seed dispersal; it is particularly important for plants
that live on hillsides or mountain slopes.

Animal-Borne Fruits and Seeds

The evolution of sweet and often highly-colored, fleshy fruits was clearly involved in the coevolution of animals and flowering plants. The majority of fruits in which much of the pericarp is fleshy—cherries, raspberries, dogwoods, grapes—are eaten by vertebrates. When such fruits are eaten by birds or mammals, the seeds the fruits contain are spread by being passed unharmed through the digestive tract or, in birds, by being regurgitated at a distance from the place where they were ingested (Figure 19–46). Sometimes, partial digestion aids the germination of the seeds by weakening their seed coats.

When fleshy fruits ripen, they undergo a series of characteristic changes, mediated by the hormone ethylene, which will be discussed in Chapter 25. Among these changes are a rise in sugar content, a softening of the fruit caused by the breakdown of pectic substances, and often a change in color from inconspicuous, leaf-like green to bright red (Figure 19–46a), yellow, blue, or black. The seeds of some plants, especially tropical ones, often have fleshy appendages, or arils, with the bright colors characteristic of fleshy fruits and, like them, are aided in their dispersal by vertebrates. The arils of yews (*Taxus*; see Figure 17–26) are not homologous to the structures called arils that occur in angiosperms, because yew arils are outgrowths of the seed, not of the fruit. Nevertheless, they play a similar role in seed dispersal.

Unripe fruits are often green or colored in such a way that they are inconspicuous among the plant's green leaves, and thus they are somewhat concealed from birds, mammals, and insects. They may also be disagreeable to the taste—such as unripe cherries (*Prunus*), which have a strong, acidic taste—thereby discouraging animals from eating them before the seeds are ripe. The changes in color that accompany ripening are the plant's "signal" that the fruit is ready to be eaten—the seeds are ripe and ready for dispersal (Figure 19–46). It is no coincidence that red is such a predominant color among ripe fruits. Red fruit is inconspicuous to insects, blending with the background of green leaves. Insects are too small to disperse the large seeds of fleshy fruits effectively, and such concealment is therefore advantageous to these plants. At the same time, the red fruits are very conspicuous to vertebrates, which then eat the fruits and may disperse their ripe seeds.

A number of angiosperms have fruits or seeds that are dispersed by adhering to fur or feathers (Figure 19–47; see also Figure 19–41b). These fruits and seeds have hooks, barbs, spines, hairs, or sticky coverings that allow them to be transported, often for great distances, attached to the bodies of animals.

Other important agents of seed dispersal in some plants are ants (Figure 19–48). Such plants have evolved a special adaptation on the exterior of their seeds, called an elaiosome, a fleshy, white appendage that contains lipids, protein, starch, sugars, and vita-

(a)

(b)

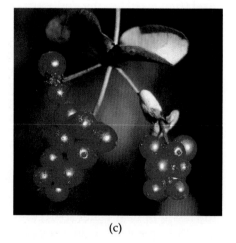

(c)

19–46

The seeds of fleshy fruits are usually dispersed by vertebrates that eat the fruits and either regurgitate the seeds or pass them as part of their feces. Examples of vertebrate-dispersed fruits are shown here. (a) Strawberries (Fragaria), an example of an aggregate fruit. The achenes are borne on the surface of a fleshy
receptacle. Immature strawberries, like the immature stages of many bird- or mammal-dispersed fruits, are green, but they become red when the seeds are mature and thus ready for dispersal. (b) The berries of many cacti, such as this prickly pear (Opuntia) growing in southern Mexico, are conspicuous at
maturity. (c) Berries of chaparral honeysuckle, Lonicera hispidula. These berries develop from inferior ovaries, and therefore have fused portions of the outer floral whorls incorporated in them. A flower of this species is shown in Figure 19–13c.

19–47
The fruits of the African plant Harpago-phytum, *a member of the sesame family* (Pedaliaceae), *are equipped with "grappling hooks," by means of which they catch in the fur on the legs of large mammals and thus are spread from place to place.*

19–48
An ant (Rhytidoponera metallica) *taking the seed of an acacia back to its nest in the Australian bushland. Note the large, yellow elaiosome.*

mins. The ants usually carry such seeds back to their nests, where the elaiosomes are consumed by other workers or larvae and the seeds are left intact. The seeds readily germinate in this location, and seedlings often become established, protected from their predators and perhaps benefiting from nutrient enrichment as well. Up to a third of the species in some plant communities, such as the herb understory communities in the deciduous forests of the central and eastern United States, are dispersed by ants in this way. These plants include such familiar species as spring beauty (*Claytonia virginiana*), dutchman's-breeches (*Dicentra cucullaria*), bloodroot (*Sanguinaria canadensis*), many species of violets (*Viola*, Figure 10–2), and *Trillium* (Figure 1–3).

Biochemical Coevolution

Also important in the evolution of angiosperms are the so-called *secondary plant products*. Once thought of as waste products, these include an array of chemically unrelated compounds, such as alkaloids, quinones, essential oils (including terpenoids), glycosides (including cyanogenic substances and saponins), flavonoids, and even raphides (needle-like crystals of calcium oxalate). The presence of certain of these compounds can characterize whole families, or groups of families, of flowering plants (Figure 19–49).

In nature, these chemicals appear to play a major role either in restricting the palatability of the plants in which they occur or in causing animals to avoid the plants altogether (Figure 19–50). When a given family of plants is characterized by a distinctive group of secondary plant substances, those plants are apt to be eaten only by insects belonging to certain families. The mustard family (Brassicaceae), for example, is characterized by the presence of mustard-oil glycosides and associated enzymes that break down these glycosides to release the pungent odors associated with cabbage, horseradish, and mustard. Plant-eating insects of most groups ignore plants of the mustard family and will not feed on them even if they are starving. However, certain groups of true bugs and beetles, and the larvae of some groups of moths, feed only on the leaves of plants of this family. The larvae of most of the members of the butterfly subfamily Pierinae (which includes the cabbage butterflies and orange-tips) also feed only on these plants. The same chemicals that act as deterrents to most groups of insect herbivores often act as feeding stimuli for these narrowly restricted feeders. For example, certain moth larvae that feed on cabbage will extrude their mouthparts and go through their characteristic feeding behavior when presented with agar or filter paper containing juices pressed from these plants.

It is clear that the ability to manufacture these chemicals and to retain them in their tissues is an important evolutionary step and gives the plants biochemical protection from most herbivores. During the

19–49
Secondary plant substances: sinigrin, from black mustard, Brassica nigra; *calactin, a cardiac glycoside, from the milkweed* Asclepias curassavica; *nicotine, from tobacco,* Nicotiana tabacum, *a member of the nightshade family; caffeine, from* Coffea arabica *of the madder family; and theobromine, a prominent alkaloid in coffee, tea* (Thea sinensis), *and cocoa* (Theobroma cacao). *Nicotine, caffeine, and theobromine are alkaloids, members of a diverse class of nitrogen-containing ring compounds that are physiologically active in vertebrates.*

SINIGRIN

CALACTIN

NICOTINE

THEOBROMINE

CAFFEINE

19–50
Poison ivy (Toxicodendron radicans) *produces a secondary plant substance, 3-pentadecanedienyl catechol, which causes an irritating rash on the skin of many people. The ability to produce this alcohol presumably evolved under the selective pressure exerted by herbivores. Fortunately, the plant is easily identifiable by its characteristic compound leaves with their three leaflets.*

course of their evolution, the members of the family Brassicaceae were doubtless protected from most herbivores because they possessed mustard-oil glycosides. From the standpoint of herbivores in general, such protected plants, because they are not already heavily utilized by other plant feeders, represent an unexploited food source for any group of insects that can tolerate or break down the poisons manufactured by the plant. The main evolutionary development of the butterfly group Pierinae probably occurred after its ancestors had acquired the ability to feed on plants of the mustard family by breaking down these toxic molecules.

Herbivorous insects that are narrowly restricted in their feeding habits to groups of plants with certain secondary plant substances are often brightly colored, which serves as a signal to their predators that they carry the noxious chemicals in their bodies and hence are unpalatable. For example, the assemblage of insects found feeding on a milkweed plant on a summer day might include bright green chrysomelid beetles, bright red cerambycid beetles and true bugs, and orange and black monarch butterflies, among others. Milkweeds (Asclepiadaceae) are richly endowed with alkaloids and cardiac glycosides, heart poisons that have potent effects in vertebrates, the main potential predators of these insects. If a bird ingests a monarch butterfly, severe gastric distress and vomiting will follow, and orange-and-black patterns like that of the monarch will be avoided by the predator in the future. Other insects, such as the viceroy butterfly (which also has an orange-and-black pattern), have evolved similar coloration and markings. Thus, they escape predation by capitalizing on their resemblance to the poisonous monarch. This phenomenon, known as **mimicry,** is ultimately dependent upon the plant's chemical defenses. Various drugs and psychedelic chemicals, such as the active ingredients in marijuana (*Cannabis sativa*) and

19–51

Some plants that produce hallucinogenic and medicinal compounds. (a) Mescaline, from the peyote cactus (Lophophora williamsii), is used ceremonially by many Native American groups of northern Mexico and the southwestern United States. (b) Tetrahydrocannabinol (THC) is the most important active molecule in marijuana (Cannabis sativa). (c) Quinine, a valuable drug used in the treatment and prevention of malaria, is derived from tropical trees and shrubs of the genus Cinchona. (d) Cocaine, a drug that has recently been abused to an unparalleled extent, is derived from coca (Erythroxylon coca), a cultivated plant of northwestern South America. A Peruvian woman is shown here harvesting the leaves of cultivated coca. The secondary substances identified in these plants presumably protect them from the depredations of insects, but they are also physiologically active in vertebrates, including humans.

the opium poppy (*Papaver somniferum*), among others, are also secondary plant compounds that in nature presumably play a role in discouraging the attacks of herbivores (Figure 19–51).

Still more complex systems are known. When the leaves of potato or tomato plants are wounded, as by the Colorado potato beetle, the concentration of proteinase inhibitors, which interfere with the digestive enzymes in the guts of the beetle, rapidly increases in the wounded tissues. Other plants manufacture molecules that resemble the hormones of insects or other predators and thus interfere with the predators' normal growth and development. One of these natural products that resembles a human hormone is a complex molecule called diosgenin, which is obtained from wild yams. Diosgenin is only two simple chemical steps

away from 16-dehydropregnenolone (16D), the main active ingredient in many oral contraceptives, and wild yams were once a major source for the manufacture of 16D. Unfortunately, these plants grow slowly, and the supply of wild yams was soon largely exhausted. Species of *Solanum* are being cultivated as alternative sources of molecules suitable for simple conversion to 16D.

As mentioned previously, pollination and fruit-dispersal systems have developed particular coevolutionary patterns in which many of the possible variants have evolved not once but several times within a particular plant family or even genus. The resulting array of forms gives the angiosperms an extremely wide variety of pollination and fruit-dispersal mechanisms. In the case of biochemical relationships, however, the evolutionary steps appear to have been large and definitive, and whole families of plants can be characterized biochemically and associated with major groups of plant-eating insects. These biochemical relationships appear to have played a key role in the early success of the angiosperms, which have a much more diverse array of secondary biochemical substances than any other group of organisms.

Summary

The relationships of angiosperms, long a subject of debate, have been investigated by constructing cladograms. Such methods have defined seed plants as one evolutionary line, or clade, and the gnetophytes as the closest living relatives of angiosperms. The further application of such methods, the incorporation of better-understood fossils into the analyses, and the application of biochemical methodology will help to define the relationships between the angiosperms and other plant groups, and within the angiosperms, more sharply in the future.

The earliest definite angiosperm remains are pollen grains from the early Cretaceous period, about 127 million years old; the earliest known flowers are about 120 million years old. The flowering plants became dominant worldwide between 80 and 90 million years ago. Possible reasons for their success include various adaptations for drought resistance, including the evolution of the deciduous habit, as well as the evolution of efficient and often specialized mechanisms for pollination and seed dispersal. The primary radiation of angiosperms took place when the southern continents were united in the great land masses of Gondwanaland, which have separated progressively during the history of the group.

Early angiosperms fall into two groups, one of which had relatively few flower parts and often unisexual flowers and the other of which had large, robust, bisexual flowers. Sepals are specialized leaves that protect the flower in bud, whereas the petals of most angiosperms have evolved from stamens that have lost their sporangia during the course of evolution. Stamens have evolved from leaflike precursors into the slender structures with terminal anthers that are characteristic of most contemporary angiosperms. Carpels are leaflike structures that have undergone folding, during the course of evolution, to enclose the ovules. In most plants, the carpels have become specialized into a swollen, basal ovary; a slender style; and a receptive, terminal stigma. The loss of individual floral whorls and fusion within and between them have led to the evolution of many specialized floral types, which are often characteristic of particular families.

Pollination by insects is basic in the angiosperms, and the first pollinating agents were probably beetles. The closing of the carpel, in an evolutionary sense, may have been a device to protect the ovules from being eaten by visiting insects. More specialized groups of insects evolved later in the history of the angiosperms, and wasps, flies, butterflies, and moths have each left their mark on the morphology of certain angiosperm flowers. The bees, however, are the most specialized and constant of flower-visiting insects and have probably had the greatest effect on the evolution of angiosperm flowers. Each group of flower-visiting animals is associated with a particular group of floral characteristics related to the animals' visual and olfactory senses. Some angiosperms have become wind-pollinated, shedding copious quantities of small, nonsticky pollen and having well-developed, often feathery stigmas that are efficient in collecting pollen from the air. Water-pollinated plants have either filamentous pollen grains that float to submerged flowers or various ways of transmitting pollen across the surface of the water.

Flowers that are regularly visited and pollinated by animals with high energy requirements, such as hummingbirds, hawkmoths, and bats, must produce large amounts of nectar. They must then protect and conceal these sources of nectar from other potential visitors with lower energy requirements, which might satiate themselves with nectar from a single flower (or from the flowers of a single plant) and therefore fail to move on to another plant of the same species to effect cross-pollination. Wind pollination can occur only when individual plants grow together in large groups, whereas insects, birds, or bats can carry pollen great distances from plant to plant.

Flower colors are determined mainly by carotenoids, which are yellow, oil-soluble pigments that occur in the chloroplasts and act as accessory pigments in photosynthesis, and by flavonoids, which are water-soluble ring compounds present in the vacuole. Anthocyanins, blue or red pigments that constitute one major class of flavonoids, are especially important in determining the colors of flowers and other plant parts.

Fruits are just as diverse as the flowers from which they are derived, and they can be classified either morphologically, in terms of their structure and development, or functionally, in terms of their methods of dispersal. Fruits are basically mature ovaries but, if additional flower parts are retained in their mature structure, are said to be accessory fruits. Simple fruits are derived from one carpel or from a group of united carpels, aggregate fruits from the free carpels of one flower, and multiple fruits from the fused carpels of several or many flowers. Dehiscent fruits split open to release the seeds, and indehiscent fruits do not.

Wind-borne fruits or seeds are light and often have wings or tufts of hairs that aid in their dispersal. The fruits of some plants expel their seeds explosively. Some seeds or fruits are borne away by water, in which case they must be buoyant and have water-resistant coats. Others are disseminated by birds or mammals and have evolved fleshy coverings that are tasty and often conspicuous to attract feeding animals. Others adhere to the coats of mammals or to feathers and are distributed in this manner. Ants disperse the seeds and fruits of many plants; such dispersal units typically have an oily appendage, an elaiosome, which the ants consume.

Biochemical coevolution has been an important aspect of the evolutionary success and diversification of the angiosperms. Certain groups of angiosperms have evolved various secondary plant substances, such as alkaloids, which protect them from most foraging herbivores. However, certain herbivores (normally those with narrow feeding habits) are able to feed on those plants and are regularly found associated with them. Potential competitors are excluded from the same plants because of their inability to handle the toxins. This pattern indicates that a stepwise pattern of coevolutionary interaction has occurred, and it appears likely that the early angiosperms also may have been protected by their ability to produce some chemicals that functioned as poisons for herbivores.

Suggestions for Further Reading

Bakker, Robert T.: "How Dinosaurs Invented Flowers," *Natural History*, November 1986, pages 30–38.

An imaginative set of hypotheses on the possible role of dinosaurs in angiosperm evolution; from the author's book The Dinosaur Heresies.

Barth, Friedrich G.: *Insects and Flowers*, Princeton University Press, Princeton, N.J., 1985.

A well-written introduction to the field of plant pollination biology, emphasizing recent discoveries in the field.

Brown, Deni: *Aroids: Plants of the Arum Family*, Timber Press, Portland, Ore., 1988.

A fine, semipopular account of the fascinating family of plants featured in the essay in this chapter.

Corbet, Sarah: "More Bees Make Better Crops," *New Scientist* 23:40–43, July 1987.

Farmers often undervalue the contributions of native bees, replacing insect-pollinated crops with wind-pollinated cereals when the bee populations begin to dwindle. The farmers then have fewer choices in what they are able to cultivate.

Cronquist, Arthur: *The Evolution and Classification of Flowering Plants*, 2nd ed., The New York Botanical Garden, Bronx, N.Y., 1988.

Excellent overall account of the features of plant families and of approaches to their classification.

Doust, Jon Lovett, and Lesley Lovett Doust (eds.): *Plant Reproductive Ecology: Patterns and Strategies*, Oxford University Press, New York, 1988.

Although advanced, this excellent book provides a rich sampling of the best of current research on plant reproductive biology. Highly recommended.

Dressler, Robert L.: *The Orchids: Natural History and Classification*, Harvard University Press, Cambridge, Mass., 1981.

A lively and interesting account of the largest plant family.

Friis, Else Marie, William G. Chloner, and Peter R. Crane: *The Origins of Angiosperms and Their Biological Consequences*, Cambridge University Press, Cambridge, England, 1987.

An outstanding review of the modern evidence bearing on this fascinating problem.

Nicklas, Karl J.: "The Aerodynamics of Wind Pollination," *The Botanical Review* 51:328–386, 1985.

A fascinating review of experiments and observations on the ways in which some angiosperms and other plants take advantage of the wind to transfer their pollen.

Richards, A.J.: *Plant Breeding Systems*, Allen and Unwin, Winchester, Mass., 1986.*

An outstanding account of reproductive biology, including pollination systems, in the angiosperms.

Robacker, David C., Bastiaan J.D. Meeuse, and Eric H. Erickson: "Floral Aroma," *BioScience* 38:390–396, 1988.

Odor-mediated relationships between flowers and insect pollinators are of key significance in pollination systems.

Stiles, Edmund W.: "Fruit for All Seasons," *Natural History*, August 1984, pages 43–53.

An examination of the dispersal of fruits by birds and mammals in the eastern United States; well written and beautifully illustrated.

Stuessy, Tod F.: *Plant Taxonomy: The Systematic Evaluation of Comparative Data*, Columbia University Press, New York, 1990.

An extensive, well-written, and well-documented account of the many lines of evidence that contribute to our classification of angiosperms.

Tanner, Ogden: "The Flowers That Afflict Us with 'A Sort of Madness,'" *Smithsonian*, November 1985, pages 168–181.

Enjoyable popular account of the orchids and their relationships with people.

Young, James A.: "Tumbleweed," *Scientific American*, March 1991, pages 82–87.

The article chronicles the introduction of tumbleweed from southern Russia; this plant became a major agricultural pest in North America more than a century ago.

* Available in paperback.

Expanding shoot of the shagbark hickory (Carya ovata). Over winter, this shoot was greatly telescoped and existed as a terminal bud. As the bud expanded, the protective bud scales (below) separated and folded back. After the shoot is fully expanded, a new terminal bud will form and pass through a dormant period before it will be capable of expanding and repeating the cycle.

CHAPTER 20

Early Development of the Plant Body

In the previous section, we traced the long evolutionary development of the angiosperms from their presumed ancestor—a relatively complex, multicellular green alga—through a series of early vascular plants whose forked axes were the forerunners of the leaves and roots of most modern vascular plants.

In this section, we shall be concerned with the end result of this evolutionary history—the flowering plant. This chapter thus begins where the story of the angiosperm life cycle ended in Chapter 18—with the seed, consisting of a seed coat, stored food, and an embryo. We will follow the formation of the embryo because it is through this process, known as embryogeny, that the vegetative parts of the plant—root, stem, and leaf—have their origin and the organization of tissues is initiated.

The Mature Embryo and Seed

The mature embryo of flowering plants consists of a stemlike axis bearing either one or two cotyledons (Figures 20–1 and 20–6). The cotyledons, sometimes referred to as the seed leaves, are the first leaves of the young sporophyte. As the names imply, the embryos of *monocotyledons* commonly have only one cotyledon and those of *dicotyledons* have two.

At opposite ends of the embryo axis are found the apical meristems of the shoot and root. As discussed previously (page 6), the apical meristems are found at the tips of all shoots and roots. Meristems are composed of meristematic cells—cells capable of repeated division. In some embryos, only an apical meristem occurs above the cotyledon or cotyledons (Figures 20–1b and 20–6a). In others, an embryonic shoot, consisting of a stemlike axis called the **epicotyl,** with one or more young leaves and an apical meristem, occurs above (*epi-*) the cotyledon(s) (Figures 20–1a and 20–6b).

441

20–1

Seeds and stages in the germination of some common dicotyledons. (a) Garden bean (Phaseolus vulgaris); seed shown open and from external edge view. (b) Castor bean (Ricinus communis); seed open, showing both flat and edge views of embryo. (c) Pea (Pisum sativum); external view of seed only.

Seed germination in both the garden bean (a) and the castor bean (b) is epigeous; that is, during germination the cotyledons are carried above ground by

the elongating hypocotyl. Note that in both of these seedlings, the elongating hypocotyl forms a hook, which then straightens out, pulling the cotyledons and plumule above ground. By contrast, seed germination in the pea (c) is hypogeous—the cotyledons remain underground. In the pea seedling, it is the epicotyl that elongates and forms a hook, which pulls the plumule above ground as it straightens out.

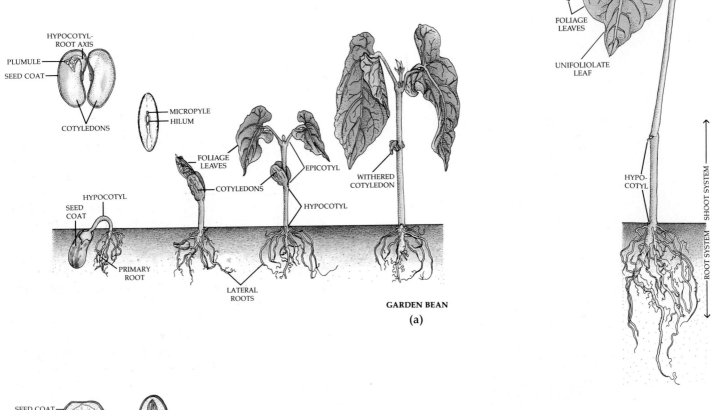

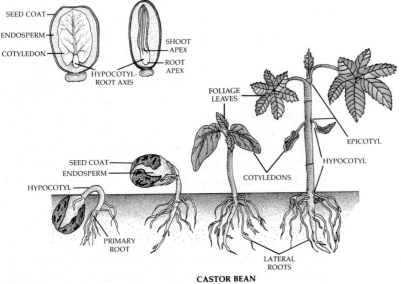

GARDEN BEAN
(a)

CASTOR BEAN
(b)

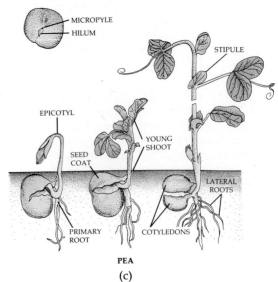

PEA
(c)

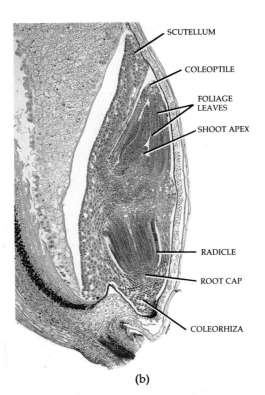

SCUTELLUM

COLEOPTILE

FOLIAGE LEAVES

SHOOT APEX

RADICLE

ROOT CAP

COLEORHIZA

(b)

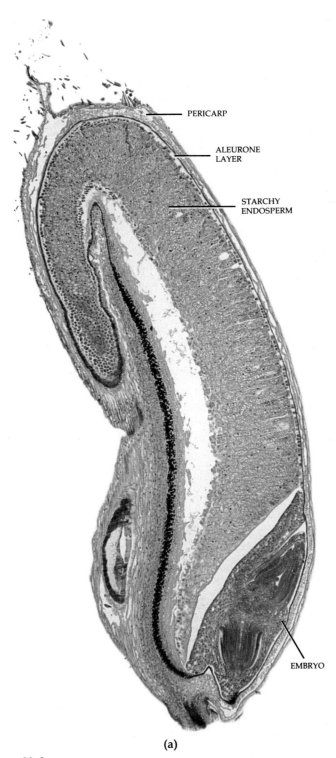

PERICARP

ALEURONE LAYER

STARCHY ENDOSPERM

EMBRYO

(a)

20–2

(a) *Longitudinal section of mature grain, or kernel, of wheat* (Triticum aestivum), *a monocot. The covering layers of the wheat kernel consist largely of the pericarp. The seed coat, which becomes fused with the pericarp, disintegrates during development of the kernel.* (b) *Detail of the mature wheat embryo.*

This embryonic shoot, the first bud, is called a **plumule.**

The stemlike axis below *(hypo-)* the cotyledons is referred to as the **hypocotyl.** At the lower end of the hypocotyl there may be an embryonic root, or **radicle,** with distinct root characteristics (Figure 20–2). In many plants, however, the lower end of the axis consists of little more than an apical meristem covered by a root cap. If a radicle cannot be distinguished in the embryo, the embryo axis below the cotyledons is called the **hypocotyl-root axis.**

In the discussion of the development of the angiosperm seed in Chapter 18, it was noted that in many dicotyledons most or all of the food-storing endosperm and the perisperm (if present) is absorbed by the developing embryo and that the embryos of such seeds develop fleshy, food-storing cotyledons. The cotyledons of most dicotyledonous embryos are fleshy and occupy the largest volume of the seed. Familiar examples of seeds lacking endosperm are sunflower, walnut, garden bean, and pea (Figure 20–1a, c). In dicots with large amounts of endosperm (for example, the castor bean), the cotyledons are thin and membranous (Figure 20–1b) and serve to absorb stored food from the endosperm during resumption of growth of the embryo.

In the monocotyledons, the single cotyledon, in addition to functioning as a food-storing or photosynthetic organ, also performs an absorbing function (Figure 20–6). Embedded in endosperm, the cotyledon absorbs food digested by enzymatic activity. The digested food is then moved by way of the cotyledon to the growing regions of the embryo. Among the most highly differentiated of monocot embryos are those of grasses (Figures 20–2 and 20–6b). When fully formed, the grass embryo possesses a massive cotyledon, the

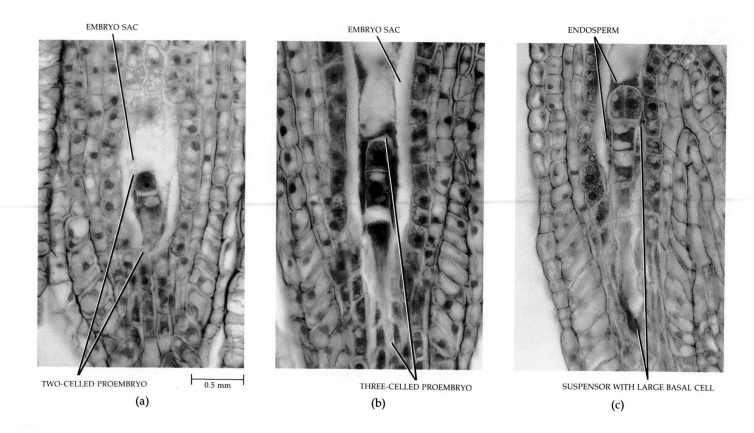

EMBRYO SAC

TWO-CELLED PROEMBRYO

0.5 mm

(a)

EMBRYO SAC

THREE-CELLED PROEMBRYO

(b)

ENDOSPERM

SUSPENSOR WITH LARGE BASAL CELL

(c)

scutellum, which is appressed to the endosperm. The scutellum, like the cotyledons of most monocots, functions in the absorption of food stored in the endosperm. The scutellum is attached to one side of the axis of the embryo, which has a radicle at its lower end and a plumule at its upper end. Both the radicle and the plumule are enclosed by sheathlike, protective structures called the **coleorhiza** and the **coleoptile,** respectively (Figures 20–2b and 20–6b).

All seeds are enclosed by a **seed coat,** which develops from the integument(s) of the ovule and provides protection for the enclosed embryo. The seed coat is usually much thinner than the integument(s) from which it is formed. The thin, dry seed coat may have a papery texture, but in many seeds it is very hard and highly impermeable to water. The micropyle (the opening in the integuments) is often visible on the seed coat as a small pore. Commonly, the micropyle is associated with a scar, called the **hilum,** which is left on the seed coat after the seed has separated from its stalk, or **funiculus** (Figure 20–1a, c).

Formation of the Embryo

The early stages of embryogeny, or embryo development, are essentially the same in dicotyledons and monocotyledons (Figures 20–3 and 20–4). Formation of the embryo begins with the division of the fertilized egg, or zygote, within the embryo sac of the ovule. In most flowering plants, the first division of the zygote is transverse with regard to the long axis of the zygote (Figures 20–3a and 20–4a). With this division, the polarity of the embryo is established: the upper (chalazal) pole is the main seat for growth of the embryo; the lower (micropylar) pole produces a stalklike **suspensor,** which anchors the embryo at the micropyle.

Through an orderly progression of divisions, the embryo eventually differentiates into a nearly spherical structure—the embryo proper—and the suspensor (Figures 20–3b through d and 20–4b through d). Before this stage is reached, the developing embryo is often referred to as the **proembryo.**

Suspensors were mentioned on pages 335 and 367 in relation to the embryos of *Selaginella* and *Pinus,* as structures that merely push the developing embryos into nutritive tissues. In the past, it was believed that the suspensors of angiosperm embryos played a similarly limited role. It now appears that the suspensors of angiosperms are also actively involved in absorption of nutrients from the endosperm. In addition, in some embryos, proteinaceous substances manufactured in the suspensor apparently are utilized by the embryo proper during periods of rapid growth.

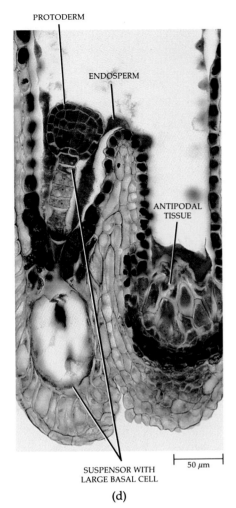

PROTODERM

ENDOSPERM

ANTIPODAL
TISSUE

SUSPENSOR WITH
LARGE BASAL CELL

50 μm

(d)

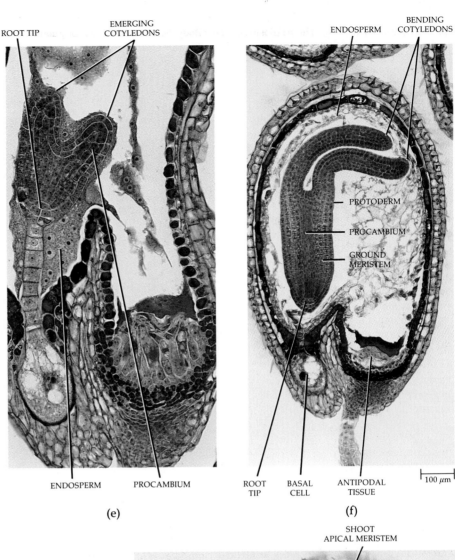

ROOT TIP

EMERGING
COTYLEDONS

ENDOSPERM

PROCAMBIUM

(e)

ENDOSPERM

BENDING
COTYLEDONS

PROTODERM

PROCAMBIUM

GROUND
MERISTEM

ROOT
TIP

BASAL
CELL

ANTIPODAL
TISSUE

100 μm

(f)

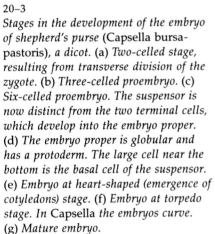

20–3

*Stages in the development of the embryo
of shepherd's purse* (Capsella bursa-
pastoris), *a dicot. (a) Two-celled stage,
resulting from transverse division of the
zygote. (b) Three-celled proembryo. (c)
Six-celled proembryo. The suspensor is
now distinct from the two terminal cells,
which develop into the embryo proper.
(d) The embryo proper is globular and
has a protoderm. The large cell near the
bottom is the basal cell of the suspensor.
(e) Embryo at heart-shaped (emergence of
cotyledons) stage. (f) Embryo at torpedo
stage. In* Capsella *the embryos curve.
(g) Mature embryo.*

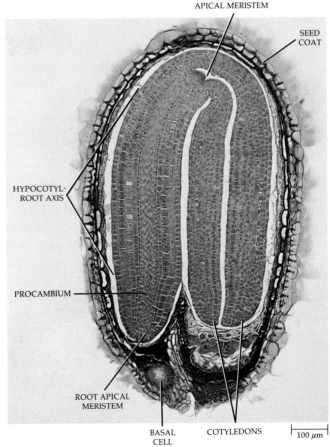

SHOOT
APICAL MERISTEM

SEED
COAT

HYPOCOTYL-
ROOT AXIS

PROCAMBIUM

ROOT APICAL
MERISTEM

BASAL
CELL

COTYLEDONS

100 μm

(g)

20–4

Some stages in the development of the embryo of arrowhead (Sagittaria), a monocot. Early stages are shown in (a) through (d). (a) The two-celled stage, resulting from transverse division of the zygote. (b) The three-celled proembryo. (c) Disregarding the large basal cell, the proembryo is now at the four-celled stage. All four of these cells, through a series of divisions, contribute to formation of the embryo proper. (d) The protoderm has been initiated at the terminal end of the embryo proper. At this stage, the suspensor consists of only two cells, one of which is the large basal cell. In (e) and (f) late stages in embryo development are shown. (e) A depression, or notch (the site of the future apical meristem of the shoot), has formed at the base of the emerging cotyledon. (f) Curving cotyledon; embryo approaching maturity. The suspensor is lacking here.

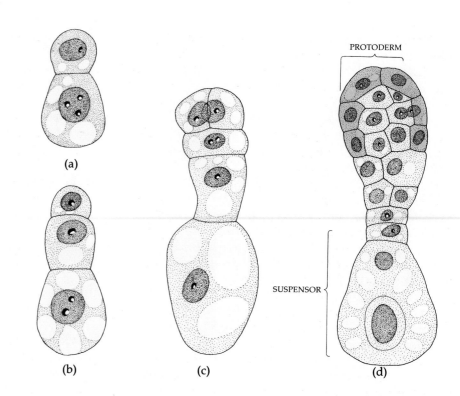

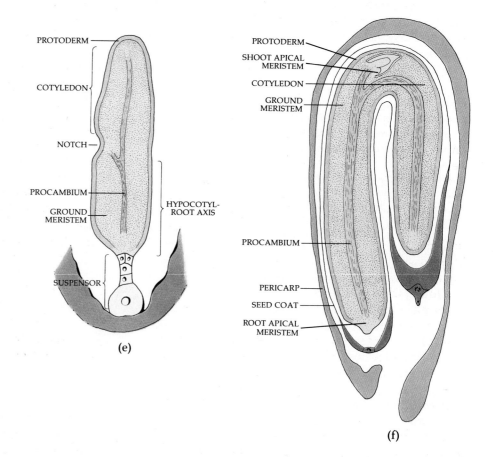

When first formed, the embryo proper consists of a mass of relatively undifferentiated cells. Soon, however, changes in the internal structure of the embryo result in the initial development of the tissue systems of the plant. The future epidermis, the **protoderm,** is formed by periclinal divisions in the outermost cells of the embryo proper (Figures 20–3d and 20–4d). (Periclinal divisions are those in which the cell plates that form between dividing cells are parallel with the surface of the plant part in which they occur.) In addition, progressive changes in the degree of vacuolation and density of the protoplasts of the cells within the embryo result in the initial distinction between **procambium** and **ground meristem.** The highly vacuolated, less dense ground meristem gives rise to the **ground tissue,** which surrounds the less vacuolated and denser procambium, the precursor of the vascular tissues, xylem and phloem. The protoderm, ground meristem, and procambium—the so-called **primary meristems**—are continuous between the cotyledons and the axis of the embryo (Figure 20–3f, g and Figure 20–4e, f).

The stage of embryo development preceding cotyledon development is often referred to as the *globular stage.* Development of the cotyledons may begin either during or after initiation of the primary meristems. With the initiation of the cotyledons, the globular embryo in dicots gradually assumes a two-lobed form. This stage of the development of the embryo in dicots is often called the *heart-shaped stage* (Figure 20–3e). The globular embryos of monocots, forming only one cotyledon, become cylindrical in shape (Figure 20–4e).

As embryo development continues, the cotyledons and axis elongate (the so-called *torpedo stage* of embryo development), and the primary meristems extend along with them (Figures 20–3f and 20–4f). During elongation, the embryo either remains straight or becomes curved. The single cotyledon of the monocot often becomes so large in comparison with the rest of the embryo that it is the dominating structure (Figure 20–6a). With continued enlargement of the embryo, the cells of the suspensor are gradually crushed.

During the early stages of embryogeny, cell division takes place throughout the young sporophyte. However, as the embryo develops, the addition of new cells gradually becomes restricted to the apical meristems of the shoot and root. In dicotyledons, the apical meristem of the shoot arises between the two cotyledons (Figure 20–3g). In monocotyledons, it arises on one side of the cotyledon and is completely surrounded by a sheathlike extension from the base of the cotyledon (Figure 20–4f). The apical meristems of shoot and root are of great importance, since these tissues are the source of virtually all of the new cells responsible for the development of the seedling and the adult plant.

Throughout the period of embryo formation, there is a continuous flow of nutrients from the parent plant to tissues of the ovule, resulting in a massive buildup of food reserves within the endosperm, perisperm, or cotyledons of the developing seed. Eventually, the stalk, or funiculus, connecting the ovule to the ovary wall separates from the ovule, and the ovule becomes a nutritionally closed system. Finally, the seed becomes desiccated as it loses water to the surrounding environment, and the seed coat hardens, encasing the embryo and stored food in "protective armor."

Requirements for Seed Germination

The growth of the embryo is usually delayed while the seed matures and is dispersed. Resumption of growth of the embryo, or **germination** of the seed, is dependent upon many factors, both external and internal. Among the external, or environmental, factors, three are especially important: water, oxygen, and temperature. In addition, small seeds, such as those of lettuce (*Lactuca sativa*) and many weeds, commonly require exposure to light for germination (see page 581).

Most mature seeds are extremely dry, normally containing only 5 to 20 percent of their total weight as water. Thus, germination is not possible until the seed imbibes the water required for metabolic activities. Enzymes already present in the seed are activated, and new ones are synthesized for the digestion and utilization of the stored foods accumulated in the cells of the seed during the period of embryo formation (see page 557). The same cells that earlier had been synthesizing enormous amounts of reserve materials now completely reverse their metabolic processes. Cell enlargement and cell division are initiated in the embryo and follow patterns characteristic of the species. Further growth requires a continuous supply of water and nutrients. As the seed imbibes water, it swells, and considerable pressure may develop within it (see the essay "Imbibition" on page 64).

During the early stages of germination, respiration may be entirely anaerobic, but as soon as the seed coat is ruptured, the seed switches to aerobic respiration, which requires oxygen. If the soil is waterlogged, the amount of oxygen available to the seed may be inadequate for aerobic respiration, and the seed will fail to grow into a seedling.

Although many seeds germinate over a fairly wide range of temperatures, they usually will not germinate below or above a certain temperature range specific for the species. The minimum temperature for many species is 0–5°C, the maximum is 45–48°C, and the optimum range is 25–30°C.

Even when external conditions are favorable, some seeds will fail to germinate. Such seeds are said to be **dormant.** The most common causes of dormancy in seeds are the physiological immaturity of the embryo and the impermeability of the seed coat to water and, sometimes, to oxygen. Some physiologically immature seeds must undergo a complex series of enzymatic and biochemical changes, collectively called **after-ripening,** before they will germinate. In temperate regions, after-ripening is triggered by the low temperatures of winter. Thus, the necessity for a period of after-ripening helps prevent germination of the seed during the inclement period of winter, when it would be unlikely to survive.

Dormancy is of great survival value to the plant. As in the example of after-ripening, it is a method of ensuring that conditions will be favorable for growth of the seedling when germination does occur. Some seeds must pass through the intestines of birds or mammals before they will germinate, resulting in wider dispersal of the plant species. Some seeds of desert species will germinate only when inhibitors in their coats are leached away by rainfall; this adaptation ensures that the seed will germinate only during those rare intervals when desert rainfall provides sufficient water for the seedling to mature. Similarly, other seeds must be cracked mechanically, as by tumbling along in the rushing water of a gravelly streambed. Still other seeds lie dormant in cones or fruits until the heat of a fire releases the seeds. The Mediterranean-type-climate vegetation of the California chaparral community, dominated by manzanita (*Arctostaphylos*), is dependent upon fire for long-term persistence at any one site because fire induces the germination of manzanita seeds (Figure 20–5). Finally, the seeds of species that live in woodland clearings depend on either the death of a canopy tree or some other disturbance to form an opening in the canopy before they will germinate. Hence, the germination strategies of plants are closely linked to the ecological problems existing in their particular habitats. (See Chapter 26, page 585, for further discussion of dormancy in seeds.)

20–5
Manzanita (Arctostaphylos viscida) *of the California chaparral community. The long-lived seeds of manzanita remain viable in the soil for years. Scarification, or breakdown of the seed coat, by fire or other means is necessary in order to break dormancy and induce germination.*

From Embryo to Adult Plant

When germination occurs, the first structure to emerge from most seeds is the radicle, or embryonic root; this enables the developing seedling to become anchored in the soil and to absorb water. As this first root, called the **primary root,** continues to grow, it develops branch roots, or **lateral roots,** and these roots in turn may give rise to additional lateral roots. In this manner, a much-branched root system develops. Commonly, the pri-

mary root in monocots is short-lived, and the root system of the adult plant develops from **adventitious roots,** which arise at the nodes (the parts of the stem at which the leaves are attached) and then produce lateral roots. ("Adventitious" refers to structures arising not at their usual sites, as roots arising from stems or leaves. Roots usually arise from other roots.)

The way in which the shoot emerges from the seed during germination varies from species to species. For example, after the root emerges from the seed of the garden bean (*Phaseolus vulgaris*), the hypocotyl elongates and becomes bent in the process (Figure 20–1a). Thus, the delicate shoot tip is protected from injury by being pulled rather than pushed through the soil. When the bend, or *hook,* as it is called, reaches the soil surface, it straightens out and pulls the cotyledons and plumule up into the air. This type of seed germination, in which the cotyledons are carried above ground level, is called *epigeous*. During germination and subsequent development of the seedling, the food stored in the cotyledons is digested and the products are transported to the growing parts of the young plant. The cotyledons gradually decrease in size, wither, and eventually drop off. By this time, the seedling has become *established*; that is, it is no longer dependent upon the stored food of the seed for its nourishment. The seedling is now a photosynthesizing, autotrophic organism.

Germination of the castor bean *(Ricinus communis)* (Figure 20–1b) is essentially similar to that of the garden bean, except that in the castor bean the stored food is found in the endosperm. As the hook straightens out, the endosperm and often the seed coat are carried upward, along with the cotyledons and plumule. During this period, the digested foods of the endosperm are absorbed by the cotyledons and are transported to the growing parts of the seedling. In both the garden bean and castor bean, the cotyledons become green upon exposure to light, but they do not have an important photosynthetic function. In some plants, such as squash *(Cucurbita maxima)*, the cotyledons become important photosynthetic organs.

In the pea *(Pisum sativum)*, the epicotyl is the structure that elongates and forms the hook. As the epicotyl straightens out, the plumule is raised above the soil surface. The cotyledons remain in the soil (Figure 20–1c), where they eventually decompose. This type of seed germination, in which the cotyledons remain underground, is called *hypogeous*.

In the large majority of monocot seeds, the stored food is found in the endosperm. In relatively simple monocot seeds, such as that of the onion *(Allium cepa)*, it is the single tubular cotyledon that emerges from the seed and forms the hook (Figure 20–6a). When the cotyledon straightens, it carries the seed coat and enclosed endosperm upward. Throughout this period, and for some time afterward, the embryo obtains much of its nourishment from the endosperm by way of the cotyledon. Furthermore, the green cotyledon in onion functions as a photosynthetic leaf, contributing significantly to the food supply of the developing seedling. Soon the plumule, enclosed within the protective, sheathlike base of the cotyledon, elongates and emerges from the cotyledon.

Our last example of seedling development is provided by corn *(Zea mays)*, a monocot that has a highly differentiated embryo (Figure 20–7). Both radicle and plumule are enclosed in sheathlike structures, the coleorhiza and the coleoptile, respectively. The coleorhiza is the first structure to grow through the pericarp (mature ovary wall) of the corn grain. (In corn, the integuments disintegrate during development of the seed and fruit, and hence the pericarp functions as a "seed

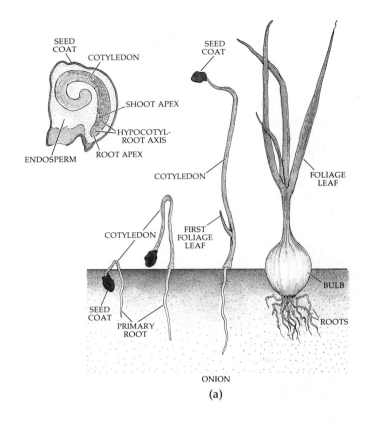

(a)

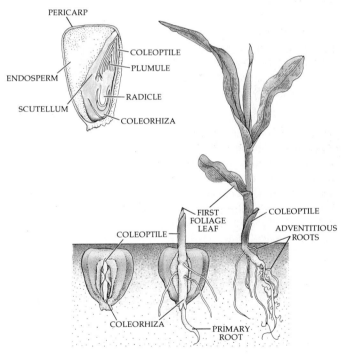

(b)

20–6
Seeds and stages in the germination of some common monocotyledons: (a) onion (Allium cepa) and (b) corn (Zea mays). Both seeds are shown in longitudinal section. Seed germination in onion is epigeous; in corn, it is hypogeous.

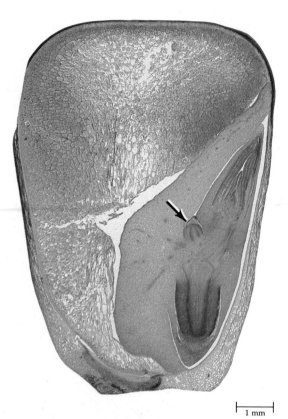

20–7
Longitudinal section of mature grain, or kernel, of corn (Zea mays). *By comparison with the corn grain diagram in Figure 20–6b, and with the photomicrographs of the wheat* (Triticum aestivum) *grain and embryo in Figure 20–2, you should be able to identify the pericarp, endosperm, and various parts of the embryo. Grain embryos commonly contain two or more seminal (seed) adventitious roots. One seminal adventitious root can be seen in this section (arrow). Although at first pointed upward, these roots become inclined downward with further growth.*

coat.") The coleorhiza is then followed by the radicle, or primary root, which elongates very rapidly and quickly penetrates the coleorhiza (Figure 20–6b). After the primary root emerges, the coleoptile is pushed upward by elongation of the first **internode,** called the *mesocotyl.* (An internode is the part of the stem between two successive nodes.) When the base of the coleoptile reaches the soil surface, its edges spread apart at the tip, and the first leaves of the plumule begin to emerge. In addition to the primary root, two or more seminal adventitious roots, which arise from the cotyledonary node, grow through the pericarp and then bend downward.

Regardless of the manner in which the shoot

Vegetative Reproduction: Some Ways and

The forms of vegetative reproduction in plants are many and varied. Some plants reproduce by means of runners, or stolons—long slender stems that grow along the surface of the soil. In the cultivated strawberry (*Fragaria ananassa*), for example, leaves, flowers, and roots are produced at every other node on the runner. Just beyond the second node, the tip of the runner turns up and becomes thickened. This thickened portion first produces adventitious roots and then a new shoot, which continues the runner.

Underground stems, or rhizomes, are also important reproductive structures, particularly in grasses and sedges. Rhizomes invade areas near the parent plant, and each node can give rise to a new flowering shoot. The noxious character of many weeds results from this type of growth pattern, and many garden plants, such as irises, are propagated almost entirely from rhizomes. Corms, bulbs, and tubers are specialized for storage and reproduction. White potatoes are propagated artificially from tuber segments, each with one or more "eyes." It is the eyes of the "seed pieces" of potato that give rise to the new plant.

The roots of some plants—for example, cherry, apple, raspberry, and blackberry—produce "suckers," or sprouts, which give rise to new plants. Commercial varieties of banana do not produce seeds and are propagated by suckers that develop from buds on underground stems. When the root of a dandelion is broken, as it may be if one attempts to pull it from the ground, each root fragment may give rise to a new plant.

In a few species, even the leaves are reproductive. One example is the house plant *Kalanchoë daigremontiana*, familiar to many people as the "maternity plant," or "mother of thousands." The common names of this plant are based on the fact that numerous plantlets arise from meristematic tissue located in notches along the margins of the leaves. The maternity plant is ordinarily propagated by means of these small plants, which, when they are mature enough, drop to the soil and take root. Another example of vegetative propagation is provided by the walking fern (*Asplenium rhizophyllum*), in which young plants form where the leaf tips touch the ground.

emerges from the seed, the activity of the apical meristem of the shoot results in the formation of an orderly sequence of leaves, nodes, and internodes. Apical meristems, which develop in the leaf axils (the upper angles between leaves and stems), produce axillary shoots, and these, in turn, may form additional axillary shoots.

The period from germination to the time the seed-

Means

(a)

(a) *The strawberry* (Fragaria ananassa) *is propagated asexually by stolons. Strawberry plants also produce flowers and reproduce sexually.* (b) *Kalanchoë daigremontiana, showing small plants that have arisen in the notches along the margins of the leaf.* (c) *A herbarium specimen of walking fern,* Asplenium rhizophyllum, *showing the way the leaves root at their tips and produce new plants. In this way, the fern is capable of forming large colonies of genetically identical plants.*

(b)

(c)

In certain plants, including some citruses, certain grasses (such as Kentucky bluegrass, *Poa pratensis*, which is discussed on page 166), and dandelions, the embryos in the seeds may be produced asexually from the parent plant. This is a kind of vegetative reproduction known as apomixis. The seeds produced in this way give rise to individuals that are genetically identical to their parents, providing another instance of asexual reproduction.

In general, asexual reproduction allows the exact replication of individuals that are particularly well suited to a certain environment or habitat. This suitability may include characteristics that are viewed as desirable in a cultivated plant or those that facilitate survival under a particular set of environmental conditions.

ling becomes established as an independent organism constitutes the most crucial phase in the life history of the plant. During this period, the plant is most susceptible to injury by a wide range of insect pests and parasitic fungi, and water stress can very rapidly prove fatal.

This type of growth of the root system and the shoot system is called vegetative growth (see the above essay). Eventually, one or more of the vegetative apical meristems of the shoot changes into a reproductive apical meristem, that is, a meristem that develops into a flower or an inflorescence (cluster of flowers). When the flowers are formed, the plant is prepared for sexual reproduction.

Wheat: Bread and Bran

Like all grasses, bread wheat *(Triticum aestivum)* is a mono-cotyledon, and its fruit—the grain or kernel—is one-seeded. The endosperm and the embryo are surrounded by covering layers composed of the pericarp and the remains of the seed coat. The endosperm represents more than 80 percent of the bulk of the wheat kernel. The outermost layer of the endosperm, called the aleurone layer, contains lipid and protein reserves. The aleurone layer surrounds the starchy endosperm, as well as the embryo.

White flour is made from the starchy endosperm. In the milling of wheat, the bran—consisting of the covering layers and the aleurone layer—is removed. The bran constitutes about 14 percent of the kernel. Actually, the bran somewhat decreases the nutritional value of the wheat kernel; because it is mostly cellulosic, it cannot be digested by humans and tends to speed the passage of food through the intestinal tract, resulting in lower absorption. The embryo ("wheat germ"), which represents about 3 percent of the kernel, is also removed, because its high oil content would reduce the length of time that the flour can be stored. The bran and wheat germ, which contain most of the vitamins found in wheat, are being used more and more for human consumption, as well as for livestock feed.

During the past few years, oat *(Avena sativa)* bran has become a popular food item among health-conscious people in many parts of the world. Research results indicate that oat bran, as part of a low-fat diet, can help reduce the amount of cholesterol in the blood. Apparently, the water-soluble

fibers found in the oat bran form a gel in the small intestine that traps cholesterol and prevents it from being reabsorbed into the bloodstream. Instead, the trapped cholesterol is excreted with other bodily wastes. Cholesterol is also reduced by water-soluble fibers found in rice *(Oryza sativa)* and barley *(Hordeum vulgare)* bran.

Summary

The seeds of flowering plants consist of an embryo, a seed coat, and stored food. When fully formed, the embryo consists basically of a hypocotyl-root axis bearing either one or two cotyledons and an apical meristem at the shoot apex and at the root apex. The cotyledons of most dicots are fleshy and contain the stored food of the seed. In other dicots and in most monocots, food is stored in the endosperm, and the cotyledons function to absorb the simpler compounds resulting from the digestion of that food. The simpler compounds are then transported to growing regions of the embryo.

During embryo development, beginning with the zygote, the shoot and root of the young plant are initiated as one continuous structure. Through an orderly progression of divisions, the embryo differentiates into a suspensor and an embryo proper, within which the primary meristems—the precursors of the epidermis, ground tissue, and vascular tissues—are formed. Development of the cotyledons may begin either during or after initiation of the primary meristems.

As the embryo develops, the addition of new cells is gradually restricted to the apical meristems. Germination of the seed—resumption of growth of the embryo—depends upon environmental factors, including water, oxygen, and temperature. Many seeds must pass through a period of dormancy before they are able to germinate.

The root is the first structure to emerge from most germinating seeds, enabling the seedling to become anchored in the soil and to absorb water. The way in which the shoot emerges from the seed varies from species to species. Many dicotyledons form a bend, or hook, in their hypocotyls or epicotyls. As the hook straightens out, the delicate shoot tip is pulled upward from the soil. Hence, in this way, the delicate shoot tip is protected from injury that might otherwise occur were the shoot tip to be pushed through the soil.

Following a period of vegetative growth, one or more apical meristems of the shoot become reproductive apical meristems, which develop into flowers or inflorescences.

CHAPTER 21

Cells and Tissues of the Plant Body

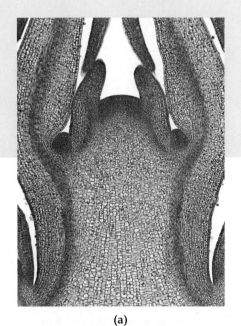

(a)

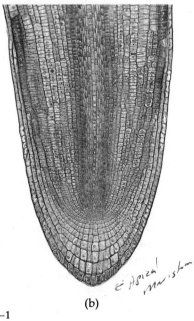

(b)

21–1

Longitudinal sections of (a) *shoot tip of lilac* (Syringa vulgaris) *and* (b) *root tip of radish* (Raphanus sativus), *showing the apical meristems. Primordia of leaves and axillary buds are present in* (a). *In* (b), *a root cap covers the apical meristem.*

Following the development of the embryo, the formation of new cells, tissues, and organs becomes restricted almost entirely to the meristems—the perpetually young tissues that play such a central role in growth. As discussed in Chapter 1, there are two main types of meristems—apical meristems and lateral meristems. The apical meristems are involved primarily with extension of the plant body; they are found at the tips of shoots and roots (Figure 21–1). This type of growth is called **primary growth,** during which primary tissues are formed. The part of the plant body composed of these tissues is called the primary plant body.

The lateral meristems, the **vascular cambium** and the **cork cambium,** produce the secondary tissues, which make up the secondary plant body. The vascular cambium produces secondary xylem and secondary phloem; the cork cambium produces mostly cork.

In both the apical and the lateral meristems, certain cells are able to divide repeatedly; after each division one of the sister cells remains in the meristem while the other becomes a new body cell. The cells that remain in the meristem are called **initials;** their sister cells are called **derivatives.** It is important to note that the initials and their derivatives are not inherently different from one another. A cell that acts as an initial apparently does so because of its preferred position in the meristem, a position that may be supplanted by a derivative, which then may serve as an initial. Commonly, the derivatives divide one or more times before they begin to differentiate into specific types of cells. Consequently, the meristem includes the initials and their immediate derivatives.

Cell division is not limited to the apical and lateral meristems. For example, the protoderm, procambium, and ground meristem, which are partly differentiated tissues, are called primary meristems because (1) they

453

21–2
Diagram illustrating some of the cell types that may originate from a meristematic cell of the procambium or the vascular cambium. The meristematic cell depicted here, with a single large vacuole, is typical of the meristematic cells of the vascular cambium.

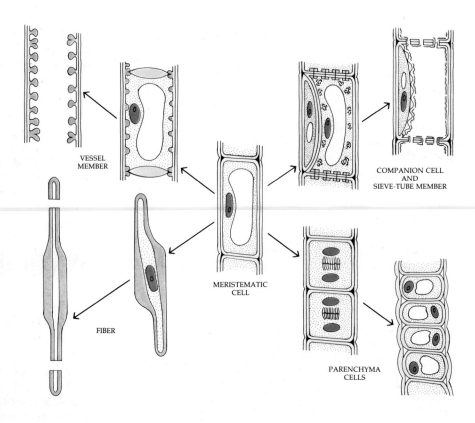

VESSEL MEMBER

COMPANION CELL AND SIEVE-TUBE MEMBER

MERISTEMATIC CELL

FIBER

PARENCHYMA CELLS

give rise to the primary tissues and (2) many of their cells remain meristematic for some time before they begin to differentiate into specific cell types. With the exception of their initiation in the embryo proper during embryogeny, the primary meristems are derived from the apical meristems.

Growth of the plant body involves both cell division and cell enlargement. The overall size of the cells increases as one progresses from younger to older meristematic tissues; eventually, the main factor involved in the increase in size of the particular region of the root, stem, or leaf is cell enlargement.

Differentiation—the process by which cells that have identical genetic constitutions become different from one another and from the meristematic cells from which they originated—often begins while the cell is still enlarging. At maturity, when differentiation is complete, some cells are living and others are dead. These living and dead cells include many different cell types (Figure 21–2). How cells that have a common origin come to be so different is generally considered one of the crucial questions of modern biology. Some factors involved in the control of cellular differentiation are discussed in Chapter 8 (pages 145 through 147) and in Chapter 25 (pages 559 through 562).

In the plant body, basic tissue patterns are established by early meristematic activity. The shape of the plant and organization of its tissues are influenced greatly by both cell division and cell enlargement. The acquisition of a particular shape or form is known as

morphogenesis (from the Greek word *morphē,* meaning "form," and the Latin word *generare,* meaning "to create").

The Tissue Systems

Botanists have long recognized that the principal tissues of vascular plants are organized into larger units in all parts of the plant. These groups of tissues are known as **tissue systems,** and their presence in root, stem, and leaf reveals both the basic similarity of the plant organs and the continuity of the plant body. The three tissue systems are: (1) the fundamental, or **ground tissue system,** (2) the **vascular tissue system,** and (3) the **dermal tissue system.** As discussed in Chapter 20, the tissue systems are initiated during the development of the embryo, where they are represented by the primary meristems—ground meristem, procambium, and protoderm, respectively.

The ground tissue system consists of the three so-called ground tissues—parenchyma, collenchyma, and sclerenchyma. Parenchyma is by far the most common of the ground tissues. The vascular tissue system consists of the two conducting tissues—xylem and phloem. The dermal tissue system is represented by the epidermis, which is the outer protective covering of the primary plant body, and later by the periderm in the secondary plant body.

Tissues and Their Component Cells

Tissues may be defined as groups of cells that are structurally and/or functionally distinct. Tissues composed of only one type of cell are called **simple tissues,** whereas those composed of two or more types of cells are called **complex tissues.** Parenchyma, collenchyma, and sclerenchyma are simple tissues; xylem, phloem, and epidermis are complex tissues.

PARENCHYMA

Parenchyma tissue is composed of **parenchyma cells.** In the primary plant body, parenchyma cells commonly occur as continuous masses—as parenchyma tissue—in the cortex of stems (Figure 21–3) and roots, in the pith of stems, in leaf mesophyll (see Figure 21–27), and in the flesh of fruits. In addition, parenchyma cells occur as vertical strands of cells in the primary and secondary vascular tissues and as horizontal strands called **rays** in the secondary vascular tissues (see Chapter 24).

Characteristically living at maturity, parenchyma cells are capable of cell division, and although their walls are commonly primary, some parenchyma cells also have secondary walls. Because of their ability to divide, parenchyma cells having only primary walls play an important role in regeneration and wound healing. It is these cells that initiate adventitious structures, such as adventitious roots on stem cuttings. Parenchyma cells are involved in such activities as photosynthesis, storage, and secretion—activities dependent upon living protoplasm. In addition, parenchyma cells may play a role in the movement of water and the transport of food substances in plants.

Transfer Cells

Considerable attention has been given to a special kind of parenchyma cell containing ingrowths of the cell wall, which often greatly increase the surface area of the plasma membrane (Figure 21–4). These paren-

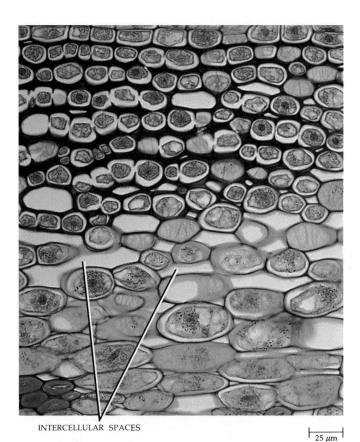

INTERCELLULAR SPACES

25 μm

21–3
Transverse section of the cortex of an elderberry (Sambucus canadensis) *stem, showing collenchyma cells with unevenly thickened walls (above) and parenchyma cells (below). Clear areas between the cells are intercellular spaces. In some of the parenchyma cells, a meshwork of lines is visible in the facing walls. The light areas within the meshwork are primary pit-fields, which are thin areas in the walls.*

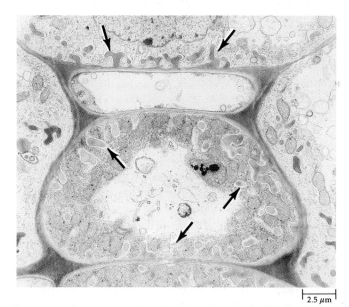

2.5 μm

21–4
Transverse section of a portion of the phloem from a small vein of a **Senecio** *leaf, showing transfer cells with their numerous wall ingrowths (arrows), bordering a relatively clear sieve element, or food-conducting cell.*

chyma cells, called **transfer cells,** are believed to play an important role in the transfer of solutes over short distances. Their presence is generally correlated with the existence of intensive solute fluxes—in either inward (uptake) or outward (secretion) directions—across the plasma membrane.

Transfer cells are exceedingly common and probably serve a similar function throughout the plant body. They occur in association with the xylem and phloem of small, or minor, veins in cotyledons and in the leaves of many herbaceous dicotyledons. Transfer cells are also associated with the xylem and phloem of leaf traces at the nodes in both dicotyledons and monocotyledons. In addition, they are found in various tissues of reproductive structures (placentae, embryo sacs, endosperm) and in various glandular structures (nectaries, salt glands, the glands of carnivorous plants). Each of these locations is a potential site of intensive short-distance solute transfer. Transfer cells can also be induced to form by external stimuli, such as by nematode infection, in a plant that does not normally develop such cells.

COLLENCHYMA

Collenchyma tissue is composed of **collenchyma cells,** which, like parenchyma cells, are living at maturity (Figures 21–3 and 21–5 through 21–7). Collenchyma tissue commonly occurs in discrete strands or as continuous cylinders beneath the epidermis in stems and petioles (leaf stalks). It is also found bordering the

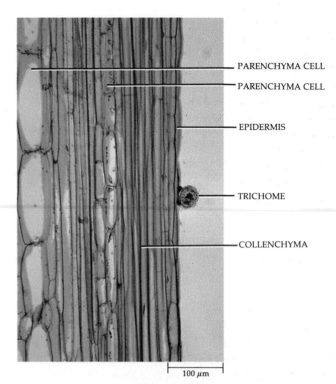

PARENCHYMA CELL
PARENCHYMA CELL
EPIDERMIS
TRICHOME
COLLENCHYMA

100 μm

21–5
Longitudinal section showing the elongated collenchyma cells in the stem of a squash (Cucurbita maxima). *A trichome is an epidermal appendage; several types of trichomes are shown in Figures 21–24 and 21–26.*

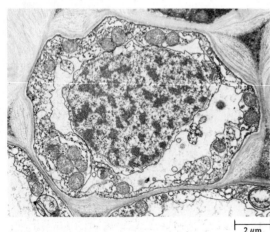

2 μm

21–6
Electron micrograph of a transverse section of a mature collenchyma cell from the filament of a wheat (Triticum aestivum) *stamen. Notice the protoplasmic contents and unevenly thickened wall of this cell.*

CELL WALL   PROTOPLAST

25 μm

21–7
Transverse section of collenchyma tissue from a petiole in rhubarb (Rheum rhaponticum). *In fresh tissue like this, the unevenly thickened collenchyma cell walls have a glistening appearance.*

veins in dicot leaves. (The "strings" on the outer surface of celery stalks, or petioles, consist almost entirely of collenchyma.) The typically elongated collenchyma cells (Figure 21–5) contain unevenly thickened, nonlignified primary walls, making these cells especially well adapted for the support of young, growing organs (see the description of the primary wall on page 34). The name *collenchyma* derives from the Greek word *kolla*, meaning "glue," which refers to the cell's characteristic thick, glistening walls in fresh tissue (Figure 21–7). The walls of collenchyma cells are readily stretched and thus offer relatively little resistance to elongation of the plant part in which they are found. In addition, because collenchyma cells are living at maturity, they can continue to develop thick, flexible walls while the organ is still elongating.

SCLERENCHYMA

Sclerenchyma tissue is composed of **sclerenchyma cells,** which may develop in any or all parts of the primary and secondary plant bodies; these cells often lack protoplasts at maturity. The term *sclerenchyma* is derived from the Greek *skleros,* meaning "hard," and the principal characteristic of sclerenchyma cells is their thick, often lignified secondary walls. Because of these walls, the sclerenchyma cells are important strengthening and supporting elements in plant parts that have ceased elongating (see the description of the secondary wall on page 35).

Two types of sclerenchyma cells are recognized: **fibers** and **sclereids.** Fibers are generally long, slender cells that commonly occur in strands or bundles (Figure 21–8). The so-called bast fibers—such as hemp, jute,

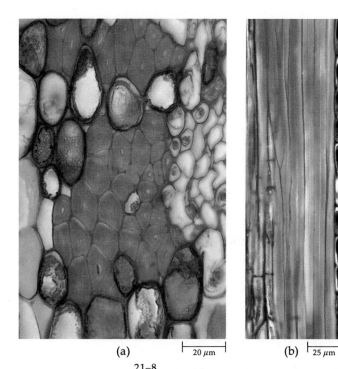

(a)   20 μm   (b)   25 μm

21–8
Primary phloem fibers from the stem of a linden, or basswood (Tilia americana), seen here in both (a) cross-sectional and (b) longitudinal views. The secondary walls of these long, thick-walled fibers contain relatively inconspicuous pits. Only a portion of the length of these fibers can be seen in (b).

and flax—are derived from the stems of dicotyledons. Other economically important fibers, such as Manila hemp, are extracted from the leaves of monocots. Sclereids are variable in shape and are often branched (Figure 21–9), but compared with most fibers, sclereids are relatively short cells. Sclereids may occur singly or in

21–9
Branched sclereid from a leaf of the water lily (Nymphaea odorata), as seen in (a) ordinary light and (b) polarized light. Numerous small, angular crystals are embedded in the wall of this sclereid.

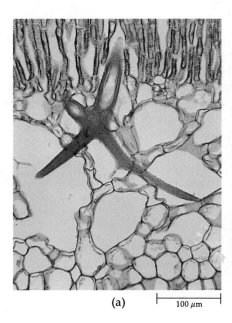
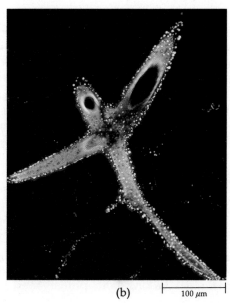

(a)   100 μm   (b)   100 μm

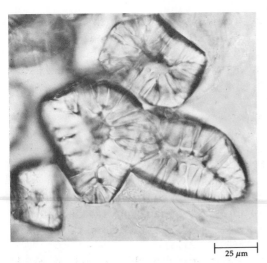

21–10

*Sclereids (stone cells) from fresh tissue
of the pear (Pyrus communis) fruit.
The secondary walls contain conspicuous simple pits, with many branches.
Branched pits such as these are called
ramiform pits. During formation of the
clusters of stone cells in the flesh of the
pear fruit, cell divisions occur concentrically around some of the sclereids formed
earlier. The newly formed cells differentiate as stone cells, adding to the cluster.*

aggregates throughout the ground tissue. They make
up the seed coats of many seeds, the shells of nuts, and
the stone (endocarp) of stone fruits, and they give pears
their characteristic gritty texture (Figure 21–10).

XYLEM

Xylem is the principal water-conducting tissue in vascular plants. It is also involved in the conduction of
minerals, in food storage, and in support. Together with
the phloem, the xylem forms a continuous system of
vascular tissue extending throughout the plant body
(Figure 21–11). Xylem may be primary or secondary in
origin. The primary xylem is derived from the procambium, and the secondary xylem is derived from the vascular cambium (see Chapter 24).

The principal conducting cells of the xylem are the
tracheary elements, of which there are two types, the
tracheids and the **vessel members,** or vessel elements.
Both are elongated cells that have secondary walls and
lack protoplasts at maturity, and both types may have
pits in their walls (Figure 21–12a through d). In addition to pits, vessel members contain *perforations*, which
are areas lacking both primary and secondary walls;

perforations are literally holes in the cell wall. Perforations generally occur on the end walls, but they may
also be found on the lateral walls. The part of the wall
bearing the perforation or perforations is called the **perforation plate** (Figure 21–13). Vessel members are
joined at their perforation plates into long, continuous
columns, or tubes, called **vessels** (Figure 21–14).

The pits of the long, tapering tracheids are concentrated on the overlapping ends of the cells. The tracheid
is the only type of water-conducting cell found in most
seedless vascular plants and gymnosperms. The xylem
of the great majority of angiosperms contains both vessel members and tracheids.

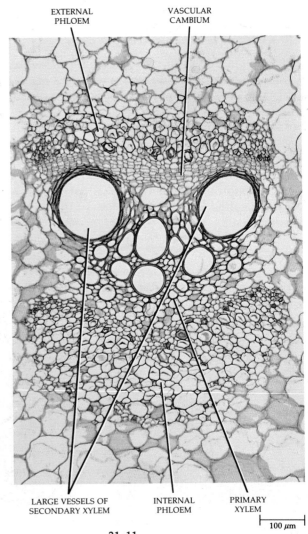

EXTERNAL
PHLOEM

VASCULAR
CAMBIUM

LARGE VESSELS OF
SECONDARY XYLEM

INTERNAL
PHLOEM

PRIMARY
XYLEM

21–11

*Transverse section of a vascular bundle
from the stem of a squash (Cucurbita
maxima). Phloem occurs on both sides of
the xylem, and a vascular cambium has
developed between the external phloem
and the xylem.*

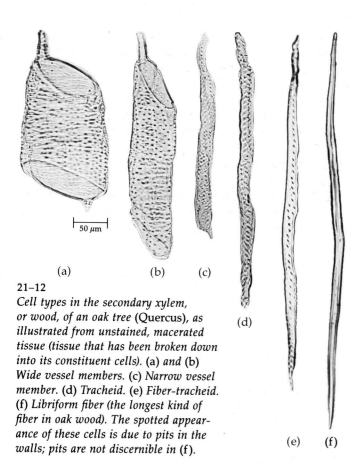

21–12
Cell types in the secondary xylem, or wood, of an oak tree (Quercus), as illustrated from unstained, macerated tissue (tissue that has been broken down into its constituent cells). (a) and (b) Wide vessel members. (c) Narrow vessel member. (d) Tracheid. (e) Fiber-tracheid. (f) Libriform fiber (the longest kind of fiber in oak wood). The spotted appearance of these cells is due to pits in the walls; pits are not discernible in (f).

The tracheid is a more primitive (less specialized) type of cell than the vessel member, which is the principal water-conducting cell in angiosperms. It is quite probable that vessel members evolved independently from tracheids in several groups of vascular plants, including the dicotyledons and monocotyledons (the two groups of angiosperms), the Gnetophyta (the vessel-containing gymnosperms), several unrelated species of ferns, and certain species of *Selaginella* (of the Lycophyta) and *Equisetum* (of the Sphenophyta). The evolution of the vessel member is an excellent example of convergent evolution, or the independent development of similar structures by unrelated or only distantly related organisms (see "Convergent Evolution" on page 518).

Vessel members are generally thought to be more efficient conductors of water than the tracheids, because water can flow relatively unimpeded from vessel member to vessel member through the perforations; however, vessel members, with their open system, are less safe for the plant than tracheids. Water flowing from tracheid to tracheid must pass through the membranes of the pit-pairs (page 36). Although the porous pit membranes offer relatively little resistance to the flow of water across them, they block even the smallest of air bubbles. Thus, air bubbles that form in a tracheid are restricted to that tracheid, and any resulting obstruction to water flow is also limited. On the other hand, air bubbles formed in a vessel member can po-

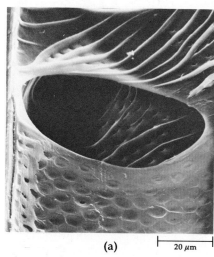

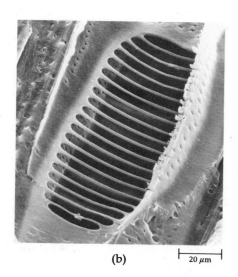

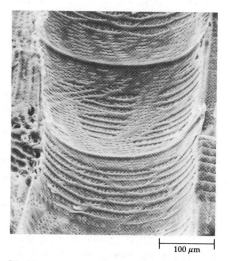

21–13
Scanning electron micrographs of the perforated end walls of vessel members from secondary xylem. (a) A simple perforation plate, with its single large opening, seen here between two vessel members in basswood, or linden (Tilia
americana). (b) The ladderlike bars of a scalariform perforation plate between vessel members of red alder (Alnus rubra). Pits can be seen in the wall below the perforation plate in (a) and in portions of the wall in (b).*

21–14
Scanning electron micrograph showing parts of three vessel members in secondary xylem of red oak (Quercus rubra). Notice the rims between vessel members.

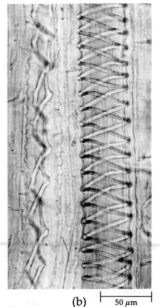

(a) ⊢ 50 μm ⊣   (b) ⊢ 50 μm ⊣

21–15

Parts of tracheary elements from the first-formed primary xylem (protoxylem) of the castor bean (Ricinus communis). (a) Annular (the ringlike shapes at left) and helical wall thickenings in partly extended elements. (b) Double helical thickenings in elements that have been extended. The element on the left has been greatly extended, and the coils of the helices have been pulled far apart.

tentially obstruct the flow of water for the entire length of the vessel. Thus, shorter vessels are safer than longer ones. Wide vessels are more efficient than narrow vessels, but wide vessels also tend to be longer.

In the secondary xylem and late-formed primary xylem (metaxylem), the secondary cell walls of the tracheids and vessel members cover the entire primary walls, except at the pits and at the perforations of the vessel members (Figure 21–12a through d). Consequently, these walls are rigid and cannot be stretched. During the period of elongation or expansion of the roots, stems, and leaves, the secondary walls of many of the first-formed tracheary elements of the early-formed primary xylem (protoxylem) are deposited in the form of rings or spirals (Figure 21–15). These thickenings, which may be annular (ringlike) or helical (spiral), make it possible for such tracheary elements to be stretched or extended, although the cells are frequently destroyed during the overall elongation of the organ. In the primary xylem, the nature of the wall thickening is greatly influenced by the amount of elongation. If little elongation occurs, pitted elements rather than extensible elements appear. On the other hand, if much elongation takes place, many elements with annular and helical thickenings will develop. Figure 21–16 illustrates some stages of differentiation of a vessel member with helical thickenings.

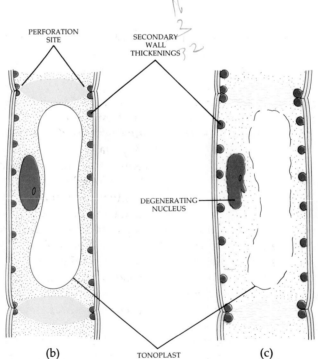

PERFORATION SITE

SECONDARY WALL THICKENINGS

PERFORATION

DEGENERATING NUCLEUS

TONOPLAST

(a)   (b)   (c)   (d)

21–16

Diagram illustrating development of a vessel member. (a) Young, highly vacuolated vessel member without secondary wall. (b) The cell has expanded laterally, secondary wall deposition has begun, and the primary wall at the perforation site has increased in thickness. (c) Secondary wall deposition has been completed, and the cell is at the stage of lysis. The nucleus is deformed, the *tonoplast is ruptured, and the wall at the perforation site is partly disintegrated. (d) The cell is now mature; it lacks a protoplast and is open at both ends.*

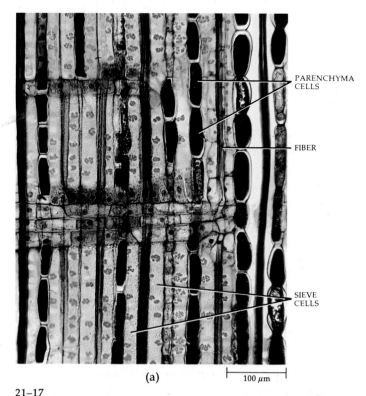

PARENCHYMA CELLS

FIBER

SIEVE CELLS

(a) ⊢ 100 μm ⊣

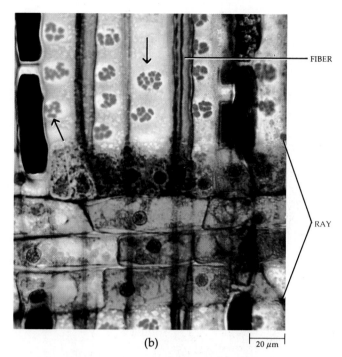

FIBER

RAY

(b) ⊢ 20 μm ⊣

21–17

(a) *Longitudinal (radial) view of secondary phloem of yew (Taxus canadensis), showing vertically oriented sieve cells, strands of parenchyma cells, and fibers.*

Parts of two horizontally oriented rays can be seen traversing the vertical cells. (b) *Detail of portion of the secondary phloem of yew, showing sieve areas, with callose*

(stained blue) on the walls of the sieve cells, and albuminous cells (see page 464), which here constitute the top row of cells in the ray.

In addition to tracheids and vessel members, xylem contains parenchyma cells that store various substances. Xylem parenchyma cells commonly occur in vertical strands. In the secondary xylem, they also are found in rays. Fibers (Figure 21–12e, f) and sclereids also occur in xylem. Many xylem fibers are living at maturity and serve a dual function of storage and support.

PHLOEM

Phloem is the principal food-conducting tissue in vascular plants (Figure 21–11). The phloem may be primary or secondary in origin, and as with primary xylem, the first-formed primary phloem (protophloem) is frequently stretched and destroyed during elongation of the organ.

The principal conducting cells of the phloem are the **sieve elements,** of which there are two types, the **sieve cells** (Figure 21–17) and the **sieve-tube members** (Figures 21–18 through 21–20). The term "sieve" refers to the clusters of pores, the **sieve areas,** through which the protoplasts of adjacent sieve elements are interconnected. In sieve cells, the pores are narrow and the sieve areas are rather uniform in structure on all walls.

Most of the sieve areas are concentrated on the overlapping ends of the long, slender sieve cells (Figure 21–17a). In sieve-tube members, the sieve areas on some walls have larger pores than those on other walls of the same cell. The part of the wall bearing the sieve areas with larger pores is called a **sieve plate** (Figures 21–18 through 21–20). Although sieve plates may occur on any wall, they generally are located on the end walls. Sieve-tube members occur end-on-end in longitudinal series called **sieve tubes.** Thus, one of the principal distinctions between the two types of sieve elements is the presence of sieve plates in sieve-tube members and their absence in sieve cells.

The sieve cell is a more primitive type of cell than the sieve-tube member. Sieve cells are the only type of food-conducting cell in most seedless vascular plants and gymnosperms, whereas only sieve-tube members occur in angiosperms.

The walls of sieve elements are generally described as primary. In cut sections of phloem tissue, the pores of the sieve areas and sieve plates are generally blocked by or lined with a wall substance called **callose,** which is a polysaccharide composed of spirally wound chains of glucose residues (Figures 21–17 and 21–18). It is now known that most, if not all, of the callose found in the pores of conducting sieve elements is deposited there in

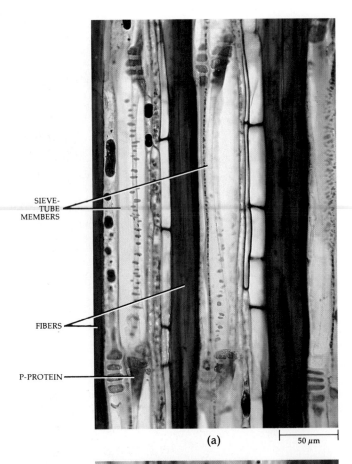

SIEVE-
TUBE
MEMBERS

FIBERS

P-PROTEIN

(a)   50 μm

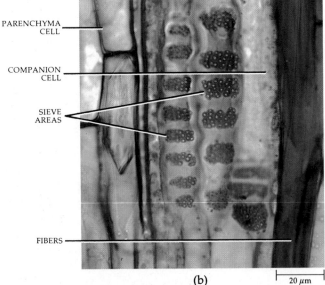

PARENCHYMA
CELL

COMPANION
CELL

SIEVE
AREAS

FIBERS

(b)   20 μm

21–18
(a) *Longitudinal (radial) view of secondary phloem of basswood (Tilia americana), showing sieve-tube members and conspicuous groups of thick-walled fibers.*
(b) *Compound sieve plates of basswood sieve-tube members. (Compound sieve plates consist of two or more sieve areas.) Each sieve area is composed of pores bordered by cylinders of callose, which are stained blue in this section.*

21–19 *(On the facing page)*
Electron micrographs of parts of mature sieve-tube members from the phloem in the stem of (a) corn (Zea mays) and (b) and (c) squash (Cucurbita maxima). (a) Longitudinal view of parts of two mature sieve-tube members and a sieve plate with open pores. The numerous round organelles with dense inclusions are plastids. The inclusions are composed of protein. The sieve-tube members of corn, like those of many monocots, lack P-protein. (b) Similar, longitudinal view of parts of two sieve-tube members of squash. The sieve-tube members of squash, like those of other dicots, contain P-protein. In these sieve-tube members, the P-protein is distributed along the wall (arrows) and the sieve-plate pores are open. At the lower left is a companion cell. (c) Face view of part of a sieve plate between two mature sieve-tube members. As in (b), the sieve-plate pores are open, lined in part with P-protein (arrows).

response to injury during preparation of the tissue for microscopy. This callose and any callose resulting from wounding (natural occurrence) is referred to as "wound callose." Callose typically is deposited at the sieve areas and sieve plates of senescing sieve elements; it is referred to as "definitive callose."

Unlike tracheary elements, sieve elements have living protoplasts at maturity (Figure 21–19). The protoplasts of mature sieve elements are unique among the living cells of the plant in that they either lack nuclei entirely or contain only remnants of them. In addition, most mature sieve elements lack a clear boundary between their cytoplasm and vacuoles. When young, the sieve element contains one or more vacuoles, each separated from the cytoplasm by a tonoplast, or vacuolar membrane. In the final stages of differentiation, the tonoplasts disappear and the distinction between cytoplasmic and vacuolar contents no longer exists. At maturity, all of the remaining components of the sieve-element protoplast are distributed along the wall; they consist of the plasma membrane, a network of smooth endoplasmic reticulum next to the plasma membrane, and some plastids and mitochondria. Ribosomes, dictyosomes, microtubules, and nuclei are lacking.

The protoplasts of the sieve-tube members of dicotyledons (and some monocotyledons) are characterized by the presence of a proteinaceous substance once called "slime" and now known as **P-protein.** (The "P" in P-protein stands for phloem.) P-protein has its origin in the young sieve-tube member in the form of discrete bodies called P-protein bodies (Figure 21–20a, c). Dur-

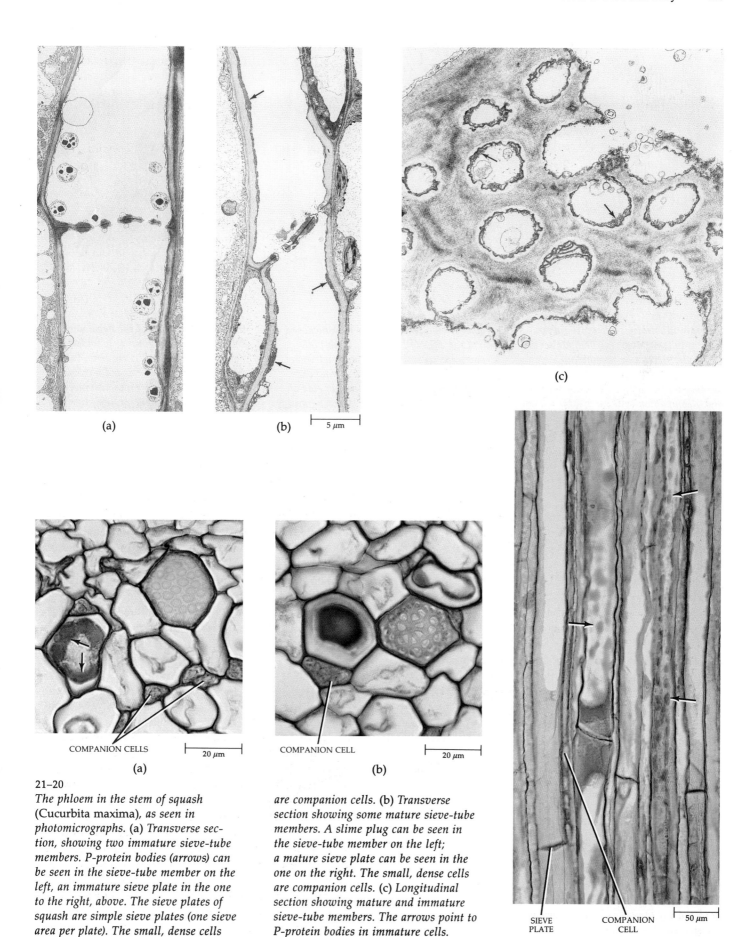

(a)

(b)   5 μm

(c)

COMPANION CELLS   20 μm

(a)

COMPANION CELL   20 μm

(b)

SIEVE   COMPANION   50 μm
PLATE   CELL

(c)

21–20

*The phloem in the stem of squash
(Cucurbita maxima), as seen in
photomicrographs. (a) Transverse section, showing two immature sieve-tube
members. P-protein bodies (arrows) can
be seen in the sieve-tube member on the
left, an immature sieve plate in the one
to the right, above. The sieve plates of
squash are simple sieve plates (one sieve
area per plate). The small, dense cells*
*are companion cells. (b) Transverse
section showing some mature sieve-tube
members. A slime plug can be seen in
the sieve-tube member on the left;
a mature sieve plate can be seen in the
one on the right. The small, dense cells
are companion cells. (c) Longitudinal
section showing mature and immature
sieve-tube members. The arrows point to
P-protein bodies in immature cells.*

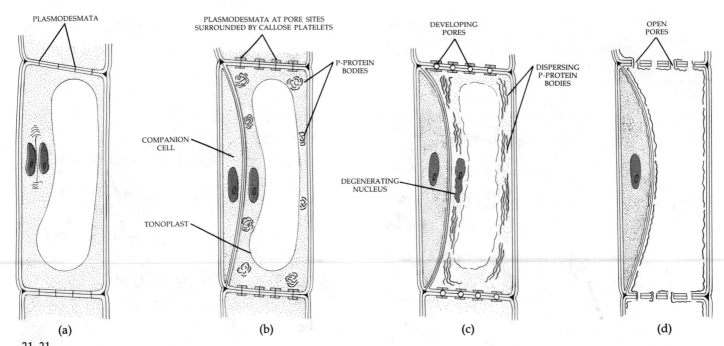

PLASMODESMATA

PLASMODESMATA AT PORE SITES
SURROUNDED BY CALLOSE PLATELETS

DEVELOPING
PORES

OPEN
PORES

P-PROTEIN
BODIES

DISPERSING
P-PROTEIN
BODIES

COMPANION
CELL

DEGENERATING
NUCLEUS

TONOPLAST

(a)

(b)

(c)

(d)

21–21
Diagram illustrating differentiation of a sieve-tube member. (a) The mother cell of the sieve-tube member undergoing division. (b) Division has resulted in formation of a young sieve-tube member and a companion cell. After division, one or more P-protein bodies arise in the cytoplasm, which is separated from the vacuole by a tonoplast. The wall of the *young sieve-tube member has thickened, and the sites of the future sieve-plate pores are represented by plasmodesmata. Each plasmodesma is now surrounded by a platelet of callose on either side of the wall. (c) The nucleus is degenerating, the tonoplast is breaking down, and the P-protein bodies are dispersing in the cytoplasm lining the wall; concomitantly,* *the plasmodesmata of the developing sieve plates are beginning to widen into pores. (d) At maturity, the sieve-tube member lacks a nucleus and a vacuole. All of the remaining protoplasmic components, including the P-protein, line the walls, and the sieve-plate pores are open. The callose platelets were removed as the pores widened.*

ing the late stages of differentiation, the P-protein bodies elongate and disperse. In cut sections of phloem tissues, "slime plugs" of P-protein are usually found near the sieve plates (Figure 21–20b). Slime plugs are not found in undisturbed cells and so are considered to result from the surging of the contents of the sieve tubes that are severed at the time the tissue is being sampled for study. In normal, mature sieve-tube members, P-protein is apparently distributed along the walls and is continuous from one sieve-tube member to the next through the pores of the sieve areas and sieve plates. The sieve-plate pores are lined with P-protein but are not plugged by it (Figure 21–19b, c). The function of P-protein has not been determined. However, some botanists believe that, together with wound callose, P-protein serves to seal the sieve-plate pores at the time of wounding, thus preventing the loss of contents from sieve tubes.

Sieve-tube members are characteristically associated with specialized parenchyma cells called **companion cells** (Figures 21–19b and 21–20), which contain all of the components commonly found in living plant cells, including a nucleus. The sieve-tube member and its associated companion cells are closely related developmentally (they are derived from the same mother cell), and they have numerous plasmodesmatal connections with one another. Figure 21–21 illustrates some stages of differentiation of a sieve-tube member with P-protein. Because of their numerous plasmodesmatal connections with the sieve-tube members and their general ultrastructural resemblance to secretory cells (high ribosome population and numerous mitochondria), it is believed that companion cells play a role in the delivery of substances to the sieve-tube members. This subject is discussed further in Chapter 28 when the mechanism of phloem transport in angiosperms is considered in detail.

The sieve cells of gymnosperms are characteristically associated with specialized parenchyma cells called **albuminous cells** (Figure 21–17b). Although generally not derived from the same mother cell as their associated sieve cells, the albuminous cells are believed to perform the same roles as companion cells. Like the companion cell, the albuminous cell contains a nucleus, in addition to other cytoplasmic components characteristic of living cells.

The sieve elements in most species apparently are short-lived, dying less than a year after their origin. This is not true, however, of all sieve elements. In the secondary phloem of the basswood, or linden tree (*Tilia americana*), some sieve elements remain alive and presumably function as conducting elements for five to ten years. Sieve elements are known to remain alive for many years in perennial monocots. In certain palms,

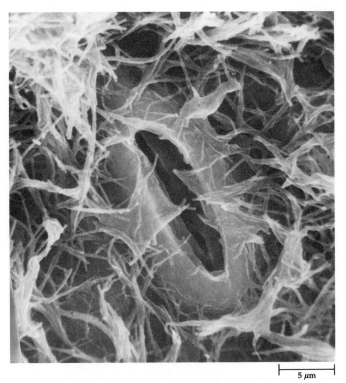

21-22
Surface view of lower epidermis of a Eucalyptus globulus *leaf, taken with a scanning electron microscope. A single stoma—flanked by two guard cells—and numerous filaments of epicuticular wax deposits can be seen here.*

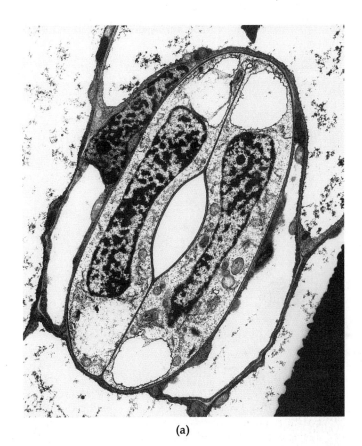

(a)

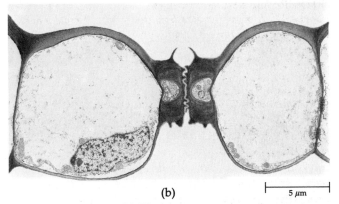

(b)

21-23
Electron micrographs of corn (Zea mays) stomata. (a) Surface view showing open stoma between guard cells, whose walls have not yet thickened, and associated subsidiary cells. (b) Transverse section through closed stoma. Each thick-walled guard cell is attached to a subsidiary cell. The interior of the leaf is below.

some sieve elements at the base of the main stem may be more than a century old. When the sieve elements die, their associated companion cells or albuminous cells also die, which is one more indication of the interdependence between sieve elements and their companion cells or albuminous cells.

Other parenchyma cells occur in the primary and secondary phloem (Figures 21–17 and 21–18). They are largely concerned with the storage of various substances. Fibers (Figures 21–17 and 21–18) and sclereids may also be present.

EPIDERMIS

The **epidermis** is the outermost layer of cells of the primary plant body, and it constitutes the dermal tissue system of leaves, floral parts, fruits, and seeds and of stems and roots until they undergo considerable secondary growth. Epidermal cells are quite variable both functionally and structurally. In addition to the relatively unspecialized ground cells, which form the bulk (groundmass) of the epidermis, the epidermis may contain **guard cells** (Figures 21–22 through 21–25), many types of appendages, or **trichomes** (Figures 21–24 and 21–26), and other kinds of specialized cells.

In most plants, the epidermis is only one layer of cells in thickness. In some plants, however, the protoderm cells divide parallel with the surface (periclinal divisions) of the plant part, and an epidermis with several layers (a multiple epidermis) is formed (Figure 21–27). A multiple epidermis is found in the leaves of such familiar house plants as the rubber plant (*Ficus elastica*) and *Peperomia*, and in the roots of some monocots, including orchids (see page 484). The multiple epidermis is believed to serve as a water-storage tissue.

465

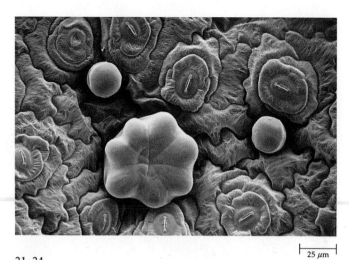

21–24
Surface view of the lower epidermis of a Swedish ivy (Plectranthus) *leaf, taken with a scanning electron microscope. Several stomata—pairs of guard cells— each flanked by two curved subsidiary cells can be seen. Also visible are two kinds of trichomes.*

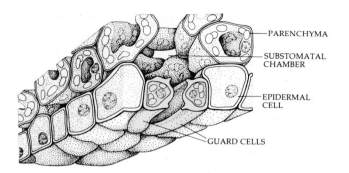

PARENCHYMA

SUBSTOMATAL
CHAMBER

EPIDERMAL
CELL

GUARD CELLS

21–25
Diagram of a mature stoma, showing its relationship to the epidermis and underlying cells. The walls next to the pores are generally thicker than those adjacent to other epidermal cells. While the stomata are being formed, they may be raised or lowered below the surface of the epidermis. Often there is a large air space, or substomatal chamber, just behind the stoma. Unlike other epidermal cells, guard cells contain chloroplasts.

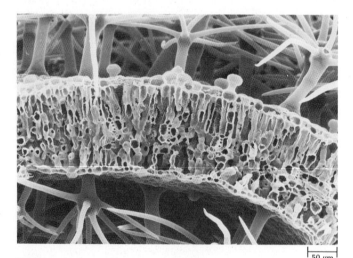

21–26
Transverse section of a common mullein (Verbascum thapsus) *leaf, taken with a scanning electron microscope. The leaves and stems of the common mullein appear densely woolly because of the presence of great numbers of highly branched trichomes such as those seen here on both the upper and lower surfaces of the leaf. Short, unbranched glandular trichomes are also present.*

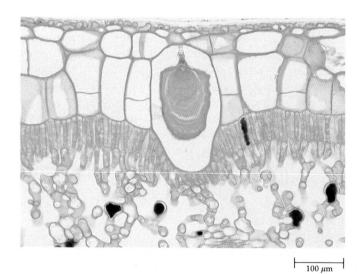

21–27
Transverse section showing upper portion of a leaf blade of the rubber plant (Ficus elastica). *Notice the thick cuticle covering the multiple epidermis of mostly large cells. The club-shaped structure in the largest epidermal cell consists mostly of calcium carbonate deposited on a cellulose stalk. The cells below the large, clear epidermal cells are mesophyll cells.*

The groundmass of epidermal cells is compactly arranged and affords considerable mechanical protection to the plant part. The walls of the epidermal cells of the aerial parts are covered with a cuticle, which minimizes water loss. The cuticle consists mainly of cutin and wax (see page 53). In many plants the wax is exuded over the surface of the cuticle, either in smooth sheets or as rods or filaments extending upward from the surface, the so-called epicuticular wax (Figure 21–22; see also Figure 3–11). It is this wax that is responsible for the whitish or bluish "bloom" on the surface of some leaves and fruits.

Interspersed among the flat, tightly packed epidermal cells are the chloroplast-containing guard cells (Figures 21–22 through 21–25), which regulate the small openings, or **stomata** (singular: stoma), in the aerial parts of the plant and hence control the movement of gases, including water vapor, into and out of those parts. (The mechanism of stomatal opening and closing is discussed in Chapter 28.) Although stomata occur on all aerial parts, they are most abundant on leaves. Stomata are often associated with epidermal cells that differ in shape from the ground cells. Such cells are termed *subsidiary cells*, or accessory cells (Figures 21–23 and 21–24).

Trichomes have a variety of functions. Root hairs facilitate the absorption of water and minerals from the soil. Recent studies of arid-land plants indicate that increase in leaf pubescence (hairiness) results in increased reflectance of solar radiation, lower leaf temperatures, and lower rates of water loss. Many "air plants" such as epiphytic bromeliads, utilize foliar trichomes for the absorption of water and minerals. In contrast, in the saltbush (*Atriplex*), salt-secreting trichomes remove salts from the leaf tissue, preventing an accumulation of toxic salts in the plant. Trichomes may provide a defense against insects. For example, in many species a positive correlation exists between hairiness and resistance to attack by insects. The hooked hairs of some species impale insects and their larvae. Glandular (secretory) hairs may provide a chemical defense.

PERIDERM

A **periderm** commonly replaces the epidermis in stems and roots having secondary growth. The periderm consists largely of protective **cork,** or *phellum*, which is nonliving and has walls that are heavily suberized at maturity, of the **cork cambium,** or *phellogen*, and of **phelloderm,** a living parenchyma tissue (Figure 21–28). The cork cambium forms cork tissue on its outer surface and phelloderm on its inner surface. The origin of the cork cambium is variable, depending on the species and plant part. The periderm is considered in detail in Chapter 24.

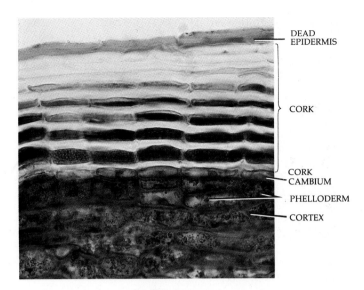

21–28
Transverse section of the periderm from the stem of apple (Malus sylvestris). The periderm shown here consists largely of cork cells, which are laid down to the outside (above) in radial rows by the cells of the cork cambium. A single layer of phelloderm cells occurs inside the cork cambium.

Summary

A summary of plant tissues and their cell types is found in Table 21–1, and the characteristics, location, and function of these types of plant cells are given in Table 21–2.

Table 21–1 *Summary of Plant Tissues and Their Cell Types*

Tissue	Cell Types
Epidermis	Ground cells; guard cells and cells forming trichomes; sclerenchyma cells
Periderm	Cork cells; cork cambium cells; parenchyma cells of the phelloderm; sclerenchyma cells
Xylem	Tracheids; vessel members; sclerenchyma cells; parenchyma cells
Phloem	Sieve cells or sieve-tube members; albuminous cells or companion cells; other parenchyma cells; sclerenchyma cells
Parenchyma	Parenchyma cells
Collenchyma	Collenchyma cells
Sclerenchyma	Fibers or sclereids

Table 21–2 *Summary of Cell Types*

Cell Type	Characteristics	Location	Function
Parenchyma	Shape: commonly polyhedral (many-sided); variable Cell wall: primary, or primary and secondary; may be lignified, suberized, or cutinized Living at maturity	Throughout the plant, as parenchyma tissue in cortex, as pith and pith rays; in xylem and phloem	Such metabolic processes as respiration, digestion, and photosynthesis; storage and conduction; wound healing and regeneration
Collenchyma	Shape: elongated Cell wall: unevenly thickened; primary only—nonlignified Living at maturity	On the periphery (beneath the epidermis) in young elongating stems; often as a cylinder of tissue or only in patches; in ribs along veins in some leaves	Support in primary plant body
Fibers	Shape: generally very long Cell wall: primary and thick secondary—often lignified Often (not always) dead at maturity	Sometimes in cortex of stems, most often associated with xylem and phloem; in leaves of monocotyledons	Support; storage
Sclereids	Shape: variable; generally shorter than fibers Cell wall: primary and thick secondary—generally lignified May be living or dead at maturity	Throughout the plant	Mechanical; protective
Tracheid	Shape: elongated and tapering Cell wall: primary and secondary; lignified; contains pits but not perforations Dead at maturity	Xylem	Chief water-conducting element in gymnosperms and seedless vascular plants; also found in angiosperms
Vessel member	Shape: elongated, generally not as long as tracheids; several vessel members end-to-end constitute a vessel Cell wall: primary and secondary; lignified; contains pits and perforations Dead at maturity	Xylem	Chief water-conducting element in angiosperms

Cell Type	Characteristics	Location	Function
Sieve cell	Shape: elongated and tapering Cell wall: primary in most species; with sieve areas; callose often associated with wall and pores Living at maturity; either lacks or contains remnants of a nucleus at maturity; lacks distinction between vacuole and cytoplasm	Phloem	Food-conducting element in gymnosperms and seedless vascular plants
Albuminous cell	Shape: generally elongated Cell wall: primary Living at maturity; associated with sieve cell, but generally not derived from same mother cell as sieve cell; has numerous plasmodesmatal connections with sieve cell	Phloem	Believed to play a role in the movement of food into and out of the sieve cell
Sieve-tube member	Shape: elongated Cell wall: primary, with sieve areas; sieve areas on end wall with much larger pores than those on side walls—this wall part is termed a sieve plate; callose often associated with walls and pores Living at maturity; either lacks a nucleus at maturity or contains only remnants of nucleus; in dicots and some monocots, contains a proteinaceous substance known as P-protein; several sieve-tube members in a vertical series constitute a sieve tube	Phloem	Food-conducting element in angiosperms
Companion cell	Shape: variable, generally elongated Cell wall: primary Living at maturity; closely associated with sieve-tube members; derived from same mother cell as sieve-tube member; has numerous plasmodesmatal connections with sieve-tube member	Phloem	Believed to play a role in the movement of food into and out of the sieve-tube member

CHAPTER 22

The Root: Structure and Development

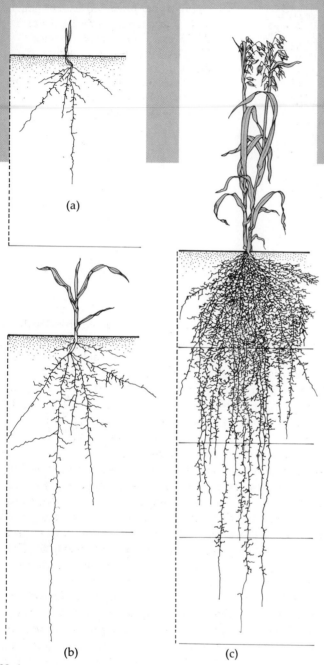

(a)

(b)   (c)

22–1

Diagrams of oat (Avena sativa) *plants, showing relative sizes of the root and shoot systems (a) 31, (b) 45, and (c) 80 days after planting. The oat plant has a fibrous root system. The roots are involved primarily with anchorage and absorption. Each of the vertical units depicted here represents 1 foot (approximately 30.5 centimeters).*

In most vascular plants, the roots constitute the underground portion of the sporophyte and are involved primarily in anchorage and absorption (Figure 22–1). Two other functions associated with roots are storage and conduction. Most roots are important storage organs, and some, such as those of the carrot, sugar beet, and sweet potato, are specifically adapted for the storage of food. Foods manufactured above ground, in photosynthesizing portions of the plant body, move through the phloem to the storage tissues of the root. This food may eventually be used by the root itself, but more often the stored food is digested and the products transported back through the phloem to the aboveground parts. In biennials (plants that complete their life cycle over a two-year period), such as the sugar beet and carrot, large food reserves accumulate in the storage regions of the root during the first year and then are used during the second year to produce flowers, fruits, and seeds. Water and minerals, or inorganic ions, absorbed by the roots move through the xylem to the aerial parts of the plant. In addition, hormones (particularly cytokinins and gibberellins) synthesized in meristematic regions of the roots are transported upward in the xylem to the aerial parts, which depend on the presence of such hormones to stimulate growth and development (see Chapter 25).

Root Systems

The first root of the plant originates in the embryo and is usually called the **primary root.** In gymnosperms and dicotyledons, this root becomes a **taproot;** it grows directly downward, giving rise to branch roots, or **lateral roots,** along the way. The older lateral roots are found nearer the base of the root (where the root and stem meet), and the younger ones are found nearer the

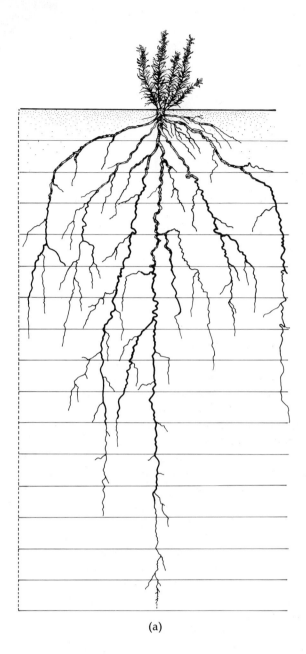

(a)

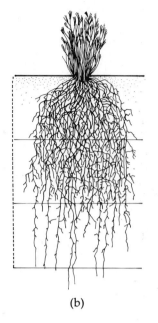

(b)

22–2
Two types of root systems, as represented by two prairie plants. (a) Taproot system of the blazing star (Liatris punctata). *(b)* Fibrous root system of wire grass (Aristida purpurea). *Each of the units depicted here represents 1 foot (approximately 30.5 centimeters). Taproot systems generally penetrate deeper into the soil than fibrous root systems.*

root tip. This type of root system—that is, one that develops from a taproot and its branches—is called a **taproot system** (Figure 22–2a).

In monocotyledons, the primary root is usually short-lived, and the root system develops from adventitious roots that arise from the stem. These adventitious roots and their lateral roots give rise to a **fibrous root system,** in which no one root is more prominent than the others (Figures 22–1 and 22–2b). Taproot systems generally penetrate deeper into the soil than fibrous root systems. The shallowness of fibrous root systems and the tenacity with which they cling to soil particles make them especially well suited as ground cover for the prevention of soil erosion.

The extent of a root system—that is, the depth to which it penetrates the soil and the distance it spreads laterally—is dependent upon several factors, including the moisture, temperature, and composition of the soil. The bulk of most so-called "feeder roots" (roots actively engaged in the uptake of water and minerals) occurs in the upper meter of soil. The majority of the feeder roots of most trees occurs in the upper 15 centimeters of soil, the part of the soil normally richest in organic matter. Some trees, such as spruces, beeches, and poplars, rarely produce deep taproots, whereas others, such as oaks and many pines, commonly produce relatively deep taproots, making such trees rather difficult to transplant. The record for depth of penetration by roots belongs to the desert shrub mesquite *(Prosopis juliflora)*. Mesquite roots were found growing at a depth of 53.3 meters at a new open-pit mine 30 kilometers southwest of Tucson, Arizona, in 1960. During the digging of the Suez Canal in Egypt, roots of *Tamarix* and *Acacia* trees were found at a depth of 30 meters. The lateral spread of tree roots is usually greater than the spread of the crown of the tree. The root systems of corn plants *(Zea mays)* often reach a depth of about 1.5 meters, with a lateral spread of about a meter on all sides of the plant. The roots of alfalfa *(Medicago sativa)* may extend to depths of up to 6 meters or more.

One of the most detailed studies on the extent of the root and shoot systems of any one plant was conducted on a 4-month-old rye plant *(Secale cereale)*. The total surface area of the root system, including root hairs, was 639 square meters, or 130 times the surface area of the shoot system. Even more amazing is the fact that the roots occupied only about 6 liters of soil.

As a plant grows, it needs to maintain a balance between the total surface area available for the manufacture of food (the photosynthesizing surface) and the surface area available for the absorption of water and minerals. In a young plant, that is, one just becoming established, the total water- and mineral-absorbing surface usually far exceeds the photosynthesizing surface. However, this relationship tends to change in favor of the photosynthesizing surface as the plant ages. Gardeners must consider these facts when transplanting plants. Even when plants are carefully transplanted, the balance between shoot and root is invariably disturbed. Most of the fine feeder roots are left behind when the plant is removed from the soil; cutting back the shoot helps to reestablish a balance between the root system and the shoot system.

Origin and Growth of Primary Tissues

The growth of many roots is apparently a continuous process that stops only under such adverse conditions as drought and low temperatures. During their growth through the soil, roots follow the path of least resistance and frequently follow spaces left by earlier roots that have died and rotted.

The tip of the root is covered by a **root cap** (Figures 22–3 through 22–5), a thimblelike mass of cells that protects the apical meristem behind it and aids the root in its penetration of the soil. As the root grows longer and the root cap is pushed forward, the cells on the periphery of the root cap are sloughed off. These sloughed-off cells and the growing root tip are covered by a slimy sheath, or **mucigel,** which lubricates the root during its passage through the soil. As quickly as root cap cells are sloughed off, new ones are added by the apical meristem. The longevity (from time of initiation to sloughing off) of root cap cells ranges from four to nine days, depending upon the length of the cap and the species.

The slimy substance forming the mucigel is a highly hydrated polysaccharide, probably a pectic substance, that is secreted by the outer root cap cells. It accumulates in dictyosome vesicles, which fuse with the plasma membrane and then release the slime into the wall. Eventually the slime passes to the outside, where it forms droplets.

In addition to protecting the apical meristem and aiding the root in its penetration of the soil, the root cap plays an important role in controlling the response of the root to gravity (gravitropism; see Chapter 26).

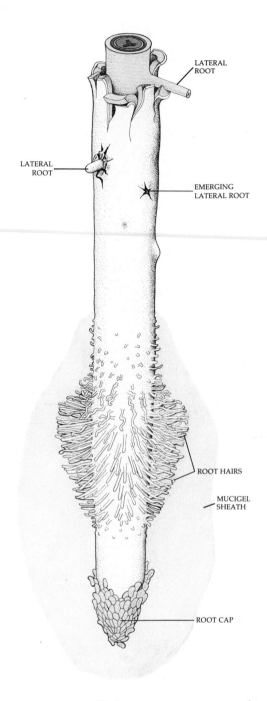

LATERAL ROOT

LATERAL ROOT

EMERGING LATERAL ROOT

ROOT HAIRS

MUCIGEL SHEATH

ROOT CAP

22–3
A portion of a dicot root, showing the spatial relationship between the root cap and region of root hairs, and (near the top) the sites of emergence of lateral roots, which arise from deep within the parent root. New root hairs arise just beyond the region of elongation at about the same rate as the older hairs die off. The root tip is covered by a mucigel sheath, which lubricates the root during its passage through the soil.

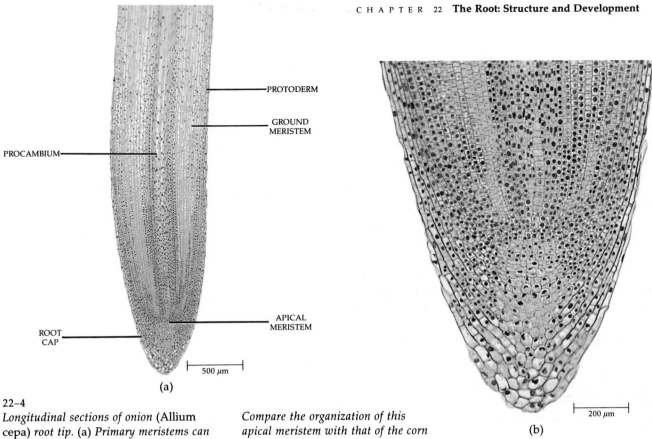

PROTODERM

GROUND
MERISTEM

PROCAMBIUM

APICAL
MERISTEM

ROOT
CAP

500 μm

(a)

200 μm

(b)

22–4

Longitudinal sections of onion (Allium cepa) root tip. (a) Primary meristems can be distinguished close to the apical meristem. (b) Detail of the apical meristem.

Compare the organization of this apical meristem with that of the corn root tip shown in Figure 22–5.

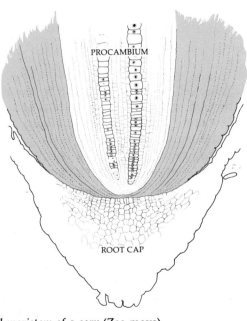

PROCAMBIUM

ROOT CAP

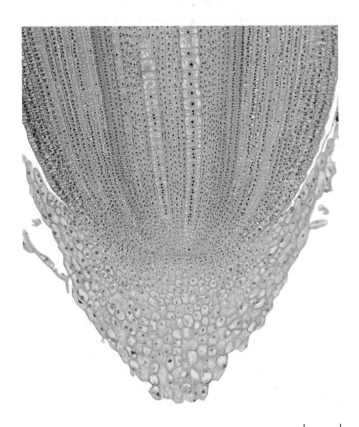

22–5

Apical meristem of a corn (Zea mays) root tip. Notice the three distinct layers of initials. The lower layer gives rise to the root cap, the middle layer to the protoderm and to the ground meristem, or cortex, and the upper layer to the procambium or vascular cylinder.

100 μm

GROWTH REGIONS OF THE ROOT

Apart from the root cap, the most striking structural feature of the root apex is the arrangement of longitudinal files of cells that emanate from the apical meristem. The apical meristem is composed of relatively small, many-sided cells—the initials and their immediate derivatives (page 453)—with dense cytoplasm and large nuclei (Figures 22–4 and 22–5). The organization and number of initiating layers in the apical meristems of roots are both quite variable.

Two main types of apical organization are found in the roots of seed plants. In one, the root cap, the vascular cylinder of xylem and phloem, and the cortex emerge from a common group of cells in the apical meristem (Figure 22–4). In the other, each region can be traced to an independent layer of cells (Figure 22–5). In the second type of organization, the epidermis has a common origin with either the root cap or the cortex.

Although the region of initials in the root apical meristems is mitotically active early in root development, divisions become infrequent in this region later in the growth of the root. Most cell division then occurs a short distance beyond the relatively quiescent initials. The relatively inactive region of the apical meristem is known as the **quiescent center** (Figure 22–6).

As indicated by the term "relatively," the quiescent center is not totally devoid of divisions under normal conditions. Moreover, the quiescent center is able to repopulate the bordering meristematic regions when they are injured. In one study, for example, the isolated quiescent centers of corn (Zea mays) grown in sterile culture were found to be able to form whole roots without first forming callus, or wound, tissue. In another study of corn roots, a striking correlation was found between the size of the quiescent center and the complexity of the primary vascular pattern of the root. These and other studies indicate that the quiescent center plays an essential role in organization and development of the root.

The distance beyond the apical meristem at which most cell division takes place varies from species to species and also within the same species, depending on the age of the root. The combination of the apical meristem and the nearby portion of root in which cell division does occur is called the **region of cell division** (Figure 22–7).

Behind the region of cell division, but not sharply delimited from it, is the **region of elongation,** which usually is only a few millimeters in length (Figure 22–7). The elongation of cells in this region results in most of the increase in length of the root. Beyond this region the root does not increase in length. Thus, growth in length of the root occurs near the root tip and results in a very limited portion of the root constantly being pushed through the soil.

├─────┤ 50 μm

22–6
Longitudinal section through the apical meristem of a corn (Zea mays) *root tip, showing the quiescent center. To prepare this autoradiograph, the root tip was supplied for one day with thymidine labeled with the radioactive hydrogen isotope, tritium (^{3}H). In the rapidly dividing cells around the quiescent center (delineated by the dashed oval), the radioactive material was quickly incorporated into the nuclear DNA. The radiation from the labeled DNA exposed a photographic emulsion coating the section, resulting in the formation of the dark grains on this autoradiograph.*

The region of elongation is followed by the **region of maturation,** in which most of the cells of the primary tissues mature (Figure 22–7). Root hairs are also produced in this region, and sometimes this part of the root is called the root-hair zone (see Figure 22–3). Of course, were root hairs to arise in the region of elongation, they soon would be sheared off.

It is important to note that there is a gradual transition from one region of the root to another. The regions are not sharply delimited from one another. Some cells begin to elongate and differentiate in the region of cell division, whereas others reach maturity in the region of elongation. For example, the first-formed elements of the phloem and xylem mature in the region of elongation and are often stretched and destroyed during elongation of the root. As can be seen in Figure 22–7, the first-formed phloem elements (protophloem sieve tubes) reach maturity nearer the root tip than do the first-formed xylem elements (protoxylem elements), a reflection of the need for substances transported in the sieve tubes for root growth.

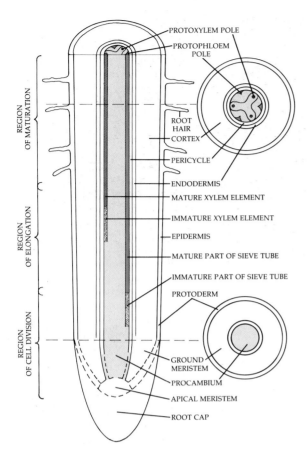

PROTOXYLEM POLE

PROTOPHLOEM POLE

ROOT HAIR

CORTEX

PERICYCLE

ENDODERMIS

MATURE XYLEM ELEMENT

IMMATURE XYLEM ELEMENT

EPIDERMIS

MATURE PART OF SIEVE TUBE

IMMATURE PART OF SIEVE TUBE

PROTODERM

GROUND MERISTEM

PROCAMBIUM

APICAL MERISTEM

ROOT CAP

REGION OF MATURATION

REGION OF ELONGATION

REGION OF CELL DIVISION

22–7
A diagram illustrating early stages in the primary development of a root tip. (Compare this figure with Figure 22–3.)

The protoderm, procambium, and ground meristem can be distinguished in very close proximity to the apical meristem (Figures 22–4 and 22–7). These are the primary meristems that differentiate into the epidermis, the primary vascular tissues, and the cortex, respectively (Chapter 20).

Primary Structure

Compared with that of the stem, the internal structure of the root is usually relatively simple. This is due in large part to the absence of leaves and the corresponding absence of nodes and internodes. Thus, the arrangement of the primary tissues shows very little difference from one level of the root to another.

The three tissue systems of the root in the primary stage of growth can be readily distinguished in both transverse (see Figures 22–8 and 22–9 on pages 476–477) and longitudinal sections (Figure 22–4). The epi-

dermis (dermal tissue system), the cortex (ground tissue system), and the vascular tissues (vascular tissue system) can be clearly distinguished from one another. In most roots, the vascular tissues form a solid cylinder (Figure 22–8), but in some they form a hollow cylinder around a pith (Figure 22–9).

EPIDERMIS

The epidermis in young roots absorbs water and minerals, and this function is facilitated by *root hairs*—tubular extensions of the epidermal cells—which greatly increase the absorbing surface of the root (Figures 22–3, 22–10 on page 477, and 22–11 on page 478). (See pages 238–241 for a discussion of the role of mycorrhizae in absorption.) As part of the previously mentioned study of a four-month-old rye plant, it was estimated that the plant contained approximately 14 billion root hairs, with an absorbing surface of 401 square meters. Placed end to end, these root hairs would extend well over 10,000 kilometers.

Root hairs are relatively short-lived and are confined largely to the region of maturation. The production of new root hairs occurs just beyond the region of elongation (Figures 22–3 and 22–7) and at about the same rate as that at which the older root hairs are dying off at the upper end of the root hair zone. As the tip of the root penetrates the soil, new root hairs develop immediately behind it, providing the root with surfaces capable of absorbing new supplies of water and minerals, or inorganic ions. (See Chapter 28 for a discussion of absorption of water and inorganic ions by roots.) Obviously, it is the new and growing roots—the feeder roots—that are primarily involved in the absorption of water and minerals. For this reason, great care must be taken by gardeners to move as much soil as possible along with the root system during transplantation. If the plant is simply "torn" from the soil, most of the feeder roots will be left behind and the plant probably will not survive.

A thin cuticle has been identified on the epidermis in the absorbing part of some roots. In other roots, the epidermal cell walls appear to contain suberin. Nevertheless, the walls of the epidermal cells offer little resistance to the passage of water and minerals into the root. The mucigel covering the surface of roots enables the roots to establish a much more intimate contact with the particles of soil and may influence the availability of ions to the roots. The mucigel has been shown to provide an environment favorable to beneficial bacteria. It may also provide the root with short-term protection from drying out (desiccation). The layer of soil bound to the root by the mucigel and root hairs and containing a variety of microorganisms and sloughed-off root cap cells is called the **rhizosphere.**

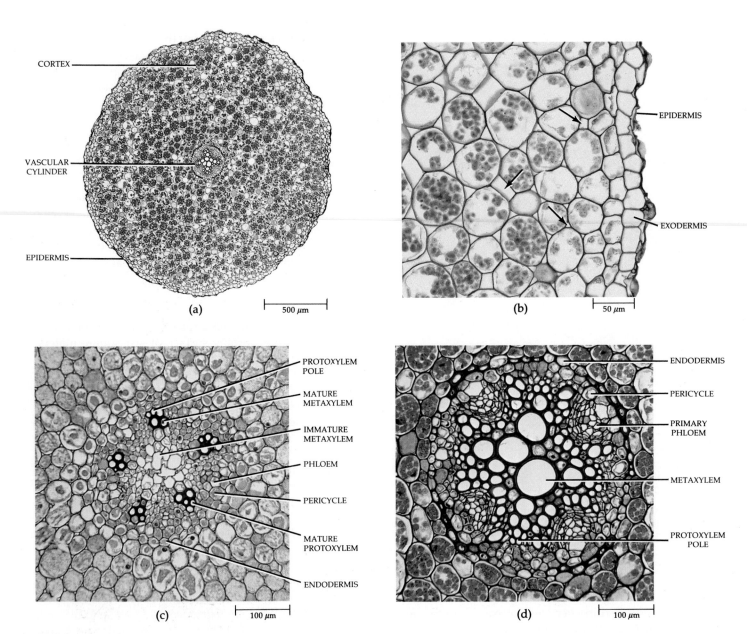

(a)

CORTEX

VASCULAR CYLINDER

EPIDERMIS

500 μm

(b)

EPIDERMIS

EXODERMIS

50 μm

(c)

PROTOXYLEM POLE

MATURE METAXYLEM

IMMATURE METAXYLEM

PHLOEM

PERICYCLE

MATURE PROTOXYLEM

ENDODERMIS

100 μm

(d)

ENDODERMIS

PERICYCLE

PRIMARY PHLOEM

METAXYLEM

PROTOXYLEM POLE

100 μm

22–8
Transverse sections of the root of a
buttercup (Ranunculus). *(a) Overall*
view of mature root. (b) Detail of outer
portion of a mature root. In this root, the
epidermis has died and the exodermis has
replaced the epidermis as the functional
surface layer. Note intercellular spaces
(arrows) among the cortical cells interior
to the compactly arranged exodermis.
(c) Detail of immature vascular cylinder.
Notice the intercellular spaces among the
cortical cells. (d) Detail of mature
vascular cylinder. Numerous starch
grains are evident in the cortical cells.
Note that in this section the root is
tetrarch (with four ridges), whereas in (c)
it is pentarch (with five ridges). Variation
in the number of ridges of primary xylem
is common along the axis of many roots.

22–9 *(On the facing page)*
Transverse sections of a corn (Zea mays)
root. (a) Overall view of mature root.
Part of a lateral root can be seen at the
lower right. The vascular cylinder, with
its pith, is quite distinct. (b) Detail of
outer portion of a mature root, showing
the epidermis with root hairs and part
of the cortex. The outer layer of cortical
cells is differentiated as a compactly
arranged exodermis. Note intercellular
spaces (arrows) among the other cortical
cells. (c) Detail of immature vascular
cylinder. (d) Detail of mature vascular
cylinder.

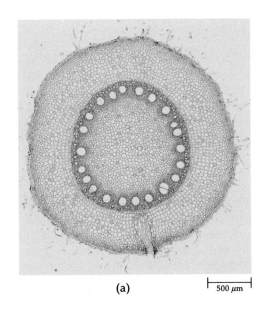

(a)

500 μm

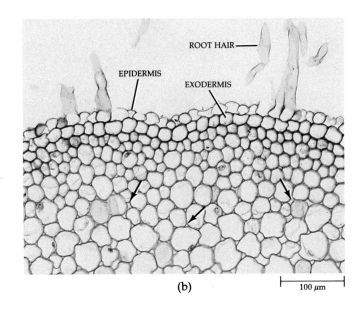

ROOT HAIR

EPIDERMIS   EXODERMIS

(b)

100 μm

ENDODERMIS   PROTOXYLEM   PRIMARY   PERICYCLE   MATURE (EARLY)
POLE   PHLOEM   METAXYLEM VESSEL

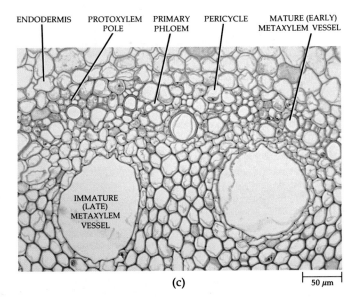

IMMATURE
(LATE)
METAXYLEM
VESSEL

(c)

50 μm

ENDODERMIS   PROTOXYLEM   PRIMARY   PERICYCLE   EARLY METAXYLEM
POLE   PHLOEM   VESSEL

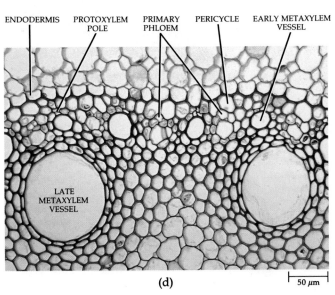

LATE
METAXYLEM
VESSEL

(d)

50 μm

22–10
A radish seedling. Note the discarded seed coat, the cotyledons, curved hypocotyl, and primary root, with numerous root hairs. Most of the uptake of water and minerals occurs through the root hairs, which form just behind the growing tip of the root.

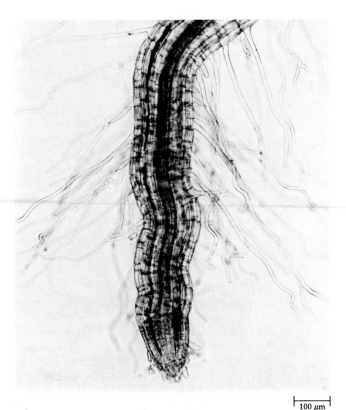

22–11

Root of a bentgrass (Agrostis tenuis) seedling, showing root hairs. Root hairs may be as much as 1.3 centimeters long and may attain their full size within hours. Each hair is comparatively short-lived, but the formation of new root hairs and the death of old ones continue as long as the root is growing.

100 μm

CORTEX

As seen in transverse section (Figure 22–8a), the cortex occupies by far the greatest area of the primary body of most roots. The cells of the cortex store starch and other substances but usually lack chloroplasts. Roots that undergo considerable amounts of secondary growth—those of gymnosperms and most dicots—shed their cortex early. In such roots, the cortical cells remain parenchymatous. In monocots, by contrast, the cortex is retained for the life of the root, and many of the cortical cells develop secondary walls that become lignified.

Regardless of the degree of differentiation, cortical tissue contains numerous intercellular spaces—air spaces essential for aeration of the cells of the root (Figures 22–8 and 22–9). The cortical cells have numerous contacts with one another, and their protoplasts are connected by plasmodesmata. Thus substances moving across the cortex may move from one protoplast to another by way of plasmodesmata, or they may follow a cell wall pathway.

Unlike the rest of the cortex, the innermost layer is compactly arranged and lacks air spaces. This layer, the **endodermis** (Figures 22–8 and 22–9), is characterized by the presence of **Casparian strips** in its anticlinal walls (the walls perpendicular to the surface of the root). The Casparian strip is a bandlike portion of the primary wall that is impregnated with a fatty substance called suberin and is sometimes lignified. The plasma membranes of endodermal cells are quite firmly attached to the Casparian strips (Figures 22–12 and 22–

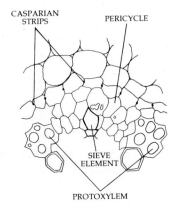

22–12

High-magnification view of a portion of an immature buttercup (Ranunculus) root, showing Casparian strips in the endodermal cells. Notice that the plasmolyzed protoplasts of the endodermal cells cling to the strips.

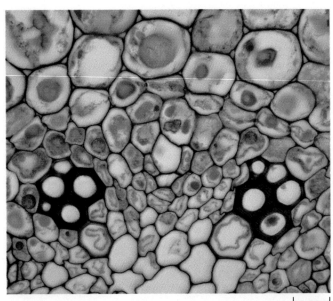

25 μm

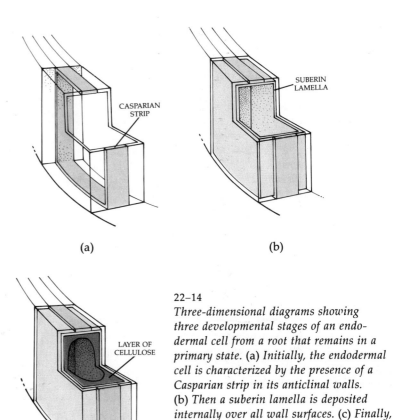

(a)   (b)

(c)

22–14
*Three-dimensional diagrams showing
three developmental stages of an endodermal cell from a root that remains in a
primary state. (a) Initially, the endodermal
cell is characterized by the presence of a
Casparian strip in its anticlinal walls.
(b) Then a suberin lamella is deposited
internally over all wall surfaces. (c) Finally,
the suberin lamella is covered by a
thick, often lignified, layer of cellulose.
The outside of the root is to the left in
all three diagrams.*

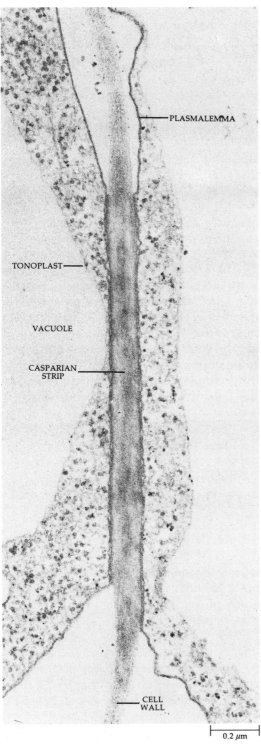

22–13
*Transverse section of a radial anticlinal
wall between two endodermal cells in the
root of field bindweed (Convolvulus
arvensis). The region of the Casparian
strip is more intensely stained than the
rest of the wall. Notice how firmly the
plasma membrane (plasmalemma) adheres
to the wall in the region of the Casparian
strip in each of these plasmolyzed cells.*

13). Inasmuch as the endodermis is compact and the
Casparian strips are impermeable to water and ions, all
substances entering and leaving the vascular cylinder
must pass through the protoplasts of the endodermal
cells. This is accomplished either by crossing the
plasma membranes of these cells or by passing through
the numerous plasmodesmata connecting the endodermal cells with the protoplasts of neighboring cells of
the cortex and vascular cylinder.

As mentioned previously, in roots that undergo secondary growth, the cortex and its endodermis are shed
early. In roots in which the cortex is retained, a suberin
lamella, consisting of alternating layers of suberin and
wax, is eventually deposited internally over all wall
surfaces of the endodermis. This is followed by the
deposition of cellulose, which may become lignified
(Figures 22–14 and 22–15). These changes in the endodermis begin opposite the phloem strands and spread
toward the protoxylem (Figure 22–8d). Opposite the
protoxylem, some of the endodermal cells may remain
thin-walled and retain their Casparian strips for a prolonged period of time. Such cells are called **passage
cells,** and in most species they eventually become suberized.

22–15
Electron micrograph showing a section through an anticlinal wall between two endodermal cells of a squash (Cucurbita pepo) root. In this late stage of differentiation of the endodermis, suberin lamellae are covered by cellulose wall layers on both sides of the primary wall. Note the alternating light and dense bands in the suberin lamellae, which are interpreted as consisting of wax and suberin, respectively.

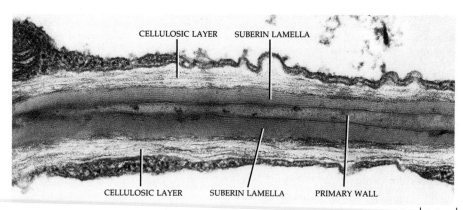

CELLULOSIC LAYER   SUBERIN LAMELLA

CELLULOSIC LAYER   SUBERIN LAMELLA   PRIMARY WALL

0.2 μm

The roots of many angiosperms have a second compact layer of cells with Casparian strips. This layer, called the **exodermis,** develops from the outermost layer or layers of cells of the cortex. Development of the Casparian strips is quickly followed by deposition of a suberin lamella and, at least in some species, by a cellulosic layer as well. The suberized cell walls of the exodermis apparently reduce water loss from the root to the soil and provide a defense against attack by microorganisms. (See Chapter 28 for further discussion of the role of the exodermis and endodermis in the movement of water and solutes across the root.)

VASCULAR CYLINDER

The vascular cylinder of the root consists of vascular tissues and one or more layers of cells, the **pericycle,** which completely surrounds the vascular tissues (Figures 22–8 and 22–9). In the young root, the pericycle is composed of parenchyma cells with primary walls, but as the root ages, the cells of the pericycle may develop secondary walls (Figure 22–9).

The pericycle plays several important roles. In most seed plants, lateral roots arise in the pericycle. In plants undergoing secondary growth, the pericycle contributes to the vascular cambium opposite the protoxylem and generally gives rise to the first cork cambium. Pericycle often proliferates—that is, gives rise to more pericycle.

The center of the vascular cylinder of most roots is occupied by a solid core of primary xylem from which ridgelike projections extend toward the pericycle (Figure 22–8). Nestled between the ridges of xylem are strands of primary phloem. Obviously, the vascular cylinder in such roots is a protostele (see page 320).

The number of ridges of primary xylem varies from species to species, sometimes even varying along the axis of a given root. If two ridges are present, the root is said to be diarch; if three are present, triarch; four, tetrarch (Figure 22–8); and if many are present, polyarch

(Figure 22–9). The first *(proto-)* xylem elements to mature in roots are located next to the pericycle, and the tips of the ridges are commonly referred to as **protoxylem poles** (Figures 22–8 and 22–9). The **metaxylem** *(meta-,* meaning "after") occupies the inner portions of the ridges and the center of the vascular cylinder and matures after the protoxylem. The roots of some monocotyledons (for example, corn) have a pith (Figure 22–9), which some botanists interpret as potential vascular tissue.

Effect of Secondary Growth on the Primary Body of the Root

As mentioned previously, secondary growth in roots and stems consists of the formation of (1) secondary vascular tissues—secondary xylem and secondary phloem—from a vascular cambium and (2) periderm, composed mostly of cork tissue, from a cork cambium. Commonly, the roots of monocots lack secondary growth and, hence, consist entirely of primary tissues. In addition, the roots of many herbaceous dicotyledons undergo little or no secondary growth and remain largely primary in composition. (See Chapter 24 for a discussion of the vascular cambium, pages 521–523, and periderm and lenticels, pages 525–527.)

In roots that exhibit secondary growth, the vascular cambium is initiated by divisions of procambial cells that remain meristematic and are located between the primary xylem and primary phloem in portions of the root that are no longer elongating. Thus, depending on the number of phloem strands present in the root, two or more independent regions of cambial activity are initiated more or less simultaneously (Figure 22–16). Soon afterward, the pericycle cells opposite the protoxylem poles also divide, and the inner sister cells resulting from those divisions contribute to the vascular cambium. Now the cambium completely surrounds the core of xylem.

As soon as it is formed, the vascular cambium opposite the phloem strands begins to produce secondary xylem toward the inside, and in the process the strands of primary phloem are displaced outward from their positions between the ridges of primary xylem. By the time the cambium opposite the protoxylem poles is actively dividing, the cambium is circular in outline and the primary phloem has been separated from the primary xylem (Figure 22–16).

By repeated divisions toward the inside and outside, secondary xylem and secondary phloem are added to the root (Figures 22–16 and 22–17). In some roots, the vascular cambium derived from the pericycle forms wide rays, whereas narrower rays are produced in other parts of the secondary vascular tissues. The rays consist of files of parenchyma cells that extend radially in the secondary xylem and secondary phloem.

With increase in width of the secondary xylem and phloem, most of the primary phloem is crushed or obliterated. Primary phloem fibers may be the only remaining distinguishable components of the primary phloem.

22–16

Root development in a woody dicot.
(a) *Early stage in primary development, showing primary meristems.* (b) *At the completion of primary growth, showing the primary tissues plus the meristematic procambium between the primary xylem and primary phloem.* (c) *Origin of vascular cambium. In the triarch root represented here, cambial activity has been initiated in three independent regions from procambium between the three primary phloem strands and the primary xylem. The pericycle cells opposite the three protoxylem poles will also contribute to the vascular cambium. Some secondary xylem has already been produced by the newly formed vascular cambium of procambial origin.* (d) *Some secondary phloem and additional secondary xylem have been formed, further separating the primary phloem from the primary xylem. A periderm has not yet formed.* (e) *After formation of additional secondary xylem and secondary phloem and of periderm.* (f) *At end of first year's growth, showing the effect of secondary growth—including periderm formation—on the primary plant body. In (d) through (f), the radiating lines represent rays.*

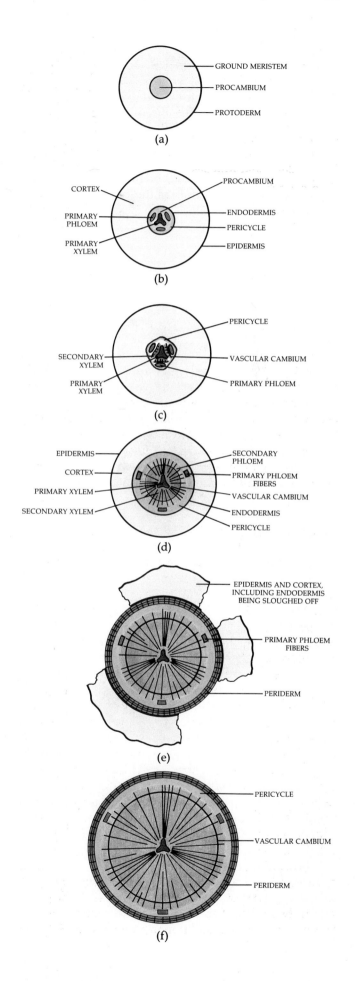

In most woody roots, periderm formation usually follows initiation of secondary xylem and phloem production and arises through division of pericycle cells. The outer sister cells of the dividing pericycle combine to form a complete cylinder of **cork cambium,** which produces **cork** toward its outer surface and **phelloderm** toward its inner surface. Collectively, these three tissues—cork, cork cambium, and phelloderm—make up the periderm. The periderm, or more specifically the cork, replaces the epidermis as the protective covering

on this portion of the root. The remaining pericycle cells may proliferate and give rise to tissue that resembles a cortex. Portions of the periderm may be differentiated as lenticels, which are spongy areas that permit the exchange of gases between the root and the soil atmosphere. (See page 527 for further discussion of lenticels.)

With the formation of the first periderm in the root, the cortex (including the endodermis) and the epidermis are isolated from the rest of the root. Being separated from the supply of water and minerals by the largely impermeable barrier of cork, the cortex and epidermis eventually die and are sloughed off (Figures 22–16 and 22–17).

At the end of the first year's growth, the following tissues are present in a woody root (from outside to inside): possibly remnants of the epidermis and cortex, periderm, pericycle, primary phloem (fibers and crushed soft-walled cells), secondary phloem, vascular cambium, secondary xylem, and primary xylem.

CORTEX   EPIDERMIS

VASCULAR
CYLINDER   200 μm

(a)

PRIMARY   PRIMARY   CORTEX
XYLEM   PHLOEM

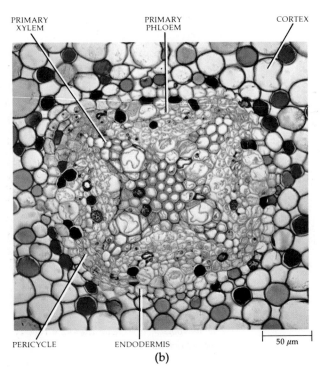

PERICYCLE   ENDODERMIS   50 μm

(b)

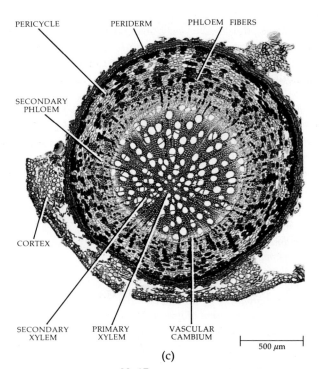

PERICYCLE   PERIDERM   PHLOEM FIBERS

SECONDARY
PHLOEM

CORTEX

SECONDARY   PRIMARY   VASCULAR
XYLEM   XYLEM   CAMBIUM   500 μm

(c)

22–17
Transverse sections of the willow (Salix) root, which becomes woody. (a) Overall view of root near completion of primary growth. (b) Detail of primary vascular cylinder. (c) Overall view of root at end of first year's growth, showing effect of secondary growth on primary plant body.

22–18
Three stages in the origin of lateral roots in a willow (Salix). *(a) One root primordium is present (below) and two others are being initiated in the region of the pericycle (arrows). The vascular cylinder is still very young. (b) Two root primordia penetrating the cortex. (c) One lateral root has reached the outside, and the other is about to break through.*

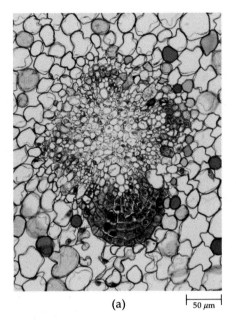

(a) $\vdash\!\!\dashv$ 50 μm

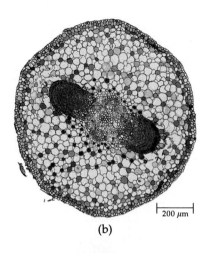

(b) $\vdash\!\!\dashv$ 200 μm

Origin of Lateral Roots

In most seed plants, lateral roots (branch roots) arise in the pericycle. Because lateral roots originate deep from within the parent root, they are said to be endogenous, meaning "originating within" (Figures 22–3 and 22–18).

Divisions in the pericycle that initiate lateral roots occur some distance beyond the region of elongation in partially or fully differentiated root tissues. In angiosperm roots, derivatives of both the pericycle and the endodermis commonly contribute to the new root primordium, although in many cases the derivatives of the endodermis are short-lived. As the young lateral root, or **root primordium,** increases in size, it pushes its way through the cortex (Figure 22–18), possibly secreting enzymes that digest some of the cortical cells lying in its path. While still very young, the root primordium develops a root cap and apical meristem, and the primary meristems appear. Initially, the vascular cylinders of lateral root and parent root are not connected to one another. The two vascular cylinders are joined later, when derivatives of intervening parenchyma cells differentiate into xylem and phloem.

(c) $\vdash\!\!\dashv$ 250 μm

Aerial Roots and Air Roots

Aerial roots are adventitious roots produced from aboveground structures. The aerial roots of some plants serve as **prop roots** for support, as in corn (Figure 22–19). When they come in contact with the soil, they branch and function also in the absorption of water and minerals. Prop roots are produced from the stems and branches of many tropical trees, such as the red mangrove (*Rhizophora mangle*), the banyan tree (*Ficus benghalensis*), and some palms. Other aerial roots, as in the ivy (*Hedera helix*), cling to the surface of objects such as walls and provide support for the climbing stem.

22–19
Prop roots of corn (Zea mays), *a type of adventitious root.*

22–20
Pneumatophores (air roots) of the white
mangrove (Laguncularia racemosa)
protruding from the mud near the base
of a tree.

(a)

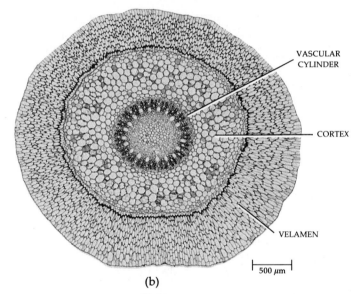

VASCULAR
CYLINDER

CORTEX

VELAMEN

500 μm

(b)

22–21
(a) *Aerial roots of an epiphytic orchid*
(Oncidium sphacelatum). (b) *Transverse*
section of an orchid root, showing the
multiple epidermis, or velamen.

Roots require oxygen for respiration, which is why most plants cannot live in soil that is inadequately drained and consequently lacks air spaces. Some trees that grow in swampy habitats develop roots, portions of which grow out of the water. Hence, the roots of such trees serve not only to anchor but also to aerate the plant. For example, the root system of the black mangrove (*Avicennia germinans*) develops negatively gravitropic extensions called **pneumatophores** (air roots), which grow upward out of the mud and so provide adequate aeration (Figure 22–20). A similar function has been attributed to the "knees" of the bald cypress (*Taxodium distichum*; see Figure 17–27), but this idea is now in doubt.

SPECIAL ADAPTATIONS

Many special adaptations of roots are found among epiphytes—plants that grow on other plants but are not parasitic on them. The epidermis of the orchid root, for example, is several layers thick (Figure 22–21) and, in some species, is the only photosynthetic organ of the plant. This multiple epidermis, called velamen, provides mechanical protection for the cortex, as well as reducing water loss. The velamen also may function in the absorption of water.

Among epiphytes, *Dischidia rafflesiana*, the "flower pot plant," has a very unusual modification. Some of its leaves are flattened, succulent structures, but others form hollow containers—the "flower pots"—that collect debris and rainwater (Figure 22–22). Ant colonies live in the pots and add to the nitrogen supply of the plant. Roots, formed at the node above the modified leaf, grow downward and into the pot, from which they absorb water and minerals.

22-22
The epiphyte Dischidia rafflesiana,
*or "flower pot plant." (a) A modified
leaf, or "pot," which collects debris and
rainwater. (b) Modified leaf cut open to
show roots that have grown down into it.*

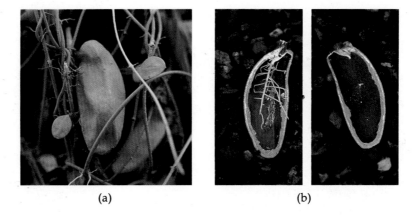

(a)   (b)

Adaptations for Food Storage

Most roots are storage organs, and in some plants the
roots are specialized for this function. Such roots are
fleshy because of an abundance of storage parenchyma, which is permeated by vascular tissue. The development of some storage roots, such as that of the
carrot *(Daucus carota)*, is essentially similar to that of
nonfleshy roots, except for a predominance of parenchyma cells in the secondary xylem and phloem of the
storage roots. The root of the sweet potato *(Ipomoea batatas)* develops in a manner similar to that of the carrot;
however, in the sweet potato, additional vascular cambium cells develop within the secondary xylem around
individual vessels or groups of vessels (Figure 22–23).
These additional cambia (plural of cambium), while
producing a few tracheary elements toward the vessels
and a few sieve tubes away from them, mainly produce

22-23
*Transverse sections of the root of a sweet
potato* (Ipomoea batatas). *(a) Overall
view. (b) Detail of xylem, showing
cambium around vessels.*

PERIDERM   VASCULAR
             CAMBIUM

PHLOEM

XYLEM

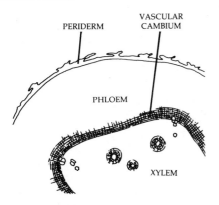

(a)   500 µm

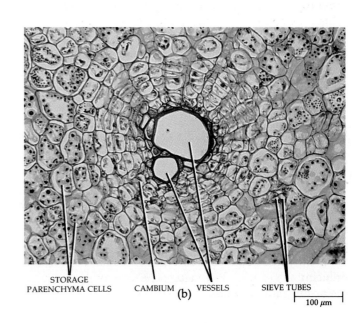

STORAGE
PARENCHYMA CELLS   CAMBIUM   VESSELS   SIEVE TUBES
                              (b)
                                                    100 µm

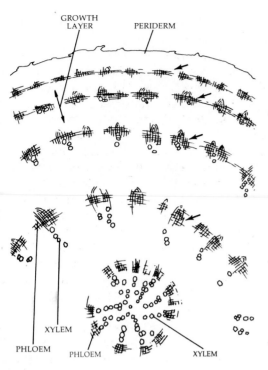

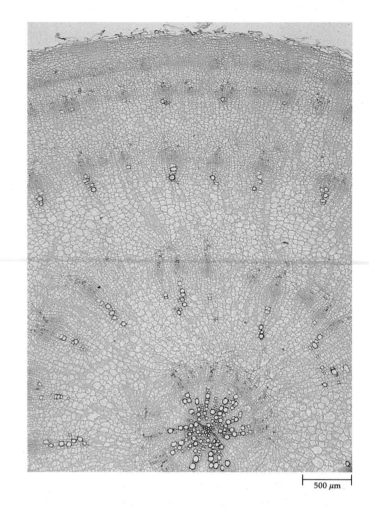

500 μm

22–24

Transverse section of a sugar beet (Beta vulgaris) root, with supernumerary cambia indicated by arrows on the diagram. The original vascular cambium produces relatively little secondary xylem and secondary phloem (in the center of the root).

storage parenchyma cells in both directions. In the sugar beet *(Beta vulgaris)*, for example, most of the increase in thickness of the root results from the development of extra cambia (supernumerary cambia) around the original vascular cambium (Figure 22–24). These concentric layers of cambia, which superficially resemble growth rings in woody roots and stems, produce parenchyma-dominated xylem toward the inside and phloem toward the outside. The upper portion of most fleshy roots actually develops from the hypocotyl.

22–25

Summary of root development in a woody dicotyledon during the first year of growth.

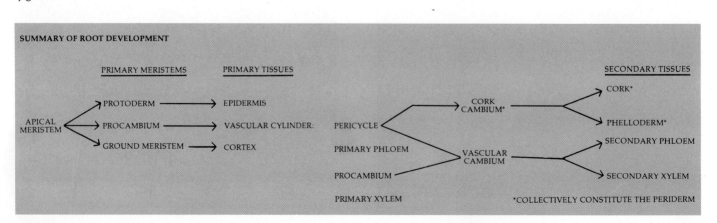

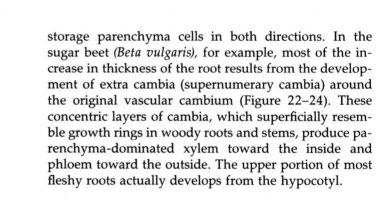

Summary

Roots are organs specialized for anchorage, absorption, storage, and conduction. Gymnosperms and dicots commonly produce taproot systems, whereas monocots usually produce fibrous root systems. The extent of the root system is dependent upon several factors, but the bulk of most feeder roots is found in the upper meter of the soil.

The apical meristems of most roots contain a quiescent center; most meristematic activity, or cell division, occurs a short distance from the apical initials. In addition to this region of cell division, two other growth regions—the region of elongation and the region of maturation—can be recognized in root tips. During primary growth, the apical meristem gives rise to the three primary meristems—protoderm, ground meristem, and procambium—which differentiate into epidermis, cortex, and vascular cylinder, respectively. In addition, the apical meristem produces the root cap, which serves to protect the meristem and aid the root in its penetration of the soil.

Many epidermal cells of the root develop root hairs, which greatly increase the absorbing surface of the root. With the exception of the endodermis, the cortex contains numerous intercellular spaces. The compactly arranged cells of the endodermis, which forms the inner boundary of the cortex, contain Casparian strips on their anticlinal walls. Consequently, all substances moving between the cortex and vascular cylinder must pass through the protoplasts of the endodermal cells. The roots of many angiosperms also have an exodermis, which forms the outer boundary of the cortex and also consists of a compact layer of cells with Casparian strips.

The vascular cylinder consists of pericycle and the primary vascular tissues, which are completely surrounded by the pericycle. The primary xylem usually occupies the center of the vascular cylinder and has radiating ridges that alternate with strands of primary phloem.

Secondary growth results in disruption of the primary body of the root as the strands of primary phloem are separated from the primary xylem through the formation of secondary xylem and secondary phloem by the vascular cambium. In roots, the vascular cambium arises partly from procambium that remains undifferentiated between the primary xylem and the primary phloem strands and partly from pericycle opposite the ridges of primary xylem. In most woody roots, the cork cambium of the first periderm originates from the pericycle. Consequently, formation of the periderm results in isolation and eventual separation of the cortex and epidermis from the rest of the root. Figure 22–25 presents a summary of root development of a woody dicotyledon, beginning with the apical meristem and ending with the secondary tissues produced during the first year's growth.

Branch roots originate in the pericycle and push their way to the outside through the cortex and epidermis.

Most roots are storage organs, and in some plants, such as the carrot, sweet potato, and sugar beet, the roots are specialized for this function. The fleshy roots have an abundance of storage parenchyma permeated by vascular tissue.

Suggested readings for this chapter appear at the end of Chapter 24.

CHAPTER 23

The Shoot: Primary Structure and Development

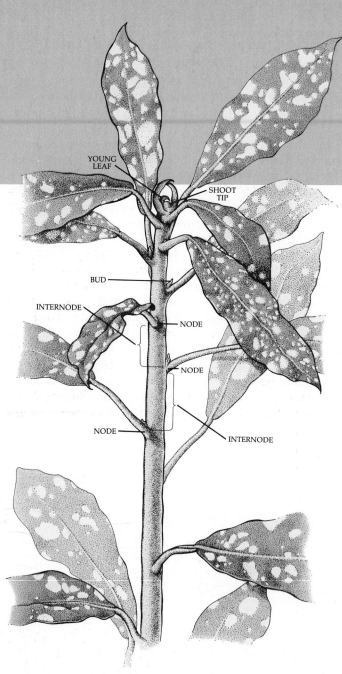

YOUNG
LEAF

SHOOT
TIP

BUD

INTERNODE

NODE

NODE

NODE

INTERNODE

23–1

A portion of a Croton *shoot. The leaves of* Croton, *a dicot, have a mottled appearance and are spirally arranged along the stem. At the apex the leaves are so close together that nodes and internodes are not distinguishable as separate regions of the stem. Growth in length of the stem between successive leaves, which are attached to the stem at the nodes, results in formation of the internodes.*

The shoot, which consists of the stem and its leaves, is initiated during development of the embryo, where it may be represented by a plumule, consisting of a stem (the epicotyl), one or more **leaf primordia** (rudimentary leaves), and an apical meristem. (The plumule can be thought of as the first bud of the plant.) With resumption of growth of the embryo during germination of the seed, new leaves develop from the apical meristem, and the stem elongates and differentiates into nodes and internodes. Gradually, **bud primordia** form in the axils of the leaves (Figures 23–1 and 23–2), and eventually they follow a sequence of growth and differentiation more or less similar to that of the first bud. This pattern is repeated many times as the shoot system of the plant is produced.

Commonly, the growing terminal bud of a shoot inhibits development of lateral buds, a phenomenon known as **apical dominance** (see Chapter 25). As the distance between shoot tip and lateral buds increases, the retarding influence of the terminal bud is lessened and the lateral buds can proceed with their development. (This is why pinching off the shoot tips, a common practice of home gardeners, results in fuller and bushier plants.)

The two principal functions associated with stems are conduction and support. Substances manufactured in the leaves are transported through the stems by way of the phloem to sites of utilization, including growing leaves, stems, and roots, as well as developing flowers, seeds, and fruits. Much of the food material is stored in parenchyma cells of roots, seeds, and fruits, but stems are also important storage organs, and some, such as the underground stems of the white potato, are specifically adapted for storage. The principal photosynthetic organs of the plant—the leaves—are supported by stems, which place the leaves in favorable positions for exposure to light. In addition, most of the plant's loss of

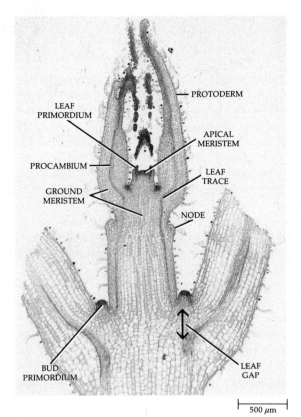

23–2
Longitudinal section of shoot tip of the common houseplant Coleus blumei, *a dicot. The leaves in* Coleus *are arranged opposite one another at the nodes, each successive pair at right angles to the previous pair; thus the leaves of the labeled node are at right angles to the plane of section. (See page 496 for discussion of leaf traces and leaf gaps.)*

water vapor occurs through the leaves (see Chapter 28). Water is conducted upward in the xylem from the roots and into the leaves via the stem.

Origin and Growth of the Primary Tissues of the Stem

The organization of the apical meristem of the shoot is more complex than that of the root. In addition to adding cells to the primary plant body, the apical meristem of the shoot is involved in the formation of leaf primordia and of the bud primordia (Figure 23–2) that develop into lateral branches. The apical meristem of the shoot also differs in its lack of a protective covering comparable to the root cap.

The vegetative shoot apex of most flowering plants has what is termed a **tunica-corpus** type of organization (Figure 23–3). The two regions—tunica and corpus—are usually distinguished by the planes of cell division that occur in them. The tunica consists of the outermost layer or layers of cells, which divide in planes perpendicular to the surface of the meristem (anticlinal divisions) and contribute primarily to surface growth. The corpus consists of a body of cells that lie beneath the tunica layers. In the corpus, the cells divide in various planes and add bulk to the developing shoot. The corpus and each layer of tunica have their own initials.

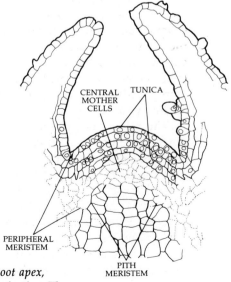

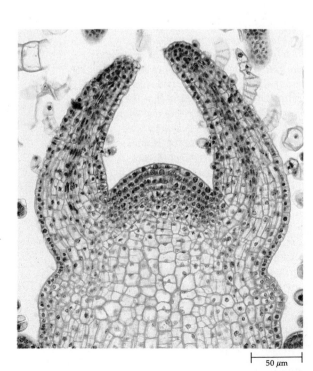

23–3
Detail of Coleus blumei *shoot apex, showing tunica-corpus organization. The zone of central mother cells roughly corresponds to the corpus.*

In the shoot apices of many angiosperms, the bulk of the corpus corresponds to an area of conspicuously vacuolated cells called the zone of **central mother cells** (Figure 23–3). This zone of vacuolated cells is surrounded by the **peripheral meristem,** which originates partly from the tunica and partly from the corpus, or central mother cell zone. Beneath the central mother cells is the **pith meristem.** Cell divisions are relatively infrequent in the central mother cell zone; by contrast, the peripheral zone is very active mitotically. The protoderm always originates from the outermost tunica layer, whereas the procambium and part of the ground meristem (the cortex and sometimes part of the pith) are derived from the peripheral meristem. The rest of the ground meristem (all or most of the pith) is formed by the pith meristem.

Although the primary tissues of the stem pass through periods of growth similar to those of the root, the stem cannot be divided along its axis into regions of cell division, elongation, and maturation as in the case of roots. When actively growing, the apical meristem of the shoot gives rise to leaf primordia in such rapid succession that nodes and internodes cannot at first be distinguished. Gradually, growth begins to occur between the levels of leaf attachment; the elongated parts of the stem take on the appearance of internodes; and the portions of the stem at which the leaves are attached become recognizable as nodes (Figure 23–4). Thus, increase in length of the stem occurs largely by internodal elongation.

Commonly, the meristematic activity causing the elongation of the internode is more intense at the base of the developing internodes than elsewhere. If elongation of the internode takes place over a prolonged period, the meristematic region at the base of the internode may be called an **intercalary meristem** (a meristematic region between two more highly differentiated regions). Certain elements of the primary xylem and primary phloem—specifically the protoxylem and protophloem—differentiate within the intercalary meristem and connect the more highly differentiated regions of the stem above and below the meristem.

23–4

Stages in growth of the terminal and lateral buds of the green ash (Fraxinus pennsylvanica). (a) *The young shoots are tightly packed in the buds and are protected by bud scales.* (b) *The terminal bud has opened just enough to reveal one of the rudimentary leaves.* (c) *With expansion of the bud, the scales fold back. Expansion of the terminal shoot is well underway. The pairs of pinnately compound leaves are still close to one another because the internodes have undergone little elongation.* (d) *Internodal elongation is well underway in both the terminal shoot and the lateral shoot (below and to the left). Note that the leaves are arranged opposite one another at the nodes but that each successive pair is at right angles to the previous pair as in the* Coleus *shoot (Figure 23–2).* (e) *Continued internodal elongation has further increased the length of both the terminal and lateral shoots.*

(a)　　(b)　　(c)

(d)　　(e)

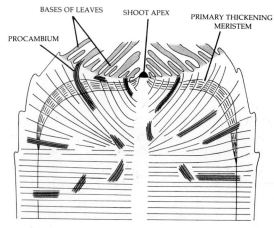

BASES OF LEAVES SHOOT APEX PRIMARY THICKENING MERISTEM

PROCAMBIUM

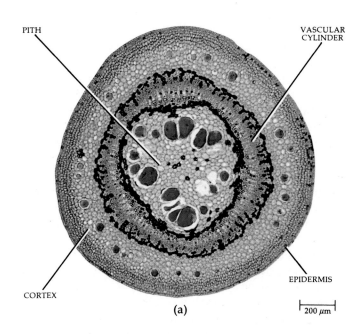

PITH

VASCULAR CYLINDER

CORTEX

EPIDERMIS

(a)

200 μm

23–5
Diagrammatic representation of the anatomy of the top, or crown, of a thick-stemmed monocot without secondary growth, such as a palm tree. Increase in thickness of such stems is due to the activity of a primary thickening meristem. The apical meristem and youngest leaf primordia are sunken below surrounding stem tissues. At the top of the stem, files of cells are arranged like a tilted stack of saucers.

Growth in thickness of the primary body of the stem involves longitudinal divisions and cell enlargement in both cortex and pith. In plants with secondary growth, this primary thickening is moderate. Monocots usually lack secondary growth, but many of them, such as the palms, have massive primary growth. This growth usually occurs close to the apical meristem (which often appears to lie in a depression at the stem apex) and is localized in a relatively narrow region near the periphery of the stem called the *primary thickening meristem* (Figure 23–5).

As in the root, the apical meristem of the shoot gives rise to the primary meristems—protoderm, ground meristem, and procambium (Figure 23–2). These primary meristems in turn develop into the epidermis, ground tissue, and primary vascular tissues, respectively.

Primary Structure of the Stem

Considerable variation exists in the primary structure of stems of seed plants, but three basic types of organization can be recognized: (1) In some conifers and dicots, the narrow, elongated procambial cells—and consequently the primary vascular tissues that develop from them—appear as a more or less continuous hollow cylinder within the ground tissue (Figure 23–6). The outer region of ground tissue is called the **cortex,** and the inner region is called the **pith.** (2) In other conifers and dicots, the primary vascular tissues develop as a cylin-

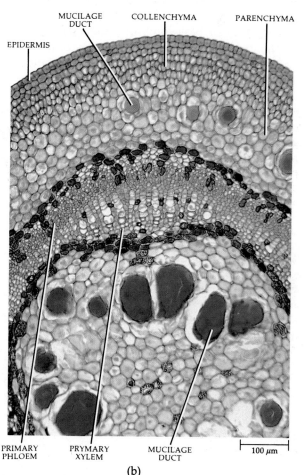

EPIDERMIS

MUCILAGE DUCT COLLENCHYMA PARENCHYMA

PRIMARY PHLOEM PRYMARY XYLEM MUCILAGE DUCT

100 μm

(b)

23–6
(a) *Transverse section of basswood (*Tilia americana*) stem in a primary stage of growth. The vascular tissues appear as a continuous hollow cylinder that divides the ground tissue into pith and cortex.*
(b) *Detail of a portion of the same basswood stem.*

der of discrete strands separated from one another by ground tissue (see Figure 23–8, page 494). The ground tissue separating the procambial strands (and later the mature vascular bundles) is continuous with the cortex and pith, and is called the **interfascicular parenchyma.** (Interfascicular means "between the bundles.") The interfascicular regions are often called **pith rays.** Narrow interfascicular regions also interconnect the cortex and pith in the first type of organization, but there such regions are inconspicuous. (3) In the stems of most monocotyledons and of some herbaceous dicotyledons, the arrangement of the procambial strands and vascular bundles is more complex. The vascular tissues do not appear as a single ring of bundles between a cortex and a pith. Instead, the vascular tissues commonly develop in more than one ring of bundles or as a system of strands scattered throughout the ground tissue. In the latter instance, the ground tissue often cannot be distinguished as cortex and pith (see Figure 23–10, page 495).

The stem of basswood, or linden (*Tilia americana*), will be used to exemplify the first type of organization. The second type of organization will be exemplified by the elderberry (*Sambucus canadensis*), alfalfa (*Medicago sativa*), and buttercup (*Ranunculus*) stems, and the third type by the stem of corn, or maize (*Zea mays*). The basswood and elderberry stems are also examples of stems that undergo much secondary growth. (They will be revisited during our discussion on secondary growth, in Chapter 24.) By contrast, the alfalfa stem undergoes relatively little secondary growth, and the stems of buttercup (a dicot) and corn (a monocot) none at all.

THE *TILIA* STEM

Figure 23–6 shows the *Tilia* stem, with its continuous hollow cylinder of primary vascular tissues. As in most stems, the epidermis is a single layer of cells covered by a cuticle. The stem epidermis generally contains far fewer stomata than the leaf epidermis.

The cortex of the *Tilia* stem consists of collenchyma and parenchyma cells. The several layers of collenchyma cells, which provide support to the young stem, form a continuous cylinder beneath the epidermis. The rest of the cortex consists of parenchyma cells that will contain chloroplasts when mature. The innermost layer of cortical cells, which have deeply colored contents, sharply delimits the cortex from the cylinder of primary vascular tissues.

In the great majority of stems, including those of *Tilia*, the primary phloem develops from the outer cells of the procambium, and the primary xylem develops from the inner ones. However, not all of the procambial cells differentiate into primary xylem and primary phloem. A single layer of cells between the primary

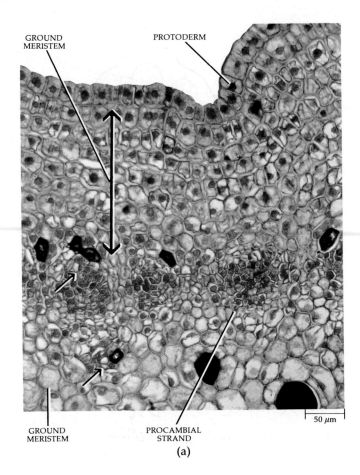

GROUND MERISTEM   PROTODERM

GROUND MERISTEM   PROCAMBIAL STRAND

50 µm

(a)

xylem and the primary phloem remains meristematic and becomes the vascular cambium. *Tilia* is also an example of a woody stem—a stem that produces much secondary xylem. After internodal elongation is completed in the *Tilia* stem, fibers develop in the primary phloem. These fibers are called **primary phloem fibers** (see Figure 24–9, page 525).

The inner boundary of the primary xylem in *Tilia* is sharply delimited by one or two layers of pith cells that have deeply colored contents. The pith is composed primarily of parenchyma cells and contains numerous large ducts, or canals, containing mucilage (a slimy carbohydrate). Similar ducts are formed in the cortex (Figure 23–6). As the cortical and pith cells increase in size, numerous intercellular spaces develop among them; these air spaces are essential for interchange of gases with the atmosphere. The cortical and pith parenchyma cells store various substances.

THE *SAMBUCUS* STEM

In the stem of *Sambucus*, the procambial strands and primary vascular bundles form a system of discrete strands around the pith. The epidermis, cortex, and pith are essentially similar in organization to those of *Tilia*, so the following discussion of the *Sambucus* stem will be used to explain in more detail the development of the primary vascular tissues of stems.

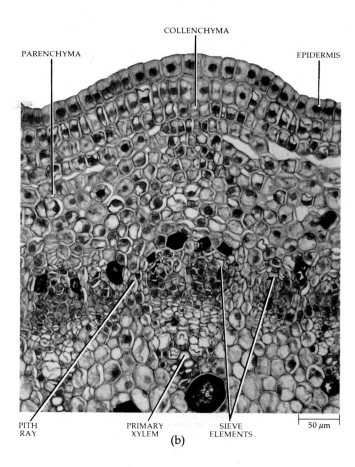

PARENCHYMA   COLLENCHYMA   EPIDERMIS

PITH RAY   PRIMARY XYLEM   SIEVE ELEMENTS   50 μm

(b)

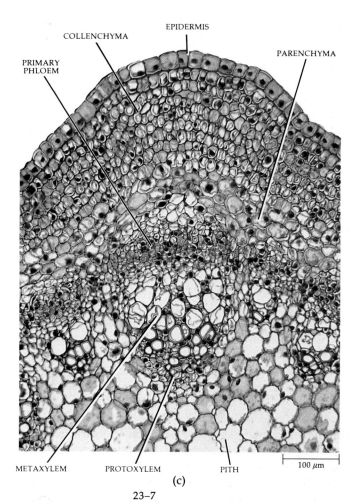

COLLENCHYMA   EPIDERMIS

PRIMARY PHLOEM   PARENCHYMA

METAXYLEM   PROTOXYLEM   PITH   100 μm

(c)

23–7

Transverse sections of the stem of the elderberry (Sambucus canadensis) in a primary stage of growth. (a) A very young stem, showing protoderm, ground meristem, and discrete procambial strands. The procambial strand on the left contains one mature sieve element (upper arrow) and one mature tracheary element (lower arrow). (b) Primary tissues further along in development. (c) Stem near completion of primary growth. Fascicular and interfascicular cambia are not yet formed. (For further stages in the growth of the elderberry stem, see Figures 24–7, 24–8, and 24–10.)

Figure 23–7a shows three procambial strands in which the primary vascular tissues have just begun to differentiate. The strand on the left is somewhat older than the two on the right and contains at least one mature sieve element and one mature tracheary element. Notice that the first mature sieve element appears in the outer part of the procambial strand (next to the cortex) and that the first mature tracheary element appears in the inner part (next to the pith). Comparing Figures 23–7a and 23–7c, we see that the more recently formed sieve elements appear closer to the center of the stem and that the xylem differentiates in the opposite direction.

The first-formed primary xylem and primary phloem elements (protoxylem and protophloem, respectively) are stretched during elongation of the internode and are frequently destroyed. As in the *Tilia* stem, fibers develop in the primary phloem after internodal elongation is completed (see Figure 24–8).

Like the stems of *Tilia*, those of *Sambucus* become woody. In *Tilia*, almost all of the vascular cambium originates from procambial cells between the primary xylem and primary phloem because the interfascicular regions are very narrow. In *Sambucus*, the interfascicular regions are relatively wide. Consequently, a substantial portion of the vascular cambium in *Sambucus* develops from the interfascicular parenchyma between the bundles.

THE *MEDICAGO* AND *RANUNCULUS* STEMS

The stems of many dicots undergo little or no secondary growth and therefore are **herbaceous,** or nonwoody (see Chapter 24). Examples of herbaceous dicot stems can be found in alfalfa *(Medicago sativa)* and in the buttercups *(Ranunculus).*

Medicago is an example of an herbaceous dicot that exhibits some secondary growth (Figure 23–8). The structure and development of the primary tissues of the *Medicago* stem are essentially similar in organization to those of *Sambucus* and other woody dicots. The vascular bundles are separated by wide interfascicular regions and surround a large pith. The vascular cambium is partly fascicular (procambial) and partly interfascicular (interfascicular parenchyma) in origin, but secondary vascular tissues are formed mainly in the vascular bundles. The interfascicular cambium generally produces only sclerenchyma cells on the xylem side.

The herbaceous stem of *Ranunculus* is an extreme example, and its vascular bundles resemble those of many monocots. The vascular bundles retain no procambium after the primary vascular tissues mature; hence, the bundles never develop a vascular cambium and lose their potential for further growth. Vascular bundles such as those of *Ranunculus* (Figure 23–9) and the monocots, in which all the procambial cells mature and the potential for further growth within the bundle is lost, are said to be "closed." Vascular bundles that do give rise to a cambium are said to be "open." In most dicots, the vascular bundles are of the open type; they produce some secondary vascular tissues.

THE *ZEA* STEM

The herbaceous stem of corn *(Zea mays)* exemplifies the stems of monocots in which the vascular bundles form a system of strands scattered throughout the ground tissue (Figure 23–10). As in other monocots, the vascular bundles of corn are closed.

Figure 23–11 shows three stages in the development of a corn vascular bundle. As in the bundles of dicot stems, the phloem develops from the outer cells of the procambial strand, and the xylem develops from the inner cells. Also, as described previously, the phloem and the xylem differentiate in opposite directions. The first-formed phloem and xylem elements (protophloem and protoxylem) are stretched and destroyed during elongation of the internode. This results in the formation of a very large air space on the xylem side of the bundle (Figure 23–11c). The mature vascular bundle contains two large vessel members (the metaxylem vessels), and the phloem (metaphloem) tissue is composed of sieve-tube members and companion cells. The entire bundle is enclosed in a sheath of sclerenchyma cells.

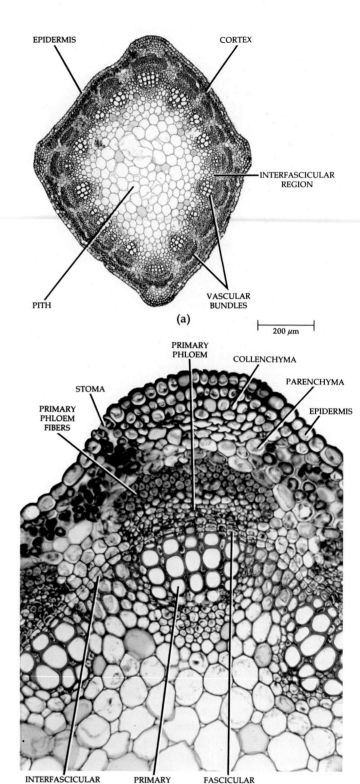

(a)

200 μm

(b)

23–8

(a) *Transverse section of stem of alfalfa* (Medicago sativa), *a dicot with discrete vascular bundles.* (b) *Detail of a portion of the same alfalfa stem.*

BUNDLE SHEATH PRIMARY PHLOEM

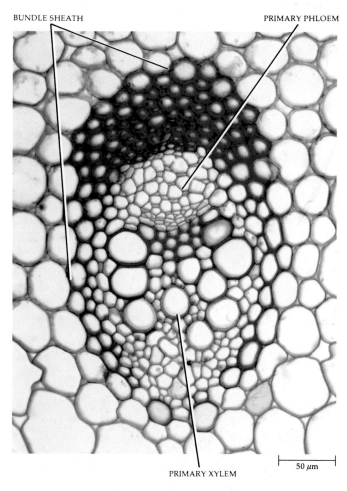

PRIMARY XYLEM

50 µm

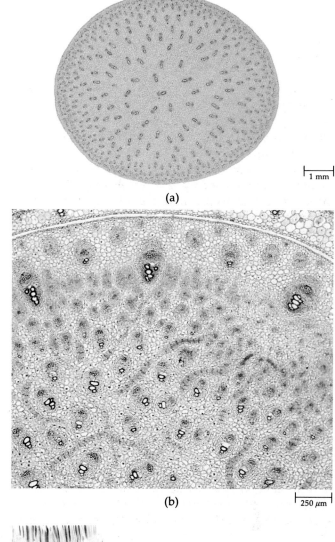

1 mm

(a)

(b)

250 µm

23–9

Transverse section of vascular bundle of the buttercup (Ranunculus), an herbaceous dicot. The vascular bundles of the buttercup are closed, that is, all of the procambial cells mature, precluding secondary growth. The primary phloem and primary xylem are surrounded by a bundle sheath of thick-walled sclerenchyma cells. Compare the vascular bundle shown here with the mature vascular bundle of corn shown in Figure 23–11c.

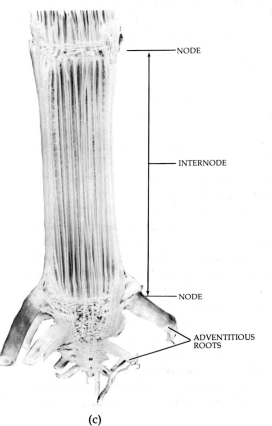

NODE

INTERNODE

NODE

ADVENTITIOUS ROOTS

23–10

Stem of corn (Zea mays). (a) Transverse section of the internodal region, showing numerous vascular bundles scattered throughout the ground tissue. (b) Transverse section of the nodal region of a young corn stem, showing horizontal procambial strands that interconnect with vertical bundles. (c) A mature stem split longitudinally; the ground tissue has been removed to expose the vascular system.

(c)

495

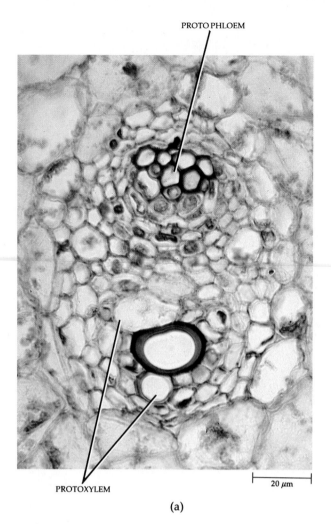

PROTOPHLOEM

PROTOXYLEM

20 μm

(a)

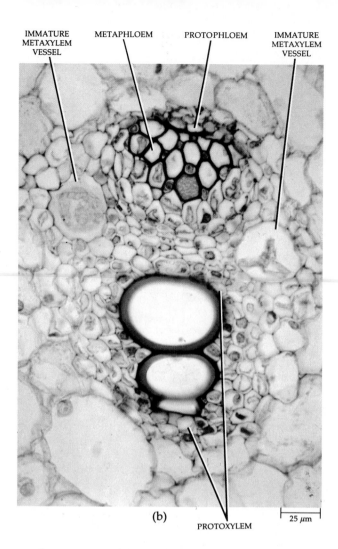

IMMATURE METAXYLEM VESSEL

METAPHLOEM

PROTOPHLOEM

IMMATURE METAXYLEM VESSEL

PROTOXYLEM

25 μm

(b)

23–11

Three stages in the differentiation of the vascular bundles of corn (Zea mays), as seen in transverse sections of the stem. (a) The protophloem elements and two protoxylem elements are mature. (b) The protophloem sieve elements are now crushed, and much of the metaphloem

is mature. Three protoxylem members are now mature, and the two metaxylem vessel members are almost fully expanded. (c) Mature vascular bundle surrounded by a sheath of thick-walled sclerenchyma cells. The metaphloem is composed entirely

of sieve-tube members and companion cells. The portion of the vascular bundle once occupied by the protoxylem elements is now a large air space. Note the wall thickenings of destroyed protoxylem elements bordering the air space.

Relation between the Vascular Tissues of the Stem and the Leaf

The pattern formed by the vascular bundles in the stem reflects the close structural and developmental relationship between the stem and its lateral appendages, the leaves. The term "shoot" serves not only as a collective term for these two vegetative organs but also as an expression of their intimate physical and developmental association.

The procambial strands of the stem arise behind the apical meristem just below the developing leaf primordia and sometimes are present below the sites of future leaf primordia even before the primordia are discernible. As the leaf primordia increase upward in length,

the procambial strands also differentiate upward within them. From its inception, the procambial system of the leaf is continuous with that of the stem.

At each node, one or more vascular bundles diverge from the cylinder of strands in the stem, cross the cortex, and enter the leaf or leaves attached at that node (Figures 23–12 and 23–13). The extensions from the vascular system in the stem toward the leaves are called **leaf traces,** and the wide gaps or regions of ground tissue in the vascular cylinder located above the level where leaf traces diverge toward the leaves are called **leaf gaps.** A leaf trace extends from its connection with a bundle in the stem—a **stem bundle**—to the level at which it enters the leaf. A single leaf may have one or more leaf traces connecting its vascular system with that in the stem. The number of internodes that leaf

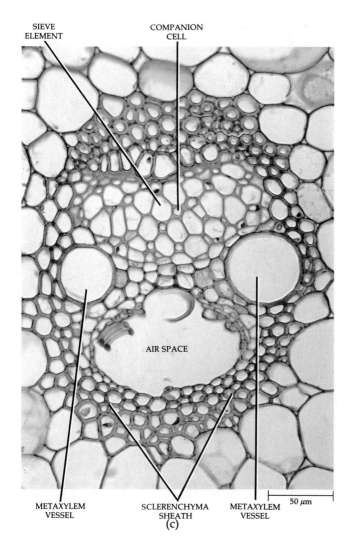

SIEVE ELEMENT COMPANION CELL

METAXYLEM VESSEL SCLERENCHYMA SHEATH METAXYLEM VESSEL

AIR SPACE

50 μm

(c)

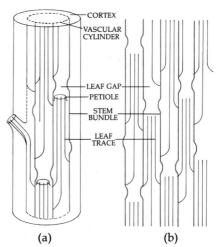

CORTEX
VASCULAR CYLINDER

LEAF GAP
PETIOLE
STEM BUNDLE
LEAF TRACE

(a)   (b)

23–12
Diagrams of the primary vascular system of a stem of the oak-leaved goosefoot (Chenopodium glaucum), *in which discrete bundles form a cylinder of interconnected strands around a pith.* (a) *A three-dimensional diagram showing the arrangement of the bundles within the stem.* (b) *Vascular system spread out in one plane. Notice the leaf traces diverging outward from the vascular cylinder and the relation of the traces to the vascular bundles of the stem. In C.* glaucum *each leaf has three leaf traces.*

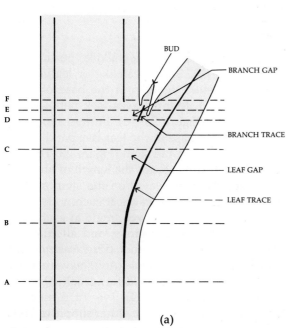

BUD
BRANCH GAP
BRANCH TRACE
LEAF GAP
LEAF TRACE

F
E
D
C
B
A

(a)

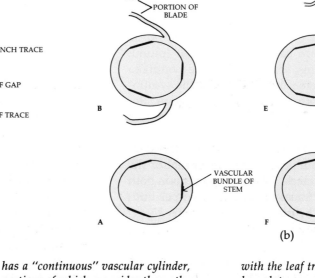

LEAF TRACE
LEAF GAP
PORTION OF BLADE

BRANCH TRACE
VASCULAR BUNDLE IN BASE OF LEAF

BRANCH TRACE
BRANCH GAP

VASCULAR BUNDLE OF STEM

BUD

C   D
B   E
A   F

(b)

23–13
Diagrams of longitudinal (a) *and transverse* (b) *sections of part of the stem of tobacco* (Nicotiana tabacum), *illustrating the relationship of the vascular systems of the leaf and the stem. The stem of tobacco* has a "continuous" vascular cylinder, portions of which are wider than others. At each node a single leaf trace diverges toward a leaf. Branch traces also occur at the nodes and are often closely associated with the leaf traces. In tobacco, two branch traces extend from the vascular system of the stem to the axillary bud, and the branch gap is continuous with the leaf gap.

497

23–14
Several examples of simple leaves. (a)
Mulberry (Morus alba). (b) Culver's-root
(Veronicastrum virginicum). (c) *Sugar*
maple (Acer saccharum). (d) Silver
maple (Acer saccharinum). (e) Red oak
(Quercus rubra). *Note the alternate*
arrangement of the leaves in mulberry,
and the whorled arrangement of those
in Culver's-root. Leaf arrangement in
the maples is distichous, and in oak it is
alternate, although only single leaves
of these trees are shown here.

traces traverse before they enter a leaf varies, so the traces vary in length.

If the stem bundles are followed either upward or downward in the stem, they will be found to be associated with several leaf traces. A stem bundle and its associated leaf traces are called a *sympodium* (plural: sympodia). In some stems, some or all of the sympodia are interconnected, whereas in others, all of the sympodia are independent units of the vascular system. Regardless, the pattern of the vascular system in the stem is a reflection of the arrangement of the leaves on the stem. Buds commonly develop in the axils of leaves, and their vascular system is connected with that of the main stem by **branch traces.** Hence, at each node both leaf traces and branch traces (commonly two per bud) diverge outward from the main stem (Figure 23–13).

Morphology of the Leaf

Leaves vary greatly in form and in internal structure. In dicots, the leaf commonly consists of an expanded por-

tion, the **blade,** or lamina, and a stalklike portion, the **petiole** (Figure 23–14). Small scalelike or leaflike appendages called *stipules* develop at the base of some leaves (Figure 23–15). Many leaves lack petioles and are said to be *sessile* (Figure 23–16). In most monocots and certain dicots, the base of the leaf is expanded into a **sheath,** which encircles the stem (Figure 23–16b). In some grasses, the sheath extends the length of an internode. The arrangement of leaves on the stem may be whorled, with three or more leaves at each node; distichous, or two-ranked, with leaves single at each node but disposed in two opposite ranks; and alternate, or helical. For example, Culver's-root (*Veronicastrum virginicum*) leaves are whorled, maple (*Acer*) leaves are distichous, and mulberry (*Morus alba*) and oak (*Quercus*) leaves are alternate (Figure 23–14).

The leaves of dicotyledons are either simple or compound. In simple leaves, the blades are not divided into distinct parts, although they may be deeply lobed (Figure 23–14). The blades of compound leaves are divided into leaflets, each usually with its own small petiole (which is called a petiolule). Two types of compound leaves can be distinguished: pinnately compound leaves and palmately compound leaves (Figure 23–17).

(a)

(b)

23–15
The pinnately compound leaf of the pea (Pisum sativum). *Notice the stipules at the base of the leaf and the slender tendrils at the tip of the leaf. In the pea leaf, the stipules are often larger than the leaflets.*

23–16
Sessile leaves (leaves without a petiole) are often found among dicots, such as Moricandia, *a member of the mustard family (a), but are particularly characteristic of grasses and other monocots. (b) In corn* (Zea mays), *a monocot, the base*

of the leaf forms a sheath around the stem. The ligule, a small flap of tissue extending upward from the sheath, is visible. The parallel arrangement of the longitudinal veins is conspicuous in the portion of the blade shown here.

23–17
Some examples of compound leaves. A palmately compound leaf is shown in (a); all the others are pinnately compound. (a) Red buckeye (Aesculus pavia). *(b)* Shagbark hickory (Carya ovata). *(c) Green ash* (Fraxinus pennsylvanica var. subintegerrima). *(d) Black locust* (Robinia pseudo-acacia). *(e) Honey locust* (Gleditsia triacanthos). *In the honey locust, each leaflet is subdivided into smaller leaflets.*

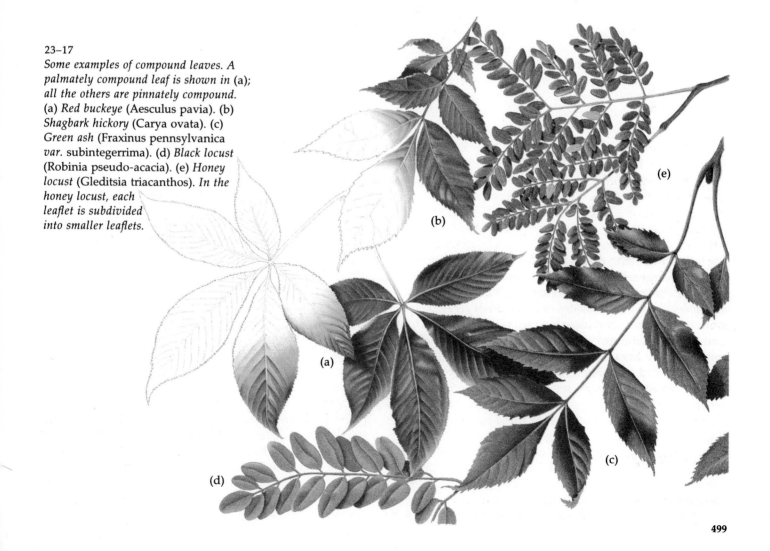

In pinnately compound leaves, the leaflets arise from either side of an axis, the **rachis,** like the pinnae of a feather. (The rachis is an extension of the petiole.) The leaflets of a palmately compound leaf diverge from the tip of the petiole, and a rachis is lacking.

Since leaflets are similar in appearance to simple leaves, it is sometimes difficult to determine whether the structure is a leaflet or a leaf. Two criteria may be used to distinguish leaflets from leaves: (1) buds are found in the axils of leaves—both simple and compound—but not in the axils of leaflets, and (2) leaves extend from the stem in various planes, whereas the leaflets of a given leaf all lie in the same plane.

Structure of the Leaf

Variations in the structure of angiosperm leaves are to a great extent related to the habitat, and the availability of water is an especially important factor affecting their form and structure. On the basis of their water requirements or adaptations, plants are commonly characterized as **mesophytes** (plants that require abundant soil water and a relatively humid atmosphere), **hydrophytes** (plants that depend on an abundant supply of moisture or grow wholly or partly submerged in water), and **xerophytes** (plants that are adapted to arid habitats). Such distinctions are not sharp, however, and leaves often exhibit a combination of features that are characteristic of different ecological types. Regardless of their varying forms, the foliage leaves of angiosperms are specialized as photosynthetic organs and, like roots and stems, consist of dermal, ground, and vascular tissue systems.

EPIDERMIS

The ground epidermal cells of the leaf, like those of the stem, are compactly arranged and covered with a cuticle that reduces water loss (see pages 53 and 467). Stomata may occur on both sides of the leaf but are usually more numerous on the lower surface (Figure 23–18). In leaves of hydrophytes that float on the surface of the water, stomata may occur in the upper epidermis only (Figure 23–19); the submerged leaves of hydrophytes usually lack stomata entirely. The leaves of xerophytes generally contain greater numbers of stomata than those of other plants. Presumably these numerous stomata permit a higher rate of gas exchange during the relatively rare periods of favorable water supply. In many xerophytes, the stomata are sunken in depressions on the lower surface of the leaf (Figures 23–20

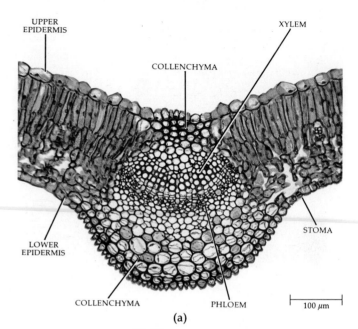

(a)

23–18
*Sections of lilac (Syringa vulgaris) leaf.
(a) A transverse section through a midrib showing the midvein. (b) A transverse section through a portion of the blade. Two small veins (minor veins) can be seen in this view. (c) This is a "paradermal section" of the leaf. Strictly speaking, a paradermal section is one cut parallel to the epidermis. In practice, such sections are more or less oblique and extend from the upper to the lower epidermis. Thus, part of the upper epidermis can be seen in the light area at the top of this micrograph, and part of the lower epidermis can be seen at the bottom. Notice the greater number of stomata in the lower epidermis, as evidenced by the number of medium-sized guard cells. (The darkly stained areas are trichomes.) The venation in lilac is netted. (d) and (e) Enlargements of portions of (c). Portions of palisade parenchyma (above) and spongy parenchyma (below) are shown in (d). A vein ending, sectioned through some tracheary elements and surrounded by a bundle sheath, can be seen at the upper right of this micrograph. A portion of the lower epidermis, with two trichomes (epidermal hairs) and several stomata, is shown in (e).*

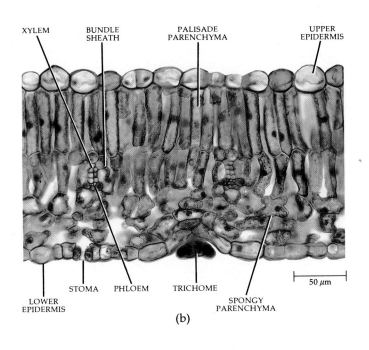

XYLEM BUNDLE SHEATH PALISADE PARENCHYMA UPPER EPIDERMIS

LOWER EPIDERMIS STOMA PHLOEM TRICHOME SPONGY PARENCHYMA

50 μm

(b)

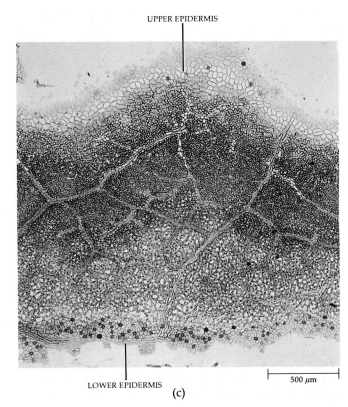

UPPER EPIDERMIS

LOWER EPIDERMIS (c)

500 μm

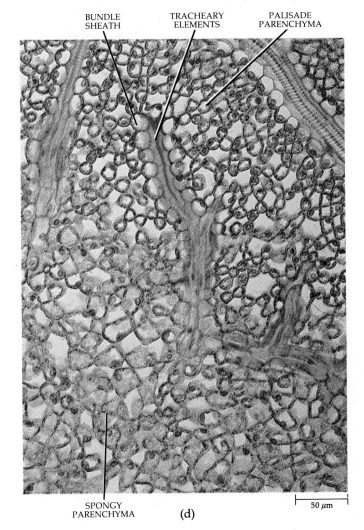

BUNDLE SHEATH TRACHEARY ELEMENTS PALISADE PARENCHYMA

SPONGY PARENCHYMA (d)

50 μm

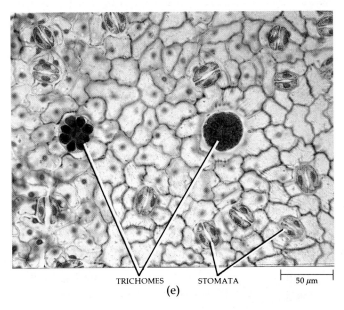

TRICHOMES STOMATA 50 μm

(e)

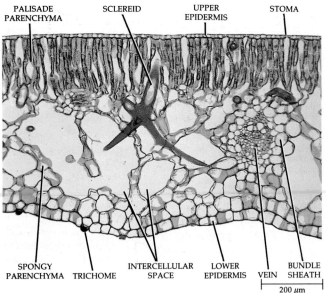

PALISADE PARENCHYMA   SCLEREID   UPPER EPIDERMIS   STOMA

SPONGY PARENCHYMA   TRICHOME   INTERCELLULAR SPACE   LOWER EPIDERMIS   VEIN   BUNDLE SHEATH

200 μm

23–19
Transverse section of the water lily
(Nymphaea odorata) *leaf, which floats
on the surface of the water and has
stomata in the upper epidermis only. As
is typical of hydrophytes, the vascular
tissue in the* Nymphaea *leaf is much
reduced, especially the xylem. The palisade parenchyma consists of several
layers of cells above the spongy parenchyma. Note the large intercellular (air)
spaces, which add buoyancy to this
floating leaf.*

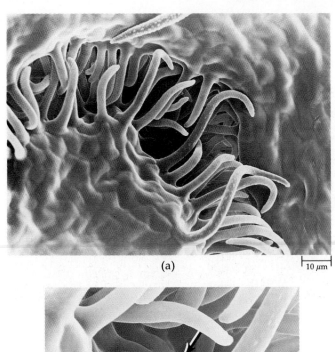

(a)

10 μm

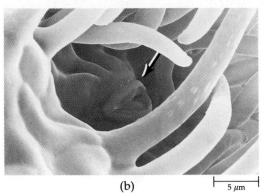

(b)

5 μm

23–21
*Scanning electron micrographs of a
stomatal crypt in the lower epidermis
of an oleander* (Nerium oleander) *leaf.*
(a) *Lower-magnification view of crypt,
showing numerous trichomes lining the
crypt.* (b) *Detail of* (a), *showing one of
the stomata (arrow), which are restricted
to the crypt.*

23–20
Transverse section of oleander (Nerium
oleander) *leaf. Oleander is a xerophyte,
and this is reflected in the structure of
the leaf. Note the very thick cuticle
covering the multiple (several-layered)
epidermis on the upper and lower surfaces
of the leaf. The stomata and trichomes
are restricted to invaginated portions
of the lower epidermis called stomatal
crypts.*

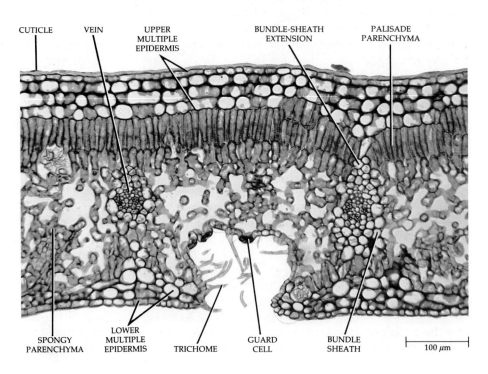

CUTICLE   VEIN   UPPER MULTIPLE EPIDERMIS   BUNDLE-SHEATH EXTENSION   PALISADE PARENCHYMA

SPONGY PARENCHYMA   LOWER MULTIPLE EPIDERMIS   TRICHOME   GUARD CELL   BUNDLE SHEATH

100 μm

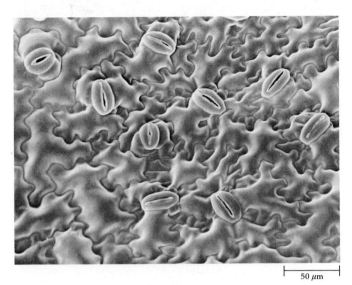

23–22
Scanning electron micrograph of a potato
(Solanum tuberosum) *leaf showing*
random arrangement of stomata typical
of the leaves of dicots. The guard cells
in potato are crescent-shaped and are
not associated with subsidiary cells.

50 µm

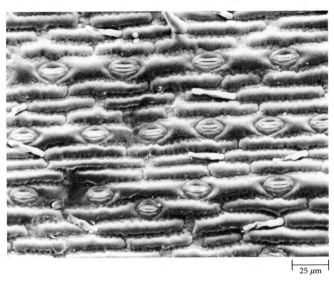

23–23
Scanning electron micrograph of a
corn (Zea mays) *leaf showing parallel*
arrangement of stomata typical of the
leaves of monocots. In corn each pair of
narrow guard cells is associated with two
subsidiary cells, one on each side of the
stoma. (See Figure 21–23, page 465.)

25 µm

and 23–21). The depressions may also contain many epidermal hairs. Together these two features may serve to reduce water loss from the leaf. Epidermal hairs, or trichomes, may occur on either or both surfaces of a leaf. Thick coats of epidermal hairs and the secreted resins of some trichomes may also retard water loss from leaves.

In the leaves of dicotyledons, the stomata are scattered and often appear to be randomly arranged (Figure 23–22); their development is mixed—that is, mature and immature stomata occur side by side in a partially developed leaf. In most monocotyledons, the stomata are arranged in rows parallel with the long axis of the leaf (Figure 23–23). Their formation begins at the tips of the leaves and progresses downward.

MESOPHYLL

It is the **mesophyll**—the ground tissue of the leaf—with its large volume of intercellular spaces and numerous chloroplasts, that is particularly specialized for photosynthesis. The intercellular spaces are connected with the outer atmosphere through the stomata, which facilitate rapid gas exchange, an important factor in

photosynthetic efficiency. In mesophytes, the mesophyll commonly is differentiated into **palisade parenchyma** and **spongy parenchyma**. The cells of the palisade tissue are columnar, with their long axes oriented at right angles to the epidermis, and the spongy parenchyma cells are irregular in shape (Figure 23–18b, d). Although the palisade parenchyma appears more compact than the spongy parenchyma, most of the vertical walls of the palisade cells are exposed to intercellular spaces, and in some leaves the palisade surface may be two to four times greater than the spongy surface. Chloroplasts are also more numerous in palisade cells than in spongy cells. Most of the photosynthesis in the leaf, therefore, apparently takes place within the palisade parenchyma.

The palisade parenchyma is usually located on the upper side of the leaf, and the spongy parenchyma is generally on the lower side (Figure 23–18). In certain plants, including many xerophytes, palisade parenchyma often occurs on both sides of the leaf. In some plants—for instance, corn (see Figure 7–20, page 113) and other grasses (see Figures 23–25 through 23–27 on page 505)—the mesophyll cells are of more or less similar shape, and a distinction between spongy and palisade parenchyma does not exist.

23–24
A cleared leaf (with the chlorophyll removed) of a cottonwood (Populus deltoides), at two successive magnifications. The cottonwood leaf has the netted venation characteristic of a dicot. The small areas of mesophyll delimited by veins are called areoles. No mesophyll cell of the leaf is far from a vein. Water and dissolved minerals are carried to the leaf through the xylem; organic molecules produced by photosynthesis in the leaf are carried out of the leaf through the phloem.

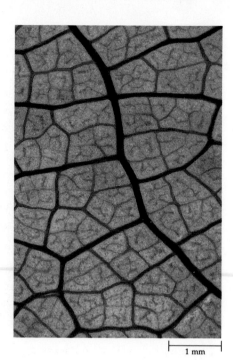

|— 10 mm —|

|— 1 mm —|

VASCULAR BUNDLES

The mesophyll of the leaf is thoroughly permeated by numerous vascular bundles, or **veins,** which are continuous with the vascular system of the stem. In most dicots, the veins are arranged in a branching pattern, with successively smaller veins branching from somewhat larger ones. This type of vein arrangement is known as **netted venation,** or reticulate venation (Figure 23–24). Often the largest vein extends along the long axis of the leaf as a midvein, which, with its associated ground tissue, makes up the so-called midrib of such leaves (Figure 23–18a). By contrast, most monocot leaves have many veins of fairly similar size that extend along the long axis of the leaf. This vein arrangement is called **parallel venation,** or striate venation (Figure 23–16b). In parallel-veined leaves, the longitudinal veins are interconnected by smaller veins, forming a complex network.

The veins contain xylem and phloem, which generally are entirely primary in origin. (The midvein and sometimes the coarser veins undergo secondary growth in some dicot leaves.) The vein endings in dicot leaves often contain only tracheary elements, although both xylem and phloem elements may extend to the ends of the vein. Commonly, the xylem occurs on the upper side of the vein, and the phloem occurs on the lower side (Figure 23–18a, b).

The small veins of the leaf that are more or less completely embedded in mesophyll tissue are called **minor veins,** while the large veins associated with ribs (protrusions on the underside of the leaf) are referred to as **major veins.** It is the minor veins that play the principal role in the collection of photosynthates from the mesophyll cells. With increasing vein size, the veins become less closely associated spatially with the meso-

phyll and increasingly embedded in nonphotosynthetic rib tissues. Hence, as the veins increase in size, their primary function changes from collection of photosynthates to transport of photosynthates out of the leaf.

The vascular tissues of the veins are rarely exposed to the intercellular spaces of the mesophyll. The large veins are surrounded by parenchyma cells that contain few chloroplasts, whereas the small veins usually are enclosed by one or more layers of compactly arranged cells that form a **bundle sheath** (Figures 23–18b, d through 23–20). The cells of the bundle sheath often resemble the mesophyll cells in which the small veins are located. The bundle sheaths extend to the ends of the veins, assuring that no part of the vascular tissue is exposed to air in the intercellular spaces and that all substances entering and leaving the vascular tissues must pass through the sheath (Figure 23–18d). Thus the bundle sheath performs a function analogous to that of the endodermis of the root.

In many leaves, the bundle sheaths are connected with either or both upper and lower epidermis by cells resembling the sheath cells (Figure 23–20). Such connections (actually, extensions of the sheaths) are called **bundle-sheath extensions.** Besides offering mechanical support to the leaf, in dicotyledons they apparently conduct water from the xylem to the epidermis.

The epidermis itself provides considerable strength to the leaf because of its compact structure and its cuticle. In addition, the larger veins of dicot leaves are often bordered by collenchyma or sclerenchyma cells, which provide support to the leaf. In monocot leaves, the veins may be bordered by fibers. Collenchyma cells and fibers may also be found along the leaf margins of dicot and monocot leaves, respectively.

Grass Leaves

After the discovery of the C_4 pathway of photosynthesis in sugarcane (see page 111), a great many studies were devoted to the comparative anatomy of grass leaves in relation to photosynthetic pathways. It was discovered that the leaves of C_3 and C_4 grasses have rather consistent anatomical differences. For example, in the leaves of C_4 grasses, the mesophyll cells and bundle-sheath cells typically form two concentric layers around the vascular bundles, as seen in transverse sections (Figure 23–25). The compactly arranged bundle-sheath cells of the C_4 grasses are very large parenchyma cells that contain many large, conspicuous chloroplasts. This concentric arrangement of the mesophyll and bundle-sheath layers in C_4 plants is referred to as *Kranz* (German for "wreath") *anatomy*.

In the leaves of C_3 grasses, by contrast, the mesophyll cells and bundle-sheath cells are not concentrically arranged. Moreover, the relatively small cells of the parenchymatous bundle sheaths in these plants have rather small chloroplasts, and at low magnifications the cells appear empty and clear. Commonly, an inner, more or less thick-walled sheath (the so-called mestome sheath) is also present in the C_3 grasses (Figure 23–26).

Another consistent structural difference between the leaves of C_3 and C_4 grasses is their interveinal distances, that is, the distances between laterally adjacent bundle sheaths. In C_4 grasses, only two to four mesophyll cells intervene between laterally adjacent bundle sheaths; in C_3 grasses, more than four (an average of 12 for the C_3 species in one study) mesophyll cells intervene between adjacent bundle sheaths.

The leaves of C_4 plants generally export photosynthates both more rapidly and more completely than leaves of C_3 plants. The reasons for these differences are unknown, but it has been suggested that the differences in physical distances between the mesophyll cells and the phloem of the vascular bundles may influence the rate of loading of photosynthates into the sieve tubes.

The epidermis of grasses is made up of a variety of cell types. Most of the epidermal cells are narrow, elongate cells. Some especially large ones, termed **bulliform cells,** occur in longitudinal rows and are believed to participate in the mechanism of folding or rolling and unfolding or unrolling of the leaves, responses resulting from changes in water potential (Figure 23–27). During excessive loss of water, the bulliform cells become flaccid and the leaf folds or rolls. The thick-walled guard cells of the stomata are associated with subsidiary cells (Figure 23–23; see also Figure 21–23, page 465).

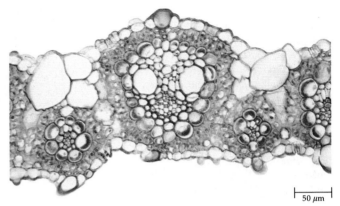

23–25
Transverse section of sugarcane (Saccharum officinarum) leaf. As is typical of C_4 grasses, the mesophyll cells are radially arranged around the bundle sheaths, which consist of large chlorophyllous cells. (See also corn, or maize, leaf section, Figure 7–20, page 113.)

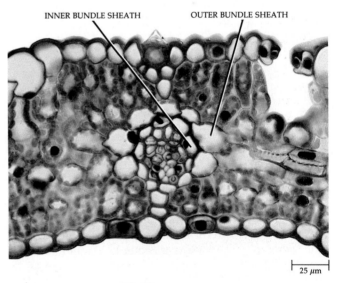

INNER BUNDLE SHEATH   OUTER BUNDLE SHEATH

23–26
Transverse section of wheat (Triticum aestivum) leaf. As is typical of C_3 grasses, the mesophyll cells are not radially arranged around the bundle sheaths. The bundle sheaths in the wheat leaf consist of two layers of cells: an outer sheath of relatively thin-walled parenchyma cells and an inner sheath, the mestome sheath, of thick-walled cells.

23-27
Transverse sections of annual bluegrass (Poa annua, a C_3 grass) leaf. Portions of folded (a) and unfolded (b) leaves including midvein. In the grass leaf, the mesophyll is not differentiated as palisade and spongy parenchyma. Strands of sclerenchyma cells commonly occur above and below the veins. The epidermis of the grass leaf contains bulliform cells—large epidermal cells thought to play a part in the folding and unfolding (rolling and unrolling) of grass leaves. In the Poa leaf shown in (a), the bulliform cells located in the upper epidermis are partly collapsed and the leaf is folded. An increase in turgor in the bulliform cells would presumably cause the leaf to unfold (b).

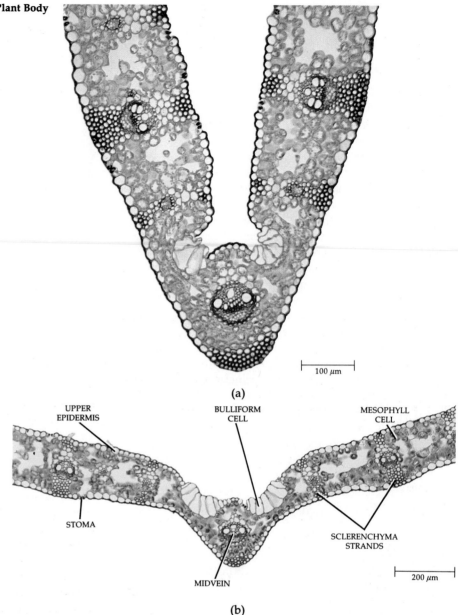

100 μm

(a)

UPPER
EPIDERMIS

BULLIFORM
CELL

MESOPHYLL
CELL

STOMA

SCLERENCHYMA
STRANDS

200 μm

MIDVEIN

(b)

Development of the Leaf

The first structural evidence of leaf initiation in most angiosperms is the appearance of periclinal divisions beneath the protoderm in the peripheral region of the shoot apex. A combination of cell enlargement and further divisions soon results in formation of a bulge, or *leaf buttress* (Figure 23–28a), as a center of mitotic activity is established at an exact time and location on the apex relative to previously initiated leaves. Either before or during buttress formation, a procambial strand appears beneath the young leaf primordium.

With continued upward growth, the leaf buttress develops into an erect, peglike structure—the *leaf primordium* (Figure 23–28b). In dicotyledons, this structure soon develops localized regions of meristematic activity on approximately opposite sides of its axis, and formation of the blade is initiated (Figure 23–29).

Apical growth of the leaf primordium is of short duration. Expansion and increase in length of the leaf occur largely by *intercalary growth*, that is, by cell division and cell enlargement throughout the blade, with cell enlargement contributing the most. Early in blade development, a certain number of layers of mesophyll cells become established. This layered pattern is maintained because of a predominance of anticlinal divisions (at right angles to the leaf surface) within the layers, although the number of layers may be increased during later development. This type of meristematic activity is attributed to what is called a *plate meristem*. Differences in rates of cell division and cell enlargement in the various layers of the blade result in the formation of numerous intercellular spaces and produce the mesophyll form characteristic of the leaf. Typically, the leaf stops growing first at the tip and last at the base. Compared with growth of the stem, the growth of

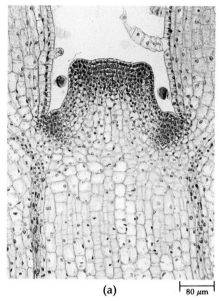

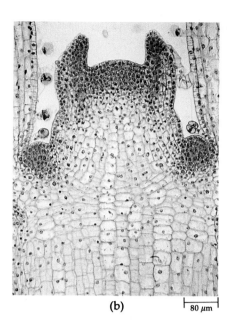

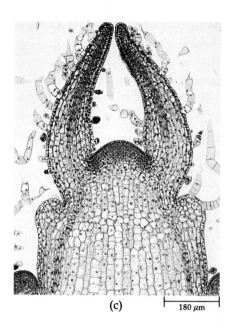

(a) 80 μm

(b) 80 μm

(c) 180 μm

23–28

Some early stages of leaf development in Coleus blumei, as seen in longitudinal sections of the shoot tip. The leaves in Coleus occur in pairs, opposite one another at the nodes (see Figure 23–2). (a) Two small bulges, or leaf buttresses, can be seen opposite one another on the flanks of the apical meristem. In addi-

tion, a bud primordium can be seen arising in the axil of each of the two young leaves, below. (b) Two erect, peglike leaf primordia have developed from the leaf buttresses. Notice the procambial strands extending upward into the leaf primordium. The bud primordia, below, are further along in development than those

in (a). As the leaf primordia elongate (c), the procambial strands, which are continuous with the leaf-trace procambium in the stem, continue to develop into the leaves. Trichomes, or epidermal hairs, develop from certain protodermal cells very early, long before the protoderm matures to become the epidermis.

MIDRIB   MIDVEIN

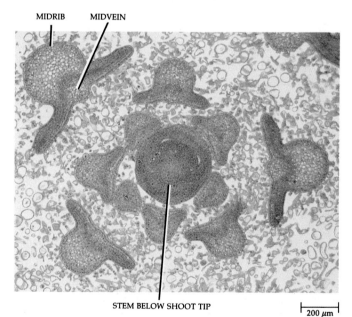

STEM BELOW SHOOT TIP

200 μm

23–29

Transverse section of developing leaves of tobacco (Nicotiana tabacum) grouped around the shoot tip, sectioned below the apical meristem. The younger leaves are nearer the axis. A leaf primordium at first lacks differentiation into midrib and blade. Some early stages in development of blade and midrib can be seen here. (Portions of numerous trichomes surround the developing leaves.)

most leaves is of short duration. The unlimited or prolonged growth of the vegetative apical meristems is described as *indeterminate*, and the restricted type of growth exhibited by the leaf and by floral apices is said to be *determinate*.

Vascular development in a dicot leaf begins with the differentiation of the procambium of the future midvein. This procambium differentiates upward into the primordium as an extension of the leaf-trace procambium (Figure 23–28c). All of the coarse, or major, veins develop upward and/or outward toward the margins of the leaf in continuity with the procambium of the midvein (Figure 23–30). The smaller, or minor, veins of the leaf are initiated at the tip of the leaf (Figure 23–30). They develop from the tip to the base of the leaf in continuity with the coarser veins. Thus, the tip of the leaf is the first part to have a complete system of veins. This course of development reflects the overall maturation of the leaf, which is from the tip to the base of the leaf.

Development of the monocot leaf differs from that of the dicot leaf in several respects. For example, in grasses such as corn and barley, development of the sheathing leaf base is apparent quite early, as growth activity spreads laterally from the flanks of the developing leaf primordium and completely encircles the shoot apex. As the primordium increases in length, it gradually acquires a hoodlike shape (Figure 23–31). Further development of the blade proceeds in a linear manner, with new cells being added by activity of a

23–30

Diagrams illustrating two stages in the development of the vascular system in the leaf of lettuce (Lactuca sativa). Whereas the major, or coarse, veins (a) develop upward in the blade, the minor, or smaller, veins (b) develop from the tip to the base of the blade. Thus, the tip of the leaf is the first region to have a complete system of veins.

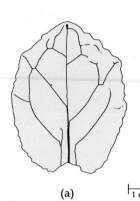

(a) |—1 mm—|

(b) |—1 mm—|

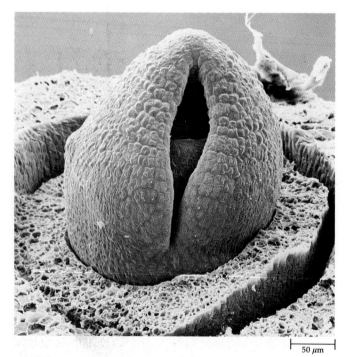

23–31

Scanning electron micrograph of developing leaf of barley (Hordeum vulgare). At this stage, the developing lamina has a hoodlike form. The shoot apex can be seen through the slitlike opening in the hood.

basal intercalary meristem. Elongation of the blade is restricted to a small zone above the region of cell division, and successively higher portions of the blade are more advanced in development. Growth of the sheath begins relatively late and lags behind the blade in development. The boundary between blade and sheath does not become distinct until the ligule, a thin projection from the top of the sheath, begins to develop (Figure 23–16b).

Development of the vascular system of a grass leaf begins with the median bundle, which in barley arises in isolation in the stem at the base of the developing leaf primordium and, as a procambial strand, develops upward into the developing leaf and downward into the stem to connect with preexisting bundles. Subsequently formed longitudinal veins arise in a similar manner on either side of the median bundle, the larger ones alternating with the smaller ones in their initiation. Transverse interconnections are initiated at the tip of the blade and progress toward the base of the blade.

SUN AND SHADE LEAVES

Environmental factors, especially light, can have substantial developmental effects on the size and thickness of leaves. In many species, leaves grown under high light intensities—the so-called *sun leaves*—are smaller and thicker than the so-called *shade leaves* that develop under low light intensities (Figure 23–32). The increased thickness of sun leaves is due mainly to a greater development of the palisade parenchyma. The vascular system of sun leaves is more extensive, and the walls of their epidermal cells are thicker than those

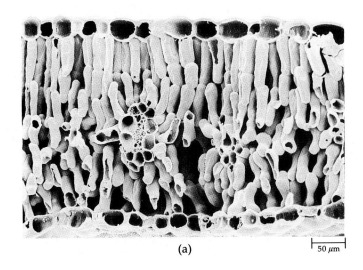

(a)   50 μm

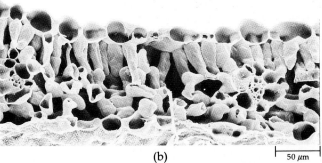

(b)   50 μm

23-32
Scanning electron micrographs showing transverse veins of (a) sun and (b) shade leaves of Thermopsis montana, *a member of the pea family (Fabaceae). Note that the sun leaf is considerably thicker than the shade leaf, a condition due mainly to a greater development of palisade parenchyma in the sun leaf.*

of shade leaves. In addition, the ratio of the internal surface area of the mesophyll to the area of the leaf blade is much higher in sun leaves. One effect of these differences is that, although both leaf types have similar photosynthetic rates at low light intensities, shade leaves are not adapted to high light intensities, and consequently they have considerably lower photosynthetic maxima under these conditions.

Because light intensities vary greatly in different parts of the crowns of trees, extreme forms of sun and shade leaves can be found there. Sun and shade leaves also occur in shrubs and in herbaceous plants. They can be induced to form by growing plants under high or low light intensities.

Leaf Abscission

In many plants, the normal separation of the leaf from the stem—the process of **abscission**—is preceded by certain structural and chemical changes near the base of the petiole. These changes result in the formation of an **abscission zone** (Figure 23-33). In woody dicotyledons, two layers can be distinguished in the abscission zone: a separation (abscission) layer and a protective layer. The separation layer consists of relatively short cells with poorly developed wall thickenings that make it structurally weak. Prior to abscission, certain reusable ions and molecules are returned to the stem, among them magnesium ions, amino acids (derived from proteins), and sugars (some derived from starch). Then, enzymes break down the cell walls in the separation layer. The cell wall changes may include weakening of the middle lamella and hydrolysis of the cellulosic walls themselves. Cell division may precede the actual

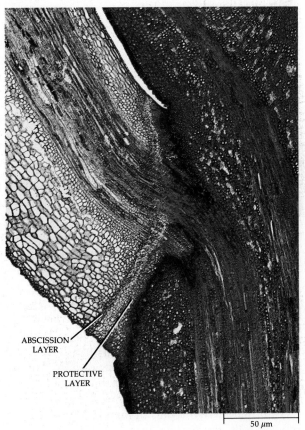

ABSCISSION LAYER

PROTECTIVE LAYER

50 μm

23-33
Abscission zone in maple (Acer) leaf, as seen in longitudinal section through the base of the petiole.

Leaf Dimorphism in Aquatic Plants

In the natural environment, the leaves of aquatic flowering plants may develop into two distinct forms. Under water they develop into narrow and often highly dissected structures (water forms), and above the surface they develop into rather ordinary-looking leaves (land forms). It is possible to force immature leaves to develop into the form atypical for a given environment by applying any one of a wide range of treatments.

During a study on the aquatic plant *Callitriche hetero-phylla*, it was found that the plant hormone gibberellic acid induced shoots growing in the air (emergent shoots) to produce water-form leaves. Abscisic acid, another plant hormone (see Chapter 25), led to the formation of land-form leaves on submerged shoots. Higher temperatures or the addition of the sugar alcohol mannitol to the water also caused submerged shoots to produce land-form leaves.

In nature, the cellular turgor pressure of the submerged (water-form) leaves is relatively high, while that of the emergent (land-form) leaves is relatively low. The lower values in the emergent leaves may be due in part to transpirational loss of water vapor via the numerous stomata on the leaf surfaces. Associated with the higher turgor pressures in developing water-form leaves are long epidermal cells in the mature leaves.

Gibberellic acid caused the cells of emergent leaves to elongate by increasing the water uptake and, hence, the turgor pressure. At maturity these leaves had all the characteristics of typical water forms, including long epidermal cells. The limited cell expansion in submerged shoots exposed to abscisic acid or high temperatures apparently was not the result of low turgor; instead, the treated cells became less yielding so that high turgor failed to promote cell expansion. Growing submerged shoots in a mannitol solution resulted in turgor pressures similar to those of the emergent controls and the production of leaves with short epidermal cells.

LAND-FORM LEAVES

WATER-FORM LEAVES

The results of these experiments suggest that the relative magnitude of cellular turgor pressure determines the final leaf size and form in *C. heterophylla*. Thus, merely the presence or absence of water surrounding the developing leaf ensures that the leaf will be suitably adapted to life above or below the water.

---

separation. If cell division occurs, the newly formed cell walls are the ones largely affected by the degradation phenomena. Beneath the separation layer, a protective layer composed of heavily suberized cells is formed, further isolating the leaf from the main body of the plant before it drops. Tyloses may form in the tracheary elements before abscission (see page 538). Eventually, the leaf is held to the plant by only a few strands of vascular tissue, which may be broken by the enlargement of parenchyma cells in the separation layer. After the leaf falls, the protective layer is recognized as a **leaf scar** on the stem (see Figure 24–17, page 530). Hormonal factors associated with abscission are discussed in Chapter 25.

Transition between Vascular Systems of the Root and the Shoot

As discussed in previous chapters, the distinction between plant organs is based primarily on the relative distribution of the vascular and ground tissues. For example, in dicot roots the vascular tissues generally form a solid cylinder surrounded by the cortex. In addition, the strands of primary phloem alternate with the radiating ridges of primary xylem. In contrast, in the stem the vascular tissues often form a cylinder of discrete strands around a pith, with the phloem on the outside of the vascular bundles and the xylem on the inside.

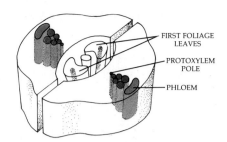

FIRST FOLIAGE LEAVES

PROTOXYLEM POLE

PHLOEM

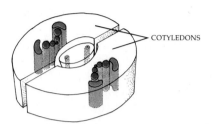

COTYLEDONS

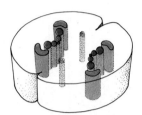

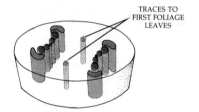

TRACES TO FIRST FOLIAGE LEAVES

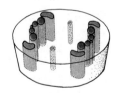

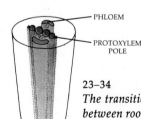

PHLOEM

PROTOXYLEM POLE

Obviously, somewhere in the primary plant body, a change from the type of structure found in the root to that found in the shoot must take place. This change is a gradual one, and the region of the plant axis through which it occurs is called the **transition region.**

As described in Chapter 20, the shoot and root are initiated as a single continuous structure during the development of the embryo. Consequently, vascular transition occurs in the axis of the embryo or young seedling. This transition is initiated during the appearance of the procambial system in the embryo and is completed with the differentiation of the variously distributed procambial tissues in the seedling. Vascular continuity between the root and shoot systems is maintained throughout the life of the plant.

The structure of the transition region can be very complex, and much variation exists in the transition regions of different kinds of plants. In most gymnosperms and dicotyledons, vascular transition occurs between the root and cotyledons. Figure 23–34 depicts a type of transition region commonly found in dicotyledons. Notice the diarch (having two protoxylem poles) structure of the root; the branching and reorientation of the primary xylem and the primary phloem, which in the upper part of the axis result in the formation of a pith; and the traces of the first leaves of the epicotyl.

Development of the Flower

The development of the flower or inflorescence terminates the meristematic activity of the vegetative shoot apex.

During the transition to flowering, the vegetative shoot apex undergoes a sequence of physiological and structural changes and is transformed into a reproductive apex. Consequently, flowering may be considered as a stage in the development of the shoot apex and of the plant as a whole. Inasmuch as the reproductive apex exhibits a determinate growth pattern, flowering in annuals indicates that the plant is approaching completion of its life cycle. By contrast, flowering in perennials may be repeated. Various environmental factors, including the length of day and the temperature, are known to be involved in the induction to flowering (see Chapter 26).

The transition from a vegetative to a floral apex is often preceded by an elongation of the internodes and the early development of lateral buds below the shoot apex. The apex itself undergoes a marked increase in mitotic activity, accompanied by changes in dimensions and organization: the relatively small apex with a tunica-corpus type of organization becomes broad and domelike.

23–34
The transition region—the connection between root and cotyledons—in the seedling of a dicotyledon with a diarch root. In the root, the primary vascular system is represented by a single cylinder of vascular tissue. In the hypocotyl-root axis, the vascular system branches and diverges into the cotyledons, and the xylem and phloem become reoriented along the hypocotyl-root axis.

Plants, Air Pollution, and Acid Rain

The plant leaf, like the human lung, can function only when it is able to exchange gases with the surrounding air. As a consequence, the leaf, like the lung, is an organ that is exceedingly susceptible to air pollution.

Air pollution has many forms. Some pollutants are particulate. The particles may be organic, such as those present in the smoke produced by burning fossil fuels and garbage, or inorganic, such as cement kiln dusts, foundry dusts, and the lead compounds released in the combustion of leaded gasoline. As a major component of smog, these particles reduce the amount of sunlight reaching the earth's surface. Such particles also have direct ill effects on plants. They may clog stomata and prevent them from functioning, or—particularly metallic particles—they may act as plant poisons.

Fluorides, which enter the air as waste products from the manufacture of phosphates, steel, aluminum, and other industrial products, act as cumulative poisons, entering the leaf through the stomata and causing collapse of leaf tissue, apparently by inhibiting enzymes concerned with cellulose synthesis. Thousands of acres of Florida citrus groves have been damaged by fluorides discharged from phosphate fertilizer factories.

When ores containing sulfur are processed, sulfur dioxide is produced:

$$2CuS + 3O_2 \longrightarrow 2CuO + 2SO_2$$

Sulfur dioxide has a disagreeable pungent-sweet taste; it is unusual among air pollutants in that it can be tasted at lower concentrations than it can be smelled. Sulfur oxides are also produced by the burning of fossil fuels containing sulfur. In moist air, sulfur oxides react with water to form droplets of sulfuric acid, a strong, corrosive acid and a component of acid rain. In several parts of the United States, virtual deserts have been created by the emission of sulfur dioxide combined with metals. As long ago as 1905, air pollution controls were instituted in areas surrounding copper smelters in Tennessee, but the surrounding area, once covered by luxuriant forest, remains barren to this day. Not only was all the vegetation killed, but the acid leached the soil of nutrients. Similarly, in a copper-smelting area of the Sacramento Valley, California, all vegetation was killed over an area of 260 square kilometers, and growth was severely affected over an additional 320 square kilometers.

The type of air pollution most familiar to Californians is photochemical smog, which is produced by sunlight acting on automobile exhausts. The Los Angeles area offers an ideal setting for the formation of photochemical smog because the life of that area is heavily geared to the use of the automobile, and the mountains to the north and east form a "basin" that prevents the dispersal of reactants. Not only are many species of plants unable to survive in the city itself (as is true in many other major cities of the world), but smog moving out of the Los Angeles basin is damaging agricultural crops and killing pine forests in mountains as far as 160 kilometers away.

One of the principal ingredients of photochemical smog is nitrogen dioxide, NO_2, which is produced by any combustion process that occurs in air (dry air is 77 percent nitrogen) and so is present in automobile exhausts. Under the influence of light, NO_2 is split into NO and atomic oxygen. The latter is extremely reactive and forms ozone, O_3, by reaction with molecular oxygen.

Similar reactions, powered by ultraviolet light, occur at the outer layers of the atmosphere, producing the ozone shield described on page 4. Ozone can also be produced by an electric spark and is the source of the "clean" smell after a lightning storm. Ozone is a highly toxic substance. In plants, it damages the thin-walled palisade cells, apparently affecting the permeability of the membranes of both the cells and their chloroplasts. Another component of photochemical smog is PAN (peroxyacetyl nitrate, $C_2H_3O_5N$). It is several times more toxic than ozone but is normally present in much lower concentrations. Photosynthesis is reduced 66 percent by a photochemical smog concentration of 0.25 part per million.

At the present time, there is much concern over the adverse effects of acid rain on the environment. Under normal conditions, rain arising in a nonpolluted area should have a pH of 5.6. With the burning of fossil fuels and the smelting of sulfide ores, the large quantities of sulfur oxides and nitrogen oxides emitted into the atmosphere react with water to form strong acids (sulfuric acid and nitric acid). Rain and snow formed in such areas have a pH of less than 5.6. The phenomenon of "acid precipitation" (meaning precipitation of pH below 5.6) is widespread today and occurs over large areas of western Europe, the eastern United States, and southeastern Canada, where the yearly average pH of precipitation ranges from 4 to 4.5. Moreover, the rain of individual storms is often much more acidic. Acid rains have been recorded in Scotland, Norway, and Iceland with pH values of 2.4, 2.7, and 3.5, respectively. The move to reduce *local* pollution problems through the building of tall stacks on smelters and industrial plants has created *regional* problems. Pollutants emanating from tall stacks are transported over long distances through the atmosphere. For example, more than 75 percent of the sulfur in rain that falls in the Scandinavian countries is believed to originate in the British Isles and central Europe.

The effect of acid rain on vegetation is not well understood but is currently being investigated throughout the Northern Hemisphere. Acid rains have been shown to decrease growth of forest trees in Sweden. In addition, simulated acid rains have been shown to injure leaves and inhibit seed germination. In just a few years the observed occurrences of damage in the forests of western Germany have increased from just a few percent to more than 50 percent.

The clearest effect of acid rain has been on fish populations, which have been virtually eliminated in acidified lakes in some parts of the world. It has been suggested that much of the toxicity to fish is actually due to increased aluminum concentrations and is not directly attributable to acidic water. Aluminum, which makes up about 5 percent of the earth's crust, is almost completely insoluble in neutral or alkaline water and thus has not been biologically available. As a result of acid rain, however, concentrations of dissolved aluminum in some lakes may increase to levels toxic to fish and other aquatic organisms. The solubility of other toxic metals, such as lead, cadmium, and mercury, also increases sharply with decreasing pH.

(a) *Sulfur dioxide injury on a blackberry (Rubus) leaf is characterized by areas of injured leaf tissue surrounded by healthy tissue.* (b) *Ozone injury on a tobacco (Nicotiana tabacum) leaf appears as stipples or flecks of dead tissue on the upper surface of the leaf. In cases of severe ozone injury, the flecks coalesce into larger lesions that are visible on both leaf surfaces.* (c) *Acid rain arises when sulfur oxides and nitrogen oxides react with water in the atmosphere to form sulfuric acid and nitric acid.*

(a)

(b)

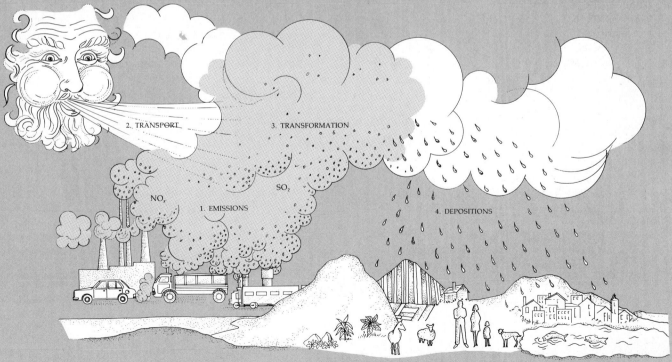
(c)

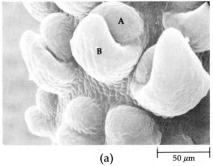

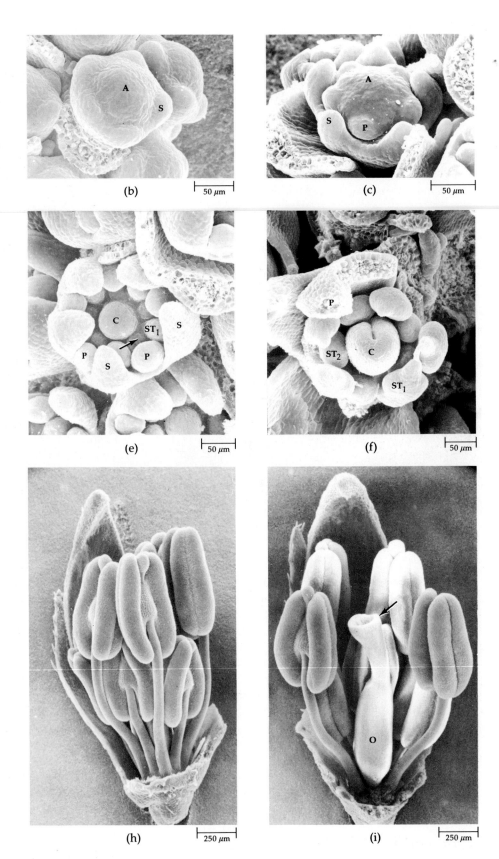

23–35

Scanning electron micrographs showing some stages in the development of the perfect flower of Neptunia pubescens, a legume with radial symmetry and whorled floral parts. Tips of the subtending bracts have been removed in (b) through (i). In (f) through (i) most sepals and petals have also been removed. (a) Floral apex (A) in axil of a bract (B). (b) Five sepal primordia (S) have been initiated around the floral apex. (c) Five petal primordia (P) have been initiated around the floral apex, alternating with the sepals (S). During their development, the sepals will form a calyx tube. (d) Five stamen primordia (arrows) have been initiated around the floral apex, alternating with the petals (P). (e) A second whorl of stamens (arrow) has been initiated, its members alternating with members of the first stamen whorl (ST_1). The carpel (C) has been initiated at the center of the floral apex. All floral parts are now present. (f) The carpel has now developed a cleft, which will form the locule of the ovary. The outer, or first, whorl of stamens (ST_1) are beginning to differentiate anthers and filaments. (ST_2 designates inner, or second, whorl of stamens.) (g) The carpel is now beginning to differentiate style and ovary. (h) Older flower with both whorls of stamens in view. (i) Older flower with some stamens removed to reveal the carpel, which has differentiated ovary (O), style, and stigma (arrow).

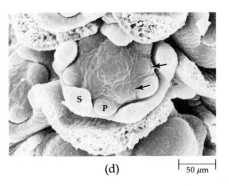

(d) 50 μm

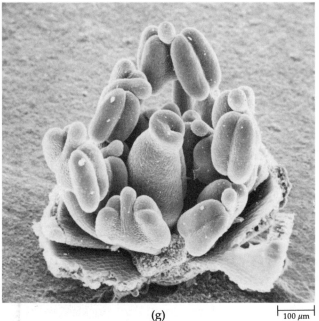

(g) 100 μm

23–36
The tendrils of grape (Vitis) *are modified stems.*

The initiation and early stages of development of the sepals, petals, stamens, and carpels are quite similar to those of leaves. Commonly, the initiation of the floral parts begins with the sepals, followed by the petals, then the stamens, and finally the carpels (Figure 23–35). This usual order of appearance of the floral parts may be modified in certain flowers, but the floral parts always have the same relative spatial relation to one another. The floral parts may remain separate during their development, or they may become united within different whorls (coalescence) and between whorls (adnation).

The basic structure of the flower and some of its variations were discussed in Chapters 18 and 19.

Stem and Leaf Modifications

Stems and leaves may undergo modifications and perform functions quite different from those commonly associated with these two components of the shoot. One of the most common modifications is the formation of **tendrils,** which aid in support. Some tendrils are modified stems. In Boston ivy *(Parthenocissus tricuspidata)* and Virginia creeper *(P. quinquefolia),* the tendrils form adhesive disks at their tips. The tendrils of the grape *(Vitis)* (Figure 23–36) are modified stems that coil around the supporting structure. In the grape, the tendrils sometimes produce small leaves or flowers.

Most tendrils are leaf modifications. In legumes, such as the garden pea *(Pisum sativum),* the tendrils constitute the terminal part of the pinnately compound leaf (Figure 23–15). One legume, the peanut *(Arachis hypogaea),* has another interesting adaptation. After fertilization takes place, the stamens and corolla of the flower fall off, and a peglike internode between ovary and receptacle, the gynophore, begins to elongate. With continued elongation, the peg bends downward and buries the developing ovary several centimeters into the ground, where the fruit ripens. If the ovary is not buried, it withers and fails to mature. While the peg elongates in the light, development of the embryo is inhibited and will resume only in the dark, when the ovary is buried. Elongation of the gynophore and development of the seed and fruit are controlled by light and darkness.

Branches that assume the form of and closely resemble foliage leaves are called **cladophylls.** The leaflike branches of asparagus *(Asparagus officinalis)* are familiar examples of cladophylls (Figure 23–37). The thick and fleshy aerial shoots ("spears") of asparagus are the edible portion of the plant. The scales found on the spears are true leaves. If the asparagus plants are allowed to continue growing, cladophylls develop in the axils of the minute, inconspicuous scales and then

23–37
The filmy branches of the common edible asparagus (Asparagus officinalis) resemble leaves. Such modified stems are called cladophylls.

23–38
The branches of the spineless cactus Epiphyllum *resemble leaves but are actually modified stems called cladophylls.*

act as photosynthetic organs. In some cacti, the branches resemble leaves (Figure 23–38). As mentioned previously, true leaves have buds in their axils. This characteristic may be used to distinguish between a leaf and a cladophyll.

In some plants, the leaves are modified as spines, which are hard and dry and nonphotosynthetic. The terms "spine" and "thorn" are frequently used interchangeably, but technically thorns are modified branches that arise in the axils of leaves (Figure 23–39). Another term commonly used interchangeably with thorn and spine is "prickle." A prickle, however, is neither a stem nor a leaf but a small, more or less slender, sharp outgrowth from the cortex and epidermis. The so-called thorns on rose stems are prickles. All three structures—spines, thorns, and prickles—may serve as

23–39
(a) *Spines, as in this* Ferocactus melocactiformis, *are modified leaves.*
(b) *Thorns are modified branches, which, as this photograph of a hawthorn (Crataegus) shows, arise in the axils of the leaves.*

(a)

(b)

defensive structures, reducing predation by herbivores (plant-eating animals). In a special plant-herbivore interaction, the spines of the bull's-horn acacias provide shelter for ants that kill other insects attempting to feed on the acacias (see page 638).

Among the most spectacular of modified or specialized leaves are those of the carnivorous plants, such as the pitcher plant, the sundew, and the Venus flytrap, which capture insects and digest them with enzymes secreted by the plant. The available nutrients are absorbed by the plant (see Chapter 27).

FOOD STORAGE

Stems, like roots, serve food-storage functions. Probably the most familiar type of specialized storage stem is the **tuber,** as exemplified by the Irish, or white, potato *(Solanum tuberosum)*. In the white potato, tubers arise at the tips of **stolons** (slender stems growing along the surface of the ground) of plants grown from seed. However, when cuttings of tubers are used for propagation, the tubers arise at the ends of long, thin **rhizomes,** or underground stems (Figure 23–40). Except for vascular tissue, almost the entire mass of the tuber inside the periderm ("skin") is storage parenchyma. The so-called "eyes" of the white potato are depressions containing groups of buds. The depression is the axil of a scalelike leaf.

A **bulb** is a large bud consisting of a small, conical stem with numerous modified leaves attached to it. The leaves are scalelike and contain thickened bases where food is stored. Adventitious roots arise from the bottom

23–40
White potato (Solanum tuberosum), *with tubers attached to a rhizome, or underground stem.*

of the stem. Familiar examples of plants with bulbs are the onion (Figure 23–41a) and the lily.

Although superficially similar to bulbs, **corms** consist primarily of thickened, fleshy stem tissue. Their leaves commonly are thin and much smaller than those of bulbs; consequently, the stored food of the corm is found within the fleshy stem. Several well-known plants, such as gladiolus (Figure 23–41b), crocus, and cyclamen, produce corms.

23–41
*Examples of modified leaves or stems.
(a) An onion* (Allium cepa) *bulb, which consists of a conical stem with scalelike, food-containing leaves attached. The leaves are the part of the onion we eat. (b) A gladiolus* (Gladiolus grandiflorus) *corm, which is a fleshy stem with small, thin leaves. (c) The fleshy storage stem of kohlrabi* (Brassica oleracea *var.* caulorapa).

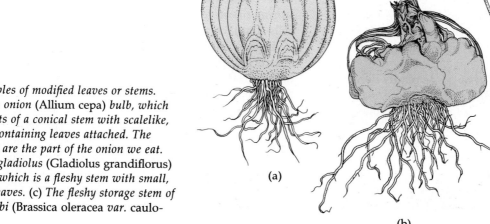

(a)

(b)

(c)

Convergent Evolution

Comparable selective forces, acting on plants growing in similar habitats but different parts of the world, often cause totally unrelated species to assume a similar appearance. The process by which this happens is known as convergent evolution.

Let us consider some of the adaptive characteristics of plants growing in desert environments—fleshy, columnar stems (which provide the capacity for water storage), protective spines, and reduced leaves. Three fundamentally different families of flowering plants—the spurge family (Euphorbiaceae), the cactus family (Cactaceae), and the milkweed family (Asclepiadaceae)—have members that have evolved in this direction. The cactuslike representatives of the spurge and milkweed families shown here evolved from leafy plants that looked quite different from one another.

Native cacti occur (with one exception) exclusively in the New World. The comparably fleshy members of the spurge and milkweed families occur mainly in desert regions in Asia and especially Africa, where they play an ecological role similar to that of the New World cacti.

Although the plants shown here—(a) *Euphorbia*, a member of the spurge family; (b) *Echinocereus*, a cactus; (c) *Hoodia*, a fleshy milkweed—have CAM photosynthesis, all three are related to and derived from plants that have C_3 photosynthesis, which indicates that the physiological adaptations involved in CAM photosynthesis also arose as a result of convergent evolution (page 116).

(a)

(b)

(c)

Kohlrabi (*Brassica oleracea* var. *caulorapa*) is one example of an edible plant with a fleshy storage stem. The short, thick stem stands above the ground and bears several leaves with very broad bases (Figure 23–41c). The common cabbage (*Brassica oleracea* var. *capitata*) is closely related to kohlrabi. The so-called "head" of cabbage consists of a short stem bearing numerous thick, overlapping leaves. In addition to a terminal bud, several well-developed axillary buds may be found within the head.

The leaf stalks, or petioles, of some plants become quite thick and fleshy. Celery (*Apium graveolens*) and rhubarb (*Rheum rhaponticum*) are familiar examples.

WATER STORAGE: SUCCULENCY

Succulent plants are plants that have juicy tissues, that is, tissues specialized for the storage of water. Most of these plants, such as the cacti of the American deserts, euphorbias of similar appearance of the African deserts (see the above essay, "Convergent Evolution"), and the century plant (*Agave*), normally grow in arid regions, where the ability to store water is necessary for their survival. The green, fleshy stems of the cacti serve as both photosynthetic and storage organs. The water-storing tissue consists of large, thin-walled parenchyma cells that lack chloroplasts.

In the century plant, the leaves are succulent. As in succulent stems, nonphotosynthetic parenchyma cells of the ground tissue constitute the water-storing tissue. Other examples of plants with succulent leaves are the ice plant (*Mesembryanthemum crystallinum*), the stonecrops (*Sedum*), and certain species of *Peperomia*. In the ice plant, large epidermal cells with appendages (trichomes), called water vesicles, which superficially resemble beads of ice, serve a water-storage function. The water-storing cells of the *Peperomia* leaf are part of a multiple (several-layered) epidermis derived by periclinal divisions of the protoderm (Figure 23–42).

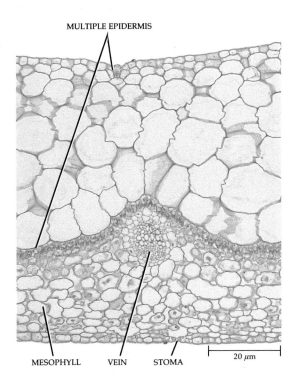

MULTIPLE EPIDERMIS

MESOPHYLL   VEIN   STOMA

20 μm

23–42
Transverse section of leaf blade of
Peperomia. *The very thick multiple*
epidermis, visible on its upper surface,
presumably functions as water-storage
tissue.

Summary

The vegetative shoot apices of most flowering plants have a tunica-corpus type of organization consisting of one or more peripheral layers of cells (the tunica) and an interior body of cells (the corpus). Although the primary tissues of the stem pass through periods of growth similar to those of the root, the stem cannot be divided into regions of cell division, elongation, and maturation in the same manner as roots. The stem increases in length largely by internodal elongation.

As in the root, the apical meristem of the shoot gives rise to protoderm, ground meristem, and procambium, which develop into the primary tissues. There are three basic patterns in the primary structure of stems, with regard to the relative distribution of ground and primary vascular tissues: the primary tissues may develop (1) as a more or less continuous hollow cylinder, (2) as a cylinder of discrete strands, or (3) as a system of strands scattered throughout the ground tissue. Regardless of the type of organization, the phloem is commonly located outside the xylem.

In dicotyledons, most leaves consist of a blade and petiole. The blades of some leaves are divided into leaflets. Stomata are commonly more numerous on the lower than the upper surface of the leaf. The ground tissue, or mesophyll, of the leaf is specialized as a photosynthetic tissue and, in mesophytes, is differentiated into palisade parenchyma and spongy parenchyma. The mesophyll is thoroughly permeated by air spaces and by veins, which are composed of xylem and phloem surrounded by a parenchymatous bundle sheath. The xylem commonly occurs on the upper side of the vein, whereas the phloem occurs on the lower side.

In most monocots, including the grasses, the leaf consists of a blade and a sheath, which encircles the stem. The leaves of C_3 and C_4 grasses have rather distinct anatomical differences. Most notable among the differences is the presence in C_4 grasses and absence in C_3 grasses of a Kranz anatomy, that is, one in which the mesophyll cells and the bundle-sheath cells are arranged in two concentric layers around the vascular bundles.

The term "shoot" serves not only as a collective term for the stem and its leaves but also as an expression of their intimate physical and developmental association. Leaves have their origin as leaf primordia in the peripheral region of the shoot apex, and their position on the stem is reflected in the pattern of the vascular system in the stem. At each node, one or more leaf traces diverge from the stem and enter the leaf or leaves at that node. The procambial strands from which the leaf traces are formed develop just below the developing leaf primordia and are sometimes present below the sites of the future leaf primordia even before the primordia are discernible.

Leaves are determinate in growth; that is, their development is of relatively short duration. However, vegetative shoot apices may exhibit unlimited, or indeterminate, growth. In many species, leaves grown under high light intensities are smaller and thicker than those grown under low light intensities. The former are called sun leaves, and the latter are called shade leaves.

In many plants, leaf abscission is preceded by formation of an abscission zone at the base of the petiole.

The change in the distribution of the vascular and ground tissues found in the root to that seen in the shoot occurs in a region of the plant axis of the embryo and young seedling called the transition region.

At flowering, in many species, the vegetative shoot apex is directly transformed into a reproductive apex.

Stems, like roots, may serve food-storage functions. Examples of fleshy stems are tubers, bulbs, and corms. Water-storing plants are known as succulents. Water-storage tissue of succulent plants is made up of large parenchyma cells. Stems or leaves or both may be succulent.

Secondary Growth in Stems

24–1

A solitary shagbark hickory (Carya ovata) in the summer condition. Plants have been able to achieve such great stature because of the ability of their roots and stems to increase in girth, that is, to undergo secondary growth. Most of the tissue produced in this manner is secondary xylem, or wood, which not only conducts water and minerals to the far reaches of the shoot but also provides great strength to the roots and stems.

In many plants (for instance, most monocotyledons and certain herbaceous dicotyledons such as *Ranunculus),* growth in a given part of the plant body ceases with maturation of the primary tissues. At the other extreme are the gymnosperms and woody dicotyledons, in which the roots and stems continue to increase in diameter in regions that are no longer elongating (Figure 24–1). This increase in thickness or girth of the plant body—termed secondary growth—results from activity of the two **lateral meristems:** the **vascular cambium** and the **cork cambium.**

Herbs, or herbaceous plants, are plants with shoots that undergo little or no secondary growth. In temperate regions either the shoot or the entire plant lives for only one season, depending on the species. Woody plants—trees and shrubs—can live for many years. At the start of each growing season, primary growth is resumed, and additional secondary tissues are added to the older plant parts through reactivation of the lateral meristems. Although most monocots lack secondary growth, some (such as the palms) may develop thick stems by primary growth alone (see page 491).

Plants are often classified according to their seasonal growth cycles as annuals, biennials, or perennials. In the **annuals**—which include many weeds, wildflowers, garden flowers, and vegetables—the entire cycle from seed to vegetative plant to flowering plant to seed again occurs within a single growing season, which may be only a few weeks in length. Only the dormant seed bridges the gap between one season of growth and the next.

In the **biennials,** two seasons are needed for the period from seed germination to seed formation. The first season of growth results in the formation of a root, a short stem, and a rosette of leaves near the soil surface. In the second growing season, flowering, fruiting, seed formation, and death occur, completing the life cycle. In temperate regions, annuals and biennials seldom become woody, although both their stems and their roots may undergo a limited amount of secondary growth.

Perennials are plants in which the vegetative structures live year after year. The herbaceous perennials pass unfavorable seasons as dormant underground roots, rhizomes, bulbs, or tubers. The woody perennials, which include vines, shrubs, and trees, survive above ground but usually stop growing during the unfavorable seasons. Woody perennials flower only when they become adult plants, which may take many years. For example, the horse chestnut, *Aesculus hippocastanum*, does not flower until it is about 25 years old. *Puya raimondii,* a large (up to 10 meters high) relative of the pineapple that is found in the Andes, takes about 150 years to flower. Many woody plants are deciduous, losing all their leaves at the same time and developing new leaves from buds when the season again becomes favorable for growth. In evergreen trees and shrubs, leaves are also lost and replaced, but not all leaves simultaneously.

The Vascular Cambium

Unlike the many-sided initials of the apical meristems, which contain dense cytoplasm and large nuclei, the meristematic cells of the vascular cambium are highly vacuolated. They exist in two forms: as vertically elongated **fusiform initials** and as horizontally elongated or squarish **ray initials.** The fusiform initials are much longer than they are wide and appear flattened or brick-shaped in transverse section. In the white pine, *Pinus strobus,* the fusiform initials average 3.2 millimeters in length; in the apple, *Malus sylvestris,* 0.53 millimeters (Figure 24–2); and in the black locust, *Robinia pseudo-acacia,* 0.17 millimeters (Figure 24–3).

Secondary xylem and secondary phloem are produced through periclinal divisions of the cambial initials and their derivatives. In other words, the cell plate that forms between the dividing cambial initials is parallel to the surface of the root or stem (Figure 24–4a). If

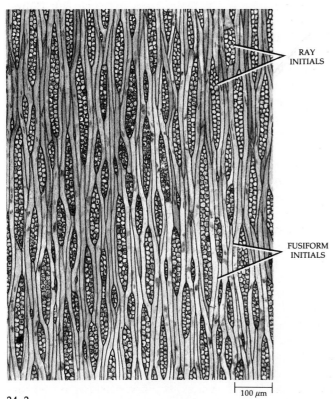

RAY INITIALS

FUSIFORM INITIALS

100 μm

24–2
Tangential view of a portion of vascular cambium of the apple (Malus sylvestris) *tree. Tangential sections are cut at right angles to the rays, so we see the rays here in transverse section. A cambium such as this, in which the fusiform initials are not arranged in horizontal tiers on tangential surfaces, is said to be nonstoried.*

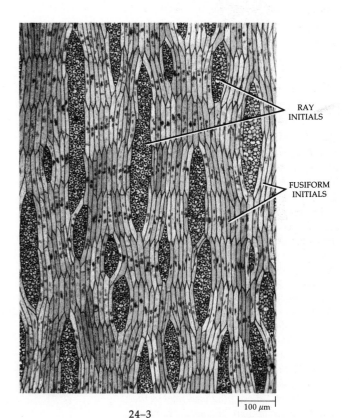

RAY INITIALS

FUSIFORM INITIALS

100 μm

24–3
Vascular cambium of the black locust (Robinia pseudo-acacia) *tree, seen here in tangential view. A cambium such as this, in which the fusiform initials are arranged in horizontal tiers on tangential surfaces, is said to be storied.*

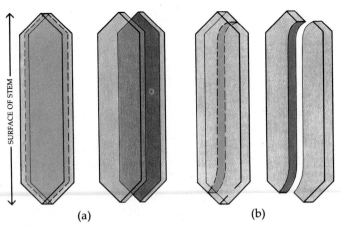

SURFACE OF STEM

(a)   (b)

24–4
*Periclinal and anticlinal divisions of
fusiform initials. (a) Periclinal divisions
are involved in the formation of secondary
xylem and secondary phloem cells, and
result in the formation of radial rows of
cells (see Figure 24–5). When an initial
divides periclinally, two cells appear,
one behind (or in front of) the other. (b)
Anticlinal divisions are involved in the
multiplication of fusiform initials. When
an initial divides anticlinally, two cells
appear side by side.*

24–5
*Diagram showing the relationship of
the vascular cambium to its derivative
tissues—secondary xylem and secondary
phloem. The darker cells are the more
recently derived ones. The vascular
cambium is made up of two types of cells
—fusiform initials and ray initials—
which give rise to the axial and radial
systems, respectively, of the secondary
vascular tissues. When the cambial
initials produce secondary xylem and
secondary phloem, they divide periclinally. Following division of an initial, one
daughter cell (the initial) remains meristematic, and the other (the derivative
of the initial) eventually develops into
one or more cells of the vascular tissue.
Cells produced toward the inner surface
of the vascular cambium become xylem
elements, and those produced toward the
outer surface become phloem elements.
The ray initials divide to form vascular
rays, which lie at right angles to the derivatives of the fusiform initials. With the
production of additional secondary xylem,
the vascular cambium and secondary
phloem are displaced in an outward direction. The diagrams (left to right) represent
successively more mature stages.*

the derivative of a cambial initial is divided off toward
the outside of the root or stem, it eventually becomes a
phloem cell; if it divides off toward the inside, it becomes a xylem cell. In this manner a long, continuous
radial file, or row, of cells is formed, extending from the
cambial initial outward to the phloem and inward to
the xylem (Figure 24–5).

The xylem and phloem cells produced by the fusiform initials have their long axes oriented vertically and
make up what is known as the **axial system** of the secondary vascular tissues. The ray initials produce horizontally oriented **ray cells,** which form the **vascular
rays** or **radial system** (Figure 24–5). The rays are composed largely of parenchyma cells and are variable in
length. The vascular rays serve as pathways for the
movement of food substances from the secondary
phloem to the secondary xylem and of water from the
secondary xylem to the secondary phloem. They also
serve as storage centers for such substances as starch,
proteins, and lipids.

In a restricted sense, the term "vascular cambium"
is used to refer only to the cambial initials, of which
there is one per radial file. However, it is often difficult,
if not impossible, to distinguish between the initials
and their immediate derivatives, which may remain
meristematic for a considerable period of time. Even in
the winter condition, when the cambium is dormant, or
inactive, several layers of undifferentiated cells that are
similar in appearance can be seen between the xylem
and phloem. Consequently, some botanists use the

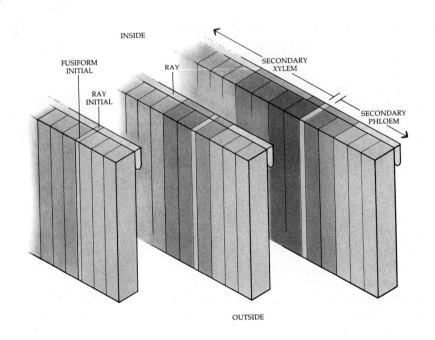

INSIDE

FUSIFORM
INITIAL

RAY INITIAL

RAY

SECONDARY
XYLEM

SECONDARY
PHLOEM

OUTSIDE

term "vascular cambium" in a broader sense to refer to the initials and their immediate derivatives, which are indistinguishable from the initials. Other botanists refer to this region of initials and derivatives as the **cambial zone.**

As the vascular cambium adds cells to the secondary xylem and the core of xylem increases in width, the cambium is displaced outward. In order to accommodate to this change, the vascular cambium undergoes an increase in circumference, which is accomplished by anticlinal divisions of the initials (Figure 24–4b). Along with an increase in the number of fusiform initials, new ray initials and rays are added, so that a fairly constant ratio of rays to fusiform cells is maintained in the secondary vascular tissues. Obviously, the developmental changes that occur in the cambium are exceedingly complex.

In temperate regions, the vascular cambium is dormant during winter and becomes reactivated in the spring. During reactivation, the cambial cells take up water, expand radially, and begin to divide periclinally. During expansion, the radial walls of the cambial cells become thinner and, as a result, the bark (all tissues to the outside of the vascular cambium) may be easily peeled or "slipped" off the stem. New growth layers, or increments, of secondary xylem and secondary phloem are laid down during the growing season. Reactivation of the vascular cambium is triggered by the expansion of the buds and resumption of their growth. The hormone auxin, produced by the developing shoots, moves downward in the stems and stimulates the resumption of cambial activity. Other factors are also involved in cambial reactivation and in continued normal growth of the cambium (see Chapter 25).

Effect of Secondary Growth on the Primary Body of the Stem

As mentioned in Chapter 23, the vascular cambium of the stem arises from the procambium that remains undifferentiated between the primary xylem and primary phloem, as well as from parenchyma of the interfascicular regions. That portion of the cambium arising within the vascular bundles is known as **fascicular cambium,** and that arising in the interfascicular regions, or pith rays, is called **interfascicular cambium.** The vascular cambium of the stem, unlike that of the root, is essentially circular in outline from its inception (Figure 24–6).

In woody stems, the production of secondary xylem and secondary phloem results in the formation of a cylinder of secondary vascular tissues, with the rays extending radially through the cylinder (Figure 24–6).

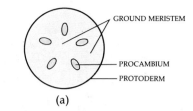

(a)

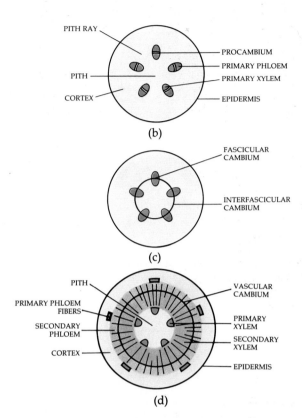

(b)

(c)

(d)

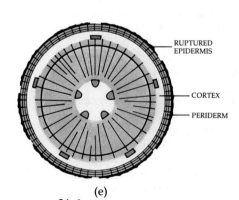

(e)

24–6
Stem development in a woody dicot. (a) Early stage in primary development, showing primary meristems. (b) At completion of primary growth. (c) Origin of vascular cambium. (d) After formation of some secondary xylem and secondary phloem. (e) At end of first year's growth, showing the effect of secondary growth—including periderm formation—on the primary plant body. In (d) and (e), the radiating lines represent rays. (Compare this with root development as depicted in Figure 22–16.)

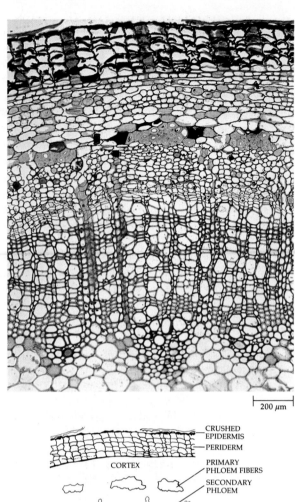

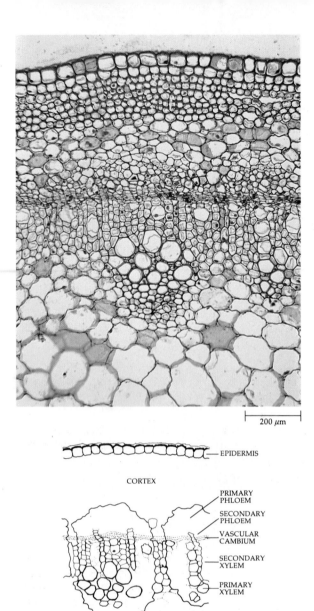

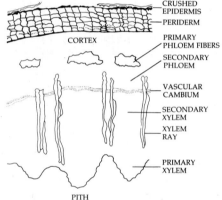

24–7
Transverse section of an elderberry (Sambucus canadensis) stem in which a small amount of secondary growth has taken place. A cork cambium has not yet been formed.

24–8
Transverse section of an elderberry (Sambucus canadensis) stem at the end of first year's growth.

Commonly, much more secondary xylem than secondary phloem is produced in the stem in any given year; this situation is also true in the root. As in the root, with secondary growth the primary phloem is pushed outward and its thin-walled cells are destroyed. Only the thick-walled primary phloem fibers, if present, remain intact (see Figure 24–8). See pages 480–482, Chapter 22, for a discussion of secondary growth in roots.

Figures 24–7 and 24–8 show an elderberry (*Sambucus canadensis*) stem in two stages of secondary growth. (See pages 492 to 493 for a description of primary growth in the *Sambucus* stem.) Only a small amount of secondary xylem and secondary phloem has

been produced in the stem of Figure 24–7. Figure 24–8 shows a stem at the end of the first year's growth. Note that considerably more secondary xylem than secondary phloem has been formed. The thick-walled cells outside the secondary phloem are primary phloem fibers.

Figure 24–9 shows one-year-old, two-year-old, and three-year-old stems of basswood (*Tilia americana*). In Chapter 23, the stem of *Tilia* was given as an example of one in which the primary tissues arise as an almost continuous hollow cylinder; thus most of the vascular cambium in the *Tilia* stem is fascicular in origin. Some of the rays in the secondary phloem of *Tilia* become

SECONDARY PHLOEM | PRIMARY PHLOEM FIBERS | EPIDERMIS | PERIDERM

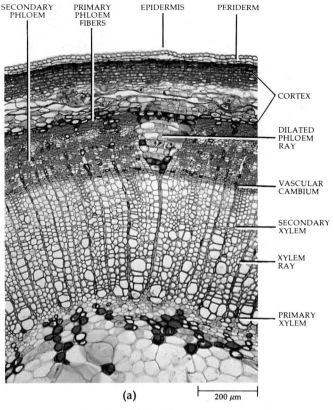

CORTEX

DILATED PHLOEM RAY

VASCULAR CAMBIUM

SECONDARY XYLEM

XYLEM RAY

PRIMARY XYLEM

(a) 200 μm

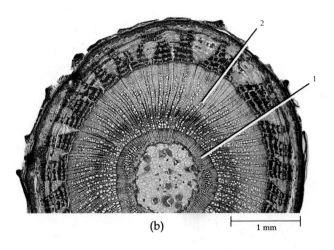

(b) 1 mm

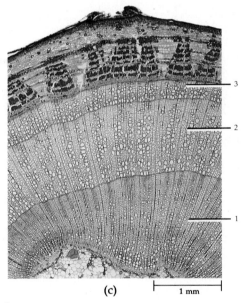

(c) 1 mm

24-9
Transverse sections of basswood (Tilia
americana) *tree stems. (a) One-year-old
stem. (b) Two-year-old stem. (c) Three-
year-old stem. Numbers indicate growth
rings in the secondary xylem.*

very wide as the stem increases in girth (Figure 24–9a).
This is one way in which the tissues outside the vascular cambium keep up with the increase in girth of the
core of xylem.

The vascular cambia and secondary tissues of root
and stem are continuous with one another. There is no
transition region in the secondary plant body as there is
in the primary plant body (see page 511).

THE PERIDERM

In most woody stems, as in most woody roots, cork formation usually follows the initiation of secondary
xylem and secondary phloem production, and the cork
tissue replaces the epidermis as the protective covering
on those portions of the plant. **Cork,** or phellem, is
formed to the outside by a **cork cambium,** or phellogen, which may also form **phelloderm** to the inside.
Together, these three tissues—cork, cork cambium, and
phelloderm—make up the **periderm** (Figure 24–10).

In the stems of most dicots and gymnosperms, the
first periderm commonly appears during the first year
of growth, most commonly originating in a layer of cortical cells immediately below the epidermis, although in
many species it originates in the epidermis.

Repeated divisions of the cork cambium result in the
formation of radial rows of compactly arranged cells,
most of which are cork cells (Figures 24–10 and 24–11).
During differentiation of the cork cells, their inner wall
surfaces are lined by suberin lamellae, consisting of
alternating layers of suberin and wax, which make the
tissue highly impermeable to water and gases. The
walls of the cork cells may also become lignified. At
maturity, the cork cells are dead.

The cells of the phelloderm are living at maturity,
lack suberin lamellae, and resemble cortical parenchyma cells. The phelloderm cells may be distinguished
from cortical cells by their inner position in the radial

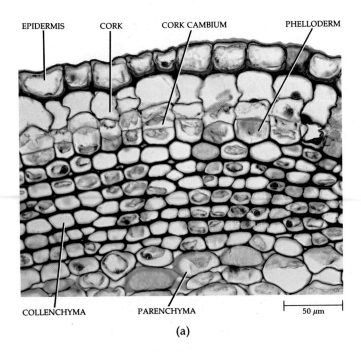

(a)

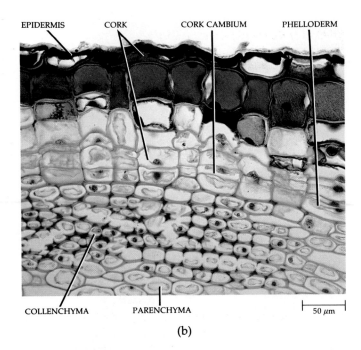

(b)

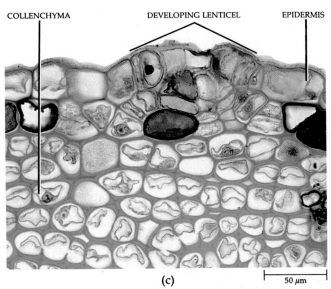

(c)

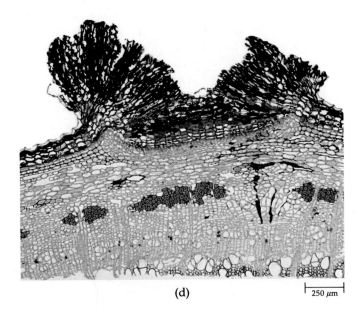

(d)

24–10

Some stages of periderm and lenticel development in elderberry (Sambucus canadensis), *as seen in transverse sections.* (a) *Newly formed periderm beneath the epidermis; collenchyma and parenchyma of cortex.* (b) *Periderm in more advanced stage of development.* (c) *Initiation of lenticel; collenchyma of cortex beneath the developing lenticel.* (d) *Well-developed lenticel. The phelloderm in Sambucus generally consists of a single layer of cells.*

rows of other periderm cells (Figure 24–11). Because the first periderm of the stem usually arises in the outer layer of cortical cells, the cortex of the stem is not sloughed off during the first year as it is in woody roots (Figures 24–6 and 24–8; compare with Figures 22–16 and 22–17c), although the epidermis dries up and peels off.

At the end of the first year's growth, the following tissues are present in the stem: remnants of the epidermis, periderm, cortex, primary phloem (fibers and crushed soft-walled cells), secondary phloem, vascular cambium, secondary xylem, primary xylem, and pith (Figure 24–6).

24–11
Lenticel of the stem of Dutchman's pipe (Aristolochia), *as seen in transverse section. Unlike that of* Sambucus, *the phelloderm of* Aristolochia *consists of several layers of cells.*

CUTICLE

EPIDERMIS

CORK

PHELLODERM

200 μm

THE LENTICELS

In the preceding discussion, it was noted that the suberized cork cells are compactly arranged and, as a tissue, present an impermeable barrier to water and gases. However, the inner tissues of the stem, like all metabolically active tissues, need to exchange gases with the surrounding air, just as the inner tissues of the root need to exchange gases with the surrounding air spaces between soil particles. In stems and roots containing periderms, this necessary gas exchange is accomplished by means of **lenticels** (Figures 24–10c, d and 24–11)— portions of the periderm with numerous intercellular spaces.

Lenticels begin to form during development of the first periderm (see Figure 24–10) and, in the stem, generally appear below a stoma or group of stomata. On the surface of the stem or root, the lenticels appear as raised circular, oval, or elongated areas (see Figure 24–17 on page 530). Lenticels are also formed on some fruits—for instance, the small dots on the surface of apples and pears are lenticels. As the roots and stems grow older, lenticels continue to develop at the bottom of cracks in the bark in newly formed periderms.

THE BARK

The terms "periderm," "cork," and "bark" are often unnecessarily confused with one another. As previously discussed, cork is one of three parts of the periderm; it is a secondary tissue that replaces the epidermis in most woody roots and stems. The term **bark** refers to all the tissues outside the vascular cambium, including the periderm when present (Figures 24–12 and 24–13). When the vascular cambium first

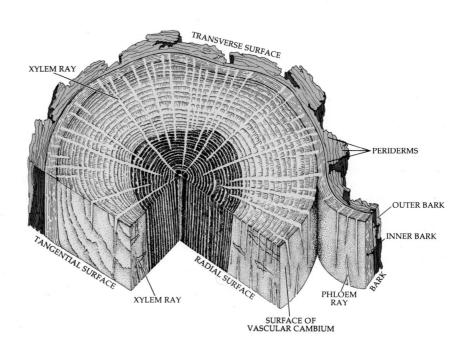

TRANSVERSE SURFACE

XYLEM RAY

PERIDERMS

OUTER BARK

INNER BARK

TANGENTIAL SURFACE

RADIAL SURFACE

BARK

XYLEM RAY

PHLOEM RAY

SURFACE OF
VASCULAR CAMBIUM

24–12
*Diagram of part of a red oak (*Quercus rubra) *stem, showing the transverse, tangential, and radial surfaces. The dark area in the center is heartwood. The lighter part of the wood is sapwood.*

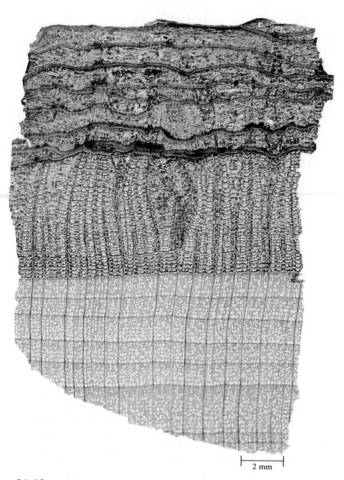

24–13
*Transverse section of the bark and some
secondary xylem from an old stem of basswood* (Tilia americana). *Several periderms can be seen traversing the mostly
brownish outer bark in the upper third of
the section. Below the outer bark is the
inner bark, which is quite distinct in
appearance from the more lightly stained
xylem in the lower third of the section.*

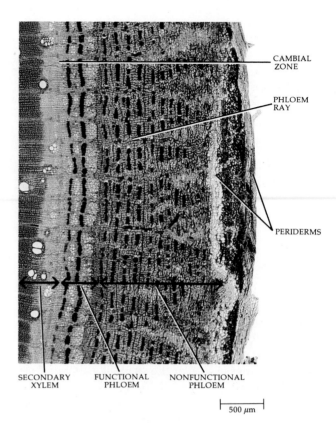

CAMBIAL
ZONE

PHLOEM
RAY

PERIDERMS

SECONDARY   FUNCTIONAL   NONFUNCTIONAL
XYLEM   PHLOEM   PHLOEM

500 μm

24–14
*Transverse section of the bark of the
black locust* (Robinia pseudo-acacia)
*stem, consisting mostly of nonfunctional
phloem.*

appears, and secondary phloem has not yet been
formed, the bark consists entirely of primary tissues. At
the end of the first year's growth, the bark includes any
primary tissues still present, the secondary phloem, the
periderm, and any dead tissues remaining outside the
periderm.

Each growing season, the vascular cambium adds
secondary phloem to the bark, as well as secondary
xylem, or wood, to the core of the stem or root. Usually
the vascular cambium produces less secondary phloem
than secondary xylem. In addition, the soft-walled cells
(sieve elements and various kinds of parenchymatous
elements) of the old secondary phloem are commonly
crushed (Figures 24–14 through 24–16). Eventually, the
old secondary phloem is separated from the rest of the

phloem by newly formed periderms. As a result, considerably less secondary phloem accumulates in the
stem or root than secondary xylem, which continues to
accumulate year after year.

As the stem or root increases in girth, considerable
stress is placed on the older tissues of the bark. In some
plants, tearing of these tissues results in the formation
of large air spaces. In many plants, the parenchyma
cells of the axial system and rays divide and enlarge; in
this manner, the old secondary phloem keeps up for a
while with the increase in circumference of the stem or
root. It was noted earlier that certain rays in the *Tilia*
stem become very wide as the stem increases in girth;
these rays are called dilated rays (Figure 24–9a).

The first-formed periderm may keep up with the increase in girth of the root or stem for several years, with
the cork cambium exhibiting periods of activity and in-
activity that may or may not correspond to the periods
of activity of the vascular cambium. In the stems of
apple (*Malus sylvestris*) and pear (*Pyrus communis*) trees,
the first cork cambium may remain active for up to 20
years. In most woody roots and stems, additional peri-

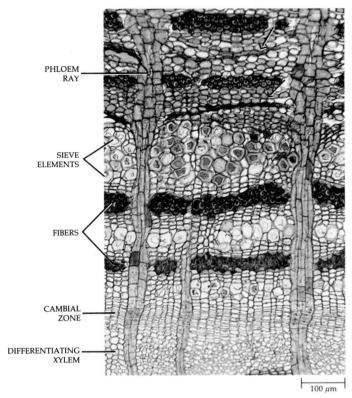

PHLOEM RAY

SIEVE ELEMENTS

FIBERS

CAMBIAL ZONE

DIFFERENTIATING XYLEM

100 μm

24–15
Transverse section of secondary phloem of the black locust (Robinia pseudo-acacia), showing mostly functional phloem. Sieve elements (indicated by arrows) of the nonfunctional phloem have collapsed.

derms are formed as the axis increases in circumference. After the first periderm, subsequently formed periderms originate deeper and deeper in the bark (Figures 24–12 and 24–13) from parenchyma cells of the phloem no longer actively engaged in the transport, or translocation, of food substances. These parenchyma cells become meristematic and form new cork cambia.

All of the tissues outside the innermost cork cambium—all of the periderms, together with any cortical and phloem tissues included among them—make up the *outer bark* (Figures 24–12 and 24–13). With maturation of the suberized cork cells, the tissues outside them are separated from the supply of water and nutrients. Hence the outer bark consists entirely of dead tissues. The living part of the bark, which is inside the innermost cork cambium and extends inward to the vascular cambium, is called the *inner bark* (Figures 24–12 and 24–13).

24–16
Radial section of the bark of the black locust (Robinia pseudo-acacia). Most of the section consists of nonfunctional phloem, in which the sieve elements are collapsed (arrows). In black locust, the functional phloem consists only of the current season's growth increment, which becomes nonfunctional in late autumn when its sieve elements die and collapse.

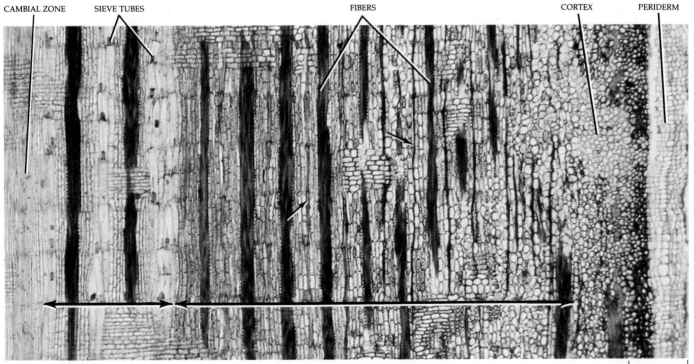

CAMBIAL ZONE   SIEVE TUBES   FIBERS   CORTEX   PERIDERM

FUNCTIONAL PHLOEM   NONFUNCTIONAL PHLOEM   200 μm

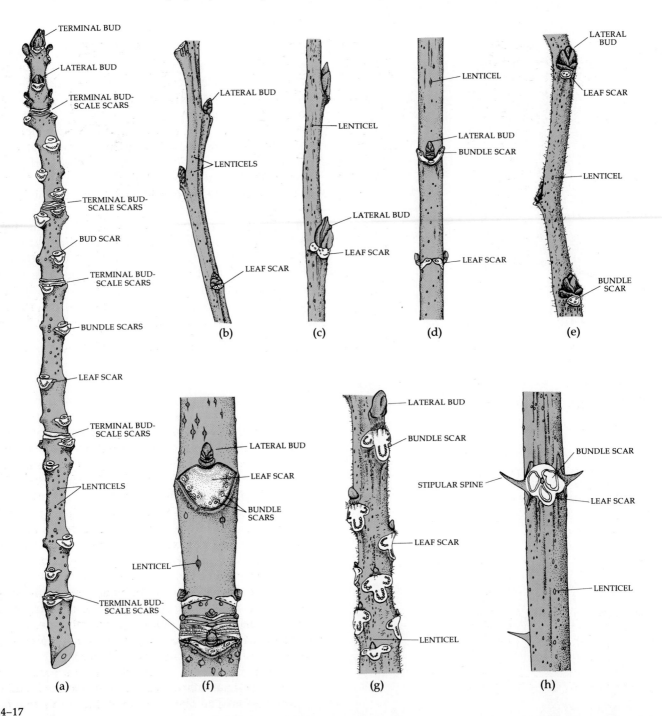

24–17

External features of woody stems. Examination of the twigs of deciduous woody plants reveals many important developmental and structural features of the stem. The most conspicuous structures on the twigs are the buds. Buds occur at the tips— the terminal buds—and in the axils of the leaves—the lateral, or axillary, buds of the twigs. In addition, accessory buds occur in some species. Commonly occurring in pairs, the accessory buds are located one on each side of an axillary bud. In some species, the accessory buds do not develop if their associated axillary bud undergoes normal development. In

others, the accessory buds give rise to flowers and the axillary bud gives rise to a leafy shoot.

After the leaves fall, leaf scars, with their bundle scars, can be seen beneath the axillary buds. The protective layer of the abscission zone produces the leaf scar. The bundle scars are the severed ends of vascular bundles that extended from the leaf traces into the petiole of the leaf, prior to abscission.

Groups of terminal bud-scale scars reveal the locations of previous terminal buds, and until they are obscured by

secondary growth, these groups of scars may be used to determine the age of portions of the stem. The portion of stem between two groups of such scars represents one year's growth. The lenticels appear as slightly raised areas on the stem. (a) *Green ash* (Fraxinus pennsylvanica *var.* subintegerrima). (b) *White oak* (Quercus alba). (c) *Basswood* (Tilia americana). (d) *Box elder* (Acer negundo). (e) *American elm* (Ulmus americana). (f) *Horse chestnut* (Aesculus hippocastanum). (g) *Butternut* (Juglans cinerea). (h) *Black locust* (Robinia pseudo-acacia).

(a)

(b)

(c)

(d)

24–18
Bark of four species of trees. (a) *Thin, peeling bark of the paper birch* (Betula papyrifera). *Horizontal lines on the surface of the bark are lenticels.* (b) *Shaggy bark of the shagbark hickory* (Carya ovata). (c) *Scaly bark of a sycamore, or buttonwood* (Platanus occidentalis). (d) *Deeply furrowed bark of the black oak* (Quercus velutina).

The manner in which new periderms are formed and the kinds of tissues isolated by them have a marked influence on the appearance of the outer surface of the bark (Figure 24–18). In some barks the newly formed periderms develop as discontinuous overlapping layers, resulting in formation of a scaly type of bark, called scale bark (Figures 24–12 and 24–13). Scale barks, for example, are found on relatively young stems of pine *(Pinus)* and pear *(Pyrus communis)* trees. Less commonly, the newly formed periderms arise as more or less continuous, concentric rings around the axis, resulting in formation of a ring bark. Grape *(Vitis)* and honeysuckle *(Lonicera)* are examples of plants with ring barks. The barks of many plants are intermediate between ring and scale barks.

Commercial cork is obtained from the bark of the cork oak, *Quercus suber,* which is native to the Mediterranean region. The first cork cambium of this tree has its origin in the epidermis, and the cork produced by it is of little commercial value. When the tree is about 20 years old, the original periderm is removed, and a new cork cambium is formed in the cortex, just a few millimeters below the site of the first one. The cork produced by the new cork cambium accumulates very rapidly and after about 10 years is thick enough to be stripped off the tree. Once again a new cork cambium arises beneath the previous one, and after about another 10 years the cork can be stripped again. This procedure may be repeated at about 10-year intervals until the tree is 150 or more years old. The spots and long dark streaks seen on the surfaces of commercial cork are lenticels.

In most woody roots and stems, very little secondary phloem is actually involved in the conduction of food. In most species, only the current year's growth increment, or growth ring, of secondary phloem is active in the long-distance transport of food through the stem. This is because the sieve elements are short-lived (see Chapter 21); most of them die by the end of the same year in which they are derived from the vascular cambium. In some plants, such as black locust (*Robinia pseudo-acacia*), the sieve elements collapse and are crushed relatively soon after they die (Figures 24–14 through 24–16).

The part of the inner bark actively engaged in the transport of food substances is called **functional phloem.** Although the sieve elements outside the functional phloem are dead, the phloem parenchyma cells (axial parenchyma) and parenchyma cells of the rays may remain alive and continue to function as storage cells for many years. This part of the inner bark is known as **nonfunctional phloem.** Only the outer bark is composed entirely of dead tissue (Figures 24–14 and 24–16).

The Wood: Secondary Xylem

Apart from use of various plant tissues as food for humans, no single plant tissue has played a more indispensable role in human survival throughout recorded history than wood, or secondary xylem (see Table 24–1, page 542). Commonly, woods are classified as either **hardwoods** or **softwoods.** The so-called hardwoods are dicot woods, and the softwoods are conifer woods. The two kinds of woods have basic structural differences, but the terms "hardwood" and "softwood" do not accurately express the degree of density (weight per unit volume) or hardness of the wood. For example, one of the lightest and softest of woods is balsa (*Ochroma lagopus*), a tropical dicot. By contrast, the woods of some conifers, such as slash pine (*Pinus elliottii*), are harder than some hardwoods. Although this chapter deals largely with secondary growth in angiosperms, it is both practical and instructive to consider conifer and dicot woods together.

C O N I F E R   W O O D

The structure of conifer wood is relatively simple compared with that of most dicots. The principal features of conifer wood are its lack of vessels (see Chapter 21 for a discussion of tracheary elements) and its relatively small amount of axial, or wood, parenchyma. Long, tapering tracheids constitute the dominant cell type in the

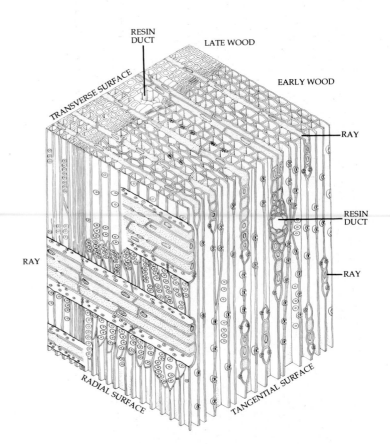

24–19
Block diagram of the secondary xylem of white pine (Pinus strobus). With the exception of the parenchyma cells associated with the resin ducts, the axial system consists entirely of tracheids. The rays are only one cell wide, except for those containing resin ducts. (Early and late wood are described on page 536.)

axial system. In certain genera, such as *Pinus*, the only parenchyma cells of the axial system are those associated with **resin ducts.** Resin ducts are relatively large intercellular spaces lined with thin-walled parenchyma cells that secrete resin into the duct. In *Pinus*, resin ducts occur in both the axial system and the rays (Figures 24–19 and 24–20). Wounding, pressure, and injuries by frost and wind can stimulate the formation of resin ducts in conifer wood, leading some investigators to suggest that all resin ducts are traumatic in origin. Resin apparently protects the plant from attack by decay-producing fungi and bark beetles.

The tracheids of conifers are characterized by large, circular, bordered pits that are most abundant on the ends of the cells, where they overlap with other tracheids (Figures 24–19 through 24–21, page 534). The pairs of pits (pit-pairs; see Chapter 2, page 36) between conifer tracheids are each characterized by the presence

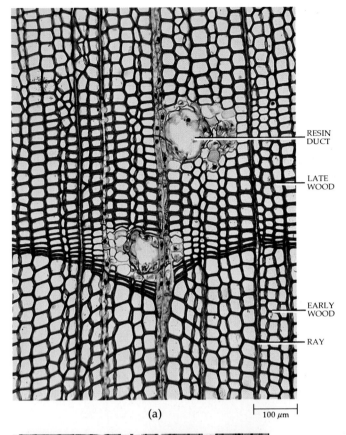

RESIN DUCT

LATE WOOD

EARLY WOOD

RAY

(a)

100 μm

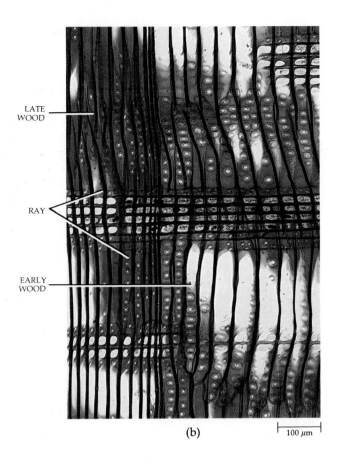

LATE WOOD

RAY

EARLY WOOD

(b)

100 μm

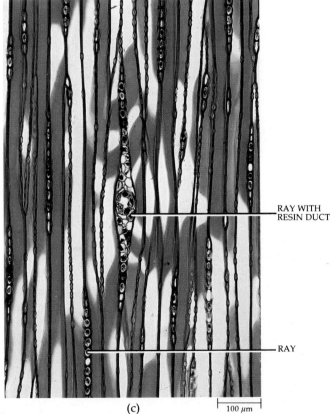

RAY WITH RESIN DUCT

RAY

(c)

100 μm

24–20

Wood of white pine (Pinus strobus), *a conifer, in* (a) *transverse,* (b) *radial, and* (c) *tangential sections. (White pine wood has a specific gravity of 0.34. See page 540.)*

of a **torus** (plural: tori). The torus is a thickened central portion of the pit membrane (Figure 24–22) and is slightly larger than the openings, or apertures, in the pit-borders (Figure 24–21). The pit membrane is flexible, and under certain conditions, the torus may block one of the apertures and prevent the movement of water or gases through the pit-pair. Although long thought to occur only in the pit membranes of the bordered pit-pairs of tracheids in certain genera of gymnosperms, tori and torus-like structures have recently been reported in the bordered pit-pairs of tracheids and vessel members in several genera of dicotyledons.

Figure 24–19 is a three-dimensional diagram of the wood of white pine *(Pinus strobus)* based on the three wood sections shown in Figure 24–20. In sections cut at right angles to the long axis of the root or stem—transverse, or cross, sections—the tracheids appear angular or squarish, and the rays can be seen in their longitudinal extent traversing the wood (Figure 24–20a). There are two kinds of longitudinal sections—radial and tangential. Radial sections are cut parallel to the rays, and in such sections, the rays appear as sheets of cells oriented at right angles to the vertically elongated tracheids of the axial system (Figures 24–20b and 24–21d). Tangential sections are cut at right angles to the rays and reveal the width and height of the rays. In *Pinus* the rays are one cell wide, except for those containing resin ducts (Figure 24–20c). Details of white pine wood are shown in Figure 24–21.

533

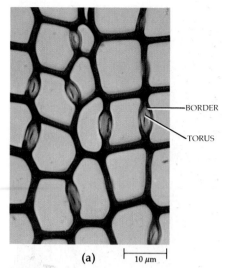

(a)

10 μm

BORDER

TORUS

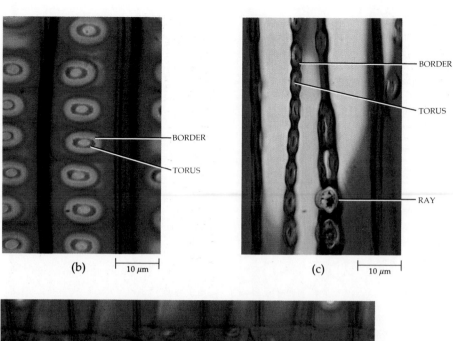

(b)

10 μm

BORDER

TORUS

(c)

10 μm

BORDER

TORUS

RAY

24–21

*Details of white pine (Pinus strobus)
wood. (a) Transverse section, showing
bordered pit-pairs on radial walls of
tracheids. (b) Radial section, showing
the face view of bordered pit-pairs in
walls of tracheids. (c) Tangential section,
showing bordered pit-pairs of tracheids.
(d) Radial section, showing ray. The rays
of pine and other conifers are composed
of ray tracheids and ray parenchyma
cells. Here ray tracheids occur at the top
and bottom of the ray and ray parenchyma cells in the middle. Notice the
bordered pits of ray tracheids. Above the
ray parenchyma are two adjacent
bordered pit-pairs.*

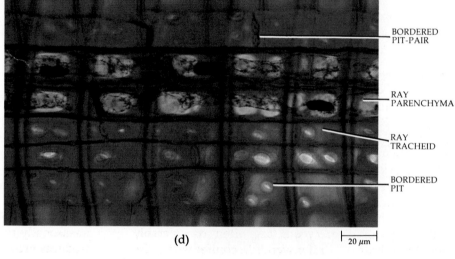

(d)

20 μm

BORDERED
PIT-PAIR

RAY
PARENCHYMA

RAY
TRACHEID

BORDERED
PIT

24–22

*Scanning electron micrograph of pit
membrane of bordered pit-pair in white
pine (Pinus strobus) tracheid. The thickened part of the membrane is the torus.
The part of the membrane surrounding
the torus is called the margo and is very
porous, allowing the movement of water
and ions from tracheid to tracheid.*

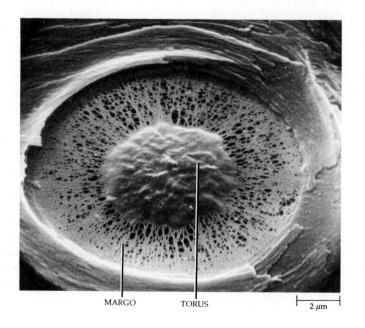

MARGO   TORUS   2 μm

24–23
Transverse sections of wood, showing growth layers. (a) *Red oak* (Quercus rubra). *The large vessels of ring-porous wood such as red oak are found in the early wood. The dark vertical lines are rays.* (b) *Tulip tree* (Liriodendron tulipifera), *a diffuse-porous wood.*

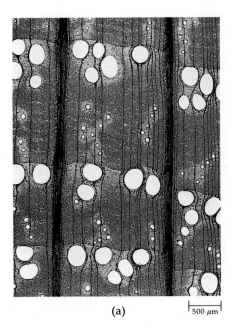

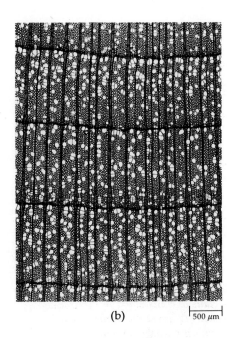

(a) |500 μm| (b) |500 μm|

DICOT WOODS

Wood structure in dicotyledons is much more varied than in conifers, owing in part to a greater number of cell types in the axial system, including vessel members, tracheids, several types of fibers, and parenchyma cells (Figures 24–23 and 24–24; see also Figures 21–12, page 459, and 24–25). The presence of vessel members, in particular, distinguishes dicot woods from conifer wood, with only a few exceptions.

The rays of dicot woods are often considerably larger than those of conifer wood. In conifer wood, the rays are predominantly one cell wide, and most range from one to 20 cells high. The rays of dicot woods range from one to many cells wide and from one to several hundred cells high. In some dicot woods, such as oak, the large rays can be seen with the unaided eye (Figure 24–12). The large rays of the red oak *(Quercus rubra)* wood illustrated in Figure 24–24c are 12 to 30 cells wide and hundreds of cells high. Besides the large rays, oak wood has numerous rays that are only one cell wide. In red oak wood, the rays make up, on average, about 21 percent of the volume of the wood. Overall, the rays of hardwoods average about 17 percent of the volume of the wood; the average for conifer wood is about 8 percent.

As in conifer wood, transverse sections of dicot woods reveal radial files of cells of both the axial and radial systems derived from the cambial initials (Figures 24–23 and 24–24a). The files may not be as orderly as in conifer wood, however, for enlargement of the vessels and elongation of fibers tend to push many of the cells out of position. The displacement of rays by vessel

members is particularly conspicuous in the transverse section of red oak wood, as shown by the rays to the left of the vessels in Figure 24–24a.

GROWTH RINGS

The periodic activity of the vascular cambium, which is a seasonally related phenomenon in temperate zones, produces growth increments, or **growth rings,** in both secondary xylem and secondary phloem (in the phloem the increments are not always readily discernible). If a growth layer represents one season's growth, it is called an **annual ring.** Abrupt changes in available water and other environmental factors may be responsible for the production of more than one growth ring in a given year; such rings are called *false annual rings.* Thus the age of a given portion of the old woody stem can be estimated by counting the growth rings, but the estimates may be inaccurate if false annual rings are included.

The width of individual growth rings may vary greatly from year to year as a function of such environmental factors as light, temperature, rainfall, available soil water, and length of the growing season. The width of a growth ring is a fairly accurate index of the rainfall of a particular year. Under favorable conditions—that is, during periods of adequate or abundant rainfall— the growth rings are wide; under unfavorable conditions, they are narrow.

In semiarid regions, where there is very little rain, the tree is a sensitive rain gauge. An excellent example

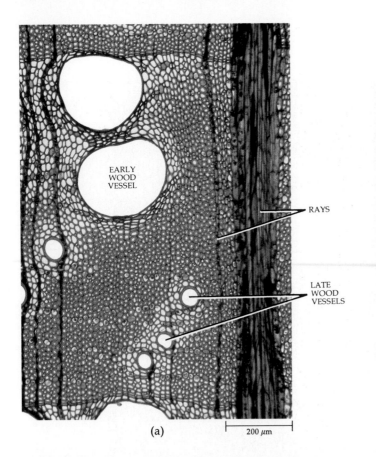

EARLY WOOD VESSEL

RAYS

LATE WOOD VESSELS

200 μm

(a)

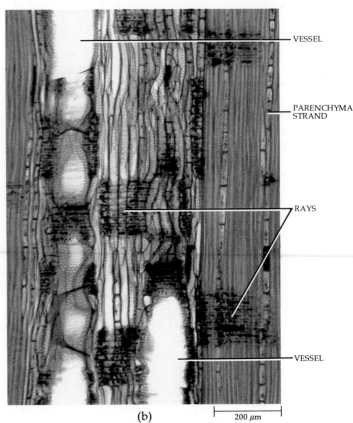

VESSEL

PARENCHYMA STRAND

RAYS

VESSEL

200 μm

(b)

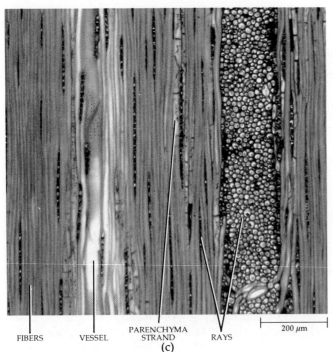

FIBERS VESSEL PARENCHYMA STRAND RAYS

200 μm

(c)

24–24
Wood of red oak (Quercus rubra) *in (a)
transverse, (b) radial, and (c) tangential
sections. (Red oak wood has a specific
gravity of 0.57.)*

of this is the bristlecone pine *(Pinus longaeva)* of the western Great Basin (Figure 24–26). Each growth ring is different, and a study of the rings tells a story that dates back thousands of years. The oldest-known living specimen of bristlecone pine is 4900 years old. Dendrochronologists—scientists who conduct historical research through the growth rings of trees—have been able to match samples of wood from living and dead trees, and in this way they have built up a continuous series of rings dating back more than 8200 years. The widths of the growth rings of bristlecone pines at the higher elevations (the upper tree line) have also been found to be closely related to temperature changes, and a record of average ring width in these trees provides a valuable guide to past temperatures and climatic conditions. For example, in the White Mountains of California, the summers were relatively warm from 3500 B.C. to 1300 B.C., and the tree line was about 150 meters above its present level. Summers were cool from 1300 B.C. to 200 B.C.

The structural basis for the visibility of growth layers in the wood is the difference in density of the wood produced early in the growing season and that produced later (Figures 24–20, 24–23, and 24–24). The **early wood** is less dense (with wider cells and proportionally thinner walls) than the **late wood** (with narrower cells and proportionally thicker walls). In a given growth layer, the change from early to late wood may be very gradual and almost imperceptible. However,

24–25
*Scanning electron micrograph of block of American elm (*Ulmus americana*) wood, showing the three faces, or surfaces, of the wood. By comparison with Figures 24–19, 24–23, and 24–24, you should be able to identify each surface. This is a semi-ring-porous wood, with the late-wood vessels arranged in wavy lines, a characteristic feature of the elms. Identify the early-wood and late-wood vessels and the rays in all three faces. The dense portion of the wood is composed largely of fibers. Axial parenchyma cells are also present but are not distinguishable at this magnification. (American elm wood has a specific gravity of 0.46.)*

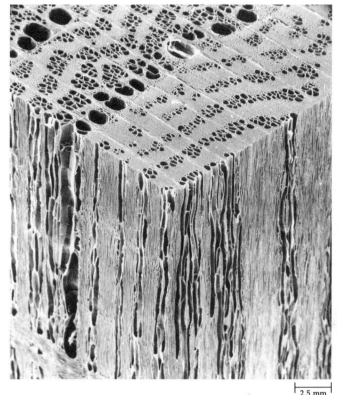

2.5 mm

(a)

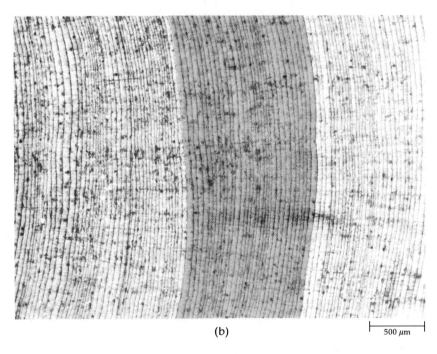

(b)

500 μm

24–26
(a) *Bristlecone pine (*Pinus longaeva*) in the White Mountains of eastern California. These pines, which grow near the timberline, are the oldest living trees; one tree has reached an age of 4900 years.* (b) *A transverse section of wood from a bristlecone pine, showing the variation in width of annual rings. This section begins approximately 6260 years ago;* the band of rings in color represents the thirty years from 4240 B.C. to 4210 B.C. The overlapping patterns of rings in dead trees have made it possible to determine relative precipitation extending back some 8200 years.

Despite its advanced age, the 4900-year-old bristlecone pine may not be the oldest living thing on earth. In fact, a ring of creosote bushes (Larrea divaricata), all apparently derived from one seed, has been estimated to be about 12,000 years old. Growing in the Mojave Desert 150 miles northeast of Los Angeles, the ring of bushes has been dubbed the "King Clone." (See also Figure 30–15.)

where the late wood of one growth layer abuts on the early wood of a following growth layer, the change is abrupt and thus clearly discernible.

In some dicot woods, size differences of the vessels, or pores, in early and late woods are quite marked, the pores of the early wood being distinctly larger than those of the late wood. (The term "pore" is used by the wood anatomist for a vessel seen in cross section.) Such woods are called **ring-porous** woods (Figures 24–23a and 24–24a). In other dicot woods, the pores are fairly uniform in distribution and size throughout the growth layer. These woods are called **diffuse-porous** woods (Figure 24–23b). In ring-porous woods, almost all the water is conducted in the outermost growth layer, at speeds about ten times greater than in diffuse-porous woods.

SAPWOOD AND HEARTWOOD

As the wood grows older and no longer serves as a conducting tissue, its parenchyma cells eventually die. Before this happens, however, the wood often undergoes visible changes, which involve the loss of reserve foods and the infiltration of the wood by various substances (such as oils, gums, resins, and tannin), which color it and sometimes make it aromatic. This often darker, nonconducting wood is called **heartwood,** while the generally lighter, conducting wood is called **sapwood** (see Figure 24–12, page 527). The proportion of sapwood to heartwood and the degree of visible difference between them vary greatly from species to species. Some trees, such as maple (Acer), birch (Betula), and ash (Fraxinus), have thick sapwoods; whereas others, such as locust (Robinia), catalpa (Catalpa), and yew (Taxus), have thin sapwoods. Still other trees, such as the poplars (Populus), willows (Salix), and firs (Abies),

have no clear distinction between sapwood and heartwood.

In many woods, tyloses are formed in the vessels when they become nonfunctional (Figure 24–27). **Tyloses** are balloonlike outgrowths from ray or axial parenchyma cells through pit cavities in the vessel wall. They may completely occlude the lumen (space bounded by the cell wall) of the vessel. Tyloses often are induced to form prematurely or unnaturally by plant pathogens and may serve as a defensive mechanism by inhibiting spread of the pathogen throughout the plant via the xylem.

REACTION WOOD

Reaction wood is a term applied to abnormal wood that develops in leaning trunks and limbs. Its formation is related to the process of straightening efforts of those plant parts. In conifers, the reaction wood develops on the underside of the leaning part and is called *compression wood*. In dicotyledons, it develops on the upper side and is called *tension wood*.

Compression wood is produced by the increased activity of the vascular cambium on the lower side of the bent stem and results in the formation of eccentric growth rings. Portions of growth rings located on the lower side are generally much wider than those located

24–27
Tyloses, balloonlike outgrowths of parenchyma cells, which partially or completely block the lumen of the vessel. (a) Transverse and (b) longitudinal sections showing tyloses in vessels of white oak (Quercus alba), *as seen with a light microscope.*

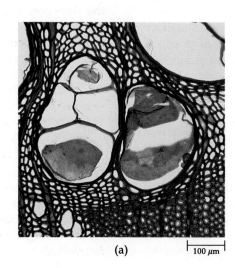

(a)   100 μm

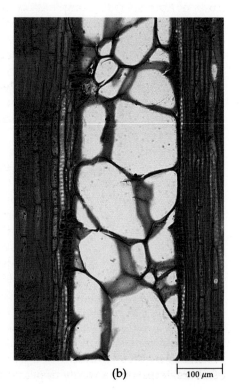

(b)   100 μm

24–28
Transverse section of the stem of a pine (Pinus sp.), showing compression wood with wider growth rings on the lower side.

24–29
Trunk of a dead white oak (Quercus alba) tree from which the bark has fallen, revealing the spiral grain of the wood.

on the upper side (Figure 24–28). Hence, compression wood causes straightening by expanding or pushing the trunk or limb upright. Compression wood has more lignin and less cellulose than normal wood, and its lengthwise shrinkage upon drying is often ten or more times as great as that of normal wood. (Normal wood usually shrinks lengthwise not more than 0.1 to 0.3 percent.) The difference in relative lengthwise shrinkage of normal and compression wood in a drying board often causes the board to twist and cup. Such wood is virtually useless except as fuel.

Tension wood is produced by the increased activity of the vascular cambium on the upper side of the stem and, as in the case of compression wood, is recognized by the presence of eccentric growth rings. To straighten the stem, the tension wood must exert a pull; hence the name "tension wood." Positive identification of tension wood requires microscopic examination of wood sections. Anatomically, the principal distinguishing feature is the presence of gelatinous fibers—fibers with little or no lignification and in which part of the secondary wall has a gelatinous appearance. Lengthwise shrinkage of tension wood rarely exceeds one percent, but boards containing it twist out of shape in drying. When sawn green, tension wood tears loose in bundles of fibers, imparting a wooly appearance to the boards.

THE GROSS FEATURES OF WOOD

The appearance of wood depends on several factors, among them color, grain, texture, and figure. Not only are some of these characteristics helpful in identification of various kinds of woods, but they are responsible for the decorative qualities of woods.

Color in wood is variable both between different kinds of wood and within a species. Color of heartwood can be important in identification of that wood; in addition, it can be responsible in part for the preferential uses of wood. The rich chocolate or purplish brown heartwood of black walnut (*Juglans nigra*) and the red-brown heartwood of black cherry (*Prunus serotina*) are examples of longtime favorites for high-quality furniture.

Grain in wood is a term used when referring to the direction of alignment of wood elements, or axial components—fibers, tracheids, parenchyma cells, and vessel members—when considered *en masse*. The alignment of the axial components is a reflection of the alignment of the fusiform initials that gave rise to them. When all the elements are oriented parallel to the longitudinal axis, the grain is said to be *straight*. When the alignment in a piece of wood does not coincide with the longitudinal axis of the piece, the wood is said to be *cross-grained*. *Spiral grain* is applied to the spiral arrangement of elements in a log or tree trunk. The log or trunk with spiral grain has a twisted appearance after the bark has been removed (Figure 24–29). If the orientation of the spiral is reversed at more or less regular intervals along a single radius, the grain is said to be *interlocked*.

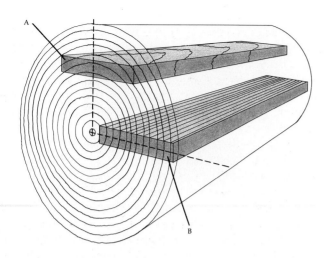

24–30

Diagram illustrating the direction in which logs are sawn to produce plain-sawn (A) and quarter-sawn (B) boards. In plain-sawn boards the growth rings are more or less parallel to the broad faces. In quarter-sawn boards, the rings are more or less at right angles to the broad faces.

Texture of wood refers to the relative size and amount of variation in size of elements within the growth increments. *Coarse* texture can result from the presence of wide bands of large vessels and broad rays, as in some ring-porous woods, and *fine* texture from the presence of small vessels and narrow rays. *Even* texture is the result of uniformity in cell dimensions—there is no perceptible difference between the early and late woods. The texture is *uneven* when distinct differences exist between early and late woods of a growth ring.

Figure refers to the patterns found on the longitudinal surfaces of wood. In a restricted sense, the term "figure" is used to refer to the more decorative woods prized in the furniture and cabinet-making industries. Figure depends on grain and texture and on the orientation of the surface that results from sawing.

Boards are sawn from logs in two ways with reference to the radius of the log (Figure 24–30). In one, the wood is cut such that the broad faces of the board roughly parallel the tangential-longitudinal plane of the log. Tangentially cut boards are said to be *flat-cut* or *plain-sawn*. The growth rings in plain-sawn boards appear as wavy bands. In the other way of sawing, a radial cut is made longitudinally through the center of the log. Radially cut boards are said to be *quarter-sawn*. In this type of stock the growth rings appear as parallel

lines extending the length of the board, with the rays crossing the growth increments at right angles. Quarter-sawn boards are preferred in many applications because the radial surfaces have more uniform wearing and finishing properties; moreover, they do not warp. Quarter-sawing is more time-consuming, however, and often more wasteful than plain-sawing.

Density and Specific Gravity of Wood

Density is the single most important indicator of the strength of wood and can be used to predict such characteristics as hardness, nailing resistance, and ease of machining. Dense woods generally shrink and swell more than light woods. In addition, the densest woods make the best fuel.

The *specific gravity* of a substance is the ratio of the weight of the substance to the weight of an equal volume of water. In the case of wood, the oven dry weight of the wood is used in figuring its specific gravity:

$$\text{specific gravity} = \frac{\text{oven dry weight of wood}}{\text{weight of the displaced volume of water}}$$

The specific gravity of dry solid wood substance (that is, oven dry cell wall material) of all plants is about 1.5. The differences in specific gravity of woods depend, therefore, upon the proportion of the wood composed of wall substance and lumen (the space bounded by the cell wall).

Fibers are especially important in determination of the specific gravity. If the fibers are thick-walled and narrow-lumened, the specific gravity tends to be high. Conversely, if the fibers are thin-walled and wide-lumened, it tends to be low. The presence of numerous thin-walled vessels also tends to lower the specific gravity.

Density is expressed as weight per unit volume, either as pounds per cubic foot (English) or as grams per cubic centimeter (metric). Water has a density of 62.4 lb/ft³, or 1 g/cm³. A wood weighing 31.2 lb/ft³, or 0.5 g/cm³, is therefore one-half as dense as water and has a specific gravity of 0.5. The *Guinness Book of World Records* lists black ironwood (*Olea capensis*) of South Africa as the heaviest (93 lb/ft³, or 1.49 g/cm³) wood and *Aeschynomene hispida* of Cuba as the lightest (2.75 lb/ft³, or 0.044 g/cm³) wood. Their respective specific gravities are 1.49 and 0.044. The specific gravities of most commercially useful woods are between 0.35 and 0.65.

See Table 24–1 on pages 542 to 543 for the uses of wood of some common North American trees.

Summary

Secondary growth (the increase in girth in regions that are no longer elongating) occurs in all gymnosperms and in most dicotyledons and involves the activity of the two lateral meristems—the vascular cambium and the cork cambium, or phellogen. Herbaceous plants may undergo little or no secondary growth, whereas woody plants—trees and shrubs—may continue to increase in thickness for many years. Figure 24–31 presents a summary of stem development in a woody plant, beginning with the apical meristem and ending with the secondary tissues produced during the first year's growth.

The vascular cambium contains two types of initials —fusiform initials and ray initials. Through periclinal divisions, the fusiform initials give rise to the components of the axial system, and the ray initials produce ray cells, which form the vascular rays, or radial system. Increase in circumference of the cambium is accomplished by anticlinal division of the initials.

The first cork cambium in most stems originates in a layer of cells immediately below the epidermis. The cork cambium produces cork toward the outside and phelloderm toward the inside. Together, the cork cambium, cork, and phelloderm constitute the periderm. Although most of the periderm consists of compactly arranged cells, isolated areas called lenticels have numerous intercellular spaces.

The bark consists of all tissues outside the vascular cambium. In old roots and stems, most of the phloem of the bark is nonfunctional. Sieve elements are short-lived, and generally only the present year's growth increment contains conducting, or functional, sieve elements. After the first periderm, subsequently formed periderms originate deeper and deeper in the bark from parenchyma cells of nonfunctional phloem.

Woods are classified as either softwoods or hardwoods. All so-called softwoods are conifers and all so-called hardwoods are dicotyledons. Compared with dicot woods, conifer woods are simple, consisting of tracheids and parenchyma cells. Some contain resin ducts. Dicot woods may contain a combination of all of the following cell types: vessel members, tracheids, several types of fibers, and parenchyma cells.

Growth layers that correspond to yearly increments of growth are called annual rings. The difference in density between the late wood of one growth increment and the early wood of the following increment makes it possible to distinguish the growth layers. Density and specific gravity are good indicators of the strength of wood.

In many plants, the nonconducting heartwood is visibly distinct from the actively conducting sapwood.

Commonly, reaction wood develops on the underside of leaning trunks and limbs of conifers and on the upper side of similar parts in dicotyledons; its formation causes straightening of the trunk or limb. Reaction wood is called compression wood in conifers and tension wood in dicots.

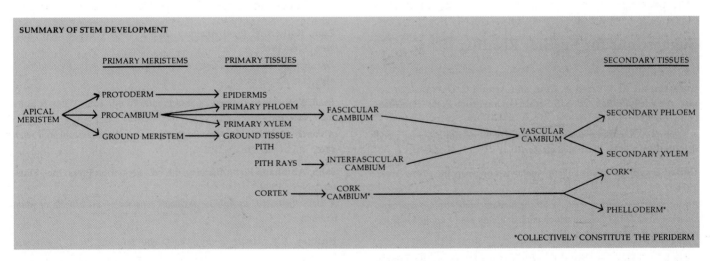

SUMMARY OF STEM DEVELOPMENT

24–31
Summary of stem development in a woody dicot during the first year of growth. (Compare this with root development, as summarized in Figure 22–25.)

Table 24-1   *Wood Uses of Some Common North American Trees*

Dicot woods

Alder, red *(Alnus rubra)*—principal hardwood of Pacific Northwest. Used for furniture, especially chairs; sash, doors, and other millwork; plywood; charcoal; important source of pulpwood.

Ash, white *(Fraxinus americana)*—handles, especially long ones (shovel, spade, rake) because of its straightness of grain, strength, moderate weight, and other qualities; almost all baseball bats; oars, paddles, tennis-racket frames, hockey sticks; kitchen cabinets; toys and woodenware.

Basswood, or linden *(Tilia americana)*—veneer for plywood used as drawer panels and other concealed furniture parts; excelsior (wood shavings used in packing); sash, doors, and other millwork; piano keys; boxes and crates; caskets and coffins.

Beech, American *(Fagus americana)*—one of three important northern hardwoods, the others being yellow birch and sugar maple. Used for planing-mill products, especially flooring; veneer; fuel wood; hardwood distillation yielding acetic acid, methanol, and other chemicals; toys and woodenware.

Birch, yellow *(Betula alleghaniensis)*—veneer; hardwood distillation; furniture; toys and woodenware; musical instruments; toothpicks.

Cherry, black *(Prunus serotina)*—a superb cabinet wood; finished furniture; printers' blocks to which electrotypes are mounted; piano actions; interior trim; paneling; handles; toys and woodenware.

Cottonwood *(Populus deltoides)*—pulpwood for high-grade paper used in books and magazines; concealed parts of furniture; excelsior; tubs and pails for food products; boxes and crates.

Elm *(Ulmus spp.)*—has interlocked grain and, consequently, is difficult to split. Used for staves and hoops; boxes and crates; veneer for fruit and vegetable containers and round cheeseboxes; bent parts in furniture; interior trim.

Hickory, bitternut *(Carya cordiformis)*—tool handles, especially for axes, picks, and sledges; ladders; furniture; woodenware; famous for smoking meats; prime fuel wood.

Locust, black *(Robinia pseudo-acacia)*—mine timbers; railroad ties; fence posts; in construction where strength and durability are of great importance.

Maple, sugar *(Acer saccharum)*—veneer; hardwood distillation; railroad ties; fuel wood; furniture; flooring, especially bowling alleys and dance floors; toys and woodenware; musical instruments.

Oak, red and white *(Quercus rubra* and *Q. alba)*—railroad ties; veneer; flooring; sash, doors, and other millwork; firewood; ship and boat building; caskets and coffins.

Suggestions for Further Reading

Core, Harold A., Wilfred A. Côté, and Arnold C. Day: *Wood Structure and Identification,* 2d ed., Syracuse University Press, Syracuse, N.Y., 1979.*

A beautifully illustrated manual of wood structure, with a key for wood identification and an illustrated glossary.

Cutler, David F.: *Applied Plant Anatomy,* Longman Inc., New York, 1978.*

An interestingly written textbook on the fundamentals of plant anatomy, showing some of the ways in which plant anatomy can be applied to solve many important everyday problems.

Cutter, Elizabeth G.: *Plant Anatomy, Part I: Cells and Tissues,* 2d ed., Addison-Wesley Publishing Co., Inc., Reading, Mass., 1978.

An introduction to plant cells and tissues, with emphasis on experimental work.

Cutter, Elizabeth G.: *Plant Anatomy, Part II: Organs,* Addison-Wesley Publishing Co., Inc., Reading, Mass., 1971.*

An introduction to the organs of the plant and possible causes underlying their development.

Esau, Katherine: *Plant Anatomy,* 2d ed., John Wiley & Sons, Inc., New York, 1965.

The standard work in the field; a well-illustrated book that considers all aspects of plant anatomy.

Esau, Katherine: *Anatomy of Seed Plants,* 2d ed., John Wiley & Sons, Inc., New York, 1977.

A shorter book than the preceding entry; an excellent textbook and reference.

Fahn, Abraham: *Plant Anatomy,* 4th ed., Pergamon Press, Inc., Elmsford, N.Y., 1990.*

A well-illustrated, up-to-date textbook considering all aspects of plant anatomy.

Hoadley, R.B.: *Understanding Wood,* The Taunton Press, Newtown, Conn., 1980.

Written by a wood technologist and woodworker, this well-illustrated book tells all about the properties of wood and how to work with it.

Metcalfe, C.R., and L. Chalk: *Anatomy of the Dicotyledons,* vol. 1, 2d ed., Clarendon Press, Oxford, England, 1979.

A systematic anatomy of leaf and stem, with a brief history of the subject.

Persimmon *(Diospyros virginiana)*—golf-club heads; boxes and crates; handles.

Sycamore, American *(Platanus occidentalis)*—has interlocked grain. Used for veneer; boxes and crates; interior trim and paneling; flour and sugar barrels; concealed parts of furniture.

Walnut, black *(Juglans nigra)*—the finest cabinet wood native to the continental United States; veneer for plywood faces used in manufacture of furniture; goes directly into high-grade tables and chairs; leading wood for gunstocks; caskets and coffins.

Yellow poplar, or tulip tree *(Liriodendron tulipifera)*—veneer for plywood used for interior finish, furniture, and cabinetwork; pulpwood; boxes and crates; sash, doors, and other millwork.

Conifer woods

Cedar, western red *(Thuja plicata)*—premier wood for shingles; poles and posts; boat-building; greenhouse construction; exterior siding; sash and doors; millwork and interior finishing; caskets and coffins.

Douglas fir *(Pseudotsuga menziesii)*—the western forests of the United States are about 50 percent Douglas fir, which furnishes more timber than any other single species grown in the United States. Used for building construction, including veneer converted largely into plywood; railroad ties; mine timbers; pulpwood; boxes and crates; ship and boat building.

Hemlock, eastern *(Tsuga canadensis)*—pulpwood; general construction; boxes and crates; sash and doors; kitchen cabinets; tannins.

Pine, loblolly *(Pinus taeda)*—interior finish; frame and sash; wainscoting, joists, and subflooring.

Pine, ponderosa *(Pinus ponderosa)*—boxes and crates; sash, doors, and other millwork; building construction; turned work (posts, balusters, porch columns); poles; toys; caskets and coffins.

Pine, slash *(Pinus elliottii)*—pulpwood; heavy timbers; railroad ties; veneer; turpentine and rosin; boxes; baskets and crates.

Pine, sugar *(Pinus lambertiana)*—boxes and crates; sash, doors, and other millwork; signs; piano keys and organ pipes.

Pine, western white *(Pinus monticola)*—matches; boxes and crates; building construction; sash, doors, and other millwork; core stock for plywood, especially tabletops.

Redwood *(Sequoia sempervirens)*—building construction of all sorts; ship and boat building; garden furniture; shingles and shakes; caskets and coffins.

Spruce, red *(Picea rubens)*—most important use is for pulpwood; sounding boards for musical instruments; paddles and oars; ladder rails; ship and boat building; boxes and crates.

Metcalfe, C.R., and L. Chalk: *Anatomy of the Dicotyledons*, vol. 2, 2d ed., Clarendon Press, Oxford, England, 1983.

This volume considers a variety of subjects, including wood structure, secretory structures, anatomy in relation to phylogeny and taxonomy, and ecological anatomy and morphology.

O'Brien, Terence, P., and Margaret E. McCully: *Plant Structure and Development*, Macmillan Publishing Company, New York, 1969.

A pictorial and physiological approach to plant structure and development.

O'Brien, Terence P., and Margaret E. McCully: *The Study of Plant Structure: Principles and Selected Methods*, Termarcarphi and Pty. Ltd., Melbourne, Australia, 1981.

A very useful book dealing with methods for the study of plant structure, ranging from fresh to variously fixed tissues studied with simple lenses, at one extreme, and the electron microscope, at the other.

Panshin, A.J., and C. De Zeeuw: *Textbook of Wood Technology*, vol. 1, 4th ed., McGraw-Hill Book Company, New York, 1980.

A technical reference on the identification, structure, and utilization of woods native to North America.

Steeves, Taylor A., and Ian M. Sussex: *Patterns in Plant Development*, 2d ed., Prentice Hall, Inc., Englewood Cliffs, N.J., 1989.

A structural approach to plant development, with emphasis on experimental and analytical data.

White, Richard A., and William C. Dickinson (eds.): *Contemporary Problems in Plant Anatomy*, Academic Press, Inc., Orlando, Fla., 1984.

A collection of papers presented at a plant anatomy symposium held at Duke University and the University of North Carolina in early 1983, dealing with problems in contemporary anatomy. The central theme of all contributions is to view the plant as a developmental, structural, and functional whole.

Zimmermann, Martin H.: *Xylem Structure and the Ascent of Sap*, Springer-Verlag, New York, 1983.

A delightfully written "idea" book on xylem structure and function by one who contributed a great deal to our understanding of functional xylem anatomy and sap movement in plants.

Zimmerman, Martin H., and Claude L. Brown: *Trees: Structure and Function*, Springer-Verlag, New York, 1975.

A discussion of how trees work, with emphasis on structure as it relates to function.

*Available in paperback.

Physiology of Seed Plants

The green, photosynthesizing leaves of the grapevine are its principal source of assimilates, and its fruits, the grapes, one of its principal sinks. The tendrils of the grapevine are modified stems, which wrap around any object with which they come in contact, a phenomenon known as thigmotropism.

CHAPTER 25

Regulating Growth and Development: The Plant Hormones

A plant, in order to grow, needs light from the sun, carbon dioxide from the air, and water and minerals, including nitrogen, from the soil. As discussed in Section 4, the plant does far more than simply increase its mass and volume as it grows. It differentiates, develops, and takes shape, forming a variety of cells, tissues, and organs. Many of the details of how these processes are regulated are not known, but it has become clear that normal development depends on the interplay of a number of internal and external factors. The principal internal factors that regulate plant growth and development are chemical, and they are the subject of this chapter. Some of the external factors—such as light, temperature, day length, and gravity—that affect plant growth are discussed in Chapter 26.

Plant hormones are organic substances that play a major role in regulating growth. Some hormones are produced in one tissue and transported to another tissue, where they produce specific physiological responses; others act within the same tissues where they are produced. Hormones are active in very small quantities. In the shoot of a pineapple plant (*Ananas comosus*), for example, there are only 6 micrograms of indoleacetic acid, a common plant hormone, per kilogram of plant material. One enterprising plant physiologist calculated that the weight of the hormone in relation to that of the shoot is comparable to the weight of a needle in 20 metric tons of hay.

The word **hormone** comes from the Greek *horman*, meaning "to set in motion." It is now clear, however, that some hormones have inhibitory influences. Therefore, rather than thinking of hormones as stimulators, it is more useful to regard them as chemical regulators. But this term also needs qualification because the response to the particular regulator depends not only on its chemical structure but on how it is "read" by the target tissue; that is, the same hormone can elicit different responses in different tissues or at different times of development in the same tissue.

545

Table 25–1   *Plant Hormones: Their Nature, Occurrence, and Effects*

Hormone	Nature	Sites of Biosynthesis	Transport	Effects
Auxin	Indole-3-acetic acid is the only known naturally occurring auxin. It is synthesized primarily from tryptophan.	Primarily in leaf primordia and young leaves and in developing seeds	IAA is transported from cell to cell and transport is unidirectional (polar).	Apical dominance; tropic responses; vascular tissue differentiation; promotion of cambial activity; induction of adventitious roots on cuttings; inhibition of leaf and fruit abscission; stimulation of ethylene synthesis; inhibition or promotion (in pineapples) of flowering; stimulation of fruit development
Cytokinins	N^6-adenine derivatives, phenyl urea compounds. Zeatin is the most common cytokinin in plants.	Primarily in root tips	Cytokinins are transported via the xylem from roots to shoots.	Cell division; promotion of shoot formation in tissue culture; delay of leaf senescence; application of cytokinin can cause release of lateral buds from apical dominance
Ethylene	The gas ethylene (C_2H_4) is synthesized from methionine. It is the only hydrocarbon with a pronounced effect on plants.	In most tissues in response to stress, especially in tissues undergoing senescence or ripening	Being a gas, ethylene moves by diffusion from its site of synthesis.	Fruit ripening (especially in climacteric fruits, such as apples, bananas, and avocados); leaf and flower senescence; leaf and fruit abscission
Abscisic acid	"Abscisic acid" is a misnomer for this compound, for it has little to do with abscission. It is synthesized from mevalonic acid.	In mature leaves, especially in response to water stress. May be synthesized in seeds.	ABA is exported from leaves in the phloem.	Stomatal closure; induction of photosynthate transport from leaves to developing seeds; induction of storage-protein synthesis in seeds; embryogenesis; may affect induction and maintenance of dormancy in seeds and buds of certain species
Gibberellins	Gibberellic acid (GA_3), a fungal product, is the most widely available. GA_1 is probably the most important gibberellin in plants. GAs are synthesized from mevalonic acid.	In young tissues of the shoot and developing seeds. It is uncertain whether synthesis also occurs in roots.	GAs are probably transported in the xylem and phloem.	Hyperelongation of shoots by stimulating both cell division and cell elongation, producing tall, as opposed to dwarf, plants; induction of seed germination; stimulation of flowering in long-day plants and biennials; regulation of production of seed enzymes in cereals

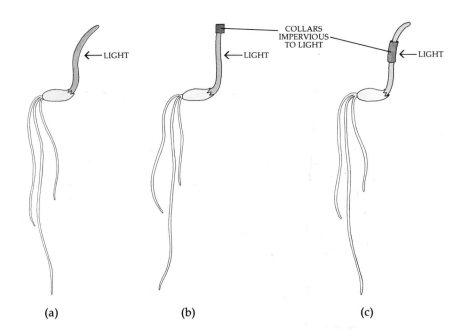

25–1

The Darwins' experiment. (a) Seedlings normally bend toward the light. (b) When the tip of a seedling was covered by a lightproof collar, this bending did not occur. (Bending did occur when the tip of a seedling was covered with a transparent collar.) (c) When a lightproof collar was placed below the tip, the characteristic light response took place. From these experiments, the Darwins concluded that, in response to light, an "influence" that causes bending is transmitted from the tip of the seedling to the area below the tip where bending normally occurs.

(a)   (b)   (c)

Five groups, or classes, of plant hormones are generally recognized: auxin, cytokinins, ethylene, abscisic acid, and gibberellins (Table 25–1). The following discussion begins with auxin because it is the first substance to have been identified as a plant hormone.

Auxin

Some of the first recorded experiments on growth-regulating substances were performed by Charles Darwin and his son Francis and were reported in *The Power of Movement in Plants*, published in 1881. The Darwins first made systematic observations of the bending or curving of plants toward light (phototropism; see page 573), using seedlings of canary grass (*Phalaris canariensis*) and oats (*Avena sativa*). They then found that if they covered the upper portion of the *coleoptile* (the sheathlike, protective structure covering the shoot of grass seedlings) with a cylinder of metal foil or a hollow tube of blackened glass (or a blackened quill) and illuminated the plant from the side, bending did not occur (Figure 25–1). If, however, the tip was enclosed in a transparent glass tube (or quill), bending occurred. From these experiments, the Darwins concluded that "when seedlings are freely exposed to a lateral light some influence is transmitted from the upper to the lower part, causing the latter to bend."

In 1926, the plant physiologist Frits W. Went succeeded in isolating this "influence" from the coleoptile tips of oat (*Avena*) seedlings. Went named this chemical substance **auxin,** from the Greek *auxein*, meaning "to increase."

As can be seen in Figure 25–2, the naturally occurring auxin, which is called indoleacetic acid (abbre-

25–2

Indoleacetic acid (IAA) is the only known naturally occurring auxin. Dichlorophenoxyacetic acid (2,4-D), a synthetic auxin, is widely used as an herbicide. Naphthaleneacetic acid (NAA), another synthetic auxin, is commonly employed to induce the formation of adventitious roots in cuttings and to reduce fruit drop in commercial crops. The synthetic auxins, unlike IAA, are not readily broken down by natural plant enzymes and microbes and so are better suited than IAA for commercial purposes.

viated IAA), closely resembles the amino acid tryptophan (see Figure 3–15c, page 56). Tryptophan is the precursor of IAA. Although four different pathways are known for IAA biosynthesis, each originates from tryptophan. Auxin is produced in the coleoptile tips of grasses and in shoot tips. Although IAA has been found in root tips, most evidence indicates that it is not produced there but is transported there via the vascular cylinder. It is synthesized in leaf primordia and young leaves and is also found in flowers, fruits, and seeds.

Shortly after the discovery of auxin and the recognition of its role in stimulating cell elongation, its inhibitory effect on the growth of lateral buds was discovered. For instance, when the shoot tip of a bean plant (*Phaseolus vulgaris*) is removed, the lateral buds begin to grow. However, when auxin is applied to the cut surface, the growth of the lateral buds is inhibited. The inhibitory influence of a shoot tip or apical bud upon the lateral buds is referred to as **apical dominance.**

AUXIN TRANSPORT

The movement of auxin in both shoots and roots is slow, only about one centimeter per hour. In addition, its transport is **polar,** or unidirectional (Figure 25–3): always toward the base (basipetal) in stems and leaves and toward the tip (acropetal) in roots. In contrast to the movement of sugars and other solutes, auxin is transported not through the conduits (sieve tubes and vessels, respectively) of the phloem and xylem but

through phloem parenchyma cells and parenchyma cells surrounding the vascular tissues. In plant parts capable of secondary growth, auxin transport also occurs through cells in the region of the vascular cambium.

Although auxin is believed to move by simple diffusion within cells, the mechanism of auxin transport is active, requiring metabolic energy for its accomplishment. In addition, auxin transport can occur in tissue whose plasmodesmata have been disrupted by plasmolysis, indicating that auxin moves from cell to cell by transport across the plasma membrane and intervening cell walls rather than primarily through plasmodesmata. One model for the mechanism of auxin transport proposes that auxin enters the cell at its apical end by diffusion across the plasma membrane in a non-ionized form (designated AH). As the AH molecules enter the cytoplasm, they become ionized to A^- and H^+, maintaining the AH gradient. At the basal end of the cell, a specific carrier protein mediates the energy-dependent efflux of auxin ions, coupled to proton efflux, across the plasma membrane. Once outside the cytoplasm and released from the carrier, the ionized auxin can acquire a proton, and the AH can enter the next cell by diffusion through the plasma membrane. In this way, the auxin moves from cell to cell along a linear file.

AUXIN AND CELL DIFFERENTIATION

The gradient of auxin caused by basipetal transport influences the differentiation of the vascular tissue in the elongating shoot. When a stem of cucumber (*Cucumis sativus*) or some other herbaceous dicotyledon is

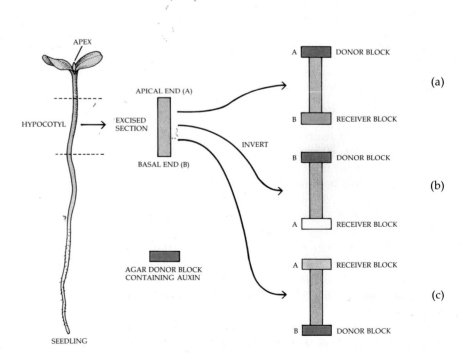

25–3
Characteristics of polar auxin transport in stems, here represented by a segment of hypocotyl from a seedling. (a) Auxin applied to the apical end of the hypocotyl is transported to the agar block at the basal end of the segment; (b) if the segment is inverted, transport does not take place; (c) transport can take place against a concentration gradient along the segment.

APEX

HYPOCOTYL

SEEDLING

APICAL END (A)

EXCISED SECTION

BASAL END (B)

INVERT

AGAR DONOR BLOCK CONTAINING AUXIN

A DONOR BLOCK

B RECEIVER BLOCK

(a)

B DONOR BLOCK

A RECEIVER BLOCK

(b)

A RECEIVER BLOCK

B DONOR BLOCK

(c)

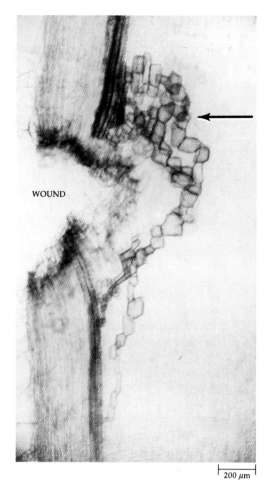

200 μm

.25–4

A longitudinal view of vascular tissue (xylem) regeneration (arrow) around a seven-day-old wound in a young internode of cucumber (Cucumis sativus) *from which the leaves and buds had been removed. IAA (0.1 percent in lanolin) was applied to the upper end of the internode immediately after wounding. The photograph shows the typical pattern of xylem differentiation induced by the basipetal polar movement of auxin.*

is wounded or when isolated cells are grown in tissue culture.) For example, in a series of experiments conducted by Ralph H. Wetmore and his coworkers, leaf buds were grafted onto the upper surface of callus blocks obtained from lilac *(Syringa)* pith. Within 10 days, the buds had burst and begun to grow. Beneath the growing buds, nodules of dividing cells arose, and then, within the nodules, xylem elements differentiated (Figure 25–5a). In a second series of experiments, agar blocks containing auxin and sucrose were substituted for buds. Now, rings of nodules containing both xylem and phloem appeared (Figure 25–5b). At a fixed concentration of auxin, an increase in the sucrose concentration resulted in an increase in the amount of phloem tissue produced. At a fixed concentration of sucrose, an increase in the auxin concentration resulted in an increase in the diameter of the ring of nodules. Apparently, both the occurrence of differentiation and its patterns were controlled by the gradient in concentration of auxin and sucrose.

In subsequent experiments with callus derived from bean *(Phaseolus vulgaris)*, it was found that when auxin alone was used, only xylem differentiation occurred. When a one-week treatment with auxin was followed by a two-week treatment with sucrose, nodules con-

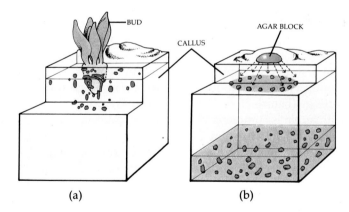

(a)   (b)

25–5

Induction of vascular tissue differentiation in blocks of lilac callus. (a) A growing leaf bud inserted into the block of callus induces the differentiation of xylem elements in nodules; (b) an agar block containing auxin and sucrose and placed on top of the callus block induces the formation of nodules containing both xylem and phloem. In (b), the culture medium also contained auxin and sucrose, and hence vascular tissue differentiation occurred in relation to both sources of these substances.

wounded in such a way as to sever and remove portions of the vascular bundles, new vascular tissues will form from cells in the pith and will connect the severed bundles. However, if the leaves and buds above the wound are removed, the formation of new cells is delayed. With the addition of IAA to the stem just above the wound, new vascular tissue begins to form (Figure 25–4). Auxin similarly plays an important role in the joining of vascular traces from developing leaves to the bundles in the stem.

Similar effects are seen in calluses. (A *callus* is a mass of undifferentiated cells that forms when a plant

taining both xylem and phloem were formed, but a one-week treatment with sucrose followed by a two-week treatment with auxin resulted in only xylem differentiation. From these results it was concluded that cell division is a prerequisite for differentiation. Apparently, the initial treatment with auxin stimulated cell division, and the subsequent treatment with sucrose, together with residual auxin, promoted differentiation.

More recent studies with cells isolated from the mesophyll of *Zinnia elegans* leaves showed a requirement for both auxin and cytokinin (see below) for the differentiation of xylem elements. Significantly, many of the isolated mesophyll cells differentiated directly into tracheary elements without preceding DNA synthesis or cytokinesis, precluding the necessity of cell division as a prerequisite for differentiation. The results of all these studies illustrate the subtlety of growth regulator function and also draw attention to the critical fact that growth regulators *never* act alone; rather, they operate in concert with other internal factors such as sugar or other regulators of plant growth.

AUXIN AND THE VASCULAR CAMBIUM

In woody plants, auxin promotes activity of the vascular cambium. With expansion of buds and resumption of their growth in the spring, auxin moves downward in the stems and stimulates the cambial cells to divide, forming secondary vascular tissue.

AUXIN AND ROOT GROWTH

The first practical application of auxin involved its promoting effect on the initiation of adventitious roots in cuttings (Figure 25–6). The practice of treating cuttings with auxin is commercially important, especially for the vegetative propagation of woody plants. (Application of a high concentration of auxin to already-growing roots, however, usually inhibits their growth.)

AUXIN AND FRUIT GROWTH

Auxin promotes the growth of many fruits. Ordinarily, if the flower is not pollinated and fertilization does not take place, the fruit will not develop. In some plants, fertilization of one egg cell is sufficient for normal fruit development, but in others, such as apples or melons, which have many seeds, several must be fertilized for the ovary wall to mature and become fleshy. By treating the female flower parts (carpels) of certain species with auxin, it is possible to produce parthenocarpic

25–6
The stalk of the African violet leaf on the left was placed in a solution containing the synthetic auxin naphthaleneacetic acid for 10 days before the picture was taken. The stalk of the leaf on the right was placed in pure water. Note the growth of adventitious roots on the stalk of the hormone-treated leaf.

fruit (from the Greek *parthenos*, meaning "maiden, virgin"), which is fruit produced without fertilization—for example, seedless tomatoes, cucumbers, and eggplants. In many or most of these seedless fruits, however, immature ovules still exist in the fruit.

Developing seeds are a source of auxin. If during development of the aggregate fruit of the strawberry (*Fragaria ananassa*) all of the seeds are removed, the receptacle will stop growing altogether. (Actually, in the strawberry, it is small achenes—one-seeded fruits—that are removed. What is recognized as the strawberry fruit is the fleshy receptacle.) If a narrow ring of seeds is left, the fruit forms a bulging girdle of growth in the area of the seeds. If auxin is applied to the deseeded receptacle, growth proceeds normally (Figure 25–7).

AUXIN AND THE CONTROL OF ABSCISSION

Abscission—the dropping of leaves or other plant parts—has been correlated with a lowered production of auxin in the leaf, among other factors; under many circumstances, abscission can be prevented by the ap-

(a) (b) (c) (d) (e)

25–7

(a) *Normal strawberry (Fragaria ananassa).* (b) *Strawberry from which all the developing "seeds" were removed.* (c) *Strawberry in which three horizontal* *rows of seeds were left.* (d) *Growth induced by one developing seed.* (e) *Growth induced by three developing seeds. If a paste containing auxin is* *applied to a strawberry from which the seeds have been removed, the strawberry grows normally.*

plication of auxin. (See discussion on leaf abscission, page 509.)

The control of abscission of leaves, flowers, and fruits is extremely important in agriculture. Auxin (and ethylene, discussed below) has been used commercially for treatment of a number of plant species. For instance, auxin prevents leaf and berry drop from evergreen holly *(Ilex aquifolium)* and therefore minimizes losses during shipment. Auxin also prevents preharvest drop of citrus fruits. On the other hand, large amounts of auxin promote fruit drop. Thus, auxin has been used for the thinning of fruit (that is, reducing the number of fruits that are retained on the tree until ripe) in the production of olives, apples, and other tree fruits.

AUXIN AND THE CONTROL OF WEEDS

Synthetic auxins such as 2,4-D have been used extensively for the control of weeds on agricultural lands. In economic terms, this is the major practical use for plant growth regulators. The mechanism by which *herbicides* kill only certain weeds is largely unknown. The selectivity of these compounds against broad-leaf weeds is due in part to the greater absorption and rates of transport of the herbicides by broad-leaf weeds than by grasses. The herbicide atrazine, which is used to eradicate weeds in corn and other crops, has been shown to kill most kinds of plants because it blocks the electron transport chain in their chloroplasts. Corn *(Zea mays),* the crop with which atrazine is used most widely, is re-

sistant to the herbicide because the corn plant contains enzymes that detoxify it. (See page 570 for a discussion of the herbicide glyphosate.) The continued use of herbicides will depend on a number of factors, including cost-effectiveness and potential or real hazards to human health.

Considerable attention has been given to Agent Orange, the herbicide most commonly used as a defoliant during the Vietnam Conflict. Agent Orange is a mixture of the *n*-butyl esters of 2,4-D and 2,4,5-T, another synthetic auxin. It also contains the dioxin 2,3,7,8-TCDD, a contaminant of 2,4,5-T that has been demonstrated to be toxic to experimental animals and to humans. In humans, 2,3,7,8-TCDD causes chloracne, a severe skin lesion that usually occurs on the head and upper body. Based on the positive evidence in animal studies, 2,3,7,8-TCDD is probably carcinogenic in humans. The manufacture and use of 2,4,5-T have been banned in the United States.

Cytokinins

In 1941, Johannes van Overbeek found that coconut *(Cocos nucifera)* milk (which is liquid endosperm) contained a potent growth factor different from anything known at that time. This factor, or factors, greatly accelerated the development of plant embryos and promoted the growth of isolated tissues and cells in the test

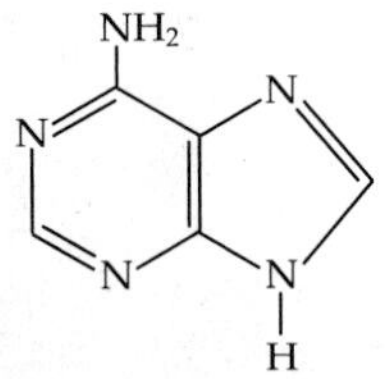

25–8

Note the similarities between the purine adenine and these four cytokinins. Kinetin and benzylamino purine (BAP) are commonly used synthetic cytokinins. Zeatin and i⁶ Ade have been isolated from plants.

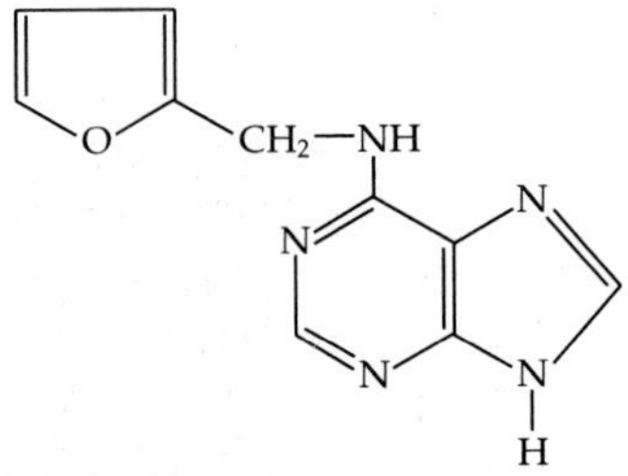

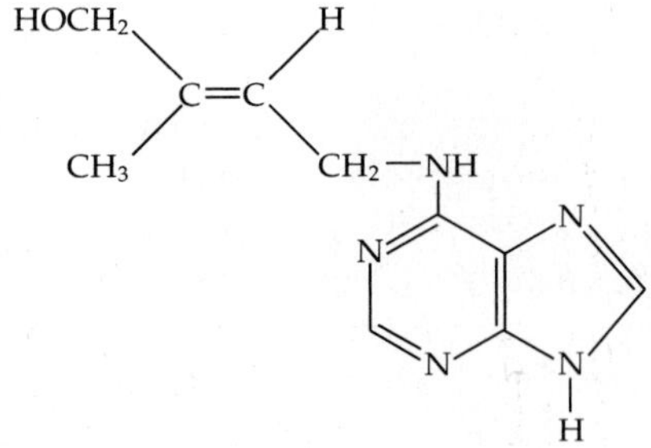

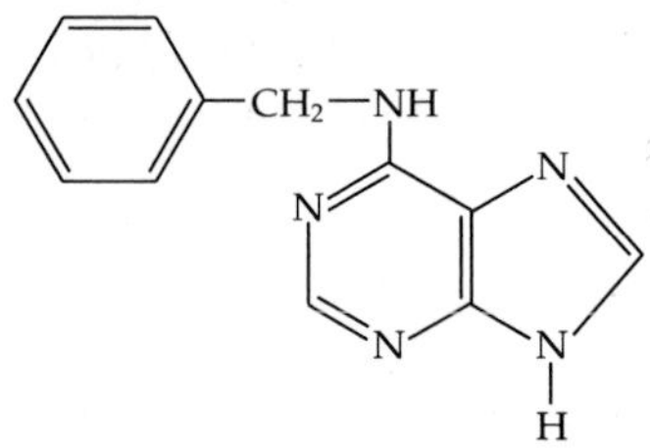

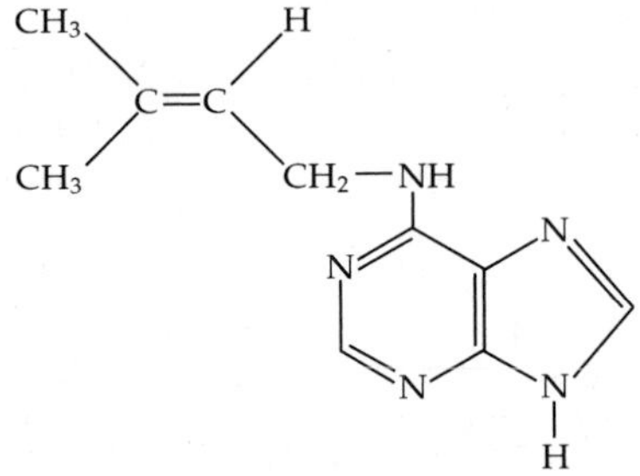

tube. Van Overbeek's discovery had two effects: it gave impetus to studies of isolated plant tissues, and it launched the search for another major group of plant growth regulators.

The basic medium for tissue culture in plant cells contains sugar, vitamins, and various salts. In the early 1950s, Folke Skoog and his coworkers showed that a stem segment of the tobacco plant (*Nicotiana tabacum*) grew initially in such a culture medium, but that its growth soon slowed or stopped. Apparently, some growth stimulus originally present in the tobacco stem became exhausted. The addition of IAA had no effect. When coconut milk was added to the medium, however, the cells began to divide and growth of the tobacco stem resumed.

Skoog and his coworkers set out to identify the growth factor in the coconut milk. After many years of effort, they succeeded in producing a thousandfold purification of a growth factor, but they could not isolate it. So, changing course, they tested a variety of purine-containing substances—largely nucleic acids—in the hope of finding a new source of the substance. This led to the discovery by Carlos D. Miller, then a postdoctoral fellow in Skoog's laboratory, that a breakdown product of DNA contained material that was highly active in promoting cell division.

Subsequently, Miller, Skoog, and their coworkers succeeded in isolating the growth factor from a DNA preparation and identifying its chemical nature. They called this substance *kinetin* and named the group of growth regulators to which it belongs the **cytokinins** because of the involvement of cytokinins in cytokinesis, or cell division. As shown in Figure 25–8, kinetin resembles the purine adenine, which was the clue that led to its discovery. Kinetin, which probably does not naturally occur in plants, has a relatively simple structure, and biochemists were soon able to synthesize a number of related compounds that behaved like cytokinins. Eventually, a natural cytokinin was isolated from kernels of corn (*Zea mays*); called *zeatin*, it is the most active of the naturally occurring cytokinins.

Cytokinins have now been isolated from many different species of seed plants, where they are found primarily in actively dividing tissues, including seeds, fruits, and leaves, and in root tips. They have also been found in bleeding sap—the sap that drips out of pruning cuts, cracks, and other wounds in many types of plants. Cytokinins also have been identified in two seedless vascular plants, a horsetail (*Equisetum arvense*) and the fern *Dryopteris crassirhizoma*.

Although practical applications for cytokinins are not as extensive as those for auxin, the former have been important in plant development research. Cytokinins are central to tissue culture methods and are extremely important for biotechnology (see page 563). Treatment of lateral buds with cytokinin often causes the buds to grow, even in the presence of auxin, thus modifying apical dominance.

CYTOKININS AND CELL DIVISION

Apparently, the undifferentiated plant cell has two courses open to it: either it can enlarge, divide, enlarge, and divide again or, without undergoing cell division, it can elongate. The cell that divides repeatedly remains essentially undifferentiated, or meristematic, whereas the elongating cell will ultimately differentiate. In studies of tobacco stem tissues, the addition of IAA to the tissue culture produces rapid cell expansion, so that giant cells are formed. Kinetin alone has little or no effect, but IAA plus kinetin results in rapid cell division, so that large numbers of relatively small, undifferentiated cells are formed. In other words, cells remain meristematic in the presence of both kinetin and IAA.

CYTOKININS AND ORGAN FORMATION IN TISSUE CULTURES

In the presence of a high concentration of auxin, callus tissue frequently gives rise to organized roots. In tobacco pith callus, the relative concentrations of auxin and kinetin determine whether roots or buds form. With higher concentrations of auxin, roots are formed; with higher concentrations of kinetin, buds are formed; and when the two are present in roughly equal concentrations, the callus continues to produce undifferentiated cells (Figure 25–9).

CYTOKININS AND LEAF SENESCENCE

In most species of plants, leaves begin to turn yellow as soon as they are removed from the plant. This yellowing, which is due to a loss of chlorophyll, can be delayed by cytokinins. For example, when excised and floated on plain water, leaves of the cocklebur (*Xanthium strumarium*) turn yellow in about ten days. With kinetin (10 milligrams per liter) present in the water, much of the green color and fresh appearance of the leaf are maintained. If an excised leaf is spotted with kinetin-containing solutions, the spots remain green while the rest of the leaf yellows. Furthermore, if the cytokinin-spotted leaf contains radioactive amino acids, it can be shown that the amino acids migrate from other parts of the leaf to the cytokinin-treated areas.

One interpretation of the senescence of detached leaves is that cytokinins are limiting in the detached leaf. This interpretation leads to an important and still unanswered question as to the site(s) of cytokinin production within plants. As mentioned previously, cytokinins are most abundant in actively dividing seeds, fruits, and leaves, and in root tips. This is not evidence, however, that cytokinins are synthesized in these organs because it is possible that cytokinins are transported there from some other sites. Based on several lines of evidence, root tips most certainly are involved in cytokinin synthesis. It is widely accepted that cytokinins synthesized in root tips are transported from them in the xylem to all other parts of the plant.

IAA CONCENTRATION (mg/liter)

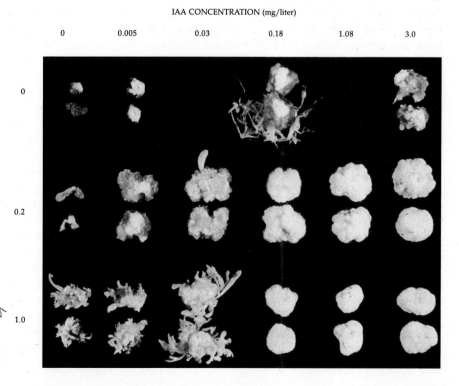

25–9
Effect of increasing IAA concentration at different kinetin levels on growth and organ formation of tobacco callus cultured on nutrient agar. Note that very little growth occurred without the addition of either IAA or kinetin (top left). Higher levels of IAA alone promoted root formation, whereas they repressed bud formation when used alone or in combination with kinetin. The higher level of kinetin was more effective than the lower level in promoting bud development, but both kinetin levels were too high for promotion of root growth.

Ethylene

Ethylene was known to have effects on plants long before the discovery of auxin. The "botanical" history of **ethylene,** a simple hydrocarbon ($H_2C = CH_2$), goes back to the 1800s, when city streets were lighted with lamps that burned illuminating gas. In Germany, illuminating gas leaking from gas mains was found to cause defoliation of shade trees along the streets.

In 1901, Dimitry Neljubov demonstrated that ethylene was the active component of illuminating gas. He noticed that exposure of pea seedlings to illuminating gas caused stems to grow in a horizontal direction. When the gaseous components of illuminating gas were individually tested for effects, all were inactive except ethylene, which was active at concentrations as low as 0.06 part per million (ppm) in air. Neljubov's findings have led to the realization that ethylene exerts a major influence on many, if not all, aspects of growth and development in plants, including fruit maturation, fruit and leaf abscission, and senescence.

The biosynthesis of ethylene begins with the amino acid methionine, which reacts with ATP to form a compound known as S-adenosylmethionine, or SAM (Figure 25–10). Next, SAM is split into two different compounds, one of which is called ACC (1-aminocyclopropane-1-carboxylic acid). Enzymes on the tonoplast then convert ACC into ethylene, CO_2, and ammonium ion. Apparently, the formation of ACC is the step that is affected by those treatments (for example, high auxin concentrations, air pollution damage, wounding) that stimulate ethylene production.

METHIONINE

S-ADENOSYLMETHIONINE

1-AMINOCYCLOPROPANE-1-CARBOXYLIC ACID

ETHYLENE

25–10
Biosynthesis of ethylene from methionine, which serves as the precursor of ethylene in all higher plant tissues. 1-Aminocyclopropane-1-carboxylic acid is the immediate precursor of ethylene.

ETHYLENE AND FRUIT RIPENING

Ripening in fruit involves a number of changes. In fleshy fruits, the chlorophyll is degraded and other pigments may form, changing the fruit color. Simultaneously, the fleshy part of the fruit softens as a result of the enzymatic digestion of pectin, the principal component of the middle lamella of the cell wall. During this same period, starches and organic acids or, as in the case of the avocado (*Persea americana*), oils are metabolized into sugars. As a consequence of these changes, fruits become conspicuous and palatable and thus attractive to animals that eat the fruit and so scatter the seed.

During the ripening of many fruits—including tomatoes, avocados, and pome fruits, such as apples and pears—there is a large increase in cellular respiration evidenced by an increased uptake of oxygen. This phase is known as the **climacteric,** and such fruits are called climacteric fruits. (Fruits that show a steady decline, or gradual ripening, such as citrus fruits, grapes, and strawberries are called nonclimacteric fruits.)

In the early 1900s, many fruit growers made a practice of improving the color and increasing the sweetness of citrus fruits by "curing" them in a room with a kerosene stove. It was believed that the heat of the stove ripened the fruits. Ambitious fruit growers, who installed more modern heating equipment, found to their sorrow that this was not the case. As subsequent experiments demonstrated, it is actually the incomplete combustion products of the kerosene that are responsible. The most active combustion product was ethylene. As little as 1 ppm of ethylene in the air will speed up the onset of the climacteric.

The effect of ethylene on fruit ripening has agricultural importance. A major use is in promoting the ripening of tomatoes that are picked green and stored in the absence of ethylene until just before marketing. It is also used to hasten ripening of walnuts and grapes.

ETHYLENE AND ABSCISSION

Ethylene promotes abscission of leaves, flowers, and fruits in a variety of plant species. In leaves, ethylene presumably triggers the enzymes that cause the cell wall dissolution associated with abscission. Ethylene is used commercially to promote fruit loosening in cherries, blackberries, grapes, and blueberries, thus making mechanical harvesting possible. It is also used as a fruit-thinning agent in commercial orchards of prunes and peaches.

ETHYLENE AND SEX EXPRESSION

Ethylene appears to play a major role in determining the sex of flowers in some monoecious plants (those plants having male and female flowers borne on the same individual). In cucurbits (family Cucurbitaceae; cucumber, squash), for example, high levels of gibberellins (see below) are associated with maleness, and treatment with ethylene changes the expression of sex to femaleness. In studies with cucumbers (*Cucumis sativus*), female buds evolved greater quantities of ethylene than male buds. In addition, cucumbers grown under short light periods, or short-day conditions, which promote femaleness, evolved more ethylene than those grown under long-day conditions (see page 579). Hence, in cucurbits, ethylene apparently participates in the regulation of sex expression and is associated with the promotion of femaleness.

Abscisic Acid

At certain times, the survival of the plant depends on its ability to restrain its growth or its reproductive activities. In 1949, Paul F. Wareing discovered that the dormant buds of ash and potatoes contain large amounts of a growth inhibitor, which he called *dormin*. During the 1960s, Frederick T. Addicott reported the discovery in leaves and fruits of a substance capable of accelerating abscission, which he called *abscisin*. Soon abscisin and dormin were found to be identical chemically. The compound is now known as **abscisic acid,** or ABA (Figure 25–11). This is an unfortunate name choice since it now appears that this substance has little role in abscission.

If a small spot of abscisic acid is placed on a leaf, the treated area yellows rapidly even though the rest of the leaf stays green, an effect that is the opposite of that of the cytokinins (see page 553). Whether this is a direct or indirect action is not known at present.

25–11
Abscisic acid, exogenous applications of which may inhibit plant growth, appears also to act as a promoter (for instance, of storage-protein synthesis in seeds).

ABSCISIC ACID AND SEED DEVELOPMENT

Abscisic acid levels increase during early seed development in many plant species. This increase in ABA stimulates the production of seed storage proteins and is also responsible for preventing premature germination. The breaking of dormancy in many seeds is correlated with declining ABA levels in the seed. In corn, there are single-gene mutants that lack the ability to make ABA. As a result, mutant embryos lack the ability to become dormant and germinate directly on the cob. Such mutants are called viviparous mutants.

ABSCISIC ACID AND WATER RELATIONS

Abscisic acid stimulates the closing of stomata in most plant species (see page 618). Since its synthesis is stimulated by water deficiency (water stress), ABA is most likely involved in the stomatal regulation of transpiration. In support of this idea, mutant plants incapable of synthesizing ABA show a wilting phenotype; that is, they are only capable of growing normally in very humid environments.

Gibberellins

The early history of gibberellin research was an exclusive product of Japanese scientists. In 1926, the same year that Went isolated auxin from oat (*Avena*) coleoptile tips, E. Kurosawa of Japan was studying a disease of rice (*Oryza sativa*) called "foolish seedling disease," in which the plants grew rapidly; were spindly, pale-col-

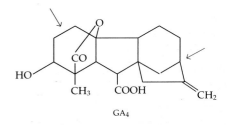

GIBBERELLIC ACID (GA₃)   GA₁₃   GA₄

25–12
Three of the more than 78 gibberellins that have been isolated from natural sources. GA₃ (gibberellic acid) is the most *abundant in fungi and biologically active in many tests. The arrows indicate where minor structural differences occur that* *distinguish the other two examples of gibberellins, GA₄ and GA₁₃.*

ored, and sickly; and tended to fall over. The cause of these symptoms, Kurosawa discovered, was a substance produced by a fungus, *Gibberella fujikuroi*, which was parasitic on seedlings.

Gibberellin was named and isolated in 1934 by the chemists T. Yabuta and Y. Sumiki. The discovery attracted little interest in the western world until after the Second World War. In 1956, J. MacMillan, in England, first successfully isolated gibberellin from a plant (the seed of the bean *Phaseolus vulgaris*). Since then, gibberellins have been identified from many species of plants, and it is now believed that they occur in all plants. They are present in various amounts in all parts of the plant, but the highest concentrations are found in immature seeds. More that 78 gibberellins now have been isolated and identified chemically. They vary slightly in structure (Figure 25–12), as well as in biological activity. The best studied of the group is GA₃ (known as gibberellic acid), which is also produced by the fungus *Gibberella fujikuroi*.

The gibberellins have dramatic effects on stem and leaf elongation in intact plants by stimulating both cell division and cell elongation.

GIBBERELLINS AND DWARF MUTANTS

The most remarkable results are seen when gibberellins are applied to dwarf mutants (Figure 25–13). Under gibberellin treatment, such plants become indistinguishable from normal tall, nonmutant plants, indicating that these mutants are unable to synthesize gibberellin and that tissue growth requires gibberellin. In corn, for example, four different types of dwarfs have been identified, each defective in a specific step of the gibberellin biosynthetic pathway. Studies of the hormone biochemistry in these mutant plants have led to a very important conclusion. Even though corn plants contain nine or more gibberellins, only the final compound in the pathway, GA₁, can cause effects directly. The other eight gibberellins must be further metabolized before a growth response is obtained.

25–13
The plant on the right was treated with gibberellin; the one on the left served as a control. The plants are Contender beans, a dwarf cultivar of the common bean (Phaseolus vulgaris).

25–14

Action of gibberellin in barley seeds. (a) Gibberellin (GA) produced by the embryo migrates into the aleurone layer, stimulating the synthesis of hydrolytic enzymes. These enzymes are released into the starchy endosperm, where they break down the endosperm reserves into sugars and amino acids, which are soluble and diffusible. The sugars and amino acids are then absorbed by the scutellum (cotyledon) and transported to the shoot and root for growth. (b) Each of these three seeds has been cut in half and the embryo removed. Forty-eight hours before the picture was taken, the seed at the lower left was treated with plain water. The seed in the center was treated with a solution of 1 part per billion of gibberellin, and the seed at the upper right was treated with 100 parts per billion of gibberellin. Digestion of the starchy storage tissue has begun to take place in the treated seeds.

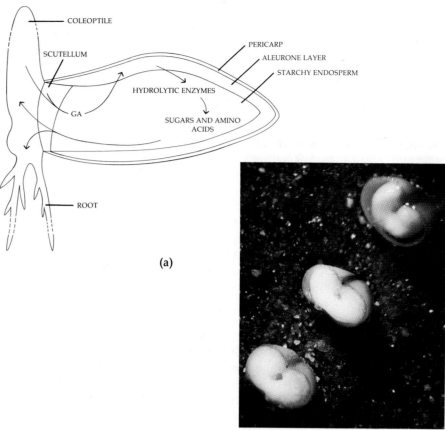

(a)

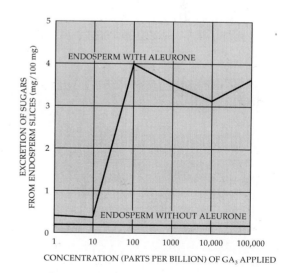

(b)

GIBBERELLINS AND SEEDS

The seeds of many plants require a period of dormancy before they will germinate. In certain plants, dormancy (see page 585) usually cannot be broken except by exposure to cold or to light. In many species, including lettuce, tobacco, and wild oats, gibberellins will substitute for the dormancy-breaking cold or light requirement and promote the growth of the embryo and the emergence of the seedling. Specifically, the gibberellins enhance cell elongation, making it possible for the root to penetrate the growth-restricting seed coat or fruit wall. This effect of gibberellin has at least one practical application. Gibberellic acid hastens seed germination and thus ensures germination uniformity for the production of the barley malt used in brewing.

In barley *(Hordeum vulgare)* and other grass seeds, a specialized layer of endosperm cells, called the aleurone (see Figure 20–2, page 443), lies just inside the seed coat. The cells in the aleurone layer are rich in protein. When the seeds begin to germinate (triggered by the uptake of water), the embryo releases gibberellins, which diffuse to the aleurone cells and stimulate them to synthesize hydrolytic enzymes. One of these enzymes is α-amylase, which hydrolyzes starch. The enzymes digest the stored food reserves of the starchy endosperm. These food reserves, in the form of sugars, amino acids, and nucleic acids, are absorbed by the scutellum and are then transported to the growing regions of the embryo (Figures 25–14 and 25–15).

25–15

The release of sugar from endosperm can be induced by gibberellin (GA₃) treatment. These data show that sugars are produced only when the aleurone layer is present. It is, in fact, the aleurone layer that is the source of the enzyme α-amylase, which digests the starches stored in the endosperm.

GIBBERELLINS AND FLOWERING

Some plants, such as cabbages *(Brassica oleracea* var. *capitata)* and carrots *(Daucus carota)*, form rosettes before flowering. (In a rosette, leaves develop but the internodes between them do not elongate.) In these plants, flowering can be induced by exposure to long days, to cold (as in the biennials), or to both. Following the appropriate exposure, the stems elongate—a phenomenon known as *bolting*—and the plants flower (Figure 25–16). Application of gibberellin to such plants causes bolting and flowering without appropriate cold or long-day exposure (Figure 25–16b). Bolting is brought about by an increase both in cell number and in cell elongation. Gibberellin can thus be used for early seed production of biennial plants.

GIBBERELLINS AND FRUIT DEVELOPMENT

Gibberellins, like auxin, can cause the development of parthenocarpic fruits, including apples, currants, cucumbers, and eggplants. In some fruits, such as mandarin oranges, almonds, and peaches, the gibberellins have been effective in the promotion of fruit development where auxin has not. The major commercial application of gibberellins, however, is in the production of table grapes. In the United States, large amounts of gibberellic acid are applied annually to the Thompson Seedless grapes, a cultivar of *Vitis vinifera*. Treatment causes larger fruit and much looser clusters (Figure 25–17).

(a)

(b)

25–16
Bolting in cabbage (Brassica oleracea
var. capitata) *occurring naturally* (a) *and induced by gibberellin treatment* (b). *The plant on the right was treated once a week for eight weeks.*

25–17
The effect of GA$_3$ on the growth of Thompson Seedless grapes, a cultivar of Vitis vinifera. *The left bunch of grapes was untreated, whereas the right bunch was treated with GA$_3$.*

The Molecular Basis of Hormone Action

The descriptions of the effects of plant hormones presented thus far have primarily involved observations of very complex morphological changes resulting from external applications of hormones to a variety of plant systems. We now turn to some of the molecular mechanisms by which these chemical regulators influence growth and development at the cellular level.

The development of organs (morphogenesis) can be described in terms of a coordinated series of cell divisions and subsequent cell enlargements. Specialization of cell types within an organ (differentiation) is the result of the selective expression of a particular set of genes within the genome of each individual cell. Clearly, the coordination of these cellular processes during development requires that individual cells communicate with each other. The concept that plant hormones help to coordinate growth and development by acting as chemical messengers between cells comes in part from numerous examples of the observable influences of plant hormones on the rate of cell division and on the rate and direction of cell expansion (Table 25–2). In addition, there is increasing evidence that all five groups of plant hormones can act either to stimulate or to repress specific genes within the nucleus, and that many observable hormone responses are the result of such differential gene expression.

HORMONAL CONTROL OF GENE EXPRESSION

The totipotency of plant cells (see page 563) is clear evidence that all of the genes present in the zygote are also present in each living cell of the adult plant. In any one cell, however, only selected genes are expressed and transcribed into mRNA and subsequently translated into proteins. The specific proteins that are produced determine the identity of the cell. It is the proteins that catalyze most of the cell's chemical reactions and that form or produce most of the structural elements within and around the cell. Thus, a cortical cell in a root and a mesophyll cell in a leaf differ from each other structurally and functionally because of differences in gene expression during the course of their development.

The molecular mechanisms by which individual genes are switched on and off in the eukaryotic nucleus are not completely understood. A number of principles are beginning to emerge, however, that are common to both plants and animals (Figure 25–18). A eukaryotic gene is composed of a coding sequence, which specifies the amino acid sequence of the gene's protein product, and regulatory sequences, regions of DNA bordering the coding sequence that play a regulatory role in gene transcription. Proteins called regulatory transcription factors can bind directly to specific DNA sequences within a regulatory sequence, activating (switching on) or repressing (switching off) that particular gene.

Plant molecular biologists are currently studying a number of plant genes that are either activated or repressed by such factors as light, environmental stress (see Chapter 26), and hormones. Regulatory sequences of DNA associated with hormone-activated genes are currently being studied using recombinant DNA technology (see pages 566 through 571). These experiments involve splicing the putative regulatory DNA sequence to the coding sequence of a foreign gene not normally activated by the hormone, then transferring this reporter gene (see page 570) back into the plant's genome. If the foreign gene is expressed (producing a detectable foreign protein) in response to a particular hormone, the specific DNA sequence involved in hormonal regulation of gene expression is thus identified.

Table 25–2 *Hormonal Influences on Basic Cellular Processes*

Hormone	Rate of Cell Division	Rate of Cell Expansion	Direction of Cell Expansion	Differentiation (Gene Expression)
Auxin	+	+	↕	+
Cytokinins	+	*		+
Ethylene	+/−	+/−	↔	+
Abscisic acid	−	−		+
Gibberellins	+	+	↕	+

KEY: + positive effect; − negative effect; * little or no effect. The vertical arrow represents longitudinal expansion; the horizontal arrow, lateral expansion.

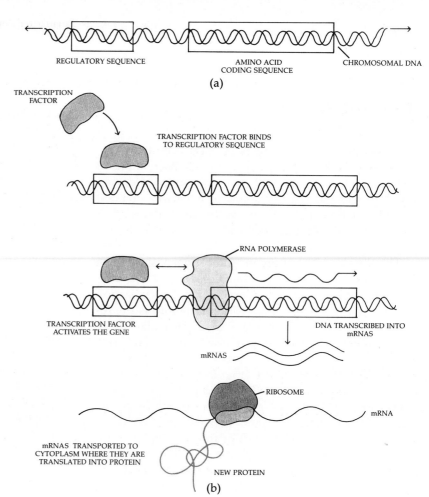

25-18

(a) *Every gene coded for in the nucleus is composed of an amino acid coding sequence and an associated regulatory sequence, which does not code for any protein but which determines whether the amino acid coding sequence will be expressed (that is, transcribed into mRNA). (b) A DNA binding protein called a transcription factor binds to the regulatory sequence. The transcription factor activates the gene; that is, RNA polymerase activity is stimulated and DNA is transcribed into mRNAs. The mRNAs are transported to the cytoplasm where they are translated into protein.*

HORMONAL REGULATION OF CELL EXPANSION

The rate at which plant cells grow depends on several factors, including their position in the plant, the cell type, and a variety of environmental influences. The rate at which an individual cell expands is controlled by (1) the amount of turgor pressure inside the cell pushing against the cell wall (see Chapter 4) and (2) the extensibility of the cell wall (Figure 25–19). Extensibility, a physical property of the wall, is a measure of how much the wall will stretch permanently when a force is applied to it. As indicated in Table 25–2, all five groups of hormones are capable of influencing the rate of cell expansion. In most cases examined, hormones affect the extensibility of the cell wall but have little direct influence on the turgor pressure. Auxin and gibberellins stimulate plant growth by increasing the extensibility of cell walls, whereas ABA and ethylene inhibit plant growth by causing a decrease in extensibility.

The mechanisms by which hormones alter the extensibility of cell walls are not well understood. Two hypotheses are currently in favor. In the *acid growth hypothesis*, hormones—particularly auxin—activate a proton-pumping enzyme in the plasma membrane. Protons are pumped from the cytoplasm into the cell wall. The resulting drop in pH is thought to cause a loosening of the cell wall structure through the breakage and reformation of noncellulosic polysaccharides, which normally cross-link the cellulose microfibrils (Figure 25–19; see also Figure 2–27). An alternative hypothesis is based on recent discoveries that auxin activates the expression of specific genes within a few minutes of application. The products of these genes are thought to influence the delivery of new wall materials in such a way as to affect cell wall extensibility. These two hypotheses are not mutually exclusive, and both may be required to explain the influence of hormones on cell expansion.

In addition to affecting the rate of cell expansion, plant hormones can also influence the *direction* of expansion. Once a cell has divided, the shape assumed by the daughter cells as they enlarge will determine the ultimate form of the developing tissue or organ. For example, many of the cells in a developing leaf tend to expand primarily in a lateral direction. This lateral expansion, in addition to the pattern of cell division, results in formation of a platelike organ. In contrast, the cells in growing stem tissues tend to expand longitudinally, resulting in the "unidirectional" growth characteristic of an elongating stem. These differences in the direction of cell expansion are apparently determined by the orientation of the cellulose microfibrils as they are deposited in the developing cell wall (Figure 25–

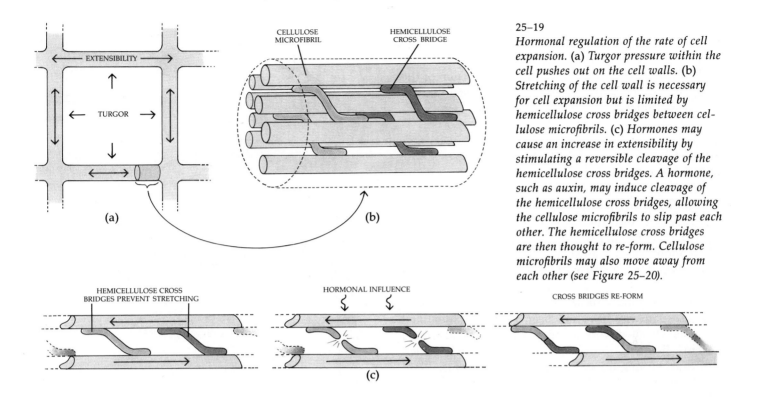

CELLULOSE MICROFIBRIL　　**HEMICELLULOSE CROSS BRIDGE**

(a)　　(b)

HEMICELLULOSE CROSS BRIDGES PREVENT STRETCHING　　*HORMONAL INFLUENCE*　　*CROSS BRIDGES RE-FORM*

(c)

25-19
Hormonal regulation of the rate of cell expansion. (a) Turgor pressure within the cell pushes out on the cell walls. (b) Stretching of the cell wall is necessary for cell expansion but is limited by hemicellulose cross bridges between cellulose microfibrils. (c) Hormones may cause an increase in extensibility by stimulating a reversible cleavage of the hemicellulose cross bridges. A hormone, such as auxin, may induce cleavage of the hemicellulose cross bridges, allowing the cellulose microfibrils to slip past each other. The hemicellulose cross bridges are then thought to re-form. Cellulose microfibrils may also move away from each other (see Figure 25-20).

20). If cellulose microfibrils are deposited in a random orientation, the cells tend to expand in all directions. If the fibrils are laid down in a primarily transverse orientation, the cells tend to expand longitudinally (just as a coiled spring is much easier to stretch in the direction perpendicular to the orientation of the coils).

The orientation of cellulose microfibrils appears to be governed by the orientation of microtubules lying just inside the plasma membrane (see page 30), and this arrangement of microtubules is influenced by hormones. Gibberellins, for example, promote a transverse arrangement of microtubules, resulting in greater longi-

25-20
Ethylene and gibberellin have opposite effects on the orientation of cellulose microfibrils as they are laid down in the developing cell wall. The orientation of cellulose microfibrils apparently is governed by the orientation of microtubules just inside the plasma membrane. Ethylene causes some reorientation of microtubules from a transverse arrangement to a longitudinal direction. Thus, when the cell expands, expansion occurs more or less equally in all dimensions because of the random arrangement of the cellulose microfibrils. Gibberellins, on the other hand, promote a transverse arrangement of microtubules. When the cell expands, it does so primarily longitudinally because the cellulose microfibrils are arranged in a transverse manner.

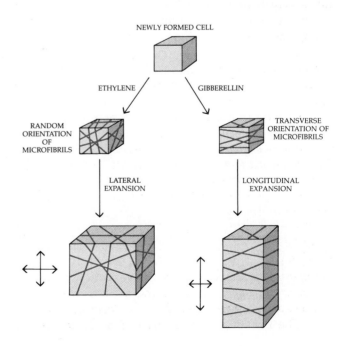

NEWLY FORMED CELL

ETHYLENE　　GIBBERELLIN

RANDOM ORIENTATION OF MICROFIBRILS　　TRANSVERSE ORIENTATION OF MICROFIBRILS

LATERAL EXPANSION　　LONGITUDINAL EXPANSION

tudinal growth, or elongation. On the other hand, in stems, ethylene treatment causes some degree of reorientation of microtubules to the longitudinal direction, which promotes a more lateral (radial) expansion of the cells. This response to ethylene results in a stem that is shorter and thicker.

HORMONE RECEPTORS AND RESPONSE PATHWAYS

In order for plant hormones to operate as chemical signals between cells, the target cells must have mechanisms for identifying the specific hormone, measuring the amount that is present, transferring this information via biochemical pathways, and converting the information into a complex set of developmental changes. The biochemical components of the hormone response pathway have not been positively identified in plants. Detailed work on microorganisms and animal systems, however, provides a number of basic principles that we might expect to be true of plants. Plant hormones may be recognized by the cell through their interaction with specific cell proteins called **receptors.** Each receptor protein would contain a hormone-binding site that is specific for a particular hormone. Binding of the hormone to the receptor would activate a particular response pathway in the cell.

Figure 25–21 illustrates a variety of ways in which a hormone-receptor complex may activate response pathways. Binding of a hormone to its receptor results in a change in conformation (shape) of the receptor protein. This conformational change alters the receptor protein, allowing the receptor to interact with other components in the cell. For example, the activated receptor may interact directly with regulatory sequences of DNA to stimulate the transcription of specific hormone-activated genes. The steroid hormones in animals operate in this way. Other types of hormone receptors are located in the plasma membrane.

25–21

Hormone response pathways. Hormones act by binding to proteins called receptors. The hormone signals are often amplified and passed down biochemical pathways by intermediary compounds called second messengers. Many responses to plant hormones are mediated through changes in gene expression within the nucleus. Specific genes are switched on or off as a result of hormone action. Hormones ultimately affect cell growth by causing changes in the architecture and chemical properties of the cell wall. See also Figure 25–18.

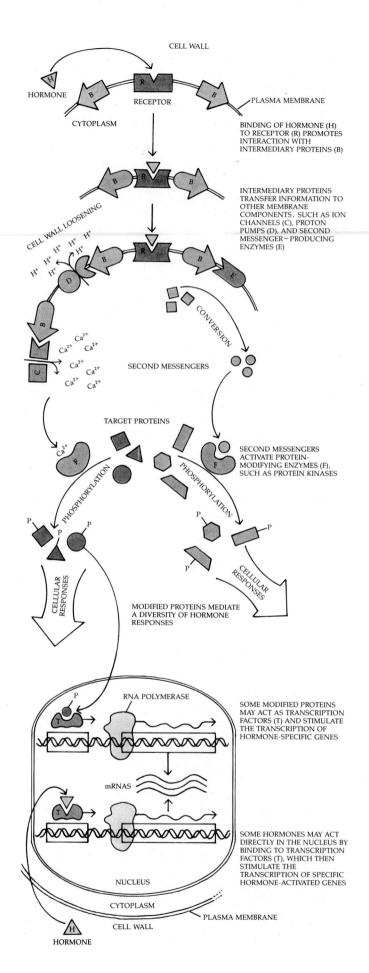

Through their interaction with other membrane proteins, some of these receptors may activate ion pumps, such as the proton pump; others may act to open ion channels in the membrane (see Chapter 4).

The calcium ion is of particular interest in hormone action. Generally, Ca^{2+} levels in the cytoplasm are very low. Hormonal stimulation of calcium ion channels results in a temporary elevation of Ca^{2+} levels. Binding of Ca^{2+} to the calcium-binding sites of certain proteins alters the activity of these proteins, much as hormones activate receptor proteins. *Protein kinases* are a class of enzymes that may be activated by Ca^{2+} or other second messengers. The protein kinases may modify "target" proteins by transferring phosphate groups onto certain amino acids of the target protein, altering its activity.

Substances, such as calcium, that mediate hormonal responses are often referred to as **second messengers** (Figure 25–21). Second messengers perform two important functions: (1) they are involved in the transfer of information from the hormone-receptor complex to the target proteins, and (2) they amplify the signal produced by the hormone. Receptor activation of a single calcium channel may result in the release of hundreds of calcium ions into the cytoplasm. Each calcium ion can in turn activate a protein kinase molecule, and each protein kinase molecule can phosphorylate many molecules of target protein. Complex response pathways involving second messengers also contribute to the diversity of possible responses to a given hormone. Different cell types may have the same plasma membrane receptor but may respond quite differently to the same hormone if they carry a different complement of protein kinases and target proteins.

Plant Biotechnology

The origins of **plant biotechnology,** the application of an array of techniques to manipulate the genetic potential of plants, can be traced back to the late 1850s and 1860s and the work of the German plant physiologists Julius von Sachs and W. Knop. It was they who demonstrated that many kinds of plants could be grown in water if provided with a few essential elements as salts dissolved in the water; in other words, plants could be grown without placing their roots in soil. By the mid-1880s, it was known that at least ten chemical elements found in plants are necessary for normal growth. Today, 16 elements are generally considered to be essential for most plants (see Chapter 27).

The use of *hydroponics* and the quest to understand the mineral nutrition of plants provided the impetus for studies on the growth of excised plant parts in nutrient solutions. Gradually, substances such as sucrose, various vitamins, and other organic substances were added to the nutrient solutions in an attempt to sustain growth. It was not until the discovery of plant hormones and an understanding of their roles in the control of plant growth and development, however, that organ and tissue culture truly became feasible.

TOTIPOTENCY

As early as 1902, the German botanist Gottlieb Haberlandt suggested that all living plant cells are **totipotent** —that is, that each cell possesses the potential to develop into an entire plant—but he was never able to accomplish this. In fact, more than half a century passed before his hypothesis was proved essentially correct. Haberlandt did not know which substances to add to the media; plant hormones had not yet been discovered.

In the late 1950s, F.C. Steward isolated small bits of phloem tissue from carrot (*Daucus carota*) root and placed them in liquid growth medium in a rotating flask. (Such pieces of tissue are called *explants*.) The medium contained sucrose and the inorganic nutrients necessary for plant growth (see Chapter 27), certain vitamins, and coconut milk, which Steward knew to be rich in plant growth compounds—although the nature of these compounds was not then understood.

In the rotating flask, individual cells continuously broke away from the growing cell mass and floated free in the medium. These individual cells were able to grow and divide. Before long, Steward observed that roots had developed in many of these new cell clumps. If left in the swirling medium, the cell clumps did not continue to differentiate, but if they were transferred to a solid medium—which was agar in these experiments —some of the clumps developed shoots. If the clumps were then transplanted to soil, the little plants leafed out, flowered, and produced seed. Similar results were obtained a few years later by V. Vasil and A.C. Hildebrandt, who used explants of tobacco (*Nicotiana*) pith from a fresh stem of a hybrid. Rather than coconut milk, the medium used by Vasil and Hildebrandt contained IAA and kinetin.

The results obtained by Steward and by Vasil and Hildebrandt indicated that at least some of the cells of the mature carrot phloem and tobacco pith contained all the genetic potential for full plant development, although this potential was not expressed. These experiments also illustrated the fact that such differentiated cells are capable of expressing portions of their previously unexpressed genetic potential to trigger those particular developmental patterns. By achieving these results, Steward and Vasil and Hildebrandt confirmed Haberlandt's hypothesis concerning totipotency.

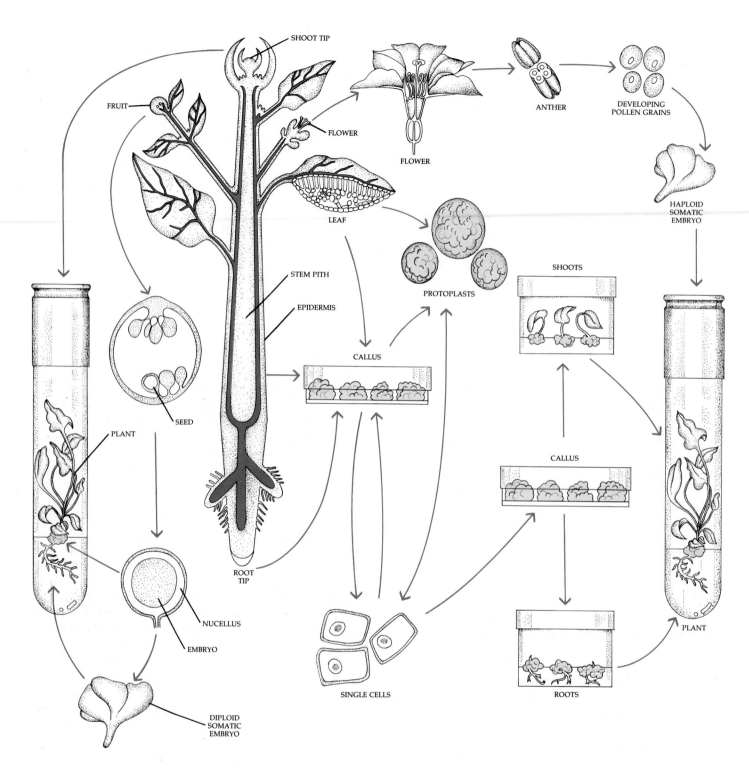

25–22

Theoretically, any plant cell—unless lacking a nucleus or enclosed by a rigid, secondary wall—is potentially capable of regenerating the organism from which it was derived and is said to be totipotent. Collections of similar cells form tissues, tissues and tissue systems are organized

into organs, and the spatial arrangement of organs constitutes the organism. Plants can be regenerated in vitro from organ explants (root and shoot tips, buds, leaf primordia, developing embryos, bud scales, and so on), tissue explants (pith,

cortex, epidermis, phloem, nucellus), cells (parenchyma, collenchyma, uni- or binucleate pollen grains), and protoplasts. The scheme above illustrates some of the pathways by which micropropagation can be achieved.

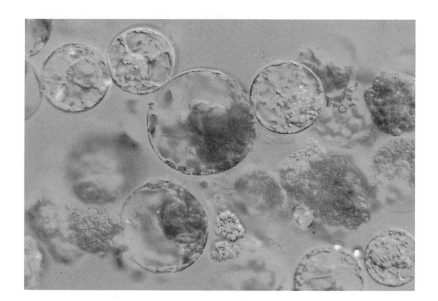

25–23
Protoplasts of a wild tobacco known as tree tobacco (Nicotiana glauca) (green) and commercial tobacco (N. tabacum) (clear). Some have fused, giving rise to hybrid cells. With appropriate treatment, the hybrid cells will regenerate cell walls, multiply, and give rise to whole plants.

ROLE OF TISSUE CULTURE IN CLONAL PROPAGATION

Plant tissue culture can be broadly defined as a collection of methods of growing large numbers of cells in a sterile and controlled environment (Figure 25–22). At present, the greatest impact of tissue culture is in the area of plant multiplication, referred to as **micropropagation,** or **clonal propagation,** since the individuals produced are genetically identical. The goal is to induce individual cells to express their totipotency.

Regeneration from Protoplasts

The most elegant use of tissue culture for biotechnology has involved plant regeneration from protoplasts that have had their walls removed by enzymatic digestion. In one procedure, known as **protoplast fusion,** protoplasts from two different plants are made to fuse together, producing a *somatic hybrid* cell. Most frequently, mesophyll cells have been used, although any living cell lacking a secondary wall can be used. Shortly after the walls have been removed and before new walls are formed, the naked protoplasts are made to fuse by the addition of an appropriate agent, such as polyethylene glycol, or by use of electrical shock (electroporation; see page 571.). The hybrid cells must then be grown on solid growth medium so that callus can develop. Provided all goes well, the callus will form somatic embryos, and then plantlets can be nurtured to grow into adult, fertile plants. Haploid hybrid plants, which are obtained through the use of anther explants or pollen, are unable to reproduce sexually and, hence, must be propagated vegetatively.

Several interspecific (between species) somatic hybrids have been produced by protoplast fusion within genera that include tobacco, petunia, potato, and carrot (Figure 25–23). Most intergeneric hybrids have been sterile. Fertile intergeneric hybrid plants have been obtained by protoplast fusion between potato and tomato, both of which are members of the Solanaceae, or nightshade, family.

Meristem Culture for the Production of Pathogen-free Plants

Apart from providing a means to produce identical copies **(clones)** of a plant, micropropagation provides a way to circumvent many plant diseases. This is due, in part, to the decontamination of the explants and the sterile conditions practiced in micropropagation, but primarily it is due to the use of meristem and shoot-tip culture techniques. Here, only very small explants of meristem and shoot tips lacking differentiated vascular tissues are cultured. Such explants are often virus-free because virus particles, which may be present in mature vascular elements below the meristems, can reach the meristematic regions of the apices only very slowly by cell-to-cell movement. The production of virus-free plants by meristem culture has greatly increased the yields of several crop plants, including potatoes and rhubarb.

GENETIC ENGINEERING

Genetic engineering (recombinant DNA) technology is providing one of the most important means by which crop plants will be improved in the future. It has a distinct advantage over recombination of plant genetic material brought about by natural and somatic hybridization because it allows individual genes to be inserted into organisms in a way that is both precise and simple. In addition, the species involved in the gene transfer do not have to be capable of hybridizing with one another.

Genetic engineering is based on the ability to cut DNA molecules precisely into specific pieces and to combine those pieces to produce new combinations (recombinant DNA). The procedure depends on the existence of **restriction enzymes** (*restriction endonucleases*) that recognize specific sequences of double-stranded DNA. These recognition sequences are typically four to six nucleotides long and are always symmetrical (that is, one strand is identical to the other strand when read in the opposite direction):

$$—GAATTC—$$
$$—CTTAAG—$$

Restriction enzymes cut DNA within or near their particular recognition sequences. Most restriction enzymes make straight cuts, but some cut through the strands a few nucleotides apart, leaving "sticky ends":

$$—G \qquad AATTC—$$
$$—CTTAA \qquad G—$$

The DNA strands at the two broken "sticky ends" of the fragment are complementary, and they can therefore pair with each other and rejoin through the action of DNA ligase (Figure 8–18). More important, these "sticky ends" can join with any other segment of DNA —from any number of sources—that has been cleaved by the same restriction enzyme and therefore has complementary "sticky ends." This property makes possible virtually unlimited recombinations of genetic material, since the ability of DNA fragments to recombine is not dependent on the original sources of the fragments.

Agrobacterium tumefaciens: A Natural Genetic Engineer

The most popular organism for the transfer of foreign genes into plants is *Agrobacterium tumefaciens*, a soil-dwelling bacterium that infects a wide range of dicotyledons, typically gaining entry through wounds. *Agrobacterium tumefaciens* induces the formation of tumors, called crown-gall tumors (Figure 25–24), on the plant by transferring a specific region, the **T-region,** or **T-DNA,** of a tumor-inducing (Ti) plasmid to the host's nuclear DNA.

Each **Ti plasmid** is a closed circle of DNA that consists of about 100 genes (Figure 25–25a). The T-DNA consists of approximately 20,000 base pairs of DNA bounded by 25 base pair repeats at each end (Figure 25–25b). The T-DNA carries a number of genes, in-

25–24
Crown galls growing on a Nicotiana glauca *stem.*

cluding one that codes for an opine-synthesizing enzyme, two that code for enzymes involved in the synthesis of auxin, and one that codes for an enzyme that catalyzes the synthesis of a cytokinin. The presence of the genes that code for the enzymes involved in hormone biosynthesis confers on the host cells the ability to grow and divide uncontrollably. The opines, unique amino acid derivatives, are used by the bacteria as sources of carbon and nitrogen. Another region of the Ti plasmid, the *vir* region, is essential for the transfer process but is not incorporated into the host's DNA.

Agrobacterium tumefaciens, with its Ti plasmid, is a powerful tool for genetic engineering of dicotyledonous plants. The tumor-promoting genes on the T-DNA can be removed and replaced by foreign genes (Figure 25–26). Infection of a plant with *Agrobacterium* containing these engineered plasmids will result in transfer of the foreign genes into the plant's genome. **Transgenic plants** (plants containing foreign genes) obtained through the use of plasmids (see page 191) will transmit the foreign genes to their progeny in a Mendelian fashion (see Chapter 8).

Considerable progress has been made in applying this method to the transfer of agriculturally useful genes. These include genes for resistance to insects, herbicides, and viruses. For example, bacterial genes for insecticidal proteins have been transferred into tomato and tobacco plants, resulting in transgenic plants that are resistant to lepidopteran larvae. Obtained from *Bacillus thuringiensis*, these genes when present result in the production of a toxin that kills larvae feeding upon the plant. The toxicity of the protein is very specific.

25–25

(a) *Diagrammatic representation of an* Agrobacterium tumefaciens *cell, showing main DNA (the bacterial "chromosome") and the Ti plasmid.* (b) *Detail of the Ti plasmid of* A. tumefaciens. O *is the gene that codes for an opine-synthesizing enzyme;* onc *is a group of three genes that code for enzymes that are involved in the biosynthesis of plant hormones;* R *represents 25-base-pair repeat sequences (only the DNA between these repeat sequences is transferred into the plant's genome);* vir *is a group of genes that control the transfer of the T-region, or T-RNA, to the host (plant) chromosome.*

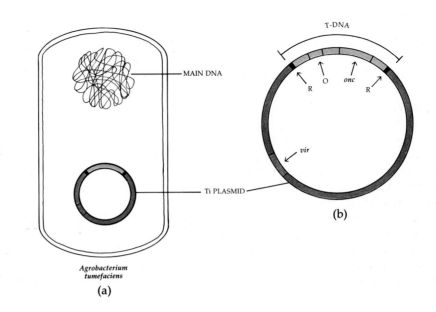

25–26

Procedure for using plasmids of Agrobacterium tumefaciens *as vectors in DNA, or gene, transfer.* (a) *A hybrid plasmid that carries only the T-DNA from a Ti plasmid is cut open with restriction enzymes and a foreign gene is inserted, creating a recombinant plasmid.* (b) *The recombinant plasmid is transferred into an* A. tumefaciens *cell that contains a Ti plasmid with its T-DNA removed, creating an engineered plasmid.* (c) *The* A. tumefaciens *containing the engineered plasmid is used to infect a plant. The* vir *region of the Ti plasmid without T-DNA controls the transfer of the foreign gene from the recombinant plasmid into the plant's chromosomes.*

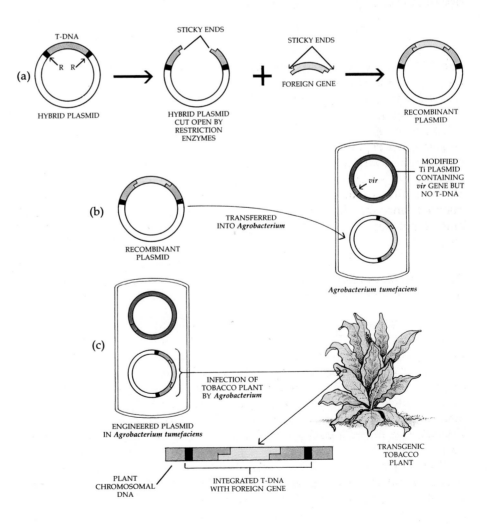

Arabidopsis thaliana: **A Weed with Promise**

Arabidopsis thaliana (family Brassicaceae), a harmless weed of no economic value, has become an invaluable experimental organism in the study of plant molecular genetics (a). Among its attributes, which make it particularly suited for both classical and molecular genetic research, are: (1) Its short generation time. Only four to six weeks are required to obtain mature plants, each of which has the potential to produce more than 10,000 seeds. (2) Its small size. *Arabidopsis* is such a small plant that literally dozens can be grown in a small pot, requiring only moist soil and fluorescent light for rapid growth. (3) Its adaptability. *Arabidopsis* plants grow well on sterile, biochemically defined media. In addition, *Arabidopsis* cells have been grown in culture and plants have been regenerated from such cells. (4) Typically, it self-fertilizes. This allows new mutations to be made homozygous with minimal effort. Many mutations have been identified in *Arabidopsis,* including visible ones useful as markers in genetic mapping. Such maps provide an approximation of the actual positions of genes on the chromosomes. (5) Its relatively small genome and small amount of repetitive DNA, which potentially simplify the task of identifying and isolating genes. (6) Its susceptibility to infection by *Agrobacterium tumefaciens* (see page 567). Successful transformations have already been achieved.

None of the crop plants share all of these traits with *Arabidopsis.* Typical crop plants have generation times of several months and require a great deal of space for growth in large numbers. In addition, those that have been used for recombinant DNA studies have large genomes and large amounts of repetitive DNA. The potential clearly exists for *Arabidopsis* to join such organisms as yeast, the fruit fly *(Drosophila),* and mice as vehicles in the molecular biologist's quest to gain insight into basic life processes. As with these other "model" organisms, plans are currently underway to characterize and completely sequence the *Arabidopsis* genome by the year 2000.

The genetic approach to solving basic problems in plant growth and development is currently being applied in major research laboratories around the world. The basic strategy involves screening large populations of plants that have been treated with mutagenic agents, which randomly cause disruptions in individual genes. The researcher looks for plants that show some alteration in behavior relative to the nonmutagenized plants. The geneticist refers to these alterations as phenotypes. Through genetic analysis it can be determined that a mutant phenotype is due to a defect in a single gene within the plant's genome. As an example, plant biologists long have recognized a series of single gene muta-

(a)

(a) *The diminutive weed* Arabidopsis thaliana, *which has become the object of investigation of hundreds of investigators world-wide.* (b) *A normal* Arabidopsis *flower, with stamens and carpels.* (c) *An agamous mutant in which the stamens and carpels have been converted into sepals and petals.*

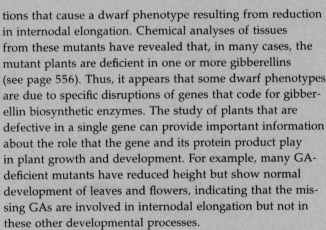

(b)

(c)

tions that cause a dwarf phenotype resulting from reduction in internodal elongation. Chemical analyses of tissues from these mutants have revealed that, in many cases, the mutant plants are deficient in one or more gibberellins (see page 556). Thus, it appears that some dwarf phenotypes are due to specific disruptions of genes that code for gibberellin biosynthetic enzymes. The study of plants that are defective in a single gene can provide important information about the role that the gene and its protein product play in plant growth and development. For example, many GA-deficient mutants have reduced height but show normal development of leaves and flowers, indicating that the missing GAs are involved in internodal elongation but not in these other developmental processes.

With a "model" organism such as *Arabidopsis*, one can take advantage of the increasingly sophisticated body of information available on the structure of the plant genome. By combining genetics and molecular biology, mutated genes can be isolated and cloned. (Mutated genes are introduced into *Escherichia coli*, using plasmids as vectors. As the bacterial cells multiply, the plasmids also replicate, producing identical segments of the gene. Such multiple copies of genes are known as clones.) Once the gene has been cloned, the nucleotide sequence can be used to determine the primary amino acid sequence of the protein coded for by the gene. In addition, the protein can be produced in unlimited quantities in *E. coli* or yeast for biochemical analysis.

One method of producing mutant genes in plants involves the use of the T-DNA of *Agrobacterium tumefaciens* as an insertional mutagen. The T-DNA of *Agrobacterium* inserts randomly into the plant's genome. If, by chance, the T-DNA is inserted into a functional gene, the gene will be disrupted and a mutant phenotype may result. In this case the mutated gene is "tagged" with the foreign T-DNA sequences, which can then be used to identify DNA clones containing the disrupted gene. Using this method, researchers working on the regulation of flower development in *Arabidopsis* have identified single gene mutations that prevent the development of specific floral organs. The agamous mutation in *Arabidopsis* blocks the development of stamens and carpels. In place of these structures, the mutant flower produces additional sepals and petals (b and c). The mutated gene was cloned and sequenced, and the derived amino acid sequence indicated that the gene coded for a transcription factor (a DNA-binding protein). The transcription factor is apparently required for activating genes that are involved in the development of reproductive structures in *Arabidopsis*.

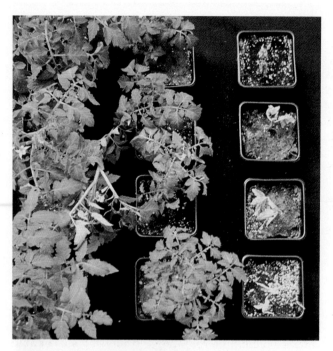

25–27
Tomato plants (middle row) engineered with the Ti plasmid of Agrobacterium tumefaciens as a vector for a bacterial gene expressing resistance to the herbicide glyphosate. The engineered plants survived spraying with 0.75 pounds of glyphosate per acre, whereas similarly treated plants without the resistance gene died (right row). The plants in the row on the left were not sprayed.

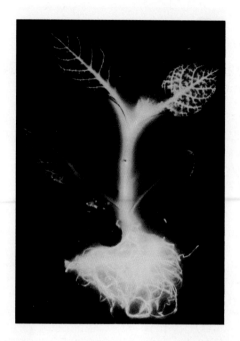

25–28
Genes for the production of the enzyme luciferase were inserted into cells isolated from a normal tobacco plant (Nicotiana tabacum), *using the Ti plasmid of* Agrobacterium tumefaciens *as a vector. After the undifferentiated callus cells developed into a whole plant, the cells that incorporated the luciferase gene into their DNA became luminescent in the presence of luciferin, ATP, and oxygen.*

Plants have been engineered for resistance to several herbicides, including glyphosate (marketed by Monsanto under the name Roundup), which works by blocking a single enzyme essential to plants for the production of aromatic amino acids. Although extremely effective and nontoxic to animals, glyphosate kills all plants, including crops. One successful approach to genetically engineering resistance into crop plants resulted from the identification of a mutant form of the target enzyme from *Salmonella* bacterium that is not blocked by glyphosate. Transfer of this mutant gene into crop plants, using the Ti plasmid, resulted in plants that were resistant to the herbicide (Figure 25–27).

Reporter Genes

One of the technical problems encountered in attempts at gene transfer is knowing whether a particular gene has actually been introduced into a new host cell and, if transferred, whether it is directing the synthesis of pro-tein. To overcome this problem, **reporter genes** have been developed, which can be transferred to the plant cell with the help of *Agrobacterium*. For gene monitoring, the reporter gene must be linked to a promoter. When the promoter is switched on, the reporter gene is activated. Two reporter genes currently in use are GUS, the *Escherichia coli β*-glucuronidase gene, and the luciferase gene, which is obtained either from fireflies (*Photinus pyralis*) or the marine bacterium *Vibrio harveyi*.

In a notable experiment, investigators at the University of California at San Diego spliced the gene for luciferase from fireflies into the DNA of tobacco cells, and plants were then regenerated from these cells. The final result was a tobacco plant that glows when it is provided with luciferin, the protein substrate of luciferase (Figure 25–28). Whenever the promoter is switched on, the luciferase gene is activated and luciferase is produced; in the presence of oxygen, luciferin plus ATP produces bioluminescence, as seen in the flash of a firefly.

Limitations to the *Agrobacterium* Gene-transfer System

Despite its great usefulness for the transfer of foreign genes into many plant species, *Agrobacterium* cannot under natural conditions transform monocotyledonous plants. The latter include, of course, the world's major cereal crops: wheat, rice, corn, barley, and sorghum. The great majority of monocots are not susceptible to *Agrobacterium*. Moreover, the tissue-culture systems and micropropagation techniques that have served so well for dicots do not work well for monocots. Considerable effort is being made to develop suitable procedures for the regeneration of plants from monocot protoplasts.

Fortunately, two new methods of gene transfer that are applicable to both monocots and dicots are providing promise. One of these, which was mentioned briefly earlier, is called **electroporation.** In this method, a high-voltage electric pulse is applied to a solution containing protoplasts and DNA. The electric shock temporarily opens pores in the plasma membrane through which the DNA passes.

In the second method, called **particle bombardment,** high-velocity microprojectiles (small tungsten particles) are used for delivery of RNA or DNA into cells. The RNA or DNA is adsorbed to the surface of the microprojectiles. This approach has an advantage over other procedures in that it can be used on intact tissues. In addition, the process does not require either cell culture or pretreatment of the recipient tissue.

Benefits and Risks of Genetic Engineering

Genetic engineering technology is providing biologists with an opportunity never before available—to transfer genetic traits between diversely different organisms. For the plant biologist this means potentially increasing crop production by introducing genes that increase the crop's resistance to various pathogens or herbicides and enhance its tolerance to various stresses. Examples of the former have already been given for the control of lepidopteran larvae and resistance to the herbicide glyphosate. A modified "ice-minus" strain of the bacterium *Pseudomonas syringa* has already been developed to reduce the susceptibility of certain crops to frost and, therefore, allow early planting. And attempts are being made to increase nitrogen availability, a limiting factor in crop production, by transferring genes responsible for nitrogen fixation into crops such as wheat and corn, which do not fix nitrogen. In addition to benefits derived from increased crop production, billions of dollars could be saved in fertilizer, insecticide, and herbicide costs.

We are reminded, however, that the benefits of genetic engineering do not come without some risks.

Some agriculturists argue that genetic engineering and biotechnology will only extend what they regard as the "invade and conquer" style of farm production and pest management at a time when more emphasis should be placed on **sustainable agriculture,** which minimizes chemical use and "focuses on working with nature, rather than trying to conquer it." It is argued that increasing the number of herbicide-resistant crops may actually encourage wider herbicide use rather than limit it and thus contribute to environmental problems. The potential also exists for genetic transfer between engineered organisms and wild organisms in the environment. The impact of such transfer would depend on the nature of the engineered trait. If, for example, it were a gene for pest resistance, it could make a hybrid weed superior to its pure wild types. Because of risks such as these, many people feel that genetic engineering technology must be carefully managed, in order to avoid serious environmental, economic, and social problems. Genetically engineered crops should be used as one element in promoting systems of sustainable agriculture. In addition, biological control and improved cultivation practices are also important elements in putting together such systems, which have a critical role to play in feeding a hungry world.

Summary

A plant hormone is a chemical produced in certain tissues that may or may not be transported to other tissues and that produces a physiological response in extremely small amounts.

Auxin is a hormone that is produced in the apical meristems of shoots and the tips of coleoptiles. Auxin travels unidirectionally toward the base of the plant, where it controls the lengthening of the shoot and the coleoptile, chiefly by promoting cell elongation. Auxin also plays a role in differentiation of vascular tissue and initiates cell division in the vascular cambium. It often inhibits growth in lateral buds, thus maintaining apical dominance. The same quantity of auxin that promotes growth in the stem inhibits growth in the main root system. Auxin promotes the formation of adventitious roots in cuttings and retards abscission in leaves, flowers, and fruits. In fruits, auxin produced by seeds or by the pollen tube stimulates growth of the ovary wall.

The cytokinins are chemically related to certain components of nucleic acids. Cytokinins act in concert with auxin to cause cell division in plant tissue culture. In tobacco pith cultures, a high concentration of auxin promotes root formation, while a high concentration of cytokinins promotes bud formation. In intact plants, cytokinins promote the growth of lateral buds, acting in opposition to the effects of auxin. Cytokinins prevent senescence in leaves by stimulating protein synthesis.

Ethylene is also a natural growth regulator in plants, where it induces a number of distinct physiological responses, such as abscission and fruit ripening.

Abscisic acid is a growth-inhibiting hormone found in dormant buds and fruits. Abscisic acid stimulates the production of seed proteins and the closing of stomata.

The gibberellins control shoot elongation, especially in dwarf plants. The application of gibberellin restores normal growth. This is also true in plants with a rosette form of growth, in which gibberellins cause bolting. Gibberellins cause seed germination in grasses. In the barley seed, the embryo releases gibberellins that cause the aleurone layer of the endosperm to produce a-amylase. This enzyme breaks down the starch stored in the endosperm, releasing sugar that nourishes the embryo and promotes germination.

At the molecular level, hormones influence developmental processes through their interaction with receptors within the plant cell. The plant hormones mediate many processes by activating or repressing sets of genes within the cell's nucleus. At the cellular level, hormones influence the rate and direction of cell expansion and the rate of cell division. These effects are mediated through complex biochemical response pathways in the cell, often involving second messengers.

Advances in hormone research and DNA biochemistry have made it possible to manipulate the genetics of plants in specific ways. Among the most important procedures in biotechnology is tissue culture. In ideal cases, tissue culture is used to obtain complete plants starting with single, genetically altered cells. The potential for single cells to develop into entire plants is called totipotency.

Genetic engineering (recombinant DNA) technology is based on the ability to cut DNA molecules precisely into specific pieces and to combine those pieces to produce new combinations. The Ti plasmid of *Agrobacterium tumefaciens*, which induces the formation of crown-gall tumors, is being used as a vector for introducing genes into plant cells. Through the use of reporter genes, it is possible to determine visually if the genes carried by the plasmid have been successfully transferred to the plant cells and are being expressed.

See the end of Chapter 26 for suggested readings.

CHAPTER 26

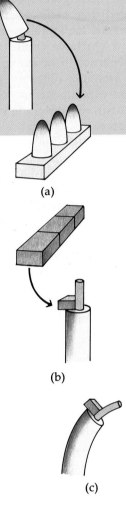

(a)

(b)

(c)

26–1
Went's experiment. (a) Went removed the coleoptile tips from seedlings and placed the tips on agar for about an hour. (b) The agar was then cut into small blocks and placed on one side of the decapitated shoots of the seedlings. (c) The seedlings, which were kept in the dark during the entire experiment, subsequently were observed to bend away from the side on which the agar block was placed. From this result, Went concluded that the "influence" that caused the seedling to bend was chemical and that it accumulated on the side away from the light.

External Factors and Plant Growth

Living things must regulate their activities in accordance with the world around them. Many animals, being mobile, can change their circumstances to some extent —foraging for food, courting a mate, and seeking or even making shelters in bad weather. A plant, by contrast, is immobilized once it sends down its first root. However, plants do have the ability to respond and make adjustments to a wide range of changes in their external environment. This ability is manifested chiefly in changing patterns of growth.

The Tropisms

A growth response involving bending, or curving, of a plant part toward or away from an external stimulus that determines the direction of movement is called a **tropism**. A response toward the stimulus is said to be *positive*; a response away from the stimulus is *negative*.

Perhaps the most familiar interaction between plants and the external world is the curving of growing shoot tips toward light (see Figure 25–1, page 547). This growth response, now known as **phototropism**, is caused by elongation—under the influence of the plant hormone auxin—of the cells on the shaded side of the tip. What role does the light play in the phototropic response? Three possible answers to this question have been suggested: (1) light decreases the auxin sensitivity of the cells on the lighted side; (2) light destroys auxin; or (3) light drives auxin to the shaded side of the growing tip.

To test these hypotheses, Winslow Briggs and his coworkers carried out a series of experiments based on earlier work by Frits Went (Figure 26–1). These investigators first showed that the same total amount of auxin is obtained from the tips of coleoptiles whether they re-

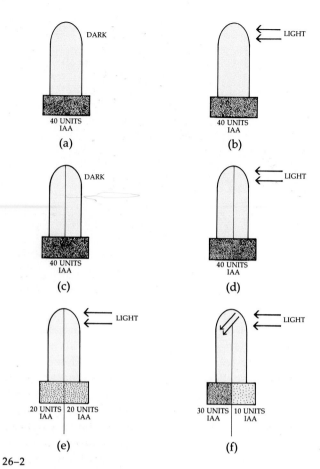

26–2

Experiments by Briggs and coworkers showing lateral displacement of IAA in unilaterally illuminated corn (Zea mays) coleoptile tips. The diffusible auxin was collected in agar blocks. The units indicate the relative amounts of IAA present in each block. (a) and (b) The same amount of IAA is collected from intact coleoptiles whether they remain in the dark or are illuminated. (c) and (d) If the coleoptiles are split in half by a thin piece of glass but remain intact, the same amount of IAA is collected as in (a) and (b). (e) If the two halves and the agar block below are completely separated, the same amounts of auxin are collected from the dark and illuminated sides. (f) When the coleoptile is not split entirely to the top, more auxin diffuses from the darkened half than from the illuminated half. These results support the idea that unilateral illumination induces a lateral movement of IAA from the illuminated to the darkened half of the coleoptile.

main in the dark or the light. However, following exposure to light on one side, the amount of auxin obtained from the shaded side is greater than the amount from the lighted side. If the tip is split and a barrier, such as a thin piece of glass, is placed between the two halves (the light and dark sides), this differential distribution of auxin is no longer observed (Figure 26–2). In other words, Briggs demonstrated that auxin migrates from the light side to the dark side. Experiments using ^{14}C-labeled indoleacetic acid (IAA) have shown that it is migration of auxin, not destruction of auxin, that accounts for the different amounts obtained from the light and dark sides. These experiments are consistent with the hypothesis that auxin redistribution is responsible for phototropic curvature.

Light of wavelengths from 400 to 500 nanometers is most effective in inducing this hormone migration in a shoot tip, indicating that a pigment absorbing blue light mediates the effect. Although the identity of the blue-light photoreceptor has not been definitely established, present evidence indicates that it is a yellow-colored pigment called a flavin.

Another familiar tropism is **gravitropism,** a response to gravity (Figure 26–3). If a seedling is placed on its side, its root will grow downward (positive gravitropism) and its shoot will grow upward (negative gravitropism). The original explanation for this mechanism involved the redistribution of auxin from the upper to the lower side. Roots are more sensitive in their response to auxin than stems, and, at high auxin concentrations, growth of roots is inhibited. The lower part of the shoot is stimulated by the auxin and grows faster than the upper side. The result is a curving upward of the stem. For the root, the relatively high level of auxin on the lower side inhibits growth; thus the root curves downward.

Experimental evidence in support of this hypothesis has been obtained for the shoot only. In many cases, particularly with dicots, bulk redistribution of auxin has not been observed. One possible explanation may be that auxin redistribution occurs between such closely associated tissues that it is difficult to measure. If, as thought, the growth rate of stems is largely determined by the extensibility of the epidermal cell layers, regulation of stem growth could involve lateral movement of auxin over the distance of only a few cell layers from the cortex to the epidermal layer. Such minor changes in auxin distribution would be difficult to detect but could have dramatic effects on growth.

The notion that auxin does play a role in gravitropic responses of stems has gained support from recent experiments. Auxin-inducible genes have been identified and cloned from a number of species. Auxin activation of transcription of some of these genes occurs only on the side of the stem showing increased growth. It is still not clear whether this transcriptional activation is due

26–3

Gravitropic responses in the shoot of a young tomato plant (Lycopersicon esculentum). *The plant in (a) was placed on its side and kept in a stationary position; the plant in (b) was placed upside down and held in position in a ring stand. Although originally straight, the stem bent and grew upward in both cases. If a plant is held horizontally and slowly rotated around its horizontal axis, no curvature (gravitropic response) will occur; that is, the plant will continue to grow horizontally. Can you explain the differences in the growth of a horizontally rotated plant and the plants shown here?*

(a)

(b)

to an increase in auxin concentration or to an increase in sensitivity of the tissue to auxin already present.

Recent studies indicate that calcium plays an important role in the gravitropic response of both shoots and roots, and that this response is mediated by the calcium-binding protein calmodulin. Calcium has been shown to move toward the upper surface of gravi-stimulated shoots before the shoots curve upward and toward the lower surface of roots before the roots curve downward. In intact corn (*Zea mays*) roots, the influence of calcium may be at the epidermis.

The perception of gravity is correlated with the sedimentation of amyloplasts (starch-containing plastids) within specific cells of the shoot and root. Such cells are often found contiguous to or surrounding the vascular bundles in shoots. In roots, however, these cells are localized in the root cap, particularly in the central column (columella) of the root cap (Figure 26–4). When a root is placed in a horizontal position, the amyloplasts, which were sedimented near the transverse walls of the vertically growing roots, slide downward and come to rest near what were previously vertically oriented walls (Figure 26–5). After several hours, the root curves downward and the amyloplasts return to their previous position along the transverse walls. How the movement of these gravity sensors (statoliths) is translated into differential growth rates has yet to be explained.

A less apparent, though common, tropism is **thigmotropism** (from the Greek *thigma*, meaning "touch"), a response to contact with a solid object. One of the

26–4

(a) *Photomicrograph of a median longitudinal section of the root cap of the primary root of the common bean,* Phaseolus vulgaris. *Arrows point to amyloplasts (starch-containing plastids) sedimented near the transverse walls in the central column (columella) cells of the root cap. (b) Electron micrograph of columella cells from a root cap similar to that shown in (a). The amyloplasts (arrows) are visible at the bottom of each cell near the transverse walls. The starch grains are clearly evident within the amyloplasts.*

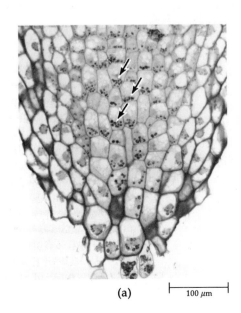

(a)   ⊢ 100 μm ⊣

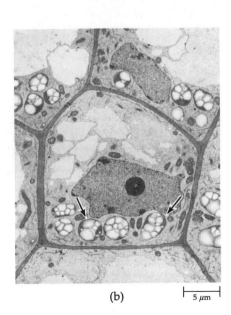

(b)   ⊢ 5 μm ⊣

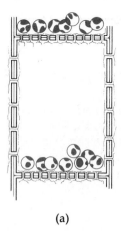
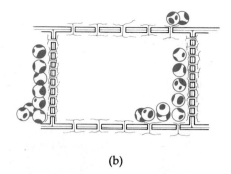
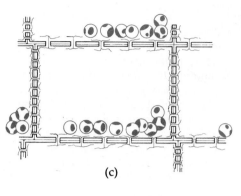

(a)   (b)   (c)

26–5

The response to gravity of amyloplasts in the columella cells of a root cap. (a) The amyloplasts are normally sedimented near the transverse walls in the root cap *of a root growing vertically downward. When the root is placed on its side, (b) and (c), the amyloplasts slide toward the normally vertical walls that are now* *parallel with the soil surface. This movement of amyloplasts plays an important role in the perception of gravity by roots.*

26–6

Tendrils of bur cucumber (Cucumis anguria). Twisting is caused by different growth rates on the inside and outside of the tendril.

most common examples of thigmotropism is seen in tendrils, which are modified leaves in some species and modified stems in others (see page 515). The tendrils wrap around any object with which they come in contact (Figure 26–6) and so enable the plant to cling and climb. The response can be rapid; a tendril may wrap around a support one or more times in less than one hour. Cells touching the support shorten slightly and those on the other side elongate.

Studies by M.J. Jaffe have shown that tendrils can store the "memory" of tactile stimulation. For example, if pea *(Pisum sativum)* tendrils are kept in the dark for three days and then rubbed they will not coil, perhaps because of the requirement for ATP. If, however, they are illuminated within two hours of the rubbing, they will show the coiling response.

Circadian Rhythms

It is a common observation that the flowers of some plants open in the morning and close at dusk; also, many leaves unfold in the sunlight and refold at night (Figure 26–7). As long ago as 1729, the French scientist Jean-Jacques de Mairan noticed that these diurnal (daily) movements continue even when the plants are kept in dim light (Figure 26–8). More recent studies have shown that photosynthesis, auxin production, and the rate of cell division also have regular daily rhythms and that these rhythms continue even when all environmental conditions are kept constant. These regular, approximately 24-hour cycles are called **circadian rhythms,** from the Latin words *circa,* meaning "approximately," and *dies,* meaning "a day." Examples of circadian rhythms are universal among eukaryotes. They appear to be absent in bacteria. Why this should be so is unknown.

(a)

(b)

26–7
Leaves of the wood sorrel (Oxalis), during the day (a) and at night (b). One hypothesis for the function of such sleep movements is that they prevent the leaves from absorbing moonlight on bright nights, thus protecting the photoperiodic phenomena discussed later in this chapter. Another hypothesis, proposed by Charles Darwin more than a century ago, is that the folding reduces heat loss from the leaves at night.

ARE THE RHYTHMS ENDOGENOUS?

Are these rhythms actually internal—that is, caused by factors entirely within the organism—or is the organism keeping itself in time with some external stimulus? Most workers agree that the rhythms are endogenous—that is, they are controlled internally. The internal timing mechanism is often referred to as the organism's **biological clock.**

SETTING THE CLOCK

Under constant environmental conditions, such as can be maintained only in a laboratory, the period of the circadian rhythm is **free-running**—that is, its natural period (usually between 21 and 27 hours) does not have to be reset at each cycle. It behaves as a self-sustained oscillator. In the natural world, the environment acts as a synchronizing agent (or "Zeitgeber," German for "time giver"). In fact, the environment is responsible for keeping a circadian rhythm in step with the daily 24-hour light-dark cycle. If a circadian rhythm of a plant were greater or less than 24 hours, the rhythm would soon get out of step with the 24-hour light-dark cycle. A phenomenon such as flowering, which generally takes place in the light period, would occur at a different time each day, including the dark period. Thus the plant must become resynchronized—that is, **entrained**—to the 24-hour day.

Entrainment is the process by which a periodic repetition of light and dark (or some other external cycle) causes a circadian rhythm to remain synchronized with the same cycle as the entraining factor. Light-dark cycles and temperature cycles are the principal factors in entrainment (Figure 26–9).

26–8
In many plants, the leaves move outward, to a position perpendicular to the stem and the sun's rays, during the day and upward, toward the stem, at night. These sleep movements can be recorded on a revolving drum using a delicately balanced pen-and-lever system attached to a leaf by a fine thread (a). Many plants, such as the common bean (Phaseolus vulgaris) shown here, will continue to exhibit these movements for several days even when kept in continuous dim light. (b) A recording of this circadian rhythm, showing its persistence under a condition of constant dim light.

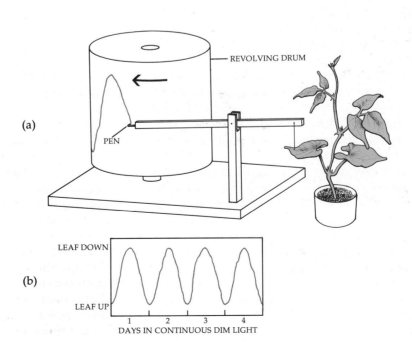

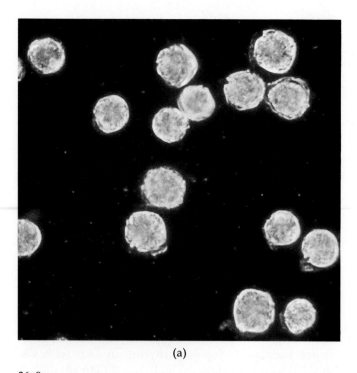

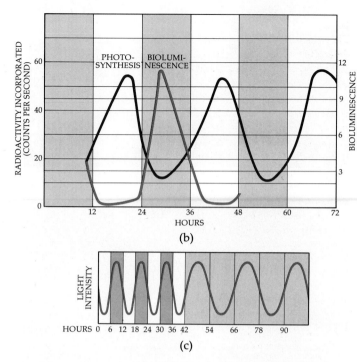

(a)

(b)

(c)

26–9

(a) *Photomicrograph of the dinoflagellate* Gonyaulax polyedra, *a single-celled marine alga.* (b) *In G.* polyedra, *three different functions follow separate circadian rhythms: bioluminescence, which reaches a peak in the middle of the night (colored curve); photosynthesis, which reaches a peak in the middle of the day (black curve); and cell division (not* shown), *which is restricted to the hours just before dawn. If Gonyaulax is kept in continuous dim light, these three functions continue to occur with the same rhythm for days and even weeks, long after a number of cell divisions have taken place.* (c) *The rhythm of bioluminescence in* Gonyaulax, *like most circadian rhythms, can be altered by modifying* the cycles of illumination. For example, if cultures of the alga are exposed to alternating light and dark periods, each 6 hours in duration, the rhythmic function will become entrained to this imposed cycle (left). If the cultures are then placed in continuous dim light, the organisms will return to their original rhythm of about 24 hours.

Another feature of interest is that these rhythms do not automatically speed up as the temperature rises, even though biochemical activities—and an internal biological clock must have a biochemical basis—take place more rapidly at high temperatures than at low ones. Some clocks run slightly faster as the temperature rises, but others go more slowly, and many are almost unchanged. Thus, the biological clock must contain within its workings a compensatory mechanism—a feedback system that allows it to adjust to temperature changes. Such a temperature-compensating feature would be very important in plants.

The primary usefulness of the biological clock is that it enables the plant or animal to respond to the changing seasons of the year by accurately measuring changing day length. Even at tropical latitudes, many plants are responsive to day length and may use this signal to synchronize flowering or other activities with such seasonal events as wet or dry periods. In this way, changes in the environment trigger responses that result in adjustments of growth, reproduction, and other activities of the organism.

Photoperiodism

The effect of day length on flowering was discovered some seventy years ago by two investigators from the U.S. Department of Agriculture, W.W. Garner and H.A. Allard, who found that neither the Maryland Mammoth variety of tobacco (*Nicotiana tabacum*) nor the Biloxi variety of soybean (*Glycine max*) would flower unless the day length was shorter than a critical number of hours. Garner and Allard called this phenomenon **photoperiodism.** Plants that flower only under certain day-length conditions are said to be *photoperiodic.* Photoperiodism is a biological response to a change in the proportions of light and dark in a 24-hour daily cycle. Although the concept of photoperiodism began with studies of plants, it has now been demonstrated in various fields of biology, including the mating behavior of such diverse animals as codling moths, spruce budworms, aphids, potato worms, fish, birds, and mammals.

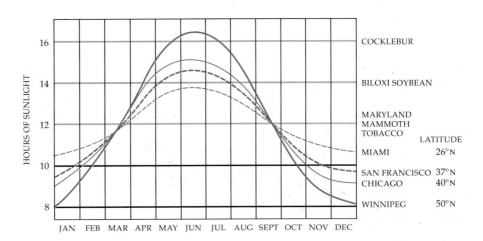

26–10
The relative length of day and night determines when plants flower. The four curves depict the annual change of day length in four North American cities at four different latitudes. The horizontal colored lines indicate the effective photoperiod of three different short-day plants. The cocklebur, for instance, requires 16 hours or less of light. In Miami, San Francisco, and Chicago, it can flower as soon as it matures, but in Winnipeg, the buds do not appear until early August, so late that the frost will probably kill the plants before the seed is set.

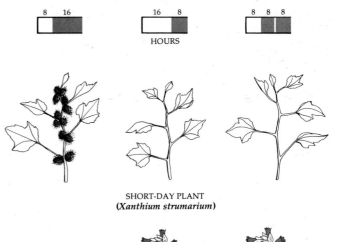

SHORT-DAY PLANT
(*Xanthium strumarium*)

LONG-DAY PLANT
(*Hyoscyamus niger*)

26–11
Short-day plants flower only when the photoperiod is less than some critical value. The common cocklebur (Xanthium strumarium), a short-day plant, requires less than 16 hours of light to flower. Long-day plants flower only when the photoperiod is longer than some critical value. The long-day plant henbane (Hyoscyamus niger) requires about 10 hours (depending on temperature) or more to flower. If the dark period is interrupted by a flash of light, Hyoscyamus will flower even when the daily light period is shorter than 10 hours. However, a pulse of light during the dark period has the opposite effect in a short-day plant—it prevents flowering. The bars at the top indicate the duration of light and dark periods in a 24-hour day.

LONG-DAY PLANTS AND SHORT-DAY PLANTS

Garner and Allard went on to test and confirm their discovery with many other species of plants. They found that plants are of three general types, which they called **short-day, long-day,** and **day-neutral.** Short-day plants flower in early spring or fall; they must have a light period *shorter* than a critical length. For instance, the common cocklebur (*Xanthium strumarium*) is induced to flower by 16 hours or less of light (Figures 26–10 and 26–11). Other short-day plants are some chrysanthemums (Figure 26–12a), poinsettias, strawberries, and primroses.

Long-day plants, which flower chiefly in the summer, will flower only if the light periods are *longer* than a critical length. Spinach, some potatoes, some wheat varieties, lettuce, and henbane (*Hyoscyamus niger*) are examples of long-day plants (Figures 26–11 and 26–12b).

Cocklebur and spinach will both bloom if exposed to 14 hours of daylight, yet only one is designated as a long-day plant. The important factor is not the absolute length of the photoperiod but whether it is longer or shorter than some critical interval. Day-neutral plants flower without respect to day length. Some examples of day-neutral plants are cucumber, sunflower, rice, corn, and garden pea.

Within individual species of plants that cover a large north-south range, different photoperiodic ecotypes (locally adapted variants of an organism) have often been observed. Thus, in many species of prairie grasses, which range from southern Canada to Texas, northern ecotypes flower before southern ones when they are grown together in a common environment. Different populations are precisely adjusted to the demands of the local day-night regimen.

(a)   (b)

26–12

Representative short-day (a) and long-day (b) plants, each grown under short-day (on left) and long-day (on right) conditions. (a) Chrysanthemum. (b) *Spinach* (Spinacia oleracea). *Note that the plants exposed to long days have* *longer stems than those exposed to short days, regardless of whether or not they flower.*

The photoperiodic response can be remarkably precise. At 22.5°C, the long-day plant henbane will flower when exposed to photoperiods of 10 hours and 20 minutes (Figure 26–11), but not with a photoperiod of 10 hours. Environmental conditions also affect photoperiodic behavior. For instance, at 28.5°C henbane requires 11.5 hours of light, whereas at 15.5°C it requires only 8.5 hours.

The response varies with different species. Some plants require only a single exposure to the critical day-night cycle, whereas others, such as spinach, require several weeks of exposure. In many plants, there is a correlation between the number of induction cycles and the rapidity of flowering or the number of flowers formed. Some plants have to reach a certain degree of maturity before they will flower, whereas others will respond to the appropriate photoperiod when they are seedlings. Some plants, when they get older, will eventually flower even if not exposed to the appropriate photoperiod, although they will flower much earlier with the proper exposure.

MEASURING THE DARK

In 1938, Karl C. Hamner and James Bonner began a study of photoperiodism, using the short-day plant cocklebur, which requires 16 hours or less of light per 24-hour cycle to flower. This plant is particularly useful for experimental purposes because a single exposure under laboratory conditions to a short-day cycle will induce flowering two weeks later, even if the plant is immediately returned to long-day conditions. The cocklebur plant can withstand a great deal of rough treatment; for example, it can survive even if all its leaves are removed. Hamner and Bonner showed that it

is the leaf blade of the cocklebur that perceives the photoperiod. A completely defoliated plant cannot be induced to flower. But if as little as one-eighth of a fully expanded leaf is left on the stem, the single short-day exposure induces flowering.

In the course of these studies, in which they tested a variety of experimental conditions, Hamner and Bonner made a crucial and totally unexpected discovery. If the period of darkness is interrupted by as little as a one-minute exposure to light from a 25-watt bulb, flowering does not occur. Interruption of the light period by darkness has absolutely no effect on flowering. Subsequent experiments with other short-day plants showed that they, too, required periods of uninterrupted darkness rather than uninterrupted light.

The most sensitive part of the dark period with respect to light interruption is in the middle of the period. If a short-day plant, such as cocklebur, is given an 8-hour light period and then an extended period of darkness, the plant passes through a stage of increasing sensitivity to light interruption lasting about 8 hours, followed by a period when light interruption has a diminishing effect. In fact, a one-minute light exposure after 16 hours of darkness stimulates flowering.

On the basis of the findings of Garner and Allard, described earlier, commercial growers of chrysanthemums had found that they could hold back blooming in short-day plants by extending the daylight with artificial light. On the basis of the new experiments by Hamner and Bonner, growers were able to delay flowering simply by switching on the light for a short period in the middle of the night.

What about long-day plants? They also measure darkness. A long-day plant that will flower if kept under light for 16 hours and dark for 8 hours will also flower on 8 hours of light and 16 hours of dark if the dark is interrupted by even a brief exposure of light.

Chemical Basis of Photoperiodism

An important clue to the mechanism of the response of plants to the relative proportions of light and darkness came from a team of research workers at the U.S. Department of Agriculture Research Station in Beltsville, Maryland. The clue came from an earlier study performed with lettuce *(Lactuca sativa)* seeds, which germinated only if they were exposed to light. This requirement is true of many small seeds, which must germinate in loose soil and near the surface in order for the seedling to emerge. In studying the light requirement of germinating lettuce seeds, earlier workers had shown that red light stimulates germination and that light of a slightly longer wavelength (far-red) inhibits germination even more effectively than does no illumination at all.

Hamner and Bonner had shown that when the dark period is interrupted by a single flash of light from an ordinary light bulb, the cocklebur will not flower. The Beltsville group, following this lead, began to experiment with light of different wavelengths, varying the intensity and duration of the flash. They found that red light with a wavelength of about 660 nanometers was most effective in preventing flowering in the cocklebur and other short-day plants and in promoting flowering in long-day plants.

Using lettuce seeds, the Beltsville group found that when a flash of red light was followed by a flash of far-red light, the seeds did not germinate. The red light most effective in inducing germination in the lettuce seeds was of the same wavelength as that involved in the flowering response—about 660 nanometers. Furthermore, the light most effective in inhibiting the effect produced by red light had a wavelength of 730 nanometers, in the far-red region. The sequence of red and far-red flashes could be repeated over and over; the number of flashes did not matter, but the nature of the final one did. If the sequence ended with a red flash, the majority of the seeds germinated. If it ended with a far-red flash, the majority did not germinate (Figure 26–13).

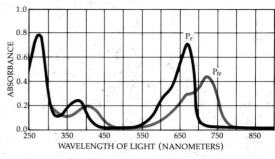

26–14
Absorption spectra of the two forms of phytochrome, P_r and P_{fr}. This difference in the absorption spectra made it possible to isolate the pigment.

DISCOVERY OF PHYTOCHROME

Plants contain a pigment that exists in two different interconvertible forms: P_r, which absorbs red light, and P_{fr}, which absorbs far-red light. When a molecule of P_r absorbs a photon of 660-nanometer light, the molecule is converted to P_{fr} in a matter of milliseconds; when a molecule of P_{fr} absorbs a photon of 730-nanometer light, it is quickly reconverted to the P_r form. These are called *photoconversion reactions.* The P_{fr} form is biologically active (that is, it will trigger a response such as seed germination), whereas P_r is inactive. The pigment molecule can thus function as a biological switch, turning responses on or off.

The lettuce-seed germination experiments are easily understood in these terms. Since P_r absorbs red light most efficiently (Figure 26–14), this light will convert a high proportion of the molecules to the P_{fr} form, thereby inducing germination. Subsequent far-red light absorbed by P_{fr} will convert essentially all of the molecules back to P_r, canceling the effect of the prior red light.

26–13
Light and the germination of lettuce seeds. (a) Seeds exposed briefly to red light; (b) seeds exposed to red light followed by far-red light; (c) seeds exposed to a sequence of red, far-red, red; (d) seeds exposed to a sequence of red, far-red, red, far-red. Whether or not the seeds germinate depends on the final wavelength in the series of exposures— red light promoting and far-red light inhibiting germination.

(a)

(b)

(c)

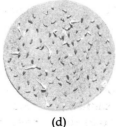

(d)

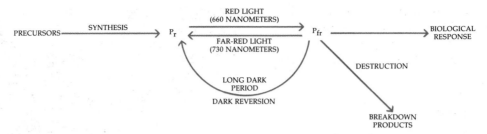

26–15
Phytochrome is continuously synthesized in the P_r form from its precursors, amino acids. P_r changes to P_{fr} when exposed to red light and accumulates in this form in dark-grown plants. P_{fr} is the active form that induces a biological response. P_{fr} is converted back to P_r by photoconversion when exposed to far-red light. In darkness, P_{fr} reverts to P_r (dark reversion) or is lost through a process termed "destruction" that occurs over several hours and probably involves hydrolysis by a protease. All three alternative pathways for P_{fr} removal provide the potential for reversing induced responses. It should be noted, however, that dark reversion has been detected only in dicots, not in monocots.

What about flowering under natural day-night cycles? Since white light contains both red and far-red wavelengths, both forms of the pigment are exposed simultaneously to photons that are efficient in promoting their photoconversion. After a few minutes in the light, then, a photoequilibrium is established in which the rates of conversion of P_r to P_{fr} and P_{fr} to P_r are equal, and the proportion of each type of phytochrome is constant (about 60 percent P_{fr} in noontime sunlight).

When plants are switched to darkness, the level of P_{fr} steadily declines over a period of several hours. If a high level of P_{fr} is regenerated by pulse irradiation with red light in the middle of the dark period (Figure 26–11), it will inhibit flowering in short-day (that is, "long-night") plants that otherwise would have flowered and promote flowering in long-day (that is, "short-night") plants that otherwise would not have flowered. In either case, the effect of the red pulse can be canceled by an immediate far-red pulse, which reconverts the P_{fr} to P_r.

In 1959, Harry A. Borthwick and his coworkers at Beltsville named this pigment **phytochrome** and presented conclusive physical evidence for its existence. The principal characteristics of the pigment, as they are presently understood, are summarized schematically in Figure 26–15.

Isolation and Characterization of Phytochrome

Phytochrome is present in plants in much smaller amounts than pigments such as chlorophyll. To detect phytochrome, one needs to use a spectrophotometer that is sensitive to extremely small changes in light absorbance. Such an instrument was not available until some seven years after the existence of phytochrome was proposed; in fact, the first use of this spectrophotometer was to detect and isolate phytochrome.

In order to avoid interference from chlorophyll, which also absorbs light of about 660 nanometers, dark-grown seedlings (in which chlorophyll has not yet developed) were chosen as the source of phytochrome. The pigment proved to be blue in color (why might you expect this color?), and it showed the characteristic P_r to P_{fr} conversion in the test tube by reversibly changing color slightly in response to red or far-red light.

The phytochrome molecule contains two distinct parts: a light-absorbing portion (the chromophore) and a large protein portion (Figure 26–16). The chromophore is much like the phycobilins that serve as accessory pigments in cyanobacteria and red algae. Genes for the protein portion have now been isolated from several species, and the amino acid sequence has been deduced from the nucleotide sequence.

The specific mechanism by which phytochrome works has not yet been established. However, it is clear that phytochrome-regulated morphogenesis results from changes in gene transcription. A number of phytochrome-activated genes have been identified and iso-

26–16
The phytochrome chromophore in its P_r form, showing its attachment to the protein part of the molecule.

lated. The regulatory regions of these genes are quite complex, with several different DNA sequences interacting with a number of DNA-binding proteins. As yet there is no evidence that phytochrome interacts directly with these genes.

Other Phytochrome Responses

Phytochrome is involved in a wide variety of plant responses. The germination of many seeds, for instance, occurs in the dark. In the seedlings, the stem elongates rapidly, pushing the shoot (or, in grasses, the coleoptile) up through the dark soil. During this early stage of growth, there is essentially no enlargement of the leaves, which would interfere with the passage of the shoot through the soil. Soil is not necessary for this growth pattern; any seedling grown in the dark will be elongated and spindly and will have small leaves. It will also be yellow to colorless because plastids do not turn green until exposed to light. Such a seedling is said to be *etiolated* (Figure 26–17).

When the seedling tip emerges into the light, the etiolated growth pattern gives way to normal plant growth. In dicots, the hook unbends, the growth rate of the stem may slow down somewhat, and leaf growth begins (see Figure 20–1). In grasses, growth of the mesocotyl (the part of the embryo axis between scutellum and coleoptile) stops, the stem elongates, and the leaves open (see Figure 20–6).

A dark-grown bean seedling that receives five minutes of red light a day, for instance, will show these light effects beginning on the fourth day. If the exposure to red light is followed by a five-minute exposure to far-red, none of the changes usually produced by the red light will appear (Figure 26–18). Similarly, in the seedlings of grains, termination of mesocotyl growth is triggered by exposure to red light, and the effect of red light is canceled by far-red.

An important feature of phytochrome in plants growing in a natural environment is the detection of shading by other plants. Radiation of wavelength below 700 nanometers is almost completely reflected or absorbed by vegetation, whereas that between 700 and 800 nanometers (in the far-red range) is largely transmitted. This causes a dramatic upward shift in the equilibrium ratio of P_r to P_{fr} (that is, more P_{fr} is converted to P_r) in shaded plants and results in a rapid increase in the rate of internodal elongation. Given the competition for light under the forest canopy, a plant's ability to sense the level of light and adjust its growth accordingly has obvious adaptive significance.

Reversible red/far-red reactions are also involved in anthocyanin formation in the apple, turnip, and cabbage; in changes in chloroplasts and other plastids; and in a tremendous variety of other plant responses in all phases of the plant life cycle.

26–17

Dark-grown seedlings, such as the bean plants on the left, are thin and pale; they have longer internodes and smaller leaves than light-grown seedlings, such as those on the right. The group of physical characteristics exhibited by these dark-grown seedlings, known as etiolation, has survival value for the seedlings because it increases their chances of reaching light before their stored energy supplies are exhausted.

26–18

All three bean plants received eight hours of light each day. The center plant was exposed to five minutes of far-red light, which promotes elongation, at the start of each dark period. The plant on the right received the same five-minute exposure to far-red light followed by a five-minute exposure to red light, which counteracted the effect of the far-red light. The plant on the left served as the control; its phytochrome is predominantly in the P_{fr} form as it enters the dark period because of its exposure to daylight.

When the existence of phytochrome was first demonstrated, its discoverers hypothesized that its behavior might explain the phenomenon of photoperiodism—that is, the reversible red/far-red reactions might be involved in the time-measuring mechanism, or biological clock. It is now generally agreed, however, that the time-measuring phenomenon of photoperiodism is not controlled by the interconversion of P_r and P_{fr} alone. A more complex explanation must be sought.

Hormonal Control of Flowering

Hamner and Bonner, in their early cocklebur experiments, showed that the leaf "perceived" the light, which caused the bud to flower. Apparently, some flowering-promoting substance is transmitted from leaf to bud. This hypothetical substance has been termed the flowering hormone, or floral stimulus.

The early experiments on the floral stimulus were carried out independently in several laboratories in the 1930s. The plant physiologist M.Kh. Chailakhyan carried out some of the earliest experiments just a few years before the first cocklebur studies. Using the short-day *Chrysanthemum indicum*, he showed that if the upper portion of the plant is defoliated and the leaves on the lower part are exposed to a short-day induction period, the plant will flower. If, however, the upper, defoliated part is kept on short days and the lower, leafy part on long days, no flowering occurs. He interpreted these results as indicating that the leaves form a hormone that moves to the stem apex and initiates flowering. Chailakhyan named this hypothetical hormone *florigen*, the "flower maker."

Further experiments showed that the flowering response will not take place if the leaf is removed immediately after photoinduction. But if left on the plant for a few hours after the induction cycle is complete, the leaf can then be removed without affecting flowering. The flowering hormone can pass through a graft from a photoinduced plant to a noninduced plant. Unlike auxin, however, which can pass through agar or non-living tissue, florigen can move from one plant tissue to another only if they are connected by living tissue. If a branch is girdled, that is, if a circular strip of bark is removed, florigen movement ceases. On the basis of these data, it was concluded that florigen moves via the phloem, the pathway followed by most organic substances in plants.

Subsequently, Anton Lang showed that several long-day plants and biennials, such as celery and cabbage, could be made to flower by treatment with gibberellin, even when the plants were grown under the wrong (that is, noninducing) photoperiod. This finding led Chailakhyan to modify his florigen hypothesis and to speculate that florigen actually consists of two classes of hormones, gibberellin and the as-yet-unidentified anthesin. According to this modified hypothesis, long-day plants produce anthesin but not gibberellin during noninducing photoperiods. Gibberellin treatments at this time can cause flowering. On the other hand, short-day plants produce gibberellin but fail to make anthesin when they are grown under noninducing conditions. The concept of florigen as a combination of gibberellin and anthesin has stimulated considerable research, but it cannot account for a critical observation: short-day plants grown under noninductive conditions (and thus, in theory, producing gibberellin) will not cause flowering of grafted long-day plants that are also under noninductive conditions (and, in theory, producing anthesin).

In some plants—the Biloxi soybean (*Glycine max*) is an example—leaves must be removed from the grafted noninduced plant or it will not flower. This observation suggests that the leaves of noninduced plants may produce an inhibitor. In fact, some investigators have concluded that there is no substance that initiates flowering but rather a substance that inhibits flowering until it is removed. Strong evidence now suggests that, at least in certain plants, both inhibitors and promoters are involved in the control of flowering.

The most convincing evidence for the existence of both flower-inducing and flower-inhibiting substances in the same plant is provided by the experimental studies conducted by Lang, Chailakhyan, and I.A. Frolova. These researchers chose three kinds of tobacco plants for their studies: the day-neutral *Nicotiana tabacum* cultivar Trabezond, the short-day tobacco cultivar Maryland Mammoth, and the long-day *Nicotiana silvestris*. They found that flower formation in the day-neutral tobacco was accelerated (compared with grafted day-neutral controls) by a graft union with the long-day plant when the grafts were kept on long days, and by a graft union with the short-day tobacco when the grafts were kept on short days. When long-day plants grafted to day-neutral plants were exposed to short days, flowering in the day-neutral receptor was greatly inhibited (Figure 26–19). By contrast, when short-day plants grafted to day-neutral plants were exposed to long days, there was little or no delay in flowering.

These results indicate that the leaves of the long-day plants are capable of producing flower-inducing substances on long days and flower-inhibiting substances on short days, and that both substances can be transferred through a graft union, an indication that both substances are transported through the plant. In the case of the short-day tobacco, the leaves apparently produce little or no flower-inhibiting substance when

26–19
Grafts of Nicotiana silvestris *(a long-day plant) onto the* Nicotiana tabacum *cultivar Trabezond (a day-neutral plant). Under long days, flower formation in the day-neutral tobacco was accelerated by the graft (left). Under short days, the day-neutral tobacco did not flower for the duration (90 to 94 days) of the experiment (right).*

exposed to long days. If such substances are produced by the short-day plant, they are much less effective in retarding flowering than those produced by the long-day plants.

All attempts to isolate either florigen or the flower inhibitor have thus far been unsuccessful.

Dormancy

Plants do not grow at the same rate all of the time. During unfavorable seasons, they limit their growth or cease to grow altogether. This ability enables plants to survive periods of water scarcity or low temperature.

Dormancy is a special condition of arrested growth. After periods of ordinary rest, growth resumes when the temperature becomes milder or when water or any other limiting factor becomes available again. A dor-

mant bud or embryo, however, can be "activated" only by certain, often quite precise environmental cues. This adaptation is of great survival importance to the plant. For example, the buds of plants expand, flowers are formed, and seeds germinate in the spring—but how do they recognize spring? If warm weather alone were enough, in many years all the plants would flower and all the seedlings would start to grow during warm autumn weather, only to be destroyed by the winter frost. The same could be said for any one of the warm spells that often punctuate the winter season. The dormant seed or bud does not respond to these apparently favorable conditions because of endogenous inhibitors, which must first be removed or neutralized before the period of dormancy can be terminated. In contrast to this "reluctance" to grow too rapidly, commercial seeds are artificially selected for their readiness to germinate promptly when they are exposed to favorable conditions, a trait that would be a great hazard for wild seeds.

DORMANCY IN SEEDS

The seeds of almost all plants growing in areas with marked seasonal temperature variations require a period of cold prior to germination. This requirement is normally satisfied by winter temperatures. The seeds of many ornamental plants have a similar cold requirement. If moist seed is exposed to a low temperature for many days (average optimum temperature and time, 5°C for 100 days), the dormancy may be broken and the seed germinated. This horticultural procedure is termed **stratification.** Many seeds require drying before they germinate (although some may be nondormant before they dry out); this prevents their germination within the moist fruit of the parent plant. Some seeds, such as lettuce seeds, require exposure to light, but others are inhibited by light.

Some seeds will not germinate in nature until they have become abraded, as by soil action. Such abrasion wears away the seed coat, permitting water or oxygen to enter the seed and, in some cases, removing the source of inhibitors. Hard seed coats that interfere with water absorption and embryo enlargement are common among legumes.

The seeds of some desert species germinate only when sufficient rain has fallen to leach away inhibitory chemicals present in the seed coat. The amount of rainfall necessary to wash off these germination inhibitors is directly related to the supply of water that the desert plant needs to become established as a seedling. Mechanical abrasion or breaking of the seed coat *(scarification)* with a knife, file, or sandpaper may allow the "hard seed" condition or inhibitor to be removed, or metabolic activity requisite to germination to be initiated. Germination may also be induced by soaking the seeds in alcohol or some other fat solvent (to dissolve waxy substances that impede entry of water) or concentrated acids. These procedures are widely used by horticulturalists. (See also Requirements for Seed Germination, page 447.)

Some seeds may remain viable for a long time in the dormant condition, enabling them to exist for many years, decades, and even centuries under favorable conditions. For example, every seed of a sacred lotus (*Nelumbo nucifera*) found in a peat deposit in Manchuria, and shown by radiocarbon dating to be some 2000 years old, germinated when the seed coats were filed to permit water to enter. Although this demonstration of endurance is impressive, even this record was reportedly broken by seeds of the arctic lupine (*Lupinus arcticus*). Some of these seeds, which were found in a frozen lemming burrow in the Yukon with animal remains estimated by carbon dating to be at least 10,000 years old, germinated within 48 hours. The age of the seeds themselves, however, was not determined.

Plant scientists have become increasingly interested in the factors involved in maintaining seed viability. Various enzyme systems may progressively fail in the stored seed, eventually leading to a complete loss of viability. Under what circumstances might viability be prolonged? Such questions are relevant to the worldwide interest in developing seed banks, with the goal of preserving the genetic characteristics of the wild and early cultivated varieties of various crop plants for use in future breeding programs. The need for such seed banks arises from the progressive replacement of older varieties with newer ones and the elimination of the replaced varieties, chiefly through the destruction of their habitats. In addition, many wild species of plants are in danger of becoming extinct, and the seeds of these plants should be preserved if possible.

0.5 mm

26–20
Longitudinal section of a dormant axillary bud of a maple (Acer). *The bud consists of an embryonic shoot enclosed by bud scales.*

DORMANCY IN BUDS

Dormancy in buds is essential to the survival of temperate herbaceous and woody perennials that are exposed to low temperatures during winter. Although dormant buds do not elongate—that is, do not exhibit observable growth—they may undergo meristematic activity during various phases of dormancy.

Bud dormancy in many trees is initiated in midsummer, long before leaf fall in autumn. The dormant bud is an embryonic shoot consisting of an apical meristem, nodes and internodes (not yet extended), and small rudimentary leaves or leaf primordia with buds or bud primordia in their axils, all enclosed by *bud scales* (Figure 26–20). The bud scales are very important because

they help prevent desiccation, restrict the movement of oxygen into the bud, and insulate the bud from heat loss. Growth inhibitors are known to accumulate in bud scales as well as in bud axes and leaves within buds. In many respects, therefore, the roles of the bud scales parallel those of the seed coat.

During and following the cessation of growth leading to the dormant condition, the plant tissues begin to undergo numerous physical and physiological changes to prepare the plant for winter, a process known as **acclimation.** Decreasing day length is the primary factor involved in the induction of dormancy in buds (Figure 26–21). Generally, more inhibitors are found in leaves

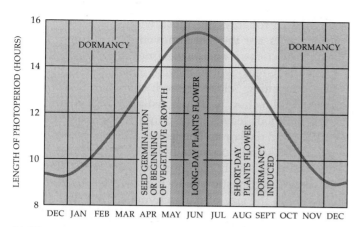

26–21
*Relationship between day length and the
developmental cycle of plants in the
North Temperate Zone.*

and buds under short-day conditions than long-day
conditions. Acclimation to cold leads to **cold hardiness**
—the ability of the plant to survive the extreme cold
and drying effects of winter weather.

As with seeds, the buds of many plant species require cold to break dormancy. If branches of flowering
trees and shrubs are cut and brought inside in autumn
they do not flower; but if the same branches are left
outside until late winter or early spring, they will bloom
in the warmer indoor temperatures. Deciduous fruit
trees, such as the apple, chestnut, and peach, cannot be
grown in climates where the winters are not cold. Similarly, bulbs such as those of tulips, hyacinths, and narcissus can be "forced," that is, made to bloom indoors
in the winter, but only if they have previously been
outside or in a cold place. As discussed in Chapter 23,
such bulbs are actually large buds in which the leaves
are modified for storage.

Cold is not required to break dormancy in all cases.
In the potato, for instance, in which the "eyes" are dormant buds, at least two months of dry storage are the
chief requirement; temperature is not a factor. In many
plants, particularly trees, the photoperiodic response
breaks winter dormancy, with the dormant buds being
the receptor organs.

Application of gibberellins sometimes breaks dormancy. For instance, gibberellin treatment of a peach
bud may induce development after the bud has been
kept for 164 hours below 8°C. Does this mean that
under normal conditions, increase in gibberellin terminates the dormancy? Not necessarily. Dormancy may
be a state of balance between growth inhibitors and
growth stimulators. Addition of any growth stimulator
(or removal of inhibitors, such as abscisic acid) may
alter the balance so that growth begins.

Cold and the Flowering Response

Cold may affect the flowering response. For example, if
winter rye *(Secale cereale)* is planted in the autumn, it
germinates during the winter and flowers the following
summer, 7 weeks after growth resumes. If it is planted
in the spring, it does not flower for 14 weeks. In 1915,
the plant physiologist Gustav Gassner discovered that
he could influence the flowering of winter rye and
other cereal plants by controlling the temperature of
the germinating seeds. He found that if the seeds of the
winter strain are kept at near-freezing (1°C) temperatures during germination, the winter rye, even when
planted in late spring, will flower the same summer it
is planted. This procedure, which came to be known
as **vernalization** (from the Latin *vernus*, meaning
"spring"), is now a common practice in agriculture.

Even after vernalization, the plant must be subjected to a suitable photoperiod, usually long days. The
vernalized winter rye behaves like a typical long-day
plant, flowering in response to the long days of summer. A similar example is seen in a biennial strain of
henbane *(Hyoscyamus niger)*. The vegetative rosette,
which culminates the first year's growth, flowers only
if it is exposed to cold. After cold exposure, it becomes a
typical long-day plant, with the same photoperiodic response as that of the annual strain.

As the above examples indicate, in some plants the
cold treatment affects the photoperiodic response.
Spinach, ordinarily a long-day plant, does not usually
flower until the days are 14 hours in length. If the spinach seeds are cold treated, however, they will flower
when the days are only 8 hours long. Similarly, cold
treatment of the clover *Trifolium subterraneum* can completely remove its dependency on day length for flowering.

In the biennial henbane and most other biennial
long-day plants that form rosettes, gibberellin treatment can substitute for the cold requirement. If gibberellin is applied, such plants will elongate rapidly and
then flower. Application of gibberellin to short-day
plants or to nonrosette long-day plants has little effect
on (or inhibits) the flowering response. However, if
gibberellin synthesis is inhibited when the plant is
being exposed to the appropriate inductive cycle, the
plant will not flower unless gibberellin is added to it.

Nastic Movements

Nastic movements are plant movements that occur in
response to a stimulus but whose direction of movement is independent of the position of the origin of the

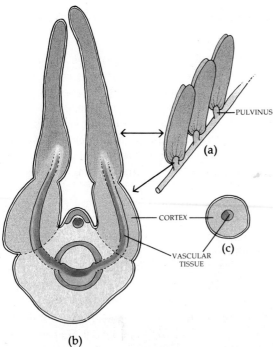

PULVINUS

(a)

CORTEX

(c)

VASCULAR
TISSUE

(b)

26–22
*Diagrammatic representation of pulvini
in* Mimosa pudica. (a) *Portion of rachis
showing three leaflets, each with a pulvinus at its base.* (b) *Transverse section
through a rachis with two leaflets in the
closed condition, showing longitudinal
view of the pulvini.* (c) *Transverse section
through a pulvinus showing the core of
the vascular tissue surrounded by a cortex,
which consists largely of thin-walled
parenchyma cells.*

stimulus. Probably the most widely occurring nastic
movements are the sleep movements. Known technically as *nyctinastic movements* (meaning "night closure"
from the Greek words *nyktos,* meaning "night," and
nastic, meaning "close-pressed; change in position in a
specified direction"), they constitute the up and down
movements of leaves in response to the daily rhythms
of light and darkness. The leaves are oriented vertically
in darkness and horizontally in the light. These movements are especially common in leguminous plants.

Most nyctinastic leaf movements result from
changes in the size of parenchyma cells in the jointlike
thickenings at the base of each leaf (or, if the leaf is
compound, at the base of each leaflet also). This thickening *(pulvinus)* is a flexible cylinder with the vascular
system concentrated in the center. Anatomically all
pulvini consist of a core of vascular tissue surrounded
by a bulky cortex of thin-walled parenchyma cells (Figure 26–22). Pulvinar movement is associated with
changes in turgor and with the concomitant contractions and expansions of the ground parenchyma on opposite sides of the pulvinus. Turgor changes in the contracting and expanding cells are brought about by the
shuttling of potassium ions between the two sides of
the pulvinus, under the control of the biological clock
and phytochrome.

Thigmonastic (seismonastic) movements are nastic
movements resulting from mechanical stimulation.
Such movements are exemplified by the familiar sensitive plant *(Mimosa pudica),* whose leaflets and sometimes entire leaves droop suddenly in response to
touch, shaking, or electrical or thermal stimulation (Figure 26–23). As with sleep movements (also exhibited

26–23
The sensitive plant (Mimosa pudica).
(a) *The normal position of leaves and
leaflets. Responses to touch are shown in*
(b). *The responses result from changes in
turgor pressure in certain cells of jointlike
thickenings (pulvini) at the base of the
leaflets. Only a single leaflet need be stimulated for the response in* (b) *to occur.*

(a)

(b)

by *M. pudica*), this response is a result of a sudden change in turgor pressure in specific cells of the pulvinus at the base of each leaflet and leaf. The loss of water from these cells follows the efflux of potassium ions from the cells into the apoplast. The accumulation of ions in the apoplast seems to be initiated by a decrease of water potential triggered by an apoplastic accumulation of sucrose unloaded from the phloem. Only a single leaflet needs to be stimulated; the stimulus then moves to other parts of the leaf and throughout the plant. Two distinct mechanisms, one electrical and the other chemical, appear to be involved with the spread of the stimulus in the sensitive plant.

The mechanism responsible for the rapid closure (within approximately 0.5 second) of the leaves of the carnivorous Venus flytrap *(Dionaea muscipula)* has generally been explained as being due to either a sudden loss of turgor pressure in the upper epidermis, which becomes flexible and causes the trap lobes to bend inward, or a sudden acid-induced wall loosening of the motor cells and their subsequent expansion. Recent studies cast doubt on both of these explanations and indicate that trap movement may be driven by turgor pressure in the layer of mesophyll cells underlying the upper epidermis. When the trap is open, the mesophyll cells are compressed and the tissue is under tension. The turgor-driven extension of the mesophyll cells, which results in trap closure, does not require a wall-loosening mechanism. The biochemical processes associated with trap closure remain obscure. They are accompanied, however, by a marked decrease in ATP levels. The touch response exhibited in the trapping mechanism of the Venus flytrap is highly specialized (Figure 26–24).

Generalized Effects of Mechanical Stimuli on Plant Growth and Development: Thigmomorphogenesis

In addition to the specialized responses of some plants —such as the sensitive plant and Venus flytrap—to touch and other mechanical stimuli, plants also respond to mechanical stimuli by altering their growth patterns, a phenomenon known as **thigmomorphogenesis.** Although botanists have long known that plants grown in a greenhouse tend to be taller and more spindly than plants grown outside, it was not until the 1970s that systematic studies undertaken by M.J. Jaffe revealed that regular rubbing or bending of stems inhibits their elongation and stimulates their radial expansion, resulting in shorter, stockier plants. Plants in a natural environment are, of course, subjected to similar stimuli, in the form of wind, raindrops, and rubbing by passing animals and machines. The swaying of plants by wind is a powerful causal factor in thigmomorphogenesis.

As we saw in Chapter 25, such growth responses are caused by changes in gene expression. Studies on *Arabidopsis thaliana* show that some of the genes whose expression is induced by touch encode proteins related to the calcium-binding protein calmodulin, suggesting a role for Ca^{2+} in mediating the growth responses. (Ca^{2+} has been implicated in the regulation of a number of plant processes, including the regulation of mitosis, polarized cell growth, cytoplasmic streaming, and both tropic and nastic movements.) *Arabidopsis* plants stimulated with touch were conspicuously shorter than the untreated plants (Figure 26–25).

26–24

Touch response in the Venus flytrap (Dionaea muscipula). *Here, an unwary fly, attracted by nectar secreted on the leaf surface, can be seen on a leaf, before and after its closure. Each leaf half is equipped with three sensitive hairs. When an insect walks on one of the leaves, it brushes against the hairs, triggering the traplike closing of the leaf. The toothed edges mesh, the leaf halves gradually squeeze shut, and the insect is pressed against digestive glands on the inner surface of the trap.*

The trapping mechanism is so specialized that it can distinguish between living prey and inanimate objects, such as pebbles and small sticks, that fall on the leaf by chance: the leaf will not close unless two of its hairs are touched in succession or one hair is touched twice.

26–25
Arabidopsis thaliana *plants six weeks of age. The plant on the left was touched twice daily; the plant on the right is the untreated control. The inhibition of growth by touch or other noninjurious mechanical stimulation resulting in reduced length and increased width is referred to as thigmomorphogenesis.*

(a)

(b)

26–26
(a) *Leaves of a lupine* (Lupinus arizonicus), *orienting to track the course of the sun. This phenomenon is commonly known as solar tracking. (b) Solar tracking by a field of sunflowers* (Helianthus annuus).

Solar Tracking

The leaves and flowers of many plants have the ability to move diurnally, orienting themselves either perpendicular or parallel to the sun's direct rays. Commonly known as *solar tracking* (Figure 26–26), this phenomenon is technically called **heliotropism** (from the Greek *helios*, meaning "sun"). Unlike stem phototropism, the leaf movement of heliotropic plants is not the result of asymmetric growth. In most cases the movements involve pulvini at the bases of leaves and/or leaflets. Some petioles appear to have pulvinal characteristics along most or all of their length. Some common plants exhibiting heliotropic leaf movements are cotton, soybeans, cowpeas, lupine, and sunflowers.

There are two types of heliotropism. In one (diaheliotropism), the movement of the leaves is such that the broad surfaces of the blades remain perpendicular to the sun's direct rays throughout the day. In such leaves, more photons are available for photosynthesis, and photosynthetic rates appear to be higher throughout the day than in nontracking leaves or leaves exhibiting the second type of heliotropism (Figure 26–27). In this second type, plants actively avoid direct sunlight during periods of drought by orienting their leaf blades parallel to the sun's rays (paraheliotropism). This orientation minimizes absorption of solar radiation rather than maximizing it (Figure 26–27), decreasing leaf temperature and transpirational water loss and enhancing survival during drought periods.

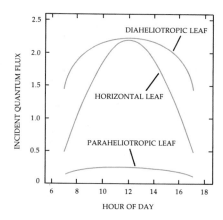

26–27
Comparison of the photosynthetically useful solar radiation between 400 and 700 nanometers incident on a diaheliotropic leaf, a nontracking (horizontal) leaf, and a paraheliotropic leaf over the course of a day.

Summary

Plants possess a variety of adaptations that enable them to detect and respond to alterations in their environment. Phototropism, or the curving of the growing shoot toward light, is an example of such an adaptation. The differential growth of the seedling is caused by lateral migration of the growth hormone auxin under the influence of light. The photoreceptor for this response is a blue-light-absorbing pigment. Gravitropism is the response by a shoot or a root to the pull of the earth's gravity. The movement of auxin to the lower surface of a horizontally oriented shoot or root may play a role in the upward curvature of the shoot and the downward curvature of the root, respectively. Calcium apparently plays an important role in the gravitropic response of both shoots and roots. The site of perception of gravity in roots appears to be in the central column of cells of the root cap. Thigmotropism is a response to contact with a solid object, such as the winding of tendrils.

Circadian rhythms are cycles of activity that recur at intervals of about 24 hours in an organism under constant environmental conditions. These rhythms are endogenous, caused not by environmental factors but by some internal timing mechanism within the organism. Such a timing mechanism, the chemical and physical nature of which is not well understood, is called the biological clock.

Photoperiodism is the response of organisms to changing 24-hour cycles of light and darkness. Such responses control the onset of flowering in many plants. Some plants, known as long-day plants, will flower only when the periods of light exceed a critical length. Other plants, short-day plants, flower only when the periods of light are less than some critical period. Day-neutral plants flower regardless of photoperiods. Experiments have shown that the dark period rather than the light period is the critical factor.

Phytochrome, a pigment commonly present in the tissues of plants, is the photoreceptor molecule for detecting transitions between light and darkness. The pigment can exist in two forms, P_r and P_{fr}. P_r absorbs red light and is thereby converted to P_{fr}. The P_{fr} absorbs far-red light and is converted to P_r. P_{fr} is the active form of the pigment; it promotes flowering in long-day plants, inhibits flowering in short-day plants, promotes germination in lettuce seeds, and promotes normal growth in seedlings.

In both long-day and short-day plants, the photoperiod is perceived in the leaves but the response takes place in the bud. Although it has not yet been isolated or identified, a chemical substance moves via the phloem from the leaves to the bud, where it induces flowering. Strong evidence suggests that, at least in certain plants, both flower-inducing and flower-inhibiting substances are involved in flowering.

Alternation of periods of growth and cessation of growth permits the plant to survive water shortages and extremes of hot or cold. Dormancy is a special condition of arrested growth in which the entire plant, or such structures as seeds or buds, do not renew growth without special environmental cues. The requirement for such cues, which include cold exposure, dryness, and a suitable photoperiod, prevents the tissue from breaking dormancy during superficially favorable conditions. Decreasing day length is the primary factor involved in the induction of dormancy in buds. Acclimation to cold leads to cold hardiness, the ability of the plant to survive the extreme cold of winter weather. Vernalization refers to the promotion of flowering in winter strains by keeping seeds at low temperatures. Hormones, cold, and light interact to modify plant responses.

Plant movements that occur in response to a stimulus, but whose direction is independent of the direction of the stimulus, are called nastic movements. Among these are the widely occurring sleep movements—the up and down movements of leaves in response to the daily rhythm of light and darkness. Nastic movements resulting from touch include the triggered closing of the carnivorous Venus flytrap. Plants may also respond to mechanical stimuli by altering their growth patterns, a phenomenon known as thigmomorphogenesis.

The leaves and flowers of certain plants follow or track the sun during the course of the day, either maximizing or minimizing absorption of solar radiation.

Suggestions for Further Reading

Becker, Wayne M., and David W. Deamer: *The World of the Cell*, 2d ed., The Benjamin/Cummings Publishing Company, Inc., Redwood City, Calif., 1991.

An outstanding general cell biology book, includes topics in bioenergetics and cellular energy metabolism; intended for undergraduates.

Bernier, G.: "The Control of Floral Evocation and Morphogenesis," *Annual Review of Plant Physiology and Plant Molecular Biology* 39:175–219, 1988.

A thorough review of all aspects of flower initiation.

Brady, John: *Biological Clocks*, Studies in Biology, No. 104, University Park Press, Baltimore, Md., 1979.*

An interesting and well-written introduction to the subject of biological clocks and their experimental study.

Davies, P.J. (ed.): *Plant Hormones and Their Role in Plant Growth and Development*, Martinus Nijhoff Publishers, Dordrecht, The Netherlands, 1987.*

An advanced textbook intended to serve as a guide to the literature for graduate-level courses in the plant hormones; written by multiple authors.

Ellmore, George S.: "The Organization and Plasticity of Plant Roots," *Scanning Electron Microscopy* III:1083–1102, 1982.

An interesting account of the plant root as a dynamic organ, describing the fluctuation of its structural and physiological components throughout its life history.

Gould, Fred: "Evolutionary Biology and Genetically Engineered Crops," *BioScience* 38:26–33, 1988.

The author considers potential evolutionary responses of indigenous organisms to genetically engineered crops that could nullify the crops' advantages, with particular attention to genetically engineered crops containing genes that code for pest-resistance.

Kendrick, Richard E., and Barry Frankland: *Phytochrome and Plant Growth*, Studies in Biology, No. 68, 2d ed., Edward Arnold Publishers, London, 1983.*

A concise, well-illustrated introduction to the molecular properties and physiology of phytochromes.

Kendrick, Richard E., and G.H.M. Kronenberg (eds.): *Photomorphogenesis in Plants*, Martinus Nijhoff Publishers, Dordrecht, The Netherlands, 1986.*

An advanced textbook with a broad interdisciplinary approach to the field of photomorphogenesis; written by multiple authors.

Lal, Rup, and Sukanya Lal (eds.): *Crop Improvement Utilizing Biotechnology*, CRC Press, Boca Raton, Florida, 1989.

A comprehensive account of virtually all aspects of plant biotechnology.

Lewis, Ricki: "Making Sense out of Antisense," *BioScience* 39:590–593, 1989.

An interesting discussion of a powerful technique that utilizes a gene's complementary sequence as a precision tool for turning off gene expression.

Raab, Mandy M., and Ross E. Koning: "How Is Floral Expansion Regulated?" *BioScience* 38:670–674, 1988.

A consideration of the role of plant hormones in the control of growth of the flower parts.

Salisbury, Frank B., and Cleon W. Ross: *Plant Physiology*, 4th ed., Wadsworth Publishing Co., Inc., Belmont, Calif., 1991.

A detailed and useful review of the entire subject.

Steeves, Taylor A., and Ian M. Sussex: *Patterns in Plant Development*, 2d ed., Prentice Hall, Inc., Englewood Cliffs, N.J., 1989.

An exceptionally well-written and well-illustrated text that should be read by all students interested in plant morphogenesis.

Stryer, Lubert: *Biochemistry*, 3d ed., W.H. Freeman and Company, San Francisco, 1988.

A good introduction to cellular energetics.

Torrey, John G.: "The Development of Plant Biotechnology," *American Scientist* 73:354–363, 1985.

An informative and interesting presentation of the history of plant biotechnology, beginning with the ideas of Justus von Liebig in the 1840s and including consideration of present-day problems and prospects for the future.

Weising, Kurt, Jeff Schell, and Günter Kahl: "Foreign Genes in Plants: Transfer, Structure, Expression, and Applications," *Annual Review of Genetics* 22:421–477, 1988.

An excellent review of plant engineering using the Ti plasmid of Agrobacterium tumefaciens, plant transformation techniques, and applications of plant genetic engineering.

Wilkins, Malcolm B. (ed.): *Advanced Plant Physiology*, Pitman Press, Bath, England, 1984.

An advanced plant physiology textbook written by multiple authors.

* Available in paperback.

CHAPTER 27

27–1
*Plants, such as the stonecrops (Sedum)
seen here flourishing on otherwise barren
cliffs high above the surf, require little
more than light, water, and certain
chemical elements to maintain life. This
scene was photographed in Redwood
National Park, California.*

Plant Nutrition and Soils

General Nutritional Requirements

Plants must obtain from the environment the specific raw materials required in the complex biochemical reactions necessary for the maintenance of their cells and for growth. In addition to light, plants require water and certain chemical elements for metabolism and growth. Much of the evolutionary development of plants has involved structural and functional specialization necessary for the efficient uptake of these raw materials and their distribution to living cells throughout the plant.

In contrast to those of animals, the nutritional demands of plants are relatively simple. Under favorable environmental conditions, most green plants can use light energy to transform CO_2 and H_2O into organic compounds for their energy source. They can also synthesize all of their required amino acids and vitamins, using inorganic nutrients drawn from the environment.

The self-sufficiency of nitrogen-fixing cyanobacteria is particularly impressive. Through the process of nitrogen fixation, they convert atmospheric nitrogen to forms of nitrogen that can be utilized in the synthesis of amino acids and proteins. In addition, of course, they carry out photosynthesis.

Plant nutrition involves the uptake from the environment of all the raw materials required for essential biochemical processes, the distribution of these materials within the plant, and their utilization in metabolism and growth.

ESSENTIAL INORGANIC NUTRIENTS

As early as 1800, chemists and plant biologists had analyzed plants and demonstrated that certain chemical elements were absorbed from the environment. Opin-

(a)

(b)

27–2

Deficiencies of one or more of the essential inorganic nutrients result in abnormal plant growth and development.

Shown here are the effects of (a) zinc and (b) boron deficiencies on a tomato plant.

In both cases, the plant on the left is a control plant.

ions differed, however, on whether the absorbed elements were impurities or constituents required for essential functions. By the mid-1880s it had been established that at least ten of the chemical elements present in plants were necessary for normal growth. In the absence of any one of these elements, plants displayed characteristic abnormalities of growth or deficiency symptoms, and often such plants did not reproduce normally. These ten elements—carbon, hydrogen, oxygen, potassium, calcium, magnesium, nitrogen, phosphorus, sulfur, and iron—were designated as essential chemical elements for plant growth. They are also referred to as essential minerals, or **essential inorganic nutrients.**

In the early 1900s, manganese was also found to be an essential element. During the next 50 years, with the aid of improved techniques for removing impurities from nutrient cultures, five additional elements—zinc, copper, chlorine, boron, and molybdenum—were determined to be essential, with chlorine being recognized only in 1954. At present, these 16 elements are generally considered to be essential for most plants (Figure 27–2).

NUTRIENT CONCENTRATIONS IN PLANTS

Chemical analyses for essential inorganic nutrients are a useful means of determining the relative amounts of various elements required for the normal growth of different plant species. Typical results of such chemical analyses are given in Table 27–1. Analyses of this sort are particularly valuable in agriculture as a guide to the nutritional well-being of plants and the need for applications of fertilizer. Potential nutritional deficiencies in livestock that consume specific plants can also be predicted by inorganic analyses.

The concentrations of specific elements in plants are known to vary over a wide range. On the basis of the usual concentrations in plants, the essential inorganic nutrients can be divided into two broad groups—macronutrients and micronutrients. In general, **macronu-**

Table 27–1 *Examples of Inorganic Analyses of Plants*

	Concentration of Element		
Element	Alfalfa	Corn*	White Oak**
Macronutrients			
Nitrogen	3.12%	2.81%	2.19%
Potassium	2.77	1.86	0.85
Calcium	1.70	0.40	0.82
Phosphorus	0.35	0.28	0.19
Magnesium	0.41	0.27	0.36
Sulfur	0.29	0.18	0.13
Micronutrients			
Iron	190 ppm†	110 ppm	126 ppm
Chlorine	8800	3100	43
Copper	9	6	8
Manganese	62	80	572
Zinc	57	27	22
Molybdenum	1.40	1.03	6.21
Boron	35	14	38
Elements Essential to Some Plants or Organisms			
Sodium	4300 ppm	127 ppm	210 ppm
Cobalt	0.21	0.16	—

* Shoot only; grain not included in analysis.

** Leaf and twig growth of current year.

†Abbreviation for part per million; ppm equals units of an element by weight per million units of oven-dried plant material; 1% equals 10,000 ppm.

trients are elements that are required in large amounts, and **micronutrients** ("trace elements") are those required in very small, or trace, amounts (Table 27–2).

Certain plant species and taxonomic groups are characterized by unusually high or low amounts of specific elements (Figure 27–3). As a result, plants growing in the same nutrient medium may differ markedly in nutrient content. Compare the analyses in Table 27–1 for corn (a monocot) and alfalfa (a dicot). Dicots generally require greater amounts of calcium and boron than do monocots.

Nutritional studies have established that some elements are essential for only limited groups of plants or for plants grown under specific environmental conditions. Legumes, such as alfalfa (*Medicago sativa*), benefit from the addition of cobalt to the culture medium. It is not the alfalfa, however, that requires cobalt but the symbiotic nitrogen-fixing bacteria growing in association with the roots of the alfalfa. As indicated in Table 27–1, sodium is present in many plants in relatively high concentrations. It has been recognized for many years that sodium can partially replace the requirement for potassium in some species. Studies have shown that, in other species, sodium is an essential element in its own right. For example, sodium appears to be required by C_4 plants and by certain halophytes (see essay, page 608). In addition, it has been demonstrated that soybean plants deprived of nickel accumulate toxic concentrations of urea in necrotic lesions (dead tissue) on the tips of their leaflets and exhibit reduced growth. Addition of nickel to the nutrient media prevents urea accumulation, necrosis, and reduction in growth. Thus nickel is essential for soybeans.

Table 27–1 includes two elements—sodium and cobalt—that are not essential for all crop plants but are essential for herbivorous animals that consume these plants. For example, if the cobalt concentration in a forage crop is less than 0.1 ppm, animals eating these plants are likely to display symptoms of cobalt deficiency.

Functions of Inorganic Nutrients in Plants

Inorganic nutrients are essential in many aspects of growth and metabolism. Table 27–2 lists the inorganic nutrients required by most plants, the forms in which they most commonly are absorbed from the environment, their usual concentration range, and some of their functions.

SPECIFIC VERSUS NONSPECIFIC FUNCTIONS

Inorganic ions affect osmosis (Chapter 4) and thus help to regulate water balance. Because several inorganic ions can serve interchangeably in this role, in many plants this particular requirement is described as *nonspecific*. On the other hand, an inorganic nutrient may function as part of an essential biological molecule; in this case the requirement is highly *specific*. An example of a specific function is the presence of magnesium in

27–3

(a) *Plants of the mustard family, such as wintercress* (Barbarea vulgaris), *use sulfur in the synthesis of the mustard oils that give the plants their sharp taste. These mustard plants were growing in Ithaca, New York. (b) Horsetails incorporate silicon into their cell walls, making the plants indigestible to most herbivores but useful, at least in colonial America, for scouring pots and pans. These bushy vegetative shoots, as well as several stalk-like fertile shoots, of the horsetail* Equisetum telemateia *were photographed in California.*

(a)

(b)

Table 27-2 *A Summary of the Functions of Inorganic Nutrients in Plants*

Element	Principal Form in Which Element is Absorbed	Usual Concentration in Healthy Plants (% or ppm of Dry Weight)	Important Functions
Macronutrients			
Carbon	CO_2	~44%	Component of organic compounds
Oxygen	H_2O or O_2	~44%	Component of organic compounds
Hydrogen	H_2O	~6%	Component of organic compounds
Nitrogen	NO_3^- or NH_4^+	1–4%	Component of amino acids, proteins, nucleotides, nucleic acids, chlorophylls, and coenzymes
Potassium	K^+	0.5–6%	Involved in osmosis and ionic balance and in opening and closing of stomata; activator of many enzymes
Calcium	Ca^{2+}	0.2–3.5%	Component of cell walls; enzyme cofactor; involved in cellular membrane permeability; component of calmodulin, a regulator of membrane and enzyme activities
Phosphorus	$H_2PO_4^-$ or HPO_4^{2-}	0.1–0.8%	Component of energy-carrying phosphate compounds (ATP and ADP), nucleic acids, several essential coenzymes, phospholipids
Magnesium	Mg^{2+}	0.1–0.8%	Part of the chlorophyll molecule; activator of many enzymes
Sulfur	SO_4^{2-}	0.05–1%	Component of some amino acids and proteins and of coenzyme A
Micronutrients			
Iron	Fe^{2+} or Fe^{3+}	25–300 ppm	Required for chlorophyll synthesis; component of cytochromes and nitrogenase
Chlorine	Cl^-	100–10,000 ppm	Involved in osmosis and ionic balance; probably essential in photosynthetic reactions that produce oxygen
Copper	Cu^{2+}	4–30 ppm	Activator or component of some enzymes
Manganese	Mn^{2+}	15–800 ppm	Activator of some enzymes; required for integrity of chloroplast membrane and for oxygen release in photosynthesis
Zinc	Zn^{2+}	15–100 ppm	Activator or component of many enzymes
Molybdenum	MoO_4^{2-}	0.1–5.0 ppm	Required for nitrogen fixation and nitrate reduction
Boron	$B(OH)_3$ or $B(OH)_4^-$	5–75 ppm	Influences Ca^{2+} utilization, nucleic acid synthesis, and membrane integrity
Elements Essential to Some Plants or Organisms			
Sodium	Na^+	Trace	Involved in osmotic and ionic balance; probably not essential for many plants; required by some desert and salt-marsh species and may be required by all plants that utilize C_4 pathway of photosynthesis
Cobalt	Co^{2+}	Trace	Required by nitrogen-fixing microorganisms

the chlorophyll molecule (see Figure 7–6, page 104). Some inorganic nutrients are essential constituents of cellular membranes, whereas others control the permeability of such membranes. Other inorganic nutrients are indispensable components of a variety of cellular enzyme systems. Still others provide a proper ionic environment in which biological reactions can occur.

Because inorganic nutrients fill such basic needs and are involved in such fundamental processes, the deficiencies affect a wide variety of structures and functions in the plant body.

Catalysts

A key role of the inorganic nutrients is their participation as catalysts in some of the enzymatic reactions of the plant cell. In some cases, they are an essential structural part (a "prosthetic group") of the enzyme. In other cases, they serve as activators or regulators of certain enzymes. Potassium, for instance, which probably affects 50 to 60 enzymes, is believed to regulate the conformation of some proteins. Changing the shape of an enzyme could, for example, expose or obstruct reaction sites.

Electron Transport

Many of the biochemical activities of cells, including photosynthesis and respiration, are oxidation-reduction reactions (see page 77). In such reactions, electrons are transferred to or from a molecule that functions as an electron acceptor or donor. The cytochromes, which contain iron (see Figure 6–9, page 93), are involved in electron transfer.

Structural and Molecular Components

Some mineral elements serve as structural components of cells, either as part of a physical structure (Figure 27–3b) or as part of the molecules involved in cellular metabolism. Calcium combines with pectic acid in the middle lamella of the plant cell wall. Phosphorus occurs in the sugar-phosphate backbone of DNA and RNA and in the phospholipids of the cellular membranes. Nitrogen is an essential component of amino acids, chlorophylls, and nucleotides. Sulfur is found in two amino acids, thus forming a component of proteins.

Osmosis

The movement of water into and out of plant cells, as discussed in Chapter 4, is largely dependent on the concentration of solute in the cells and in the surrounding medium. The uptake of ions by a plant cell thus may result in the entry of water into the cell. The increased turgor pressure results in expansion of the immature cell, which is the chief cause of cellular growth, and in the maintenance of turgor in the mature cell (see page 66). This is an example of conversion of energy from one form to another by a living system; the chemical (ATP) energy expended in the active uptake of ions by the plant cell is translated into the physical energy of water movement.

Effects on Cell Permeability

Calcium has a direct effect on the physical properties of cellular membranes. When there is a calcium deficiency, membranes seem to lose their integrity, and solutes within the membranes or cells leak out.

The Soil

Soil is the primary nutrient medium for plants. Soils must provide plants not only with physical support but also with adequate inorganic nutrients at all times, as well as with adequate water and a suitable gaseous environment for the root systems. Understanding the origins of soils and their chemical and physical properties in relation to plant growth requirements is critical in planning for the nutrition of field crops.

WEATHERING OF THE EARTH'S CRUST

The inorganic nutrients utilized by plants are derived from the atmosphere and from the weathering of rocks in the crust of the earth. The earth is composed of 92 naturally occurring elements, which are often found in the form of minerals. **Minerals** are naturally occurring inorganic compounds that are usually composed of two or more elements in definite proportions by weight (see Appendix A). Quartz (SiO_2) and calcite ($CaCO_3$) are examples of minerals.

Most rocks consist of several different minerals and are divided into three groups based on origin and formation. *Igneous* rocks, such as granite, are derived directly from molten material and, for the most part, were originally formed when the earth cooled and solidified. As a result of weathering, igneous rocks and other kinds of rocks may be broken down into soluble and insoluble components. After being transported by water, wind, or glaciers, these components form new deposits —usually in water—that in time become cemented and solidified into *sedimentary* rocks, such as shale, sandstone, and limestone. Although sedimentary rocks form only about 5 percent of the earth's crust, they are of great importance because they occur extensively at or

27–4
The fibrous root systems of grasses bind and anchor prairie soil in place.

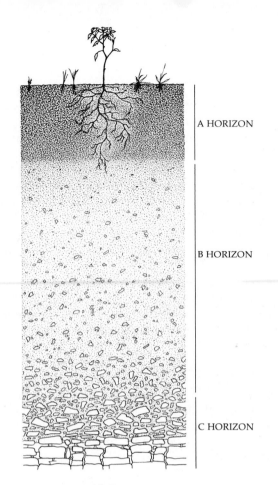

A HORIZON

B HORIZON

C HORIZON

27–5
The three primary horizons, or soil layers, in a typical soil. Each horizon is described in the text.

near the surface. Under the extreme heat and pressure deep within the earth, sedimentary and igneous rock may be transformed into a third type of rock—*metamorphic*. In this way, quartzite is formed from sandstone, slate from shale, and marble from limestone.

Weathering processes, involving the physical disintegration and chemical decomposition of minerals and rocks at or near the earth's surface, produce the inorganic materials from which soils are formed. Weathering involves freezing and thawing or heating and cooling, which cause substances in the rocks to expand and contract, splitting rocks apart. Water and wind often carry the rock fragments great distances, exerting a scouring action that breaks and wears the fragmented rock into even smaller particles. Water enters between the particles, and soluble materials dissolve in the water. Water, in combination with carbon dioxide and air impurities such as sulfur dioxide and oxides of nitrogen, forms dilute acids that help to dissolve materials that are less soluble in pure water. Soil formation may occur at the site of weathering, or the parent materials may be transported elsewhere by gravity, wind, water, or glaciers.

Soils also contain organic materials. If light and temperature conditions permit, bacteria, fungi, algae, lichens, and bryophytes and small vascular plants gain a foothold on or among the weathered rocks and minerals. Growing roots also split rocks, and their disintegrating bodies and those of the animals associated with them add to the accumulating organic material. Finally, larger plants move in, anchoring the soil in place with their root systems (Figure 27–4), and a new community begins.

Examining a vertical section of soil (Figure 27–5), one can see variations in the color, the amount of living and dead organic matter, the porosity, the structure, and the extent of weathering. These variations generally result in a succession of rather distinct layers that soil scientists refer to as **horizons.** At least three horizons are recognized.

The A horizon (sometimes called the "topsoil") is the upper region, that of the greatest physical, chemical, and biological activity. The A horizon contains the greatest portion of the soil's organic material, both living and dead, such as large amounts of dead and decaying leaves and other plant parts, insects and other small arthropods, earthworms, protists, nematodes, and decomposer organisms (Figure 27–6).

The B horizon is a region of deposition. Iron oxide, clay particles, and small amounts of organic matter are among the materials carried from the A horizon to the B horizon by water percolating, or moving down, through the soil. The B horizon contains much less organic material and is less weathered than the A horizon above it.

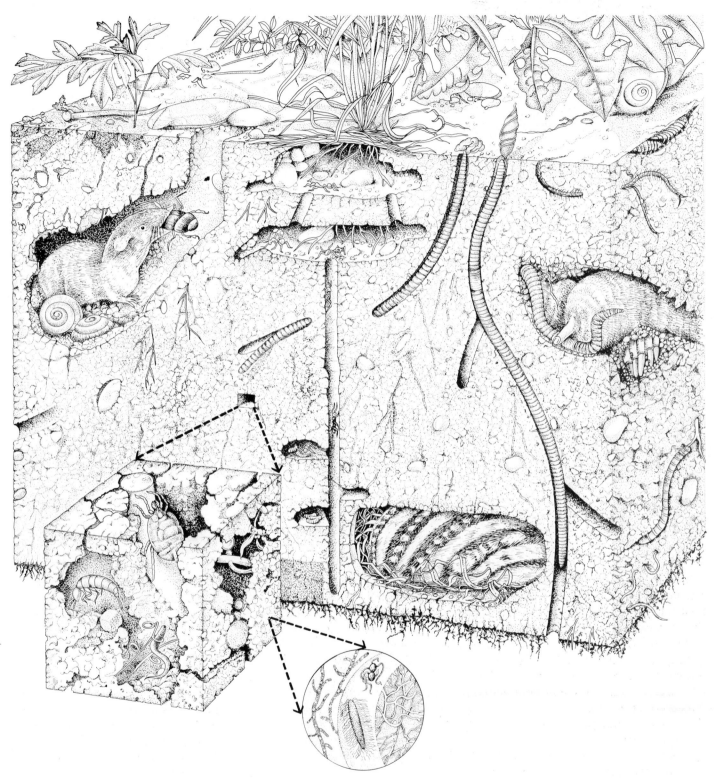

27–6

Plants share the soil with a vast number of living organisms, ranging from microbes to small mammals such as moles, shrews, and ground squirrels. Multitudes of burrowing creatures—most notably ants and earthworms—aerate the soil and improve its ability to absorb water. Called by Aristotle "the intestines of the earth," earth-worms refine the soil by processing it through their gut. The refined soil is then deposited on the soil surface in the form of castings. In a single year, the combined activities of earthworms may produce as much as 500 metric tons of castings per hectare. The castings are very fertile, containing five times the nitrogen content of the surrounding soil, seven times the phosphorus, 11 times the potassium, three times the magnesium, and twice the calcium. Bacteria and fungi are the principal decomposers of the organic matter in soils.

The Water Cycle

The earth's supply of water is stable and is used over and over again. Most of the water (98 percent) is present in oceans, lakes, and streams. Of the remaining 2 percent, some is frozen in polar ice and glaciers, some is found in the soil, some is in the atmosphere as water vapor, and some is in the bodies of living organisms.

Sunshine evaporates water from the oceans, lakes, and streams, from the moist soil surfaces, and from the bodies of living organisms, drawing the water back up into the atmosphere, from which it falls again as rain. This constant movement of water from the earth into the atmosphere and back again is known as the water cycle. The water cycle is driven by solar energy.

Some of the water that falls on the land percolates down through the soil until it reaches a zone of saturation. In the zone of saturation, all holes and cracks in the rock are filled with water. Below the zone of saturation is solid rock through which the water cannot penetrate. The upper surface of this zone of saturation is known as the water table.

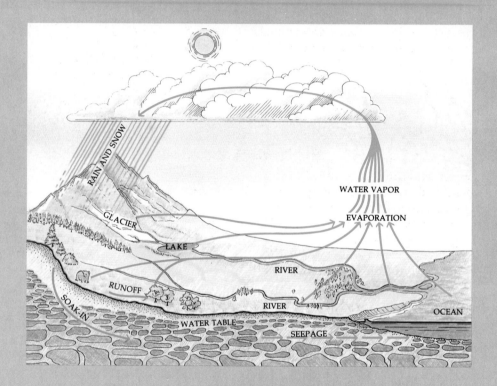

The C horizon is composed of the broken-down and weathered rocks and minerals from which the true soil in the upper horizons is formed.

SOIL COMPOSITION

Soils consist of solid matter and pore space (the space around the soil particles). Different proportions of air and water occupy the pore space, depending on prevailing moisture conditions. The soil water is present primarily as a film on the surfaces of soil particles.

The fragments of rocks and minerals in the soil vary in size from sand grains, which can be seen easily with the naked eye, to clay particles too small to be seen even under the low power of a light microscope. The following classification is one scheme for categorizing soil particles (also known as soil separates) according to size:

Separate	Diameter (in micrometers)
Coarse sand	200–2000
Fine sand	20–200
Silt	2–20
Clay	Less than 2

Soils contain a mixture of particles of different sizes and are divided into textural classes according to the proportions of different particles present in the mixture. For example, soils that contain 35 percent or less clay and 45 percent or more sand are sandy; those containing 40 percent or less clay and 40 percent or more silt are silty. **Loam soils** contain sand, silt, and clay in proportions that result in ideal agricultural soils. This mixture is ideal for two reasons. The coarser soil particles add better drainage, while the finer soil particles have high nutrient-retention capabilities.

The solid matter of soils consists of both inorganic and organic materials, with the proportions varying greatly in different soils. The organic component includes the remains of organisms in various stages of decomposition, a highly decomposed fraction known as **humus,** and a wide range of living plants and animals. Structures as large as tree roots may be included, but the living phase is dominated by fungi, bacteria, and other microorganisms.

SOILS AND WATER

Approximately 50 percent of the total soil volume is represented by pore space, which is occupied by varying proportions of air and water, depending on moisture conditions. When no more than half the pore space is occupied by water, adequate oxygen is available for root growth and other biological activity.

Following a heavy rain or irrigation, soils retain a certain amount of water and remain moist even after gravity has removed loosely bound water. If the fragments that make up a soil are large, the pores and spaces between them will be large; water will drain through the soil rapidly, and relatively little will be available for plant growth in the A and B horizons. Because of their finer pores and the attractive forces that exist between water molecules and colloidal-size clay particles, clay soils are able to hold a much greater amount of water against the action of gravity. Thus, clay soils may retain three to six times more water than a comparable volume of sand; hence, soils with more clay can hold more water that is available to plants. The percentage of water that a soil can hold against the action of gravity is called its **field moisture capacity.**

If a plant is allowed to grow indefinitely in a sample of soil and no water is added, the plant will eventually not be able to absorb water rapidly enough to meet its needs, and it will droop and wilt. When wilting is severe, plants fail to recover even when placed in a humid chamber. The percentage of water remaining in a soil when such irreversible wilting occurs is called the **permanent wilting percentage** of that soil.

Figure 27–7 shows the relationship between the soil water content and the potential at which water is held by sandy, loam, and clay soils. The forces that retain water in the soil can be expressed in the same terms (in

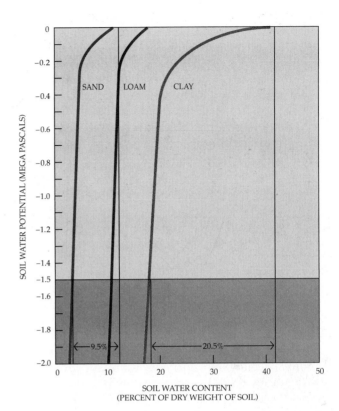

27–7
Relationship between soil water content and soil water potential in sandy, loam, and clay soils. (The curves are plotted on a linear scale.) Note that, whereas water available to plants in a sandy soil is only about 9.5 percent of the dry weight of sand, that available to plants in a clay soil is considerably greater, about 20.5 percent of the dry weight of clay. Soil with a water potential of −1.5 megapascals is considered to be at permanent wilting percentage.

this case, water potential) as the forces for water uptake in cells and tissues (see Chapter 4, pages 62–64). The soil water potential decreases gradually with a decrease in the soil moisture below field moisture capacity. Soil scientists have agreed to consider soil with a water potential of −1.5 megapascals to be at permanent wilting percentage.

CATION EXCHANGE

The inorganic nutrients taken in through the roots of plants are present in the soil solution as ions. Most metals form positively charged ions, that is, cations, such as Ca^{2+}, K^+, and Na^+. Clay particles provide a reservoir of such cations for the plant—at various points on the crystalline lattice of the clay there is an excess of negative charge, where cations can be bound and thus held against the leaching action of percolating soil water.

The cations bound in this way to the clay particles can be replaced by other cations (in a process called **cation exchange**) and then released into the soil solution, where they become available for plant growth. This is one reason why clay particles are an essential component of productive soils.

The principal negatively charged ions, or anions, found in soil are NO_3^-, SO_4^{2-}, HCO_3^-, and OH^-. Anions are leached out of the soil more rapidly than cations because anions do not attach to clay particles. An exception is phosphate, which is retained against leaching because it forms insoluble precipitates. Phosphate is specifically adsorbed by, or held on the surface of, compounds containing iron, aluminum, and calcium.

The acidity or alkalinity of soil is related to the availability of inorganic nutrients for plant growth. Soils vary widely in pH, and many plants have a narrow range of tolerance on this scale. In alkaline soils, some cations are precipitated, and such elements as iron, manganese, copper, and zinc may thereby become unavailable to plants. Mycorrhizae (see Chapter 12) are especially important in the absorption and transfer of phosphorus in most plants, but these structures have also been implicated in the increased absorption of manganese, copper, and zinc by plants.

Nutrient Cycles

The bulk of the organic matter of soil is made up of dead leaves and other plant material, together with the decomposing bodies of animals. This organic debris is mixed with the inorganic particles of the soil, and in this mixture live astonishing numbers of small organisms that spend all or part of their lives beneath the soil surface. A single teaspoon of soil may contain 5 billion bacteria, 20 million small filamentous fungi, and 1 million protists. The soil animals and microorganisms (Figure 27–6) break down the organic matter, releasing its inorganic nutrients, which can be reutilized by the plants. Thus, except for those nutrients that leach out of the soil, are carried away by streams and rivers, and eventually precipitate in the ocean, substances that are taken from the soil are constantly returned to it. Both macronutrients and micronutrients are recycled constantly through plant and animal bodies, returned to the soil, broken down, and taken up into plants again. Each element has a different cycle, involving many different organisms and different enzyme systems.

The end results are the same, however—a significant amount of the element is constantly returned to the soil and is available for plant use.

Nitrogen and the Nitrogen Cycle

The nitrogen in soil is derived from the earth's atmosphere. Although the atmosphere is 78 percent nitrogen, most living things cannot use atmospheric nitrogen to make proteins and other organic substances. Unlike carbon and oxygen, nitrogen is chemically unreactive. The highly specialized capacity for converting atmospheric nitrogen to a form that can be used by cells—a process known as nitrogen fixation—is limited to a few species of bacteria.

On a worldwide basis, usable nitrogen is the major limiting nutrient in crop plant growth. The processes by which nitrogen is circulated from the atmosphere through plants and the soil by the action of living organisms are illustrated in Figure 27–8.

NITROGEN FIXATION

If the nitrogen that is removed from the soil were not steadily replaced, virtually all life on this planet would slowly disappear. Nitrogen is replenished in the soil primarily by nitrogen fixation. Much lesser amounts are added via precipitation.

Nitrogen fixation is the process by which dinitrogen (N_2) is reduced to ammonium (NH_4^+) and made available for amination reactions (the process by which ammonium ions are transferred to carbon-containing compounds to produce amino acids and other nitrogen-containing organic compounds). Nitrogen fixation, which can be carried out only by certain bacteria, is a process on which all living organisms are now dependent, just as most organisms are ultimately dependent on photosynthesis as the source of their energy.

The enzyme that catalyzes the fixation of nitrogen is called *nitrogenase*. Nitrogenase contains molybdenum, iron, and sulfide prosthetic groups, and therefore these elements are essential for nitrogen fixation. Nitrogenase also uses large amounts of ATP as an energy source, making nitrogen fixation an expensive metabolic process.

Of the various classes of nitrogen-fixing organisms, the symbiotic bacteria are by far the most important in terms of total amounts of nitrogen fixed. The most common of the nitrogen-fixing bacteria are *Rhizobium* and *Bradyrhizobium*, both of which invade the roots of legumes, such as alfalfa (*Medicago sativa*), clovers (*Trifolium*), peas (*Pisum sativum*), soybean (*Glycine max*), and beans (*Phaseolus*) (Figure 27–9).

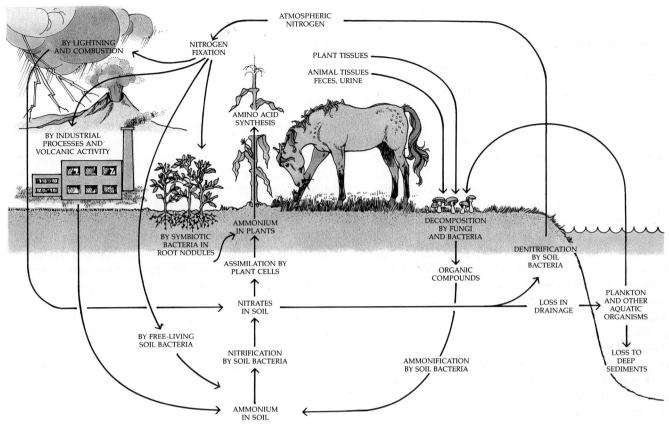

27–8

The nitrogen cycle in a terrestrial ecosystem. The primary reservoir of nitrogen is the atmosphere, where it makes up 78 percent of dry air. Only a few microorganisms, some symbiotic and others free-living, are capable of fixing *nitrogen gas into inorganic compounds that can be used by plants in the synthesis of amino acids and other nitrogenous compounds.*

The nitrogen cycle in aquatic ecosystems is similar to the terrestrial *cycle. Although the specific organisms differ, the processes are essentially the same, with specific groups of bacteria carrying out the various chemical reactions on which the cycle depends.*

27–9

(a) *Nitrogen-fixing nodules on the roots of a soybean* (Glycine max) *plant. These nodules are the result of a symbiotic relationship between the root cells of this leguminous plant and the bacterium* Bradyrhizobium. (b) *Scanning electron micrograph of nitrogen-fixing nodules on white clover* (Trifolium repens).

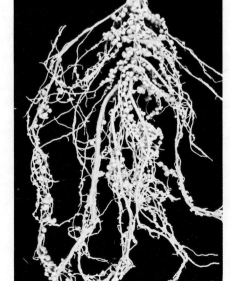

(a)

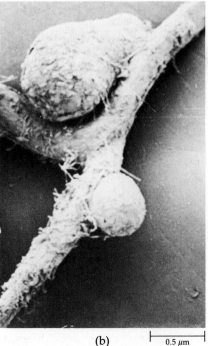

(b)

0.5 μm

The beneficial effects on the soil of growing leguminous plants have been recognized for centuries. Theophrastus, who lived in the third century B.C., wrote that the Greeks used crops of broad beans (*Vicia faba*) to enrich the soils. Where leguminous plants are grown, some of the "extra" nitrogen may be released into the soil, becoming available for other plants. In modern agriculture, it is common practice to rotate a nonleguminous crop, such as corn (*Zea mays*), with a leguminous crop, such as alfalfa, or corn with soybeans and then wheat. The leguminous plants are then either harvested for hay, leaving behind the nitrogen-rich roots, or, better still, they are simply plowed under. A crop of alfalfa that is plowed back into the soil may add as much as 300 to 350 kilograms of nitrogen per hectare of soil. As a conservative estimate, 150 to 200 million metric tons of fixed nitrogen are added to the earth's surface each year by such biological systems.

An industrial process for the chemical fixation of atmospheric nitrogen was developed in 1914. Since that time, the commercial production of fixed nitrogen has increased steadily to the current level of approximately 50 million metric tons per year. Most of this nitrogen is used in agricultural fertilizers. Industrial fixation, unfortunately, is accomplished at a high energy cost in terms of fossil fuels.

Small amounts of nitrogen can also be fixed by lightning or by internal combustion engines (such as that found in the automobile), but, instead of forming ammonia, various oxides of nitrogen are produced by these reactions. These oxides dissolve in rainwater forming nitric acid (one of the major constituents of acid rain) and other nitrogen oxides (see essay, page 512). Measurements at an experimental station in England over a five-year period showed that the rainwater brought down 7.1 kilograms of nitrogen per hectare each year.

Nitrogen-Fixing Symbioses

Rhizobium and *Bradyrhizobium* bacteria, commonly called rhizobia (Figure 27–10), enter the root hairs of leguminous plants when the plants are still seedlings. Establishment of the nitrogen-fixing symbiosis between *Bradyrhizobium japonicum* and soybean (*Glycine max*) begins with the attachment of rhizobia to emerging root hairs (Figure 27–11a), which typically develop into tightly curled structures, entrapping the rhizobia (Figure 27–11b). Invasion of the root hairs and underlying cortical cells by the rhizobia occurs via **infection threads,** tubular structures formed by progressive inward growth of the root-hair cell walls from the sites of penetration. A single root hair may be penetrated by several rhizobia and, hence, may contain several infection threads (Figure 27–11b). The bacterial symbiont also induces cell division in localized regions of the cortex, which it then enters through growth and branching of the infection thread (Figures 27–11c and 27–12). Release of the rhizobia from the infection thread into envelopes derived from the root-hair plasma membrane and continued proliferation of both **bacteroids** (the name given the now-enlarged nitrogen-fixing rhizobia, which are shown in Figures 27–10b and 27–14) and cortical cells of the root result in the formation of tumorlike growths known as **nodules** (Figure 27–9). The manner of infection and nodule formation in the roots of other legumes is apparently similar to that for soybean.

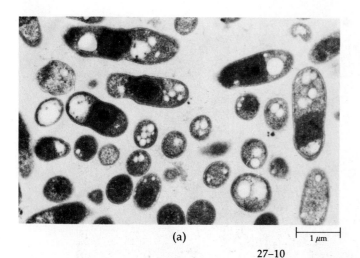

(a)   1 μm

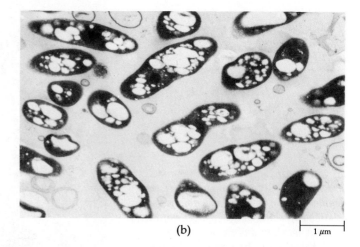

(b)   1 μm

27–10

Rhizobium. (a) *Bacteria in their free-living form.* (b) *The bacteroid form that* *the rhizobia assume when they enter root cells.*

27–11

Early events in the infection of soybean by Bradyrhizobium japonicum. *(a) Scanning electron micrograph showing rhizobia (arrows) attached to recently emerged root hair. (b) Differential-interference contrast photomicrograph showing a short, curled root hair containing multiple infection threads (arrows). (c) Transverse section of an infected root showing early development of the nodule meristem. At this stage, most of the cell divisions are anticlinal (at right angles to the surface) across the cortex. In the nodule meristem adjacent to the infected root hair, cells have divided in other planes as well.*

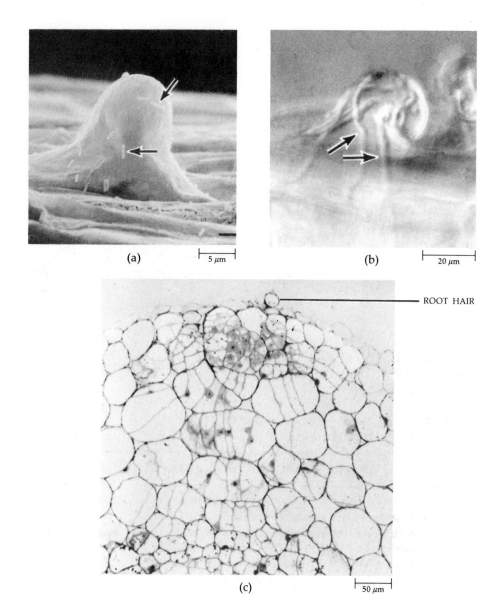

(a) |— 5 μm

(b) |— 20 μm

ROOT HAIR

(c) |— 50 μm

27–12

Electron micrograph of a branched infection thread containing bacteroids in an infected nodule cell of soybean.

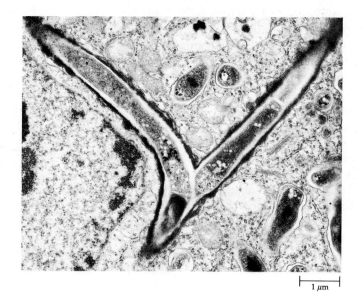

|— 1 μm

27–13

Section through a mature soybean root nodule. Vacuolate uninfected cells can be seen among the darkly stained infected cells in the central zone of the nodule. The nodule cortex, containing vascular bundles and a layer of darkly stained sclerenchyma cells, surrounds the central zone. A variable oxygen-diffusion barrier is thought to be present in the innermost layers of the cortex.

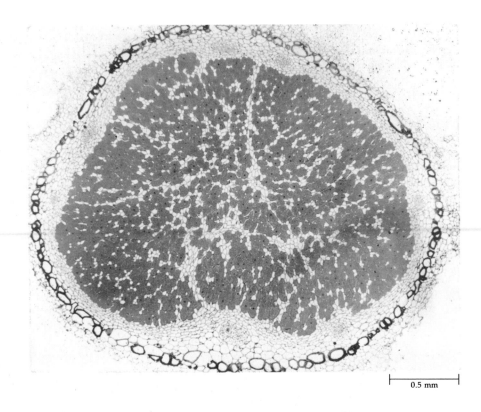

0.5 mm

The root nodules of legumes consist of a relatively narrow cortex, which surrounds a large central zone containing both bacteroid-infected and uninfected cells (Figure 27–13). Vascular bundles, which radiate from the point of attachment of the nodule to the root, occur in the inner cortex. In soybean, both infected and uninfected cell types play a role in the production of ureides (derivatives of urea), the end product of nitrogen fixation that is exported from the nodule to the plant (Figure 27–14). Other legumes produce other nitrogenous compounds, the amino acid asparagine being one of the most common.

In the symbiotic association between rhizobia bacteria and legumes, the bacteria provide the plant with a form of nitrogen that it can use to make protein, while the plant provides the bacteria with both an energy source (carbon compounds) and a highly regulated oxygen environment. The O_2 concentration in the bacteria-infected cells must be carefully regulated since O_2 is a potent irreversible inhibitor of the nitrogenase enzyme and yet is required for aerobic respiration to meet the large ATP demands of the nitrogenase as well as other metabolic activities in both the bacteria and the plant cells. Studies have shown that the infected-cell O_2 concentration is maintained at an extremely low level, approximately 10^{-4} times the O_2 concentration in the root environment. This is accomplished by a variable barrier to oxygen diffusion that is thought to be located in the nodule inner cortex (Figure 27–13). The nodule appears to have the ability to regulate the proportion of air- to water-filled intercellular spaces in the inner cortex, and since O_2 is less soluble and diffuses more slowly in water than in air, an increase in water-filled spaces restricts O_2 diffusion into the central zone of the nodule. In fact, the O_2 concentration in the central zone is so low that the infected cells produce large quantities of an O_2-binding protein called *leghemoglobin*. This protein picks up the oxygen at the surface of each infected cell and facilitates its diffusion to the bacteroids within the cell. Hence, the cortical diffusion barrier is primarily responsible for maintaining a low O_2 concentration in the central zone of the nodule, while leghemoglobin works at the low O_2 concentration to supply the bacteroids with the O_2 needed for respiration.

The symbiosis between a species of *Rhizobium* or *Bradyrhizobium* and a legume is quite specific; for example, bacteria that invade and induce nodule formation in clover (*Trifolium*) roots will not induce nodules on the roots of soybeans (*Glycine*). The bacteria and legume may recognize each other through a plant protein (*lectin*) found on the root surface. The lectin binds to a polysaccharide on the surface of an effective *Rhizobium* or *Bradyrhizobium* cell but not to the surface polysaccharides of ineffective bacteria. In studies aimed at developing new effective associations between *Rhizobium* and nonlegumes, the *Rhizobium*-host plant specificity barrier was removed by treating the roots with enzymes that degraded the cell walls at the tips of emerging root hairs. In this way, nodular structures were induced on roots of wheat (*Triticum aestivum*) and rice (*Oryza sativa*) and of the dicotyledonous nonlegume oilseed rape (*Brassica napus*). Whether nodular structures effective in nitrogen fixation can be induced on these nonlegumes remains to be determined.

The interaction between the symbiotic partners involves an intricate exchange of molecular signals that regulate the expression of genes essential for infection and nodule involvement. Initiation of the symbiotic relationship begins when flavonoids secreted by the legume bind to and activate the bacterial nodulation (nod) gene known as nodD. The nodD gene product then induces the expression of other nod genes, whose products are required for such processes as root-hair curling and cell wall degradation. In addition, other bacterial nod gene products activate the plant genes that encode plant cell proteins called nodulins, which are essential for cortical cell division and for growth and function of the nodule.

The legumes are by far the largest group of plants that enters into a nitrogen-fixing partnership with symbiotic bacteria. There are, however, a few nitrogen-fixing symbioses that involve plants other than legumes.

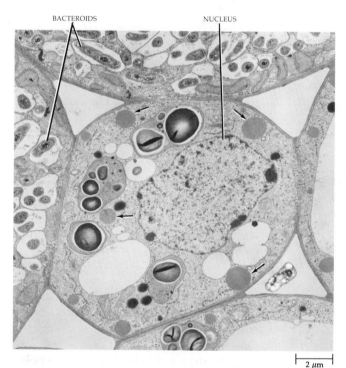

BACTEROIDS   NUCLEUS

$\vdash$ 2 μm $\dashv$

27–14

Electron micrograph of portion of nodule of soybean showing an uninfected cell (middle) adjacent to infected cells (top and left). The infected cells contain many bacteroids. Note the numerous peroxisomes (arrows) in the uninfected cell. The enzymes involved in the final steps of ureide formation apparently are located in these cellular components. The volume of mitochondria is four times greater in infected cells than in uninfected ones— apparently a reflection of the enormous demand of the bacteroids for ATP.

Alder trees (Alnus), for example, form nodules that are induced by and contain nitrogen-fixing actinomycetes, rather than Rhizobium or Bradyrhizobium. Sweet gale (Myrica gale), sweet fern (Comptonia), and mountain lilacs (Ceanothus) also form symbiotic associations with actinomycetes (see Chapter 11).

Another symbiotic relationship is of considerable practical interest in certain parts of the world. Azolla is a small floating water fern, and Anabaena is a nitrogen-fixing cyanobacterium that lives in the cavities of the Azolla fronds (Figure 27–15, page 608). The Azolla-Anabaena symbiosis is unique among nitrogen-fixing symbioses in that the relationship is sustained throughout the life cycle of the host. Azolla infected with Anabaena may contribute as much as 50 kilograms of nitrogen per hectare. In the Far East, for example, heavy growths of Azolla-Anabaena are permitted to develop on rice paddies (see also Figure 11–16b on page 198). The rice plants eventually shade out the Azolla and as the fern dies, nitrogen is released for use by the rice plants.

Free-Living Nitrogen-Fixing Microorganisms

Nonsymbiotic bacteria of the genera Azotobacter and Clostridium are both able to fix nitrogen. Azotobacter is aerobic, and Clostridium is anaerobic; both are common saprophytic soil bacteria. It is estimated that they probably add about 7 kilograms of nitrogen to a hectare of soil per year. Another important group includes many photosynthetic bacteria, such as cyanobacteria.

AMMONIFICATION

Much of the soil nitrogen is derived from dead organic materials in the form of complex organic compounds such as proteins, amino acids, nucleic acids, and nucleotides. These nitrogenous compounds are usually rapidly decomposed into simpler compounds by soil-dwelling saprobic bacteria and various fungi, which incorporate the nitrogen into amino acids and proteins and release excess nitrogen in the form of ammonium ions (NH_4^+) by a process known as **ammonification.** In alkaline media, the nitrogen may be converted to ammonia gas (NH_3), but this conversion usually occurs only during the decomposition of large amounts of nitrogen-rich material, as in a manure pile or a compost heap that has contact with the atmosphere. Within soil, the ammonia produced by ammonification is dissolved in the soil water, where it combines with protons to form the ammonium ion. In some ecosystems, the ammonium ion is not rapidly oxidized (see Nitrification on page 610) but remains in the soil. Plants growing in these soils are able to take up the NH_4^+ and use it in the synthesis of plant protein.

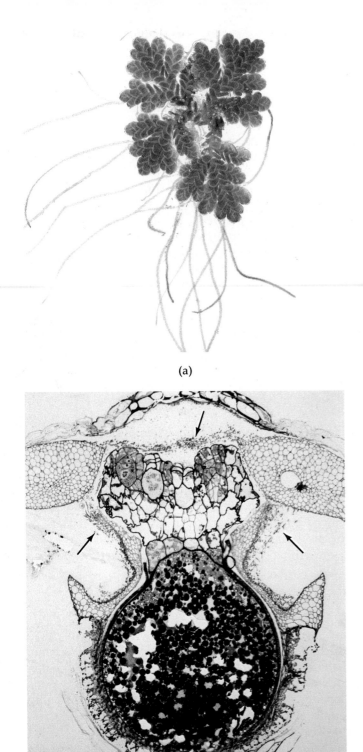

(a)

(b)

27–15
(a) Azolla filiculonides, *a water fern that grows in symbiotic association with the cyanobacterium* Anabaena. *(b) Filaments of* Anabaena *can be seen (arrows) associated with a female gametophyte (megagametophyte) that has developed from a germinated megaspore of Azolla.*

Halophytes: A Future Resource?

Unlike most animals, most plants do not require sodium and, moreover, cannot survive in brackish waters or saline soils. In such environments, the solution surrounding the roots often has a higher solute concentration than the cells of the plants, causing water to move out of the roots by osmosis. Even if the plant is able to absorb water, it faces additional problems from the high level of sodium ions. If the plant takes up water and excludes sodium ions, the solution surrounding the roots becomes even saltier, increasing the likelihood of water loss through the roots. The salt may even become so concentrated that it forms a crust around the roots, effectively blocking the supply of water to them. Another problem is that sodium ions may enter the plant in preference to potassium ions, depriving the plant of an essential nutrient as well as inhibiting some enzyme systems.

Some plants—known as halophytes—can grow in saline environments such as deserts, salt marshes, and coastal areas. All of these plants have evolved mechanisms for dealing with high sodium concentrations, and for some of them sodium appears to be a required nutrient. The adaptations of halophytes vary. In many halophytes, a sodium-potassium pump seems to play a major role in maintaining a low sodium concentration within the cells while simultaneously ensuring that a sufficient supply of potassium ions enters the plant. In some species, the pump operates primarily in the root cells, pumping sodium back to the environment and potassium into the root. The presence

(a) Atriplex *(saltbush) is one of several halophytes being evaluated as potential crop plants. (b) The surface of a leaf of* Atriplex. *Salt is pumped from the leaf tissues through narrow stalk cells into the large, expandable bladder cells.*

of calcium ions (Ca^{2+}) in the soil solution is thought to be essential for the effective functioning of this mechanism.

Other halophytes take in sodium through the roots but then either secrete it or isolate it from the living cytoplasm of the plant body. In *Salicornia* (pickleweed), a sodium-potassium pump (or a variant of it) operates in the vacuolar membranes (tonoplasts) of leaf cells. Sodium ions enter the cells but are immediately pumped into vacuoles and isolated from the cytoplasm. In such plants, the solute concentration of the vacuoles is higher than that of the environment, establishing the necessary osmotic potential for the movement of water into the roots. In other genera, the salt is pumped into the intercellular spaces of the leaves and then secreted from the plant. In *Distichlis palmeri* (Palmer's grass), the salt exudes through specialized cells (not the stomata) onto the surface of the leaf. In *Atriplex* (saltbush), it is concentrated by special salt glands and pumped into bladders. The bladders expand as the salt accumulates, finally bursting. Rain or the passing tide washes the salt away.

Halophytes are of current interest not only because of the light they may shed on the osmoregulatory mechanisms of plants but also because of their potential as crop plants. In a world with an ever-increasing need for food, vast areas are unsuitable for agricultural purposes because of the salinity of the soil. For example, there are over 30,000 kilometers of desert coastline and about 400 billion hectares of desert with potential water supplies that are too salty for crop plants. Moreover, each year about 200,000 hectares of irrigated crop land becomes so salty that further agriculture is impossible. When arid land is heavily irrigated, as in large areas of the western United States, salts from the irrigation waters accumulate in the soil. This accumulation occurs because, in both evaporation from the soil and transpiration from plants, essentially pure water is given off, leaving all solutes behind. Over many years, the salt concentration of the soil increases, eventually reaching levels that cannot be tolerated by most plants. It has been suggested that the ancient civilizations of the Near East ultimately fell because their heavily irrigated land became so salty that food could no longer be grown on it.

One way to extend the life of irrigated crop lands and to bring presently barren areas into agricultural use would be to breed salt tolerance into conventional crop plants. Thus far, however, such efforts have met with little success. Scientists at the University of Arizona's Environmental Research Laboratory are taking what appears to be a more promising approach. They have gathered halophytes from all over the world and are engaged in an extensive research program to determine optimal growing conditions, potential yields, and the nutritional value and palatability of the seeds and vegetative parts of the various species. Their results suggest that a number of halophyte species have great potential for use in livestock feed and, quite possibly, for human consumption as well.

(a)

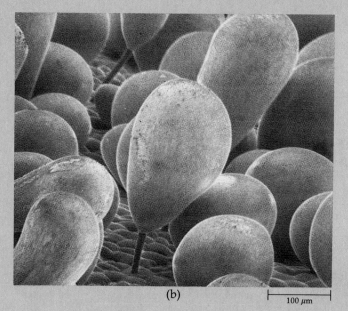

(b)

100 μm

(a)

(b)

27–16

Carnivorous plants entrap their prey in a variety of ways. (a) The common bladderwort (Utricularia vulgaris) is a free-floating aquatic plant. The traps are tiny, flattened, pear-shaped bladders. Each bladder has a mouth guarded by a hanging door. The tripping mechanism consists of four stiff bristles near the lower free edge of the door. When a small animal brushes against these bristles, the hairs distort the lower edge of the door, causing it to spring open. Water then rushes into the bladder, carrying the *animal inside, and the door snaps shut behind it. A range of enzymes secreted by the internal wall of the bladder and by the residential bacterial population digest the animal, and released minerals and organic compounds are taken up through the cell walls of the trap. The undigested exoskeletons remain within the bladders. (b) The sundew (Drosera rotundifolia) is a tiny plant, often only a few centimeters across, with club-shaped hairs on the upper surface of its leaves. The tips of these glandular hairs secrete a clear,* *sticky liquid, or mucilage, that attracts insects such as the damselfly shown here. When an insect is caught in the mucilage, the hairs bend inward until the leaf finally curves around the insect. The hairs are known to secrete at least six enzymes, which, together with bacteria-produced enzymes, especially chitinase, digest the insect. Nutrients released from the prey into the mucilage are resorbed by the same glands that secreted the digestive enzymes.*

NITRIFICATION

Several species of bacteria common in soils are able to oxidize ammonia or ammonium ions. The oxidation of ammonia, or **nitrification,** is an energy-yielding process, and the energy released in the process is used by these bacteria to reduce carbon dioxide in much the same way that photosynthetic autotrophs use light energy in the reduction of carbon dioxide. Such organisms are known as chemosynthetic autotrophs (as distinct from photosynthetic autotrophs). The chemosynthetic nitrifying bacterium *Nitrosomonas* is primarily responsible for oxidation of ammonia to nitrite ions (NO_2^-):

$$2NH_3 + 3O_2 \longrightarrow 2NO_2^- + 2H^+ + 2H_2O$$

Nitrite is toxic to plants, but it rarely accumulates in the soil. *Nitrobacter*, another genus of bacteria, oxidizes the nitrite to form nitrate ions (NO_3^-), again with a release of energy:

$$2NO_2^- + O_2 \longrightarrow 2NO_3^-$$

Because of nitrification, nitrate is the form in which almost all nitrogen is absorbed by plants. Most nitrogen fertilizer used commercially contains either ammonium ions (NH_4^+) or urea, which breaks down into NH_4^+ in soils. The NH_4^+ is converted to NO_3^- by nitrification.

A few species of plants are able to use animal proteins directly as a nitrogen source. These carnivorous plants (Figure 27–16) have special adaptations that are used to lure and trap insects and other very small animals. The plants digest the trapped organisms, absorbing the nitrogenous compounds they contain as well as other organic compounds and minerals, such as potassium and phosphate. Most of the carnivores of the plant world are found in bogs, a habitat that is usually quite acidic and thus not favorable for the growth of nitrifying bacteria.

ASSIMILATION OF NITROGEN

Once nitrate ions enter a plant cell, they are reduced to ammonium ions. This reduction process requires energy, in contrast to nitrification, which involves the oxidation of NH_4^+ and releases energy. The ammonium ions formed by reduction are transferred to carbon-containing compounds to produce amino acids and other nitrogen-containing organic compounds. This process is known as **amination.** The incorporation of nitrogen into organic compounds takes place largely in young, growing root cells. The initial stages in the me-

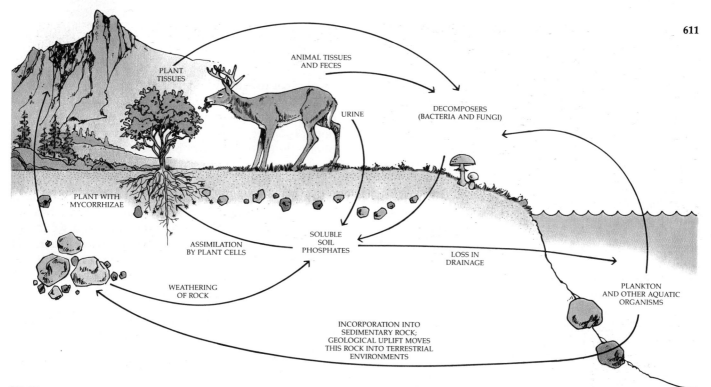

27–17

The phosphorus cycle in a terrestrial ecosystem. Phosphorus is essential to all living organisms as a component of energy-carrier molecules, such as ATP, and also of the nucleotides of DNA and RNA. Like other minerals, it is released from dead tissues by the activities of decomposers, taken up from soil by

plants, and cycled through the ecosystem.

The phosphorus cycle in an aquatic ecosystem involves different organisms but is, in most respects, similar to the terrestrial cycle shown here. However, in aquatic ecosystems a considerable amount of phosphorus is incorporated into the shells and skeletons of aquatic

organisms. This phosphorus, along with phosphates that precipitate out of the water, is subsequently incorporated into sedimentary rock. Such rock, returned to the land surface as a result of geological uplift, is the primary terrestrial reservoir of phosphorus.

tabolism of nitrogen also appear to occur in the root; almost all the nitrogen ascending the stem in the xylem is already in the form of organic substances, largely molecules of various amino acids.

NITROGEN LOSS

The nitrogen-containing compounds of green plants are returned to the soil with the death of the plants (or of the animals that have eaten the plants) and are reprocessed by soil organisms. Nitrate dissolved in the soil water is then taken up by plant roots and reconverted to organic compounds. In the course of this cycle, a certain amount of nitrogen is always "lost," in the sense that it becomes unavailable to the plants in specific ecosystems.

A main source of nitrogen loss from an ecosystem is the removal of plants from the soil. Soils under cultivation often show a steady decline of nitrogen content. Nitrogen may also be lost when topsoil is carried off by soil erosion or when ground cover is destroyed by fire. Nitrogen is also removed by leaching; nitrates and nitrites, both of which are anions, are particularly susceptible to being washed from the root zone by water percolating through the soil.

Under anaerobic conditions, nitrate is often reduced to volatile forms of nitrogen, such as nitrogen gas (N_2) and nitrous oxide (N_2O), which then return to the atmosphere. This process of reduction, called **denitrification,** is carried out by numerous microorganisms. The low oxygen conditions necessary for denitrification have long been perceived as characteristic of waterlogged soils and such habitats as swamps and marshes. Scientists now recognize that these conditions commonly exist within soil aggregates even in the absence of excessive water. Consequently, denitrification is virtually a universal process in soils. A fresh supply of readily decomposable organic matter provides the energy source required by denitrifying bacteria and, if other conditions are appropriate, promotes denitrification. Lack of an energy source allows nitrate concentrations to build up to excessive levels in groundwater.

The Phosphorus Cycle

The phosphorus cycle (Figure 27–17) seems simpler than the nitrogen cycle because there are fewer steps, and because particular steps are not dependent on specific groups of microorganisms. The phosphorus cycle also differs from the nitrogen cycle in that the earth's crust, rather than the atmosphere, is the primary reservoir of phosphorus. As previously discussed, the weathering of rocks and minerals over long periods of time is the source of most of the phosphorus in the soil solution.

Compared with the amount of nitrogen, the amount of phosphorus required by plants is relatively small (Tables 27–1 and 27–2). Nevertheless, of all the elements for which the earth's crust is the primary reservoir, phosphorus is the most likely to limit plant growth. In Australia, for example, where the soils are extremely weathered and deficient in phosphorus, the distribution and limits of native plant communities are often determined by the available soil phosphate.

Phosphorus circulates from plants to animals and is returned to the soil in organic forms in residues and wastes; these organic forms of phosphorus are converted to inorganic phosphate, and thus the phosphorus again becomes available to plants (Figure 27–17).

Through erosion, pollution, and loss in drainage water, large amounts of phosphorus are discharged into rivers and streams. This phosphorus eventually reaches the oceans, where it is deposited in sediments as precipitates and in the remains of organisms. In the past, the use of guano (deposits of seabird feces) as agricultural fertilizer returned some of the ocean phosphorus to terrestrial ecosystems. However, most of the phosphorus in deep-sea sediments will become available only as a result of major geological uplifts. To counter this loss, deposits of phosphate rock are being mined on a large scale for use as agricultural fertilizer.

Human Impact on Nutrient Cycles

Normal functioning of the phosphorus, nitrogen, and other nutrient cycles requires the orderly transfer of elements between steps in a cycle in order to prevent buildup or depletion of nutrients at any one stage. Over millions of years, organisms have been provided with needed quantities of essential inorganic nutrients as a result of the normal functioning of such cycles. However, in recent years, the need to adequately feed an exponentially growing human population has had drastic effects on some cycles, and in time this may lead to harmful accumulations and depletions of nutrients at specific stages of one or more nutrient cycles. For example, crop removal and increased soil erosion have accelerated phosphorus loss from soils. Also, sewage effluents have often been discharged into streams, the phosphorus being transported (and lost) to the oceans rather than being recycled.

Normal functioning of the nitrogen cycle involves a balance between fixation processes, which remove nitrogen from the atmospheric reservoir, and denitrification processes, which return nitrogen to the atmosphere. Recently there has been massive introduction of fixed nitrogen (nitrates) into the environment through the extensive use of commercial fertilizers. Because nitrates are readily leached from soil, increased nitrogen pollution of groundwater, lakes, and streams has resulted. To add to this problem, marshes and wetlands, the primary sites of denitrification, are being destroyed at an alarming rate through their conversion to building sites, agricultural land, and dump sites.

A method has been developed to reduce the loss of soil nitrates by leaching. It involves the application of an organic chemical that selectively inhibits the nitrifying bacterium *Nitrosomonas* for a limited period of time and then is broken down in the soil. This practice helps retain fertilizer nitrogen in the form of ammonium until the nitrogen can be absorbed by plants. Unfortunately, unreliable economic returns from use of this chemical have militated against its widespread use.

Soils and Agriculture

In natural situations, the elements present in the soil recirculate and so become available again for plant growth. As discussed previously, negatively charged clay particles and organic matter are able to bind such positively charged ions as Ca^{2+}, Na^+, and K^+. These ions are then displaced from the particles by other cations (cation exchange) and absorbed by roots from the soil solution. In general, the cations that are required by plants are present in large amounts in fertile soils, and the amounts removed by a single crop are small. However, when a series of crops is grown on a particular field and the nutrients are continuously removed from the cycle as the crops are harvested, some of these cations (most commonly potassium) may become depleted to such an extent that fertilizers containing the missing element must be added.

Programs for supplementing nutrient supplies for agricultural and horticultural crops should be based on a bookkeeping approach that equates amounts of nutrients required to produce a specific crop and the nutrients available to the plants from all sources. Often the soil and plant residues cannot supply the required nutrients, and supplemental amounts must be provided in applications of commercial fertilizers, organic residues such as manure or compost, or a combination of the two.

Nitrogen, phosphorus, and potassium are the three elements that are commonly included in commercial fertilizers. Fertilizers are usually labeled with a formula that indicates the percentage of each of these elements. A 10-5-5 fertilizer, for example, is one that contains 10 percent nitrogen (N), 5 percent phosphorus (reported as phosphorus pentoxide, P_2O_5), and 5 percent potassium (reported as potassium oxide, K_2O). This method of reporting phosphorus and potassium contents of fertilizer is a historical artifact dating from the days when analytical chemists reported all of their analyses as oxides.

Other essential inorganic nutrients, although required in very small amounts, can sometimes become limiting factors in soils on which crops are grown. Experience has shown that the most common deficiencies are those for sulfur, magnesium, zinc, and boron.

Plant Nutrition Research

Research on the inorganic nutrients essential for crop plants—particularly on the quantities of those nutrients required for optimal crop yields and on the capacities of various soils to provide the nutrients—has been of great practical value in agriculture and horticulture. Because of the steady increase in worldwide food needs, this type of research undoubtedly will continue to be essential.

SOIL DEFICIENCIES AND TOXICITIES

Modification and manipulation of soils by addition of nutrients in fertilizers, by raising of the pH with lime, or by removal of excess salts by leaching with water may not be the only means of improving and maintaining crop production in below-optimum soils. By using the knowledge and techniques of plant breeding and plant nutrition, it may be possible to select and develop cultivars of crop species that are better adapted for growth in nutrient-deficient environments. The validity of this research approach is confirmed by the occurrence of wild plants in nutritional environments that are very different from the average soil environments in which crop plants are grown; examples are acidic *Sphagnum* bogs, in which the pH may be less than 4.0, and mine tailings, which often contain high concentrations of potentially toxic metals, such as zinc.

In an attempt to develop bean *(Phaseolus vulgaris)* strains that are tolerant of potassium-deficiency stress, strains were collected from around the world and grown in a nutrient solution containing suboptimal potassium levels. All other nutrients were at optimal concentrations. The extremes in bean growth under the imposed potassium-deficiency stress are shown in Figure 27–18. Emanuel Epstein and his associates have

27–18
Comparison of potassium-deficiency symptoms in strains of the bean (Phaseolus vulgaris). In the bean strain on the left (strain 66), the plant is tolerant of a deficiency level of potassium, whereas the plant on the right (strain 63) is clearly a nontolerant strain.

had extraordinary success in isolating barley *(Hordeum vulgare)* strains tolerant of high salt concentrations. The most tolerant strains grew reasonably well even when irrigated with sea water rather than fresh water.

EFFICIENCY OF NITROGEN FIXATION

Manipulation of biological nitrogen fixation also offers tremendous potential for improved efficiency in nitrogen utilization. One aspect of research in this area is concerned with improving the efficiency of the association of legumes with either *Rhizobium* or *Bradyrhizobium*, for example, through the genetic screening of both legumes and bacteria. Such screening could identify combinations that would result in increased fixation in specific environments. This could result from a greater photosynthetic efficiency in legumes, so that more carbohydrate is available for bacterial nitrogen fixation and growth. However, nitrogen fixation requires considerable energy, and so any increase in fixation might be at the expense of shoot productivity.

A second research approach is to develop additional and more effective associations of free-living nitrogen-fixing bacteria and higher plants. In Brazil in the early 1970s, several types of nitrogen-fixing bacteria were found growing in association with the roots of certain tropical grasses; for example, the grass *Digitaria* was found to support populations of the bacterium *Azospirillum*. Similar associations with some of the world's

Compost

Composting, a practice as old as agriculture itself, has attracted increased interest as a means of utilizing organic wastes by converting them to fertilizer. The starting product is any collection of organic matter—leaves, kitchen garbage, animal manure, straw, lawn clippings, sewage sludge, sawdust—and the population of bacteria and other microorganisms normally present. The only other requirements are oxygen and moisture. Grinding of the organic matter is not essential, but it provides a greater surface area for microbial attack and thus speeds the process.

In a compost heap, microbial growth accelerates quite rapidly, generating heat, much of which is retained because the outer layers of organic matter act as insulation. In a large heap (2 meters × 2 meters × 1.5 meters, for instance), the interior temperature rises to about 70°C; in small heaps, it usually reaches 40°C. As the temperature rises, the population of decomposers changes, with thermophilic and thermotolerant forms replacing the organisms previously present. As the original forms die, their organic matter also becomes part of the product. A useful side effect of the temperature increase is that most of the common pathogenic bacteria that may have been present, for example, in sewage sludge, are destroyed, as are cysts, eggs, and other immature forms of plant and animal parasites.

With the passage of time, changes in pH also occur in a compost heap. The initial pH value is usually slightly acidic (about 6.0), which is comparable to the pH of most plant fluids. During the early stages of decomposition, the production of organic acids causes a further acidification, with the pH decreasing to about 4.5 to 5.0. However, as the temperature rises, the pH also increases; the composted material eventually levels off at slightly alkaline values (7.5 to 8.5).

An important factor in composting (as in any biological growth process) is the ratio of carbon to nitrogen. A ratio of about 30 to 1 (by weight) is optimal. If the carbon ratio is higher, microbial growth slows. If the nitrogen ratio is higher, some escapes as ammonia. If the compost materials are quite acidic, limestone (calcium carbonate) may be added to balance the pH; however, if too much is added, it will increase the nitrogen loss.

Studies involving municipal compost piles in Berkeley, California, demonstrated that if large piles were kept moist and aerated, composting could be completed in as little as

two weeks. Generally, though, three months or more during the winter is needed to complete the process. If compost is added to the soil before the composting process is complete, it may temporarily deplete the soil of soluble nitrogen.

Because it greatly reduces the bulk of plant wastes, composting can be a very useful means of waste disposal. In Scarsdale, New York, for example, leaves composted in a municipal site were reduced to one-fifth their original volume. At the same time, they formed a useful soil conditioner, improving both the aeration and water-holding capacity. Chemical analyses indicate, however, that a rich compost commonly contains, in dry weight, only about 1.5 to 3.5 percent nitrogen, 0.5 to 1.0 percent phosphorus, and 1.0 to 2.0 percent potassium, far less than commercial fertilizers. Yet, unlike commercial fertilizers, compost can be a source of nearly all the elements known to be needed by plants. Compost provides a continuous balance of nutrients, releasing them gradually as it continues to decompose in the soil.

Today, the main driving force behind composting is the increasing cost of waste disposal and the increasing difficulty of finding suitable disposal sites—sites that will not negatively affect the environment or pollute our waters. Unfortunately, composting is a long way from becoming economically viable as a substitute for manufacturing fertilizers for commercial agriculture.

major food crops, such as corn *(Zea mays)* and sugarcane *(Saccharum)*, have since been reported. Although the practical benefits to agriculture could be tremendous, the effectiveness of associations of nitrogen-fixing bacteria and grasses such as corn is still uncertain.

Probably the most exciting research approach is to be found in genetic engineering: the genetic modifications and transfer of the genes necessary for nitrogen fixation from one organism to another (page 566).

EFFECTS OF POLLUTION

The toxic effects of various inorganic agents discharged into the environment as pollutants are also of current research interest. Crop plants, for example, may be adversely affected by heavy metals such as copper and cadmium. Aquatic ecosystems, in particular, have been damaged because they have become common disposal sites for industrial and municipal wastes. The introduction of nitrogen and phosphorus—the primary eutrophication nutrients—into freshwater ecosystems has resulted in massive growths of algae and aquatic flowering plants, thus seriously reducing the recreational value of affected lakes and streams.

Both terrestrial and, particularly, aquatic ecosystems are subject to widespread environmental damage because of acid rain. Acid rain results from the interaction of sulfur dioxide and oxides of nitrogen—derived principally from the combustion of fossil fuels—with atmospheric moisture to form sulfuric and nitric acids. These acids impart a high degree of acidity to the rainfall. Acid rain can have adverse effects on plants, the weathering of rocks and minerals, the solubilities of potentially toxic metals in the environment, and even human health. (See essay, pages 512 and 513.)

Summary

A total of 16 inorganic nutrients are required by most plants for normal growth. Of these, carbon, hydrogen, and oxygen are derived from air and water. The rest are absorbed by roots in the form of ions. These 16 elements are categorized as either macronutrients or micronutrients, depending on the amounts in which they are required. The macronutrients are carbon, oxygen, hydrogen, nitrogen, potassium, calcium, phosphorus, magnesium, and sulfur. The micronutrients are iron, chlorine, copper, manganese, zinc, molybdenum, and boron. Some inorganic nutrients, such as sodium and cobalt, are essential only for specific organisms.

Inorganic nutrients perform a number of important roles in cells. They regulate osmosis and affect cell permeability. Some also serve as structural components of cells, as components of critical metabolic compounds, and as activators and components of enzymes.

The chemical and physical properties of soils are critical in determining the capability of the soils to provide the inorganic nutrients, water, and other conditions necessary for maximum crop plant production. The weathering of rocks and minerals supplies the inorganic component of soils. All of the inorganic nutrients except nitrogen are derived from weathering processes. In addition, soils contain organic matter and pore space occupied by varying proportions of water and gases. Under agricultural conditions, nitrogen, phosphorus, and potassium are the nutrients most often limiting to plant growth and most frequently added to soils in fertilizers.

Each essential inorganic nutrient is circulated, in a complex cycle, among organisms and between organisms and the environment. The circulation of nitrogen through the soil, through the bodies of plants and animals, and back to the soil again is known as the nitrogen cycle.

Nitrogen is replenished in the soil primarily by nitrogen fixation, the process by which N_2 is reduced to NH_3 (NH_4^+) and made available for assimilation into amino acids and other organic nitrogenous compounds. Biological nitrogen fixation is carried out only by bacteria. These include bacteria *(Rhizobium* and *Bradyrhizobium)* that are symbionts of leguminous plants, free-living bacteria, and actinomycetes in symbiotic relationship with a few genera of nonleguminous plants. In agriculture, plants are removed from the soil and nitrogen and other elements are not recycled, as they are in nature; therefore, they must be replenished in either organic form or inorganic form.

Nitrogen also reaches the soil in the form of organic material of plant and animal origin. These substances are decomposed by soil organisms. Ammonification— the release of NH_4^+ ions from nitrogen-containing compounds—is carried out by soil bacteria and fungi. Nitrification is the oxidation of ammonia or ammonium ions to form nitrites and nitrates; one type of bacterium is responsible for the oxidation of ammonia to nitrite, and another for the oxidation of nitrite to nitrate. Nitrogen enters plants almost entirely in the form of nitrates. Nitrogen is lost from the soil by crop removal, erosion, fire, leaching, and the action of denitrifying bacteria.

See the end of Chapter 28 for suggested readings.

CHAPTER 28

The Movement of Water and Solutes in Plants

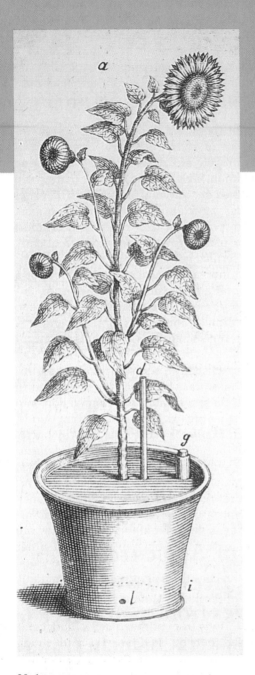

28–1
Diagram of the sunflower plant from Stephen Hales's description of his experiments on the movement of water through the plant body. Hales found that most of the water "imbibed" by the plant was lost by "perspiration."

The capacity of a plant to transport both organic and inorganic nutrients and water throughout the plant body is critical in determining the ultimate structure and function of its component parts, as well as the development and form of the plant as a whole. In the first part of this chapter, we examine the movement of water and solutes through the plant body from the soil to the aerial plant parts. Then, toward the end of the chapter, we consider the movement of solutes and water from the sites of photosynthesis to the nonphotosynthetic parts of the plant body. We begin with a description of transpiration because this process is a major determining factor in the movement of water through the plant body.

Movement of Water and Inorganic Nutrients through the Plant Body

TRANSPIRATION

In the early eighteenth century, Stephen Hales, an English physician, noted that plants "imbibe" a much greater amount of water than animals do. He calculated that one sunflower plant, bulk for bulk, "imbibes" and "perspires" 17 times more water than a human every 24 hours (Figure 28–1). Indeed, the total quantity of water absorbed by any plant is enormous—far greater than that used by any animal of comparable weight. An animal uses less water because much of its water is recirculated through its body over and over again, in the form (in vertebrates) of blood plasma and other fluids. In plants, nearly 99 percent of the water taken in by the roots is released by the plant into the air as water vapor (Table 28–1). This process, known as **transpira-**

Table 28–1 *Water Loss by Transpiration in One Plant in a Single Growing Season*

Plant	Water Loss (liters)
Cowpea (*Vigna sinensis*)	49
Potato (*Solanum tuberosum*)	95
Wheat (*Triticum aestivum*)	95
Tomato (*Lycopersicon esculentum*)	125
Corn (*Zea mays*)	206

After J.F. Ferry, *Fundamentals of Plant Physiology*, Macmillan Publishing Company, New York, 1959.

tion, may involve any aboveground part of the plant body; leaves, however, are by far the principal organs of transpiration.

Why do plants lose such large quantities of water to transpiration? This question can be answered by considering the requirements for the chief function of the leaf, photosynthesis—the source of all the food for the entire plant body. The energy necessary for photosynthesis comes from sunlight. Therefore, for maximum photosynthesis, a plant must spread a maximum surface to the sunlight, creating a large transpiring surface at the same time. But sunlight is only one of the requirements for photosynthesis; the chloroplasts also need carbon dioxide. Under most circumstances, carbon dioxide is readily available in the air surrounding the plant, but in order for carbon dioxide to enter a plant cell, which it does by diffusion, it must go into solution because the plasma membrane is nearly impervious to the gaseous form of carbon dioxide. Hence, the gas must come into contact with a moist cell surface. However, wherever water is exposed to air, evaporation occurs. In other words, the uptake of carbon dioxide for photosynthesis and the loss of water by transpiration are inextricably bound together in the life of the green plant.

REGULATION OF TRANSPIRATION

Transpiration—sometimes called an "unavoidable evil" —can be extremely injurious to a plant. Excessive transpiration (water loss exceeding water uptake) retards the growth of many plants and kills many others by dehydration. Despite their long evolutionary history, plants have not developed a structure favorable to the entrance of the CO_2 essential in photosynthesis but unfavorable to the loss of water vapor by transpiration. However, a number of special adaptations minimize water loss while optimizing carbon dioxide gain.

Cuticle and Stomata

Leaves are covered by a cuticle that makes their surfaces largely impervious to both water and carbon dioxide. Only a small fraction of the water transpired by plants is lost through this protective outer coating, and another small fraction is lost through the lenticels in the bark. By far the largest amount of water transpired by a vascular plant is lost through the stomata (Figure 28–2). Stomatal transpiration involves two steps: (1) evaporation of water from cell wall surfaces bordering the intercellular spaces, or air spaces, of the leaf, and (2) diffusion of the resultant water vapor from the intercellular spaces into the atmosphere by way of the stomata (see Figure 28–19).

Stomata are small openings in the epidermis, each surrounded by two guard cells, which can change their shape to bring about the opening and closing of the pores. Stomata are also found in young stems, but they are far more abundant in leaves. The number of stomata may be quite large; for example, there are approximately 12,000 stomata per square centimeter of leaf surface in tobacco (*Nicotiana tabacum*) leaves. The stomata lead into a honeycomb of air spaces that surround the thin-walled mesophyll cells within the leaf. The air in these spaces—which make up 15 to 40 percent of the total volume of the leaf—is saturated with water vapor that has evaporated from the damp surfaces of the mesophyll cells. Although the stomatal openings account for only about 1 percent of the total leaf surface, more than 90 percent of the water transpired by the plant is lost through the stomata. The rest is lost through the cuticle.

Closing of stomata not only prevents loss of water vapor from the leaf, but, as mentioned, it also prevents entry of carbon dioxide into the leaf. A certain amount of carbon dioxide, however, is produced by the plant during respiration, and as long as light is available, this carbon dioxide can be used to sustain a very low level of photosynthesis even when the stomata are closed.

The Mechanism of Stomatal Movements

Stomatal movements result from changes in turgor pressure within the guard cells. Opening occurs when solutes are actively accumulated in these cells. The accumulation of solutes (and resultant decrease in guard cell water potential) causes osmotic movement of water into the guard cells and a buildup of turgor pressure in excess of that in the surrounding epidermal cells. Stomatal closing is brought about by the reverse process: with a decline in guard cell solutes (and resultant increase in guard cell water potential), water moves out of the guard cells and the turgor pressure decreases. Thus, turgor is maintained or lost due to the passive os-

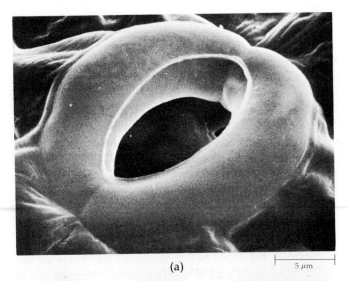

(a)

5 μm

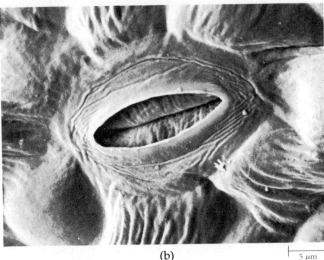

(b)

5 μm

28–2
Scanning electron micrographs showing
(a) open stoma in epidermis of cucumber
(Cucumis sativus) leaf and (b) closed
stoma in epidermis of parsley (Apium
petroselinum) leaf. The stomata lead into
a honeycomb of air spaces that surround
the thin-walled, photosynthetic mesophyll
cells within the leaf. The air spaces are
saturated with water vapor that has evapo-
rated from the surfaces of the mesophyll
cells.

motic movement of water into or out of the cells along a gradient of water potential.

The major solute responsible for these gradients in water potential is the potassium ion (K^+). This ion has been found in the guard cells of open stomata of more than 50 species, including CAM plants whose stomata open at night (see Chapter 7). Techniques for estimating potassium levels within a single guard cell show that the K^+ concentration rises when the stomata open and drops when the stomata close. The surrounding cells provide the required reservoir of potassium ions. The gradient of potassium between the guard cells and surrounding cells changes significantly and is accompanied by the osmotic flow of water and resultant turgor changes (Figure 28–3).

Transported with the K^+ ions are the negatively charged ions (anions) needed to counter the positive charge. Two anions implicated in this role are chloride and malate (an ion with two $-COO^-$ groups). Evidence now indicates that guard cell chloroplasts fix CO_2 photosynthetically and that sugar from guard cell photosynthesis can contribute to the solute buildup required for stomatal opening.

The structure of the guard cell walls plays a crucial role in stomatal movements. During expansion of the paired guard cells, two physical constraints cause the guard cells to bend and thus to open the pore. One of these constraints is the radial orientation of the cellulose microfibrils in the guard cell walls (Figure 28–4a). This **radial micellation** allows the guard cells to lengthen while preventing them from expanding laterally. The second constraint is found at the ends of the guard cells, where they are attached to one another. This common wall remains almost constant in length during opening and closing of the stomata. Consequently, increase in turgor pressure causes the outer (dorsal) walls of the guard cells to move outward relative to their common walls. As this happens, the radial micellation transmits this movement to the wall bordering the pore (the ventral wall), and the pore opens. Figure 28–4b through d depicts the results of some experiments with balloons that have been used in support of the role of radial micellation in stomatal movement.

Factors Affecting Stomatal Movements

A number of environmental factors affect stomatal opening and closing, water loss being the major influence. When the turgor of a leaf drops below a certain critical point, which varies with different species, the stomatal opening becomes smaller. The effect of water loss overrides other factors affecting the stomata, but stomatal changes can occur independently of overall water gain or loss by the plant. The most conspicuous example is found in the many species in which the sto-

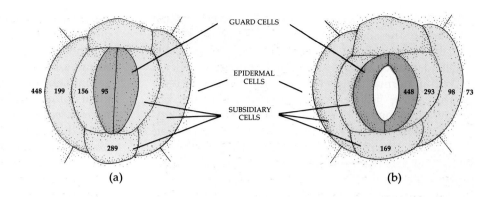

28–3

Quantitative changes in potassium concentrations across the stomatal complex (guard cells and subsidiary cells) and neighboring epidermal cells in the dayflower (Commelina communis) *leaf. The guard cells of the dayflower leaf are associated with six subsidiary cells, four lateral and two terminal. (a) Vacuolar potassium concentrations in the various cells when the stomata are closed. (b) Vacuolar potassium concentrations in the various cells when the stomata are open. Potassium-sensitive microelectrodes were used to determine the potassium content of the individual cells. The K^+ values are expressed in mM (millimoles per milliliter).*

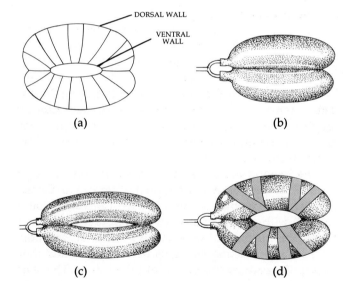

28–4

(a) Diagram of a pair of guard cells. The radial lines indicate the radial arrangement of cellulose microfibrils in the guard cell walls. (b) through (d) Diagrams of models (balloons) used to study the effect of radial micellation on the opening of stomata. (b) Two partially inflated balloons glued together near their ends. (c) The same balloons at higher pressure (fully inflated). A narrow slit is visible. (d) A pair of fully inflated balloons after bands of tape have been added to simulate radial micellation. The opening is much greater than in (c).

mata open regularly in the morning and close in the evening, even though there may be no changes in the amount of water available to the plant.

In many plants, there is a marked increase in the level of abscisic acid (ABA) during periods of water stress. When "fed" or applied to leaves, ABA causes stomatal closure within a few minutes; moreover, the effect of ABA on stomatal movement is readily reversible. Solute (K^+) loss from guard cells begins when ABA of mesophyll origin arrives at the stomata, signaling the stomata that the mesophyll cells are experiencing water stress.

Other environmental factors that affect stomatal movement include *carbon dioxide concentration, light,* and *temperature*. In most species, an increase in CO_2 concentration causes the stomata to close. The magnitude of this response to CO_2 varies greatly from species to species and with the degree of water stress a given plant has undergone or is undergoing. In corn (*Zea mays*), the stomata may respond to changes in CO_2 in a matter of seconds. The site for sensing the level of CO_2 is located within the guard cells.

In most species, the stomata open in the light and close in the dark. This can be explained in part by the photosynthetic utilization of CO_2, which brings about a reduction in the CO_2 level within the leaf. Light may, however, have a more direct effect on stomata. Blue light has long been known to stimulate stomatal opening independently of CO_2. For example, guard cell protoplasts of onion (*Allium cepa*) swell in the presence of K^+ when illuminated with blue light. The blue-absorbing pigment (a flavin or a flavoprotein located in the tonoplast and possibly the plasma membrane) promotes K^+ uptake by the guard cells. The blue light response is involved in stomatal opening in the early morning and in stomatal responses to sunflecks, spots of light that penetrate the leaf canopy and range in duration from a few seconds to minutes.

Evidence also exists for the presence of red-light-stimulated stomatal opening mediated by the guard cell chloroplasts. Apparently the chloroplasts supply the

ATP that fuels proton pumping at the guard cell plasma membrane, the site of active K^+ uptake. It is believed that the guard cell chloroplasts are involved with stomatal adaptation to sun, shade, and temperature.

Within normal ranges (10°–25°C), changes in temperature have little effect on stomatal behavior, but temperatures higher than 30° to 35°C can lead to stomatal closure. The closing can be prevented, however, by holding the plant in air without carbon dioxide, which suggests that temperature changes work primarily by affecting the concentration of carbon dioxide in the leaf. An increase in temperature results in an increase in respiration and a concomitant increase in the concentration of intercellular carbon dioxide, which may actually be the cause of stomatal closure in response to heat. Many plants in hot climates close their stomata regularly at midday, apparently because of the effect of temperature on carbon dioxide accumulation and because of dehydration of leaves as water loss by transpiration exceeds water uptake by absorption.

Stomata not only respond to environmental factors but also exhibit daily rhythms of opening and closing that appear to be controlled from within the plant— that is, they also exhibit circadian rhythms (see page 576).

The stomata of most plants are open during the day and closed at night, but this is not true of all plants. A wide variety of succulents—including cacti, the pineapple (*Ananas comosus*), and members of the stonecrop family (Crassulaceae)—open their stomata at night, when conditions are least favorable to transpiration. The Crassulacean acid metabolism (CAM) characteristic of such plants has a pathway for carbon flow not substantially different from that of C_4 plants, as discussed in Chapter 7. At night, when their stomata are open, CAM plants take in carbon dioxide and convert it to organic acids. During the day, when their stomata are closed, the carbon dioxide is released from these organic acids for use in photosynthesis.

FACTORS AFFECTING THE RATE OF TRANSPIRATION

Although stomatal opening and closing are the major plant factors affecting the rate of transpiration, a number of other factors both in the environment and in the plant itself influence transpiration. One of the most important of these is *temperature*. The rate of water evaporation doubles for every 10°C rise in temperature. However, because evaporation cools the leaf surface, its temperature does not rise as rapidly as that of the surrounding air. As noted previously, stomata close when temperatures exceed 30° to 35°C.

Humidity is also important. Water is lost much more slowly into air already laden with water vapor. Leaves of plants growing in shady forests, where the humidity is generally high, typically spread large, luxuriant leaf surfaces because for these plants the main problem is getting enough light, not losing water. In contrast, plants of grasslands or other exposed areas often have narrow leaves characterized by relatively little leaf surface, thick cuticles, and sunken stomata. Grassland plants obtain all the light they can use but are constantly in danger of excessive water loss.

Air currents also affect the rate of transpiration. A breeze cools your skin on a hot day because it blows away the water vapor that has accumulated near the skin surface and so accelerates the rate of evaporation of water from your body. Similarly, wind blows away the water vapor from leaf surfaces. Sometimes, if the air is very humid, wind may decrease transpiration by cooling the leaf, but a dry breeze will greatly increase evaporation. Leaves of plants that grow in exposed, windy areas are often hairy; the hairs are believed to protect the leaf surface from wind action and so slow the rate of transpiration by stabilizing the boundary layer of air over the leaf surface.

WATER TRANSPORT: THE COHESION-TENSION MECHANISM

Water enters the plant by the roots and is given off, in large quantities, by the leaves. How does the water get from one place to another, often over large vertical distances? This question has intrigued many generations of botanists.

The general pathway that the water follows in its ascent has been clearly identified. One can trace this pathway simply by placing a cut stem in water that is colored with any harmless dye (preferably, the stem should be cut under the water to prevent air from entering the conducting elements of the xylem) and then tracing the path of the liquid into the leaves. The stain quite clearly delineates the conducting elements of the xylem. Experiments using radioactive isotopes confirm that the isotope and, presumably, the water do indeed travel by way of vessel elements (or tracheids) in the xylem. In the experiment shown in Figure 28–5, care had to be taken to separate the xylem from the phloem. Earlier experiments in which this separation was not made produced ambiguous results because there is a great deal of lateral movement from the xylem into the phloem. This lateral movement, however, as the experiment shows, is not necessary for the overall movement of water and minerals from soil to leaf.

Now that we know the path water takes, the next question we ask is, "How does the water move?" Logic suggests two possibilities: it can be pushed from the

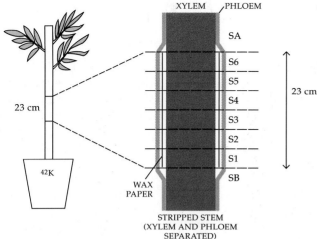

	STEM SEGMENT	ppm ^{42}K IN XYLEM	ppm ^{42}K IN PHLOEM
ABOVE STRIP	SA	47	53
STRIPPED SECTION	S6	119	11.6
	S5	122	0.9
	S4	112	0.7
	S3	98	0.3
	S2	108	0.3
	S1	113	20
BELOW STRIP	SB	58	84

28-5
Radioactive potassium (^{42}K) added to the soil water shows that the xylem is the channel for the upward movement of both water and inorganic ions. Wax paper was inserted between the xylem and the phloem to prevent lateral transport of the isotope. The relative amounts of ^{42}K detected in each segment of the stem are given in the table.

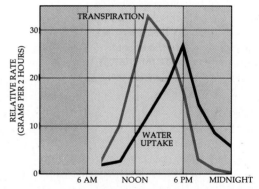

28-6
Measurements of water movement in ash (Fraxinus) trees show that a rise in water uptake follows a rise in transpiration. These data suggest that the loss of water generates the force needed for its uptake.

bottom or pulled from the top. As we shall see, the first of these possibilities is not the answer. Briefly, root pressure (see page 626) does not exist in all plants, and in those plants in which it is present, it is not sufficient to push water to the top of a tall tree. Moreover, the simple experiment just described (that involving the cut stem) rules out root pressure as a crucial factor. So we are left with the hypothesis that water is pulled up through the plant body, and this hypothesis is correct according to all present evidence.

When water evaporates from the cell wall surfaces bordering the intercellular spaces in the interior of a leaf during transpiration, it is replaced by water from within the cell. This water diffuses across the plasma membrane, which is freely permeable to water but not to the solutes of the cell. As a result, the concentration of solutes within the cell increases and the water potential of the cell decreases. A gradient of water potential then becomes established between this cell and adjacent, more saturated cells. These cells, in turn, gain water from other cells until eventually this chain of events reaches a vein and exerts a "pull," or tension, on the water of the xylem. Because of the extraordinary cohesiveness of water molecules, this tension is transmitted all the way down the stem to the roots, so that water is withdrawn from the roots, pulled up the xylem, and distributed to the cells that are losing water vapor to the atmosphere (Figure 28–6). This water loss makes the water potential of the roots more negative and increases their capacity to extract water from the soil. Hence, the lowered water potential in the leaves, brought about by transpiration and/or by the use of water in the leaves, results in a gradient of water potential from the leaves to the soil solution at the surface of the roots. This gradient of water potential provides the driving force for the movement of the water along the soil-plant-atmosphere continuum.

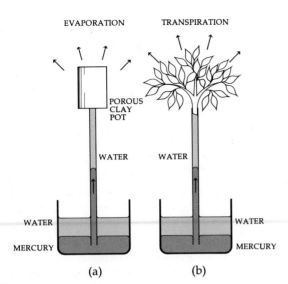

28–7

(a) *A simple physical system that demonstrates the cohesion-tension theory. A porous clay pot is filled with water and attached to the end of a long, narrow glass tube that is also filled with water. The water-filled tube is placed with its lower end below the surface of a volume of mercury in a beaker. Water evaporates from the pores in the pot and is replaced by water "pulled up" through the tube in a continuous column. As the water rises, mercury is pulled up into the tube to replace it. (b) Transpiration from leaves results in sufficient water loss to create a similar negative pressure.*

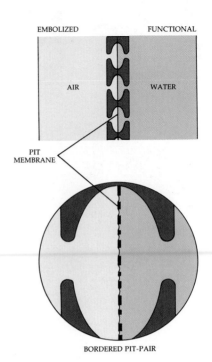

28–8

(a) *Diagram showing bordered pit-pairs between tracheary elements, one of which is embolized (filled with air) and thus nonfunctional. (b) Detail of a pit membrane. The small size of the pores in the pit membrane helps to prevent air from spreading between tracheary elements. The bordered pit-pairs diagrammed here do not have a torus, the impermeable central thickening typical of the pit membranes of conifer tracheids and found in the pit membranes between vessel members of some dicotyledons.*

This theory of water movement is known as the **cohesion-tension theory** because it depends on the cohesiveness of water, the property of water that permits it to withstand tension (Figure 28–7). However, the theory might also be known as the *cohesion–adhesion–tension theory* because adhesion of the water molecules to the walls of the tracheids and vessels of the xylem and to the cell walls of the leaf and root cells is just as important for the rise of water as are cohesion and tension. The cell walls along which the water moves have evolved as a very effective water-attracting surface, taking maximal advantage of water's adhesiveness and thus providing a situation in which cohesiveness is readily expressed.

The cohesion of water in the xylem is increased by the filtering effect of the roots, which remove fine particles that can act as nuclei for bubble formation, and by the small diameter of the xylem conduits—vessels and tracheids—through which the water moves. Despite the filtering provided by the roots, bubble formation does occur, and it is a normal occurrence in many

trees. *Cavitation* (rupture of the water columns) and subsequent *embolism* (filling of vessels and/or tracheids with air) are the bane of the cohesion-tension mechanism. (Embolized tracheary elements cannot conduct water; see also the discussion on page 459.) Fortunately, the small size of the pores in the pit membranes of the bordered pit-pairs between adjacent conduits helps to isolate a bubble in a single vessel or tracheid (Figure 28–8).

There is no doubt that the tensile strength of water is great enough to prevent the pulling apart of adjacent water molecules under the tension required to move water up the xylem of tall trees. For example, it has been demonstrated that a column of water in a fine capillary tube is capable of withstanding a tension of -26.4 megapascals; the estimated tension required to move water to the top of a tall redwood (*Sequoia sempervirens*) is only about -2.0 megapascals. Calculated values for the tensile strength of water are very high: between 130 and 150 megapascals, or possibly as high as 1500 megapascals!

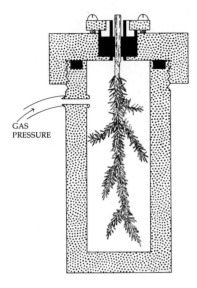

28–9
*Measuring water tension in the xylem.
A branch whose xylem tension the investigator wishes to measure is cut off and
placed in a pressure "bomb." When the
branch is cut, some of the xylem sap—
which was under tension before the
branch was cut—recedes into the xylem
below the cut surface. Pressure is raised
in the bomb until sap emerges from the
cut end of the stem. If one assumes that
equal pressure is required to force the
sap in either direction, the positive
pressure needed to force out the sap is,
ideally, equal to the tension that existed
in the branch before it was severed.*

How can the cohesion-tension theory be tested?
One way to test this theory directly is by measuring the
tension of water within the xylem. When a twig is cut
from a transpiring tree, the columns of water in the
vessels abruptly recede below the cut surfaces. By
mounting the twig in a pressure chamber, as shown in
Figure 28–9, it is possible to apply pressure to the
leaves until the curved upper surfaces of the water columns appear (when magnified) at the cut surface of the
twig. The magnitude of the pressure required to return
the water to the cut surface is approximately equal to
the magnitude of the tension under which the water
existed in the twig before its excision. Results obtained
by this method are entirely consistent with the predictions of the cohesion-tension theory.

A second set of data that is in accord with the cohesion-tension theory indicates that the movement of
water begins at the top of the tree. The velocity of sap
flow in various parts of the tree has been measured by
an ingenious method involving a small heating element
to warm the xylem contents for a few seconds and a
sensitive thermocouple to detect the moment at which
the heated xylem sap moves past a specific point (Figure 28–10). As shown in the graph, in the morning the
sap begins to flow first in the twigs, as tension arises
close to the leaves, and later in the trunk. In the evening, the flow diminishes first in the twigs, as water
loss from the leaves diminishes, and later in the trunk.
Trees with wide vessels (200 to 400 micrometers in diameter) showed midday peak velocities of 16 to 45
meters per hour (measured at breast height), while
those with narrow vessels (50 to 150 micrometers in diameter) had slower midday peak velocities, ranging
from 1 to 6 meters per hour.

A third set of supporting data comes from measurements of minute changes in the diameter of the tree
trunk (Figure 28–11). The shrinking of the trunk is interpreted by some investigators as occurring because of
the negative pressures in the water passages of the
xylem. Presumably, the water molecules adhering to
the sides of the vessels pull the sides inward. When
transpiration begins in the morning, first the upper part

28–10
*A method for measuring velocity of sap
flow. A small heating element inserted
into the xylem heats the ascending sap
for a few seconds. A thermocouple above
the heating element records the passing
wave of heat. The experimenter times the
interval between these two events. As
shown in the graph, in the morning the
sap begins to increase its velocity of flow
first in the twigs and then in the trunk.
In the evening, velocity diminishes first
in the twigs and then in the trunk.*

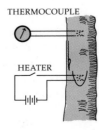

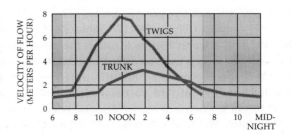

28–11
A dendrometer (left) records small daily fluctuations in the diameter of a tree trunk at two different heights. As shown in the graph, in the morning, shrinkage occurs in the upper trunk slightly before it occurs in the lower trunk. These data suggest that transpiration from the leaves "pulls" water out of the trunk before it can be replenished from the roots. The shaded strips signify nighttime.

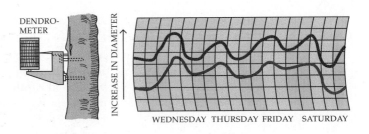

of the trunk shrinks, as water is pulled out of the xylem before it can be replenished from the roots; then the lower part shrinks. Later in the day, as the transpiration rate decreases, the upper trunk expands before the lower trunk does. There is some evidence that changes in trunk diameter are due in part or largely to shrinkage and expansion of bark tissues as water moves laterally to and from the xylem in response to changing pressures in the xylem.

Note that the energy for evaporation of water molecules—and thus for movement of water and inorganic nutrients through the plant body—is supplied not by the plant but directly by the sun. Note also that movement is possible because of the extraordinary cohesive and adhesive properties of water, to which the plant is so exquisitely adapted.

The cohesion-tension theory is sometimes called the "transpiration-pull theory." This is unfortunate because "transpiration-pull" implies that transpiration is essential for water to move to leaves. Although transpiration may increase the rate at which water moves, any use of water in leaves produces forces causing movement of water to them.

WATER UPTAKE BY ROOTS

The root system serves to anchor the plant in the soil and, above all, to meet the tremendous water requirements of the leaves resulting from transpiration. Most of the water that a plant takes from the soil enters through the younger parts of the root (see Chapter 22). Absorption takes place directly through the epidermis of younger roots. The root hairs, located several millimeters above the root tip, provide an enormous surface area for absorption (Figure 28–12; Table 28–2). From the root hairs, the water moves through the cortex (the outer layer(s) of which may be differentiated as an exodermis), through the endodermis (the innermost layer of cortical cells), and into the vascular cylinder. Once in the conducting elements of the xylem, the water moves upward through the root and stem and into the leaves,

Table 28–2   Density of Root Hairs on the Root Surface in Three Plant Species

Plant	Root Hair Density (per square centimeter)
Loblolly pine (*Pinus taeda*)	217
Black locust (*Robinia pseudo-acacia*)	520
Rye (*Secale cereale*)	2500

After J.F. Ferry, *Fundamentals of Plant Physiology*, Macmillan Publishing Company, New York, 1959.

from which most of it is lost to the atmosphere by transpiration. Hence, the soil-plant-atmosphere pathway can be viewed as a continuum of water movement.

The pathway followed by water across the root depends in large part on the degree of differentiation of the various tissues that make up the root. In each of the tissues, water may follow one or more of three possible pathways: (1) **apoplastic** (via the cell walls), (2) **symplastic** (from protoplast to protoplast via plasmodesmata), and/or (3) **transcellular** (from cell to cell, passing from vacuole to vacuole) (Figure 28–13). For example, in a root without an exodermis, water can move apoplastically as far as the endodermis. At the endodermis, however, water is forced to traverse the plasma membranes and protoplasts of the tightly packed endodermal cells because of the presence of the water-impermeable Casparian strips in the radial and transverse walls of these cells (see page 478). In roots having an exodermis, by contrast, the Casparian strips in the radial and transverse walls of the compactly arranged exodermal cells preclude apoplastic movement of water across that cell layer. Water could follow either a symplastic or transcellular pathway across such roots. If, however, the outer tangential walls of the exodermal cells possess suberin lamellae, movement across that surface may be limited to the symplast, and the movement of water across such roots could be entirely symplastic.

28–12

(a) *Primary root of a radish (Raphanus sativus) seedling, showing root hairs. (b) Root hairs surrounded by soil particles; each particle is covered by a layer of adhered water.*

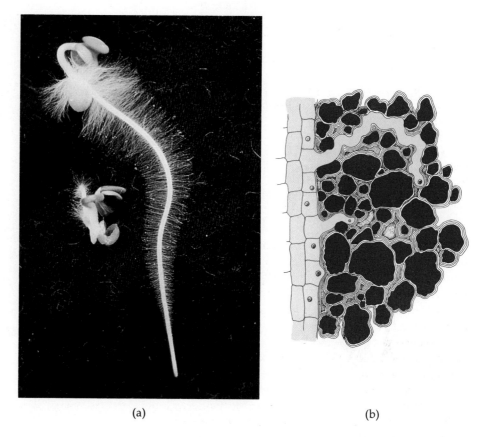

(a)   (b)

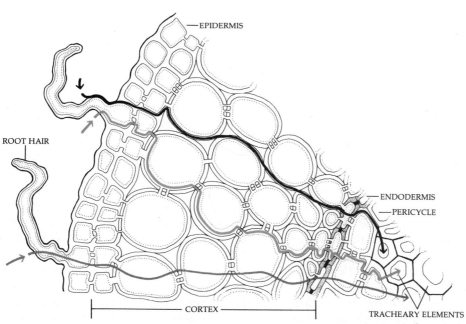

EPIDERMIS

ROOT HAIR

ENDODERMIS

PERICYCLE

CORTEX

TRACHEARY ELEMENTS

28–13

Possible pathways for the movement of water from the soil, across the epidermis and cortex, and into the tracheary elements, or water-conducting elements, of the root: apoplastic (black line), symplastic (blue line), and transcellular (red line). The root depicted here lacks an exodermis (a subepidermal layer of cells with Casparian strips). *Note that water following an apoplastic pathway is forced by the Casparian strips of the endodermal cells to cross the plasma membranes and protoplasts of the endodermal cells on its way to the xylem. Having crossed the plasma membrane on the inner surface of the endodermis, the water may* once again enter the apoplastic pathway *and make its way into the lumina of the tracheary elements. Inorganic ions are actively absorbed by the epidermal cells and then follow a symplastic pathway across the cortex and into parenchyma cells, from which they may be secreted into the tracheary elements.*

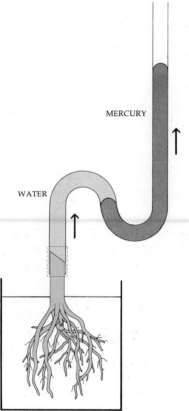

28–14
*Demonstration of root pressure in the cut
stump of a plant. Uptake of water by the
plant's roots causes the mercury to rise
in the column. Pressures of 0.3 to 0.5
megapascal have been demonstrated by
this method.*

28–15
*Guttation droplets at the margins of
a leaf of lady's mantle (Achemilla
vulgaris) also demonstrate the presence
of root pressure. These droplets are not
condensation from water vapor in the
surrounding air; rather, they are forced
out of the leaf through openings in
structures called hydathodes at the tips
along the margins of the leaf (see Figure
28–16).*

Root Pressure and Guttation

The driving force for movement of water across the
root is the difference in water potential between the soil
solution at the surface of the root and the fluid contents
of the xylem (the "xylem sap"). When transpiration is
very slow or absent, as it is at night, the gradient of
water potential is brought about by the secretion of ions
into the xylem. Because the vascular tissue of the root is
surrounded by the endodermis, ions do not tend to leak
back out of the xylem. Therefore, the water potential of
the xylem becomes more negative, and water moves
into the xylem by osmosis through the surrounding
cells. In this manner, a positive pressure, called **root
pressure,** is created, and it forces both water and dissolved ions up the xylem (Figure 28–14).

Dewlike droplets of water at the tips of grass and
other leaves in the early morning demonstrate the effects of root pressure (Figure 28–15). These droplets are
not dew—which is water that has condensed from the
air—but come from within the leaf by a process known
as **guttation** (from the Latin *gutta*, meaning "drop").
They exude through openings—commonly stomata that
lack the capacity to close and open—in special structures called hydathodes, which occur at the tips and
margins of leaves (Figure 28–16). The water of guttation is literally forced out of the leaves by root pressure.

Root pressure is least effective during the day, when
the movement of water through the plant is the fastest,
and the pressure never becomes high enough to force
water to the top of a tall tree. Moreover, many plants,
including conifers such as pine, do not develop root
pressure. Thus, root pressure can perhaps be regarded
as a by-product of the mechanism of pumping ions into
the xylem and as a subsidiary means of moving water
into the shoot under special conditions.

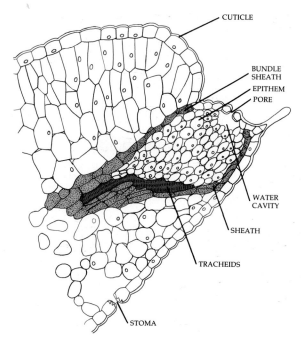

28–16
Longitudinal view of a hydathode of the leaf of Saxifraga lingulata. *The hydathode consists of the terminal tracheids of a bundle end, thin-walled parenchyma (the epithem) with numerous intercellular spaces, and epidermal pores. The tracheids are in direct contact with epithem. The epidermal pores commonly are stomata that lack the capacity to close and open.*

Passive Water Absorption

During periods of high transpiration rates, ions accumulated in the xylem of the root are swept away in the **transpiration stream,** and the amount of osmotic movement across the endodermis decreases. At such times, the roots become passive absorbing surfaces through which water is pulled by bulk flow generated in the transpiring shoots. Some investigators believe that almost all of the absorption of water by the roots of transpiring plants occurs in this passive manner.

During periods of rapid transpiration, water may be removed from around the root hairs so quickly that the local soil becomes depleted of water; water will then move from some distance away toward the root hairs through fine pores in the soil. In general, however, the roots come into contact with additional water by growing, although roots will not grow in dry soil. For example, under normal conditions, roots of apple trees grow

an average of about 3 to 9 millimeters a day; roots of prairie grasses may grow more than 13 millimeters a day; and the main roots of corn plants average 52 to 63 millimeters a day. The results of such rapid growth can be remarkable: a four-month-old rye *(Secale cereale)* plant has over 10,000 kilometers of roots and many billions of root hairs.

UPTAKE OF INORGANIC NUTRIENTS

The uptake, or absorption, of inorganic ions takes place through the epidermis of younger roots. Current evidence suggests that the major pathway followed by ions from the epidermis to the endodermis of the root is symplastic. Ion uptake by the symplastic route begins at the plasma membrane of the epidermal cells. The ions then move from the epidermal cell protoplasts to the first layer of cortical cells (possibly an exodermis) through the plasmodesmata in the epidermal-cortical cell walls (see Figure 28–13). Radial movement of ions continues in the cortical symplast—from protoplast to protoplast via plasmodesmata—through the endodermis and into the parenchyma cells of the vascular cylinder by diffusion, possibly aided by cytoplasmic streaming within the individual cells.

Nutrient uptake from soil by most seed plants is greatly enhanced by naturally occurring mycorrhizal fungi associated with the root systems of the plants (see page 238). Mycorrhizae are especially important in the absorption and transfer of phosphorus, but the increased absorption of zinc, manganese, and copper has also been demonstrated. These nutrients are relatively immobile in soil, and depletion zones for them quickly develop around the root and root hairs. The hyphal network of mycorrhizae extends several centimeters out from colonized roots, thus exploiting a large volume of soil more efficiently.

How the ions enter the mature vessels (or tracheids) of the xylem from the parenchyma cells of the vascular cylinder has been the subject of considerable debate. At one time it was suggested that the ions leak passively from the parenchyma cells into the vessels, but there is now substantial evidence that the ions are secreted either directly into the vessels from the parenchyma cells or indirectly via apoplastic channels that extend from the parenchyma cells to the vessels.

The mineral composition of root cells is far different from that of the medium in which the plant grows. For example, in one study, cells of pea *(Pisum sativum)* roots had a K^+ ion concentration 75 times greater than that of the nutrient solution. Another study showed that the vacuoles of rutabaga *(Brassica napus* var. *napobrassica)* cells contained 10,000 times more potassium than the external solution.

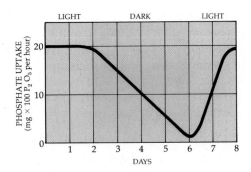

28–17
The rate of phosphate absorption (measured as P_2O_5; see page 613) by corn (Zea mays) plants fell to near zero after four days of continuous darkness. The rate began to rise again when the plants were reilluminated. These and other data indicate that mineral ion uptake in plants is a process that requires energy.

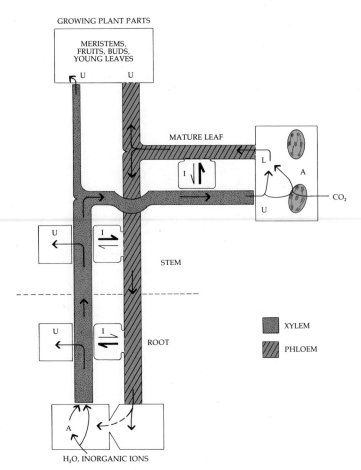

28–18
Diagram showing basic elements in the circulation of water, inorganic ions, and assimilates in the plant. Water and inorganic ions taken up by the root move upward in the xylem in the transpiration stream. Some move laterally into tissues of the root and stem, while others are transported to growing plant parts and mature leaves. In the leaves, substantial amounts of water and inorganic ions are transferred to the phloem and are exported with sucrose in the assimilate stream. The growing plant parts, which are relatively ineffective in acquiring water through transpiration, receive much of their nutrients and water via the phloem. Both water and solutes entering the roots in the phloem may be transferred to the xylem and recirculated in the transpiration stream. The symbol A designates sites specialized for the absorption and assimilation of raw materials from the environment. L and U designate sites of loading and unloading, respectively, and I designates principal points of interchange between the xylem and phloem.

Since substances do not diffuse against a concentration gradient, it is clear that minerals are absorbed by **active transport** (see page 70). Indeed, the uptake of minerals is known to be an energy-requiring process. For instance, if roots are deprived of oxygen, or poisoned so that respiration is curtailed, mineral uptake is drastically decreased. Also, if a plant is deprived of light, it will cease to absorb salts after its carbohydrate reserves are depleted (Figure 28–17) and will finally release them back into the soil solution. Hence, ion transport from the soil to the vessels of the xylem requires two active, carrier-mediated membrane events: (1) uptake at the plasma membrane of the epidermal cells and (2) secretion into the vessels at the plasma membrane of the parenchyma cells bordering the vessels.

TRANSPORT OF INORGANIC NUTRIENTS

Once secreted into the xylem vessels (or tracheids), the inorganic ions are rapidly transported upward and throughout the plant in the transpiration stream. Some ions move laterally from the xylem into surrounding tissues of the roots and stems, while others are transported into the leaves (Figure 28–18).

Much less is known about the pathways followed by the ions in leaves than about those in roots. Within the leaf, the ions are transported along with the water in the leaf apoplast, that is, in the cell walls. Some ions may remain in the transpiration stream and reach the main regions of water loss—the stomata and other epidermal cells. Most eventually enter the protoplasts of the leaf cells, probably by carrier-mediated transport mechanisms similar to those in roots. The ions may then move symplastically to other parts of the leaf, including the phloem. Inorganic ions may also be absorbed in small amounts through the leaves; consequently, fertilization by direct application of micronutrients to the foliage has become standard agricultural practice for some crop plants.

Substantial amounts of the inorganic ions that are imported into the leaves through the xylem are exchanged with the phloem of the leaf veins and exported from the leaf together with sucrose in the assimilate, or translocation, stream (Figure 28–18; see also the discussion of translocation that follows). For instance, in the annual white lupine (*Lupinus albus*), transport in the phloem accounted for more than 80 percent of the fruit's vascular intake of nitrogen and sulfur, and 70 to 80 percent of its phosphorus, potassium, magnesium, and zinc intake. The uptake of such inorganic ions by developing fruits is undoubtedly coupled to the flow of sucrose in the phloem.

Recycling may occur in the plant as nutrients reaching the roots in the descending assimilate stream of the phloem are transferred to the ascending transpiration stream of the xylem (Figure 28–18). Only those ions that can move in the phloem—*phloem-mobile ions*—can be exported from the leaves to any great extent. For example, K^+, Cl^-, and HPO_4^{2-} are readily exported from leaves, whereas Ca^{2+} is not. Solutes such as calcium are said to be *phloem-immobile*.

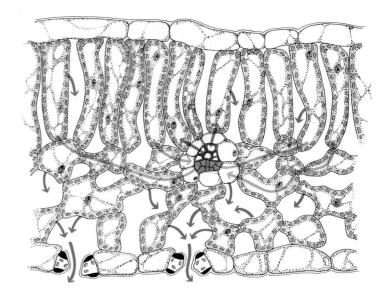

28–19
Diagram of leaf showing the pathways followed by water molecules of the transpiration stream as they move from the xylem of a minor vein to the mesophyll cells, evaporate from the surface of the mesophyll cell walls, and then diffuse out of the leaf through an open stoma (blue arrows).

Also shown are the pathways followed by sugar molecules manufactured during photosynthesis as they move from mesophyll cells to the phloem of the same vein and enter the assimilate stream. The sugar molecules manufactured in the palisade parenchyma cells are believed to move to the spongy parenchyma cells and then laterally to the phloem via the spongy cells (gold arrows).

Translocation: The Movement of Substances through the Phloem

As discussed in Chapters 21 through 23, the xylem and phloem together form a continuous vascular system that penetrates practically every part of the plant. Whereas water and inorganic solutes ascend the plant in the transpiration stream of the xylem, the sugars manufactured during photosynthesis move out of the leaf in the **assimilate stream** of the phloem (Figure 28–19) to sites where they are utilized, such as growing shoot and root tips, and to sites of storage, such as fruits, seeds, and the storage parenchyma of stems and roots (see Figure 28–18).

Assimilate movement is said to follow a source-to-sink pattern. The principal **sources** of assimilate solutes are the photosynthesizing leaves, but storage tissues may also serve as important sources. All plant parts un-

able to meet their own nutritional needs may act as **sinks**, that is, importers of assimilates. Thus, storage tissues act as sinks when they are importing assimilates and as sources when they are exporting assimilates.

Source-sink relations may be relatively simple and direct, as in some young seedlings, where cotyledons containing reserve food often represent the major source and growing roots represent the major sink. In older plants, the upper, most recently formed mature leaves commonly export assimilates primarily toward the shoot tip, the lower leaves export assimilates primarily to the roots, and those in between export assimi-

(a) (b)

28–20
Diagrams of a plant in (a) the vegetative stage and (b) the fruiting stage. The arrows indicate the direction of assimilate transport in each stage.

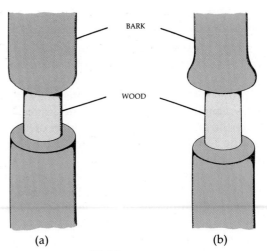

(a) (b)

28–21
As early as the seventeenth century, Marcello Malpighi of Italy noticed that, when a ring of bark was removed from a stem (a), the tissues above the ring became swollen (b). He correctly interpreted this phenomenon as new growth of wood and bark tissues stimulated by an accumulation of food moving down from the leaves and intercepted at the ring. Malpighi studied the effect of ringing at different times of the year and found that no swelling occurred during the winter months.

lates in both directions (Figure 28–20a). This pattern of assimilate distribution is markedly altered during the change from vegetative to reproductive growth. Developing fruits are highly competitive sinks that monopolize assimilates from the nearest leaves and frequently from distant ones as well, often causing a marked decline of vegetative growth (Figure 28–20b).

EVIDENCE OF SUGAR TRANSPORT IN PHLOEM

Early evidence supporting the role of the phloem in assimilate transport came from observations of trees from which a complete ring of bark had been removed. As noted in Chapter 24, the bark in older stems is composed largely of phloem. When a photosynthesizing tree is ringed, or "girdled," in this manner, the bark above the ring becomes swollen, indicating the formation of new wood and bark tissues stimulated by an accumulation of assimilates moving downward in the phloem from the photosynthesizing leaves (Figure 28–21).

Much convincing evidence of the role of the phloem in assimilate transport has been obtained with radioactive tracers. Experiments with radioactive assimilates (such as ^{14}C-labeled sucrose) have not only confirmed the movement of such substances in the phloem but have also shown conclusively that sugars are transported in the sieve tubes (see the essay on page 633).

APHIDS IN PHLOEM RESEARCH

Much valuable information on movement of substances in the phloem has come from studies utilizing aphids—small insects that suck the juices of plants. Most species of aphids are phloem-feeders. These aphids insert their modified mouth parts, or stylets, into a stem or leaf, extending them until the tips of the stylets puncture a conducting sieve tube (Figure 28–22). The turgor pressure of the sieve tube then forces the sieve-tube sap through the aphid's digestive tract and out through its posterior end as droplets of "honeydew." If feeding aphids are anesthetized (to prevent them from withdrawing their stylets from the sieve

(a)

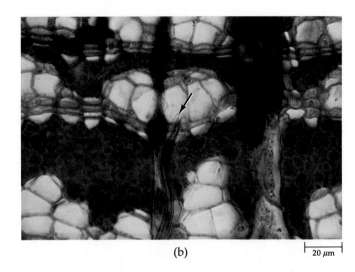

(b)

20 μm

28–22
(a) *Aphid* (Longistigma caryae) *feeding on a basswood* (Tilia americana) *stem. A droplet of "honeydew" can be seen* *emerging from the aphid.* (b) *A photomicrograph showing part of the modified mouth parts (stylets) of the aphid in a* *sieve tube of the secondary phloem of the basswood stem. An arrow points to the tips of the stylets.*

tube) and severed from their stylets, exudation of sieve-tube sap often continues from the cut stylets for many hours. The exudate can be collected with a micropipette. Analyses of exudates obtained in this manner reveal that sieve-tube sap contains 10 to 25 percent dry matter, 90 percent or more of which is sugar— *mainly sucrose* in most plants. Low concentrations (less than 1 percent) of amino acids and other nitrogen-containing substances are also present.

Data obtained from studies utilizing aphids and radioactive tracers indicate that the rates of longitudinal movement of assimilates in the phloem are remarkably fast. For example, sieve-tube sap moved at a rate of about 100 centimeters per hour at the sites of the stylet tips.

THE MECHANISM OF PHLOEM TRANSPORT: PRESSURE FLOW

Several mechanisms have been proposed to explain assimilate transport in the sieve tubes of the phloem. Probably the earliest explanation was that of diffusion, followed by that of cytoplasmic streaming. Normal diffusion and cytoplasmic streaming of the kind found in plant cells were largely abandoned as possible translocation mechanisms when it became known that the velocities of assimilate transport (typically 50 to 100 centimeters per hour) are far too great for either of these phenomena to account for long-distance transport via sieve tubes.

Alternative hypotheses have been advanced to explain the mechanism of phloem transport, but only one, the pressure-flow hypothesis, satisfactorily accounts for practically all of the data obtained in experimental and structural studies on phloem.

Originally proposed in 1927 by the German plant physiologist Ernst Münch, and since modified, the pressure-flow hypothesis is the simplest and the most widely accepted explanation for long-distance assimilate transport in sieve tubes. It is the simplest explanation because it depends only on osmosis as the driving force for assimilate transport.

Briefly stated, the **pressure-flow hypothesis** asserts that assimilates are transported from sources to sinks along a gradient of turgor pressure developed osmotically. For example, sucrose produced by photosynthesis in a leaf is actively secreted into the sieve tubes of the minor veins (Figure 28–23). This active process, called **phloem loading**, decreases the water potential in the sieve tube and causes water entering the leaf in the transpiration stream to move into the sieve tube by osmosis. With the movement of water into the sieve tube at this source, the sucrose is carried passively by the water to a sink, such as growing tissue or a storage root, where the sucrose is removed *(unloaded)* from the sieve tube. The removal of sucrose results in an increased water potential in the sieve tube at the sink and subsequent movement of water out of the sieve tube there. The sucrose may be either utilized in growth or respiration or stored at the sink, but most of the water returns to the xylem and is recirculated in the transpiration stream.

28–23
Diagram of osmotically generated pressure-flow mechanism. Yellow dots represent sugar molecules. Sugar is actively loaded into the sieve tube at the source. With the increased concentration of sugar, the water potential is decreased, and water from the xylem enters the sieve tube by osmosis. Sugar is removed (unloaded) at the sink, and the sugar concentration falls; as a result, the water potential is increased, and water leaves the sieve tube. With the movement of water into the sieve tube at the source and out of it at the sink, the sugar molecules are carried passively by the water along the concentration gradient between source and sink. Note that the sieve tube between source and sink is bounded by a differentially permeable membrane, the plasma membrane. Consequently, water enters and leaves the sieve tube not only at the source and sink, but all along the pathway. Evidence indicates that few if any of the original water molecules entering the sieve tube at the source end up in the sink, because they are exchanged with other water molecules that enter the sieve tube from the phloem apoplast along the pathway.

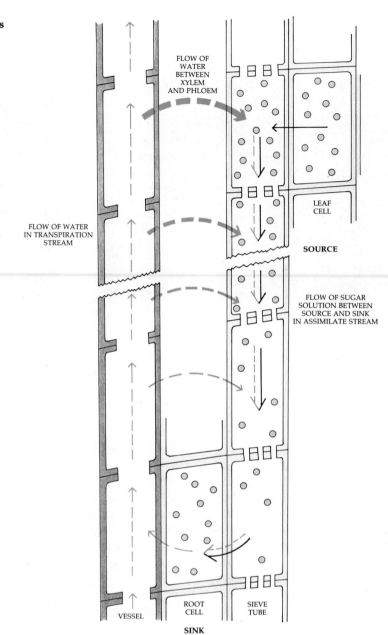

Note that the pressure-flow hypothesis casts the sieve tubes in a passive role in the movement of the sugar solution through them. Active transport is also involved in the pressure-flow mechanism; however, active transport is not directly involved with the long-distance transport through the sieve tubes but rather with loading and possibly unloading of sugars and other substances into and out of the sieve tubes at the sources and sinks. Considerable evidence indicates that the driving force for sucrose accumulation (phloem loading) at the source is provided by a proton pump that is energized by ATP and mediated by an ATPase at the plasma membrane, and involves a sucrose-proton cotransport (symport) system (see page 70). The metabolic energy required for the loading and unloading may be expended entirely by companion cells or parenchyma cells bordering the sieve tubes, rather than by the sieve tubes. Until recently, it was assumed that loading occurred across the plasma membrane of the companion cell, which then transferred the sugar to its associated sieve tube via the many plasmodesmatal connections in their common wall. It now appears, however, that some sieve tubes are capable of loading themselves, the site of active transport being their own plasma membranes. Whatever the case, the mature sieve tube probably is dependent upon its companion cell for much or most of its energy needs. In addition, the companion cells may maintain the enucleate (nucleus lacking) sieve-tube members through the transfer of information molecules.

Phloem loading is a selective process. As mentioned previously, sucrose is by far the most common sugar transported; in addition, all of the sugars found in sieve-tube sap are nonreducing sugars. Certain amino acids and ions also are selectively loaded into the phloem.

Radioactive Tracers and Autoradiography in Plant Research

Radioactive tracers can be used in a number of ways to study the synthesis, transport, and use of materials within plants. Initially, in any tracer study, the radioactive isotope must be incorporated into the plant. Radioactive carbon, for example, will be taken up by a plant if its leaves are exposed to ^{14}C-labeled carbon dioxide, or radioactive phosphorus will be taken up by a plant if the roots are exposed to a solution containing ^{32}P-labeled phosphate ions.

The length of time the plant must be exposed to the radioactive material is determined by the information that the investigators hope to obtain. For example, in studies designed to determine the time required for carbon dioxide to be incorporated into the various products of photosynthesis, a sequence of exposure times would be used. In studies focusing on the location of a particular product formed by the metabolic processes that involve the radioactive substance, the length of exposure would depend on the time

required for the chemical reactions under study to occur.

In whole-plant autoradiography, the plant is quick-frozen and freeze-dried after exposure to the radioactive substance. The plant is then flattened and pressed against a sheet of X-ray film. Radiation given off by the isotope exposes the film adjacent to the portions of the plant in which the tracer is located. By comparing the flattened plant with the developed film, investigators can determine the location of the radioactive substance within the plant.

In tissue autoradiography (microautoradiography), the freeze-dried plant tissues are embedded in paraffin, resin, or a similar material. Next, they are sliced into very thin sections, which are mounted on microscope slides. The tissue sections are then placed in contact with a photographic emulsion or film. Comparison, under a microscope, of the developed film and the underlying tissue section reveals the exact location of the radioactive substance in the plant tissues.

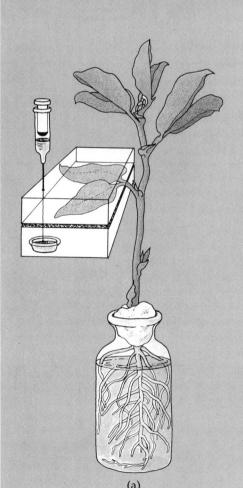

(a)

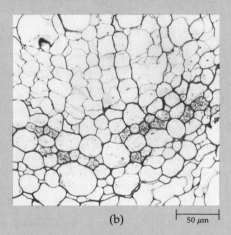

(b)   50 μm

(a) *Two leaflets of a broad bean plant* (Vicia faba) *were enclosed in a Plexiglas container and exposed to* $^{14}CO_2$ *and light for 35 minutes. During that time,* $^{14}CO_2$ *was incorporated into sugars, which were then transported to other parts of the plant. A cross section (b) and a longitudinal section (c) from the stem were placed in contact with autoradiographic film for 32 days. When the film was developed and compared with the underlying tissue sections, it was apparent that the radioactivity (visible as dark grains on the film) was confined almost entirely to the sieve tubes.*

(c)   50 μm

Summary

In plants, nearly 99 percent of the water taken in by the roots is given off into the air as water vapor. This process is called transpiration. Most of the water transpired by vascular plants is lost through the stomata in the leaves.

The rate of transpiration is affected by such factors as carbon dioxide concentration in the intercellular spaces (and, conversely, exposure of the leaf to atmospheric carbon dioxide), light, temperature, atmospheric humidity, air currents, and availability of soil water. Most of these factors have an effect on responses of stomata. Stomatal opening and closing are controlled by changes in turgor of the guard cells; these turgor changes are closely correlated with changes in the potassium ion level within the guard cells. Abscisic acid and both blue and red light also play roles in stomatal movement. The stomata open when the guard cells become turgid and close when they become flaccid.

Uptake of water takes place largely through the root hairs, which provide an enormous surface area for water uptake. In some plants, the uptake of water from the soil results in the buildup of positive pressure, or root pressure. This osmotic uptake depends on transport of inorganic ions from the soil into the xylem by the living cells of the root and can result in guttation, a process in which liquid water is forced out through special structures (hydathodes) in the tips or margins of the leaves. The pathway followed by water across roots may be apoplastic, symplastic, or transcellular; however, apoplastic movement is blocked at the endodermis by the Casparian strips. The water must pass through the plasma membrane and protoplast of the endodermal cells on its way to the xylem.

From the roots, water moves to the leaves through the xylem. The current and widely accepted theory of water movement to the top of tall plants through the xylem is the cohesion-tension theory. Water within the vessels is under tension because the water molecules cling together (cohere) in continuous columns pulled by evaporation from above. Water has sufficient tensile strength to withstand such tension when contained in small-diameter conduits. Other supporting evidence includes observations that water in the xylem is under tension, that water movement in trees begins in the topmost branches, and that the trunk of a tree shrinks slightly at the beginning of water movement.

Inorganic nutrients become available to plants in soil solution in the form of ions. Plants employ metabolic energy to concentrate the ions they require. Ion transport from the soil to the vessels of the xylem requires two active, carrier-mediated events: uptake at the plasma membrane of epidermal cells and secretion into the vessels at the plasma membrane of parenchyma cells bordering the vessels. Inorganic ions follow a mostly symplastic pathway from the epidermis to the xylem. Nutrient uptake from the soil by most seed plants is greatly enhanced by mycorrhizal fungi.

Research on movement of substances in the phloem has been greatly aided by the use of aphids and radioactive tracers. Sieve-tube sap contains sugar (mainly sucrose), small quantities of nitrogenous substances, and phloem-mobile ions. Rates of longitudinal movement of substances in the phloem greatly exceed the normal rate of diffusion of sucrose in water—the rates typically range from 50 to 100 centimeters per hour.

According to the pressure-flow hypothesis, assimilates move from sources to sinks along turgor-pressure gradients developed osmotically. Sugars are actively loaded into the sieve tube at a source and unloaded from it at a sink, resulting in mass flow of solution in the sieve tube.

Suggestions for Further Reading

Allen, Michael F.: *The Ecology of Mycorrhizae*, Cambridge University Press, New York, 1991.

A valuable review of a fascinating topic.

Baker, David A., and John L. Hall (eds.): *Solute Transport in Plant Cells and Tissues*, Longman Scientific & Technical, Harlow, Essex, England, and John Wiley & Sons, Inc., New York, 1988.

This multiauthored book is intended to introduce the reader to the processes involved in the transport of solutes by plant cells and tissues, fungi, and algae as these processes are currently perceived. Among the topics covered are membrane transport, movements of ions across roots, phloem transport, and mechanisms of ion accumulation in guard cells.

Baker, David A., and John L. Hall (eds.): *Transport of Photoassimilates*, Longman Scientific & Technical, Harlow, Essex, England, and John Wiley & Sons, Inc., New York, 1989.

A multiauthored book aimed at presenting to advanced undergraduate and postgraduate students the major conceptual developments in the study of the transport of photoassimilates in plants, beginning with the transport of photoassimilates within photosynthetic cells and ending with the compartmentation and partitioning of photoassimilates within the plant.

Bem, Robyn: *Everyone's Guide to Home Composting*, Van Nostrand Reinhold Company, New York, 1978.*

A comprehensive, easy-to-read manual.

Bonnemain, Jean L., Serge Delrot, William J. Lucas, and Jack Dainty: *Recent Advances in Phloem Transport and Assimilate Compartmentation*, Quest Editions, Presses Académiques, Nantes Cedex, France, 1991.

The proceedings of the 4th International Conference on Phloem Transport, considering structural, physiological, biochemical, and molecular aspects.

Brill, Winston J.: "Nitrogen Fixation: Basic to Applied," *American Scientist* 67:458–466, 1979.

An excellent review of a complex subject presented in an easy-to-understand and readable style.

Campbell, Stu: *The Gardener's Guide to Composting,* revised ed., Storey Publishing, Pownal, Vt., 1990.*

A concise presentation on all you need to know about composting.

Campbell, Stu: *The Mulch Book: A Complete Guide for Gardeners,* Storey Publishing, Pownal, Vt., 1991.*

A concise presentation on all you need to know about mulching.

Cronshaw, James, William J. Lucas, and Robert T. Giaquinta (eds.): *Phloem Transport,* Alan R. Liss, Inc., New York, 1986.

The proceedings of the 3rd International Conference on Phloem Transport, considering membrane transport of sugars, phloem unloading and sink metabolism, production and allocation of photoassimilates, and the relationship between photoassimilate partitioning and crop yield.

Epstein, Emanuel: *Mineral Nutrition of Plants: Principles and Perspectives,* John Wiley & Sons, Inc., New York, 1972.

A thorough coverage of the whole field of mineral nutrition, well illustrated and written by one of the leading students of the subject.

Gregory, P.J., J.V. Lake, and D.A. Rose (eds.): *Root Development and Function,* Society for Experimental Biology, Seminar Series 30, Press Syndicate of the University of Cambridge, Cambridge, England, 1987.

A volume based on review papers considering the effects of environmental factors on root development and function at four levels of organization: cell, tissue, single plant, and plant community.

Gresshoff, Peter M. (ed.): *Molecular Biology of Symbiotic Nitrogen Fixation,* CRC Press, Boca Raton, Fla., 1990.

A multiauthored book summarizing the knowledge of the day, which is undergoing rapid expansion. Although aimed at the research worker in the field of nitrogen fixation, it is also intended to provide the backbone for a graduate course on the subject.

Gresshoff, Peter M., L. Evans Roth, Gary Stacey, and William E. Newton: *Nitrogen Fixation: Achievements and Objectives,* Chapman and Hall, New York, 1990.

A collection of outstanding papers on all aspects of this very active field of research, from the 8th International Congress on Nitrogen Fixation.

Harley, J.L.: *The Biology of Mycorrhiza,* Leonard Hill, London, 1969.

A book dealing with all aspects of mycorrhizae.

Kramer, Paul J.: *Water Relations of Plants,* Academic Press, San Diego, Calif., 1983.

A comprehensive text on all important aspects of plant–water relations.

Kramer, Paul J., and Theodore T. Kozlowski: *Physiology of Woody Plants,* Academic Press, New York, 1979.

A well-written, comprehensive review of the entire subject.

Lewis, Owen A.M.: *Plants and Nitrogen,* Studies in Biology, No. 166, Edward Arnold Publishers, London, 1986.*

A concise introduction to the association that exists between plants and nitrogen. Ecological, physiological, and biochemical aspects are combined to provide the reader with an overall view of nitrogen in the biosphere.

Lüttge, Ulrich, and Noe Higinbotham: *Transport in Plants,* Springer-Verlag, New York, 1979.

A comprehensive review of the various processes by which inorganic and organic substances are transported in plants, from the level of cellular organelles to long-distance movements in trees.

Marks, C.G., and T.T. Kozlowski (eds.): *Ectomycorrhizae: Their Ecology and Physiology,* Academic Press, New York, 1973.

A collection of articles by a number of scientists working on mycorrhizae.

Marschner, H.: *Mineral Nutrition of Higher Plants,* Academic Press, New York, 1986.

A comprehensive account of the subject.

Martin, E. Stephen, Maria E. Donkin, and R. Andrew Stevens: *Stomata,* Studies in Biology, No. 155, Edward Arnold Publishers, London, 1983.*

A concise introduction to many aspects of the mechanism of stomatal movements.

Mengel, K., and E.A. Kirkby: *Principles of Plant Nutrition,* 3d ed., International Potash Institute, Worblaufen-Bern, Switzerland, 1982.

A textbook for students interested in plant science and crop production; integrates soil science, plant physiology, and biochemistry.

Salisbury, Frank B., and Cleon W. Ross (eds.): *Plant Physiology,* 4th ed., Wadsworth Publishing Co, Inc., Belmont, Calif., 1991.

A detailed and useful review of the entire subject.

Steward, F.C. (ed.): *Plant Physiology, A Treatise, Vol. IX: Water and Solutes in Plants,* F.C. Steward, James F. Sutcliffe, and John E. Dale (coeds.), Academic Press, New York, 1986.

A multiauthored book dealing with many aspects of water and solute transport in plants, including transpiration and water balance, physiology of stomata, and phloem transport.

Zimmermann, Martin H.: *Xylem Structure and the Ascent of Sap,* Springer-Verlag, New York, 1983.

A delightfully written "idea" book on xylem structure and function by one who contributed a great deal to our understanding of functional xylem anatomy and sap movement in plants.

Zimmermann, Martin H., and Claud L. Brown: *Trees: Structure and Function,* Springer-Verlag, New York, 1975.

This book is devoted to those aspects of tree physiology that are peculiar to tall woody plants. The emphasis is on function, and the book includes material not found in general physiology texts.

* Available in paperback.

SECTION 6 · Ecology and the Human Prospect

Hairy Solomon's seal (Polygonatum pubescens), *a common herb of moist woods, thickets, and roadsides from Canada to South Carolina. With its leaves arranged along an arching stem,* Solomon's seal is well-adapted for light capture in shady conditions. On steep slopes, herbs with arching stems have an apparent advantage over herbs with erect stems (see page 676).

CHAPTER 29

The Dynamics of Communities and Ecosystems

Ecology is the study of the interactions of organisms with one another and with all other components of their environment. As a science, ecology attempts to explain why particular plants and animals can be found living together in one area and not in others; why there are so many organisms of one sort and so few of another; what changes one might expect the interactions among them to produce in a particular area; and how ecosystems function, with particular reference to the flow of energy, the use of organic compounds, and the cycling of chemical elements.

A few definitions are necessary before we proceed. At the individual level, ecologists ask how the individual organism functions in and interacts with its environment. Individuals exist in **populations,** groups of individuals, usually of one species, that occur together at a given place. At the population level, ecologists ask what determined the abundance and distribution of individuals of a particular species.

Communities, in turn, consist of populations; they include the groups of plants, animals, and other organisms that live in a particular area. Although we can speak of an herb community or a community of vertebrate animals, the word "community" when unmodified is generally taken to mean *all* of the organisms that occur together at a particular place. An **ecosystem** includes not only these organisms but also their physical environment and their interactions, both with other organisms and with the environment. In this chapter, the properties of populations and individuals will be discussed in the context of interactions between organisms in communities and ecosystems. Once those interactions have been treated, we shall move on to consider properties of communities and ecosystems as a whole, centering our discussion on nutrient cycling, trophic levels, and the development of communities and ecosystems through time.

Biomes are large terrestrial complexes of communities of living organisms characterized by distinctive growth forms that have evolved in relation to a particular kind of climate. Examples include forests, grasslands, and deserts. The principal biomes are discussed in Chapter 30.

Interactions between Organisms

From the preceding discussion, it should be evident that interaction is central to the field of ecology. No organism living in a community—whether that community be a patch of woodland, a pasture, a pond, or a coral reef—exists in isolation. Each organism participates in a number of interactions, both with other organisms and with the nonliving components of the environment. We shall organize our discussion around three major kinds of interactions between species of organisms: mutualism, competition, and plant-herbivore (and plant-pathogen) interactions.

MUTUALISM

Mutualism is a biological interaction in which the growth, survival, and/or reproduction of both interacting species are enhanced. In many mutualisms, neither partner can survive without the other, particularly when competition from other plants and predation are taken into account. We have already come across several examples of mutualism in earlier chapters. Lichens are an outstanding example (Chapter 12). Others are the relationship between legumes and the nitrogen-fixing bacteria that live in nodules on the legume roots (see Chapter 27, pages 602–607) and some of the closely linked pollination relationships discussed in Chapter 19, such as that between the yucca moth and the yucca plant. Two additional examples of mutualism will now be discussed in greater detail.

Mycorrhizae

One of the most interesting and ecologically significant examples of mutualism concerns the interaction between fungi and plants. As discussed in Chapters 12 and 27, the roots of most vascular plants are associated with fungi, forming compound structures known as mycorrhizae. These fungi play a vital role in the absorption of phosphorus and other essential nutrients; without the fungi, the normal growth of the plants would be impossible. Mycorrhizae appear to have played a crucial role in the establishment of early plants on land.

In many vascular plants, nonmycorrhizal individuals are rarely encountered under natural conditions, even though growth may be possible without fungi if nutrients are abundant. Most vascular plants are dual organisms in the same sense that lichens are dual organisms, although the relationship is not obvious above ground. As University of Wisconsin soil scientist S.A. Wilde stated, "A tree removed from the soil is only a part of the whole plant, a part surgically separated from its . . . absorptive and digestive organ."

The fungi that form mycorrhizal associations in most plants are zygomycetes; as discussed in Chapter 12, the associations are called endomycorrhizae, and they are characteristic of a majority of species of herbs, shrubs, and trees. In some groups of conifers and dicots —mainly trees—the associations are mostly with basidiomycetes but also with certain ascomycetes; such associations are called ectomycorrhizae. Some of them are highly specific, with one species of fungus forming ectomycorrhizal associations only with a particular species or group of related species of vascular plants. For example, the pore fungus *Boletus elegans* is known to associate only with larch (*Larix*), a conifer. Other fungi, such as *Cenococcum geophilum*, have been discovered living in ectomycorrhizal association with forest trees of more than a dozen genera. Ectomycorrhizae are particularly characteristic of relatively pure stands of trees growing at high latitudes in the Northern Hemisphere or at high elevations, two places where slow decomposition rates may make nutrients particularly difficult to obtain.

Ants and Acacias

The most intricate examples of mutualism occur in the tropics, where many more kinds of organisms are found than in temperate regions. For example, trees and shrubs of the genus *Acacia* occur widely in tropical and subtropical regions. The interaction between certain species of *Acacia* in the lowlands of Mexico and Central America and the ants that inhabit their thorns (actually, stipular "thorns," or spines) provides a remarkable example of the complexity that plant-animal interactions can entail. The particular relationship between the bull's-horn acacias and the inhabitants of their thorns, ants of the genus *Pseudomyrmex*, is a good example (Figure 29–1).

The bull's-horn acacias have a pair of greatly swollen thorns more than 2 centimeters long at the base of each leaf. Nectaries occur on the petioles, and small nutritive organs known as Beltian bodies are located at the tip of each leaflet. The ants live inside the hollow thorns and obtain sugars from the nectaries and fats and proteins by eating the Beltian bodies. The acacias grow extremely rapidly and are particularly common in disturbed areas, where the competition between rapidly growing colonizing plants is often intense.

Thomas Belt first described the relationship between *Pseudomyrmex* and the bull's-horn acacias in his book,

(a)

(b)

(c)

(d)

29–1

Ants and acacias. (a) A queen ant cuts an entrance into a thorn on a seedling bull's-horn acacia (Acacia cornigera). She will hollow out the thorn and raise her first brood inside it. (b) The tip of an acacia leaf. The orange structures at the tips of the leaflets are Beltian bodies. They are a source of food for the ants. (c) A worker ant (Pseudomyrmex ferruginea) *drinking from a nectary at the base of a petiole of one of the compound leaves. (d) Warriors in a battle for* possession of an acacia. Obtaining all of their food from the acacia, the ants, in turn, kill most other insects that attempt to feed on it and girdle all plants that come into contact with it.

The Naturalist in Nicaragua (1874). Following his observations, there was a prolonged controversy about whether the presence of the ants actually benefited the acacia plants. This question was finally and definitively solved in 1964 by Daniel Janzen. Janzen found that the worker ants, which swarm over the surface of the plant, bite and sting animals of all sizes that contact the plant, thus protecting it from the activities of herbivores and ensuring a home for themselves. Moreover, whenever the branches of another plant touch an inhabited acacia tree, the ants girdle the other plant's bark, destroying the invading branches and producing a tunnel to the light through the rapidly growing tropical vegetation within which the acacia grows.

When Janzen removed the ants from a plant by poisoning them or clipping off the portions of the plant that contained ants or when an acacia was naturally unoccupied by ants (a situation that occurs rarely), the plant's growth was extremely slow, and the plant usually died after a few months as a result of insect damage and shading by other plants. Plants inhabited by ants grew very rapidly, soon reaching 6 meters or more in height and overtopping the other second-growth vegetation. These ants make their nests only in these particular acacias and are completely dependent on the acacias' nectaries and Beltian bodies for food. Thus the ant-acacia system is as much a dual biological entity as, for example, a lichen. One element usually cannot survive without the other in the community in which it occurs.

COMPETITION

Competition is defined as interaction between members of the same population or of two or more populations in order to obtain a mutually required resource available in limited supply. In competition, one species interferes with another species enough to keep the second species from gaining access to the resource.

Unlike animals, which ingest food, green plants are dependent on a single process, photosynthesis, to obtain their energy. Competition in plants is therefore manifested largely in terms of a "struggle for light." Those plants whose photosynthetic rate, growth form, and pattern of allocation of energy to leaves, roots, and stems result in the highest rate of growth in a particular environment will often have an advantage in this struggle. Those plants with a lower growth rate must either be able to photosynthesize at low light intensities, have a high growth rate relative to competitors elsewhere, or die. The evolution of plants with C_4 and CAM photosynthesis (pages 116 to 117) can be seen in this light: the physiological properties of such species enable them to dominate climatic areas where their C_3 ancestors would have either perished or else been competitively excluded by more rapidly growing C_4 or CAM species.

Differences in height, leaf arrangement, crown shape, and allocation to roots versus leaves affect a plant's growth rate relative to potential competitors.

(a)   (b)

29–2

Lullington Heath National Nature
Reserve, East Sussex, England, an area
of chalk grassland, before (a) *and after*
(b) *the elimination of rabbits by the viral*

disease myxomatosis. The first photograph
was taken in 1954, the second in 1978.
In an effort to restore the diversity of
herbaceous plants, the authorities have

introduced grazing programs for sheep
and horses over much of the reserve,
which has an area of 62 hectares.

Different combinations of these traits are likely to maximize growth in different environments; no single combination can produce the best competitor in all environments. As a consequence, vegetation in different climatic areas is dominated by plants with different growth forms; within a given area (indeed, within a single community, such as a temperate forest or prairie), differences in growth form, photosynthetic physiology, and allocation may permit species to coexist by granting each a competitive advantage in different microenvironments.

The ways in which individual plants are able to enhance their overall growth and thus compete for light, water, and mineral nutrients largely determine their success in different habitats. An understanding of such factors is therefore the key to increasing yield in extreme environments, such as those that are relatively poor in nutrients or water, or are heavily shaded. These relationships are important in agriculture, therefore. Knowledge of these factors will also provide the information necessary to predict the performance of individual plant species and communities in a world that is rapidly changing, as by the global warming that is resulting from increased levels of CO_2 and certain other atmospheric gases, largely produced by human activities.

Competition between plant species can be visualized more easily under experimental conditions than in nature. From observations of the growth of organisms in simple environments, it has been deduced that two species with similar environmental requirements cannot coexist indefinitely in the same habitat. This is a simplified version of what Garrett Hardin has called **the principle of competitive exclusion.** If the environment is complex, as it often is in nature, various organisms may relate to it in different ways, in effect subdividing the environment and achieving competitive success in different microenvironments. These various organisms may then continue to coexist indefinitely.

For example, Engelmann spruce (*Picea engelmannii*) and subalpine fir (*Abies lasiocarpa*) coexist and form the dominant tree community in the subalpine zone of the central and northern Rocky Mountains. Careful field measurements suggest that these two tree species are maintained as codominants in this area because the greater longevity and size of the spruce are balanced by the faster growth in height and more flexible seedling-establishment requirements of the fir. Spruce seedlings are found primarily in forest gaps or associated with a canopy of firs, whereas fir seedlings are found more commonly in the forest. In other words, fir seedlings outcompete spruce in shaded understories (that is, they survive), whereas spruce seedlings outcompete fir in sunnier sites, by virtue of higher growth rates there and lower sensitivity to drought. Constant disturbance from storms, flooding, and avalanches, among other factors, and the relatively short lifespan of fir prevent it from taking over the entire landscape as succession proceeds. Patterns of this sort are frequent in different plant communities, but they are often not at all obvious on simple inspection. The different requirements of the species involved remove them from direct competition and allow them to coexist indefinitely.

If populations of coexisting species are kept at low levels, competitive exclusion may be avoided even though the species do, in principle, compete for the same limiting resources. Such a situation occurred in the chalk grasslands of England when the grasses were kept closely cropped by rabbits and many kinds of flowering plants were able to grow in this habitat. Diversity is enhanced in such situations when competitive dominant species—grasses and other tall herbs in this case—are attacked preferentially. This allowed many other kinds of shorter grasses and other herbs to coexist. However, the situation changed drastically earlier in this century, when a severe epidemic of myxomatosis, a disease caused by a virus, drastically reduced the population of rabbits. After this decline, the grass cover of the chalk soils became deeper and more dense, and many of the formerly abundant species of flowering plants became rare (Figure 29–2). Similar effects are often seen when comparing grazed and ungrazed pastures or grasslands, and they also arise in areas where natural disasters, such as hurricanes, are common. Thus, as we might expect, the diversity of species is greater along a wave-lashed coast, where disturbance is continuous, than in a more stable environment.

Clonal reproduction, which is important in many plants, sometimes makes it difficult to define the limits of individual plants in nature. This, in turn, may make it difficult to evaluate the competitive interactions of specific genotypes within a species. Genetically identical individuals may be widely dispersed but still occur in widely separated but similar environments within a single complex habitat. This is characteristic not only of rhizomatous perennials, such as many grasses and sedges, but also of dandelions and other plants in which the seeds are produced asexually and contain embryos that are genetically identical to their parents. It is characteristic also of those plants, such as white clover (*Trifolium repens*), in which an original plant simply grows apart into distinct individuals.

Depletion of a shared resource, such as light or water, is not the only mechanism by which plants may compete. In some competitive interactions, one (or both) of the competing organisms produces chemical substances that inhibit either the growth of members of its own species, resulting in increased spacing of like individuals, or the growth of other species. For example, the fungus *Penicillium chrysogenum*, which grows on organic substrates such as seeds, produces significant quantities of penicillin in nature. Penicillin inhibits the growth of gram-positive bacteria, which might otherwise compete directly with the fungus for the same nutrients. However, bacteria that produce penicillinases (enzymes that break down penicillin), such as *Bacillus cereus*, often replace the *Penicillium*.

Analogous relationships among plants are grouped under the general heading of **allelopathy**. An excellent example of this phenomenon is provided by the effects of black walnuts (*Juglans nigra*) on other plants, which are very sparse under the walnut trees. Tomatoes (*Lycopersicon esculentum*) and alfalfa (*Medicago sativa*) wilt when grown near black walnuts, and their seedlings die if their roots contact walnut roots. Similarly, white pine (*Pinus strobus*) and black locust (*Robinia pseodoacacia*) are often killed by black walnuts growing in their vicinity, especially in poorly drained soils, where the toxins leached from the walnuts apparently accumulate. The bare zones around sage (*Salvia leucophylla*) shrubs in California (Figure 29–3) may also occur in part because of allelopathic effects, although the birds and rodents that the shrubs shelter exert an important influence on the patterns observed (Figure 29–4).

Allelopathic effects are likewise being applied in agriculture. For example, a strip planted with sorghum will have two to four times fewer weeds the following

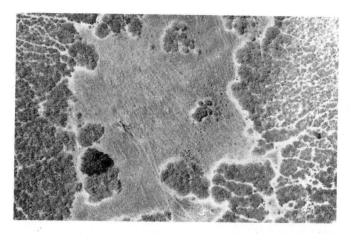

29–3
Purple sage (Salvia leucophylla) *shrubs produce terpenes that evaporate and spread through the air, ultimately reaching the soil, where they inhibit the growth of other plants. In this aerial view, the bare zone around the individual* Salvia *shrubs and colonies can be seen plainly. Immediately around the shrubs is a completely bare zone, and then a zone of inhibited grassland with a few stunted annual herbs.*

year than other strips. The sorghum plants evidently leave behind allelopathic compounds in the soil that depress the growth of weeds.

29–4
When mammals such as mice are excluded from the bare zone, annual plants grow luxuriantly right up to the edge of the shrubs in areas where they are normally absent. Such results, reported by Bruce Bartholomew of the California Academy of Sciences, indicate that the effect of these mammals would be sufficient to account for the bare zone, regardless of whether the plants are producing toxic chemicals.

PLANT-HERBIVORE AND PLANT-PATHOGEN INTERACTIONS

We have already considered one important area of plant-herbivore interactions involving flowering plants —the relationships that exist between flowers and their visitors, and between fruits and their dispersers (Chapter 19). These specialized interactions arose, during the course of evolution, from the more general relationships between plants and the animals that consume them. Relationships between plants and pathogenic organisms, especially fungi and bacteria, are similar in their effects.

The important role of plant-herbivore and plant-pathogen interactions in determining the structure of natural communities is not always evident. For example, vast areas of Australia were at one time covered with spiny clumps of prickly-pear cactus (Opuntia), a plant that was introduced from Latin America. Fertile lands became useless for grazing, and the economy of great stretches of the interior was severely threatened. Today, the cactus has been nearly eliminated by a cactus moth (Cactoblastis cactorum) discovered in South America and deliberately introduced into Australia to control the cactus. The larvae of this moth destroy the cactus plants by eating them. The moth, once abundant in Australia, can scarcely be found today, even by a careful inspection of the few remaining cactus clumps; yet there is no doubt that it continues to exert a controlling influence over the populations of this plant in Australia (Figure 29–5).

The effects of herbivores on plants are profound, both in the short term and in the long term. Herbivores control the reproductive potential of plants by destroying their photosynthetic surfaces, their food-storage organs, or their reproductive structures. As discussed in Chapter 19, these interactions have led, over the course of time, to the evolution by plants of a wide variety of chemical defenses—in the form of molecules commonly referred to as "secondary plant products." The ability of plants to produce toxic chemicals and to retain them in their tissues gives the plants a tremendous competitive advantage. Indeed, these chemicals are apparently the most important factors in controlling herbivorous insects in nature. This advantage is analogous to the advantage achieved by the production of thorns or tough, leathery leaves, which obviously protect plants from grazing. Scientists working to improve the resistance of crops to herbivores are focusing much of their effort on these chemicals.

In the sea, many seaweeds have evolved comparable defenses. These marine algae are consumed by many different kinds of herbivores, including fishes, sea urchins, mollusks, and other animals; nearly all of the biomass is ultimately consumed in some habitats. To escape these herbivores, some seaweeds grow in cracks and holes or in other habitats that the herbivores do not visit. Palatable seaweeds may gain protection by growing intermixed with chemically protected ones. Many seaweeds produce chemical defenses that render them distasteful to herbivores, whereas others (for instance, corralline red algae) may be too tough to consume. In other words, the array of defenses against herbivores that is employed by seaweeds is virtually as extensive as that which occurs among flowering plants on land.

29–5

(a) *Dense prickly-pear cactus* (Opuntia inermis) *growing in a mixed scrub forest in Queensland, Australia, in October 1926;* (b) *the same forest in October 1929, after the cacti were destroyed by the deliberately introduced South American moth* Cactoblastis cactorum. *First introduced in May 1925, the larvae of this moth destroyed the cacti on more than 120 million hectares of rangeland.*

(a)   (b)

Pesticides and Ecosystems

Approximately half a billion metric tons of pesticides and herbicides are produced annually for application to crops in the United States alone. Of this enormous total, it has been estimated that only approximately 1 percent actually reaches the target organisms. Most of the remainder either reaches soil, water, or nontarget organisms in the same ecosystem or spreads into neighboring ecosystems, where it may have important detrimental effects. This residue may result in the elimination of certain species, for example, and the loss of those species may affect the functioning of the ecosystem as a whole or lead indirectly to the elimination of other species. The abundance of important decomposers, such as earthworms and other soil organisms, may be greatly reduced by pesticides and herbicides, so that the ecosystem as a whole ceases to function normally. The intensity of such effects depends on the toxicity of the chemicals and on their persistence in the environment.

One of the problems associated with some chemical pollutants is that they tend to be concentrated as they pass up through food chains, reaching their highest concentrations in the top predators (see pages 646 to 648). For example, chlorinated hydrocarbons, such as DDT (now outlawed in the United States for use but not for manufacture and export, unfortunately, and not in most other industrialized countries), become concentrated in the tissues of predatory birds and cause the shells of their eggs to be abnormally thin. The shell of such an egg is likely to break before the chick is ready to hatch, causing its death. In addition, many strains of insects, bacteria, and fungi have evolved that are resistant to the pesticides designed to control them. Of the estimated 2000 species of major insect pests, about a quarter have already evolved strains that are resistant to one or more insecticides. Similarly, a number of species of weeds have evolved resistance to herbicides.

Other effects are less direct. For example, predator species that naturally control populations of pest species may be eliminated by pesticide poisoning, leading to outbreaks of the very pest organisms that the chemicals were employed to control in the first place. As with most human activities, the net effect of the application of pesticides and herbicides is to lessen the diversity of the ecosystems affected. Yet productive modern agriculture depends, to a large extent, on the application of these useful substances.

Because the effects of pesticides and herbicides are often so drastic, however, scientists are actively searching for less damaging methods of improving agricultural yields. These include selective breeding to produce crops that are resistant to pests (the methods of genetic engineering described in Chapter 25 will be especially useful in achieving this goal), increased research to develop pesticides and herbicides that are less toxic and less persistent than those currently used, and integrated pest-management systems, involving combinations of control measures, including the encouragement of predators and diseases of pest species as well as the judicious application of pesticides.

Plant-herbivore and plant-pathogen interactions may be quite complex. For example, pea plants (*Pisum sativum*) are largely protected from parasitic fungi by a substance called pisatin, which the plants produce. Many strains of the important parasitic fungus *Fusarium*, however, have enzymes called monooxygenases, which convert pisatin into a less toxic compound; such fungi then have the ability to attack peas. Humans also utilize monooxygenases to detoxify certain chemicals that would otherwise be harmful to the body. In such ways, "chemical warfare" between plants and their herbivores is continuously being waged.

The protective chemicals that plants produce are often not only distasteful, but may display still other features that deter herbivores. Chromenes, for example, can interfere with insect juvenile hormone (essential to an insect's life cycle) and thus can act as true insecticides. A Mexican sneezeweed (*Helenium* sp.) produces helanalin, which functions as a powerful insect repellent. Pyrethrum is another natural insecticide produced commercially from a species of *Chrysanthemum*. Even the waxy surfaces of leaves, which are difficult to digest, may be important in retarding attacks by insects and fungi.

When infected with fungi or bacteria, plants often defend themselves by producing natural antibiotics called **phytoalexins** (page 33). These are lipidlike compounds whose synthesis can also be stimulated by leaf damage. They appear to be produced in response to the presence of specific carbohydrate molecules, called **elicitors,** that are present in fungal and bacterial cell walls. The elicitors, which are released from the fungal or bacterial cell walls by enzymes present in the plants being attacked, diffuse through the plant cells more or less like hormones. Ultimately the elicitors bind to specific receptors on the plasma membranes of the plant cells, bringing about metabolic changes that result in the production of the phytoalexins. In principle, it

(a)

(b)

29–6

(a) *The monarch butterfly* (Danaus plexippus) *obtains cardiac glycosides from the plants of the milkweed family* (Asclepiadaceae), *upon which its larvae* (b) *feed. As a result, the monarch is unpalatable to birds and other vertebrates. It "advertises" this fact by bright orange-and-black adult coloration and by larvae that are conspicuously banded with white, yellow, and black. Even the eggs of the monarch, which are bright yellow and conspicuous, contain enough cardiac glycosides to be protected.*

should be possible to spray elicitors onto crops before the crops become infected and thereby protect them from fungal and bacterial pests. Such a process would be analogous to vaccination in humans and domestic animals. One possible problem, however, is that the energetic cost to the plant of producing large quantities of phytoalexins might lower the plant's ultimate yield more than would the fungal or bacterial infection. A natural advantage of the phytoalexins as defensive substances is that the plant does not need to expend the energy necessary to produce them unless it is actually attacked. Nevertheless, understanding phytoalexin production in plants is of considerable importance for crop protection, and several synthetic elicitors have already been produced and tested. Manipulating the genetic basis of resistance, now possible through the methods of genetic engineering (described in Chapter 25), also offers new possibilities for enhancing crop resistance that do not carry a high energy cost.

Just as some plants produce phytoalexins, others produce tannins and other phenolic compounds, and these compounds seem to play a similar role in nature. Tannins are generally static defenses, always present in the plant parts where they occur. In some instances, however, they may be marshalled by the plant when it is attacked. For example, when gypsy moths *(Lymantria dispar)* attack and defoliate oak trees *(Quercus* spp.), the trees produce new leaves that are much higher in tannins and other phenolic compounds than normal. The new leaves produced under such conditions are also tougher, and they contain less water than those they replace. Indeed, the differences are great enough that larvae feeding on the new leaves experience reduced growth, and further outbreaks of gypsy moths are diminished in intensity. The tannins apparently interfere with digestion in the insects by combining with other plant proteins, making them indigestible. Similar effects may be common in other plants as well. For example, when snowshoe hares heavily browse some trees and shrubs, such as paper birch *(Betula papyrifera),* these plants produce new shoots that are much richer in resins and phenolic compounds than the earlier shoots.

The secondary plant products ingested by herbivores may, in turn, play a role in the animals' ecological relationships with other animals. For example, some insects store these poisons within their tissues and are thereby protected from their predators (Figure 29–6). In addition, some sex attractants in insects are derived from the plants on which they feed.

Viewed as a whole, the relationships within a community are incredibly complex. Organisms that coexist within a community often have evolved together. Within the community, they affect one another in an endless variety of ways, a few of which scientists are just beginning to understand.

Nutrient Cycling

With this section, we turn to the overall properties of ecosystems. As mentioned earlier, ecosystems have the property of regulating the flow of energy, originally derived from the sun, and regulating the cycling of nutrients. Aspects of these properties will be explored in the remaining pages of this chapter.

In terms of its nutrient supply, an ecosystem is more or less self-sustaining. One of the more important reasons for this autonomy is the continuous cycling of chemical elements between organisms and the environment. The pathways of some of these essential elements, known as nutrient cycles, were discussed in Chapter 27. Ideally, nothing is lost, so that the pool of nutrients is continually renewed and continually available for the growth of organisms. The rate of flow from the nonliving pool to the organisms and back again, the amount of available material in the nonliving pool, and the form of this pool differ among nutrients and from habitat to habitat.

By considering the interaction of organisms, it is easy to see that competition for nutrients can easily occur. In some cases, this competition has become so intense that organisms benefit from a mutualistic interaction, such as the formation of mycorrhizae, in which one partner assumes the responsibility of providing nutrients for the other.

RECYCLING IN A FOREST ECOSYSTEM

Studies of a deciduous forest ecosystem in the Hubbard Brook Experimental Forest of the White Mountain National Forest of New Hampshire have shown that the plants of this community play a major role in the retention of nutrient elements. The investigators first established a procedure for determining the mineral budget—input and output, or "gain" and "loss"—of different areas in the forest. By analyzing the nutrient content of rain and snow, they were able to estimate atmospheric input, and by constructing concrete weirs that channeled the water flowing out of selected areas, they were able to calculate output (Figure 29–7). A particular advantage of this site was that a granitic bedrock is present just below the soil surface, so that very little material leaches downward; that is, the soil water percolates only a short distance. In addition, few nutrients are added by dissolution of the highly resistant bedrock.

The investigators discovered that the natural forest was extremely efficient in conserving its mineral elements. For example, annual net loss of calcium from the ecosystem was 9.2 kilograms per hectare. This represents only about 0.3 percent of the calcium in the sys-

29–7
Weir in the Hubbard Brook Experimental Forest in New Hampshire. Water from each of six experimental ecosystems was channeled through a weir (such as this one, built where the water leaves the watershed) and was analyzed for chemical elements. The trees and shrubs in the watershed behind this weir have been cut down. The experiments showed that deforestation disrupted the tight cycling of nutrients by various living components of the ecosystem and greatly increased the loss of nutrient elements from that system.

tem. In the case of nitrogen, the ecosystem was actually accumulating this element at a rate of about 2 kilograms per hectare per year. There was a similar, though somewhat smaller, net gain of potassium in the system.

At Hubbard Brook, the biological regulation of element cycling was tested by the following experiment. In the winter of 1965–1966, all trees, saplings, and shrubs in one 15.6-hectare area in one small watershed of the forest were cut down. No organic materials were removed, however, and the soil was undisturbed. During the following spring, the area was sprayed with a herbicide to inhibit regrowth. During the four months from June to September 1966, the runoff of water from the area was four times greater than in previous years. Net loss of calcium was 20 times higher than in the undisturbed forest, and potassium loss was 21 times higher. The most severe disturbance was seen in the nitrogen cycle. The tissues of dead plants and animals continued to be decomposed to ammonia or ammonium ions, which were then acted upon by nitrifying bacteria to produce nitrates, the form in which nitrogen is usually assimilated by plants. However, no plants were present, and the nitrate ions were not held in the soil. The net loss of nitrogen averaged 120 kilograms per hectare per year from 1966 to 1968. As a side effect, the stream that drained the area, polluted with nitrates, supported an algal bloom. The nitrate concentration of the stream increased to levels above those established by the U.S. Public Health Service as safe for drinking water.

Nitrogen is not always lost so readily from disturbed forest ecosystems. In the Hubbard Brook experiment, vegetation regrowth was prevented by herbicides so no uptake was possible; therefore, these results represent the maximum possible loss. The actual local rates of nitrogen loss depend on the details of the nitrogen cycle in the particular forest before the disturbance takes place. For example, if the microorganisms in the ecosystem have a high demand for the nitrogen released after forest clear-cutting, forest disturbance would not result in losses of the magnitude measured at Hubbard Brook.

29–8
A food chain. A three-toed box turtle (Terrapene carolina triunguis) *feeds on a snail that has fed on the mushrooms that are decomposing organic matter in the soil.*

Trophic Levels

In addition to its *physical* (or nonliving) components, each ecosystem includes two classes of *biotic* (or living) components—autotrophs and heterotrophs. Autotrophs are mainly photosynthetic organisms, which are able to use light energy to manufacture their own food; they are known as **primary producers.** Autotrophs consist of green plants, algae, and autotrophic bacteria. Because heterotrophs cannot manufacture their own food, they must use the organic molecules made by the autotrophs. Several feeding levels, or **trophic levels,** occur among the heterotrophs: **primary consumers,** or **herbivores**—animals that feed on living plants; **secondary consumers,** or parasites and carnivores—animals that feed on animals; and **decomposers**—fungi, bacteria, and various small animals that break down organic matter stored in the bodies of other organisms. All three levels are present in most ecosystems.

In a given ecosystem, organisms from each of the trophic levels make up what is called a **food chain** (Figure 29–8). The relationships between the organisms in a food chain regulate the flow of energy and nutrients through the ecosystem. The length and complexity of such food chains vary a great deal. Usually an organism has more than one source of food and is itself preyed on by more than one kind of organism. Under most circumstances, it is more nearly correct to speak of a **food web** (Figure 29–9). The complexity of trophic relationships has a number of important implications for the overall properties of that ecosystem.

THE FLOW OF ENERGY

In an ecosystem, the flow of energy typically begins with solar energy that is captured in photosynthesis and used to make carbohydrate molecules. Energy does not flow in a cycle in ecosystems; rather, it flows through autotrophs (usually photosynthetic organisms; that is, plants, algae, and some bacteria) to consumers (animals and heterotrophic protists) and then to decomposers. At each step, the great majority of the energy is dissipated in the form of heat energy, which eventually is returned to space as infrared radiation.

A very large amount of **biomass** is produced on earth each year. ("Biomass" is a convenient shorthand term for organic matter.) It is currently estimated that the earth's annual production of biomass amounts to about 200 billion metric tons. Despite this enormous figure, photosynthetic organisms are not very efficient in converting the sun's energy into organic compounds. Generally, less than 1 percent of the light that falls on a plant is utilized in biosynthesis (see Chapter 5). How-

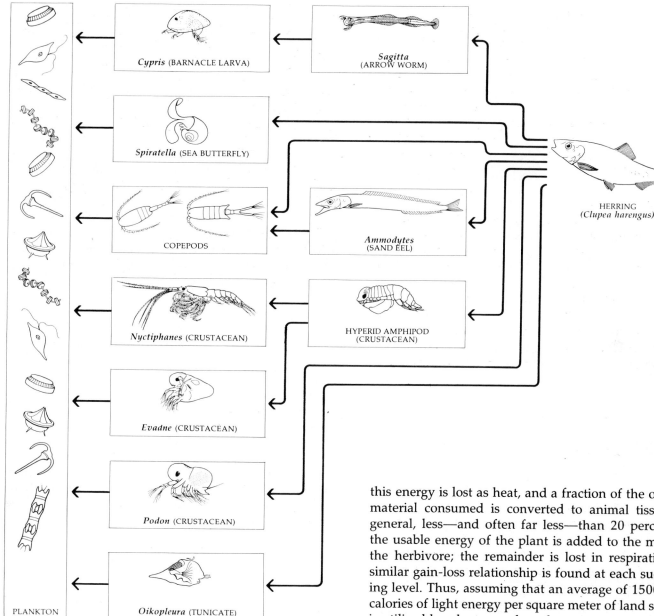

29–9

The feeding relationships of the herring
(Clupea harengus) *in the North Atlantic,*
illustrating some of the complexities
and interrelationships of a food web.

ever, particularly productive stands of vegetation, and some aquatic systems, may convert up to 3 percent of the annual incident solar radiation into chemical energy.

When the organic material produced by plants is consumed by an herbivore, energy is released. Most of

this energy is lost as heat, and a fraction of the organic material consumed is converted to animal tissue. In general, less—and often far less—than 20 percent of the usable energy of the plant is added to the mass of the herbivore; the remainder is lost in respiration. A similar gain-loss relationship is found at each succeeding level. Thus, assuming that an average of 1500 kilocalories of light energy per square meter of land surface is utilized by plants per day, about 1 percent (as noted above), or 15 kilocalories, is converted to plant material. Of this amount, perhaps 10 percent, or 1.5 kilocalories, is incorporated into the bodies of the herbivores that eat the plants, and 10 percent of that, or about 0.15 kilocalorie, is incorporated into the bodies of the carnivores that prey on the herbivores.

To give a concrete example, Lamont Cole of Cornell University, in his studies of Lake Cayuga, near the Cornell campus in New York State, calculated that, for every 1000 kilocalories of light energy utilized by algae in the lake, about 150 kilocalories are reconstituted as small aquatic animals. In addition, of these 150 kilocalories, 30 kilocalories are reconstituted as smelt (a small fish). If we were to eat these smelt, we would gain about 6 kilocalories from the original 1000 kilocalories used by the algae. But if trout eat the smelt and we then eat the trout, we gain only about 1.2 kilocalories from

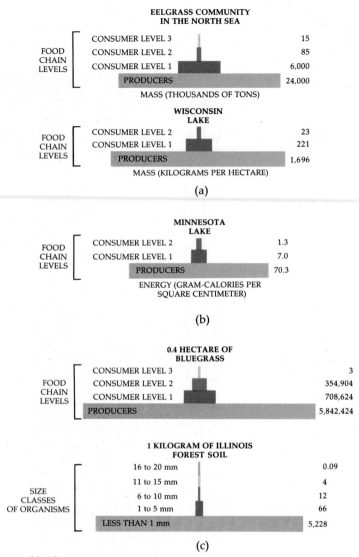

EELGRASS COMMUNITY
IN THE NORTH SEA

FOOD CHAIN LEVELS

CONSUMER LEVEL 3	15
CONSUMER LEVEL 2	85
CONSUMER LEVEL 1	6,000
PRODUCERS	24,000

MASS (THOUSANDS OF TONS)

WISCONSIN LAKE

FOOD CHAIN LEVELS

CONSUMER LEVEL 2	23
CONSUMER LEVEL 1	221
PRODUCERS	1,696

MASS (KILOGRAMS PER HECTARE)

(a)

MINNESOTA LAKE

FOOD CHAIN LEVELS

CONSUMER LEVEL 2	1.3
CONSUMER LEVEL 1	7.0
PRODUCERS	70.3

ENERGY (GRAM-CALORIES PER SQUARE CENTIMETER)

(b)

0.4 HECTARE OF BLUEGRASS

FOOD CHAIN LEVELS

CONSUMER LEVEL 3	3
CONSUMER LEVEL 2	354,904
CONSUMER LEVEL 1	708,624
PRODUCERS	5,842,424

1 KILOGRAM OF ILLINOIS FOREST SOIL

SIZE CLASSES OF ORGANISMS

16 to 20 mm	0.09
11 to 15 mm	4
6 to 10 mm	12
1 to 5 mm	66
LESS THAN 1 mm	5,228

(c)

29–10
Pyramids of mass (a), energy (b), and numbers of organisms (c) in various communities. A relatively small amount of mass or energy is transferred to each successively higher level. The units of measurement are listed at the bottom of each figure, with the quantities opposite each level, at the right of the figure.

the original 1000 kilocalories. Smelt are much more abundant, and constitute a much larger biomass, in Lake Cayuga than the trout. Thus, more of the original energy is available to us if we eat smelt rather than the trout that feed on the smelt; yet trout are considered a delicacy, smelt a much less desirable food for humans. Under conditions of starvation, people cannot afford the tenfold loss of energy that occurs when plants are fed to animals; they must then become herbivores to obtain as much food as possible.

Food chains are generally limited to three or four links; the amount of energy remaining at the end of a longer food chain is so small that few organisms can be supported by it. Body size also plays a role in the structure of food chains. For example, an animal constituting one link generally has to be large enough to capture prey from the previous, lower link on the food chain (though small insects do feed on large trees!). In the end, most of the biomass in an ecosystem is used by decomposers, such as fungi and bacteria.

Owing to the relationships just discussed, the total energy at successively higher trophic levels in an ecosystem generally decreases sharply, setting up the sort of relationship described by the expression "pyramid of energy" (Figure 29–10). This relationship may not hold if there is rapid turnover among the primary producers, such as algae in a lake; the turnover rate then becomes the controlling factor, and the total energy present at that level may be relatively small at any one time. A similar relationship is characteristic of mass, leading to the existence of "pyramids of mass." In general, there are also far more individuals at the lower levels than at the higher levels, which leads to a "pyramid of numbers." It also follows that, if all the organisms in an ecosystem are divided into size classes, the small animals will be far more numerous than the large ones.

A practical aspect of the flow of energy through ecosystems concerns human efforts to develop renewable sources of energy from plants, a process known as biological energy conversion. Plantings of fast-growing trees and other plants may provide a major, renewable energy source in the future, and they may constitute one of the most environmentally benign ways of efficiently capturing solar energy. Even now, scientists estimate that the waste materials available each year after the harvest of crop and forest products could provide an amount of energy equivalent to 1 percent of the gasoline consumed annually in the United States, or 4 percent of the nation's annual consumption of electrical energy. The potential is limited, however, by the energy costs of harvesting the material. In developing countries, more than 2 billion people rely on biomass for virtually all their cooking, heating, and lighting, so that the ability of plants to convert energy from the sun to a form useful to humans is a matter of central importance on a global scale.

Development of Communities and Ecosystems

SUCCESSION

Some plant communities appear to remain the same year after year, whereas others change rapidly. In the latter case, there is a progression of changes in the community called **succession.** A cleared woodlot is rapidly colonized by the remaining trees in the vicinity; in a similar fashion, a pasture eventually gives way to a forest. Analogous series of events occur in naturally disturbed areas, such as lakes, floodplain forests, or steep hillsides. Natural disturbances (for instance, floods, windstorms, earthquakes, landslides, fires) are a pervasive feature of all ecosystems, even "pristine" wilderness areas that are little disturbed by humans. The process of succession is both continuous and worldwide in scope.

Succession occurs at a variable rate in all temporarily disturbed areas. Some ponds, for example, fill with aquatic plant remains and debris; emergent vegetation builds soil, while sediment washes in and contributes to filling in of the lake; the site is taken over by meadow; moisture-loving shrubs may become established; and finally the forest characteristic of the region develops in the meadow that formed earlier where there had been a pond (Figure 29–11). In another example, rocks

(a)   (b)

(c)   (d)

29–11
(a) *Emerging vegetation grows along the edge of a pond.* (b) *Aquatic plants with floating leaves, such as water lily* (Nymphaea odorata), *grow across the* surface of a pond and eventually choke out bottom-dwelling plants. (c) *Water hyacinths* (Eichhornia crassipes) *play a similar role in warmer climates.* (d) *Marsh grasses, sedges, and cattails* (Typha spp.) *growing on an old pond bed continue the process of succession.*

29–12
An early stage of succession. Lichens have begun to break down the rocks, while ferns and bryophytes are accumulating soil in a small crevice.

29–13
Seedling trees of balsam fir (Abies balsamea) growing under and replacing quaking aspen (Populus tremuloides) in northern Arizona—a stage in forest succession leading to a climax community of white spruce (Picea glauca) and balsam fir.

weather and break down because of freezing and thawing and other physical factors, and the process is sometimes expedited by the action of lichens, which secrete chemicals that erode the rocks directly, and by mosses, which expand when wet, continually breaking off little flakes of rock (Figure 29–12). Soil accumulates around the bases of the lichens and mosses, and flowering plant seedlings eventually may become established. Their roots penetrate cracks, breaking the rocks down further. Eventually, perhaps after many centuries, the rock may be completely reduced to a component of soil, which also includes organic matter from the generations of organisms that have grown in it. The soil will ultimately be occupied by forests or other types of vegetation characteristic of the region. In the early stages of succession, plants with symbionts that have the ability to fix nitrogen may be prominent, at least in certain areas. Another example of succession is shown in Figure 29–13.

The process of succession is often not completely unidirectional, particularly in its later stages. It is generally simple to observe the early stages of succession and very difficult to interpret the later ones. In more mature communities, reversals in the expected direction of succession often occur, and the ultimate outcome may be heavily influenced by the nature of the adjoining communities.

The creation and refilling of **gaps** created by natural disturbances are processes that play a key role in the process of succession and in the maintenance of species diversity in various plant communities. The gaps that occur when trees fall in a forest, for example, provide opportunities for many plant species with relatively high light requirements to flourish in such areas. In temperate North America, a number of species that characteristically occur in gaps, such as black cherries and blackberries, have fleshy fruits eaten by birds. The birds digest the fleshy parts of these fruits and drop the seeds in new gaps, which are thus colonized efficiently. Such pioneer species often have multilayered, diffuse crowns; light, nondurable wood; and high rates of growth under sunny conditions. The dominant, late successional trees in the same forest often have very different characteristics, such as more tightly packed crowns, heavy but durable wood, and slow rates of growth under sunny conditions; however, they can

29-14
(a) *When fire sweeps through a forest, recolonization—with regeneration from nearby unburned stands of vegetation—is initiated. Some plants produce sprouts from the base, others seed abundantly on the burned area. In one group of pines, the closed-cone (serotinous) pines, the cones do not open to release their seeds until they have been exposed to fire.* (b) *Sugar pines* (Pinus lambertiana) *in Yosemite National Park in California. With the prevention of forest fires by human beings, sugar pines are being replaced by other trees, such as white fir* (Abies concolor), *the trees seen here growing at the base of the sugar pines.*

(a)

(b)

grow in the shade and live longer than the pioneers. Such differences between tree species play a major role in determining their local success and thus the structure of the mature forest.

One of the most significant forms of natural disturbance affecting plant communities is fire. For example, when European settlers first arrived in California, they found a magnificent forest of sugar pine *(Pinus lambertiana)* along much of the length of the Sierra Nevada. Although conservationists tried to preserve some of this forest in national parks and forests, many of the stands of pines were eventually replaced by other trees, such as white fir *(Abies concolor)* and incense cedar *(Calocedrus decurrens)*. Why did this change take place?

Sugar pine was a member of a successional stage in the forests of this region that was maintained by periodic fires. These fires were greatly reduced in number and scope after the influx of settlers to the area. Without lightning-set fires of low intensity periodically racing through the groves, a thick growth of brush and smaller trees grew up, evidently creating conditions so shady that sugar pine seedlings could not compete effectively. Only a policy of letting the occasional fires that occur burn, or one of controlled burning, can preserve the remaining groves of sugar pine in the open form that most people find so attractive (Figure 29–14b). Similar relationships are found in all vegetation types in which fires periodically burn, either naturally or caused by humans.

The Great Yellowstone Fire

It was an exceptionally dry summer in 1988, with the hottest, driest conditions since the Dust Bowl of the 1930s. But the park managers of Yellowstone National Park were confident that major fires could not happen. Yet in July of that year, Yellowstone broke into flames. Soon new fires were breaking out in every section of the park; they continued to spread throughout August and September. Park concessions were endangered. Tourists had to be evacuated. A holocaust had begun. What had gone wrong?

Yellowstone Park was established in 1872 in order to preserve its natural beauty, and that included protecting it from fires. Eventually, however, scientists and managers realized that there was a gradual decrease in certain species, as seen in photographs taken through the years. This evidence, along with the existence of charred stumps, strongly suggested that fire was a normal part of the Yellowstone ecosystem. In 1972, park officials initiated a program in which they allowed natural fires in remote areas to burn without interference. The program had subsequently been extended throughout the park to all but 5 percent of the total area—places where people were living. This policy was widely viewed as a success.

But by October 1988, 4300 square kilometers lay charred, and nowhere in the park could you get a vantage point where there was no visible fire damage. Dry winds, gusting to more than 100 kilometers per hour, had frustrated the efforts of more than 9000 firefighters to control the spread of the flames. Was this havoc the result of buildup of abnormal amounts of biomass in the trees and shrubs? Could it have been prevented by controlled burning?

Fire scars among the annual growth rings of lodgepole pines *(Pinus contorta)* indicated that any one spot in Yellowstone had burned naturally about every 20 to 25 years, with a wide range of variation. Lodgepole pine, like other pines, is adapted to fire. Its cones open only in intense heat, and its seeds regenerate best following fires, benefit when the light-blocking mature trees are killed, and thrive on the nutrients released from burned pine needles. The beautiful, even-aged lodgepole pine stands that are such a characteristic feature of Yellowstone National Park are a direct consequence and reflection of periodical catastrophic fires.

At higher elevations, with cooler climates, mostly small fires had burned almost every decade. In areas where there were no significant fires, the lodgepole pines were eventually replaced by other tree species, such as subalpine fir *(Abies lasiocarpa)* and Engelmann spruce *(Picea engelmannii)*. These species usually succeed lodgepole pine, starting about 150 years after a fire, and become dominant during the following century. As the original lodgepole pines die, however, the forest canopy becomes uneven, with many gaps, and is highly flammable. Fires that occur at early stages of succession are usually small, but as the forest becomes more mature, fires are extensive and highly destructive.

In Yellowstone, fire usually intervenes and the climax fir-spruce communities are very rare. The extensive fires of 1988 seem to have been similar to a series of fires that occurred around 1700 and perhaps should also be viewed as a natural feature of this area. Major efforts to suppress fires in the Yellowstone area may merely have delayed this natural cycle for a few years.

At any event, the great fire of 1988 appears to have had no major lasting effect on the ecosystem. Lush meadows feed healthy buffalo and deer. Colorful lupines help to replace the soil's depleted nitrogen. Pine seedlings thrive on the sunny flats and slopes. Lodgepole pines grow slowly in Yellowstone. In 30 years, these seedlings will be 2 to 4 meters tall; perhaps in 80 or 100 years, the forests will begin to resemble those that burned in 1988. In two or three centuries, the stage may have been set for a repeat of the events that occurred in the late summer of that year. Meanwhile, curiosity about the recovery of the Yellowstone ecosystem from the great fire has proved a great stimulus to tourism, and people have been visiting the area in record numbers.

In management terms, the continuation of a program in which lightning-set fires are allowed to burn is clearly an appropriate strategy for maintaining the natural ecosystems of Yellowstone, but the total area—like that of all protected areas—is so limited that fires cannot be allowed to burn completely freely. They affect too many interests both within the protected areas and beyond their boundaries to be treated with complete passivity. Scientists and managers currently are studying the possibility of burning areas near human habitation periodically to reduce the amount of natural fuel. The struggle between the interests of humans and the continuance of nature continues. In this context, fire is clearly seen as an example of a natural, pervasive influence affecting an apparently pristine ecosystem.

(a) *Satellite image of Yellowstone area at the height of the fire activity, September 8, 1988. Red denotes the area within the burn perimeter, and green identifies unaffected vegetation. Yellowstone Lake is the dark blue shape on the right, just below the center. (b) Regeneration near Elk Park in Yellowstone, a year after the fire.*

(b)

(a)

(b)

29–15

The most recent severe eruption of Mount St. Helens in Washington State took place on May 18, 1980. (a) What had been productive forests of Douglas fir (Pseudotsuga menziesii), *western hemlock* (Tsuga heterophylla), *and firs* (Abies amabilis *and A.* procera) *were blown down by the lateral blast. (b) Over an area of more than 60 square kilometers, a deep layer of volcanic debris was deposited. (c) During the summer of 1981, many perennial plants, like the fireweed* (Epilobium angustifolium) *shown here, resprouted from ash-covered clearcuts on steep slopes, where erosion removed the overlying ash. Although there is blown-down forest in the background, the area of lush regrowth is an old clearcut where the fireweed had become well established prior to the eruption. (d) Douglas fir and many other plants with wind-dispersed seeds recolonized areas that had been so deeply covered by volcanic debris that the plants beneath were killed. This photograph was taken in 1984.*

Following natural disasters, recolonization produces similar successional changes. For example, in August 1883, a violent volcanic eruption destroyed half of the island of Krakatau, in the Java Straits about 40 kilometers from Java, Indonesia. The remaining half of the island was covered by a layer of pumice and ash more than 31 meters thick. Neighboring islands were also buried, and the entire assemblage of plants and animals on these islands was wiped out. Soon afterward, however, the recolonization of Krakatau began, and the expected number (based on the number originally occupying the area) of about 30 species of land and freshwater birds was reached within about 30 years. Recolonization by plants also proceeded rapidly, with a total of more than 270 species being recorded for the island of Krakatau by 1934.

In the state of Washington, the violent eruption of Mount St. Helens on May 18, 1980, sent a massive avalanche of volcanic debris from the top and north side of the mountain into the North Toutle River Valley. Within 15 minutes, more than 61,000 hectares of forest and recreation land were devastated by the lateral blast, which blew down forests on about 21,000 hectares and killed trees and other plants but left them standing on another 9700 hectares. In addition, the nine-hour eruption covered the whole area with up to 0.5 meter of ash, pumice, and rock pulverized by the blast (Figure 29–15).

Life began to reappear on the slopes affected by the eruption almost immediately, however, with plants sprouting up through the volcanic debris the following spring. Much of this debris soon eroded off, and wind-dispersed seeds and fruits blew back into the area. Such dispersal was especially important in areas that had been buried by debris avalanches, which were so deep that they killed the plants buried beneath them, and on the volcanic flows. Many small animals also survived, both underground and in lakes and streams, and terres-

(c)

(d)

trial vertebrates soon moved back into the area. The eruptive periods of Mount St. Helens have been separated by only 100 to 500 years in the last 35 centuries, and both the blast itself and the recovery from its effects that scientists were able to observe in the early 1980s were typical of natural phenomena that periodically affect living systems in volcanic areas.

A general property of successional processes is that, if climatic factors remain constant, the process of succession should ultimately slow and nearly stop, and terminate in a **climax community.** Such a community is self-perpetuating; its characteristics relate to the specific climatic conditions under which it is produced. A climax community, however, is an ideal concept that mainly serves as a reference point against which to measure community change. In reality, climatic conditions often change, natural disturbances such as hurricanes and landslides occur, and animals modify the nature of the changing communities. The communities that develop at a particular place reflect a balance of many different environmental factors.

As human beings become more and more numerous, and their impact on ecosystems correspondingly profound, the science of **restoration ecology,** which attempts to understand the process of succession better and use its principles to reestablish natural communities, will become more and more important. It is often not a simple matter to re-create natural communities once they have been destroyed; yet the process is one of great significance to an increasingly overcrowded world.

Although the eruption of Mount St. Helens has provided a recent and very dramatic example of natural disturbance and the ensuing early stages of succession, these phenomena are typical of all communities and occur throughout the world. Disturbance and succession are two important factors that account for the full extent of the diversity of life on earth.

Summary

Biomes are large terrestrial complexes of living organisms that are characterized by distinctive vegetation and climate. Specific growth features are characteristic of the dominant plants of each biome. Ecosystems are self-sustaining systems that include the aggregation of living organisms together with the nonliving (physical) elements of the environment with which they interact. Communities consist of all the organisms that live in a particular area, or of a subset of these organisms that is defined for a particular purpose.

Some of the relationships that occur in communities can be grouped under three headings: mutualism, competition, and plant-herbivore (and plant-pathogen) relationships. In mutualism, two populations interact to the benefit of both. Examples include lichens, mycorrhizal associations between fungi and the roots of plants, and the relationships between flowering plants and their pollinators and fruit- and seed-dispersers. In the bull's-horn acacias of Latin America, the thorns are inhabited by specialized ants, which obtain their food from the plants and protect them from most herbivores and from competition with other plants.

Competitive interactions are found between most kinds of plants that grow together and between most individual plants also. The principle of competitive exclusion states that, when two kinds of organisms occurring together compete for the same limiting resources, ultimately only one of them will survive in that area. One of the most important kinds of competition is competition for light; often plants with the highest growth rate relative to other species in a particular environment will be the most successful competitors there. Plants have also evolved chemical weapons with which to compete aggressively with nearby plants, and such allelopathic relationships can also affect community composition. Finally, plants frequently compete for the services of pollinators and seed dispersers; such competition can have a profound impact on the reproduction and long-term abundance of different species.

Plants counter the effects of herbivores, which limit the reproductive potential of the plants, through the evolution of spines, tough leaves, and similar structures or structural alterations, and, most important, chemical defenses. An insect or other herbivore that has overcome a plant's chemical defenses not only has a new and often largely untapped food resource at its disposal but may also utilize the toxic substances produced by the plant to gain a degree of protection from its own predators.

An ecosystem consists of nonliving elements and two different kinds of living elements—autotrophs and heterotrophs. Among the heterotrophs are the primary consumers, or herbivores; the secondary consumers, or carnivores and parasites; and the decomposers. The organisms found at these levels are members of food chains or food webs.

The properties of ecosystems have been studied especially well experimentally at Hubbard Brook, in New Hampshire, where it has been shown that undisturbed natural communities control the cycling of nutrients but that the control tends to be lost when the ecosystem is disturbed. Energy flows through ecosystems, with 1 percent or less of the incident solar energy converted into chemical energy by green plants. When these plants are consumed, less than 20 percent of their potential energy is stored at the next trophic level; a similar degree of efficiency characterizes transfers farther up the food chain. The amounts of energy remaining after several transfers are so small that food chains are rarely more than three or four links long. In most ecosystems, more energy, biomass, and individuals occur at lower trophic levels, giving rise to the phenomena known as pyramids of energy, mass, and numbers.

Succession occurs in naturally open areas, such as lakes, ponds, or meadows in a forested region, and after an area has been denuded by artificial or natural means. In the course of succession, the kinds of plants and animals in the area change continuously, some being characteristic only of the early stages of succession, for example. The creation and refilling of gaps created by natural disturbances play a key role in the process of succession and in the maintenance of species diversity in various forest communities. The pioneer species that occur when the gaps are formed grow rapidly under sunny conditions and have other characteristics different from those of the trees that dominate the mature forest. Eventually succession may result in the production of a climax community, which reproduces itself indefinitely unless there are major environmental changes.

Fire plays a very important role in the dynamics of many ecosystems, as in the maintenance of sugar pine forests in the Sierra Nevada of California. Succession after volcanic eruptions, such as that of Krakatau in Java in 1883 or Mount St. Helens in Washington State in 1980, provides a spectacular example of the process, and these areas have been studied extensively.

CHAPTER 30

The Biomes

(a)

(b)

30–1
Regions with mild winters and long, dry summers are often dominated by spiny shrubs with broad, thick evergreen leaves. This vegetation, known formally as Mediterranean scrub, has been given a variety of local names. (a) In North America, it is known as chaparral, as seen here in southern California. (b) In Mediterranean regions, it is known as maquis, as seen here on the Greek island of Corfu. Although the plants of these two areas are mostly unrelated, they closely resemble one another in their growth patterns and characteristic appearance.

You are already familiar with many of the vast biomes of the earth—the tundra, with its herds of caribou; the temperate deciduous forests, leafless in winter but bursting into life in the spring; and the deserts, with their cacti and other succulent and spiny plants, for example. Defined in ecological terms, a **biome** is a terrestrial, climatically controlled set of ecosystems that are characterized by distinctive vegetation and among which there is an exchange of water, nutrients, gases, and biological components, including people. The plants and animals that occur in particular biomes have characteristic growth forms and other adaptations that have evolved in relation to particular climates. It is because of these common growth forms that we can recognize biomes such as savannas or deserts wherever they occur, even though the individual organisms from a biome in one region are often completely different from those of the same biome in another region, having achieved similar characteristics as a result of parallel evolution. A single biome occupies large areas of land surface and usually occurs on more than one continent (Figure 30–1). Biomes can be classified in a number of different ways, but the categories that are discussed in this book (Figure 30–2) provide a basis for an overall account of the vegetation of the world.

Life on the Land

Land plants face a variety of problems. In most regions, they are subjected to periodic drought and to rapid diurnal and seasonal changes in temperature. They must survive during seasons unfavorable to their growth, they must often grow on substrates of unfavorable mineral composition, and they are subject to the action of gravity (which affects organisms on land much more strongly than it does aquatic organisms).

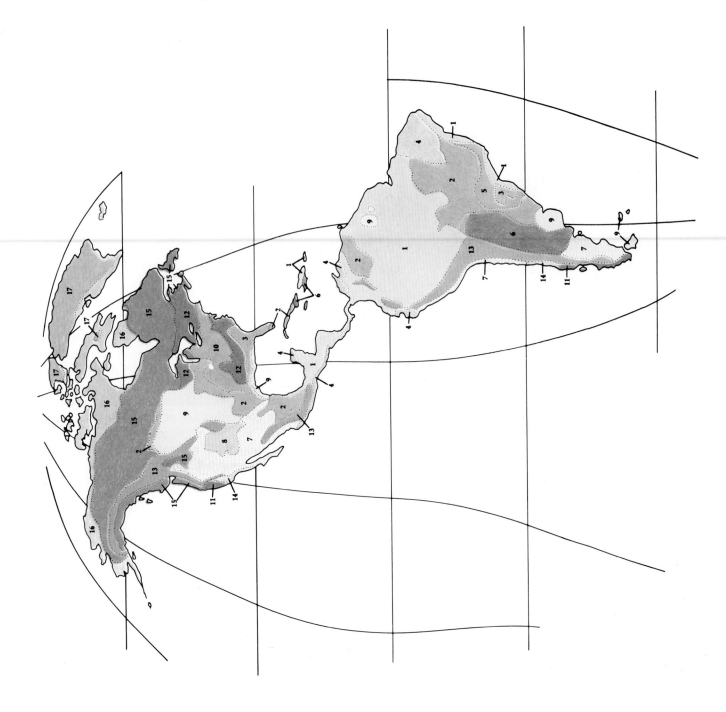

30–2
The distribution of biomes throughout the world. The biomes recognized are indicated in the legend. This map was specially prepared for this book by A.W. Küchler of the University of Kansas. Because of the global coverage of the map, its scale is relatively small and its content generalized. For this reason, the various biomes shown are not always highly uniform; all of them include considerable variations in vegetation. The boundaries between the biomes may be sharp, but more often they are blurred, consisting frequently of broad zones of transition from one type of biome to another.

1	RAINFORESTS
2	SAVANNAS
3	SUBTROPICAL MIXED FORESTS
4	MONSOON FORESTS
5	TROPICAL MIXED FORESTS
6	SOUTHERN WOODLAND AND SCRUB
7	DESERTS AND SEMIDESERTS
8	JUNIPER SAVANNA
9	GRASSLANDS
10	TEMPERATE DECIDUOUS FORESTS
11	MIXED WEST-COAST FORESTS
12	TEMPERATE MIXED FORESTS
13	ALPINE TUNDRA AND MOUNTAIN FORESTS
14	MEDITERRANEAN SCRUB
15	TAIGA
16	ARCTIC TUNDRA
17	ICE DESERT

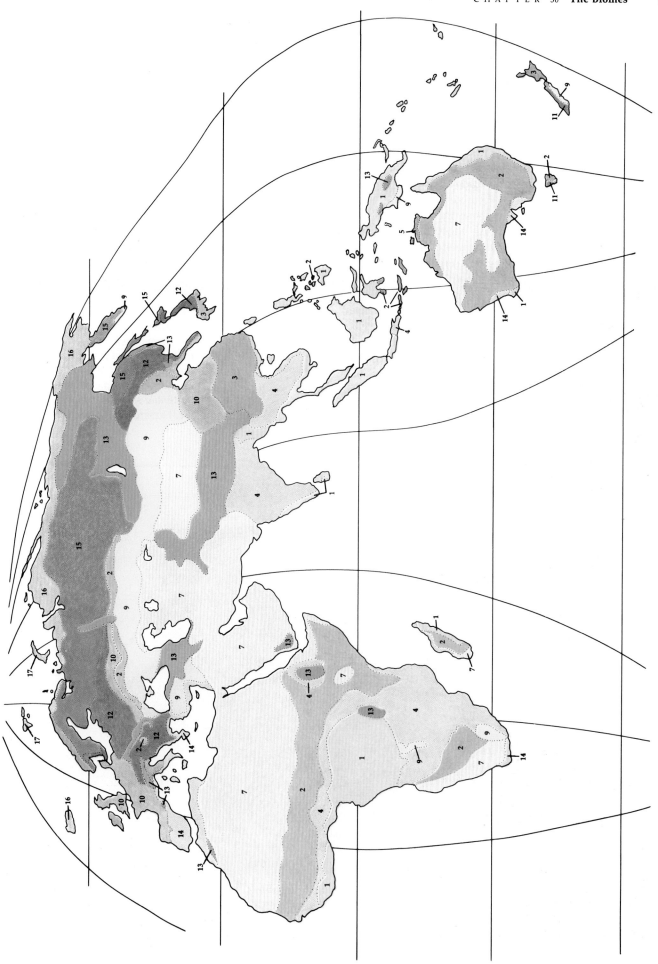

Living on land has some advantages, however; oxygen is more uniformly distributed in air than it is in water, and carbon dioxide is more readily available.

The distribution of biomes results from three kinds of physical factors: (1) the distribution of heat from the sun and the relative seasonality of different portions of the earth; (2) global patterns of air circulation (Figure 30–3) , particularly the directions in which the prevailing moisture-laden winds blow; and (3) such geological factors as the distribution of mountains and their height and orientation. All of these factors interact to produce the varying patterns of vegetation over the face of the earth and underlie differences in productivity in different geographical regions (Figure 30–4).

The earth's land areas are discontinuous, and this separation of land masses has an important effect on the distribution of organisms. In addition, there are often sharp differences in precipitation, substrate, climate, and other factors between different places on land. Such differences mean that the distribution of any particular kind of land organism is likely to be much more limited than that of an organism of similar size and motility in the sea (Figure 30–5).

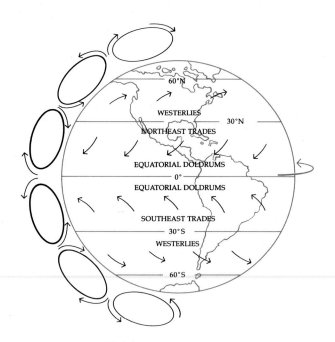

30–3

The earth's surface is covered by belts of air currents, which determine the major patterns of wind and rainfall. Air rising at the equator loses moisture in the form of rain, and falling air at latitudes of 30° north and south is responsible for the great deserts found at these latitudes.

30–4

The earth's biological productivity, as reconstructed from three years of satellite data by NASA. Rainforests and other highly productive areas are dark green, *deserts yellow. More than two-thirds of all biological productivity occurs on land, the remainder in the sea. The concentration of phytoplankton, which* *consists of photosynthetic protists, is represented by a scale that runs from red (highest productivity) to orange, yellow, green, and blue (lowest productivity).*

30–5
This fern, the spleenwort Asplenium viride, *is growing in rock crevices at Sierra Buttes, Sierra County, northern California. It occurs in similar habitats all around the Northern Hemisphere—across both North America and Eurasia—but in North America it occurs no closer to Sierra Buttes than central Washington and northeastern Nevada, both nearly 1000 kilometers away. Throughout its range,* A. viride *grows on cliffs at scattered localities, reaching new places by means of its wind-dispersed spores. Sierra Buttes, however, is far from the edge of the fern's more continuous range, and its occurrence there provides an excellent example of the disjunct and dispersed nature of the ranges of plants and animals in general.*

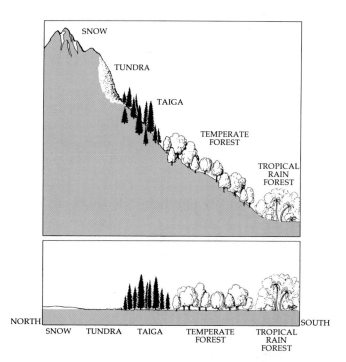

30–6
We can see a similar sequence of plant communities whether we travel north for hundreds of kilometers or merely ascend a tall mountain. This relationship between altitude and latitude was first pointed out by Alexander von Humboldt.

LOCAL MODIFICATIONS IN CLIMATE

Plants and plant communities vary with the latitude of their habitat. For instance, the mean atmospheric temperature decreases with increasing latitude. In terms of annual averages, the warmest atmospheric temperature is found not at the equator but at 10° north latitude; only during January is the equator the warmest circle of latitude. In July, the warmest parallel of latitude is 20° north. The Northern Hemisphere as a whole has a higher annual mean temperature than the Southern Hemisphere because of the much larger land surfaces in the north. The annual mean temperature at the equator is 26.3°C; that at 40° north latitude is 14°C; and that at 40° south latitude is 12.4°C.

Just as one encounters lower temperatures in traveling north or south from the equator, one encounters them also when ascending to higher elevations. In gen-

eral, a change in mean atmospheric temperature corresponding to an increase in latitude of 1° occurs with each increase in elevation of approximately 100 meters. This relationship has important consequences for the distribution of land organisms. For example, plants and animals characteristic of Arctic regions may approach or even reach the equator at high elevations, particularly in mountain ranges trending north and south (Figure 30–6).

There are, however, important differences between high-latitude habitats and high-altitude habitats. In the mountains, the air is clearer and the solar radiation is more intense. Most of the water vapor in the atmosphere—which plays a major role in preventing heat from radiating away from the earth at night—occurs below 2000 meters. Consequently, nights are often much cooler in the mountains than at lower elevations at the same latitude. Moreover, in the Arctic and Antarctic, day length varies considerably, from 24 hours of light in the summer in the extreme north and south to 24 hours of darkness in the winter, whereas at the equator day length varies little from 12 hours per day all through the year. Temperature variation is likewise much greater away from the equator. Those "Arctic" or "Antarctic" organisms that extend their ranges toward the equator in the mountains must make physiological adjustments for such differences between high-latitude (Arctic and Antarctic) and high-altitude (alpine) environments.

There are often pronounced temperature variations from slope to slope on a particular mountain. In the middle latitudes of the Northern Hemisphere, for example, the southern and western slopes of mountains are often drier and warmer than the northern and eastern ones, because the northern slopes receive no direct sunlight and the eastern slopes receive sunlight only in the mornings, when it is cooler and the solar radiation is less intense. Clouds that form in the mornings often dissipate by the afternoon, enhancing this effect. Because of the relationships just described, the driest slope of a mountain in the Northern Hemisphere is usually that oriented toward the southwest. (Which would you expect to be the driest slope of a mountain in the Southern Hemisphere?) For the same reasons, mosses and lichens often grow, or are better developed, on the northeastern side of trees in the Northern Hemisphere, leading to a well-known folk saying about how to identify the points of the compass when trying to find one's way in the woods. The relationship is valid, however, only in sufficiently moist and well-lighted situations.

As a result of the complex interactions between varying local conditions, the distribution of biomes is not what we would expect if the earth were a perfect sphere. In addition, major shifts in vegetation types have occurred in the past, profoundly affecting the sort of vegetation that we see today. In the following pages, some examples are provided of the ways in which climate, topography, and soil type interact to control the distributions of the biomes. We shall discuss the 17 biomes shown on the map in Figure 30–2 under the following categories: rainforests; savannas and deciduous tropical forests (including savannas, subtropical mixed forests, monsoon forests, tropical mixed forests, and southern woodland and scrub); deserts (including deserts and semideserts, and juniper savanna); grasslands; temperate deciduous forests; temperate mixed and coniferous forests (including mixed west-coast forests, temperate mixed forests, and alpine tundra and mountain forests); mediterranean scrub; taiga; and Arctic tundra (including Arctic tundra and ice desert).

(a)

30–7
Tropical rainforest. (a) The diversity of the trees in the forest, which may reach several hundred species per hectare, is revealed when individual trees burst into bloom, like this guayacan tree (Tabebuia chrysantha) *in Panama. (b) A buttressed tree in a lowland forest in Thailand. (c) Stinkhorn fungus* (Dictyophora duplicata). *(d) Dense growth of epiphytes on a tree in Colombia. (e) An orchid growing on the trunk of a mango in Ecuador. (f) Lianas, such as this hummingbird-pollinated, red-flowered passion flower* (Passiflora coccinea), *are abundant in the tropical rainforest (g).*

Rainforests

More species of plants and animals live in tropical rainforests than in all the rest of the biomes of the world combined. Neither water nor temperature is a limiting factor during any part of the year. Although there are many species, there are usually few individuals per species; a species of tree may be represented by only one individual per hectare. Not only is there a large number of species in the tropical rainforests, but their interrelationships are more complex than those of plants and animals in any other biome (Figure 30–7).

(b)

(c)

(d)

(e)

(f)

(g)

Alexander von Humboldt

Alexander von Humboldt (1769–1859) was perhaps the greatest scientific traveler who ever lived and was certainly one of the greatest writers and scientists of his era. A native of Germany, Humboldt ranged widely across the interior of Latin America from 1799 to 1804 and climbed some of its highest mountains. Exploring the region between Ecuador and central Mexico, Humboldt was the first to recognize the incredible diversity of tropical life and, consequently, the first to realize just how vast a number of species of plants and animals there must be in the world.

In his travels, Humboldt was impressed with the fact that plants tend to occur in repeatable groups, or communities, and that wherever there are similar conditions—relating to climate, soil, or biological interactions—similar groupings of plants appear. He also discovered a second major principle—the relationship between altitude and latitude. He found that

Tropical rainforests are dominated by broad-leaved evergreen trees. Little light penetrates to the floor of these forests, and the rainfall is generally between 200 and 400 centimeters per year. There is little accumulation of organic debris; decomposers rapidly break down leaves, stems, and the bodies of animals in the constantly warm and moist climate. Nutrients released from this breakdown are quickly absorbed by mycorrhizal roots or are leached from the soil by the rain. Although there may be a notable variation in precipitation from month to month, there is no pronounced dry season.

Generally, plants in the tropical rainforest have not evolved particular mechanisms that would permit them to survive unfavorable seasons of drought or cold. Nearly all the plants are woody, and woody vines are abundant. There is a large flora of **epiphytes,** which grow on the branches of other plants in the illuminated zone far above the forest floor. Epiphytes, which have no direct contact with the forest floor, obtain water from the humid air of the canopy, as well as directly from rain, and they obtain minerals from dust and the surfaces of the plants on which they grow. Along with epiphytes and climbing vines, many kinds of animals live in the treetops; this is the area of the tropical rainforest in which animal life is most abundant and diverse.

Because so little light reaches the forest floor, very few herbaceous plants grow there, and those that do are found mostly in gaps in the forest canopy created by treefalls. Many plants in the tropical rainforest are trees; they are usually larger than those found in temperate forests, often reaching 40 to 60 meters in height. Furthermore, the tree species are diverse; there are seldom fewer than 40 species per hectare. This is in sharp contrast with temperate zone forests, where there are rarely more than a few tree species per hectare. Nevertheless, the trees of the tropical rainforest are remarkably homogeneous in appearance. Generally, they branch only near the crown. Because their roots are usually shallow, they often have buttresses at the base of the trunk that provide a firm, broad anchorage. Their leaves are medium-sized, leathery, untoothed and unlobed, and dark green; their bark is thin and smooth; and their flowers are generally inconspicuous and greenish or whitish in color. Such forests often form several layers of foliage, the lower layers consisting of seedlings of the taller species, together with a relatively small number of lower-growing species.

Tropical rainforests are well developed in three major areas of the world. The largest is in the Amazon Basin of South America, with extensions into coastal Brazil, Central America, and eastern Mexico, as well as some of the islands in the West Indies. In Africa, there is a large area of rainforest in the Zaïre Basin, with an extension along the west coast of Liberia. The third area

climbing a mountain in the tropics is analogous to traveling farther north (or south) from the equator. Humboldt illustrated this point with his well-known diagram of the zones of vegetation on Mount Chimborazo in Ecuador, which he climbed. On this mountain, he reached the highest elevation on record attained by any human being up to that date.

On leaving Latin America in 1804, Humboldt visited the United States for eight weeks. He spent three of these weeks as Thomas Jefferson's guest at Monticello, talking over many matters of mutual interest. It is thought that Humboldt's enthusiasm for learning about new lands encouraged Jefferson's own great scheme for the exploration of the western United States. Thus, it is fitting that Humboldt's name is commemorated in the names of several counties, mountain ranges, and rivers in the American West.

of rainforest extends from Sri Lanka and eastern India to Thailand, the Philippines, the large islands of Malaysia, and a narrow strip along the northeastern coast of Australia (Figure 30–2). In the North American region, tropical rainforest exists only in southern Mexico and on Puerto Rico and the northern side of Hispaniola in the West Indies (Figure 30–8). Temperate rainforests, dominated by broad-leaved evergreen trees like those in the tropics, occur along the eastern side of Australia, in Tasmania, and in parts of New Zealand, as shown in Figure 30–2.

The tropical rainforest now forms about half of the forested area of the earth, but it is in the process of being destroyed systematically by human activities. The rapidly expanding human population in the tropics, coupled with traditional patterns of land ownership, have made the usual practices of tropical agriculture—clear-cutting of the forest, followed by short-term cultivation—immensely destructive when carried out on such a wide scale. Many kinds of plants, animals, and microorganisms will become extinct in the course of this destruction, as we shall discuss in the next chapter.

One reason for the rapidity with which tropical forests disintegrate under human pressure has to do with the nature of tropical soils. Many of these soils are conditioned by high and constant temperatures and abundant rainfall and are relatively infertile. A majority of

many mineral nutrients are locked up in the plants themselves, not in the soil (which is the case in many temperate communities). More than half of the soils of the tropics are acidic and deficient in calcium, phosphorus, potassium, magnesium, and other nutrients. Furthermore, the phosphorus in such soils tends to combine with iron or aluminum to form insoluble compounds that are not available for plant growth, and the soils often contain toxic levels of aluminum ions. Plant roots tend to spread out in a thin layer no more than a few centimeters in depth, and they rapidly transfer the nutrients released by the decomposition of fallen leaves and branches back into the living plants. Below this thin layer of topsoil, which is easily destroyed during the process of clearing away the trees, there is virtually no organic matter. For all of these reasons, the cultivation of tropical soils presents formidable problems, even though nutrients may be available in abundance right after the soils have been cleared and crops may be good for two or three years. Despite these relationships, the tropical forests of the world are being cut and burned at an ever-increasing rate, mainly to produce fields that become completely useless to agriculture within a few years. It is estimated that, by early in the next century, most of the tropical rainforests will have disappeared, with the exception of those in the western Amazon Basin and central Africa.

30–8
Tropical rainforest in North America. Such forest also occurs over part of Hispaniola and Puerto Rico in the West Indies.

30–9

A savanna in Kenya, with zebras, a giraffe, and an impala. The transitional nature of this biome, relative to the characteristic vegetation of the tropical rainforest biome and the desert biome, is evident in the grasses, shrubs, and short trees seen here. The trees in the background are acacias.

Savannas and Deciduous Tropical Forests

Savannas consist of grassland with scattered broad-leaved deciduous and evergreen trees, which may occur singly or in groves. Some savannas are dominated by trees, often thickly so, and others are dominated by shrubs. Savannas occupy the region between the evergreen tropical rainforest biome and the desert biome (Figure 30–9). Savanna trees are generally deciduous and lose their leaves in the dry season. Savannas cover large areas of East Africa (Figure 30–2), but they are also found on the margins of rainforests everywhere, where the rainfall is seasonal and limiting. Savannas also occur in the region between the prairies and temperate deciduous forests and in the region between prairies and taiga in North America and over much of eastern Mexico, Cuba, and southernmost Florida (Figure 30–10).

Savannas usually have much lower annual rainfall than tropical rainforests—frequently in the range of 90 to 150 centimeters a year. There is also a wider range in average monthly temperatures, owing to the seasonal drought and the sparse covering of vegetation. The thorn forest (cerrado) that covers wide areas in Brazil belongs to the savanna biome. Savannas and similar seasonally deciduous tropical and subtropical plant communities grade into tropical rainforests as the rainfall becomes higher and its seasonal distribution more

even, and corridors of tropical rainforest (called gallery forest) extend out into the savannas along rivers. At their poleward margins, savannas and related plant communities grade into deserts.

In savannas, because of the scattered distribution of trees, the ground is generally well illuminated, and perennial herbs (mostly grasses) are common. Bulbous plants, which are able to withstand periodic burning, are abundant. Because of the dense cover of perennial herbs made possible by the abundant seasonal rainfall, there are few annual herbs. Epiphytes are also rare.

Savanna trees often have thick bark. They are well branched, but they are seldom more than 15 meters tall. Nearly all are deciduous; they lose their leaves at the start of the dry season and flower when they are leafless. Their leaves are generally smaller than those of the evergreen trees of the rainforest, and so they lose less water by transpiration.

In Southeast Asia, there are extensive areas of monsoon forest (Figure 30–2), which resembles savanna in being seasonally dry. In the monsoon forest, however, the density of trees is much higher than it is in savannas, and the trees and shrubs are mainly broad-leaved and deciduous. Monsoon forests are one particularly widespread kind of seasonally deciduous forest, and a number of others are found in different

30–10
*Distribution of savannas in North
America.*

30–11
*Distribution of subtropical mixed forests
in North America.*

parts of the tropics. Monsoon forests are also found in
the northern Yucatán Peninsula in Mexico and in eastern Brazil (Figure 30–2). In the areas where such forests
occur, there is high precipitation during portions of the
year, when the moisture-bearing monsoon wind blows
steadily off the ocean, but there is also a well-defined
dry season during which the trees lose their leaves.

Other kinds of deciduous and partly deciduous forests occur in the tropics and subtropics, in areas where
there is a pronounced dry season. Most of Florida and
extensive areas throughout the southeastern United
States are covered by subtropical mixed forests (Figure
30–11). In these forests, pines (Figure 30–12) and other
evergreen trees are mixed with deciduous trees; the
precipitation occurs mainly in the summer. Tropical
mixed forests, in which evergreen trees and shrubs are
better represented than in subtropical mixed forests,
occur locally in eastern and southern Brazil and northern Australia (Figure 30–2). Related communities—
subtropical and tropical mixed forests and woodland
and scrub—are found in Argentina (Figure 30–2).

The existence of the savannas (Figure 30–9) and
monsoon forests of Africa and other regions depends to
a large extent on periodic burning. The winter is long
and hot, and people frequently burn the plain to encourage the growth of young grass to support domestic
herds of both grazing animals and game animals.

30–12
Slash pine (Pinus elliottii) *is one of the
widespread evergreen tree species that
occur in the subtropical mixed forests
of the southeastern United States.*

Deserts

The great deserts of the world are all located in the zones of atmospheric high pressure that flank the tropics at about 30° north latitude and 30° south latitude, and they extend poleward in the interior of the large continents (Figure 30–2). Many deserts receive less than 20 centimeters of rainfall per year. In the Atacama Desert of coastal Peru and northern Chile, the average rainfall is less than 2 centimeters per year. Extensive deserts are located in North Africa and in the southern part of the African continent, where the Namib Desert is inhabited by some of the world's most extraordinary organisms, including *Welwitschia* (see Figure 17–36 on page 378). Other deserts occur in the Near East, in western North America and western South America, and in Australia. The Sahara, which extends all the way from the Atlantic coast of Africa to the Arabian peninsula, is the largest desert in the world. Australia's desert, which covers some 44 percent of the continent, is the next largest. Less than 5 percent of North America is desert (Figure 30–13).

The temperatures in many deserts are very high; summer temperatures of more than 36°C are common in some deserts. Other deserts are cooler. In the Great Basin Desert of western North America, which lies between the Sierra Nevada–Cascade Mountain system and the Rocky Mountains, there are only a few weeks of high temperatures each year. In general, however, the water vapor in desert air, even when relatively abundant, fails to condense because of the high temperatures. Because the cover of vegetation is generally sparse, heat normally radiates from deserts rapidly at night, creating large differences in temperature in the course of a 24-hour period.

The annual distribution of rainfall in deserts generally reflects that of the adjacent areas. On the equatorial side, it rains in the summer; on the poleward side, it rains in the winter. Between the two, as in the lowlands of Arizona, there may be two annual peaks of precipitation. As a result, in such an area there are generally two periods of active plant growth—one in the winter and one in the summer—and different plants are active in each period. In general, the patterns of activity of desert plants reflect the origins of the plants. Characteristically, plant species that have migrated into the desert from areas where they grew actively in the winter continue doing so in the desert. Similarly, plant species that have migrated into the deserts from areas where they grew during periods of summer rainfall continue to grow actively in the summer in the desert, provided that there is a sufficient amount of summer rainfall to make this possible.

Annual plants are better represented, both in number and in kind, in the deserts and semiarid regions

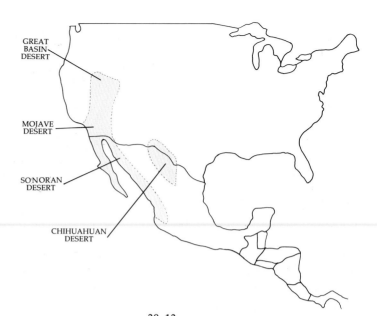

30–13
*North American deserts. The Sonoran Desert stretches from southern California to western Arizona and south into Mexico. Some of its characteristic plants are shown in Figure 30–14a through c. North of the Sonoran Desert is the Mojave Desert; one of its characteristic plants is the Joshua tree (*Yucca brevifolia*; Figure 30–14d). The Mojave Desert includes Death Valley, the lowest point on the continent (90 meters below sea level). Death Valley is only 130 kilometers from Mount Whitney, which, with an elevation of more than 4000 meters, is the highest point in the contiguous United States. The Mojave Desert blends into the Great Basin Desert, a cold desert that is bounded by the Sierra Nevada to the west and the Rocky Mountains to the east. Vast stretches of the Great Basin Desert are dominated by sagebrush (*Artemisia tridentata*; Figure 30–14e) and rabbitbush (*Chrysothamnus*). To the east of the Sonoran Desert is the Chihuahuan Desert.*

of the world than anywhere else. Because of the erratic supply of water, perennial herbs do not succeed well in deserts and semideserts, and there is no dense covering of perennials to inhibit the growth of annuals. Annuals, with their very active growth, can germinate and complete their life cycles in the open areas during the limited periods when water is available. The seeds of these plants can survive in the soil during the long periods of drought, which sometimes extend over many years. Then, when sufficient water is available to induce their

(a)

(b)

(c)

30–14

Desert plants. (a) Community of saguaro cactus (Carnegiea gigantea) *and barrel cactus* (Ferocactus) *in the Sonoran Desert of southern Arizona. (b) Washington palms* (Washingtonia filifera) *growing along a creek in Palm Canyon, in the Sonoran Desert of southern California. (c) Coastal desert of Baja California, Mexico, which receives much of its moisture from fog, dominated by boojum trees* (Fouquieria columnaris, *formerly called* Idria) *and agaves* (Agave shawii). *(d) Joshua trees* (Yucca brevifolia) *at Joshua Tree National Monument in the Mojave Desert of southern California. (e) Great Basin Desert, dominated by sagebrush* (Artemisia tridentata), *in the John Day Basin of eastern Oregon.*

(d)

(e)

germination and support their growth to flowering, the seeds can germinate rapidly.

The relatively few perennial herbs that grow in the desert are often bulbous and dormant for much of the year. Most of the taller plants either are succulents—such as the cacti, euphorbias, and other characteristic desert plants (Figure 30–14)—or else have small leaves that either are leathery or are shed during unfavorable seasons (Figure 30–15). Usually the leaves have a thicker cuticle and fewer stomata than those of plants in less arid regions. Many desert plants have green stems, rich in chlorophyll, which may contribute large amounts of photosynthate to the plants; such plants often lack leaves, as do most species of cacti. Many of the succulent plants have adopted CAM photosynthesis (see page 116) and absorb carbon dioxide only at night; their stomata remain closed during the day. C_4 photosynthesis is likewise more common among the plants of deserts and other seasonally dry, warm habitats than it is elsewhere (see page 114).

30–15
*One of the most characteristic plants of
the Mojave, Sonoran, and Chihuahuan
deserts of North America is the creosote
bush (Larrea divaricata). Creosote
bushes in the Mojave Desert may form
circular or elliptical clones because of
new branch production at the periphery
of stem crowns and the segmentation and
death of older stem segments, resulting
in a ring of satellite bushes around a
central bare area. The latter usually accumulates a mound of sand, which may
reach a depth of about a half meter and
may function as a water-storage medium,
thus contributing to the survival of
creosote bushes in periods of drought.
Some clones may attain extreme ages: the
King Clone, shown here, is estimated to
be nearly 12,000 years old. Such ancient
clones started from seeds that germinated
near the end of the last glacial expansion.*

30–16
Juniper savanna in North America.

The temperatures at which the maximum photosynthetic rate occurs in desert plants are often much higher
than the corresponding temperatures for the plants of
other biomes. For example, in *Tidestromia oblongifolia*, a
C_4 perennial herb common in the Death Valley region
of California and Nevada, the maximum photosynthetic rate is achieved at temperatures between 45° and
50°C in full sunlight in midsummer. In addition, the
leaves of many desert plants are small, the reduction in
surface area being another adaptation to conserving
water. The leaves may also be oriented in such a way as
to minimize heat absorption, or they may be covered
with a dense coat of hairs for the same reason. The
trees and shrubs of deserts either have wide-ranging
roots that effectively absorb the periodic rainfall or are
restricted to washes and arroyos, along which underground water is concentrated.

The juniper savannas that are such a familiar sight
throughout the interior of the western United States
(Figures 30–16 and 30–17) occur in areas of cold desert,
but generally at higher elevations. During the pluvial
periods of the Pleistocene epoch, at the times of maximum expansion of the continental glaciers, such plant
communities moved down into lowland areas that have
subsequently become, for the most part, deserts. Because the preservation of plant materials is so good in
the dry desert air, the shifting patterns of vegetation in
a particular area can be traced in the dried leaves and
stems in old nests of pack rats.

How Does a Cactus Function?

The barrel cactus *Ferocactus acanthodes*, which grows along the northwestern edge of the Sonoran Desert in southern California, gets its name because it is shaped like a barrel. It looks as if its stem has been folded like a fan, and it is covered with spines like a porcupine. Why should this desert plant present such an unusual and formidable appearance?

When the barrel cactus is full of water, the folds swell and are barely visible, but when the plant dries, the folds are deep and the stem is able to contract without crushing the cells. This ridge-and-valley folding of the stem has other advantages, too. Deep inside the valleys between the folds are the stomata. The spines help to break up the wind currents, and thus the valleys serve as protected retreats where dry desert winds cannot easily reach to carry moist air away from the vicinity of the stomatal chambers. These formidable spines also help to protect the cactus from the rodents and birds that are in constant search of water, even going so far as to steal it from a succulent stem. Park Nobel, of the University of California, Los Angeles, found that barrel cacti stored enough water in their succulent stems to permit them to open their stomata for about 40 days after the soil became too dry to furnish them with any additional water. Under such conditions, many of the fine roots are sloughed off, to prevent water loss to the soil. After seven

months of drought, stomatal activity ceased, and the osmotic potential of the stem was more than double the value it had been during wet periods, in spite of the ability of the stem to fold and shrink. When rain finally returned, the shallow root systems (mean depth of only about 8 centimeters) took in water so rapidly that the stomata were fully functional again within 24 hours of rainfall.

But these are not the only mechanisms the barrel cactus uses for conserving water. Like many other desert plants, the barrel cactus exhibits Crassulacean acid metabolism (CAM). It opens its stomata only at night and so faces cooler air, which can hold less water than warmer air; consequently, the plant loses less water to the atmosphere through transpiration. Its ratio of mass of water transpired to mass of CO_2 fixed is only about 70 for the entire year, considerably lower than for a typical C_3 plant, which requires larger quantities of water to fix an equivalent amount of carbon but has a higher maximum rate of photosynthesis.

Since seedlings of the barrel cactus cannot tolerate extremely high temperatures and prolonged drought like their parents can, they survive only in certain years and in protected microhabitats. By the age of 26 years, these plants have usually grown to only about 34 centimeters tall, adding about 10 percent to the mass of their stems each year.

30–17

(a) *Juniper (Juniperus osteosperma) savanna in the Great Basin near Wellington, Utah.* (b) *Cold winters are characteristic of the pinyon-juniper savannas and woodlands of the Great Basin, as shown here south of Moab, Utah.*

(a)

(b)

Grasslands

Grasslands form the zone that lies between deserts and temperate deciduous forests, occurring where the amount of rainfall is intermediate between the amounts characteristic of those two biomes. Included in grasslands are a wide variety of plant communities; some intergrade with savannas, others with deserts, and still others with temperate deciduous forests. As the amount of precipitation decreases, the major grasslands of the world often grade into deserts toward the equator. The more fertile grasslands receive as much as 100 centimeters of precipitation per year; they grade into temperate deciduous forests, where the moisture supply is even more abundant. Unlike savannas, grasslands generally lack trees, except along streams, and are characterized by cold winters.

Fires play a critical role in the development of grasslands. Toward the deserts, there is too little biomass to sustain the spread of fires; toward moister areas, such as temperate deciduous forests, trees often grow rapidly enough to escape the influence of ground fires, at least on fertile sites. Fire selects against plants with permanent aboveground parts and favors grasses, sedges, and other herbs that have basal meristems that are not easily destroyed by fire or by grazers. When disturbed, grasslands have often changed into either forests or deserts. Thus, the grasslands that once stretched from southern Arizona to western Texas, which were encountered by early European settlers, were largely overgrazed and converted to deserts during the past century. The failure of farmers and graziers to manage grasslands properly led to the "dust bowl" disasters of the central United States in the 1930s (Figure 30–18).

Grasslands have traditionally been heavily exploited for agriculture, both as pasture and, when plowed up, for cultivation. Much effort has been devoted to learning how to convert other biomes into grassland and, once the conversion has been made, how to maintain them in this condition. The most productive soils for temperate agriculture are found in areas formerly occupied by tall-grass prairies.

Grasslands generally occur over large areas in the interior portions of the continents (Figure 30–2). In North America, there is a transition from the more desertlike, western, short-grass prairie (the Great Plains), through the moister, richer, tall-grass prairie (the Corn Belt), to the eastern temperate deciduous forest (Figures 30–19 and 30–20). Grasslands become progressively drier with increasing distance from the Atlantic Ocean and the Gulf of Mexico, which are the major sources of moisture-bearing winds in the eastern half of the North American continent. Farther north, grasslands become moister again as evaporation decreases at relatively cool temperatures. "Typical" short-grass prairie occurs under near-drought conditions (Figure 30–21); in moister years, taller grasses (but not the very tall ones of the tall-grass prairie) tend to predominate. In many forested areas in the central and eastern United States, particularly where the soil is shallow, there are open areas, called glades, that are typically occupied largely by prairie plants (Figure 30–22).

30–18

The prairie soils were once so bound together with the roots of grasses that they could not be cultivated until adequate plows were developed. But once the plants were removed by overgrazing or careless cultivation, prairie soils rapidly deteriorated and were blown away by the wind. This photograph of a badly eroded farmyard, taken in Oklahoma in 1937, vividly recalls the "dust bowl" conditions that led many thousands of people to migrate away from the central United States. John Steinbeck's novel The Grapes of Wrath *was based on the experiences of these migrants.*

30–19

Distribution of the grasslands of North America. Aside from the main areas of prairie in the center of the continent, there are disjunct areas of grassland in and around the Central Valley of California and along the Gulf Coast of southeastern Texas and Louisiana. The soils in all of these areas are fertile, of good structure, and excellent for sustained agriculture.

30–20

(a) Tall-grass prairie in the Flint Hills of eastern Kansas. This beautiful area has been proposed as the site of a prairie national park. (b) Hill prairie near Alton, Illinois, dominated by big bluestem (Andropogon gerardi). *(c) The rich hill prairies are almost entirely under cultivation, as in this Illinois scene. Soybeans* (Glycine max) *are being grown in the field.*

(a)

(b)

(c)

(a)

30-21

(a) *Healthy short-grass prairie in northeastern Colorado, with a young pronghorn antelope* (Antilocapra americana). *The bluish clumps of vegetation are fringed sagebrush* (Artemisia frigida). *(b) Overgrazed short-grass prairie in northeastern Colorado.*

(b)

30-22

Coreopsis *dominating a glade in temperate deciduous forest near St. Louis, Missouri. Such openings in the forest, which usually occur in areas of shallow soil, are often dominated by prairie plants that form continuous stands beyond the borders of the forest.*

Perennial bunchgrasses and sod-forming grasses are dominant, but other perennial herbs are common. Although the growth of grassland plants is seasonal, there is little room for the development of annual herbs, and these are essentially absent from this biome. Occasionally, annuals and introduced weeds become established in disturbed areas, such as around the burrows of animals, near buildings, and along roadsides and railroads.

All of the great grasslands of the world were once inhabited by herds of grazing mammals associated with large predators. Such herds were widespread in the periods of maximum expansion of the glaciers during the Pleistocene epoch, when they were widely hunted by our ancestors. Many of these grazing mammals, such as the American bison, have since been hunted almost to extinction. They survive today mainly in refuges, having given way to herds of domestic animals and cultivated fields.

Temperate Deciduous Forests

Temperate deciduous forests are almost absent in the Southern Hemisphere, but they are represented on all the major land masses of the north (Figure 30-2). This type of forest reaches its best development in areas with warm summers and relatively cool winters (Figures 30-23 and 30-24). Annual precipitation generally ranges from about 75 to 250 centimeters and is either

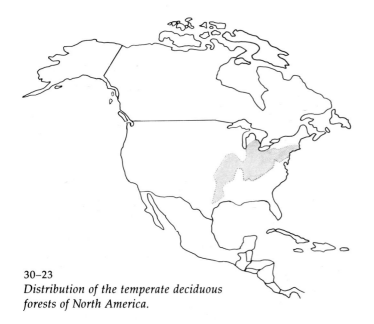

30–23
Distribution of the temperate deciduous forests of North America.

distributed evenly throughout the year or concentrated somewhat during the summer months. The deciduous (leaf-dropping) habit may be related to the unavailability of water during much of the winter, which is a consequence of soil temperatures below the freezing point. These conditions delay the movement of water to the roots. In the colder regions occupied by the temperate deciduous forest, the soils are frozen for many months. In effect, the winters in regions occupied by temperate deciduous forest are analogous to the hot, dry seasons that occur in savannas–deciduous tropical forests. The unavailability of water in both biomes is associated with the seasonal loss of leaves in trees and shrubs.

The ecology of temperate deciduous forests differs from that of evergreen forests in the nature of the annual cycle of growth (Figure 30–25). In winter, the trees are leafless and their metabolic activity is greatly reduced. In spring, a variety of herbaceous plants bursts

(a)

(b)

(c)

(d)

30–24
The deciduous forests of the southern Appalachians are among the richest temperate-zone forests in the world, containing many kinds of hardwoods.
(a) *Mixed forest with hemlocks* (Tsuga canadensis) *in Tennessee.* (b) *A deep cove, with rich deciduous forest.* (c) *In fall, the leaves of the maples and eastern sourwoods* (Oxydendrum) *turn a beautiful scarlet.* (d) *Several species of* Trillium *are among the herbaceous perennials that occur in these forests.*

Competing for Light

Forest herbs show a remarkable diversity in leaf height and arrangement. Some species, such as rattlesnake plantain (a), bear foliage at or near ground level. Others, such as trillium (see Figures 1–3b and 30–24d) and bloodroot, hold their leaves 20 to 50 centimeters above the ground. Yet others, such as jewelweed, spread their crowns a meter or more above the ground. Most forest herbs have umbrella-like canopies, but some have several layers of leaves along an erect stem, and a few have tightly packed foliage along an arching stem. What is the significance of these variations, and how are they related to the distribution of individual species?

Recent studies by Thomas Givnish, of the University of Wisconsin at Madison, have shed light on some of these issues. According to his analysis, leaf height is likely to affect an herb's growth relative to its competitors—and hence its competitive ability—primarily through its effect on the way the herb captures light and allocates energy. Taller herbs, which capture more light than shorter ones, spend more of their annual energy income from photosynthesis to build unproductive stems at the expense of productive leaves. Herbs should evolve taller and taller canopies until the growth advantage each obtains from increased height is balanced by the growth loss caused by reduced energy allocation to leaves. In sparsely covered areas, there is little advantage to growing tall and "wasting" a large amount of energy on stems, and so short plants are favored. In densely covered areas, taller plants are favored because they are better able to capture light. Indeed, there is evidence to support these predictions. Tall species—such as jewelweed, stinging nettles, and many asters and goldenrods (b)—are usually found in dense stands under partially open canopies, particularly on moist, fertile floodplains. Short species—such as rattlesnake plantain and striped wintergreen—are found in areas of sparse herb cover, especially in dry, infertile oak forests.

An intriguing variation among forest herbs involves an arching growth form. Solomon's seal and similar species with leaves packed along arching stems (see page 636) are less efficient than umbrella-like herbs, in that they require more stem tissue to hold the same amount of leaf mass a given height above level ground. On steep slopes, however, such plants gain an advantage because they orient downhill; as a result, erect plants rooted downslope require longer stems to hold their leaves in the same position. Again, as predicted, in virgin forests, dominance by arching herbs increases dramatically on slopes steeper than approximately 20°.

These findings illustrate the general principle that different environments favor different growth forms and patterns of energy allocation: adaptation and competitive success are strongly dependent on the conditions in which the plant grows.

(a)

(b)

30–25

The four seasons in a temperate deciduous forest in Illinois. The trees leaf out early in spring and begin to manufacture food; they lose their leaves in autumn and enter an essentially dormant condition, in which they pass the unfavorable growing conditions of winter. Many herbs grow under the trees (see Figure 30–24d), and a number of them flower very early in spring, before the tree leaves reach full size and shade the forest floor. In spring, most of the trees produce abundant pollen, which is carried by the wind.

forth in profusion on the well-illuminated forest floor (Figure 30–24d). Some species (the **spring ephemerals**) have leaves that spring fully formed from bulbs or rhizomes and complete their period of activity in sunlight before the trees leaf out overhead. Others (for example, **early** and **late summer species,** and **evergreen species**) emerge more slowly and sustain photosynthetic activity longer and later during the shady conditions prevailing in summer. Summer-active species generally have leaves that are broader but thinner in cross section

677

than those active only in spring and usually have smaller storage organs than species with shorter growing periods. Forest herbs vary in height from a few centimeters to more than one meter; variation in leaf height appears to be related to the typical density of the competing foliage that different species encounter in their microenvironments (see page 676). Most of the species that ripen their seeds in spring are dispersed by ants (pages 434 to 435), which are active when few other dispersers are present; most of the species that ripen seeds in the fall are dispersed by birds, at a season that coincides with the height of fall migration. Very few annual plants occur in deciduous forests, probably because the dense plant cover in the moist understory puts seedlings at a strong competitive disadvantage.

The soils in regions occupied by temperate deciduous forest are usually acid, and they tend to contain relatively low amounts of nutrients. For these reasons, these soils are easily depleted of nutrients after the forests are cleared, and they may become highly infertile following years of intensive cultivation. Prairie soils are far more fertile and have characteristics much more favorable to sustained cultivation.

One of the outstanding characteristics of the temperate deciduous forest is that the plants found in each of its three main regions in the Northern Hemisphere are similar to one another, often representing related species of particular genera. For example, the herbaceous plants of the deciduous forests of Japan and east Asia generally resemble those of eastern North America more closely than either group resembles those of western North America, where temperate deciduous forests mostly disappeared millions of years ago.

Temperate Mixed and Coniferous Forests

Bordering the deciduous forests on the north are mixed forests, in which conifers form an important element along with the deciduous trees. Such temperate mixed forests (Figures 30–26 and 30–27) are characteristic of the Great Lakes–Saint Lawrence River region, much of the southeastern United States, eastern Europe, the northern and eastern border regions of Manchuria and adjacent Siberia, eastern Korea, and northern Japan (Figure 30–2). Temperate mixed forests sometimes occur in areas with colder winters than the areas where temperate deciduous forests occur. Temperate mixed forests are also partly characteristic of regions with more nutrient-poor soils or with less seasonal environments. Temperate mixed forests represent a broad transitional zone between the temperate deciduous forests to the south and the taiga to the north.

30–26
Distribution of temperate mixed forests in North America.

30–27
Temperate mixed forest in southern Ontario, with the evergreen conifers appearing dark green and the deciduous trees showing their fall colors.

Most of the deciduous trees and associated herbs were eliminated in western North America during the latter half of the Cenozoic era as the amount of summer rainfall was greatly reduced. In their place now grow the mountain coniferous forests and west-coast forests of western North America (Figure 30–29), which contain such trees as the coast redwood (*Sequoia sempervirens;* Figure 30–30), the big tree or giant sequoia (*Sequoiadendron giganteum),* the Douglas fir (*Pseudotsuga menziesii),* and the sugar pine (*Pinus lambertiana;* see Figure 29–14b), all of which were much more widely distributed in the past (see the essay "Jobs versus Owls" on page 682). Similar kinds of vegetation are found in areas with similar climatic characteristics in Scandinavia, central Europe, the Pyrenees, the Caucasus, the Urals, southern Tibet, and the Himalayas northward to eastern Siberia. Vegetation types that resemble these are also found in disjunct areas in western South America, central New Guinea, southwestern New Zealand, the southern Arabian Peninsula, Ethiopia, and the mountains of Central Africa (Figure 30–2). At higher elevations in these regions are found the open grassland communities called alpine tundra (Figure 30–28), which are intermixed with mountain forests from the Brooks Range in southern Alaska southward into the Rocky Mountains, the Cascades, and the Sierra Nevada (Figure 30–29).

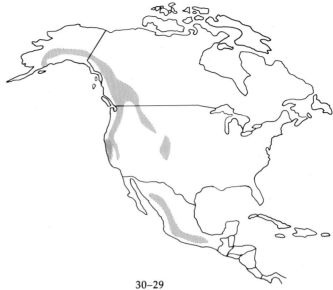

30–29
Distribution of mixed west-coast forests (pink), alpine tundra, and mountain forests (blue) in North America. The alpine tundra occupies limited areas at high elevations within the area of mountain forests.

30–30
Redwoods (Sequoia sempervirens) *are a prominent feature of the mixed west-coast forests of California. Watered by the frequent fogs of the region during the dry summers and protected from freezing temperatures by their proximity to the ocean, redwoods often form spectacular groves, many of which, like the one shown here in Muir Woods near San Francisco, are now protected in parks and reserves.*

30–28
Alpine tundra on the Olympic Peninsula in Washington. Such alpine tundra is comparable in many respects with the Arctic tundra found hundreds of miles to the north. Here, however, forested slopes are found within a hundred meters or so of the alpine meadows.

Mediterranean Scrub

Highly distinctive scrub communities have evolved from mixed deciduous–evergreen forests in areas with mediterranean climates, which are characterized by cool, moist winters and hot, dry summers (Figure 30–31). Such climates are found along the shores of the Mediterranean Sea, over a large part of California (and for a short distance to the north and south of that state), in central Chile, in southwestern Africa, and along portions of the coast of southern and southwestern Australia (Figure 30–2). The plants in these areas—often evergreen or summer-deciduous trees and shrubs—have relatively short growing seasons that are restricted to the cool part of the year, when moisture is relatively abundant. They may lock up nutrients efficiently in their evergreen leaves. In mediterranean climates, the luxuriant growth of spring is followed by drought and dormancy during the summer.

Fire is a prominent ecological factor in mediterranean-type vegetation, and fires apparently occurred regularly in such plant communities long before they were invaded by human beings. Today, fire can be a serious problem in such areas as southern California, where homes extend far up into the chaparral. *Chaparral* is the word used in California and elsewhere in western North America for evergreen, often spiny shrubs that typically form dense thickets. Similar vegetation types have evolved elsewhere, particularly in areas characterized by mediterranean climates, and have been given different names locally; the equivalent vegetation around the Mediterranean Sea is called *maquis* (Figure 30–1); in Chile, *matorral;* in South Africa, *fynbos.*

As is true of the deserts of the world, each area of mediterranean climate is isolated from the others, and each has evolved its own distinctive assemblage of plants and animals. The degree of ecological convergence is high, however, and the chaparral of California is quite similar in appearance to the matorral of Chile or the maquis of the Mediterranean, even though the plants are mostly unrelated. Seasonal drought enhances the importance of edaphic (soil-related) and biotic variation, and small differences in precipitation often have profound effects on the vegetation and animal life present in the area. Hence, these areas often have high proportions of extremely local species of plants and animals, many of them now in great danger of extinction. In their modern form, these areas have already been profoundly changed by people; much of their vegetation now occurs in a condition that is very different from what it was before people occupied the areas with their grazing animals—for example, today's areas might be inhabited by more shrubs and fewer trees than formerly, or by more spiny and poisonous plants.

30–31

Mediterranean scrub, locally called chaparral, in the mountains near Los Angeles, California. This sort of plant community, consisting of broad-leaved, drought-resistant, and often spiny evergreen shrubs, occurs only in limited areas of the world, but it has evolved independently on five continents in areas having mediterranean (summer-dry) climates.

30–32
The dense, tall forest that grows on the rainy slopes and flats of the Olympic Peninsula in Washington is the southern expression of the taiga, or northern coniferous forest. Epiphytic mosses, liverworts, Selaginella, and lichens often grow luxuriantly on the trees, as seen here.

30–33
The northern taiga, which covers hundreds of thousands of square kilometers in the cooler part of the North Temperate Zone, is dominated by white spruce (Picea glauca) and larch (Larix), which are smaller than shown here in the northern part of their range. This photograph was taken in northern Manitoba, Canada.

Taiga

This northern coniferous forest, or boreal forest, is characterized by a persistent cover of snow in the winter and by a severe climate toward the north. In the southern reaches of the taiga, the trees are taller and more luxuriant, often reaching 75 meters or more in height (Figure 30–32), but in its main, northern area, the trees are shorter, and thousands of square kilometers are covered by this uniform forest, with relatively few species of plants and animals (Figure 30–33). *Taiga* is the Russian word for vegetation of the sort found in this biome, which extends over much of Russia, Scandinavia, and northern North America (Figure 30–34; see also Figure 30–2). It is evergreen almost everywhere except in large areas of northeastern Siberia, where larches *(Larix)*, which are deciduous conifers, are dominant.

Taiga occurs in the interior of large continental masses at the appropriate latitudes and extends along the Pacific coast south to California. In such regions, extreme temperatures range from −50° to 35°C. Taiga is flanked on the south by montane forests (as in west-

30–34
The distribution of taiga in North America.

Jobs versus Owls

Some of the world's most magnificent forests grow along the western side of the Cascade Mountains in Washington, Oregon, and northern California. In the Olympic Peninsula of Washington, these forests are truly temperate rainforests, with more than 400 centimeters of rain per year; other forests, more distant from the coast and father south, are drier. In ancient forests, also known as old-growth forests, the huge trees are more than 250 years old; some of them reach heights of nearly 100 meters and diameters of approximately 2 meters. These forests are biologically diverse, representing a late successional stage with well-developed understories of herbs, ferns, mosses, and other plants.

Originally, such ancient forests probably covered more than 76,000 square kilometers in the Pacific states. Estimates of the area now remaining vary widely because of disagreements about what constitutes an ancient forest, but it is not more than one-third of the original area and may be as little as one-sixth. Furthermore, the forests that are left tend to be highly fragmented, with the penetration of sun, wind, and other factors that do not affect undisturbed forest causing the loss of species and leading to diminished genetic variability.

An enormous controversy erupted in Oregon and Washington in 1989 when the U.S. Forest Service proposed taking more than 1500 square kilometers of national forest out of production in order to protect the northern spotted owl, which had recently been declared an endangered species by the U.S. Fish and Wildlife Service. This set-aside would reduce the available timber supply by about 5 percent and result in the loss of about 3000 jobs; it is projected to be of sufficient size to protect about 1300 pairs of owls.

In a broader sense, however, the controversy in the Pacific Northwest does not concern owls versus jobs at all; it concerns regional values and the way in which we choose to treat the remnants of ancient forest, centuries old and fundamentally important to the appearance and sound ecological functioning of the region. Economically, tourism is even more important than lumbering in the Pacific Northwest, but the benefits of tourism accrue more to city-dwellers than to the inhabitants of rural towns, who may be directly dependent on the timber industry for their livelihoods. If all the ancient trees are cut—the only remaining ones, for all practical purposes, are on federal lands—the jobs will still run out in a few years, and the transition to sustainable forestry, or to other forms of employment, must then be made in any case. Although relatively few species of plants and animals seem to be absolutely restricted to ancient forests, these forests do protect many kinds of organisms other than the owls.

Ancient forests of Douglas fir (Pseudotsuga menziesii) at the H.J. Andrews Experimental Forest in the western Cascade Range of Oregon.

As is true for ecological problems in general, people must find new ways to determine their overall priorities and not necessarily chop down every tree in an effort to "protect" jobs over the short term. A country that has the will to do so can find better ways to take care of its people and encourage the development of jobs than to use irreplaceable natural resources. Meanwhile, the high prices paid by Japan for unsawn logs from the Pacific Northwest will continue to ensure the rapid destruction of the ancient forests, and the owl will remain as a symbol of the deeper regional values that are being lost permanently.

ern North America), deciduous forests, savannas, or grassland, depending on the amount of precipitation in the region. Because continental masses do not occur at the appropriate latitudes in the Southern Hemisphere, taiga is absent there (Figure 30–2). Owing to the influence of the prevailing westerlies blowing over relatively warm ocean currents between 40° and 50° north latitude, the western portions of North America and Eurasia are characterized by milder climates than their eastern portions. Consequently, taiga is found somewhat farther north toward the Pacific coast than it is along the Atlantic coast (Figure 30–34), and the same is true of the individual distributions of many kinds of plants and animals that inhabit the biome. The northern limits of taiga are ultimately determined by the severity of the Arctic climate, reaching a limit where the maximum monthly temperature is approximately 10°C.

In the extensive northern reaches of the taiga, most of the precipitation falls in the summer; the cold winter air in these regions has a very low moisture content. The annual total precipitation usually amounts to less than 30 centimeters. The rate of evaporation is low, however, and lakes, bogs, and marshes are common (Figure 30–33). More than three-quarters of the northern reaches of the taiga is underlain by permanent ice, or **permafrost,** usually within less than a meter of the surface; the permafrost tends to trap surface water and form lakes. Fires are common in the taiga, and they result in generally warmer, more productive sites for at least 10 to 20 years afterward, owing to the local melting of the permafrost. In general, taiga soils are highly acidic, very low in nutrients, and poorly suited for agriculture.

Species of a few genera of trees are common in the northern taiga, including spruce (Picea), larch (Larix), fir (Abies), and poplar (Populus). Among the more common shrubs are dewberries (Rubus), Labrador tea (Rhododendron), willows (Salix), birches (Betula), and alders (Alnus). Occasionally, in warm, dry areas, groves of pine (Pinus) are found. The members of all these genera of trees and shrubs are ectomycorrhizal, and they occur in dense stands consisting of only one or a very few species. Perennial herbs are common; mosses and lichens are especially prevalent, often forming luxuriant masses. Annual plants are essentially absent.

At its northern limits, taiga grades unevenly into tundra. In both these biomes, the days are long during the relatively short growing season; north of the Arctic Circle, the sun does not set during at least part of the summer. Because of the seasonally abundant light and favorable temperatures, cool-season cultivated plants, such as cabbages (Brassica oleracea var. capitata), may grow rapidly in cleared areas in the taiga, attaining large sizes in a remarkably short period of time. Yet the infertile, highly leached soils of the taiga are inimical to most forms of agriculture.

Arctic Tundra

Arctic tundra is a treeless biome that extends to the farthest northern limits of plant growth (Figure 30–35). It occupies an enormous area: fully one-fifth of the earth's land surface (Figure 30–2). Arctic tundra is best developed in the Northern Hemisphere and is mostly found north of the Arctic Circle, although it extends farther south along the eastern sides of the continents than along the western sides (Figure 30–36). The Arctic tundra essentially constitutes one huge band across Eurasia and North America, with alpine tundra, more closely related to the adjacent mountain forests, extending southward in the mountains (Figure 30–28; see also Figure 30–6). Some species of plants that occur in the Arctic tundra have wide circumpolar ranges.

The entire region occupied by Arctic tundra is underlain by permafrost. The soils are acidic to neutral, low in nutrients, and generally poor for agriculture. Even though precipitation is usually less than 25 centimeters per year, much of it is held near the surface by underlying permafrost, and the ground is usually wet. Fixed nitrogen is generally in very short supply, and there are only a few species of legumes and other plants with symbiotic bacteria that fix atmospheric nitrogen. The evaporation rate is low because of the relatively high moisture content of the air and low temperatures. In contrast, some tundra areas are so dry as to constitute true polar deserts. Most of the land north of 75° north latitude is a desert or semidesert, with few plants taller than 5 centimeters. This is a result of the very low precipitation in both summer and winter and the very cold winter temperatures.

For plant growth to occur at all, the mean temperature must be above freezing for at least one month of the year. The growing season (the time from one killing frost to the next) in many areas of Arctic tundra is less than two months. Several genera of low shrubs, including birch (Betula), willow (Salix), blueberry (Vaccinium), and Labrador tea (Rhododendron), are common in the tundra, and there are a number of genera of perennial herbs but only one native annual species (Koenigia islandica). Many of the plants that grow in the Arctic tundra, including a number of the grasses and sedges, are evergreen, thus enabling them to initiate photosynthesis soon after adequate light, moisture, and temperature are available. Plant height is controlled primarily by snow depth in winter; large woody plants are absent also because of their relatively high allocation to unproductive stem tissue, which cannot be supported given the short, cool growing season. A number of tundra plants have relatively large, showy flowers, the production of which costs the plants considerable quantities of energy. Such flowers hold energy-rich rewards for their pollinators—necessary rewards, given

(a)

(b)

30–35
(a) *Arctic wet coastal tundra near Prudhoe Bay, Alaska, in late summer. The reddish brown plants are the grass* Arctophila fulva, *the green ones are the sedge* Carex aquatilis. *Note that the water table is above the surface because of the presence of permafrost; such conditions are characteristic of the Arctic tundra.* (b) *Arctic tundra at Barrow, Alaska. A clone of the sedge* Eriophrum angustifolium *growing out into a solid patch of another species of the same genus, E. scheuchzeri. Vegetative reproduction, like that in E. angustifolium illustrated here, is characteristic of many tundra plants.*

the low temperatures that prevail at high latitudes. Vegetative propagation is characteristic of many of the perennials, and this may be correlated with the uncertainties of setting seed during the brief Arctic summer. Much of the biomass of tundra plants—from half to as much as 98 percent—is underground, consisting not only of roots but also of rhizomes and other kinds of underground stems.

North of the Arctic tundra is ice desert, where physical conditions are even more extreme and vegetation is absent. Ice desert is characteristic of the interior of Greenland, Svalbard off Norway, and Novaya Zemlya off the north coast of Siberia. Much of Antarctica, not mapped in Figure 30–2, is also covered by ice.

Summary

Tropical rainforest, in which neither water nor low temperature is a limiting factor, is by far the richest biome in terms of number of species. The trees are evergreen and characterized by medium-sized leathery leaves. A poorly developed layer of herbs grows on the forest floor, but there are many vines and epiphytes at higher levels. Tropical soils are often acidic and very poor in nutrients; such soils lose their fertility rapidly when the forest is cleared.

Mostly tropical and subtropical communities that are characterized by a seasonal drought are termed savannas, subtropical mixed forests, monsoon forests, tropical mixed forests, and southern woodland and

30–36
The distribution of Arctic tundra in North America.

scrub. The trees and shrubs of these communities are wholly or partly deciduous, shedding their leaves during times of drought. Herbaceous perennials are common. Savannas also occur between the prairies and temperate deciduous forests and between the prairies and the taiga in the United States and Canada. Subtropical mixed forests cover most of Florida and the Coastal Plain of the southeastern United States. In these forests, evergreen trees, such as pines, grow intermixed with deciduous trees. Away from the equator, communities of these sorts grade into deserts and semideserts, which are characterized by low precipitation and often by high daytime temperatures during at least part of the year. Succulent plants and annual herbs are common in deserts.

Grasslands intergrade with savannas, deserts, and temperate forests; they are characterized by a general lack of trees, except along streams. The most productive soils for temperate agriculture are grassland soils.

In the temperate deciduous forests, most of the trees lose their leaves during the cold (usually snowy) winters, when moisture may be unavailable for growth. Many genera are common to the temperate deciduous forests of eastern North America and eastern Asia. Temperate deciduous forests are bordered by temperate mixed and coniferous forests to the north, in which conifers play an important role.

Distinctive scrub communities, called chaparral in North America and maquis in the Mediterranean region, have evolved in the five widely separated areas of the world with a mediterranean climate—a dry summer and a cool, rainy winter growing season. Such communities occur in western North and South America, around the Mediterranean, in the Cape Region of South Africa, and in southwestern Australia.

The taiga is a vast northern coniferous forest that extends in unbroken bands across Eurasia and North America and down the Pacific Coast to northern California. In its southern reaches, the taiga is dominated by tall trees with a lush growth of bryophytes and lichens; northward, it consists of vast monotonous stretches of forest with very few tree species. North of the taiga is the tundra, a treeless region that also extends around the Northern Hemisphere, mostly above the Arctic Circle, in a band that is broken only by bodies of water. Both the northern reaches of the taiga and all of the tundra are underlain by permafrost. Because of this permafrost, and especially because of the low rates of evapotranspiration, tundra and taiga soils are relatively moist and highly leached of nutrients.

CHAPTER 31

Plants and People

31–1

Hunter-gatherers in the Congo Forest of Central Africa. The man has just caught the turtle; the woman is carrying the family dog. People who must obtain their food in this way cannot build cities.

Our own species, *Homo sapiens*, has been in existence for at least 500,000 years and has been abundant for approximately 150,000 years. Like all other living organisms, we represent the product of at least 3.5 billion years of evolution. Our immediate antecedents, members of the genus *Australopithecus*, first appeared no less than 5 million years ago. At about that time, they apparently diverged within Africa from the evolutionary line that gave rise to the chimpanzees and gorillas, our closest living relatives. The members of the genus *Australopithecus* were relatively small apes that often walked on the ground on two legs.

Larger, tool-using human beings—members of the genus *Homo*—first appeared about 2 million years ago. They almost certainly evolved from members of the genus *Australopithecus*, but they had much larger brains, apparently associated with their ability to use tools and reinforced by it. Early members of the genus *Homo* probably subsisted mainly by gathering food (picking fruits, seeds, and nuts; harvesting edible shoots and leaves; and digging roots), scavenging dead animals, and occasionally hunting. They learned how to use fire no less than 1.4 million years ago. Their style of life seems to have resembled that of some contemporary groups of humans (Figure 31–1). Our species, *Homo sapiens*, appeared in Africa by about 500,000 years ago and in Eurasia by about 250,000 years ago.

About 34,000 years ago, the powerfully built, short, stocky Neanderthal people, who had been abundant in Europe and western Asia, disappeared completely. They were replaced by human beings who were essentially just like us. From that time onward, our ancestors made increasingly complex tools of stone and also of bone, ivory, and antler, materials that had not been used earlier. These people were excellent hunters, preying on the herds of large animals with which they shared their environment. They also began to create often magnificent ritual paintings on the walls of caves. The foundations of modern society had been laid.

The Agricultural Revolution

THE BEGINNINGS OF AGRICULTURE

The modern human beings who replaced the Neanderthals soon migrated over the entire surface of the globe. They colonized Siberia soon after they appeared in Europe and western Asia, and they reached North America by 12,000 to 13,000 years ago. Their migration eastward took place during one of the cold periods of the Pleistocene epoch, when savannas, with their large herds of grazing mammals, were widespread. As these people migrated, they seem to have been responsible for the extinction of many species of these animals. At any rate, extensive hunting by humans, together with major changes in climate, occurred at the same time that these animals were disappearing from many parts of the world.

About 18,000 years ago, the glaciers began to retreat, just as they had done 18 or 20 times before during the preceding two million years. Forests migrated northward across Eurasia and North America, while grasslands became less extensive and the large animals associated with them dwindled in number. Probably no more than 5 million humans existed throughout the world, and they gradually began to utilize new sources of food. Some of them lived along the seacoasts, where animals that could be used as sources of food were locally abundant; others, however, began to cultivate plants, thus gaining a new, relatively secure source of food.

The first deliberate planting of seeds was probably the logical consequence of a simple series of events. For example, the wild cereals (grain-producing members of the grass family, Poaceae) are weeds, ecologically speaking; that is, they grow readily on open or disturbed areas, patches of bare land where there are few other plants to compete with them. People who gathered these grains regularly might have spilled some of them accidentally near their campsites, or planted them deliberately, and thus created a more dependable source of food. When this sequence of events was initiated, cultivation began (Figure 31–2). In places where wild grains and legumes were abundant and readily gathered, humans would have remained for long periods of time, eventually learning how to increase their yields by saving and planting their seeds, by protecting their crops from mice, birds, and other pests, and by watering and fertilizing them.

Under the influence of cultivation, the characteristics of the domesticated crops would have changed gradually as people selected more seeds from plants with certain specific characteristics because these characteristics made the plants easier to gather, store, or use. For example, the stalk (rachis) breaks readily in the

31–2
Harvesting (above) and winnowing (below) of wheat (Triticum) *in Tunisia, North Africa. Similar small-scale cultivation of wheat has been taking place around the Mediterranean basin for more than 10,000 years.*

wild wheats and their relatives, scattering the ripe seeds. In the cultivated species of wheat, the rachis is tough and holds the seeds until they are harvested. Seeds held in this way would not be dispersed well in nature, but they can be gathered easily by humans for food and replanting. As this selection process is continued, a crop plant steadily becomes more and more dependent on the humans who cultivate it, just as the humans become more and more dependent on the plant.

AGRICULTURE IN THE OLD WORLD

The domestication of plants and animals began about 11,000 years ago in the area known as the "Fertile Crescent" of the eastern Mediterranean, in lands that extend through parts of what are now Lebanon, Syria,

687

Turkey, Iraq, and Iran. In this region, barley *(Hordeum vulgare)* and wheat *(Triticum)* were apparently the first plants to be brought into cultivation, with lentils *(Lens culinaris)* and peas *(Pisum sativum)* soon following (Figure 31–3). Additional plants that were domesticated early in this area were chickpeas, or garbanzos *(Cicer arietinum)*, vetch *(Vicia* spp.), olives *(Olea europaea)*, dates *(Phoenix dactylifera)*, pomegranates *(Punica granatum)*, and grapes *(Vitis vinifera)*. Wine made from the grapes and beer brewed from the grains were used from very early times. Flax *(Linum usitatissimum)* was also cultivated very early, probably both as a source of food (the seeds are still eaten today in Ethiopia) and as a source of fiber for weaving cloth.

Among the first cultivated plants, the cereals provided a rich source of carbohydrates, and the legumes provided an abundant source of proteins. The seeds of legumes are among the richest in proteins of all plant parts, and their proteins, in turn, often are rich in the particular amino acids that are poorly represented in the cereals. It is not surprising, therefore, that legumes have been cultivated, along with cereals, from the very beginnings of agriculture in various parts of the world. Of all the protein consumed by human beings worldwide, plants contribute about 70 percent, animals about 30 percent. Only 18 percent of the total plant protein comes from legumes; about 70 percent comes from cereals, even though they contain smaller percentages of protein. Despite this relationship, legume proteins are very important in human diets, and there seem to be great opportunities for increasing the quality—in terms of their amino acid composition—of the legumes that we consume.

The cultivation of plants became more and more organized with the passage of time. For example, specialized implements associated with the harvesting and processing of grains, including flint sickle blades, grinding stones, and stone mortars and pestles, were in use more than 10,000 years ago. About 8000 years ago, humans began to make pottery vessels for storing grain. At the same time that plants were first being cultivated in the Near East, various animals—including dogs (which may have been the first animals kept regularly by humans), goats, sheep, cattle, and pigs—were also domesticated. Horses were domesticated later in southwestern Europe, cats in Egypt, and chickens in Southeast Asia, but all of these animals spread rapidly throughout the world.

Everywhere that they were kept, grazing animals ate the plants that were available to them, whether cultivated or not; produced wool, hides, milk, cheese, and eggs; and could themselves be eaten by their owners. As human beings increased in number, they built up herds of grazing animals so large that they began to destroy their own pastures, causing widespread ecological destruction (Figure 31–4). Much of the Near East and

31–3
Two of the first plants to be cultivated in the Near East were barley, Hordeum vulgare *(above), and peas,* Pisum sativum *(below).*

31–4
Herds of domesticated animals, like these karakul sheep in Afghanistan, devastated large areas of the eastern Mediterranean region as their numbers increased. In many areas, only spiny or poisonous plants survived, while previously fertile fields were converted to desert.

31–5

*Soybeans, Glycine max, have been
a major crop in the United States for
only about 50 years. They are one of the
richest sources of nutrients among food
plants, the seeds consisting of 40 to 45
percent protein and 18 percent fats and
oils. In the Orient, soybeans are used to
make bean curd and soy sauce, among
other products. Soybeans can be grown
most successfully in temperate regions,
and the United States produces more
than half of the world's crop. Like many
legumes, they harbor nitrogen-fixing
bacteria in nodules on their roots, by
means of which they are able to obtain
nitrogen for their own growth and to
enrich the soil. Soybeans are often
rotated with corn in the United States,
primarily to interrupt the life cycles of
the important nematode and insect pests
that attack these two major crops.*

other arid areas around the Mediterranean Sea are still badly overgrazed, and deserts continue to spread, as they have from the initial formation of large herds of domestic animals, up to the present.

At any rate, humans now had steady sources of food, in the form of domesticated plants and animals, and they were able to organize villages, starting about 10,000 years ago, and eventually towns, about 4000 years later. Land that was productive, and on which humans could live permanently, could be owned, accumulated, and passed on to their descendants. In this way, the world became divided into semipermanent groups of haves and have-nots, just as it is today.

The agriculture that had originated in the Near East spread northwestward, extending over much of Europe and reaching Britain by about 4000 B.C. At the same time, agriculture was developed independently in other parts of the world. There is evidence that agriculture may have been practiced in the subtropical Yellow River area of China nearly as long ago as it was in the Near East. Several genera of cereal grasses known as millets were cultivated for their grain, and eventually rice (*Oryza sativa*), now one of the most valuable cereals in the world, was added to them. Later, rice displaced the millets as a crop throughout much of their earlier range. Soybeans (*Glycine max*) have been cultivated in China for at least 3100 years (Figure 31–5).

In other parts of subtropical Asia, an agriculture based on rice and different legumes and root crops was developed. There is some archaeological evidence of the cultivation of rice in Thailand about 10,000 years ago, but much further research is needed before these dates can be established with certainty. The humid, rainy conditions of the tropics tend to destroy most of the evidence with which an archaeologist might trace these ancient events. Such animals as water buffaloes, camels, and chickens were domesticated in Asia long ago, and they became important elements in the systems of cultivation that were practiced there (Figure 31–6).

Eventually, plants such as the mango (*Mangifera indica*) and the various kinds of citrus (*Citrus* spp.) were brought into cultivation in tropical Asia, and others, such as rice and soybeans, began to be cultivated farther north. Taro (*Colocasia esculenta*) is a very important food plant in tropical Asia, where it is grown for its starchy tubers; *Xanthosoma* is a related food plant of the New World tropics. *Colocasia* and other similar genera, including *Xanthosoma*, are the source of poi, a staple starch food of the Pacific Islands, including Hawaii, where these plants were brought by Polynesian settlers about 1500 years ago.

Bananas (*Musa* × *paradisiaca*) are among the most important domesticated plants to come from tropical Asia; their fruits are a staple food throughout the tropical parts of the world. Starchy varieties of bananas,

31–6

Rice, Oryza sativa, *provides half of the food consumed by some 1.6 billion people and more than a quarter of the food consumed by another 400 million. It has been cultivated for at least 6000 years, and it is currently grown on 145 million hectares, about 11 percent of the world's* arable land. Rice is the source of several kinds of alcoholic beverages, including sake, a kind of wine that is a traditional drink in Japan. When rice is grown wet, in paddy fields, fish are often grown in the flooded fields and harvested along with the rice. Since its establishment in 1962, the International Rice Research Institute, in the Philippines, has made important contributions to the improvement of this crop. Seen here are (a) rice terraces and (b) water buffalo being used to cultivate rice in Bali, Indonesia.

known as plantains, are much more important as a source of food in tropical countries, where two-thirds of the total banana crop is consumed, than are the sweet varieties, which are more familiar in temperate regions. Wild bananas have large, hard seeds, but the cultivated varieties, like many kinds of cultivated citrus fruits, are seedless. The banana reached Africa about 2000 years ago, and it was brought to the New World soon after the voyages of Columbus.

There was also early domestication of food plants in Africa, but, again, we have little direct evidence of the timing of its origins. At any event, there is a gap of at least 5000 years between the origins of agriculture in the Fertile Crescent and its appearance at the southern tip of the African continent. Such grains as sorghum (*Sorghum* spp.) and various kinds of millet (*Pennisetum* spp. and *Panicum* spp.); a number of kinds of vegetables, including okra (*Hibiscus esculentus*); several kinds of root crops, but notably yams (*Dioscorea* spp.); and a species of cotton (*Gossypium*)—all were initially brought into cultivation in Africa. The various species of cotton are widespread as wild plants, mainly in seasonally arid but mild climates, and they are obviously useful; the long hairs on their seeds can easily be woven into cloth (Figure 31–7). Fragments of cotton cloth 4500 years old have been found in India. Cotton seeds are also used as a source of oil, and the seed meal from which the oil has been expressed is used for feeding animals. Coffee (*Coffea arabica*) is another crop of African origin. It was brought into cultivation much later than the other plants mentioned here, but it is now a very important commercial crop in the tropics.

31–7

Cotton, Gossypium, *one of the most useful of plants cultivated for fiber, seems to have been domesticated independently in Africa and/or India (the same species of cotton is cultivated in both areas), Mexico (another species), and western South America (yet a third species). Cotton has been cultivated for thousands of years for cloth and has also become an important source of edible oil during the past century. The cotton that is grown widely throughout the world today is a polyploid from the New World; the diploid species of the Old World are more locally cultivated.*

AGRICULTURE IN THE NEW WORLD

A parallel development of agriculture took place in North and South America. No domesticated plants appear to have been brought by human beings from the Old World to the New World prior to 1492. Dogs were certainly brought by people migrating to North America across the Bering Straits but were the only kind of domestic animal that these humans brought with them. This attests to the great value of dogs for protection, hunting, herding, and, commonly, as a source of meat. It may also be related to the very early domestication of dogs.

The plants that were brought into cultivation in the New World were different from those that were first cultivated in the Old World. In place of wheat, barley, and rice, there was corn (*Zea mays;* Figure 31–8), and in place of lentils, peas, and chickpeas, the inhabitants of the New World cultivated kidney beans (*Phaseolus vulgaris*), lima beans (*Phaseolus lunatus*), and peanuts (*Arachis hypogaea*), among other legumes. Among the other important crop plants of Mexico were cotton (*Gossypium* spp.), chili peppers (*Capsicum* spp.), tomatoes (*Lycopersicon* spp.), tobacco (*Nicotiana tabacum*), cacao (*Theobroma cacao*, which yields cocoa, the major ingredient in chocolate), pineapple (*Ananas comosus*), pumpkins and squashes (*Cucurbita* spp.), and avocados (*Persea americana*). Cotton was domesticated independently in the New and in the Old World, with different species involved in the different centers of domestication. Its cultivation by humans in Mexico dates back at least 4000 years, and cotton has been cultivated even longer in Peru. The New World cottons are polyploid, and they are the source of nearly all the cotton that is cultivated throughout the world today; in contrast, the Old World cottons are diploid. After the diverse cultivated plants of the New World were discovered by Europeans following the voyages of Columbus, many of them were brought into cultivation in Europe, spreading from there to the rest of the world. Not only were some of these plants totally new to Europeans, but others, such as the New World species of cotton, were better than the forms already in cultivation in Eurasia and soon replaced them.

The earliest evidence of the existence of domesticated plants in Mexico dates from about 9000 years ago, but extensive agriculture appears not to have become well established until considerably later. The evidence thus suggests that the domestication of plants began later in the New World than in Eurasia. Many of the crops that were originally domesticated in Mexico eventually became widespread from Central America and Mexico northward into Canada. Similar crops were also widely grown throughout the lowlands and middle elevations of South America. In fact, it is possible that agriculture was developed independently in Mexico

31–8
Corn, or maize, Zea mays, *the most important crop plant in the United States. Originally used mainly to feed people, corn is now one of the world's chief foods for domestic animals. In the United States, some 80 percent of the crop is consumed by animals. At the time of Columbus, corn was cultivated from southern Canada to southern South America. Five main types are recognized: popcorn, flint corns, flour corns, dent corns, and sweet corn. Dent corns, in which there is a dent in each kernel, are primarily responsible for the productivity of the U.S. corn belt and are mainly used as animal food.*

and in Peru, although the question cannot be settled with the evidence that is now available. The earliest evidence of agriculture in Peru is almost as ancient as that from Mexico, and some domestic plants, such as tomatoes and peanuts, may well have been brought from South America to Mexico by humans.

In the northern Andes of South America, a distinctive kind of agriculture was developed (Figure 31–9), based on tuberous crops such as potatoes (*Solanum tuberosum* and related species) and seed crops such as quinoa (*Chenopodium quinoa*) and lupines (*Lupinus* spp., of the family Fabaceae). Potatoes were cultivated throughout the highlands of South America at the time of Columbus, but they did not reach Central America or Mexico until they were taken there by the Spaniards. They have been one of the most important foods in Europe for two centuries, yielding more than twice as many calories per hectare as wheat.

Other crops were first brought into cultivation outside these main centers. For example, the sunflower (*Helianthus annuus*) was domesticated by Native Americans living in what is now the United States (Figure

(a)

(b)

(c)

(d)

31–9

A distinctive form of agriculture was developed at high elevations in the Andes of South America. (a) Cultivated fields in the mountains of northwestern Argentina. (b) A potato (Solanum tuberosum) *field in Ecuador. Only a limited selection of the available genetic diversity of potatoes has been used for the improvement of the cultivated crop, one of the most important in the world. (c) Three of the four major root crops that are cultivated in the Andes, shown here for sale in the market at Tarma, Peru—potatoes* (S. tuberosum), *añu* (Tropaeolum tuberosum), *and ullucu* (Ullucus tuberosus). Ullucu, *which can be grown at higher elevations than potatoes, forms large, nutritious tubers; it might well become a useful crop elsewhere in the world. The fourth root crop that is common in the Andes is oca* (Oxalis tuberosa), *which is cultivated to a limited extent in New Zealand and elsewhere. The tubers of all four of these plants are naturally freeze-dried by Andean farmers, after which they can be easily stored for later consumption. (d) Quinoa* (Chenopodium quinoa), *a grain of the pigweed family (Cheno-podiaceae), is an important crop in the Andes—here shown in cultivation in northern Chile—and is now being tested for more widespread cultivation.*

The Origin of Corn

Ears of corn differ so greatly from those of its ancestor that for many years the wild progenitor species was not recognized. We now know, however, that corn (*Zea mays* subspecies *mays*) is simply the domesticated form of a large wild grass, the southern Mexican annual teosinte (*Zea mays* subspecies *parviglumis*). Teosinte plants have hundreds of small, narrow, two-rowed ears, each made up of 5 to 12 fruit cases (each permanently enclosing one grain) that disarticulate and fall free at maturity. Because the fruit cases are woody and not separable from the grains, the flour (meal) that results from grinding them is unpalatable. The several species of teosinte grow from southern Chihuahua, Mexico, to Nicaragua. They are all capable of forming hybrids with corn and may do so spontaneously when corn and teosinte grow together (the *maíz de coyote*, or wild corn, as the hybrids have been called).

Corn is known only as a cultivated plant and could not survive on its own in the wild. Not only are its large naked grains vulnerable, but they are permanently attached to a central axis (the "corn cob") and permanently protected by many overlapping leaf sheaths (the "husks") that would not permit dissemination even if the grains were free. In short, the corn ear is a well-packaged, high-yielding, and easily harvested agricultural artifact, with all its principal characteristics selected by the early farmers.

The domestication of corn, which began in southern Mexico more than 7000 years ago, obviously involved making the teosinte grains accessible as food, as well as increasing grain and ear size and harvestability. As a result of human selection for these characteristics, corn came to differ from teosinte in much the same way as the un-branched and monocephalic (single-headed) cultivated sunflower differs from its highly branched and many-flowered wild ancestor. Both corn and the cultivated sunflower have all the reproductive resources of a branch concentrated in one, gigantic, many-seeded, terminal structure, the corn ear or the sunflower, which results in easier harvesting and higher yield.

But corn evolution, it turns out, had its own peculiar twist. A wild teosinte plant looks like a gigantic candelabrum, with the main stem and each of the several long leafy lateral branches terminated by a staminate "tassel," while the many tiny female ears are hidden within the leaf sheaths. In corn, only the main stem retains its tassel, while each of the one or two very short lateral branches are terminated by a gigantic, solitary, husk-covered ear. In other words, the same (homologous) position occupied by a tassel (male) in teosinte is occupied in corn by an ear (female). Hugh Iltis of the University of Wisconsin, Madison, has therefore suggested that the ear of corn could not have been derived from an ear of teosinte but was instead derived from the terminal portion of a teosinte tassel, in effect from an originally male (staminate) structure. One drastic effect of this sex change has been the telescoping and thickening of the inflorescence axis (the cob) and of the lateral branch on which it sits (the shank). Dramatic sexual abnormalities (for example, "tassel seed," male "tails" on the ear), which are very common in corn, are less so in teosinte and are either the result of genetic mutation or are easily induced in some individuals by special environmental conditions—for example, cold nights, short days, and hormonal weed-killer. (Such drastic changes might also have been related to genetic transposition in corn, studies of which earned Barbara McClintock the Nobel Prize.) The abnormality most likely to have initially triggered human selection must have been the occurrence on teosinte tassels of such "tassel seed" grains, which, not imprisoned within woody fruitcases as in teosinte ears and hence naked, were usable as food right from the start. Amerindians, having settled in Mexico and already knowledgeable in the elements of agriculture, noticed this new food source, saved and planted its seeds, and, over centuries of selection, invented from a basically useless weed the wonder crop we call corn or maize.

An exciting new development in the story of corn evolution has been the discovery of a new species of wild perennial teosinte, *Z. diploperennis*. This rare plant was first found in 1977 in southwestern Mexico by Rafael Guzmán, then an undergraduate student at the University of Guadalajara, and was recognized in 1978 as a new species by Hugh Iltis, John Doebley, and their associates in Madison. Interfertile with annual corn, it carries the genes for resistance to seven of the nine major viruses that infect corn in the United States; for five of these, no other source of resistance is known. The economic implications are obvious when one considers the world-wide gross value of corn—nearly $60 billion in 1991 alone. Virus-resistant types of corn are now grown in South Africa following the incorporation of *Z. diploperennis* genes. By using these genes, plant breeders have also tried to develop a semiperennial corn, which might be useful in home gardens of subtropical regions and which has been tested in Argentina. *Zea diploperennis* occurs naturally in only three small areas, totaling a little more than two hundred hectares, in the cloud forests of the Sierra de Manantlán, a mountain range between Guadalajara and Puerto Vallarta. There it might easily have been destroyed by spreading cultivation and cattle ranching without ever having been made known to science. To protect this one species, and the associated rich flora and fauna (which include ocelots, jaguars, and mountain lions), a large biosphere nature reserve and research laboratory has been established by the University of Guadalajara.

31–10
Sunflowers, Helianthus annuus, *an important crop because of the oil that is obtained from its seeds, were first domesticated at least 3000 years ago in what is now the central United States.*

31–10). Another very important crop plant of the New World, manioc (also called cassava or yuca), was domesticated in the drier areas of South America but is now cultivated on a major scale throughout the tropics (Figure 31–11).

Although the familiar white, or Irish, potato is a member of the nightshade family (Solanaceae), the sweet potato (*Ipomoea batatas*), as you might know if you have ever seen a field of it in flower, is a member of the morning-glory family (Convolvulaceae) and is therefore not closely related to the white potato. At the time of the voyages of Columbus, the sweet potato was widely cultivated in Central and South America, but it was also widespread on some Pacific islands, as far as New Zealand and Hawaii, to which it was apparently carried by human beings in the course of their early voyages. Subsequent to the time of Columbus, the sweet potato became a very important crop throughout most of Africa and tropical Asia.

Very few animals were domesticated in the New World. The so-called Muscovy duck (which, despite its name, does not come from Moscow), turkeys, guinea pigs, llamas, and alpacas are among the few to have originated there. Towns and eventually large cities were organized in the New World, just as they had been earlier throughout Eurasia, wherever the agricultural systems were well enough developed to support them. Crops were extensively cultivated around such centers, but there were no large herds of domestic animals comparable to those that had become so prominent throughout Europe and Asia.

When Europeans colonized the Western Hemisphere, however, they brought their herds with them. In time, these herds brought about the same sort of widespread ecological devastation in parts of the New World as that which had taken place millennia earlier in the Near East and other parts of Eurasia. Many natural plant communities cannot readily be converted to pastures. For example, the widespread clearing of moist tropical forests to make pastures has been enormously destructive everywhere that it has been attempted. For

31–11
Manioc, Manihot esculenta, *one of the most important root crops throughout the tropics. Tapioca is produced from the starch extracted from the roots of this plant. Some of the cultivated strains, the bitter maniocs, contain poisonous cyanide compounds, which must be removed before the tuber is eaten. (a) Manioc cultivated in a forest clearing in southern Venezuela. (b) A Tirió woman in southern Surinam peeling a manioc root to prepare it for cooking.*

(a)

(b)

the most part, pastures in such areas are productive only in the short term, until the nutrients in the soil have been exhausted, and they must often be abandoned after no more than ten or fifteen years.

SPICES AND HERBS

The substances that are produced by plants primarily to defend themselves from insects and other herbivores were discussed in Chapter 19. These substances contribute to the flavors, odors, and tastes of many plants and thus add features that have been utilized by human beings since prehistoric times. Some of these substances make plants poisonous to humans and other animals, but others add characteristics that humans find desirable.

Spices—strongly flavored parts of plants, usually rich in essential oils—may be derived from the roots, bark, seeds, fruits, or buds. **Herbs**, on the other hand, are usually the leaves of nonwoody plants, although laurel, or bay, leaf and a few other condiments derived from trees or shrubs are considered herbs also. In practice, herbs and spices intergrade completely. Both have traditionally been used by humans to flavor food, especially if that food has become stale or somewhat spoiled.

Spices and herbs have been used widely in cooking for as long as we have records; the search for spices played a major role in the great Portuguese, Dutch, and English voyages that began in the thirteenth century and eventually led to Europeans' discovery of the Western Hemisphere. The most important spices came from the tropics of Asia; they were responsible for the voyages and for a great deal of warfare as well. By the third century B.C., caravans of camels—often requiring two years for the trip—were carrying spices from tropical Asia back to the civilizations of the Mediterranean region. Included among these spices were cinnamon (the bark of *Cinnamomum zeylanicum*), black pepper (the dried, ground fruits of *Piper nigrum*; Figure 31–12), cloves (the dried flower buds of *Eugenia aromatica*), cardamom (the seeds of *Elettaria cardamomum*), ginger (the rhizomes of *Zingiber officinale*), and nutmeg and mace (the seeds and the dried outer seed coverings of *Myristica fragrans*; Figure 31–13). When the Romans learned that, by taking advantage of the seasonal shifts of the monsoon winds, they could reach India by sea from Aden, they shortened the trip to about a year—but it was still a highly dangerous and uncertain enterprise. A smaller number of additional spices, including vanilla (the dried, fermented seedpods of the orchid *Vanilla planifolia*; see Figure 19–17), red peppers (*Capsicum* spp.), and allspice (the berries of *Pimenta officinalis*), came from the New World tropics after the voyages of Columbus.

31–12
Black pepper (Piper nigrum) *has been known for thousands of years as an important spice.*

In Europe and the Mediterranean region generally, there were many different kinds of native herbs, ones that were more familiar locally and, perhaps for that reason, not as highly prized as some of the spices that were available only from distant lands. Especially prominent among these herbs are members of the mint family (Menthaceae). These include thyme (*Thymus* spp.), mint (*Mentha* spp.), basil (*Ocimum vulgare*), oregano (*Origanum vulgare*), and sage (*Salvia* spp.). Also important were members of the parsley family (Apiaceae), including parsley (*Petroselinum crispum*), dill (*Anethum graveolens*), caraway (*Carum carvi*), fennel (*Foeniculum vulgare*), coriander (*Coriandrum sativum*), and anise (*Anethum graveolens*). Some members of this family (pars-

31–13
Nutmeg (Myristica fragrans) *is one of the most important traditional spices from tropical Asia. The spice nutmeg is derived from the ground seeds, whereas the spice mace comes from the fleshy seed coatings, seen here as bands of red tissue.*

ley, for example) are grown primarily for their leaves, and some (such as caraway) are grown for their seeds, but many (such as dill and coriander) are valued for both.

Tarragon (*Artemisia dracunculus*) is an herb that consists of the leaves of a plant belonging to the same genus as wormwood and the sagebrush of the western United States and Canada. Mustard (*Brassica nigra*) seed, which can be ground into the spice we call mustard, likewise comes from a plant native to Eurasia. Bay leaves, traditionally taken from the tree *Laurus nobilis* of the Mediterranean region but now also often taken from *Umbellularia californica* of California and Oregon, are another herb, one that is derived from temperate members of the largely tropical laurel family (Lauraceae). Saffron, popular in the Near East and adjacent regions, consists of the dried stigmas of *Crocus sativus*, a small, bulbous plant of the iris family (Iridaceae). The stigmas are gathered laboriously by hand, which ac-

counts for the extremely high price of saffron and the fact that it is so prized—as much for its color as for its taste.

Coffee, *Coffea arabica* (Figure 31–14), and tea, *Camellia sinensis*, provide the two most important beverages in the world; both are consumed primarily because of the stimulating alkaloid, caffeine, that they contain. Coffee is made from seeds of the coffee plant that have been dried, roasted, and ground, whereas tea is prepared from the dried leafy shoots of the tea plant. Coffee, as mentioned earlier, was domesticated in the mountains of northeastern Africa, whereas tea was first cultivated in the mountains of subtropical Asia; both are now widespread crops throughout the warm regions of the world. Coffee now provides the livelihood for some 25 million people around the world, and it provides a major source of income for the 50 tropical nations that export it. A third of the world's supply of coffee comes from Brazil.

WORLDWIDE AGRICULTURE

For the last 500 years, the important crops have been carried throughout the world and cultivated wherever they grow best. The major grains—wheat, rice, and corn—are grown everywhere that the climate will permit. Plants unknown in Europe before the voyages of Columbus, including corn, tomatoes, and *Capsicum* peppers, are now cultivated throughout the world. Sunflowers, first domesticated in the area that is now the United States, now produce more than half of their total worldwide yield in Russia. Sunflowers are also displacing the traditional olives as a source of oil in many parts of Spain and elsewhere in the Mediterranean region. Throughout the world, the sunflower is now being extensively grown for its oil, and it is second only to the soybean among plants grown for this purpose.

Some tropical crops have also become widespread. For example, rubber (derived from several species of trees of the genus *Hevea*, of the family Euphorbiaceae) was brought into cultivation on a commercial scale about 150 years ago. The main area of rubber production is in tropical Asia. For rubber, as for many other crops, cultivation away from the areas where the plants are native seems to be favored; the plants are then often free of the pests and diseases that attack them in their native lands. Oil palms (*Elaeis guineensis*) are native to West Africa but are grown now in all tropical regions. Although they have been grown on a commercial scale for only about 75 years, oil palms are among the most important cash crops of the tropics today. Among the others are coffee and bananas, both widespread. Cacao, which was first semidomesticated in tropical Mexico and Central America, is now most important as a crop in West Africa (Figure 31–15). Sugar

31–14

Coffee, Coffea arabica, *is an important cash crop throughout the tropics of the world. It is a member of the madder family (Rubiaceae), as is cinchona (Cinchona),* which produces the medically important alkaloid quinine. Rubiaceae is one of the larger families of flowering plants, with about 6000 species, mainly tropical.

31–15
Cacao, Theobroma cacao, *the source of chocolate and cocoa. The fruits, shown here, contain several large seeds, or "beans." Cacao was first domesticated in Mexico, where chocolate was a prized drink among the Aztecs; at times, the beans were used as currency.*

31–16
Sugar beet, Beta vulgaris. *The sugar beet is simply a variety of the ordinary beet—selected from strains grown earlier for fodder and not those cultivated as a root crop—in which the sucrose content has been increased by selection from about 2 percent to more than 20 percent. Beets were domesticated in Europe, where their leaves have long been used for food; Swiss chard is another variety of beet, one that is grown for its edible leaves. For about 300 years, beets have also been used as a source of sugar that has been competitive with sugar cane, which must be grown in the tropics and imported to the countries of the developed world. In 1984, U.S. production of raw sugar from beets amounted to more than 2.5 million metric tons, or about a third of the total sugar consumed domestically.*

cane *(Saccharum officinale)* was domesticated in New Guinea and adjacent regions, whereas sugar beets were developed from other cultivated members of their species in Europe. Yams *(Dioscorea spp.)* are an important tropical root crop. A number of species of yams are cultivated throughout the tropics, some of them from West Africa, others from Southeast Asia, and a few—less important—are cultivated in Latin America. The better yams have now spread throughout the tropics, and they provide a staple food over wide areas. Manioc *(Manihot spp.)*—the source of tapioca—is another very important tropical root crop. Originally from seasonally dry tropical areas in South America, manioc is now grown abundantly in Africa and Asia (Figure 31–11).

Another of the most important cultivated plants of the tropics is the coconut palm *(Cocos nucifera)*, which seems to have originated in the western Pacific–tropical Asian region but was widespread in the western and central Pacific Ocean area before the European voyages of exploration; natural stands of coconuts are rare in the eastern Pacific, with a few known from Central America. The vast range of the coconut may be the result of natural dispersal of the fruits floating in the sea, rather than by human intervention. Each tree produces about 50 to 100 nuts each year, and they are a rich source of protein, oils, and carbohydrates. Coconut shells, leaves, husk fibers, and trunks are used to make many useful items, including clothing, buildings, and utensils; it is the solid and milky endosperm that we eat.

Modern agriculture has become highly mechanized in temperate regions and also in some parts of the tropics. It has also become highly specialized, with just six kinds of plants—wheat, rice, corn, potatoes, sweet potatoes, and manioc—directly or indirectly (that is, after having been fed to animals) providing more than 80 percent of the total calories consumed by human beings. These plants are rich in carbohydrates, but they do not provide a balanced diet. They are usually eaten with legumes, such as common beans, peas, lentils, peanuts, or soybeans, which are rich in protein, and with leafy vegetables, such as lettuce, cabbage, spinach *(Spinacia oleracea)*, and chard, which are abundant sources of vitamins and minerals. Such plants as sunflowers and olives provide fats, which are also necessary in the human diet.

In addition to the six major food crops, there are eight others of considerable importance to human beings: sugar cane, sugar beets (Figure 31–16), common beans, soybeans, barley, sorghum, coconuts, and bananas. Taken together, these 14 kinds of crop plants constitute the great majority of those widely cultivated as sources of food.

There are great regional differences in the human diet. For example, rice (Figure 31-6) provides more than three-quarters of the diet in many parts of Asia, and wheat (Figure 31-17) is equally dominant in parts of North America and Europe. Corn can be grown successfully in many parts of the world only with supplementary irrigation, which is usually not necessary for wheat. Extending the productive ranges of these major grains and finding additional crop species are tasks of the greatest importance to the human species, as we shall see.

31-17
Bread wheat, Triticum aestivum, *cultivated with modern techniques. First domesticated in the Near East, wheat has become the most widely grown crop in the world today. Along with barley, which is now largely used for animal food and as the source of malt for making beer, wheat was probably one of the first two plants to be cultivated. Because of the special properties of some of its proteins, wheat is used more extensively than any other cereal for making bread. Wheat proteins form a sticky substance called gluten that facilitates the handling of the dough and helps to hold the bread together.*

The Growth of Human Populations

The roughly 5 million humans who lived 11,000 years ago were already the most widely distributed large land mammal in the world. Subsequently, however, with the development of agriculture, this number has grown at an accelerating pace.

A striking characteristic of groups of humans who make their living by hunting is that they strictly limit their own numbers. A woman on the move cannot carry more than one infant along with her household baggage, minimal though that baggage may be. When simple means of birth control—often simply abstention from sex—are not effective, such a woman may resort to abortion or, more commonly, infanticide. In addition, there is a high natural mortality in these populations, particularly among the very young, the old, the ill, the disabled, and women in childbirth. As a result of these factors, populations that are dependent on hunting tend to remain small. In addition, there is not much incentive for specialization of knowledge or skills; those basic skills on which individual survival depends are of the greatest importance.

Once most groups of humans became sedentary, there was no longer the same urgent need to limit the number of births, and children may have become more of an asset to their families than previously, helping with agricultural and other chores. People could live much more densely on the land than ever before. For hunting and food-gathering economies, an area of 5 square kilometers, on the average, is required to provide enough food for one family to eat. In crop-based economies, only a small fraction of that amount of land is required. In the cities and towns that agricultural productivity made possible, human knowledge became increasingly specialized. Because the efforts of a few people could produce enough food for everyone, patterns of life became more and more diversified. People became merchants, artisans, bankers, scholars, poets—all the rich mixture of which a modern community is composed. The development of agriculture set human society on its modern course.

As a consequence of the development of agriculture, the numbers of human beings grew to about 130 million, distributed all over the earth, by the beginning of the Christian era. Over a period of about 8000 years, the human population had increased about 25-fold. By 1650, the world population had reached 500 million, with many people living in urban centers (Figure 31-18). The development of science and technology had begun, as had the process of industrialization, bringing about further profound changes in the lives of human beings and their relationships with the natural world. The human birth rate has remained essentially constant throughout the world from the seventeenth century

31–18
London was one of the thriving cities of seventeenth-century Europe, its influence reaching out to distant parts of the world as the global population increased, attaining a level of 1 billion people by 1850.

31–19
These impoverished people, living on the outskirts of Tegucigalpa, Honduras, represent the condition of a majority of the world's population. Their prospects for the future depend directly on limiting population growth, incorporating the poor into the global economy, and finding new and improved methods for productive agriculture in the tropics and subtropics.

onward, but the death rate has decreased dramatically on a regional basis, with the result that the population overall has grown at an unprecedented rate. In the twentieth century, the birth rate itself has dropped in developed countries.

By 1992, there were about 5.4 billion humans on our planet; this compares with 2.5 billion in 1950. The global population has doubled in less than 35 years. In 1992, about 23 percent of the people lived in the developed world (the United States, Canada, Europe, Russia, Australia, and New Zealand), some 21 percent in China, and the remainder in the less developed world, mostly consisting of countries that are tropical or subtropical, at least in part.

For the planet as a whole, the population is growing at about 1.8 percent per year. This means that about 180 humans are being added to the world population every minute—more than 260,000 each day, or 96 million every year. A high proportion—usually about 38 to 44 percent—of the humans living in the less developed countries are under 15 years of age; the comparable percentage for developed countries is about 20 to 24 percent. Such young people have not yet reached the age at which they usually bear children. Consequently, population growth in developing countries cannot soon be brought under control, even though government policy and individual choice often favor such a trend. It is predicted that there will be approximately 6.3 billion people on earth by the year 2000, with a billion people

—equalling the population of the entire globe 300 years ago, at the start of the Industrial Revolution—being added during the 1990s alone. More than 95 percent of this growth will take place in developing countries.

If worldwide efforts to promote family planning continue to be pursued strongly, the earth's population may stabilize in about a century at somewhere between 10 billion and 14 billion people. In the context of the next 100 years, the next few decades are likely to be one of the most difficult periods ever faced by the human race (Figure 31–19). In 1990, the World Bank estimated that 1.1 billion people—one out of every five of us—were living in absolute poverty. These people were unable to obtain food, shelter, or clothing dependably. Some 500 million people—nearly one out of every ten—were receiving less than 80 percent of the daily intake of food calories recommended by the United Nations; their minds and bodies are literally wasting away.

Humans are presently estimated to be consuming some 40 percent of the total net photosynthetic productivity on land, and that productivity has been greatly reduced by our burning and clearing in the past. Although we have achieved a 2.6-fold increase in world grain production since 1950, this increase has been accomplished at the expense of no less than 20 percent of the topsoil on the world's cultivated lands—lost at the rate of some 24 billion metric tons per year. As we saw in Chapter 30, we lack the agricultural technology to convert most of the lands in the tropics to sustainable productivity. Most of the world's lands that can be cultivated using available techniques are already in cultivation. Despite this, the rapidly growing majority of the world's population that lives in the tropics must somehow be fed. The job cannot be done by exporting surpluses from the more productive lands of the developed world. In 1983, for example, just 8 percent of the food consumed in the developing world came from the developed world; yet that food constituted more than half of the total food that was exported from the developed countries that year. This pattern has continued to the present. The solution to the problem of feeding the world's population must be found in the regions where most people live—the tropics and subtropics.

The Food and Agriculture Organization (FAO) of the United Nations has estimated that an increase in world food production of 60 percent will be needed by the year 2000 if the world's population is to be fed adequately. Realistically, there appears to be little hope of attaining this goal, even though we must try to approach it as closely as possible. In some tropical areas, such as Africa south of the Sahara, food production per capita has actually been falling. Recently, even the *total* food production has declined in this vast area, whose population is well over 500 million and rapidly growing. For U.S. citizens, who spend on the average less than a fifth of their personal income on food, the increasing cost of food already occasions serious concern. For those in developing nations, who may spend 80 to 90 percent of their income on food, it can be a death sentence. Indeed, in such countries as Bangladesh and Haiti and in such regions as East Africa, humans are dying in increasing numbers because of the lack of food. How can the situation be improved?

Agriculture in the Future

THE PRESENT SITUATION

The first significant advance that led to a massive increase in agricultural productivity was the development of irrigation (Figure 31–20). The necessity of providing water to crops has always been so evident that irrigation was practiced in the Near East as early as 7000 years ago, and it was developed independently in Mexico at least 3000 years ago. During the past two centuries, increasingly specialized and efficient machinery has been developed for agriculture, and great increases in productivity have been achieved. Fertilizers have been widely used, with their production drawing heav-

31–20
This irrigated cotton field in Texas is representative of modern, intensive agriculture. Irrigation may pose severe environmental problems over the long run, however, especially if it is coupled with the intensive use of pesticides and herbicides. More pesticides are used on cotton than on any other crop in the world.

ily on the use of fossil fuels. One of the great problems worldwide is how to utilize the gains in productivity that are made possible by increased mechanization, irrigation, and fertilization without simultaneously displacing millions of workers. Throughout much of the developing world, well over three-quarters of the people are directly engaged in food production, as compared with fewer than 3 percent in the United States in the late twentieth century.

Research has already done a great deal to improve agriculture. In the United States, the land-grant college system and the associated state agricultural experiment stations have made major contributions in this area. Nevertheless, many problems remain. The energy cost of producing crops in the United States and other developed countries is very high. Modern agriculture likewise depends on an elaborate distribution system, which is expensive in terms of energy and is easily disrupted. A significant proportion of each crop, the exact amount depending on the region and the year, is lost to insects and other pests. In many regions, an additional significant proportion is lost after harvest, either to spoilage or to insects, rats, mice, and other pests. Water is increasingly expensive in many areas, and the quality of local water supplies is often degraded by runoff from fields to which fertilizers and pesticides have been applied. Soil erosion is a problem everywhere, and it increases with the intensity of the agriculture (Figure 31–21). Many efforts are under way to achieve improvements in crop productivity, in the protection of crops from pests, and in the efficiency with which crops use water. Each of these will be discussed in this chapter.

31–21

Use of the no-tillage cropping system, which combines ancient and modern agricultural practices, has been increasing rapidly. By the year 2000, as much as 65 percent of the acreage of crops grown in the United States may be grown under no-tillage conditions. Soil erosion is virtually eliminated in this system, as shown in the corn planted without tillage into a killed cover crop of clover (right). Compare with the evident erosion in corn planted conventionally (left). This photograph was taken soon after a spring rainstorm. When no-tillage methods are used, the energy input into corn and soybean production is reduced by 7 and 18 percent, respectively, and the crop yields are as high as or higher than those obtained by conventional plowing and disking methods.

31–22

Norman Borlaug, who was awarded the Nobel Peace Prize in 1970. Borlaug was the leader of a research project sponsored by the Rockefeller Foundation under which new strains of wheat were developed at the International Center for the Improvement of Maize and Wheat (CIMMYT) in Mexico. Widely planted, these new strains changed the status of Mexico from that of a wheat importer when the program began in 1944 to that of an exporter by 1964.

IMPROVING THE QUALITY OF CROPS

The most promising approach to alleviating the world's food problem seems to lie in the further development of existing crops grown on land that is already in cultivation. Most land suitable for plant agriculture is already in cultivation, and increasing the supplies of water, fertilizers, and other chemicals for crops is not economically feasible in many parts of the world. Thus the improvement of existing crops is of exceptional importance. Such development entails not only improving the yield of these crops but also improving the quantity of proteins and other nutrients that they contain. The *quality* of the protein in food plants is also of the greatest importance to human nutrition: animals, including human beings, must be able to obtain from food the right amounts of all the essential amino acids—the ones that they cannot manufacture themselves. Eight of the 20 amino acids required by human beings must be obtained from food; the other 12 can be manufactured

in the human body. Plants that have been selected for an improved protein content, however, inevitably have higher requirements for nitrogen and other nutrients than do their less modified ancestors. For this reason, improved crops cannot always be grown on the marginal lands where they would be especially useful.

The quality of crops can also be improved in many respects other than yield and protein composition and quantity. Newly developed crop varieties may be more resistant to disease, as when they contain secondary metabolites that their herbivores find distasteful; more interesting in form, shape, or color (such as apples of a brighter red); more amenable to storage and shipping (such as tomatoes that stack better in boxes); or improved in other features that are important for the crop in question.

For decades, the plant breeder, patiently sifting the available genetic diversity and producing hybrids of better promise, has produced thousands of improved strains of our major crops (Figure 31–22). Typically, thousands of hybrids are made and evaluated in order to find a few that represent genuine improvements on those already in wide cultivation. For example, the production of corn per hectare in the United States was increased about eightfold between the 1930s and the 1980s, even though very little of the genetic diversity of this remarkable plant was utilized in the process.

Hybrid Corn

The increase in the production of corn was made possible largely because of the introduction of hybrid corn seed. Inbred lines of corn (themselves originally of hybrid origin) are used as parents. When they are crossed, the result is seed that produces very vigorous hybrid plants. The strains to be crossed are grown in alternating rows, with the tassels (staminate inflorescences) being removed from one row of plants by hand, so that all of the seeds set on those plants will be of hybrid origin. Through a careful selection of the best inbred lines, vigorous strains of hybrid corn that are suitable for cultivation at any given locality can be produced. Because of the uniform characteristics of the hybrid plants, they are easier to harvest, and they uniformly produce much higher yields than nonhybrid individuals. Less than 1 percent of the corn grown in the United States in 1935 was hybrid corn, but now virtually all of it is. Far less labor is now needed to produce much greater yields per hectare.

Progress at the International Plant-Breeding Centers

Extensive efforts have been made during the past few decades to increase the yields of wheat and other grains, particularly in warmer regions. These efforts have led to dramatic improvements in crop yields, largely at international crop-improvement centers located in subtropical areas. When the improved strains of wheat, corn, and rice produced at these centers were grown in such countries as Mexico, India, and Pakistan, they made possible the pattern of improved agricultural productivity that has been called the Green Revolution. The techniques of breeding, fertilizing, and irrigation that have been developed as a part of the Green Revolution have been applied in many countries of the developing world.

Any grain requires excellent conditions to produce high yields, and fertilization, mechanization, and irrigation have all been necessary components of the successes of the Green Revolution. Only relatively wealthy landowners have been able to cultivate the new crop strains, and in many areas the net effect has been to accelerate the consolidation of farmlands into a few large holdings by the very wealthy. Such consolidation has not necessarily provided either jobs or food for the majority of the local peoples.

31–23
Triticale (Triticosecale), *a modern polyploid hybrid between wheat and rye, which combines the high-quality yield of wheat with the ruggedness of rye.*

Triticale

The traditional methods of the plant breeder can sometimes lead to surprising results. For example, hybrids of wheat *(Triticum)* with rye *(Secale)*, which are called **triticale** (*Triticosecale* is the scientific name), have become a crop of increasing importance in several areas and hold much promise for the future (Figure 31–23). Triticale hybrids arose following the spontaneous doubling of the chromosome number in a sterile hybrid of wheat and rye and were later produced artificially. In the mid-1950s J.G. O'Mara of Iowa State University produced such hybrids using the chemical colchicine, which arrests the formation of the cell plate and thus allows the chromosome number of the treated plant cells to double.

Unfortunately, triticale is only partly fertile, and its endosperm often develops poorly. It combines the high yield and quality of wheat with the winter hardiness and disease resistance of rye, but the best combinations of these features are still being sought by plant breeders. Improved strains of triticale are gaining in popularity in the 1990s and are already grown on more than 1 million hectares in over 30 countries around the world. Most triticale is used for animal feed, but it is being consumed in increasing amounts by humans also.

Preserving and Using the Genetic Diversity of Crops

Intensive programs of breeding and selection have tended to narrow the genetic variability of crop plants.

31–24

Southern leaf blight of corn, a disease that is caused by the fungus Helminthosporium maydis.

(a)

(b)

31–25

(a) *U.S. Department of Agriculture seed bank at Fort Collins, Colorado, showing seeds being sorted out and sealed for long-term storage. Approximately 200,000 germ plasm lines are stored at this facility.*
(b) *The seed potato farm at Three Lakes, Wisconsin, is the national repository for strains of potatoes.*

Much artificial selection, for obvious reasons, has been concerned with yield, and sometimes disease resistance has been lost among the highly uniform members of a progeny that has been selected strongly for increased yield. Overall, individual crop plants have tended to become more and more uniform, as particular traits have been stressed more strongly than others, and these plants have become more vulnerable to attack by diseases and pests. In 1970, for example, the fungus that causes southern leaf blight of corn, *Helminthosporium maydis*, destroyed approximately 15 percent of the U.S. corn crop—a loss of approximately 1 billion dollars (Figure 31–24). These losses were apparently related to the appearance of a new race of the fungus that was highly destructive to some of the major strains of corn that were used extensively in the production of hybrid corn seed. The non-nuclear genetic factors in many of the commercially important strains of corn were identical, having been introduced through the methods used to produce hybrid corn when identical carpellate parents were employed repeatedly.

In order to help guard against such losses, it is necessary to locate and to preserve distinct strains of our important crops, because these strains—even though their overall characteristics may not be attractive economically—may contain genes useful in the continuing fight against pests and diseases (Figure 31–25). Since the beginnings of agriculture, huge reserves of variability have accumulated in all crop plants by the processes of mutation, hybridization, artificial selection, and adaptation to a wide range of conditions. For such crops

as wheat, potatoes, and corn, there are literally thousands of known strains. In addition, even more genetic variability exists among the wild relatives of the cultivated crops, often, however, in areas where it is being lost to advancing civilization. The problem lies in finding, preserving, and using the genetic variability of cultivated plants and their wild relatives before they are lost.

As an example of the role of genetic diversity in the history of, and the prospects for, a single crop plant, we shall consider the potato. There are more than 60 species of potatoes, most of them never cultivated, and thousands of different strains of cultivated potatoes are known (Figure 31–9). Despite this, most of our cultivated potatoes are descended from a very few strains that were brought to Europe in the late sixteenth century. This genetic uniformity led directly to the Irish potato famine of 1846 and 1847, in which the potato crop was nearly wiped out by a blight caused by the mold *Phytophthora infestans* (see Chapter 13). Within three years, the population of Ireland fell from 8.5 million to 6.5 million, with about one out of ten people either starving to death or dying of diseases associated with starvation and an additional one out of five emigrating. The subsequent breeding of strains of potatoes resistant to the blight restored the status of the plant as a crop in Ireland and elsewhere. Looking to the future, the potential for further developing potatoes as a crop through the use of additional cultivated and wild strains is enormous.

Striking examples of the potential of obtaining additional genetic material from the wild have been achieved by tomato breeders. The collection of strains of tomatoes, mostly carried out in recent years by Charles Rick and his associates at the University of California, Davis, has led to effective control of many of the important diseases of tomatoes, such as the rots caused by the deuteromycetes *Fusarium* and *Verticillium* and by several viral diseases. The nutritive value of tomatoes has been greatly enhanced, and their ability to tolerate saline or other basically unfavorable conditions has been increased greatly through the systematic collection, analysis, and use of strains of wild tomatoes in breeding programs.

Who Owns Crop Genetic Diversity?

A serious question has arisen concerning the ownership of the genetic resources of the world's food crops, nearly all of which are located in developing countries. Some of the nations of the developing world are concerned that their plant genes are being taken freely to industrialized countries, incorporated into food crops in breeding programs there, and then sold back to them at a profit. On the other hand, virtually everyone believes that access to the genetic diversity of plants ought to be possible at low cost. At a meeting of the Food and Agriculture Organization (FAO) of the United Nations, held in Rome in 1983, an intergovernmental committee was set up to oversee the conservation and use of plant genetic resources. It was decided at the same FAO meeting that individual countries should be allowed to withhold valuable strains of crops from free international exchange in order to benefit directly from their use and to obtain profit from their sale to other countries.

NEW CROPS

In addition to the plant species already widely cultivated, there are many wild plants, or plants that are cultivated only locally, that could, if brought into more

general cultivation, make important contributions to the world economy. For example, as mentioned earlier, we still derive more than 80 percent of our calories from only six kinds of angiosperms out of the approximately 235,000 species that exist. Only about 3000 angiosperm species have ever been cultivated for food, and the great majority of those either are no longer used or are used very locally. Only about 150 plant species have ever been cultivated widely.

Many plants other than this limited number, however—especially those that have been used before but have been completely abandoned or seen as having less importance—may prove to be very useful. Some of these species are still in cultivation in different areas of the world (Figure 31–26). Although we are accustomed to thinking of plants primarily as important sources of food, they also produce oils, drugs, pesticides, perfumes, and many other products that are important to our modern industrial society. We have come to regard the production of such substances from plants as a rather archaic method that has now been fully sup-

31–26

The winged bean, Psophocarpus tetragonolobus, *a legume, is a traditional crop in New Guinea, the Philippines, Indonesia, Thailand, Burma, Malaysia, and adjacent areas, used primarily as a green vegetable. Recently introduced in many other areas throughout the tropics, the winged bean shows great promise as a crop because of the high protein content of its seeds, which rivals that of soybeans. All parts of the plant—pods, seeds, leaves, and fleshy tubers—can be eaten.*

31–27
Jojoba, Simmondsia chinensis, *an important crop that is increasingly being planted throughout the arid areas of the world as a source of waxes with special lubricating characteristics and other useful properties.*

31–28
Guayule, Parthenium argentatum, *a desert shrub that produces natural rubber. First harvested from natural populations, guayule is now being cultivated widely.*

planted by chemical synthesis in commercial laboratories. Their production by plants, however, requires no energy other than that of the sun: it occurs naturally. As our sources of nonrenewable energy near exhaustion and prices increase, it has become increasingly important to find ways to produce complex chemical molecules less expensively. Further, the vast majority of plants have never been examined or tested to determine their usefulness.

A few examples of crops that have recently been developed will demonstrate the great potential that exists in nature. Jojoba (pronounced "ho-*ho*-ba"), *Simmondsia chinensis,* despite its Latin name, is a shrub native to the deserts of northwestern Mexico and the adjacent United States (Figure 31–27). The large seeds of jojoba contain about 50 percent liquid wax, a substance that has impressive industrial potential. Wax of this sort is indispensable as a lubricant where extreme pressures are developed, as in the gears of heavy machinery and in automobile transmissions. It is difficult to produce synthetic liquid wax commercially, and the endangered sperm whale is the only alternative natural source. Jojoba wax is also used in cosmetics and as a food additive, and other applications are still being discovered for this unusual substance. The plant thrives in hot deserts that are not suitable for the cultivation of most other crops, and some varieties are even highly salt-tolerant. Crops of jojoba are being produced in arid areas throughout the world, and they may make an important contribution to the economic development of these areas, especially by helping to provide employment for local peoples in the cultivation of the crop.

Jojoba might also be able to help stop the spread of deserts, since it can grow in sandy soil when the rainfall is as little as 7.5 centimeters per year. In some areas, plantations of jojoba might make a useful contribution to stabilizing the soil. Such considerations are of prime importance, in view of the fact that the Sahara, for example, is spreading southward at the rate of about 5 kilometers per year and thus limiting the potential for food production of areas that are not now desert. At present, Saudi Arabia, Kuwait, Egypt, Morocco, Ecuador, and Nigeria are experimenting with jojoba for this purpose.

Another interesting new crop is guayule, *Parthenium argentatum* (Figure 31–28). A member of the daisy, or sunflower, family (Asteraceae), guayule is closely related to ragweed (*Ambrosia* spp.). It is a low shrub native to northern Mexico and the southwestern United States. As much as 20 percent of the living plant body may consist of rubber.

Synthetic rubber has replaced natural rubber for many purposes and now constitutes about two-thirds of the world supply. It is manufactured from petroleum and is therefore not renewable, as is rubber produced by plants. Almost all natural rubber is now obtained from rubber trees, members of the genus *Hevea* of the spurge family (Euphorbiaceae). Although they are native to the Amazon Basin of South America, rubber trees are cultivated most successfully in tropical Asia. Guayule provides a promising alternative that can be cultivated in the desert and that has the potential of greatly increasing the world rubber supply. Wild plants of guayule have been used as a source of rubber for

(a)

(b)

31–29

The development of new crops, or new strains of presently cultivated crops, that can succeed at relatively high salt concentrations is important in many areas of the world, particularly those that are arid or semiarid. (a) Eggplants (Solanum melongena) *cultivated in the Arava Valley of Israel using drip irrigation (water reaches the plants through plastic tubes and is greatly conserved in this way) with highly saline water (1800 parts per million). Fifteen years ago, this amount of salinity was thought to be incompatible with commercial crop production. (b) At an experimental site on the Mediterranean coast of Israel, a highly nutritious fodder shrub,* Atriplex nummularia, *is grown with 100 percent seawater. Yields are similar to those achieved with alfalfa, but the leaves and stems of the* Atriplex *are highly salty and thus not as valuable for fodder as alfalfa. Research being conducted by D. Pasternak and his co-workers at the Ben-Gurion University of the Negev aims at overcoming the problems of seawater irrigation, and this type of irrigation may one day make possible the cultivation of huge areas that are now coastal deserts.*

nearly a century, and cultivated fields in the United States produced more than 1.3 million kilograms of rubber during World War II, when the rubber supply was cut off by the Japanese. Although guayule received relatively little attention during the four decades following the end of the war, it is under active investigation again.

A third crop that might well be grown more widely is the grain amaranths (various species of *Amaranthus*), plants that have been grown for food for thousands of years in Latin America, but mainly on a relatively small scale. The grain amaranths have recently emerged as an excellent example of a potential new food crop. Weedy varieties of *Amaranthus* are called "pigweed" in English-speaking countries, but in pre-Columbian times the seeds of these plants were one of the basic foods of the New World—almost as important as corn and beans. Some 20,000 metric tons of seeds were sent annually to Tenochtitlan (present-day Mexico City) in tribute to the chief ruler of the Aztecs. The Spanish conquistadors forbade the Mexican people to use amaranth because of its association with pagan rituals and human sacrifice, and it survived as a crop only in very small, local areas. The protein content of amaranth seeds is as high as that of the cereals, and, furthermore, amaranth protein contains large amounts of the amino acid lysine, making it nutritionally complementary to the cereals. (Many cereals are low in lysine, which is an essential nutrient for human beings.) The leaves of some races of amaranth can also be used as nutritious vegetables. In view of all these virtues, it is not surprising that the cultivation of amaranth is spreading throughout the world.

One important area of investigation in the search for valuable new crops focuses on salt-tolerant species. Increasingly, intensive agriculture is being practiced in the arid and semiarid areas of the world, primarily in response to the demands of the ever-increasing human population. Such practices have placed enormous demands on very limited local water supplies, which have become increasingly brackish as the water is used, reused, and polluted with fertilizer from adjacent cultivated areas. In addition, there are many parts of the world, especially near seashores, where the soil and the local water supplies are naturally saline. Unproductive when farmed using traditional agricultural methods, such areas might be brought under cultivation if the right kinds of plants could be found (Figure 31–29).

Finding suitable plants for saline areas entails not only seeking out entirely new crop plants but also breeding salt-tolerance into traditional crops (see the essay on pages 608–609). In tomatoes, for example, a wild species that grows on the sea-cliffs of the Galápagos Islands and is therefore very salt-resistant, *Lycopersicon cheesmanii*, has been used as a source of salt-tolerance in new hybrids with the commonly cultivated tomato, *Lycopersicon esculentum.* The selection of these hybrids was carried out in a medium having half the salinity of seawater. Researchers have succeeded in selecting hybrid individuals that will complete their development in water of that salinity. Salt-tolerant strains of barley have been developed using similar methods.

Drugs from Plants

In addition to their other uses, plants are important sources of drugs. In fact, about a quarter of the prescriptions written in the United States contain at least one product that has been derived from a plant. For millennia, humans have been using plants for medicinal purposes. Botany, in fact, was traditionally regarded as a branch of medicine, and it has only been in the past 150 years or so that there have been professional botanists, as distinct from physicians. There has not, however, been any comprehensive effort to identify and bring into use previously unexploited secondary plant products of the sort discussed in Chapter 19.

One of the reasons that plants will continue to be of importance as sources of drugs, despite the ease with which many drugs can be synthesized in the laboratory, is that plants manufacture the drugs inexpensively, without the provision of additional energy. Furthermore, the structure of some molecules—such as the steroids that make up the basic structure of cortisone and the hormones used in birth-control pills—is so complex that, although the molecules can be synthesized chemically, the products made from them are prohibitively expensive (Figure 31–30). For this reason, antifertility pills and cortisone were manufactured in the past principally from substances extracted from the roots of wild yams *(Dioscorea)*, obtained largely from Mexico. When these sources were virtually exhausted, other plants, including *Solanum aviculare*, a member of

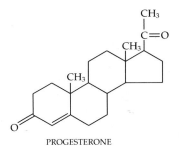

PROGESTERONE

31–30
The steroid progesterone is the precursor of human male and female sex hormones. It is closely related in structure to cholesterol, another abundant human steroid, and to cortisol, which is marketed as cortisone and used to treat inflammation. Other steroids are found in plants, and they are abundant in yams (Dioscorea), from which they are extracted and used as the starting material for the synthesis of the active ingredient in birth-control pills. Steroids are extremely active physiologically in vertebrates.

the same genus as the nightshades and potatoes, were developed into crops for the same purpose.

In addition to considerations of cost, the marvelous variety of substances that plants produce is important: a source of new products that is seemingly inexhaustible (Figure 31–31). One of the ways in which scientists have learned about potential new drugs is to study the medical uses to which plants are put by rural or indigenous peoples (Figure 31–32). For example, the contraceptive properties of Mexican yams were "discovered" in this way.

31–31
Periwinkle, Catharanthus roseus, *the natural source of the drugs vinblastine and vincristine. These drugs, which were developed in the 1960s, are highly effective against certain cancers. Vinblastine is typically used to treat Hodgkin's disease, a form of lymphoma, and vincristine is used in cases of acute leukemia. Before the development of vinblastine, a person with Hodgkin's disease had a one-in-five chance of survival; now, the odds have been increased to nine in ten. This periwinkle is widespread throughout the warmer regions of the earth, but it is native only to Madagascar, an island where only a small proportion of the natural vegetation remains undisturbed.*

31–32

Mark Plotkin, a scientist from Conservation International, collecting herbarium specimens of medicinal plants with the advice of a Wayana medicine man, southeastern Surinam. Although the study of the medicinal uses of forest plants has led to the discovery of important drugs such as D-tubocurarine chloride (used as a muscle relaxant in open-heart surgery) and ipecac (employed in the treatment of amoebic dysentery), the opportunities for gaining such knowledge are rapidly being lost as tribal cultures disappear and whole groups of people lose their traditional life styles. Valuable knowledge that has accumulated through thousands of years of trial and error is being forgotten in a very short period of time. Much of this information has been transmitted orally, and written records do not exist.

As we expand our search for useful plants, we should keep in mind the rapid loss of plant species that is related to (1) the rapid growth of the human population; (2) poverty, especially in the tropics, where about two-thirds of the world's species of plants occur; and (3) our incomplete understanding of how to construct productive agricultural systems in the tropics. With the complete destruction of undisturbed tropical forests, which seems almost certain to occur within a century, many kinds of plants, animals, and microorganisms will become extinct. Because our knowledge of plants, especially those of the tropics, is so rudimentary, we are faced with the prospect of losing many of them before we even learn of their existence, much less have the opportunity to examine them to see if they might be of some use to us. The examination of wild plants for potential human use must be accelerated, and promising species must be preserved in seed banks, in cultivation, or, preferably, in natural reserves.

GENETIC ENGINEERING

One of the most important ways in which crops will be improved is by the methods of genetic engineering. Molecular biologists have learned how to transfer foreign genes into plant cells. As we saw in Chapter 9, natural hybridization, which brings about the recombination of genetic material, plays an important role in the ongoing process of plant evolution. Similarly, the plant breeder has utilized the techniques of hybridization to recombine genetic material with the aim of developing crop plants with improved characteristics. What is new about the methods of genetic engineering is that they allow individual genes to be inserted into organisms in a way that is both precise and simple. Characteristics of interest can therefore be enhanced directly, with far less necessity for backcrossing and selection of the progeny than has been necessary in the past. Genetic engineering is discussed further in Chapter 25.

THE FUTURE: AN INTEGRATED APPROACH

The problem of how to ease the burdens of hunger and extreme poverty that are borne by at least a fifth of the world's people remains severe. The Green Revolution must of course go forward, but, at the same time, we must recognize that the broader solutions to this problem are social, political, and ethical. They involve not only the growth of food and its distribution but also the creation of jobs whereby the poor can earn the means to buy food. They involve not only limiting population growth but also raising the living standards of the poor to tolerable levels. No matter how striking the advances of agricultural science might be, they will never be adequate to eliminate hunger in a world in which the population is growing rapidly. In countries with large numbers of people living in poverty, there must be an organizational system that facilitates the introduction of new agricultural practices. There must be sources of fertilizers, pesticides, equipment, credit, and water that are available to all; farmers must be able to sell their products, and they must be able to get these products to market in order to do so. Genetic engineering, coupled with an increased knowledge of plant physiology, will certainly produce superior new crops, but will the billion poor farmers of the Third World have access to them?

Plants also must be developed more fully as sources of nonfood products, such as medicines, other chemicals, and energy. Our major crop plants were mainly brought into widespread cultivation thousands of years ago, and many additional ones could be of great use if we were to identify them, work out their requirements for cultivation, and put them into production. However, many observers believe that, as vegetation is being destroyed rapidly throughout the world, 15 to 20 percent of the total number of plant species may become extinct during the next 30 years. Such a loss (of up to 50,000 plant species) would tragically and unnecessarily limit our options, and it must be reduced as much as possible. Certainly, the prospect of a loss of this magnitude makes the search for useful plants and their conservation seem especially urgent.

A broad knowledge of plant biology has become increasingly relevant to achieving solutions to some of society's most crucial problems. The stabilization of the human population must come first, but then our attention must turn both to the amelioration of poverty and malnutrition and to finding ways in which to provide food for the people of all countries. All the principles of plant growth and development must be brought into play in improving the practice of agriculture throughout the world, in making it possible for our planet to support unprecedented numbers of people in dignity and relative prosperity. Doing this will tax human imagination and ingenuity to the limit, but the task is of immense importance. We must make every effort to accomplish it.

Summary

The human race evolved in Africa. The genus *Australopithecus* existed there at least 5 million years ago. Our genus, *Homo*, apparently evolved from *Australopithecus* about 2 million years ago, and our species, *Homo sapiens*, has existed for at least 500,000 years.

Starting about 11,000 years ago, in the Fertile Crescent—an area that extends from Lebanon and Syria through Iraq to Iran—humans began to cultivate such plants as barley, lentils, wheat, and peas. In cultivating and caring for these crops, the early farmers changed the characteristics of the plants, so that they were more nutritious and easier to harvest, and differed in other ways from their wild relatives. Agriculture spread from this center across Europe, reaching Britain by about 6000 years ago. It also seems to have spread southward through Africa, although agriculture may have originated there independently in one or more centers. Many crops were first brought into domestication in Africa, including yams, okra, coffee, and cotton, which was also domesticated independently in the New World and perhaps in Asia. In Asia, agriculture based on staples such as rice and soybeans—and, farther south, citrus, mangos, taro, bananas, and other crops—was developed.

Domestic animals, starting with the dog, were an important feature of agriculture in the Old World from the earliest times. Herds of sheep, goats, cattle, horses, and hogs were ecologically destructive to many of the semiarid areas of the Old World, especially as the animals increased in number, but they were also important sources of food. Other animals, such as water buffaloes, camels, chickens, and elephants, were domesticated as agriculture spread. When the grazing animals that were so important in the Old World were introduced into Latin America following the voyages of Columbus, they proved to be enormously destructive in many habitats, including tropical forests.

Agriculture was developed independently in the New World. It began as much as 9000 years ago in Mexico and Peru. Dogs were brought to the New World by people migrating from Asia, but apparently no other domestic animals or plants were introduced in this way. Columbus and those who followed him found a virtual cornucopia of new crops to bring back to the Old World. These crops included corn, kidney beans, lima beans, tomatoes, tobacco, chili peppers, potatoes, sweet potatoes, pumpkins and squashes, avocados, cacao, and the major cultivated species of cotton.

Spices and herbs are plants prized for their flavors and odors. Spices, which are strongly flavored plants usually rich in essential oils, may be derived from the roots, bark, seeds, fruits, or buds of the plants, whereas herbs are usually the leaves of nonwoody plants. Cinnamon, black pepper, and cloves are examples of spices; mint, dill, and tarragon are herbs.

For the last 500 years, the important crops have been cultivated throughout the world. Wheat, rice, and corn, which provide 60 percent of the calories we consume, are cultivated wherever they will grow, and a limited number of other plants have achieved worldwide commercial status.

The human population has grown from an estimated 5 million people at the time when agriculture was first developed to about 5.4 billion people in 1992. The world population is growing very rapidly, and more than 95 percent of the growth is taking place in developing countries, where the rural poor constitute about 40 percent of the population. As a result of this growth and of widespread poverty, and because relatively little has been done to develop agricultural practices suitable for tropical regions, the tropics are being devastated ecologically, with up to 20 percent of the world's species likely to be lost over the next 30 years.

The world food supply can be improved by the traditional methods of plant breeding and selection, by the cultivation of new crops, and by the methods of genetic engineering. Genetic diversity is necessary to allow crops to be modified to meet the different requirements of growing successfully in different regions; modern commercial agriculture tends to reduce genetic diversity. Among the more promising of recently developed new crops are such food plants as triticale and the grain amaranths. Also promising are such industrial crops as jojoba, valuable as a source of liquid wax for lubrication, and guayule, a source of rubber. Many drugs are also derived from plants, and without doubt numerous additional ones await discovery. The widespread destruction of habitats in the tropics, however, is threatening the existence of many potentially useful plant species before they can be identified and exploited.

Suggestions for Further Reading

Abrahamson, Warren G. (ed.): *Plant-Animal Interactions,* McGraw-Hill Book Company, New York, 1988.

An excellent collection of chapters that accurately convey contemporary knowledge about these complex systems; highly recommended.

Ahmadjian, V., and S. Paracer: *Symbiosis: An Introduction to Biological Associations,* University Press of New England, Hanover, N.H., 1986.

An outstanding treatment of all aspects of symbiotic relationships between organisms.

Barbour, Michael G., and W. Dwight Billings (eds.): *North American Terrestrial Vegetation,* Cambridge University Press, New York, 1988.

This book outlines the major vegetation types of North America in the light of contemporary information.

Barbour, Michael G., Jack H. Burk, and Wanna D. Pitts: *Terrestrial Plant Ecology,* 2d ed., Benjamin/Cummings Publishing Company, Inc., Menlo Park, Calif., 1987.

A lucid and comprehensive introduction to this exciting field.

BioScience: "Tree Death as an Ecological Process," a series of articles in *BioScience* 37:550–610, 1987.

These articles deftly integrate plant population biology with community and ecosystem functioning, shedding much light on many facets of plant ecology.

BioScience: "How Plants Cope: Plant Physiological Ecology," a series of articles in *BioScience* 38:18–67, 1988.

These highly recommended articles provide an overview of research opportunities in the physiological ecology of plants.

BioScience: "Fire Impact on Yellowstone," a series of articles in *BioScience* 39:667–722, 1989.

These articles analyze the causes of the Yellowstone fires of 1988 and the process of recovery from them.

Caufield, Catherine: *In the Rainforest,* Alfred A. Knopf, Inc., New York, 1985.

A beautifully written account of the troubles of tropical lands and their effect upon us all.

Chabot, B.F., and H.A. Mooney (eds.): *Physiological Ecology of North American Plant Communities,* Chapman and Hall, New York, 1985.

A good introduction to the mechanisms that enable plants to occupy the diverse environments of North America.

Colinvaux, Paul A.: *Ecology,* John Wiley & Sons, Inc., New York, 1986.

A balanced treatment of ecology, with special emphasis on ecosystems.

Conniff, R.: "Yellowstone's 'Rebirth' amid the Ashes Is Not Neat or Simple, but It's Real," *Smithsonian,* September 1989, pages 36–49.

The complicated process of regeneration in Yellowstone National Park is well under way.

Ehrlich, Paul R., and J. Roughgarden: *The Science of Ecology,* Macmillan Publishing Company, New York, 1987.

A broad overview of ecology, well set in an evolutionary context; highly recommended.

Forman, Richard T.T., and Michael Godron: *Landscape Ecology,* John Wiley & Sons, New York, 1986.

A good introduction to landscapes as an ecological phenomenon, a rapidly growing field of ecology.

Forsyth, A.: *Portraits of the Rainforest,* Camden House, Camden East, Ontario, 1990.

A beautifully illustrated and extraordinarily interesting book on the rainforest.

Givnish, Thomas J. (ed.): *On the Economy of Plant Form and Function,* Cambridge University Press, New York, 1986.

An excellent collection of papers on the ways in which the form and function of plants is related to their environment.

Heiser, Charles B., Jr.: *Seed to Civilization: The Story of Food,* 2d ed., W.H. Freeman and Company, San Francisco, 1981.

A concise account of the plants that feed people.

Heiser, Charles B., Jr.: *Of Plants and People,* University of Oklahoma Press, Norman, Okla., 1985.

A lively account of the relationships between people and some very interesting economically important plants.

Hinman, C.W.: "Potential New Crops," *Scientific American,* July 1986, pages 33–37.

A useful review of current efforts to identify new crops.

Horn, Henry S.: *The Adaptive Geometry of Trees,* Princeton University Press, Princeton, N.J., 1971.*

This interesting short book will give you a good idea of why trees are shaped as they are, opening new insights into their structure and form.

Hunter, A.E., and L.W. Aarssen: "Plants Helping Plants," *BioScience* 38:34–40, January 1988.

Although we normally think of the relationships between different kinds of plants as predominantly competitive, they can affect both kinds of plants positively.

Jeffrey, David W.: *Soil–Plant Relationships: An Ecological Approach,* Timber Press, Portland, Ore., 1987.*

A balanced approach to all aspects of the relationship between plants and soils.

MacQueen, Hilary: "Biotechnology Puts the Squeeze on Plants," *New Scientist*, 14 April 1988, pages 50–53.

This article explores the ways in which secondary plant products might be obtained more efficiently, as from cell cultures.

McIntyre, Loren: "Humboldt's Way," *National Geographic* 168:318–351, 1985.

Excellent article on Alexander von Humboldt, inspired nineteenth-century explorer who laid the foundations of biogeography.

Myers, Norman: *A Wealth of Wild Species*, Westview Press, Boulder, Colo., 1983.

A lucid account of the value of plants, animals, fungi, and microorganisms to people.

National Research Council: *Lost Crops of the Incas*, National Academy Press, Washington, D.C., 1990.

Based on its evaluations of Andean crops, the committee presents a good example of the ways in which new kinds of economically useful plants can be used to feed people.

Nobel, Park S.: *Biophysical Plant Physiology and Ecology*, W.H. Freeman and Company, New York, 1983.

An excellent account of the physical interactions between plants and their environment; highly recommended.

Norse, Elliott A.: *Ancient Forests of the Pacific Northwest*, The Island Press, Washington, D.C., 1990.

A balanced account of the controversy surrounding the destruction of ancient forests in the Pacific Northwest.

Odum, Eugene P.: *Ecology and Our Endangered Life-Support Systems*, Sinauer Associates, Inc., Sunderland, Mass., 1989.*

This book outlines the principles of ecology, the ways in which we are violating them, and the consequences.

Pain, Stephanie: "The Case of the Stolen Slippers," *New Scientist*, 24 June 1989, pages 48–53.

An interesting discussion of orchid smuggling, one of the reasons that many of these beautiful plants are disappearing so rapidly in nature.

Parfit, Michael: "The Dust Bowl," *Smithsonian*, June 1989, pages 44–57.

Half a century ago, parts of the Great Plains of North America blew away, and the question now is: Could it be happening all over again?

Romme, W.H., and D.G. Despain: "The Yellowstone Fires," *Scientific American*, November 1989, pages 37–46.

This excellent article analyzes the reasons for the extent of the Yellowstone fire and the ecological processes that are important for understanding it.

Rosenthal, Gerald A.: "The Chemical Defenses of Higher Plants," *Scientific American*, January 1986, pages 94–99.

An outstanding essay on the chemical defenses of flowering plants and the ways in which animals overcome or exploit them.

Ross, David A.: *Introduction to Oceanography*, 4th ed., Prentice-Hall, Englewood Cliffs, N.J., 1988.

An excellent treatment of the biological, chemical, physical, and geological aspects of—as well as the human uses of—the resources of the world oceans.

Simpson, Beryl B., and Molly Conner-Ogorzaly: *Economic Botany: Plants in Our World*, McGraw-Hill Book Company, New York, 1986.

An excellent and well-illustrated account of useful plants, presented according to their uses.

Urban, Dean L., Robert V. O'Neill, and Herman H. Shugart, Jr.: "Landscape Ecology," *BioScience* 37:119–127, 1987.

A hierarchical perspective on landscapes can help scientists understand spatial patterns.

Vankat, J.L.: *The Natural Vegetation of North America: An Introduction*, John Wiley & Sons, Inc., New York, 1979.

A valuable resource for information regarding the vegetation of the biomes of North America.

Vitousek, Peter M., Paul R. Ehrlich, Anne H. Ehrlich, and Pamela A. Matson: "Human Appropriation of the Products of Photosynthesis," *BioScience* 36:368–373, 1986.

About 25 percent of potential terrestrial net primary productivity is used directly by humans.

Waring, Richard H., and Willian H. Schlesinger: *Forest Ecosystems: Concepts and Management*, Academic Press, Inc., Orlando, Fla., 1985.*

An excellent digest of all aspects of forest ecology, outlining the relationship between forests and the rest of the global ecosystem.

Weiner, Jonathan: *The Next One Hundred Years: Shaping the Future of Our Living Earth*, Bantam Books, New York, 1990.

Well-written account of the ecological problems that confront us as we approach the next century.

Wilson, E.O. (ed.): *Biodiversity*, National Academy Press, Washington, D.C., 1988.*

Excellent overall account of biodiversity throughout the world, with essays by many of its leading students.

* Available in paperback.

Fundamentals of Chemistry

Atoms

All matter is composed of **atoms** (from the Greek *atomos*, meaning "indivisible"), which are the smallest complete units of **elements.** There are 92 naturally occurring elements, and each is unique in the structure of its atoms. Each type of atom has a characteristic number of **protons**—positively charged particles—in its nucleus (center). The number of protons ranges from the lightest element—hydrogen—which has 1 proton, to the heaviest—uranium—which has 92. The **atomic number** of an element represents the number of protons in the nucleus of one atom (Table A–1). Outside the nucleus of the atom are **electrons**—negatively charged particles—that are attracted by the positive charges of the protons. The way electrons are arranged in an atom determines the chemical properties of that atom, and chemical reactions involve changes in the number and the distribution of an atom's electrons.

Table A–1 *Atomic Structure of Some Familiar Elements*

Element	Symbol	Nucleus		Number of Electrons
		Number of Protons	Number of Neutrons*	
Hydrogen	H	1	0	1
Helium	He	2	2	2
Carbon	C	6	6	6
Nitrogen	N	7	7	7
Oxygen	O	8	8	8
Sodium	Na	11	12	11
Phosphorus	P	15	16	15
Sulfur	S	16	16	16
Chlorine	Cl	17	18	17
Potassium	K	19	20	19
Calcium	Ca	20	20	20

*In most common isotope.

Atoms also contain **neutrons,** which are uncharged particles of about the same weight as protons (see Table A–1). The **atomic weight** of an element is essentially equal to the number of protons and neutrons in the nucleus of one atom. By comparison, electrons are so light that their weight is usually disregarded. For instance, when you weigh yourself, only about 30 grams, or about 1 ounce, of your total weight is made up of electrons.

ISOTOPES

Not all atoms of the same element have the same atomic weight. These different kinds of atoms are known as **isotopes.** They have the same number of protons—and hence the same atomic number—but different numbers of neutrons. For example, the common form of hydrogen, with its one proton, has an atomic weight of 1 and is symbolized as ^1H, or simply H. Deuterium, ^2H, is an isotope of hydrogen that contains one proton and one neutron, and so it has an atomic weight of 2. Tritium, ^3H, a third isotope of hydrogen, has one proton and two neutrons; it has an atomic weight of 3 (Figure A–1). Like many (but not all) of the less common isotopes, tritium is **radioactive,** which means that its nucleus is unstable and emits energy as it changes into a more stable form. Both deuterium and tritium have nearly the same chemical properties as the more common isotope of hydrogen (^1H), and either can substitute for it in chemical reactions. However, if a hydrogen (^1H) atom gains a proton along with two neutrons, it is no longer hydrogen but helium. It now has an atomic number of 2 and an atomic weight of 4 (Figure A–1). The fusion of hydrogen nuclei (which are termed protons) to form helium is the source of energy at the heart of the sun and also provides the terrible destructive force of the hydrogen bomb.

Many naturally occurring isotopes are radioactive.

All the heavier elements—atoms that have 84 or more protons in their nucleus—are unstable and, therefore, radioactive. All radioactive isotopes emit nuclear particles at a rate proportional to the number of atoms present; they are said to undergo radioactive "decay" as they change to another element. The rate of decay is measured in terms of half-life: the **half-life** of a radioactive isotope is defined as the time taken for half the atoms in a sample to change into another isotope or into a stable element. Because the half-life of an element is constant, it is possible to calculate the fraction of decay that will occur for a given isotope over a given period of time.

Half-lives vary widely, depending on the isotope. The radioactive nitrogen isotope ^{13}N has a half-life of only 10 minutes; tritium has a half-life of 12.25 years. The most common isotope of uranium (^{238}U) has a half-life of 4.5 billion years. The uranium atom decays through a series of isotopes and is eventually transformed to an isotope of lead (^{206}Pb).

Isotopes play a number of important roles in biological research. One use is in dating the age of fossils or the rocks in which fossils are found. For example, the proportion of ^{238}U to ^{206}Pb in a given rock sample is a good indication of how long ago that rock was formed. (The lead formed as a result of the decay of uranium is not the same as the lead commonly present in the original rock, ^{204}Pb.) Isotopes are also used as radioactive tracers. The use of radioactive carbon dioxide ($^{14}CO_2$) has played an important role in enabling plant physiologists to trace the path of carbon in photosynthesis, as described in Chapter 7 (see also "Radiocarbon Dating," page 53). A third use of isotopes is in autoradiography, a technique in which a sample of material containing a radioactive isotope is placed on a sheet of photographic film. Energy emitted from the isotope leaves traces on the film and so reveals the exact location of the isotope within the specimen; Figure 22–6, page 474, is an example of an autoradiograph (see also "Radioactive Tracers and Autoradiography in Plant Research," page 633).

A-1

Diagrams of atoms of hydrogen, deuterium, tritium, and helium. Note that hydrogen, deuterium, and tritium each have only a single proton and a single electron, so they are chemically similar even though they differ in the number of neutrons in their nuclei. Helium, which has one more proton and one more electron, is chemically very different from hydrogen and its isotopes.

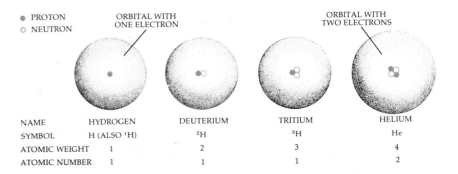

	PROTON				
	NEUTRON				
	ORBITAL WITH ONE ELECTRON			ORBITAL WITH TWO ELECTRONS	
NAME		HYDROGEN	DEUTERIUM	TRITIUM	HELIUM
SYMBOL		H (ALSO ^1H)	^2H	^3H	He
ATOMIC WEIGHT		1	2	3	4
ATOMIC NUMBER		1	1	1	2

ELECTRONS AND ORBITALS

As early as 400 to 500 B.C., Greek philosophers suggested that matter cannot be forever divided into smaller and smaller parts. However, the modern concept of the atom as the fundamental unit of the chemical elements is less than 200 years old, and our ideas about its structure have undergone many changes in that time. These ideas, or hypotheses, have usually been presented in the form of models, as are many scientific hypotheses.

The earliest model was of an indivisible atom, resembling a billiard ball. When it was realized that electrons could be removed from the atom, the billiard-ball model gave way to the plum-pudding model, in which the atom was represented as a solid, positively charged mass with negatively charged particles—electrons—embedded in it. Subsequently, however, physicists found that an atom is in fact mostly open space. The distance from electron to nucleus, experiments indicated, is about 100,000 times the diameter of the nucleus; the electrons are so exceedingly small that the space is almost entirely empty. The physicist Niels Böhr proposed a type of planetary model in which the electrons in the atom were depicted as moving in definite orbits around the nucleus, with a specific energy associated with each orbit (Figure A–2).

The current model of electron configurations is quite different from all previous ones; it reflects our increased knowledge of the behavior of electrons. According to this model, the electron moves unpredictably around the nucleus, and its position at any given moment cannot be known with certainty. For convenience, its pattern of motion is defined as the volume of space in which the electron can be found 90 percent of the time. This volume is known as the electron's **orbital.** Each orbital can hold a maximum of two electrons. In this model, however, the electrons in the atom have definite energies, or energy levels.

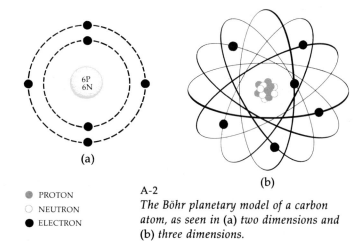

(a)

- ● PROTON
- ○ NEUTRON
- ● ELECTRON

A-2

The Böhr planetary model of a carbon atom, as seen in (a) *two dimensions and* (b) *three dimensions.*

The energy levels in the atom are roughly arranged into **shells,** each of which is composed of one or more **subshells.** Each subshell contains one or more variously shaped **orbitals.** The first two electrons occupy a single spherical orbital. Thus, 90 percent of the time, hydrogen's single electron moves about the nucleus within a single spherical orbital, as do the two electrons of helium. This single spherical orbital, with its maximum of two electrons, makes up the first shell.

The second shell is composed of two subshells and four orbitals, each of which can hold two electrons. The first subshell has a single spherical orbital, and the second subshell has three dumbbell-shaped orbitals. The axes of these three orbitals are perpendicular to one another (Figure A–3). The spherical orbital is filled first, followed by the dumbbell-shaped ones. The second shell can hold a total of eight electrons. Atoms with electrons in as many as seven shells are known.

Atoms tend to stabilize, or complete their energy levels, and the chemical behavior of atoms is governed by this tendency. For instance, helium (atomic number 2), neon (atomic number 10), and argon (atomic number 18) have completely filled outer energy levels and so tend to be unreactive; they are called the "noble" gases because of this apparent "disdain" for reacting with other elements. Atoms of hydrogen (atomic number 1), lithium (atomic number 3), sodium

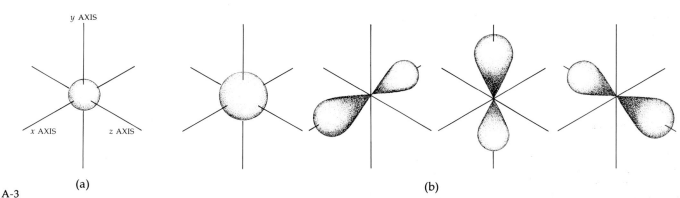

A-3

Orbital models. (a) *The two electrons in the first shell of an atom occupy a single spherical orbital.* (b) *The second shell has four orbitals, each containing up to* two electrons. One of these orbitals is spherical and the other three are dumbbell-shaped. The axes of the dumbbell-shaped orbitals (indicated by the lines) are perpendicular to one another. The nucleus is at the intersection of the axes.

(atomic number 11), and potassium (atomic number 19) have a single electron in their outermost energy level, and they tend to lose this electron. As a consequence of such a loss, each has one more proton than electron and therefore acquires a positive charge: H^+, Li^+, Na^+, and K^+. By contrast, fluorine and chlorine, with atomic numbers of 9 and 17, respectively, tend to gain an electron in order to complete an outer energy level and so become negatively charged: F^- and Cl^-. Similarly, an atom with two electrons in its outer energy level may lose both of them, thus acquiring a double positive charge. For example, magnesium (atomic number 12) and calcium (atomic number 20) become Mg^{2+} and Ca^{2+}. Such charged atoms are known as **ions.** Positively charged ions are called **cations,** and negatively charged ions are called **anions.**

Ions make up less than 1 percent of the weight of most living matter, but they play particularly crucial roles. For instance, K^+ is the principal positively charged ion in most cells, and many essential biological reactions cannot proceed in its absence. Both Na^+ and K^+ are involved in the active transport of sugar and amino acids across the plasma membrane in many plant and animal cells. The calcium ion, Ca^{2+}, has a direct effect on the physical properties of the membrane; Mg^{2+} forms a part of chlorophyll—the molecule that traps radiant energy from the sun.

ELECTRONS AND ENERGY

As was noted previously, electrons, which are negatively charged, are attracted to the atomic nucleus because of the positive charge of the protons. The orbital occupied by an electron is related to the amount of energy of the electron; this energy is in the form of potential energy. The following analogy may be useful: a boulder on flat ground may be said to have no potential energy. If you push it up a hill, you give it energy—potential energy. So long as it sits on the top of the hill, it neither gains nor loses energy. If it rolls down the hill, however, it loses its potential energy as it rolls back toward its original position on level ground.

The electron is like the boulder in that an input of energy can raise it to a higher energy level—to a position farther away from the nucleus. As long as it remains at this higher level, it possesses this added energy. Also, the electron tends to go to its lowest possible energy level, just as the boulder rolls downhill (Figure A–4).

In a given atom, the first spherical orbital is the lowest energy level. The four orbitals of the second level are occupied by electrons with more energy, and so on. Energy is needed to move a negatively charged electron farther away from the positively charged nucleus, just as energy is needed to push a boulder to the top of a

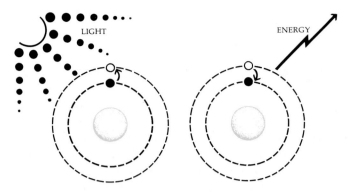

A-4
When an input of energy—such as light energy—boosts an electron to a higher energy level, the electron gains additional potential energy. This energy is released when the electron returns to its previous energy level.

hill. However, unlike the boulder on the hill, the electron cannot be pushed part way up. With an input of energy, electrons may move from a lower energy level to any one of several higher ones, but they cannot move to someplace in between. For an electron to move from one level to another, the atom must absorb a discrete packet of energy, known as a quantum of energy, which contains precisely the amount of energy needed for the transition.

Electronegativity

The atomic nuclei of different elements have various degrees of attraction for electrons. The strength of the attraction depends on the number of protons in the nucleus, the number of electrons, and their proximity to the nucleus. The affinity of an atom for electrons is called **electronegativity.** Electronegativity is expressed on a scale of 0 to 4. Helium and the other unreactive noble gases have electronegativities of 0. At the other end of the scale is fluorine, which has an electronegativity of 4. The value for oxygen, the next most electronegative element, is 3.5. Electronegativity values for certain elements are given in Table A–2.

When an electron moves from an atom that is less

Table A–2   *Electronegativity of Some Common Elements*

Oxygen (O)	3.5
Nitrogen (N)	3.0
Chlorine (Cl)	3.0
Carbon (C)	2.5
Sulfur (S)	2.5
Hydrogen (H)	2.1
Phosphorus (P)	2.1
Sodium (Na)	0.9

electronegative to one that is more electronegative, it moves "downhill"—energetically speaking—and energy is released, just as energy is released when a boulder rolls downhill.

In the cells of photosynthesizing green plants and algae, the radiant energy of sunlight raises electrons to a higher energy level. In the course of a series of electron-transfer reactions (which are described in Section 1), the electrons are passed downhill, and the radiant energy of sunlight is changed to the chemical energy on which nearly all life on earth depends.

Bonds and Molecules

Atoms may be held together by forces known as **chemical bonds**; an assembly of atoms held together by chemical bonds is called a **molecule.** When atoms interact to form a bond, only electrons in their outer shells (termed the valence shells) are involved. There are two general types of chemical bonds: ionic bonds and covalent bonds.

IONIC BONDS

Positive and negative ions attract one another. Bonds involving the mutual attraction of ions of opposite charge are known as **ionic bonds,** or *ionic interactions.* The ionic bond is a very common type of bond in inorganic molecules. Thus, the sodium ion (Na^+)—with its single positive charge—is attracted to the chloride ion (Cl^-)—with its single negative charge (Figure A–5). The calcium ion (Ca^{2+}), with its double positive charge, can attract and hold two Cl^- ions, forming $CaCl_2$—the subscript 2 indicates that two atoms of chlorine are present for each atom of calcium.

The combining power of an element is called its **valence.** The valence is determined by the number of electrons that an atom can gain, or lose, or share. The valences of Na^+ and Cl^- are 1, and the valence of Ca^{2+} is 2. Thus, Na^+ combines with Cl^- in a ratio of 1 to 1, and Ca^{2+} combines with Cl^- in a ratio of 1 to 2.

COVALENT BONDS

Another way for an atom to complete its outer energy level is by *sharing* electrons with another atom. Bonds formed by shared pairs of electrons are known as **covalent bonds.** Covalent bonds figure prominently in organic chemistry, which is essentially the chemistry of carbon-containing compounds. The simplest covalent bond is that formed between hydrogen atoms in the hydrogen molecule (Figure A–6). In a covalent bond, the shared pair of electrons forms a new orbital—or molecular orbital—that encompasses both atoms. In

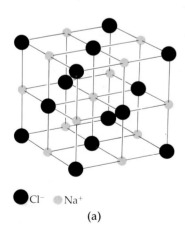

$\bullet$ Cl$^-$ $\bullet$ Na$^+$

(a)

(b)

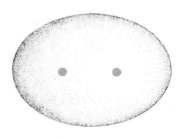

A-5
Sodium (Na), which has only one electron in its outer energy level, becomes more stable if it loses the electron. Chlorine (Cl), which has seven electrons in its outer energy level, becomes more stable if it gains one. When sodium and chlorine interact, sodium loses its single electron and chlorine gains it; following this transaction, sodium has a positive charge (Na$^+$) and chlorine has a negative charge *one (Cl$^-$). Such charged atoms are called ions. (a) Oppositely charged ions attract one another. Table salt is crystalline NaCl, a latticework of alternating Na$^+$ and Cl$^-$ ions held together by their opposite charges. Such bonds between oppositely charged ions are known as ionic bonds. (b) The regularity of the latticework is reflected in the structure of salt crystals.*

A-6
In a molecule of hydrogen, each atom shares its single first-orbital electron with the other atom. As a result, both atoms effectively have a filled first orbital, containing two electrons—a highly stable configuration. The orbital shown here—a molecular orbital—indicates the volume in which the two electrons can be found 90 percent of the time.

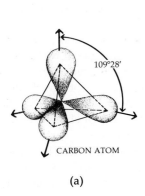

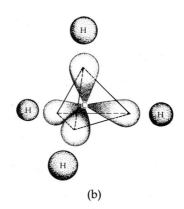

CARBON ATOM

(a)

METHANE (CH₄)

(b)

(c)

A-7

(a) *When carbon forms covalent bonds with four other atoms, new orbitals are formed. These new orbitals, which are all the same shape, are oriented toward the four corners of a tetrahedron. Thus, the* *four orbitals are spatially separated as much as possible.* (b) *When a carbon atom reacts with hydrogen to form a methane molecule, the four unpaired electrons of the carbon atom form pairs* *with each of the unpaired electrons in four hydrogen atoms.* (c) *Each pair of electrons moves in a new molecular orbital. The molecule has the shape of a regular tetrahedron.*

such a bond, each electron spends part of its time around one nucleus and part of its time around the other.

Carbon, which has an atomic number of 6, needs to share four electrons to achieve a filled and therefore stable outer energy level; and so carbon is able to form covalent bonds with as many as four other atoms. When these bonds are formed, the electron pairs assume new orbitals. These molecular orbitals are distributed symmetrically in space, forming a regular tetrahedron (Figure A–7). The capacity of carbon to form four covalent bonds is crucial to its central role in the chemistry of living systems.

Double and Triple Bonds

There are various ways in which atoms can participate in covalent bonds and satisfy their valence requirements. Oxygen, with two unpaired electrons in its outer electron orbitals, has a valence of 2. Carbon, as noted above, has a valence of 4. Carbon and oxygen can thus form a simple compound, carbon dioxide (CO_2), in which each oxygen atom shares two pairs of electrons with a central carbon atom. Such bonds in which two atoms are held together by two pairs of electrons (four electrons) are called double bonds. They are symbolized in a structural formula by two lines connecting the atomic symbols: O=C=O. Carbon atoms can form double or even triple bonds (in which three pairs of electrons are shared) with each other as well as with other atoms.

Electrons shared in double and triple bonds form orbitals that differ in shape from the orbitals filled by single electron pairs. For instance, when four bonds satisfy the electron requirements of carbon, they are directed toward the four corners of a tetrahedron, as in

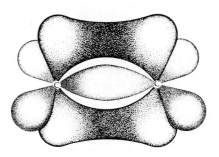

A-8

An orbital model showing the carbon-carbon double bond of ethylene, $CH_2{=}CH_2$. One pair of electrons occupies the inner orbital between the two carbon atoms. The other pair of electrons occupies the outer orbital, which has two phases—one above the plane of the two nuclei and one below. This creates a rigid bond, about which the atoms cannot rotate. The two smaller orbitals extending from each carbon atom contain one electron each and form molecular orbitals with each hydrogen atom.

Figure A–7. When two bonds are replaced by a double bond, the remaining single bonds form the arms of a Y with the double bond as its leg (Figure A–8). When two double bonds are made by a single carbon atom, as in carbon dioxide, the three bonded atoms lie in a straight line.

Single bonds are flexible, allowing atoms to rotate in relation to one another. Double and triple bonds hold the atoms relatively rigid in relation to one another.

The presence of double bonds in a molecule can make a significant difference in its properties. For example, both fats and oils are composed of covalently bonded carbon and hydrogen atoms. In oils some of the bonds between carbon atoms are double, and in fats all of the carbon-carbon bonds are single.

Polar Covalent Bonds

The electrons in covalent bonds are not always shared equally between the two atoms involved. Some atoms have a greater attractive force for the electrons than others—that is, they are more electronegative. Because of this differential attraction, the shared electrons tend to spend more time around the more electronegative atom. As a result, the more electronegative atom in the molecule will have a slightly negative charge, and the less electronegative atom will have a slightly positive one, since its nuclear charge is not entirely neutralized (Figure A–9). The polar properties of some covalent bonds have very important consequences for living things. For example, many of the special properties of water, upon which life depends, derive largely from its polar nature, as explained below.

Ionic, covalent, and polar covalent bonds may be considered different versions of the same type of bond. The differences depend on the differences in electronegativity among the combining atoms. In a wholly nonpolar covalent bond, the electrons are shared equally; such bonds can exist only between identical atoms, as in H_2, Cl_2, O_2, and N_2. In polar covalent bonds, electrons are shared unequally. In ionic bonds, there is no electron sharing; an electrostatic attraction between the negatively charged and positively charged ions results from their having gained or lost electrons.

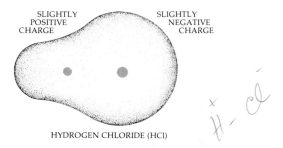

SLIGHTLY POSITIVE CHARGE

SLIGHTLY NEGATIVE CHARGE

HYDROGEN CHLORIDE (HCl)

A-9

In a polar molecule, such as hydrogen chloride (HCl), the shared electrons tend to spend more time around the more electronegative atom—in this case, the chlorine atom. As a result, chlorine has a slightly negative charge, and the less electronegative atom (hydrogen) has a slightly positive charge.

ATOMIC AND MOLECULAR WEIGHTS

The atomic weight of any element is an average value for the naturally occurring mixture of isotopes of that element relative to the common isotope of carbon (^{12}C) which has an atomic weight of 12. Hypothetically, an element exactly twice as heavy as carbon would have an atomic weight of 24; an element one-half as heavy would have an atomic weight of 6.

The **molecular weight** of a substance is the sum of the atomic weights of all the atoms in one molecule. For example, the molecular weight of carbon dioxide, CO_2, is $12 + 16 + 16$, or 44. Similarly, the molecular weight of a protein (see Chapter 3) is the sum of the atomic weights of all of the atoms of all of the constituent amino acids.

Atoms and molecules are measured in amounts called **moles.** One mole of any substance contains the same number of particles (atoms, ions, or molecules) as 1 mole of any other substance. This number, 6.022×10^{23}, is known as **Avogadro's number.** Thus 1 mole of water contains 6.022×10^{23} molecules of water, and 1 mole of glucose contains 6.022×10^{23} molecules of glucose. One mole of a substance weighs an amount, in grams, that is numerically equal to its atomic weight (if an element) or molecular weight (if a molecule). Thus 1 mole of water (H_2O) weighs 18 grams, and 1 mole of sodium weighs 23 grams. The mole is useful for defining quantities involved in chemical reactions.

In order to make water, for instance, one would combine 2 moles of hydrogen (4 grams) with 1 mole of oxygen (32 grams); in other words, four hydrogen atoms for every two oxygen atoms. Two moles of water would be produced, each weighing 18 grams. Similarly, to make table salt (NaCl), one would combine 1 mole of sodium (about 23 grams) and 1 mole of chlorine (about 35.5 grams).

FUNCTIONAL GROUPS

Sometimes clusters of atoms joined by covalent bonds tend to react together as a group, known as a **functional group.** The specific chemical properties of organic molecules are determined primarily by their functional groups. The $-OH$ group is an example ($-OH$, the functional group, is called hydroxyl; OH^-, the ion, is called hydroxide). When one hydrogen and one oxygen are bonded covalently, one electron on the oxygen is left over for sharing. A compound formed when a hydroxyl group replaces one or more of the hydrogens in a hydrocarbon (a compound consisting only of carbon and hydrogen) is known as an alcohol.

Thus, when one hydrogen atom in methane (CH_4) is replaced by a hydroxyl group, the compound becomes methanol (CH_3OH), a pleasant-smelling, poisonous al-

Table A–3 *Some Functional Groups that Play Important Roles in Organic Compounds*

Group	Name
—OH	Hydroxyl
—NH$_2$	Amino
—C—CH$_3$ (with =O below C)	Acetyl
—C—C—C— (with O and =O above middle C)	Keto
—C—OH (with =O below C)	Carboxyl
—P—OH (with OH above, =O below)	Phosphate

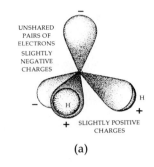

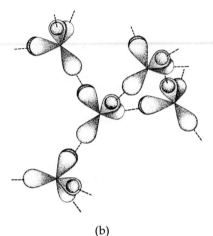

(a)

(b)

A-10

(a) *An orbital diagram of a single water molecule. Note the four-cornered distribution of positive and negative charges.* (b) *As a result of these slight positive and negative charges, each water molecule can form hydrogen bonds (indicated by dashed lines) with four other water molecules.*

cohol noted for its ability to cause blindness and death. Ethane similarly becomes ethanol (C$_2$H$_5$OH), which is present in all alcoholic beverages. Glycerol, C$_3$H$_5$(OH)$_3$, is an alcohol that contains three hydroxyl groups.

The carboxyl group (—COOH) is a functional group that gives a compound the properties of an acid. Table A–3 shows some functional groups of considerable biological importance.

Water and the Hydrogen Bond

Water is composed of very small molecules held together by the mutual attraction of positively and negatively charged atoms. Each water molecule is made up of two atoms of hydrogen and one atom of oxygen, held together by two covalent bonds.

The water molecule as a whole is neutral in charge, having an equal number of electrons and protons. However, because oxygen is more electronegative than hydrogen, the water molecule is polar. The paired electrons in the outer orbitals spend more time around the oxygen nucleus than around the hydrogen nuclei. Consequently, the region near the oxygen nucleus has two weakly negative zones, and each of the regions near the hydrogen nuclei has a weakly positive zone. Thus, the water molecule—in terms of its polarity—is four-cornered, with two positively charged corners and two negatively charged ones (Figure A–10a).

When any of these charged regions comes close to the oppositely charged region of another water molecule, the force of the attraction forms a bond between

them—a **hydrogen bond.** Hydrogen bonds are not only found in water but can form between any hydrogen atom that is covalently bonded to an electronegative atom—usually oxygen or nitrogen—and the electronegative atom of another molecule. In liquid water, hydrogen bonds form between the negative "corners" of one water molecule and the positive "corners" of another. Thus, every water molecule can establish hydrogen bonds with four other water molecules. Liquid water is made up of water molecules bound together in this way, as shown in Figure A–10b.

Any single hydrogen bond is relatively weak and has an exceedingly short lifetime; on an average, such bonds last about 1/100,000,000,000th of a second (10^{-12} seconds). But, as one is broken, another is formed. As a group, hydrogen bonds have considerable strength, making water both liquid and stable under ordinary conditions of pressure and temperature.

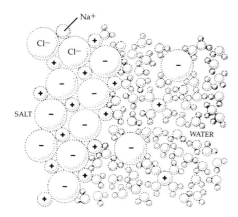

A-11
Because of the polarity of water molecules, water can serve as a solvent for polar atoms or molecules. This diagram shows sodium chloride (NaCl) dissolving in water as the water molecules cluster around the individual ions, separating them from each other. Substances that dissociate into ions in solution produce solutions that can conduct electricity. Such solutions are called electrolytes.

Water As a Solvent

Many substances within living systems are found in solution. A **solution** is a uniform mixture of the molecules of two or more substances. The substance present in the greatest amount—usually a liquid—is called the **solvent;** the substances present in lesser amounts are called **solutes.** The polarity of the water molecule is responsible for the capacity of water to act as a solvent for polar or charged substances. The polar water molecules tend to separate substances such as NaCl into their constituent ions. Then, as shown in Figure A–11, the water molecules cluster around and segregate the charged ions.

Many of the molecules important in living systems, such as glucose, also have polar areas. Such regions of partial positive or negative charges arise in the neighborhood of covalently bonded atoms of unequal electronegativity. Thus, these molecules attract water molecules and are dissolved in the water.

Molecules that readily dissolve in water are called **hydrophilic** ("water-loving"). Such molecules slip into aqueous solution easily because their partially charged regions attract water molecules and thus compete with the attraction between the water molecules themselves.

Molecules that lack polar regions, such as fats, tend to be insoluble in water. The hydrogen bonding between the water molecules acts as a force to exclude the nonpolar molecules. As a result of this exclusion, nonpolar molecules tend to cluster together in water, just as droplets of fats tend to coalesce on the surface of chicken soup, for example. Such molecules are said to be **hydrophobic** ("water-fearing").

Acids and Bases

Acids taste sour, like sour milk, citrus fruits, and vinegar. Bases taste flat, like milk of magnesia, and feel slippery and soapy in solution.

To define "acid" and "base" in chemical terms, it is easiest to begin by looking at water. Water consists of two atoms of hydrogen and one of oxygen held together by covalent bonds. Water molecules also have a slight tendency to ionize—separate into H^+ and OH^- ions. (The H^+ ions tend to combine with other H_2O molecules to produce the hydronium ion, H_3O^+, but we shall omit the hydronium ion from our consideration in this discussion.) In any given volume of pure water, a very small but constant number of water molecules will be dissociated into ions. The number is constant because the tendency of water to dissociate is exactly offset by the tendency of the ions to reunite; thus, even as some water molecules are ionizing, an equal number of others are re-forming, a state known as dynamic equilibrium.

In pure water, the number of H^+ ions exactly equals the number of OH^- ions. This is necessarily the case, because neither ion can be formed without the other when only H_2O molecules are present. A solution acquires the properties we recognize as acidic when the number of H^+ ions exceeds the number of OH^- ions; conversely, a solution is basic when the concentration of OH^- ions exceeds that of H^+ ions. There is always an inverse relationship between the concentration of H^+ and OH^- ions; when H^+ is high, OH^- is low, and vice versa. This is because the product of their concentrations is a constant. Thus, at $25°C$,

$$[H^+][OH^-] = 1 \times 10^{-14} \text{ (moles per liter)}^2$$

We now can define our terms chemically:

1. An **acid** is a substance that donates H^+ ions to a solution. Because an H^+ ion is a proton, acids can also be defined as **proton donors.**

2. A **base** is a substance that decreases the number of H^+ ions, or protons. More specifically, a base is a **proton acceptor.** The OH^- ion is a base because it can accept a proton and thus be neutralized:

$$H^+ + OH^- \longrightarrow H_2O$$

3. In any acid-base reaction, there is always a proton donor and a proton acceptor.

STRONG AND WEAK ACIDS AND BASES

Hydrochloric acid (HCl) is a common strong acid, meaning that it tends to be almost completely ionized in an aqueous solution into H^+ and Cl^- ions. Sodium hydroxide (NaOH) is a common strong base; in an aqueous solution, it exists entirely as Na^+ and OH^- ions. Weak acids and weak bases are those that ionize only slightly. Compounds that contain the carboxyl group ($-COOH$) are often weak acids because the hydrogen atom may partially dissociate from the carboxyl group to yield a proton:

$$R-COOH \rightleftharpoons R-COO^- + H^+$$

In this reaction, R represents any chemical structure that may be attached to a carboxyl group.

Compounds that contain the amino group ($-NH_2$) act as weak bases, because the amino group has a weak tendency to accept a hydrogen ion:

$$R-NH_2 + H^+ \rightleftharpoons R-NH_3^+$$

THE pH SCALE

Chemists define degrees of acidity by means of the pH scale. In the expression "pH," the "p" stands for "power" and the "H" stands for the concentration of hydrogen ions.

In a liter of pure water, 0.0000001 mole of hydrogen ions can be detected. For convenience, this is written in terms of a power of ten, 10^{-7}, and in terms of the pH scale, it is simply referred to as pH 7 (see Table A-4). At pH 7, the concentrations of free H^+ and OH^- ions are exactly the same, and pure water is thus "neutral." Any pH below 7 is acidic, and any pH above 7 is basic. The lower the pH number, the higher the concentration of hydrogen ions. Thus, pH 2 means 10^{-2} mole of hydrogen ions per liter of water, or 0.01 mole per liter—a much higher concentration than 0.0000001. A difference of one pH unit represents a tenfold difference in the concentration of hydrogen ions.

Lemon juice has a pH of about 2, as do the stomach contents of humans and other animals. Orange juice has a pH of about 3; human blood has a pH of 7.4. The best soil pH for most plants is about 6.4, but alkaline soils have a pH between 7 and 9, and peat bogs may have a pH as low as 3.

Chemical Reactions

All chemical reactions involve the breaking of bonds and the formation of new bonds. According to present concepts, atoms or molecules react with one another only when they collide with sufficient force to overcome the initial forces of repulsion. The force required varies with the nature of the atoms or molecules; the more stable their initial state, the more forceful the collision must be for a reaction to occur. In any given group of atoms, it is likely that some proportion is moving with sufficient energy for a reaction to occur, but often this proportion is so small that the reaction, for all practical purposes, does not take place.

Table A-4 *The pH Scale*

	Concentration of H^+ Ions (moles per liter)		pH	Concentration of OH^- Ions (moles per liter)	
Acidic	1.0	10^0	0	10^{-14}	
	0.1	10^{-1}	1	10^{-13}	
	0.01	10^{-2}	2	10^{-12}	
	0.001	10^{-3}	3	10^{-11}	
	0.0001	10^{-4}	4	10^{-10}	
	0.00001	10^{-5}	5	10^{-9}	
	0.000001	10^{-6}	6	10^{-8}	
Neutral	0.0000001	10^{-7}	7	10^{-7}	
Basic		10^{-8}	8	10^{-6}	0.000001
		10^{-9}	9	10^{-5}	0.00001
		10^{-10}	10	10^{-4}	0.0001
		10^{-11}	11	10^{-3}	0.001
		10^{-12}	12	10^{-2}	0.01
		10^{-13}	13	10^{-1}	0.1
		10^{-14}	14	10^0	1.0

A-12

Chemical reactions require an initial input of energy—called the activation energy—in order to take place. An uncatalyzed reaction requires more "input," or activation energy, than does a catalyzed one, such as an enzymatic reaction. The lower activation energy in the presence of the catalyst is often within the range of energy possessed by the molecules, and so the reaction can occur with little or no added energy. Note, however, that the overall energy change from the initial state to the final state is the same with or without the catalyst.

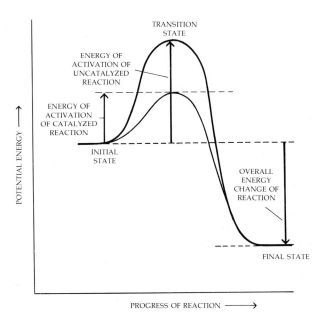

Reaction rates can be increased by increasing the likelihood of forceful collisions. One way to do this is to raise the temperature, thereby increasing the average velocity at which the atoms or molecules move and so increasing their likelihood of colliding with sufficient force. Sometimes, as in the oxidation of methane to carbon dioxide and water, a spark is all that is needed. Once the reaction begins, it liberates heat that is transferred to the other CH_4 molecules until all are moving rapidly enough to react with oxygen almost simultaneously with explosive force. Driving a reaction by heat is a method commonly used in chemical laboratories, as well as in industry.

A second way to increase the rate of a reaction is to increase the concentration of reacting molecules. Chemists working in research laboratories or in industry usually work with pure chemicals in high concentrations. By this means, the chemical reactions of interest are not only speeded up, but are also easier to control.

A third way to increase a reaction rate is to use catalysts. Catalysts lower the **energy of activation** of a reaction, that is, the minimum energy that must be available in a collision for a reaction to be initiated (Figure A–12). Metals, such as iron, nickel, and platinum, are commonly used as catalysts in industrial laboratories. Certain molecules apparently tend to cluster on the surfaces of such metals, thus increasing the likelihood of the necessary close encounters between the reactants. Although they participate in the reactions, catalysts are not used up, so they can be used over and over again. Enzymes are the catalysts of the chemical reactions that take place in living cells (see Chapter 5, pages 79–84).

TYPES OF REACTIONS

Chemical reactions can be classified into a few general types. One type, a combination reaction, can be represented by the expression:

$$A + B \longrightarrow AB$$

An example of this sort of reaction is the combination of hydrogen gas with oxygen gas to produce water:

$$2H_2 + O_2 \longrightarrow 2H_2O$$

A reaction may also take the form of a dissociation:

$$AB \longrightarrow A + B$$

For example, the equation for the formation of water can be reversed to show the breakdown of water into its component elements:

$$2H_2O \longrightarrow 2H_2 + O_2$$

This means that water molecules yield hydrogen and oxygen gases.

A reaction may also involve an exchange, taking the form:

$$AB + CD \longrightarrow AD + CB$$

An example of such a reaction is the combination of hydrochloric acid and sodium hydroxide to form table salt and water:

$$NaOH + HCl \longrightarrow NaCl + H_2O$$

CHEMICAL EQUILIBRIUM

Some chemical reactions can go in either direction, as discussed previously. When net change ceases, the reaction is said to be at equilibrium. In the reaction

$$A + B \rightleftharpoons C + D$$

the point of equilibrium is reached when as many molecules of C and D are being converted to molecules of A and B as molecules of A and B are being converted to molecules of C and D.

The concentration of reactants does *not* have to equal the concentration of products in order for equilibrium to be established; only the *rates* of the forward and reverse reactions must be the same. Consider the reaction above. The different lengths of the arrows indicate that more C + D is present at equilibrium than A + B. If only A and B molecules are present initially, the reaction occurs at first to the right, with A and B molecules converting into C and D molecules. Figure A–13 shows the relative changes in concentration as the reaction continues. As C and D accumulate, the rate of the reverse reaction increases, and at the same time, the rate of the forward reaction decreases because of the decreasing concentrations of A and B. At some point (after six minutes in the example in Figure A–13), the rates of the forward and reverse reactions equalize, and no further changes in concentration take place. The proportions of A + B and C + D will remain the same, but there will always be more C and D molecules.

THE ENERGY FACTOR

Each bond within a chemical compound has a characteristic energy content, or **bond energy** (Figure A–14); the higher the bond energy, the stronger the chemical bond. The total bond energy of a compound can be defined as the amount of energy required to break that compound into its constituent atoms. All chemical reactions involve the rearrangement of bonds; hence, they are accompanied by changes in energy. Most chemical reactions result in the loss of energy to the surroundings.

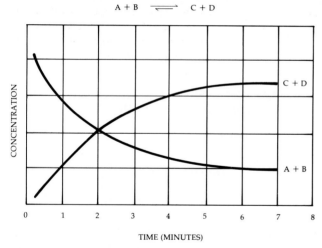

A-13

A graph of the changes in concentration of products and reactants in a reversible reaction. At first, only molecules of A and B are present. The reaction begins when A and B start to yield products C and D. At the end of two minutes, the concentrations of A + B and C + D are equal. As the reaction proceeds, the concentration of C + D will continue to increase to the point of chemical equilibrium (at about six minutes) and will thereafter remain greater than the concentration of A + B. At equilibrium, the rates of forward and reverse reactions are the same.

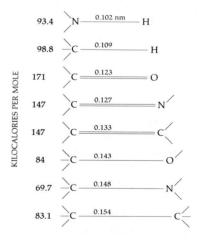

A-14

A chemical bond is a force holding atoms together. The strength of a bond is measured in terms of the energy required to break it. The figures at the left indicate the number of kilocalories per mole needed to break the bonds between the pairs of atoms shown. The lines connecting the atoms represent the bonds, and the figures above them represent the characteristic center-to-center distances between the atoms, in nanometers (10^{-9} meters). Double lines indicate double bonds, which hold the atoms closer together and thus are stronger.

Consider, for example, the burning (oxidation) of methane, represented by the following equation:

$$CH_4 + 2O_2 \longrightarrow CO_2 + 2H_2O$$

This reaction can be set in motion by a spark (which is what causes explosions in coal mines), and then energy is released in the form of heat. The amount of energy released (that is, 213 kilocalories per mole of methane) can be measured quite precisely (Figure A–15). This release of energy can be expressed by a simple equation: $\Delta H = -213$ kcal/mole. The Greek letter delta (Δ) represents "change" and H represents "heat content." In general, the change in heat content is approximately equal to the change in potential energy. The minus sign indicates that energy has been released. (A calorie is defined as the amount of heat necessary to raise the temperature of 1 gram of water by 1°C; 1000 calories = 1 kilocalorie. In expressing the heat- or energy-producing value in a food that is oxidized, the calorie unit ordinarily used is actually a "nutritional calorie" or "large Calorie"—that is, a kilocalorie.)

Similarly, changes in energy occur in the chemical reactions that take place in living systems. However, living systems have evolved ways to minimize the activation energy required for molecules to react, as well as ways to convert a portion of the energy released in some reactions into more usable forms than heat (see Chapter 5).

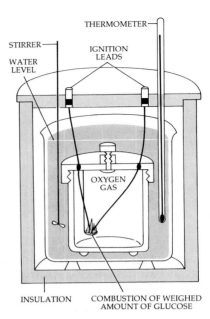

A-15
A calorimeter. A known quantity of glucose or some other material is ignited electrically. As it burns, the increase in the temperature of a known amount of water is measured. Based on the specific heat of water (the heat in calories required to raise the temperature of one gram of water one degree Celsius) one can then calculate the number of calories released by the burning of the sample.

ENDERGONIC AND EXERGONIC REACTIONS

A reaction that requires a net input of energy is said to be **endergonic** (from the Greek *endon*, meaning "within," and *ergon*, meaning "work"), because energy must be put into it. A reaction that liberates energy is an **exergonic** reaction. An endergonic reaction can be thought of as an "uphill" reaction and an exergonic reaction as a "downhill" reaction. Only exergonic (downhill) reactions can proceed spontaneously.

Endergonic reactions do not occur by themselves; they must be coupled to some downhill process in such a way that the energy released by the downhill process can be used for the uphill process. Thus, the energy released in the downhill process must be at least slightly greater than that required for the uphill process. In living systems, the uphill and downhill processes are often accomplished in very small stages, so that large amounts of energy are not required or released all at once.

Metric Table

	Fundamental Unit	Quantity	Numerical Value	Symbol	English Equivalent
Area		hectare	10,000 m²	ha	2.471 acres
Length	meter			m	39.37 inches
		kilometer	1000 (10³) m	km	0.62137 mile
		centimeter	0.01 (10⁻²) m	cm	0.3937 inch
		millimeter	0.001 (10⁻³) m	mm	
		micrometer	0.000001 (10⁻⁶) m	μm	
		nanometer	0.000000001 (10⁻⁹) m	nm	
		angstrom	0.0000000001 (10⁻¹⁰) m	Å	
Mass	gram			g	0.03527 ounce
		kilogram	1000 g	kg	2.2 pounds
		milligram	0.001 g	mg	
		microgram	0.000001 g	μg	
Time	second			sec	
		millisecond	0.001 sec	msec	
		microsecond	0.000001 sec	μsec	
Volume (solids)	cubic meter			m³	35.314 cubic feet
		cubic centimeter	0.000001 m³	cm³	0.061 cubic inch
		cubic millimeter	0.000000001 m³	mm³	
Volume (liquids)	liter			l	1.06 quarts
		milliliter	0.001 liter	ml	
		microliter	0.000001 liter	μl	

Temperature Conversion Scale

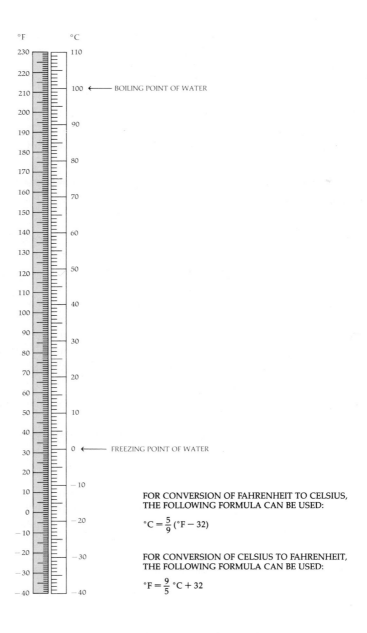

°F   °C

230   110
220
210   100 ← BOILING POINT OF WATER
200
190   90
180
170   80
160
150   70
140   60
130
120   50
110
100   40
90
80   30
70   20
60
50   10
40
30   0 ← FREEZING POINT OF WATER
20
10   −10
0
−10   −20
−20   −30
−30
−40   −40

FOR CONVERSION OF FAHRENHEIT TO CELSIUS,
THE FOLLOWING FORMULA CAN BE USED:

$$°C = \frac{5}{9}(°F - 32)$$

FOR CONVERSION OF CELSIUS TO FAHRENHEIT,
THE FOLLOWING FORMULA CAN BE USED:

$$°F = \frac{9}{5}°C + 32$$

Classification of Organisms

There are several alternative ways to classify organisms. The one presented here follows the overall scheme described in Chapter 10, in which organisms are divided into six major groups, or kingdoms: Archaebacteria, Eubacteria, Protista, Animalia, Fungi, and Plantae. The chief taxonomic categories are kingdom, division (phylum), class, order, family, genus, species.

The classification that follows includes the divisions of Protista, except those considered Protozoa, as well as the Fungi and Plantae. Certain classes given prominence in this book are also included, but the listings are far from complete. The number of species given for each group is the estimated number of living species that have been described and named. Only groups that include living species are described. Viruses are not included in this appendix, but are discussed in Chapter 11.

Kingdom ARCHAEBACTERIA

Archaebacteria are prokaryotic cells that lack a nuclear envelope, plastids and 9-plus-2 flagella. Bacteria are unicellular but sometimes aggregate into filaments or other superficially multicellular bodies. Their predominant mode of nutrition is absorption, but one group of genera obtains its energy by metabolizing sulfur, and another genus, *Halobacterium*, does so through the operation of a proton pump. Many archaebacteria are methanogens, generators of methane. Reproduction is asexual, by fission; genetic recombination has not been observed. They are diverse morphologically, being motile flagellated or nonmotile rods, cocci, and spirilla. Archaebacteria differ fundamentally from Eubacteria in the base sequences in RNAs, lack of muramic acid in their cell walls, and lipid composition of their plasma membranes. There are fewer than 100 named species.

Kingdom EUBACTERIA

Like Archaebacteria, Eubacteria lack a nuclear envelope, plastids, and mitochondria and other membrane-bound organelles, and 9-plus-2 flagella; they are unicellular, but many form aggregates. Their predominant mode of nutrition is absorption, but some groups are photosynthetic or chemosynthetic. Reproduction is predominantly asexual, by fission or budding, but portions of DNA molecules may also be exchanged between cells under certain circumstances. They are motile by simple flagella, or by gliding, or they may be nonmotile.

About 2500 species of eubacteria are recognized at present, but that is probably only a small fraction of the actual number. The recognition of species is not comparable with that in eukaryotes and is based largely upon metabolic features. One group, the class Rickettsiae—very small bacteria—occurs widely as parasites in arthropods and may contain tens of thousands of species, depending upon the classification criteria used; they have not been included in the estimate given here.

Scientists have begun recently to divide Eubacteria into phyla, and about 20 are recognized; there is not yet agreement about the limits of these groups. Among these, cyanobacteria are an ancient group that is abundant and important ecologically. Formerly and misleadingly called "blue-green algae," cyanobacteria have a photosynthesis that is based on chlorophyll *a*. Cyanobacteria, like the red algae, also have accessory pigments called phycobilins. Many cyanobacteria can fix atmospheric nitrogen, often in specialized cells called heterocysts. Some cyanobacteria form complex filaments or other colonies. Although some 7500 species of cyanobacteria have been described, a more reasonable estimate puts the number of these specialized bacteria at about 200 distinct nonsymbiotic species.

Kingdom FUNGI

Eukaryotic multicellular or rarely unicellular organisms in which the nuclei occur in a basically continuous mycelium; this mycelium becomes septate in certain groups and at certain stages of the life cycle. Fungi are heterotrophic; they obtain their nutrition by absorption. Members of all three divisions form important symbiotic relationships with the roots of plants, called mycorrhizae. Reproductive cycles typically include both sexual and asexual phases. There are about 100,000 valid species of fungi to which names have been given, and many more will eventually be found. Some have been named two or more times; this is particularly so for fungi that may be classified both as ascomycetes and as members of the deuteromycetes.

Division

Division Zygomycota: Terrestrial fungi with the hyphae septate only during the formation of reproductive bodies; chitin predominant in the cell walls. The class includes about 765 described species, of which about 100 occur as components of the endomycorrhizae that are found in about 80 percent of all vascular plants.

Division Ascomycota: Terrestrial and aquatic fungi with the hyphae septate but the septa perforated; complete septa cut off the reproductive bodies, such as spores or gametangia. Chitin is predominant in the cell walls. Sexual reproduction involves the formation of a characteristic cell—the ascus—in which meiosis takes place and within which ascospores are formed. The hyphae in many ascomycetes are packed together into complex "bodies" known as ascomata. Most yeasts are unicellular ascomycetes that reproduce asexually; about a quarter of the genera are actually basidiomycetes. There are about 30,000 species of ascomycetes, in addition to some 17,000 additional species of deuteromycetes (members of the artificial group Deuteromycota), in which sexual stages do not occur or are not known.

Lichens: The lichens are symbiotic associations between ascomycetes and certain genera of green algae or cyanobacteria that multiply within their densely packed hyphae. (There are also about a dozen species of basidiomycetes that form associations with algae, but they are closely related to free-living basidiomycetes and do not resemble other lichen-forming fungi.) There are about 20,000 described species of lichens.

Division Basidiomycota: Terrestrial fungi with the hyphae septate but the septa perforated; complete septa cut off reproductive bodies, such as spores. Chitin is predominant in the cell walls. Sexual reproduction involves formation of basidia, in which meiosis takes place and on which the basidiospores are borne. Basidiomycetes are dikaryotic during most of their life cycle, and there is often complex differentiation of "tissues" within their basidiomata. They are components of most ectomycorrhizae. There are some 25,000 described species.

Class Hymenomycetes: Basidiomycetes that produce basidiospores in a hymenium exposed on a basidioma; the mushrooms, coral fungi, and shelf, or bracket, fungi. The basidia are either aseptate (internally undivided) or septate.

Class Gasteromycetes: Basidiomycetes that produce basidiospores inside basidiocarps, where they are completely enclosed for at least part of their development; puffballs, earthstars, stinkhorns, and their relatives.

Class Teliomycetes: Basidiomycetes that do not form basidiocarps and have septate basidia; the rusts and smuts.

Kingdom PROTISTA

Eukaryotic unicellular or multicellular organisms. Their modes of nutrition include ingestion, photosynthesis, and absorption. True sexuality is present in most divisions. They move by means of 9-plus-2 flagella or are nonmotile. Fungi, plants, and animals are specialized multicellular groups derived from Protista. The divisions treated in this book are categorized as heterotrophic protists (water molds, slime molds, and chytrids; the first four divisions listed below) and autotrophic protists (the algae). The characteristics of the divisions of Protisa are outlined in Table 13–1 (page 247).

Division

Division Oomycota: Water molds and related organisms. Aquatic or terrestrial organisms with motile cells characteristic at certain stages of their life cycle. The flagella are two in number—one tinsel and one whiplash. Their cell walls are composed of cellulose or celluloselike polymers, and they store their food as glycogen. There are about 475 species.

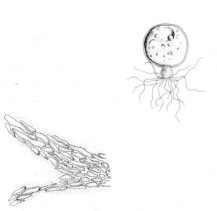

Division Chytridiomycota: Chytrids. Aquatic heterotrophic organisms with motile cells characteristic at certain stages in their life cycle. The motile cells have a single, posterior, whiplash flagellum. Their cell walls are composed of chitin, but other polymers may also be present, and they store their food as glycogen. There are about 750 species.

Division Acrasiomycota: Cellular slime molds. Heterotrophic organisms that exist as separate amoebas (called myxamoebas). Eventually, the myxamoebas swarm together to form a pseudoplasmodium, within which they retain their individual identities. Ultimately, the pseudoplasmodium differentiates into a fruiting body. Sexual reproduction involves structures known as macrocysts. Pairs of amoebas first fuse, forming zygotes. Subsequently, these zygotes attract and then engulf nearby amoebas. The principal mode of nutrition is by ingestion. There are several genera, with about 65 known species.

Division Myxomycota: Plasmodial slime molds. Heterotrophic amoeboid organisms that form a multinucleate plasmodium that creeps along as a mass and eventually differentiates into sporangia, each of which is multinucleate and eventually gives rise to many spores. Sexual reproduction is occasionally observed. The predominant mode of nutrition is by ingestion. There are about 450 species.

Division Chrysophyta: Chrysophytes. Autotrophic organisms that possess chlorophyll *a*, chlorophyll *c*, and carotenoids, including fucoxanthin. Food is stored as the carbohydrate chrysolaminarin. The cell walls are absent, or consist of cellulose, with silica scales in some, and silica in diatoms. There are about 6650 living species.

Class Chrysophyceae: Golden algae. A diverse group of mostly unicellular organisms, most of which are flagellated; others lack flagella, and some are amoeboid. Many golden algae lack a clearly defined cell wall but have silica scales or skeletal structures. There are about 500 species.

Class Xanthophyceae: Yellow-green, mainly unicellular algae that have chlorophylls *a* and *c* but lack fucoxanthin. Most yellow-green algae are nonmotile, although some are amoeboid or flagellated. About 550 species.

Class Bacillariophyceae: Diatoms. Chrysophyta with double siliceous shells, the two halves of which fit together like a pillbox. Diatoms have chlorophylls *a* and *c*, as well as fucoxanthin. They are sometimes motile by the secretion of mucilage fibrils along a specialized groove, called the raphe. There are about 5600 living species, plus a large number of extinct ones.

Division Pyrrhophyta: Dinoflagellates. Autotrophic organisms possessing chlorophylls *a* and *c* and carotenoids. Food is stored as starch. Cell walls contain cellulose. This division contains some 1100 species, mostly biflagellated organisms. These all have lateral flagella, one of which beats in a groove that encircles the organism. Sexual reproduction is generally isogamous, but anisogamy is also present. The mitosis of dinoflagellates is unique. Many—in a form called zooxanthellae—are symbiotic in marine animals, and they make important contributions to the productivity of coral reefs.

Division Euglenophyta: The euglenoids. About a third of the approximately 40 genera of euglenoids have chloroplasts, with chlorophylls *a* and *b* and carotenoids; the others are heterotrophic and essentially resemble members of the phylum Zoomastigina, within which they would probably best be included. They store food as paramylon, an unusual carbohydrate. Euglenoids usually have a single apical flagellum of the tinsel variety and a contractile vacuole. The flexible pellicle is rich in proteins. Sexual reproduction is unknown. There are some 800 species, most of which occur in fresh water.

Division Rhodophyta: Red algae. Primarily marine algae characterized by the presence of chlorophyll *a* and phycobilins. Their carbohydrate food reserve is starch, and the cell walls are composed of cellulose or pectic materials, with calcium carbonate in many. No motile cells are present at any stage in the complex life cycle. The vegetative body is built up of closely packed filaments in a gelatinous matrix and is not differentiated into roots, leaves, and stems. It lacks specialized conducting cells. There are some 4000 species.

Division Phaeophyta: Brown algae. Multicellular, nearly entirely marine algae characterized by the presence of chlorophyll *a*, chlorophyll *c*, and fucoxanthin. The carbohydrate food reserve is laminarin, and the cell walls have a cellulose matrix with alginic acids in it. Motile cells are biflagellated, with one forward flagellum of the tinsel type and one trailing flagellum of the whiplash type. A considerable amount of differentiation is found in some of the kelps (some of the large brown algae of the order Laminariales), with specialized conducting cells for transporting the products of photosynthesis to the dimly lighted regions of the body. There is, however, no differentiation into roots, leaves, and stems, as in the vascular plants. There are about 1500 species.

Division Chlorophyta: Green algae. Unicellular or multicellular photosynthetic organisms characterized by the presence of chlorophyll *a*, chlorophyll *b*, and various carotenoids. The carbohydrate food reserve is starch; only green algae and plants, which are clearly descended from the green algae, store their reserve food inside their plastids. The cell walls of green algae are formed of polysaccharides, sometimes cellulose. Motile cells have two lateral or apical whiplash flagella. True multicellular genera do not exhibit complex patterns of differentiation. Multicellularity has arisen at least three times, and quite possibly more often. There are about 7000 known species.

Class Charophyceae: Unicellular, few-celled, filamentous, or parenchymatous green algae in which cell division involves a phragmoplast—a system of microtubules perpendicular to the plane of cell division. The nuclear envelope breaks down during the course of mitosis. Motile cells, if present, are asymmetrical and possess two flagella that are lateral or subapical and extend laterally at right angles from the cell. Sexual reproduction always involves the formation of a dormant zygote and zygotic meiosis. Certain members of this class resemble plants more closely than do any other organisms. Predominantly occur in fresh water.

Class Ulvophyceae: Cell division like that of the Charophyceae; the nuclear envelope persists during mitosis. Motile cells, if present, are symmetrical and possess two, four, or many flagella that are apical and directed forward. Sexual reproduction often involves alternation of generations and sporic meiosis, and dormant zygotes are rare. Predominantly marine algae.

Class Chlorophyceae: Those green algae in which the unique mode of cell division involves a phycoplast—a system of microtubules parallel to the plane of cell division. The nuclear envelope persists throughout mitosis, and chromosome division occurs within it. Motile cells, if present, are symmetrical and possess two, four, or many flagella that are apical and directed forward. Sexual reproduction always involves the formation of a dormant zygote and zygotic meiosis. Predominantly occur in fresh water.

Kingdom PLANTAE

The plants are autotrophic (some are derived heterotrophs), multicellular organisms possessing advanced tissue differentiation. All plants have an alternation of generations, in which the diploid phase (sporophyte) includes an embryo, and the haploid phase (gametophyte) produces gametes by mitosis. Their photosynthetic pigments and food reserves are similar to those of the green algae. Plants are primarily terrestrial.

Division

Division Hepatophyta: The liverworts. Hepatophyta and the two following divisions, all of which constitute the bryophytes, have multicellular gametangia with a sterile jacket layer; their sperm are biflagellated. In all three divisions, most photosynthesis is carried out in the gametophyte, upon which the sporophyte is dependent. Liverworts lack specialized conducting tissue (possibly a few exceptions), a cuticle, and stomata; they are the simplest of all living plants. The gametophytes are thallose or leafy, and the rhizoids are single-celled. There are about 6000 species.

Division Anthocerophyta: The hornworts. Hornworts have thallose gametophytes; the sporophyte grows from a basal intercalary meristem for as long as conditions are favorable. Stomata are present on the sporophyte; there is no specialized conducting tissue. There are about 100 species.

Division Bryophyta: The mosses. Mosses have leafy gametophytes; the sporophytes have complex patterns of dehiscence. Specialized conducting tissue is often present in both gametophytes and sporophytes. Rhizoids are multicellular. Stomata are present on the sporophytes. There are about 9000 species.

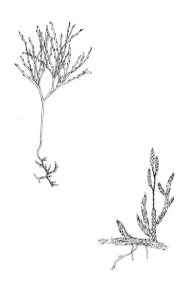

Division Psilotophyta: Psilopsids. Homosporous vascular plants, with two genera, one of which has leaflike appendages on the stem; both genera have extremely simple sporophytes, with no differentiation between root and shoot. Motile sperm. There are several species.

Division Lycophyta: The lycophytes. Homosporous and heterosporous vascular plants characterized by the presence of microphylls; the lycophytes are extremely diverse in appearance. All lycophytes have motile sperm. There are 10 to 15 genera, with about 1000 species.

Division Sphenophyta: The horsetails. A single genus of homosporous vascular plants, *Equisetum,* with jointed stems marked by conspicuous nodes and elevated siliceous ribs. Sporangia are borne in a strobilus at the apex of the stem. Leaves are scalelike. Sperm are motile. There are 15 living species of horsetails.

Division Pterophyta: The ferns. Mostly homosporous, although some are heterosporous. All possess a megaphyll. The gametophyte is more or less free-living and usually photosynthetic. Multicellular gametangia and free-swimming sperm are present. There are about 11,000 species.

Division Coniferophyta: The conifers. Gymnosperms with active cambial growth and simple leaves; ovules and seeds exposed; sperm nonflagellated. The most familiar group of gymnosperms. There are some 50 genera, with about 550 species.

Division Cycadophyta: The cycads. Gymnosperms with sluggish cambial growth and pinnately compound, palmlike or fernlike leaves; ovules and seeds exposed. The sperm are flagellated and motile but are carried to the vicinity of the ovule in a pollen tube. There are 10 genera, with about 100 species.

Division Ginkgophyta: Ginkgo. Gymnosperm with considerable cambial growth and fan-shaped leaves with open dichotomous venation; ovules and seeds exposed; seed coats fleshy. Sperm are carried to the vicinity of the ovule in a pollen tube but are flagellated and motile. There is only one species.

Division Gnetophyta: Gnetophytes. Gymnosperms with many angiospermlike features, such as vessels; the gnetophytes are the only gymnosperms in which vessels occur. Motile sperm are absent. There are three very distinctive genera, with about 70 species.

Division Anthophyta: The flowering plants. Seed plants in which ovules are enclosed in a carpel and seeds are borne within fruits. The angiosperms are extremely diverse vegetatively but are characterized by the flower, which is basically insect-pollinated. Other modes of pollination, such as wind pollination, have been derived in a number of different lines. The gametophytes are much reduced, with the female gametophyte often consisting of only seven cells at maturity. Double fertilization involving the two sperm of the mature microgametophyte gives rise to the zygote (sperm and egg) and to the primary endosperm nucleus (sperm and polar nuclei); the former becomes the embryo and the latter becomes a special nutritive tissue, called the endosperm. There are about 235,000 species.

Class Monocotyledones: The monocots. Flower parts are usually in threes; leaf venation is usually parallel; primary vascular bundles in the stem are scattered; true secondary growth is not present; there is one cotyledon. There are about 65,000 species.

Class Dicotyledones: The dicots. Flower parts are usually in fours or fives; leaf venation is usually netlike; primary vascular bundles in the stem are in a ring; many with a vascular cambium and true secondary growth; there are two cotyledons. There are about 170,000 species.

Some other common terms used to describe major groups of plants deserve mention here. In systems in which the algae and fungi are regarded as plants, they are often grouped as a subkingdom, Thallophyta, the thallophytes: organisms with no highly differentiated tissues, such as root, stem, or leaf, and no vascular tissues (xylem and phloem). The bryophytes and vascular plants are then grouped into a second subkingdom, Embryophyta, in which the zygote develops into a multicellular embryo still encased in an archegonium or an embryo sac. All embryophytes are marked by an alternation of heteromorphic generations.

Although they are no longer used in formal schemes of classification, terms such as "algae," "thallophytes," "vascular plants," and "gymnosperms" are still sometimes useful in an informal sense. An even earlier scheme divided all plants into "phanerogams," for those with flowers, and "cryptogams," for those lacking flowers; these terms are still occasionally seen.

APPENDIX D

Geologic Eras

Era*	Period*	Epoch*	Life Forms	Climates and Major Physical Events
CENOZOIC (65)	Quaternary (2)	Recent (0.01) Pleistocene (2)	Age of human beings. Extinction of many large mammals and birds.	Fluctuating cold to mild. More than two dozen glacial advances and retreats; final uplift of many mountain ranges.
	Tertiary (65)	Pliocene (5.1)	Aridity, formation of deserts; climates diversify. First appearance of man-apes.	Cooler. Much uplift and mountain building; widespread glaciation in Northern Hemisphere. Uplift of Panama joins North and South America.
		Miocene (24.6)	Spread of grasslands as forests contract. Grazing animals, apes.	Moderate. Extensive glaciation begins again in Southern Hemisphere.
		Oligocene (38)	Browsing mammals, monkey-like primates; many modern genera of plants evolve.	Rise of Alps and Himalayas. South America separates from Antarctica. Volcanoes in Rocky Mountains.
		Eocene (54.9)	Extensive radiation of mammals and birds; initial formation of grasslands.	Mild to very tropical. Australia separates from Antarctica; India collides with Asia.
		Paleocene (65)	Early insectivorous mammals and primates.	Mild to cool. Wide, shallow continental seas largely disappeared.
MESOZOIC (248)	Cretaceous (144)		Angiosperms and many groups of insects appear and become dominant. Age of reptiles. Extinction of dinosaurs at end of period.	Climate tropical to subtropical throughout. Africa and South America separate.
	Jurassic (213)		Gymnosperms, especially cycads. Birds appear.	Mild. Continents low, with large areas covered by seas.
	Triassic (248)		Forests of gymnosperms and ferns. First dinosaurs and first mammals.	Continents mountainous, joined into a supercontinent. Large areas arid.
PALEOZOIC (590)	Permian (286)		Origin of conifers, cycads, and ginkgos; earlier forest types wane. Reptiles diversify.	Extensive glaciation in Southern Hemisphere. Appalachians uplifted. Worldwide aridity.
	Carboniferous (360) Pennsylvanian (320) Mississippian (360)		Amphibians appear on land; forests appear and become dominant. Origin of insects and reptiles. Age of amphibians.	Warm, with little seasonal variation; lands low, swampy, with formation of coal deposits.
	Devonian (408)		Age of fishes. Rise of land plants; extinction of primitive vascular plants.	Sea over most of the land, with mountains locally.
	Silurian (438)		Period starts with major extinction event. First fossil plants. First jawed fishes.	Mild. Continents generally flat.
	Ordovician (505)		Period starts with first major extinction event. Oldest fossil crustaceans. Diversification of mollusks. Possible invasion of land by plants. First fungi.	Mild. Shallow seas, continents generally flat; seas cover much of present United States.
	Cambrian (590)		Evolution of external skeletons in animals. Explosive evolution of phyla and divisions. Evolution of chordates.	Mild. Extensive seas spilling over present continents.
PRECAMBRIAN (4500)			Origin of life (at least 3.5 billion years ago). Origin of eukaryotes, at least 1.5 billion years ago. Multicellular animals by 700 million years ago.	Formation of earth's crust and beginning of continental movements.

* A number following the name of a geologic time division indicates the age (in millions of years) at which it began.

Glossary

Å: *See* ångstrom.

a- [Gk. *a-*, not, without]: Prefix, negates the succeeding part of the word; "an-" before a vowel.

abscisic acid [L. *abscissus*, to cut off]: A plant hormone that brings about dormancy in buds, maintains dormancy in seeds, and brings about stomatal closing, among other effects.

abscission (ăb·sizh'ŭn): The dropping off of leaves, flowers, fruits, or other plant parts, usually following the formation of an abscission zone.

abscission zone: The area at the base of a leaf, flower, or fruit, or other plant part containing tissues that play a role in the separation of a plant part from the plant body.

absorption spectrum: The spectrum of light waves absorbed by a particular pigment.

accessory bud: A bud generally located above or on either side of the main axillary bud.

accessory cell: *See* subsidiary cell.

accessory pigment: A pigment that captures light energy and transfers it to chlorophyll *a*.

acclimation: The process by which numerous physical and physiological processes prepare a plant for winter.

achene: A simple, dry, one-seeded indehiscent fruit in which the seed coat is not adherent to the pericarp.

acid: A substance that dissociates in water, releasing hydrogen ions (H^+) and thus causing a relative increase in the concentration of these ions; having a pH in solution of less than 7; a proton donor; the opposite of "base."

acid growth hypothesis: The hypothesis that acidification of the cell wall leads to hydrolysis of restraining bonds within the wall and, consequently, to cell elongation driven by the turgor pressure of the wall.

actinomorphic [Gk. *aktis*, ray of light, + *morphe*, form]: Pertaining to a type of flower that can be divided into two equal halves in more than one longitudinal plane; also called radially symmetrical or regular; *see* zygomorphic.

action spectrum: The spectrum of light waves that elicits a particular reaction.

active transport: Energy-requiring transport of a solute across a membrane in the direction of increasing concentration (against the concentration gradient).

ad- [L. *ad-*, toward, to]: Prefix meaning "toward" or "to."

adaptation [L. *adaptare*, to fit]: A peculiarity of structure, physiology, or behavior that aids in fitting an organism to its environment.

adaptive radiation: The evolution from one kind of organism to several divergent forms, each specialized to fit a distinct and diverse way of life.

adenine (ăd'e·nēn): A purine base present in DNA, RNA, and nucleotide derivatives, such as ADP and ATP.

adenosine triphosphate (ATP): A nucleotide consisting of adenine, ribose sugar, and three phosphate groups; the major source of usable chemical energy in metabolism. On hydrolysis, ATP loses one phosphate to become adenosine diphosphate (ADP), releasing usable energy.

adhesion [L. *adhaerere*, to stick to]: The sticking together of unlike objects or materials.

adnate [L. *adnatus*, grown together]: Said of fused unlike parts, as stamens and petals; *see also* connate.

ADP: *See* adenosine triphosphate.

adsorption [L. *ad-*, to, + *sorbere*, to suck in]: The adhesion of a liquid, gaseous, or dissolved substance to a solid, resulting in a higher concentration of the substance.

adventitious [L. *adventicius*, not properly belonging to]: Referring to a structure arising from an unusual place, such as buds at other places than leaf axils, or roots growing from stems or leaves.

aeciospore (ē'sĭ·o·spor) [Gk. *aikia*, injury, + *spora*, seed]: A binucleate spore of rust fungi; produced in an aecium.

aecium, *pl.* **aecia:** In rust fungi, a cuplike structure in which aeciospores are produced.

aerobic [Gk. *aer*, air, + *bios*, life]: Requiring free oxygen.

aerobic respiration: *See* respiration.

after-ripening: Term applied to the metabolic changes that must occur in some dormant seeds before germination can occur.

agar: A gelatinous substance derived from certain red algae; used as a solidifying agent in the preparation of nutrient media for growing microorganisms.

aggregate fruit: A fruit developing from the several separate carpels of a single flower.

akinete: A vegetative cell that is transformed into a thick-walled resistant spore in cyanobacteria.

albuminous cell: Certain ray and axial parenchyma cells in gymnosperm phloem that are closely associated with the sieve cells both morphologically and physiologically.

aleurone [Gk. *aleuron*, flour]: A proteinaceous material, usually in the form of small granules, occurring in the outermost cell layer of the endosperm of wheat and other grains.

alga, *pl.* **algae** (ăl′ga, ăl′je): Traditional term for a series of unrelated groups of photosynthetic eukaryotic organisms lacking multicellular sex organs (except for the charophytes); the "blue-green algae," or cyanobacteria, are one of the groups of photosynthetic bacteria.

alkali [Arabic *algili*, the ashes of the plant saltwort]: A substance with marked basic properties.

alkaline: Pertaining to substances that release hydroxyl ions (OH^-) in water, having a pH greater than 7.

alkaloids: Nitrogen-containing ring compounds produced by plants that are physiologically active in vertebrates. Many alkaloids have a bitter taste and some are poisonous; examples are nicotine, caffeine, and strychnine.

allele (ă·lēl′) [Gk. *allelon*, of one another, + *morphe*, form]: One of the two or more alternative forms of a gene.

allelopathy [Gk. *allelon*, of each other, + *pathos*, suffering]: The inhibition of one species of plant by chemicals produced by another plant.

allosteric interaction [Gk. *allos*, other, + *steros*, shape]: A change in the shape of a protein resulting from the binding to the protein of a nonsubstrate molecule; in its new shape, the protein typically has different properties.

alternate: Referring to bud or leaf arrangement in which there is one bud or one leaf at a node.

alternation of generations: A reproductive cycle in which a haploid (*n*) phase, the gametophyte, produces gametes, which, after fusion in pairs to form a zygote, germinate, producing a diploid (2*n*) phase, the sporophyte. Spores produced by meiotic division from the sporophyte give rise to new gametophytes, completing the cycle.

amino acids [Gk. *Ammon*, referring to the Egyptian sun god, near whose temple ammonium salts were first prepared from camel dung]: Nitrogen-containing organic acids, the units, or "building blocks," from which protein molecules are built.

ammonification: Decomposition of amino acids and other nitrogen-containing organic compounds, resulting in the production of ammonia (NH_3) and ammonium ions (NH_4^+).

amoeboid [Gk. *amoibe*, change]: Moving or eating by means of pseudopodia (temporary cytoplasmic protrusions from the cell body).

amphi- [Gk. *amphi-*, on both sides]: Prefix meaning "on both sides," "both," or "of both kinds."

amylase (ăm′ĭ·lās): An enzyme that breaks down starch into smaller units.

amyloplast: A leucoplast (colorless plastid) that forms starch grains.

an- [Gk. *an-*, not, without]: Prefix, equivalent to "a-," meaning "not" or "without"; used before vowels and "h."

anabolism [Gk. *ana-*, up,+ *-bolism* (as in metabolism)]: The constructive part of metabolism; the total chemical reactions involved in biosynthesis.

anaerobic [Gk. *an-*, without, + *aer*, air, + *bios*, life]: Any process that can occur without oxygen, or an organism that can live without oxygen; strict anaerobes cannot survive in the presence of oxygen.

analogous [Gk. *analogos*, proportionate]: Applied to structures similar in function but different in evolutionary origin, such as the phyllodes of an Australian *Acacia* and the leaves of an oak.

anaphase [Gk. *ana*, away, + *phasis*, form]: A stage in mitosis in which the chromatids of each chromosome separate and move to opposite poles; similar stages in meiosis in which chromatids or paired chromosomes move apart.

anatomy: The study of the internal structure of organisms; morphology is the study of their external structure.

andro- [Gk. *andros*, man]: Prefix meaning "male."

androecium [Gk. *andros*, man, + *oikos*, house]: (1) The floral whorl that comprises the stamens; (2) in leafy liverworts, a packetlike swelling containing the antheridia.

angiosperm [Gk. *angion*, a vessel, + *sperma*, a seed]: Literally, a seed borne in a vessel (carpel); thus one of a group of plants whose seeds are borne within a mature ovary (fruit).

ångstrom [after A. J. Ångstrom, a Swedish physicist, 1814–74]: A unit of length equal to 10^{-10} meter; abbreviated Å.

anion [Gk. *anienae*, to go up]: A negatively charged ion.

anisogamy [Gk. *aniso*, unequal, + *gamos*, marriage]: The condition of having dissimilar motile gametes.

annual [L. *annulus*, year]: A plant in which the life cycle is completed in a single growing season.

annual ring: In wood, the growth layer formed during a single year; *see also* growth layer.

annulus [L. *anus*, ring]: In ferns, a row of specialized cells in a sporangium; in gill fungi, the remnant of the inner veil forming a ring on the stalk.

anterior: Situated before or toward the front.

anther [Gk. *anthos*, flower]: The pollen-bearing portion of a stamen.

antheridiophore [Gk. *anthos*, flower, + *phoros*, bearing]: In some liverworts, a stalk that bears antheridia.

antheridium: A sperm-producing organ, which may be multicellular or unicellular.

anthocyanin [Gk. *anthos*, flower, + *kyanos*, dark blue]: A water-soluble blue or red pigment found in the cell sap.

antibiotic [Gk. *anti*, against or opposite, + *biotikos*, pertaining to life]: Natural organic substances that retard or prevent the growth of organisms; generally used to designate substances formed by microorganisms that prevent growth of other microorganisms.

anticlinal: Perpendicular to the surface.

antipodals: Three (sometimes more) cells of the mature embryo sac, located at the end opposite the micropyle.

apical dominance: The influence exerted by a terminal bud in suppressing the growth of lateral buds.

apical meristem: The meristem at the tip of the root or shoot in a vascular plant.

apomixis [Gk. *apo*, separate, away from, + *mixis*, mingling]: Reproduction without meiosis or fertilization; vegetative reproduction.

apoplast [Gk. *apo*, away from, + *plastos*, molded]: The cell wall continuum of a plant or organ; the movement of substances via the cell walls is called apoplastic movement or transport.

apothecium [Gk. *apotheke*, storehouse]: A cup-shaped or saucer-shaped open ascoma.

arch-, archeo- [(Gk. *arche, archos,* beginning]: Prefix meaning "first," "main," or "earliest."

archegoniophore [Gk. *archegonos,* the first of a race, + *phoros,* bearing]: In some liverworts, a stalk that bears archegonia.

archegonium, *pl.* **archegonia:** A multicellular organ in which a single egg is produced; found in the bryophytes and some vascular plants.

aril (ăr′ĭl) [L. *arillus,* grape, seed]: An accessory seed covering, often formed by an outgrowth at the base of the ovule; often brightly colored, which may aid in dispersal by attracting animals that eat it and, in the process, carry the seed away from the parent plant.

artifact [L. *ars,* art, + *facere,* to make]: A product that exists because of an extraneous, especially human, agency and does not occur in nature.

artificial selection: The breeding of selected organisms to produce strains with desired characteristics.

ascogenous hyphae [Gk. *askos,* bladder, + *genous,* producing]: Hyphae containing paired haploid male and female nuclei; they develop from an ascogonium and eventually give rise to asci.

ascogonium: The oogonium or female gametangium of the ascomycetes.

ascoma, *pl.* **ascomata:** A multicellular structure in ascomycetes lined with specialized cells called asci, in which nuclear fusion and meiosis occur. Ascomata may be open or closed.

ascospore: A spore produced within an ascus; found in ascomycetes.

ascus, *pl.* **asci:** A specialized cell, characteristic of the ascomycetes, in which two haploid nuclei fuse to produce a zygote that immediately divides by meiosis; at maturity, an ascus contains ascospores.

aseptate [Gk. *a-,* not, + L. *septum,* fence]: Nonseptate; lacking cross walls.

asexual reproduction: Any reproductive process, such as fission or budding, that does not involve the union of gametes.

atom [Gk. *atomos,* indivisible]: The smallest unit into which a chemical element can be divided and still retain its characteristic properties.

atomic nucleus: The central core of an atom, containing protons and neutrons, around which electrons orbit.

atomic number: The number of protons in the nucleus of an atom.

atomic weight: The weight of a representative atom of an element relative to the weight of an atom of carbon ^{12}C, which has been assigned the value 12.

ATP: *See* adenosine triphosphate.

ATP synthase: An enzyme complex that forms ATP from ADP and phosphate during oxidative phosphorylation in the inner mitochondrial membrane.

auto- [Gk. *autos,* self, same]: Prefix, meaning "same" or "self-same."

autoecious [Gk. *auto,* self, + *oikia,* dwelling]: In some rust fungi, completing the life cycle on a single species of host plant.

autoradiograph: A photographic print made by a radioactive substance acting upon a sensitive photographic film.

autotroph [Gk. *autos,* self, + *trophos,* feeder]: An organism that is able to synthesize the nutritive substances it requires from inorganic substances in its environment; *see also* heterotroph.

auxin [Gk. *auxein,* to increase]: A plant growth-regulating substance; controls cell elongation, among other effects.

axial system: In secondary xylem and secondary phloem; the term applied collectively to cells derived from fusiform cambial initials. The long axes of these cells are oriented parallel with the main axis of the root or stem. Also called longitudinal system and vertical system.

axil [Gk. *axilla,* armpit]: The upper angle between a twig or leaf and the stem from which it grows.

axillary: Term applied to buds or branches occurring in the axil of a leaf.

bacillus, *pl.* **bacilli** (ba·sĭl′ŭs) [L. *baculum,* rod]: A rod-shaped bacterium.

backcross: The crossing of a hybrid with one of its parents or with a genetically equivalent organism; a cross between an individual whose genes are to be tested and one that is homozygous for all of the recessive genes involved in the experiment.

bacteriophage [Gk. *bakterion,* little rod, + *phagein,* to eat]: A virus that parasitizes bacterial cells.

bacterium, *pl.* **bacteria:** A prokaryotic organism.

bark: A nontechnical term applied to all tissues outside the vascular cambium in a woody stem; *see also* inner bark and outer bark.

basal body: A self-reproducing, cylinder-shaped cytoplasmic organelle from which cilia or flagella arise; identical in structure to the centriole, which is involved in mitosis and meiosis in most animals and protists.

base: A substance that dissociates in water, causing a decrease in the concentration of hydrogen ions (H^+), often by releasing hydroxyl ions (OH^-); bases have a pH in solution of more than 7; the opposite of "acid."

basidioma, *pl.* **basidiomata:** A multicellular structure, characteristic of the basidiomycetes, within which basidia are formed.

basidiospore: A spore of the basidiomycetes, produced within and borne on a basidium following nuclear fusion and meiosis.

basidium, *pl.* **basidia:** A specialized reproductive cell of the basidiomycetes, often club-shaped, in which nuclear fusion and meiosis occur.

berry: A simple fleshy fruit that includes a fleshy ovary wall and one or more carpels and seeds; examples are the fruits of grapes, tomatoes, and bananas.

bi- [L. *bis,* double, two]: Prefix meaning "two," "twice," or "having two points."

biennial: A plant that normally requires two growing seasons to complete its life cycle, flowering and fruiting in its second year.

bilaterally symmetrical: *See* zygomorphic.

biological clock [Gk. *bios,* life, + *logos,* discourse]: The internal timing mechanism that governs the innate biological rhythms of organisms.

biomass: Total dry weight of all organisms in a particular population, sample, or area.

biome: A complex of terrestrial communities of very wide extent, characterized by its climate and soil; the largest ecological unit.

biosphere: The zone of air, land, and water at the surface of the earth that is occupied by organisms.

biotechnology: The practical application of advances in hormone research and DNA biochemistry to manipulate the genetics of plants.

biotic: Relating to life.

bisexual flower: A flower that has at least one functional stamen and one functional carpel.

bivalent [L. *bis,* twice, + *valere,* to be strong]: A pair of synapsed homologous chromosomes.

blade: The broad, expanded part of a leaf; the lamina.

body cell: Vegetative or somatic cell; *see also* spermatogenous cell.

bordered pit: A pit in which the secondary wall arches over the pit membrane.

bract: A modified, usually reduced leaflike structure.

branch root: *See* lateral root.

bryophytes (bri'o·fits): The members of the divisions of nonvascular plants; the mosses, hornworts, and liverworts.

bud: (1) An embryonic shoot, often protected by young leaves; (2) a vegetative outgrowth of yeasts and some bacteria as a means of asexual reproduction.

bulb: A short underground stem covered by enlarged and fleshy leaf bases containing stored food.

bulk flow: The overall movement of water or some other liquid induced by gravity, pressure, or an interplay of both.

bulliform cell: A large epidermal cell present, with other such cells, in longitudinal rows in grass leaves; also called motor cell. Believed to be involved with the mechanism of rolling and unrolling of the leaves.

bundle scar: Scar or mark left on leaf scar by vascular bundles broken at the time of leaf fall, or abscission.

bundle sheath: Layer or layers of cells surrounding a vascular bundle; may consist of parenchyma or sclerenchyma cells, or both. Bundle sheaths are especially characteristic of C_4 plants.

bundle-sheath extension: A group of cells extending from a bundle sheath of a vein in the leaf mesophyll to either or both upper and lower epidermis; may consist of parenchyma, collenchyma, or sclerenchyma.

C_3 pathway: *See* Calvin cycle.

C_4 pathway: The set of reactions through which carbon dioxide is fixed to a compound known as phosphoenolpyruvate (PEP) to yield oxaloacetate, a four-carbon compound.

callose: A complex branched carbohydrate that is a common wall constituent associated with the sieve areas of sieve elements; may develop in reaction to injury in sieve elements and parenchyma cells.

callus [L. *callos*, hard skin]: Undifferentiated tissue; a term used in tissue culture, grafting, and wound healing.

calorie [L. *calor*, heat]: The amount of energy in the form of heat required to raise the temperature of one gram of water 1°C. In making metabolic measurements, the kilocalorie (kcal), or Calorie—the amount of heat required to raise the temperature of one kilogram of water 1°C—is generally used.

Calvin cycle: The series of enzymatically mediated photosynthetic reactions during which carbon dioxide is reduced to 3-phosphoglyceraldehyde and the carbon dioxide acceptor, ribulose 1,5-bisphosphate, is regenerated. For every six molecules of carbon dioxide entering the cycle, a net gain of two molecules of glyceraldehyde 3-phosphate results.

calyptra [Gk. *kalyptra*, covering for the head]: The hood or cap that partially or entirely covers the capsule of some species of mosses; it is formed from the expanded archegonial wall.

calyx (kā'lĭks) [Gk. *kalyx*, a husk, cup]: The sepals collectively; the outermost flower whorl.

CAM: *See* Crassulacean acid metabolism.

cambial zone: A region of thin-walled, undifferentiated meristematic cells between the secondary xylem and secondary phloem; consists of cambial initials and their recent derivatives.

cambium [L. *cambiare*, to exchange]: A meristem that gives rise to parallel rows of cells; commonly applied to the vascular cambium and the cork cambium, or phellogen.

capsid: The protein coat of a virus particle.

capsule: (1) In angiosperms, a dehiscent, dry fruit that develops from two or more carpels; (2) a slimy layer around the cells of certain bacteria; (3) the sporangium of bryophytes.

carbohydrate [L. *carbo*, ember, + *hydro*, water]: An organic compound consisting of a chain of carbon atoms to which hydrogen and oxygen are attached in a 2:1 ratio; examples are sugars, starch, glycogen, and cellulose.

carbon cycle: Worldwide circulation and utilization of carbon atoms.

carbon fixation: The conversion of CO_2 into organic compounds during photosynthesis.

carnivorous: Feeding upon animals, as opposed to plants (herbivorous); also refers to plants that are able to utilize proteins obtained from trapped animals, chiefly insects.

carotene (kăr'o·tēn) [L. *carota*, carota, carrot]: A yellow or orange pigment belonging to the carotenoid group.

carotenoids (kă·rŏt'e·noids): A class of fat-soluble pigments that includes the carotenes (yellow and orange pigments) and the xanthophylls (yellow pigments); found in chloroplasts and chromoplasts of plants. Carotenoids act as accessory pigments in photosynthesis.

carpel [Gk. *karpos*, fruit]: One of the members of the gynoecium, or inner floral whorl; each carpel encloses one or more ovules. One or more carpels forms a gynoecium.

carpellate: Pertaining to a flower with one or more carpels but no functional stamens; also called pistillate.

carpogonium [Gk. *karpos*, fruit, + *gonos*, offspring]: In red algae, the female gametangium.

carposporangium [Gk. *karpos*, fruit, + *spora*, seed, + *angeion*, vessel]: In red algae, a carpospore-containing cell.

carpospore: In red algae, the single diploid protoplast found within a carposporangium.

carriers: Transport proteins that bind specific solutes and undergo conformational change in order to transport the solute across the membrane.

caryopsis [Gk. *karyon*, a nut, + *opsis*, appearance]: Simple, dry, one-seeded indehiscent fruit, with the pericarp firmly united all around the seed coat; a grain characteristic of the grasses (family Poaceae).

Casparian strip [Robert Caspary, a German botanist]: A bandlike region of primary wall containing suberin and lignin; found in anticlinal—radial and transverse—walls of endodermal and exodermal cells.

catabolism [Gk. *katabole*, throwing down]: Collectively, the chemical reactions resulting in the breakdown of complex materials and involving the release of energy.

catalyst [Gk. *katalysis*, dissolution]: A substance that accelerates the rate of a chemical reaction but is not used up in the reaction: enzymes are catalysts.

cation [Gk. *katienai*, to go down]: A positively charged ion.

catkin: A spikelike inflorescence of unisexual flowers; found only in woody plants.

cDNA: *See* complementary DNA.

cell [L. *cella*, small room]: The structural unit of organisms; in plants, cells consist of the cell wall and the protoplast.

cell division: The division of a cell and its contents into two roughly equal parts.

cell plate: The structure that forms at the equator of the spindle in the dividing cells of plants and a few green algae during early telophase. The cell plate, when mature, becomes the middle lamella.

cell sap: The fluid contents of the vacuole.

cellulase: An enzyme that hydrolyzes cellulose.

cellulose: A carbohydrate; the chief component of the cell wall in plants and some protists; an insoluble complex carbohydrate formed of microfibrils of glucose molecules attached end to end.

cell wall: The rigid outermost layer of the cells found in plants, some protists, and most bacteria.

central mother cells: Relatively large vacuolate cells in a subsurface position in apical meristems of shoots.

centriole [Gk. *kentron*, center, + L. *-olus*, little one]: A cytoplasmic organelle found outside the nuclear envelope, and identical in structure to a basal body; centrioles generally found in the cells of most eukaryotes other than fungi, red algae, and the nonflagellated cells of plants. Centrioles divide and organize spindle fibers during mitosis and meiosis.

centromere [Gk. *kentron*, center, + *meros*, a part]: Region of constriction of chromosome that holds sister chromatids together.

chalaza [Gk. *chalaza*, small tubercle]: The region of an ovule or seed where the funiculus unites with the integuments and the nucellus.

channel proteins: Transport proteins that form water-filled pores that extend across cellular membranes; when open, the channel proteins allow specific solutes to pass through them.

chemical potential: The activity or free energy of a substance; it is dependent upon the rate of motion of the average molecule and the concentration of the molecules.

chemical reaction: The making or breaking of chemical bonds between atoms or molecules.

chemiosmotic coupling: Coupling of ATP synthesis to electron transport via an electrochemical H^+ gradient across a membrane.

chemoautotrophic: Refers to bacteria that are able to manufacture their own basic foods by using the energy released by specific inorganic reactions; *see* autotroph.

chiasma (kī·ăz′ma) [Gk. *chiasma*, a cross]: The X-shaped figure formed by the meeting of two nonsister chromatids of homologous chromosomes; the site of crossing-over.

chitin (kī′tin) [Gk. *chiton*, tunic]: A tough, resistant, nitrogen-containing polysaccharide forming the cell walls of certain fungi, the exoskeleton of arthropods, and the epidermal cuticle of other surface structures of certain other protists and animals.

chlor- [Gk. *chloros*, green]: Prefix meaning "green."

chlorenchyma: Parenchyma cells that contain chloroplasts.

chlorophyll [Gk. *chloros*, green, + *phyllon*, leaf]: The green pigment of plant cells, which is the receptor of light energy in photosynthesis; also found in algae and photosynthetic bacteria.

chloroplast: A plastid in which chlorophylls are contained; the site of photosynthesis. Chloroplasts occur in plants and algae.

chlorosis: Loss or reduced development of chlorophyll.

chroma- [Gk. *chroma*, color]: Prefix meaning "color."

chromatid [Gk. *chroma*, color, + L. *-id*, daughters of]: One of the two daughter strands of a duplicated chromosome, which are joined at the centromere.

chromatin: The deeply staining complex of DNA and proteins that forms eukaryotic chromosomes.

chromatophore [Gk. *chroma*, color, + *phorus*, a bearer]: In some bac-

teria, a discrete vesicle delimited by a single membrane and containing photosynthetic pigments.

chromoplast: A plastid containing pigments other than chlorophyll, usually yellow and orange carotenoid pigments.

chromosome [Gk. *chroma*, color, + *soma*, body]: The structure that carries the genes. Eukaryotic chromosomes are visualized as threads or rods of chromatin, which appear in contracted form during mitosis and meiosis, and otherwise are enclosed in a nucleus; bacterial chromosomes consist only of a closed circle of DNA.

chrysolaminarin: The storage product of the chrysophytes and diatoms.

cilium, *pl.* **cilia** (sĭl′ĭ·ŭm) [L. *cilium*, eyelash]: A short, hairlike flagellum, usually numerous and arranged in rows.

circadian rhythms [L. *circa*, about, + *dies*, a day]: Regular rhythms of growth and activity that occur on an approximately 24-hour basis.

circinate vernation [L. *circinare*, to make round, + *vernare*, to flourish]: As in ferns, the coiled arrangement of leaves and leaflets in the bud; such an arrangement uncoils gradually as the leaf develops further.

cisterna, *pl.* **cisternae** [L. *cistern*, a reservoir]: A flattened or saclike portion of the endoplasmic reticulum or a dictyosome (Golgi body).

cladistics: System of arranging organisms following an analysis of their primitive and advanced features so that their phylogenetic relationships will be reflected accurately.

cladophyll [Gk. *klados*, shoot, + *phyllon*, leaf]: A branch resembling a foliage leaf.

clamp connection: In the basidiomycetes, a lateral connection between adjacent cells of a dikaryotic hypha; ensures that each cell of the hypha will contain two dissimilar nuclei.

class: A taxonomic category between division and order in rank. A class contains one or more orders, and belongs to a particular division.

cleistothecium [Gk. *kleistos*, closed, + *thekion*, small receptacle]: A closed, spherical ascoma.

climax community: The final stage in a successional series; its nature is determined largely by the climate and soil of the region.

cline: A graded series of changes in some characteristics within a species, often correlated with a gradual change in climate or another geographical factor.

clone [Gk. *klon*, twig]: A population of cells or individuals derived by asexual division from a single cell or individual; one of the members of such a population.

cloning: Producing a cell line or culture all of whose members are characterized by a specific DNA sequence; a key element in genetic engineering.

closed bundle: A vascular bundle in which a cambium does not develop.

coalescence [L. *coalescere*, to grow together]: The union of floral parts of the same whorl, as petals to petals.

coccus, *pl.* **cocci** (kŏk′ŭs) [Gk. *kokkos*, a berry]: A spherical bacterium.

codon (kō′dŏn): Sequence of three adjacent nucleotides in a molecule of DNA or mRNA that form the code for a single amino acid, or for the termination of a polypeptide chain.

coenocytic (se·nō·sĭ′tic) [Gk. *koinos*, shared in common, + *kytos*, a hollow vessel]: A term used to describe an organism or part of an organism that is multinucleate, the nuclei not separated by walls or membranes; also called siphonaceous, siphonous, or syncytial.

coenzyme: An organic molecule, or nonprotein organic cofactor, that plays an accessory role in enzyme-catalyzed processes, often by acting as a donor or acceptor of electrons; NAD$^+$ and FAD are common coenzymes.

cofactor: One or more nonprotein components required by enzymes in order to function; many cofactors are metal ions, while others are called coenzymes.

cohesion [L. *cohaerrere*, to stick together]: The mutual attraction of molecules of the same substance.

cold hardiness: The ability of a plant to survive the extreme cold and drying effects of winter weather.

coleoptile (kō′le·op′till) [Gk. *koleos*, sheath, + *ptilon*, feather]: The sheath enclosing the apical meristem and leaf primordia of the grass embryo; often interpreted as the first leaf.

coleorhiza (kō′le·o·rī′za) [Gk. *koleos*, sheath, + *rhiza*, root]: The sheath enclosing the radicle in the grass embryo.

collenchyma [Gk. *kolla*, glue]: A supporting tissue composed of collenchyma cells; often found in regions of primary growth in stems and in some leaves.

collenchyma cell: Elongated living cell with irregularly thickened primary cell wall.

colloid (kŏl′oid): A permanent suspension of fine particles.

community: All the organisms inhabiting a common environment and interacting with one another.

companion cell: A specialized parenchyma cell associated with a sieve-tube member in angiosperm phloem and arising from the same mother cell as the sieve-tube member.

competition: Interaction between members of the same population or of two or more populations to obtain a resource that both require and which is available in limited supply.

complementary DNA (cDNA): A single-stranded molecule of DNA that has been synthesized from an mRNA template by reverse transcription.

complete flower: A flower having four whorls of floral parts—sepals, petals, stamens, and carpels.

compound: A combination of atoms in a definite ratio, held together by chemical bonds.

compound leaf: A leaf whose blade is divided into several distinct leaflets.

compression wood: The reaction wood of conifers; develops on the lower sides of leaning trunks or limbs.

concentration gradient: The concentration difference of a substance per unit distance.

cone: *See* strobilus.

conidiophore: Hypha on which one or more conidia are produced.

conidium, *pl.* **conidia** [Gk. *konis*, dust]: An asexual fungal spore not contained within a sporangium; it may be produced singly or in chains; most conidia are multinucleate.

conifer: A cone-bearing tree.

conjugation: The temporary fusion of pairs of bacteria, protozoa, and certain algae and fungi during which genetic material is transferred between the two individuals.

conjugation tube: A tube formed during the process of conjugation to facilitate the transfer of genetic material.

connate (kŏn′āt): Said of similar parts that are united or fused, as petals fused in a corolla tube; *see also* adnate.

consumer: In ecology, an organism that derives its food from another organism.

continuous variation: Variation in traits to which a number of different genes contribute; the variation often exhibits a "normal" or bell-shaped distribution.

contractile vacuole: A clear, fluid-filled vacuole in some groups of protists that takes up water within the cell and then contracts, expelling its contents from the cell.

convergent evolution [L. *convergere*, to turn together]: The independent development of similar structures in organisms that are not directly related; often found in organisms living in similar environments.

cork, or **phellem:** A secondary tissue produced by a cork cambium; made up of polygonal cells, nonliving at maturity, with walls infiltrated with suberin, a waxy or fatty material resistant to the passage of gases and water vapor; the outer part of the periderm.

cork cambium, or **phellogen:** The lateral meristem that forms the periderm, producing cork (phellem) toward the surface (outside) of the plant and phelloderm toward the inside; common in stems and roots of gymnosperms and dicots.

corm: A thickened underground stem, upright in position, in which food is accumulated, usually in the form of starch.

corolla [L. *corona*, crown]: The petals collectively; usually the conspicuously colored flower whorl.

corolla tube: A tubelike structure resulting from the fusion of the petals along their edges.

cortex: Ground-tissue region of a stem or root bounded externally by the epidermis and internally by the vascular system; a primary-tissue region.

cotransport: Membrane transport in which the transfer of one solute depends on the simultaneous or sequential transfer of a second solute.

cotyledon (kŏt′i·lē′dŭn) [Gk. *kotyledon*, cup-shaped hollow]: Seed leaf; generally stores food in dicotyledons and absorbs food in monocotyledons.

coupled reactions: reactions in which energy-requiring chemical reactions are linked to energy-releasing reactions.

covalent bond: A chemical bond formed between atoms as a result of the sharing of two electrons.

Crassulacean acid metabolism, or **CAM:** A variant of the C$_4$ pathway; phosphoenolpyruvate fixes CO$_2$ in C$_4$ compounds at night and then, during the day time, the fixed CO$_2$ is transferred to the ribulose bisphosphate of the Calvin cycle within the same cell. Characteristic of most succulent plants, such as cacti.

cristae, *sing.* **crista:** The enfoldings of the inner mitochondrial membrane, which form a series of crests or ridges containing the electron-transport chains involved in ATP formation.

crop rotation: The practice of growing different crops in regular succession to aid in the control of insects and diseases, to increase soil fertility, and to decrease erosion.

cross-fertilization: The fusion of gametes formed by different individuals; the opposite of self-fertilization.

crossing-over: The exchange of corresponding segments of genetic material between the chromatids of homologous chromosomes at meiosis.

cross-pollination: The transfer of pollen from the anther of one plant to the stigma of a flower of another plant.

cross section: *See* transverse section.

cryptogam: An archaic term for all organisms except the flowering plants (phanerogams), animals, and heterotrophic protists.

cultivar: A variety of plant found only under cultivation.

cuticle: Waxy or fatty layer on outer wall of epidermal cells, formed of cutin and wax.

cutin [L. *cutis*, skin]: Fatty substance deposited in many plant cell walls and on outer surface of epidermal cell walls, where it forms a layer known as the cuticle.

cyclic electron flow: In chloroplasts, the light-induced flow of electrons originating from and returning to photosystem I.

cyclosis (si·klō′sis) [Gk. *kyklosis*, circulation]: The streaming of cytoplasm within a cell.

-cyte, cyto- [Gk. *kytos*, hollow vessel, container]: Suffix or prefix meaning "pertaining to the cell."

cytochrome [Gk. *kytos*, container, + *chroma*, color]: Heme proteins serving as electron carriers in respiration and photosynthesis.

cytokinesis: [Gk. *kytos*, vessel, + *kinesis*, motion]: Division of the cytoplasm of a cell following nuclear division.

cytokinin [Gk. *kytos*, hollow vessel, + *kinesis*, motion]: A class of plant hormones that promote cell division, among other effects.

cytology: The study of cell structure and function.

cytoplasm: The living matter of a cell, exclusive of the nucleus; the protoplasm.

cytoplasmic ground substance: The least differentiated part of the cytoplasm, as seen with the electron microscope, and the part surrounding the nucleus and various organelles; also called hyaloplasm.

cytosine: One of the four pyrimidine bases found in the nucleic acids DNA and RNA.

cytoskeleton: The flexible network within cells, composed of microtubules and microfilaments, or actin filaments.

dark reactions: In photosynthetic cells, the light-independent enzymatic reactions concerned with the synthesis of glucose from CO_2, ATP, and NADPH; "light-independent reactions" is the preferred term.

day-neutral plants: Plants that flower without regard to day length.

de- [L. *de-*, away from, down, off]: Prefix meaning "away from," "down," or "off"; for example, dehydration means removal of water.

deciduous [L. *decidere*, to fall off]: Shedding leaves at a certain season.

decomposers: Organisms (bacteria, fungi, heterotrophic protists) in an ecosystem that break down organic material into smaller molecules that then are recirculated.

dehiscence [L. *de*, down, + *hiscere*, split open]: The opening of an anther, fruit, or other structure, which permits the escape of reproductive bodies contained within.

denitrification: The conversion of nitrate to gaseous nitrogen; carried out by a few genera of free-living soil bacteria.

deoxyribonucleic acid (DNA): Carrier of genetic information in cells; composed of chains of phosphate, sugar molecules (deoxyribose), and purines and pyrimidines; capable of self-replication as well as determining RNA synthesis.

deoxyribose [L. *deoxy*, loss of oxygen, + *ribose*, a kind of sugar]: A five-carbon sugar with one less atom of oxygen than ribose; a component of deoxyribonucleic acid.

dermal tissue system: The epidermis or the periderm.

desmotubule [Gk. *desmos*, to bind, + L. *tubulus*, small tube]: The tubule traversing a plasmodesmatal canal and uniting the endoplasmic reticulum of the two adjacent cells.

determinate growth: Growth of limited duration, characteristic of floral meristems and of leaves.

deuterium: Heavy hydrogen; a hydrogen atom, the nucleus of which contains one proton and one neutron. (The nucleus of most hydrogen atoms consists of only one proton.)

dichotomy: The division or forking of an axis into two branches.

dicotyledon: One of the two classes of angiosperms, Dicotyledones; often abbreviated as dicot. Dicotyledons are characterized by having two cotyledons, net-veined leaves, and flower parts usually in multiples of fours or fives.

dictyosome: In eukaryotes, a group of flat, disk-shaped sacs that are often branched into tubules at their margins; serve as collecting and packaging centers for the cell and concerned with secretory activities; also called Golgi bodies. The term "Golgi apparatus" is used to refer collectively to all the dictyosomes, or Golgi bodies, of a given cell.

differentially permeable membrane: Cellular membrane that permits the passage of water but blocks the passage of most dissolved substances.

differentiation: A developmental process by which a relatively unspecialized cell undergoes a progressive change to a more specialized cell; the specialization of cells and tissues for particular functions during development.

diffuse-porous wood: A wood in which the pores, or vessels, are fairly uniformly distributed throughout the growth layers or in which the size of pores changes only slightly from early wood to late wood.

diffusion [L. *diffundere*, to pour out]: The net movement of suspended or dissolved particles from a more concentrated region to a less concentrated region as a result of the random movement of individual molecules; the process tends to distribute such particles uniformly throughout a medium.

digestion: The conversion of complex, usually insoluble foods into simple, usually soluble forms by means of enzymatic action.

dikaryon (Gk. *di*, two, + *karyon*, a nut): In fungi, mycelium with paired nuclei, each usually derived from a different parent.

dikaryotic: In fungi, having pairs of nuclei within cells or compartments.

dimorphism [Gk. *di*, two + *morphe*, form]: The condition of having two distinct forms, such as sterile and fertile leaves in ferns, or sterile and fertile shoots in horsetails.

dioecious [Gk. *di*, two + *oikos*, house]: Unisexual; having the male and female (or staminate and ovulate) elements on different individuals of the same species.

diploid: Having two sets of chromosomes; the $2n$ (diploid) chromosome number is characteristic of the sporophyte generation.

disaccharide: A carbohydrate formed of two simple sugar molecules linked by a covalent bond; sucrose is an example.

disk flowers: The actinomorphic, tubular flowers in Asteraceae; contrasted with flattened, zygomorphic ray flowers. In many Asteraceae, the disk flowers occur in the center of the inflorescence, the ray flowers around the margins.

distal: Situated away from or far from point of reference (usually the main part of body); opposite of proximal.

division: One of the major kinds of groups used by botanists in classifying organisms, the equivalent of phyla in animals and hetrotrophic protists; kingdoms are divided into divisions (phyla).

DNA: *See* deoxyribonucleic acid.

dominant allele: One allele is said to be dominant with respect to an alternative allele if the homozygote for the dominant allele is indistinguishable phenotypically from the heterozygote; the other allele is said to be recessive.

dormancy [L. *dormire*, to sleep]: A special condition of arrested growth in which the plant and such plant parts as buds and seeds do not begin to grow without special environmental cues. The requirement for such cues, which include cold exposure and a suitable photoperiod, prevents the breaking of dormancy during superficially favorable growing conditions.

double fertilization: The fusion of the egg and sperm (resulting in a 2n fertilized egg, the zygote) and the simultaneous fusion of the second male gamete with the polar nuclei (resulting in a 3n primary endosperm nucleus); a unique characteristic of all angiosperms.

doubling rate: The length of time required for a population of a given size to double in number.

drupe [Gk. *dryppa*, overripe olive]: A simple, fleshy fruit, derived from a single carpel, usually one-seeded, in which the inner fruit coat is hard and may be adherent to the seed.

druse: A compound, more or less spherical crystal with many component crystals projecting from its surface; composed of calcium oxalate.

early wood: The first-formed wood of a growth increment; it contains larger cells and is less dense than the subsequently formed late wood: replaces the term "spring wood."

eco- [Gk. *oikos*, house]: Prefix meaning "house" or "home."

ecology: The study of the interactions of organisms with their physical environment and with one other.

ecosystem: A major interacting system that involves both living organisms and their physical environment.

ecotype [Gk. *oikos*, house, + L. *typus*, image]: A locally adapted variant of an organism, differing genetically from other ecotypes.

ectoplast: *See* plasma membrane.

edaphic [Gk. *edaphos*, ground, soil]: Pertaining to the soil.

egg: A nonmotile female gamete, usually larger than a male gamete of the same species.

egg apparatus: The egg cell and synergids located at the micropylar end of the female gametophyte, or embryo sac, of angiosperms.

elater [Gk. *elater*, driver]: (1) An elongated, spindle-shaped, sterile cell in the sporangium of a liverwort sporophyte (aids in spore dispersal); (2) clubbed, hygroscopic bands attached to the spores of the horsetails.

electrolyte: A substance that dissociates into ions in aqueous solution and so makes possible the conduction of an electric current through the solution.

electron: A subatomic particle with a negative electric charge equal in magnitude to the positive charge of the proton, but with a mass of 1/1837 of that of the proton. Electrons orbit the atom's positively charged nucleus and determine the atom's chemical properties.

electron-dense: In electron microscopy, not permitting the passage of electrons and so appearing dark.

element: A substance composed of only one kind of atom; one of more than 100 distinct natural or synthetic types of matter that, singly or in combination, compose virtually all materials of the universe.

embryo [Gk. *en*, in, + *bryein*, to swell]: A young sporophytic plant, before the start of a period of rapid growth (germination in seed plants).

embryogeny: The formation of the embryo.

embryo sac: The female gametophyte of angiosperms, generally an eight-nucleate, seven-celled structure; the seven cells are the egg cell, two synergids and three antipodals (each with a single nucleus), and the central cell (with two nuclei).

endergonic: Describing a chemical reaction that requires energy to proceed; the opposite of "exergonic."

endo- [Gk. *endo*, within]: Prefix meaning "within."

endocarp [Gk. *endo*, within, + *karpos*, fruit]: The innermost layer of the mature ovary wall, or pericarp.

endocytosis [Gk. *endon*, within, + *kytos*, hollow vessel]: The uptake of material into cells by means of invagination of the plasma membrane; if solid material is involved, the process is called phagocytosis; if dissolved material is involved, it is called pinocytosis.

endodermis [Gk. *endon*, within, + *derma*, skin]: A single layer of cells forming a sheath around the vascular region in roots and some stems; the endodermal cells are characterized by a Casparian strip within radial and transverse walls. In roots and stems of seed plants, the endodermis is the innermost layer of the cortex.

endogenous [Gk. *endon*, within, + *genos*, race, kind]: Arising from deep-seated tissues, as in the case of lateral roots.

endomembrane system: Collectively, the cellular membranes that form a continuum (plasma membrane, tonoplast, endoplasmic reticulum, Golgi bodies, and nuclear envelope).

endoplasmic reticulum: A complex, three-dimensional membrane system of indefinite extent present in eukaryotic cells, dividing the cytoplasm into compartments and channels. Those portions that are densely coated with ribosomes are called rough endoplasmic reticulum, and other portions with fewer ribosomes are called smooth endoplasmic reticulum.

endosperm [Gk. *endon*, within, + *sperma*, seed]: A tissue, containing stored food, that develops from the union of a male nucleus and the polar nuclei of the central cell; it is digested by the growing sporophyte either before or after the maturation of the seed; found only in angiosperms.

entrainment: The process by which a periodic repetition of light and dark, or some other external cycle, causes a circadian rhythm to remain synchronized with the same cycle as the modifying, or entraining, factor.

entropy: A measure of the randomness or disorder of a system.

enzyme: A protein that is capable of speeding up specific chemical reactions by lowering the required activation energy, but is unaltered itself in the process; a biological catalyst.

epi- [Gk. *epi*, upon]: Prefix meaning "upon" or "above."

epicotyl: The upper portion of the axis of an embryo or seedling, above the cotyledons (seed leaves) and below the next leaf or leaves.

epidermis: The outermost layer of cells of the leaf and of young stems and roots; primary in origin.

epigeous [Gk. *epi*, upon, + *ge*, the earth]: Type of seed germination in which the cotyledons are carried above ground level.

epigyny [Gk. *epi*, upon, + *gyne*, woman]: A pattern of floral organization in which the sepals, petals, and stamens apparently grow from the top of the ovary; the opposite of hypogyny.

epiphyte (ĕp′i·fīt): An organism that grows upon another organism but is not parasitic on it.

epistatic [Gk. *epistasis*, a stopping]: Term used to describe a gene the action of which modifies the phenotypic expression of a gene at another locus.

ergastic substance: A passive product of the protoplast; a storage or waste product.

ethylene: A simple hydrocarbon that is a plant hormone involved in the ripening of fruit; $H_2C = CH_2$.

etiolation (e′tĭ·o·lā′shŭn) [Fr. *etioler*, to blanch]: A condition involving increased stem elongation, poor leaf development, and lack of chlorophyll; found in plants growing in the dark or with a greatly reduced amount of light.

etioplast: Plastid of a plant grown in the dark and containing a prolamellar body.

eukaryote [Gk. *eu,* good, + *karyon,* kernel]: A cell that has a membrane-bound nucleus, membrane-bound organelles, and chromosomes in which the DNA is associated with proteins; an organism composed of such cells. Plants, animals, fungi, and protists are the four kingdoms of eukaryotes.

eusporangium: A sporangium that arises from several initial cells and, before maturation, forms a wall of more than one layer of cells.

eustele [Gk. *eu-,* good, + *stele,* pillar]: A stele in which the primary vascular tissues are arranged in discrete strands around a pith: typical of gymnosperms and angiosperms.

evolution: The derivation of progressively more complex forms of life from simple ancestors; Darwin proposed that natural selection is the principal mechanism by which evolution takes place.

exergonic [L. *ex,* out, + Gk. *ergon,* work]: Energy-yielding, as in a chemical reaction; applied to a "downhill" process.

exine: The outer wall layer of a spore or pollen grain.

exocarp [Gk. *exo,* without, + *karpos,* fruit]: The outermost layer of the mature ovary wall, or pericarp.

exocytosis [Gk. *ex,* out of, + *kytos,* vessel]: A cellular process in which particulate matter or dissolved substances are enclosed in a vesicle and transported to the cell surface; there, the membrane of the vesicle fuses with the plasma membrane, expelling the vesicle's contents to the outside.

exodermis: The outer layer, one or more cells in depth, of the cortex in some roots; these cells are characterized by Casparian strips within the radial and transverse walls. Following development of Casparian strips, a suberin lamella is deposited on all walls of the exodermis.

exon [Gr. *exo,* outside]: A segment of DNA that is both transcribed into RNA and translated into protein; exons are characteristic of eukaryotes. Contrasts with intron.

eyespot: Also stigma; a small, pigmented structure in flagellate unicellular organisms that is sensitive to light.

F$_1$: First filial generation. The offspring resulting from a cross. F$_2$ and F$_3$ are the second and third generations resulting from such a cross.

facilitated diffusion: Passive transport with the assistance of carrier proteins.

family: A taxonomic group between order and genus in rank; the ending of family names in animals and heterotrophic protists is *-idae;* in all other organisms it is *-aceae.* A family contains one or more genera, and each family belongs to an order.

fascicle (fås′ï•k′l) [L. *fasciculus,* a small bundle]: A bundle of pine leaves or other needlelike leaves of gymnosperms; an obsolete term for a vascular bundle.

fascicular cambium: The vascular cambium originating within a vascular bundle, or fascicle.

fats: A molecule composed of glycerol and three fatty acid molecules; the proportion of oxygen to carbon is much less in fats than it is in carbohydrates. Fats in the liquid state are called oils.

feedback inhibition: Control mechanism whereby an increase in the concentration of some molecule inhibits the further synthesis of that molecule.

fermentation: The extraction of energy from organic compounds without the involvement of oxygen.

ferredoxin: Electron-transferring proteins of high iron content; some are involved in photosynthesis.

fertilization: The fusion of two gamete nuclei to form a diploid zygote.

fiber: An elongated, tapering, generally thick-walled sclerenchyma cell of vascular plants; its walls may or may not be lignified; it may or may not have a living protoplast at maturity.

fibril: Submicroscopic threads composed of cellulose molecules, which constitute the form in which cellulose occurs in the cell wall.

field moisture capacity: The percentage of water a particular soil will hold against the action of gravity.

filament: (1) The stalk of a stamen; (2) a term used to describe the threadlike bodies or segments of certain algae or fungi.

fission: Asexual reproduction involving the division of a single-celled individual into two new single-celled individuals of equal size. Also applied to the division of plastids.

fitness: The genetic contribution of an organism to future generations, relative to the contributions of organisms living in the same environment that have different genotypes.

flagellum, *pl.* **flagella** [L. *flagellum,* whip]: A long threadlike organelle that protrudes from the surface of a cell. The flagella of bacteria are capable of rotary motion and consist of a single protein fiber each; eukaryotic flagella, which are used in locomotion and feeding, consist of an array of microtubules with a characteristic internal 9 + 2 microtubule structure and are capable of a vibratory, but not rotary, motion. A cilium is a small eukaryotic flagellum.

flavoprotein: A dehydrogenase that contains a flavin and often a metal and plays a major role in oxidation; abbreviated FP.

floral tube: A cup or tube formed by the fusion of the basal parts of the sepals, petals, and stamens; floral tubes are often found in plants that have an inferior ovary.

floret: One of the small flowers that make up the composite inflorescence or the spike of the grasses.

florigen [L. *flor-,* flower, + Gk. *-genes,* producer]: A hypothetical plant hormone that promotes flowering.

flower: The reproductive structure of angiosperms; a complete flower includes calyx, corolla, androecium (stamens), and gynoecium (carpels), but all flowers contain at least one stamen or one carpel.

fluid-mosaic model: Model of membrane structure, with the membrane composed of a lipid bilayer in which globular proteins are embedded.

follicle [L. *folliculus,* small ball]: A dry, dehiscent simple fruit derived from a single carpel and opening along one side.

food chain, food web: A chain of organisms existing in any natural community such that each link in the chain feeds on the one below and is eaten by the one above; there are seldom more than six links in a chain, with autotrophs on the bottom and the largest carnivores at the top.

fossil [L. *fossils,* dug up]: The remains, impressions, or traces of an organism that has been preserved in rocks found in the earth's crust.

fossil fuels: The altered remains of once-living organisms that are burned to release energy; oil, gas, and coal.

FP: *See* flavoprotein.

free energy: Energy available to do work.

frond: The leaf of a fern; any large, divided leaf.

fruit: In angiosperms, a mature, ripened ovary (or group of ovaries), containing the seeds, together with any adjacent parts that may be fused with it at maturity; sometimes applied informally, and misleadingly, as in "fruiting body," to the reproductive structures of other kinds of organisms.

fucoxanthin (fū′kŏ·zăn′thĭn) [Gk. *phykos*, seaweed, + *xanthos*, yellowish-brown]: A brownish carotenoid found in brown algae and chrysophytes.

fundamental tissue system: *See* ground tissue system.

funiculus [L. *funiculus*, small rope or cord]: The stalk of the ovule.

fusiform initials [L. *fusus*, spindle]: The vertically elongated cells in the vascular cambium that give rise to the cells of the axial system in the secondary xylem and secondary phloem.

gametangium, *pl.* **gametangia** [Gk. *gamein*, to marry, + L. *tangere*, to touch]: A cell or organ in which gametes are formed.

gamete [Gk. *gamete*, wife]: A haploid reproductive cell; gametes fuse in pairs, forming zygotes, which are diploid.

gametophore [Gk. *gamein*, to marry, + *phoros* bearing]: In the bryophytes, a fertile stalk that bears gametangia.

gametophyte: In plants, which have an alternation of generations, the haploid (*n*), gamete-producing phase.

gel: A mixture of substances having a semisolid or solid constitution.

gemma, *pl.* **gemmae** (jĕm′ă) [L. *gemma*, bud]: A small mass of vegetative tissue; an outgrowth of the thallus, for example, in liverworts or certain fungi; it can develop into an entire new plant.

gene: A unit of heredity; a sequence of DNA nucleotides that codes for a protein, tRNA, or rRNA molecule, or regulates the transcription of such a sequence.

gene frequency: The relative occurrence of a particular allele in a population.

generative cell: (1) In many gymnosperms, the cell of the male gametophyte that divides to form the sterile and spermatogenous cells: (2) in angiosperms, the cell of the male gametophyte that divides to form two sperm.

genetic code: The system of nucleotide triplets (codons) in DNA and RNA that dictates the amino acid sequence in proteins; except for three "stop" signals, each codon specifies one of 20 amino acids.

genetic recombination: The occurrence of gene combinations in the progeny that are different from the combinations present in the parents.

genotype: The genetic constitution, latent or expressed, of an organism, as contrasted with the phenotype; the sum total of all the genes present in an individual.

genus, *pl.* **genera:** The taxonomic group between family and species in rank; genera include one or more species.

geotropism: *See* gravitropism.

GERL: The Golgi associated–endoplasmic reticulum–lysosomal complex, which may be involved with lysosome or vacuole formation in some cells.

germination [L. *germinare*, to sprout]: The beginning or resumption of growth by a spore, seed, bud, or other structure.

gibberellins (jĭb·ĕ·rĕ′lĭns) [*Gibberella*, a genus of fungi]: A group of growth hormones, the best known effect of which is to increase the elongation of plant stems.

gill: The plates on the underside of the cap in basidiomycetes.

girdling: The removal from a woody stem of a ring of bark extending inward to the cambium; also called ringing.

glucose: A common six-carbon sugar ($C_6H_{12}O_6$); the most common monosaccharide in most organisms.

glycerol: A three-carbon molecule with three hydroxyl groups attached; glycerol molecules combine with fatty acids to form fats or oils.

glycogen [Gk. *glykys*, sweet, + *gen*, of a kind]: A carbohydrate similar to starch that serves as the reserve food in bacteria, fungi, and most organisms other than plants.

glycolysis: The anaerobic breakdown of glucose to form two molecules of pyruvic acid, resulting in the net liberation of two molecules of ATP.

glyoxylate cycle: A variant of the Krebs cycle; present in bacteria and some plant cells, for net conversion of acetate into succinate and, eventually, new carbohydrate.

glyoxysome: A microbody containing enzymes necessary for the conversion of fats into carbohydrates; glyoxysomes play an important role during the germination of seeds.

Golgi body (gôl′jē): *See* dictyosome.

grafting: A union of different individuals in which a portion, called the scion, of one individual is inserted into a root or stem, called the stock, of the other individual.

grain: *See* caryopsis; also a term used to refer to the alignment of wood elements.

grana, *sing.* **granum:** Structures within chloroplasts, seen as green granules with a light microscope and as a series of stacked thylakoids with an electron microscope; the grana contain the chlorophylls and carotenoids and are the sites of the light reactions of photosynthesis.

gravitropism [L. *gravis*, heavy, + Gk. *tropes*, turning]: The response of a shoot or root to the pull of the earth's gravity; also called geotropism.

ground meristem [Gk. *meristos*, divisible]: The primary meristem, or meristematic tissue, that gives rise to the ground tissues.

ground tissue system: All tissues other than the epidermis (or periderm) and the vascular tissues; also called fundamental tissue system.

growth layer: A layer of growth in the secondary xylem or secondary phloem; *see also* annual ring.

growth ring: A growth layer in the secondary xylem or secondary phloem, as seen in transverse section; may be called a growth increment, especially where seen in other than transverse section.

guanine [Sp. from Quechua, *huanu*, dung]: A purine base found in DNA and RNA. Its name is based on the fact that guanine is abundant in the form of a white crystalline base in guano and other animal excrements.

guard cells: Pairs of specialized epidermal cells surrounding a pore, or stoma; changes in the turgor of a pair of guard cells cause opening and closing of the pore.

guttation [L. *gutta*, a drop]: The exudation of liquid water from leaves, caused by root pressure.

gymnosperm [Gk. *gymnos*, naked, + *sperma*, seed]: A seed plant with seeds not enclosed in an ovary; the conifers are the most familiar group.

gynoecium [Gk. *gyne*, woman, + *oikos*, house]: The aggregate of carpels in the flower of a seed plant.

habit [L. *habitus*, condition, character]: Characteristic form or appearance of an organism.

habitat [L. *habitare*, to inhabit]: The environment of an organism; the place where it is usually found.

haploid [Gk. *haploos*, single]: Having only one set of chromosomes (*n*), in contrast to diploid (2*n*).

hardwood: A name commonly applied to the wood of a dicot tree.

Hardy-Weinberg law: The mathematical expression of the relationship between the relative frequencies of two or more alleles in a population; it demonstrates that the frequencies of alleles and genotypes will remain constant in a random-mating population in the absence of inbreeding, selection, or other evolutionary forces.

haustorium, *pl.* **haustoria** [L. *haustus*, from *haurire*, to drink, draw]: A projection of fungal hypha that functions as a penetrating and absorbing organ.

heartwood: Nonliving and commonly dark-colored wood in which no water transport occurs; it is surrounded by sapwood.

heliotropism [Gk. *helios*, sun]: *See* solar tracking.

hemicellulose (hĕm'ĭ·sĕl'u·lŏs): A polysaccharide resembling cellulose but more soluble and less ordered; found particularly in cell walls.

herb [L. *herba*, grass]: A nonwoody seed plant with a relatively short-lived aerial portion.

herbaceous: An adjective referring to nonwoody plants.

herbarium: A collection of dried and pressed plant specimens.

herbivorous: Feeding upon plants.

heredity [L. *heredis*, heir]: The transmission of characteristics from parent to offspring through the gametes.

hermaphrodite [Gk. for Hermes and Aphrodite]: An organism possessing both male and female reproductive organs.

hetero- [Gk. *heteros*, different]: Prefix meaning "other" or "different."

heterocyst [Gk. *heteros*, different, + *cystis*, a bag]: A kind of large, transparent, thick-walled, nitrogen-fixing cell that forms in the filaments of certain cyanobacteria.

heteroecious (hĕt'er·ē'shŭs) [Gk. *heteros*, different, + *oikos*, house]: As in some rust fungi, requiring two different host species to complete the life cycle.

heterogamy [Gk. *heteros*, other, + *gamos*, union or reproduction]: Reproduction involving two types of gametes.

heterokaryotic [Gk. *heteros*, other, + *karyon*, kernel]: In fungi, having two or more genetically distinct types of nuclei within the same mycelium.

heteromorphic [Gk. *heteros*, different, + *morphe*, form]: A term used to describe a life history in which the haploid and diploid generations are dissimilar in form.

heterosis [Gk. *heterosis*, alteration]: Hybrid vigor, the superiority of the hybrid over either parent in any measurable character.

heterosporous: Having two kinds of spores, designated as microspores and megaspores.

heterothallic [Gk. *heteros*, different, + *thallus*, sprout]: A term used to describe a species, the haploid individuals of which are self-sterile or self-incompatible; two compatible strains or individuals are required for sexual reproduction to take place.

heterotroph [Gk. *heteros*, other, + *trophos*, feeder]: An organism that cannot manufacture organic compounds and so must feed on organic materials that have originated in other plants and animals; *see also* autotroph.

heterozygous: Having two different alleles at the same locus on homologous chromosomes.

Hill reaction: The oxygen evolution and photoreduction of an artifi-

cial electron acceptor by a chloroplast preparation in the absence of carbon dioxide.

hilum [L. *hilum*, a trifle]: (1) Scar left on seed after separation of seed from funiculus; (2) the part of a starch grain around which the starch is laid down in more or less concentric layers.

histone: The group of five basic proteins associated with the chromosomes of all eukaryotic cells.

holdfast: (1) Basal part of a multicellular alga that attaches it to a solid object; may be unicellular or composed of a mass of tissue; (2) cuplike structures at the tips of some tendrils, by means of which they become attached.

homeo-, homo- [Gk. *homos*, same, similar]: Prefix meaning "similar" or "same."

homeostasis (hŏ'me·ō·stā'sĭs) [Gk. *homos*, similar, + *stasis*, standing]: The maintaining of a relatively stable internal physiological environment within an organism, or a steady-state equilibrium in a population or ecosystem. Homeostasis usually involves feedback mechanisms.

homokaryotic [Gk. *homos*, same, + *karyon*, kernel]: In fungi, having nuclei with the same genetic makeup within a mycelium.

homologous chromosomes: Chromosomes that associate in pairs in the first stage of meiosis; each member of the pair is derived from a different parent. Homologous chromosomes are also called homologues.

homology [Gk. *homologia*, agreement]: A condition indicative of the same phylogenetic, or evolutionary, origin, but not necessarily the same in present structure and or function.

homosporous: Having only one kind of spore.

homothallic [Gk. *homos*, same, + *thallus*, sprout]: A term used to describe a species in which the individuals are self-fertile.

homozygous: Having identical alleles at the same locus on homologous chromosomes.

hormogonium, *pl.* **hormogonia:** A portion of a filament of a cyanobacterium that becomes detached and grows into a new filament.

hormone [Gk. *hormaein*, to excite]: A chemical substance produced usually in minute amounts in one part of an organism, from which it is transported to another part of that organism on which it has a specific effect.

host: An organism on or in which a parasite lives.

humus: Decomposing organic matter in the soil.

hyaloplasm: *See* cytoplasmic ground substance.

hybrid: Offspring of two parents that differ in one or more heritable characteristics; offspring of two different varieties or of two different species.

hybridization: The formation of offspring between unlike parents.

hybrid vigor: *See* heterosis.

hydrocarbon [Gk. *hydro*, water, + L. *carbo*, charcoal]: An organic compound that consists only of hydrogen and carbon atoms.

hydrogen bond: A weak bond between a hydrogen atom attached to one oxygen or nitrogen atom and another oxygen or nitrogen atom.

hydrolysis [Gk. *hydro*, water, + *lysis*, loosening]: Splitting of one molecule into two by addition of the H^+ and OH^- ions of water.

hydrophyte [Gk. *hydro*, water, + *phyton*, a plant]: A plant that depends on an abundant supply of moisture or that grows wholly or partly submerged in water.

hydroxyl group: An OH^- group; a negatively charged ion formed by the dissassociation of a water molecule.

hymenium [Gk. *hymen*, a membrane]: The layer of asci on an ascoma, or of basidia on a basidioma, together with any associated sterile hyphae.

hyper- [Gk. *hyper*, above, over]: Prefix meaning "above" or "over."

hypertonic: Refers to a solution that has a concentration of solute particles high enough to gain water across a permeable membrane from another solution.

hypha, pl. **hyphae** [Gk. *hyphe*, web]: A single tubular filament of a fungus, oomycete, or chytrid; the hyphae together comprise the mycelium.

hypo- [Gk. *hypo*, less than]: Prefix meaning "under" or "less."

hypocotyl: The portion of an embryo or seedling situated between the cotyledons and the radicle.

hypocotyl-root axis: The embryo axis below the cotyledon or cotyledons, consisting of the hypocotyl and the apical meristem of the root or the radicle.

hypodermis [Gk. *hypo*, under, + *derma*, skin]: One or more layers of cells beneath the epidermis, which are distinct from the underlying cortical or mesophyll cells.

hypogeous [Gk. *hypo*, under, + *ge*, the earth]: Type of seed germination in which the cotyledons remain underground.

hypogyny [Gk. *hypo*, under, + *gyne*, woman]: Floral organization in which the sepals, petals, and stamens are attached to the receptacle below the ovary; the opposite is epigyny.

hypothesis [Gk. *hypo*, under, + *tithenai*, to put]: A temporary working explanation or supposition based on accumulated facts and suggesting some general principle or relation of cause and effect; a postulated solution to a scientific problem that must be tested by experimentation and, if disproved or shown to be unlikely, is discarded.

hypotonic: Refers to a solution that has a concentration of solutes low enough to lose water across a permeable membrane to another solution.

IAA: *See* indoleacetic acid.

imbibition (im·bi·bish′ŭn): Adsorption of water and swelling of colloidal materials because of the adsorption of water molecules onto the internal surfaces of the materials.

imperfect flower: A flower lacking either stamens or carpels.

imperfect fungi: The deuteromycetes, or conidial fungi, which reproduce only asexually, or in which the sexual cycle has not been observed; most deuteromycetes are ascomycetes.

inbreeding: The breeding of closely related plants or animals: in plants, it is usually brought about by repeated self-pollination.

incomplete flower: A flower lacking one or more of the four kinds of floral parts, that is, lacking sepals, petals, stamens, or carpels.

indehiscent (in′de·his′ĕnt): Remaining closed at maturity, as are many fruits (samaras, for example).

independent assortment: *See* Mendel's second law.

indeterminate growth: Unrestricted or unlimited growth, as with a vegetative apical meristem that produces an unrestricted number of lateral organs indefinitely.

indoleacetic acid, or **IAA:** Naturally occurring auxin, a kind of plant hormone.

indusium, pl. **indusia** (in·dū′zi·ŭm) [L. *indusium*, a woman's undergarment]: Membranous growth of the epidermis of a fern leaf that covers a sorus.

inferior ovary: An ovary that is completely or partially attached to the calyx; the other floral whorls appear to arise from the ovary's top.

inflorescence: A flower cluster, with a definite arrangement of flowers.

initial: (1) In the meristem, a cell that remains within the meristem indefinitely and at the same time, by division, adds cells to the plant body; (2) a meristematic cell that eventually differentiates into a mature, more specialized cell or element

inner bark: In older trees, the living part of the bark; the bark inside the innermost periderm.

integument: The outermost layer or layers of tissue enveloping the nucellus of an ovule; develops into the seed coat.

inter- [L. *inter*, between]: Prefix meaning "between," or "in the midst of."

intercalary [L. *intercalare*, to insert]: Descriptive of meristematic tissue or growth not restricted to the apex of an organ; that is, growth in the regions of the nodes.

interfascicular cambium: The vascular cambium arising between the fascicles, or vascular bundles, from interfascicular parenchyma.

interfascicular region: Tissue region between vascular bundles in stem; also called pith ray.

internode: The region of a stem between two successive nodes.

interphase: The period between two mitotic or meiotic cycles; the cell grows and its DNA replicates during interphase.

intine: The inner wall layer of a spore or pollen grain.

intra- [L. *intra*, within]: Prefix meaning "within."

intron [L. *intra*, within]: A portion of mRNA as transcribed from eukaryotic DNA that is removed by enzymes before the mRNA is translated into protein. *See* exon.

ion: An atom or molecule that has lost or gained one or more electrons, and thus becomes positively or negatively charged.

irregular flower: A flower in which one or more members of at least one whorl differ in form from other members of the same whorl.

iso- [Gk. *isos*, equal]: Prefix meaning "equal"; like "homeo-" or "homo-."

isogamy: A type of sexual reproduction in which the gametes (or gametangia) are alike in size; found in some algae and fungi.

isomer [Gk. *isos*, equal, + *meros*, part]: One of a group of compounds identical in atomic composition but differing in structural arrangement; for example, glucose and fructose.

isomorphic [Gk. *isos*, equal, + *morphe*, form]: Identical in form.

isotonic: Having the same osmotic concentration.

isotope: One of several possible forms of a chemical element that differ from other forms in the number of neutrons in the atomic nucleus, but not in chemical properties.

karyogamy [Gk. *karyon*, kernel, + *gamos*, marriage]: The union of two nuclei following plasmogamy.

kelp: A common name for any of the larger members of the order Laminariales of the brown algae.

kinetin [Gk. *kinetikos*, causing motion]: A purine that probably does not occur in nature but that acts as a cytokinin in plants.

kinetochore [Gk. *kinetikos*, putting in motion, + *chorus*, chorus]: Specialized protein complexes, which develop on each centromere, and to which spindle fibers are attached during mitosis or meiosis.

kingdom: One of the six chief taxonomic categories; for example, Eubacteria or Plantae.

Kranz anatomy [Ger. *Kranz*, wreath]: The wreathlike arrangement of mesophyll cells around a layer of large bundle-sheath cells, forming two concentric layers around the vascular bundle; typically found in the leaves of C_4 plants.

Krebs cycle: The series of reactions that results in the oxidation of pyruvic acid to hydrogen atoms, electrons, and carbon dioxide. The electrons, passed along electron-carrier molecules, then go through the oxidative phosphorylation and terminal oxidation processes. Also called the tricarboxylic acid cycle or TCA cycle.

lamella (la·mël′a) [L. *lamella*, thin metal plate]: Layer of cellular membranes, particularly photosynthetic, chlorophyll-containing membranes; *see also* middle lamella.

lamina: the blade of a leaf.

laminarin: One of the principal storage products of the brown algae; a polymer of glucose.

lateral meristems: Meristems that give rise to secondary tissue; the vascular cambium and cork cambium.

lateral root: A root that arises from another, older root; also called a branch root, or secondary root, if the older root is the primary root.

late wood: The last part of the growth increment formed in the growing season; it contains smaller cells and is denser than the early wood; replaces the term "summer wood."

leaching: The downward movement and drainage of minerals, or inorganic ions, from the soil by percolating water.

leaf buttress: A lateral protrusion below the apical meristem; represents the initial stage in the development of a leaf primordium.

leaf gap: Region of parenchyma tissue in the primary vascular cylinder above the point of departure of the leaf trace or traces.

leaflet: One of the parts of a compound leaf.

leaf primordium [L. *primordium*, beginning]: A lateral outgrowth from the apical meristem that will eventually become a leaf.

leaf scar: A scar left on a twig when a leaf falls.

leaf trace: That part of a vascular bundle extending from the base of the leaf to its connection with a vascular bundle in the stem.

legume [L. *legumem*, leguminous plant]: (1) A member of the Fabaceae, the pea or bean family: (2) a type of dry simple fruit that is derived from one carpel and opens along both sides.

lenticels (lĕn′tĭ·sĕls) [L. *lenticella*, a small window]: Spongy areas in the cork surfaces of stem, roots, and other plant parts that allow interchange of gases between internal tissues and the atmosphere through the periderm; occur in vascular plants.

leptosporangium: A sporangium that arises from a single initial cell and whose wall is composed of a single layer of cells.

leucoplast (lū′kō·plăst) [Gk. *leuko*, white, + *plasein*, to form]: A colorless plastid; leucoplasts are commonly centers of starch formation.

liana [F. *liane*, from *lier*, to bind]: A large, woody vine that climbs upon other plants.

life cycle: The entire sequence of phases in the growth and development of any organism from time of zygote formation to gamete formation.

light reactions: The reactions of photosynthesis that require light and cannot occur in the dark; "light-dependent reactions" is the preferred term.

lignin: One of the most important constituents of the secondary wall of vascular plants, although not all secondary walls contain lignin; after cellulose, lignin is the most abundant plant polymer.

ligule [L. *ligula*, small tongue]: A minute outgrowth or appendage at the base of the leaves of grasses and those of certain lycophytes.

linkage: The tendency for certain genes to be inherited together owing to the fact that they are located on the same chromosome.

lipid [Gk. *lipos*, fat]: One of a large variety of nonpolar organic molecules that are insoluble in water (which is polar) but dissolve readily in nonpolar organic solvents; lipids include fats, oils, steroids, phospholipids, and carotenoids.

locule (lŏk′ŭl) [L. *loculus*, small chamber]: A cavity within a sporangium or a cavity of the ovary in which ovules occur.

locus, *pl.* **loci:** The position on a chromosome occupied by a particular gene.

long-day plants: Plants that must be exposed to light periods longer than some critical length for flowering to occur; they flower in spring or summer.

lumen [L. *lumen*, light, an opening for light]: The space bounded by the plant cell wall; the transparent space of endoplasmic reticulum.

lysis [Gk. *lysis*, a loosening]: A process of disintegration or cell destruction.

lysogenic bacteria: Bacteria-carrying viruses (phages) that eventually break loose from the bacterial chromosome and set up an active cycle of infection, producing lysis in their bacterial hosts.

lysosome [Gk. *lysis*, loosening, + *soma*, body]: An organelle, bounded by a single membrane, and containing hydrolytic enzymes that are released when the organelle ruptures and are capable of breaking down proteins and other complex macromolecules.

macrocyst: A flattened, irregular structure, encircled by a thin membrane, in which zygotes are formed during the life cycle of cellular slime molds.

macrofibril: An aggregation of microfibrils, visible with the light microscope.

macromolecule [Gk. *makros*, large]: A molecule of very high molecular weight; refers specifically to proteins, nucleic acids, polysaccharides, and complexes of these.

macronutrients [Gk. *makros*, large, + L. *nutrire*, to nourish]: Inorganic chemical elements required in large amounts for plant growth, such as nitrogen, potassium, calcium, phosphorus, magnesium, and sulfur.

maltase: An enzyme that hydrolyzes maltose to glucose.

mannitol: One of the storage molecules of the brown algae; an alcohol.

marginal meristem: The meristem located along the margin of a leaf primordium and forming the blade.

mating type: A particular genetically defined strain of an organism that is incapable of sexual reproduction with another member of the same strain but capable of such reproduction with members of other strains of the same organism.

mega- [Gk. *megas*, large]: Prefix meaning "large."

megagametophyte [Gk. *megas*, large, + *gamos*, marriage, + *phyton*, plant]: In heterosporous plants, the female gametophyte; located within the ovule of seed plants.

megaphyll [Gk. *megas*, large, + *phyllon*, leaf]: A generally large leaf with several to many veins; its leaf trace is associated with a leaf gap; in contrast to microphyll.

megasporangium, *pl.* **megasporangia:** A sporangium in which megaspores are produced; *see* nucellus.

megaspore: In heterosporus plants, a haploid (*n*) spore that develops into a female gametophyte; in most groups, megaspores are larger than microspores.

megaspore mother cell: A diploid cell in which meiosis will occur, resulting in the production of four megaspores; also called a megasporocyte.

megasporocyte: *See* megaspore mother cell.

megasporophyll: A leaf or leaflike structure bearing a megasporangium.

meiosis (mī·ō′sĭs) [Gk. *meioun*, to make smaller]: The two successive nuclear divisions in which the chromosome number is reduced from diploid (2*n*) to haploid (*n*) and segregation of the genes occurs; gametes or spores (in organisms with an alternation of generations) may be produced as a result of meiosis.

Mendel's first law: The factors for a pair of alternate characteristics are separate and only one may be carried in a particular gamete (genetic segregation).

Mendel's second law: The inheritance of one pair of characteristics is independent of the simultaneous inheritance of other traits, such characteristics assorting independently as though there were no others present (later modified by the discovery of linkage).

meristem [Gk. *merizein*, to divide]: The undifferentiated plant tissue from which new cells arise.

meso- [Gk. *mesos*, middle]: Prefix meaning "middle."

mesocarp [Gk. *mesos*, middle, + *karpos*, fruit]: The middle layer of the mature ovary wall, or pericarp, between the exocarp and endocarp.

mesophyll: The ground tissue (parenchyma) of a leaf, located between the layers of epidermis; mesophyll cells generally contain chloroplasts.

mesophyte [Gk. *mesos*, middle, + *phyton*, a plant]: A plant that requires abundant soil water and a relatively humid atmosphere.

messenger RNA (mRNA): The class of RNA that carries genetic information from the gene to the ribosomes, where it is translated into protein.

metabolism [Gk. *metabole*, change]: The sum of all chemical processes occurring within a living cell or organism.

metaphase: The stage of mitosis or meiosis during which the chromosomes lie in the equatorial plane of the spindle.

metaxylem [Gk. *meta*, after]: The part of the primary xylem that differentiates after the protoxylem; the metaxylem reaches maturity after the portion of the plant part in which it is located has finished elongating.

micro- [Gk. *mikros*, small]: Prefix meaning "small."

microbody: An organelle bounded by a single membrane and containing a variety of enzymes; generally associated with the endoplasmic reticulum. Peroxisomes and glyoxysomes are kinds of microbodies.

microfibril: A threadlike component of the cell wall, composed of cellulose molecules, visible only with the electron microscope.

microfilament [Gk. *mikros*, small + L. *filum*, a thread]: A long filament, 5 to 7 nanometers thick, composed of actin; plays a causative role in cytoplasmic streaming; also F-actin, actin filament.

microgametophyte [Gk. *mikros*, small, + *gamos*, marriage, + *phyton*, plant]: In heterosporous plants, the male gametophyte.

micrometer: A unit of microscopic measurement convenient for describing cellular dimensions; 1/1000 of a millimeter; its symbol is μm.

micronutrients [Gk. *mikros*, small, + L. *nutrire*, to nourish]: Inorganic chemical elements required only in very small, or trace, amounts for plant growth, such as iron, chlorine, copper, manganese, zinc, molybdenum, and boron.

microphyll [Gk. *mikros*, small, + *phyllon*, leaf]: A small leaf with one vein and one leaf trace not associated with a leaf gap; in contrast to megaphyll. Microphylls are characteristic of the lycophytes.

micropyle: In the ovules of seed plants, the opening in the integuments through which the pollen tube usually enters.

microsporangium: A sporangium within which microspores are formed.

microspore: In heterosporous plants, a spore that develops into a male gametophyte.

microspore mother cell: A cell in which meiosis will occur, resulting in four microspores; in seed plants, often called a pollen mother cell; also called a microsporocyte.

microsporocyte: *See* microspore mother cell.

microsporophyll: A leaflike organ bearing one or more microsporangia.

microtubule [Gk. *mikros*, small, + L. *tubulus*, little pipe]: Narrow (about 25 nanometers in diameter), elongate, nonmembranous tubule of indefinite length; microtubules occur in the cells of eukaryotes. Microtubules move the chromosomes in cell division and provide the internal structure of cilia and flagella.

middle lamella: The layer of intercellular material, rich in pectic compounds, cementing together the primary walls of adjacent cells.

mimicry [Gk. *mimos*, mime]: The superficial resemblance in form, color, or behavior of certain organisms (mimics) to other more powerful or more protected ones (models), resulting in protection, concealment, or some other advantage for the mimic.

mineral: A naturally occurring chemical element or inorganic compound.

mitochondrion, *pl.* **mitochondria** [Gk. *mitos*, thread, + *chondrion*, small grain]: A double-membrane-bound organelle found in eukaryotic cells; contains the enzymes of the Krebs cycle and the electron-transport chain; the major source of ATP in nonphotosynthetic cells.

mitosis (mītō′sĭs) [Gk. *mitos*, thread]: A process during which the duplicated chromosomes divide longitudinally and the daughter chromosomes then separate to form two genetically identical daughter nuclei; usually accompanied by cytokinesis.

mole: The name for a gram molecule. The number of particles in 1 mole of any substance; always equal to Avogadro's number: 6.022×10^{23}.

molecular weight: The relative weight of a molecule when the weight of the most frequent kind of carbon atom is taken as 12; the sum of the relative weights of the atoms in a molecule.

molecule: Smallest possible unit of a compound, consisting of two or more atoms.

mono- [Gk. *monos*, single]: Prefix meaning "one" or "single."

monocotyledon: A plant whose embryo has one cotyledon; one of the two great classes of angiosperms, Monocotyledones; often abbreviated as monocot.

monoecious (mō·nē′shŭs) [Gk. *monos*, single, + *oikos*, house]: Having the anthers and carpels produced in separate flowers on the same individual.

monokaryotic [Gk. *monos*, one, + *karyon*, kernel]: In fungi, having a single haploid nucleus within one cell or compartment.

monosaccharide [Gk. *monos*, single, + *sakcaron*, sugar]: A simple sugar, such as five-carbon and six-carbon sugars, that cannot be disassociated into smaller sugar particles.

-morph, morph- [Gk. *morphe*, form]: Suffix or prefix meaning "form."

morphogenesis: The development of form.

morphology [Gk. *morphe*, form, + *logos*, discourse]: The study of form and its development.

mRNA: *See* messenger RNA.

mucigel: A slime-sheath covering the surface of many roots.

multigene family: A collection of related genes on a chromosome; most eukaryotic genes appear to be members of multigene families.

multiple epidermis: A tissue composed of several cell layers derived from the protoderm; only the outer layer assumes characteristics of a typical epidermis.

multiple fruit: A cluster of matured ovaries produced by a cluster of flowers, as in the pineapple.

mutagen [L. *mutare*, to change, + Gk. *genaio*, to produce]: An agent that increases the mutation rate.

mutant: A mutated gene or an organism carrying a gene that has undergone a mutation.

mutation: An inheritable change of a gene from one allelic form to another.

mutualism: The living together of two or more organisms in an association that is mutually advantageous.

myc-, myco- [Gk. *mykes*, fungus]: Prefix meaning "pertaining to fungi."

mycelium [Gk. *mykes*, fungus]: The mass of hyphae forming the body of a fungus, oomycete, or chytrid.

mycology: The study of fungi.

mycorrhiza, *pl.* **mycorrhizae:** A symbiotic association between certain fungi and plant roots; characteristic of most vascular plants.

NAD⁺: *See* nicotinamide adenine dinucleotide.

NADP⁺: *See* nicotinamide adenine dinucleotide phosphate.

nanoplankton: (năn′o·plăngk′tŏn) [Gk. *nanos*, dwarf, + *planktos*, wandering]: Plankton with dimensions of less than 70 to 75 micrometers.

nastic movement: A plant movement that occurs in response to a stimulus, but whose direction is independent of the direction of the stimulus.

natural selection: The differential reproduction of genotypes based on their genetic constitution.

nectary [Gk. *nektar*, the drink of the gods]: In angiosperms, a gland that secretes nectar, a sugary fluid that attracts animals to plants.

netted venation: The arrangement of veins in the leaf blade that resembles a net; characteristic of dicot leaves; also called reticulate venation.

neutron [L. *neuter*, neither]: An uncharged particle with a mass slightly greater than that of a proton, found in the atomic nucleus of all elements except hydrogen, in which the nucleus consists of a single proton.

niche: The role played by a particular species in its ecosystem.

nicotinamide adenine dinucleotide (NAD⁺): A coenzyme that functions as an electron acceptor in many of the oxidation reactions of respiration.

nicotinamide adenine dinucleotide phosphate (NADP⁺): A coenzyme that functions as an electron acceptor in many of the reduction reactions of biosynthesis; similar in structure to NAD⁺ except that it contains an extra phosphate group.

nitrification: The oxidation of ammonium ions or ammonia to nitrate; a process carried out by a specific free-living soil bacterium.

nitrogen fixation: The incorporation of atmospheric nitrogen into nitrogen compounds; carried out by certain free-living and symbiotic bacteria.

nitrogen-fixing bacteria: Soil bacteria that convert atmospheric nitrogen into nitrogen compounds.

nitrogenous base: A nitrogen-containing molecule having basic properties (tendency to acquire an H atom); a purine or pyrimidine; one of the building blocks of nucleic acids.

node [L. *nodus*, knot]:The part of a stem where one or more leaves are attached; *see* internode.

nodules: Enlargements or swellings on the roots of legumes and certain other plants inhabited by symbiotic nitrogen-fixing bacteria.

noncyclic electron flow: The light-induced flow of electrons from water to NADP⁺ in oxygen-evolving photosynthesis; it involves both photosystems I and II.

nonseptate: *See* aseptate.

nucellus (nu·sĕl′ŭs) [L. *nucella*, a small nut]: Tissue composing the chief part of the young ovule, in which the embryo sac develops; equivalent to a megasporangium.

nuclear envelope: The double membrane surrounding the nucleus of a cell.

nucleic acid: An organic acid consisting of joined nucleotide complexes; the two types are deoxyribonucleic acid (DNA) and ribonucleic acid (RNA).

nucleolar organizer region: A special area on a certain chromosome associated with the formation of the nucleolus.

nucleolus, *pl.* **nucleoli,** (nu·klē′ō·lŭs) [L. *nucleolus*, a small nucleus]: A small, spherical body found in the nucleus of eukaryotic cells, which is composed chiefly of rRNA in the process of being transcribed from copies of rRNA genes; the site of production of ribosomes.

nucleoplasm: The ground substance of a nucleus.

nucleotide: A single unit of nucleic acid, composed of a phosphate, a five-carbon sugar (either ribose or deoxyribose), and a purine or a pyrimidine.

nucleus, *pl.* **nuclei:** (1) A specialized body within the eukaryotic cell bounded by a double membrane and containing the chromosomes; (2) the central part of an atom of a chemical element.

nut: A dry, indehiscent, hard, one-seeded simple fruit, generally produced from a gynoecium of more than one fused carpel.

obligate anaerobe: An organism that is metabolically active only in the absence of oxygen.

-oid [Gk. *oid*, like, resembling]: Suffix meaning "like" or "similar to."

Okazaki fragments (after R. Okazaki, Japanese geneticist): In DNA replication, the discontinuous segments in which the 3′ to 5′ strand (the lagging strand) of the DNA double helix is synthesized; typically 1000 to 2000 nucleotides long in prokaryotes, and 100 to 200 nucleotides long in eukaryotes.

ontogeny [Gk. *on*, being, + *genesis*, origin]: The development, or life history, of all or part of an individual organism.

oo- [Gk. *oion*, egg]: Prefix meaning "egg."

oogamy: Sexual reproduction in which one of the gametes (the egg) is large and nonmotile, and the other gamete (the sperm) is smaller and motile.

oogonium (ō'o·gō'nĭ·ŭm): A unicellular female sex organ that contains one or several eggs.

oospore: The thick-walled zygote characteristic of the oomycetes.

open bundle: A vascular bundle in which a vascular cambium develops.

operator: A site of gene regulation; a sequence of nucleotides recognized by a repressor protein.

operculum (o·pûr'ku·lŭm) [L. *operculum*, lid]: In mosses, the lid of the sporangium.

operator: A segment of DNA that interacts with a repressor protein to regulate the transcription of the structural genes of an operon.

operon [L. *opus, operis*, work]: In the bacterial chromosome, a segment of DNA consisting of a promoter, an operator, and a group of adjacent structural genes; the structural genes, which code for products related to a particular biochemical pathway, are transcribed onto a single mRNA molecule, and their transcription is regulated by a single repressor protein.

opposite: Term applied to buds or leaves occurring in pairs at a node.

order: A category of classification between the rank of class and family; classes contain one or more orders, and orders in turn are composed of one or more families.

organ: A structure composed of different tissues, such as root, stem, leaf, or flower parts.

organelle (ôr'găn·ĕl) [Gk. *organella*, a small tool]: A specialized, membrane-bound part of a cell.

organic: Pertaining to living organisms in general, to compounds formed by living organisms, and to the chemistry of compounds containing carbon.

organism: Any individual living creature, either unicellular or multicellular.

osmosis (ŏs·mō'sĭs) [Gk. *osmos*, impulse or thrust]: The diffusion of water, or any solvent, across a differentially permeable membrane; in the absence of other forces, the movement of water during osmosis will always be from a region of greater water potential to one of lesser water potential.

osmotic potential: The change in free energy or chemical potential of water produced by solutes; carries a negative (minus) sign; also called solute potential.

osmotic pressure: The potential pressure that can be developed by a solution separated from pure water by a differentially permeable or semipermeable membrane; it is an index of the solute concentration of the solution.

outcrossing: Cross-pollination between individuals of the same species.

outer bark: In older trees, the dead part of the bark; the innermost periderm and all tissues outside it; also called rhytidome.

ovary [L. *ovum*, an egg]: The enlarged basal portion of a carpel or of a gynoecium composed of fused carpels; a mature ovary, sometimes with other adherent parts, is a fruit.

ovule (ō·vūl) [L. *ovulum*, a little egg]: A structure in seed plants containing the female gametophyte with egg cell, all being surrounded by the nucellus and one or two integuments; when mature, an ovule becomes a seed.

ovuliferous scale: In certain conifers, the appendage or scalelike shoot to which the ovule is attached.

oxidation: The loss of an electron by an atom or molecule. Oxidation and reduction (gain of an electron) take place simultaneously, because an electron that is lost by one atom is accepted by another. Oxidation-reduction reactions are an important means of energy transfer in living systems.

oxidative phosphorylation: The formation of ATP from ADP and inorganic phosphate; oxidative phosphorylation takes place in the electron-transport chain of the mitochondrion.

pairing of chromosomes: Side-by-side association of homologous chromosomes.

paleobotany [Gk. *palaios*, old]: The study of fossil plants.

palisade parenchyma: A leaf tissue composed of columnar chloroplast-bearing parenchyma cells with their long axes at right angles to the leaf surface.

panicle (păn'ĭk'l) [L. *panicula*, tuft]: An inflorescence, the main axis of which is branched, and whose branches bear loose flower clusters.

para- [Gk. *para*, beside]: Prefix, meaning "beside."

paradermal section [Gk. *para*, beside, + *derma*, skin]: Section cut parallel with the surface of a flat structure, such as a leaf.

parallel evolution: The development of similar structures having similar functions in two or more evolutionary lines as a result of the same kinds of selective pressures.

parallel venation: The pattern of venation in which the principal veins of the leaf are parallel or nearly so; characteristic of monocots.

paramylon: The storage molecule of euglenoids.

paraphysis, *pl.* **paraphyses** [Gk. *para*, beside, + *physis*, growth]: As in certain fungi and brown algae, a sterile filament growing among the reproductive cells in the fruiting body.

parasexual cycle: The fusion and segregation of heterokaryotic haploid nuclei in certain fungi to produce recombinant nuclei.

parasite: An organism that lives on or in an organism of a different species and derives nutrients from it; the association is beneficial to the parasite and harmful to the host.

parenchyma (pa·rĕng'kĭ·ma) [Gk. *para*, beside, + *en*, in, + *chein*, to pour]: A tissue composed of parenchyma cells.

parenchyma cell: Living, generally thin-walled cell of variable size and form; the most abundant kind of cell in plants.

parthenocarpy [Gk. *parthenos*, virgin, + *karpos*, fruit]: The development of fruit without fertilization; parthenocarpic fruits are usually seedless.

passage cell: Endodermal cell of root that retains thin wall and Casparian strip when other associated endodermal cells develop thick secondary walls.

passive transport: Non-energy-requiring transport of a solute across a membrane down the concentration or electrochemical gradient by either simple diffusion or facilitated diffusion.

pathogen [Gk. *pathos*, suffering, + *genesis*, beginning]: An organism that causes a disease.

pathogenic: Disease-causing.

pathology: The study of plant or animal diseases, their effects on the organism, and their treatment.

pectin: A complex organic compound present in the intercellular layer and primary wall of plant cell walls; the basis of fruit jellies.

pedicel (pĕd'ĭ·sĕl): The stalk of an individual flower in an inflorescence.

peduncle (pe·dŭng'k'l): The stalk of an inflorescence or of a solitary flower.

pentose phosphate cycle: The pathway of oxidation of glucose 6-phosphate to yield pentose phosphates.

peptide: Two or more amino acids linked by peptide bonds.

peptide bond: The type of bond formed when two amino acid units are joined end to end by the removal of a molecule of water; the bonds always form between the carboxyl (—COOH) group of one amino acid and the amino (—NH$_2$) group of the next amino acid.

perennial [L. *per*, through, + *annuus*, a year]: A plant that persists and produces reproductive structures year after year.

perfect flower: A flower having both stamens and carpels; hermaphroditic flower.

perfect stage: That phase of the life history of a fungus that includes sexual fusion and the spores associated with such fusions.

perforation plate: Part of the wall of a vessel member that is perforated.

peri- [Gk. *peri*, around]: Prefix meaning "around" or "about."

perianth (pĕr'ĭ·ănth) [Gk. *peri*, around, + *anthos*, flower]: (1) The petals and sepals taken together; (2) in leafy liverworts, a tubular sheath surrounding an archegonium, and later, the developing sporophyte.

pericarp [Gk. *peri*, around, + *karpos*, fruit]: The fruit wall, which develops from the mature ovary wall.

periclinal: Parallel with the surface.

pericycle [Gk. *peri*, around, + *kykos*, circle]: A tissue, characteristic of roots that is bounded externally by the endodermis and internally by the phloem.

periderm [Gk. *peri*, around, + *derma*, skin]: Outer protective tissue that replaces epidermis when it is destroyed during secondary growth; includes cork, cork cambium, and phelloderm.

perigyny [Gk. *peri*, around, + *gyne*, female]: A form of floral organization in which the sepals, petals, and stamens are attached to the margin of a cup-shaped extension of the receptacle; superficially, the sepals, petals, and stamens appear to be attached to the ovary.

perisperm [Gk. *peri*, around, + *sperma*, seed]: Food-storing tissue derived from the nucellus that occurs in the seeds of some flowering plants.

peristome (pĕr'ĭ·stōm) [Gk. *peri*, around, + *stoma*, a mouth]: In mosses, a fringe of teeth around the opening of the sporangium.

perithecium: A spherical or flask-shaped ascoma.

permanent wilting percentage: The percentage of water remaining in a soil when a plant fails to recover from wilting even if placed in a humid chamber.

permeable [L. *permeare*, to pass through]: Usually applied to membranes through which liquid substances may diffuse.

peroxisome: A microbody that plays an important role in glycolic acid metabolism associated with photosynthesis; the site of photorespiration.

petal: A flower part, usually conspicuously colored; one of the units of the corolla.

petiole: The stalk of a leaf.

pH: A symbol denoting the relative concentration of hydrogen ions in a solution; pH values run from 0 to 14, and the lower the value the more acidic a solution, that is, the more hydrogen ions it contains; pH 7 is neutral, less than 7 is acidic, and more than 7 is alkaline.

phage: *See* bacteriophage.

phagocytosis: *See* endocytosis.

phellem: *See* cork.

phelloderm (fĕl'o·dûrm) [Gk. *phellos*, cork, + *derma*, skin]: A tissue formed inwardly by the cork cambium, opposite the cork; inner part of the periderm.

phellogen (fĕl'lo·jĕn): *See* cork cambium.

phenotype: The physical appearance of an organism; the phenotype results from the interaction between the genetic constitution (genotype) of the organism and its environment.

phloem (flō'ĕm) [Gk. *phloos*, bark]: The food-conducting tissue of vascular plants, which is composed of sieve elements, various kinds of parenchyma cells, fibers, and sclereids.

phloem loading: The process by which substances (primarily sugars) are actively secreted into the sieve tubes.

phosphate: A compound formed from phosphoric acid by replacement of one or more hydrogen atoms.

phospholipids: A phosphorylated lipid; similar in structure to a fat, but with only two fatty acids attached to the glycerol backbone, with the third space occupied by a phosphorus-containing molecule; important components of cellular membranes.

phosphorylation (fŏs'fo·rĭl·ā'shŭn) [Gk. *phosphoros*, bringing light]: A reaction in which phosphate is added to a compound; e.g., the formation of ATP from ADP and inorganic phosphate.

photo-, -photic [Gk. *photos*, light]: Prefix or suffix meaning "light."

photolysis: The light-dependent oxidative splitting of water molecules that takes place in photosystem II of the light reactions of photosynthesis.

photon [Gk. *photos*, light]: The elementary particle of light.

photoperiodism: Response to duration and timing of day and night; a mechanism evolved by organisms for measuring seasonal time.

photophosphorylation [Gk. *photos*, light, + *phosphoros*, bringing light]: The formation of ATP in the chloroplast during photosynthesis.

photorespiration: The light-dependent production of glycolic acid in chloroplasts and its subsequent oxidation in peroxisomes.

photosynthesis [Gk. *photos*, light, + *syn*, together + *tithenai*, to place]: The conversion of light energy to chemical energy; the production of carbohydrates from carbon dioxide and water in the presence of chlorophyll by using light energy.

photosystem: A discrete unit of organization of chlorophyll and other pigment molecules embedded in the thylakoids of chloroplasts and involved with the light-requiring reactions of photosynthesis.

phototropism [Gk. *photos*, light, + *trope*, turning]: Growth in which the direction of the light is the determining factor, as the growth of a plant toward a light source; turning or bending in response to light.

phragmoplast: A spindle-shaped system of fibrils, which arises between two daughter nuclei at telophase and within which the cell plate is formed during cell division, or cytokinesis. The fibrils of the phragmoplast are composed of microtubules. Phragmoplasts are found in all green algae except the members of the class Chlorophyceae and in plants.

phycobilins: A group of water-soluble accessory pigments, including phycocyanins and phycoerythrins, which occur in the red algae and cyanobacteria.

phycology [Gk. *phykos*, seaweed]: The study of algae.

phycoplast: A system of microtubules that develops between the two daughter nuclei parallel to the plane of cell division. Phycoplasts occur only in green algae of the class Chlorophyceae.

phyllo-, phyll- [Gk. *phyllon*, leaf]: Prefix meaning "leaf."

phyllode (fĭl'ōd): A flat, expanded, photosynthetic petiole or stem; phyllodes occur in certain genera of vascular plants, where they replace leaf blades in photosynthetic function.

phylogeny [Gk. *phylon*, race, tribe]: Evolutionary relationships among organisms; the developmental history of a group of organisms.

phylum, *pl.* **phyla** [Gk. *phylon*, race, tribe]: A unit of classification equivalent to "division," used for animals and heterotrophic protists.

physiology: The study of the activities and processes of living organisms.

phyto-, -phyte [Gk. *phyton*, plant]: Prefix or suffix meaning "plant."

phytoalexin: [Gk. *phyton*, plant, + *alexein*, to ward off, protect]: A chemical sustance produced by a plant to combat infection by a pathogen (fungi or bacteria).

phytochrome: A phycobilinlike pigment found in the cytoplasm of plants and a few green algae that is associated with the absorption of light; photoreceptor for red and far-red light; involved in a number of timing processes, such as flowering, dormancy, leaf formation, and seed germination.

phytoplankton: Autotrophic plankton.

pigment: A substance that absorbs light, often selectively.

pileus [L. *pileus*, a cap]: The caplike part of mushroom basidiomata and of certain ascomata.

pinna, *pl.* **pinnae** (pĭn'a) [L. *pinna*, feather]: A primary division, or leaflet, of a compound leaf or frond; may be divided into pinnules.

pinocytosis: *See* endocytosis.

pistil [L. *pistillum*, pestle]: A term sometimes used to refer to an individual carpel or a group of fused carpels.

pistillate: *See* carpellate.

pit: A recessed cavity in a cell wall where the secondary wall does not form.

pit membrane: The middle lamella and two primary cell walls between two pits.

pit-pair: Two opposite pits plus the pit membrane.

pith: The ground tissue occupying the center of the stem or root within the vascular cylinder; usually consists of parenchyma.

pith ray: *See* interfascicular region.

placenta, *pl.* **placentae** [L. *placenta*, cake]: The part of the ovary wall to which the ovules or seeds are attached.

placentation: The manner of ovule attachment within the ovary.

plankton [Gk. *planktos*, wandering]: Free-floating, mostly microscopic, aquatic organisms.

plaque: Clear area in a sheet of cells resulting from the killing or lysis of contiguous cells by viruses.

-plasma, plasmo-, -plast [Gk. *plasma*, form, mold]: Prefix or suffix meaning "formed," or "molded"; examples are protoplasm, "first-molded" (living matter), and chloroplast, "green-formed."

plasma membrane or **plasmalemma:** Outer boundary of the proto-

plast, next to the cell wall; consists of a single membrane; also called cell membrane, and ectoplast.

plasmid: A relatively small fragment of DNA that can exist free in the cytoplasm of a bacterium and can be integrated into and then replicated with a chromosome. Plasmids make up about 5 percent of the DNA of many bacteria, but are rare in eukaryotes.

plasmodesma, *pl.* **plasmodesmata** [Gk. *plasma*, form, + *desma*, bond]: The minute cytoplasmic threads that extend through openings in cell walls and connect the protoplasts of adjacent living cells.

plasmodium (plăz·mo'di·ŭm): Stage in life cycle of myxomycetes (plasmodial slime molds); a multinucleate mass of protoplasm surrounded by a membrane.

plasmogamy [Gk. *plasma*, form, + *gamos*, marriage]: Union of the protoplasts of gametes that is not accompanied by union of their nuclei.

plasmolysis (plăz·mŏl'ĭ·sĭs) [Gk. *plasma*, form, + *lysis*, a loosening]: The separation of the protoplast from the cell wall because of the removal of water from the protoplast by osmosis.

plastid (plăs'tĭd): Organelle in the cells of certain groups of eukaryotes that is the site of such activities as food manufacture and storage; plastids are bounded by two membranes.

pleiotropy [Gk. *pleros*, more, + *trope*, a turning]: The capacity of a gene to affect more than one phenotypic characteristic.

plumule [L. *plumula*, a small feather]: The first bud of an embryo; the portion of the young shoot above the cotyledons.

pneumatophores [Gk. *pneuma*, breath, + *-phoros*, carrying]: Negatively geotropic extensions of the root systems of some trees growing in swampy habitats; they grow upward and out of the water and probably function to assure adequate aeration.

point mutation: An alternation in one of the nucleotides in a chromosomal DNA molecule.

polar molecule: A molecule with positively and negatively charged ends.

polar nuclei: Two nuclei (usually), one derived from each end (pole) of the embryo sac, which become centrally located; they fuse with a male nucleus to form the primary ($3n$) endosperm nucleus.

pollen [L. *pollen*, fine dust]: A collective term for pollen grains.

pollen grain: A microspore containing a mature or immature microgametophyte (male gametophyte); pollen grains occur in seed plants.

pollen mother cell: *See* microspore mother cell.

pollen sac: A cavity in the anther that contains the pollen grains.

pollen tube: A tube formed after germination of the pollen grain; carries the male gametes into the ovule.

pollination: In angiosperms, the transfer of pollen from an anther to a stigma. In gymnosperms, the transfer of pollen from a pollen-producing cone directly to an ovule.

poly- [Gk. *polys*, many]: Prefix meaning "many."

polyembryony: Having more than one embryo within the developing seed.

polygenic inheritance: The inheritance of quantitative characteristics determined by the combined effects of multiple genes.

polymer (pŏl'ĭ·mer): A large molecule composed of many similar molecular subunits.

polymerization: The chemical union of monomers such as glucose or nucleotides to form polymers such as starch or nucleic acid.

polynucleotide: A single-stranded DNA or RNA molecule.

polypeptide: A molecule composed of amino acids linked together by peptide bonds, not as complex as a protein.

polyploid (pŏl'ĭ·ploid): Referring to an organism, tissue, or cell with more than two complete sets of chromosomes.

polysaccharide: A polymer composed of many monosaccharide units joined in a long chain, such as glycogen, starch, and cellulose.

polysome or **polyribosome:** An aggregation of ribosomes actively involved in the translation of the same RNA molecule, one after another.

pome (pŏm) [Fr. *pomme*, apple]: A simple fleshy fruit, the outer portion of which is formed by the floral parts that surround the ovary and expand with the growing fruit; found only in one subfamily of the Rosaceae (apples, pears, quince, pyracantha, and so on).

population: Any group of individuals, usually of a single species, occupying a given area at the same time.

P-protein: Phloem-protein; a proteinaceous substance found in cells of angiosperm phloem, especially in sieve-tube members; also called slime.

preprophase band: A ringlike band of microtubules, found just beneath the plasma membrane, that delimits the equatorial plane of the future mitotic spindle of a cell preparing to divide.

primary endosperm nucleus: The result of the fusion of a sperm nucleus and the two polar nuclei.

primary growth: In plants, growth originating in the apical meristems of shoots and roots, as contrasted with secondary growth.

primary meristem or **primary meristematic tissue** (měr'ĭ-ste·măt'ĭk): A tissue derived from the apical meristem; of three kinds: protoderm, procambium, and ground meristem.

primary pit-field: Thin area in a primary cell wall through which plasmodesmata pass, although plasmodesmata may occur elsewhere in the wall as well.

primary plant body: The part of the plant body arising from the apical meristems and their derivative meristematic tissues; composed entirely of primary tissues.

primary root: The first root of the plant, developing in continuation of the root tip or radicle of the embryo; in gymnosperms and dicots it becomes the tap root.

primary thickening meristem: In many monocotyledons, the meristem responsible for the increase in thickness of the shoot axis.

primary tissues: Cells derived from the apical meristems and primary meristematic tissues of root and shoot; as opposed to secondary tissues derived from a cambium; primary growth results in an increase in length.

primary wall: The wall layer deposited during the period of cell expansion.

primordium, *pl.* **primordia** [L. *primus*, first, + *ordiri*, to begin to weave]: A cell or organ in its earliest stage of differentiation.

pro- [Gk. *pro*, before]: Prefix meaning "before" or "prior to."

procambium (pro·kăm'bĭ·ŭm) [L. *pro*, before, + *cambiare*, to exchange]: A primary meristematic tissue that gives rise to primary vascular tissues.

proembryo: Embryo in early stages of development, before the embryo proper and suspensor become distinct.

prokaryote [Gk. *pro*, before, + *karyon*, kernel]: A cell lacking a membrane-bound nucleus and membrane-bound organelles; a bacterium.

prolamellar body: Semicrystalline body found in plastids arrested in development by the absence of light.

promoter: A specific segment of DNA to which RNA polymerase attaches to initiate transcription of mRNA from an operon.

prophage (prō'fāj'): Noninfectious phage units linked with the bacterial chromosome which multiply with the growing and dividing bacteria but do not bring about their lysis. Prophage is a stage in the life cycle of a temperate phage.

prophase [Gk. *pro*, before, + *phasis*, form]: An early stage in nuclear division, characterized by the shortening and thickening of the chromosomes and their movement to the metaphase plate.

proplastid: A minute self-reproducing body in the cytoplasm from which a plastid develops.

prop roots: Adventitious roots arising from the stem above soil level and helping to support the plant: common in many monocots, for example, corn (*Zea mays*).

prosthetic group: A heat-stable metal ion or an inorganic group (other than an amino acid) that is bound to a protein and serves as its active group.

protease (prō'te·ās): An enzyme that digests protein by the hydrolysis of peptide bonds; proteases are also called peptidases.

protein [Gk. *proteios*, primary]: A complex organic compound composed of many (100 or more) amino acids joined by peptide bonds.

prothallial cell [Gk. *pro*, before, + *thallos*, sprout]: The sterile cell or cells found in the male gametophytes, or microgametophytes, of vascular plants other than angiosperms; believed to be remnants of the vegetative tissue of the male gametophyte.

prothallus (pro·thăl'ŭs): In homosporous vascular plants, such as ferns, the more or less independent, photosynthetic gametophyte; also called the prothallium.

protist: A member of the kingdom Protista.

proto- [Gk. *protos*, first]: Prefix meaning "first," for example, Protozoa, "first animals."

protoderm [Gk. *protos*, first, + *derma*, skin]: Primary meristematic tissue that gives rise to epidermis.

proton: A subatomic, or elementary, particle, with a single positive charge equal in magnitude to the charge of an electron and a mass of 1; the basic component of every atomic nucleus.

protonema, *pl.* **protonemata** [Gk. *protos*, first, + *nema*, a thread]: The first stage in development of the gametophyte of mosses and certain liverworts; protonemata may be filamentous or platelike.

protoplasm: A general term for the living substance of all cells, but not their organelles.

protoplast: The protoplasm of an individual cell; in plants, the unit of protoplasm inside the cell wall.

protostele [Gk. *protos*, first, + *stele*, pillar]: The simplest type of stele, consisting of a solid column of vascular tissue.

protoxylem: The first part of the primary xylem, which matures during elongation of the plant part in which it is found.

proximal (prŏk'sĭ·măl) [L. *proximus*, near]: Situated near the point of reference, usually the main part of a body or the point of attachment; opposite of distal.

pseudo- [Gk. *pseudes*, false]: Prefix meaning "false."

pseudoplasmodium: A multicellular mass of individual amoeboid cells, representing the aggregate phase in the cellular slime molds.

pulvinus, *pl.* **pulvini:** Jointlike thickening at the base of the petiole of a leaf or petiolule of a leaflet, and having a role in the movements of the leaf or leaflet.

pumps: Transport proteins driven by either chemical energy (ATP) or light energy; in plant and fungal cells, they typically are proton pumps.

punctuated equilibrium: A model of evolutionary change that proposes that there are long periods of little or no change punctuated with brief intervals of rapid change.

purine (pū'rēn): The larger of the two kinds of nucleotide base found in DNA and RNA; a nitrogenous base with a double-ring structure, such as adenine or guanine.

pyramid of energy: Energy relationships among various feeding levels involved in a particular food chain; autotrophs (at the base of the pyramid) represent the greatest amount of available energy; herbivores are next; then primary carnivores; secondary carnivores; and so forth. Similar pyramids of mass, size, and number also occur in natural communities.

pyrenoid [Gk. *pyren*, the stone of a fruit, + *oides*, like]: Differentiated regions of the chloroplast that are centers of starch formation in green algae and hornworts.

pyrimidine: The smaller of the two kinds of nucleotide base found in DNA and RNA: a nitrogenous base with a single-ring structure, such as cytosine, thymine, or uracil.

quantasome [L. *quantus*, how much, + [Gk. *soma*, body]: Granules located on the inner surfaces of chloroplast lamellae; believed to be involved in the light-dependent reactions of photosynthesis.

quantum: The ultimate unit of light energy.

quiescent center: The relatively inactive initial region in the apical meristem of a root.

raceme [L. *racemus*, bunch of grapes]: An indeterminate inflorescence in which the main axis is elongated but the flowers are borne on pedicels that are about equal in length.

rachis (rā'kĭs) [Gk. *rachis*, a backbone]: Main axis of a spike; the axis of a fern leaf (frond), from which the pinnae arise; in compound leaves, the extension of the petiole corresponding to the midrib of an entire leaf.

radially symmetrical: *See* actinomorphic.

radial section: A longitudinal section cut parallel to the radius of a cylindrical body, such as a root or stem; in the case of secondary xylem, or wood, and secondary phloem, parallel to the rays.

radial system: In secondary xylem and secondary phloem, the term applied to all the rays, the cells of which are derived from ray initials; also called the horizontal system, or ray system.

radicle [L. *radix*, root]: The embryonic root.

radioisotope: An unstable isotope of an element that decays or disintegrates spontaneously, emitting radiation; also called a radioactive isotope.

raphe [Gk. *raphe*, seam]: (1) Ridge on seeds, formed by the stalk of the ovule, in those seeds in which the stalk is sharply bent at the base of the ovule; (2) groove on the shell, or frustule, of a diatom.

raphides (răf'ĭ-dēz) [Gk. *rhaphis*, a needle]: Fine, sharp, needlelike crystals of calcium oxalate found in the vacuoles of many plant cells.

ray flowers: *See* disk flowers.

ray initial: An initial in the vascular cambium that gives rise to the ray cells of secondary xylem and secondary phloem.

reaction center: The chlorophyll molecule of a photosystem capable of using energy in the photochemical reaction.

reaction wood: Abnormal wood that develops in leaning trunks and limbs; *see also* compression wood and tension wood.

receptacle: That part of the axis of a flower stalk that bears the floral organs.

recessive: Describing a gene whose phenotypic expression is masked in the heterozygote by a dominant allele; heterozygotes are phenotypically indistinguishable from dominant homozygotes.

recombinant DNA: DNA formed either naturally or in the laboratory by the joining of segments of DNA from different sources.

reduction [L. *reductio*, a bringing back; originally "bringing back" a metal from its oxide]: Gain of an electron by an atom; reduction takes place simultaneously with oxidation (the loss of an electron by an atom), because an electron that is lost by one atom is accepted by another.

regular: *See* actinomorphic.

regulator gene: A gene that prevents or represses the activity of the structural genes in an operon.

replicate: Produce a facsimile or a very close copy; used to indicate the production of a second molecule of DNA exactly like the first molecule or of a sister chromatid.

replication fork: In DNA synthesis, the Y-shaped structure formed at the point where the two strands of the original molecule are being separated and the complementary strands are being synthesized.

repressor: A protein that regulates DNA transcription; this occurs because RNA polymerase is prevented from attaching to the promoter and transcribing the gene. *See* operator.

resin duct: A tubelike intercellular space lined with resin-secreting cells (epithelial cells) and containing resin.

respiration: An intracellular process in which molecules, particularly pyruvate in the Krebs cycle, are oxidized with the release of energy. The complete breakdown of sugar or other organic compounds to carbon dioxide and water is termed aerobic respiration, although the first steps of this process are anaerobic.

restriction enzymes: Enzymes that cleave the DNA double helix at specific nucleotide sequences.

reticulate venation: *See* netted venation.

reverse transcription: The process by which an RNA molecule is used as a template to make a single-stranded copy of DNA.

rhizobia [Gk. *rhiza*, root, + *bios*, life]: Bacteria of the genera *Rhizobium* or *Bradyrhizobium*, which may be involved with leguminous plants in a symbiotic relationship that results in nitrogen fixation.

rhizoids [Gk. *rhiza*, root]: (1) Branched rootlike extensions of fungi and algae that absorb water, food, and nutrients; (2) root-hairlike structures in liverworts, mosses, and some vascular plants, which occur on free-living gametophytes.

rhizome: A more or less horizontal underground stem.

ribonucleic acid (RNA): Type of nucleic acid formed on chromosomal DNA and involved in protein synthesis; composed of chains of phosphate, sugar molecules (ribose), and purines and pyrimidines; RNA is the genetic material of many kinds of viruses.

ribose: A five-carbon sugar; a component of RNA.

ribosome: A small particle composed of protein and RNA; the site of protein synthesis.

ring-porous wood: A wood in which the pores, or vessels, of the early wood are distinctly larger than those of the late wood, forming a well-defined ring in cross sections of the wood.

RNA *See* ribonucleic acid.

root: The usually descending axis of a plant, normally below ground, which serves to anchor the plant and to absorb and conduct water and minerals into it.

root cap: A thimblelike mass of cells that covers and protects the growing tip of a root.

root hairs: Tubular outgrowths of epidermal cells of the root in the zone of maturation.

root pressure: The pressure developed in roots as the result of osmosis, which causes guttation of water from leaves and exudation from cut stumps.

Rubisco: RuBP carboxylase, the enzyme that catalyzes initial reaction of the Calvin cycle, involving the fixation of carbon dioxide to ribulose 1,5-bisphosphate (RuBP).

runner: *See* stolon.

samara: Simple, dry, one-seeded or two-seeded indehiscent fruit with pericarp-bearing, winglike outgrowths.

sap: (1) A name applied to the fluid contents of the xylem or the sieve elements of the phloem; (2) the fluid contents of the vacuole are called cell sap.

saprobe [Gk. *sapros*, rotten, + *bios*, life]: An organism that secures its food directly from nonliving organic matter.

sapwood: Outer part of the wood of stem or trunk, usually distinguised from the heartwood by its lighter color, in which active conduction of water takes place.

satellite DNA: A short nucleotide sequence repeated in tandem fashion many thousands of times; this region of the chromosome has a distinctive base composition and is not transcribed.

savanna: Grassland containing scattered trees.

scarification: The process of cutting or softening a seed coat to hasten germination.

schizo- [Gk. *schizein*, to split]: Prefix meaning "split."

schizocarp (skĭz′o·kärp): Dry simple fruit with two or more united carpels that split apart at maturity.

sclereid [Gk. *skleros*, hard]: A sclerenchyma cell with a thick, lignified secondary wall having many pits. Sclereids are variable in form but typically not very long; they may or may not be living at maturity.

sclerenchyma (skle·rĕng′kĭ·má) [Gk. *skleros*, hard, + L. *enchyma*, infusion]: A supporting tissue composed of sclerenchyma cells, including fibers and sclereids.

sclerenchyma cell: Cell of variable form and size with more or less thick, often lignified, secondary walls; may or may not be living at maturity; includes fibers and sclereids.

scutellum (sku·tĕl′ŭm) [L. *scutella*, a small shield]: The single cotyledon of a grass embryo, specialized for absorption of the endosperm.

secondary growth: In plants, growth derived from secondary or lateral meristems, the vascular and cork cambiums; secondary growth results in an increase in girth, and is contrasted with primary growth, which results in an increase in length.

secondary plant body: The part of the plant body produced by the vascular cambium and the cork cambium; consists of secondary xylem, secondary phloem, and periderm.

secondary root: *See* lateral root.

secondary tissues: Tissues produced by the vascular cambium and cork cambium.

secondary wall: Innermost layer of the cell wall, formed in certain cells after cell elongation has ceased; secondary walls have a highly organized microfibrillar structure.

seed: A structure formed by the maturation of the ovule of seed plants following fertilization.

seed coat: The outer layer of the seed, developed from the integuments of the ovule.

seedling: A young sporophyte, which develops from a germinating seed.

segregation: The separation of the chromosomes (and genes) from different parents at meiosis; Mendel's first law.

semipermeable membrane: A membrane that is permeable to water but not to solutes.

sepal (se′păl) [L. *sepalum*, a covering]: One of the outermost flower structures, a unit of the calyx; sepals usually enclose the other flower parts in the bud.

septate [L. *septum*, fence]: Divided by cross walls into cells or compartments.

sessile (sĕs′ĭl) [L. *sessilis*, of or fit for sitting, low, dwarfed]: Attached directly by the base; referring to a leaf lacking a petiole or to a flower or fruit lacking a pedicel.

seta, *pl.* **setae** (sē′ta) [L. *seta*, bristle]: In bryophytes, the stalk that supports the capsule, if present; part of the sporophyte.

sexual reproduction: The fusion of gametes followed by meiosis and recombination at some point in the life cycle.

sheath: (1) The base of a leaf that wraps around the stem, as in grasses; (2) a tissue layer surrounding another tissue, such as a bundle sheath.

shoot: The above-ground portions, such as the stem and leaves, of a vascular plant.

short-day plants: Plants that must be exposed to light periods shorter than some critical length for flowering to occur; they usually flower in autumn.

shrub: A perennial woody plant of relatively low stature, typically with several stems arising from or near the ground.

sieve area: A portion of the sieve-element wall containing clusters of pores through which the protoplasts of adjacent sieve elements are interconnected.

sieve cell: A long, slender sieve element with relatively unspecialized sieve areas and with tapering end walls that lack sieve plates; found in the phloem of gymnosperms and seedless vascular plants.

sieve element: The cell of the phloem that is involved in the long-distance transport of food substances; further classified into sieve cells and sieve-tube members.

sieve plate: The part of the wall of sieve-tube members bearing one or more highly differentiated sieve areas.

sieve tube: A series of sieve-tube members arranged end-to-end and interconnected by sieve plates.

sieve-tube member: One of the component cells of a sieve tube; found primarily in flowering plants and typically associated with a companion cell; also called sieve-tube element.

silique [L. *siliqua*, pod]: The fruit characteristic of the mustard family; two-celled, the valves splitting from the bottom and leaving the placentae with the false partition stretched between. A small, compressed silique is called a silicle.

simple fruit: A fruit derived from one carpel or several united carpels.

simple leaf: An undivided leaf; as opposed to a compound leaf.

simple pit: A pit not surrounded by an overarching border of secondary wall; as opposed to a bordered pit.

siphonaceous, siphonous [Gk. *siphon*, a tube, pipe]: In algae, multinucleate cells without cross walls; coenocytic.

siphonostele [Gk. *siphon*, pipe, + *stele*, pillar]: A type of stele containing a hollow cylinder of vascular tissue surrounding a pith.

slime: See P-protein.

softwood: A name commonly applied to the wood of a conifer.

solar tracking: The ability of the leaves and flowers of many plants to move diurnally, orienting themselves either perpendicular or parallel to the sun's direct rays; also called heliotropism.

solute: A molecule dissolved in a solution.

solute potential: *See* osmotic potential.

solution: Usually liquid, in which the molecules of the dissolved substance, the solute (e.g., sugar), are dispersed between the molecules of the solvent (e.g., water).

somatic cells [Gk. *soma*, body]: All cells except the gametes and the cells from which the gametes develop.

soredium, *pl.* **soredia** [Gk. *soros*, heap]: A specialized reproductive unit of lichens, consisting of a few cyanobacterial or green algal cells surrounded by fungal hyphae.

sorus, *pl.* **sori** (sō'rŭs) [Gk. *soros*, heap]: A group or cluster of sporangia or spores.

specialized: (1) Of organisms, having special adaptations to a particular habitat or mode of life; (2) of cells, having particular functions.

species, *pl.* **species** [L. kind, sort]: A kind of organism; species are designated by binomial names written in italics.

specific epithet: The second part of a species name; for example, *mays* of *Zea mays*, which is corn.

specificity: Uniqueness, as in proteins in given organisms or enzymes in given reactions.

sperm: A mature male gamete, usually motile and smaller than the female gamete.

spermatangium, *pl.* **spermatangia** (spûr'ma·tăn'ji·ŭm) [Gk. *sperma*, sperm, + L. *tangere*, to touch]: In the red algae, the structure that produces spermatia.

spermatium, *pl.* **spermatia** [Gk. *sperma*, sperm]: In the red algae and some fungi, a minute, nonmotile male gamete.

spermatogenous cell: The cell of the male gametophyte, or pollen grain, of gymnosperms, which divides mitotically to form two sperm.

spermatophyte [Gk. *sperma*, seed, + *phyton*, plant]: A seed plant.

spermogonium, *pl.* **spermogonia** (spûr'ma·gō'ni·ŭm) [Gk. *sperma*, sperm, + *gonos*, offspring]: In the rust fungi, the structure that produces spermatia.

spherosome: Single, membrane-bound spherical structures in the cytoplasm of plant cells, many of which contain mostly lipids and apparently are centers of lipid synthesis and accumulation.

spike [L. *spica*, head of grain]: An indeterminate inflorescence in which the main axis is elongated and the flowers are sessile.

spikelet: The unit of inflorescence in grasses; a small group of grass flowers.

spindle fibers: A group of microtubules that extend from the centromeres of the chromosomes to the poles of the spindle or from pole to pole in a dividing cell.

spine: A hard, sharp-pointed structure; usually a modified leaf, or part of a leaf.

spirillum, *pl.* **spirilli** [L. *spira*, coil]: A long coiled or spiral bacterium.

spongy parenchyma: A leaf tissue composed of loosely arranged, chloroplast-bearing cells.

sporangiophore (spo·răn'ji·o·fōr') [Gk. *spora*, seed, + *pherein*, to carry]: A branch bearing one or more sporangia.

sporangium, *pl.* **sporangia** (spo·răn'ji·ŭm) [Gk. *spora*, seed, + *angeion*, a vessel]: A hollow unicellular or multicellular structure in which spores are produced.

spore: A reproductive cell, usually unicellular, capable of developing into an adult without fusion with another cell.

spore mother cell: A diploid (2*n*) cell that undergoes meiosis and produces (usually) four haploid cells (spores) or four haploid nuclei.

sporophyll (spō'ro·fĭl): A modified leaf or leaflike organ that bears sporangia; applied to the stamens and carpels of angiosperms, fertile fronds of ferns, and other similar structures.

sporophyte (spōr'ro·fĭt): The spore-producing, diploid (2*n*) phase in a life cycle characterized by alternation of generations.

sporopollenin: The tough substance of which the exine, or outer wall, of spores and pollen grains is composed.

stalk cell: See sterile cell.

stamen (stā'měn) [L. *stamen*, thread]: The part of the flower producing the pollen, composed (usually) of anther and filament; collectively, the stamens make up the androecium.

staminate (stăm'ĭ·nat): Pertaining to a flower having stamens but no functional carpels.

starch [M.E. *sterchen*, to stiffen]: A complex insoluble carbohydrate, the chief food storage substance of plants; composed of a thousand or more glucose units.

statoliths [Gk. *statos*, stationary, + *lithos*, stone]: Gravity sensors; starch grains or other bodies in the cytoplasm.

stele (stē'le) [Gk. *stele*, a pillar]: The central cylinder, inside the cortex, of roots and stems of vascular plants.

stem: The part of the axis of vascular plants that is above ground, as well as anatomically similar portions below ground, such as rhizomes or corms.

stem bundle: Vascular bundle belonging to the stem.

sterigma, *pl.* **sterigmata** [Gk. *sterigma*, a prop]: A small, slender protuberance of a basidium, bearing a basidiospore.

sterile cell: One of two cells produced by division of the generative cell in developing pollen grains of gymnosperms; it is not a gamete, and eventually degenerates.

stigma: (1) The region of a carpel that serves as a receptive surface for pollen grains and on which they germinate; (2) a light-sensitive eyespot, found in some kinds of algae.

stipe: A supporting stalk, such as the stalk of a gill fungus or the leaf stalk of a fern.

stipule (stĭp'ŭl): An appendage, often leaflike, that occurs on either side of the basal part of a leaf, or encircles the stem, in many kinds of flowering plants.

stolon (stō'lŏn) [L. *stolo*, shoot]: A stem that grows horizontally along the ground surface and may form adventitious roots, such as the runners of a strawberry plant.

stoma, *pl.* **stomata** (stō'ma) [Gk. *stoma*, mouth]: A minute opening bordered by guard cells in the epidermis of leaves and stems through which gases pass; also used to refer to the entire stomatal apparatus—the guard cells plus their included pore.

stratification: The process of exposing seeds to low temperatures for an extended period before attempting to germinate them at warm temperatures.

strobilus, *pl.* **strobili** (strōb'ĭ·lŭs) [Gk. *strobilos*, a cone]: A reproductive structure consisting of a number of modified leaves (sporophylls) or ovule-bearing scales grouped terminally on a stem; a cone. Strobili occur in many kinds of gymnosperms, lycophytes, and sphenophytes.

stroma [Gk. *stroma*, anything spread out]: The ground substance of plastids.

structural gene: Any gene that codes for a protein; in distinction to regulatory genes.

style [Gk. *stylos*, column]: A slender column of tissue that arises from the top of the ovary and through which the pollen tube grows.

sub- [L. *sub*, under, below]: Prefix meaning "under" or "below"; for example, subepidermal, "underneath the epidermis."

suberin (su'ber·in) [L. *suber*, the cork oak]: Fatty material found in the cell walls of cork tissue and in the Casparian strip of the endodermis.

subsidiary cell: An epidermal cell morphologically distinct from other epidermal cells and associated with a pair of guard cells; also called an accessory cell.

subspecies: The primary taxonomic subdivision of a species. Varieties are used as equivalent to subspecies by some botanists, or subspecies may be divided into varieties.

substrate [L. *substatus*, strewn under]: The foundation to which an organism is attached; the substance acted on by an enzyme.

substrate phosphorylation: Phosphorylation—the formation of ATP from ADP and inorganic phosphate—that takes place during glycolysis.

succession: In ecology, the orderly progression of changes in community composition that occurs during the development of vegetation in any area, from initial colonization to the attainment of the climax typical of a particular geographic area.

succulent: A plant with fleshy, water-storing stems or leaves.

sucker: A sprout produced by the roots of some plants and that gives rise to a new plant; erect sprouts that occur from the base of stems.

sucrase (su'kras): An enzyme that hydrolyzes sucrose into glucose and fructose; also called invertase.

sucrose (su'kros): A disaccharide (glucose plus fructose) found in many plants; the primary form in which sugar produced by photosynthesis is translocated.

superior ovary: An ovary that is free and separate from the calyx.

suspension: A heterogeneous dispersion in which the dispersed phase consists of solid particles sufficiently large that they will settle out of the fluid dispersion medium under the influence of gravity.

suspensor: A structure at the base of the embryo in many vascular plants. In some plants, it pushes the embryo into nutrient-rich tissue of the female gametophyte.

symbiosis (sim'bi·o'sis) [Gk. *syn*, together with, + *bios*, life]: The living together in close association of two or more dissimilar organisms; includes parasitism (in which the association is harmful to one of the organisms) and mutualism (in which the association is advantageous to both).

symplast [Gk. *syn*, together with, + *plastos*, molded]: The interconnected protoplasts and their plasmodesmata; the movement of substances in the symplast is called symplastic movement, or symplastic transport.

sympodium, *pl.* **sympodia:** A stem bundle and its associated leaf traces.

syn-, sym- [Gk. *syn*, together with]: Prefix meaning "together."

synapsis [Gk. *synapsis*, a contract or union]: The pairing of homologous chromosomes that occurs prior to the first meiotic division; crossing-over occurs during synapsis.

synergids (si·nur'jids): Two short-lived cells lying close to the egg in the mature embryo sac of the ovule of flowering plants.

syngamy [Gk. *syn*, together with, + *gamos*, marriage]: *See* fertilization.

synthesis: The formation of a more complex substance from simpler ones.

systematics: Scientific study of the kinds and diversity of organisms and of the relationships between them.

taiga: The northern coniferous forest.

tandem clusters: Multiple copies of the same gene lying side by side in a series.

tangential section: A longitudinal section cut at right angles to the radius of a cylindrical structure, such as a root or stem; in the case of secondary xylem, or wood, and secondary phloem, at right angles to the rays.

tapetum (ta·pe'tum) [Gk. *tapes*, a carpet]: Nutritive tissue in the sporangium, particularly an anther.

taproot: The primary root of a plant formed in direct continuation with the root tip or radicle of the embryo: forms a stout, tapering main root from which arise smaller, lateral branches.

taxon: General term for any one of the taxonomic categories, such as species, class, order, or division.

taxonomy [Gk. *taxis*, arrangement, + *nomos*, law]: The science of the classification of organisms.

teliospore (te'li·o·spor'): In the rust fungi, a thick-walled spore in which karyogamy and meiosis occur and from which basidia develop.

telium, *pl.* **telia:** In the rust fungi, the structure that produces teliospores.

telophase: The last stage in mitosis and meiosis, during which the chromosomes become reorganized into two new nuclei.

template: A pattern or mold guiding the formation of a negative or complement; a term applied especially to DNA duplication, which is explained in terms of a template hypothesis.

tendril [L. *tendere*, to extend]: A modified leaf or part of a leaf or stem modified into a slender coiling structure that aids in support of the stems; tendrils occur only in some angiosperms.

tension wood: The reaction wood of dicots; develops on the upper side of leaning trunks and limbs.

tepal: One of the units of a perianth that is not differentiated into sepals and petals.

test cross: A cross of a dominant with a homozygous recessive; used to determine whether the dominant is homozygous or heterozygous.

tetrad (tet'rad): A group of four spores formed from a spore mother cell by meiosis.

tetraploid (tet'ra·ploid) [Gk. *tetra*, four, + *ploos*, fold]: Twice the usual, or diploid (2n), number of chromosomes (that is, 4n).

tetrasporangium, *pl.* **tetrasporangia** [Gk. *tetra*, four, + *spora*, seed, + *angeion*, vessel]: In certain red algae, a sporangium in which meiosis occurs, resulting in the production of tetraspores.

tetraspore [Gk. *tetra*, four, + *spora*, seed]: In certain red algae, the four spores formed by meiotic division in the tetrasporangium of a spore mother cell.

tetrasporophyte [Gk. *tetra*, four, + *spora*, seed, + *phyton*, plant]: In certain red algae, a diploid individual that produces tetrasporangia.

texture: Of wood, refers to the relative size and amount of variation in size of elements within the growth rings.

thallophyte: A term previously used to designate fungi and algae collectively, now largely abandoned.

thallus (thăl′ŭs) [Gk. *thallos*, a sprout]: A type of body that is undifferentiated into root, stem, or leaf; the word *thallus* was used commonly when fungi and algae were considered to be plants, to distinguish their simple construction, and that of certain gametophytes, from the differentiated bodies of plant sporophytes and the elaborate gametophytes of the bryophytes.

theory [Gk. *theorein*, to look at]: A well-tested hypothesis; one unlikely to be rejected by further evidence.

thermodynamics [Gk. *therme*, heat, + *dynamis*, power]: The study of energy exchanges, using heat as the most convenient form of measurement of energy. The first law of thermodynamics states that in all processes, the total energy of the universe remains constant. The second law of thermodynamics states that the entropy, or degree of randomness, tends to decrease.

thorn: A hard, woody, pointed branch.

thigmotropism [Gk. *thigma*, touch]: A response to contact with a solid object.

thylakoid [Gk. *thylakos*, sac, + *oides*, like]: A saclike membranous structure in cyanobacteria and the chloroplasts of eukaryotic organisms; in chloroplasts, stacks of thylakoids form the grana; chlorophylls are found within the thylakoids.

thymine: A pyrimidine occurring in DNA but not in RNA; *see also* uracil.

tissue: A group of similar cells organized into a structural and functional unit.

tissue culture: A technique for maintaining fragments of plant or animal tissue alive in a medium after removal from the organism.

tissue system: A tissue or group of tissues organized into a structural and functional unit in a plant or plant organ. There are three tissue systems: dermal, vascular, and ground, or fundamental.

tonoplast [Gk. *tonos*, stretching, tension, + *plastos*, formed, molded]: The cytoplasmic membrane surrounding the vacuole in plant cells; also called vacuolar membrane.

torus, *pl.* **tori:** The central thickened part of the pit-membrane in the bordered pits of conifers and some other gymnosperms.

totipotent: Said of tissues or cells that are capable of developing into any structure of the mature plant.

tracheary element: The general term for a water-conducting cell in vascular plants; tracheids and vessel members.

tracheid (trā′ke·ĭd): An elongated, thick-walled conducting and supporting cell of xylem. It has tapering ends and pitted walls without perforations, as contrasted with a vessel member. Found in nearly all vascular plants.

transcription: The enzyme-catalyzed assembly of an RNA molecule complementary to a strand of DNA.

transduction: The transfer of genes from one organism to another by a virus.

transfer cell: Specialized parenchyma cell with wall in-growths that increase the surface area of the plasma membrane; apparently functions in the short-distance transfer of solutes.

transfer RNA (tRNA): Low-molecular-weight RNA that becomes attached to an amino acid and guides it to the correct position on the ribosome for protein synthesis; there is at least one tRNA molecule for each amino acid.

transformation: The transfer of naked DNA from one organism to another; also called "gene transfer." Transposons are often used as vectors in transformation when it is carried out in the laboratory.

transgenic organism: An organism whose genome contains DNA—from the same or a different species—that has been modified by the methods of genetic engineering.

transition region: The region in the primary plant body showing transitional characteristics between structures of root and shoot.

translation: The assembly of a protein on the ribosomes; mRNA is used to direct the order of the amino acids.

translocation: (1) In plants, the long-distance transport of water, minerals, or food; most often used to refer to food transport; (2) in genetics, the interchange of chromosome segments between nonhomologous chromosomes.

transmembrane proteins: Globular proteins that traverse the lipid bilayer of cellular membranes. Some extend across the lipid bilayer as a single alpha helix and others as multiple alpha helices.

transpiration [Fr. *transpirer*, to perspire]: The loss of water vapor by plant parts; most transpiration occurs through stomata.

transport protein: A specific membrane protein responsible for transferring solutes across membranes; grouped into three broad classes: pumps, carriers, and channels.

transposon [L. *transponere*, to change the position of something]: A DNA sequence that carries one or more genes and is flanked by sequences of bases that confer the ability to move from one DNA molecule to another; an element capable of transposition, which is the changing of a chromosomal location.

transverse section: A section cut perpendicular, or at right angles, to the longitudinal axis of a plant part.

tree: A perennial woody plant generally with a single stem (trunk).

tricarboxylic acid cycle or **TCA cycle:** *See* Krebs cycle.

trichogyne [Gk. *trichos*, a hair, + *gyne*, female]: In the red algae and certain ascomycetes and basidiomycetes, a receptive protuberance of the female gametangium for the conveyance of spermatia.

trichome [Gk. *trichos*, hair]: An outgrowth of the epidermis, such as a hair, scale, and water vesicle.

triose [Gk. *tries*, three, + *ose*, suffix indicating a carbohydrate]: Any three-carbon sugar.

triple fusion: In angiosperms, the fusion of the second male gamete, or sperm, with the polar nuclei, resulting in formation of a primary endosperm nucleus, which is triploid ($3n$) in most groups.

triploid [Gk. *triploos*, triple]: Having three complete chromosome sets per cell ($3n$).

tritium: A radioactive isotope of hydrogen, ^{3}H. The nucleus of a tritium atom contains one proton and two neutrons, whereas the more common hydrogen nucleus consists only of a proton.

tRNA: *See* transfer RNA.

-troph, tropho- [Gk. *trophos*, feeder]: Suffix or prefix meaning "feeder," "feeding," or "nourishing"; for example, autotrophic, "self-nourishing."

trophic level: A step in the movement of energy through an ecosystem, represented by a particular set of organisms.

tropism [Gk. *trope*, a turning]: A response to an external stimulus in which the direction of the movement is usually determined by the direction from which the most intense stimulus comes.

tube cell: In male gametophytes, or pollen grains, of seed plants, the cell that develops into the pollen tube.

tuber [L. *tuber*, swelling]: An enlarged, short, fleshy underground stem, such as that of the potato.

tundra: A treeless circumpolar region, best developed in the Northern Hemisphere and most found north of the Arctic Circle.

tunica-corpus: The organization of the shoot apex of most angiosperms and a few gymnosperms, consisting of one or more peripheral layers of cells (the tunica layers) and an interior (the corpus). The tunica layers undergo surface growth (by anticlinal divisions), and the corpus undergoes volume growth (by divisions in all planes).

turgid (tûr'jĭd) [L. *turgidus,* a swollen]: Swollen, distended, referring to a cell that is firm due to water uptake.

turgor pressure [L. *turgor,* a swelling]: The pressure within the cell resulting from the movement of water into the cell.

tylose [Gk. *tylos,* a lump]: A balloonlike outgrowth from a ray or axial parenchyma cell through the pit in a vessel wall and into the lumen of the vessel.

type specimen: Usually a dried plant specimen housed in an herbarium; selected by a taxonomist to serve as the basis for comparison with other specimens in determining whether they are members of the same species or not.

umbel (ŭm'bĕl) [L. *umbella,* sunshade]: An inflorescence, the individual pedicels of which all arise from the apex of the peduncle.

unicellular: Composed of a single cell.

unisexual: Usually applied to a flower lacking either stamens or carpels; a perianth may be present or absent.

unit membrane: A visually definable, three-layered membrane, consisting of two dark layers separated by a lighter layer.

uracil (ū'ra·sil): A pyrimidine found in RNA but not in DNA; *see also* thymine.

uredinium, *pl.* **uredinia** [L. *uredo,* a blight]: In rust fungi, the structure that produces urediniospores.

urediniospore [L. *uredo,* a blight, + *spora,* spore]: In rust fungi, a reddish, binucleate spore produced in summer.

vacuolar membrane: *See* tonoplast.

vacuole [L. *vaccus,* empty]: A space or cavity within the cytoplasm filled with a watery fluid, the cell sap; part of the lysosomal compartment of the cell.

variation: The differences that occur within the offspring of a particular species.

variety: A group of plants or animals of less than species rank; some botanists view varieties as equivalent to subspecies, and others consider them divisions of subspecies.

vascular [L. *vasculum,* a small vessel]: Pertains to any plant tissue or region consisting of or giving rise to conducting tissue; e.g., xylem, phloem, vascular cambium.

vascular bundle: A strand of tissue containing primary xylem and primary phloem (and procambium if still present) and frequently enclosed by a bundle sheath of parenchyma or fibers.

vascular cambium: A cylindrical sheath of meristematic cells, the division of which produces secondary phloem and secondary xylem.

vascular rays: Ribbonlike sheets of parenchyma that extend radially through the wood, across the cambium, and into the secondary phloem; they are always produced by the vascular cambium.

vascular system: All the vascular tissues in their specific arrangement in a plant or plant organ.

vector [L., a bearer, carrier, from *vehere,* to carry]: (1) A pathogen that carries a disease from one organism to another; (2) in genetics, any virus or plasmid DNA into which a gene is integrated and subsequently transferred into a cell.

vegetative: Of, relating to, or involving propagation by asexual processes; also referring to nonreproductive plant parts.

vegetative reproduction: (1) In seed plants, reproduction by means other than by seeds; apomixis; (2) in other organisms, reproduction by vegetative spores, fragmentation, or division of the somatic body. Unless a mutation occurs, each daughter cell or individual is genetically identical with its parent.

vein: A vascular bundle forming a part of the framework of the conducting and supporting tissue of a leaf or other expanded organ.

velamen [L. *velumen,* fleece]: A multiple epidermis covering the aerial roots of some orchids and aroids; also occurs on some terrestrial roots.

venation: Arrangement of veins in leaf blade.

venter [L. *venter,* belly]: The enlarged basal portion of an archegonium containing the egg.

vernalization [L. *vernalis,* spring]: The induction of flowering by cold treatment.

vessel [L. *vasculum,* a small vessel]: A tubelike structure of the xylem composed of elongate cells (vessel members) placed end to end and connected by perforations. Its function is to conduct water and minerals through the plant body. Found in nearly all angiosperms and a few other vascular plants (e.g., gnetophytes).

vessel member: One of the cells composing a vessel; also called vessel element.

viable [L., *vita,* life]: Able to live.

volva [L. *volva,* a wrapper]: A cuplike structure at the base of the stalk of certain mushrooms.

wall pressure: The pressure of the cell wall exerted against the turgid protoplast; opposite and equal to the turgor pressure.

water potential: The algebraic sum of the solute potential and the pressure potential, or wall pressure; the potential energy of water.

water vesicle: An enlarged epidermal cell in which water is stored; a type of trichome.

weed [O.E. *weod,* used at least since the year 888 in its present meaning]: Generally an herbaceous plant not valued for use or beauty, growing wild, and regarded as using ground or hindering the growth of useful vegetation.

whorl: A circle of leaves or of flower parts.

wild type: In genetics, the phenotype or genotype that is characteristic of the majority of individuals of a species in a natural environment.

wood: Secondary xylem.

xanthophyll: (zăn'thō·fil) [Gk. *xanthos,* yellowish-brown, + *phyllon,* leaf]: A yellow chloroplast pigment; a member of the carotenoid class.

xerophyte [Gk. *xeros,* dry, + *phyton,* a plant]: A plant that has adapted to arid habitats.

xylem [Gk. *xylon,* wood]: A complex vascular tissue through which most of the water and minerals of a plant are conducted; characterized by the presence of tracheary elements.

zeatin (zē'a·tin): Plant hormone; a natural cytokinin isolated from corn.

zoosporangium: A sporangium bearing zoospores.

zoospore (zō'o·spŏr): A motile spore, found among algae, oomycetes, and chytrids.

zygomorphic [Gk. *zygo*, pair, + *morphe*, form]: A type of flower capable of being divided into two symmetrical halves only by a single longitudinal plane passing through the axis; also called bilaterally symmetrical.

zygospore: A thick-walled, resistant spore that develops from a zygote, resulting from the fusion of isogametes.

zygote (zī'gōt) [Gk. *zygotos*, paired together]: The diploid (2*n*) cell resulting from the fusion of male and female gametes.

Illustration Acknowledgments

All photographs not credited herein are by Ray F. Evert. All section openers are by Rhonda Nass.

Chapter 1

Opener, p. xx, Rhonda Nass; **1.1** © Biological Photo Service; **1.2** Richard E. Dickerson, "Chemical Evolution and the Origin of Life," *Scientific American*, vol. 239(3), pages 70–86, 1978; **1.3(a)** © Walter H. Hodge/Peter Arnold, Inc.; **(b)** © Larry West/Bruce Coleman, Inc.; **1.4** © Dianne Edwards; **1.5** © Doug Wechsler/ Animals Animals; **1.7** After W. Troll, *Vergleichende Morphologie der Höheren Pflanzen*, vol. 1, pt. 1, Verlag von Gebrüder Borntraeger, Berlin, 1937; **1.8(a)** © Dr. Anne La Bastille/Photo Researchers, Inc.; **(b)** © John Johnson/Earth Scenes; **(c)** © Stephen P. Parker/Photo Researchers, Inc.; **(d)** © Tom Bean/DRK Photo; **(e)** © Stephen J. Krasemann/DRK Photo; **(f)** © Gregory G. Dimijian, M.D./Photo Researchers, Inc.; **1.9** From H. Curtis and N. Sue Barnes, *Biology*, 5th ed., Worth Publishers, New York, 1989

Chapter 2

2.1 © A. Ryter; **2.2** (photo) © M. A. Walsh; **2.5(a)** © D. Branton; **2.8** © Myron C. Ledbetter; **2.9** © Damian Neuberger; **2.10** (photo) © R. R. Dute; **2.11** After W. W. Thompson and J. M. Whatley, *Annual Review of Plant Physiology*, vol. 31, pages 375–394, 1980; **2.12** © David Stetler; **2.13** © K. Esau; **2.14** © K. Esau; **2.15** © Mary Alice Webb; **Page 27** From Curtis and Barnes, *op. cit.*, 1989; after B. Alberts, D. Bray, J. Lewis, M. Raff, K. Roberts, and J. D. Watson, *Molecular Biology of the Cell*, 2nd ed., Garland Publishing, Inc., New York, 1989; **2.17(a)** © R. R. Dute; **(b)** © R. R. Dute; **2.18** After D. J. Morré and H. H. Mollenhauer, in *Dynamic Aspects of Plant Ultrastructure*, A. W. Robards (Ed.), McGraw-Hill Book Company, New York, 1974; **2.20(b)** M. U. Parthasarathy et al., *American Journal of Botany*, vol. 72, pages 1318–1323, 1985; **2.21** M. Kruatrachue and R. Evert, *American Journal of Botany*, vol. 64, pages 310–325, 1977; **2.22** © R. R. Powers; **2.23(a)** © Lewis Tilney, from Alberts et al., *op. cit.*, 1989; **Page 33** After Richard E. Williamson, *Plant Physiology*, vol. 82, pp. 631–634, 1986 and Alberts et al., *op. cit.*, 1989, Figure 11–33, page 632; **2.24** © Brian Wells and Keith Roberts, in Alberts et al., *op. cit.*, 1989, page 1140; **2.26** After K. Esau, *Anatomy of Seed Plants*, 2nd ed., John Wiley and Sons, Inc., New York, 1977; **2.27** After Alberts et al., *op cit.*, 1989; **2.29** After R. D. Preston, in Robards, *op. cit.*, 1974; **2.34** W. T. Jackson, *Physiologia Plantarum*, vol. 20, pages 20–29, 1967; **Page 41** **(a,b)** S. M. Wick and J. Duniec, *Protoplasma*, vol. 122, pages 45–55, 1984; **(c,d)** S. M. Wick, R. W. Seagull, M. Osborn, K. Weber, and B.E.S. Gunning, *Journal of Cell Biology*, vol. 89, pages 685–690, 1981; **2.36(b)** © M. J. Schibler; **2.37** © R. R. Dute; **2.38** © J. Cronshaw; **2.39** P. K. Hepler, *Protoplasma*, vol. 111, pages 121–133, 1982

Chapter 3

3.1 © Keith Kent/Peter Arnold, Inc.; **3.4(a-c)** After A. L. Lehninger, *Biochemistry*, 2nd ed., Worth Publishers, Inc., New York, 1975; © courtesy Shirley Baty, from Curtis and Barnes, *op. cit.*, 1989; **3.4(d)** © L. M. Biedler; **3.8** M. Kruatrachue and R. Evert, *American Journal of Botany*, vol. 64, pages 310–325, 1977; **3.10** After S. H. Crowdy, in *Systemic Fungicides*, O.B.E. Marsh (Ed.), Longman, Inc., New York, 1977; **3.11** © B. E. Juniper; **3.17** Courtesy Shirley Baty, from Curtis and Barnes, *op. cit.*, 1989; **3.18** Courtesy Shirley Baty, from Curtis and Barnes, *op. cit.*, 1989; adapted from R. E. Dickerson and I. Geis, *The Structure and Action of Proteins*, W. A. Benjamin, Inc., 1969, © 1969 by Dickerson and Geis; **3.19** After Gerald Karp, *Cell Biology*, McGraw-Hill Book Company, New York, 1979

Chapter 4

4.2 © Fredrick J. Dodd/Peter Arnold, Inc.; **4.3** Courtesy Shirley Baty, from Curtis and Barnes, *op. cit.*, 1989; **4.4** After Lehninger, *op. cit.*, 1975; **4.8** From Alberts et al., *op. cit.*, 1989, Figure 6–14, page 284; **4.12** © K. B. Raper; **4.13** © David G. Robinson; **4.15** E. B. Tucker, *Protoplasma*, vol. 113, pages 193–201, 1982

Chapter 5

5.1 © David R. Frazier/Photo Researchers, Inc.; **5.3** © Larry Ulrich/DRK Photo; **5.5** © Thomas A. Steitz; **5.6** From G. Schneider, Y. Lindqvist, and T. Lundqvist, *Journal of Molecular Biology*, vol. 211, pages 989–1008, 1990, drawing by Ulla Uhlin; **5.7** Redrawn from W. M. Becker and D. W. Deamer, *The World of the Cell*, 2nd ed., Benjamin/Cummings, Redwood City, Calif., 1991; **5.9** After Curtis and Barnes, *op. cit.*, 1989

Chapter 6

6.2 M. E. Nuttall, DuPont Experimental Station, Wilmington, Del., from Raskin et al., *Science*, vol. 237, pages 1601–1602, 1987; **6.6** After A. J. Vander, J. N. Sherman, and D. S. Luciano, *Human Physiology*, 5th ed., McGraw-Hill Book Company, New York, 1990; **6.7** After Lehninger, *op. cit.*, 1975; **6.10** After Lehninger, *op. cit.*, 1975; **6.13(a)** From Becker and Deamer, *op. cit.*, 1991; **6.13(b)** © John N. Telford; **Page 97** © M. P. Price/Bruce Coleman, Inc.; **6.14** After P. C. Hinckle and R. E. McCarty, *Scientific American*, vol. 238, pages 104–123, 1978; **6.15 6.16(b)** © Grant Heilman Photography

Chapter 7

7.2 © Paul W. Johnson /Biological Photo Service; **7.3** © Colin Milkins/Oxford Scientific Films; **7.4** From Peter Gray, *Psychology,*

Worth Publishers, New York, 1991; **7.5** © Linda Graham; **7.7** Prepared by Govindjee; **7.10** © D. Branton; **7.15** S. G. Pallardy and T. T. Kozlowski, *New Phytologist*, vol. 85, pages 363–368, 1980; **7.23** After J. A. Teeri and L. G. Stowe, *Oecologia*, vol. 23, pages 1–12, 1976

Chapter 8

8.1 © V. Orel, The Moravian Museum, Brno, Czechoslovakia; **8.2(a,b)** © Don Kyhos; **8.3(a)** © John Bova/Photo Researchers, Inc.; **(b)** © Arnold Sparrow; **8.4** © P. B. Moens; **8.5** © B. John; **8.6** © G. Ostergren; **8.7** (photo) © W. Tai; **8.14(a)** © C.G.G.J. van Steenis; **(b)** © Warren L. Wagner, Smithsonian, Washington, D.C.; **8.15(a)** J. D. Watson, *The Double Helix*, Antheneum, New York, 1968; **(b)** W. Etkin, *BioScience*, vol. 23, pages 652–653, 1973, as modified by R. A. Kellin and J. R. Gear, *BioScience*, vol. 30, pages 110–111, 1980; **8.20** O. L. Miller, Jr., B. A. Hamkalo, and C. A. Thomas, Jr., *Science*, vol. 169, pages 392–395, 1970; **8.17, 8.18, 8.19** Courtesy Shirley Baty, from Curtis and Barnes, *op. cit.*, 1989; **8.23(b)** © Sung Hou-Kim; **8.26** © Hans Ris; **8.28, 8.29** Courtesy Shirley Baty, from Curtis and Barnes, *op. cit.*, 1989; **8.30** Redrawn from Curtis and Barnes, *op. cit.*, 1989

Chapter 9

9.1 The Royal College of Surgeons of England; **9.5** After S. Carlquist, *Island Life*, The Natural History Press, Garden Press, Garden City, New York, 1965; **9.6(a)** © Heather Angel/BioFotos; **(b)** © Robert Ornduff, University of California, Berkeley; **9.7** After S. Ross-Craig, *Drawings of British Plants*, part IV, G. Bell and Sons, Ltd., London, 1950; **9.8** B. Crandall, Carnegie Institution of Washington, Publication 540; **9.10(a,b)** © M. A. Nobs; **(c,d)** © L. R. Heckard and C. S. Webber, Jepson Herbarium, University of California, Berkeley; **9.11** © D. Myrick, Jepson Herbarium, University of California, Berkeley; **9.12(a)** © Thase Daniel/Bruce Coleman, Inc.; **Page 164** After R. H. Robichaux; **Page 165** © Gerald D. Carr; **9.13(a)** © Heather Angel; **(b)** © Heather Angel/Biofotos; **(c-e)** © C. J. Marchant

Chapter 10

10.1 The Bettmann Archive; **10.2(a)** © Larry West; **(b)** © C. S. Webber; **(c)** © Imagery; **Page 174(a)** © G. R. Roberts; **(b)** © David Thompson/Oxford Scientific Films; **10.3(a)** L. V. Leak, *Journal of Ultrastructural Research*, vol. 21, pages 61–74, 1967; **(b)** © K. Esau; **10.5(a,c)** © E. V. Gravé; **(b)** © H. Forest; **10.7(a)** © R. M. Meadows/Peter Arnold, Inc.; **(b)** © L. E. Graham; **(c)** © Kim Taylor/Bruce Coleman, Inc.; **(d)** © Runk and Schoenberger/Grant Heilman Photography; **(e)** © E. V. Gravé; **10.8(a)** © Photo Researchers, Inc.; **(b)** © E. S. Ross; **(c)** © Larry West; **(d)** © K. Sandved; **(e)** © John A. Lynch/Photo NATS; **10.9(a,d)** © R. Carr/Bruce Coleman, Inc.; **(b)** © J. W. Perry; **(c)** © E. S. Ross/Bruce Coleman, Inc.; **(e)** © J. Shaw/Bruce Coleman, Inc.; **(f-h)** © J. Dermid; **(i)** © Steve Solum 1985/Bruce Coleman, Inc.; **(j)** © E. Beals

Chapter 11

11.1 © Huntington Potter and David Dressler, Harvard Medical School; **11.2** © T. D. Pugh and E. H. Newcomb; **11.3(a,b)** © D. Greenwood; **(c)** © J. L. Pate; **11.4(a)** © Hans Reichenbach; **(b)** © MSU Instructional Media Center/R. Hammerschmidt; **11.5** D. A. Cuppels and A. Kelman, *Phytopathology*, vol. 70, pages 1110–1115, 1980; **11.6** © C. C. Brinton, Jr., and John Carnahan; **11.7** A. Berkaloff, J. Bourquet, P. Favard, and M. Guinnebault, *Introduction à la Biologie: Biologie et Physiologie Cellulaire*, Hermann Collection, Paris, 1967; **11.8** After G. N. Agrios, *Plant Pathology*, 2nd ed., Academic Press, Inc., New York, 1978; **11.9(top)** © Fred Bavendam/Peter Arnold, Inc.; **11.9(bottom)** After M. R. Walter, *American Scientist*, vol. 65, pages 563–571, 1977; **11.10(a)** © E. V. Gravé; **(c)** © Winton Patnode/Photo Researchers, Inc.; **(d)** © E. J. Ordal; **11.11** © R. D. Warmbrodt; **11.12** © Johnson 1984/Biological Photo Service; **11.13** © R. Davis, U.S. Department of

Agriculture; **11.14(a)** © M. V. Parthasarathy; **(b)** © Henry Donselman; **11.15(a,c)** © John G. Torrey, Cabot Foundation, Harvard University; **(b)** © BioPhoto Associates/Science Source/Photo Researchers, Inc.; **11.16(a)** N. J. Lang, *Journal of Psychology*, vol. 1, pages 127–134, 1965; **(b)** © Robert and Linda Mitchell; **11.17** © W. A. Niering; **11.18** After G. N. Agrios, *Plant Pathology*, 2nd ed., Academic Press, Inc., New York, 1978; **11.19** © L. D. Simon; **11.20** © G. A. de Zoeten and G. Gaard; **11.21** After K. Namba, D. L. D. Casper, and G. J. Stubbs, *Science*, vol. 227, pages 773–776, 1985; **11.22** J. D. Almeida and A. F. Howatson, *Journal of Cell Biology*, vol. 16, pages 616–620, 1963; **11.23(a)** © G. Gaard and R. W. Fulton; **(b,c)** © G. Gaard and G. A. deZoeten; **11.24(a)** © K. Maramorosch; **(b)** © E. Shikata; **(c)** © E. Shikata and K. Maramorosch; **11.25** © T. White; **11.26** © Th. Koller and J. M. Sogo, Swiss Federal Institute of Technology, Zurich

Chapter 12

12.1 © R. M. Meadows 1989/Peter Arnold, Inc.; **12.2** © Charles Marden Fitch/Taurus Photos; **12.3** © E. S. Ross; **12.5** M. D. Coffey, B. A. Palevitz, and P. J. Allen, *The Canadian Journal of Botany*, vol. 50, pages 231–240, 1972; **12.6** © R. J. Howard; **Page 214** John Hogdin, from A.H.R. Buller, *Researches on Fungi*, vol. 6, Longman, Inc., New York; **12.9(a)** © Jeff Lepore, 1986/Photo Researchers, Inc.; **(b)** © Rod Planck/Photo Researchers, Inc.; **(c)** © G. J. Breckon **12.11(a,b)** © J. C. Pendland and D. G. Boucias; **12.12(a)** © C. Bracker; **(b)** © D. S. Neuberger; **(c)** © Bryce Kendrick; **12.14(a)** © E. V. Gravé; **12.15(a)** © A. McClenaghan/Photo Researchers, Inc.; **(b)** © John Durham/Photo Researchers; **12.16** © G. L. Barron; **Page 221** © John Webster, University of Exeter; **Page 223(a,b)** © G. L. Barron, University of Guelph; **(c)** © N. Allin and G. L. Barron, University of Guelph; **12.17** © E. Imre Friedmann; **12.18** © E. S. Ross; **12.19(a)** © Meredith Blackwell, Louisiana State University; **(b)** © Robert A. Ross; **(c)** © Jeff Foott; **12.20(a,b)** © E. S. Ross; **(c)** © Larry West; **12.22** © R. L. Chapman; **12.23(a,b)** © V. Ahmadjian; **(c)** © V. Ahmadjian and B. J. Jacobs; **12.24(b)** © E. S. Ross; **12.26** © David J. McLaughlin, University of Minnesota, St. Paul; **12.27, 12.28(b)** © Haisheng Lee and David J. McLaughlin, University of Minnesota, St. Paul; **12.29(a)** © Brian Enting/Photo Researchers, Inc.; **(b)** © Tom Branch 1976/Photo Researchers, Inc.; **(c)** © Peter Katsaros/Photo Researchers, Inc.; **(d)** © J. W. Perry; **12.31** © E. S. Ross; **12.32** © Walter Hodge/Peter Arnold; **12.33(a,b)** © R. Gordon Wasson, Botanical Museum of Harvard University; **12.34(a)** © Jane Burton/Bruce Coleman, Inc.; **(b)** © Hartmut Noeller/Peter Arnold, Inc.; **(c,d)** © Jeff Lepore 1986/Photo Researchers, Inc.; **12.36** © S. A. Wilde; **12.37** © Bryce Kendrick; **12.38** © B. Zak, U.S. Forest Service; **12.39** © R. D. Warmbrodt; **12.40** © Thomas N. Taylor, Ohio State University

Chapter 13

13.1 © G. Vidal and T. D. Ford, *Precambrian Research* (in press); **13.2** © D. P. Wilson/Science Source/Photo Reseachers, Inc.; **13.3(a)** © L. Wingren; **(b)** L. E. Graham and J. M. Graham, *Transactions of the American Microscopical Society*, vol. 99, pages 160–166, 1980; **13.4(a)** © A. W. Barksdale; **(b)** A. W. Barksdale, *Mycologia*, vol. 55, pages 493–501, 1963; **Page 250 (a,b)** © A. W. Barksdale; **(c)** A. W. Barksdale, *op. cit.*, 1963; **13.6** After J. H. Niederhauser and W. C. Cobb, *Scientific American*, vol. 200, pages 100–112, 1959; **13.7** © John W. Taylor; **13.9(a-c)** © K. B. Raper; **(d-g)** © David Scharf/Peter Arnold, Inc.; **13.10** © K. B. Raper; **13.11** © D. S. Neuberger; **13.13(c)** © Victor Duran; **(d)** © Ed Reschke/Peter Arnold, Inc.; **13.14(a)** © G. A. Fryxell; **(b)** © Elizabeth Vuerick, Scripps Institute of Oceanography, University of California, San Diego, La Jolla, Calif.; **(c)** © William Balch, RSMAS/MBF, University of Miami; **13.15(a)** © M. I. Walker/Photo Researchers, Inc.; **(b)** © F. Rossi; **(c)** © Dr. Ann Smith/SPL/Photo Researchers, Inc.; **(d)** © Biophoto Associates/Science Source/Photo Reseachers, Inc.; **13.17** © D. S. Neuberger; **Page 263 (a)** © Robert E. Pelham/Bruce Coleman, Inc.; **(b)** © Florida Department of Natural Resources, Bureau of Marine Research; **13.19(a,b)** © D. P. Wilson/Science Source/Photo Researchers, Inc.; **(c)** © Florida

Department of Natural Resources; **13.20** Robert F. Sisson ©
National Geographic Society; **Page 265** © D. F. Kubai and H. Ris;
13.21(a) © Biophoto Associates/Photo Researchers, Inc.; **(b)**
© D. Longanecker

Chapter 14

14.1 © Antonio Laped; **14.2** © Anne Wertheim/Animals
Animals; **14.3(a)** © L. E. Graham; **(b)** R. G. Shealth, J. A.
Hellebust, and T. Sawa, *Psychologia*, vol. 20, pages 22–31, 1981;
14.4(a) © Kim Taylor/Bruce Coleman, Inc.; **(b,c)** © Ronald
Hoham, Colgate University; **14.5(a)** © D. P. Wilson/Eric and
David Hosking Photography; **(b)** © E. S. Ross; **(c)** © R. C.
Carpenter; **(d)** © Jean Baxter/Photo NATS; **14.6(a,b)** © M. Littler
and D. Littler, Smithsonian Institution; **(c)** © C. Chulamanis, M.
Littler, and D. Littler; **14.7(a)** C. M. Pueschel and K. M. Cole,
American Journal of Botany, vol. 69, pages 703–720, 1982; **14.9(a)**
© G. R. Roberts; **(b,c)** © D. P. Wilson/Eric and David Hosking
Photography; © D. P. Wilson/Eric and David Hosking Photo-
graphy **14.10(b)** © Oxford Scientific Films; **14.11** © C. J.
O'Kelly; **14.12(a,b)** © Ray Evert and John West; **14.15(a–f)** ©
Ronald Hoham; **Page 281 (a)** © Bob Evans/Peter Arnold, Inc.;
(b) © W. H. Hodge/Peter Arnold, Inc.; **(c)** © Kelco
Communications; **14.16** After G. L. Floyd; **14.17** After K. R.
Mattox and K. D. Stewart, in *Systematics of the Green Algae*, D.E.G.
Irvine and D. M. John (Eds.), 1984; **14.18** © L. E. Graham;
14.19(a–d) © M. I. Walker/Science Source/Photo Researchers;
14.20(a–c) © Alan J. Brook; **(d)** © Spike Walker; **14.21(a,b)** ©
L. E. Graham; **14.22** © K. J. Niklas; **14.23(a)** © W. H. Amos/
Bruce Coleman, Inc.; **14.24(a)** © James Graham; **(b,c)** K. Esser,
Cryptograms, Cambridge University Press, Cambridge, 1982; **(d)**
© E. S. Ross; **14.26** © D. P. Wilson/Eric and David Hosking
Photography **14.28(a)** © Robert A. Ross; **(b)** © Grant Heilman
Photography; **(c)** © L. R. Hoffman; **14.29** © V. Paul;
14.30 (photo) © G. E. Palade; **14.33** © J. Robert Waaland/Biologi-
cal Photo Service; **14.34(a)** © Linda E. Graham; **(b)** H. J.
Marchant and J. D. Pickett-Heaps, *Australian Journal of Biological
Science*, vol. 25, pages 1199–1213, 1972; **14.35(a–d)** © L. E.
Graham; **14.36** © David L. Kirk, Washington University, St.
Louis; **14.37(a)** © Gary Floyd; **14.38** After R. F. Skagel, R. J.
Bandoni, G. E. Rouse, W. B. Schofield, J. R. Stein, and T.M.C.
Taylor, *An Evolutionary Survey of the Plant Kingdom*, Wadsworth
Publishing Company, Inc., Belmont, Calif., 1966; **14.39** © L. E.
Graham; **14.40(a,b)** © Larry Hoffman

Chapter 15

15.1 © T. S. Elias; **15.2(a,b)** W. Remy, *Science*, vol. 219, pages
1625–1627, 1982; **15.3(a)** © D. S. Neuberger; **15.4(a)** After G. M.
Smith, *Cryptogamic Botany*, vol. 2, in *Bryophytes and Pteridophytes*,
2nd ed., McGraw-Hill Book Company, New York, 1955; **15.4(b)** ©
R. E. Magill, Botanical Research Institute, Pretoria; **15.6(a,b)** © Dr.
G. J. Chafaris/Dr. E. R. Degginger; **15.9(a)** © Harold Taylor AB
IPP/Oxford Scientific Films; **15.10(a,b)** J. J. Engel, *Fieldiana: Botany*
(New Series), vol. 3, pages 1–229, 1980; **15.10(c)** J. J. Engel; **Page
206** After C. T. Ingold, *Spore Discharge in Land Plants*, Clarendon
Press, Oxford, 1939; **15.11** © K. B. Sandved; **15.12(a)** © Robert
A. Ross; **(d)** © D. S. Neuberger; **15.14** © D. S. Neuberger;
15.15 © Rod Plank/Photo Researchers, Inc.; **15.16(a)** © A. E.
Staffan; **(b)** © E. S. Ross; **15.17(a)** © D. S. Neuberger; **15.18(a–c)**
C. Hebant, *Journal of the Hattori Botanical Laboratory*, vol. 39, pages
235–254, 1975; **15.20(a)** After Ingold, *op. cit.*, 1939; **15.20(b)** © R.
Magill, Botanical Research Institute, Pretoria; **15.21(a)** © Larry
West; **15.22(a)** © M.C.F. Proctor; **15.21** After Skagel et al., *op. cit.*,
1966; **15.21** After Ingold, *op. cit.*, 1939

Chapter 16

16.1 © Dr. Jeremy Burgess/Science Photo Library/Photo Researchers, Inc.; **16.2** © Kristine Rasmussen and Stuart Naquin;
16.3 After A. S. Foster and E. M. Gifford, Jr., *Comparative Morphology of Vascular Plants*, 2nd ed., W. H. Freeman & Company,
Publishers, New York, 1974; **16.4(a,d)** After K. K. Namboodiri and
C. Beck, *American Journal of Botany*, vol. 55, pages 464–472,

1968; **(b,c)** After K. Esau, *Plant Anatomy*, 2nd ed., John Wiley &
Sons, Inc., New York, 1965; **16.6** After Smith, *op. cit.*, 1955;
16.8 After H. P. Banks, *Evolution and Plants of the Past*, Wadsworth
Publishing Company, Inc., Belmont, Calif., 1970; **16.9(a)** After J.
Walton, *Fossil Plants*, MacMillan Publishing Company, New York,
1940; **16.9(b)** After J. Walton, *Phytomorphology*, vol. 14, pages
155–160, 1964; **16.9(c)** After F. M. Heuber, *International Symposium
on the Devonian System*, vol. 2, D. H. Oswald (Ed.), Alberta Society
of Petroleum Geologists, Calgary, Alberta, Canada, 1968; **16.10** ©
Field Museum of Natural History; **16.12(a)** R. L. Peterson, M. J.
Howarth, and D. P. Whittier, *Canadian Journal of Botany*, vol. 59,
pages 711–720, 1981; **(b)** © Heather Angel; **16.13(a)** © D.
Cameron; **(b)** © R. Schmid; **16.15** Specimen provided by Ripon
Microslides, Ripon, Wisconsin; **16.16** © D. S. Neuberger;
16.18(a) © David Johnson, Big Bend National Park; **(b)** © D. S.
Neuberger; **(c)** © Fletcher & Baylis/Photo Researchers, Inc.;
16.20 © W. H. Wagner; **16.21** After Kristine Rasmussen and Stuart
Naquin; **16.23(a)** © R. Carr; **(b)** © E. S. Ross; **16.27(a)** © James
L. Castner; **(b,c)** © W. H. Wagner; **(d)** © Murry Johnson; **(e)** ©
W. H. Wagner; **(f)** © David Johnson; **16.29(a)** © W. H. Wagner;
(b) © Joseph M. Beitel; **Page 346 (a)** After M. Hirmer, *Handbuch
der Paläobotanik*, vol. 1, Druck and Verlag von R. Oldenbourg,
Munich and Berlin, 1927; **(b)** After W. N. Stewart and T.
Delevoryas, *Botanical Review*, vol. 22, pages 45–80, 1956; **(c)** After
Banks, *op. cit.*, 1970; **16.31** © M. dos Passos, 1978/Photo
Researchers, Inc.; **16.32** © C. Neidorf; **16.34** © D. Farrar;
16.36(a) © D. S. Neuberger; **(b)** © W. H. Wagner

Chapter 17

17.1 Kristine Rasmussen and Stuart Naquin; **17.3 (a,b)** After J. M.
Pettitt and C. B. Beck, *Contributions from the Museum of Paleontology*,
University of Michigan Press, vol. 22, pages 139–154, 1968; **(c)** J.
M. Pettitt and C. B. Beck, *Science*, vol. 156, pages 1727–1729,
1967; **17.4** © Charles B. Beck; **17.5** After S. E. Scheckler, *Ameri-
can Journal of Botany*, vol. 62, pages 923–934, 1975; **17.6, 17.7**
Kristine Rasmussen and Stuart Naquin; **17.8, 17.9(a)** © J. Dermid;
17.10 © E. S. Ross; **17.11** © J. Kummerow; **17.14** © B. Haley;
17.16(c) © G. J. Breckon; **17.19** © N. Fox-Davies/Bruce Coleman
Ltd.; **17.22(a)** © W. H. Hodge/Peter Arnold, Inc.; **(b)** © E. S.
Ross; **17.23, 17.24** © H. H. Iltis; **17.25** © Grant Heilman
Photography; **17.26(a)** © Larry West; **(b)** © J. Burton/Bruce
Coleman, Inc.; **17.27** © Gene Ahrens/Bruce Coleman, Inc.;
17.28 © Sichuan Institute of Biology; **17.29** © (photo) Carolina
Biological Supply Company; **17.30** © D. A. Steingraeber;
17.31(a) © Knut Norstog; **(b)** © D. T. Hendricks and E. S. Ross;
17.32 © Knut Norstog; **17.33(a)** © J. W. Perry; **(b)** © Runk and
Schoenberger/Grant Heilman Photography; **17.34(a)** © F.S.P.
Ng; **(b,c)** © G. Davidse; **17.35(a)** © E. S. Ross; **(b)** © J. W.
Perry; **(c)** © K. J. Niklas; **(d)** © J. W. Perry; **17.36(a)** © C. H.
Bornman; **(b,c)** © E. S. Ross

Chapter 18

18.1 © E. S. Ross; **18.2** © W. P. Armstrong; **18.3(a,b)** © G. J.
Breckon; **(c)** © E. S. Ross; **18.4(a)** © D. S. Neuberger; **(b)** ©
E. S. Ross; **(c)** © E. R. Degginger/Earth Scenes; **18.5(a,c)** © E. S.
Ross; **(b)** © R. Carr; **18.8(a,c)** © Larry West; **(b)** © J. H.
Gerard; **(c)** © Larry West; **(d)** © Grant Heilman Photography;
(e) © J. H. Gerard; **18.10** © E. S. Ross; **18.12(a)** © Runk and
Schoenberger/Grant Heilman Photography; **18.13(a)** © Larry
West; **(b)** Specimen provided by Rudolf Schmid; **18.15(a)** © P.
Echlin; **(b,c)** © J. Heslop-Harrison and Y. Heslop-Harrison; **(d)** ©
J. Mais; **18.22** © Peter Hoch; **18.23** © James L. Castner; **Page 398**
© J. Heslop-Harrison; **18.24** © Charles Webber

Chapter 19

19.1(a) © E. S. Ross; **19.1(b)** © D. L. Dilcher; **19.2** © E. Dorf;
19.3, 19.4, 19.5 After Peter Crane; **19.7** © D. W. Taylor and L. J.
Hickey; **19.8** © E. M. Friis; **19.10** © Mike Andrews/ Animals
Animals/Earth Scenes; **19.11** © J. Castner; **19.12** © David L.
Dilcher (Reconstructions by Megan Rohn in consultation with D.
Dilcher); **19.13(a,c)** © E. S. Ross; **(b)** © J. W. Perry; **19.14(b,c)** ©

E. S. Ross;   **(d)** © A. Sabarese;   **19.15(a)** © E. S. Ross;   **19.16** © J.A.L. Cooke;   **19.17** © W. H. Hodge;   **19.18(a)** © Larry West; **(b)** © T. J. Hawkeswood; **Page 416   (a)** © J. Bogner;   **(b)** © ERM Group; **Page 417   (c)** © Larry West;   **(d)** © J. Dermid;   **(e)** © J. Bogner;   **19.19, 19.20, 19.21** © E. S. Ross;   **19.22** © Larry West; **19.23** © T. Eisner;   **19.24, 19.25** © E. S. Ross;   **19.26(a,c)** © E. S. Ross;   **(b)** © L. B. Thein;   **19.27** © E. S. Ross;   **19.28** © M.P.L. Fogden/Bruce Coleman, Inc.;   **19.29** © R. A. Tyrell;   **19.30** © Oxford Scientific Films;   **19.31** © E. S. Ross;   **19.32** © D. J. Howell;   **19.33(a,b)** © T. Hovland/Grant Heilman Photography; **(c)** © E. S. Ross;   **(d)** © J. F. Skvarla, University of Oklahoma; **19.34(b,c)** After D. B. Swingle, *A Textbook of Systematic Botany*, McGraw-Hill Book Company, New York, 1946;   **19.35** © D. S. Neuberger;   **19.36(a)** John M. Pettitt and R. Bruce Knox, *Scientific American*, vol. 244, pages 134–144, 1981;   **19.36(b)** © Sean Morris/Oxford Scientific Films;   **19.38(a)** © James L. Castner; **(b)** After Skagel et al., *op. cit.*, 1966;   **19.39(a)** © E. S. Ross; **(b)** After Skagel et al., *op. cit.*, 1966;   **(c)** After L. Berson, *Plant Classification*, D. C. Heath and Company, Boston, 1957;   **19.40(a)** © E. S. Ross;   **(b,c)** © K. B. Sandved;   **19.41(a)** After Skagel et al., *op. cit.*, 1966;   **(b)** © E. S. Ross;   **19.43** © E. S. Ross;   **19.44** © Larry Atkinson/Mobridge Tribune;   **19.45(a)** © E. S. Ross;   **(b)** © U.S. Forest Service;   **19.46** © E. S. Ross;   **19.47** © E. S. Ross; **19.48** © Andrew J. Beattie;   **19.50** © Robert and Linda Mitchell; **19.51(a,c,d)** © Timothy Plowman;   **(b)** © E. S. Ross

Chapter 20

20.4 After A. S. Foster and E. M. Gifford, Jr., *Comparative Morphology of Vascular Plants*, 2nd ed., W. H. Freeman & Company, Publishers, New York, 1974;   **20.5** © Tom McHugh/Photo Researchers, Inc.; **Page 451   (a)** © G. I. Bennard/Oxford Scientific Films;   **(b,c)** © Heather Angel; **Page 452** © Werner H. Muller/Peter Arnold, Inc.

Chapter 21

21.6 M. C. Ledbetter and K. B. Porter, *Introduction to the Fine Structure of Plant Cells*, Springer-Verlag, Inc., New York, 1970; **21.13(a,b)** H. A. Core, W. A. Cote, and A. C. Day, *Wood: Structure and Identification*, 2nd ed., Syracuse University Press, Syracuse, N.Y., 1979;   **21.14** © I. B. Sachs, Forest Products Laboratory, U.S.D.A.;   **21.16, 21.21** After Esau, *op. cit.*, 1977;   **21.22** © J. S. Pereira;   **21.24** © Randall Brand;   **21.25** After Esau, *op. cit.*, 1977;   **21.26** © T. Vogelman and G. Martin

Chapter 22

22.1, 22.2 After J. E. Weaver, *Root Development of Field Crops*, McGraw-Hill Book Company, New York, 1926;   **22.6** © F.A.L. Clowes;   **22.10** © Robert Mitchell/Earth Scenes;   **22.13** H. T. Bonnett, Jr., *Journal of Cell Biology*, vol. 37, pages 199–205, 1968;   **22.14** After W. Braune, A. Leman, and H. Taubert, *Pflanzenanatomisches Praktikum*, VEB Gustav Fischer Verlag, Jena, 1967; **22.15** Robert Warmbrodt, *New Phytologist*, 1986, 102, pages 175–192;   **22.19** © E. R. Degginger/Bruce Coleman, Inc.;   **22.20** © Robert and Linda Mitchell;   **22.22(a,b)** © D. A. Steingraeber

Chapter 23

23.5 After D. A. DeMason, *American Journal of Botany*, vol. 70, pages 955–962, 1983;   **23.10(c)** © W. Eschrich;   **23.12** After A. Fahn, *Plant Anatomy*, 2nd ed., Pergamon Press, Inc., Elmsford, N.Y., 1974;   **23.14** © Rhonda Nass/Ampersand;   **23.16(a)** © J. W. Perry;   **23.17** © Rhonda Nass/Ampersand;   **23.22** © Michele McCauley;   **23.24(a,b)** © W. A. Russin;   **23.30** © Daniel J. Barta;   **23.31** J. M. Dannenhoffer, © R. Evert;   **23.32(a,b)** © Raymon Donahue and Greg Martin; **Page 510** After P. A. Deschamp and T. J. Cooke, *Science*, vol. 219, pages 505–507, 1983; **Page 513   (a)** © H. C. Jones, Tennessee Valley Authority;   **(b)** J. S. Jacobson and A. C. Hill (Eds.), *Recognition of Air Pollution Injury to Vegetation: A Pictorial Atlas*, Air Pollution Control Association, Pittsburgh, 1970.;   **(c)** After T. H. Maugh, II, *Science*, vol. 226, pages 1408–1410, 1984;   **23.35(a–i)** © Shirley C. Tucker;   **23.36** © J. W.

Perry;   **23.37** © Norman Owen Tomalin;   **23.38** © D. A. Steingraeber;   **23.39(a,b)** © J. W. Perry;   **23.40** © G. R. Roberts; **Page 518   (a–c)** © E. S. Ross

Chapter 24

24.22 © I. B Sachs, Forest Products Laboratory, U.S.D.A.; **24.25** Core, Côté, and Day, *op. cit.*, 1979;   **24.26(a)** © Galen Rowell 1985/Peter Arnold, Inc.;   **(b)** © C. W. Ferguson, Laboratory of Tree-Ring Research, University of Arizona;   **24.28** © Regis Miller; **24.30** After R. B. Hoadley, *Understanding Wood*, The Taunton Press, Newton, Conn., 1980

Chapter 25

25.4 © Roni Aloni, Tel Aviv University;   **25.6** © Runk and Schoenberger/Grant Heilman Photography;   **25.7(a)** © Biological Photo Service;   **(b–e)** J. P. Nitsch, *American Journal of Botany*, vol. 37, pages 211–215, 1950;   **25.9** © Folke Skoog, University of Wisconsin;   **25.13** © S. W. Wittwer;   **25.14(a)** After M. B. Wilkins (Ed.), *Advanced Plant Physiology*, Pitman Publishing Ltd., London, 1984;   **25.14(b)** © J. E. Varner;   **25.15** J. van Overbeck, *Science*, vol. 152, pages 721–731, 1966;   **25.16(a)** © Carolina Biological Supply Co.;   **(b)** © S. W. Wittwer;   **25.17(a,b)** © Abbott Laboratories;   **25.23** © Philip Harrington;   **25.22** After art by Hilleshög, Laboratory for Cell and Tissue Culture, Research Division, Landskrona, Sweden;   **25.24** © Eugene W. Nester; **Page 568   (a)** S. Ross-Craig, *Drawings of British Plants*, Part III, *Cruciferae*, Bell, London, 1949;   **(b,c)** © Elliot Meyerowitz, California Institute of Technology;   **25.27** © Luca Comai/Calgene;   **25.28** © Keith Wood, University of California, San Diego

Chapter 26

26.3(a,b) © Dept. of Botany, University of Wisconsin;   **26.4(a,b)** J. S. Ranson and R. Moore, *American Journal of Botany*, vol. 70, pages 1048–1056, 1983;   **26.5** After B. E. Juniper, *Annual Review of Plant Physiology*, vol. 27, pages 385–406, 1976;   **26.6** © Stephen P. Parker/Photo Researchers, Inc.;   **26.7(a,b)** © Jack Dermid; **26.8** After A. W. Galston, *The Green Plant*, Prentice-Hall, Inc., Englewood Cliffs, New Jersey, 1968;   **26.9(a)** © Biophoto Associates/Science Source/Photo Researchers, Inc.;   **(b,c)** After B. Sweeney, *Rhythmic Phenomena in Plants*, Academic Press, Inc., New York, 1969;   **26.10** After A. W. Naylor, *Scientific American*, vol. 186, pages 49–56, 1952;   **26.11** After P. M. Ray, *The Living Plant*, Holt, Rinehart, & Winston, Inc., New York, 1963;   **26.12(a,b)** © Department of Botany, University of Wisconsin;   **26.13(a–d)** © U.S. Department of Agriculture;   **26.17** © Breck P. Kent/Earth Scenes; **26.18** © U.S. Department of Agriculture;   **26.19** A. Lang, M.Kh. Chailakhyan, and I. A. Frolova, *Proceedings of the National Academy of Sciences*, vol. 74, pages 2412–2416, 1977;   **26.21** After Naylor, *op. cit.*, 1952;   **26.23(a)** © Robert L. Dunne/Bruce Coleman, Inc.;   **(b)** © Bruce Coleman, Inc.;   **26.24(a,b)** © Runk and Schoenberger/Grant Heilman Photography;   **26.25** © Janet Braam;   **26.26(a)** © J. Ehleringer and I. Forseth, University of Utah;   **(b)** © Gene Ahrens/Bruce Coleman, Inc.;   **26.27** After Ehleringer and I. Forseth, *Science*, vol. 210, pages 1094–1098, 1980

Chapter 27

27.1 © Roger Archibald/Earth Scenes;   **27.2(a,b)** © Department of Botany, University of Wisconsin;   **27.3(a)** © Donald Specker/Earth Scenes;   **(b)** © Biological Photo Service;   **27.4** © Eric Crichton/ Bruce Coleman, Ltd.;   **27.5** J. Bonner and A. W. Galston, *Principles of Plant Physiology*, W. H. Freeman & Company, Publishers, New York, 1952;   **27.6** After B. Gibbons, *National Geographic*, vol. 166, pages 350–388, 1984 (Ned M. Seidler, artist); **Page 600** Courtesy Shirley Baty, from Curtis and Barnes, *op. cit.*, 1989;   **27.7** After F. C. Salisbury and Ross, *Plant Physiology*, Wadsworth Publishing Co., Belmont, Calif., 1985;   **27.9(a)** © The Nitrogen Company, Inc.;   **(b)** © Oxford University Botany School;   **27.10(a,b)** © R. R. Herbert, R. D. Holsten, and R.W.F. Hardy, E. I. duPont de Nemours Company;   **27.11(a–c)** © B. F. Turgeon and W. D. Bauer, *Canadian Journal of Botany*, vol. 60, pages 152–161, 1982;   **27.12** © R. R.

Herbert, R. D. Holsten, and R.W.F. Hardy, E. I. duPont de Nemours Company; **27.13** J.M.L. Selker and E. H. Newcomb, *Planta*, vol. 165, pages 446–454, 1985; **27.14** E. H. Newcomb, Sh. R. Tandon, and R. R. Kowal, *Protoplasma*, vol. 125, pages 1–12, 1985; **27.15(b)** © H. E. Calvert; **Page 609** **(a)** © Grant Heilman Photography; **(b)** J. H. Troughton and L. Donaldson, *Probing Plant Structures*, McGraw-Hill Book Company, New York, 1972; **27.16(a)** © Dwight Kuhn/Bruce Coleman, Inc.; **(b)** © L. West/ Bruce Coleman, Inc.; **27.17** From Curtis, *op. cit.*, 1989; **27.18** © G. C. Gerloff; **Page 614** © Foster/Bruce Coleman, Inc.

Chapter 28

28.2(a) © L. M Beidler; **(b)** © J. H. Troughton; **28.3** After M. G. Penny and D.J.F. Bowling, *Planta*, vol. 119, pages 17–25, 1974; **28.4** After D. E. Aylor, J. Y. Parlange, and A. D. Krikorian, *American Journal of Botany*, vol. 60, pages 163–171, 1973; **28.5** After M. Richardson, *Translocation in Plants*, Edward Arnold Publishers, Ltd., London, 1968; **28.6** After A. C. Leopold, *Growth and Development*, McGraw-Hill Book Company, New York, 1964; **28.7** After Richardson, *op. cit.*, 1968; **28.9** After P. F. Scholander, H. T. Hammel, E. D. Bradstreet, and E. A. Hemmingsen, *Science*, vol. 148, pages 339–246, 1965; **28.10, 28.11** After M. H. Zimmermann, *Scientific American*, vol. 208, pages 132–142, 1963; **28.12(a)** © G. R. Roberts; **28.14** After Richardson, *op. cit.*, 1968; **28.15** © H. Reinhard/Bruce Coleman, Inc.; **28.16** After E. Hausermann and A. Frey-Wyssling, *Protoplasma*, vol. 57, pages 37–80, 1963; **28.18** After J. S. Pate, *Transport in Plants*, I, *Phloem Transport*, M. H. Zimmermann and J. A. Milburn (Eds.) , Springer-Verlag Berlin, Inc., 1975; **28.21** After Malpighii, *Opera Posthuma*, London, 1675; **28.22(a,b)** © M. H. Zimmerman; **Page 633** **(b,c)** © E. Fritz

Chapter 29

29.1(a-d) © Daniel Janzen; **29.2(a,b)** © D. H. Harvey; **29.3** © C. H. Muller; **29.4** © B. Bartholomew; **29.5(a,b)** © J. Mann, Australian Department of Lands; **29.6(a)** © Dwight Kuhn/Bruce Coleman, Inc.; **(b)** © Frans Lanting; **29.7** © G. E. Likens; **29.9** After J. Phillipson, *Ecological Energetics*, Edward Arnold Publishers, Ltd., London, 1966; **29.11(a)** © Fred Bauendam/Peter Arnold, Inc.; **(b)** © Wendell Metzen/Peter Arnold, Inc.; **(c)** © James H. Carmichael/Bruce Coleman, Inc.; **(d)** © L. West/Bruce Coleman, Inc.; **29.12** © Jane Burton/Bruce Coleman, Inc.; **29.13** © Bruce Coleman, Inc.; **29.14(a)** © J. Dermid; **(b)** © E. S.

Ross; **Page 652** **(a)** © U.S. Geological Survey, Eros Data Center Survey; **Page 653** **(b)** © Jeff Henry/Peter Arnold, Inc.; **29.15(a)** © Larry Nielsen/Peter Arnold, Inc.; **(b)** © David K. Yamaguchi; **(c)** © P. Frenzen, U.S.F.S.; **(d)** © David K. Yamaguchi

Chapter 30

30.1(a) © Jack Wilburn/Earth Scenes; **(b)** © Ardea Photographics; **30.2** After W. W. Küchler; **30.3** Courtesy Shirley Baty, from Curtis and Barnes, *op. cit.*, 1989; **30.4** © Dr. Gene Feldman/ NASA/Goddard Space Flight Center; **30.5** © C. D. MacNeill; **30.6** After Curtis, *op. cit.*, 1989; **30.7(a)** © C. S. Rettenmeyer; **(b,c,f)** © E. S. Ross; **(d)** © P. Raven; **(e)** © C. W. Rettenmeyer; **(g)** © 1983 James P. Blair, National Geographic Society; **Page 665** The Bettmann Archive; **30.9** © J. Van Wormer/Bruce Coleman, Inc.; **30.12** © J. Dermid; **30.14(a-e)** © E. S. Ross; **30.15** © F. C. Vasck; **30.17(a,b)** © J. Reveal; **30.18** © Soil Conservation Service; **30.20(a)** © H. E. Eversmeyer; **(b,c)** © J. H. Gerard; **30.21(a)** © M. Travis; **(b)** © J. F. Dodd; **Page 676** **(a)** © Kevin Byron/Bruce Coleman, Inc.; **(b)** © John Shaw/Bruce Coleman, Inc.; **Page 682** © Jerry Franklin, University of Washington

Chapter 31

31.1 James P. Blair, © 1983, National Geographic Society; **31.2(a,b)** © E. S. Ross; **31.3(a,b)** © E. S. Ross; **31.4** © E. S. Ross; **31.5** © W. H. Hodge; **31.6(a)** James P. Blair, © 1983, National Geographic Society; **(b)** © Susan Pierres/Peter Arnold, Inc.; **31.7** © Robert and Linda Mitchell; **31.8** © G. R. Roberts; **31.9(a)** © E. Zardini; **(b)** © C. B. Heiser, Jr.; **(c)** © A. Gentry; **(d)** © M. K. Arroyo; **31.10** © E. S. Ross; **31.11(a)** © C. F. Jordan; **(b)** © M. J. Plotkin; **31.12** © M. J. Plotkin; **31.13** © K. B. Sandved; **31.14** © W. H. Hodge/Peter Arnold, Inc.; **31.15** © E. S. Ross; **31.16** © W. H. Hodge/Peter Arnold, Inc.; **31.17** © Harvey Lloyd/Peter Arnold, Inc.; **31.18** © New York Public Library Picture Collection; **31.20** © R. Abernathy; **31.21** © C. A. Black; **31.22** © AP/Wide World Photos; **31.23** © W. H. Hodge; **31.24** © Agricultural Research Service, U.S.D.A.; **31.25(a)** © Agricultural Research Service, U.S.D.A.; **(b)** © University of Wisconsin; **31.26** © N. Vietmeyer, National Academy of Sciences; **31.27** © Robert and Linda Mitchell; **31.28** © 1982 Angelina Lax/Photo Researchers, Inc.; **31.29(a,b)** © J. Aronson; **31.30** After Lehninger, *op. cit.*, 1982; **31.31, 31.32** © M. J. Plotkin

Index

Numbers in **boldface** indicate figures and tables.

A (aminoacyl) site, 143, **144,** 145
ABA (*see* abscisic acid)
Abies, 369, 538, 683 (*see also* firs)
    amabilis, **654**
    balsamea, **369, 650**
    concolor, **651**
    lasiocarpa, 640, 653
    procera, **654**
abortion, 698
abscisic acid (ABA), 510 **546, 555,** 560, 572
    and dormancy, 555, 587
    influence on cell division, 559
    influence on cell expansion, 510, 560
    and seed development, 555
    and stomatal movements, 555
    and water relations, 555
abscisin, 555
abscission, 509, 510
    and auxin, 550, 551
    and ethylene, 555
    zone, **509, 530**
absolute temperature, 77
absorption
    of food by cotyledons, 443, 444, 452
    as nutritive mode, 255
    by roots, of inorganic nutrients, 471, 475, 627, 628
    by roots, of water, 467, 471, 475, 624–627
absorption spectrum, 103, **104, 105,** 581
Acacia (acacia), **8, 435,** 471, 666
    -ant symbiosis, 517, 638–639
    bull's-horn, 638–639
    cornigera, **639**
ACC (1-aminocyclopropane-1-carboxylic acid), **554**
accessory pigments, 104, 105, (*see also* carotenoids; chlorophylls; phycobilins)
acclimation, 586, 587
Acer, 432, 498, **509,** 538, **586**
    negundo, **530**
    saccharinum, **498**
    saccharum, 498, **542**
Aceraceae, 432
Acetabularia, **288**
acetaldehyde, 98
acetyl CoA, **91,** 92, 96, 98
acetyl group (CH$_3$CO), 91, 92, 720
achene, 432, **434,** 550
Achlya, 248, 249
    ambisexualis, **248,** 250

Achnanthes, 258
acid growth hypothesis, 560, **561**
acid rain, 512, 513, 615
acids, 721, 722
    fatty, 52, **53,** 91
    hydrochloric (HCl), 721
acorn, 427, 432
Acrasiomycota (acrasiomycetes) **178,** 179, 245, 246, **247,** 252–255, 730 (*see also* slime molds)
    life cycles of, **254**
acritarchs, **244,** 264
actin, 30, **33,** 46
actin filaments, 30, **31, 33,** 46
actin-myosin interactions, 33
actinomycetes, 188, **189,** 190, 197, 607
action spectrum, 103, **104, 105**
activation energy, 76, 83, 722
active (membrane) transport, **68,** 70, 73, 628, 632
    primary and secondary, **70**
active site, **79, 80**
adaptation, 6, 407, 415, 419
    mutation and, 135
adaptive radiation, 160, 164–165, 406–407
Addicott, Frederick T., 555
adenine, **60,** 61
    in ATP, **83**
    in DNA, 136, 137, **138,** 139
    in NAD, **81**
adenosine
    diphosphate (*see* ADP)
    monophosphate (*see* AMP)
    triphosphate (*see* ATP)
adenovirus, **202**
adhesion (adhesiveness), 622
Adiantum, **345, 348**
adnation, 387, 412
ADP, 61, 87–98, 101, **108,** 109, 117
adventitious structures, 455
aecia, aeciospores, **237,** 238
Aedes, **422**
Aeschynomene hispida, 540
Aesculus
    hippocastanum, **389,** 521, **530**
    pavia, **499**
aethalium, 255, **257**
aflatoxins, 209, 222
African violet, **550**
after-ripening, 448
Agallia constricta, **203**
agar/agarose, 270, 281, **548, 549,** 573, 574
Agaricus
    bisporus, 173, **174,** 234
    campestris, 234, **425,** 518, 669
Agave (agave), **425,** 518, 669
    shawii, **669**

Agent Orange, 551
agomanous mutant, 569
Agoseris, **412, 432**
agricultural revolution, 687–698
agriculture, 10, 551, 554, 571, 602, 604, 609, 612, 629, 641
    in the future, 700–710
    new developments in, 565–571, 608, 609, 700–709
    productivity, 571, 665, 700
    sustainable, 571, **673,** 678, 700
    technology, 700
    and world food crisis, 571
Agrobacterium, **199,** 200, 566, 569, 571
    tumefaciens, 566, **567,** 568, 569, **570,** 572
Agropyron, **426**
    cristatum, **126–127**
Agrostis tenuis, 116, 155–156, **478**
A horizon (topsoil), 598
AIDS, 200, 222
air chambers of liverworts, **301**
air currents and transpiration, 620
air plants, 467
air pollution, 512, 513
air (intercellular) spaces, **330, 455, 476, 477,** 478, 492, 503, 504, 527, 609, 617, 621
akinetes, 194, **195**
alanine, **56**
Albizzia polyphylla, **430–431**
albuminous cell, **461, 464, 469**
alcohol(s), 227, 248, 719, 720
    ethyl (ethanol), 97, **98,** 99
    from fermentation, 97, **98,** 99
alcohol dehydrogenase, **98**
alders, 197, 607, 683
    red, **549**
aldolase, **88**
aleurone layer, **443,** 452, 557
Alexopoulos, C. J., 249
alfalfa, 199, 471, **494, 594,** 595, 602, 641
algae, 173, 179, 257–297, 615 (*see also by name of group*)
    accessory pigments of, **247,** 257, 262
    bioluminescence of, 264, **587**
    blooms of, **258**
    blue-green (*see* cyanobacteria)
    brown (*see* brown algae)
    cell walls of, **247,** 257, 258, 262, 270, 280
    chlorophylls of, 105, **247,** 257
    classification of, 282–283
    diatoms (*see* diatoms)
    economic importance of, 263, 268, 274, 276, 281
    evolution of, 24, 270, 282
    food reserves of, **247,** 257, 262, 265, 269–270, 280

    as food source, 281, 291, 292
    freshwater, 247, 257
    golden, (chrysophyceae), **32,** 257–258, 259, 262
    green (*see* green algae)
    life cycles of, 260, 270, **277, 278, 287, 290,** 291
    marine, **115, 247,** 257, 258, 642
    multicellular, 11
    phytoplanktonic, 245, 258, 292
    red (*see* red algae)
    snow, 279
    soil, 186, 291
    symbiotic, 259, 264, 265, 272, 279, 294
    terrestrial, **247**
    unicellular, 11, 257, 291
    yellow-green, (xanthophyceae) 261–262
algin (alginic acid), **247,** 276
alginates, 281
alkaloids, 220, 435, **436,** 696
Allard, H. A., 578–580
allele(s), 130
    deleterious, 154
    dominant/recessive, 130, 131, 132, 152, **153**
    frequency, 152, **153**
    homozygous/heterozygous, 130, 131
    independent assortment of, 131–132
    mutation, 134
allelopathy, 641
allergens, 397
Allium, 41
    cepa, **19,** 41, **449,** 473, 517, 619
    porrum, **239**
allolactose, 147
Allomyces,
    arbusculus, 252
    life cycle of, **184, 253**
allosteric effector, 147
allspice, 695
almonds, 558
Alnus, 197, 517, 683 (*see also* alders)
    rubra, **197, 459**
Alopecurus pratensis, 398
alpha-amylase, 557
alpha-glucose, 49
alpha helix, 57, **58,** 80
    and transmembrane proteins, **67**
alternation of generations, 183, 184
    in algae, 253, 274, 276, 283
    in ancestor of plants, 298
    in bryophytes, 298, **302–303**
    in chytrids, 252, **253**
    heteromorphic, 184, 253, 274, 276, **277,** 298, **302–303,** 322, **328**
    isomorphic, 184, **253,** 274, **277,** 286
    in vascular plants, 253, 298

aluminum (Al), 513
Amanita,
    muscaria, **3**, 232
    virosa, 234
Amaranthus, 706
    retroflexus, **99**
Ambrosia, 640, 705
    psilostachya, 389
Amaryllis belladonna (pink-flowered
    amaryllis), 423
amination, 602, 610
amino acid(s), 55, **56**, 509, **596** (*see also by name*)
    breakdown of, 91
    coding sequence, 560
    and cytokinins, 553
    and genetic code, 140, 141, **142**, 559, 560
    in phloem, 276, 631, 632
    in phytochrome synthesis, **582**
    as protein components, 55–59
    in protein synthesis, 140–142, **144**, 145
    in seeds, 557
    structure, 55, **56**
    synthesis, 593, **596**, 610, 611
aminoacyl tRNAs, 144, 145
aminoacyl-tRNA synthetases, 142, **148**
1-aminocyclopropane-1-carboxylic acid
    (ACC), **554**
amino group (-NH$_2$), 55, **56**, **720**, 722
ammonia/ammonium (NH$_3$/NH$_4^+$), 2, 198,
    253, **596**, 607, 610, 612
    oxidation of, 610
ammonification, 198, 607
amoebas, **71**
    of cellular slime molds, 71, 252, **254**, 255
    of plasmodial slime molds, **256**, 257
amoebic dysentery, **708**
amoeboid organisms, 71, 244, 246, 252,
    257–258
amoeboid plastid, **23**
Amorphophallus titanum, **134**, 417
AMP, 83
Amphibolis, **428**
amylase, 59, 557
amylopectin, **50**, 248, 270
amyloplasts, **22**, **23**, 50, 52, 575
amylose, **50**, 270
Anabaena, 194, **195**, 197, 353, 607, 608
    azollae, **198**, 353
anabolism, 78
anaerobes, **179**, 607
    facultative, 188
    obligate, 188
anaerobic conditions, 2
analogy, 403
analysis, 11
Andreaea, 315–316
    rothii, **315**
Andreaeidae, 308, 315–316
Andreaeobryum, 316
androecium, **305**, 383, 410
Andropogon gerardi, **673**
Anethum graveolens, 695
Aneurophyton, **359**
angiosperms (flowering plants), 358, 359,
    380–400 (*see also* flowers)
    adaptations of, 406
    ancestors of, 401–404, 415
    aquatic, **428**
    embryogeny, 441–447
    evolution of, 401–439
    evolutionary radiation of, 406–407
    evolutionary trends in, 410–411
    and gymnosperms compared, 393
    life cycle of *glycine max*, **394–395**
    life cycles of, 388–395
    origin of, 405–406
    parasitic, 382, **383**
    phylogenetic relationships of, 401–404

saprophytic, 382, **383**
    specialized familes of, 411–414
    vascular tissue of, 458–465
Angophora woodsiana, **415**
Animalia (animal) kingdom, 11, 173, 176,
    180
animals, 573
    cells, 26, 97
    dependency on plants, 9–10
    distribution of (*see specific biomes*)
    domestication of, 687, 688–689, 694
    grazing, 688
    nutrition, 180, 593
    and photoperiodism, 578
    as pollinators (*see under* pollination)
    reproduction of, 180
    respiration in, 87
    as fruit and seed distributors, 434–435,
        554
    soil, 597–602
    sugar transport in, 49
    water use by, 616
    as zoo plankton, 245
anions (negatively charged ions), 602, 716
    (*see also specific ions*)
anise, 695
anisogamy, 264, 286, **291**
annual rings, 535
annuals, 6, 160, 511,520, 521
annular scars, 295, **296**
annulus, 343, **344**, 351
anther, 383, **384**, 389, 395, 514, **515**
antheridia, 299
    of algae, **262**, 276, **277**, **278**, **279**, 285,
        295, 296
    of ancestor of plants, 299
    of bryophytes, 300, 301, **303**, 313
    of *Equisetum*, 339, **341**
    of ferns, 345, 349, **351**
    of fungi, 216
    of liverworts, 301, **305**
    of Lycopodiaceae, 331, **333**
    of *Marchantia*, **300**, **303**, 304
    of oomycetes, 248, **249**, 250
    of *Psilotum*, 327, **329**
    of seedless vascular plants, 323, **333**
    of *Selaginella*, 335
    of true moss, 309, **310**, 313
antheridiol, 250
antheridiophore, 301, **303**
Anthoceros, **307**, 308
Anthocerophyta, **178**, 299, 307–308, 733 (*see
    also* bryophytes; hornworts)
anthocyanins, 25, 26, 428–429, 583
Anthophyta, **174**, **178**, **183**, 358, 380–400,
    733 (*see also* angiosperms)
anthracnose diseases, 221
anthrax, 190
Anthreptes collarii, **424**
Anthurium, 416
antibiotics, **643**
    and bacteria, 190, 191
    fungal origin of, 209, 222
antibodies, fluorescent, 41
anticlinal divisions, 343, 489, **522**, 523, 605
anticodon, 142, **143**, 144, 145, **148**
antigens, 397
Antilocapra americana, **674**
antiparallel strands, **138**, 139
antipodals, 392, **393**
antiports, 70
ants, **3**, 157, 424, 434, 435, 517, 678 (*see also
    Acacia*)
apex (apices)
    floral, 511, **514**, **515**
    root, 453, 472–474
    shoot, 453, 489, **490**
    vegetative, 451, 511
aphids, 204, **631**
    in phloem research, 630, 631
Apiaceae, 173, **387**, 401, 418, 432, 695

apical cell, 272, 274, 276, 301, 309, **343**
apical dominance, 488, 548, 552
apical meristems, 6, 302, 441, 447, 453
    of embryo, 367, **443**, **445**, **446**, 447
    of liverworts, 301
    of pines, 362, 363, 367
    reproductive (floral), 451
    of root, 367, **453**, **473**, **474**, **475**, 483
    root and shoot compared, 453, 489
    of shoot, 321, 367, 447, **453**, **488–491**,
        507
    vegetative, 451, 511
apical pore, **284**
Apis mellifera, **418**
Apium
    graveolens, 518
    petroselinum, 618
apoenzyme, 81
apogamy, 352
apomixis, 166, 451
apoplast, 72
apoplastic transport (movement, pathway),
    72, 624, **625**
apothecia, **215**, 216
apomixis, 451
apple, 388, 430, 450, **467**, **521**, 528, 550, 587
    diseases of, **188**, 250
appressoria, 227
aquatic plants, 510, 607, **608**
    organisms, 245, 248, 250
Aquilegia canadensis, 424
Arabidopsis, 569
    thaliana, 568, 569, 589, **590**
Araceae, 87, **381**, 416–417
Arachis hypogaea, 515, 691
arachnids, 210
Araucaria
    araucana, 369
    heterophylla, 369, **402**
Araucariaceae, **402**
Araucarites longifolia, **402**
arbuscles, 239–240
Arceuthobium, **433**
Archaeanthus, 401, 430
    linnenbergeri, 408, **409**
Archaebacteria, 11, 175, 176, **178–179**, 186,
    188, 191–192, 195, 728 (*see also*
    bacteria)
Archaeopteris, 360
    macilenta, 360
archegonia,
    of ancestors of plants, 299
    of bryophytes, 300, 301, **303**, 313
    of *Equisetum*, 339, **341**
    of ferns, 345, 349, **351**
    of gymnosperms, 323, 361
    of liverworts, 301, **305**
    of Lycopodiaceae, 331, **333**
    of *Marchantia*, **300**, **302**, **303**
    of pine, 367, **371**
    of *Psilotum*, 327, **329**
    of seedless vascular plants, 323, 333
    of *Selaginella*, 335
    of true moss, 309, **310**, 311, **313**
archegoniophore, 301, **303**
Archeosperma arnoldii, **358**
Arctic tundra, 8, 683–684
    biome map of, **684**
    ice desert, 684
    distribution of, 683
    soils of, 683
    vegetation characteristics of, 683–684
Arctium, **431**
Arctophila fulva, **684**
Arctostaphylos, 161, 448
Arcyria nutans, **257**
areoles, **504**
arginine, **56**, 122
Argyroxiphium, 164
    grayanum, 164
    sandwicense, 164, **165**
aril, 369, **372**, 434
Aristolochia, 527
Aristotle, 99, 599

aroid family, **381**, **416–417**
aromatic compounds, 429
arrowroot, **446**
arrow-woods, 402
arroyos, 670
Artemisia
    biennis, 172
    dracunculus, 696
    frigida, **674**
    tridentata, **668**, **669**
Arthrobotrys dactyloides, 223
arthropod(s), 214
    exoskeleton, 210
    as soil components, 598, **599**
ascidians, 196
Asclepiadaceae, **418**, 436, 518, **644** (*see also*
    milkweed)
Asclepias, **431**, 432
    curassavica, 436
Asclera ruficornis, 415
ascocarp (*see* ascoma)
ascospores, 215, **216**, 218
Ascodesmis nigricans, **217**
ascogonium, **216**
ascoma (-ta), **215**, 216, **217**, 226
Ascomycota (ascomycetes), **178**, 211, 212,
    215–219, 220, 222, 223, 241, 729
    evolution of, 211
    life cycle of, 215–216, 218
ascus (-i), 215, 216, 218
asexual reproduction, 450, 451
    in algae, 258, 259, 264, 265, 283, 284,
        286, 290, 291, 293, 295
    in bryophytes, 304, 311
    in diatoms, **260– 261**
    in euglenoids, 265
    and evolution, 156–157
    in fungi, 212, 213, **216**, 220
    in hybrids, 166
    in vascular plants, 157, 158, 161, 450,
        451
    in protists, 248–249, 255
    in seedless vascular plants, **163**
ash, **431**, 432, 538, 621
    green, 490, **499**, **530**
    white, **542**
asparagine, **56**, 606
asparagus, 515, **516**
Asparagus officinalis, 515, **516**
aspartate (aspartic acid), **56**, 111, **112**
aspen, quaking, 650
Aspergillus, **219**, 220, 222
    flavus, 222
    fumigatus, 219
    nidulans, 221
    oryzae, 221, 222
    parasiticus, 222
    soyae, 221
aspirin, 416
Asplenium
    rhizophyllum, 450, **451**
    septentrionale, **342**
    viride, 661
assimilate transport, 629–633
asters, 676
Asteraceae (Compositae, composites), 164,
    173, **389**, 390, 398, 413–414, **431**,
    432, 705
athlete's foot, 222
atmosphere, 47, 115, 600
    early, 2
atom, 713–717 (*see also specific elements*)
    models of, 714, 715, 716
    naturally occurring, 47, 713
    in oxidation-reduction, 723, 725
    structure of, 714–717
atomic number, 713
atomic weight, 714, 719
ATP, 61, 83, 84
    in active transport, 68, 70, 597, 632
    and bioluminescence, 97
    in biosynthesis, 554
    in glycolysis, fermentation, 87–90, 97,
        98

in nitrogen fixation, 602
in photosynthesis, 101, 108–110, 112
in respiration, 24, 92
in touch response, 576, 589
ATPase, 68, 70, 95, 632
ATP synthase, **95, 96, 108**
atrazine, 551
Atriplex, 467, 609
  nummularia, **706**
aureomycin, 190
Auricularia auricula, **231**
Australopithecus, 686
autoecious parasite, 235, 238
autoradiography, **474**, 633, 714
autotrophs, 245
  chemosynthetic, 610
  photosynthetic, 3, 4, 11, **244**, 246,
    257–266
auxiliary cell, **273**, 274
auxin, **546**, 547–551 (*see also* IAA; 2, 4-D; 2,
    4, 5-T; 2, 3, 7, 8-TCDD; NAA)
  and the control of abscission, 550, 551
  and apical dominance, 548
  and cell differentiation, 548–550
  and cell division, **559**
  and cell elongation, 36, 66, 548
  and cell expansion, **559**, 560, 561
  and cell turgor, 66, 560
  and circadian rhythms, 576
  commercial uses of, 550, 551
  -cytokinin interactions, 550, 553
  and ethylene, 551
  and fruit growth, 550, **551**
  and gene expression, 559, 560
  -gibberellin interactions, 560, 561
  and gravitropism, 574
  inhibitory effects of, 548, 551
  -inducible genes, 574
  and leaf development, 348
  and phototropism, 547, 573, 574
  and plasma membrane, 560
  and root growth, 550, **553**
  synthetic, **547**, 551
  in tissue culture, 552, 553, 563
  transport, 548
  and vascular cambium, 523, 550
auxospores, **260–261**
Avena, 547
  sativa, 114, 470, 547
avocado, 251, 554, 691
Avogadro's number, 719
Axelrod, Daniel, 406
axial cell, **273**
axial core, **124**
axial system, 522, 528, 532, 535
axil, leaf, **7**, 450, 488, 500, 516
Azolla, 197, **198**, 353, 607
  filiculonides, 608
Azospirillum, 613
Azotobacter, 607

bacilli, 188
Bacillariophyceae, 257, 258–261 (*see also*
    algae)
  cereus, 641
  thuringiensis, 566
bacteria, 11, 16, 173, 175, 178–179,
    186–200, 244, 475 (*see also by name*)
  aerobic, 177
  anaerobic, 177, 188, 191
  autotrophic, 188, 190, 191
  capsule of, 187
  cell wall of, 179, 187, 188, 190, 193
  chemosynthetic autotrophs, 188, 190,
    198, 610
  chlorophyll in, 105, **107, 187**, 192, 195,
    196
  chromosomes of, 135, 146–147, 186,
    187, 190, 191
  classification of, 187–188, 190, 200,
    728–729
  commercial uses of, 190

conjugation of, **189**, 191
DNA of, 135, 175, **186**, 187, 190–191
as decomposers, 188, **189**, 190, 196 (*see
    also* decomposers)
ecological role of, 188, 190
endospores of, 190
in evolution, 4, 24, 191, 192
and fermentation, 97
filamentous, 188, 190, 193, 197
flagella, 188, **189**
function of pigments in, 105
gene regulation in, 146–147
in genetic engineering, 191, 566, **567**,
    568, **570**, 571
genetic recombination in, 175–176, 183,
    188, 191
gliding, **189**, 193
gram-positive and negative, 187, 198,
    222, 641
green sulfur, 105, 195, 196
habitats of, 190, 192, 193
halophilic, 191, **192**
heterotrophic, 191, 245
and human diseases, 190
major groups of, 190–200, 728–729
membranes of, **187**, 192
metabolism of, 97, 190, 191
methanogenic, **179**, 191, 192
movement of, 178, **189**, 193
mutations in, 135
nitrifying, denitrifying, 198, 610, 611
nitrogen-fixing, 190, 191, 193, **195**,
    196–198, 593, 595, 602–607
and the operon, 146–147
pathogenic, 188, 190, 195, 198–200
photosynthetic, 100, 105, **187**, 188,
    195–196
pili, 188, **189**
pigments of, 187, 191, 192, 196
and plant diseases, 188, 190, 195, **196**,
    198–200
and protein synthesis, 145–147, **148**
purple, 105, 107
purple nonsulfur, 177, 195, 196
purple sulfur, 100, 195, 196
reproductive techniques of, 190, 194
ribosomes of, **143, 148**, 175, 187
saprobic, 188
saprophytic, 188, 607 (*see also by name*)
sheath, 187, 192
storage products of, 192
structure of, 187–188
in sulfur cycle, 190, 191
symbiotic, 194–195, **197**, 604–607
thermophilic, 191
bacteriochlorophyll, 105, **107**
bacteriology, 11
bacteriophage, 201, 205
bacteriopheophytin, **107**
bacteriorhodopsin, 191
bacteroids, 604, **605**, 606, **607**
Baldwin, Bruce, 164
balsa, 532
banana(s), 168, **188**, 200, **381**, 424, 425, 429,
    689, 690, 696, 698
banyan tree, 483
BAP (6-benzylamino purine), **552**
bar (unit of pressure), 63
Barbarea vulgaris, 595
barberry, **236–237**, 238
bare zones, 641
bark, 364, 527–532, 630
  types of, 531
Barksdale, Alma, 250
barley, 452, 557, 613, 688, 698
  foxtail, **183**
  seeds, **557**
barley malt, 557
Bartholomew, Bruce, **641**
basal bodies, 32, **266**, 270, **280**, 282
basal cell, **445**, 446
basidioma(ta), **229**, 231, 233, 234
Basidiomycota (basidiomycetes), **174, 178,**

208, 211, 212, 218, 219, 220, 223,
    228–238, 241, 730
evolution of, 211
life cycles of, 229, **236–237**
basidiospores, basidia, 229, **230**, 232, 233,
    234, **235, 236**, 238'
basil, 695
basswood, 457, **491**, 492, 524, **525, 528, 530,
    542, 631**, (*see also* linden)
Bateson, William, 133
Batrachospermum, **271**, 272, 274
  moniliforme, **269, 271**
  siridotia, **271**
bay leaf, 696
Beagle, H.M.S., 150
bean, 199, 430, **442**, 448, 449, 548, 549, 556,
    575, 577, 583, 602, 697, 698 (*see also
    Phaseolus*)
  broad, **7**, 604, 633
  castor, **442**, 449, **460**
  curd, 221
  kidney, 691
  lima, 691
  winged, **704**
beech(es)
  American, **542**
  family, 240
  southern, 240, **408**
beer, brewing, 209, 218–219, 557, 688, **698**
bees, 396, 416, 418–422
  carpenter, **163**
  honey-, **381**
bee's purple, **420**, 429
beetles, 87, 155, 234, 374, 401, **415**, 416,
    417, 418, 435, 436, 437
beets, 429
beggar-ticks, 412
bell-shaped curve, 128, **129**
Belt, Thomas, 638–639
Beltian bodies, 638, **639**
6-benzylamino purine (BAP), **552**
Berberis, 238
berry, 429–430, **434**
beta-carotene, **105**
betacyanins (betalains), 429
beta glucose, 51
β-glucuronidase gene (GUS), 570
beta pleated sheets, 57, **58**, 80
Beta vulgaris, **486**, 697
  stomata, **6**
Betula, 538, 683
  alleghaniensis, **542**
  papyrifera, **427, 531**, 644
B horizon of soil, 598
bicarbonate(s), 112
Bidens, 412
biennials, 470, 520, 521, 558, 584, 587
big tree, 372, 679
binomial system of nomenclature, 172
binomial theorem, 152
binucleate cells, **265**
biochemical coevolution, 435–438
biochemical methods, 404
biological classification, **174**
biological clocks, 577, 578, 587, 588 (*see also*
    circadian rhythms)
biological control of pests, 214, 252
biological time clock, **10**
bioluminescence, **97**, 570, **578**
bioluminescent algea, **264**
biomass, 646, 648
biomes, 6, 657–685 (*see also under specific
    types*)
  Arctic tundra, 683–684
  and cacti, 671
  categories of, 662
  competition for light in, 676
  defined, 657
  deserts, 668–671
  distribution of, 660, **658–659**
  and distribution of organisms, 660
  and earth's biological productivity, 660
  grasslands, 672, 674
  and jobs versus owls, 682

maps of, **665, 667, 668, 670, 673, 675,
    678, 679, 681, 684**
mediterranean scrub, 680
and patterns of wind and rainfall, **660**
rainforests, 662–665
and relationship between altitude and
    latitude, **661**, 664
savannas and deciduous tropical forests,
    666–667
taiga, 681, 683
temperate deciduous forests, 645,
    674–675, 677–678
temperate mixed and coniferous forests,
    678–679
biosphere, 3, 10, 99, 208, 268
biosynthetic pathways, 83, 84, 562
biotechnology, 201, 563–571
  and crop improvement, 565, 566, 570
  gene transfer in, 565, 566, **567**
  tissue culture in, 564, 565
biotic components in ecosystems, 646
biotic variation, 680
birch(es), 396, 397, 538, 683
  paper, **8, 427, 531**, 644
  yellow, **542**
bird-of-paradise, **424**
birds, in fruit and seed dispersal, 448
bird's-nest fungi, 234, **235**
birth control, 698
birth-control drugs, 707
bivalents (chromosomes), 124, 125, 126, **127,**
    132
blackberries, 166, **228**, 513, 650
black-eyed Susan, 413
Blackman, F. F., 101
bladder
  air (algal), **275**, 276
  salt, 609
bladderwort, 610
blade (algal), 274, **275**, 276, **277**
blazing star, **471**
blights, 199
  chestnut, 215
  fire, **188**, 199
bloodroot, 435, 676
blooms
  of algae, **258**, 646
  of cyanobacteria, 193
  of dinoflagellates, 263
bluebell(s), 171, **385**
blueberries, 683
blue-green algae (*see* cyanobacteria)
body cell, spermatogenous, 367
bogs, 610, 613
Böhr, Niels, 715
bolete, **232**, 233
Boletus elegans, 638
bolting, **558**
Bombus, 421
bonds (chemical), 77, 78, 83, 84, 717–719,
    724 (*see also organic compounds*)
  carbon-hydrogen, 52, 718
  conjugated, **104**
  covalent, 84, 717–**718**, 719
  disulfide, 57
  double, 52, 53, **718**, 719
  energy, 83, 84, 724, 725
  high-energy, 84
  hydrogen, 57, **58**, 59, **137, 138**, 139, **143,**
    720
  ionic, **717**
  peptide, 57, **58, 144**, 145, **148**
  polar covalent, **719**
  phosphate, 83, 84, 89
  radiation effects on, 103
  triple, 718
Bonnemaisonia hamifera, **271**
Bonner, James, 580, 581, 584
boojum trees, **669**
Bordeaux mixture, 250
bordered pit, 36, 532, 533, **534**
Borlaug, Norman, **701**
boron (B), 594, **596**
  ions (B(OH)$_4^-$), **596**

Borthwick, Harry A., 582
botany, science of, 10–12, 136
Botrychium, 343, 344, 345
    parallelum, 344
botulism, 190
Bougainvillea, 429
boxelder, **530**
Brachythecium, **314**
bracket fungi, 232
bracts, of flower, **427**
Bradyrhizobium, 196, 197, 602, **603**, 604, 606
    japonicum, 604, **605**
bran, 452
branch gap, **497**
branch traces, **497**, 498
Brassica
    napus var. *napobrassica*, 627
    nigra, 436, 696
    oleracea, 166
    oleracea var. *capitata*, 518, 558, 683
    oleracea var. *caulorapa*, **517**, 518
    rapa, **431**
Brassicaceae, 166, 238, 398, 430, **431**, 435, 568
bread, baking, 209, 218–219
bread wheat, 167
breeding (*see also* plant breeding)
    cross-, **153**
    inter-, **153**, 160
    random, **153**
    selective, 643, 701
Briggs, Winslow, 573, 574
bristlecone pine, 535, 536, **537**
British soldiers (lichens), **225, 228**
Bromeliaceae, **225**
bromeliads, 467
brown algae, **5**, 247, 248, 257, 269–270, 274–279 (*see also* algae)
    conducting tissue in, 270, **276**
    economic uses of, 274, 276, 281
    life cycles of, 270, **277**
Bruchidae, bruchids, 155
Bryidae, 308, 309–314
Bryophyta (*see also* mosses),
    life cycles of, **178, 183**, 299, 308–316, 733
bryophytes, 298–316, 322 (*see also* hornworts, liverworts, mosses)
    in biological succession, **650**
buckeye, red, **499**
buckwheat family, 432
bud(s), **7**, 500, **530**
    accessory, **530**
    dormant, 555, 585, 586, 587
    growth of, **490**, 523, 552
    and hormones, 523, 547, 552, 553, 555
    lateral (axillary), **490**, 511, **530**
    in mosses, 309, **313**, 314
    primordia, 488, **489, 507**
    scales, **490**, 586
    terminal, 440, **490, 530**
    winter, 586
budding, in yeast, 218
bud-scale scars, 530
bugs, true, 435, 436
bulbs, **417, 449**, 450, 587
bulk flow, 62, 63, 631, 632
bulliform cell, 505, **506**
bundle scar, **530**
bundle sheath
    in leaf, **501, 502,** 504, 505
    in stem, 494, **495, 496,** 497
bundle-sheath cells, **16, 17,** 54, 112, **113, 114,** 504, 505
bundle-sheath extension, **502,** 504
burdock, **431**
butter-and-eggs, **385,** 432
buttercup, 476, 494, **495**
    family, 432
butterflies, 421–423, 436
    cabbage, 435
    copper, **422**
    monarch, 436, **644**
    orange tips, 435

viceroy, 436
butternut, 530
button stage, 232
buttressed trees, **662–663,** 664

cabbage, 166, 435, 518, 558, 683, 697
cacao, 250, **414,** 691, 696, **697**
Cactaceae, 116, 518
Cactoblastis cactorum, 642
cactus, 116, **419,** 424, 429, **516,** 518, 620, 669 (*see also by name*)
    barrel, **669,** 671
    Christmas, **203**
    fishhook, **183**
    organ-pipe, **425**
    peyote, **637**
    prickly-pear, **434,** 642
    saguaro, 4, 198, **382, 425, 669**
    spineless, **516**
cadmium, 513, 615
caffeine, 436, 696
calactin, 436
calamites (*Calamites*), 338, 346, **356**
calcium (Ca/Ca²⁺), 74, 589, 594, 595, **596,** 597, 609, 629, **713,** 716
    -binding protein, 575
    carbonate (calcite), **247, 258,** 270, **466,** 597
    cycling, 645
    deposits (stromalites), **193**
    and hormones, 575
    -inducible genes, 574
    oxalate, 25, 435
    as second messenger, 563
Callitriche heterophylla, 510
Callixylon newberryi, 359
callose, **276,** 461, 462, **464**
callus, 549
    culture from, 549, 550, 553, **564,** 565
calmodulin, 575, 589, 596
Calocedrus decurrens, **183, 651**
Caloplaca sacicola, 224
caloric, 76, 77
calories
    from ATP hydrolysis, 84
    defined, 725
    from glucose oxidation, 76, 77, 86, 96
calorimeter, 76, **725**
Calothrix, **194, 195**
Caltha palustris, **420,** 429
Calvin, Melvin, 110
Calvin cycle, 110, 111
Calypte anna, **423**
calyptra, 300, **302,** 311, **312, 328**
calyx, 383, 387, 408, **432, 514, 515**
cambial cells, **52**
cambial initials, 521, 522
cambial zone, 523, **528, 529**
cambium
    cork, 6, 320, 453, 467, 480, 482, 520, 525, **526,** 528
    fascicular, 494, 523, 531
    interfascicular, 494,523
    of *Isoetes,* 335
    reactivation of, 523, 550
    storied, nonstoried, **521**
    supernumerary, 486
    vascular, **6,** 320 (*see also* vascular cambium)
Cambrian period, 270
Camellia sinensis, 696
cAMP, 253, 256
CAM photosynthesis, 116,117, 518, 620
CAM plants, 116, 117, 518, 618
    stomatal behavior in, 117, 620
cancer, 12, 134, 222, **707,** (*see also* tumors, plant)
Candida albicans, 222
Cannabis sativa, 386, 436, **437**
cap (*see also* annular scars)
    of mushroom, 232
Capsella bursa-pastoris, **444, 445**
Capsicum, 691, 695, 696

frutescens, **203**
capsid, 200, **202,** 204
capsules (*see also* sporangia)
    of angiosperms, 430, **431,** 433
    bacterial, 187
    of bryophtes, 301, **304, 305, 306,** 310, 311, **312, 314, 315,** 316
    of ferns, **343,** 344
caraway, 695, 696
carbohydrates, 48–52, **247,** 248, 265, 688 (*see also by name*)
    in carbon cycle, 115
    formation of, 99, 109–112
    in human diet, 688, 697
    in membrane structure, 57
    oxidation of, 86–98
carbon (C), 2, 47, 594, **596**
    atomic structure and bonding of, 715, 718, 719
    -containing compounds, 47, 48
    fixation of, 101, 109, 111, 112, 116
    radioactive (¹⁴C), 53, **110,** 630, 633
carbon cycle, 115
carbon dating, 53, 586
carbon dioxide (CO₂)
    in atmosphere, 2, 12, 115
    in CAM plants, 116, 117
    in carbon cycle, 115
    in C₃ pathway, 110
    in C₄ pathway, 111, 112
    from fermentation, 97, 98, 218
    fixation of, 109–112
    and greenhouse effect, 12
    from Krebs cycle, 91, 92
    in photosynthesis, 53, 100, 101, 617
    radioactive, 714
    and stomatal movements, 619
carbonic anhydrase, 79
Carboniferous period, 212, 322, 323, 330, 338, 343, 345, 346, 347, 361, 374
    swamp forest, **324, 356–357,** 401
carboxyl group (-COOH), 52, **53,** 720
carcinogenic substances, 209, 374
cardamom, 695
Carex aquatilis, **684**
carinal canals, 338, **339**
Carnegiea gigantea, **198, 382, 669**
carnivores, 268
carnivorous plants, 517, 589, 610
carotenes, 105
carotenoids, 20, 22, 105, **187,** 192, **247,** 257, 262, 265, 269, 274, **279,** 298, **389,** 428–429
carpels, 383, **384,** 410, **514,** 515, 569
carpogonial branch, 273
carpogonium, carposporangium, carpospores, carposporophyte, **271, 273,** 274
Carr, Gerald, 164
carrageenan, 270, **271,** 281
carrier-mediated diffusion, **68**
carrier proteins, 68
carrot, 470, 485, 558, 563
Carum carvi, 695
Carya
    cordiformis, **542**
    ovata, 440, **499, 520, 531**
caryopsis, 432
Casparian strip, 54, 478–480, 624, **625**
cassava, 694
castor bean, **442, 449, 460**
catabolism, 78, 84
Catalpa, 538
catalysts, 59, 79, **722,** 723
caterpillar, **217**
Catharanthus roseus, **707**
cation (positively charged ions; divalent and monovalent), 601, 602 (*see also specific ions*)
    exchange, 601, 602
catkins, **384**
catnip, 172
cattail, 382, **649**
Cattleya, **413**

caulimoviruses, 201, 204
cavitation, 622
Ceanothus, 161, **162,** 197, 607
    cuneatus, **162**
    gloriosus, **162**
    masonii, **162**
    ramulosus, **162**
    sonomensis, **162**
cedar
    incense, **183,** 651
    western red, **543**
celery, 518
cell(s), 15, 18 (*see also* specific cell types)
    chemical components of, 47–61
    cycle, 38–44
    and diffusion, 63–66
    division (*see* cell division; cytokinesis; meiosis; mitosis)
    eukaryotic, **4,** 15–46 (*see also* eukaryotes)
    eukaryotic and prokaryotic compared, 16, 121
    fossils, 1–2, 4
    microscopy of, 26, 27
    prokaryotic, 16 (*see also* bacteria; prokaryotes)
    sap, 25
    structure, 15–46 (*see also* specific types)
    theory, 18
    and totipotency, 563
    types in plants, 455–469
cell differentiation, 454, **460, 464**
    and hormones, 548–550, 559–563
cell division, 38–44 (*see also* cytokinesis; meiosis; mitosis)
    in algae, 259, **280, 282, 283,** 284, **290**
    in embryo, 444–447
    and hormones, 553, 559
    in plant growth, 453, 454, 474, 489–491, 506–508, 521–523
    in root nodule, 604, **605**
    in slime molds, 252
cell elongation, 36, 37
    and hormones, 510, 559–562
    in plant growth, 474, 510
cell enlargement, 25, 36, 37, 510
cell expansion, 510, 559, 588
cell membrane (*see* plasma membrane)
cell permeability, 65, 66, 68, 597
cell-plate formation, 43–45
    in algae, 270, **280, 282,** 283, 294, **295**
    in bryophytes and vascular plants, 298
cellular membranes, 16–18, 62, 66–68, 597
cellular respiration, 24, 62, **75,** 86, 87, 115
cellular slime molds (*see* slime molds)
cellulose, 33–35, 50, 51 (*see also* cell wall)
    and bacteria, 191
    in cellular slime molds, 252, 253, **254,** 255
    in cell wall structure, 33–35, **247,** 270, **280,** 298, 618, **619**
    in cuticle, 54
    in dinoflagellates, **262**
    enzymatic digestion of, 50
    in pollen grains, 388
    in root, 478–480
    in sclereids, 34
    synthesis, 37
cell wall, 17, 33–38
    of acritarchs, **244**
    algal, **247,** 257, 270, 272, 280, 281, 285, 288
    bacterial, 187, 188, 191
    calcareous, 270
    of chytrids, **247,** 251
    components of, 33, 34, 50, 53, 54, 270, 288, 290, 597
    of chrysophytes, **247,** 257
    of diatoms, **247**
    of dinoflagellates, **247,** 264
    of euglenoids, **247**
    formation of, 29, 30, 43, 44, **127,** 392
    fungal, 210, 211, 248
    growth of, 36, 37

of guard cells, 618, **619**
movement across/along, 622
mucilaginous compounds in, 472
of oomycetes, **247**, 248
and plant hormones, 560, 562
of plant tissues, 298, 455–462, 478–480,
    509, 510
pressure, 66
primary, 34, 35, 44, **45**, 457, 460, 461
secondary, 35, 36, 455, 457–460
of slime molds, **247**, 252
Cenococcum geophilum, 638
Center for the Improvementof maize and
    wheat (CIMMYT), **701**
Cenozoic era, 679
central cell, 392
central dogma of genetics, 141
central mother cells, **489**, 490
central pore (fungal), 211
centriole, 42, 248, 252, 255, 270
flagellar, **280**
centromere, 39, 41, **42**, 124, 125, **126**, **127**,
    132
centrosome, 41, 42
century plant, 518
Cephaleuros virescens, **227**
Cephalozia, **306**
Cerambycidae, 401
Ceratium, **262**
    triops, **264**
Cercis canadensis, **25**
cereals/grains, 235, **443**, 449, 450, 688
cerrado, 666
CFCs, 12
Chaetophorales, 294
Chailakhyan, M. Kh., 584
chalaza, chalazal pole, **391**, 392, 444
chalcone, 429
Chaney, Ralph, 374
channel proteins, 68, 70
chaparral, 448, **657**, 680
pine-oak, **8**
Chara, 33, **285**
Charales, 284, 285
chard, 697
charge, electrical, 714–717
Charophyceae, **280**, 282, 283–285
    cell division in, **280**, 282
cheese, 190, 209, 221
chemical defenses, 33, 642
chemical messengers (*see* hormones)
chemical mutagens, 134
chemical reactions, 76–80, 83, 84, 713,
    722–725
chemical reactivity, 714–717
chemiosmotic coupling, 95, **96**, **108**
chemosynthetic autotrophs (chemoauto-
    trophs), 188, 190, 198, 610
Chenopodiaceae, **692**
chenopodiales (centrospermae), 429
Chenopodium
    glaucum, **497**
    quinoa, 691, **692**
chernobyl nuclear power plant, 227
cherry, **387**, 430, 434, 450, 539, **542**
    black, 539, **542**, 650
chestnut, 587
    horse, 521, **530**
chiasma (-ata), 124, **125**, 126, **127**, 133
chicken pox, 200
chickory, 412
chickpeas (garbanzos), 688, 691
Chimaphila umbellata, **440–441**
chitin, 52, 181, 210, **247**, 248, 251
chitinase, 610
Chlamydomonas, 175, 279, **280**, 289, **293**, 294
    life cycle of, **184**, **290**, 291
    nivalis, **279**
chloracne, 551
chloramphenicol, 21
Chloranthaceae, 408
Chlorella, **246**, **269**, 292, 294
chlorine/chloride (Cl/Cl⁻), 594, **596**, 629,
    713

Chlorococcum echinozygotum, **291**
chlorofluorocarbons (CFCs), 12
Chloromonas, 279
    brevispina, **279**
    granulosa, **279**
Chlorophyceae, **280**, 282, 289–296
    cell division in, **280**
chlorophyll(s), 1, 20, 103–105, 107, 159,
    246, **596**
    a, 104, 105, 107, **187**, 192, 195, 196,
        245, **247**, 257, 258, 265, 269,
        274, 280, 298
    absorption spectra of, 103, 104, 105, 582
    b, 104, 105, **187**, 196, **247**, 265, 269, 280,
        298
    in bacteria, **187**
    c, 105, **247**, 257, 258, 262, 269, 274
    chlorobium, 105
    structure of, **104**
Chlorophyta, **178**, **180**, **247**, 269–270,
    279–297, 732
chloroplast, **16**, **17**, 20–22, **99**, **106**
    in algae, 21, 196, 246, 257, 258, 259,
        262, 269–270, 274, 280, **283**,
        290
    in bryophytes, 298, 311
    in chrysophytes, 257, 258, 259, 262, 274
    in C₃, C₄ plants, 111, 112, 505
    in euglenoids, 265
    evolutionary origin of, 24, 177, 192,
        245–246, 269, 274
    functions of, 20, 22, 107–109
    genetic material of, 21
    in guard cells, 618
    in photosynthesis, 20, **99**, 104, 106, 107,
        110, 503, 505
    and phytochrome, 583
    and ribosomes, 145
    symbiotic, 245–246, 262, 269
chocolate, **414**, 691, **697**
cholera, 190
cholesterol, 67, 452, **707**
Chondromyces crocatus, **189**
Chondrus crispus, **271**
Chorizon of soil, **598**, 600
Chorthippus parallelus, **125**
chromatid
    in meiosis, 124, 125, **126**, **127**
    in mitosis, 39–41, **42**
    sister, 125
chromatin, 19, 39, 122
chromatophores, 187
chromenes, 643
chromomeres, 122, **123**
chromophore, 582
chromoplasts, 22, **23**
chromosome(s), 16, 19, 122–128 (*see also*
    DNA)
    and amino acids, 122
    bacterial, **186**, 190, 191, 265
    bivalents, 122, 123, 124
    and centromere, 125
    and chiasmata, 124, **125**, **127**
    and chromomeres, 122, **123**
    crossing-over in, 124, 133
    daughter, **39**, 40, 42, 43, 125
    of dinoflagellates, 265
    and DNA, 16, 19, 122, **123**
    of euglenoids, 265
    eukaryotic, 122, 123, 176, 248, 265
    fungal, 248
    and histones, 122, **123**, 176
    homologous, 122, 124, 125, **126**, **127**,
        130, 133
    independent assortment of, 131
    in mitosis, 38–43
    and nucleosomes, 122, **123**
    number, 19, 20, 122, 126, 135, 345, 392
    of oomycetes, 248
    and polyploidy, 166–168
    and proteins, 16, 19, 122
    segregation of, 131, 166
    sets, 122, 135, 166, 167, 168
    of slime molds, 255

structure of, 122–123
    and synaptonemal complex, 124, 125
chrysanthemum, 156, 168, 204, 579, **580**
Chrysanthemum indicum, 584, 643
chrysolaminarin, **247**, 248, 257, 258
Chrysophyceae, 257–258 (*see also* algae)
Chrysophyta (chrysophytes), **154**, **161**, 232,
    178, **180**, **247**, 248, 257–262, 731 (*see*
    also diatoms; golden algae;
    life cycle of, 260,–261
Chrysothamnus, 668
chytrids (Chytridiomycota), **178**, 179, 183,
    245, 246, **247**, 251–253, 730
    life cycle of, 252, **253**
Chytridium confervae, 252
Cicer arietinum, 688
Cicuta maculata, 385
cider, 219
ciguatera, 263
ciliates, 244
cilium (-a), 31 (*see also*, flagellum)
Cinchona (cinchona), **437**, **696**
Cinnamomum zeylanicum, 695
cinnamon, 695
circadian rhythms, 576–578m 620 (*see also*
    biological clocks)
    endogenous, 577
    entrainment, 577
    free-running period, 577
    Zeitgeber, 577
circinate vernation, 348
Cirsium pastoris, 412
cis-aconitate (*cis*-aconitic acid), **91**
cisternae, 28, 29
citrate (citric acid), 91, 222
citrus, 25, 155, 250, 551, 689, 690
    canker, 200
    industry, 512, 554
clade, 403, 404
cladistics, 404
cladogram, **403**, **404**
Cladonia
    cristatella, **225**, **228**
    subtenuis, **225**
Cladophora, **286**
cladophylls, 515, 516
Cladosporium herbarum, 209
clamp connections, 220, **231**
Clarkia, **399**
    cyklindrica, **399**
    heterandra, **399**
Clasmatocolea
    humilis, **305**
    puccionana, **305**
class, classification by, 168, 173
classification, binomial system of, 172
    of living things, 11, 171–185
    of major groups of organisms, 175–176,
        282
    of plants, 11, 181, 348, 349, 728
clathrin, 71
Clausen, Jens, 158
Clavariaceae, **181**
Clvibacter, **199**
Claviceps purpurea, **221**
clay, 600, 601
Claytonia virginiana, 435
clear zone, 39, 40
clear cutting, 665
cleistothecium 216, **217**
climacteric, 554
climate, 535, 536, **537**, 660–662
clinical variation, 158–159
clines, 159
clonal reproduction, 565, 641
clones, 565, 569
cloned gene, 569
Clostridium, 607
    tetani, **188**
clover, 155, **421**, 587, 602, **603**, 606, 641
cloves, 695
club moss(es), **183**, 308, **319**, 323, 330–331,
    332–333
Clupea harengus, **647**

CoA, 90–92
coal age plants, 346–347
coal deposits, 346
coalescence, 386, 412, 515
coastal regions, 4, 5
coated pits, 71
coated vesicles, 71
cobalt (Co/Co²⁺), **594**, 595, **596**
coca, **437**
cocaine, **437**
cocci, 188
coccolithophorids, 258
cocklebur, 553, 579, **580**, 584
cocoa, 436, 691, **697**
coconut, 204, 430, 433, 698,
    milk, **430**, 551, 552, 563
    palm, **196**, **430**
Cocos nucifera, **196**, **381**, **430**, 551, 697
coding sequence, 559
Codium, **288**
    magnum, 279, 288
codominants, 640
codon(s), 141, **142**, 144, 145, **148**
    initiation, 144
    termination (stop), 141, 144, 145
    stop signals, **142**
Coelomomyces, 252
coenocytic organisms, 211, 248, **249**, 251,
    262, 270, 283, 286, 288, 292
coenzyme, 81, 82, **596**
coenzyme A, 90–92
coenzyme Q, **94**
coevolution
    of animals and angiosperms, 434
    of flowers and pollinators, 415, 421,
        425, 437
    of fruits and dispersal agents, 432, 437
cofactors, 81, **596**
Coffea arabica, 436, 690, 696
coffee, 436, 690, 696
cohesion (cohesiveness), 64, 622
cohesion (adhesion) tension mechanism,
    620–624
Colacium, 265
colchicine, 166, 702
cold
    and dormancy, 585–587
    and flowering response, 587
    and germination, 585
    hardiness, 586, 587
colds (viral), 200, **202**
Cole, Lamont, 647
Coleochaetales, 284
Coleochaete, 270, 280, **284**, 285, 307, 318
coleoptile, **443**, 444, 449, 450, 547, **557**, 573,
    574
coleorhiza, **443**, 444, 449, 450
Coleus, 489
    blumei, **489**, **507**
collenchyma, 456, 457, **468**, 491, 492, **493**,
    494, **500**, 504
Colocasia esculenta, 689
colonies
    algal, 258, 265
color of wood, 539
columbines, **424**, 430
    red, 424
columella, 575
column of flower, 413
combustion, 115
Commelina communis, 619
communities, 6, 637–656(*see also* biomes;
    ecosystems)
    climax, **650**, 653, 655
    defined, 637
    development of, 649–655
    evolution of, 8–10
    recolonization/regeneration in, 651–655
comoviruses, 204
companion cell, **462**, **463**, 464, **497**, 632
competition, 3, 8, 160, 638, 639–641
    defined, 639
    for light, 676
competitive exclusion, principle of, 640, 641

Composite, 173
composites, 398, **412**
compost, composting, 607, 614
compression wood, 538, 539
concentration gradient, 64, 68, **632**
conceptacles, **278, 279**
condensation (reaction), 48, **49**
conducting tissue 6, 299, **308**, 317, 458–465
    (*see also specific tissues*)
    algal, 270, **276**
    bryophytes and vascular plants
        compared, 299
cones (strobili), 346, 347
    of *Abies*, 369
    of *Cypressus*, 369
    of *Cycas*, 375
    of *Encephalartos*, **375**
    of *Ephedra*, **377**
    of *Juniperus*, **372**
    of *Larix*, 369
    of lycophyte trees, 346
    of Lycopodiaceae, 331, **332**
    microsporangiate, **364, 372, 374, 377, 378**
    ovulate, 364, 365, **366, 367, 369, 372, 374, 375, 378**
    of *Pinus*, **363, 364, 365, 366, 367, 370**
    of *Taxus*, **372**
    of *Welwitschia*, **378**
    of *Zamia*, **374**
cone scales (*see* ovuliferous scales)
conidiogenous cells, 212, 218, **219**
conidium, conidiophore, 215, **216, 217, 219,** 220
conifers (Coniferophyta), 626, **178, 183,** 347, 357, 358, 361–374, 426, 626, 733
Coniochaeta, **217**
conjugation pilus, **189**
conjugation tubes of algae, 283, 290
connexons, 72
conservation of mass, 75, 76
conservation of plants, 586
consumers
    primary, 646, **647, 648** (*see also*
        herbivores)
    secondary, 268, 646, **647, 648** (*see also*
        parasites)
contraceptives, oral, 437
convergent evolution, 404, 432, 459, 518
Convoluta roscoffensis, 294
Convolvulaceae, **383,** 694
Cooksonia, **4,** 318, 325
copepods, 252
copper (Cu/Cu²⁺), 239, 594, 596, 615
copper sulfate, 250
Coprinus, **233**
    cinereus, **230**
coral fungus, 181, 232
Corallinaceae (coralline algae), 270, **271, 272,** 274
coral reef(s), 246, 264, 272
cordaites (Cordaitales), 346, 347, 361
Cordaites, **347, 357,** 358
Coreopsis, **674**
corepressor, 147
coriander, 695, 696
Coriandrum sativum, 695
cork, 6, 54, **465, 467,** 482, 525, **526, 527**
    commercial, 531
corms, 52, **335,** 450, 517
corn, **16, 17, 21, 113,** 114, **120,** 173, 386, 397, **426, 449, 450,** 571, **617, 628, 689,** 693, 696, 697, 698 (*see also Zea mays*)
    annual, 693
    belt, 672
    Black Mexican, **129**
    cob, 693
    cultivation of, **691**
    domesticated, 693
    dwarf, 556
    ear, 693
    genetics, 128, **129**
    herbicide resistance of, 551
    husks, 693

hybrid, 702
    leaf blight, **703**
    origin of, 693
    smut, **228**
    tassel, 693
    wild, 693
Cornus florida, 221
corolla, 383, 408, 410
cortex, **319,** 325, 326
    of lichen, **226**
    of root, **7, 330, 476, 477,** 478–480, **481, 482,** 624, **625,** 627
    of stem, **7, 330, 334, 339, 345, 455,** 492, **494, 523, 524, 526, 529**
cortisol, **707**
cortisone, 707
corymb, **384**
cotransport systems, 70, 632
cotton, 168, **411,** 690, 691
cottonwoods, 110, 240, **504, 542**
cotyledons, **7, 367, 371, 395, 396,** 441–447, 449, 450, 510, 511
coupled reactions, 83
cowpea, 590, 617
C₃ photosynthesis, 110, 111, 505
C₄ photosynthesis, 111, 112, 505, **596**
C₃ plants, 111, 112, 114, 116, 505
C₄ plants, 111, 112, 114, 116, 505
    efficiency of, 112, 114, 116
    evolution of, 114, 116
Crane, Peter, 402
Crassulaceae, 116, 620
Crassulacean acid metabolism (CAM), 116, 117, 518, 620
Crataegus, 166, **516**
crested wheat grass, **126–127**
creosote bushes, 537
Cretaceous period, 259, 358, 361, 372, 401, **402, 405, 406, 407,** 410, 430
Crick, Francis, 136, **137,** 138, 139
cristae, 23, **86,** 90, **96**
Crocus sativus, 696
crops, agricultural, 167, 168, 209, 565–571, 602, 612–615
    improvement of, 565, 566, 570
    new, 566, 570
cross(es), genetic, 129
    artificial, **162**
    crossbreeding, **129, 153**
    dihybrid, 129
    monohybrid, 129
    test-, **130,** 131
crossing-over, 124, 125, 126, **127,** 133, 134
cross pollination, **129,** 166
cross wall (*see* septum)
Croton, **488**
crown gall(s), 200, 566
crozier, **216,** 218
Crucibulum laeve, **235**
crucifers, 199, 398
crustaceans, 210, 245
crustose algae, **272**
crypt, **502**
cryptococcosis, 219
Chryphonectria parasitica, 215
Cryptothecodinium cohnii, **265**
crystalline bodies (diatoms), 261
crystals, 25, 201, **457**
crystal violet, 187
C-terminus, 57
cucumber, 548, **549,** 550, 555
Cucumis sativus, 548, **549,** 555, 618
Cucurbita, 691
    maxima, **456, 458, 463**
    pepo, 480
Cucurbitaceae, 555
cucurbits, 199, 555
cultivars, 558, 584, 613
culture techniques, 414, 549, 563–565
Culver's-root, **498**
cup (fungal), **232**
Cupressus, 369
    goveniana, **369**
cupules, seed, 358

Curculionidae, 374
curd, **689**
currants, 558
Cuscuta, 382
    salina, **383**
cuticle, 5, 53, **54,** 299, 308, 317, 325, 326, 338, 363, 467, 475, 492, 500, **502,** 504, 617
cutin, 34, 53, **54,** 317
cyanidin, 429
cyanobacteria (blue-green algae), 179, 187, 192–195
    accessory pigments in, 192
    in carbon cycle, 115
    chlorophylls of, 192
    and chloroplast evolution, 24, 192, 195, 269
    ecological importance, 192, 353
    evolution of, 192
    as functional chloroplasts, 195
    and glycogen, 192
    membranes of, 192
    nitrogen-fixing, 193, **195,** 197–198, 308, 374, 593, 607
    nonsymbiotic, 192
    and photosynthesis, 105
    planktonic, 193, 245
    reproduction in, 194
    ribosomes of, 192
    symbiotic, 194–195, 197–198, 223, 308
        (*see also* lichens)
cyanogenic substances, 435
Cyathea, 342, 343
    australis, **327**
cycadeoids (Cycadeoidophyta), 361, 374, 404
cycads (Cycadophyta), 178, 357, 358, 361, 374, **375, 396,** 733
Cycas siamensis, **375**
cycle(s)
    carbon, 115
    of elements, 9
    nitrogen, 602, **603**
    nutrient, 602, **603,** 611, 612
    phosphorus, 611, 612
    seasonal growth, 520, 521
    water, **600**
cyclic adenosine monophosphate (cAMP), 253
cyclic electron flow, 109
cyclic photophosphorylation, 109, 194
cyclin, 39
cyclosis, 18
cyclosporine, 209
Cyclotella meneghiniana, 259
Cymbidium, **183**
Cyperaceae, 238
cypress, 369
    bald, 372, **373, 484**
    Gowen, 369
Cypripedium calceolus, 170
cypsela, **431, 432**
Cyrtomium falcatum, **349**
cysteine, **56,** 57, 93
Cystopteris bulbifera, **183**
cytochromes, 93, 108, 109, **596, 597**
cytokinesis, 38, 43, 44, **45,** 124, **127,** 286, **295,** 552 (*see also* cell division)
cytokinin(s), 546, 551–553
    -auxin interactions, 553
    and cell division, 553, 559
    and leaf senescence, 553
    and organ formation, 553
    in roots, 470, **546**
    structure of, **552**
    synthetic, **552**
    and tissue culture, 553, 563
cytology, 11
cytoplasm, 17, 18, **20,** 25 (*see also specific components of cytoplasm*)
cytoplasmic sleeve (annulus), 72
cytoplasmic streaming, 18, 33, 627, 631
cytosine, **60,** 61, 136, 137, **138,** 139
cytoskeleton, 29, 30
cytosol, 18

2,4-D (2,4-dichlorophenoxyacetic acid), **547,** 551
Dactylis glomerata, 155, 398
daisies, 422
daisy family, 705
Danaus plexippus, **644**
dandelion, 171, **183, 412, 432**
darkness, and flowering, 580
dark reactions, (*see* light-independent reactions)
Darwin, Charles, **150,** 151–152, 160, 401, 407, 547
Darwin, Francis, 547
Dasycladus, **288**
dates, 429, 688
dating methods, 53, 586
Daucus carota, 34, **485,** 558, 563
Dawsonia superba, **311**
dayflower, 619
day length
    and biological clock, 578
    and bud dormancy, 586, 587
    and flowering, 579, 580, 587
daylilies, 168
day-neutral plants, 579, 584
DDT, 643
decarboxylation, 97, 112
deciduous forests, 8, **658–659**
    temperate, 674–678
    tropical, 666–667
deciduous trees, 8, 369, 344, 521, **530,** 587
decomposers, 115, 188, **189, 190,** 196, 208, 598, **599,** 602, 646, 648
defense mechanisms, 467, 516, 517, 538
Deisenhofer, Johann, 107
dehiscence (*see also* spore dispersal)
    of anther, **389, 390,** 392
    of spore capsule, 308, **315,** 316
16-dehydropregnenolone (16D), 437
deletions, 134
delphinidin, **429**
de Mairan, Jean-Jacques, 576
denaturation, 59, 82
dendrochronology, 536
dendrometer, **624**
denitrification, 198, 611
Dennstaedtia punctilobula, **348**
dent corns, **691**
deoxyribonucleic acid (*see* DNA)
deoxyribonucleotides, **60,** 61
deoxyribose sugar, **60,** 61, 137, **138,** 139
derivatives
    of cambial initials, 521– 523
    of meristematic cells, 453, 474, 489–491
dermal tissue system, 319, 320, 454, 465, 467, 475
dermatophytes, 222
desert(s), 8, 668–671
    agriculture, 608
    biome maps of, **668, 670**
    deserts and semideserts, 668–670
    distribution of, 668
    juniper savanna, 670–671
    photosynthetic rates in, 669–670
    plants of, 518
    rainfall in, 668
    temperatures in, 668
    vegetation characteristics of, 668–669
desmids, 283, **284**
desmotubule, 37, **44,** 72
desiccation, 475
destroying angel, 234
Desulfovibrio, 190
deuterium, 714
Deuteromycota, 219–222
    and day length, 587
    of embryo, 441–451
    of flower, 511, 514, 515
    of leaf, 506–508
    regulation by hormones, 545–563
    of root, 472–475, **481**
    of stem, 489–491, 521–523
Devonian period, **285,** 299, 318, 319, 320,

323, **324–325,** 326, 330, 338, 347, 358, 359, 360, 361, 401
deVries, Hugo, **134**
dewberries, 683
diaheliotropism, 590
diatomaceous earth, 259
diatoms, **245, 247,** 257, 258–261 (*see also* algae)
    cell wall of, **247,** 257, 258
    centric, **259,** 260–261
    life cycle of, **260–261**
    motility in, 261
    pennate, **180, 259,** 261
Dicentra cucullaria, 435
2,4-dichlorophenoxyacetic acid (2,4-D), 547, 551
dichogamy, 396
Dicksonia, **345**
dicot(s) (Dicotyledones), 116, **178, 183,** 380, **382,** 412–413, 441, **442,** 462
    herbaceous, 456, 492, 494
    and monocots compared, 380, **382,** 388, 393, 443, 444, 448–451, 462
Dictyophora duplicata, **662–663**
dictyosomes, **20,** 28, 29, 43, 472
Dictyostelium
    discoideum, 253, **254,** 256
    life cycles of, **254**
    mucoroides, **254,** 255
Dieffenbachia, 87
Diener, Theodor, 205
differentiation, 454, 559
    in algae, 294
    cellular, 294
    in embryo, 444–447
    and hormones, 559–563
    in leaf tissue, 506–508
    physiological, 159
    in root tissue, 472–475, **476, 477**
    in stem tissues, 489–496
    of vascular tissues, **460, 464**
diffusion, 63–66, 617, 627, 631
    facilitated, 68
digestive glands, 610
Digitalis purpurea, **420**
Digitaria, 613
    sanguinalis, 116
dihybrid(s), 131
dihybrid cross, 129, 131
dihydroxyacetone phosphate, 88
Dilcher, David, **409**
dill, 695–696
Dinobryon, 258
dinoflagellates (Pyrrhophyta), **244, 245,** 246, **247,** 248, 257, 262–265, 289, **578** (*see also* algae)
    cell wall of, **262**
    as symbionts, 264, 272
dinucleotide, 61, 81
dioecious plants, **378,** 386, 396, 427
Dionaea muscipula, 589
Dioscorea, 690, 697, 707
diosgenin, 437
Diospyros, **37**
    virginiana, **543**
dioxin, 551
dipeptide, 144
Diphasiastrum, 331
    complenatum, **330**
    digitatum, **183**
diphtheria, 190
Diplococcus pneumoniae, **188**
diploid
    cells, 19, 20, 126, 130
    generation, 181
    number, 19, 20, 122
    organisms, 122, 124, 130, 135, 166, 183–184, **278**
diploidy, evolution of, 183–185
diplophase, **256**
disaccharides, 48, 49
Dischidia rafflesiana, 484, **485**
Discula, 221
disease(s) (*see also by name*)

bacterial, 187, **189,** 190, 195, 198–200, 222
    fungal, 168, 209, **219,** 221
    by oomycetes, 248, 250–251
    viral, 200, 201, 202–204, 640
    viroid, 200, 204–205
disk flower, 412, 413
Distephanus speculum, **258**
disturbance, 641, 645, 650
    natural, 649, 651, 652–653, 654–655
disulfide bonds, 57
diurnal movements, 576–578
division, classification by, 173
DNA, **60,** 61, 136–140, 552, 597
    antiparallel strand of, **138,** 139
    in bacteria (prokaryotes), 16, 175, **186,** 187, 190, 195
    base pairing, **137, 138, 139**
    binding protein, 569, 583
    building blocks of, 59–61
    in chloroplasts, 21, 245
    in chromosomes, 16, 19, 38, 122
    circular, 135, 175, 187, 190
    complementary strand of, 139, **140**
    composition of, 136
    and cytokinins, 552
    double helix, 137–138, **140**
    of eukaryotes and prokaryotes compared, 16, 175, 176
    in genetic engineering, 565–571
    and hormone-activated genes, 559
    in humans, 138
    lagging strand of, 139
    leading strand of, 139
    ligase, **140,** 566
    linker, 122, **123**
    in mitochondria, **23,** 24
    nucleotides of, **138,** 139, 140
    Okazaki fragments of, 139, **140**
    polymerases, 139, **140**
    recombinant, 565–571
    replication of, 38, 138–140
    replication fork, 139, **140**
    sequences, 559, 583
    spacer, 122, **123**
    structure of, 136–138, 139
    as template, 139, **140, 141, 148**
    viral, 138, 200, 201, 202–204, **205**
    Watson-Crick model, 136, **137,** 139
Doctrine of Signatures, 300
dodder, 383
Dodecatheon pauciflorum, **385**
Doebley, John, 693
dogwood, 221, 418, 434
dolipore, 230
dominance, incomplete, 131
dominant/recessive characteristics (phenotypes), 129, 130
dormancy, 318, 555, 557, 585–587, 680
    breaking of, 557, 586, 587
    of buds, 555, 585, 586, 587
    of cambium, 552
    of seeds, 448, 557, 585, 586
    of sporangia, **252**
    of zygosporangium, 286, 291
dormin, 555
double fertilization, 378, 392, **393**
double helix, 137–138, **140**
Drepanophycus, 318
Drosophila, 568
    melanogaster, 568
drought, 590
drugs, 190, 704, 707–708
drupe, **381,** 429, 430
druses, **20,** 25
Dryopteris
    crassirhizoma, 552
    marginalis, **348**
Dubautia, 164
    reticulata, **165**
    scabra, **165**
duckweeds, 380, **381**
Durvillea antarctica, **275**
dust bowl, 672

Dutch elm disease, 215
dutchman's breeches, 435
Dutchman's pipe, 527
dwarf phenotype, 568
dwarf (mutants) plants, 556
dynamic equilibrium, 64, 721
3' and 5' ends, **138, 139, 140, 143**

earth, 1, 2
    age, 1
    crust, 597, 599, 600, 612
earthstars, 234, **235**
earthworms, **599,** 643
Echinocereus, **419,** 518
ecology, 11, 637–656
    bacterial, 188, 190
    defined, 11, 637
    of protista, 245
    restoration, 655
ecosystem(s), 8–9, 637–656
    aquatic, 615
    biotic components in, 646
    cycling of nutrients in, 602, 645
    defined, 637
    development of, 649–655
    flow of energy in, 645
    lichens in, 227
    management of, 653
    and pesticides, 643
    physical components in, 646
ecotype(s), 158
ecotypic variation, 158–159
Ectocarpus, 274, **275,** 276
    siliculosus, 275
ectomycorrhizae, 239, 240–241, 638
edaphic variation, 680
effector, 147
egg apparatus, 392
egg cells
    of algae, **273,** 274, 276, **277, 278, 279,** 285, **293,** 295–296
    of angiosperms, 392
    of byrophytes, 300, 302, **313**
    of diatoms (egg nucleus), **260**
    of gymnosperms, 367, **371**
    of oomycetes, 248, **249,** 250
    of seedless vascular plants, **329,** 331, **333, 335, 337,** 339, **341,** 351
eggplant, 199, 200, 550, **706**
egg sac, 278
Eichhornia crassipes, 649
einstein, 109
Einstein, Albert, 103
Elaeis guineensis, 696
elaiosomes, 434, **435**
Elaphoglossum, 342
elaters, 302, **304, 306,** 308, **317, 339, 341**
elder, 418
elderberry, **455,** 492, 493, 524–527
electrical impulses in plants, 589
electrochemical gradient, 68, 70, 95, 96
electrodes (injection and receiver), **72,** 73
electrolytes, 721, 722
electromagnetic spectrum, **102**
electron(s)
    acceptor/donor, 77, 78, 81, 94, 107, 109, 597
    in atoms, 713–719
    carriers, 92, 93, 101, 109
    in photosynthesis, 103, 104, 107–109, 196
    transfer, 92–95, 597
electronegativity, 716, 717
electron flow, photosynthetic, 1, 107–109
electron microscope, 26, 27
electron transport chain, 87, 92–95
electroporation, 565, 571
elements, 47, 597, 713 (*see also specific types*)
    cycling of, 9, 602
    essential, 594, **596**
    of living matter, 2
elephants, 151
Elettaria cardamomum, 695

elicitors, 643
elm, **431, 542**
    American, **530, 537**
Elodea, 18, **66, 101**
elongation (of polypeptide), 144
embolism, 622
embryo (young sporophyte), 6, 323, 357–358 (*see also* embryogeny)
    angiosperm, **395,** 441–451
    asexually produced, 166
    in bryophytes, 298, 300, **302, 312**
    corn, 449, 450
    development in arrowroot, **446**
    development in shepherd's purse, 444, 445, 447
    dicot, 441–444
    dormancy of, 448
    of *Equisetum,* 339, **340**
    and evolution of seed, 323, 357–358
    of ferns, 345, **350**
    formation of, 444–447
    grain, 443, 449, 450, 557
    of gymnosperms, 361
    of Lycopodiaceae, 331, **332**
    monocot, 441–444
    of pine, 367
    of plants, 298–299
    of *Psilotum,* 328
    sac, 390, **391,** 392, 395, 444, 445
    of seed plants, 357–358
    of *Selaginella,* **336**
    vascular transition in, 510, 511
embryogenesis, in biotechnology, 563–565
embryogeny
    in angiosperms, 393, 441–452
    in pine, 367
Emiliania huxleyi, **258**
Encephalartos
    altensteinii, 3
    ferox, 375
endangered species, 170, 680, 682
endergonic reaction, 76, 77, 78, 725
endocarp, 393, 430, 433, 458
endocytosis, 70, 71
endodermal cells, 54, **479, 480**
endodermis
    in pine leaf, 363
    in root, **475, 476, 477,** 478, 479, **480,** 624, **625,** 627
    in stem, **334, 339, 345**
endogenous rhythms, 577
endomembrane system, **17,** 29
endomycorrhizae, 239–240, 299, 318, 323, 326, 638
endophytes, 220–221
endoplasmic reticulum (ER), **19, 20,** 28, **31,** 44
    and meiosis, 125
    and ribosomes, 145
endosperm, **37,** 393, **395, 442,** 443–445, **449,** 452, 551, 557, 702
    liquid, 697
endospores, 190
endothermic reaction, 77
energy (*see also* ATP; cellular respiration; phosphorylation; photosynthesis)
    activation, 83, 723, **725**
    for active transport, **628**
    biological flow of, 74–84
    in biosynthesis, 77–79
    bond, 52, 83, 84, 724
    chemical, 1, 48, 52, 63, 74, 75, 99, 104
    conversions, 1, 74–77, 99, 104
    in ecosystems, 9
    electrical, 2, 63, 76, 102
    for evaporation, 624
    -exchange molecules, 61
    flow in ecosystem, 646–648
    from glucose oxidation, 76–78, 86, 96
    heat, 74–76, 87
    kinetic, 74–76
    levels of electrons, 103, 104, 715, 716
    levels in electron transport, 92–95, 107–109
    light, 1, 74, **75,** 100, 101, 103, 106–109

"loss," 74, 75
mechanical, 63, 74, 76
nonrenewable sources, 705
in nitrogen cycle, 602
potential, 62, 63, 76, 716
release of in respiration, 86, 96
renewable sources of, 11
solar (radiant), 74, 75, 99 (see also sun)
and thermodynamic laws, 74–77
transfer, 86, 89, 96
yield in photosynthesis, 109
Englemann, T. W., **104**
enolase, **89**
Entogonia, **259**
Entomophthora muscae, **181**
Entomophthorales, 214
entrainment, 577, 578
entropy, **76**, 77
envelope, viral, 200, 202
environment
and circadian rhythms, 576–578
disturbances to, 160, 161
and dormancy, 447, 585
influence on selection, 154
modification, adaptation to, 154–160
and stomatal movements, 618–628
enzymatic pathways, 82
enzyme(s), 59, 78–83, 597
and abscission, 509, 510
activity, regulation of, 82, 83
bacterial, 190
biosynthesis, 146
digestive, 26, 27, 447, 517, 554, 610
in DNA synthesis, 139, **140**
in ethylene synthesis, 554
glycolytic, 87, 88
hydrolytic, 33, 59, 557
inducible, **146**
in Krebs cycle, **91**
in membranes, 67, 68, 70
in mitochondria, 90, **96**
nitrogen-fixing, **197**, 602
in photosynthesis, 110–112
in prokaryotes, **197**
in protein synthesis, 139, 140, **141**, 142
repressible, 146
restriction (endonucleases), 565–567
in roots, 483
in seeds, 443, 447, 557, 586
specificity, 79
structure, 59, 79, 80, 597
synthesis, 557
eoplast, **23**
Ephedra, 375, 377, 378, 392
trifurca, **377**
viridis, **377**
ephemerals, spring, 677
epicotyl, 441–443, 449, 488, 511
epidermis, 5, **319**, 320, 325, 454, 465–467
of leaf, **7**, **22**, 363, 465, 466, 500–503, 505, **506**
multiple, 465, **466**, 484, **502**, 519
of root, 475, **476**, **477**, **478**, **482**, 484, 624, **625**
of stem, **7**, 338, 339, **345**, 491, 492, **493**, **494**, **524**, **525**, **526**
epigyny, 387, 412
Epilobium, 432
angustifolium, **396**
Epiphyllum, **516**
epiphytes, 330, 343, 484, **662–663**, 664, 681
epistasis, 135–136, 154
epithem, **627**
Epstein, Emanuel, 613
equatorial plane (in meiotic/mitotic division), 39–40, 125, **126**, **127**, 132
equilibrium, 64, 77, **724**
Equisetites, 338
Equisetum, 320, 323, 338–341, 346, 348, 377, **595**
arvense, **317**, 552
X *ferrissii*, **163**
hyemale, **28**, **42**, 163
laevigatum, 163

life cycle of, **340–341**
sylvaticum, **183**
ergastic substances, 30, 31
ergot, ergotism, **221**
Ericaceae, 241
Eriophrum
angustifolium, **684**
scheuchzeri, **684**
erosion, soil, 471, 611, 612
Erwinia, **199**
amylovora, **188**, 199
carnegieana, **198**
Erysiphe aggregata, **217**
erythrocyte, **123**
Erythroxylon coca, **437**
Escherichia coli, **16**, 55, **71**, **186**, **188**, 195, 219
conjugation in, **189**
and DNA studies, **136**
in genetic engineering, 570
and gene transcription studies, 145–146
ribosomes of, **143**
viral infection of, **201**
Eschscholzia californica, **421**
Essay on the Principles of Population, 151
essential inorganic nutrients, 593, 594, **596**
essential oils, 435, 695
ethanol (ethyl alcohol), 97, **98**, 209, 210, 218, 719
ethylene, **546**, 554, 555, **718**
and abscission, 555
and auxin, 554
biosynthesis of, 554
and cell division, **559**
and cell expansion, **559**, 560–562
commercial use of, 554, 555
and fruits, 434, 554
and gene expression, 559
and sex expression, 555
etiolation, **583**
etioplast(s), **23**
Euastrum armatum, **284**
Eubacteria, 11, 175, 176, 177, **178–179**, 186–188, **189**, 190–191, 192–200, 728–729 (see also bacteria)
Eucalyptus, 161, 240, 250, 380, 424
cloeziana, **54**
jacksonii, **380**
Eucheuma, 281
euchromatin, 122
Eudorina, **293**, 294
Eugenia aromatica, 695
Euglena, 173, 265, 266
euglenoids (*Euglenophyta*), **178**, 246, **247**, 248, 731 (see also algae)
nonphotosynthetic, 265
eukaryotes, 4, 11, 16, 135, 576
classification of, 175–176
DNA and chromosomes of, 19, 20
evolution of, 16, 24, 244, 246
origin of multicellularity of, 176
and prokaryotes compared, 4, 16, 175–176
eukaryotic cell, 15–46
Euphorbia, 518
pulcherrima, **424**
Euphorbiaceae, 518, 696, 705
euphorbias, 518, 669
Eupoecila australasinae, **415**
eusporangia (see sporangia, of ferns)
eusporangiate ferns (see ferns)
eustele, 320, 321, 322, 360
eutrophication, 615
evaporation, 600, 617, 620, 621, **622**, 624
evergreen species, 677
evergreens, 521
evolution, 1–5, **10**, 150–169
angiosperm, 378, 380, 401–439
of autotrophs, 3, 4
of bacteria, 1–4
of bryophytes, 298–300
of cells, 3
of chloroplast, 24
of communities, 6, 8–10
convergent, 320, 403, 459, 518

of diploidy, 183–185
of eukaryotes, 4, **10**, 18, 24, 176–177
of ferns, 326, 345, 347, 359
of flower, 408–429
of flower parts, 410
of fruits, 429–438
of fungi, 208, 251
of gametophytes of vascular plants, 318, 323
and genetic variability, 152
of glycolytic sequence, 87
of gymnosperms, 357, 358, 359–361, 378
of heterospory, 323
of hornworts, 300
and hybridization, 160–163, 166–168
of liverworts, 299–300, 301
macro-, 168
mechanisms of, 151, 152, 161, 168
micro-, 168
of mitochondrion, 24
of mosses, 300
of multicellularity, 10
and mutations, 135
by natural selection, 150, 151, 152
parallel, 377, 657
of photosynthesis, 1, 3, 4, 114, 116, 117, 518
of plants, 5–6, 244, 285, 298–299, 318, 323
process of, 150–169
of prokaryotes, 1–4, **10**
of protista, 245
of reproduction, 270
of seed, 318, 323, 357–358, 360
of sporophyte, 318
of stomata, 317
theory of, 15, 151, 160, 168
of vascular plants, 317–319, 321, **324–325**, 330, 357–361
of vascular tissues, 317, 319, 320, 321, 406, 459
evolutionary relationships, 176, 177, 402
exergonic reaction, 48, 76, 77, 78, 725
exine, 388, **389**, 390, 428
exocarp, 393
exocytosis, 70, 71
exodermis, 480, 624
exons, 140, 141, **142**
exoskeletons, 52, 210
explants
meristem, 565
organ, **564**
tissue, 563
Exuviaella, **262**
"eyes" of white potato, 450, 517
eyespot (stigma), 289, 292, 294

F_1, F_2, F_3 (see generation, filial)
fMet, **144**, 145
Fabaceae, 155, 173, 430, **509**, 691 (see also legumes)
facilitated diffusion, 68
FAD/FADH$_2$, 92, 93, 94, 96
Fagaceae, 240, **386**
Fagus americana, **542**
fairy rings, 233, **234**
false truffles, 234
family, classification by, 168, 173
fasciation, **199**
fascicles, 362, **363**
fats, 52, 53, 91, 718, 719
in human diet, 697
fatty acids, 52, 53, 91
saturated/unsaturated, 52, 53
female nuclei, 216
fennel, 695
fermentation, 86
alcoholic, 97, 98
lactate, 97, 98
of wine and/or beer, 97, 98
ferns, 117, 317, 320, 322, 323, **324**, **340**, 342–353, 359, 401, 459

adder's tongues, 344
in biological succession, **650**
bracken, 348
bulblet, **183**
of Carboniferous period, 346, 347
climbing, 343
eusporangiate, 343, 344–345, 347
evergreen wood, 348
grape, 344
grass, 62
heterosporous, 344, 352–353
homosporous, 344, 345, 349
leaf cells, **30**
leptosporangiate, 343, 344, 345, 348–352
life cycle of leptosporangiate, **350–351**
maidenhair, 345, 348
ostrich, 348
seed, **324**, 346, **347**, 356–357, 361
spleenwort, **661**
tree, 342, 343, 345, 347, 356–357
walking, 450, **451**
water, 197–198, 344, 345, 352–353
Ferocactus melocactiformis, 516, **669**
acanthodes, **671**
ferredoxins, **108**, 109
Fertile Crescent, 687, 690
fertilization, 122, 124, 128, **129**, 357
in algae, **273**, 277, 278, 279, 283, 287
in angiosperms, 392–393, 395 (see also pollination)
in bryophytes, **302–303**, 312–313
in chytrids, **253**
cross-, 339
in diatoms, 260–261
and fruit development, 550
in fungi, **213**
in gymnosperms, 361, **370–371**
in oomycetes, **249**
in pine, 367
in seedless vascular plants, 329, 331, **333**, **337**, **340**, **351**, 361
in true mosses, **312–313**
fertilization tubes, 248, 249
fertilizers, 281, 571, 604, 610, 612, 613, 614, 700
Festuca arundinacea, 220
fibers, 457, 468, 504
bast, 457
gelatinous, 539
libriform, 459
in phloem, **457**, **461**, **462**, 481, **482**, 492, 493, **494**
in xylem, 459, 461, 535, 539, 540
fiber-tracheid, 459
fibrils, cellulose, 30, 33, 34, **35**, **36**
Ficus
benghalensis, 483
elastica, 465, **466**
fiddleheads, 348, **350**
field moisture capacity, 601
figure, of wood, 540
filament (of flower), 383, **384**, 456, **514**
Filicales, 344, 345, 348–352
finches, Galapagos, 160
fire, 448, **651**, **652–653**, 672, 680
in Yellowstone Park, 652–653
fire bug, 378
fireflies, 97, 570
fireweed, **396**, 419, 432
fir(s), 361, 538, **654**, 683
balsam, **369**, 650
Douglas, 369, **543**, **654**, **655**, 679, 682
subalpine, 640, 653
white, **651**
fish, 245, 248, 513
Fissidens, **298**
fission, 190
"fit" (see natural selection)
flagellum(a), 31, 32
of algae, 24, **32**, **247**, 257, 259, 270, **280**, **282**, 283, 286
of chrysophytes, **247**, 270

of chytrids, 245, 246, **247**, 251
of diatoms, **247**, 258, **260**
of dinoflagellates, **247**, 262, 263
emergent, 266
of euglenoids, 247, 265–266
of eukaryotes, 176
nonemergent, 266
of oomycetes, 245, 246, **247**
of prokaryotes, 188, **189**
of slime molds, **247**, **256**
structure of, 31, 32
tinsel, 31, **32**, 246, 248, 259, 270
of water molds, 245
whiplash, 31, **32**, 246, 248, 251, 256, 270, 289
flatworm, 294
flavin, 574, 619
flavin adenine dinucleotide (FAD), 92, 93, 94, 96
flavin mononucleotide (FMN), **92**, 94
flavonoids, 428–429, 435, 607
flavonols, 429
flavoprotein, 619
flax, **211**, 457, 688
Fleming, Sir Alexander, 222
flies, 87, **181**, 234, **235**, 238, 252, 416, 417, 418–422, 568
flint corns, **691**
flora(s), **3**, 358
floral stimulus, 584
florets, **427**
Florey, Howard, 222
floridean starch, 247, 248, 269, 270
florigen, 584
flour, 452
flour corns, **691**
flower(s), 383–388 (see also angiosperms)
arrangement of parts, 386–387
bisexual, 408, 410, 415
buds, **384**
carpellate (ovulate), 386, 396, 428
cluster(s) (see inflorescence)
colors of, 22, 25, 410, 428–429
complete, 386
definition of, 383
development of, 511, **514**, 515, 569
of dicots, **382**
epigynous, 387, 412
evolution of, 380, 408–429
and hormones, 555, 558
hypogynous, 383, **384**, 387
imperfect, 386
incomplete, 386
irregular (bilateral; zygomorphic), 388
levels of insertion of parts of, 387
of monocots, **382**
odors of, 410, 416, 418, 423, 425
parts of, **129**, 383, **384**, 386, 408, 410, **514**, **515**
perfect, 386
perigynous, 387
pistillate, 386
placentation of, 386
regular (radial; actinomorphic), **382**, 388
saprophytic, 413
staminate, 386, 396, 428
symmetry, **382**, 387–388
terminology, 383, 386–388
underground, 413, **414**
unisexual, 408, 410
vasculature of, 410
and viruses, 204
flowering, 511, 515, 558
and cold treatment, 587
and daylength, 578–580
hormonal control of, 558, 584
inducers and inhibitors of, 587
and light, 581
photoinduction of, 581
and photoperiod, 578–580
and phytochrome, 581, 582
plants (see angiosperms)
time of, 160
flower pot plant, 484, **485**

fluid mosaic model of membranes, 66, 67
fluorescein dye, 72
fluorescence, 103, 104
microscopy, 41
fluorides, 512
fly, **181**
fly agaric, **3**, **232**
FMN, 92, 94
Foeniculum vulgare, **695**
fog, **669**, 679
follicle, 430, **431**
Fontinalis, 309
food bodies, 418
food chain, web, 643, **646**, **647**, 648
food-conducting cells, tissues, 5, 270, **276**, 300, 461–465 (see also leptoids; phloem)
food reserves, **247**, 298, (see also by type)
food storage, 25, 48, 52, 455, 470, 629
in leaf, 517, 518
in root, 485, 486
in seed, 367, 443, 447, 448, 449
in stem, **52**, 488, 517, 518
food (assimilate) transport, 629–633
foolish seedling disease, 555, 556
of bryophyte, 301, **302**, 308, **313**
of seedless vascular plants, 326, **328**, **332**, **336**, **340**, **350**, 352
Foraminifera (marine protozoa), 259
foreign genes, 566, **567**
forest(s), (see also biomes)
and acid rain, 512, 513
ancient, 682
boreal, 681
fires, 448
gallery, 666
old-growth, 682
forestry, sustainable, 682
fossil fuels, 115, 346, 512, 513, 604, 700
fossils, 152, **325**, **401**
of algae, 270, 272, **285**
of ancestors of plants, 299
of angiosperms, 358, **402**
of bacteria, **2**, 186
dating of, 53, 714
of diatoms, 259
earliest known, 1, **2**, 4
of eukaryotic cells, 4, **244**
of flowers, 323, **405**, **406**, 408, **409**
of fungi, 211
of gymnosperms, 372, **402**
of *Metasequoia* (redwoods), **373**, 374
of multicellular organisms, 5
of mycorrhizae, 241
of photosynthetic organisms, 3
of plants, **4**, 241, 299
pollen, 390, 405, 407
of prokaryotes, **2**
seed, 361
of vascular plants, **4**, 320, 321, 323, 325
wood, 359
Fouquieria columnaris, **669**
foxglove, **420**
Fragaria, 158, **434**
× *ananassa*, 450, **451**, 550, **551**
fragmentation, 283, 304, 311
Frankia, **197**
Franklin, Rosalind, 136
Fraxinus, **431**, 538, 621
americana, **542**
pennsylvanica var. *subintegerrima*, **490**, **499**, **530**
free-energy change, 77, 96, **97**, 109
freeze-drying, 633
freeze-fracture, -etch preparation, **19**, **106**
frequency (-ies)
of alleles, 152, 154, 166
genotypic, 152, 154
of polyploids, 166
friction, 75
Friis, Else Marie, **406**
Fritschiella, **280**, 294, **295**
Frolova, I. A., 584

fructans, 50
fructose, 49, 50, 84
1,6-bisphosphate, 88
6-phosphate, 88
fruit(s), 388, 454
accessory, 429
aggregate, 429, **434**
climacteric, 554
color, 22, 25, 26, 554
definition, 429
dehiscent, 430, 431
dispersal of, 429–438
dry, 393
fleshy, 393, 429, 430
fly, 568
growth and development of, 550, **551**, 558, 630
and hormones, 550, **551**, 554, 558
indehiscent, 430, **431**, 432
multiple, 429
parthenocarpic, 429, 550, 558
ripening, 22, 515, 554
simple, 429
size of, **155**
fruit drop, 550, **551**, 555
fruitlets, 429
Frullania, **305**, **306**
frustules, 258
Fucales, 275
fuchsia, 424
fucose, **51**
fucoxanthin, **247**, 257, 258, 262, 274 (see also carotenoids; xanthophylls)
Fucus, 276
life cycle of, **184**, **278–279**
vesiculosus, 275
fumarate (fumaric acid), **91**
functional groups (chemical), 719, **720**
fundamental tissue system, 454
fungal mantle, 240, 241
Fungi (fungi), Kingdom, **174**, 176, **178**, 208
fungi, **3**, 10, 52, 70, 97, **98**, 135, **181**, 208–243, **244**, 729 (see also by name)
cell walls of, 181, 210, 211
classification of, 175
coenocytic, 211, 212, 213
conidial, 219–222
as decomposers, 186, 208, 228, 598
destructiveness of, 209
economic importance of, 209, 214, 215, 221, 234
endophytic, 220–221
evolution of, 208, 211–212, 251
and fermentation, 97, **98**, 210, 218
and genetic engineering, 218
habitats of, 209, 212
life cycles of, **213**, **216**, 218, **229**, **236–237**
major groups, 212–237
marine, 209
mycorrhizal, 209, 214, 238–241, **383**, 638
nutrition of, 181, 210
parasitic, 210, 211, 212, 214, 555, 556
pathogenic, 209, 212, **219**, 220, 555, 556 (see also rusts; smuts)
and phototropism, 214
and plant diseases, 209, 215, 221, **228**, 235
predaceous, 223
reproductive techniques of, 181, 212
saprobic, 210, 212, 218
in soil, 210
spores of, 211 (see also specific spore types)
structure of, 210
symbiotic, 209–210, 214, 221, 223–228, 329, 333 (see also lichens; mycorrhizae)
toxins of, 209
Fungi Imperfecti (see Deuteromycota)
fungus roots (see mycorrhizae)
funiculus, 390, **391**, 432
furrowing, of cell, **280**, 282, 283
Fusarium, 643, 704

oxysporum, 222
fusion cell, 218
fynbos, 680

GA-deficient mutants, 569
GA$_3$, GA$_4$, GA$_7$ (gibberellins), 546, 556
galactons, 247
galactose, **146**, 270
β-galactosidase, **146**, 147
α-galacturonic acid, 51
Galapagos Islands, 150, **151**
finches, 160
tortoise, 151, 160
gametangia (see also antheridia; archegonia)
of algae, **260–261**, 276, **285**, **287**
of ancestors of plants, 298
of bryophytes, 298, **300**, 310
of chytrids, **252–253**
of fungi, **213**
plurilocular, 276
of vascular plants, 298
gamete(s), 20, 31, 183, **184** (see also eggs; sperm)
of algae, 259, 274, **287**, **290**, 291
of angiosperms, **395**
of chytrids, **252–253**
defined, 124
of diatoms, **260–261**
evolution of, 6
formation of, 130
of fungi, 212, 213
in genetics, **153**
nonmotile, 291
of oomycetes, **249**
of slime molds, 256
gametophores, 301 (see also antheridiophores; archegoniophores)
gametophyte(s), 184, 318
of algae, 272, **273**, 276, **277**, **287**
of ancestors of plants, 298, **299**
of angiosperms, 388, **395** (see also pollen)
bisexual, 308, 322, 326, 327, 331, **333**, 339, 349
of bryophytes, 298, 299, 300–301, **302–303**, **313**
of chytrids, **252–253**
of *Equisetum*, 339, **340**, **341**
of ferns, 323, 345, 349, **350–351**
of granite mosses, 315
of gymnosperms, 361, 375
of hornworts, 307, 308
of liverworts, 300, 301, **302–303**, 304
of Lycopodiaceae, **333**
of *Marchantia*, **302–303**
of peat mosses, 314, **315**
of pines, 365
of *Polytrichum*, **312–313**
of *Psilotum*, 326, **328–329**, 345
of seed plants, 323
of *Selaginella*, 335, **337**
of true mosses, 309, 310, 311, **313**
unisexual, 308, 310, 323, 331
of vascular plants, 298, 299, 319, **333**
vascular tissue in, 326
gamma rays, **102**
gangrene, 221
Ganoderma applanatum, **232**
gap junctions, 72
gaps created by disturbances, 650, 664
garbanzos, 688
Garner, W. W., 578, 579, 580
gas vacuoles, 193
Gassner, Gustav, 587
Gasteromycetes, 231, 234–235
gastrointestinal illness, 263
gating, 69
gazelles, Thomson's, **8**
Geastrum saccutum, **235**
geminiviruses, 201, 204
gemma cup, **303**, 304, **305**
gemmae, **303**, 304, **305**, 311, 352
gene(s) (see also alleles; chromosomes; DNA)

bacterial, **141**, 142, 145–146
coding and noncoding, 140
dominant/recessive, 152
eukaryotic, 559
expression, 555, 559–561, 565–570
flow, 157
foreign, 559
frequency, 152, **153**
in genetic engineering, 565–571
linkage, 132–133, 154
mutant, 135
nature of, 152
nodulation (nod) gene, 607
nonallelic, 136
pool, 152
regulator, 146–147
reporter, 559
sequencing, 568, 569
structural, 146–147
transcription, 140–142, 145–147, 574
transfer, 539, 565–570
transposition, **120**
generation, filial (F₁, F₂, F₃), 129, 130, 131, 132, 133, 152, **153**
generative cell, **365, 371**, 390, **395**
generic name, 172
genetic(s), 11, 121–149
in biotechnology, 565–570
changes and mutations (*see* mutation)
code, 134, 140–142
crosses, 129–133
engineering, 12, 200, 204, 218, 565–571, 643, 644, 708
equilibrium, 152
isolation, 157
maps, 133, 568
material, 126, 136, 139, 188 (*see also* DNA)
Mendelian, 129–133
and natural selection, 154
races, 158
recombination, 156, **157**, 173, 187, 220, 256
self-incompatibility, 397, 398, 427
variability, 152, 157
genome, 186, 200, 204
genotype, 130, **153**, 166
genotypic ratio(s), **130, 131**, 133, **153**
genus (genera), 171
origins of, 168
geologic eras, 735
geotropism (*see* gravitropism)
GERL, 27
germination (seed), 55, 64, **442**, 447–451
and cold treatment, 585
epigeous, **442**, 448, 449
and hormones, 555, 557
hypogeous, **442**, 448, 449
and light, 581
and phytochrome, 583
requirements for, 447, 448
germ tube, 251
giant cell, **254**
giant sequoia, 679
giant tortoises, 151, 160
Gibberella fujikuroi, 556
acuminata, **211**
gibberellic acid (GA₃), 510, **546**, 556, 557, 560, 561, 569
gibberellins, **546**, 555–558, 559 (*see also* gibberellic acid)
and abscisic acid, 560
and auxins, 560
biosynthetic enzymes, 569
and cell division, 556, 559
and cell elongation, 510, 556, 559, 560–562
commercial use of, 557, 558
and dormancy, 557, 587
and dwarf mutants, 556
and ethylene, 555, 560–562
and flowering, 558, 584
and fruit development, 558
mechanism of action of, 559–562

in roots, 470
and seeds, 557
and sex expression, 555
structure of, **556**
Gibbs, Josiah Willard, 77
gill fungi, 234
gills, 232, **233**
ginger, 695
Ginkgo biloba, 357, 374–375, **376**, 396
Ginkgophyta (maidenhair tree, ginkgo), **178**, 358, 361, 374–375, 733
giraffe, 8
girdling, 639
Givnish, Thomas, 676
glades, 674
gladiolus, 517
Gladiolus grandiflorus, **517**
glands, 609, 610
gleba, 234
Gleditsia triacanthos, **499**
global warming, 639
Glomus, 214
versiforme, **239**
glucose, 49, 50, **51**, 86, **146**, 270
fermentation of, 97, 98
oxidation of, 76–78, 86, 96, 97 (*see also* cellular respiration; glycolysis; Krebs cycle)
6-phosphate, 88
as product of photosynthesis, 110, 111
glumes, **427**
glutamic acid, **56**
glutamine, **56**
gluten, **698**
glyceraldehyde, **49**
3-phosphate, 88, 110, 111
3-phosphate dehydrogenase, 89
glycerate 1, 3-bisphosphate (1,3-bisphosphoglycerate), 89
glycerate 2-phosphate (2-phosphoglycerate), 89
glycerate 3-phosphate (3-phosphoglycerate), 89
glycerate 3-phosphate kinase, 89
glycerol, 52, **53**, 264, 720
glycine, **56**
Glycine, 606
life cycle of, **394–395**
max, **22**, 394–395, 578, 584, 602, **603, 673, 689**
glycocalyx, 187
glycogen, 50, 192, 210, **247**, 270 (*see also* starch, cyanophycean)
glycolate (glycolic acid), 114
glycolate oxidase, 282, 285
glycolysis, 87–90
glycoprotein, 29, 34, 37, 290, 294
glycosides, 435
cardiac, **436, 644**
mustard oil, 435, 436
glyoxysomes, 25
glyphosate, 570
Gnetophyta (gnetophytes), **178**, 358, 361, 375–378, 733
and angiosperms compared, 377–378, 392
Gnetum, 375, **376**, 377
golden algae, **32**, 257–258 (*see also* algae)
goldenrod, 159, 676
Golgi apparatus, bodies, 28, 29, 43, 44 (*see also* dictyosomes)
Golgi associated-endoplasmic reticulum-lysosomal complex (GERL), 26, 27
Gondwanaland, 407
Gonium, 292, **293**, 294
gonorrhea, 190
Gonyaulax
polyedra, 264, **578**
tamarensis, 263
goosefoot, 429
oak-leaved, **497**
Gossypium, **411**, 690, 691
G phases of cell cycle, 38
grafting, 584, **585**

grain (kernel), 427, 432, 443, 444, 449, 450, 687, 689, 696
grain (pigweed), **692**
grain, of wood, 539
grain amaranths, 706
gram atom, molecule, 719
grana, 21, **99, 106**
granite, 597
grape, 98, 250, 429, 433, 515, 531, 558, 688
family, **383**
Grapes of Wrath, The, 672
grasses, 155–156, **381**, 382, 397, 398, **427**, 432, 498, 505, **506**, 507, 508, **649, 684**, 687 (*see also specific types*)
annual blue, **506**
bent, **478**
big bluestem, **673**
C₃, C₄, 114, 505, **506**
canary, 547
crab-, 114
eel, **428**
embryos of, 443, 444
foxtail, 398
Kentucky blue-, 114, 155, 166
orchard, 155, 398
prairie, 471, 579, **598**, 627
rye, 221, 398
salt-marsh, 166, **167**
tall fescue, 220
tropical, 613, 615
wire, **471**
grasshopper chromosomes, **125**
grasslands, 640–641, 672–674
agriculture in, 672
biome, map of, **673**
chalk, 640–641
communities in, 672
distribution of, 672
grazing in, 672, 674
productivity of, 672
rainfall in, 672
role of fire in, 672
short-grass prairie, 672, **674**
soils of, 672, 678
tall-grass prairie, 672, **673**
vegetation, characteristics of, 672
gravitropism, 472, 574, 575
gravity, 472, 574
grazing, **8**, 155, 220, 640–641, **674**, 680, 688, 689, 694
green algae, 175, **180, 244, 247**, 265, 269–270, 279–297 (*see also* algae)
as ancestors of plants, 280, 284, 285, 298, 299, 318
cell division in, 280, 282
classification of, 282
economic uses of, 281, 291, 292
food reserves of, 248, 269
life cycles of, 270
and plants compared, **269**, 270, 280, 282, 284, 291, 298–299, 309
siphonous, 270, 283, **288, 289**
symbiotic, 223, **246**, 279, 288, 294 (*see also* lichens)
volvocine line of, 292–294
greenhouse effect, 12
Green Revolution, 702, 708
greensword, 164
Griffonia simplicifolia, **430–431**
ground meristem, 447, 453, 475, 491, **492**
ground substance, 16, 17, 145
ground tissue system, 319, 320, 363, 447, 454, 475, 491
growth, 6, 66, 545 (*see also* meristems)
bud, **490**, 523, 550
deficiency symptoms of, 594, 595
determinate, 362, 383, 507, 511
embryonic, 444–451
"forced," 587
increments, layers, 535–538
indeterminate, 362, 507, 511
inhibitors, 545, 548, 550, 555, 560
intercalary, 506

nutritional requirements for, 593–595, **596** (*see also* nutrition)
and phytochrome, 583
primary, 6, 320, 453
regulators, 545–565 (*see also by name*)
reproductive, 451, 511, 514, 515, 630
rings, 535–538, 540
in roots, 472–475, 480–482
secondary to, 6, 320, 364, 480–482
in stems, 489–491, 520–526
vegetative, 451, 511, 630
GTP, 144
guanine, **60**, 61, 136, **138**, 139
guanosine triphosphate (GTP), 144
guard cells (stomata), 5, **6**, 311, 465–467, 500–503, **506**, 617–620
guayacan tree, **662**
guayule, **705**, 706
gums, 30
guttation, 626
Guzmán, Rafael, 693
Gymnodinium
breve, 263
catenella, 263
costatum, **262**
neglectum, 262
gymnosperms, 324, 357–379 (*see also* conifers; Coniferophyta; cycads; *Ephedra*; *Ginkgo*; *Gnetum*; *Welwitschia*)
of Carboniferous period, 347
origin and evolution, 346, 347
parasitic, 382
of Triassic period, 241
vascular tissue of, 459, 461
gynoecium, 383, **384**

Habenaria elegans, **422**
Haberlandt, Gottlieb, 563
habitat(s), 157, 168
adaptation to, 154–159, 161, 164–165
of bacteria, 190, 192
of bryophytes, 300, 301, 305, 308
destruction of, 166
of fungi, 209
of lichens, 223, 224, 227
of protists, **247**, 248–249, 257, 259, 262, 265, 268–269, 279, 289
of seedless vascular plants, 323, 326, 330, 331, 338, 343, 344, 345, 349, 375, 377, 380
Haemanthus katherinae, 40
hair cells, algal, **284**
hairs, 465 (*see also* trichomes)
branched, **466**
glandular, **466**, 610
leaf, 164, 348, **466, 501, 502**, 503, **507**
root, **472**, 475, **477, 478**, 604, **605**, 607, 624, **625**
sensitive, 589
staminal, 72
Hales, Stephen, 616
half-life, 714
Halictidae, **419**
Halimeda, 288, **289**
hallucinogens, 234, **437**
Halobacterium halobium, 191
halophiles, 191
halophytes, 595, 608, 609
Hammamelis, 433
Hamner, Karl C., 580, 581, 584
haploid (*see also* meiosis)
cells, 124
generation, 181
number, 20, 122
organisms, 124, 135, 183
haplophase, **256**
Hardin, Garrett, 640
hardwoods, 532, 535 (*see also by name*)
Hardy, G.H., 152
Hardy-Weinberg law, 152–154
Harpagophytum, **435**
Harris tweed, 224

Hartig net, 240
Hatch-Slack pathway, 111
haustorium (-ia), 211, 227, 375, 383
Hawaiian islands, 164–165
Hawaiian silversword, 164–165
Hawaiian tarweeds, 164–165, 168
hawthorns, 166, **516**
hay fever, **389**, 397
hazelnuts, **215**, 432
head (*see also* inflorescence), **384**
heartwood, **527**, 538
heat
    denaturation by, 59, 82
    in flowers, 416–417
heat energy, 74–76, 87
heather family, 241
Hedera helix, 483
helanalin, 643
Helenium, 643
Helianthus annuus, **412, 490,** 691, **694**
helicases, 139, **140**
Heliconia, **47**
heliotropism, 590
helium, **713**
helix
    alpha, 57, **58**, 67
    double, 122, 137–138, **140**
Helminthosporium maydis, **703**
heme, 93
hemicellulose, 34, **35**, 280
Hemitrichia serpula, **257**
Hemizonia pungens, **123**
hemlock, 369, **675**
    eastern, 543
    western, **240, 654**
hemp, 386, 457
henbane, 579, 587
hepatica, round-lobed, **382**, 415
Hepatica americana, 382, 415
hepatics, 300
Hepatophyta, **178, 183**, 299, 300–307, 732
    (*see also* bryophytes; liverworts)
hepatitis, 200
herbaceous plants, 456, 520
    annual, 160, 511, 520
    biennial, 520, 521
    forest, 676
    perennial, 158, 160, 511, 521, 586
herbicide resistance, 566
herbicides, 547, 551, 566, 570, 643
herbivores, 516, 517, 571, 595, 646 (*see also*
    animals; mammals; plant-herbivore
    interactions)
herbs, 382, 695–696
Hercium coralloides, **232**
hereditary information, 19
    mechanism of, 152
    molecular basis of, 121
    physical basis of, 121
heredity, 12
herpes, 200
Herpothallon sanguineum, **181**
heterochromatin, 122, 135
heterocysts, 193, **194, 195**, 198
heteroecious parasite, 235
heterokaryosis, 220
heterospory, 322–323, 346
    in angiosperms, 357, 388
    in ferns, 344, 349
    in gymnosperms, 357, 388
    in *Isoetes*, 335
    in progymnosperms, 360
    in *Selaginella*, 331, **336–337**
heterothallic organisms, 213, 215, 245, 248,
    250
heterotrophic protists, 244–246, 248–257,
    259, 262, 265
heterotrophs, 3, 188, 210, 323
heterozygote(s), 131, 152, **153**
heterozygous state, 131
Hevea, 696, 705
hexokinase, 88
hexoses, **49**
hibiscus, 424

Hibiscus esculentus, 690
Hickey, L.J., 405
hickory
    bitternut, **542**
    shagbark, **440, 499, 520, 531**
hierarchy, 403
Hiesey, William, 158
Hildebrandt, A.C., 563
hilum, in seeds, **442, 444**
histidine, **56**
histones, 19, 122, **123**, 176
Hodgkin's disease, **707**
holdfast, 274, **275, 276, 277**, 286, **287, 295**
holly, 551
homologs (chromosomes), 122, 124, 135
homology, 402–404, 693
Homo, 686
Homo sapiens, 686
homospory, 322–323, 325, 326, 346
    in *Equisetum*, 338
    in ferns, 348, 349
    in Lycopodiaceae, 330, **333**
    in progymnosperms, 360
homothallic organisms, 213, 215, 248
homozygote, 152, **153**, 154
homozygous state, 131
honeydew, 630, 631
honey guides, 419, **420**
honeysuckle, 531
    chaparral, **411, 434**
Hoodia, **518**
hook, fungal, 218
Hooke, Robert, 18
Hordeum
    jubatum, **183**
    vulgare, 86, 452, 557, 613, 688
horizons of soil, 598, 600
hormonal control of gene expression, 559
hormonal regulation of cell expansion, 560,
    561
    hormogonia, 193
hormone(s), 470, 545–547 (*see also by name*)
    action, molecular basis of, 559–563
    and control of gene expression, 559
    and control of sexuality, 250
    defined, 545
    influences on basic cellular processes,
    559
    plant, 545–565, 574, 575, 584
    receptors and response pathways, 562,
    563
    -receptor complex, 562
    and regulation of cell expansion, 560,
    561
    and tissue culture, 552, 553, 563–565
hornworts, 283, 299, 307–308 (*see also*
    bryophytes)
horse chestnut, 389
horseradish, 435
horsetails, **28, 42,** 163, **183,** 317, 320, 323,
    326, 338–341, 552, **595**
    of Carboniferous period, 346, 347, **356**
    life cycle of, **340–341**
horticulture, 585, 612
hot spot, 165
Hoya carnosa, 117
Hubbard Brook Experimental Station, 645
Huber, Robert, 107
human(s), 171
    and agriculture, 167, 687
    chromosomes of, 126, 128
    culture and society, 10, 686, 687, 698,
    708–709
    DNA, 134, **136**
    evolution, 10, 686
    hormones, 707
    impact on environment, 12, 155, 166,
    665, 699
    impact on nutrient cycles, 612
    and the Industrial Revolution, 699
    population growth, 698–700, 708, 709
humidity and transpiration, 620
hummingbirds, **163, 225, 412, 423, 424,**
    662–663

hunter-gatherers, **686**
Huperzia lucidula, **331**
hyacinths, 587
hydathodes, 626, **627**
hybridization
    by genetic engineering, 565
    interspecific, 173, 565
hybrids, 160, 161, **162**
    crop, 167
    fertile, 160, 161, 164
    plasmid, 567
    from protoplast fusion, 565–567
    sterile, 160, 161, 166, 167
hydration, 64
hydrocarbons, chlorinated, 643
Hydrodictyon, 292
    reticulatum, 292
hydrogen (H_2), 2, 594, **596, 713**
    gas (H_2), 2
    isotope(s), 714
    in living tissues, 47, **48**
    molecular structure, 717
    in organic compounds, 47, **48**
    in water molecule, 720
hydrogen bonds, **137, 138**, 139, **143**
hydrogen chloride, **719**
hydrogen sulfide (H_2S), 2
    in bacterial metabolism, 100, 190, 191,
    194
hydroids, 309, **311**
hydrolysis, 48, **49**, 52, 86, 91
hydronium ions (H_3O^+), 721
hydrophilic/hydrophobic interactions,
    molecules, 52, **55**, 67, 721
hydrophytes, 500, **502**
hydroponics, 563
hydrostatic pressure, 63 (*see also* turgor)
hydroxyl groups, **51**, 719, 720
hydroxyl ions, 719
hydroxyproline, 290
hygroscopic wall thickenings, 304
hymenium (hymenial layer), 216, **230**, 231,
    232, 234
Hymenomycetes, **174, 229,** 231, 232–234,
    238
    life cycle of, **229**
Hyoscyamus niger, 579, 587
hypanthium, 387
hyphae, 209, 245 (*see also* conidiophore;
    haustoria; rhizoids)
    ascogenous, 216, 218
    coenocytic, 211, 212, 213, 246, **249**
    dikaryotic, 216, 218, 235
    monokaryotic, 216
    of protists, 246
    receptive, **237**, 238
    septate, 211, 215
hypocotyl, **371**, 443, 448, 449
hypocotyl-root axis, 367, 374, **395, 442**, 443,
    445, 446, 449, 511
hypodermis, 363
hypogyny, 383, **384**, 387
hystrichospheres, 264

IAA (indoleacetic acid), 546, 547, 548, 549,
    552, 553, 563
i⁶Ade (⁶N-isopentenyladenine), **547**
ice plant, 518
Idria, **669**
igneous rocks, 597
Ilex aquifolium, 551
Iltis, Hugh, 693
imbibition, 64, 66, 226, 447
immune reactions, 209
immune systems, 222, 397
immunofluorescence microscopy, **41**
impala, 666
Impatiens, 433
incompatibility response, 3
incomplete dominance, 131
independent assortment of genes (alleles),
    132
Indian pipe, 383

indoleacetic acid (*see* IAA)
induced-fit hypothesis, 80
inducer, 146, 147
induction
    floral, 511, 515, 578, 584
indusia, 348, **349**
Industrial Revolution, 699
infanticide, 698
infection(s) (*see* diseases)
infection thread, 70, 604, **605**
inflorescences, **134, 376,** 383, **384**
influenza, 200
infrared radiation, **102**
Ingenhousz, Jan, 100
ingestion, as nutritive mode, 262
inheritance, 128, 129 (*see also* heredity)
    polygenic, 133
    molecular basis of, 136
    blending, 152
initials, **343, 344,** 453
    apical, **473, 474,** 489, 490
    cambial, 521–523
    fusiform, 521–523
    ray, 521–523
initiation
    codon, 144
    complex, 144, 145
inky cap, **230**
inorganic nutrients, 593, 594, **596**
    uptake and movement in plants, 471,
    627–629
insecticide resistance, 566
insecticides, 571, 643
insects, 210 (*see also under* pollination)
    in bryophyte sperm dispersal, 309
    and carnivorous plants, 517, 589, 610
    and "copulation," 420
    as disease carriers, 204
    herbivorous, 155
    juvenile hormones, 643
    mouthparts, 630, 631
    as pollinators, 22, 157, 374
    as soil components, **599**
integral (membrane) proteins, 67, 145
integuments, 357, 358
    of angiosperms, 390, **395**
    of pines, 365, 367, **371**
intercalary meristem, 490
intercellular (air) spaces, **455, 476, 477, 478,**
    492, 503, 504, 527, 609, 617, 621
interfascicular regions, 492, 497
intergeneric hybrids, 565
intermatrical spores, 239
International Code of Botanical Nomenclature,
    172
International Rice Research Institute, **690**
internodal elongation, 490, 492, 493, 511,
    515
internodes, **7,** 338, 340, 383, 450, 488, 490,
    495, 496, 498
interphase
    meiotic, 124, 125
    mitotic, 38
intine, 388
introns, 140, 141, **142**
inversions, chromosomal, 135
ion channels, 68, 69
ion pumps, 563
ion(s), 68, **596,** 601, 602, 715, 716, 717, 721
    (*see also specific ions*)
    and active transport, 68, 70, 628
    channels, 69
    as cofactors, 81, **596**
    functions of, **596**
    as inorganic nutrients, **596**
    movement of, 601, 602
    phloem-immobile, phloem-mobile, 629
    in plant tissues, 594, **596,** 627
    in soil, 601, 602
    transport in plant, 25, 627–629
ipecac, **708**
Ipomoea batatas, 485, **694**
Iridaceae, 696
irises, 156, 382, 696

iron (Fe/Fe²⁺/Fe³⁺), 594, **596**
   in electron carriers, 93
   oxide, 598
iron-porphyrin, 93
iron-sulfur proteins, 93, 108
ironwood, black, 540
irrigation, 700
isidia, 226
isocitrate (isocitric acid), **91**
Isoetes, 332, 335, 338, 346
   muricata, **52**
   storkii, **335**
Isoetaceae, 335, 338
isogametes, 286
isogamy, 259, 264, **291**, 292, 294
isoleucine, **56**
isomorphic (appearance), 274, 276
⁶N-isopentenyladenine (i⁶Ade), **552**
   radioactive, 53, 586, **621**, 630, 633
isthmus (desmids), 283
ivy, 156, **466**, 483, 515

jacket layer (sterile), 270, 299, 300, **351**
Jacob, Francois, 146
Jaffe, M. I., 576, 589
Janzen, Daniel, 155, 639
Jefferson, Thomas, 665
jelly fungus, 232, **233**, 235
jewelweed, 676
jojoba, **705**
Joshua tree, **668, 669**
Juglans
   cinerea, 530
   nigra, 539, **543**, 641
Juniperus (junipers), 369, 396, 671
   communis, **372**
   osteosperma, 671
jute, 457

Kalanchoë
   daigremontiana, 117, 450, **451**
kapok, 425
karyogamy, **216**, 218, **229, 236,** 238, **256,** 290
Keck, David, 158
kelp, **5**, 268, 270, 274, **275,** 276, **281** (*see also* brown algae)
Kentucky bluegrass, 116, 155, 166, 451
keratin, 222
kernel, **443, 450**
   wheat, **133**
α-ketoglutarate (α-ketoglutaric acid), **91**
keto group, **720**
kinases, 84
   protein, 563
kinetic energy, 74–76
kinetin, 552, 553
King Clone, **670**
kingdom, (*see also* by name)
   classification by, 11, 173, 175
   schemes of, 175, 176
Klebsormidium, **280**
Knop, W., 563
Koenigia islandica, 683
kohlrabi, 517
kombu, 281
Kranz anatomy, 112, **113,** 505
Krakatau, 654
Krebs, Sir Hans, 91
Krebs cycle, 86, 87, 90, 91, 92
Küchler, A. W., 658
Kuehneola uredinis, **228**
Kurosawa, E., 555
Kyhos, Donald, 164

Labrador tea, 683
lac operon, 146–147
lactate (lactic acid), 97
Lactobacillus
   acidophilus, **179**
   biennis, 172
lactose, 146, 190

(*lac*) system, 145–146
Lactuca sativa, 44, 447, **508,** 581
lady's slipper, yellow, 170
lagging strand, 139, **140**
Laguncularia racemosa, **484**
Lamarck, Jean Baptiste de, 150
Laminaria, 274, 275, 276
   life cycle of, 277
Laminariales, 274 (*see also* kelps)
laminarin, **247,** 248, 257, 270
landing platform (flower), 419
Lang, Anton, 584
larch, 638, **681,** 683
   European, **369**
Larix, 638, **681,** 683
   decidua, **369**
Larrea divaricata, **670**
late wood, 535, 536, 538
Lathyrus odoratus, 133
Latin names, 171
latitude, 579
Lauraceae, 696
laurel, 696
Laurus nobilis, 696
leaching, 601, 602, 611, 612
lead (Pb), 155, 512
leading strand, 139, **140**
leaf, leaves, 5, 319, 489–510, 628, 629
   abscission, 509, 510, 550, 551
   adaptations/modifications, 484, **485,** 515–519
   and air pollution, 512, 513
   of angiosperms, 489–510
   of aquatic plants, **502,** 510
   arrangements, 498
   axil, **7,** 450, **488,** 500, 516
   blade (lamina), 322, 344, 348, 498, 506–508, 580
   of bryophytes, 305, **315**
   buds, **7, 490**
   buttress, 506
   cells, **16, 17,** 24, **31,** 66
   coloration, 26, 429
   composting, 614
   compound, 498, **499,** 500
   of C₃, C₄ plants, 112, **113,** 505, **506**
   of cycads, 374
   of desert plants, 518
   development of, 348, 506–508
   of dicots, 456, 498–500, 503, 504
   dimorphism, 510
   of *Ephedra*, 377
   epidermis of, **22,** 464, 466, 500–503, 505
   of *Equisetum*, 338, **340**
   evolution of, 319, 321, **322**
   of ferns, 343, 344, **349**
   floating, 500, **502**
   food storage in, 517, 518
   function, 488, 489, 503
   gaps, 320, 321, 322, 345, **489,** 496–498
   gas exchange by, 503, 512, 513
   of *Ginkgo*, **376**
   of *Gnetum*, **376**
   of grasses, 498, **499, 503,** 505, 620
   hairs (trichomes), 164, 465, **466, 501, 502,** 503, **507**
   height and competition, 676
   heliotropism of, 590
   and hormones, 550, 551, 553, 555
   of hydrophyte, 500, **502**
   of *Isoetes*, 335
   juvenile, **362**
   of Lycopodiaceae, **319**
   mesophyll of, **7,** 363, 503, 504, 505, **506, 629**
   of mesophyte, 500, 503
   of monocots, 456, 498, **499,** 503, 504, 505, **506**
   morphology of, 498–500
   movements of, 505, **506,** 576, **577,** 587–589, 590, **610**
   and nitrogen fixation, 197
   photoperiodic responses in, 580

photosynthesis in, 112, **113,** 503, 505
   of pine, 362, 363
   primordium (primordia), 321, 488, 506–508
   pubescence (hairiness), 467, 620
   scalelike, **338,** 362, **377**
   scar, 509, 510, 530
   of *Selaginella*, 331, **336**
   senescence, 553
   sessile, 498, **499**
   shade, sun, 158, 508, 509
   shape and size, 164
   sheath, 498, **499**
   simple, 498
   in soil, 598
   stalk (petiole), 343, 348, **456,** 498, 509, 518
   -stem relation, 496–498
   structure, 500–506
   succulent, 117, 518, **519**
   traces, 320, 322, 456, 484, 496–498
   trifoliolate, **394**
   unifoliolate, **394**
   vascular tissue of, 504, 629
   veins (*see* veins; venation)
   water storage in, 518, **519**
   of *Welwitschia*, 377, **378**
   of xerophyte, 500, **502**
leafhopper, 201, **203,** 204
leaflets, 498, **499,** 500
leaf stalk (*see* petiole(s))
lectins, 606
leeks, 239
leghemoglobin, 606
legumes (leguminous plants), 155, 398, 430, **514,** 515, 587, 595, 602, **603,** 606, 687, 688, 689, 691, 697, **704**
   cultivation of, 604–608
Leguminosae, 173, 430
lemma, 427
Lemnaceae, 381
Lemna gibba, **381**
Lens culinaris, 688
lenticels, 526, 527, **530,** 531
lentils, 430, 688, 691, 697
Lentinus edodes, 234
Lepidodendron, **346, 356**
lepidopteran larvae, 566
leptoids, 309, **311**
leptosporangia (*see* sporangia, of ferns)
leptosporangiate ferns, (*see* ferns)
lettuce, **44,** 447, 579, 508, 581, 697
   wild, 172
leucine, **56, 142**
leucoplasts, 22
leukemia, **707**
Liatris punctata, **471**
lichen acids, 227
lichen substances, 227
lichens, **181,** 209, 223–228, 729
   algal or cyanobacterial component of, 194, 223, 224, 226
   in biological succession, 224, **650**
   biology of, 226–227
   commercial uses of, 224
   controlled parasitism of, 228
   crustose, **224, 225**
   as dyes, 224, 227
   evolution of, 227
   foliose, **225**
   fruticose, **225**
   fungal component of, 223, 227–228
   goldeye, **225**
   marine, 224
   as medicines, 224
   and pollution, 227
   reproduction in, 226
   relationship between components of, 227–228
   secondary metabolites of, 227
   structure of, **226**
   water content of, 226

licmophora flasellatta, **259**
lid (*see* operculum)
life
   and energy, 1
   origin of, 1
   and shores, 4–5
   primitive, 5
   transition to land, 5
life cycle(s), 322 (*see also* specific group or organism)
   duration of, 160
   types of, 183, **184**
ligase, **140,** 566
light, 1, 2, 101–103 (*see also* energy; solar radiation)
   absorption of, 103, 104
   blue, 574, 619
   fitness of, 103
   and flowering, 578–580
   and leaf development, 508, 509
   nature of, 101–103
   in photosynthesis, 103–109, 590, **591,** 717
   perception, 289
   plant response to, 547, 573, 574
   and plastid formation, 22, 23
   properties of, 101–103
   red, far-red, 581–584
   and secondary growth, 535
   and seed germination, 447, 581, 585
   and stomatal movements, 619
   wavelengths of, 101–103, 581, 582
light-dependent reactions, 107–109
light-independent reactions, 109–112
light microscope, 26, 27
lignin, 34, 36, 318, 320, 457, 478, 525, 539
lignin-like substances, 280, 299
ligule, 331, 335, **336,** 499
lilac, 47, **453,** 500, 501, 549, 607
lilies, 382, **389,** 418, **422,** 517
   African blood, 40
   arum, **416–417**
   calla, **134,** 381
   Easter, 397
   water, 382, 408, 410, **457, 502,** 649
Lilium, **124,** 389, **390, 391, 393**
   henryi, **384**
   longiflorum, **389,** 397
lime(s), 250
limestone, 598, 614
Linaria vulgaris, **385,** 432
linden, **457, 459, 462,** 464, 530 (*see also* basswood)
Lindsaea, 342
linkage, 132–133, 154
Linnaeus, Carl, **172,** 173
Linum usitatissimum, **211,** 688
lip cells, 343, 344
lip (of flower), **413**
lipid(s), **48,** 52–55
   in cell walls, 53, 54
   droplets, 21, **22,** 31
   in fungi, 211
   as membrane component, 66–68
   in protists, 248
   in seeds, 452
   synthesis, 28
lipid bilayer, 66–68
lipopolysaccharide, 187
Liriodendron tulipifera, **543**
Lithocarpus densiflora, **386**
liverworts, **183,** 299, 300–307, 309 (*see also* bryophytes)
   asexual reproduction in, 304
   leafy, 305, 307
   life cycle of, **302–303**
   sexual reproduction in, **302–303**
   spore discharge in, 306
   thallose, 301–305, 307
loam, 601
Lobaria verrucosa, **226**
lockjaw, **188**
locule, 386, **514**
locus (-i) (on chromosome), 130

incompatibility, 398
locust
    black, 521, **528, 529, 530**, 532, **542**, 624, 641
    honey, **499**
lodicules, **427**
London plane tree, **161**
long-day plants, 579, 580, 584, 587
Longistigma caryae, **631**
Lonicera, 531
    hispidula, **411, 433**
Lophophora williamsii, **437**
lotus, **411**, 586
LSD, 234
luciferase, 570
luciferase gene, 570
luciferin, 570
lumen
    of endoplasmic reticulum, 145
lupine(s), **385**, 419, 590
    arctic, 586, 691
    white, 629
Lupinus, 691
    albus, 629
    arcticus, 586
    arizonicus, 590
    diffusus, **385**
Lycaena gorgon, **422**
Lycogala, 257
Lycoperdon ericetorum, **235**
Lycopersicon, 691
    cheesmanii, **706**
    esculentum, 575, **617**, 641, 707
Lycophyta (lycophytes), **178**, 183, **318**, 320, 321, 322, 323, 326, 330–338, 346, 347, 348, 401, 733 (*see also Isoetes;*
    Lycopodiaceae; Lycopodium;
    Selaginella)
    of Carboniferous period, **324**, 346, **356–357**
    life cycles of, **332–333, 336–337**
Lycopodiaceae, 323, 330–333, **340**, 355
Lycopodiella, 331
Lycopodium, 330, 331, 346 (*see also*
    Lycopodiaceae)
    clavatum, 319, 331, **332–333**
    life cycle of, **332–333**
Lyell, Charles, 151
Lygodium, 343
Lymantria dispar, 644
lymphoma, **707**
Lysichiton americanum, 417
lysine, **56**, 122, 706
lysosomes, lysosomal activity, 26, 27

macaroni, 167
mace, **695**
MacMillan, J., 556
macrocyst, **254**, 255
Macrocystis, 274, 276, 281
    integrifolia, **276**
    pyrifera, **281**
macrofibrils, cellulose, 34, **35**
macromolecules (*see molecular biology*)
macronutrients, 594, 595, **596**
madder(s), 436
madder family, **696**
Madiinae, 164
magnesium (Mg/Mg²⁺), **509**, 594, **596**
    in chlorophyll molecule, **104**, 595, **596**, 597
    as cofactor, 81
Magnolia (magnolia), 408, 418, 429, 430
    grandiflora, **409**
    southern, **409**
maidenhair tree, 374, **376**
maiz de coyote, 693
maize, **120**, 386, 397, **691** (*see also* corn; *Zea mays*)
malaria, **437**
malate (malic acid), 25, **91, 92**, 111, 112, 116, 618
male nuclei, 216, 248, **249**, 274

Malloch, D.W., 241
Mallow, 410, **411**
Malpighi, Marcello, **630**
malt (barley), **698**
Malthus, Thomas, 151
Malus sylvestris, **388, 467**, 521, 528
mammals (*see also by name*)
    soil-dwelling, 599
Mammillaria microcarpa, **183**
manganese (Mn/Mn²⁺), 239, 594, **596**
    in photosynthesis, 108
Mangifera indica, 689
mangoes, 425, **662–663**, 689
mangrove
    black, 484
    white, **484**
Manihot, 697
    esculenta, **694**
Manila hemp, 457
manioc, **694**, 697
mannitol, **247**, 248, 257, 270, 276, 510
mannose, 248
manzanitas, 161, 448
maple(s), 432, **498, 509**, 538, **586, 675**
    red, **8**, 26
    silver, **498**
    sugar, **498**, 542
maquis, **657**, 680
Marasmius oreades, **234**
Marattiales, 344–345, 347
marble, 598
Marchantia, **183, 300, 301, 304, 305, 306**
    life cycle of, **302–303**
margo, **534**
María Sabina, 234
marijuana, 436, **437**
marshes, 611, 612
marsh marigold, **420**, 429
Marsilea, 352
    polycarpa, **352**
Marsileales, 344, 352, 353
maternity plant (mother-of-thousands), 117, 450, **451**
mating strains/types, **213, 229**, 238, **283, 290**, 291
Matteuccia struthiopteris, **348**
mattoral, 680
Maxwell, James Clerk, 101
McClintock, Barbara, **120**, 693
meadows, 245
measles, 200
measurement, units of, **21**
mechanical stimulation (of plant growth), 589
Medicago, 494
    sativa, 471, 494, 595, 602, 641
medicinal compounds, plants, **437, 708**
medicine and botany, 10, 11, 707
medicine man, Wayana, **708**
mediterranean scrub, **8, 657**, 680
    distribution of, 680
    vegetation characteristics of, 680
    role of fire in, 680
    grazing in, 680
    rainfall in, 680
medulla (lichen), **226**
Medullosa, **356–357**
    noei, 347
megagametogenesis, 390, 392
megagametophytes, 323, 335, **336, 337**, 357, 367, **371**, 388, 390, 391, **395**, 608
megapapascal (MPa), 63
megaphylls, 321, **322**, 338, 345, 348, **350, 352**, 358, 383
megasporangia, 323, 331, 335, **336, 352–353**, 357, 358, 364, 365, 367, 390
megaspores, 323, 335, **337**, 357, 358, 365, 367, **371**, 390, 391, **395**
    functional, **371**
megasporocytes (megaspore mother cells), 365, **366, 370**, 389, **391, 395**
megasporogenesis, 390, 392
megasporophylls, 331, 335, **336**, 383
meiosis, 121, 122, 124–128

in algae, **273**, 274, **275**, 276, **277, 278, 287, 290**
in angiosperms, 388, 390, **395**
in chytrids, 251, **252–253**
in diatoms, **260–261**
in eukaryotes, 248
    and evolution, 318
    first division, 124–125
in fungi, 212, 213, **216**, 218, **229, 236**
    gametic, 183, **184, 261**, 270, **278–279**, 283
in gymnosperms, 365, 367
in liverworts, **302–303**
    and mitosis contrasted, 125, 126, **127, 128**
in oomycetes, **249**
    and polyploidy, 166, 167
    second division, 125–126
in seedless vascular plants, **328, 332, 336, 340**, 344, **350**
in slime molds, 255, **256**
    sporic, 183, **184**, 270, 276, **277**, 283, 286
in true mosses, **312**
    zygotic, 183, **184**, 212, 270, 282, 283, 286, 291, 292, 295, 318
Melampsora lini, **211**
Melilotus alba, **203**
melon, 410, 550
membrane(s), 16
    cellular, 16, 62, 65, 66–68
    chloroplast, 20, 21
    of endoplasmic reticulum, 28
    mitochondrial, 23, 90, 93, **96**
    mobility of, 29
    permeability of, 65
    photosynthetic, 21, 106, **108**
    plasma (*see plasma membrane*)
    structure of, 54, 55, 66–68
    synthesis of, 28, 29
    transport across, 65, 68–73, 628, 632
    turnover of, 29
    unit, 16
    vacuolar (tonoplast), 18, 25
Mendel, Gregor, 121, **122**, 129, 131, 152
    experiments by, 129, **130**
Mendelian genetics, 129–133
Mendelian ratio, **131, 153**
Mendel's first law, 131
Mendel's second law, 132
Mentha, 695
Menthaceae, 695
meristem(s), 6, 447, 453 (*see also* apical, cambium, ground meristems; pro-cambium; vascular cambium)
    culture, 204, 414, 565
    in hornworts, 308
    intercalary, 274, 490
    lateral, 6, 320, 453, 520
    peripheral, 490
    pith, 490
    primary, 447, 453, 455, 475, 483, 491
    primary thickening, 491
    of root nodule, **603, 604, 605**, 606
    of *Welwitschia*, 377
mermaid's wine glass, **288**
Mertensia virginica, **385**
mescaline, **234, 437**
Mesembryanthemum crystallinum, 518
mesocarp, 393
mesocotyl, 450
mesophyll (*see* leaf, mesophyll)
mesophyll cells, 20, 111, 112, **113, 466**
mesophytes, 500, 503
Mesozoic era, 323, 347, 374, 401
mesquite, 471
messenger RNA (*see* mRNA)
mestome sheath, 505
metabolism, 78, 79
metals, 102, 723
    toxicity of, 512, 513, 595, 613
metamorphic rocks, 598
metaphase
    meiotic, 125, **126–127**
    mitotic, 40, 41, 125
    plate, **126**

metaphloem, **494**
Metasequoia, 372, **373**, 374
    glyptostroboides, **373**
metaxylem, 460, **476, 477**, 480, **493, 496, 497**
methane, 2, 12, 191, **718**
methanogens, 191
Methanosarcina, 191
methionine, **56**, 144, 554, (*see also* fMet)
metric table, **726**
mice, 568, **641**
micelles, 33, 34, **35**
Michel, Hartmut, 107
Micrasterias thomasiana, **284**
microautoradiography, 633
microbodies, 25
microcyst, 256
microelectrodes, **619**
microenvironments, 639
microfibrils, 30, 33, 34, **35**, 270, 560, 561 (*see also* cell wall)
microfilaments (actin filaments), 30
microgametogenesis, 388, 390
microgametophytes, 323, 335, **337**, 388, 390, 392
microhabitats, 671
Micromonadophyceae, 282
micronutrients, **594, 595, 596**
microorganisms, 50, 598, **599**
microphyll, **318, 319**, 321, **322**, 330, 331, **332**, 335, **336**, 338, 346, 347
micropropagation, 565
micropylar fluid, 367
micropylar pole, 444
micropyle, 357, 358, 365, 367, **371**, 390, **391**, 392, **395**, 444
microsporangia, 323, 331, 335, **336**, 347, **352–353**, 364, **365**, 370, 383, 388
microspores, 323, 335, **337**, 365, **371**, 388, **395**
microsporocytes (microspore mother cells), 365, **370**, 388, **389, 395**
microsporogenesis, 388, 390
microsporophylls, 331, 335, **336**, 365, 370, 383
microtubules, 30, **59**
    and cell cycle, 44, **45**
    and cell-plate formation, 43, 44, **45**
    and cell wall formation, 30, 36, 37, 560, 561
    in cytokinesis, 43, 44, **45**
    in flagella and cilia, 32, **280**
    kinetochore, 41, 42
    in meiosis, 125
    in mitosis, 38–43
    polar, 41
    and spindle fibers (*see* spindle fibers, apparatus)
middle lamella, 34, **36**, 43, 44, 509, 554
midrib, midvein, **500, 504, 507**
    of bryophytes, 305, 309
migraine, 221
migration, 154
    of slime molds, 253, **254**
Miki, Shigeru, 372
mildew (*see also* mold)
    downy, 250
    powdery, 215
milk, **188**, 190
milkweed, 430, **431**, 432, 436, 518
    family, **418**, 518, **644**
Miller, Carlos O., 552
Miller, Stanley, **2**
millets, 689, 690
mimicry, 436
Mimosa pudica, 588, 589
Mimulus cardinalis, **423**
mineral(s), 4 (*see also* nutrients; soils)
    absorption by roots, 467, 470, 471, 475, 627, 628
    cycles, 645
    in enzyme catalysis, 597
    in human diet, 697
    kingdom, 173
    transport, 467

mint, 695
mint family, 410, 695
Miocene epoch, 372
miso, 221
Mississippian period, 360
mistletoe, 382
   dwarf, **433**
Mitchell, Peter, 95
mitochondrion (-ia), **20**, 23, 24, 31, 86, **607**
   evolutionary origin of, 24, 177
   and ribosomes, 145
   structure and functions of, 23, 24, 86, **96**
mitosis, 38–43, 156 (*see also* cell division)
   in algae, 265, 284, 286, **290**
   and bacteria, 265
   in chytrids, 251
   and colchicine, 166
   in dinoflagellates, 265
   in euglenoids, 265
   in eukaryotes, 248
   in fungi, 212, 265
   and meiosis contrasted, 125, 126, **127, 128**
   in oomycetes, 248
   and polyploidy, 166
   in protozoa, 265
   in slime molds, 252, 255, 256
moa plant, **327**
molds, 245 (*see also* water molds)
   black, **212**
   blue, 251
   bread, 215
mole (gram molecular weight), 76, 77, 719
molecular biology, 11, 19, 142, 201, 559, 569
molecular weight, 719
molecules, 717–720
   diffusion of, 63–65
   and energy exchange, 83, 84
   energy storage in, 48, 86, 87
   excited, 103, 107
   organic, 2, 3, 47–61
mollusks, 288
molybdenum (Mo), 594, **596**, 602
   ions (MoO_4^{2-}), **596**
monkeyflower, scarlet, **423**
monkey-puzzle tree, 369
monocots (Monocotyledones), **174**, 178, **183**, 380, **381**, 412, 413–414, 441, 443, 444, **446**, 449, 456, 462, 471, 491, 520, 595
   and dicots compared, 380, **382**, 388, 393, 441, 443, 444, 456, 470, 471, **594**, 595
   perennial, 464, 520
Monod, Jacques, 146
monoecious plants, 386, 396, 427, 555
monohybrid cross, 129
monomers, 48
monooxygenases, 643
monosaccharides, 48
Monotropa uniflora, **383**
Monstera, 87
   deliciosa, 416
Morchella, **217**
   esculenta, **215**
morel, 215, **217**
Moricandia, **499**
morning glory family, **383**, 694
morphogenesis, 454, 559
morphology, 10
Morus alba, 498
mosaic diseases, 201, 202, **203**, 204
mosquitoes, 252, 421, **422**
mosses, **298**, 299, 308–316 (*see also* bryophytes)
   aquatic, 309
   asexual reproduction in, 311
   club, **183**, 308, 317, **319**, 323, 330–331, **332–333**
   cushiony, 310
   distribution of, 308
   economic uses of, 314, 315
   feathery, 310
   fir, 331

granite (rock), 308, 315–316
   growth forms of, 310
   Irish, **271**, 308
   life cycle of true moss, **312–313**
   peat, **183**, 308, 314–315, 346
   reindeer, 224, **225**, 308
   sea, 308
   sexual reproduction in, **312–313**
   Spanish, **225**, 308, **373**
   tree club, 331
   true, 308, 309–314
moths, 421–423, 642, 644
   hawk-, 423
   larvae of, 435
   yucca, **423**
motility
   of algae, 24, 31, 32, 282, 291
   of bacteria, 178, **189**, 193
   of chytrids, 248, 251
   of diatoms, 261
   of dinoflagellates, 262
   of euglenoids, 265, 266
   of oomycetes, 246, 248
   of slime molds, 253, **254**, 255
Mount St. Helen, **654–655**
mountain lilacs, 162, **162**, 197
M-phase-promoting factor (MPF), 39
mRNA, 125, 140 (*see also* RNA)
   and protein synthesis, 140, **148**
   transcription, 122, 140, **148**
mucigel, 472
mucilage, 492, **610**
   duct, **491**, 492
mucilaginous matrix, 270, 272
mucilaginous substances, 276, 281, **284**, 308
Mucor, 222
mulberry, 498
mullein, **466**
multicellularity, **5**, **10**, 38, 124, 176, 183, 208, 244, 268, 269
   in prokaryotes, 187
multilayered structure (MLS), 283, 285
multinucleate cells (*see* coenocytic organisms)
mumps, 200
Münch, Ernst, 631
muramic acid, 179, 188
Musa × *paradisiaca*, **381**, 689
muscle,
   contraction, 253
   relaxant, **708**
Muscovy duck, 694
mushroom(s), **181**, 228–229, 231
   edible, 173, **174**, 232
   life cycle of, **229**
   oyster, **223**
   poisonous, **3**, 232, 234
mustard, 435, 696
   black, 436
   family, 166, 238, 430, **431**, 435, 436, **499**, 595
mutagenic agents, 568
mutagens, 134
mutant, 144
   dwarf, 556
mutated genes, 568, 569
mutation(s), 134–135, 154, 638
   early studies of, 134
   and evolution, 135
   point, 134
   types of, 134–135
mutualism, 223, 238, 638–639
mycelium(a), **208**, 210, 211, **229**, 230, **249**
   dikaryotic, 229, 230–231
   heterokaryotic, 229, 231
   homokaryotic, 231
   monokaryotic, 229, 230–231
   septate, 211, 230
Mycena, 181
   lux-coeli, 97
mycologist, 210, 232, 244
mycology, 11
Mycoplasma (mycoplasmas), 191, 195
mycoplasmalike organisms (MLOs), **196**
mycorrhizae, 209, 214, **215**, 299, 602, 627

fossil, 241, 318
   vesicular-arbuscular (V/A), 239–240, 241
mycorrhizal fungus, 627, 638
   of angiosperms, 382
   of ferns, 345
   of Lycopodiaceae, 323, 331, **333**
   of plants, 299
   of *Psilotum*, 326
Mydaea urbana, **235**
myosin, 253
Myrica, 197
   gale, 607
Myristica fragrans, **695**
myxamoeba, 252, 255
myxobacteria, 189
myxomatosis, 640
Myxomycota (myxomycetes), 178, 179, 245, 246, **247**, 255–257, 730–731 (*see also* slime molds)
   life cycle, **256**

NAA (naphthalenacetic acid), **547**
NAD⁺/NADH, 61, 81, 87–90, 91–95, 96, 97, 98
NADP⁺/NADPH, 101, 109, 110, 111
nanoplankton, 245, 257, 258
naphthalenacetic acid (NAA), **547**
narcissus, 587
nastic movements, 587–589
natural gas, 191
Naturalist in Nicaragua, The, 639
natural history, 150
natural selection, 135, 150, 151, 152
Neanderthals, 686–687
neck canal cells, 300, **312–313**
necroses, 199, 204
necrosis, 595
nectar, 238, 378, 415, 418, 419, 421, 424, 425
nectaries, 410, 423, 638, **639**
nectarine, 173
needles (*see* leaves, of pines)
Neljubov, Dimitry, 554
Nelumbo
   lutea, **411**
   nucifera, 586
nematodes, 204, **223**
neomycin, 190
Nepeta cataria, 172
Neptunia pubescens, **514, 515**
Nereocystis, 276
Nerium oleander, **502**
nettle, stinging, 676
neurological illness, 263
Neurospora, 135, 215
neutron, 715, 716
Newton, Sir Isaac, 101
N-formyl methionine (fMet), 144
niacin, 81
nickel, 595
Nicotiana, 423
   glauca, **565, 566**
   silvestris, 584
   tabacum, **24**, 436, **497**, 513, 507, 552, **565**, **570**, 578, 584, 691
nicotinamide, 81
nicotinamide adenine dinucleotide (*see* NAD⁺/NADH)
nicotine, 436
nightshade family, 436, 565, 694
Nilson-Ehle, H., 133
Nitella, 33
nitrate ions (NO_3^-) and nitrite ions (NO_2^-), 4, 196, 198, **596**, 602, 610, 611
nitric acid, 512, 615
nitrification, 198, 610
Nitrobacter, Nitrococcus, Nitrosomonas, 198, 610
nitrogen (N), 2, 47, 48, 597, 602, 611
   assimilation of, 610, 611
   in bacterial metabolism, 196, 602, 604, 606, 607, 610

cycling of, 602, **603**, 645
   fixation, 602–607
   gas (N_2), 2
   isotope (¹³N), 714
   loss, 611
   and mycorrhizae, 241
   in plant nutrition, 241, 593, 594, **596**
nitrogenase, 194, 197, 602, 606
nitrogen cycle, 196–198, 602, **603**
nitrogen dioxide, 512, 598, 615
nitrogen fixation
   biological, 190, 191, 193, **195**, 196–198, 227, 430, 571, 593, 595, 602, 603, 606, 607
   efficiency of, 613, 615
   industrial, 604
   by lightning, 604
nitrogen-fixing bacteria, 190, 191, 193, **195**, 196–198, 308, 604, 606, 607
nitrogen-fixing genes, 571, 607
nitrogenous bases, **60**, 61, 81, 136, 137, 138, 139 (*see also* by type)
nitrogenous compounds, **596**
nitrogen oxides, 12
Nitrosococcus nitrosus, **188**
Nitrosomonas, 612
Nobel, Park, 671
noble gases, 715, 716
Noctiluca scintillans, **264**
nodal structure, **495**, 497
nodes, 285, **321**, 338, **340**, 346, 386, 450, **488**, 497
nodD gene, 607
nodules (*see* root nodules)
nodulins, 607
Nomuraea rileyi, **217**
noncyclic electron flow, photophosphorylation, 107–, 109
non-heme iron proteins, 93
nonpolar molecules, 56
Norfolk Island pine, 369
nori, 281
Nostoc, **179**, 198, 223, 227, 308
   commune, **194**
Nothofagus, 240
   menziesii, 408
no-tillage cropping, 701
N-terminus, 57
nucellus, 357, 358, 361, 365, **366**, 367, **371**
nuclear division, 122, 124, 128 (*see also* meiosis)
nuclear envelope, 19, 28, 39, 40, **42**, 43, 125, 126, 176, 212, 252, 265, 280, 282, 284, 286, 294
nuclear pores, 19
nucleating site, 30
nucleic acids, 48, 59–61, 552 (*see also specific compounds*)
   viral, 200
nucleoid, **16**, 21, **23**, 24, 187
nucleolar organizer regions, 20
nucleolus, **16**, **17**, 20, **39**, 40, 43, 125
nucleoplasm, 19
nucleoside, 81
nucleotide(s), 59–61, 81 (*see also specific compounds*)
   in ATP, 61
   DNA sequences of, **138**, 139, 140, 146, 245
   in nucleic acids, 59–61, 136, **137**, **138**
   and operators, 146
   origin of replication of, 139
   and promoters, 140, 146
   RNA sequences of, **140**, 141, 142, 175
   sequences of, 582
   and terminators, 140
nucleus, atomic, 713–717
nucleus, of cell, **16**, **17**, 19, 20, 176
   dinokaryotic, 265
   diploid and hapolid, 20
   division of, 38–43 (*see also* meiosis; mitosis)
   fungal, 212
   interphase, 38

and ribosomes, 145
Nudaria, **281**
nudibranch, 288
nutcrackers, 368
nutmeg, **695**
nutrient cycling, 645–646
  in a forest ecosystem, 645
nutrients (inorganic) (*see also* soils)
  concentrations in plants, 594, 595
  cycles, 602, 603, 611, 645–646
  cycles and humans, 612
  essential, 593, 594, **596**
  functions of, 595–597
  movement in plants, 627–629 (*see also* xylem)
  source of, 597
nutrition (plant), 593–615
  general requirements of, 593–595, **596**
  and mycorrhizae, 209, 238–241
  research, 613, 615
  and soils, 597–602
nutrition, modes of, 176, 257, 382
nutritive tissue, 367
nuts, 432
nyctinastic movements, 588
Nymphaea odorata, **381, 457, 502, 649**

oak(s), 161, 171, **215**, 240, 386, 396, 427, 459, 498, 535, 644
  black, 160, 161, **531**
  cork, 531
  red, **459**, 498, **527**, 535, **536, 542**
  scarlet, 160, 161
  tan-bark, **386**
  true, 386
  white, **530, 538, 542, 594**
oak-leaved goosefoot, **497**
oat(s), 114, 452, **470**, 547
oca, **692**
oceans, 2, 4, 115
Ochroma lagopus, 532
Ocimum vulgare, 695
octamers, 122
Oedemeridae, **415**
Oedogoniales, 294
Oedogonium, 295, **296**
  cardiacum, **296**
  foveolatum, **296**
Oenothera, 423
  biennis, 172
  glazioviana, **134**
oil droplets (*see* lipids)
oils, 52, 53, 390, 704
  in human diet, 697
  olive, 696
  sunflower seed, **694**
okra, 690
old man's beard, **225**
Olea europaea, 688
oleander, **502**
olefinic hydrocarbons, 276
oleic acid, **53**
oligosaccharins, 33
olive(s), 430, 688, 696, 697
Oliver, F.W., 347
O'Mara, J.G., 702
Onagraceae, 399
Oncidium sphacelatum, **484**
onion, 19, **41**, 199, 250, 449, **473, 517**, 619
Onoclea sensibilis, **348**
oogamy, 248, 259, **262**, 284, **291**, 294, 295–296, 298, 322
oogoniol, 250
oogonium (-ia)
  of algae, **262**, 276, **277, 278, 279**, 285, 295
  of oomycetes, 248, **249**, 250
Oomycota (oomycetes) **178**, 179, **184**, 245, 246, **247**, 248–251, 730
  economic importance of, 250–251
  life cycle of, **249**
oospore, 248, **249**
opaline silica, 258

opals, 338
operator (nucleotide sequence), 146
operculum, 311, **312, 314, 315**
operon, 146–147
Ophioglossales, 344–345
Ophioglossum, **344**, 345
  reticulatum, 345
Ophiostoma ulmi, 215
Ophrys, 420, **421**
opines, 566
opine-synthesizing enzyme, 566, **567**
Opuntia, **434, 642**
  inermis, **642**
orbitals, electron, **714, 715, 716**
orchard grass, 155
orchid(s), 382, **413**, 420, **421, 422**, 424, 432, **662–663**
  Cymbidium, **183**
  epiphytic, 484
  lady-slipper, **170, 413**
  mycorrhizae of, 241
  propagation of, 414
  saprophytic, 413
  underground, 413, **414**
  vanilla, 414, 695
Orchidaceae, 170, 241, 412, 413–414
order, classification by, 173
Ordovician period, 212, 299, 318
oregano, 695
organelle, 17, 18 (*see also specific types*)
organic acids, 554
organic compounds, **2**, 3, 47–61, **596**, 607, 719 (*see also by name*)
  formation of, 99, 101
  oxidation of, 77, 78
organic matter, 598, 599, 614
organisms
  classification of, 178–181
  interactions between, 638–645
  major groups of, 175–176
  multicellular, 176
  origin of, 168
  unicellular, 176
organs, of plants, 7 (*see also* leaf; root; stem)
organ transplants, 209, 222
Origanum vulgare, 695
origin of replication (nucleotides), 139
Origin of Species, On the, 151
Orthotrichum, **314**
Oryza sativa, 114, **381**, 452, 555, 689, **690**,
Oscillatoria, **179, 194**
osmoregulatory mechanisms, 609
osmosis, 65, 66, 596, 608, 617, 626, 631, **632**
osmotic potential (pressure), 65, 609
outcrossing, 166, 173, 349, 364, 396–397, 399, 425, 427
ovary, **384**, 386, **387, 388**, 390, **395**, 410, 444, **514, 515**
  development into fruit, 393, 550, **551**
  inferior, **381**, 387
  superior, 387
ovule(s), **129**, 357, 358, 444
  of angiosperms, 383, 386, 388, **391, 395**, 410
  of *Ginkgo*, 375
  of gymnosperms, 361, **370**
  of pines, 365
ovuliferous scale (cone scale), 365, 367, **370, 371**
owl, northern spotted, 682
Oxalis, 577
  tuberosa, **692**
oxaloacetate (oxaloacetic acid), 91, 92, 111, **112, 113**
oxidation, 77, 78
  of glucose, 76–78, 86–95
  of pyruvate, 90–92
  -reduction reactions, 77, 78, 90, 597, 723, 724
oxidative phosphorylation, 93, 95, 96
Oxydendron, 675
oxygen (O), 594, **596, 713, 716** (*see also* oxidation)

in cellular respiration, 90–95
  and fruit development, 554
  gas (O_2), 2, 4, **10**
  heavy isotope of, 100
  as inhibitor of nitrogenase, 606
  in living tissues, 47, **48**
  in organic compounds, 47, **48**
  and photosynthesis, 4, 100, **101**, 107, 108
  and seed germination, 447
  in water molecule, 720
Oxyria digyna, 159
ozone, 4, 12, 512, 513
  injury, 513

P_{680}, P_{700}, 107, 108, 109
P (peptidyl) site, 143, **144**, 145
"Pac-Man" model (for anaphase chromosome movement), 42
palea, **427**
paleobotany, 11, 347, 372, 374
Paleozoic era, 270, 330, 338, 359
Palhinhea, 331
Palmaria, **272**
palmettos, **402**
palmitic acid, **53**, 91
palms, 374, **381**, 382, 464, 465, 491, 520
  coconut, **196, 381**, 697
  fan, **402**
  oil, 696
  sago, 374
  Washington, **402**
PAN (peroxyacetyl nitrate), 512
Pandorina, 292, **293**, 294
panicle, **384**
Panicum, 690
pansy, 168, **173**
pantothenic acid, 90, 91
Papaver, somniferum, **431**, 437 (*see also* poppy)
Papaveraceae, 430, **431**
papayas, 250
pappus, 412, **432**
paraheliotropism, 590
paramylon, **247**, 248, 265, **266**
paraphyses, **310**
Parasitaxus ustus, 382
parasites,
  angiosperm, 382, **383**, 433
  bacterial, 190, 198–200
  fungal, 210, 211, 212, 214, 235, 555, **556**
  gymnosperm, 382
  protista, 245, 248, 251
parenchyma, **28**, 31, **42**, 320, 455, 456, **468**, 492, 493, 628
  in algae, 274, 294
  axial, 532, 535
  interfascicular, 492, 493
  of *Isoetes*, 335
  palisade, **501, 502**, 503, 508, 512, 629
  in phloem, **461, 462**, 464, 465, 528, 532
  of pine leaf, 363
  of pulvinus, 588
  ray, 455, 522
  spongy, **501, 502**, 503, 629
  storage, 485, 486, 517, 532
  wood (xylem), **461**, 532, 535
parenthesomes, 230
Parka, 270
  decipiens, **285**
Parmelia perforata, **225**
parsley, 695
  family, 173, 401, 418, 432, **618**, 695–696
Parthenium argentatum, **705**
parthenocarpic fruit, 550, 558
Parthenocissus quinquefolia, 515
particle bombardment, 571
particle theory of light, 103
passage cells, 479
Passiflora coccinea, **662–663**
passion flower, 424, **662–663**

Pasternak, D., **706**
pasture plants and grazing, 155, 156
patch-clamp recording, 69
pathogen(s), (*see also* diseases; plant diseases)
  animal, 190, 218, 219
  plant, 538
pea(s), **130, 442**, 449, **499**, 515, 576, 579, 602, 627, 643, 688, 691, 697
  family, 155, 173, 410, 430, **509**
  garden, **129**, 131, 132, **430–431**
  Mendel's experiments on, 129–132
peach, 173, 430, 587
peanut(s) 200, 515, 691, 697
pear, 34, 430, 458, 528, 531
  fire blight, **188**
peat mosses, 314–315 (*see also Sphagnum*)
pectic acid, **51**, 597
pectic compounds (substances), 34, 35, 44, 50, 280, 472
pectin, 50, 388, 554
Pedaliaceae, **435**
pedicel, 383, **384**
peduncle, 383
pegs
  fungal, 227
  lichen, 232
pelargonidin, **429**
pellicle, **247**, 266
penicillin, 209, **219**, 641
penicillinase(s), 641
Penicillium, **219**, 220, 221
  camembertii, 221
  chrysogenum, 641
  notatum, **219**
  roquefortii, 221
Pennisetum, 690
Pennsylvanian period, 346
Penstemon
  centranthifolius, **163**
  grinnellii, **163**
  spectabilis, **163**
3-pentadecanedienyl catechol, **436**
pentoses, **49**
Peperomia, 465, 519
pepper(s), **203**, 405, 696
  black, **695**
  red, 695
  chile, 691
peptide bond, 57, **144**, 145, **148**
peptide transferase, 145
peptidoglycan, 187
perennials, 6, 158, 160, 511, 521, 586
perforation plate, 458
  scalariform, **459**
  simple, **459**
perforations, in vessel members, 458, 459, 460
perfumes, 704
perianth, 383, **384**, 408, 410
  in liverworts, **305**
pericarp, 393, 430, 434, **443, 446**, 449, **450**
  in algae, **273**
periclinal division, 343, 447, 465, 506, 521, 522
pericycle, **345, 475, 476, 477, 478**, 480, **481**, **482**, 483, **625**
periderm, 320, 346, 364, 454, 467, 480, **481**, 482, **523, 524**, 525, 526, 527
peridinin, **247**, 262
perigyny, 387
peripheral (membrane) proteins, 67
periphyses, 238
perisperm, 447
peristome, 311, **312, 314**
perithecium, 216, **217**
periwinkle, **707**
permafrost, **8**, 683
permanent wilting percentage, 601
Permian period, 322, 323, 346, 347, 357, 362, 375
Peronospora hyoscyami, 251
Peronosporales, 249
peroxisomes, 25, 282, 285, 607
  and ribosomes, 145

peroxyacetyl nitrate (PAN), 512
Persea americana, 554, 691
persimmon, **37**, 543
pesticides, 643, 700, 704
pest management systems, integrated, 643
petals, 383, **384**, 410, **514**, 515
petiole(s), **7**, **456**, 498, 509, 518 (*see also* leaf,
  leaves)
petiolule, 498
Petroselinum crispum, 695
pH, 82, 83, 614
  of acid rain, 512, 513, 615
  of bogs, 315
  gradient, 70
  scale, **722**
Phaeophyta, **5**, **178**, 180, **247**, 274–279, 731
  (*see also* brown algae)
phages, 201 (*see also* bacteriophage)
phagocytosis, 70
Phalaris canariensis, 547
Phallus impudicus, **235**
phase contrast microscopy, **40**
Phaseolus, 602
  lunatus, 691
  vulgaris, **442**, 448, 449, 549, 550, 556,
    575, **577**, 613, 691
phellem, 525 (*see also* cork)
phelloderm, 467, 525, 526
phellogen, 467, 525, 527
phenolic compounds, 227, 644
phenotype, **129**, **130**, 131, 135, 151
  and environment, 154, 158
phenotypic ratio(s), **130**, **131**, **132**, **153**
phenylalanine, 56
Philodendron, 87, **381**, 416
  scandens, 416
phlegmarius 330, 331
phloem, 6, 317, **319**, 320, 325, **327**, 330, **339**,
  345, 454, **458**, 461–465, 486,
  629–632, 620, **621**
  in brown algae, 270, **276**
  cells of, **467**, **469**
  development of, **475**, 494, **496**, **497**
  disease in, 195, 204
  florigen transport in, 584
  functional, **528**, **529**, 532
  loading, 631, 632
  nonfunctional, **528**, **529**, 532
  and plant hormones, **546**, 548–550
  primary, 320, 338, 461, 465, 480, 481,
    482, 492, 493, **494**, **495**
  secondary, 320, 453, **461**, **462**, 464, 465,
    480, 481, **482**, 521, 522, 523,
    524, **525**, **528**, **529**
  secondary, of pine, **364**
  secondary, of progymnosperms, 359
  transport (translocation), 71–73,
    629–633
  and xylem interchange, 629
phlox, 410
Phoenix dactylifera, 688
phosphate, 54, **55**, 61, **596**, 601, 602, **720**
  absorption, **628**
  in DNA, **137**, **138**
  inorganic (P_i), 83, 84, 89, 110, 612
  in oxidative phosphorylation, 93, 95, 96
  and mycorrhizae, 240
phosphoenolpyruvate (PEP), 89, 111, 112
phosphoenolpyruvate carboxylase, 111, 112,
  114, 116
phosphofructokinase, 88
phosphoglucoisomerase, 88
3-phosphoglycerate (PGA), 110, **111**
phosphoglyceromutase, **89**
phosphoglycolate (phosphoglycolic acid), 114
phospholipid bilayer, 54, **55**
phospholipids, 54, **55**, 66–68, 597
phosphorus (P), **48**, 594, **596**, 597, 611, 612,
  713
  cycle, **611**
  ions ($H_2PO_4^-$; HPO_4^{2-}), **596**, 629 (*see
    also* phosphate)
  and mycorrhizae, 238
  pentoxide (P_2O_5), 613

in plant nutrition, 238, 594, **596**, 597
  radioactive (^{32}P), 633
phosphorylation, 84, **596**
  oxidative, 93, 95, 96
  photo-, 108, 109
  substrate-level, 89
Photinus pyralis, 570
photoconversion reactions, 581–582
photoelectric effect, 102
photolysis, 107, **108**
photons, 103, 106, 109
photoperiodism, 578–584
  in animals, 578
  chemical basis of, 581–584
  and darkness, 580
  and dormancy, 586, 587
  and flowering, 579–584
  and phytochrome, 581, 582
  and temperature, 580, 587
  and vernalization, 587
photophosphorylation, 108, 109
photoreceptor, 289
photorespiration, 112, 114
photosynthates, 504, 505
photosynthesis, 1, **75**, 78, 99–119, 248, 472,
  590, **596**, 617
  and air pollution, 512
  and algae, 291
  in bacteria, 107, 186, 188, 196
  in cacti, 518
  CAM, 116, 117, 338, 518, 618, 639, 671
  in carbon cycle, 115
  and chloroplasts, 20, 21, 99, 106, 108,
    113, **114**
  and circadian rhythm, 576, **578**
  in C_3 plants, 112, 116, 117, 505
  in C_4 plants, 112, **113**, 114, 116, 117,
    505, **596**, 639
  in desert plants, 518
  in ecosystems, 9
  efficiency of, 112, 114, 116, 159, **272**,
    318
  evolution of, 3, 4, 114, 116, 518
  general equation for, 100
  in *Isoetes*, 338
  in lichens, 226
  and stomatal movements, 619
  and temperature, 101, 117
  and transpiration, 617
photosynthetic organisms (*see also*
  autotrophs)
  requirements for, **5**
  multicellular, **5**
photosystem(s), 106, 107, 109
  II, 248
phototropism, 214, 547, 573, 574
phragmoplast, **38**, 41, 43, 44, **45**, **280**, 282,
  283, 284, 285, 298
phycobilins, 105, 192, 196, **247**, 269, 270
phycocyanin, 192
phycoerythrin, 192
phycologist, 245, 282
phycology, 11
phycoplast, 280, 282, 289, 294, 295
phylogenetic analysis, 404
phylogenetic relationships, **177**, 401–404
phylum, 173
Physarum, **180**
physical components in ecosystems, 646
phytoalexins, 33, 643, 644
phytochrome, 581–584
  activated genes, 582, 583
  in algae, 280
  dark reversion of, 581, 582
  photoconversion, 581
  and photoperiodism, 581, 582
phytohormones, 546
Phytophthora, 250, 251
  cinnamomi, **250**
  infestans, **251**, 704
phytoplankton, 245, 258, 292
Picea, 369, 687
  engelmannii, 640, 653
  glauca, **650**, 681

rubens, 543
pickles, 190
pickleweed, 609
Pierinae, 435, 436
pigment(s), 104, 105 (*see also* by name)
  absorption spectrum of, 104, 105
  accessory, 104, 105, 192, 196, 246, 298
  antenna, 106
  in bacteria, 187, 191, 192
  body, 289
  of flowers, 428–429
  P_{680}, P_{700}, 107, 108, 109
  and photoperiodism, 581, 582
  photosynthetic (*see* accessory pigments;
    carotenoids; chlorophylls;
    phycobilins)
  in plastids, 20–22
  P_r, P_{fr}, 581–584
  and stomatal movements, 619
  in vacuole, 25, 26
  visual, 289
pigweeds, 99, **692**, 706
pileus, 232
pili, 188, **189**
Pilobolus, **214**
Pimenta officinalis, 695
Pinaceae, 240, 369
pine(s), 157, 171, 361, 362–368, 397, 531,
  543, **651**, 683 (*see also Pinus*)
  bristlecone, **363**, 535, 536, **537**
  closed-cone, **651**
  digger, 366
  eastern white, **366**
  family, 240
  life cycle of, **370–371**
  limber, 368
  loblolly, **543**, 624
  lodgepole, **367**, 653
  longleaf, **362**
  Monterey, **363**, 364
  Norfolk Island, 369, **402**
  nuts, **366**
  pinyon, 362, **366**, 368
  ponderosa, 543
  red, **366**
  slash, 532, **543**, 667
  serotinous, **651**
  sugar, **183**, 366, 543, **651**, 679
  western white, 543
  white, **239**, 521, 532–534, 641
  whitebark, 368
  yellow, 366
pineapple, 117, **225**, 250, 429, 545, 620, 691
pines, running, 331
pinnae, 348
Pinnularia, **261**
pinnae, 348
pinocytosis, 70, 71
Pinus, **240**, 362–368, 531, 532, 683
  albicaulis, 368
  contorta, **367**, 653
  edulis, **362**, 366
  elliottii, 532, **543**, 667
  flexilis, 368
  lambertiana, **183**, 366, **543**, **651**, 679
  life cycle of, **370–371**
  longaeva, **363**, 535, 536, **537**
  monticola, **543**
  palustris, 362
  ponderosa, **366**, **543**
  radiata, **363**, **364**
  resinosa, **366**
  sabiniana, **366**
  strobus, **239**, **366**, 521, 532–534, 641
  taeda, **543**
pinyon-juniper savannas, woodlands, **671**
pioneer species, 650
Piperaceae, 405, 408
Piper nigrum, **695**
Pirozynski, K. A., 241
pisatin, 643
pistil, 383
Pisum sativum, 129, **430–431**, **442**, 449, 458,
  499, 515, 576, 579, 643, 688

pit, 36, 458, 459, 532, 533
  apeture, 533
  bordered, 36, 532, **534**
  cavity, **36**
  membrane, 36, 533
  ramiform, **458**
  simple, 36, **458**
pit connections, **272**
pit-fields, primary, 35
pith, 320, **321**
  of roots, 475, **476**, **477**
  of stems, 338, 347, 360, 490, 491, 492,
    493, 494
pit-pairs, 36, 459, **534**
placenta, 386, 390
placentation, 386
Placobranchus ocellatus, 288
Plagiogyria, **342**
plane tree, **161**
plankton, 193, **244**, 245, 269 (*see also*
  nanoplankton; phytoplankton;
  zooplankton)
plant(s), 175, 181
  and adaptations to life on land, 657,
    660–662
  and air pollution, 512, 513
  anatomy, 11
  ancestors of, 298
  and animals compared, 573, 593, 616,
    617
  biology, 10–12
  biotechnology, 563–571
  body, 319–322, 453, 454
  C_3, C_4, 114, 116, 117, 505
  CAM, 116, 117, 518
  classification of, 11, 172–173, 181
  dark grown, 583
  distributions of, 157–160
  domestication of, 687–694
  in ecosystems, 6, 8–10
  and green algae compared, 269
  -herbivore interactions, 518
  insect-pollinated, 157
  morphology, 10
  nutrition and soils, 593–615
  physiology, 10
  and people, 686–711
  reproduction, 270
  succulent, 116, 518, **519**, 620
  tissues, 454–468
  woody, 520–523
Plantae (plant), kingdom, 11, 173, **174**, 176,
  178, 181–183, 244, 732 (*see also*
  bryophytes; vascular plants)
plantains, 690
plant body, 6
plant breeding, 151, 235, 701, 702, 704
plant cells, 17–38
  and animal cells compared, 33
  differentiation of, 454
  division of, 38–44
  types of, **467**, **468**, **469**
plant diseases
  bacterial, **188**, 190, 195
  viral, 201–204
  viroid, 204–205
plant-herbivore interactions, 638, 642–644
plantlets, 450
plant-pathogen interactions, 638, 642–644
plant pathogens
  bacterial, 198–200
  oomycetes, 250
plasma membrane(s) (plasmalemma), 18, 62,
  65
  and active transport, 70, 628
  of algae, 270, 290, 294
  bacterial, 187, 190, 191, 195
  in cytokinesis, 44, **127**
  of dinoflagellates, **262**
  and endo-, exocytosis, 70, 71
  of euglenoids, 266
  growth of, 29
  permeability, 65, 632
  in secretion, 29

of slime molds, 253, 255
    water movement across, 65, 624, **625**, **632**
plasmid, 135, **186**, 191, 569
    recombinant, **567**
    T region, **567**
    Ti, 566, **567**, 569, 570
plasmodesmata, 28, 37, 43, 44, 71–73, 176, 187, 193, 204, 294, **464**, 478, 624, 627
plasmodesmatal canal, 37
plasmodial slime molds (see slime molds)
plasmodiocarp, 255, 257
plasmodium(-a), 180, 246, 255, **256**, 257
plasmogamy, 216, 229, **237**, 238, **256**, 290
plasmolysis, 66
Plasmopara viticola, 249–250
plastid(s), 20–23 (see also amyloplasts; chloroplasts; leucoplasts)
    developmental cycle of, 23
plastocyanin, 108, **109**
plastoquinone, **108**
Platanus, 161
    X *hybrida*, 161
    occidentalis, 161, **531**, 543
    orientalis, 161
Plectranthus, 466
pleiotropy, 135, 154
Pleistocene epoch, 670, 674, 687
Pleurastrophyceae, 282, 294
Pleurotus ostreatus, **223**
Pliny, 235
Plotkin, Mark, **708**
plum, 430
plumule, 443, 444, 449, 450, 488
pneumatophores, 484
pneumonia,
    bacterial, **188**, 190
Poa
    annua, **506**
    pratensis, 155, 166, 451
Poaceae, **174**, 432, 687 (see also grasses)
pod (fruits), **395**
poi, 689
poinsettias, **424**, 579
poison ivy, **436**
polar bears, 192
polar desert (see Arctic tundra)
polar molecules, 55, **56**, 719, 721, 722
polar nuclei, 391, 392, **393**, 395
polar rings, 270
pollen (grains)
    of angiosperms, 388, **389**, 390, 395
    dicot, monocot, 388, **426**
    dispersal, 389
    filiform, 428
    as food source, 390
    formation of, **127**, 388, 390
    fossil, 390
    germination of, 390, 392, **395**
    of gymnosperms, 361
    mother cell, 125
    of pine, **364**, **365**, 367, **371**
    and spores compared, 390
pollen baskets, brushes, combs, **418**
pollen sacs, 383, 388, **389**, 395, 421
pollen tubes,
    of angiosperms, 388, 392, **395**
    of gymnosperms, 361, 367, **371**, 375
pollination, **129**, 392–393
    of angiosperms, 388, 398, 410, 414–428
    by bats, 425
    by birds, **163**, 423–425
    cross-, **129**, **157**, 396, 398, 415
    direct, 410
    of gymnosperms, 361, 375, 378, 399, 415
    "hot," 416–417
    by insects, 22, 160, **163**, 410, 415, 416–417, 418–428
    mechanisms of, 406
    of pines, 367
    self-, 130, **157**, 166, 399
    water-, 428

by wind, 157, 160, 426–427
pollinium, 413, 420, **421**, **422**
pollutants, 512, 513
pollution, 512, 513, 612, 615
    and bryophytes, 308
    and *Ginkgo*, 375
    and lichens, 227, 308
polyembryony, 361, 367
polygenic inheritance, 133, 154
Polygonaceae, 432
Polygonatum, **392**
    pubescens, **636**
polymerases (DNA/RNA), 139, **140**, **141**, **146**, 147
polymers, 48, 50, 55, 270
polynomials, 172
polynucleotide chains, **137**
polypeptide(s), 57, 143, **144**, 202
    elongation of chain, 144
    targeting and sorting, 145
polyploids, 166–168
polyploidy, 135, 161, 166–168
Polypodium
    life cycle of, **350–351**
    polypodioides, **342**
polyribosome(s), **145**
polysaccharides, 33, 34, 50, **247**, 248, 257, 265 (see also by name)
    in bacteria, 192
    in fungi, 210
    in cell walls, 33, 34, 50, 270, 461
Polysiphonia, life cycle, **273**, 274
polysomes, **19**, **20**, 27, **141**, 145
Polytrichum
    juniperinum, **310**
    life cycle of, **312–313**
    piliferum, **309**
pome, 429, 430, 554
pomegranates, 688
popcorn, **691**
poplars, 240, 432, 538, 683
    yellow, **543**
poppy, **431**
    California, **418**, **421**
    family, 430, **431**
    opium, 437
population(s)
    arctic, alpine, 159
    defined, 637
    diploid, 152, **153**, 166
    divergence of, 157–160
    genetic variability of, 152
    groups, 173
    and Hardy-Weinberg equilibrium, 152–154
    human, 151
    isolation of, 157
    natural, 154–157
    natural, changes in, 155–156
    natural selection within, 154, 155, 157
    size of, 154
Populus, 538, 683
    deltoides, **504**, 542
    tremuloides, **650**
pores
    in bryophytes, 301, **315**
    epidermal, 617
    fungal, **232**, 233
    in pollen grains, **389**, 390, 392
    in vascular tissues, 309
    of wood, 538
Porolithon craspedium, **271**
Porphyra, 272, 281
Porphyridium, 270
porphyrin ring, 98, **104**
portulaca family, 429
position effects, 134
Postelsia palmiformis, **180**
potassium (K/K+), 47, 594, **596**, 597, 613, 621, 627, **713**, 715, 716
    cycling, 645
    deficiency, 613
    and leaf movements, 589
    oxide (K₂O), 613

radioactive (⁴²K), 621
    and stomatal movements, **596**, 618, **619**
potato, 50, 204, 437, **503**, 565, 579, 587, **617**, 691, 697
    blight, **251**
    cultivation of, 168, **692**, 703
    eye (bud), 517, 587
    famine, Irish, 251, 704
    ring rot, **199**
    scab, **199**
    soft rot, 199
    spindle tuber disease (PSTV), 204–205
    sweet, 485, 486, 694, 697
    white (Irish), 517, 694
    wilt, **188**, 200
Potentilla glandulosa, ecotypes of, 158–159
Power of Movement in Plants, The, 547
P-protein (slime), 462, **463**, 464
P-protein (slime) bodies, 462, **463**, 464
Precambrian period, 272
preprophase band, 39, 41, **45**
pressure-flow mechanism (hypothesis), 631, 632
prickles, 516, 517
Priestley, Joseph, 100
primary consumers, 646
primary endosperm nucleus, 392, 393, **395**
primary growth, 6, 320, 453
    in roots, 472–475
    in stems, 489–491
primary pit-fields, 35, 36, **455**
primary plant body, 320, 453, 455
primary producers, production, 646
primary structure
    comparison in root and stem, 475
    of root, 475–480
    of stem, 491–496
primary tissues, 320, 343
    of root, 472–480
    of stem, 489–496
primary vascular tissues, 458–465, 475, 480, 491, 492
primordium (-ia)
    bud, 488, **489**, **507**
    leaf, 490, 506–508
    root, 483
primroses, **399**, 579
    evening, **134**, 172, 423
principle of independent assortment, 132
principle of segregation, 131
prism, 102
procambial strands, 491, 492, 494, **495**, 496, **507**
procambium, **445**, **446**, 447, 453, 454, **473**, 475, 481, 491, 493, 507
Prochloron, **187**, 195, 196, 269
productivity, ecosystem, 9, 245, 246, 257, 264, 268, 269, 272, 353
proembryo, **371**, 444, **445**, 446
progesterone, **707**
progymnosperm (Progymnospermophyta), 321, 323, **324**, 326, 358, 359–361, 401
prokaryotes, 4, 186–200 (see also bacteria; cyanobacteria)
    classification of, 175–176
    and eukaryotes compared, 4, 16, 175–176
prolamellar bodies, 22, 23
proline, **56**
prometaphase, 40, 41
promoter (nucleotide sequence), 140, 146–147
propagation, 158, 161, 163, 450, 451, 517, **564**, 565 (see also asexual and sexual reproduction)
prophase
    meiotic, 124, 125, **126–127**, 133
    mitotic, 39, 40
proplastids, 22, 23
Prosopis juliflora, 471
prosthetic groups, 82, 597
Proteaceae, 238
protein, 55–59, **596** (see also by name)

amino acid sequences in, 140
    binding, **140**
    biosynthesis, 21, 22, 24, 28
    carrier, 68, 70
    as cell components, 55
    as cell-wall component, **247**, 266
    coat (viruses), 176
    enzymes as, 59
    globular, 57, **58**, 59, 79
    histone, 19, 122, **123**, 176
    in human diet, 688, 697, **698**, 706
    initiator, 139
    integral, 67
    iron-sulfur, 82, 93, 108, 109
    as membrane component, 57, 66–68, 70
    molecular weights, 57
    non histone, 122
    peripheral, 67
    plant, 55
    repressor, 146–147
    and ribosomes, 143
    in seeds, 25, 55, 452
    structure of, 55–59, 597
    synthesis, 125, 140–145, **147**, **148**
    targeting and sorting, 145
    transport, 28–29, 70
    viral, 200
proteinase inhibitors, 437
prothallial cells, 335, **365**, 371
prothallus, 349, **351**
Protista (protists), 11, **178**, 244, 730 (see also algae; chytrids; slime molds; water molds)
    autotrophic, 11, 246
    in evolution, **177**, 244
    heterotrophic, 244, 245, 246, 248
    parasitic, 245
    photosynthetic, **177**, 246
    saprobic, 245
Protista kingdom (protists), 176, 244
protoderm, **445**, **446**, 447, 453, 454, 465, 475, **489**, 491, **492**, 506, **507**
protofilaments, 30
Protolepidodendron, 318
proton, 713–716
    acceptors, donors, 721
    gradient, 93, 95, 108
    pump, 68, **69**, 108, 191, 560, 563, 632
protonema, 309, **313**, 314
protopectin, **51**
protophloem, 461, 475, **476**, 490, 493, 494, **496**
protoplasm, 17
protoplast, 17
    fusion, 565
    of mature sieve element, 462
    in plant regeneration, 565
    of slime mold, 256
protostele, 320, 321, 480
    of Lycopodiaceae, **319**, **330**
    of progymnosperms, 359
    of *Psilotum*, **327**
    of *Selaginella*, 331, **334**
protoxylem, 460, 475, **476**, **477**, 479, 480, 490, 493, 494, **496**, 511
protozoa, 11, 175, 179, 186, 244, 245, **246**, 259
protozoologist, 245
Prunella vulgaris, **156**
prunes, 555
Prunus, **387**, 434
    persica var. *nectarina*, 173
    persica var. *persica*, 173
    serotina, 539, **542**
Psaronius, 345, 347, **356–357**
Pseudolycopodiella, 331
Pseudomonas, **198**, **199**
    marginalis, **189**
    solanacearum, **188**, 199
Pseudomyrmex, 638, **639**
pseudoplasmodium (-ia), 253, **254**, 255
pseudopodium
    of granite moss, 315
    of peat moss, 314

Pseudotrebouxia, 223
Pseudotsuga, 369
  menziesii, **453, 654, 655**, 679, **682**
Psilocybe mexicana, 234
psilocybin, **234**
Psilophyta (psilophytes), **178**, 320, 322, **325**, 326–329, **340**, 345, 733 (*see also Psilotum; Tmesipteris*)
Psilophyton, 318, **325**
  princeps, **325**
Psilotum, 323, 326
  life cycle of, **328–329**
  nudum, **327**
Psophocarpus tetragonolobus, **704**
psychedelic drugs, 436, 437
Pteridium aquilinum, **348**
Pteridospermales, 346
Pteridospermophyta, 361
Pterophyta, **178, 183**, 323, 342–353, 733 (*see also* ferns)
Puccinia graminis, 235–238
  life cycle of, 236–237
puffballs, **181**, 228, 231, 234, **235**
pulvinus, 588, 589
pumpkins, 691
pumps, 68, 70, 91, 608, 609 (*see also by name*)
punctuated equilibrium model of evolution, 168
Punica granatum, **688**
Punnett, Reginald Crundall, 132
Punnett square, 132
purines, **60**, 61, **136, 137, 138**
Puya raimondii, 521
pyramids of energy, mass, numbers, 648
pyrenoid, **266**, 283, 284, 286, 290, 291, 292, 307
pyrethrum, 643
pyridoxal phosphate, 82
pyrimidines, **60**, 61, **136, 137, 138**
Pyrococcus, 191
pyrophosphate bridge, 81
Pyrrhophyta, **178, 247**, 262–265, 731 (*see also* dinoflagellates)
Pyrus communis, **34, 458, 528**, 531
pyruvate (pyruvic acid), 86, 87, 89, 90, 96, 97, **98**, 111, 112, **113**
pyruvate decarboxylase, **98**
pyruvate kinase, **89**

Q (electron acceptor), 107, **108**
quantum, 103
quartz (SiO_2), 597
quartzite, 598
Quercus, 161, 171, **386**, 459, 498, 535, 644
  alba, **530, 538, 542**
  coccinea, 160
  rubra, **459, 498, 527**, 535, **536, 542**
  suber, 531
  velutina, 160, **531**
quiescent center, 474
quillworts, 335, 338
quinine, **437, 696**
quinoa, 691, **692**
quinones, 93, 94, 108, 435

rabbitbush, **668**
rabbits, 640
rabies, 200
raceme, **384**
races, 166, 173
rachilla, **427**
rachis, 343, 348, 500, **588**, 687
radial files, 522, 535
radial micellation, 618, **619**
radial section (surface), **527, 529**, 532–537
radial system, 522, 535
radiation, 101 (*see also* light; sun)
  infrared, 102
  ionizing, 102, 103
  ultraviolet, 2, 4, 102
radicle, 443, **449**, 450

radioactive dating method, 53, 586, 714
radioactive substances and lichen growth, 225
radioactive tracers, **621, 628**, 633, 714
radioautography (*see* autoradiography)
radio waves, 102
radish, 166, 453, **477, 625**
Rafflesia, 382
Rafflesia arnoldii, **383**
ragweed, 396, 397, 705
  western, **389**
rain, 615 (*see also* precipitation)
  acid, 512, 513, 615
  in fruit, seed dispersal, 433
  and plant growth, 535, 536, **537**
  and seed germination, 585
rainforests, **8**, 662–665
  agriculture in, 665
  biome map of, **665**
  distribution of, 664–665
  and human activity, 665
  interrelationships in, 662
  rainfall in, 662
  soils of, 665
  species diversity of, 662
  temperate, 665, 682
  tropical, 662–665
  vegetation characteristics of, 664
Ranunculaceae, **387**, 432
Ranunculus, **476, 478, 495**
Raper, John R., 250
rapeseed, 397
Raphanus sativus, 166, **453, 625**
raphe, **261**
raphides, **25**, 435
raspberry, 429, 434
rattlesnake plantain, 676
ray(s), 455, 461, 481, 521, 522, 524, 525, **527**
  of conifer wood, 364, 532–534
  of dicot wood, 535, **536**
  dilated, 524, 525
  phloem, 364
  xylem, 364
ray cells, 492
ray flowers, 412, 413
ray parenchyma, 534
ray tracheid, 534
reactants, 79
reaction center, 106, 107
receptacle (algal), **278, 279**
receptacle (flower), 383, **384**, 550
receptors (membrane), 562
recombinant DNA technology (genetic engineering), 559, 565–571
recombinant plasmid, **567**
recombination, genetic, **153**
recycling (*see also* cycles)
  bacteria and, 188
red algae, **180, 192**, 247, 248, 269–274 (*see also* algae)
  economic uses of, 281
  life cycle of, 270, 272, **273**
red blood cells, **123**
red tides, **263**, 264
red tingle, **380**
redox reactions, 78
reduction, 77, 78
reduction division (*see* nuclear division)
redwood(s), 361, **543**, 622
  coast, 372, **679**
  dawn, 372, **373**, 374
  family, **373**
regeneration, 35, 455
  in tissue culture, **564, 565**
region
  of cell division, 474, **475**
  of elongation, **474, 475**
  internodal and nodal, **495**
  of maturation, 474, **475**
regulatory sequence, 559
regulatory transcription factors, 559
release factor, **144**
reporter genes, 570
repressor, 147

reproduction (*see also subentries* life cycle)
  asexual (*see* asexual reproduction)
  by cell division, 38
  nonrandom, 154
  rate of, 151
  sexual (*see* sexual reproduction)
  vegetative, 450, 451
reproductive cells, 6
reproductive isolation, 157, 160
reservoir, in *Euglena*, 266
resin(s), 30, 532
resin duct, 363, 532, **533**
respiration (cellular), 24, **75**, 86, 87, 115
restriction enzymes (restriction endonucleases), 566
restriction point (start), 38
  in flowers, 416–417
  in fruits, 554
  rates of, 159
respiratory illness, 263
  in roots, 484
  in seeds, 447
resurrection plant, 331, **334**
retinal, 105
Rheum rhaponticum, **456**, 518
Rhizanthella, **414**
rhizine, **226**
rhizobia, 196, 197, **199**, 430, 604
Rhizobium, 70, 196, 197, **199**, 430, 602, **604**, 606, 607
rhizoids
  algal, 288, 294, **295**
  of bryophytes, 300, **301, 303**, 309, **313**
  of chytrids, 251, **252**
  fungal, 211, 213
  of vascular plants, 326, **329, 333, 337**, **340**, 345, 349, **350, 351**
rhizomes, 157, 167, 325, 346, 450, 517
  of *Equisetum*, 338, 340
  of ferns, 344, **345**, 348, **351**
  of Lycopodiaceae 330, **332**
  of *Psilotum*, 326, **328**
  of *Selaginella*, **334**
  of *Trillium*, **3**
Rhizophora mangle, 483
Rhizopoda, 252, 257
Rhizopus, **209**, 222
  life cycle of, **213**
  stolonifer, **212, 213–214**
Rhodococcus, **199**
Rhododendron, 683
Rhodophyta, **178, 180**, 247, 269–274, 730 (*see also* red algae)
Rhodopseudomonas viridis, 107
rhodopsin, 289
Rhodospirillum rubrum, **80**
Rhopalothrix, 374
rhubarb, 204, **456**, 518
Rhynia, 318, 325
  major, **325**
Rhynie chert, **299**
Rhyniophyta (rhyniophytes), 323, 324–325, 326, 401
Rhytiodpononera metallica, **435**
rib, of leaf, **500**, 504
D-ribitol, 227
riboflavin, **92**
ribonucleic acid (*see* RNA)
ribonucleotide, **60**
ribose sugar, 49, **60**, 61, **81**, 83
ribosomes, 17, 21, 27, **141, 144**, 145
  of eukaryotes, **143**
  free, 145
  of prokaryotes, **143, 148**, 187, 192, 195
ribulose 1,5-bisphosphate (RuBP), 110, **111**, 112, **113**, 116
ribulose bisphosphate carboxylase, 110, **111**, 112
ribulose bisphosphate carboxylase/oxygenase (Rubisco), **80**, 112
  structure in *Rhodospirillum rubrum*, **80**

rice, **381**, 542, 555, 579, 607, 689, **690**, 696–698
  and cyanobacteria, 197, **198**, 353
Ricinus communis, **442**, 449, 460
Rick, Charles, 704
ringworm, 222
RNA, 60, 61, 597
  in bacteria, 141, **148**, 175, 179, 188, 191, 195, 196
  in eukaryotes, 141
  and genetic code, 140
  messenger (mRNA), 140–142, 144–145, **148**
  nucleotides of, **140**, 141, **143**, 175
  primary transcript of, 141
  in protein synthesis, 140, 141, 144–145
  in protists, 251
  ribosomal (rRNA), **19, 20**, 140, 143, 175, 179
  splicing of, 141
  synthesis of, 125, 140–141
  transcription of, 140, **141**, 142, 145–147, **148**
  transfer (tRNA), 140, 142–143, **148**
  translation of, 142, 144–145, **148**
  viral, 176, 200, 201, 202, **203**, 204
  of viroid, 204–205
RNA polymerase, 140, **141**, 146, 147, 560
RNA primase, **140**
RNA primer, **140**
Robichaux, Robert, 164, **165**
Robinia, 538
  pseudo-acacia, **499**, 521, **529, 530**, 532, **542**, 617, **624**
rocks
  types of, 597, 598
  weathering of, 597–600, 611, 612
rockweed, **275**
Rohn, Meagan, 409
root(s), 5, **7**, 319, 320, 326, 470–486, 608, 624, **625**, 627 (*see also* vascular plants)
  abnormalities of, 156
  adaptations, 484–486
  adventitious, 330, **332, 336**, 338, **340**, 345, 347, **350, 351**, 449, 450, 471, 483, 517, 550
  aerial, 483, 484
  of angiosperms, 470–486
  apical meristem of, **7**, 447, **473, 474**, 475, 483
  and auxins, 548, 550
  branch (*see* lateral)
  cap, 443, 472, **473, 475**, 483, 575
  cells, **41, 44**
  and cytokinins, 470, 552, 553
  development of, 472–475, **481, 486**
  of dicots, 470, **472**, 478
  embryonic, **442, 449**
  feeder, 471, 475
  of ferns, 343, 345
  food storage in, 470, 485, 486
  functions of, 5, 470, 624, 627
  of grasses, 471, **598**
  gravitropism of, 472, 574, 575
  growth regions of, 474, 475
  of gymnosperms, 470, 478
  hairs, 241, 467, **472**, 474, 475, **477**, 604, **605**, 624
  initiation of, 483
  of *Isoetes*, **335**
  lateral (branch, secondary), **7, 442**, 448, 470, 471, 483
  of Lycopodiaceae, **319, 330**
  of monocots, 471, 478, 480
  and mycorrhizal fungi (*see* mycorrhizae)
  and nitrogen assimilation, 610, 611
  nodules of legumes, 196–197, 604–607
  and nutrient uptake, 472, 593, 627, 628
  origin (evolutionary), 318, 321
  periderm of, **481, 482**
  pressure, 626, **627**
  primary, **442**, 448, **449**, 470, 471, **625**
  primary structure of, 475–480

primordium, 483
prop, 483
secondary growth of, 478, 480–482
secondary structure of, 481, 482
of *Selaginella*, 331
and stem compared, 475
structure, 475–480, 481
systems, 6, 7, 319, 347, 448, **470**, 471, **598**
tip, **41**, **445**, **446**, **473**, 475, 548
transition to shoot, 510, 511
tree, 471
vascular cambium of, 480, 481, 482
vascular tissue of, **476**, **477**, 480, 481, 482
woody, 481
root-hair (region) zone, 472, 474
Rosa, 172
Rosaceae (rose family), **387**, 430
rose, 1, 171, 418
rosette, 558, 587
Rosmarinus officinalis, **418**
rots, 199, **232**
Roundup, 570
roundworm (*see* nematode)
rRNA, 140, 142, 143 (*see also* RNA)
rubber, 250, 696, **705**, 706
rubber plant, 465, **466**
Rubiaceae, **696**
Rubus, 166, 513, 683
RuBP (ribulose 1,5-bisphosphate), 110, **111**, 112, **113**, 114, 116
RuBP carboxylase, 110, 112, 283
RuBP carboxylase/oxygenase, 112, 283
ruminants, 50
runner, 450, 451
rusts, 211, **228**, 231, 235
life cycle of, 235, **236–237**, 252
rutabaga, 627
Rutaceae, **155**
rye, 168, 221, 471, 475, **624**, 702
winter, 587

Sabalites montana, **402**
Saccharomyces cerevisiae, **218**, 219, 222
carlsbergensis, 219
Saccharum
officinale, 697
officinarum, **505**
sacred lotus, 586
sacred mushroom, 234
safflower, 168
saffron, 696
sage, 695
sagebrush, **668**, **669**, 696
fringed, **674**
Sagittaria, 446
saguaro, **8**, **198**
St. Anthony's fire, 221
sake, 219, 222, **690**
Salicaceae, 240, 432
Salicornia, 609
salicylic acid, 416
saline environments, 608
Salix, **482**, **483**, 532, 614, 683 (*see also* willows)
salmon, DNA in, **136**
Salmonella, 570
Salsola, **433**
salt (NaCl), 191, **192**, **717**, **721**
saltbushes, 609
salt marsh, 166, **167**, 608
salt secretion, 467
salt tolerance, 608, 609, 706–707
Salvia, 695
leucophylla, 641
Salvinia, 352, **353**
Salviniales, 344, 352–353
SAM (*S*-adenosylmethionine), 554
samara, **431** 432
Sambucus canadensis, **455**, 492, 493, 524, 526
sand, 600
sandstone, 597
sandy, 601

Sanguinaria canadensis, 435
Sansevieria, **25**
zeylanica, 117
sap (*see also* cytoplasm)
bleeding, 552
cell, 25
flow rate in phloem, 631
flow rate in xylem, 623
sieve-tube, 631
xylem, 623
saponins, 435
saprobes, 188, 245, 248, 251 (*see also* bacteria; fungi; protista)
Saprolegnia, 248, 249
life cycle of, **249**
sapwood, **527** 538
Sarcoscypha coccinea, **215**
Sargassum, 274, **275**, 276
saturated, unsaturated fatty acids, 52
saturation, zone of, 600
sauerkraut, 190
savannas and deciduous tropical forests, **8**, 666–667
biome map of, **667**
distribution of, 666–667
monsoon forests, 666–667
rainfall in, 666
savannas, **8**, 666–667
southern woodland and scrub, 667
subtropical mixed forests, 667
tropical mixed forests, 667
vegetation characteristics of, 666
Saxifragaceae, **406**
Saxifraga lingulata, **627**
scalelike outgrowth, **328**, **329**
scales
of algae, **282**
of ferns, 346
of liverworts, 301
scanning electron microscope, 26, 27
scarlet cup, **215**
schizocarp, 432
Schizosaccharomyces octosporus, **218**
Schleiden, Matthias, 18
Schwann, Theodor, 18
scientific names, 172, 173
sclereids, **34**, 457, 458, 461, 465, **468**
sclerenchyma, 457, 458, **495**, **497**, **506**
Scleroderma aurantium, **181**
sclerotia, 256
Scott, D. H., 347
scouring rushes, 338, 595
scutellum, 443, 444, **449**, 557
sea, 4–5
sea lettuce, 286, **287**
sea-nymph, **428**
sea otters, 268
sea palm, **180**
sea slug, 288
seasons, 585
and biological clock, 578
seaweed, 268, 270, 274, 642 (*see also* brown and red algae)
Sebdenia polydactyla, **180**
Secale, 168, 702
cereale, 114, 221, 398, 471, 587, **624**
second messengers, 563
secondary consumers, 646
secondary growth, 6, 320, 453, 467, 480–482, 520–543
effect on primary plant body, 480–482, 523–532
secondary plant body, 320, 453
secondary products (metabolites/substances), 25, 31, 222, 227, 288, 435, **436**, **437**, 438, 642, 644, 701, 707
secondary plant body, 320, 453
secretion, 28, 29, 467, 472
secretory vesicles, 28, 29
sedges, 238, **649**, 684
sedimentary rocks, 597, 598
Sedum, 518
seed(s), 357–358, 520
of angiosperms, **129**, 383, 388, **395**, 441, 443, 444

and apomixis, 166
and auxin, 550, **551**
and beetles, 155
banks, 586, **703**
coat, **7**, 357, 358, 367, **371**, 393, **395**, 444, 447, 448, 458, 585, 586
cold-treated, 585
corn, **449**, 450
of cycads, 375
cytokinins in, 552
development of, 393, 441–447
dispersal, 154, **157**, 368, 406, 412–413, 429–438, 448
dormancy of, 448, 585, 586
of *Ephedra*, 377
evolution of, 155, 357–358
ferns, 357, 359
germination (*see* germination (seed)) and gibberellins, 557
of *Ginkgo*, 375, **376**
of *Gnetum*, **376**
of grasses (grains), 443, 444, 449, 450, 557
of gymnosperms, 361, **371**
of pines, 367, **368**
poisonous, 155
-scale complexes, 365
size of, 155
stored food in, 155, 358, 392, 443, 447
viability of, 586
seed beetles, 155
seed leaves, 367
seedless vascular plants, **163**, 317–354
seedling, **362**, 442, 448, 449, 450, 451, 629
dark-grown, 583
effects of light on, 547, 583
establishment of, 450, 451
etiolated, 583
and gravitropism, 574
"seed pieces," 450
seed plants, 6, 323, **356–357**
evolution of, 357–361, 401–439
segregation, principle of, 131, 132
seismonastic (thigmonastic) movements, 588
Selaginella, 320, **321**, 331, **334**, 335, 346, 681
lepidophylla, 331, **334**
life cycle of, **336–337**
rupestris, **334**
willdenovii, **334**
Selaginellaceae, 331, 334–335, **336–337**
selection
advantages in, 407, 415
artificial, 151, 687
natural, 150, 151, 152, 154, 168
responses to, 154–157, 160
self-fertilization, 129
self-pollinate, 130, 132
senescence
and cytokinins, 553
and ethylene, 554
sensitive plant, 588
sepals, 383, **384**, 410, **514**, 515
separation layer, 509, 510 (*see also* abscission zone)
septum (a), fungal, 211, 212, 215, 230
Sequoia, **679**
sempervirens, 361, 372, **679**
Sequoiadendron giganteum, 372, 679
serine, **56**
sesame family, **435**
Setcreasea purpurea, **72**
seta (stalk), 301, **302**, 310, **311**, **312–313**, 314, 316
sex attractants, 644
sex expression and ethylene, 555
sex hormones/inducers, 252, 294
sex organs (*see* antheridia; gametangia; oogonia)
sexual reproduction, 121, 122
in algae, 259, 264, 282–283, 286, 290, **291**
in chytrids, **252–253**
in diatoms, **260**
in eukaryotes, 176, 183

and evolution, 156–157
in fungi, 213, 216, 218, 220
oogamous, 248
in oomycetes, 248–249
in plants, **157**
in slime molds, **254**, 256
sexuality and hormones, 250
shagbark hickory, **499**, **520**, 531
shale, 598
shaman, **234**
sheath
bacterial, 187, 192, 193
bundle, **495**, **497**, **501**, **502**, 504, 505
gelatinous or mucilaginous, 187, 192, 193, **198**
leaf, 498, **499**
sheep, karakul, **688**
sheep liver, DNA in, **136**
shelf fungi, 228, 231, 232, 233
shells, of diatoms, 257, 258, 259, **260**
shepherd's purse, **444**, **445**
shiitake, 234
shoot
of angiosperms, 488, 496, 498
apex, **328**, **332**, **336**, **340**, **350**, **371**, 489–491
apical meristem of, 441, **445**, 447, 488
axillary (lateral), 450, 488
determinate, 383
embryonic, **7**
emergence during germination, 448–451, 488
of *Equisetum*, **317**, **338**, **340**
fertile, **317**, **338**, **340**
gravitropism in, 574, 575
and hormones, 546
origin (evolutionary), 318
reproductive (floral), 451, 511
and root compared, 471, 472, 489
system, **7**, 319, 451, 471, 472
tip, **446**, **453**, 488, **489**
transition to root, 510, 511
vegetative, **338**, **340**, 451, 490, 511
shooting star, **385**
short-day plants, 579, 580
short shoots, 362, 363
shoyu, 221
shrubs, 161, 382, 520
sieve area, 461, **462**
sieve cell, 364, 461, 464, **469**
sieve element, 270, 276, 309, **311**, 320, 335, **345**, 461, 464, 465, 493, **529**
sieve plate, 461, **462**, **463**
in algae, 276
sieve-plate pore, 195, **196**, 461, **462**, **463**
sieve tube, 461, **529**
development of, **475**
sap, 631
transport in, 631, 632
sieve-tube member, 461, **462**, **463**, 464, **469**
development of, **464**
Sigillaria, **356–357**
signal peptidase, 145
signal sequence (signal peptide), 145
silica, **247**, 257, 258
siliceous compounds, 338, 595
silique, 430, **431**
silk, **426**
Silphium, **390**
silt, 600
silty, 601
Silurian period, **4**, 212, 270, 299, 318–319, 320, 323, **324–325**
silversword, 164
Silvianthernum suecicum, **406**
Simmondsia chinensis, **705**
simple pit, 36, **458**
sinigrin, 436
siphonostele, 320, **321**, 322
of ferns, **345**
siphonous algae, 270, 283, **288**, **289**
sirenin, **252**
sisal, 425
six-kingdom system, 176, 178–181

Skoog, Folke, 552
skunk cabbage, 87, 416
    eastern, 87, **417**
    western, **417**
slate, 598
sleep movements in leaves, 576, 577, 587, 588
slime (P-protein), 462, **463**, 464
slime (P-protein) bodies, 462, **463**, 464
slime mold(s), **189**, **247** (*see also* Acrasiomycota; Myxomycota)
    cellular, 179, 245, **247**, 252–255
    life cycles of, **254**, **256**
    nutrition of, 245
    phagocytosis by, 70
    plasmodial, 179, **180**, 245, **247**, 255–257
    slime (P-protein) plug, **463**, 464
slime sheath, **254**, 255
slug (cellular slime mold), 253, **254**, 255
smallpox, 200
smog, 512
smuts, 228, 231, 235
snapdragon, 131
snapdragon family, 410
sneeze-weed, 643
snow algae, 279
snowshoe hares, 644
sodium (Na/Na⁺), 47, **594**, 595, **596**, 608, **713**, **716**, 717
    chloride (NaCl), **717**, 721
    concentration gradient, 64, 68
    hydroxide (NaOH), 721
    -potassium pump, 608, 609
soft rot, **189**, 199
softwoods, 532
soil(s), 597–602
    air in, 601
    algae, 186
    bacteria, 186, 190, 598, **599**, 602
    and biomes, (*see by type*)
    cellular slime molds in, 252
    classification of, 600, 601
    composition of, 600, 601
    deficiencies and toxicities, 613, 665
    depletion of, 612, 613
    erosion, 471, 611, 612, 699, 700, **701**
    formation, 597, 598, 600
    fungi, 186, 208, 210, 598, **599**, 602
    horizons of, 598, 599
    inorganic (mineral) components of, 597, 598, 600, 601, 602
    microorganisms, organisms, 186, 190, 208, 210, 252, 598, **599**, 600, 607
    of mining area, 155–156
    and mycorrhizae, 238–241
    nitrogen in, 602, 607 (*see also* nitrogen cycle)
    organic components, 598, **599**, 601
    particles (separates), 600, 601
    pH, 601, 602, 613
    plant-atmosphere continuum, 621
    and plant nutrition, 593–615
    pore space of, 601
    prairie (grassland), **598**
    recycling of nutrients in, 602
    and roots, 471, 598, **625**
    and water, 601
Solanaceae, 565, 694
Solanum, 437
    aviculare, 707
    melongena, **706**
    tuberosum, 50, **54**, **503** 517, **617**, 691, **692**
solar
    energy, 74 (*see also* sun)
    radiation, 164 (*see also* light)
    tracking, 590
Solidago virgaurea, 159
Solomon's seal, 392, **636**, 676
solutes, solutions, solvents, 65, 720, 721
solute transfer, 455, 456
somatic hybrid cell, 565
D-sorbitol, 227
soredia, 226

sorghum, 114, 641, 690, 698
Sorghum, 690
    vulgare, 114
sori (-us),
    of ferns, 348, **349**, 352, **353**, 356
    of fungi, 235
sorrel, wood, **577**
source-to-sink movement, 629, 631, 632
sourwoods, eastern, **675**
soybean, 22, 221, 250, 590, 595, 605–607, **673**, **689**, 696, 697, 698
    Biloxi, 578, 584
soy paste (miso), 221
soy sauce (shoyu), 221, **689**
spadix, 416
Spartina, 166, **167**
    alterniflora, 166, **167**
    anglica, **167**
    maritima, 166, **167**
    X *townsendii*, 167
spathe, 416
special creation theory, 150
species, 166, 173
    apomictic, 166
    classification by, 168, 172
    clusters, 160, 161
    definition of, 160, 173
    formation/origin of, 160, 161, 163
    name, 172
Species Plantarum, 172
specific epithet, 172
specific gravity, 540
spectrophotometer, 582
spectrum
    electromagnetic, 101, **102**
    of visible (white) light, 102
S-phase activator, 38
S-phase of cell cycle, 38
sperm
    of algae, 274, 276, **277**, **278**, **279**, 285, **293**, 295–296
    of angiosperms, 388, **390**, **392**, **395**
    of bryophytes, **299**, 300, **303**, 313
    of conifers, 361
    of cycads, 361, 374, **375**
    of *Equisetum*, 339, **341**
    of ferns, 349, **351**
    fossils, **299**
    of *Ginkgo*, 361, **375**
    of gnetophytes, 361
    of gymnosperms, 361
    of liverworts, 301
    of Lycopodiaceae, 331, **333**
    of *Marchantia*, **303**
    of pine, 367, **371**
    of plants, 280, 282, **299**
    of *Psilotum*, 326, **329**
    of seedless vascular plants, 323, **333**, **337**
    of *Selaginella*, 335, **337**
    of true mosses, 309, **313**
spermatangia, **273**, 274
spermatia, 237, 238, **273**, 274
spermatogenous cell (body cell), **351**, 367, **371**
    in bryophytes, 300
spermatogenous tissue, **329**, **333**, **337**, **341**
spermogonia, 237, 238
sperm packet, **278**
Sphacelia typhina, 220–221
Sphagnidae, 308, 314–315
Sphagnum, **183**, 314, **315**
Sphenophyta (sphenophytes), **178**, **183**, 320, 322, 323, **325**, 338–341, 346, 401, 733 (*see also Equisetum*; horsetails)
spices, 695–696
spike (flower), **384**
spikelet, **427**
spinach, 23, 579, **580**, 587, 697
Spinacia oleracea, 23, 579, **580**, 697
spindle, 125, 166
    in fungi, 212, 218
    in dinoflagellates, 265
    nonpersistant, **280**, 295
    persistant, **280**, 282, 283, 286

spindle fibers, apparatus, 40–42, 43, **45**
    and colchicine, 166
    in meiosis, 125
    in mitosis, 166, **280**, 283
spindle pole bodies, 212
spine fungi, 232
spines, 516, 517, 638, 671
spines (fungal), 232
spiraea, 418
Spiranthes, **422**
spirilli, 188
Spirogyra, **283**
Spiroplasma citri, 195
spiroplasmas, **195**
splash cups, 309
sporangia, (*see also* capsule; megasporangia; microsporangia)
    of algae, 275, 276, **277**, 287
    of *Allomyces*, 253
    of ancestors of plants, 299
    asexual, 252
    of bryophytes, 301, **302–303**, 304, 307, 308, 310, **312–313**
    of chytrids, 252–253
    of early vascular plants, 4, **325**, 326, 360
    of *Equisetum*, 317, **338**, 340
    of ferns, 343–344, 345, **348**, **349**, 350, **351**
    of fungi, 212, 213–214
    of granite mosses, 315
    of hornworts, 307, 308
    of liverworts, 301, **302–303**, 306
    of Lycopodiaceae, 330, **331**, 332–333
    of oomycetes, 248, **249**, 251
    plurilocular, 276
    of *Psilotum*, 327, **328**, 329
    of *Selaginella*, 331
    sexual, 252
    of slime molds, 255, **256**
    of true moss, 310, **312–313**
    unilocular, 276, **277**
sporangiophore, 346
    of *Equisetum*, 338, **340**
    of fungi, 213
    of oomycetes, **251**
spore dispersal
    in *Equisetum*, 317, 339, 340
    in ferns, 344, 349
    in fungi, 211, **214**, 234, 235
    in hornworts, **307**, 308
    in liverworts, 301, **306**
    in mosses, 311, **314**, **315**, 316
    in plants, 3, 317
spore mother cells, 328, **343**, 344
spore(s), 122, 183, **184**, 317, 358 (*see also* ascospores; basidiospores; conidia; megaspores; microspores; zoospores; zygospores)
    of algae, 276, **287**
    asexual, 212, 214, 215, **216**, **217**, 246
    of bacteria, 194
    of bryophytes, 299, **302–303**, 304, 311, **312–313**, 314, **315**
    of cellular slime molds, 252–**254**, 255
    defined 124
    diploid, 349, 352
    dormant, 213
    of eusporangia, 343, 344
    of fungi, 211, **213**, 214–217
    haploid, **213**, 276
    of hornworts, 307, 308
    of leptosporangia, 343, 344
    of liverworts, **302–303**
    motile, 246, 248, 276
    nonmotile, 212
    of oomycetes, 248
    of plants, 299, **328–329**, 351
    of plasmodial slime molds, 255, **256**
    resistant, 251
    resting, 248
    of vascular plants, 317, 326, **329**, 333, 339, **340**, **351**, 352
sporocarp, 352–353
sporogenous cells, 343, 388

sporogenous tissue, **302**, **312**, **328**, **332**, **336**, **340**, **350**
sporophylls (*see also* megasporophylls; microsporophylls)
    of angiosperms, 383
    of early vascular plants, 346
    of gymnosperms, 361
    of Lycopodiaceae, 330, 331, **323–333**, 335
    of *Selaginella*, 331
sporophyte, 184, 185, 318
    of algae, 276, **277**, 287
    of ancestor of plants, 299
    of angiosperms, 441, 447, 470
    of bryophytes, 299, 300, **302–303**, 304, **312–313**
    of chytrids, 253
    of *Equisetum*, **338**, **339**, **340**
    of ferns, 345, 349, **350–351**, 352
    of granite mosses, 315, 316
    of hornworts, 307, 308
    of *Isoetes*, 335
    of liverworts, 301, **305**
    of Lycopodiaceae, 319, 331, 332
    of *Marchantia*, **302–303**, 304
    of peat mosses, 314, 315
    of pines, **370**
    of plants, 318
    of *Psilotum*, 326, **328**
    of *Selaginella*, **336**
    of true mosses, 310, **311**, **312–313**
    of vascular plants, 299, 319
sporopollenin, 298, 388, **389**, 390
spring beauty, 435
spruce, 361, 369, 683
    Englemann, 640, 653
    red, 543
    white, **650**, **681**
spur shoots, 374
spurge family, 518, 705
squash, 199, **456**, **458**, 691
stalk
    of bryophytes, 301, **305**
    of fungus, 232
    of ovule, 390
    of slime molds, **254**, 255
stalk (rachis), 687
stalkcell (*see* sterile cell)
stamens, 383, **384**, 410, **514**, 515, 569
Stanley, Wendell, 201
Stapelia schinzii, 418
Staphylinidae, 417
Staphylococcus, 222
starch, 50, 111, 390
    in algae, **247**, 248, 262, 269, 270, 280
    breakdown of, 86, 557
    in fruits, 554
    floridean, **247**, 248, 269, 270
    grains, **50**, **476**
    in plastids, **22**, **52**, 99, 248, 269, 298
    in protists, **247**, 262, 269
    in seeds, 557
statoliths, 575
staurastrum brasiliense, **284**
stearic acid, 53
Steinbeck, John, 672
stele(s), 320, 347
stem, 5, 6, **7**, 319, 320, 489–498
    apical meristem of, **7**, 489–491
    bundle, 496, **497**, 498
    of cactus, 516
    development, 489–491, 492, 493, 523, 541
    of dicot, 456, 491, 494, **495**, **523**
    elongation, 490
    of *Equisetum*, 163, 338, **339**
    fertile, **338**, 339
    fleshy, 518
    food storage in, 517, 518
    functions of, 488, 489
    and herbaceous dicot, 494
    and hormones, 556
    -leaf relation, 496–498
    of *Lycopodium*, 319, **330**

modifications, 515–518
of monocot, 492, 494, **495**
periderm of, 467, 525, 526
of *Pinus*, 364
primary structure of, 491–496, **523**
of *Psilotum*, 327
and root compared, 475
secondary growth of, 490, 491, 494, 520–532
secondary structure of, **523**, 526
of *Selaginella*, 331, **334**
succulent, 518, **519**
vascular cambium of, 492, 493, 494, 521–523
vascular tissue of, 491–498, 523–525, 527, 528
vegetative, **338**, 339
water storage in, **8, 382**, 518, **519**, 671
woody, 492, 523, 525, 526, 527–529, **530**, 531, 535 (*see also* wood)
Stemonitis splendens, **257**
stenocereus thurberi, **425**
sterigma (-ata), **230**, 232, 233, 238
sterile bract, 365, **371**
sterile cell, 365, **371**
steroid(s), 707
sterols, 67
Steward, F. C., 563
Stigeoclonium, 294, **295**
stigma (eyespot), **266**, 289
stigma (flower), **381, 384**, 386, **395**, 410, **514, 515**
stigmatic tissue, 392
stimulus, response to
in diatoms, 261
stinkhorns, 228, 234, **235, 662–663**
stipe
algal, 274, **275, 276, 277**
fungal, 232
stipular thorns, 638
stipules, 498, **499**
stolon, **157,** 213, 517, 451
in fungi, 213
stoma (-ata), 5, **6,** 109, **110,** 299, 317, 325, 326, 363, 465, 466, 467, 500–503, 617
of CAM plants, 117, 618, 620
of C₄ plants, 116
development of, 503
gas exchange through, 617
of hornworts, 300, 307–308
in leaf epidermis, **465, 466,** 500–503, 617, **618**
of mosses, 300, 311, 314, **315**
movement of, 5, 555, 617–620
in stem epidermis, 417, 492, **494,** and transpiration, 510, 616–620
stomatal crypt, **502**
stone cells, **34, 458**
stonecrops, 116, 518, 620
stoneworts, 270, 285
storage products
in animals, 211, 248
in bacteria, 192
in cytoplasm, 248, 257, **269**
in fungi, 210, 211, 248
in plastids, 248, **269,** 280, 298
in protista, **247,** 248, 262, 264, 265, 269–270
strands
antiparallel, **138,** 139
complementary, 139, **140, 141**
inactive DNA, **141**
lagging, 139
leading, 139
stratification, of seeds, 585
Straw, Richard, **162**
strawberry, 158, **209,** 250, 429, **434,** 450, 451, 550, 551, 579
streams, 611, 612, 615
Strelitzia reginae, 424
Streptococcus, lactis, **188**
Streptomyces, **199**
streptomycin, 190

Strigula elegans, **227**
strobili 317, 331, **332, 336, 338, 340,** 361, 366
(*see also* cones)
stroma (matrix)
of mitochondria, **23,** 90, **96**
of plastids, 20, **99, 106,** 110
stromatolites, **193**
structural gene, 146–147
style, **384,** 386, 410, **514, 515**
suberin, 34, 53, 54, 467, 478, 510, 525
lamellae, 54, 479, 480
subsidiary cell, **465, 466,** 467, 503, 505, **619**
subsoil, 598 (*see also* B horizon)
subspecies, 173
substomatal chamber, **466**
substrate, 79
succession, 224, 649–655, 682
succinate (succinic acid), **91**
succulents, 117, 518, **519,** 620, 669
sucrase, 59
sucrose, 49, 111, **628,** 631
sugar(s), 48, 49, 248 (*see also* by name)
in fruits, 554
in leaves, 509
nonreducing, 632
in nucleic acids, 58–61
from photosynthesis, 99, 111, 629
and plant hormones, 549, 550
in seeds, 557
storage of, 50
transport in phloem, 629–632
sugar beet, 470, 486, **697**
sugarcane, 114, 168, 505, 615, 696–697
suckers, 450
sulfur (S), **48,** 57, 594, **596,** 713
in bacterial metabolism, 190, 191, 196
oxides of, 512
sulfur dioxide, 227, 308, 512, 513, 598, 615
sulfuric acid, as pollutant, 512, 615
sumac, 14
Sumiki, Y. 556
summer species, early and late, 677
sun, 1–3, 49, 74, **75,** 99, 103, 600 (*see also* energy; solar)
sunbird, collared, **424**
sundew, 610
sunflower, **412,** 579, 590, 616, 693, **694,** 696, 697
family, 164, 172, 173, **389,** 410, 705
supernumerary cambia, 486
supporting cell, **273**
suspensor, 335, **332, 336,** 371, 444, **445, 446**
cells, 367
swamps, 611
sweet clover, **203**
sweet corn, **691**
sweet fern, 607
sweet gale, 607
sweet pea, 133
sweet potato, 485, 694, 697
Swiss chard, **697**
sycamore, **161,** 171, **531**
American, **531, 543**
family, 408
symbiotic associations, 24, 246 (*see also* lichens; mycorrhizae)
of bacteria, 194, 604–607
of chloroplasts, 245–246
of cyanobacteria, 194
of dinoflagellates, 246, 272
and evolution, **177**
of green algae, **246**
symmetry (*see* flower, symmetry)
symplast, 71, 72, 624
symplastic transport (movement, pathway), 71, 72, 624, 627
Symplocarpus foetidus, 417
sympodium, 498
symport, 70, 632
synapsis, 124
synaptonemal complex, **124**
synergids, 392, 393
syngamy, 122 (*see also* fertilization)
Synura petersenii, **32**

syphilis, 190
Syringa vulgaris, **453, 500, 501,** 549
Syrphus, 181
systematics, 11
Szent-Gyorgyi, Albert, 1

2, 4, 5–T, 551
2, 3, 7, 8–TCDD, 551
Tabebuia chrysantha, **662**
taiga, 681, 683
biome map of, 681
distribution of, 681
rainfall in, 683
soils of, 683
vegetation characteristics of, 683
tangential section (surface), **527, 532,** 533, **534, 536**
tannins, 30, **31,** 644
tapeta, 343, 344, 388, **389**
tapioca, **694,** 697
taproot, 362, 471
Taraxacum, **412,** 432
officinale, **183**
target proteins, 563
taro, 689
tarragon, 696
tarweeds, subtribe Madiinae, **123,** 164
Taxaceae, 369, **372**
Taxodiaceae, 372, **373**
Taxodium, 372
distichum, **373,** 484
taxonomic categories, 168
taxonomy, 11
Taxus, **372,** 433
canadensis, **461**
Taylor, D.W., 405
tea, 436, 696
teeth (*see* peristome)
Tegeticula yucasella, **423**
telegraph plant, 390
teleomorph, 220
telia, **236–238**
Teliomycetes, **228,** 231, 232, 235–238
teliospores, **236,** 238
telophase
meiotic, 125, **126–127**
mitotic, 38, **39, 40,** 43
Teloschistes chrysophthalmus, **225**
tempeh, 221
temperate deciduous forests, **8,** 645, 674–675, 677–678
biome map of, **675**
distribution of, 674
growth cycle in, 675
rainfall in, 674
soils of, 675
vegetation characteristics of, 675
temperate mixed and coniferous forests, 678–679
alpine tundra and mountain forest, 679
biome maps of, **678, 679**
distribution of, 678
mixed west-coast forests, 679
soils of, 678
temperate mixed forests, 678
vegetation characteristics of, 679
temperate regions, 520, 521, 535
temperature(s)
and biological clocks, 577
cycles, 577
and dormancy, 585
and enzymes, 82
and evaporation, 617
and flowering, 587
and fruit ripening, 554
and leaf development, 510
and photosynthesis, 101, 117
and root growth, 471
and secondary growth, 533, 536
and seed germination, 447, 585, 586, 586
and stomatal movement, 619
temperature conversion scale, **727**

tendrils, **499,** 515
and touch responses, 576
tension (negative pressure), 622, 623
tension wood, 538, 539
teosinte
annual, 693
wild perennial, 693
tepals, 384
termination, 144–145
codon, 144
termites, 50
terpenes, 641
terpenoids, 270, 435
Terrapene carolina triunguis, **646**
Tertiary period, 361, 372, 415
testcross (backcross), **130,** 131
tetanus, 190
tetracycline, 190
tetrads
of spores, **371,** 395
tetrahydrocannabinol (THC), **437**
Tetraselmis convolutae, 294
tetrasporangia, tetraspores, **273,** 274
tetrasporophyte, **273,** 274
texture of wood, 540
thallus
of algae, 286
of liverworts, 301
Thaxter, Roland, **225**
Thea sinensis, **436**
theca, 262
Theobroma cacao, 436, 691, **697**
theobromine, 436
Theophrastus, 604
thermocouple, 623
thermodynamics, laws of, 74–77, 87
thermogenicity, 87
thermonuclear reactions, 74
thermophilic organisms, 614
Thermoplasma, 191–192, 195
Thermopsis montana, **509**
thermotolerant organisms, 614
thigmomorphogenesis, 589, 590
thigmonastic movements, 588
thigmotropism, 575, 576
Thiobacillus, 190
Thiothrix, **194**
thistle, **412**
thorns, 516, 517
of *Acacia,* 517, 638–639
threonine, 56
thrush, 219, 222
Thuidium delicatulum, **310**
Thuja plicata, **543**
thylakoids, 21, **99, 106,** 108
in bacteria, 192
thyme, 695
thymine, **60,** 61, 136, **138,** 139
Thymus, 695
Tidestromia oblongifolia, 670
Tilia americana, **457, 459, 462,** 464, **491,** 492, 524, 525, **528, 530,** 542, 631
Tillandsia usneoides, **373**
Ti plasmid, 566, 567
tissue culture, 552, 553, 563–565
tissue(s), 320, 454–469 (*see also* by name)
autoradiography, 633
complex, simple, 455
development of, 447, 454, 475
embryonic, 6 (*see also* meristems)
explants, 563
tissue systems, 320, 447, 454, 475
Tmesipteris, 326
lanceolata, **327**
parva, **327**
TMV (tobacco mosaic virus), 202
toadstools, 228, 232
tobacco, 24, 168, 200, 250, 251, 423, 436, **497, 507,** 513, 552, **553,** 584, 691
Maryland Mammoth, 578, 584
Trabezond, 584
tobacco mosaic virus, 201, 202, **203,** (*see also* TMV)

tofu, 221
Tolypocladium inflatum, 209
tomato(es), 429, 437, 550, 554, **575, 594, 617,** 641
   cultivation of, 691, 706–707
   diseases of, **188,** 199, 200, 250
tonoplast (vacuolar membrane), 18, **24,** 66, **460,** 462, **464,** 609
tooth fungi, 232
topoisomerases, **140**
topsoil, 598
Torrey, John, **197**
tortoises, 151, 160
torus(-i), 532, 533, **534**
totipotency, 559, 563
touch-me-not, 433
touch responses (thigmonastic movements), 588, 589
Toxicodendron radicans, **436**
toxic substances, 155, 193, 222, 227, 262, 289, **348, 372,** 436, 641, 642, 643
toxins, 220, 263, 374 (*see also* secondary substances)
trabeculae, **334**
trace elements, 595
tracheary elements, 309, 320, 458, 459, 460, **492, 493, 501, 504, 622**
   evolution of, 320, 325, 459
tracheids, 299, 320, 335, **345,** 359, 363, 364, 458, 459, 461, **468,** 532–534, 622
traits
   dominant/recessive, **130**
   true-breeding, 129
transcellular pathway, 624
transcription, 140–142, **148**
   factor, 560, 569
   regulation of, 145–147
transduction, 191
transfer cells, 455, 456
transformation (bacterial), 191
transfusion tissue, 363
transgenic plants, 566
transition region, 510, 511
transition vesicles, 29
translation, 142, 144–145, **148**
   stages of, 144
translocation, gene, 135
translocation, in phloem, 629–632
transmembrane proteins, 66, 67
transmission electron microscope, 26, 27
transmitting tissue, 392
transpiration, 510, 616–620
transpiration-pull theory, 624
transpiration stream, 620, 621, 627
transplanting, 472
transport
   active, 25, **68,** 70, 628, 632
   between cells, 37, 38
   apoplastic, 71, 72, 624, 628, 629
   passive, 68
   of plant hormones, **546**
   proteins, 68
   symplastic, 71, 72, 624, 627, 628, 629
transposons, 135
transverse section (surface), **527, 528, 532, 533, 535, 536**
Trebouxia, 223, **228**
tree(s), 382, 520, 521, 536 (*see also* by name; forests; gymnosperms)
   dormancy in, 587
   mycorrhizal associations in, **239,** 240
   ringing (girdling) of, 630
   root systems of, 471
   water movement in, 622
Trentepohlia, 223, 279
Triassic period, **241,** 404
tricarboxylic acid (TCA) cycle, 91, 92
Trichodesmium, 193, 197
trichogyne, 216, **273,** 274
Trichomanes, 349
trichomes, **456,** 465, **466,** 467, **501, 502,** 503, **507,** 518
trichomycetes, 214
trichothecenes, 222

Tridachnidae, 264
Trifolium, 602, 606
   repens, 155, **421, 603,** 641
   subterraneum, 587
triglycerides, 52, 53, 54
Trillium, 435, **675,** 676
   erectum, 123
   grandiflorum, **3**
Triloboxylon ashlandicum, 359
Trimerophyta (trimerophytes), 318, 323, **324–325,** 326, 359, 401
Trimerophyton, **325,** 326
triple fusion, 392, **393**
triticale, 168, 702
Triticosecale, 168, 702
Triticum, 167, 168, 687, 688, 702
   aestivum, **125,** 167, **443,** 452, **456, 505,** 617, **698**
tritium (³H), **474**
tRNA, 140–142 (*see also* RNA)
   acceptor end of, **143**
   translation of, 142–145
Tropaeolum tuberosum, **692**
trophic levels, 646–648
tropical environments, 666–667 (*see also* savannas and deciduous tropical forests)
tropism(s), 573–576 (*see also* gravitropism; phototropism)
truffles, 215, 241
   black, **215**
   false, 234
tryptophan, 56, 147, 548
tryptophan (*trp*) operon, 147
Tsang Wang, 372, 374
Tsuga, 369
   canadensis, 543, **675**
   heterophylla, **240, 654**
tube cell, **365, 371,** 390, **395**
tube nucleus, 371, **395**
tuber(s), 450, 517, **692**
tuberculosis, 190
Tuber melanosporum, **215**
D-tubocurarine chloride, 708
tubulin, 30, 59
Tulare apple mosaic virus, 202, 203
tulips, **204,** 587
tulip tree, **535,** 543
tumbleweeds, 432, 433
tundra (*see* Arctic tundra)
tunica-corpus, 489, 490, 511
Turesson, Göte, 158
turgor, turgor pressure, 25, 66, 164, 560, 597
   in aquatic plants, 510
   and leaf movements, 505, **506,** 588, 589
   in sieve tubes, 631, 632
   and stomatal movement, 617, 618
turtle, box, **646**
twigs, **530**
tyloses, 538
type specimen, 172
Typha, **649**
typhoid fever, 190
tyrosine, 56

ubiquinol, **94**
ubiquinone, **94**
ullucu, 692
Ullucus tuberosus, **692**
Ulmus, **431,** 542
   americana, **530,** 542
Ulothrix, **295**
   Zonata, 286, **287,** 294
ultraviolet radiation, 2, 4, 12, 102
   and bees, 419, **420,** 429
   and flowers, 428
   and mutations, 134
Ulva, 281, 286, **287**
   life cycle of, **287**
Ulvophyceae, **280,** 282, 286–289, **295**
   cell division in, **280,** 282
umbels, **384**

Umbelliferae, 173
Umbellularia californica, 696
uniport, 70
United Nations, 699
   Food and Agricultural Organization (FAO), 700
uracil, **60,** 61, 140, **141**
uranium, 714
urea, 595
Uredinales, 235
uredinia, urediniospores, **236–237,** 238
ureide, 606
Usnea, **225**
Ustilaginales, 235
Ustilago maydis, **228**
uterine contractions, 221
Utricularia vulgaris, 610

Vaccinium, 683
vacuolar membrane (*see* tonoplast)
vacuole(s), 176
   contractile, 65, **254, 266,** 290
   gas, 193
   perialgal, 246
   plant, 18, **20,** 24, 25–27, **31,** 66, 462, 609
valence, 717
valine, **56**
Vallisneria, **428**
valves (diatoms), 258, **260**
van Helmont, Jan Baptista, 99
vanilla, vanillin, **414,** 695
Vanilla, 414
   planifolia, **414,** 695
van Niel, C. B., 100, 196
van Overbeek, Johannes, 551
variety, classification by, 173
vascular bundle(s), 112, **113, 458,** 492, 494, **495, 496, 497**
   closed, open, 494
   differentiation of, **496, 497**
vascular cambium, 6, 320, 347, 453, 521–523, 535, 538, 539
   and auxin, 523, 550
   bifacial, 359
   of *Botrychium,* 343
   of conifers, **364**
   of cycads, 374
   in root, 480, 481, 482, 485
   in stem, **458,** 492, 493, 494, 523, **524,** 525
vascular cylinder
   of root, **476, 477,** 480, **482**
   of stem, 491, 492, 497
vascular plants, 3, 6, 7, 317–355, 356–379, 380–400
   CAM, 116
   early, 318
   evolution of, 357–361, 401–439
   mycorrhizae, 238–241
   phylogenetic relationships of, 401–404
   seedless, 317–355
vascular rays, 522
vascular strand
   of *Equisetum,* **339**
   of flowers, 410
vascular systems, 6, **7,** 319–320, 629
vascular tissue(s), 320, 454, 458–465, 475 (*see also* by name)
   induction of, by hormones, 548–550
   primary, 320
   secondary, 320, 522, 523
Vasil, V., 563
Vaucheria, **262**
vectors, 201, 204, 309, 393, 399 (*see also* by type)
vegetation types (*see* biomes)
vegetative reproduction, **163,** 166, 450, 451
vein(s), 322, 363, **500–502,** 504, 505, **506**
   major, 504
   minor, 456, **500, 501,** 504, 629
velamen, 484
venation, **499,** 504

venter, 300, **302, 312, 329,** 331, 333, **341**
Ventricaria (Valonia), **288**
Venus flytrap, 589
Verbascum thapsus, **466**
vernalization, 587
Veronicastrum virginicum, 498
Verrucaria serpuloides, 224
Verticilium, 704
vesicles
   of dinoflagellates, **262**
   of mycorrhizae, 239
   in nitrogen fixation, **197**
   secretory, 29
   transition, 29
   in transport, 71
vesicular-arbuscular (V/A) mycorrhizae, 239–240
vessel members (elements), 320, 458, 459, 460, 535, 588, 620
   development of, **460, 494, 496, 497**
vessels, 458, **536,** 622, 623
   in Gnetophyta, 320, 377
vetches, 688
Vibrio harveyi, 570
Viburnum marginatum, **402**
Vicia, 688
   faba, **7,** 604, 633
Vietnam Conflict, 551
Vigna sinensis, **617**
vinblastine, **707**
vincristine, **707**
vine(s), 521
   woody, 664
Viola, 172, 435
   quercetorum, **173**
   rostrata, **173**
   tricolor var. *hortensis,* **173**
violets, 157, 172, 435
   long-spurred, **173**
   yellow-flowered, **173**
Virchow, Rudolf, 18
Virginia creeper, 515
viroids, 200, 204–205
virology, 11
vir region, 566, **567**
virus(es), 11, 176, 200–205
   animal, 201
   bacterial, 201 (*see also* bacteriophage)
   commercial uses of, 204
   DNA, 176, 200, **201,** 202
   human cold, 202, **202**
   influenza, 200
   mosaic, 201, 202, **203,** 204
   nature of, 201
   origin of, 205
   plant, 201, 202–204
   polio, 200
   protein coat, 176
   replication of, 201, 202
   RNA, 176, 200, 201, 202, **203,** 204
   structure of, 201, 202
   transmission of, 204
   wound tumor, 203
Vitaceae, 383
vitamin(s), 81, 214, 452, 593, 697
   A, 105
   B-complex, 90, 91
   B₂, 92
Vitis, 515, 531
   vinifera, 558, 688
Vittaria, 349
Vittaria guineensis, **62**
viviparous mutants, 555
volcanoes, 2
volva, 232
volvocine line, 292
Volvox, **180,** 292, 293–294
   carteri, **293,** 294
von Frisch, Karl, 419
von Humboldt, Alexander, **661,** 664–665
von Sachs, Julius, 563
voodoo lily, 87
Vorticella, 246
Voyage of the Beagle, The, 150

wake-robin, **123**
Wald, George, 103
Wallace, Alfred Russel, 151, 407
Wallace's line, 407
wall pressure, 66
walnut, black, 539, **543**, 641
Wareing, Paul F., 555
washes, 670
Washingtonia filifera, **669**
wasps, **163**, 418–421
water
    absorption of, by roots, 467, 470, 471,
        475, 624–627
    balance, 595
    characteristics of, 64, 622, 720, 721
    as component of cell sap, 25
    as component of living matter, 47
    conducting cells, tissues, 300, 301, 458,
        459, 504 (*see also* hydroids;
        xylem)
    cycle, 600
    of guttation, 626
    loss of (*see* water loss)
    molecular structure, properties, 720, 721
    movement of, 62–66
    movement of, in plants, 5, 455, 456,
        522, 538, 597
    photolysis of, 107, 108
    and photosynthesis, 4, 100, **101**, 107,
        108
    plant requirement for, 5, 447, 472, 500,
        616, 617
    potential, 62, 63, 76, 621
    pressure, 63
    from respiration, 86, 94, 95
    and soils, 598, 600, 601
    as a solvent, 721
    stress, 450, 451, 555, 621–623
    table, 600
    tensile strength of, 622
    transport, 620–624
    vapor, 2, 617
    vesicle, 518
water dispersal of fruits and seeds, 433
waterfall, 63
water felt, **262**
water hemlock, **385**
water hyacinth, **649**
water loss, 53, 116, 617 (*see also*
        transpiration)
    protection against, 467, 500, 555, 590,
        617–620
watermelon, 199
water molds, 179, 244, 245, **247**, 248–249
        (*see also* Oomycota)
    life cycle of, **249**
water net, **292**
Water-soluble fibers, 452
water stress, 164
Watson, James D., 136, **137**, 138, 139
Watson-Crick model, 136, **137**, 139
wavelengths, 102 (*see also* light)

wave theory of light, 101–103
wax deposits, **54**, **465**, 467
waxes, 34, 53, 317, **705**
wax myrtle, 197
weathering of rocks, 597, 598, 600, 611, 612
weeds, 166, 674, 687
    control of, 551, 570
weevils, 374
Weinberg, G., 152
Welwitschia, 375, 377–378, 668
    mirabilis, **378**
Went, Frits W., 547, 555, 573
Wetmore, R. H., 549
whales, 245
Whatley, J. M., 23
wheat, 114, 168, **443**, 452, **456**, **505**, 566,
        567, **617**, **687**, 688, 696, 697, 698,
        702
    bran, 452
    germ, 452
    hybrid, 167
    kernels, **133**
    polyploid, 167
wheat rusts,
    life cycle of, **236–237**
whiteflies, 201
whooping cough, 190
Wilde, S. A., 638
Wilkesia, 164
Wilkesia gymnoxiphium, **165**
Wilkins, Maurice, 136
willow(s), 240, 386, 396, 416, 427, 432, **482**,
        483, 538, 683
wilt, bacterial (disease), **188**, 199
wilting, 66, 601
wind dispersal
    of fruits and seeds, 432–433
wine, 97, **98**, 209, 218–219, 250, 688, **690**
wintergreen, **410–411**
    striped, 676
witch hazel, 433
Woese, Carl R., 191, 195
Wolffia, **381**
    borealis, **381**
wood(s), 36, 64, **459**, **520**, 532–540
    commercial uses of, 542, 543
    of conifer, 532–534, 538, 539
    density of, 540
    of dicot, 532, 535, **536**
    diffuse-porous, **535**, 538
    early, late, **532**, **533**, 536, **537**, 538
    features and identification of, 539
    methods of sawing, 540
    ring-porous, **535**, 538, 540
    semi-ring-porous, **537**
    shrinkage of, 538–539
    specific gravity of, 540
woody plants, in biomes
    and hybridization, 161, **162**
work, 74, 75
World Bank, 699
wormwood, 172, 696

wound healing and repair, 35, 455, 464
wound tumor virus, **203**

Xanthomonas, 198, **199**
    campestris, 199, 200
Xanthium strumarium, 553, 579
xanthophylls, 105, 274
Xanthophyceae, 261–262 (*see also* algae)
Xanthosoma, 689
xerophytes, 500, **502**
X-ray diffraction studies, 136
X-rays, 101, 134
xylem, 6, 317, **319**, 320, 325, 326, **327**, **330**,
        339, **345**, 454, 458–461, **467**, 485,
        486
    bacterial infections in, 199
    of calamites, 346
    of early vascular plants, 346
    of gymnosperms, **363**
    and ion transport, 628, 629
    and nutrient transport, 628, 629
    and phloem interchange, **628**, 629
    and plant hormones, 548–550
    primary, 320, 338, 458, 460, **476**, **477**,
        480, **491**, 492, 493, **494**, **495**
    primary development of, 475, **496**, **497**
    of progymnosperms, 359
    secondary, 36, 320, 346, 347, 453, 458,
        460, 532–543
    secondary, of pine, 364
    and water transport, 72, 620–624
xyloglucan, 51
xylose, 51

Yabuta, T., 556
yams, 171, 437, 690, 697, 707
yeast(s), 209, 210, 215, 218–219
yellow-green algae, 261–262 (*see also* algae)
yew(s), 369, **372**, 434, **461**, 538
yogurt, 190
yucca, 694
Yucca
    brevifolia, **668**, **669**
yucca flower, **423**
yucca moth, **423**

Zamia, 374
Zamia pumila, **374**, **375**
Zantedeschia, 381
Zanthoxylum
    ailanthoides, 155
    dipetala, 155
Zea diploperennis, 693
Zea mays (corn)
    classification of, 173, **174**
    cultivation of, **691**
    diseases of, 195

    ear, **120**, **129**
    flowers, **426**
    hormones of, 552
    leaf, **16**, **17**, **37**, **465**, **499**, **503**, 618
    photosynthesis in, **21**, 113, 114
    root, 471, **473**, **477**, **483**
    seedling, 449
    stem, **494**, **495**, **496**, **497**
    subspecies *mays*, 693
    subspecies *parviglumis*, 693
    water loss in, **617**
zeatin, 552
zeaxanthin, 105
zebra, 666
Zebrina, **22**
Zigadenus fremontii, **422**
zinc, 155, 239, 594, **596**, 613
Zingiber officinale, 695
zone of saturation, 600
zoologists, 11, 173, 245
zoomastigotes (Zoomastigina), 246, 265
zooplankton, 245
zoosporangium, **249**
zoospores
    of algae, 258, 276, **277**, **287**, **290**, **291**,
        292, 295
    of chytrids, 24*̂*, 251, **252–253**
    of dinoflagellates, 264
    of oomycetes, 246, 248, **249**, 250, 251
zooxanthellae, 264, 272
Zostera, 326
Zosterophyllophyta (zosterophyllophytes),
        323, **324–325**, 326, 330
Zosterophyllum, 318, **325**, 326
Zygocactus truncatus, **203**
Zygomycota (zygomycetes), **178**, **181**, 211,
        212–214, 220, 222, 239, 241, 299,
        326, 729
    evolution of, 211–212
    life cycle of, **213**
zygosporangium, **212**, **213**
zygospore(s)
    of algae, 282, 283, 284, 286, **290**, **291**
    of fungi, **212**, 213
zygotes, 183, **184**
    of algae, **244**, 273, 276, **277**, 278,
        282–283, **284**, **287**, **290**, **291**,
        295–**296**
    of angiosperms, 392, 393, 444, 446
    of bryophytes, **302**, **312–313**
    of cellular slime molds, **254**, 255
    of chytrids, **252–253**
    definition, 124
    of diatoms, **260–261**
    of dinoflagellates, 264
    division of, 183, **184**
    of fungi, 213, 218
    of gymnosperms, **371**
    of plasmodial slime molds, 256
    of oomycetes, 248, **249**
    of seedless vascular plants, **329**, 331,
        333, **336**, **340**, **351**

2-HOUR RESERVE

This item is due at the
Limited Loan Counter on or before
the latest time/date stamped below.

FEB 16 '99 -7 45 PM

FEB 16 '99 -7 05 PM

MAR 02 '99 -8 00 PM

MAR 02 '99 -7 00 PM

APR 16 '99 -1 45 PM

APR 16 '99 -12 45 PM

MAY 12 '99 -6 10 PM

MAY 12 '99 -5 55 PM

MAY 13 '99 -9 45 PM

MAY 13 '99 -8 45 PM

MAY 14 '99 -2 30 PM

MAY 14 '99 -1 20 PM

MAY 15 '99 -7 15 PM

MAY 16 '99 -11 40 AM

MAY 6 '99 -1 45 PM

when the library is closed.

MAY 16 '99 -2 05 PM